CONVERSIONS BETWEEN U.S. CUSTOMARY UNITS AND SI UNITS (Continued)

U.S. Customary unit		Times conversion factor		Equals SI unit	
		Accurate	Practical		
Moment of inertia (area)					
inch to fourth power	in.⁴	416,231	416,000	millimeter to fourth power	mm⁴
inch to fourth power	in.⁴	0.416231×10^{-6}	0.416×10^{-6}	meter to fourth power	m⁴
Moment of inertia (mass)					
slug foot squared	slug-ft²	1.35582	1.36	kilogram meter squared	kg·m²
Power					
foot-pound per second	ft-lb/s	1.35582	1.36	watt (J/s or N·m/s)	W
foot-pound per minute	ft-lb/min	0.0225970	0.0226	watt	W
horsepower (550 ft-lb/s)	hp	745.701	746	watt	W
Pressure; stress					
pound per square foot	psf	47.8803	47.9	pascal (N/m²)	Pa
pound per square inch	psi	6894.76	6890	pascal	Pa
kip per square foot	ksf	47.8803	47.9	kilopascal	kPa
kip per square inch	ksi	6.89476	6.89	megapascal	MPa
Section modulus					
inch to third power	in.³	16,387.1	16,400	millimeter to third power	mm³
inch to third power	in.³	16.3871×10^{-6}	16.4×10^{-6}	meter to third power	m³
Velocity (linear)					
foot per second	ft/s	0.3048*	0.305	meter per second	m/s
inch per second	in./s	0.0254*	0.0254	meter per second	m/s
mile per hour	mph	0.44704*	0.447	meter per second	m/s
mile per hour	mph	1.609344*	1.61	kilometer per hour	km/h
Volume					
cubic foot	ft³	0.0283168	0.0283	cubic meter	m³
cubic inch	in.³	16.3871×10^{-6}	16.4×10^{-6}	cubic meter	m³
cubic inch	in.³	16.3871	16.4	cubic centimeter (cc)	cm³
gallon (231 in.³)	gal.	3.78541	3.79	liter	L
gallon (231 in.³)	gal.	0.00378541	0.00379	cubic meter	m³

*An asterisk denotes an exact conversion factor

Note: **To convert from SI units to USCS units,** *divide* **by the conversion factor**

Temperature Conversion Formulas

$$T(°C) = \frac{5}{9}[T(°F) - 32] = T(K) - 273.15$$

$$T(K) = \frac{5}{9}[T(°F) - 32] + 273.15 = T(°C) + 273.15$$

$$T(°F) = \frac{9}{5}T(°C) + 32 = \frac{9}{5}T(K) - 459.67$$

Traffic and Highway Engineering

FIFTH EDITION

SI Edition

Nicholas J. Garber
Lester A. Hoel

University of Virginia

SI Edition prepared by

K. Ramachandra Rao

Indian Institute of Technology, Delhi

CENGAGE
Learning

Australia • Brazil • Japan • Korea • Mexico • Singapore • Spain • United Kingdom • United States

CENGAGE
Learning®

Traffic and Highway Engineering, Fifth Edition, SI Edition
Nicholas J. Garber, Lester A. Hoel
SI Edition prepared by: K. Ramachandra Rao

Publisher: Timothy Anderson

Senior Developmental Editor: Hilda Gowans

Senior Editorial Assistant: Tanya Altieri

Senior Content Manager: D. Jean Buttrom

Production Director: Sharon Smith

Media Assistant: Ashley Kaupert

Rights Acquisition Director: Audrey Pettengill

Rights Acquisition Specialist,
    Text and Image: Amber Hosea

Text and Image Researcher: Kristiina Paul

Manufacturing Planner: Doug Wilke

Copyeditor: Patricia Daly

Proofreader: Pamela Ehn

Indexer: Shelly Gerger-Knechtl

Compositor: Integra

Senior Art Director: Michelle Kunkler

Internal Designer: Carmela Pereira

Cover Designer: Rose Alcorn

Cover Image: © Ron Niebrugge/Alamy

Library of Congress Control Number: 2013948641

ISBN-13: 978-1-133-60708-3
ISBN-10: 1-133-60708-X

Cengage Learning
200 First Stamford Place, Suite 400
Stamford, CT 06902
USA

Cengage Learning is a leading provider of customized learning solutions with office locations around the globe, including Singapore, the United Kingdom, Australia, Mexico, Brazil, and Japan. Locate your local office at: **international. cengage.com/region.**

Cengage Learning products are represented in Canada by
Nelson Education Ltd.

For your course and learning solutions, visit **www.cengage.com/engineering.**

Purchase any of our products at your local college store or at our preferred online store **www.cengagebrain.com.**

Unless otherwise noted, all items © Cengage Learning

Printed in the United States of America
1 2 3 4 5 6 7 18 17 16 15 14

This book is dedicated to our wives,
Ada and Unni
and to our daughters,
Alison, Elaine, and Valerie
and
Julie, Lisa, and Sonja
With appreciation for the support, help, and encouragement that we received
during the years that were devoted to writing this textbook.

Donated by

Spartans.4.Spartans (S.4.S)

S.4.S is a student organization that solicits for textbooks donations and makes them available to students through the MSU Library over subsequent semesters.

Get involved by:

- *Donating a textbook*
- *Joining the volunteer team*
- *Spreading the word – tell a friend!*

Contact us

Email:
spartans.4.spartans2017@gmail.com
Facebook: *Spartans4Spartans*
Tel: *+1 (517) 899 1261*

Contents

Preface

PURPOSE IN WRITING AND REVISING THIS TEXTBOOK

The purpose of *Traffic and Highway Engineering,* Fifth Edition, is to serve as a resource textbook for students in engineering programs where courses in transportation, highway, or traffic engineering are offered. In most cases, these courses are usually taught in the third or fourth year but may also be covered at the graduate level.

Another purpose of this book is to serve as a reference for transportation engineers who are in practice or are preparing for a professional engineering examination.

The initial motivation for writing this textbook, which was first published in 1988, was many years of teaching highway and traffic engineering using textbooks that were primarily descriptive and lacked examples that illustrated the concepts presented. We also noted that none were comprehensive in dealing with all aspects of the subject and that some were written with transportation engineering titles but lacked specific focus. We also saw the need to demonstrate the challenges of the field and to explain the solid quantitative foundations that underlie the practice of transportation engineering. We wanted to select a mode that is ubiquitous and of worldwide application and one that students had contact with on a daily basis. Accordingly, we decided to focus on motor vehicle transportation and the highways that are an essential partner for this mode to exist. Our experience and instincts proved correct as the book became known and widely used.

The objectives of this textbook are: (1) To be a contemporary and complete text in highway and traffic engineering that can be used both at the undergraduate and at the graduate level for courses that emphasize highway and traffic engineering topics and (2) To serve as a reference for engineers in the highway and traffic field and as a study guide for use in preparing for the professional engineering license exam, review courses, and preparation for graduate comprehensive exams in transportation engineering.

The Fourth Edition of this textbook was published in 2009 and in the ensuing years there have been significant changes to the highway transportation literature that mandated a major revision. Professors from transportation programs at twenty-one major universities reviewed various editions of the book and their comments and suggestions have been incorporated into the Fifth Edition.

The book is appropriate for a transportation curriculum or as an introductory transportation course because it provides an opportunity to present material that is not only useful to engineering students who may pursue careers in or related to transportation engineering, but is also interesting and challenging to those who intend to work in other areas. Furthermore, this book can serve as a reference for practicing transportation engineers and for use by students in graduate courses. Thus, the textbook provides a way for students to get into the area of transportation engineering, develop a feel for what it is about, and thereby experience the challenges of the profession

MAJOR ORGANIZING FEATURES OF THE TEXT

The scope of transportation engineering is broad and covers many modes and disciplines. Accordingly, several approaches have been used to introduce this topic to students.

One approach is to cover all transportation modes-air, highway, pipeline, public, rail, and water, presented in an overview course. This approach ensures comprehensive coverage but tends to be superficial with uneven coverage of some modes and can be lacking in depth.

A second approach is to present the subject of transportation by generic elements, such as vehicle and guideway characteristics, capacity analysis, planning, design, safety, human factors, administration, finance, system models, information technology, operations, and so forth. This approach is appealing because each of the modes is considered within a common context and the similarities between various modes are emphasized. Our textbook, *Transportation Infrastructure Engineering: A Multi-Modal Integration,* is based on this concept.

A third approach is to select a single mode and cover the relevant disciplines to provide a comprehensive treatment focused on that mode. Our book follows this approached by emphasizing the subject of traffic and highway engineering, which is a major area within civil engineering. It is a topic that appeals to students because they can relate directly to problems created by motor vehicle travel and is useful to professionals employed by federal, state and local agencies as well as private consulting and construction organizations.

Each chapter presents material that will help students understand the basis for transportation, its importance, and the extent to which transportation pervades our daily lives. The text also provides information about the basic areas in which transportation engineers work: traffic operations and management, planning, design, construction, and maintenance. Thus, this book has been categorized into five parts

- Part 1: Introduction to the profession, its history, systems, and organizations
- Part 2: Traffic Operations
- Part 3: Transportation Planning
- Part 4: Location, Geometrics, and Drainage
- Part 5: Materials and Pavements.

The topical division of the book organizes the material so that it may be used in one or more separate courses.

For a single course in transportation engineering, which is usually offered in the third year where the emphasis is on traffic and highway aspects, we recommend that material from Parts 1, 2, and 3 (Chapters 1-13) be covered.

For a course in highway engineering, where the emphasis is on highway location, design, materials, and pavements, we recommend that material from Parts 2, 4, and 5 (Chapters 3 and 14-21) be used.

A single introductory course in transportation facilities design could include Chapters 1, 2, 3, 14, 15, 16, 19, and 21.

The book also is appropriate for use in a two-semester sequence in transportation engineering in which traffic engineering and planning (Chapters 3-13) would be covered in the first course, and highway design (Chapters 14-21) would be covered in the second course.

The principal features of this textbook are:

- Comprehensive treatment of the subject.
- Extensive use of figures and tables.
- Numbering of subsections for easy reference.
- Completed examples in each chapter that illustrate the concepts presented.
- Representative homework problems at the end of each chapter
- References and additional readings at the end of each chapter

CHANGES TO THE NEW EDITION

In addition to responding to reviewer comments on the Fourth Edition and updating each chapter, substantial changes were made in several chapters due to the availability of new editions of the following professional publications:

- *A Policy on Geometric Design of Highways and Streets,* 6th Edition, 2011, American Association of State Highway and Transportation Officials.
- *HCM 2010 Highway Capacity Manual,* Transportation Research Board
- *Highway Safety Manual,* 1st edition, American Association of State Highway and Transportation Officials, Washington, D.C., 2010.
- *Manual on Uniform Traffic Control Devices (MUTCD),* 2009 Edition, U.S. Department of Transportation, Federal Highway Administration
- *Roadway Design Guide,* 4th Edition 2011, American Association of Highway and Transportation
- *Transportation Planning Handbook,* 3rd Edition, Institute of Transportation Engineers

New Learning Objectives have been added for each chapter, and the Problem Sets have been thoroughly revised and updated to match the new content in the book.

ANCILLARIES TO ACCOMPANY THE TEXT

An Instructor's Solutions Manual is provided with each problem completely solved. All figures and tables in the text are provided as PowerPoint slides. Also LectureBuilder PowerPoint slides are provided for all equations and examples so that instructors may easily and quickly build their own lectures. A digital version of the ISM and both sets of PPT slides are available for instructors through registration at www.cengagebrain.com

MINDTAP ONLINE COURSE AND READER

In addition to the print version, this textbook will also be available online through Mind-Tap, a personalized learning program. Students who purchase the MindTap version will have access to the book's MindTap Reader and will be able to complete homework and assessment material online, through their desktop, laptop, or iPad. If your class is using a Learning Management System (such as Blackboard, Moodle, or Angel) for tracking course content, assignments, and grading, you can seamlessly access the MindTap suite of content and assessments for this course.

In MindTap, instructors can:

- Personalize the Learning Path to match the course syllabus by rearranging content, hiding sections, or appending original material to the textbook content
- Connect a Learning Management System portal to the online course and Reader
- Customize online assessments and assignments
- Track student progress and comprehension with the Progress app
- Promote student engagement through interactivity and exercises

Additionally, students can listen to the text through ReadSpeaker, take notes and highlight content for easy reference, and check their understanding of the material.

ACKNOWLEDGMENTS

The success of our textbook has been a source of great satisfaction, because we believe that it has contributed to the better understanding of highway transportation in all its dimensions. We wish to thank our colleagues and their students for selecting this book for use in transportation courses taught in colleges and universities throughout the United States and abroad and for the many suggestions received during the preparation of all five editions.

The authors are indebted to many individuals who assisted in reviewing various chapters and drafts of the original manuscript and succeeding editions. We especially wish to thank the following individuals for their helpful comments and suggestions:

Maher Alghazzawi, Edward Beimborn, Rakim Benekohal, David Boyce, Stephen Brich, Chase Buchannan, Bernard Carlson, Christian Davis, Michael Demetsky, Brian Diefenderfer, Stacey Diefenderfer, Conrad Dudek, Lily Elefteriadou, Thomas Freeman, Ron Gallagher, Per Garder, Alan Gesford, Richard Gunther, Jiwan Gupta, Jerome Hall, Kathleen Hancock, Marvin Hilton, Jotin Khisty, Lydia Kostyniak, Michael Kyte, Feng-Bor Lin, Qun Liu, Yuan Lu, Tom Maze, Catherine McGhee, Kenneth McGhee, Richard McGinnis, Carl Monismith, Thomas Nelson, Ken O'Connell, Jack Page, Emelinda Parentela, Brian Park, Mofreh Saleh, Mitsuru Saito, Anthony Saka, Gerald Seeley, Robert Smith, Hamid Soleymani, James Stoner, Ed Sullivan, James Taylor, Egons Tons, Erol Tutumluer, Joseph Vidunas, Joseph Wattleworth, Peter Weiss, W James Wilde, F. Andrew Wolfe, Hugh Woo, Lewis Woodsen, Robert Wortman, Shaw Yu, Yihua Ziong and Michael Zmuda.

In the preparation of the Fifth Edition and earlier editions as well, we received reviews, comments and suggestions on individual chapters from several of our colleagues who have special expertise in the topics covered. We are most grateful for their willingness to devote this effort and for their help in validating and augmenting these chapters. They are: Richard Boaz, Michael Fontaine, Arkopal Goswami, Winston Lung, John Miller, Adel Sadek and Rod Turochy.

We also received a significant number of helpful comments from the reviewers of the fourth edition. We wish to thank them for their insightful comments and helpful suggestions many of which have been incorporated into this book. They are: Montasir Abbas, Virginia Tech, Mashrur Chowdhury, Clemson University, Shauna Hallmark, Iowa State University, David S. Hurwitz, Oregon State University, Wesley Marshall, University of

Colorado, Sam Owusu-Ababio, University of Wisconsin-Platteville, Kevan Shafizadeh, California State University, Sacramento, Anuj Sharma, University of Nebraska, Lincoln, Edward Smaglik, Northern Arizona University, Claudia Mara Dias Wilson, New Mexico Institute of Mining and Technology.

The many organizations cited herein that permitted us to include material from their publications deserve special mention because, without their support, our book would not have been a reality. We also wish to thank our editor Hilda Gowans, for her help and guidance in the preparation of this edition, and Rose Kernan of RPK Editorial Services for her Production skills.

Nicholas J. Garber
Lester A. Hoel

Preface to the SI Edition

This edition of **Traffic and Highway Engineering** has been adapted to incorporate the International System of Units (*Le Systeme Internationale d'Unites* or SI) throughout the book, wherever possible.

LE SYSTEME INTERNATIONAL D'UNITES

The United State Customary System (USCS) of units uses FPS (foot-pound-second) units (also called English or Imperial units). SI units are primarily the units of the MKS (meter-kilogram-second) system. However, CGS (centimeter-gram-second) units are often accepted as SI units, especially in textbooks.

USING SI UNITS IN THIS BOOK

In this book, we have used both MKS and CGS units. USCS units or FPS units used in the US Edition of the book have been converted to SI units throughout the text and problems, wherever possible. However, in the case of data sourced from handbooks, government standards, and product manuals, it is not only extremely difficult to convert all values to SI, it also encroaches on the intellectual property of the source. Some data in figures, tables, example, and references, therefore, remain in FPS units.

To solve problems that require the use of sourced data, the sourced values can be converted from FPS units to SI units before they are to be used in a calculation. To obtain standardized quantities and manufacturers' data in SI units, or country-specific codes and regulations, readers may need to contact the appropriate government agencies or authorities in their countries/regions.

INSTRUCTOR RESOURCES

A Printed Instructor's Solutions Manual in SI units is available on request. A digital version of the ISM, and PowerPoint slides of both figures, tables and images and examples and equations from the SI text are available for instructors registering on the book website.

Feedback from users of this SI Edition will be greatly appreciated and will help us improve subsequent editions.

The Publishers

About the Authors

Nicholas J. Garber is the Henry L. Kinnier Emeritus Professor of Civil Engineering at the University of Virginia and served as chairman of the Department from 1996 to 2002. Before joining the University of Virginia, Dr. Garber was Professor of Civil Engineering in the faculty of Engineering of the University of Sierra Leone, where he was also the Dean of the faculty of Engineering. He taught at the State University of New York at Buffalo, where he played an important role in the development of the graduate program in Transportation Engineering. Dr. Garber worked as a Design Engineer for consulting engineering firms in London between 1961 and 1964 and as an Area Engineer and Assistant Resident Engineer in Sierra Leone between 1964 and 1967.

Dr. Garber received the degree of Bachelor of Science (B.S.) in Civil Engineering from the University of London and the Masters (M.S.) and Doctoral (Ph.D.) degrees from Carnegie-Mellon University.

Dr. Garber's research is in the areas of Traffic Operations and Highway Safety. He has been the principal investigator for many federal-, state-, and private-agency-sponsored research projects. He is the author of over 120 refereed publications and reports. He is a co-author of the textbook *Transportation Infrastructure Engineering: A Multi-Modal Integration,* Thomson/Nelson, 2007.

Dr. Garber served as a member of the Executive Committee of the Transportation Research Board (TRB) and served for many years as chair of the TRB Committee on Traffic Safety in Maintenance and Construction Operations, currently the Committee on Work Zone Traffic Control. He has served as a member of several TRB Policy Studies on speed management, size and weight of large trucks, transportation of hazardous materials, and research priorities and coordination in highway infrastructure and operations safety. He also served as a member of the TRB Oversight Committee for the Strategic Highway Research Program II (SHRP II). Dr. Garber also has served as a member of several other national committees of the American Society of Civil Engineers (ASCE) and The Institute of Transportation Engineers (ITE). He also served as a member of the Editorial Board of the ASCE *Journal of Transportation Engineering*.

Dr. Garber is a member of the National Academy of Engineering. He is a recipient of many awards, including the TRB D. Grant Mickle Award, the ITE Edmund R. Ricker Transportation Safety Award, and the American Roads and Transportation Builders (ARTBA) S. S. Steinberg Outstanding Educator Award. He is listed in *Who's Who* in Science and Engineering and *Who's Who* in the world.

Dr. Garber is a Distinguished member of the American Society of Civil Engineers, a Fellow of the Institute of Transportation Engineers, a Fellow of the Institution of Civil

Engineers of the United Kingdom, a member of the American Society for Engineering Education, and a member of Chi Epsilon.

Lester A. Hoel is the L. A. Lacy Distinguished Professor of Engineering Emeritus, at the University of Virginia. He held the Hamilton Professorship in Civil Engineering from 1974 to 1999. From 1974 to 1989 he was Chairman of the Department of Civil Engineering and from 2002 to 2009 was Director of the Center for Transportation Studies. Previously, he was Professor of Civil Engineering and Associate Director, Transportation Research Institute at Carnegie-Mellon University. He has been a registered professional engineer in California, Pennsylvania, and Virginia. His degrees are: BCE from the City College of New York, MCE from the Polytechnic Institute of New York, and the Doctorate in Engineering from the University of California at Berkeley.

Dr. Hoel's area of expertise is the management, planning, and design of surface transportation infrastructure with emphasis on highway and transit systems. He is an author of over 150 publications and was co-editor (with G.E. Gray) of the textbook *Public Transportation,* and co-author (with N.J. Garber and A.W. Sadek) of the textbook *Transportation Infrastructure Engineering: A Multi-Modal* Integration and Lead Editor of the Textbook, Intermodal Transportation: Moving Freight in A Global Economy.

Dr. Hoel is a member of the National Academy of Engineering, a Distinguished Member of the American Society of Civil Engineers, a Fellow of the Institute of Transportation Engineers, a member of the American Society for Engineering Education and the Norwegian Academy of Technical Sciences. As a student, he was elected to the national honor societies Chi Epsilon, Tau Beta Pi, and Sigma Xi. He was a member of the Executive Committee of the Transportation Research Board (TRB) from 1981 to 1989 and from 1995 to 2004 and served as its Chairman in 1986. He was an ex-officio member of the National Research Council (NRC) Governing Board of the National Academies and the Transportation Research Board Division Chairman for NRC Oversight from 1995 to 2004. In that capacity, he was responsible for oversight of the NRC review process for all TRB policy studies produced during that period. He served as the Chairman of two congressionally mandated policy studies. He also has served on TRB technical committees and NCHRP/TCRP panels. He was a member of the TRB Transit Research Analysis Committee, whose purpose is to advise the Federal Transit Administration on its research program, and was a member of the National Research Council Report Review Committee, in which he oversees the review process for policy studies prepared by the National Research Council of the National Academies.

He is a recipient of the American Society of Civil Engineers' Huber Research Prize, the Transportation Research Board Pyke Johnson Award, the Highway Users Federation Stanley Gustafson Leadership Award, the TRB W.N. Carey, Jr. Distinguished Service Award, the ASCE Frank Masters Transportation Engineering Award, the ASCE James Laurie Prize, the Virginia Society of Professional Engineers Service Award, the Institute of Transportation Engineers' Wilbur Smith Distinguished Educator Award, the American Road and Transportation Builders S. S. Steinberg Outstanding Educator Award, and the Council of University Transportation Centers Distinguished Professor Award. He is listed in *Who's Who in America* and *Who's Who in the World*. He resides in Saint Helena, California.

Dr. Hoel has served as president of the Council of University Transportation Centers and on the ASCE accreditation board for engineering and technology. He was chairman of the Board of Regents of the Eno Transportation Foundation Leadership Center and served on its Board of Advisors. He is Senior Editor of the *Journal of Transportation of the Institute of Transportation Engineers* and has served on the editorial boards of transportation journals, including *Transportation Research, Journal of Advanced Transportation, Journal of Socio-Economic Planning Sciences, Journal of Specialized Transportation, Computer-Aided Civil and Infrastructure Engineering,* and *Urban Resources.*

P A R T   1

Introduction

Transportation is essential for a nation's development and growth. In both the public and private sector, opportunities for engineering careers in transportation are exciting and rewarding. Elements are constantly being added to the world's highway, rail, airport, and mass transit systems, and new techniques are being applied for operating and maintaining the systems safely and economically. Many organizations and agencies exist to plan, design, build, operate, and maintain the nation's transportation system.

CHAPTER 1

The Profession of Transportation

For as long as the human race has existed, transportation has played a significant role by facilitating trade, commerce, conquest, and social interaction while consuming a considerable portion of time and resources. The primary need for transportation has been economic, involving personal travel in search of food or work and travel for the exchange of goods and commodities; in addition, travel has been spurred by exploration, a quest for personal fulfillment, and the desire to improve a society or a nation. The movement of people and goods, which is the basis of transportation, always has been undertaken to accomplish these basic objectives or tasks, which require transfer from one location to another. For example, a farmer must transport produce to market, a doctor must see a patient in the office or in the hospital, and a salesperson must visit clients located throughout a territory. Every day, millions of people leave their homes and travel to a workplace—be it a factory, office, classroom, or distant city.

CHAPTER OBJECTIVES:

- Explain the importance of transportation in a modern and developed society.
- Become familiar with the critical issues in transportation.
- Understand how transportation technology has evolved over time.
- Discuss the principal technical areas and employment opportunities in transportation and highway engineering.
- Identify and discuss the challenges faced by transportation engineers in the twenty-first century.

1.1 IMPORTANCE OF TRANSPORTATION

Tapping natural resources and markets and maintaining a competitive edge over other regions and nations are linked closely to the quality of the transportation system. The speed, cost, and capacity of available transportation have a significant impact on

3

the economic vitality of an area and the ability to make maximum use of its natural resources. Examination of most developed and industrialized societies indicates that they have been noted for high-quality transportation systems and services. Nations with well-developed maritime systems (such as the British Empire in the 1900s) once ruled vast colonies located around the globe. In more modern times, countries with advanced transportation systems—such as the United States, Canada, and countries in Asia and Europe—are leaders in industry and commerce. Without the ability to transport manufactured goods and raw materials and without technical know-how, a country is unable to maximize the comparative advantage it may have in the form of natural or human resources. Countries that lack an abundance of natural resources rely heavily on transportation in order to import raw materials and export manufactured products.

1.1.1 Transportation and Economic Growth

Good transportation, in and of itself, will not assure success in the marketplace, as the availability of transportation is a necessary but insufficient condition for economic growth. However, the absence of supportive transportation services will serve to limit or hinder the potential for a nation or region to achieve its economic potential. Thus, if a society expects to develop and grow, it must have a strong internal transportation system consisting of good roads, rail systems, as well as excellent linkages to the rest of the world by sea and air. Transportation demand is a byproduct derived from the needs and desires of people to travel or to transfer their goods from one place to another. Transportation is a necessary condition for human interaction and economic competitiveness.

The availability of transportation facilities can strongly influence the growth and development of a region or nation. Good transportation permits the specialization of industry or commerce, reduces costs for raw materials or manufactured goods, and increases competition between regions, thus resulting in reduced prices and greater choices for the consumer. Transportation is also a necessary element of government services, such as delivering mail, providing national defense, and assisting U.S. territories. Throughout history, transportation systems (such as those that existed in the Roman Empire and those that now exist in the United States) were developed and built to ensure economic development and efficient mobilization in the event of national emergencies.

1.1.2 Social Costs and Benefits of Transportation

The improvement of a region's economic position by virtue of improved transportation does not come without costs. Building vast transportation systems requires enormous resources of energy, material, and land. In major cities, transportation can consume as much as half of all the land area. An aerial view of any major metropolis will reveal vast acreage used for railroad terminals, airports, parking lots, and freeways. Transportation has other negative effects as well. Travel is not without danger; every mode of transportation brings to mind some major disaster—be it the sinking of the *Titanic*, the explosion of the zeppelin *Hindenburg*, the infrequent but dramatic passenger air crashes, and frequent highway crashes. In addition, transportation can create noise, spoil the natural beauty of an area, change the environment, pollute air and water, and consume energy resources.

Society has indicated a willingness to accept some risk and changes to the natural environment in order to gain the advantages that result from constructing new transportation systems. Society also values many social benefits brought about by good transportation. Providing medical and other services to rural areas and enabling people to socialize who live some distance apart are just two examples of the benefits that transportation provides.

A major task for the modern transportation engineer is to balance society's need for reasonably safe and efficient transportation with the costs involved. Thus, the most efficient and cost-effective system is created while assuring that the environment is not compromised or destroyed. In carrying out this task, the transportation engineer must work closely with the public and elected officials and needs to be aware of modern engineering practices to ensure that the highest quality transportation systems are built consistent with available funds and accepted social policy.

1.1.3 Transportation in the United States

Is transportation very important? Why should you study the subject and perhaps consider transportation as a professional career? Many "gee whiz" statistics can be cited to convince the reader that transportation is vital to a nation, but before we do so, consider how transportation impacts people's daily lives.

Perusal of a local or national newspaper will inevitably produce one or more articles on transportation. The story might involve a traffic fatality, a road construction project, the price of gasoline, trends in purchases of motor vehicles, traffic enforcement and road conditions, new laws (such as cell phone use while driving), motor vehicle license requirements, neighborhood protests regarding road widening or extensions, proposals to increase road user fees or gasoline taxes to pay for maintenance and construction projects, the need for public transit services, or the debate over "sprawl" versus "smart growth." The enormity of transportation can be demonstrated by calculating the amount of land consumed for transportation facilities, such as sidewalks, parking lots, roads, driveways, shoulders, and bike paths, which in some cases can exceed 50 percent of the land area.

The examples cited suggest that transportation issues are largely perceived at local and state levels where people live. Mayors and governors are elected based on their promises to improve transportation without raising taxes. At the national level, transportation does not reach the "top 10" concerns, and transportation is rarely mentioned in a presidential address or national debate. At this level, issues of defense, health care, immigration, voting rights, taxes, and international relations take center stage. While many Americans probably know the name of the Secretary of State or Defense, few could answer the question, "Who is the Secretary of Transportation?"

The Executive Committee of the Transportation Research Board of the National Academies periodically develops a list of "critical issues" in transportation, which are posted on the committee's Web site. Among the issues identified are the following: (1) congestion; (2) emergency preparedness, response, and mitigation, such as vulnerability to terrorist strikes and natural disasters; (3) energy, environment, and climate change; (4) finance and equity; (5) safety; (6) twentieth-century institutions mismatched to twenty-first-century missions; and (7) human and intellectual capital as reflected in the inadequate investment in innovation. Each issue suggests the importance of transportation and the priorities of concern to the transportation professional community.

The importance of transportation in the United States can also be illustrated by citing statistics that demonstrate its national and worldwide influence. For example, data furnished by agencies such as the Bureau of Transportation Statistics of the U.S. Department of Transportation, the Federal Highway Administration, and the U.S. Bureau of Labor Statistics provide major indices, such as the following: approximately 18 percent of U.S. household expenditure is related to transportation. Regarding energy consumption, transportation accounts for about 28 to 29 percent of total energy consumption in the United States, of which approximately 95 percent of the energy utilized for propelling transport vehicles is derived from petroleum products. The extent of U.S. travel is summarized as by the almost 90 percent of the driving age population in the United States that possess a license to operate a motor vehicle and the 19,000 – 24,000 km per person travelled annually in the United States during the past decade. Transportation infrastructure is vast; for example, in the United States there are almost 6.43 million km of paved roadway, of which 1.2 million km are used for intercity travel and 75,000 km are interstate highways. In addition, there are approximately 225,300 km of freight railroads, 5300 public use airports, 41,850 km of navigable channels, and 580,000 km of oil and gas pipelines. These statistics demonstrate that transportation will continue to play a key role in the economy even as it shifts from manufacturing to a focus on services, which is the largest and fastest growing sector in the U.S. economy.

1.2 TRANSPORTATION HISTORY

The story of transportation in the United States has been the subject of many books; the story covers a 300-year period and includes the development of many modes of transportation. Among the principal topics are travel by foot and horseback, automobile and truck travel, development of roads and highways, the building of canals and inland waterways, expansion of the West, construction of railroads, the use of public transportation (such as bus and metro systems in cities), and the development of air transportation. A summary of the historical highlights of transportation development is shown in Table 1.1.

1.2.1 An Overview of U.S. Transportation History

In its formative years, the United States was primarily rural, with a population of about 4 million in the late 1700s. Only about 200,000 people, or 5 percent of the population, lived in cities; the remainder inhabited rural areas and small communities. That pattern remained until the early 1900s. During the twentieth century, the urban population continued to increase such that at present over 75 percent of the U.S. population lives in urban or suburban areas. Large cities have been declining in population, and increases have occurred in suburban and rural areas. These changes have a significant impact on the need for highway transportation.

Early Road Building and Planning

During the eighteenth century, travel was by horseback or in animal-drawn vehicles on dirt roads. As the nation expanded westward, roads were built to accommodate the settlers. In 1794, the Lancaster Turnpike, the first toll road, was built to connect the Pennsylvania cities of Lancaster and Philadelphia. The nineteenth century brought further expansion of U.S. territorial boundaries, and the population increased from 3 million to 76 million. Transportation continued to expand with the nation. The remainder of the nineteenth century saw considerable activity, particularly in canal and railroad building.

The Canal Boom

An era of canal construction began in the 1820s when the Erie Canal was completed in 1825 and other inland waterways were constructed. Beginning in the 1830s, this efficient means

Table 1.1   Significant Events in Transportation History

1794: First toll road, the Lancaster Turnpike, is completed.

1807: Robert Fulton demonstrates a steamboat on the Hudson River. Within several years, steamboats are operating along the East Coast, on the Great Lakes, and on many major rivers.

1808: Secretary of Treasury Albert Gallatin recommends a federal transportation plan to Congress, but it is not adopted.

1825: Erie Canal is completed.

1830: Operations begin on Baltimore and Ohio Railroad, first railroad constructed for general transportation purposes.

1838: Steamship service on the Atlantic Ocean begins.

1857: First passenger elevator in the United States begins operation, presaging high-density urban development.

1865: First successful petroleum pipeline is laid between a producing field and a railroad terminal point in western Pennsylvania.

1866: Bicycles are introduced in the United States.

1869: Completion of first transcontinental railroad.

1887: First daily railroad service from coast to coast.

1888: Frank Sprague introduces the first regular electric streetcar service in Richmond, Va.

1903: The Wright brothers fly first airplane 37 m at Kitty Hawk, N.C.

1914: Panama Canal opens for traffic.

1915–18: Inland waters and U.S. merchant fleet play prominent roles in World War I freight movement.

1916: Interurban electric-rail mileage reaches a peak of 25,000 km.

1919: U.S. Navy and Coast Guard crew cross the Atlantic in a flying boat.

1927: Charles Lindbergh flies solo from New York to Paris.

1956: Construction of the 68,400-km Interstate and Defense Highway System begins.

1959: St. Lawrence Seaway is completed, opening the nation's fourth seacoast.

1961: Manned spaceflight begins.

1967: U.S. Department of Transportation is established.

1969: Men land on moon and return.

1972: San Francisco's Bay Area Rapid Transit System is completed.

1981: Space shuttle *Columbia* orbits and lands safely.

1991: The Interstate highway system is essentially complete.

1992: Intelligent transportation systems (ITS) usher in a new era of research and development in transportation.

1995: A 259,100-km National Highway System (NHS) is approved.

1998: Electric vehicles are introduced as an alternative to internal combustion engines.

2000: A new millennium ushers in a transportation–information technology revolution.

2005: Energy-efficient autos as hybrid vehicles gain in popularity and ethanol production increases.

2011: Global warming and climate change become an emerging issue in planning, design, and emergency preparedness aspects of highway transportation.

---

of transporting goods was replaced by the railroads, which were being developed at the same time. By 1840, the number of kilometers of canals and railroads was approximately equal (5150 km), but railroads, which could be constructed almost anywhere in this vast, undeveloped land at a much lower cost, superseded canals as a form of intercity transportation. Thus, after a short-lived period of intense activity, the era of canal construction came to an end.

The Railroad Era

The railroad was the emerging mode of transportation during the second half of the nineteenth century, as railway lines were spanning the entire continent. Railroads dominated intercity passenger and freight transportation from the late 1800s to the early 1920s. Railroad passenger transportation enjoyed a resurgence during World War II but has steadily declined since then, owing to the competitiveness of the automobile. Freight rail was consolidated and remains viable. Railroad mileage reached its peak of about 426 500 km by 1915.

Transportation in Cities

Each decade has seen continuous population growth within cities, and with it, the demand for improvements in urban transportation systems. Urban transportation began with horse-drawn carriages on city streets; these later traveled on steel tracks. They were succeeded by cable cars, electric streetcars, underground electrified railroads, and bus transportation. City travel by public transit has been replaced largely by the use of automobiles on urban highways, although rail rapid transit and light rail systems have been built in many large and medium-sized cities since the 1970s.

The Automobile and Interstate Highways

The invention and development of the automobile created a revolution in transportation in the United States during the twentieth century. No facet of American life has been untouched by this invention; the automobile (together with the airplane) has changed the way we travel within and between cities. Only four automobiles were produced in 1895. By 1901, there were 8000 registered vehicles, and by 1910, there were over 450,000 cars and trucks. Between 1900 and 1910, 89,500 km of surfaced roads were constructed, but major highway-building programs did not begin in earnest until the late 1920s. By 1920, more people traveled by private automobile than by rail transportation. By 1930, 23 million passenger cars and 3 million trucks were registered. In 1956, Congress authorized a 68,400 km interstate highway network, which is now completed.

The Birth of Aviation

Aviation was in its infancy at the beginning of the twentieth century with the Wright brothers' first flight, which took place in 1903. Both World Wars I and II were catalysts in the development of air transportation. The carrying of mail by air provided a reason for government support of this new industry. Commercial airline passenger service began to grow, and by the mid-1930s, coast-to-coast service was available. After World War II, the expansion of air transportation was phenomenal. The technological breakthroughs that developed during the war (coupled with the training of pilots) created a new industry that replaced both ocean-going steamships and passenger railroads.

1.2.2 Evolution of America's Highways

To commemorate the 200th anniversary of the signing of the Declaration of Independence in 1776, the Federal Highway Administration published a landmark commemorative volume titled *America's Highways* that described the evolution of the federal government's involvement in roads, which culminated with the establishment of the U.S. Bureau of Public Roads and its successor, the Federal Highway Administration. The book follows the major milestones in highway transportation, beginning with the colonial period and early settlement, when roads were unpaved and nearly impassable, with few bridges to span streams and rivers, and horse paths were unsuited for wheeled vehicles. It concludes with the growth of motor vehicle transportation in the twentieth century and its impact on highway transportation. The following sections summarize this evolutionary journey.

Turnpikes and Canals

In the eighteenth and nineteenth centuries, surface transportation improvements were focused on improving both roads and inland waterways, as together they comprised the internal network of transportation for a new nation led by its first president, George Washington, who had been elected in 1789. Federal interest and support for "internal improvements" was limited, as these functions were seen as the purview of states. The earliest attempt by a state government to develop a plan to build roads and canals was in Pennsylvania, when the legislature authorized private companies to build and maintain roads and canals and collect tolls that would cover costs and yield a profit for its investors—a practice still prevalent in the nineteenth and early twentieth centuries. (This method of financing was rejected for the U.S. Interstate Highway System, but recent trends have been moving toward this earlier financing model as states are turning to the private sector for "partners" to own, build, and maintain state highways.)

In 1791, the Philadelphia and Lancaster Turnpike Road Company was formed, having been granted a charter to build a 100-km highway that would include a 6-m hard surface and a 15-m right-of-way with grades not to exceed 7 percent. The road, designed by the well-known Scottish road builder John Loudon McAdam, was completed in three years and served the travel needs of horse-drawn coaches and freight-carrying wagons. The road served as a model for similar toll roads constructed in East Coast states connecting cities and towns. The Lancaster "Pike" was so named because the toll gate was similar to a pivoted lancer's pike. It proved to be a success, yielding up to a 15 percent annual return on investment. It was later extended across the state to Pittsburgh.

Spurred on by the success of the Philadelphia and Lancaster Road, a "turnpike building frenzy" ensued with construction in Connecticut, New York, Maryland, and Virginia. By 1850, thousands of miles of turnpikes were in existence. Not many were as successful as the Lancaster Pike, and eventually there were failures due to low traffic demand and competition from canals and railroads. (Similar experiences were noted in the twentieth century, when many toll roads went bankrupt due to competition from free roads and other modes.)

The 1800s were a dark period for roads because other modes were dominant and vehicle technology had not changed since the time of the Roman Empire. Accordingly, animal and wind power continued to be the means of propulsion. Since the United States had an extensive system of rivers and lakes, it was logical that water navigation was a priority, and the building of canals would be a natural enhancement. Then in 1830, the "iron horse" appeared on the scene, and for the next 100 years, the railroad would dominate. Railroads initially appeared in Europe and were horse drawn. They, too, were regarded as "public highways" and had little to offer other than serving as short extensions from quarries to rivers as roads and canals were already in place.

The most extensive and successful of all canal projects was the Erie Canal, a 590 km connection between the Hudson River in New York and Lake Erie, Pennsylvania. Construction began in 1817 and was completed in 1825. It had a trapezoidal cross section 12.5 m wide at the top and 8.5 m at the bottom, and it had a uniform depth of 1.2 m. The canal ascended and descended a height of 206 m through 83 separate locks. Eighteen aqueducts spanned rivers, and numerous bridges connected roads on opposite sides of the canal. Since the profession of civil engineering had yet to be established, this project became known as the first school of civil engineering in the United States. The canal was profitable, which convinced other states to undertake similar projects. However, most were not "money-makers" for their investors, and eventually canals became largely extinct.

A National Plan and a National Road

In the eighteenth and nineteenth centuries, sources of revenue for highways and canals included tolls, state and local taxes, and donated labor, while the federal government played a small (albeit important) role. The first act of Congress to support road building occurred in 1796. This authorized Colonel Ebenezer Zane (1741–1811) to build a 515-km post road (called Zane's Trace) through the northwest territory (now Ohio) between Wheeling, Virginia (which became West Virginia in 1863) and Limestone (now Maysville), Kentucky. The road was primitive, following Native American trails, but it was to serve as a mail route and later was widened for wagon travel. It became part of the National Road in 1825. The federal government did not pay for the road but permitted Colonel Zane to purchase selected tracts of land where the road crossed three major rivers. Unfortunately, this small beginning of federal involvement in early road development was to have little influence on future events.

During the administration of President Thomas Jefferson (1801–1809), two events of major significance occurred that had an impact on road and canal building. The first was the completion of the Gallatin Report on internal improvements, and the second was the authorization of the Cumberland Road.

Secretary of the Treasury Albert Gallatin, at the request of the U.S. Senate, prepared the first national transportation inventory in 1807. The report, titled *Roads, Canals, Harbors and Rivers,* was submitted to Congress on April 4, 1808. The document contained a detailed program of "internal improvements" intended to increase the wealth of this undeveloped nation, as had occurred in France and England. The proposed 10-year program contained projects totaling $20 million and was to be financed by the federal government. This bold plan was fiercely debated in Congress but was not completed in time to be acted upon by President Jefferson. Rather, the bill reached the desk of President James Madison (1809–1817), who vetoed it on the grounds that direct federal support for internal improvements was unconstitutional as these matters were to be dealt with by the states. Gallatin earlier had proposed that the states use a portion of federal land sales for building roads, and some states did adopt this funding mechanism.

The Cumberland Road (later known as the National Road) is the first example of federal aid for a major road project in the United States. On March 29,1806, President Thomas Jefferson signed a bill authorizing the construction of a 200-km road from Cumberland, Maryland, to Wheeling, Virginia, on the Ohio River. Road construction began in 1811, and the project was completed in 1818. In 1820, Congress appropriated additional funds to extend the road to the banks of the Mississippi River. Appropriations continued until 1838, and construction ceased in 1840 at Vandalia, Illinois. The National Road, now about 1200 km in length, was poised to open the western territories for settlement. However, this was not to be, because the federal government ceded the road to those states through which it traversed, and soon after, railroads were constructed—further sealing the National Road's fate.

The Demise of Federal Support for Roads

Another blow for federal support for road building was struck by President Andrew Jackson (1829–1837), who vetoed a bill that would have allowed the Secretary of the Treasury to purchase $150,000 in shares to help build a 100-km turnpike from Maysville to Louisville in Kentucky. His veto was based on the continuing belief by U.S. presidents that since "internal improvements" were not specified in the Constitution as a federal responsibility, he could not sign the bill. Jackson's decision effectively ended attempts to secure federal funds for roads. The Maysville Turnpike was eventually completed with the support of state and private funds and was used as a mail route by the government.

In subsequent years, with the exception of the National Road, the responsibility for building toll roads fell to the states and private investors. Military roads were built during this period, most of them in the territories, and consisted primarily of clearings for wagon wheels. The total mileage of military roads was about 41,845, and they often served as the sole routes available to settlers moving westward. Following the Civil War (1860–1865), there was a reversal in federal policy that provided significant support to a new and emerging technology that would open the West to development and span the continent. The railroad era was about to begin.

Steamboats and Railroads

A "golden age" of transportation was to emerge in the nineteenth century, thanks to the genius of James Watt (1736–1819), a Scottish inventor and engineer who, with his partner Matthew Bolton (1728–1809), perfected and produced the steam engine. Steam engines originally were used to pump water from tin and copper mines and for spinning and weaving. Later, they were adapted to propel marine vessels and steam locomotives. The introduction of the first successful steamboat in 1807 is credited to Robert Fulton (1765–1815), who, with his partner Robert Livingston (1746–1813), used a 20-horsepower Watt and Bolton steam engine to propel a 40.5-m-long passenger vessel. The *Clermont* (Figure 1.1) left New York City for Albany, New York, on a 240-km journey up the Hudson River and arrived safely after 30 hours. This demonstration proved the viability of steamboat travel on rivers and lakes, and thus steamboats became instrumental in opening the West for settlement during the first half of the nineteenth century. By 1859, 2000 steamboats plied the Mississippi and its tributaries. The federal government subsidized inland waterway transportation, primarily

Figure 1.1   The *Clermont*—1807

SOURCE: Courtesy of the Library of Congress, LC-USZ62-110382 on the Ohio and Mississippi Rivers, the Great Lakes, and coastal ports. However, by 1850, it was increasingly clear that the railroad was the dominant mode compared to waterways and turnpikes because it was faster, cheaper, and more adaptable.

During the 10 years prior to the Civil War, railroad construction was widespread, with entrepreneurs and financiers seeking to gain fortunes while meeting a growing demand for rail connections between towns, villages, lakes, and seaports. To encourage railroad expansion westward, the federal government provided land grants to railroads totaling over 1.5 million ha. Rail lines were built without a system-wide plan, and the result was a plethora of unconnected short lines with varying track gauges. Later, many of these lines would form the basis for a system-wide network connecting major cities, all with a common track gauge of 1.435 m. At the time of the Civil War, two-thirds of all railroad mileage was in the Northern states—an advantage that proved increasingly significant as the war progressed. After hostilities ended, railroads expanded rapidly (Figure 1.2), paralleling rivers and canals and heading inland and westward. Fierce competition ensued between steam packet ships and the railroads in a brutal and unregulated environment

Figure 1.2 Workmen Repairing Railroad Track—1895

SOURCE: Jackson, William Henry, 1843–1942. World's Transportation Commission photograph collection, Library of Congress, LC-W7-637

leading to the demise of waterways for moving freight and the dominance of railroads for moving freight and passengers. Additional land grants totaling 14.8 million ha were given to 50 railroads to encourage expansion westward. Eventually, four transcontinental railroads were completed (the first in 1869), with all liberally subsidized by generous federal land grants. This frenzy of railroad construction during the last half of the nineteenth century had produced approximately 418,400 km of railroad track as the nation entered the twentieth century.

The Automobile and Resurgence of Highways

Highway transportation remained primitive and unchanged during the nineteenth century, as railroads dominated the landscape. Long-distance freight and stagecoach companies had been driven out of business, and toll road revenues continued to decline. Even though this "dark age" of roads seemed to be unending, over 2.4 million km of rural roads were built—most composed of natural soil or stones that could be muddy in rainy seasons and dusty in dry ones. Rural roads were paid for and maintained by local citizens through property taxes, land donations, and donated labor. In cities and towns, the transportation situation was considerably better, as streets were paved with granite blocks, and public transit was introduced by 1880. Electric or cable streetcars (Figure 1.3) were common by the turn of the century.

The introduction of bicycles in the United States occurred as early as 1817. Bicycle transportation did not become practical for the general public until the introduction of a "safety bicycle," which used two wheels of equal size and pneumatic tires. Bicycle riding became a popular pastime, and many "wheel clubs" were formed followed by a national organization called the League of American Wheelmen. This organization is still in existence as the League of American Bicyclists, and reflects a persistent and increasing demand for improved bicycle facilities. (Figure 1.4.)

To their dismay, the new bicycle owners were soon to discover that a ride into the country was nearly impossible to complete due to the poor quality of roads, many of which were rutted, uneven, and lacked bridge links over streams and rivers. Thus was formed the first "highway lobby" seeking to influence the building of "good roads." A Good Roads Association was formed in 1891 in Missouri, with similar organizations to follow in other states. Ironically, an ally in this movement were the railroads themselves, whose representatives believed that if roads were improved, access to rail stations would be easier, thus increasing their market. Rail cars were outfitted with exhibits to demonstrate the benefits of "good roads" and how they should be built. These trains traveled throughout the nation, stopping at cities, towns, and villages and convincing citizens and politicians alike that it was time for the nation to begin investing in roads. Thus, "good road" trains roamed the nation proclaiming the benefits of a transport mode that by the end of the twentieth century would contribute to rail travel's own demise.

The introduction of a successful and practical gasoline-powered vehicle was the result of inventions by Gottlieb Daimler in 1885 and Karl Benz in 1886 and sparked a fury of innovation that culminated in a vehicle design that could be mass produced. The Ford Model T transformed the automobile market from that of a "rich man's toy" to "everyman's transport." The Ford Motor Company, led by Henry Ford, began to mass-produce cars selling for $950, and production of this model (available in all colors as long as they were black) totaled 15.5 million by the time production ceased in 1927 (Figure 1.5). Not to be outdone, other manufacturers followed suit, and an orgy of auto building began such that by 1921 there were 10.5 million registered vehicles in the United States.

Figure 1.3   Cable Car in Tacoma, Washington—1906

SOURCE: Courtesy of the Library of Congress, LC-USZ6-173

The next 50 years would witness a transformation in highways, from largely unpaved rural roads to an impressive network of rural and urban highways, despite an economic depression (1929–1939) and World War II (1941–1945). However, along with the mobility offered by automobiles (and later trucks) came traffic congestion, traffic fatalities, and diminished environmental quality. In 1956, highway construction entered a new era with the authorization of a 68,400 km National Interstate Highway System, which when completed at the end of the twentieth century would total 76,900 km and change the way people lived and traveled. Thus, the highway revolution (which began with the

Figure 1.4   Bicycling on the Golden Gate Bridge

SOURCE: Moreno Novello/Shutterstock.com

invention of the internal combustion engine in 1885 and its mass production in 1908) coupled with the introduction of "heavier than air" flight in 1903 dominated travel and reduced the role of rail and water transportation.

Looking ahead, can we expect that things will remain as they have in the past or will history be the prologue for future changes in transportation?

1.3 TRANSPORTATION EMPLOYMENT

Employment opportunities in transportation exist in the United States as well as in many other countries throughout the world. The principal areas of this field are logistics and supply-chain management, vehicle design and manufacture, provision of services related to transportation, and planning, design, construction, operations, and management of the infrastructure required if vehicles are to function as intended.

1.3.1 Logistics and Supply-Chain Management

The physical-distribution aspect of transportation, known as business logistics or physical-distribution management, is concerned with the movement and storage of freight between the primary source of raw materials and the location of the finished manufactured product. Logistics is the process of planning, implementing, and controlling the efficient and effective flow and storage of goods, services, and related information from origination

Figure 1.5 Parked Automobiles—1920

SOURCE: Library of Congress, Prints & Photographs Division, Theodor Horydczak Collection, LC-H823-V01-004

to consumption as required by the customer. An expansion of the logistics concept is called supply-chain management: a process that coordinates the product, information, and cash flows to maximize consumption satisfaction and minimize organization costs.

1.3.2 Vehicle Design and Transportation Services

Vehicle design and manufacture is a major industry in the United States and involves the application of mechanical, electrical, and aerospace engineering skills as well as those of technically trained mechanics and workers in other trades.

The service sector provides jobs for vehicle drivers, maintenance people, flight attendants, train conductors, and other necessary support personnel. Other professionals, such as lawyers, economists, social scientists, and ecologists, also work in the transportation fields when their skills are required to draft legislation, to facilitate right-of-way acquisition, or to study and measure the impacts of transportation on the economy, society, and the environment.

1.3.3 Transportation Infrastructure Services

Although a transportation system requires many skills and provides a wide variety of job opportunities, the primary opportunities for civil engineers are in the area of transportation infrastructure. A transportation engineer is the professional who is concerned with the planning, design, construction, operations, and management of a transportation system (Figure 1.6). Transportation professionals must make critical decisions about the system that will affect the thousands of people who use it. The work depends on the results of experience and research and is challenging and ever changing as new needs emerge and new technologies replace those of the past. The challenge of the transportation engineering profession is to assist society in selecting the appropriate transportation system consistent with society's economic development, resources, and goals, and to construct and manage the system in a safe and efficient manner. It is the engineer's responsibility to ensure that the system functions efficiently from an economic point of view, and that it meets external requirements concerning energy, air quality, safety, congestion, noise, and land use.

1.3.4 Specialties within Transportation Infrastructure Engineering

Transportation engineers are typically employed by the agency responsible for building and maintaining a transportation system, including federal, state, or local government, railroads, or transit authorities. They also work for consulting firms that help carry out the planning and engineering tasks for these organizations. During the past century, transportation engineers have been employed to build the nation's railroads, the interstate highway system, and rapid transit systems in major cities, airports, and turnpikes. Each decade has seen a new national need for improved transportation services.

In the twenty-first century, there will be increased emphasis on rehabilitating the highway system, including its surfaces and bridges, as well as on devising a means to ensure improved safety and utilization of the existing system through traffic control, information technology, and systems management. Highway construction will be required, particularly in suburban areas. Building of roads, highways, airports, and transit systems is likely to accelerate in less-developed countries, and the transportation engineer will be called on to furnish the services necessary to plan, design, build, and operate highway systems throughout the world. Each of the specialties within the transportation infrastructure engineering field is described next.

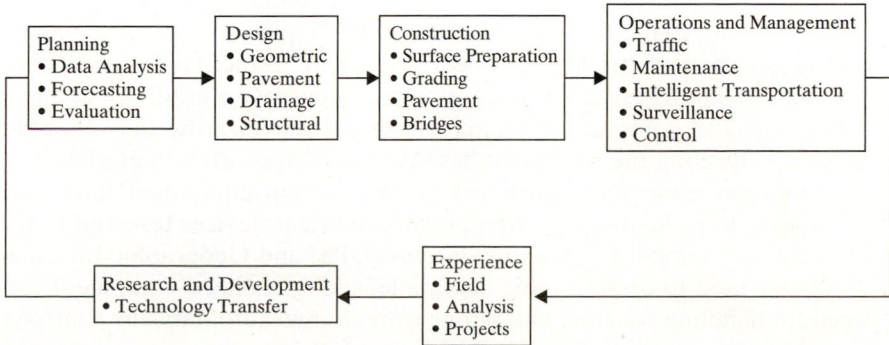

Figure 1.6 The Profession of Transportation Engineering

Transportation Planning

Transportation planning deals with the selection of projects for design and construction. The transportation planner begins by defining the problem, gathering and analyzing data, and evaluating various alternative solutions. Also involved in the process are forecasts of future traffic; estimates of the impact of the facility on land use, the environment, and the community; and determination of the benefits and costs that will result if the project is built. The transportation planner investigates the physical feasibility of a project and makes comparisons among various alternatives to determine which one will accomplish the task at the lowest cost—consistent with other criteria and constraints.

A transportation planner must be familiar with engineering economics and other means of evaluating alternative systems; be knowledgeable in statistics and data-gathering techniques, as well as in computer applications for data analysis and travel forecasting; and be able to communicate with the public and policy makers.

Transportation Infrastructure Design

Transportation design involves the specification of all features of the transportation system to assure that it will function smoothly, efficiently, and in accord with physical laws of nature. The design process results in a set of detailed plans that can be used for estimating the facility costs and for carrying out its construction. For highway design, the process involves the selection of dimensions for all geometrical features, such as the longitudinal profile, vertical curves and elevations, the highway cross section, pavement widths, shoulders, rights-of-way, drainage ditches, and fencing. The design processes also include the pavement and structural requirements for base courses and the concrete or asphalt surface material. Highway design also includes bridges and drainage structures as well as provision for traffic control devices, roadside rest areas, and landscaping. The highway designer must be proficient in civil engineering subjects (such as soil mechanics, hydraulics, land surveying, pavement design, and structural design) and is concerned primarily with the geometric layout of the road, its cross section, paving materials, roadway thickness, and traffic-control devices. Special appurtenances (such as highway bridges and drainage structures) are usually designed by specialists in these areas.

Highway Construction

Highway construction involves all aspects of the building process, beginning with clearing of the native soil, preparation of the surface, placement of the pavement material, and preparation of the final roadway for use by traffic. Highways initially were built with manual labor assisted by horse-drawn equipment for grading and moving materials. Today, modern construction equipment is used for clearing the site, grading the subgrade, compacting the pavement base courses, transporting materials, and placing the final highway pavement. Advances in construction equipment have made the rapid building of large highway sections possible. Nuclear devices test compaction of soil and base courses, Global Positioning Systems (GPS) and Geographic Information Systems (GIS) are used to establish line and grade, and specialized equipment has been developed for handling concrete and bridge work. Large, automatically controlled mix plants have been constructed, and new techniques for improving durability of structures and for substituting of scarce materials have been developed.

Traffic Operations and Management

The operation of the nation's highway system is the responsibility of the traffic engineer. Traffic engineering involves the integration of vehicle, driver, bicyclist, and pedestrian characteristics to improve the safety and capacity of streets and highways. All aspects of the transportation system are included after the street or highway has been constructed and opened for operation. Among the elements of concern are traffic accident analyses, parking and loading, design of terminal facilities, traffic signs, markings, signals, speed regulation, and highway lighting. The traffic engineer works to improve traffic flow and safety, using engineering methods and information technology to make decisions that are supported by enforcement and education. Traffic engineers work directly for municipalities, county governments, and private consulting firms.

Maintenance Operations and Management

Highway maintenance involves all the work necessary to ensure that the highway system is kept in proper working order. Maintenance includes patching, repair, and other actions necessary to ensure that the roadway pavement is at a desired level of serviceability. Maintenance management involves recordkeeping and data analysis regarding work activities, project needs, and maintenance activities to assure that the project is carried out in the most efficient and economical manner. Scheduling work crews, replacing worn or damaged signs, and repairing damaged roadway sections are important elements of maintenance management. The work of the civil engineer in the area of maintenance involves the redesign of existing highway sections, economic evaluation of maintenance programs, testing of new products, and scheduling of personnel to minimize delay and cost. The maintenance engineer must also maintain an inventory of traffic signs and markings and ensure that they are in good condition.

1.3.5 Professional Challenges in Transportation Engineering

What will be the challenges for the transportation engineer whose career can be expected to encompass the better part of the twenty-first century? How can these challenges be addressed, and what are the necessary attributes and skills? The answers to these questions have many facets and will require continued renewal of knowledge and experience through life-long learning and participation in professional society activities. Since transportation is a complex, multidimensional, and interactive system, the transportation engineer will need an arsenal of resources to respond to the many situations that can be expected. This section identifies some of the likely major challenges and suggests the kinds of skills and abilities that should prove valuable.

The principal challenge will be to meet the expectation of the public that transportation will be efficient, effective, long lasting, and safe. Meeting this expectation is no small feat and requires extensive knowledge and experience regarding human behavior, systems performance, and technology. Transportation systems are not produced on an assembly line, and they cannot be easily discarded for something better. When introduced into service, careful integration within an existing environment is required. Transportation projects are unique; each project is "one of a kind" and require many years to complete—for example, 50 years were devoted to construction of the Interstate Highway System. A typical highway project requires 5 to 20 years from start to finish.

Transportation engineers are required to possess a long-term vision of the future. They must remain steadfast, patient, and persistent in guiding a transportation project to completion. The transportation engineer works in an environment where change is gradual and sometimes imperceptible. To illustrate using an example from history, a major milestone in transportation occurred when the Wright brothers assembled a heavier-than-air machine in 1903 and demonstrated that it could fly under its own power. Almost a quarter of a century transpired before this "flying machine" transported a single person across the Atlantic from New York to Paris.

A related challenge for transportation engineers is to understand how innovation and new technology can be used to transport people and goods in new and different ways. Were it not for the inventive spirit of transportation pioneers, such as Robert Fulton in 1807 and Henry Ford in 1903, today's modern transportation systems would not exist. Yet another innovation with profound consequences for public transportation was the use of iron wheels on iron rails. Prior to this innovation, the "old way" relied on wheels in contact with the pavement, which was usually uneven, or more likely, nonexistent. With this new innovation—the street railway—travel became faster, smoother, and more comfortable—with less energy required to pull the trams, first by horses, then by steam engines and cables, and ultimately by electric motors. The new idea had successfully competed with the "old way" and beaten it at its own game.

Technological innovation can be expected to accelerate in the twenty-first century. Thus, the challenge for the transportation engineer will be to distinguish between technology with solutions that are looking for a problem and solutions that successfully compete with the old way. To illustrate, monorail transit has been promoted as a futuristic answer to urban congestion. The results have proven otherwise, and consequently, monorail transit remains a perennial answer to tomorrow's transportation problems. In contrast, when railroads appeared in the 1830s, canals were the dominant mode. Many of the early railroads were built parallel to canal towpaths and soon demonstrated their superiority. Modern examples of two emerging technologies are automated highways and automated toll collection. Automated highways with driverless vehicles have been proposed as a way to increase speed and capacity while increasing safety. Automated toll collection has been successfully introduced as a substitute for manual collection.

Another important challenge for transportation engineers is to be a good steward of people's investment in transportation. Doing so requires attention to designs that are appropriate, cost effective, and economical. Attention to maintenance is essential. Civil engineering is the professional discipline that seeks to harness nature in a cost-effective manner for the benefit of society. In the twenty-first century, with limited budgets and competition for the use of public funds, the existing system must be managed while prudently adding capacity.

Transportation engineers produce a product that is highly visible. The system is paid for by taxes and user fees and serves every segment of society. Accordingly, another major challenge is to deal with the public and their representatives at the local, state, and national level. To adequately respond to citizen and political concerns, the transportation engineer must be a technical expert and have the requisite communication skills for clearly explaining ideas and concepts. Communication involves two elements: speaking and writing. Speaking skills are gained by preparation, practice, and experience. Writing skills can be improved through practice in class assignments and personal communications, as well as by reading well-written books and articles.

Transportation engineers also must learn to be good listeners, especially under trying and tense conditions. They are often confronted by opponents who voice objections because of their belief that proposals may negatively impact their lives or property.

Transportation engineers face hostile audiences more often than supportive ones. Citizens who are not affected by a project show little interest or support. No one wants highway noise and congestion near their home and, consequently, opponents are sometimes referred to as NIMBYs, meaning, "not in my back yard." Such groups deserve a fair and complete hearing, and attempts at mitigation should be made when possible. The key challenge for transportation engineers is to convincingly and honestly communicate the project's need, importance, and design location.

A constant challenge for transportation engineers is to deal ethically and fairly with everyone involved in a transportation project, including contractors, suppliers, real estate developers, funding agencies, construction and maintenance workers, and fellow employees. Ethics is a broad topic that can be summed up simply as "doing what is right within the context of the situation." To be ethical, one must be honest, trustworthy, dependable, fair, even-handed, loyal, and compassionate. Ethical persons refrain from slander, fraud, and malicious activities. For example, a transportation engineer who withholds an alternate plan/design concept for inspection by the public because it is not favored by the agency in charge of constraction is acting unethically. Ethics involves sharing credit for work done as a team effort and not taking credit for the work of others. It involves treating subordinates and colleagues with respect and without bias.

When planning transportation projects, ethics implies evaluating cost and benefits in an objective manner and disclosing the underlying assumptions of the analysis. Ethics requires that transportation engineers wrestle with thorny issues, such as who is being served with improved mobility and safety. For example: Do rail systems or high-occupancy toll (HOT) lanes benefit suburban commuters who own automobiles while neglecting the inner-city poor, who cannot afford to purchase an automobile and depend solely on public transit for their economic survival? Should highway funds be used to reduce truck traffic and congestion on interstate highways or should resources be devoted to local and rural tourist routes near historic and scenic destinations?

In summary, if transportation engineers are to meet the challenges of the twenty-first century, they will require technical knowledge and judgment as well as emotional intelligence. Good technical judgment is based on personal behavior and work experience and includes persistence, high performance standards, emphasis on quality, a sense of priorities, and adaptability to change. Emotional intelligence reflects the ability to deal with others by being supportive and helpful, avoiding malicious behavior, and treating others as we would treat ourselves.

1.4 SUMMARY

Transportation is an essential element in the economic development of a society. Without good transportation, a nation or region cannot achieve the maximum use of its natural resources or the maximum productivity of its people. Progress in transportation is not without its costs, both in human lives and environmental damage, and it is the responsibility of the transportation engineer working with the public to develop high-quality transportation consistent with available funds and social policy and to minimize damage. Transportation is a significant element in our national life, accounting for about 18 percent of household expenditure and employing over 10 percent of the workforce.

The history of transportation illustrates that the way people move is affected by technology, cost, and demand. The past 200 years have seen the development of several modes of transportation: waterways, railroads, highway, and air. Each mode has been dominant during one period of time; several have been replaced or have lost market share when a new mode emerged that provided a higher level of service at a competitive price.

The specialties in transportation engineering are planning, design, construction, traffic management and operations, and maintenance. Planning involves the selection of projects for design and construction; design involves the specification of all features of the transportation project; construction involves all aspects of the building process; traffic management and operations involves studies to improve capacity and safety; and maintenance involves all work necessary to ensure that the highway system is kept in proper working order.

Engineering students have exciting career opportunities in transportation. In the past, transportation engineers planned and built the nation's railroads, highways, mass transit systems, airports, and pipelines. In the coming decades, additional system elements will be required, as will efforts toward maintaining and operating in a safe and economical manner the vast system that is already in place. New systems, such as magnetically levitated high-speed trains or Intelligent Transportation Systems (ITS), will also challenge the transportation engineer in the future.

PROBLEMS

1-1   To illustrate the importance of transportation in our national life, identify a transportation-related article that appears in a local or national newspaper. Discuss the issue involved and explain why the article is newsworthy.

1-2   Arrange an interview with a transportation professional in your city or state (that is, someone working for a consulting firm, city, county or state transportation department, transit or rail agency). Inquire about the job he or she performs, why he or she entered the profession, and what he or she sees as the future challenges in the field.

1-3   Keep a diary of all trips you make for a period of three to five days. Record the purpose of the trip, how you traveled, the approximate distance traveled, and the trip time. What conclusions can you draw from the data?

1-4   Identify one significant transportation event that occurred in your city or state. Discuss the significance of this event.

1-5   Describe how transportation influenced the initial settlement and subsequent development of your home city or state.

1-6   Describe your state's transportation infrastructure. Include both passenger and freight transportation.

1-7   What is the total number of km of public roadways in your state? What percentage of the highway system mileage is maintained by the state government (as opposed to local and federal government)? What percentage of the total public highway system in your state is comprised of interstate highways?

1-8   Estimate the number of personal motor vehicles in your city or state. What is the total number of km driven each year? How much revenue is raised per vehicle per year for each 1 cent/gallon tax? Assume that the average vehicle achieves 10.6 km per litre in fuel economy.

1-9   How many railroad trains pass through your city each week? What percentage of these are passenger trains?

1-10   Compare the extent of the U.S. railroad system of today with that of 100 years ago. What changes have occurred and what factors have led to these changes?

1-11   What proportion of American household expenditures is associated with transportation, and what is the breakdown of these expenditures by category (such as ownership, fuel, maintenance, etc.)? Estimate the proportion of your monthly budget that is spent on transportation.

1-12 Identify an ITS project or application that is underway in your home state. Describe the project, its purpose, and the way it is operated.

1-13 Most Departments of Transportation incorporate at least five major transportation engineering subspecialties within their organization. List and briefly indicate at least three tasks under each specialty.

1-14 List four major detrimental effects that are directly related to the construction and use of our highway transportation system.

1-15 Cite four statistics that demonstrate the importance of transportation in the United States.

1-16 A state has a population of 17 million people and an average ownership of 1.1 cars per person, each driven an average of 19,310 km/yr, at an average fuel economy of 10.2 km/l of gasoline (mpg). Officials estimate that an additional $60 million per year in revenue will be required to improve the state's highway system, and they have proposed an increase in the gasoline tax to meet this need. Determine the required tax in cents per gallon.

1-17 Select a single event in Table 1.1 and explain why this is a significant achievement in the history of transportation.

1-18 Name and describe the first successful turnpike effort in the newly independent United States of America.

1-19 What mode of transportation was the primary contributor to the demise of road construction in the United States in the early nineteenth century, and what advantages did the new mode offer?

1-20 What mode of transportation succeeded the mode noted in Problem 1-19, and what advantages did it offer?

1-21 What was the interest of the U.S government in supporting expansion of railroads in the mid-nineteenth century, and how did the government provide support?

1-22 Public expectations for the transportation system continue to increase. What is the principal challenge faced by the transportation engineer in meeting these expectations? What fields of knowledge beyond traditional transportation engineering are needed?

REFERENCES

America's Highways: 1776–1976, U.S. Department of Transportation, Federal Highway Administration, Washington, D.C., 1976.

McNichol, Dan, *The Roads that Built America*, Barnes & Noble, 2002.

U.S. Department of Transportation, www.dot.gov, Research and Innovative Technology (Bureau of Transportation Statistics) and Federal Highway Administration, Washington, D.C., 2011.

Transportation Research Board of the National Academies, History Committee, http://transportationhistorycommittee.blogspot.com.

CHAPTER 2

Transportation Systems and Organizations

The transportation system in a developed nation is an aggregation of vehicles, guideways, terminal facilities, and control systems that move freight and passengers. These systems are usually operated according to established procedures and schedules in the air, on land, and on water. The set of physical facilities, control systems, and operating procedures referred to as the nation's transportation system is not a system in the sense that each of its components is part of a grand plan or was developed in a conscious manner to meet a set of specified regional or national goals and objectives. Rather, the system has evolved over a period of time and is the result of many independent actions taken by the private and public sectors, which act in their own or in the public's interest.

Each day, decisions are made that affect the way transportation services are used. The decisions of a firm to ship its freight by rail or truck, of an investor to start a new airline, of a consumer to purchase an automobile, of a state or municipal government to build a new highway or airport, of Congress to deny support to a new aircraft, and of a federal transportation agency to approve truck safety standards are just a few examples of how transportation services evolve and a transportation system takes shape.

CHAPTER OBJECTIVES:

- Understand how transportation systems are created and developed.
- Discuss the comparative advantages of both passenger and freight transportation modes.
- Explain the use of a supply-demand curve to determine the effect of tolls on travel demand.
- Become familiar with public transportation modes, their capacity, and their service attributes.
- Understand how highway projects are developed and their sources of funds.
- Become familiar with the U.S. national highway system and the organizations and associations involved in transportation.

2.1 DEVELOPING A TRANSPORTATION SYSTEM

Over the course of a nation's history, attempts are made to develop a coherent transportation system, usually with little success. A transportation plan for the United States was proposed by Secretary of the Treasury Gallatin in 1808, but this and similar attempts have had little impact on the overall structure of the U.S. transportation system. As stated in the *TRNews* special issue on the fiftieth anniversary of the Interstate highway system, engineers and planners failed to recognize or account for the impact of this immense national system on other transportation modes or on its effect on urbanization and sprawl. The creation of the U.S. Department of Transportation (DOT) in 1967 had the beneficial effect of focusing national transportation activities and policies within one cabinet-level agency. In turn, many states followed by forming transportation departments from existing highway agencies.

The Interstate Commerce Commission (ICC), created in 1887 to regulate the railroads, was given additional powers in 1940 to regulate water, highway, and rail modes, preserving the inherent advantages of each and promoting safe, economic, and efficient service. The intent of Congress was to develop, coordinate, and preserve a national transportation system; however, the inability to implement vague and often contradictory policy guidelines coupled with the extensive use of congressionally mandated projects, known as earmarks, has not helped to achieve the results implied by national policy. More recently, regulatory reform has been introduced, earmarks have been eliminated, and transportation carriers are developing new and innovative ways of providing services.

2.1.1 Comparative Advantages of Transportation Modes

The transportation system that evolves in a developed nation may not be as economically efficient as one that is developed in a more analytical fashion, but it is one in which each of the modes provides unique advantages for transporting the nation's freight and passengers. A business trip across the country may involve travel by taxi, airplane or rail, and auto; transportation of freight often requires trucks for pick-up and delivery and railroads, barges, or motor carriers for long-distance hauling.

Each mode has inherent advantages of cost, travel time, convenience, and flexibility that make it "right for the job" under a certain set of circumstances. The automobile is considered to be a reliable, comfortable, flexible, and ubiquitous form of personal transportation for many people. However, when distances are great and time is at a premium, air transportation will be selected—supplemented by the auto for local travel. If cost is important and time is not at a premium or if an auto is not available, then intercity bus or rail may be used.

Selecting a mode to haul freight follows a similar approach. Trucks have the advantages of flexibility and the ability to provide door-to-door service. They can carry a variety of parcel sizes and usually can pick up and deliver to meet the customer's schedule. Waterways can ship heavy commodities at low cost, but only at slow speeds and between points on a river or canal. Railroads can haul a wide variety of commodities between any two points, but usually require truck transportation to deliver the goods to a freight terminal or to their final destination. In each instance, a shipper must decide whether the cost and time advantages are such that the goods should be shipped by truck alone or by a combination of truck, waterway, and rail.

Many industries have been trying to reduce their parts and supplies inventories, preferring to transport them from the factory when needed rather than stockpiling them in a warehouse. This practice has meant shifting transportation modes from rail to truck. Rail shipments are usually made once or twice a week in carload lots, whereas truck deliveries can be made in smaller amounts and on a daily basis, depending on demand. In this instance, lower rail-freight rates do not compete with truck flexibility, since the overall result of selecting trucking is a cost reduction for the industry. There is a trend toward intermodalism, which has combined the capabilities of both modes.

Example 2.1 Selecting a Transportation Mode

An individual is planning to take a trip between the downtown area of two cities, A and B, which are 650 km apart. There are three options available:

Travel by air. This trip will involve driving to the airport near city A, parking, waiting at the terminal, flying to airport B, walking to a taxi stand, and taking a taxi to the final destination.

Travel by auto. This trip will involve driving 650 km through several congested areas, parking in the downtown area, and walking to the final destination.

Travel by rail. This trip will involve taking a cab to the railroad station in city A, a direct rail connection to the downtown area in city B, and a short walk to the final destination.

Since this is a business trip, the person making the trip is willing to pay up to $25 for each hour of travel time reduced by a competing mode. (For example, if one mode is two hours faster than another, the traveler is willing to pay $50 more to use the faster mode.) After examining all direct costs involved in making the trip by air, auto, or rail (including parking, fuel, fares, tips, and taxi charges) the traveler concludes that the trip by air will cost $250 with a total travel time of five hours, the trip by auto will cost $200 with a total travel time of eight hours and the trip by rail will cost $150 with a total travel time of 12 hours.

Which mode is selected based on travel time and cost factors alone?

What other factors might be considered by the traveler in making a final selection?

Solution: Since travel time is valued at $25/hr, the following costs would be incurred:

Air: $250 + 25(5) = \$375$
Auto: $200 + 25(8) = \$400$
Rail: $150 + 25(12) = \$450$

In this instance, the air alternate reflects the lowest cost and is the selected mode. However, the traveler may have other reasons to select another alternative. These may include the following considerations.

Safety. While each of these modes is safe, the traveler may feel "safer" in one mode than another. For example, rail may be preferred because of concerns regarding air safety issues.

Reliability. If it is very important to attend the meeting, the traveler may select the mode that will provide the highest probability of an on-time arrival. If the drive involves travel through work zones and heavily congested areas, rail or air would be preferred. If potential air delays are likely due to congestion, flight cancellations, or inclement weather, another mode may be preferred.

Convenience. The number of departures and arrivals provided by each mode could be a factor. For example, if the railroad provides only two trains/day and the airline has six flights/day, the traveler may prefer to go by air.

2.1.2 Interaction of Supply and Demand

The transportation system that exists at any point in time is the product of two factors that act on each other. These are (1) the state of the economy, which produces the demand for transportation; and (2) the extent and quality of the system that is currently in place, which constitutes the supply of transportation facilities and services. In periods of high unemployment or rising fuel costs, the demand for transportation tends to decrease. On the other hand, if a new transportation mode is introduced that is significantly less costly when compared with existing modes, the demand for the new mode will increase, decreasing demand for the existing modes.

These ideas can be illustrated in graphic terms by considering two curves, one describing the demand for transportation at a particular point in time, and the other describing how the available transportation service or supply is affected by the volume of traffic that uses that system.

The curve in Figure 2.1 shows how demand in terms of traffic volume could vary with cost. The curve is representative of a given state of the economy and of the present population. As is evident, if the transportation cost per km, C, decreases, then, since more people will use it at a lower cost, the volume, V, will increase. In Figure 2.1, when the traffic volume/day is 6000, the cost is \$0.75/km. If cost is decreased to \$0.50/km, the volume/day increases to 8000. In other words, this curve provides an estimate of the demand for transportation under a given set of economic and social conditions.

Demand can occur only if transportation services are available between the desired points. Consider a situation where the demand shown in Figure 2.1 represents the desire to travel between the mainland of Florida and an inaccessible island that is located off the coast, as shown in Figure 2.2.

If a bridge is built, people will use it, but the amount of traffic will depend on cost. The cost to cross the bridge will depend on the bridge toll and the travel time for cars and trucks. If only a few vehicles cross, little time is lost waiting at a tollbooth or in congested traffic. However, as more and more cars and trucks use the bridge, the time required to cross will increase unless automated tollbooths are installed. Lines will be long at the tollbooth; there might also be traffic congestion at the other end. The curve in Figure 2.3 illustrates how the cost of using the bridge could increase as the volume of traffic increases, assuming that the toll is \$0.25/km. In this figure, if the volume is

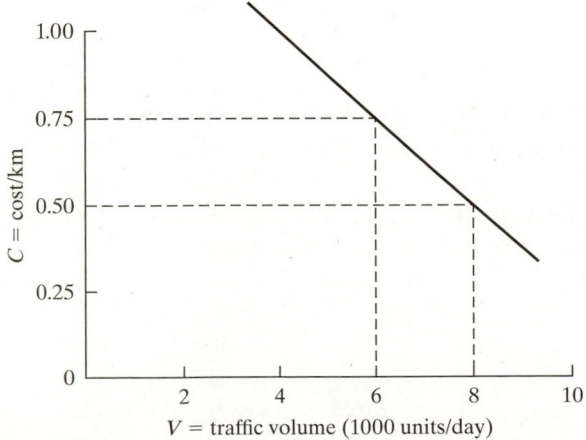

Figure 2.1 Relationship between Transportation Demand and Cost

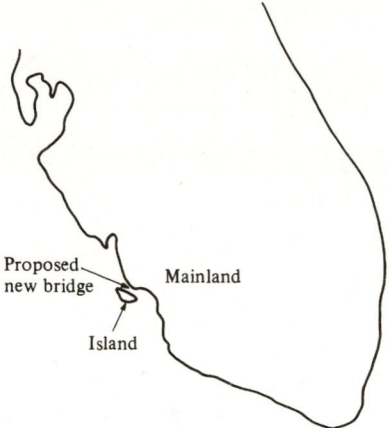

Figure 2.2   Location of a New Bridge between the Mainland and an Island

less than 2000 units/day, there is no delay due to traffic congestion. However, as traffic volumes increase beyond 2000 units/day, delays occur and the travel time increases. Since "time is money," the increased time has been converted to $/km. If 4000 units/day use the bridge, the cost is $0.50 km; at 6000 units/day, the cost is $0.75/km.

The two curves (Figures 2.1 and 2.3) determine what volume (V) can be expected to use the bridge. This value will be found where the demand curve intersects the supply curve as shown in Figure 2.4, because any other value of V will create a shift in demand either upward or downward, until the equilibrium point is reached. If the volume increased beyond the equilibrium point, cost would go up and demand would drop. Likewise, if the volume dropped below equilibrium, cost would go down and demand would increase. Thus, in both instances equilibrium is achieved. In this example, the number of units crossing the bridge would be 6000 units/day. The traffic volume could be raised or lowered by changing the toll—an example of congestion pricing.

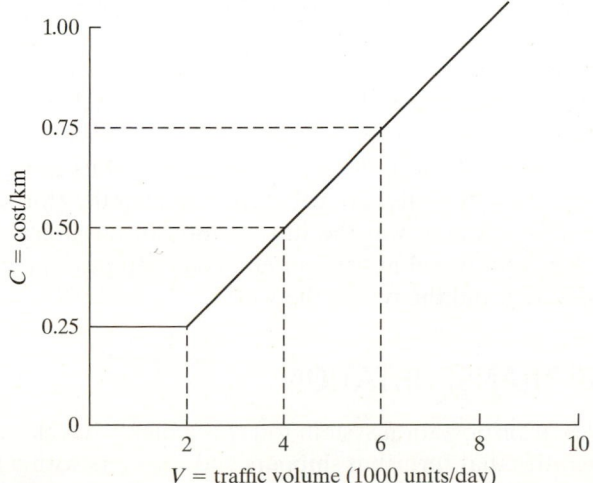

Figure 2.3   Relationship between Transportation Supply and Cost

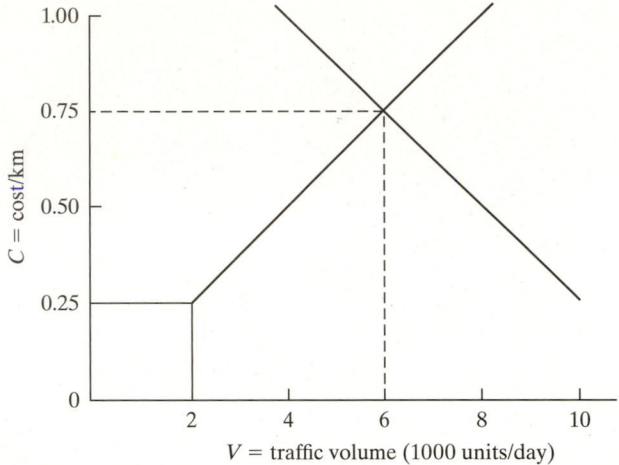

Figure 2.4   Equilibrium Volume for Traffic Crossing a Bridge

2.1.3 Forces That Change the Transportation System

At any point in time, the nation's transportation system is in a state of equilibrium as expressed by the traffic carried (or market share) for each mode and the levels of service provided (expressed as travel attributes such as time, cost, frequency, and comfort). This equilibrium is the result of market forces (state of the economy, competition, costs, and prices of service), government actions (regulation, subsidy, and promotion), and transportation technology (speed, capacity, range, and reliability). As these forces shift over time, the transportation system changes as well, creating a new set of market shares (levels of demand) and a revised transportation system. For this reason, the nation's transportation system is in a constant state of flux, causing short-term changes due to immediate revisions in levels of service (such as raising the tolls on a bridge or increasing the gasoline tax) and long-term changes in lifestyles and land-use patterns (such as moving to the suburbs after a highway is built or converting auto production from large to small cars).

If gasoline prices were to increase significantly, there could be a measurable shift of long-haul freight from truck to rail. In the long run, if petroleum prices remained high, there might be shifts to coal or electricity or to more fuel-efficient trucks and autos.

Government actions also influence transportation equilibrium. For example, the federal government's decision to build the national interstate system affected the truck-rail balance in favor of truck transportation. It also encouraged long-distance travel by auto and was a factor in the decline of intercity bus service to small communities.

Technology has also contributed to substantial shifts in transportation equilibrium. A dramatic example was the introduction of jet aircraft, which essentially eliminated passenger train travel in the United States and passenger steamship travel between the United States and the rest of the world.

2.2 MODES OF TRANSPORTATION

The U.S. transportation system today is a highly developed, complex network of modes and facilities that furnishes shippers and travelers with a wide range of choices in terms of services provided. Each mode offers a unique set of service characteristics in terms of travel time, frequency, comfort, reliability, convenience, and safety. The term *level of*

Example 2.2 Using a Supply-Demand Curve to Compute the Bridge Cost/Vehicle (Toll) That Will Maximize Total Revenue

A toll bridge carries 5000 veh/day. The current cost (toll) is 150 cents. When the cost (toll) is increased by 25 cents, traffic volume decreases by 500 veh/day. Determine the cost/veh (new toll) that should be charged such that revenue is maximized. How much additional revenue will be received?

Solution: Let x = the cost increase in cents.

Assuming a linear relation between traffic volume and cost, the expression for V is

$$V = 5000 - x/25(500)$$

The new cost/veh (toll) is the original cost plus the cost increase

$$T = 150 + x$$

The revenue produced is the product of total cost/veh (toll) and vehicle volume:

$$\begin{aligned} R &= (V)(T) \\ &= \{5000 - x/25(500)\}(150 + x) \\ &= (5000 - 20x)(150 + x) = 750{,}000 - 3000x + 5000x - 20x^2 \\ &= 750{,}000 + 2000x - 20x^2 \end{aligned}$$

For maximum value of x, compute the first derivative and set equal to zero:

$$dR/dt = 2000 - 40x = 0$$
$$x = 50 \text{ cents}$$

The total cost/veh (new toll) is the current total cost (toll) plus the toll increase. Thus the total cost/veh toll for maximum revenue = 150 + 50 = 200 cents or $2.00 The additional revenue, AR, is

$$\begin{aligned} AR &= (V_{max})(T_{max}) - (V_{current})(T_{current}) \\ &= \{(5000 - (50/25)(500)\}\{2\} - (5000)(1.50) \\ &= (4000)(2) - 7500 \\ &= 8000 - 7500 \\ &= \$500 \end{aligned}$$

---

service is used to describe the relative values of these attributes. The traveler or shipper must compare the level of service offered with the cost in order to make tradeoffs and mode selection. Furthermore, a shipper or traveler can decide to use a public carrier or to use private (or personal) transportation. For example, a manufacturer can ship goods through a trucking firm or with company trucks; a homeowner who has been relocated can hire a household moving company or rent a truck; and a commuter can elect to ride the bus to work or drive a car. Each of these decisions involves a complex set of factors that require tradeoffs between cost and service.

2.2.1 Freight and Passenger Traffic

The principal modes of intercity freight transportation are highways, railroads, water, and pipelines. Traffic carried by each mode, expressed as tonne-kms or passenger-kms, has varied considerably in the past 70 years. The most current information regarding modal market share is available from the Bureau of Transportation Statistics (BTS) Web site. Changes in tonne-kms carried from 1960 through 2008 are illustrated in Figure 2.5.

Class I Railroads, which accounted for about 0.87 tonne kms tonne-kms of freight traffic in 1960, carried about 2.63 tonne kms tonne-kms by the year 2008. Oil pipelines have increased their share of freight traffic from 0.3 tonne kms tonne-kms in 1960 but have remained relatively constant at about 0.87 tonne kms tonne kms since 1980. Domestic water transportation has increased its tonne-kms between 1965 and 1980 and has declined since then. Intercity trucking has steadily increased each year, from 0.44 tonne kms tonne-kms in 1960 to about 1.9 tonne kms tonne-kms in 2003. Air freight is an important carrier for high-value goods, but it is insignificant on a tonne-km basis.

The four principal carriers for freight movement (rail, truck, pipeline, and water) account for varying proportions of total number of tonne-kms of freight. The railroads' share is highest on a tonne-km basis, but it has been reduced significantly due to competition from truck and pipeline. Overall the railroads have lost traffic due to the advances in truck technology and pipeline distribution. Government policies that supported highway and waterway improvements are also a factor. Subsequent to World War II, long-haul trucking was possible because the U.S. highway system was developed. As petroleum became more widely used, construction of a network of pipelines for distribution throughout the nation was carried out by the oil industry.

During the past 50 years, the railroad's share of revenue has decreased while that of trucking has increased to about 80 percent of the total. These trends reflect two factors: the increased dominance of trucking in freight transportation and the higher ton-mile rates that are charged by trucking firms compared with the railroads. Although trucks move fewer tonne-kms than does rail, the value of the goods moved by truck comprises about 75 percent of the total value of all goods moved in the United States.

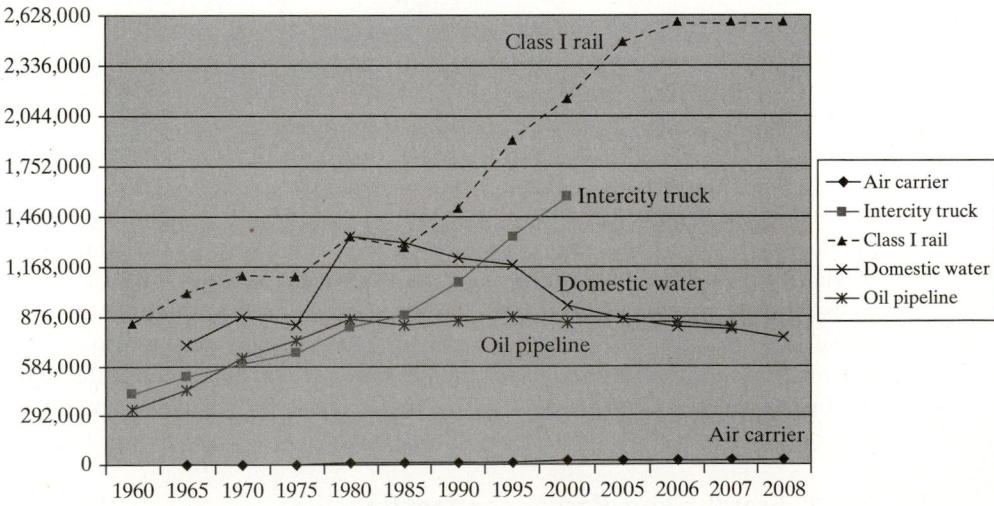

Figure 2.5   U.S. Tonne-Kms of Freight (Millions)

SOURCE: U.S. Department of Transportation, Research and Innovative Technology Administration, Bureau of Transportation Statistics, *National Transportation Statistics*, Table 1-46a.

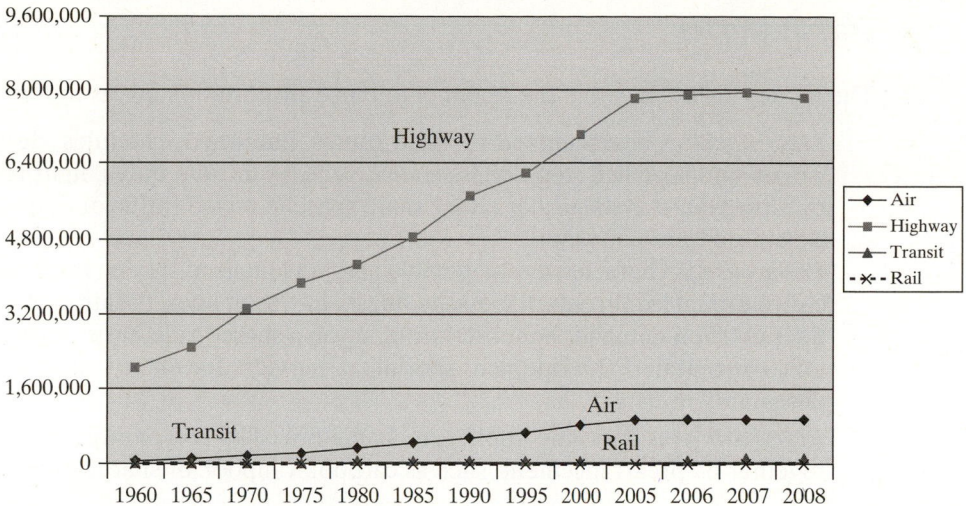

Figure 2.6   U.S. Passenger–Kms (Millions)

SOURCE: U.S. Department of Transportation, Research and Innovative Technology Administration, Bureau of Transportation Statistics, *National Transportation Statistics*, Table 1-37.

The distribution of passenger transportation is much different from that for freight: one mode—the automobile—accounts for the highest number of all domestic intercity passenger-kms traveled in the United States, as illustrated in Figure 2.6. With the exception of the World War II years (when auto use declined), passenger-kms by automobile, small trucks, and vans have accounted for as much as 90 percent of all passenger-kms traveled. The remaining modes—air, bus, and rail—shared a market representing about one-quarter of the total, with air being the dominant mode and intercity bus, private air carriers, and rail representing 1 percent or less of the total.

Of the four transportation carriers for intercity passenger movement, two—air and auto—are dominant, representing 98 percent of all intercity passenger-kms. If the public modes (rail, bus, and air) are considered separately from the auto, a dramatic shift in passenger demand is evident. The largest increase has occurred in air transportation, which represents over 90 percent of all intercity passenger-kms traveled using public modes. In cities, buses are the major public transit mode, with the exception of larger urban areas that have rapid rail systems. Intercity bus transportation has declined as a percentage of the total number of passenger kms. Buses now serve a market consisting primarily of passengers who cannot afford to fly or drive. As air fares have decreased with deregulation, bus travel has declined. Most passenger rail traffic is concentrated in the corridor between Washington, D.C., New York, and Boston.

2.2.2 Public Transportation

Public transportation is a generic term used to describe the family of transit services available to urban and rural residents. Thus, it is not a single mode but a variety of traditional and innovative services, which should complement each other to provide system-wide mobility.

Transit Modes

The modes included within the realm of public transportation are as follows:

Mass transit is characterized by fixed routes, published schedules, designated networks, and specified stops. Mass-transit vehicles include buses, light rail (trolleys), or rapid transit systems that either share space in mixed traffic or operate on grade-separated rights-of-way.

Paratransit is characterized by flexible and personalized service intended to replace conventional fixed-route, fixed-schedule mass-transit lines. Paratransit is available to the public on demand, by subscription, or on a shared-ride basis. Examples include taxi, car rental, dial-a-ride, and specialized services for elderly, medical, and other designated users.

Ridesharing (as the name implies) is characterized by two or more persons traveling together by prearrangement, such as carpool, vanpool, or shared-ride taxi.

Transit Capacity and Level of Service

A basic attribute of any transit mode is its carrying capacity, defined as the number of vehicles or persons that pass a given point in a specified time (usually an hour). The numerical value of carrying capacity (usually referred simply as capacity) is dependent on two variables: (1) the number of vehicles that pass a point at a given time and (2) the number of passengers within each vehicle. For example, if, for a given lane along a section of highway, there are 60 buses that pass by in an hour (or one per minute) and each bus carries 50 seated passengers, then the carrying capacity of this highway lane is 60 buses/ln/hr or $(50) \times (60) = 3000$ passengers/ln/hr.

Carrying capacity is influenced by (1) the "spacing" in seconds between each vehicle (called the headway) and (2) the "comfort factor" experienced by passengers (called the level of service). Thus, carrying capacity can be increased in two ways: (1) reduce the headway or (2) increase the number of passengers per vehicle. In the bus capacity example, the headway was 60 seconds and the level of service was that all passengers had a seat. Time spacing between buses could possibly be reduced, but there are limits to lowering headway values: these limits are dictated by safe distance requirements between vehicles and/or the time spent at transit stops and terminals (called dwelltime). Similarly, passenger loading could be increased by allowing standees, but this would decrease the comfort level for passengers. Were the bus equipped with computer tables and a refreshment area (thus offering a higher level of service), fewer passengers could be accommodated, resulting in a lower carrying capacity but a higher level of service.

Accordingly, when reporting transit capacity, it is important to specify the units as either vehicles or passengers/hour and the corresponding level of service in terms of passengers/vehicle. Public transit is often compared with the automobile when issues of carrying capacity are involved, as it is commonly believed that transit capacity is superior to auto capacity. As will be discussed in greater detail in Chapters 9 and 10, the capacity of a single lane of passenger vehicles is approximately 2000 vehicles/hour, representing a headway of 1.8 seconds. Since most cars have at least five seats, the person capacity of a highway lane could be as great as $(5) \times (2000) = 10,000$ persons/hour. Capacities of this magnitude never have been achieved, since most cars carry only one person and the average car occupancy is about 1.5. Why is this so? Have you ever driven in a car carrying five people? It may not be a pleasant experience, which could be why carpooling is not very popular. Given the opportunity, most people choose to drive alone or with just one other person.

Travelers usually consider many factors other than simply the in-vehicle level of service, and they don't really consider how they can contribute to increasing "carrying capacity." If drivers were to optimize the carrying capacity of a highway, they would all drive at a speed that they would choose as discussed in Chapter 6. Other major considerations in selecting the travel mode include reliability, punctuality, cost, travel time, and safety. Transit systems that receive "high marks" for the out-of-vehicle level-of-service factors are typically the ones that use exclusive lanes or tracks with no interference from other vehicles or pedestrians and have adequate capacity at station stops and terminals. Thus, rapid-transit services (whether bus or fixed guideway) are the superior mode but are more costly to build and maintain and require high volumes of demand to be feasible.

The carrying capacity of a highway lane and selected transit modes are listed in Table 2.1 in terms of vehicles/hour. The number of passengers/hour is so varied due to assumptions regarding level of service that it is less meaningful, but the reader is encouraged to try various values to see the results. The values shown are of necessity within a range and approximate. However, they are useful in that they provide perspective for engineers, planners, and public officials in selecting a mode that will accommodate anticipated demand.

The Role and Future of Public Transportation

Public transportation is an important element of the total transportation services provided within large and small metropolitan areas. A major advantage of public transportation is that it can provide high-capacity, energy-efficient movement in densely traveled corridors. It also serves medium- and low-density areas by offering an option for auto owners who do not wish to drive and an essential service to those without access to automobiles, such as school children, senior citizens, single-auto families, and others who may be economically or physically disadvantaged.

For most of the twentieth century, public transportation was provided by the private sector. However, increases in auto ownership, shifts in living patterns to low-density suburbs, and the relocation of industry and commerce away from the central city, along with changes in lifestyle have resulted in a steady decline in transit ridership. Since the early 1960s, most transit services have been provided by the public sector. Income from fares no longer represents the principal source of revenue, and over a 25- to 30-year period, the proportion of funds for transit provided by

Table 2.1   Capacity of Urban Transportation Modes

Mode	Capacity (veh /hr)
Automobiles on a freeway	2000–2200
Buses in mixed traffic	60–120
Bus rapid transit	150–300 (single or articulated)
Light rail in mixed traffic	15–30
Light rail transit	30–60 (single or multiple cars)
Heavy rail transit	20–30 (10-car trains)

federal, state, and local governments has increased steadily. While it generally is believed that highways and motor transport will play a dominant role in providing personal transportation in the beginning decades of the twenty-first century, there are many unforeseen changes that could alter the balance between public and private transportation. Among these are rising fuel prices and a slow-growth economy. Some could contribute to a decline in transit ridership while others might cause transit to become stronger, and for the remainder, there would be little or no effect. The potential changes that could influence transit usage are as follows:

Factors Bad for Transit

- Growth of suburbs
- Industry and employment moving from the central city
- Increased suburb-to-suburb commuting
- Migration of the population to the south and west
- Loss of population in "frost-belt" cities
- Growth in private vehicle ownership
- Increased diversity in vehicle types such as SUVs, pickup trucks, and RVs
- High cost per km to construct fixed-rail transit lines
- High labor costs

Factors Good for Transit

- Emphasis by the federal government on air quality
- Higher prices of gasoline
- Depletion of energy resources
- Trends toward higher-density living
- Legislation to encourage "livable cities" and "smart growth"
- Location of mega-centers in suburbs
- Need for airport access and circulation within airports
- Increased number of seniors who cannot or choose not to drive

Factors Neutral for Transit

- Increases in telecommuting may require less travel to a work site.
- Internet shopping and e-commerce could reduce shopping trips.
- Changes in work schedules to accommodate childcare could increase trip chaining.
- Staggering work hours, flex-time, and four-day work weeks reduce peak-hour congestion.
- Aging population, most of whom are not transit users, may continue to drive.
- Increased popularity in walking and biking could be a substitute for transit riding.

Thus, the future of public transportation appears to be one of stability and modest growth. In the 1990s and into the twenty-first century, public support and financing for transit increased. Political leadership and the citizenry generally agree that transit is essential to the quality of life and is a means to reduce traffic congestion. Furthermore, there remains a significant proportion of the population who must rely primarily on transit because they do not have access to an automobile or are disabled. As with all modes of transportation, change is incremental. If the factors favorable to transit should prevail, then, over time, transit will continue to improve service, expand its network, and attract an increasing number of riders.

2.2.3 Highway Transportation

Highway transportation is the dominant mode in passenger travel and one of the principal freight modes. The U.S. highway system comprises approximately 6.3 million km of highways, ranging from high-capacity and multilane freeways to urban streets and unpaved rural roads. Although the total number of roadway km has not increased greatly, the quality of roadway surfaces and their capacity has improved due to increased investment in maintenance and construction.

The Highway Project Development Process

In order for a new highway project to become a reality, a process is followed that includes the topics outlined in the previous chapter, including planning, evaluation, design, right-of-way acquisition, and construction. When the project is completed, highway operations, management, and maintenance are needed. In most cases, the state transportation agency is responsible for overall coordination; public and business involvement; interaction with local, state, and national agencies; and project management. Proposed projects require approval from a policy board that is mandated to set the project agenda typically in terms of a Transportation Improvement Program that typically spans 3 to 6 years and specifies the work program for the entire state.

A considerable amount of information is provided to the general public through the Internet by state DOTs regarding the status of highway projects. The process that must be followed "from start to finish" is typified by information provided by the Virginia and Wisconsin DOTs and other states on their Web sites. According to the Virginia Department of Transportation, the road-building process includes

1. Planning (6 to 24 months)
2. Design (15 to 24 months)
3. Environment (9 to 36 months)
4. Right-of-Way (6 to 21 months)
5. Construction (12 to 36 months)

Thus, total project time can range from 4 to 12 years, depending on the physical characteristics, scope, and community support for the project. Throughout the process, the Commonwealth Transportation Board (CTB), a citizen panel appointed by the Governor, updates the Six Year Improvement Program (SYIP), approves the project location prior to land acquisition and final design, and awards the contract to the lowest qualified bidder.

The Wisconsin Department of Transportation follows a six-step process that includes the following phases.

Step 1. **Select Project.** Highway projects are selected based on a variety of criteria. These include public concerns, traffic crash data, pavement and bridge condition, traffic volume and trends, and forecasts of future growth. A highway's concept takes shape through a series of public meetings. Citizens are encouraged to ask questions about the type of improvement being considered and how it will improve the quality of the transportation system. A preliminary project list is submitted by each transportation district for consideration in the statewide program. If a project is selected to be funded, it becomes part of the state's Six-Year Highway Improvement Program. A major highway project is more complex and costly, involving either a capacity expansion of over eight km or the creation of a highway on a new location at least four km long. Major highway projects must undergo an extensive environmental review, including public involvement and approval by the Transportation Projects Commission, the State Legislature, and the Governor.

Step 2.   Investigate Alternates. After a project is selected for inclusion, the Six-Year Highway Improvement Program project design alternatives are identified, using citizen input. Each alternative is analyzed and assessed based on cost and its impact on people, businesses, farmlands, wetlands, endangered species, historic structures, artifacts, and landfills. Recommended improvements are presented at public meetings, where detailed information about the impacts is provided. Changes can be made that address public concerns.

Step 3.   Obtain Final Approvals. Information about acquisitions from farm operations is furnished to the Department of Agriculture, Trade and Consumer Protection (DATCP), where it is reviewed to determine if an Agricultural Impact Statement (AIS) is required. When DATCP prepares the AIS, copies are sent to the affected farm operations, legislators, and the public. The AIS is commonly included in the Environment Impact Statements (EIS).

Citizen input is incorporated during the development of the EIS. While an official advertised public hearing is held to obtain citizen comments on the Draft EIS, the public can provide comments about the project at any time during this phase.

Environmental documents for federal projects are prepared for the approval of the Federal Highway Administration (FHWA). Similarly, projects that are state funded are approved by the Department of Transportation.

Step 4.   Develop Project Design. The specific project route and related details are finalized. Affected property owners are contacted to discuss land purchases and relocation plans. Every effort is made to ensure that offering prices reflect "just compensation" for the property. Local businesses affected by the project receive assistance during the construction phase intended to mitigate the economic effect of the project. Among the services that may be stipulated during construction are the following:

(1) Maintaining access for customers, employees, and service vehicles
(2) Posting highway signs that inform motorists of business district locations
(3) Permitting the temporary posting of signs in the highway right-of-way to reassure customers that businesses are accessible
(4) Assisting businesses to identify and inform motorists of alternate routes during construction.

Access management is analyzed to ensure that the existing highway system continues to perform with acceptable efficiency and safety. The following engineering documents are prepared: (1) a plan that depicts the physical layout of the project, (2) specifications that define how each item in the plan is to be built, and (3) engineering estimates that list the expected cost of each item of work.

Step 5.   Prepare for Construction. The DOT Bureau of Highway Construction reviews each plan, specification, and estimate package and prepares a document suitable to be used by a contractor in preparing a bid. All land required for the project must have been purchased so that the project site can be prepared for construction. The construction schedule is coordinated with utility companies in the event that replacement of sewer, gas, power, or phone lines are required. Citizens receive prior notification should there be a need to disrupt utility service.

Request for project bids are advertised and those received within a specified period are checked for completeness and accuracy. All bids awarded are forwarded to the Governor for signature. With contractor contracts approved and signed, construction usually begins within 30 to 45 days—weather permitting.

Step 6.   **Construct the Project.** A pre-construction meeting is held with the contractor, local utilities, Department of Natural Resources, and local government officials. The DOT coordinates with property owners and local businesses to ensure that all prior commitments to landowners, such as access to homes and businesses, are fulfilled. DOT officials also meet with the contractors throughout the construction period. Citizens are kept informed of construction progress through meetings and news releases sent to local media. Web sites, project newsletters, and brochures are developed for larger projects.

When requested by local officials, completion of a major project may be celebrated with a ribbon cutting. The new roadway is then operated with the stipulation that safety and traffic flow will be inspected every two years to monitor the road's pavement performance.

Sources of Funds for Highways

Highway users are the primary source of funds to build and maintain the nation's roads. Taxes are paid when purchasing fuel, and excise taxes (which are taxes targeted to specific goods or activities) are paid on items such as sales of trucks, batteries, and tires. In addition, motorists pay state and local license and registration fees, highway and parking tolls, and traffic fines. Fuel taxes represent the largest segment of highway user fees (almost two-thirds of the total) while other user fees accounted for almost 30 percent, with tolls less than 10 percent.

In recent years, there has been a growing concern regarding the viability of the gas tax to continue as a sustainable revenue source. Since the inception of this tax in Oregon in 1919, gasoline that is sold at the pump has been taxed on a cents/litre basis. This method proved to be successful as long as gasoline consumption continued its upward climb. The fuel tax has been less effective when the price/litre increased with no corresponding increase in the tax, as would have been the case had gasoline tax been a sales tax based on a percentage of the total cost. Furthermore, politicians are loath to increase a tax that is so visible to the consumer.

The revenue from the gas tax is vulnerable to changes in the economy that reduce gasoline consumption. Among contributing factors leading to less fuel consumption are (1) crude oil shortages, (2) improved automobile efficiency, (3) use of alternative energy sources, and (4) shifts to public transportation to achieve environmental and energy conservation goals. To further exacerbate the problem, highway fuel taxes have been diverted to support other worthwhile programs, such as transit. These taxes are also used to subsidize fuel additives like ethanol. Highway taxes have been used for general revenue purposes and to help reduce state budget deficits. The net effect of the trend toward lower highway revenue from the gas tax has been a diminution of the state's ability to build infrastructure, which is further compounded by maintenance needs—both routine (i.e., snow removal) and periodic (i.e., bridge and pavement repair).

In order to strengthen the nation's highway finance system, the Transportation Research Board published a policy study report in 2005, *The Fuel Tax and Alternatives for Transportation Funding*. The Committee's charge was to examine current practices and trends in finance of roads and public transit and to evaluate options for a transition to an alternative finance arrangement. The principal recommendations were as follows:

1. Maintain and reinforce the existing user-fee finance system.
2. Expand the use of tolls and test road use metering.
3. Provide stable, broad-based support for transit.
4. Evaluate the impact of finance arrangements on transportation system performance.

New concepts of highway finance are emerging. Were the technology available so that vehicles could be charged on the basis of a per vehicle (or ton) km driven, then cost could be allocated directly on the basis of units consumed, as is the case for consumption of water, natural gas, and electricity. Electronic tolling represents a movement in that direction. The use of "value pricing" is being tested and considered as a means to charge for services rendered as well as "variable pricing" that recognizes changes in demand.

The Federal Highway System

The federal-aid system, which includes the interstate and other federal-aid routes, consists of a network of roads totaling approximately 1,500,000 km. These roads are classified as rural or urban and as arterials or collectors. Urban roads are located in cities of 25,000 or more, and they serve areas of high-density land development. Urban roads are used primarily for commuting and shopping trips. Rural roads are located outside of cities and serve as links between population centers.

Arterial roads are intended primarily to serve travel between areas and provide improved mobility. Collector roads and minor arterials serve a dual purpose of mobility and access. Local roads serve primarily as access routes and for short trips. Arterial roads have higher speeds and capacities, whereas local roads, which serve abutting land uses, and collector roads are slower and not intended for through traffic. A typical highway trip begins on a local road, continues onto a collector, and then to a major arterial. This hierarchy of road classification is useful in allocating funds and establishing design standards. Typically, local and collector roads are paid for by property owners and through local taxes, whereas collectors and arterials are paid for jointly by state and federal funds.

The interstate highway system consists of approximately 76,900 km of limited-access roads. This system, which is essentially complete, is illustrated in Figure 2.7. It connects major metropolitan areas and industrial centers by direct routes, connects the U.S. highway system with major roads in Canada and Mexico, and provides a dependable highway network to serve in national emergencies. Each year, about 25300 route-km of the national highway system become 20 years old, which is their design life, and a considerable portion of the system must undergo pavement repair and resurfacing, bridge maintenance, and replacement. Ninety percent of the funds to build the interstate system were provided by the federal government, while 10 percent were provided by the state. The completion of the system was originally planned for 1975, but delays resulting from controversies, primarily in the urban sections of the system, caused the completion date to be extended into the 1990s.

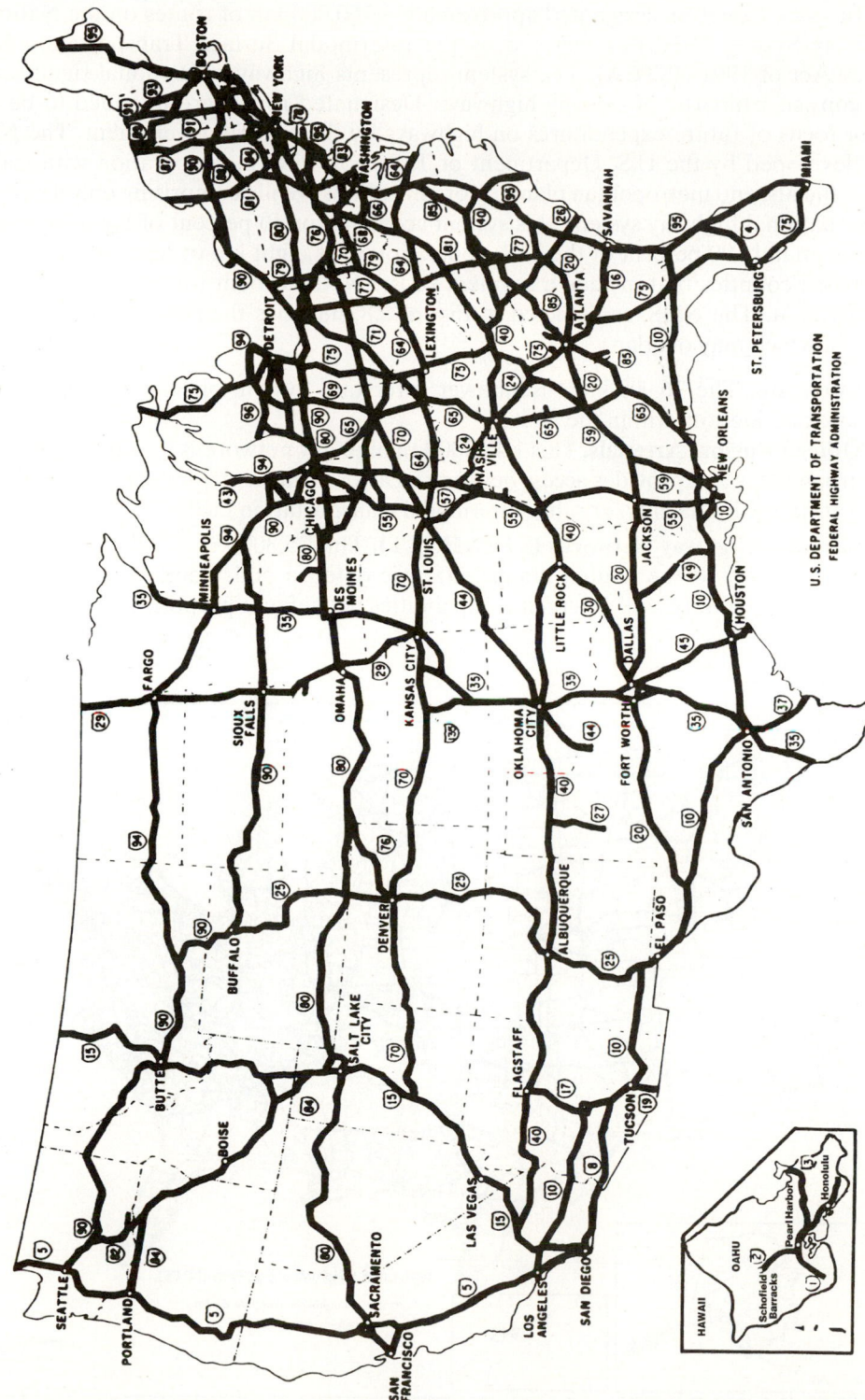

Figure 2.7   Dwight D. Eisenhower System of Interstate and Defense Highways

SOURCE: Federal Highway Administration, Washington, D.C., 1992.

In 1995, Congress designated approximately 260,000 km of routes on the National Highway System (NHS) as required by the Intermodal Surface Transportation Efficiency Act of 1991 (ISTEA). The system represents highways of national significance and consists primarily of existing highways. Designated routes are intended to be the major focus of future expenditures on highways by the federal government. The NHS was developed by the U.S. Department of Transportation in cooperation with states, local officials, and metropolitan planning organizations. While comprising only 4 percent of the national highway system, this system accounts for 40 percent of highway traffic. Approximately 90 percent of the U.S. population lives eight km or less from the NHS, and those counties that include a segment of the NHS account for 99 percent of U.S. employment. The NHS, as depicted in Figure 2.8, includes the following subsystems, some of which may overlap.

Interstate. The 76,900-km Eisenhower Interstate System of Highways retains its separate identity within the NHS.

Other Principal Arterials. This 1,48,000-km highway network is located in rural and urban areas and provides access between an arterial and a major port, airport, public transportation facility, or other intermodal transportation facility.

Strategic Highway Networks (STRAHNET). This 25,300-km network of highways is important to the United States' strategic defense policy and provides defense access, continuity, and emergency capabilities for defense purposes.

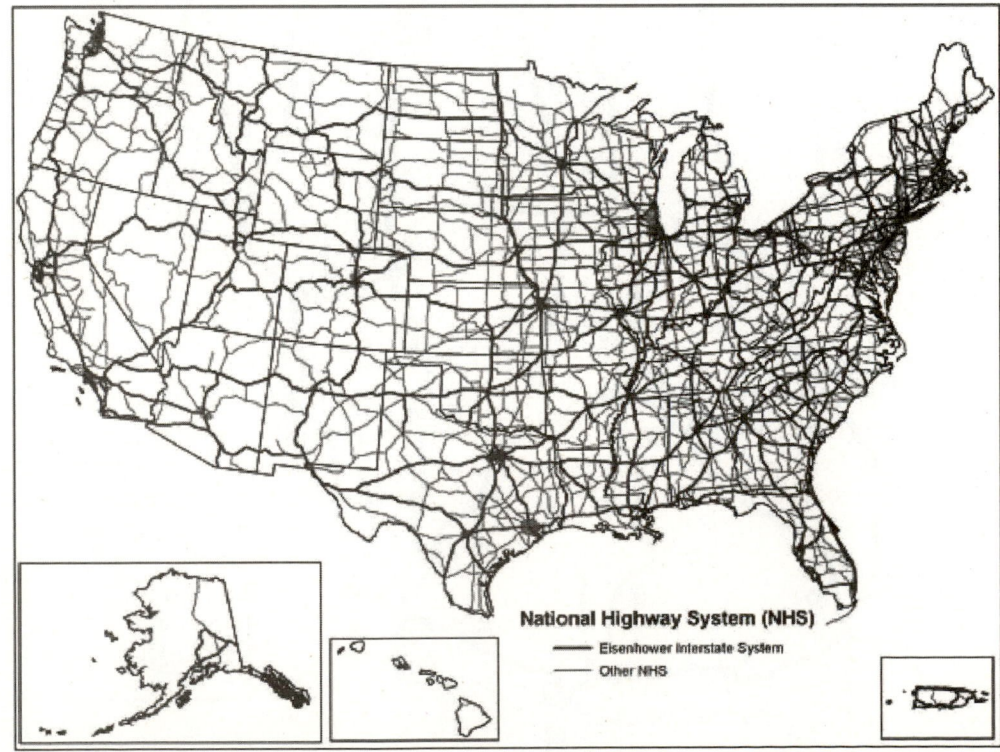

Figure 2.8   National Highway System

SOURCE: U.S. Department of Transportation, Federal Highway Administration, 2007.

Major Strategic Highway Network Connectors. This 3050-km highway element provides access between major military installations and highways that are part of the Strategic Highway Network.

Intermodal Connectors. Highways that provide access between major intermodal facilities and other subsystems that comprise the National Highway System.

Intercity bus transportation services have benefited from the interstate highway system. Buses have provided extensive nationwide coverage by connecting as many as 15,000 cities and towns with bus routes. Private companies usually provide intercity bus service, the principal one being Greyhound, which operates at a profit with little or no support from state or federal governments. Bus transportation is a highly energy-efficient mode, averaging 128 seat-km/litre. Buses are also very safe. Their accident rate of 7.45 fatalities/100 billion passenger-kms is over 100 times better than that of automobiles. In spite of its positive characteristics of safety and energy efficiency, the riding public generally views bus travel unfavorably. Buses are slower and less convenient than other modes and often terminate in downtown stations that are located in the less attractive parts of cities. Other factors, such as lack of through ticketing, comfortable seats, and system-wide information, which the riding public is accustomed to receiving when traveling by air, reinforce the overall negative image of intercity bus transportation.

Truck Transportation

The motor-carrier industry is very diverse in terms of size, ownership, and use. Most vehicles registered as trucks are less than 4500 kg in gross weight, and over half of these are used for personal transportation. Only about 9 percent are heavy trucks used for over-the-road intercity freight. Many states consider recreational vehicles, mobile homes, and vans as trucks. Approximately 20 percent of trucks are used in agriculture; 10 percent are used by utilities and service industries to carry tools and people rather than freight. Personal truck ownership varies by region and is more prevalent in the mountain and western states than in the mid-Atlantic and eastern states.

Trucks are manufactured by the major auto companies as well as by specialty companies. The configurations of trucks are diverse; a truck is usually produced to an owner's specifications. Trucks are classified by gross vehicle weight. The heaviest trucks, weighing over 11,800 kg, are used widely in intercity freight; lighter ones transport goods and services for shorter distances. Until 1982, truck size and weight were regulated by states, with little consistency and uniformity. Trucks were required to conform to standards for height, width, length, gross weight, weight per axle, and number of axles, depending on the state. The absence of national standards for size and weight reflected the variety of truck dimensions that existed. An interstate truck shipment that met requirements in one state could be overloaded in another along the same route. A trucker would then carry a minimal load, use a circuitous route, change shipments at state lines, or travel illegally. With the passage of the Surface Transportation Act of 1982, states were required to permit trucks pulling 8.5-m trailers on interstate highways and other principal roads. The law also permits the use of 14.63-m semi-trailers and 2.6-m-wide trucks carrying up to 36,300 kg.

The trucking industry has been called a giant composed of midgets because, although it is a major force in the U.S. transportation picture, it is very fragmented and diverse. Regulation of the trucking industry has been limited to those carriers that transport under contract (known as *for hire),* with the exception of carriers operating within a single state or in a specified commercial zone or those carrying exempt agricultural products. Private carriers, which transport their own company's goods or products, are not economically regulated but must meet federal and state safety requirements. The for-hire portion of the trucking industry accounts for about 45 percent of intercity tonne-kms and includes approximately 15,000 separate carriers. With deregulation of various transportation industries, the trucking industry has become more competitive.

The industry also has benefited significantly from the improvements in the U.S. highway transportation system. The diversity of equipment and service types is evidence of the flexibility furnished by truck transportation. Continued growth of this industry will depend on how well it integrates with other modes, on the future availability of energy, on limitations on sizes and weights of trucks, on highway speed limits, and on safety regulations.

2.3 TRANSPORTATION ORGANIZATIONS

The operation of the vast network of transportation services in the United States is carried out by a variety of organizations. Each has a special function to perform and serves to create a network of individuals who, working together, furnish the present-day transportation systems and services. The following sections will describe some of the organizations and associations involved in transportation. The list is illustrative only and is intended to show the wide range of organizations active in the transportation field. The following seven categories, described briefly in the sections that follow, outline the basic purposes and functions that these organizations serve:

1. Private companies available for hire to transport people and goods
2. Regulatory agencies that monitor the behavior of transportation companies in areas such as pricing of services and safety
3. Federal agencies such as the Department of Transportation and the Department of Commerce, which, as part of the executive branch, are responsible for carrying out legislation dealing with transportation at the national level
4. State and local agencies and authorities responsible for the planning, design, construction, and maintenance of transportation facilities such as roads and airports
5. Trade associations, each of which represents the interests of a particular transportation activity, such as railroads or intercity buses, and which serve these groups by furnishing data and information, by representing them at congressional hearings, and by furnishing a means for discussing mutual concerns
6. Professional societies composed of individuals who may be employed by any of the transportation organizations but who have a common professional bond and benefit from meeting with colleagues at national conventions or in specialized committees to share the results of their work, learn about the experience of others, and advance the profession through specialized committee activities
7. Organizations of transportation users who wish to influence the legislative process and furnish its members with useful travel information

Other means of exchanging information about transportation include professional and research journals, reports and studies, and university research and training programs.

2.3.1 Private Transportation Companies

Transportation by water, air, rail, highway, or pipeline is furnished either privately or on a for-hire basis. Private transportation (such as automobiles or company-owned trucks) must conform to safety and traffic regulations. For-hire transportation (regulated until recently by the government) is classified as common carriers (available to any user), contract carriers (available by contract to particular market segments), and exempt (for-hire carriers that are exempt from regulation). Examples of private transportation companies are Greyhound, Smith Transfer, United Airlines, and Yellow Cab, to name a few.

2.3.2 Regulatory Agencies

Common carriers have been regulated by the government since the late 1800s, when abuses by the railroads created a climate of distrust toward the railroad "robber barons," who used their monopoly powers to grant preferential treatment to favored customers or charged high rates to those who were without alternative routes or services. The Interstate Commerce Commission (ICC) was formed to make certain that the public received dependable service at reasonable rates without discrimination. It was empowered to control the raising and lowering of rates, to require that the carriers had adequate equipment and maintained sufficient routes and schedules, and to certify the entry of a new carrier and control the exit of any certified carrier. Today, these concerns are less valid because the shipper has alternatives other than shipping by rail, and the entry and exit restrictions placed on companies no longer tend to favor the status quo or limit innovation and opportunities for new service. For these reasons, the ICC, which had regulated railroads since 1887 and trucking since 1935, was abolished; now private carriers can operate in a more independent economic environment that should result in lower costs and improved services. Similarly, the Civil Aeronautics Board (CAB), which regulated the airline industry starting in 1938, is no longer empowered to certify routes and fares of domestic airline companies; it was phased out in 1985. Other regulatory agencies are the Federal Maritime Commission, which regulates U.S. and foreign vessels operating in international commerce, and the Federal Energy Regulatory Commission, which regulates certain oil and natural gas pipelines.

2.3.3 Federal Agencies

Because transportation pervades our economy, each agency within the executive branch of the federal government is involved in some aspect of transportation. For example, the Department of State develops policy recommendations concerning international aviation and maritime transportation, and the Department of Defense, through the Army Corps of Engineers, constructs and maintains river and harbor improvements and administers laws protecting navigable waterways.

The Department of Transportation is the principal assistant to the President of the United States in all matters relevant to federal transportation programs. Within the department are the 10 administrations, illustrated in Figure 2.9, that deal with programs for highways, aviation, railroads, transit, motor carrier safety, highway traffic safety, research and innovative technology, maritime, pipelines, hazardous materials safety, and the St. Lawrence Seaway Development Corporation. The department also deals with special problems such as public and consumer affairs, civil rights, and international affairs.

The U.S. Congress, in which the Senate and the House of Representatives represent the people, has jurisdiction over transportation activities through the budget and legislative process. Two committees in the Senate are concerned with transportation: the Commerce, Science, and Transportation Committee and the Committee on Environment and Public Works. The two House committees are Commerce, and Transportation and

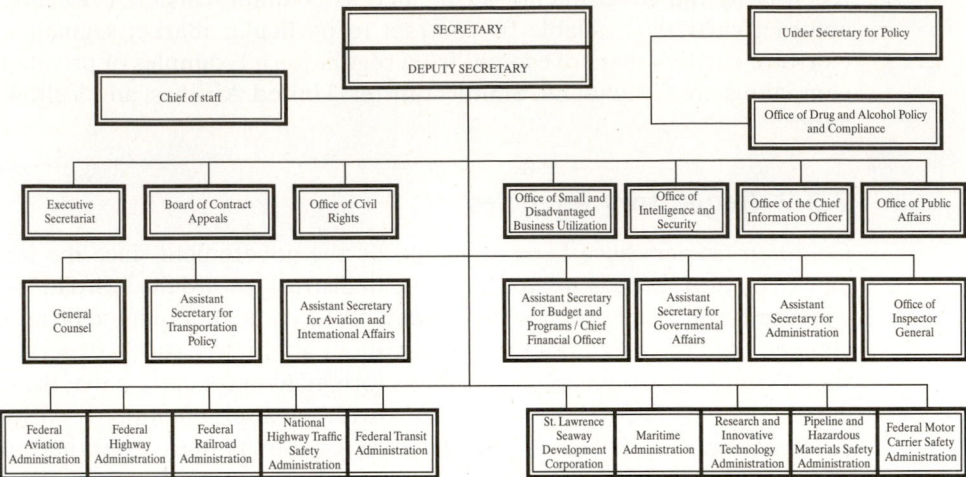

Figure 2.9   Organization Chart of the U.S. Department of Transportation 2007

SOURCE: U.S. Department of Transportation, Washington, D.C.

Infrastructure. The Appropriations Committees in both the Senate and the House have transportation subcommittees.

2.3.4 State and Local Agencies and Authorities

Each of the 50 states has its own highway or transportation department that is responsible for planning, building, operating, and maintaining its highway system and for administering funds and programs in other modes such as rail, transit, air, and water. These departments also may be responsible for driver licensing and motor vehicle registration, policing, and inspection. The organization and functions of these departments vary considerably from state to state, but since highway programs are a direct state responsibility and other modes are either privately owned and operated (such as air and rail) or of greater concern at a local level (such as public transit), highway matters tend to predominate at the state level. In many states, the responsibilities of state transportation departments closely parallel those of the U.S. Department of Transportation.

Local agencies are responsible for carrying out specific transportation functions within a prescribed geographic area. Larger cities may have a regional transportation authority that operates the bus and rapid transit lines, and many communities have a separate traffic department responsible for operating the street system, signing, traffic signal timing, and parking controls. Local roads are often the responsibility of county or township agencies, which vary considerably in the quality of engineering staff and equipment. The jurisdiction and administration of the roads in a state can create severe economic and management problems if the responsibility for road improvements is fragmented and politicized.

2.3.5 Trade Associations

Americans are joiners; for each occupation or business involved in transportation, there is likely to be an organization that represents its interests. These associations are an attempt to present an industry-wide front in matters of common interest. They also promote and develop new procedures and techniques to enhance the marketability of their products and to provide an opportunity for information exchange.

Examples of modally oriented organizations are the Association of American Railroads (AAR), the American Road and Transportation Builders Association (ARTBA), the American Public Transit Association (APTA), and the American Bus Association (ABA).

In addition to carrying out traditional lobbying efforts, trade associations collect and disseminate industry data, perform research and development, provide self-checking for safety matters, publish technical manuals, provide continuing career education, and provide electronic-data exchange between carriers and shippers.

Examples of trade associations that improve product performance and marketability are the Asphalt Institute (AI) and the American Concrete Pavement Association. These groups publish data and technical manuals used by design engineers and maintain technical staffs for special consultations. These and many similar associations attempt to keep abreast of changing conditions and to provide communications links for their members and between the many governmental agencies that use or regulate their products.

2.3.6 Professional Societies

Professional societies are composed of individuals with common interests in transportation. Their purposes are to exchange ideas, to develop recommendations for design and operating procedures, and to keep their memberships informed about new developments in transportation practice. Membership in professional organizations is essential for any professional who wishes to stay current in his or her field.

Examples of professional societies are the Women's Transportation Seminar (WTS) and the Institute of Transportation Engineers (ITE), which represent professionals who work in companies or agencies as transportation managers, planners, or engineers. Members of the American Association of State Highway and Transportation Officials (AASHTO) are representatives of state highway and transportation departments and of the Federal Highway Administration. AASHTO produces manuals, specifications standards, and current practices in highway design, which form the basis for practices throughout the country.

The Transportation Research Board (TRB) of the National Academies is a division of the National Research Council and is responsible for encouraging research in transportation and disseminating the results to the professional community. It operates through a technical committee structure composed of knowledgeable practitioners who assist in defining research needs, review and sponsor technical sessions, and conduct workshops and conferences. The Transportation Research Board receives support from state transportation departments, federal transportation administrations, trade associations, transportation companies, and individual memberships. The TRB annual meeting was created to provide a venue to share results of research and to foster information exchange. Attendees represent the United States and countries throughout the world.

2.3.7 Users of Transport Services

The transportation user has a direct role in the process of effecting change in transportation services and is represented by several transportation groups or associations. The American Automobile Association (AAA) has a wide membership of highway users and serves its members as a travel advisory service, lobbying organization, and insurance agency. The American Railway Passenger Association, the Bicycle Federation of America, and other similar consumer groups have attempted to influence legislation to improve transportation services such as requiring passenger-nonsmoking sections or building bike lanes. Also, environmental groups have been influential in ensuring that highway planning includes consideration of air quality, land use, wetlands, and noise. In general, however, the average transportation consumer plays a minor role in the transportation decision process and is usually unaware of the way in which transportation services came about or are supported. Quite often, the only direct involvement of the consumer is through citizen participation in the selection process for new transportation facilities in the community.

2.4 SUMMARY

The transportation system in a developed nation consists of a network of modes that have evolved over many years. The system consists of vehicles, guideways, terminal facilities, and control systems; these operate according to established procedures and schedules in the air, on land, and on water. The system also requires interaction with the user, the operator, and the environment. The systems that are in place reflect the multitude of decisions made by shippers, carriers, government, individual travelers, and affected nonusers concerning the investment in or the use of transportation. The transportation system that has evolved has produced a variety of modes that complement each other. Intercity passenger travel often involves auto and air modes; intercity freight travel involves pipeline, water, rail, and trucking. Urban passenger travel involves auto or public transit; urban freight is primarily by truck.

The nation's transportation system can be considered to be in a state of equilibrium at any given point in time as a result of market forces, government actions, and transportation technology. As these change over time, the transportation system also will be modified. During recent decades, changes in gasoline prices, regulation by government, and new technology have affected the relative importance of each mode. The passenger or shipper thinks of each mode in terms of the level of service provided. Each mode offers a unique set of service characteristics at a given price: travel time, frequency, comfort, convenience, reliability, and safety. The traveler or shipper selects the mode based on how these attributes are valued.

The principal carriers of freight are rail, truck, pipeline, and water. Passenger transportation is by auto, air, rail, and bus. Highway transportation is the dominant mode in passenger travel. Trucks carry most freight in urban areas and are a principal mode in intercity travel. The United States highway system comprises 6.27 million km of roadway. The Interstate system, consisting of 77,000 km of limited-access roads, represents the backbone of the nation's highway network.

A wide range of organizations and agencies provide the resources to plan, design, build, operate, and maintain the nation's transportation system. These include private companies that furnish transportation; regulatory agencies that monitor the safety and service quality provided; federal, state, and local agencies that provide funds to build roads and airports and carry out legislation dealing with transportation at a national level; trade associations that represent the interests of a particular group of transportation providers or suppliers; professional organizations; and transportation user groups.

PROBLEMS

2-1   How would your typical day be changed without the availability of your principal mode of transportation? Consider both personal transportation as well as goods and services that you rely on.

2-2   What are the central problems in your state concerning one of the following: (a) air transportation, (b) railroads, (c) water transportation, (d) highways, or (e) public transportation? (To answer this question, obtain a copy of the governor's plan for transportation in your state or contact a key official in the transportation department.)

2-3   A bridge has been constructed between the mainland and an island. The total cost (excluding tolls) to travel across the bridge is expressed as $C = 50 + 0.5V$, where V is the number of veh/h and C is the cost/veh in cents. The demand for travel across the bridge is $V = 2500 - 10C$.

   (a) Determine the volume of traffic across the bridge.
   (b) If a toll of 25 cents is added, what is the volume across the bridge? What volume would be expected with a 50 cent increase?
   (c) A tollbooth is to be added, thus reducing the travel time to cross the bridge. The new cost function is $C = 50 + 0.2V$. Determine the volume of traffic that would cross the bridge.
   (d) Determine the toll to yield the highest revenue for demand and supply function in part (a), and the associated demand and revenue.

2-4   A toll bridge carries 6000 veh/day. The current toll is $3.50/vehicle. Studies have shown that for each increase in toll of 50 cents, the traffic volume will decrease by 500 veh/day. It is desired to increase the toll to a point where revenue will be maximized.

   (a) Write the expression for travel demand on the bridge, related to toll increase and current volume.
   (b) Determine the toll charge to maximize revenues.
   (c) Determine traffic in veh/day after the toll increase.
   (d) Determine the total revenue increase with the new toll.

2-5   Officials are considering to increasing the toll on a bridge that now carries 4500 veh/day. The current toll is $1.25/veh. Experience has shown that the daily traffic volume will decrease by 400 veh/day for each 25 cent increase in toll. Therefore, if x is the increase in toll in cents/veh, the volume equation for veh/day is $V = 4500 - 400(x/25)$, and the new toll/veh would be $T = 125 + x$. In order to maximize revenues, what would the new toll charge be per vehicle and what would the traffic in veh/day be after the toll increase?

2-6   A large manufacturer uses two factors to decide whether to use truck or rail for movement of its products to market: cost and total travel time. The manufacturer uses a utility formula that rates each mode. The formula is $U = 6C + 14T$, where C is cost ($/ton) and T is time (hr). For a given shipment of goods, a trucking firm can deliver in 12 hr and charges $30/ton, whereas a railroad charges $22/ton and can deliver in 16 hr.

   (a) Which mode should the shipper select?
   (b) What other factors should the shipper take into account in making a decision? (Discuss at least two.)

2-7   An individual is planning to take an 950-km trip between two large cities. Three possibilities exist: air, rail, or auto. The person is willing to pay $25 for every hour saved in making the trip. The trip by air costs $450 and travel time is 6 hr, by rail the cost is $400 and travel time is 10 hr, and by auto the cost is $250 and travel time is 15 hr.

   (a) Which mode is the best choice?
   (b) What factors other than cost might influence the decision regarding which mode to use?

2-8   Name the two key influences on transit system carrying capacity.

2-9   What factors affect the long-term viability of fuel taxes as a stable source of revenue to fund highway system improvements?

2-10   What emerging concepts for financing highway improvements are currently being explored?

2-11   Describe the organization and function of your state highway/transportation department.

2-12   What are the major activities performed by the highway/transportation department in your state as described by the organization chart and other information furnished on your state DOT's Web site?

2-13   Consult with the U.S. Department of Transportation Web site and identify the name and location of highways in your state that are included as part of the National Highway System.

2-14   List three transportation organizations located in your state. What services do they provide?

2-15   Obtain a copy of a *Transportation Research Record: Journal of the Transportation Research Board* (published by the Transportation Research Board).

(a) Select one article and write a short summary of its contents.
(b) Describe the technical area of transportation covered by this article.

2-16   Write out the names of the organizations represented by these acronyms and, for each organization, briefly note the type of organization, its purpose, and its members and constituency.

AAA   AAR   AASHTO   APTA
ARTBA   FHWA   TRB

2-17   List the seven categories of transportation organizations and cite one example of each.

2-18   Explain the role of AASHTO with respect to state highway/transportation agencies.

2-19   What are the four principal modes for moving freight? Which of these modes carries the largest share of tonne-kms? Which carries the lowest?

2-20   What are the four principal modes for moving people? Which of these modes accounts for the largest share of passenger-kms? Which mode accounts for the lowest?

2-21   (a) List four major factors that will determine the future of public transportation in the United States.
(b) Indicate if the factor is positive, neutral, or negative to the success of transit.

2-22   What are the advantages and disadvantages of using intercity bus transportation?

REFERENCES

Black, Alan, *Urban Mass Transportation Planning.* McGraw-Hill, Inc., 1995.

Coyle, John J., Novack, Robert A., Gibson, B., and Bardi, Edward J., *Transportation: A Supply Chain Perspective,* 6th Edition. Thomson-South Western Cengage Learning, 2011.

Hoel, L. A., Garber, N. J., and Sadek, A. S., *Transportation Infrastructure Engineering: A Multi-Modal Integration.* Thomson-Nelson, 2008.

Hoel, L. A., Meyer, M. D., and Guiliano, G., Editors. *Intermodal Transportation: Moving Freight in a Global Economy.* Eno Transportation Foundation, 2011.

National Transportation Statistics. U.S. Department of Transportation Research and Innovative Technology Administration, Bureau of Transportation Statistics, 2011.

TRNews: *Special Issue Commemorating the Interstate's 50th Anniversary.* Transportation Research Board, National Academies, May–June 2006.

Transportation Research Board of the National Academies, *The Fuel Tax and Alternatives for Transportation Funding.* Special Report 285, Washington, D.C., 2005.

PART 2

Traffic Operations

The traffic or highway engineer must understand not only the basic characteristics of the driver, the vehicle, and the roadway, but how each interacts with the others. Information obtained through traffic engineering studies serves to identify relevant characteristics and define related problems. Traffic flow is of fundamental importance in developing and designing strategies for intersection control, rural highways, and freeway segments.

CHAPTER 3

Characteristics of the Driver, the Pedestrian, the Bicyclist, the Vehicle, and the Road

The four main components of the highway mode of transportation are the driver, the pedestrian, the vehicle, and the road. The bicycle is also becoming an important component in the design of urban highways and streets. To provide efficient and safe highway transportation, a knowledge of the characteristics and limitations of each of these components is essential. It is also important to be aware of the interrelationships among these components in order to determine the effects, if any, that they have on each other. Their characteristics are also of primary importance when traffic engineering measures, such as traffic-control devices, are to be used in the highway mode. Knowing average limitations may not always be adequate; it may be necessary to obtain information on the full range of limitations. Consider, for example, the wide range of drivers' ages in the United States, which usually begins at 16 and can exceed 85.

Highway statistics provided by the Federal Highway Administration indicate that the number of drivers age 70 years and older with a valid license increased from 8.8 million in 1980 to 21.9 million in 2009, representing an increase of about 149 percent of drivers in this age group since 1980. Sight and hearing vary considerably across age groups, with the ability to hear and see usually decreasing after age 65. In addition, these can vary even among individuals of the same age group.

Similarly, a wide range of vehicles, from compact cars to articulated trucks, is being designed. The maximum acceleration, turning radii, and ability to climb grades vary considerably among different vehicles. The road therefore must be designed to accommodate a wide range of vehicle characteristics while allowing use by drivers, cyclists, and pedestrians with a wide range of physical and psychological characteristics.

This chapter discusses the main components of the highway mode and demonstrates their importance and their use in the design and operation of highway facilities.

CHAPTER OBJECTIVES:

- Become familiar with the main components of the highway mode.
- Understand the relationships among the different components of the highway mode.

- Understand the human response process.
- Become familiar with the characteristics of the driver, the vehicle, and the highway.
- Understand how the driver and vehicle characteristics influence the geometric design of the highway.

3.1 DRIVER CHARACTERISTICS

One problem that faces traffic and transportation engineers when they consider driver characteristics in the course of design is the varying skills and perceptual abilities of drivers on the highway, as demonstrated by a wide range of abilities to hear, see, evaluate, and react to information. Studies have shown that these abilities may also vary in an individual under different conditions, such as the influence of alcohol, fatigue, and the time of day. Therefore, it is important that criteria used for design purposes be compatible with the capabilities and limitations of most drivers on the highway. A principal concern is that the engineer must balance the tradeoffs between accommodating the abilities of as many drivers as possible with the potential cost implications of designing for drivers with abilities far below normal. However, the use of an average value, such as mean reaction time, may not be adequate for a large number of drivers. Both the 85th percentile and the 95th percentile have been used to select design criteria; in general, the higher the chosen percentile, the wider the range covered.

3.1.1 The Human Response Process

Actions taken by drivers on a road result from their evaluation of and reaction to information they obtain from certain stimuli that they see or hear. However, evaluation and reaction must be carried out within a very short time, as the information being received along the highways is continually changing. Most of the information received by a driver is visual, implying that the ability to see is of fundamental importance in the driving task. It is therefore important that highway and traffic engineers have some fundamental knowledge of visual perception as well as of hearing perception.

Visual Reception

The principal characteristics of the eye are visual acuity, peripheral vision, color vision, glare vision and recovery, and depth perception.

Visual Acuity.   Visual acuity is the ability to see fine details of an object. It reflects the quality of an individual's sight along his/her direct line of vision. Visual acuity is usually measured by using the Snellen eye chart, which is based on the usual standard length of an eye exam room (6 m, normal visual acuity is taken as 6/6, which indicates that an individual with a 6/6 visual acuity can distinguish a letter that is subtended by an angle of 5' of arc ($\frac{1}{12}$ of a degree) at the eye. It can be represented by the visual angle, which is the angle that a viewed object subtends at the eye, given in degrees of arc and given as

$$\varphi = 2 \arctan\left(\frac{H}{2D}\right) \tag{3.1}$$

where
- H = height of the target letter or symbol
- D = distance from the eye to target in the same units as H
- φ = angle of arc subtended at the eye

For example, given the ability to resolve a pattern detail with a visual acuity of 6/6, at a distance 6 m from the object, the height of the object can be determined as shown below:

Angle of arc with 20/20 vision = 5' = 0.0833 degrees

At 6 m away,

$$0.0833 = 2 \arctan (H/(2 \times 6))$$
$$\tan (0.0833/2) = H/12$$
$$H = 8.7 \text{ mm} = 0.0087 \text{ m}$$

This value of 8.7 mm therefore represents the height of a 6/6 letter in the Snellen acuity eye chart. Note that a 6/6 visual acuity does not necessarily indicate perfect vision, as one can have a better acuity then 6/6 vision. The lower the bottom number in the acuity ratio, the better the acuity. For example, an acuity vision of 6/4 is better than that of 6/6. For example, if a driver with an acuity vision of 6/6 can read a sign 9 m away, the same sign can be read at a distance of 12 m by a another driver with an acuity vision of 6/4; $(6/4) \times 8 = 12$ m. This means that if a driver with a visual acuity of 20/20 is just able to distinguish a sign at a distance d ft from the sign, then a driver with a visual acuity of 6/8 can see the same sign at a distance of $(6/8) \times d$ ft, or one with a visual acuity of 6/4 can see the same sign at $(6/4) \times d$. Two types of visual acuity are of importance in traffic and highway emergencies: static and dynamic visual acuity. The driver's ability to identify an object when both the object and the driver are stationary depends on his or her static acuity. Factors that affect static acuity include background brightness, contrast, and time. Static acuity increases with an increase in illumination up to a background brightness of about 32 candles (cd)/sq m and then remains constant even with an increase in illumination. When other visual factors are held constant at an acceptable level, the optimal time required for identification of an object with no relative movement is between 0.5 and 1.0 seconds.

The driver's ability to clearly detect relatively moving objects, not necessarily in his or her direct line of vision, depends on the driver's dynamic visual acuity. Most people have clear vision within a conical angle of 3 to 5 degrees and fairly clear vision within a conical angle of 10 to 12 degrees. Vision beyond this range is usually blurred. This is important when the location of traffic information devices is considered. Drivers will see clearly those devices that are within the 12-degree cone, but objects outside this cone will be blurred.

Peripheral Vision.   Peripheral vision is the ability of people to see objects beyond the cone of clearest vision. For example, a driver sees a vehicle approaching from his/her side because of peripheral vision. Although objects can be seen within this zone, details and color are not clear. The cone for peripheral vision could be one subtending up to 160 degrees; this value is affected by the speed of the vehicle. Age also influences peripheral vision. For instance, at about age 60, significant reductions can occur in a person's peripheral vision.

Color Vision.   Color vision is the ability to differentiate one color from another, but deficiency in this ability, usually referred to as *color blindness,* is not of great significance in highway driving because other ways of recognizing traffic information devices (e.g., shape) can compensate for it. Combinations of black and white and black and yellow have been shown to be those to which the eye is most sensitive and are therefore commonly used in warning and regulatory traffic signs.

Glare Vision and Recovery.   There are two types of glare vision: direct and specular. Rowland and others have indicated that *direct glare* occurs when relatively bright light appears in the individual's field of vision, like headlights shining in a driver's eyes, and *specular glare* occurs when the image reflected by the relatively bright light appears in the field of vision, like the image of the sun reflected off a windshield. Both types of glare result in a decrease of visibility and cause discomfort to the eyes. It is also known that age has a significant effect on the sensitivity to glare; at about age 40, a significant increase occurs in a person's sensitivity to glare.

The time required by a person to recover from the effects of glare after passing the light source is known as *glare recovery*. Studies have shown that this time is about 3 seconds when moving from dark to light and can be 6 seconds or more when moving from light to dark. Glare vision is of great importance during night driving for older people, who see much more poorly at night. This phenomenon should be taken into account in the design and location of street lighting so that glare effects are reduced to a minimum.

Glare effects can be minimized by reducing luminaire brightness and by increasing the background brightness in a driver's field of view. Specific actions taken to achieve this in lighting design include using higher mounting heights, positioning lighting supports farther away from the highway, and restricting the light from the luminaire to obtain minimum interference with the visibility of the driver. Likewise, glare shields or vegetated medians can reduce glare effects due to headlights.

Depth Perception.   Depth perception is the ability to see objects in three dimensions and estimate speed and distance. It is particularly important on two-lane highways during passing maneuvers or turning maneuvers, when crashes may result from a lack of proper judgment of speed and distance.

The ability of the human eye to differentiate between objects is fundamental to this phenomenon. It should be noted, however, that the human eye is not very good at estimating absolute values of speed, distance, size, and acceleration. This is why traffic-control devices are standard in size, shape, and color. Standardization not only aids in distance estimation but also helps the color-blind driver to identify signs.

Hearing Perception

The ear receives sound stimuli, which is important to drivers only when warning sounds, usually given out by emergency vehicles, are to be detected. Loss of some hearing ability is not a serious problem since it normally can be corrected by a hearing aid.

Example 3.1   Distance Away from Object for Different Visual Acuities

A driver with a visual acuity of 6/10 can just decipher a sign at a distance 6 m from a sign. Determine the maximum distance from the sign at which drivers with the following visual acuities will be able to see the same sign:

(i) 6/4
(ii) 6/15

Solution:

- **Determine distances for different visual acuities**
  - Solve for a distance that the sign can be seen for visual acuity of 6/6
    The distance $d = (10/6) \times 6 = 10$ m
  - For visual acuity of 6/4
    The distance $d = (6/4) \times 10 = 15$ m
  - For visual acuity of 6/15
    The distance $d = (6/15) \times 10 = 4$ m

3.2 PERCEPTION–REACTION PROCESS

The process through which a driver, cyclist, or pedestrian evaluates and reacts to a stimulus can be divided into four subprocesses:

1. *Perception:* The driver sees a control device, warning sign, vehicle, or object on the road.
2. *Identification:* The driver identifies the object or control device and thus understands the stimulus.
3. *Emotion:* The driver decides what action to take in response to the stimulus; for example, to step on the brake pedal, to pass, to swerve, or to change lanes.
4. *Reaction* or *volition:* The driver actually executes the action decided on during the emotion subprocess.

Time elapses during each of these subprocesses. The time that elapses from the start of perception to the end of reaction is the total time required for perception, identification, emotion, and volition, sometimes referred to as PIEV time or (more commonly) as perception-reaction time.

Perception-reaction time is an important factor in the determination of braking distances, which in turn dictates the minimum sight distance required on a highway and the length of the yellow phase at a signalized intersection. Perception-reaction time varies among individuals and may, in fact, vary for the same person as the occasion changes. These changes in perception-reaction time depend on how complicated the situation is, driver training, the existing environmental conditions, age, gender, whether the person is tired or under the influence of drugs and/or alcohol, and whether the stimulus is expected or unexpected.

Several studies have been conducted to determine perception-reaction times of drivers and pedestrians. These studies have resulted in a wide range of values and indicate that "driver expectation" is the most important factor that influences the value. These studies noted that the 85th-percentile time for drivers range from about 1.60 seconds to about 7.8 seconds for unexpected information and from about 1.0 second to about 5.1 seconds for expected information. The reaction time selected for design purposes should, however, be large enough to include reaction times for most drivers using the highways. Recommendations made by AASHTO stipulate a perception-reaction time of 2.5 seconds for stopping-sight distances. This encompasses the decision times for about 90 percent of drivers under most highway conditions. Note, however, that a perception-reaction time of 2.5 seconds may not be adequate for unexpected conditions or for some very complex conditions, such as those at multiphase at-grade intersections and ramp terminals. For example, when signals are unexpected, reaction times can increase by 35 percent.

Example 3.2 Distance Traveled during Perception-Reaction Time

A driver with a perception-reaction time of 2.5 sec is driving at 105 km/h when she observes that an accident has blocked the road ahead. Determine the distance the vehicle would move before the driver could activate the brakes. The vehicle will continue to move at 105 km/h during the perception-reaction time of 2.5 sec.

Solution:

- Convert km/h to m/s:

$$105 \text{ km/h} = \left(105 \times \frac{1000}{3600}\right) \text{m/s} = 105 \times 0.278 = 29.2 \text{ m/s}$$

- Find the distance traveled:

$$D = vt$$
$$= 29.2 \times 2.5$$
$$= 72.9 \text{ m}$$

where v = velocity and t = time.

3.3 OLDER DRIVERS' CHARACTERISTICS

Growth progression of the United States population as published by the Administration on Aging indicates that by 2030, the population of persons aged 60+ in the United States will be about 17.1 million, representing about 19.3 percent of the population, compared to 12.9 percent in 2009. This in turn will result in 20 percent of U.S. drivers being 65 years or older. Also, as noted above, as one grows older, his or her sensory, cognitive, and physical functioning abilities decline, which can result in older drivers being less safe than their younger counterparts and having a higher probability of being injured when involved in a crash. Older drivers seem to be depending more on the automobile for meeting their transportation needs, so it is vital that traffic and highway engineers consider these diminished characteristics of older drivers in making decisions about highway design and operational characteristics that are influenced by human characteristics. Specific declining abilities of older drivers include reduced visual acuity, reduced ability to see at night, and reduced flexibility and motion range. This group also suffers from narrower visual fields, greater sensitivity to glare, higher reaction times, and reduced muscle strength, which may result in the older driver having a higher crash risk. For example, reduction in visual acuity results in older drivers being less capable to discern letters on road signs, while greater sensitivity to glare results in older drivers taking a much longer time in recovering from the disabling effect of glare and perhaps failing to respond to roadway signs or roadway obstacles, particularly at work zones.

3.4 PEDESTRIAN CHARACTERISTICS

Pedestrian characteristics relevant to traffic and highway engineering practice include those of the driver, discussed in the preceding sections. In addition, other pedestrian characteristics may influence the design and location of pedestrian-control devices.

Such control devices include special pedestrian signals, safety zones and islands at intersections, pedestrian underpasses, elevated walkways, and crosswalks. Apart from visual and hearing characteristics, walking characteristics play a major part in the design of some of these controls. For example, the design of an all-red phase, which permits pedestrians to cross an intersection with heavy traffic, requires knowledge of the walking speeds of pedestrians. Observations of pedestrian movements have indicated that walking speeds vary between 0.9 and 2.4 m/s. Significant differences have also been observed for different age groups and between male and female walking speeds. At intersections, the mean male walking speed has been determined to be 1.5 m/s, and for females, 1.4 m/s. A more conservative value of 4.0 ft/sec is normally used for design purposes. However, studies have shown that the average walking speed depends on the population of elderly pedestrians. For example, the average walking speed is 1.22 m/s when the percentage of elderly pedestrians is 20 percent or lower, but decreases to 0.9 m/s when the percentage of elderly pedestrians is higher than 20 percent. This factor therefore should be taken into consideration for the design of intersection pedestrian signals at locations where a high number of older pedestrians is expected. For example, it has been recommended that a walking speed of 0.88 m/s should be used for pedestrian clearance at intersections where pedestrians 65 years and older are predominant. Consideration also should be given to the characteristics of handicapped pedestrians, such as the blind. For example, many states are now incorporating accessible pedestrian signals (APSs), which are devices that transmit information on pedestrian signal timing in a nonvisual format such as audible tones, speech messages, and /or vibrating surfaces. An example of an APS is a device that gives audible information to blind pedestrians on the WALK and DON'T WALK intervals at signalized intersections. Studies have shown that crashes involving blind pedestrians can be reduced by installing special signals. Ramps are also now being provided at intersection curbs to facilitate the crossing of the intersection by the occupant of a wheelchair. Also, consideration should be given to the relatively lower average walking speed of the physically challenged pedestrian, which can vary from a low of 0.6–1.12 m/s. These factors have led to the *Manual of Uniform Traffic Control Devices* recommending that a pedestrian walking speed of 1.07 m/s rather than 1.2 m/s be used in the design of pedestrian intervals and signal phases.

3.5 BICYCLISTS AND BICYCLES CHARACTERISTICS

Bicycles are now an important component of the highway mode, especially for highways located in urban areas. It is therefore essential that highway and traffic engineers understand the characteristics of bicycles and bicyclists. The basic human factors discussed for the automobile driver also apply to the bicyclist, particularly with respect to perception and reaction. However, unlike the automobile driver, the bicyclist is not only the driver of the bicycle, but he/she also provides the power to move the bicycle. The bicycle and the bicyclist therefore unite to form a system, thus requiring that both be considered jointly.

Three classes of bicyclists (A, B, and C) based on skill level and comfort have been identified in the *Guide for the Development of Bicycle Facilities* by AASHTO. Experienced or advanced bicyclists are within class A, while less experienced bicyclists are within class B, and children riding on their own or with parents are classified as C. Class A bicyclists typically consider the bicycle as a motor vehicle and can comfortably ride in traffic. Class B bicyclists prefer to ride on neighborhood streets and are more comfortable on designated bicycle facilities, such as bicycle paths. Class C bicyclists use mainly residential streets that provide access to schools, recreational facilities, and stores.

In designing urban roads and streets, it is useful to consider the feasibility of incorporating bicycle facilities that will accommodate class B and class C bicyclists.

A study by the Toole Design Group for the National Cooperative Highway Research Program, Transportation Research Board of the National Academies also suggested two classes of bicyclists based on skill and comfort: "experienced and confident" and "casual and less confident." "Experienced and confident" bicyclists can ride comfortably on nearly all types of bicycle facilities and may include groups such as commuters, long-distance riders, and bicyclists, who are regular participants of long bicycle trips organized by bicycle clubs. Bicyclists in the "casual and less confident" class include the majority of bicyclists, who may ride frequently for different trip purposes; recreational riders; those who are comfortable riding only on bicycle facilities on low-volume streets when conditions are favorable; and those for whom the necessary mode of transportation is the bicycle. The study also suggests that a well-designed, physically visible network of bicycle facilities is necessary if this group is to select the bicycle as the mode of transportation.

The bicycle, like the automobile, also has certain characteristics that are unique. For example, studies have suggested that for experienced and confident riders, speeds vary from 32 km/h on level terrain to 72 km/h on steep downgrades, and for casual and less confident riders, speeds are usually between 12 to 20 km/h.

3.6 VEHICLE CHARACTERISTICS

Criteria for the geometric design of highways are partly based on the static, kinematic, and dynamic characteristics of vehicles. Static characteristics include the weight and size of the vehicle, while kinematic characteristics involve the motion of the vehicle without considering the forces that cause the motion. Dynamic characteristics involve the forces that cause the motion of the vehicle. Since nearly all highways carry both passenger-automobile and truck traffic, it is essential that design criteria take into account the characteristics of different types of vehicles. A thorough knowledge of these characteristics will aid the highway and/or traffic engineer in designing highways and traffic-control systems that allow the safe and smooth operation of a moving vehicle, particularly during the basic maneuvers of passing, stopping, and turning. Therefore, designing a highway involves the selection of a *design vehicle,* whose characteristics will encompass those of nearly all vehicles expected to use the highway. The characteristics of the design vehicle are then used to determine criteria for geometric design, intersection design, and sight-distance requirements.

3.6.1 Static Characteristics

The size of the design vehicle for a highway is an important factor in the determination of design standards for several physical components of the highway. These include lane width, shoulder width, length and width of parking bays, and lengths of vertical curves. The axle weights of the vehicles expected on the highway are important when pavement depths and maximum grades are being determined.

For many years, each state prescribed by law the size and weight limits for trucks using its highways, and in some cases local authorities also imposed more severe

Table 3.1 Range of State Limits on Vehicle Lengths by Type and Maximum Weight of Vehicle

Type	Allowable Lengths (ft)
Bus	35–60
Single truck	35–60
Trailer, semi/full	35–48
Semitrailer	55–85
Truck trailer	55–85
Tractor semitrailer trailer	55–85
Truck trailer trailer	65–80
Tractor semitrailer, trailer, trailer	60–105

Type	Allowable Weights (lb)
Single axle	20,000–22,400
Tandem axle	34,000–44,000
State maximum gross vehicle weight	80,000–129,000
Interstate maximum gross vehicle weight	80,000–117,000

SOURCE: Adapted from *Vehicle Sizes and Weights Manual*, J. J. Keller & Associates, Inc.®, July 2010.

restrictions on some roads. Table 3.1 shows some features of static characteristics for which limits were prescribed. A range of maximum allowable values is given for each feature.

Since the passage of the Surface Transportation Assistance Act of 1982, the maximum allowable truck sizes and weights on interstate and other qualifying federal-aided highways are at most:

- 36,500 kg gross weight, with axle loads of up to 9100 kg for single axles and 15,400 kg for tandem (double) axles
- 2.60 m width for all trucks
- 14.6 m length for semitrailers and trailers
- 8.5 m length for each twin trailer

(*Note:* Those states that had higher weight limits before this law was enacted are allowed to retain them for intrastate travel.)

The federal regulations also stipulate that the overall maximum gross weight for a group of two or more consecutive axles should be determined from the "Bridge Formula" given in Eq. 3.2.

$$W = 230\left[\frac{LN}{N-1} + 23N + 36 \right] \tag{3.2}$$

where

W = overall gross weight (calculated to the nearest 230 kg)
L = the extreme spacing of any group of two or more consecutive axles (m)
N = number of axles in the group under consideration

The regulations also stipulate that a gross load of 15,400 kg may be carried by two consecutive sets of tandem axles if the overall distance between the first and last axles of the consecutive sets is 11 m or more. Equation 3.2 is used to determine whether a loaded truck satisfies the federal regulations for different axle configurations. Three conditions should be satisfied for a loaded truck to comply with the Bridge Formula.

(i)   Gross weight should not exceed 36,500 kg.
(ii)   Total weight on any single axle should not exceed 9100 kg.
(iii)   Total weight on any set of tandem axles should not exceed 15,400 kg.

Example 3.3 Estimating Allowable Gross Weight of a Truck

A 5-axle truck traveling on an interstate highway has the following axle characteristics:

Distance between the front single axle and the front set of tandem axles = 6 m
Distance between front single axle and the back set of tandem axles = 20 m
Distance between the first set of tandem axles and the back set of tandem axles = 15 m
Load carried by front set of single axles = 4600 kg
Load carried by each axle of the tandem axles = 7360 kg

Determine whether this truck satisfies federal weight regulations.

Solution: Overall gross weight = (2 sets of tandem axles × 2 axles/tandem × 7360 kg/axle + 4600 kg on front single axle = 34,040 kg < 36,800 lb.

Although the overall gross weight is less than the maximum allowable of 36,800 lb, the allowable gross weight based on the axle configuration should be checked.
Use Eq. 3.2. For the front set of tandem axles to the second set of tandem axles:

$$W = 230\left[\frac{LN}{N-1} + 23N + 36\right]$$

$$W = 230\left[\frac{15 \times 4}{4-1} + 23 \times 4 + 36\right]$$

$$= 34{,}040\,\text{kg}$$

Total Applied Load = 7360 × 4 = 29,400 kg < 34,040 kg, so this is satisfied.
For the front single axle to the first set of tandem axles:

$$W = 230\left[\frac{6 \times 3}{3-1} + (23 \times 3) + 36\right] = 26{,}220\text{ kg}$$

Total Applied Load = 4600 + 7360 × 2 = 19320 kg < 26,220 kg, so this is satisfied.
For the front single axle to the second set of tandem axles:

$$230\left[\frac{20 \times 5}{5-1} + (23 \times 5) + 36\right] = 40{,}480\text{ kg}$$

However, max load allowable is 34,040 kg
Total Applied Load = (7360 × 4 + 4600) kg = 34,040 kg < 36,800. The truck therefore satisfies the federal truck weight regulations.

States are no longer allowed to set limits on overall truck length. As stated earlier, the static characteristics of vehicles expected to use the highway are factors that influence the selection of design criteria for the highway. It is therefore necessary that all vehicles be classified so that representative static characteristics for all vehicles within a particular class can be provided for design purposes. AASHTO has selected four general classes of vehicles: passenger cars, buses, trucks, and recreational vehicles. Included in the passenger-car class are sport/utility vehicles, minivans, vans, and pick-up trucks. Included in the bus class are intercity motor coaches, city transit buses, conventional school buses, large school buses, and articulated buses. Within the class of trucks are single-unit trucks, intermediate-semitrailers (WB-40), interstate semitrailers (WB-62, and WB-67), double bottom–semitrailer/trailer (WB67-D), triple-semitrailer/trailers (WB-100T), and turnpike double-semitrailer/trailers (WB-109D). Within the class of recreational vehicles are motor homes, cars with camper trailers, cars with boat trailers, and motor homes, boat trailers, and farm tractors. A total of 19 different design vehicles have been selected to represent the different categories of vehicles within all four classes. Table 3.2 shows the physical dimensions for each of these design vehicles, and Figure 3.1 shows examples of different types of trucks.

AASHTO also has suggested the following guidelines for selecting a design vehicle:

- For a parking lot or series of parking lots, a passenger car may be used.
- For intersections on residential streets and park roads, a single-unit truck could be considered.
- For the design of intersections of state highways and city streets that serve bus traffic but with relatively few large trucks, a city transit bus may be used.
- For the design of intersections of highways with low-volume county and township/local roads with Average Annual Daily Traffic (AADT; see Chapter 4 for definition) of 400 or less, a large school bus with a capacity of 84 passengers or a conventional bus with a capacity of 65 passengers may be used. The selection of the bus type depends on the expected frequency of each of the buses on the facility.
- For intersections of freeway ramp terminals and arterial highways, and for intersections of state highways and industrial streets with high traffic volumes, or with large truck access to local streets, the WB-65 or 67 may be used.

In carrying out the design of any of the intersections referred to above, the minimum turning radius for the selected design vehicle traveling at a speed of 10 mph should be provided. Minimum turning radii at low speeds (10 mi/h = 16 km/h or less) are dependent mainly on the size of the vehicle. The turning-radii requirements for single-unit (SU-9 [SU-30]) trucks and the WB-20 (WB-67) design vehicles are given in Figures 3.2 and 3.3, respectively. The turning-radii requirements for other vehicles can be found in AASHTO's Policy on Geometric Design of Highways and Streets. These turning paths were selected by conducting a study of the turning paths of scale models of the representative vehicles of each class. It should be emphasized, however, that the minimum turning radii shown in Figures 3.2 and 3.3 are for turns taken at speeds less than 10 mi/h = 16 km/h. When turns are made at higher speeds, the lengths of the transition curves are increased, so radii greater than the specified minimum are required. These requirements will be described later.

3.6.2 Kinematic Characteristics

The primary element among kinematic characteristics is the acceleration capability of the vehicle. Acceleration capability is important in several traffic operations, such as passing maneuvers and gap acceptance. Also, the dimensioning of highway features such as

Table 3.2 Design Vehicle Dimension

| | | U.S. Customary — Dimensions (ft) | | | | | | | | | | | |
| | | Overall | | | Overhang | | | | | | | | Typical Kingpin to Center of Rear Axle |
Design Vehicle Type	Symbol	Height	Width	Length	Front	Rear	WB_1	WB_2	S	T	WB_3	WB_4	
Passenger Car	P	4.25	7	19	3	5	11	—	—	—	—	—	—
Single-Unit Truck	SU-30	11-13.5	8.0	30	4	6	20	—	—	—	—	—	—
Buses													
Intercity Bus	BUS-40	12.0	8.5	40.5	6.3	9.0[a]	25.3	3.7	—	—	—	—	—
(Motor Coaches)	BUS-45	12.0	8.5	45.5	6.2	9.0[a]	28.5	4.0	—	—	—	—	—
City Transit Bus	CITY-BUS	10.5	8.5	40	7	8	25.0	—	—	—	—	—	—
Conventional School Bus (65 pass.)	S-BUS 36	10.5	8.0	35.8	2.5	12	21.3	—	—	—	—	—	—
Large School Bus (84 pass.)	S-BUS 40	10.5	8.0	40	7	13	20	—	—	—	—	—	—
Articulated Bus	A-BUS	11.0	8.5	60	8.6	10	22.0	19.4	6.2[b]	13.2[b]	—	—	—
Trucks													
Intermediate Semitrailer	WB-40	13.5	8.0	45.5	3	4.5[a]	12.5	25.5	—	—	—	—	25.5
Intermediate Semitrailer	WB-50	13.5	8.5	55	3	4.5[a]	14.6	35.4	—	—	—	—	37.5
Interstate Semitrailer	WB-62*	13.5	8.5	69.0	4	4.5[a]	19.5	41.0	—	—	—	—	41.0
Interstate Semitrailer	WB-67**	13.5	8.5	73.5	4	4.5[a]	19.5	45.5	—	—	—	—	45.5
"Double-Bottom"–Semitrailer/Trailer	WB-67D	13.5	8.5	72.3	2.3	3	11.0	23.0	3.0[c]	7.0[c]	22.5	—	23.0
Triple-Semitrailer/Trailer	WB-100T	13.5	8.5	104.8	2.3	3	11.0	22.5	3.0[d]	7.0[d]	22.5	22.5	22.5
Turnpike Double-Semitrailer/Trailer	WB-109D*	13.5	8.5	114	2.3	4.5[e]	12.2	40.0	4.5[e]	10.0[e]	40.0	—	40.5
Recreational Vehicles													
Motor Home	MH	12	8	30	4	6	20	—	—	—	—	—	—
Car and Camper Trailer	P/T	10	8	48.7	3	12	11	—	5	17.7	—	—	—
Car and Boat Trailer	P/B	—	8	42	3	8	11	—	5	15	—	—	—
Motor Home and Boat Trailer	MH/B	12	8	53	4	8	20	—	6	15	—	—	—
Farm Tractor[f]	TR	10	8-10	16[g]	—	—	10	9	3	6.5	—	—	—

* = Design vehicle with 48-ft trailer as adopted in 1982 Surface Transportation Assistance Act (STAA).

** = Design vehicle with 53-ft trailer as grandfathered in with 1982 Surface Transportation Assistance Act (STAA).

a = Length of the overhang from the back axle of the tandem axle assembly.

b = Combined dimension is 19.4 ft and articulating section is 4 ft wide.

c = Combined dimension is typically 10.0 ft.

d = Combined dimension is typically 10.0 ft.

e = Combined dimension is typically 12.5 ft.

f = Dimensions are for a 150–200 hp tractor excluding any wagon length.

g = To obtain the total length of tractor and one wagon, add 18.5 ft to tractor length. Wagon length is measured from front of drawbar to rear of wagon, and drawbar is 6.5 ft long.

Notes:
- WB_1, WB_2, WB_3, and WB_4 are the effective vehicle wheelbases, or distances between axle groups, starting at the front and working toward the back of each unit.
- S is the distance from the rear effective axle to the hitch point or point of articulation.
- T is the distance from the hitch point or point of articulation measured back to the center of the next axle or center of tandem axle assembly.

SOURCE: Based on *A Policy on Geometric Design of Highways and Streets*, 2011, AASHTO, Washington, D.C.

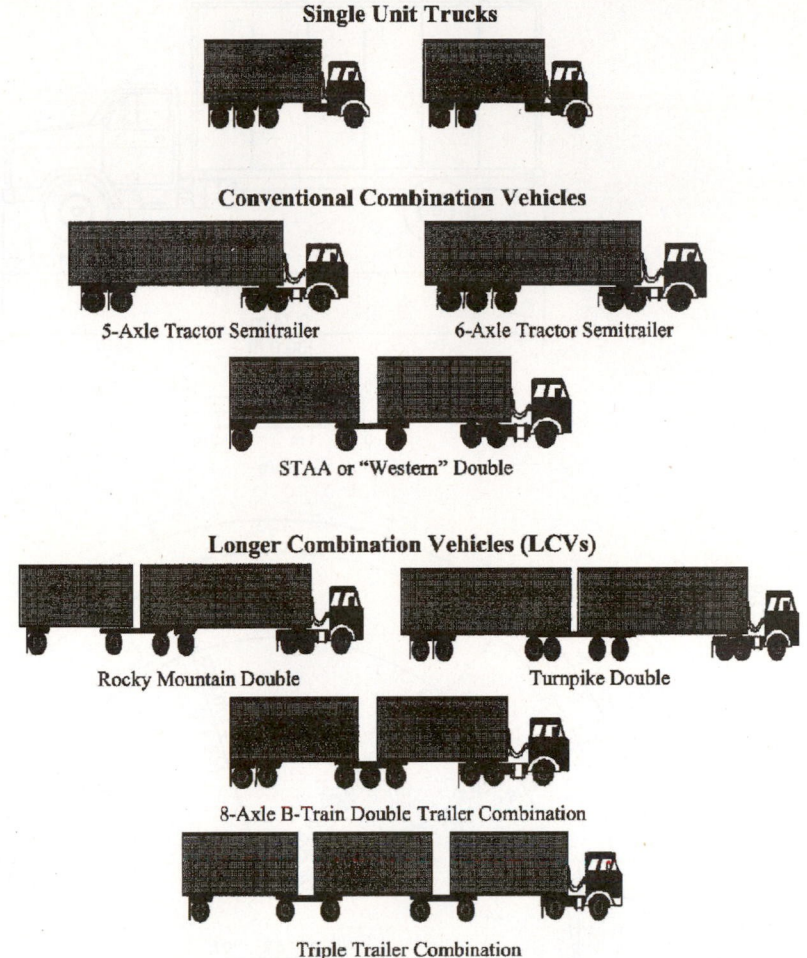

Figure 3.1 Examples of Different Types of Trucks

freeway ramps and passing lanes is often governed by acceleration rates. Acceleration is also important in determining the forces that cause motion. Therefore, a study of the kinematic characteristics of the vehicle primarily involves a study of how acceleration rates influence the elements of motion, such as velocity and distance. We therefore review in this section the mathematical relationships among acceleration, velocity, distance, and time.

Let us consider a vehicle moving along a straight line from point o to point m, a distance x in a reference plane T. The position vector of the vehicle after time t may be expressed as

$$r_{om} = x\hat{i} \tag{3.3}$$

where

$\quad r_{om}$ = position vector for m in T
$\quad \hat{i}$ = a unit vector parallel to line om
$\quad x$ = distance along the straight line

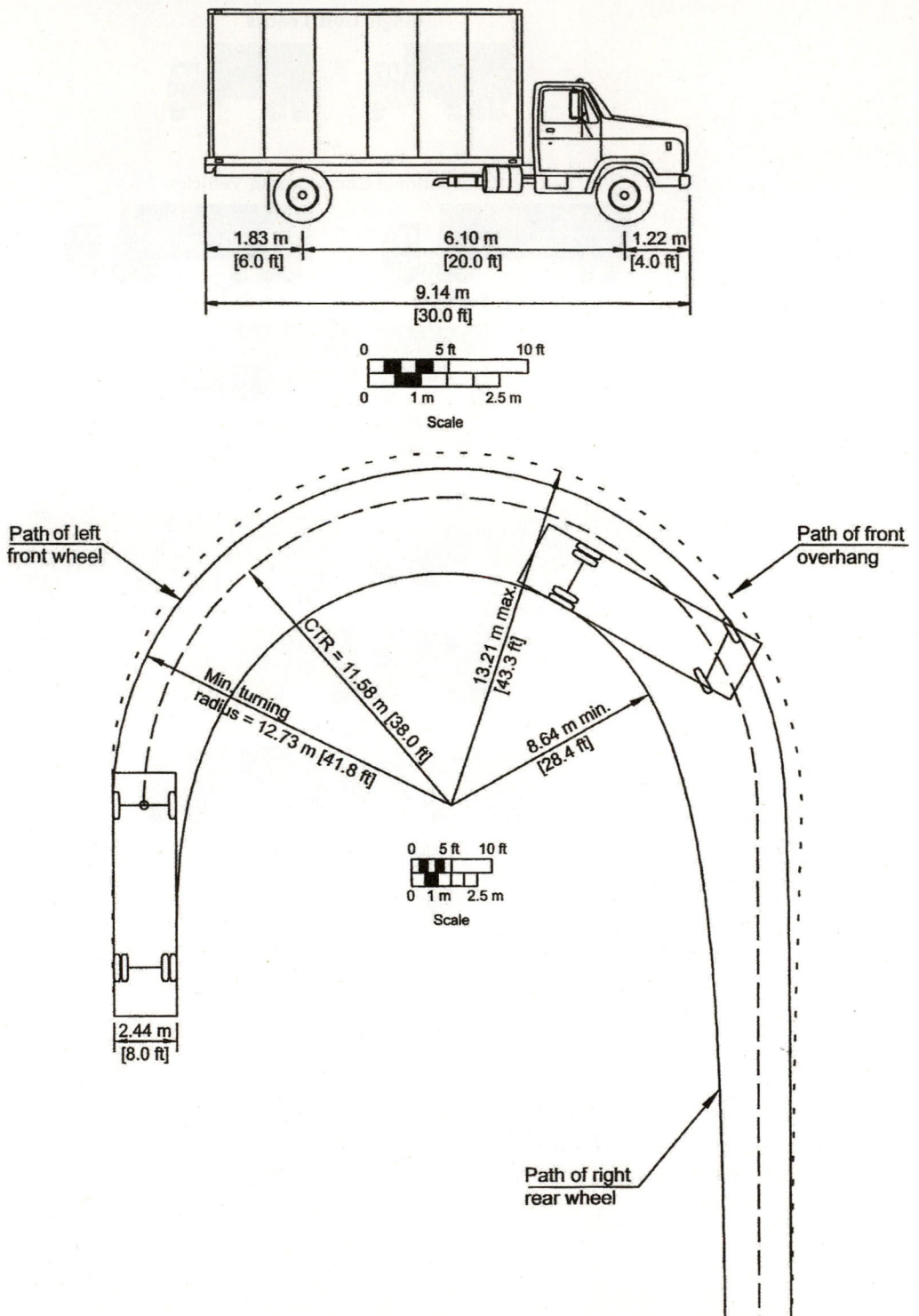

Figure 3.2 Minimum Turning Path for Single Unit (SU-9 [SU-30]) Design Vehicle

SOURCE: From Texas State Department of Highways and Public Transportation, reprinted in *A Policy on Geometric Design of Highways and Streets*, 2004, AASHTO, Washington, D.C. Used by permission.

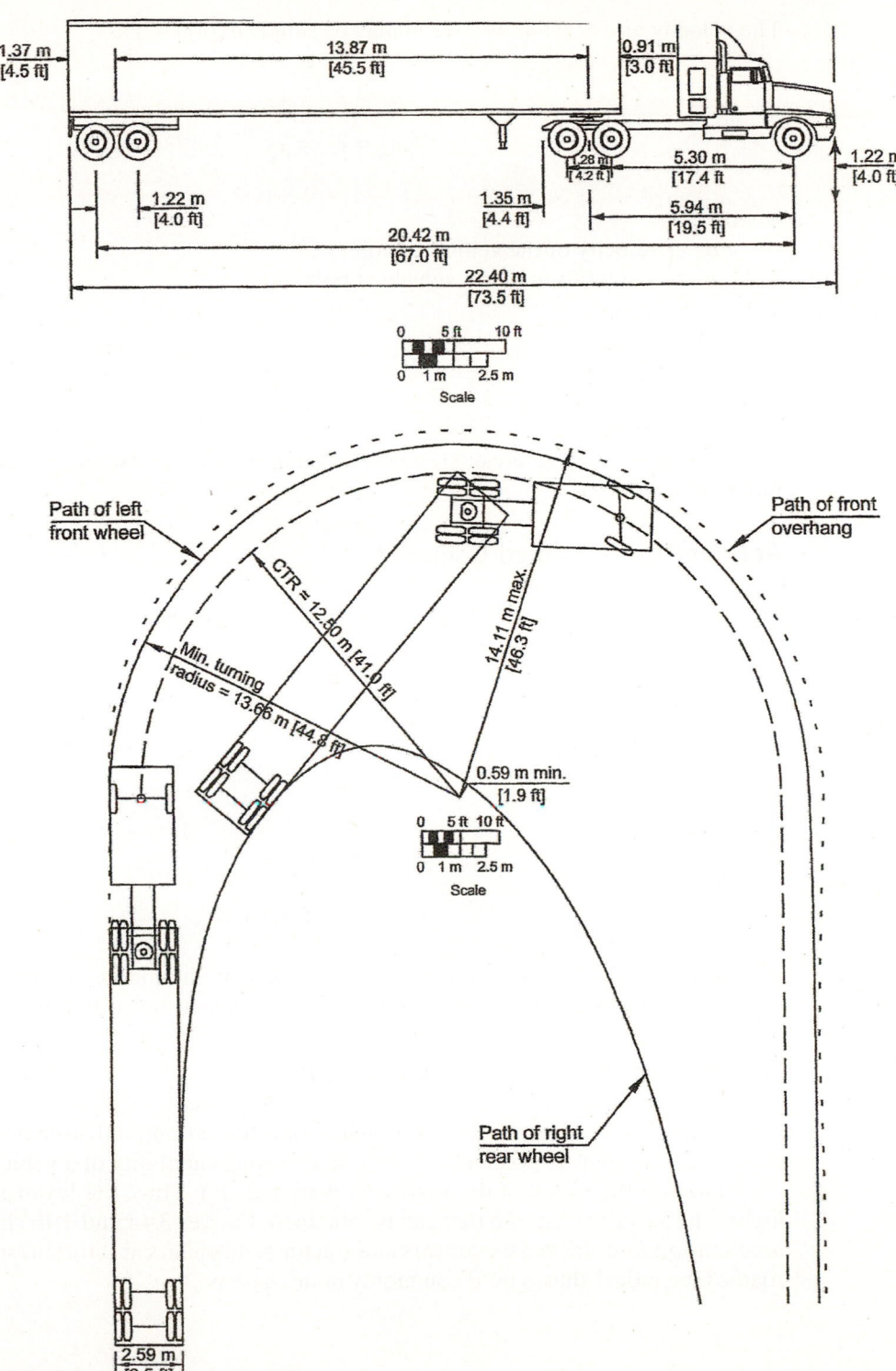

Figure 3.3 Minimum Turning Path for Interstate Semitrailer (WB-20 [WB-67]) Design Vehicle

SOURCE: From *A Policy on Geometric Design of Highways and Streets*, 2004, AASHTO, Washington, D.C. Used by permission.

The velocity and acceleration for m may be simply expressed as

$$u_m = \dot{r}_{om} = \dot{x}_{\hat{i}} \tag{3.4}$$

$$a_m = \ddot{r}_{om} = \ddot{x}_{\hat{i}} \tag{3.5}$$

where

u_m = velocity of the vehicle at point m

a_m = acceleration of the vehicle at point m

$$\dot{x} = \frac{dx}{dt}$$

$$\ddot{x} = \frac{d_2 x}{dt^2}$$

Two cases are of interest: (1) acceleration is assumed constant; (2) acceleration is a function of velocity.

Acceleration Assumed Constant

When the acceleration of the vehicle is assumed to be constant,

$$\ddot{x}_i = a$$

$$\frac{d\dot{x}}{dt} = a \tag{3.6}$$

$$\dot{x} = at + C_1 \tag{3.7}$$

$$x = \frac{1}{2}at^2 + C_1 t + C_2 \tag{3.8}$$

The constants d, C_1, and C_2, are determined either by the initial conditions on velocity and position or by using known positions of the vehicle at two different times.

Acceleration as a Function of Velocity

The assumption of constant acceleration has some limitations and often does not realistically represent vehicle acceleration. The accelerating capability of a vehicle at any time t is related to the speed of the vehicle at that time (u_t). Thus, the lower the speed, the higher the acceleration rate that can be obtained. Figures 3.4a and 3.4b show maximum acceleration rates for passenger cars and tractor-semitrailers at different speeds on level roads. One model that is used commonly in this case is

$$\frac{du_t}{dt} = \alpha - \beta u_t \tag{3.9}$$

where α and β are constants.

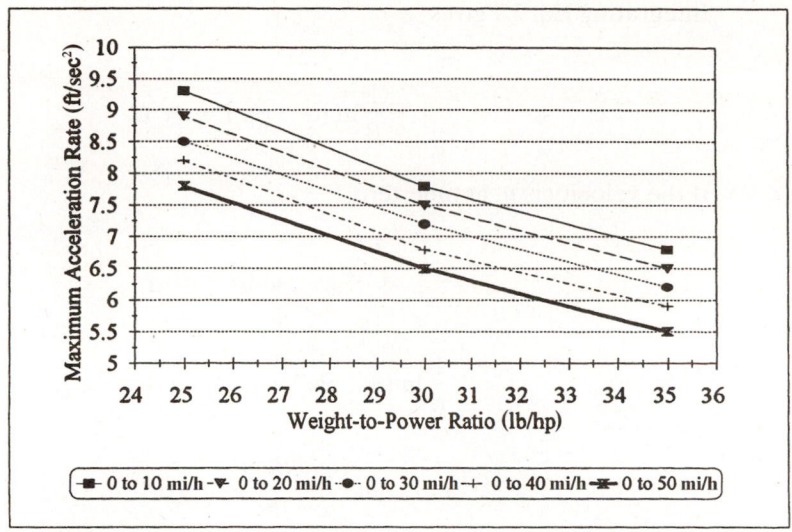

(a) Passenger Cars

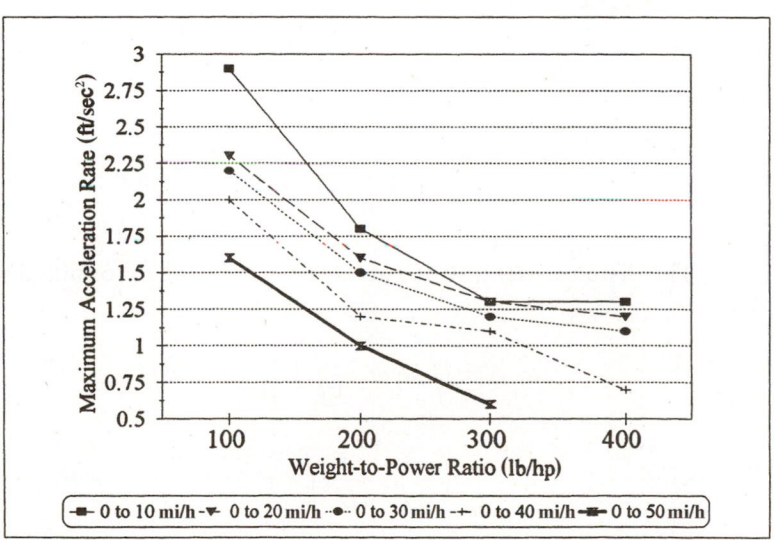

(b) Tractor-Semitrailers

Figure 3.4 Acceleration Capabilities of Passenger Cars and Tractor-Semitrailers on Level Roads

SOURCE: Modified from *Traffic Engineering Handbook*, 5th ed., © 2007 Institute of Transportation Engineers, Washington, 1099 14th Street, NW, Suite 300 West, Washington, D.C., 20005 USA. www.ite.org. Used by permission.

In this model, the maximum acceleration rate that can be achieved is theoretically α, which means that α has units of acceleration as its unit. The term βu_t also should have units of acceleration as its unit, which means that β has the inverse of time (for example, $\sec^{-1}$) as its unit.

Integrating Eq. 3.9 gives

$$-\frac{1}{\beta}\ln(\alpha - \beta u_t) = t + C$$

If the velocity is u_0 at time zero,

$$C = -\frac{1}{\beta}\ln(\alpha - \beta u_0)$$

$$-\frac{1}{\beta}\ln(\alpha - \beta u_t) = t - \frac{1}{\beta}\ln(\alpha - \beta u_0)$$

$$\ln\frac{(\alpha - \beta u_0)}{\alpha - \beta u_t} = -\beta t$$

$$t = \frac{1}{\beta}\ln\frac{(\alpha - \beta u_0)}{\alpha - \beta u_t} \tag{3.10}$$

$$\alpha - \beta u_t = (\alpha - \beta u_0)e^{-\beta t}$$

$$u_t = \frac{\alpha}{\beta}(1 - e^{-\beta t}) + u_0 e^{-\beta t} \tag{3.11}$$

The distance $x(t)$ traveled at any time t may be determined by integrating Eq. 3.11.

$$x = \int_0^t u_t\, dt = \int_0^t \left[\frac{\alpha}{\beta}(1 - e^{-\beta t}) + u_0 e^{-\beta t}\right] dx$$

$$= \left(\frac{\alpha}{\beta}\right)t - \frac{\alpha}{\beta^2}(1 - e^{-\beta t}) + \frac{u_0}{\beta}(1 - e^{-\beta t}) \tag{3.12}$$

Example 3.4 Time to Achieve a Given Speed and Distance Traveled

The acceleration of a vehicle can be represented by the following equation:

$$\frac{du}{dt} = 1.0 - 0.025u$$

where u is the vehicle speed in m/s. If the maximum allowable speed of an urban interstate highway is 88.5 km/h at the vicinity of a signal-controlled ramp, determine the minimum total length of the ramp and acceleration lane from the signal stop line to the freeway that will allow a driver starting at the stop line to achieve the allowable speed on the interstate when her vehicle enters the freeway.

Solution:

- Convert 88.5 km/h to m/s
    $88.5 \times (5/18) = 24.6$ m/s
- Determine time to accelerate to 88.5 km/h.

Since the vehicle starts at the stop line, the original velocity is zero.

- Use Eq. 3.10 to determine time it takes the driver to achieve the speed of 88.5 km/h:

$$t = \frac{1}{\beta} \ln \left(\frac{(\alpha - \beta u_0)}{(\alpha - \beta u_t)} \right)$$

$$t = \frac{1}{0.025} \ln \left(\frac{(1.0 - 0.025 \times 0)}{(1.0 - 0.025 \times 24.6)} \right)$$

Note that u_0 is 0 as the vehicle starts from a stop line.

$$t = 38.18 \, \text{sec}$$

- Use Eq. 3.12 to determine distance required to achieve speed of 55 mph.

$$x = \left(\frac{\alpha}{\beta} \right) t - \frac{\alpha}{\beta^2} \left(1 - e^{-\beta t} \right) \left(\text{since } u_0 \text{ is zero} \right)$$

$$x = \left(\frac{1}{0.025} \right) 38.18 - \frac{1.0}{0.025^2} \left(1 - e^{-0.025 \times 38.18} \right)$$

$$= 1527.2 - 1600 \times (1 - 0.385) = 543.2 \, \text{m}$$

The total length of the ramp and the acceleration lane should be at least 543.2 m.

3.6.3 Dynamic Characteristics

Several forces act on a vehicle while it is in motion: air resistance, grade resistance, rolling resistance, and curve resistance. The extents to which these forces affect the operation of the vehicle are discussed in this section.

Air Resistance

A vehicle in motion has to overcome the resistance of the air in front of it as well as the force due to the frictional action of the air around it. The force required to overcome these is known as the *air resistance* and is related to the cross-sectional area of the vehicle in a direction perpendicular to the direction of motion and to the square of the speed of the vehicle. Claffey has shown that this force can be estimated from the formula

$$R_a = 0.5 \, \frac{(0.077pC_DAu^2)}{g} \tag{3.13}$$

where

R_a = air resistance force (N)

p = density of air (1.0567 kg/m³) at sea level; less at higher elevations

C_D = aerodynamic drag coefficient (drag coefficients for most passenger cars vary from about 0.2 to 0.75. For example, the Hummer H2 has a CD of 0.57, while the Mercedes E220 has a CD of 0.24. Current average value for passenger cars is 0.4; for trucks, this value ranges from 0.5 to 0.8, but a typical value is 0.5)

A = frontal cross-sectional area (m)

u = vehicle speed (m/h)

g = acceleration of gravity (9.81 m/sec²)

Grade Resistance

When a vehicle moves up a grade, a component of the weight of the vehicle acts downward, along the plane of the highway. This creates a force acting in a direction opposite that of the motion. When the vehicle is traveling downgrade, the force acts in the direction of travel. This force is the grade resistance. A vehicle traveling up a grade will therefore tend to lose speed unless an accelerating force is applied. The speed achieved at any point along the grade for a given rate of acceleration will depend on the grade. Figure 3.5 shows the relationships between speed achieved and distance traveled on different grades by a typical heavy truck of 200 lb/hp during maximum acceleration. (*Note:* grade resistance = weight × grade, in decimal.)

Rolling Resistance

There are forces within the vehicle itself that offer resistance to motion. These forces are due mainly to frictional effects on moving parts of the vehicle, but they also include the frictional slip between the pavement surface and the tires. The sum effect of these forces on motion is known as *rolling resistance*. The rolling resistance depends on the speed of the vehicle and the type of pavement. Rolling forces are relatively lower on smooth pavements than on rough pavements.

The rolling resistance force for passenger cars on a smooth pavement can be determined from the relation

$$R_r = (C_{rs} + 0.077C_{rv}u^2)W \tag{3.14}$$

where

R = rolling resistance force (N)

C_{rs} = constant (typically 0.012 for passenger cars)

C_{rv} = constant (typically 7×10^{-6} sec²/m² for passenger cars)

u = vehicle speed (km/h)

W = gross vehicle weight (N)

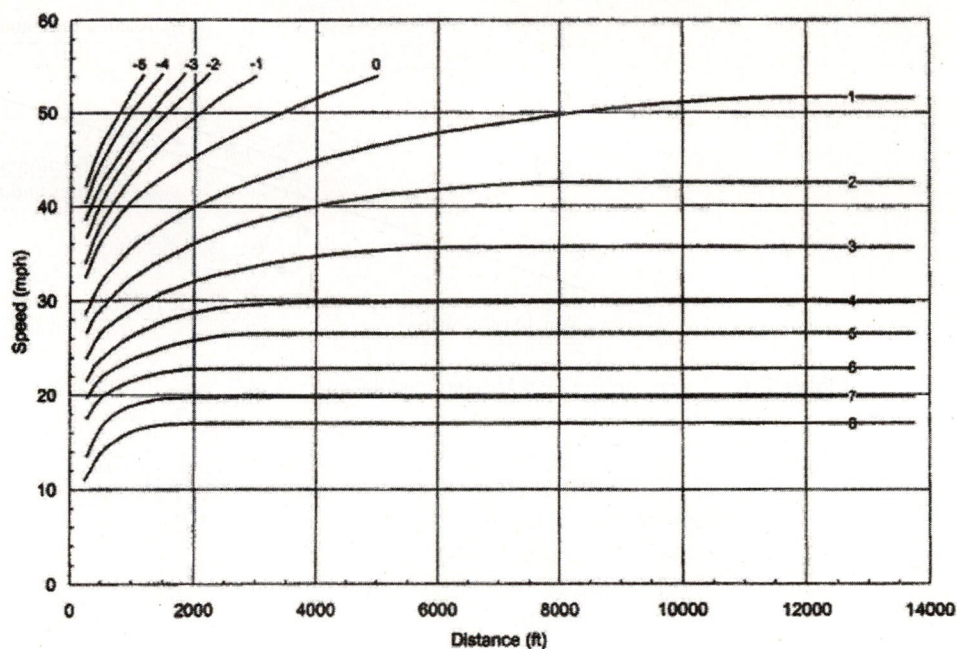

Figure 3.5   Speed-Distance Curves for Acceleration of a Typical Heavy Truck of 120 kg/kw [200 lb/hp] on Upgrades and Downgrades

SOURCE: From *A Policy on Geometric Design of Highways and Streets*, 2004, AASHTO, Washington, D.C. Used by permission.

For trucks, the rolling resistance can be obtained from

$$R_r = (C_a + 0.278C_b u)W \tag{3.15}$$

where

R_r = rolling resistance force (N)
C_a = constant (typically 0.02445 for trucks)
C_b = constant (typically 0.00147 for trucks)
u = vehicle speed (km/h)
W = gross vehicle weight (N)

Curve Resistance

When a passenger car is maneuvered to take a curve, external forces act on the front wheels of the vehicle. These forces have components that have a retarding effect on the forward motion of the vehicle. The sum effect of these components constitutes the curve resistance. This resistance depends on the radius of the curve, the gross weight of the vehicle, and the velocity at which the vehicle is moving. It can be determined as

$$R_c = 0.5 \frac{(0.077u^2 W)}{gR} \tag{3.16}$$

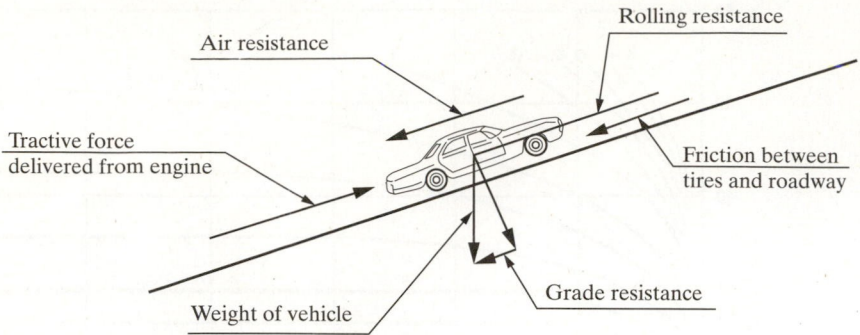

Figure 3.6   Forces Acting on a Moving Vehicle

where

R_c = curve resistance (N)
u = vehicle speed (km/h)
W = gross vehicle weight (N)
g = acceleration of gravity (9.81 m/sec^2)
R = radius of curvature (m)

Power Requirements

Power is the rate at which work is done. It is usually expressed in horsepower (a U.S. unit of measure), where 1 horsepower is 746 N-m/s or W. The performance capability of a vehicle is measured in terms of the horsepower the engine can produce to overcome air, grade, curve, and friction resistance forces and put the vehicle in motion. Figure 3.6 shows how these forces act on the moving vehicle. The engine power required to overcome resistive forces is

$$P = \frac{0.278\,Ru}{760} \qquad (3.17)$$

where

P = horsepower delivered (hp)
R = sum of resistance to motion (N)
u = speed of vehicle (km/h)

Example 3.5   Vehicle Horsepower Required to Overcome Resistance Forces

Determine the horsepower produced by a passenger car traveling at a speed of 105 km/h on a straight road of 5 percent grade with a smooth pavement. Assume the weight of the car is 18 kN and the cross-sectional area of the car is 3.8 m^2.

Solution: The force produced by the car should be at least equal to the sum of the acting resistance forces.

$$R = \text{(air resistance)} + \text{(rolling resistance)} + \text{(upgrade resistance)}$$

Note: There is no curve resistance since the road is straight.

- Use Eq. 3.13 to determine air resistance.

$$R_a = 0.5\left(\frac{0.077pC_DAu^2}{g}\right)$$

$$= 0.5\frac{0.077 \times 1.277 \times 0.4 \times 3.8 \times 105 \times 105}{1}$$

$$= 823 \text{ N}$$

- Use Eq. 3.14 to determine rolling resistance.

$$R_r = (C_{rs} + 0.077C_{rv}u^2)(18000)$$
$$= (0.012 + 0.077 \times 0.65 \times 10^{-6} \times 105 \times 105)(18000)$$
$$= (0.012 + 0.006)(18000)$$
$$= 0.018 \times 18000$$
$$= 324\text{N}$$

$$\text{Grade resistance} = 18000 \times \frac{5}{100} = 900\text{N}$$

- Determine total resistance.

$$R = R_a + R_r + \text{grade resistance} = 823 + 324 + 900 = 2047\text{N}$$

- Use Eq. 3.17 to determine horsepower produced.

$$P = \frac{0.278\,Ru}{760}$$

$$= \frac{0.278 \times 3047 \times 105}{760}$$

$$= 78.6 \text{ hp}$$

Braking Distance

The action of the forces (shown in Figure 3.6) on the moving vehicle and the effect of perception-reaction time are used to determine important parameters related to the dynamic characteristics of the vehicles. These include the braking distance of a vehicle and the minimum radius of a circular curve required for a vehicle traveling around a curve with speed u, where $u > 16$ km/h. Also, relationships among elements, such as the acceleration, the coefficient of friction between the tire and the pavement, the distance above ground of the vehicle's center of gravity, and the track width of the vehicle, could be developed by analyzing the action of these forces.

Braking.   Consider a vehicle traveling downhill with an initial velocity of u, in km/h, as shown in Figure 3.7.

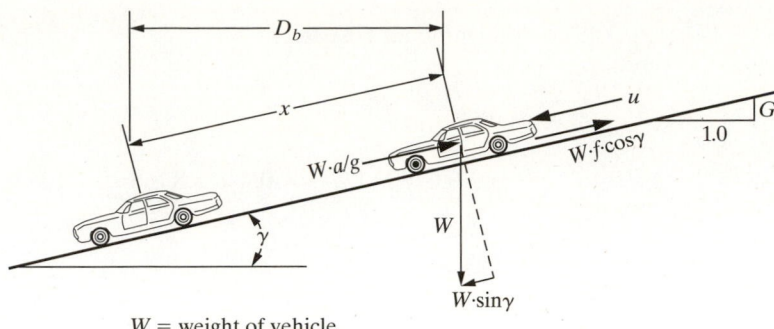

W = weight of vehicle
f = coefficient of friction
g = acceleration of gravity
a = vehicle acceleration
u = speed when brakes applied
D_b = braking distance
γ = angle of incline
G = tan γ (% grade/100)
x = distance traveled by the vehicle along the road during braking

Figure 3.7   Forces Acting on a Vehicle Braking on a Downgrade

Let

W = weight of the vehicle
f = coefficient of friction between the tires and the road pavement
γ = angle between the grade and the horizontal
a = deceleration of the vehicle when the brakes are applied
D_b = horizontal component of distance traveled during braking (that is, from time brakes are applied to time the vehicle comes to rest)

Note that the distance referred to as the braking distance is the horizontal distance and not the inclined distance x. The reason is that measurements of distances in surveying are horizontal and, therefore, distance measurements in highway design are always made with respect to the horizontal plane. Since the braking distance is an input in the design of highway curves, the horizontal component of the distance x is used.

Frictional force on the vehicle = $Wf \cos \gamma$

The force acting on the vehicle due to deceleration is Wa/g, where g is acceleration due to gravity. The component of the weight of the vehicle is $W \sin \gamma$. Substituting into $\Sigma f = ma$, we obtain

$$W \sin \gamma - Wf \cos \gamma = \frac{Wa}{g} \tag{3.18}$$

The deceleration that brings the vehicle to a stationary position can be found in terms of the initial velocity u as $a = u^2/2x$ (assuming uniform deceleration), where x is the distance traveled in the plane of the grade during braking. Equation 3.16 can then be written as

$$W \sin \gamma - Wf \cos \gamma = \frac{Wu^2}{2gx} \tag{3.19}$$

However, $D_b = x \cos \gamma$, and we therefore obtain

$$\frac{Wu^2}{2gD_b} \cos \gamma = Wf \cos \gamma - W \sin \gamma$$

giving

$$\frac{u^2}{2gD_b} = f - \tan \gamma$$

and

$$D_b = \frac{u^2}{2g(f - \tan \gamma)} \tag{3.20}$$

Note, however, that tan y is the grade G of the incline (that is, percent of grade/100), as shown in Figure 3.7.

Equation 3.20 can therefore be written as

$$D_b = \frac{u^2}{2g(f - G)} \tag{3.21}$$

If g is taken as 9.81 m/s^2 and u is expressed in km/h, Eq. 3.19 becomes

$$D_b = \frac{u^2}{254(f - G)} \tag{3.22}$$

and D_b is given in m.

A similar equation could be developed for a vehicle traveling uphill, in which case the following equation is obtained.

$$D_b = \frac{u^2}{254(f + G)} \tag{3.23}$$

A general equation for the braking distance can therefore be written as

$$D_b = \frac{u^2}{254(f \pm G)} \tag{3.24}$$

where the plus sign is for vehicles traveling uphill, the minus sign is for vehicles traveling downhill, and G is the absolute value of tan γ.

AASHTO represents the friction coefficient as a/g and notes that to ensure that the pavement will have and maintain these coefficients, it should be designed to meet the criteria established in the AASHTO *Guidelines for Skid Resistance Pavement Design*. These include guidelines on the selection, quality, and testing of aggregates. AASHTO also recommends that a deceleration rate of 3.41 m/sec^2 be used, as this is a comfortable deceleration rate for most drivers. This rate is further justified because many studies have shown that when most drivers need to stop in an emergency, the rate of deceleration is greater than 4.51 m/sec². Equation 3.24 then becomes

$$D_b = \frac{u^2}{254\left(\dfrac{a}{g} \pm G\right)} \tag{3.25}$$

Similarly, it can be shown that the horizontal distance traveled in reducing the speed of a vehicle from u to u_2 in km/h during a braking maneuver is given by

$$D_b = \frac{u_1^2 - u_2^2}{254\left(\dfrac{a}{g} \pm G\right)} \tag{3.26}$$

The distance traveled by a vehicle between the time the driver observes an object in the vehicle's path and the time the vehicle actually comes to rest is longer than the braking distance, since it includes the distance traveled during perception-reaction time. This distance is referred to in this text as the *stopping sight distance S* and is given as

$$S(m) = 0.278ut + \frac{u^2}{254\left(\dfrac{a}{g} \pm G\right)} \tag{3.27}$$

where the first term in Eq. 3.27 is the distance traveled during the perception-reaction time t (seconds) and u is the velocity in mi/h at which the vehicle was traveling when the brakes were applied.

Example 3.6 Determining Braking Distance

A student trying to test the braking ability of her car determined that she needed 5.64 more to stop her car when driving downhill on a road segment of 5 percent grade than when driving downhill at the same speed along another segment of 3 percent grade. Determine the speed at which the student conducted her test and the braking distance on the 5 percent grade if the student is traveling at the test speed in the uphill direction.

Solution:

- Let x = downhill braking distance on 5 percent grade.
- $(x - 5.64)$ m = downhill braking distance on 3 percent grade.
- Use Eq. 3.25.

$$D_b = \frac{u^2}{254\left(\dfrac{a}{g} - G\right)}$$

$$x = \frac{u^2}{254\left(\dfrac{a}{g} - 0.05\right)}$$

$$x - 5.64 = \frac{u^2}{254\left(\dfrac{a}{g} - 0.03\right)}$$

$$5.64 = \frac{u^2}{254\left(\dfrac{a}{g} - 0.05\right)} - \frac{u^2}{254\left(\dfrac{a}{g} - 0.03\right)}$$

Determine u, the test velocity. Using $a = 3.41$ m/sec^2 and $g = 32$,

$$\frac{a}{g} = \frac{3.41}{9.81} = 0.35$$

$$\frac{u^2}{254(0.35 - 0.05)} - \frac{u^2}{254(0.35 - 0.03)} = 5.64$$

which gives

$$0.0131u^2 - 0.0123u^2 = 5.64$$

from which we obtain $u = 83.96$ km/h. The test velocity is therefore 51.4 mi/h.

- Determine braking distance on uphill grade (use Eq. 3.23).

$$D_b = \frac{83.96^2}{254(0.35 + 0.05)} = 69.38 \text{ m}$$

Example 3.7 Exit Ramp Deceleration Distance

A motorist traveling at 105 km/h on an expressway intends to leave the expressway using an exit ramp with a maximum speed of 56 km/h. At what point on the expressway should the motorist step on her brakes in order to reduce her speed to the maximum allowable on the ramp just before entering the ramp, if this section of the expressway has a downgrade of 3 percent?

Solution: Use Eq. 3.26.

$$D_b = \frac{u_1^2 - u_2^2}{254\left(\dfrac{a}{g} - 0.03\right)}$$

$$a/g = 3.41/9.81 = 0.35$$

$$D_b = \frac{105^2 - 56^2}{254(0.35 - 0.03)} = 97.06 \text{ m}$$

The brakes should be applied at least 97.06 m from the ramp.

Example 3.8 Distance Required to Stop for an Obstacle on the Roadway

A motorist traveling at 88 km/h down a grade of 5 percent on a highway observes a crash ahead of him, involving an overturned truck that is completely blocking the road. If the motorist was able to stop his vehicle 9 m from the overturned truck, what was his distance from the truck when he first observed the crash? Assume perception-reaction time = 2.5 sec.

Solution:

- Use Eq. 3.27 to obtain the stopping distance.

$$\frac{a}{g} = 0.35$$

$$S = 0.278ut + \frac{u^2}{254(0.35 - 0.05)}$$

$$= 0.278 \times 88 \times 2.5 + \frac{88^2}{254 \times 0.30}$$

$$= 61.16 + 101.63$$

$$= 162.78 \text{ m}$$

- Find the distance of the motorist when he first observed the crash.

$$S + 9 = 170.46 \text{ m}$$

Estimate of Velocities.   It is sometimes necessary to estimate the speed of a vehicle just before it is involved in a crash. This may be done by using the braking-distance equations if skid marks can be seen on the pavement. The steps taken in making the speed estimate are as follows:

Step 1.   Measure the length of the skid marks for each tire and determine the average. The result is assumed to be the braking distance D_b of the vehicle.

Step 2.   Determine the coefficient of friction by performing trial runs at the site under similar weather conditions, using vehicles whose tires are in a state similar to that of the tires of the vehicle involved in the accident. This is done by driving the vehicle at a known speed u_k and measuring the distance traveled D_k while braking the vehicle to rest. The coefficient of friction f_k can then be estimated by using Eq. 3.24.

$$f_k = \frac{u_k^2}{254D_k} \pm G$$

Alternatively, a value of 0.35 for a/g can be used for f_k.

Step 3.   Use the value of f_k obtained in step 2 or the assumed value of 0.35 to estimate the unknown velocity u_u just prior to impact; that is, the velocity at which the vehicle was traveling just before observing the crash. This is done by using Eq. 3.26.

If it can be assumed that the application of the brakes reduced the velocity u_u to zero, then u_u may be obtained from

$$D_b = \frac{u_u^2}{254(f_k \pm G)}$$

or

$$D_b = \frac{u_u^2}{254\left(\dfrac{u_k^2}{254D_k} \mp G \pm G\right)} = \left(\frac{u_u^2}{u_k^2}\right)D_k \qquad (3.28)$$

giving

$$u_u = \left(\frac{D_b}{D_k}\right)^{1/2} u_k \qquad (3.29)$$

However, if the vehicle involved in the accident was traveling at speed u_1 when the impact took place and the speed u_1 is known, then, using Eq. 3.24, the unknown speed u_u just prior to the impact may be obtained from

$$D_b = \frac{u_u^2 - u_1^2}{254\left(\dfrac{u_k^2}{254D_k} \mp G \pm G\right)} = \left(\frac{u_u^2 - u_1^2}{u_k^2}\right)D_k$$

giving

$$u_u = \left(\frac{D_b}{D_k}u_k^2 + u_1^2\right)^{1/2} \qquad (3.30)$$

Note that the unknown velocity just before the impact, obtained from either Eq. 3.29 or Eq. 3.30, is only an estimate, but it always will be a conservative estimate because it always will be less than the actual speed at which the vehicle was traveling before impact. The reason is that some reduction of speed usually takes place before skidding commences, and, in using Eq. 3.29, the assumption that the initial velocity u_u is reduced to zero is always incorrect. The lengths of the measured skid marks do not reflect these factors.

Example 3.9 Estimating the Speed of a Vehicle from Skid Marks

In an attempt to estimate the speed of a vehicle just before it hit a traffic signal pole, a traffic engineer measured the length of the skid marks made by the vehicle and performed trial runs at the site to obtain an estimate of the coefficient of friction. Determine the estimated unknown velocity of the vehicle when the brakes are applied if the following data were obtained.

Length of skid marks = 178 m, 180 m, 177 m, and 181 m
Speed of trial run = 48 km/h
Distance traveled during trial run = 90 m

Examination of the vehicle just after the crash indicated that the speed of impact was 56 km/h.

Solution:

$$\text{Average length of skid marks} = \frac{178 + 180 + 177 + 181}{4}$$

$$= 179\,\text{m}$$

(This is assumed to be the braking distance D_b.) Use Eq. 3.30 to determine the unknown velocity.

$$u_u = \left(\frac{D_b}{D_k}u_k^2 + u_1^2\right)^{1/2}$$

$$= \left(\frac{179}{90}48^2 + 56^2\right)^{1/2}$$

$$= (4582.4 + 3136)^{1/2}$$

$$= 87.85\,\text{km/h}$$

Minimum Radius of a Circular Curve

When a vehicle is moving around a circular curve, there is an inward radial force acting on the vehicle, usually referred to as the centrifugal force. There is also an outward radial force acting toward the center of curvature as a result of the centripetal acceleration. In order to balance the effect of the centripetal acceleration, the cross-slope of the road is inclined downward toward the center of the curve. The inclination of the roadway toward the center of the curve is known as *superelevation*. The centripetal acceleration depends on the component of the vehicle's weight along the inclined surface of the road and the side friction between the tires and the roadway. The action of these forces on a vehicle moving around a circular curve is shown in Figure 3.8.

The minimum radius of a circular curve R for a vehicle traveling at u mi/h can be determined by considering the equilibrium of the vehicle with respect to its moving up or down the incline. If α is the angle of inclination of the highway, the component of the weight down the incline is $W \sin \alpha$, and the frictional force also acting down the incline is $Wf \cos \alpha$. The centrifugal force F_c is

$$F_c = \frac{Wa_c}{g}$$

where

a_c = acceleration for curvilinear motion = u^2/R (R = radius of the curve)
W = weight of the vehicle
g = acceleration of gravity

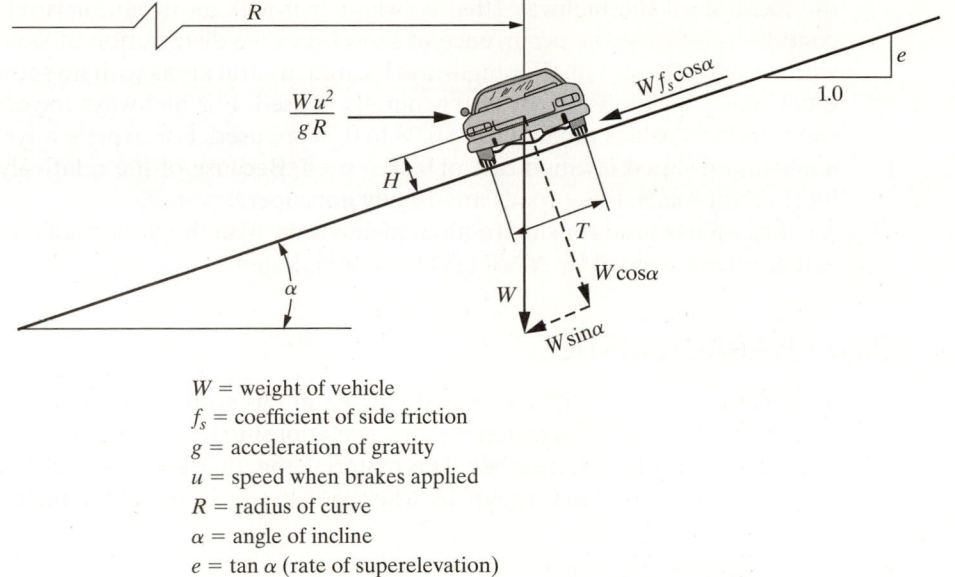

W = weight of vehicle
f_s = coefficient of side friction
g = acceleration of gravity
u = speed when brakes applied
R = radius of curve
α = angle of incline
e = tan α (rate of superelevation)
T = track width
H = height of center of gravity

Figure 3.8   Forces Acting on a Vehicle Traveling on a Horizontal Curve Section of a Road

When the vehicle is in equilibrium with respect to the incline (that is, the vehicle moves forward but neither up nor down the incline), we may equate the three relevant forces and obtain

$$\frac{Wu^2}{gR} \cos \alpha = W \sin \alpha + Wf_s \cos \alpha \qquad (3.31)$$

where f_s = coefficient of side friction and $(u^2/g) = R(\tan \alpha + f_s)$. This gives

$$R = \frac{u^2}{g(\tan \alpha + f_s)} \qquad (3.32)$$

In addition, tan α, the tangent of the angle of inclination of the roadway, is known as the *rate of superelevation e.* Equation 3.32 can therefore be written as

$$R = \frac{u^2}{g(e + f_s)} \qquad (3.33)$$

Again, if g is taken as 9.81 m/sec^2 and u is measured in km/h, the minimum radius R is given in meters as

$$R = \frac{u^2}{127(e + f_s)} \qquad (3.34)$$

Equation 3.34 shows that to reduce R for a given velocity, either e or f_s or both should be increased.

There are, however, stipulated maximum values that should be used for either e or f_s. Several factors control the maximum value for the rate of superelevation. These include the location of the highway (that is, whether it is in an urban or rural area), weather conditions (such as the occurrence of snow), and the distribution of slow-moving traffic within the traffic stream. For highways located in rural areas with no snow or ice, a maximum superelevation rate of 0.10 generally is used. For highways located in areas with snow and ice, values ranging from 0.08 to 0.10 are used. For expressways in urban areas, a maximum superelevation rate of 0.08 is used. Because of the relatively low speeds on local urban roads, these roads are usually not superelevated.

The values used for side friction, mainly vary with the design speed. Table 3.3 gives values recommended by AASHTO for use in design.

3.7 ROAD CHARACTERISTICS

The characteristics of the highway discussed in this section are related to stopping and passing because these have a more direct relationship to the characteristics of the driver and the vehicle discussed earlier. This material, together with other characteristics of the highway, will be used in Chapter 15, where geometric design of the highway is discussed.

3.7.1 Sight Distance

Sight distance is the length of the roadway a driver can see ahead at any particular time. The sight distance available at each point of the highway must be such that, when a driver is traveling at the highway's design speed, adequate time is given after an object is observed in the vehicle's path to make the necessary evasive maneuvers without colliding with the object. The three types of sight distance are (1) stopping sight distance, (2) decision sight distance, and (3) passing sight distance.

Stopping Sight Distance

The *stopping sight distance* (SSD), for design purposes, is usually taken as the minimum sight distance required for a driver to stop a vehicle after seeing an object in the vehicle's path without hitting that object. This distance is the sum of the distance traveled during perception-reaction time and the distance traveled during braking. The SSD for a vehicle traveling at u km/h is therefore the same as the stopping distance given in Eq. 3.27. The SSD is therefore

$$\text{SSD} = 0.278 \, ut + \frac{u^2}{30\left(\dfrac{a}{g} \pm G\right)} \tag{3.35}$$

Table 3.3   Coefficient of Side Friction Factors Assumed for Design

Design Speed (km/h)	Coefficients of Side Friction, f_s
48	0.2
64	0.16
80	0.14
96	0.12
112	0.1

SOURCE: Based on *A Policy on Geometric Design of Highways and Streets*, 2011, AASHTO, Washington, D.C.

Table 3.4   SSDs for Different Design Speeds on Level Roads

	U.S. Customary Stopping Sight Distance			
Design Speed (km/h)	*Brake Reaction Distance (m)*	*Braking Distance on Level (m)*	*Calculated (m)*	*Design*
24	16.53	6.48	23.01	24
32	22.05	11.52	33.57	34.5
40	27.57	18	45.57	46.5
48	33.09	25.92	59.01	60
56	38.58	35.28	73.86	75
64	44.1	46.08	90.18	91.5
72	49.62	58.32	107.94	108
80	55.14	72	127.14	127.5
88	60.63	87.09	147.72	148.5
96	66.15	103.65	169.8	171
104	71.67	121.65	193.32	193.5
112	77.19	141.09	218.28	219
120	82.68	161.97	244.65	246
128	88.2	184.29	272.49	273

Note: Brake reaction distance predicated on a time of 2.5 sec; deceleration rate of 11.2 ft/sec^2 used to determine calculated sight distance.
SOURCE: Based on *A Policy on Geometric Design of Highways and Streets*, 2011, AASHTO, Washington, D.C.

It is essential that highways be designed such that sight distance along the highway is at least equal to the SSD. Table 3.4 shows SSDs for different design speeds. The SSD requirements dictate the minimum lengths of vertical curves and minimum radii for horizontal curves that should be designed for any given highway. It should be noted that the values given for SSD in Table 3.4 are for horizontal alignment and the grade is zero. On upgrades, the SSDs are shorter; on downgrades, they are longer (see Eq. 3.35).

Decision Sight Distance

The SSDs given in Table 3.4 are usually adequate for ordinary conditions, when the stimulus is expected by the driver. However, when the stimulus is unexpected or when it is necessary for the driver to make unusual maneuvers, longer SSDs are usually required, since the perception-reaction time is much longer. This longer sight distance is the *decision sight distance (DSD);* it is defined by AASHTO as the "distance required for a driver to detect an unexpected or otherwise difficult-to-perceive information source or hazard in a roadway environment that may be visually cluttered, recognize the hazard of its threat potential, select an appropriate speed and path, and initiate and complete the required safety maneuvers safely and efficiently."

The DSDs depend on the type of maneuver required to avoid the hazard on the road, and also on whether the road is located in a rural or urban area. Table 3.5 gives AASHTO's recommended decision sight distance values for different avoidance maneuvers; these values can be used for design.

Table 3.5   Decision Sight Distances for Different Design Speeds and Avoidance Maneuvers

Design Speed (km/h)	U.S. Customary Decision Sight Distance (m)				
	Avoidance Maneuver				
	A	B	C	D	E
48	66	147	135	160.5	186
56	82.5	177	157.5	187.5	216
64	99	207	180	214.5	247.5
72	118.5	240	202.5	240	279
80	139.5	273	225	267	309
88	160.5	309	259.5	294	340.5
96	183	345	297	337.5	384
104	208.5	382.5	315	366	409.5
112	234	423	331.5	382.5	433.5
120	262.5	463.5	354	409.5	463.5
128	291	505.5	378	436.5	495

Note: Brake reaction distance predicted on a time of 2.5 sec; deceleration rate of 11.2 ft/sec^2 used to determine calculated sight distance.
Avoidance Maneuver *A:* Stop on rural road—$t = 3.0$ sec
Avoidance Maneuver *B:* Stop on urban road—$t = 9.1$ sec
Avoidance Maneuver *C:* Speed/path/direction change on rural road—t varies between 10.2 and 11.2 sec
Avoidance Maneuver *D:* Speed/path/direction change on suburban road—t varies between 12.1 and 12.9 sec
Avoidance Maneuver *E:* Speed/path/direction change on urban road—t varies between 14.0 and 14.5 sec
SOURCE: Based on *A Policy on Geometric Design of Highways and Streets*, 2011, AASHTO, Washington, D.C.

Passing Sight Distance

The *passing sight distance* is the minimum sight distance required on a two-lane, two-way highway that will permit a driver to complete a passing maneuver without colliding with an opposing vehicle and without cutting off the passed vehicle. The passing sight distance will also allow the driver to successfully abort the passing maneuver (that is, return to the right lane behind the vehicle being passed) if he or she so desires. In determining minimum passing sight distances for design purposes, only single passes (that is, a single vehicle passing a single vehicle) are considered. Although it is possible for multiple passing maneuvers to occur (that is, more than one vehicle pass or are passed in one maneuver), it is not practical for minimum design criteria to be based on them.

In order to determine the minimum passing sight distance, certain assumptions are made regarding the movements of the vehicles involved in a passing maneuver:

1. The passing and opposing vehicles travel at the same speed that represents the design speed of the road.
2. The speed of the vehicle being passed (impeder) does not vary during the passing maneuver and it is less than that of the passing vehicle by 19.2 km/h.
3. The passing vehicle is capable of accelerating to the passing speed (19.2 km/h higher than that of the vehicle being passed) within a distance of about 40 percent of the distance covered during the passing maneuver.
4. The lengths of the vehicle being passed and that of the passing vehicle are that for the passenger car design vehicle (i.e., 5.7 m).

Table 3.6   AASHTO-Recommended Values for Passing Sight Distances

| Design Speed (km/h) | Assumed Speeds (km/h) | | Passing Sight Distance (ft) |
	Passed (impeder) Vehicle	Passing Vehicle	
32	12.8	32	120
40	20.8	40	135
48	28.8	48	150
56	36.8	56	165
64	44.8	64	180
72	52.8	72	210
80	60.8	80	240
88	68.8	88	270
96	76.8	96	300
104	84.8	104	330
112	92.8	112	360
	100.8	120	390
	108.8	128	420

SOURCE: Based on *A Policy on Geometric Design of Highways and Streets*, 2011, AASHTO, Washington, D.C.

5. The perception-reaction time of the driver of the passing vehicle is 1.0 sec for aborting the pass.
6. The same deceleration rate of 3.36 m/sec^2 used for developing stopping sight distances is used for aborting a passing maneuver.
7. A time headway of 1 sec exists between the passing and impeder vehicles at the end of an aborted or completed pass.
8. At the point where the passing vehicle returns to its normal lane, the time headway between the passing and impeder vehicles is 1 sec.

Table 3.6 gives AASHTO-recommended values for passing sight distances for different design speeds.

3.8 SUMMARY

The highway or traffic engineer needs to study and understand the fundamental elements that are important in the design of traffic-control systems. This chapter has presented concisely the basic characteristics of the driver, pedestrian, bicyclist, vehicle, and the road that should be known and understood by transportation and/or traffic engineers.

The most important characteristic of the driver is the driver response process, which consists of four subprocesses: perception, identification, emotion, and reaction or volition. Each of these subprocesses requires time to complete, the sum of which is known as the perception-reaction time of the driver. The actual distance a vehicle travels before coming to rest is the sum of the distance traveled during the perception-reaction time of the driver and the distance traveled during the actual braking maneuver. Perception-reaction times vary from one person to another, but the recommended value for design is 2.5 sec. The static, kinematic, and dynamic characteristics of the vehicle are also important because they are used to determine minimum radii of horizontal curves for low speeds (*u* <10 mi/h), the acceleration and deceleration capabilities of the vehicle (through which distance traveled and velocities attained can be determined), and the resistance forces that act on the moving vehicle. The characteristic of the road that has a direct relationship to the characteristics of the driver is the sight distance on the road. Two types of sight distances are considered to be part of the characteristics of the road: the stopping sight distance, which is normally taken as the minimum sight distance required for a driver to stop a vehicle after seeing an object in the vehicle's path without hitting that object, and the passing sight distance, which is the minimum sight distance required on a two-lane, two-way highway that will permit a driver to complete a passing maneuver without colliding with an opposing vehicle and without cutting off the passed vehicle.

Although these characteristics are presented in terms of the highway mode, several of these are also used for other modes. For example, the driver and pedestrian characteristics also apply to other modes, such as motorcycles, in which vehicles are manually driven and some possibility exists for interaction between the vehicle and pedestrians. It should be emphasized again that because of the wide range of capabilities among drivers and pedestrians, the use of average limitations of drivers and pedestrians in developing guidelines for design may result in the design of facilities that will not satisfy a significant percentage of the people using the facility. High-percentile values (such as 85th- or 95th-percentile values) are therefore normally used for design purposes.

PROBLEMS

3-1 Briefly describe the five characteristics of visual reception relevant to transportation engineering.

3-2 Briefly describe the two types of visual acuity.

3-3 **(a)** What color combinations are used for regulatory signs (e.g., speed limit signs) and for general warning signs (e.g., advance railroad crossing signs)?
(b) Why are these combinations used?

3-4 Briefly describe the changes that occur in the abilities of older drivers (age 65 and over) that influence their driving capability.

3-5 Determine your average walking speed. Compare your results with that of the recommended walking speed in the MUTCD. Which value is more conservative and why?

3-6 Describe the three types of vehicle characteristics.

3-7 Determine the maximum allowable overall gross weight of the WB-67 design vehicle.

3-8 The design speed of a multilane highway is 96 km/h. Determine (a) the minimum stopping sight distance that should be provided for a level roadway, and (b) the minimum stopping sight distance that should be provided for a roadway with a maximum grade of 7 percent.

  Note: The term $\frac{a}{g}$ in the appropriate equation is typically rounded to 0.35 in calculations. Assume perception reaction time = 2.5 sec.

3-9 The acceleration of a vehicle can be expressed as:

$$\frac{du}{dt} = 1.1 - 0.06u$$

If the vehicle speed, *u,* is 9 m/sec at time T$_0$, determine:

(a) Distance traveled when the vehicle has accelerated to 13.5 m/sec
(b) Time for vehicle to attain the speed of 13.5 m/sec
(c) Acceleration after 4 sec

3-10 The gap between two consecutive automobiles (distance between the back of a vehicle and the front of the following vehicle) is 19.5 m. At a certain time the front vehicle is traveling at 64 km/h and the following vehicle at 48 km/h. If both vehicles start accelerating at the same time, determine the gap between the two vehicles after 10 sec if the acceleration of the vehicles can be assumed to take the following forms:

$$\frac{du}{dt} = 1.02 - 0.07u_t \text{ (leading vehicle)}$$

$$\frac{du}{dt} = 0.99 - 0.065u_t \text{ (following vehicle)}$$

where u_t is the vehicle speed in m/sec.

3-11   The driver of a vehicle on a level road determined that she could increase her speed from rest to 80 km/h in 12.8 sec and from rest to 104 km/h in 19.8 sec. If it can be assumed that the acceleration of the vehicle takes the form:

$$\frac{du}{dt} = \alpha - \beta u_t$$

determine the maximum acceleration of the vehicle.

3-12   If the vehicle in Problem 3-11 is traveling at a speed of 64 km/h, how long will it take after the driver starts accelerating for the vehicle to achieve a speed of 72 km/h?

3-13   Determine the horsepower developed by a passenger car traveling at a speed of 80 km/h on an upgrade of 5 percent with a smooth pavement. The weight of the car is 15.87 kN and the cross-sectional area of the car is 3.6 m².

3-14   Repeat Problem 3-13 for a 108.84 kN truck with a cross-sectional area of 9 m² and coefficient of drag of 0.5 traveling at 80 km/h.

3-15   A 11.34 kN passenger vehicle originally traveling on a straight and level road gets onto a section of the road with a horizontal curve of radius = 255 m. If the vehicle was originally traveling at 88 km/h, determine (a) the additional horsepower on the curve the vehicle must produce to maintain the original speed, (b) the total resistance force on the vehicle as it traverses the horizontal curve, and (c) the total horsepower. Assume that the vehicle is traveling at sea level and has a front cross-sectional area of 2.7 m².

3-16   A horizontal curve is to be designed for a section of a highway having a design speed of 80 km/h.

(a) If the physical conditions restrict the radius of the curve to 120 m, what value is required for the superelevation at this curve?
(b) Is this a good design?

3-17   Determine the minimum radius of a horizontal curve required for a highway if the design speed is 96 km/h and the superelevation rate is 0.08.

3-18   The existing posted speed limit on a section of highway is 88 km/h and studies have shown that that the current 85th percentile speed is 104 km/h. If the posted speed limit is to be increased to the current 85th percentile speed, what should be the increase in the radius of a curve that is just adequate for the new posted speed limit? Assume a superelevation rate of 0.08 for the existing curve and for the redesigned curve.

3-19   The radius of a horizontal curve on an existing highway was field-measured to be 243 m. The pavement on this two-lane highway is 6.6 m wide, and the elevation difference between the inside and outside of the curve is 0.43 m. The posted speed limit on the road is 96 km/h. Is this a hazardous location? If so, why? What action will you recommend to correct the situation?

3-20   A section of highway has a superelevation of 0.05 and a curve with a radius of only 91.5 m. What is the maximum safe speed at this section of the highway?

3-21   What is the distance required to stop an average passenger car when brakes are applied on a 3.3 percent downgrade if that vehicle was originally traveling at 56 km/h?

3-22   A driver on a level two-lane highway observes a truck completely blocking the highway. The driver was able to stop her vehicle only 9 m from the truck. If the driver was driving at 80 km/h, how far was she from the truck when she first observed it (assume perception-reaction time is 1.5 sec)? How far was she from the truck at the moment the brakes were applied (use $a/g = 0.35$)?

3-23   A temporary diversion has been constructed on a highway of +4 percent grade due to major repairs that are being undertaken on a bridge. The maximum speed allowed on the diversion is 32 km/h. Determine the minimum distance from the diversion that a road sign should be located informing drivers of the temporary change on the highway. Assume that a driver can read a road sign within his or her area of vision at a distance of 9 m for each inch of letter height.

        Speed limit on highway = 104 km/h
        Letter height of road sign = 200 mm
        Perception-reaction time = 2.5 sec

3-24   Repeat Problem 3-23 for a highway with a downgrade of −3.5 percent and for which the speed allowed on the diversion is 24 km/h. Assume that a driver can read a road sign within his or her area of vision at a distance of 12 m for each inch of letter height.

3-25   An elevated expressway goes through an urban area and crosses a local street as shown in Figure 3.9. The partial cloverleaf exit ramp is on a 2 percent downgrade, and all vehicles leaving the expressway must stop at the intersection with the local street. Determine (a) minimum ramp radius and (b) length of the ramp for the following conditions:

        Maximum speed on expressway = 96 km/h
        Distance between exit sign and exit ramp = 78 m
        Letter height of road sign = 75 mm
        Perception-reaction time = 2.5 sec
        Maximum superelevation = 0.08
        Expressway grade = 0%

Assume that a driver can read a road sign within his or her area of vision at a distance of 15 m for each inch of letter height, and the driver sees the stop sign immediately on entering the ramp.

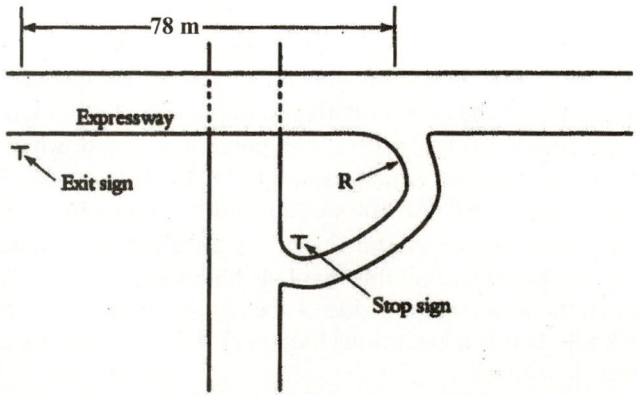

Figure 3.9   Layout of Elevated Expressway Ramp, and Local Street for Problem 3-25

REFERENCES

A Policy on Geometric Design of Highways and Streets. American Association of State Highway and Transportation Officials, Washington, D.C., 2011.

Bridge Formula Weights. Report No. FHWA-HOP-06-105, U.S. Department of Transportation, Federal Highway Administration, August 2006.

Fambro, D. B., et al., *Determination of Stopping Sight Distances.* NCHRP Report 400, Transportation Research Board, National Research Council, Washington, D.C., 1997.

Garber, N. J., and M. Saito, *Centerline Pavement Markings on Two-Lane Mountainous Highways.* Research Report No. VHTRC 84-R8, Virginia Highway and Transportation Research Council, Charlottesville, VA, March 1983.

Garber, N. J. and S. Subramanyan, "Feasibility of Developing Congestion Mitigation Strategies that Incorporate Crash Risk. A Case Study: Hampton Roads Area." Virginia Transportation Research Council, 2001.

Guide for the Development of Bicycle Facilities, 4th edition, American Association of State Highway and Transportation Officials, Washington, D.C., 2012.

Harwood, D. W., *Traffic and Operating Characteristics.* Institute of Transportation Engineers, Washington, D.C., 1992.

Highway Capacity Manual Transportation Research Board, National Research Council, Washington, D.C., 2000.

Highway Statistics 2009. Washington D.C., Federal Highway Administration, 2011.

Manual on Uniform Traffic Control Devices. U.S. Department of Transportation, Federal Highway Administration, Washington, D.C., 2009.

Peripheral Vision Horizon Display (PHVD). Proceedings of a conference held at NASA Ames Research Center, Dryden Flight Research, March 15–16, 1983, National Aeronautics and Space Administration, Scientific and Technical Information Branch, 1984.

Rouphail, N., J. Hummer, P. Allen, and J. Milazzo. *Recommended Procedures for Chapter 13, Pedestrians, of the Highway Capacity Manual.* Report FFHWA-RD-98-107, FHWA, U.S. Department of Transportation, Washington, D.C., 1998.

Review of Truck Characteristics as Factors in Roadway Design, NCHRP Report 505, Transportation Research Board, Washington, D.C., 2003.

Revising the AASHTO Guide for the Department of Bicycle Facilities Final Report, the National Cooperative Highway Research Program, Transportation Research Board, Washington, D.C., 2010.

CHAPTER 4

Traffic Engineering Studies

The availability of highway transportation has provided several advantages that contribute to a high standard of living. However, several problems related to the highway mode of transportation exist. These problems include highway-related crashes, parking difficulties, congestion, and delay. To reduce the negative impact of highways, it is necessary to adequately collect information that describes the extent of the problems and identifies their locations. Such information is usually collected by organizing and conducting traffic surveys and studies. This chapter introduces the reader to the different traffic engineering studies that are conducted to collect traffic data. Brief descriptions of the methods of collecting and analyzing the data are also included.

Traffic studies may be grouped into three main categories: (1) inventories, (2) administrative studies, and (3) dynamic studies. Inventories provide a list or graphic display of existing information, such as street widths, parking spaces, transit routes, traffic regulations, and so forth. Some inventories—for example, available parking spaces and traffic regulations—change frequently and therefore require periodic updating; others, such as street widths, do not.

Administrative studies use existing engineering records, available in government agencies and departments. This information is used to prepare an inventory of the relevant data. Inventories may be recorded in files but are usually recorded in automated data processing (ADP) systems. Administrative studies include the results of surveys, which may involve field measurements and/or aerial photography.

Dynamic traffic studies involve the collection of data under operational conditions and include studies of speed, traffic volume, travel time and delay, parking, and crashes. Since dynamic studies are carried out by the traffic engineer to evaluate current conditions and develop solutions, they are described in detail in this chapter.

CHAPTER OBJECTIVES:

- Understand the importance of the different categories of traffic studies.
- Learn the different methods for collecting spot speed data.

- Learn the different methods for presenting and analyzing spot speed data.
- Learn the different methods for collecting and analyzing traffic volume data.
- Become familiar with the different methods for collecting and analyzing travel time and delay data.
- Become familiar with the methods for collecting parking data.
- Become familiar with the different methods for analyzing parking data.

4.1 SPOT SPEED STUDIES

Spot speed studies are conducted to estimate the distribution of speeds of vehicles in a stream of traffic at a particular location on a highway. The speed of a vehicle is defined as the rate of movement of the vehicle; it is usually expressed in miles per hour (km/h) or kilometers per hour (km/h). A spot speed study is carried out by recording the speeds of a sample of vehicles at a specified location. Speed characteristics identified by such a study will be valid only for the traffic and environmental conditions that exist at the time of the study, and for segments that have similar traffic and geometric characteristics similar to the study location. For example, data obtained from a spot speed study may not represent the speed characteristics of a segment of road that includes the location of the study and locations of interrupted flows such as signalized intersections. Speed characteristics determined from a spot speed study may be used to

- Establish parameters for traffic operation and control, such as speed zones, speed limits (85th-percentile speed is commonly used as the speed limit on a road), and passing restrictions.
- Evaluate the effectiveness of traffic control devices, such as variable message signs at work zones.
- Monitor the effect of speed enforcement programs, such as the use of drone radar and the use of differential speed limits for passenger cars and trucks.
- Evaluate and or determine the adequacy of highway geometric characteristics, such as radii of horizontal curves and lengths of vertical curves.
- Evaluate the effect of speed on highway safety through the analysis of crash data for different speed characteristics.
- Determine speed trends.
- Determine whether complaints about speeding are valid.

4.1.1 Locations for Spot Speed Studies

The following locations generally are used for the different applications listed:

1. Locations that represent different traffic conditions on a highway or highways are used for *basic data collection.*
2. Mid-blocks of urban highways and straight, level sections of rural highways are sites for *speed trend analyses.*
3. Any location may be used for the solution of a *specific traffic engineering problem.*

When spot speed studies are being conducted, it is important that unbiased data be obtained. This requires that drivers be unaware that such a study is being conducted. Equipment used, therefore, should be concealed from the driver, and observers conducting the study should be inconspicuous. Since the speeds recorded eventually will be subjected to statistical analysis, it is important that a statistically adequate number of vehicle speeds be recorded.

4.1.2 Time of Day and Duration of Spot Speed Studies

The time of day for conducting a speed study depends on the purpose of the study. In general, when the purpose of the study is to establish posted speed limits, to observe speed trends, or to collect basic data, it is recommended that the study be conducted when traffic is free-flowing, usually during off-peak hours. However, when a speed study is conducted in response to citizen complaints, it is useful if the time period selected for the study reflects the nature of the complaints.

The duration of the study should be such that the minimum number of vehicle speeds required for statistical analysis is recorded. Typically, the duration is at least 1 hour and the sample size is at least 30 vehicles, although more data will provide a better indication of the speed distribution and overall mean.

4.1.3 Sample Size for Spot Speed Studies

The calculated mean (or average) speed is used to represent the true mean value of all vehicle speeds at that location. The accuracy of this assumption depends on the number of vehicles in the sample. The larger the sample size, the greater the probability that the estimated mean is not significantly different from the true mean. It is therefore necessary to select a sample size that will give an estimated mean within acceptable error limits. Statistical procedures are used to determine this minimum sample size. Before discussing these procedures, it is first necessary to define certain significant values that are needed to describe speed characteristics:

1. **Average Speed**, which is the arithmetic mean of all observed vehicle speeds (which is the sum of all spot speeds divided by the number of recorded speeds). It is given as

$$\bar{u} = \frac{\Sigma f_i u_i}{\Sigma f_i} \tag{4.1}$$

where

$\bar{u}$ = arithmetic mean
f_i = number of observations in each speed group
u_i = midvalue for the ith speed group

The formula also can be written as

$$\bar{u} = \frac{\Sigma u_i}{N}$$

where

u_i = speed of the ith vehicle
N = number of observed values

2. **Median Speed**, which is the speed at the middle value in a series of spot speeds that are arranged in ascending order. Fifty percent of the speed values will be greater than the median; 50 percent will be less than the median.
3. **Modal Speed**, which is the speed value that occurs most frequently in a sample of spot speeds.
4. The **ith-percentile Spot Speed,** which is the spot speed value below which i percent of the vehicles travel; for example, 85th-percentile spot speed is the speed at or

below which 85 percent of the vehicles travel and above which 15 percent of the vehicles travel.

5. **Pace**, which is the range of speed—usually taken at the 10-km/h speed range that has the largest number of observations. For example, if a set of speed data includes speeds between 30 and 60 km/h, the speed intervals will be 30 to 40 km/h, 40 to 50 km/h, and 50 to 60 km/h, assuming a range of 10 km/h. The pace is 40 to 50 km/h if this range of speed has the highest number of observations.

6. **Standard Deviation of Speeds**, which is a measure of the spread of the individual speeds. It is estimated as

$$S = \sqrt{\frac{\Sigma(u_j - \bar{u})^2}{N - 1}} \tag{4.2}$$

where

S = standard deviation
$\bar{u}$ = arithmetic mean
u_j = jth observation
N = number of observations

However, speed data are frequently presented in classes where each class consists of a range of speeds. The standard deviation is computed for such cases as

$$S = \sqrt{\frac{\Sigma f_i(u_i - \bar{u})^2}{N - 1}} \tag{4.3}$$

where

u_i = midvalue of speed class i
f_i = frequency of speed class i

Probability theory is used to determine the sample sizes for traffic engineering studies. Although a detailed discussion of these procedures is beyond the scope of this book, the simplest and most commonly used procedures are presented. Interested readers can find an in-depth treatment of the topic in publications listed in the References at the end of this chapter.

The minimum sample size depends on the precision level desired. The precision level is defined as the degree of confidence that the sampling error of a produced estimate will fall within a desired fixed range. Thus, for a precision level of 90-10, there is a 90 percent probability (confidence level) that the error of an estimate will not be greater than 10 percent of its true value. The confidence level is commonly given in terms of the level of significance (α), where $\alpha = (100 - $ confidence level). The commonly used confidence level for speed counts is 95 percent.

The basic assumption made in determining the minimum sample size for speed studies is that the normal distribution describes the speed distribution over a given section of highway. The normal distribution is given as

$$f(x) = \frac{1}{\sigma\sqrt{2\pi}} e^{-(x-\mu)^2/2\sigma^2} \text{ for } -\infty < x < \infty \tag{4.4}$$

where

μ = true mean of the population
σ = true standard deviation
σ^2 = true variance

The properties of the normal distribution are then used to determine the minimum sample size for an acceptable error d of the estimated speed. The following basic properties are used (see Figure 4.1):

1. The normal distribution is symmetrical about the mean.
2. The total area under the normal distribution curve is equal to 1 or 100 percent.
3. The area under the curve between $\mu + \sigma$ and $\mu - \sigma$ is 0.6827.
4. The area under the curve between $\mu + 1.96\sigma$ and $\mu - 1.96\sigma$ is 0.9500.
5. The area under the curve between $\mu + 2\sigma$ and $\mu - 2\sigma$ is 0.9545.
6. The area under the curve between $\mu + 3\sigma$ and $\mu - 3\sigma$ is 0.9971.
7. The area under the curve between $\mu + \infty$ and $\mu - \infty$ is 1.0000.

The last five properties are used to draw specific conclusions about speed data. For example, if it can be assumed that the true mean of the speeds in a section of highways is 80 km/h and the true standard deviation is 7 km/h, it can be concluded that 95 percent of all vehicle speeds will be between $(80 - 1.96 \times 7) = 66.3$ km/h and $(80 + 1.96 \times 7) = 93.7$ km/h. Similarly, if a vehicle is selected at random, there is a 95 percent chance that its speed is between 66.3 and 93.7 km/h. The properties of the normal distribution have been used to develop an equation relating the sample size to the number of standard deviations corresponding to a particular confidence level, the limits of tolerable error, and the standard deviation. The formula is

$$N = \left(\frac{Z\sigma}{d}\right)^2 \qquad\qquad (4.5)$$

where

N = minimum sample size
Z = number of standard deviations corresponding to the required confidence level
    = 1.96 for 95% confidence level (Table 4.1)
σ = standard deviation (km/h)
d = limit of acceptable error in the average speed estimate (km/h)

The standard deviation may be estimated from previous data or a small sample size may be used.

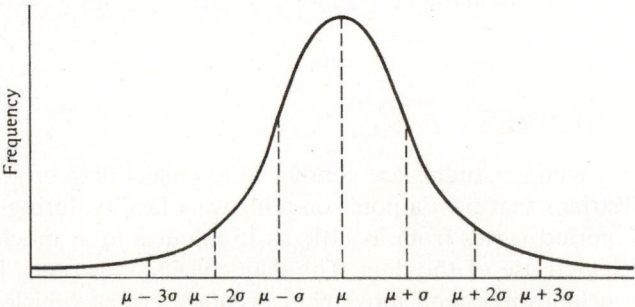

Figure 4.1   Shape of the Normal Distribution

Table 4.1   Constant Corresponding to Level of Confidence

Confidence Level (%)	Constant Z
68.3	1.00
86.6	1.50
90.0	1.64
95.0	1.96
95.5	2.00
98.8	2.50
99.0	2.58
99.7	3.00

Example 4.1 Determining Spot Speed Sample Size

As part of a class project, a group of students collected a total of 120 spot speed samples at a location and determined from this data that the standard variation of the speeds was ±10 km/h. If the project required that the confidence level be 95% and the limit of acceptable error was ±2.5 km/h, determine whether these students satisfied the project requirement.

Solution: Use Eq. 4.5 to determine the minimum sample size to satisfy the project requirements.

$$N = \left(\frac{Z\sigma}{d}\right)^2$$

where

Z = 1.96 (from Table 4.1)
σ = ±10
d = 2.5

$$N = \left(\frac{1.96 \times 10}{2.5}\right)^2$$
$$= 61.45$$

Therefore, the minimum number of spot speeds collected to satisfy the project requirement is 62. Since the students collected 120 samples, they satisfied the project requirements.

4.2 VOLUME STUDIES

Traffic volume studies are conducted to collect data on the number of vehicles and/or pedestrians that pass a point on a highway facility during a specified time period. This time period varies from as little as 15 minutes to as much as a year depending on the anticipated use of the data. The data collected also may be put into subclasses, which may include directional movement, occupancy rates, vehicle classification. When selecting locations for volume counts, it is important to note whether upstream bottlenecks could be limiting flow into the location where the count is being made. If an upstream bottleneck exists, the observed traffic volume may be less than the demand that would ideally like to use the roadway. Traffic volume studies are usually conducted when certain volume characteristics are needed, some of which follow:

1. **Annual Average Daily Traffic** (AADT) is the average of 24-hour counts collected every day of the year. AADTs are used in several traffic and transportation analyses for
   a. Estimation of highway user revenues
   b. Computation of crash rates in terms of number of crashes per 100 million vehicle kms traveled
   c. Establishment of traffic volume trends
   d. Evaluation of the economic feasibility of highway projects
   e. Development of freeway and major arterial street systems
   f. Development of improvement and maintenance programs
2. **Average Daily Traffic** (ADT) is the average of 24-hour counts collected over a number of days greater than one but less than a year. ADTs may be used for
   a. Planning of highway activities
   b. Measurement of current demand
   c. Evaluation of existing traffic flow
3. **Peak Hour Volume** (PHV) is the maximum number of vehicles that pass a point on a highway during a period of 60 consecutive minutes. PHVs are used for
   a. Functional classification of highways
   b. Design of the geometric characteristics of a highway; for example, number of lanes, intersection signalization, or channelization
   c. Capacity analysis
   d. Development of programs related to traffic operations; for example, one-way street systems or traffic routing
   e. Development of parking regulations
4. **Vehicle Classification** (VC) records volume with respect to the type of vehicle for example, passenger cars, two-axle trucks, or three-axle trucks. VC is used in
   a. Design of geometric characteristics, with particular reference to turning-radii requirements, maximum grades, lane widths, and so forth
   b. Capacity analyses, with respect to passenger-car equivalents of trucks
   c. Adjustment of traffic counts obtained by machines
   d. Structural design of highway pavements, bridges, and so forth
5. **Vehicle Kms of Travel** (VKT) is a measure of travel along a section of road. It is the product of the traffic volume (that is, average weekday volume or ADT) and the length of roadway in miles to which the volume is applicable. VKTs are used mainly as a base for allocating resources for maintenance and improvement of highways.

4.3 METHODS FOR CONDUCTING SPOT SPEED AND VOLUME STUDIES

The methods for conducting spot speed and volume studies can generally be divided into two main categories: intrusive and nonintrusive. The intrusive methods use sensors that are placed in or on the road and connected to counters located on the side of the road. Nonintrusive methods use remote observational techniques.

4.3.1 Intrusive Methods

Intrusive methods and counters used for spot speed studies can in general be used as intrusive methods for volume counts. These include pneumatic road tubes, piezoelectric sensors, and inductive loops.

Pneumatic road tubes are rubber tubes laid perpendicular to the direction of travel, connected to an electronic traffic counter/classifier. They can be used to collect data on speeds, volumes, and vehicle classification. Figure 4.2 shows an example setup of a surface detector using pneumatic road tubes. When a moving vehicle passes over the tube, an air pulse is transmitted through the tube to the counter. When used for speed measurements, two tubes are placed at a known distance apart across the lane/lanes. An air pulse is transmitted when the front wheels of a moving vehicle pass over the first tube; shortly afterward a second impulse is recorded when the front wheels pass over the second tube. The time elapsed between the two impulses and the distance between the tubes are used to compute the speed of the vehicle. When data are being collected on two or more lanes, the distance between the two tubes is important as the longer the distance, the larger the area between the tubes (error zone), and therefore the higher the probability of two vehicles being in the area at the same time or a vehicle straddling both lanes (both of which would result in erroneous data). To counteract this problem, a high-precision counter can be used, with road tubes at a very close spacing; or multiple counters can be used, with road tubes knotted at the lane lines to separate air pulses from different lanes from one another. Road tubes are typically used primarily for short-term studies. The tubes may fail if exposed to traffic for long periods (greater than one week), especially if there is a high volume of heavy trucks. Furthermore, research has shown that tubes can have a small impact on driver speeds since they are visible to oncoming traffic. Figure 4.3 shows the PEEK JR 161 portable loop traffic counter, which can be used with pneumatic road tubes. It is powered by a self-contained battery and consists of a solid-state loop detector and a six-digit electronic counter.

A *piezoelectric sensor* for traffic data collection consists of a mineral-based powder that surrounds a solid copper wire at the center of the cable and serves as the dielectric between the solid copper wire and the copper tube (the latter of which acts as the outer conductor). The principle used in the operation of this sensor is that mechanical energy is converted to potential energy, which results in a potential difference between the electrodes. This process creates an analog signal that is proportional to the pressure exerted

Figure 4.2 An Example of a Sensor Setup of a Surface Detector Using Pneumatic Road Tubes

SOURCE: Photograph by Lewis Woodson, Virginia Transportation Research Council, Charlottesville, VA.

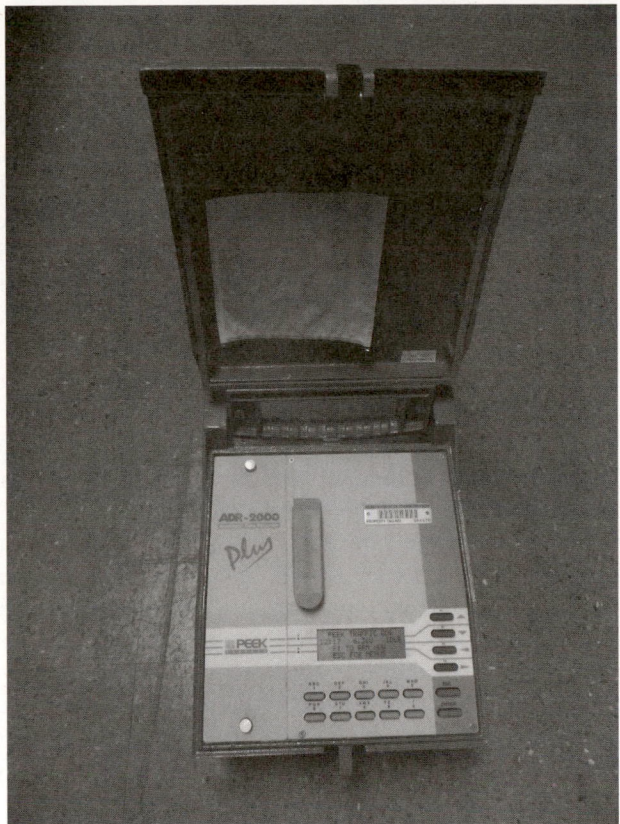

Figure 4.3   PEEK JR 161 Portable Loop Traffic Counter

SOURCE: Photograph by Lewis Woodson, Virginia Transportation Research Council, Charlottesville, VA.

by a tire when it passes over the sensor. The difference in the signals created as different axles pass the sensor is used to differentiate individual axles. Depending on the installation method, piezoelectric sensors can be used for either short- or long-term studies.

An *inductive loop* consists mainly of a single loop or several rectangular loops of insulated wire buried under the roadway surface and connected by a cable to a pull box, which is connected by another cable to a central controller cabinet that houses an electronics unit. The system serves as the detector of a resonant circuit. It operates on the principle that a disturbance in the electrical field is created when a motor vehicle passes across it. This causes a change in potential that is amplified, resulting in an impulse being sent to the counter. Inductive loops are primarily used for permanent installations since they must be installed in the pavement. Temporary inductive loops can be installed on the pavement surface with tape, but they are often more labor intensive to install than road tubes and will fail faster than permanent installations of loops.

4.3.2 NonIntrusive Methods

Usually, nonintrusive methods do not require the placing of sensors on or into the road pavement. These sensors are usually located above the road, either at the center or on the side of the road. Nonintrusive methods include manual, microwave radar, video image processing, ultrasonic, and passive acoustic technologies.

Manual methods are not now commonly used, except for turning movements at intersections. Therefore, these methods are not discussed here except for counts at intersections.

Counting turning movements at intersections involves the use one or more persons recording observed vehicles using a counter. With this type of counter, both the turning movements at the intersection and the types of vehicles can be recorded. Note that, in general, the inclusion of pickups and light trucks that have four tires and are in the category of passenger cars does not create any significant deficiencies in the data collected, since the performance characteristics of these vehicles are similar to those of passenger cars. In some instances, however, a more detailed breakdown of commercial vehicles may be required that would necessitate the collection of data according to number of axles and/or weight. However, the degree of truck classification usually depends on the anticipated use of the data collected.

Figure 4.4 shows the TDC-12 electronic manual counter, which may be used to conduct manual traffic volume counts at an intersection. The TDC-12 electronic manual counter, produced by Jamar Technologies, is powered by four AA batteries, which can be supplemented by an external power supply to extend the life of the batteries. Several buttons are provided, each of which can be used to record volume data for different movements and different types of vehicles. The data for each movement can be recorded in 1-, 5-, 15-, 30-, or 60-minute intervals, although the default value is 15 minutes. The recorded data can be viewed as data collection proceeds by using either the status screen, which indicates the current time and amount of time left in the interval, or a TAB key, which shows totals for each of the primary movements. The stored data can be extracted manually or transferred to a computer through a serial port. An associated software program (PETRAPro) can be used to read, edit, store, or print a variety of reports and graphs. Figure 4.4 shows the hook-up of the equipment to a PC for data transmittal.

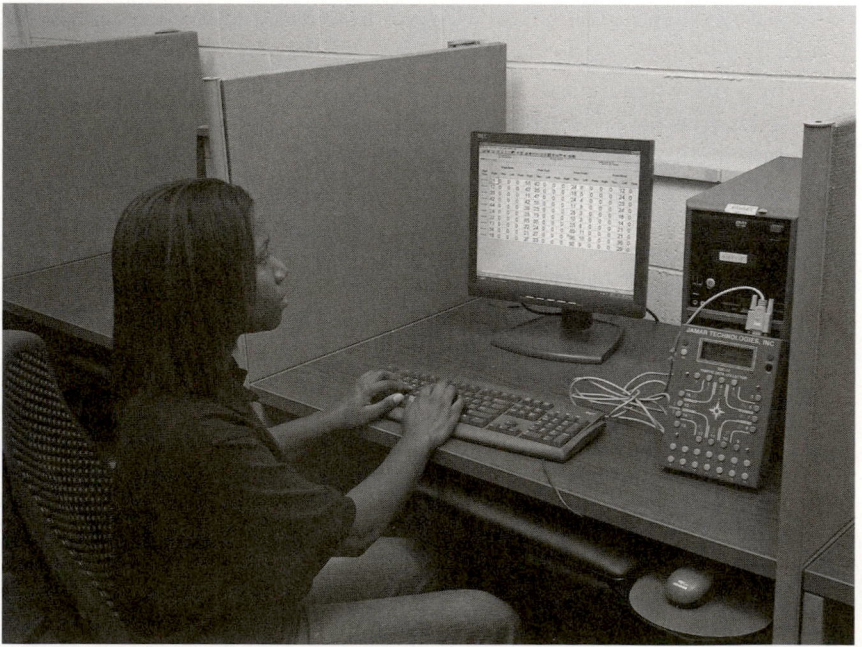

Figure 4.4 Jamar Traffic Data Collector TDC-12 Hooked to a Computer

SOURCE: Photograph by Nicholas J. Garber

The main disadvantages of the manual count method are that (1) it is labor intensive and therefore can be expensive, (2) it is subject to the limitations of human factors, and (3) it cannot be used for long periods of counting.

Microwave radar-based traffic sensors work on the principle that when a signal is transmitted onto a moving vehicle, the change in frequency between the transmitted signal and the reflected signal is proportional to the speed of the moving vehicle. The difference between the frequency of the transmitted signal and that of the reflected signal is measured by the equipment and then converted to speed in km/h. Radar-based speed studies can be done in two ways. First, an operator can manually collect speeds using a radar gun similar to the ones used by the police. Second, side-fire radar sensors can be installed beside or above the road to collect data in an automated manner. When placed above the road, they can measure the speeds of vehicles traveling toward or away from the equipment on a single lane. When placed at the side of the road, in a direction perpendicular to the flow of traffic, they can be used to obtain data on several lanes. When a radar unit is positioned on the side of the road to capture data in a direction parallel to the direction of travel, care must be taken to reduce the angle between the direction of the moving vehicle and the line joining the center of the transmitter and the vehicle. The value of the speed recorded depends on that angle. If the angle is not zero, an error related to the cosine of that angle is introduced, resulting in a lower speed than that which would have been recorded if the angle had been zero. However, this error is not very large, because the cosines of small angles are not much less than one. When radar is used to collect speed data in an automated manner, an advantage is that it is inconspicuous and unlikely to influence driver speeds. When collecting data using a hand-held radar unit, operators must take care to ensure that they are inconspicuous so that they do not influence traffic speeds.

Figure 4.5 shows a Wavetronix SmartSensor IQ radar-based traffic sensor. This sensor can be deployed either in the forward-looking mode, as shown in Figure 4.5a, or in the side-fire mode, as illustrated in Figure 4.5b. When deployed in the forward mode, a speed trap or Doppler system is used, and in the side mode a frequency modulated continuous wave (FMCW) system is used.

Video image processing, sometimes referred to as machine vision and commonly referred to today as a video detection system, usually consists of one or more cameras, a computer that can digitize and process imagery, and a software package that can convert these images to information on traffic flow in real time, such as speed, volume, queues, and headways. Autoscope is one example of equipment that uses video image processing. It consists of one or more electronic cameras, which overlook a large section of the roadway, and a microprocessor. The electronic camera receives the images from the road; the microprocessor determines the vehicle's presence or passage. This information is then used to determine the traffic characteristics in real time. Figure 4.6a schematically illustrates the configuration of the autoscope. It has a significant advantage over loops in that it can detect traffic in many locations within the camera's field of view. The locations to be monitored are selected by the user through interactive graphics, a process that normally takes only a few minutes. This flexibility is achieved by placing electronic detector lines along or across the roadway lanes on the monitor showing the traffic. The detector lines are therefore not fixed on the roadway because they are not physically located on the roadway but are placed on the monitor. A detection signal, which is similar to that produced by loops, is generated whenever a vehicle crosses the detector lines, indicating the presence or passage of the vehicle. The autoscope is therefore a wireless detector with one or more cameras that can replace many loops, thereby providing a wide-area detection system. The device therefore can be installed without

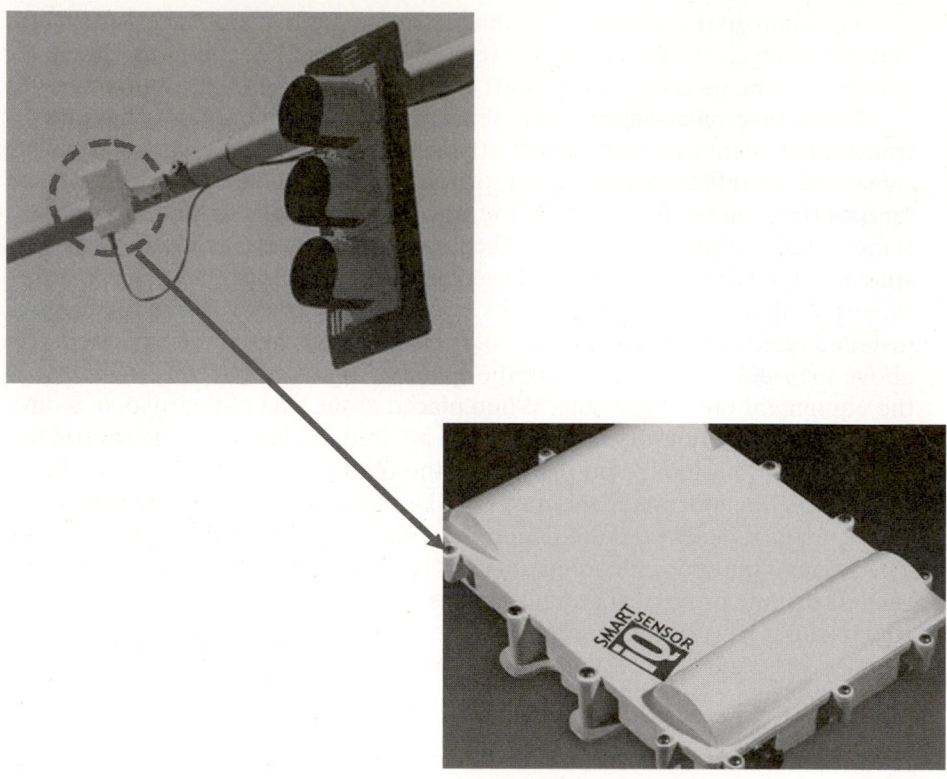

(a) Wavetronix SmartSensor IQ Deployed in the Forward Looking Mode

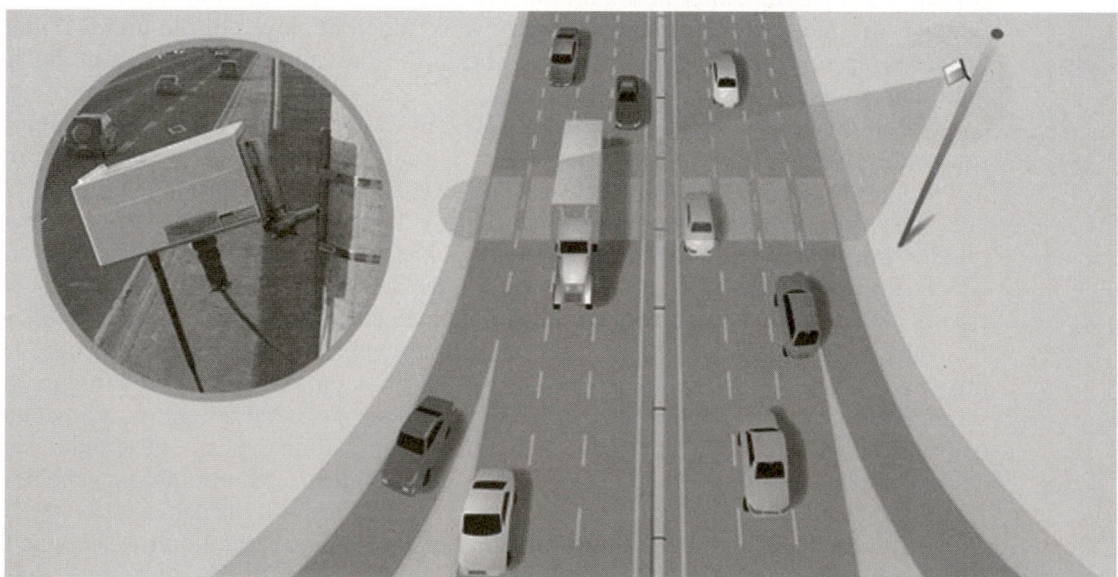

(b) Wavetronix SmartSensor IQ Deployed in the Side-Fire Mode

Figure 4.5   Wavetronix SmartSensor IQ Radar-Based Traffic Sensor

SOURCE: Courtesy EIS Traffic/ISS Canada Ltd

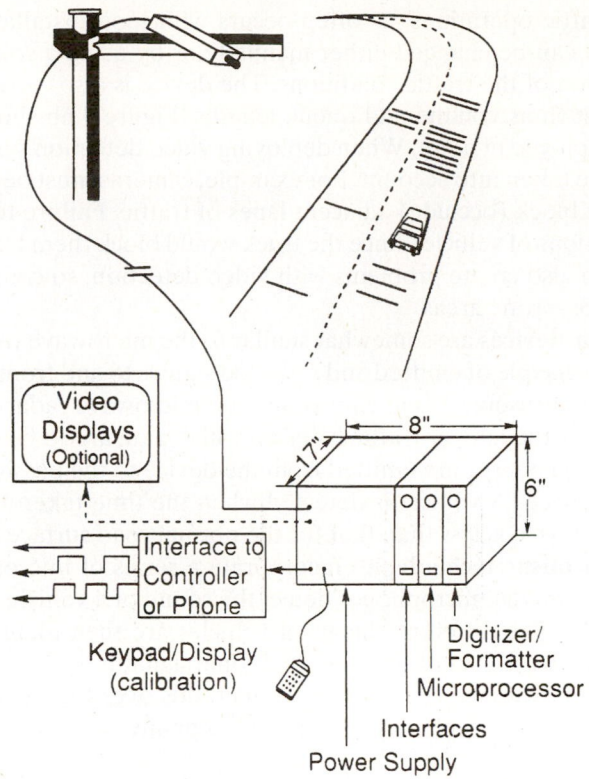

(a) Schematic Illustration of the Autoscope

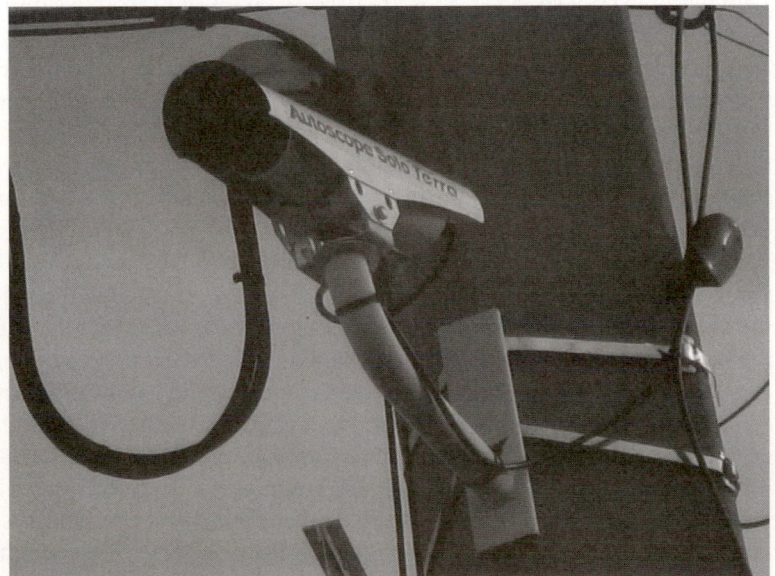

(b) The Autoscope Deployed

Figure 4.6   The Autoscope

SOURCE: Alan Blanchard Jr.

disrupting traffic operations, as often occurs with loop installation, and the detection configuration can be changed either manually or by using a software routine that provides a function of the traffic conditions. The device is also capable of extracting traffic parameters, such as volume and queue lengths. Figure 4.6b shows a photograph of an autoscope deployed at a site. When deploying video detection systems, several considerations must be taken into account. For example, cameras must be positioned so that large trucks do not block (occlude) adjacent lanes of traffic. Failure to do this could result in missed detections of vehicles since the truck would block them from the camera's field of view. Fog can also create problems with video detection, so care should be taken when it is used in fog-prone areas.

Ultrasonic devices are somewhat similar to the microwave radar devices in that they work on the principle of emitted and reflected signals to and from moving vehicles. However, they use ultrasonic sound energy and not microwave radar. These devices are usually placed over the lane or traffic lanes on which data are to be collected, and pulses of ultrasonic sound energy are emitted from the device to the passing vehicle and reflected back to the device. A vehicle is detected when the time taken for the sound energy to return to the device is less than that for the normal road surface background.

Passive acoustic technologies incorporate a series of microphones pointed toward the traffic stream; the microphones detect the sound of a vehicle as it passes through the detection zone. The different classes of vehicles are then identified by comparing the sound detected with a set of programmed sonic signatures.

The great advantage of nonintrusive methods over the use of intrusive methods is that it is not necessary to physically install loops or any other type of detector on the road.

4.4 PRESENTATION AND ANALYSIS OF SPOT SPEED DATA

The data collected in spot speed studies are usually taken only from a sample of vehicles using the section of the highway on which the study is conducted, but these data are used to determine the speed characteristics of the whole population of vehicles traveling on the study site. It is therefore necessary to use statistical methods in analyzing these data. Several characteristics are usually determined from the analysis of the data. Some of them can be calculated directly from the data; others can be determined from a graphical representation. Thus, the data must be presented in a form suitable for the specific analysis to be carried out.

The presentation format most commonly used is the frequency distribution table. The first step in the preparation of a frequency distribution table is the selection of the number of classes—that is, the number of velocity ranges—into which the data are to be fitted. The number of classes chosen is usually between 8 and 20, depending on the data collected. One technique that can be used to determine the number of classes is to first determine the range for a class size of 8 and then for a class size of 20. Finding the difference between the maximum and minimum speeds in the data and dividing this number first by 8 and then by 20 gives the maximum and minimum ranges in each class. A convenient range for each class is then selected and the number of classes determined. Usually the midvalue of each class range is taken as the speed value for that class. The data also can be presented in the form of a frequency histogram, or as a cumulative frequency distribution curve. The frequency histogram is a chart showing the midvalue for each class as the abscissa and the observed frequency for the corresponding class as the ordinate. The frequency distribution curve shows a plot of the frequency cumulative percentage against the upper limit of each corresponding speed class.

Example 4.2 Determining Speed Characteristics from a Set of Speed Data

Table 4.2 shows the data collected on a rural highway in Virginia during a speed study. Develop the frequency histogram and the frequency distribution of the data and determine

1. The arithmetic mean speed
2. The standard deviation
3. The median speed
4. The pace
5. The mode or modal speed
6. The 85th-percentile speed

Solution: The speeds range from 34.8 to 65.0 km/h, giving a speed range of 30.2. For eight classes, the range per class is 3.75 km/h; for 20 classes, the range per class is 1.51 km/h. It is convenient to choose a range of 2 km/h per class, which will give 16 classes. A frequency distribution table can then be prepared, as shown in Table 4.3, in which the speed classes are listed in column 1 and the midvalues are in column 2. The number of observations for each class is listed in column 3; the cumulative percentages of all observations are listed in column 6.

Figure 4.7 shows the frequency histogram for the data shown in Table 4.3. The values in columns 2 and 3 of Table 4.3 are used to draw the frequency histogram,

Table 4.2   Speed Data Obtained on a Rural Highway

Car No.	Speed (km/h)	Car No.	Speed (km/h)	Car No.	Speed (km/h)	Car No.	Speed (km/h)
1	35.1	23	46.1	45	47.8	67	56.0
2	44.0	24	54.2	46	47.1	68	49.1
3	45.8	25	52.3	47	34.8	69	49.2
4	44.3	26	57.3	48	52.4	70	56.4
5	36.3	27	46.8	49	49.1	71	48.5
6	54.0	28	57.8	50	37.1	72	45.4
7	42.1	29	36.8	51	65.0	73	48.6
8	50.1	30	55.8	52	49.5	74	52.0
9	51.8	31	43.3	53	52.2	75	49.8
10	50.8	32	55.3	54	48.4	76	63.4
11	38.3	33	39.0	55	42.8	77	60.1
12	44.6	34	53.7	56	49.5	78	48.8
13	45.2	35	40.8	57	48.6	79	52.1
14	41.1	36	54.5	58	41.2	80	48.7
15	55.1	37	51.6	59	48.0	81	61.8
16	50.2	38	51.7	60	58.0	82	56.6
17	54.3	39	50.3	61	49.0	83	48.2
18	45.4	40	59.8	62	41.8	84	62.1
19	55.2	41	40.3	63	48.3	85	53.3
20	45.7	42	55.1	64	45.9	86	53.4
21	54.1	43	45.0	65	44.7		
22	54.0	44	48.3	66	49.5		

Table 4.3   Frequency Distribution Table for Set of Speed Data

1	*2*	*3*	*4*	*5*	*6*	*7*
Speed Class (km/h)	*Class Midvalue, u_i*	*Class Frequency (Number of Observations in Class), f_i*	*$f_i u_i$*	*Percentage of Observations in Class*	*Cumulative Percentage of All Observations*	*$f(u_i - \bar{u})^2$*
34–35.9	35.0	2	70	2.3	2.30	420.50
36–37.9	37.0	3	111	3.5	5.80	468.75
38–39.9	39.0	2	78	2.3	8.10	220.50
40–41.9	41.0	5	205	5.8	13.90	361.25
42–43.9	43.0	3	129	3.5	17.40	126.75
44–45.9	45.0	11	495	12.8	30.20	222.75
46–47.9	47.0	4	188	4.7	34.90	25.00
48–49.9	49.0	18	882	21.0	55.90	9.00
50–51.9	51.0	7	357	8.1	64.0	15.75
52–53.9	53.0	8	424	9.3	73.3	98.00
54–55.9	55.0	11	605	12.8	86.1	332.75
56–57.9	57.0	5	285	5.8	91.9	281.25
58–59.9	59.0	2	118	2.3	94.2	180.50
60–61.9	61.0	2	122	2.3	96.5	264.50
62–63.9	63.0	2	126	2.3	98.8	364.50
64–65.9	65.0	1	65	1.2	100.0	240.25
Totals		86	4260			3632.00

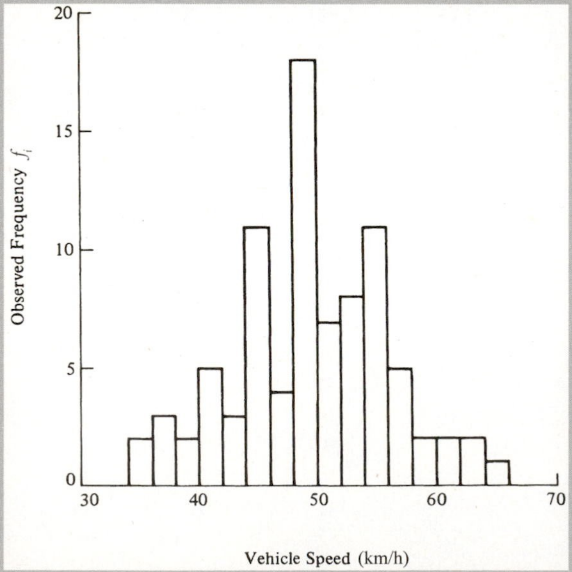

Figure 4.7   Histogram of Observed Vehicles' Speeds

where the abscissa represents the speeds and the ordinate the observed frequency in each class.

Figure 4.8 shows the frequency distribution curve for the data given. In this case, a curve showing percentage of observations against speed is drawn by plotting values from column 5 of Table 4.3 against the corresponding values in column 2. The total area under this curve is one or 100 percent.

Figure 4.9 shows the cumulative frequency distribution curve for the data given. In this case, the cumulative percentages in column 6 of Table 4.3 are plotted against the upper limit of each corresponding speed class. This curve, therefore, gives the percentage of vehicles that are traveling at or below a given speed.

The characteristics of the data can now be given in terms of the formula defined at the beginning of this section.

- The arithmetic mean speed is computed from Eq. 4.1.

$$\bar{u} = \frac{\sum f_i u_i}{\sum f_i}$$

$$\sum f_i = 86$$

$$\sum f_i u_i = 4260$$

$$\bar{u} = \frac{4260}{86} = 49.5 \, \text{km/h}$$

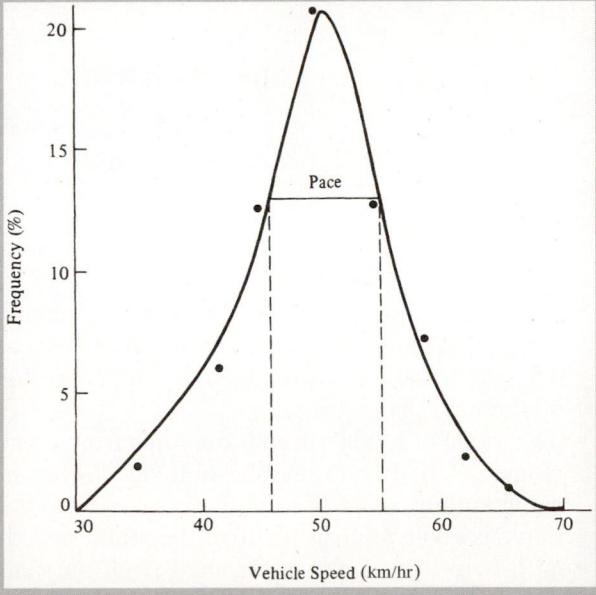

Figure 4.8   Frequency Distribution

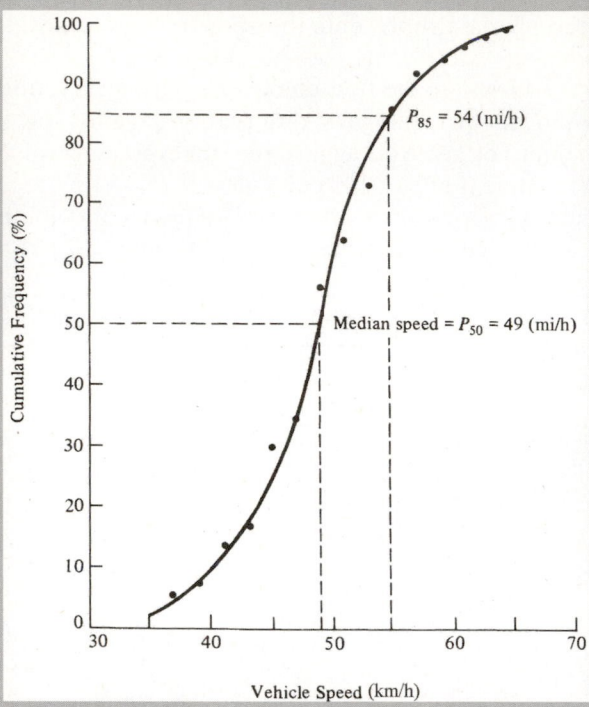

Figure 4.9   Cumulative Distribution

- The standard deviation is computed from Eq. 4.2.

$$S = \sqrt{\frac{\sum (u_i - \overline{u})^2}{N - 1}}$$

$$\sum (u_i - \overline{u})^2 = 3632$$

$$(N - 1) = \sum f_i - 1 = 85$$

$$S^2 = \frac{3632}{85} = 42.73$$

$$S = \pm 6.5 \, \text{km/h}$$

- The median speed is obtained from the cumulative frequency distribution curve (Figure 4.9) as 49 km/h, the 50th-percentile speed.
- The pace is obtained from the frequency distribution curve (Figure 4.8) as 45 to 55 km/h.
- The mode or modal speed is obtained from the frequency histogram as 49 km/h (Figure 4.7). It also may be obtained from the frequency distribution curve shown in Figure 4.8, where the speed corresponding to the highest point on the curve is taken as an estimate of the modal speed.
- 85th-percentile speed is obtained from the cumulative frequency distribution curve as 54 km/h (Figure 4.9).

Note that not all spot speed sets of data exhibit a true normal distribution. For example, the frequency distribution for the set of data shown in Table 4.4 is shown in Figure 4.10, which somewhat deviates from a true normal distribution. However, certain statistical characteristics, such as the 85th percentile speed, can be determined as shown in Figure 4.11.

Table 4.4   Speed Data at a Site in Virginia

Speed Bin (km/h)	Frequency
<50	15
50 – 51.9	11
52 – 53.9	17
54 – 55.9	16
56 – 57.9	13
58 – 59.9	27
60 – 61.9	46
62 – 63.9	56
64 – 65.9	74
66 – 67.9	68
68 – 69.9	67
70 – 71.9	32
72 – 73.9	19
74 – 75.9	6
76 – 77.9	2
78 – 79.9	0
≥80	2

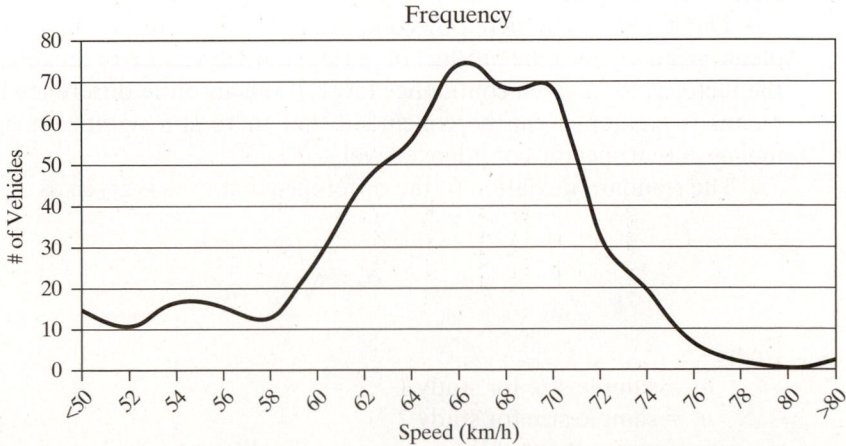

Figure 4.10   Frequency Distribution of Speed Data at a Location in Virginia

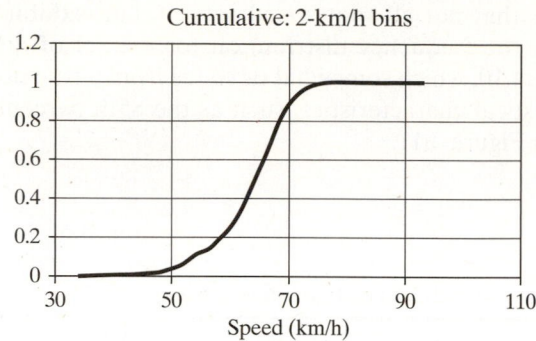

Figure 4.11   Cumulative Frequency Distribution of Speed Data at a Location on a Highway in Virginia

4.4.1 Other Forms of Presentation and Analysis of Speed Data

Certain applications of speed study data may require a more complicated presentation and analysis of the speed data. For example, if the speed data are to be used in research on traffic flow theories, it may be necessary for the speed data to be fitted into a suitable theoretical frequency distribution, such as the normal distribution or the gamma distribution. This is done first by assuming that the data fit a given distribution and then by testing this assumption using one of the methods of hypothesis testing, such as chi-square analysis. If the test suggests that the assumption can be accepted, specific parameters of the distribution can be found using the speed data. The properties of this distribution are then used to describe the speed characteristics, and any form of mathematical computation can be carried out using the distribution. Detailed discussion of hypothesis testing is beyond the scope of this book, but interested readers will find additional information in any book on statistical methods for engineers.

4.4.2 Statistical Comparison of Mean Speeds

There are several statistical tests that can be used to determine whether there is a significant difference between the mean speeds obtained from two spot speed studies. The tests briefly discussed are the *Z-test* and the *t-test* for comparison of two means.

The Z-test is conducted by comparing the absolute difference between the sample mean speeds against the product of the standard deviation of the difference in means and the factor Z for a given confidence level. If the absolute difference between the sample means is greater, it can be concluded that there is a significant difference in sample means at that specific confidence level.

The standard deviation of the difference in means is given as

$$S_d = \sqrt{\frac{S_1^2}{n_1} + \frac{S_2^2}{n_2}} \tag{4.6}$$

where

   n_1 = sample size for study 1
   n_2 = sample size for study 2
   S_d = square root of the variance of the difference in means
   S_1^2 = variance about the mean for study 1
   S_2^2 = variance about the mean for study 2

If $\bar{u}_1$ = mean speed of study 1, $\bar{u}_2$ = mean speed of study 2, and $|\bar{u}_1 - \bar{u}_2| > ZS_d$, where $|\bar{u}_1 - \bar{u}_2|$ is the absolute value of the difference in means, it can be concluded that the mean speeds are significantly different at the confidence level corresponding to Z. This analysis assumes that $\bar{u}_1$ and $\bar{u}_2$ are estimated means from the same distribution. Since it is typical to use the 95 percent confidence level in traffic engineering studies, the conclusion will, therefore, be based on whether $|\bar{u}_1 - \bar{u}_2|$ is greater than $1.96S_d$.

Example 4.3 Significant Differences in Average Spot Speeds

Speed data were collected at a section of highway during and after utility maintenance work. The speed characteristics are given as $\bar{u}_1, S_1$ and $\bar{u}_2, S_2$ as shown below. Determine whether there was any significant difference between the average speed at the 95% confidence level.

$$\bar{u}_1 = 35.5 \text{ km/h} \qquad \bar{u}_2 = 38.7 \text{ km/h}$$
$$S_1 = 7.5 \text{ km/h} \qquad S_2 = 7.4 \text{ km/h}$$
$$n_1 = 250 \qquad n_2 = 280$$

Solution:

- Use Eq. 4.6.

$$S_d = \sqrt{\frac{S_1^2}{n_1} + \frac{S_2^2}{n_2}}$$

$$= \sqrt{\frac{(7.5)^2}{250} + \frac{(7.4)^2}{280}} = 0.65$$

- Find the difference in means.

$$38.7 - 35.5 = 3.2 \text{ km/h}$$
$$3.2 > (1.96)(0.65)$$
$$3.2 > 1.3 \text{ km/h}$$

It can be concluded that the difference in mean speeds is significant at the 95 percent confidence level.

The t-test is one of many tests used for *hypothesis testing,* a procedure used to make inferences on populations based on samples of these populations. The t-test is preferred over the Z-test when sample sizes are less than 25 to 30. In hypothesis testing, the analyst first assumes that the means of two or more independent samples are equal. This assumption is referred to as the null hypothesis. The alternative hypothesis nullifies this assumption.

Consider the case where the mean speeds obtained from two spot speed studies are $\bar{\mu}_1$ and $\bar{\mu}_2$, respectively; the t-test can be used to determine whether there is a significant difference between the average number of crashes for the before and after periods. In this case, the null hypothesis is usually written as

$$H_0: \mu_1 = \mu_2$$

Depending on the statement of the problem, the alternative hypothesis can take one of the following forms:

$$H_1: \overline{\mu}_1 > \overline{\mu}_2 \quad \text{(one tail)}$$

$$H_1: \overline{\mu}_1 < \overline{\mu}_2 \quad \text{(one tail)}$$

$$H_1: \overline{\mu}_1 \neq \overline{\mu}_2 \quad \text{(two tail)}$$

The estimated means and/or variances of the populations obtained from the data sets are used to test the hypothesis by computing the test statistic T, which is then compared with a similar value t obtained from the theoretical distribution (which is shown in Appendix A). The T value is computed from Eq. 4.7.

$$T = \frac{\overline{\mu}_1 - \overline{\mu}_2}{S_p \sqrt{\dfrac{1}{n_1} + \dfrac{1}{n_2}}} \tag{4.7}$$

where

$\overline{\mu}_1$ and $\overline{\mu}_2$ = sample means
n_1 and n_2 = sample sizes
S_p = square root of the pooled variance given by

$$S_p^2 = \frac{(n_1 - 1)S_1^2 + (n_2 - 1)S_2^2}{n_1 + n_2 - 2} \tag{4.8}$$

where

S_1 and S_2 = standard deviations of the populations

The theoretical value of t depends on the degrees of freedom and the level of significance α used for the test, and whether the test is for a one or two-tailed test. The degree of freedom for the distribution is $(n_1 + n_2 - 2)$. The level of significance is the probability of rejecting the null hypothesis when it is true and is usually referred to as the type 1 error. The most commonly used value in traffic studies is 5 percent, although 10 percent is sometimes used. When the test is one-tailed, the t value selected is for α, and when it is two-tailed, the t value selected is $\alpha/2$. The region of rejection of the null hypothesis is as follows:

If H_1 is	Then reject H_0
$\mu_1 > \mu_2$ (one-tailed test)	$T > t_\alpha$
$\mu_1 < \mu_2$ (one-tailed test)	$T < t_\alpha$
$\mu_1 \neq \mu_2$ (two-tailed test)	$|T| > t_{\alpha/2}$

Appendix A gives theoretical t_α values for different levels of significance and degrees of freedom. The main criterion attributed to the use of the Z and t-tests is that the test assumes a normal distribution for data being tested. However, large sample sizes tend to diffuse this deficiency.

Example 4.4

In response to several complaints from residents that vehicle speeds on a neighborhood street were very high, the traffic engineer decided to post a speed limit of 35 kmph on the street. She collected speed data before and after posting the speed limit signs and obtained the results shown below. Using the *t*-test and a 95 percent significance level, determine whether it can be concluded that the posting of the speed limit signs resulted in a significant reduction in the mean speed of vehicles on the street.

Number of observations before posting of the speed limit signs = 65
Number of observations after posting of the speed limit signs = 70
Estimated mean speed before posting of the speed limit = 41.3 kmph (μ_1)
Estimated mean speed after posting of the speed limit = 36.5 kmph (μ_2)
Estimated standard deviation of speeds before posting of the speed limit = ±7.8 kmph (S_1)
Estimated standard deviation of speeds after posting of the speed limit = ±8.1 kmph (S_2)

Solution:

- Determine square root of the pooled variance (use Eq. 4.8).

$$S_p^2 = \frac{(n_1 - 1)S_1^2 + (n_2 - 1)S_2^2}{n_1 + n_2 - 2}$$

$$S_p^2 = \frac{(65 - 1)7.8^2 + (70 - 1)8.1^2}{65 + 70 - 2} = \frac{3893.76 + 4527.09}{133}$$

$$= 63.31$$

$$S_p = 7.96$$

- Compute test statistic T (use Eq. 4.7).

$$T = \frac{\overline{\mu}_1 - \overline{\mu}_2}{S_p\sqrt{\frac{1}{n_1} + \frac{1}{n_2}}} = T = \frac{41.3 - 36.5}{7.96\sqrt{\frac{1}{65} + \frac{1}{70}}}$$

$$= 3.50$$

- Determine whether μ_1 is significantly higher than μ_2.

Degree of freedom = 65 + 70 − 2 = 133

Theoretical $t_\alpha = 1.656$

Since 3.50 > 1.656, then $T > t_\alpha$ and the null hypothesis should be rejected.

It can therefore be concluded that posting the speed signs has resulted in a significant reduction in the mean speed on the street at a significance level of 0.05.

4.5 TYPES OF VOLUME COUNTS AND ANALYSIS OF VOLUME DATA

Different types of traffic counts are carried out, depending on the anticipated use of the data to be collected. These different types will now be discussed briefly.

4.5.1 Cordon Counts

When information is required on vehicle accumulation within an area such as the central business district (CBD) of a city, particularly during a specific time, a cordon count is undertaken. The area for which the data are required is cordoned off by an imaginary closed loop; the area enclosed within this loop is defined as the *cordon area.* Figure 4.12 shows such an area where the CBD of a city is enclosed by the imaginary loop ABCDA. The intersection of each street crossing the cordon line is taken as a count station; volume counts of vehicles and/or persons entering and leaving the cordon area are taken. The information obtained from such a count is useful for planning parking facilities, updating and evaluating traffic operational techniques, and making long-range plans for freeway and arterial street systems.

4.5.2 Screen Line Counts

In screen line counts, the study area is divided into large sections by running imaginary lines, known as screen lines, across it. In some cases, natural and synthetic barriers, such as rivers or railway tracks, are used as screen lines. Traffic counts are then taken at each point where a road crosses the screen line. It is usual for the screen lines to be designed or chosen such that they are not crossed more than once by the same street. Collection of data at these screen line stations at regular intervals facilitates the detection of variations in the traffic volume and traffic flow direction due to changes in the land-use pattern of the area.

4.5.3 Intersection Counts

Intersection counts are taken to determine vehicle classifications, through movements and turning movements at intersections. These data are used mainly in determining phase lengths and cycle times for signalized intersections, in the design of

Figure 4.12   Example of Station Locations for a Cordon Count

channelization at intersections, in capacity and level of servive analysis, and in the general design of improvements to intersections.

4.5.4 Pedestrian Volume Counts

Volume counts of pedestrians are made at locations such as subway stations, mid-blocks, and crosswalks. The counts are usually taken at these locations when the evaluation of existing or proposed pedestrian facilities is to be undertaken. Such facilities may include pedestrian overpasses or underpasses.

Pedestrian counts can be made using the TDC-12 electronic manual counter described earlier and shown in Figure 4.4. The locations at which pedestrian counts are taken also include intersections, along sidewalks, and mid-block crossings. These counts can be used for crash analysis, capacity analysis, and determining minimum signal timings at signalized intersections.

4.5.5 Periodic Volume Counts

In order to obtain certain traffic volume data, such as Annual Average Daily Traffic (AADT), it is necessary to obtain data continuously. However, it is not feasible to collect continuous data on all roads because of the cost involved. To make reasonable estimates of annual traffic volume characteristics on an area-wide basis, different types of periodic counts, with count durations ranging from 15 minutes to continuous, are conducted; the data from these different periodic counts are used to determine values that are then employed in the estimation of annual traffic characteristics. The periodic counts usually conducted are continuous, control, or coverage counts.

Continuous Counts

These counts are taken continuously using mechanical or electronic counters. Stations at which continuous counts are taken are known as permanent count stations. In selecting permanent count stations, the highways within the study area must first be properly classified. Each class should consist of highway links with similar traffic patterns and characteristics. A highway link is defined for traffic count purposes as a homogeneous section that has the same traffic characteristics, such as AADT and daily, weekly, and seasonal variations in traffic volumes at each point. Broad classification systems for major roads may include freeways, expressways, and major arterials. For minor roads, classifications may include residential, commercial, and industrial streets.

Control Counts

These counts are taken at stations known as control count stations, which are strategically located so that representative samples of traffic volume can be taken on each type of highway or street in an area-wide traffic counting program. The data obtained from control counts are used to determine seasonal and monthly variations of traffic characteristics so that expansion factors can be determined. These expansion factors are used to determine year-round average values from short counts.

Control counts can be divided into major and minor control counts. Major control counts are taken monthly, with 24-hour directional counts taken on at least two to three days during the week (Tuesday, Wednesday, and Thursday) and also on Saturday and Sunday to obtain information on weekend volumes. It is usual to locate at least one major control-count station on every major street. The data collected give information regarding hourly, monthly, and seasonal variations of traffic characteristics. Minor control counts are five-day weekday counts taken every other month on minor roads.

Coverage Counts

These counts are used to estimate ADT, using expansion factors developed from control counts. The study area is usually divided into zones that have similar traffic characteristics. At least one coverage count station is located in each zone. A 24-hour nondirectional weekday count is taken at least once every four years at each coverage station. The data indicate changes in area-wide traffic characteristics.

4.5.6 Presentation and Analysis of Traffic Volume

The data collected from traffic volume counts may be presented in one of several ways, depending on the type of count conducted and the primary use of the data. Descriptions of some of the conventional data presentation techniques follow.

Traffic Flow Maps

These maps show traffic volumes on individual routes. The volume of traffic on each route is represented by the width of a band, which is drawn in proportion to the traffic volume it represents, providing a graphic representation of the different volumes that facilitates easy visualization of the relative volumes of traffic on different routes. When flows are significantly different in opposite directions on a particular street or highway, it is advisable to provide a separate band for each direction. In order to increase the usefulness of such maps, the numerical value represented by each band is listed near the band. Figure 4.13 shows a typical traffic flow map.

Intersection Summary Sheets

These sheets are graphic representations of the volume and directions of all traffic movements through the intersection. These volumes can be either ADTs or PHVs, depending on the use of the data. Figure 4.14 shows a typical intersection summary sheet, displaying peak-hour traffic through the intersection.

Time–Based Distribution Charts

These charts show the hourly, daily, monthly, or annual variations in traffic volume in an area or on a particular highway. Each volume is usually given as a percentage of the average volume. Figure 4.15 shows typical charts for monthly, daily, and hourly variations.

Summary Tables

These tables give a summary of traffic volume data such as PHV, vehicle classification (VC), and ADT in tabular form. Table 4.5 is a typical summary table.

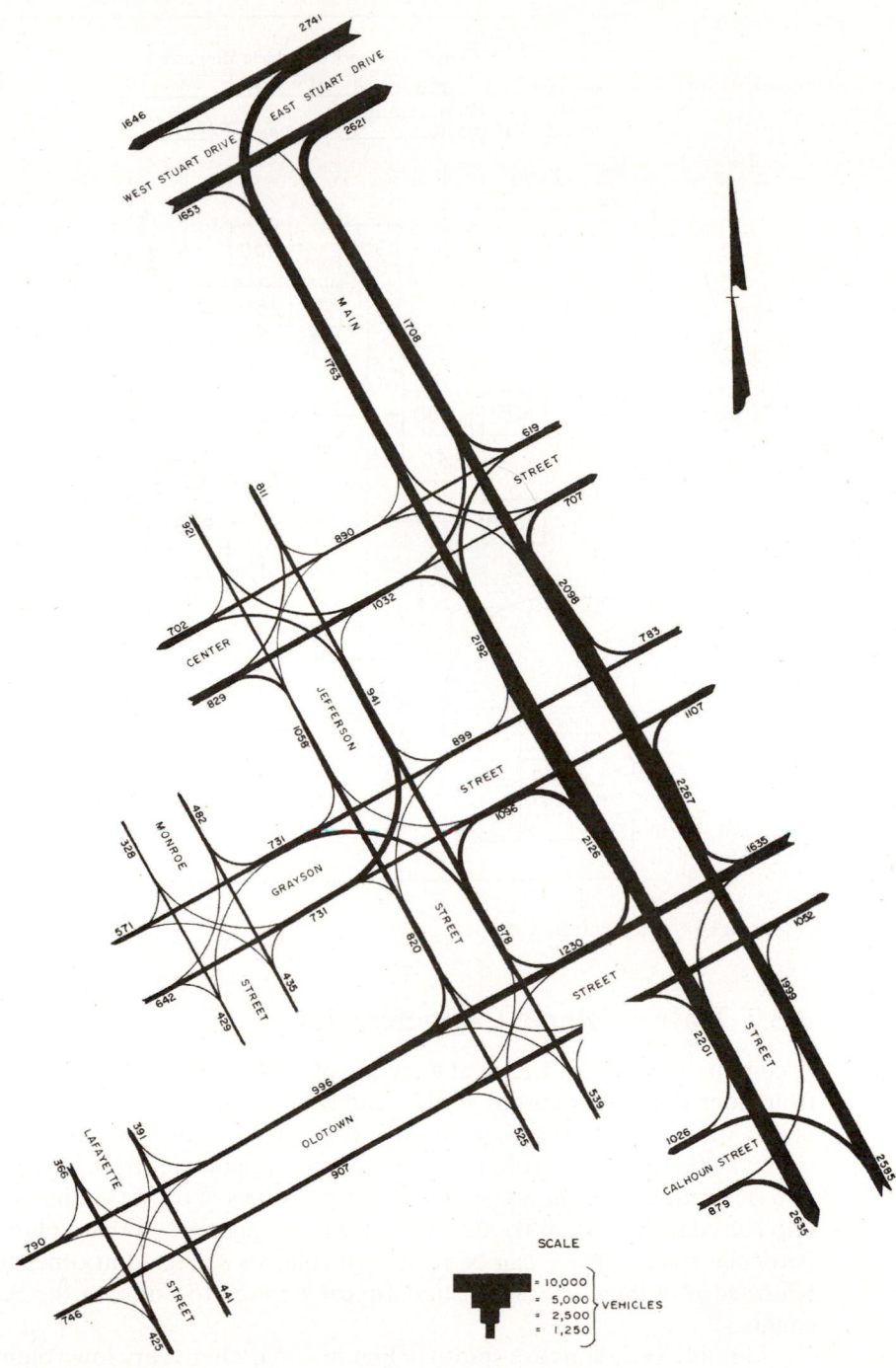

Figure 4.13   Example of a Traffic Flow Map

SOURCE: Galax Traffic Study Report, Virginia Department of Transportation, Richmond, VA., 1965.

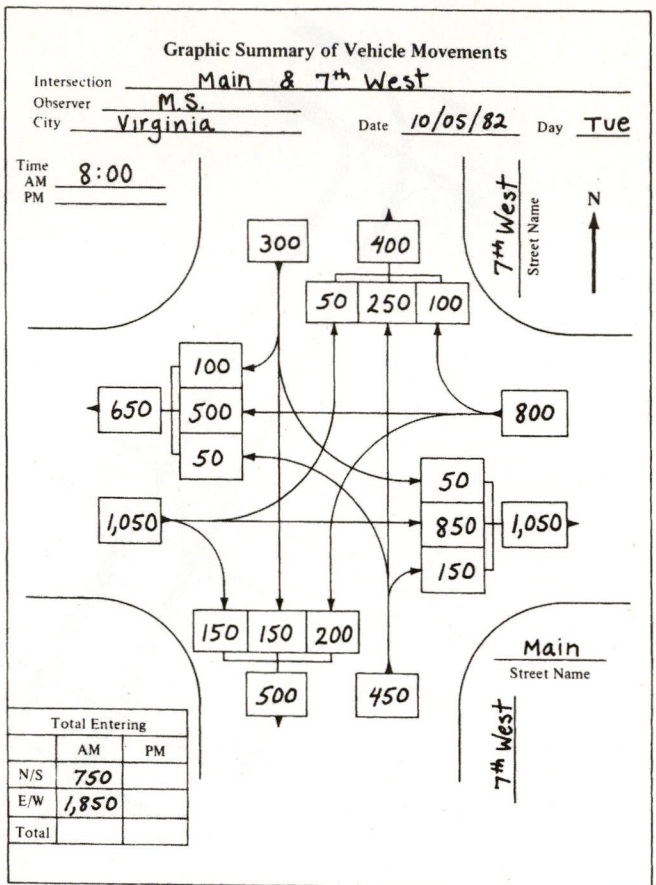

Figure 4.14   Intersection Summary Sheet

4.5.7 Traffic Volume Characteristics

A continuous count of traffic at a section of a road will show that traffic volume varies from hour to hour, from day to day, and from month to month. However, the regular observation of traffic volumes over the years has identified certain characteristics showing that although traffic volume at a section of a road varies from time to time, this variation is repetitive and rhythmic. These characteristics of traffic volumes are usually taken into consideration when traffic counts are being planned so that volumes collected at a particular time or place can be related to volumes collected at other times and places. Knowledge of these characteristics also can be used to estimate the accuracy of traffic counts.

Monthly variations are shown in Figure 4.15a, where very low volumes are observed during January and February, mainly because of the winter weather, and the peak volume is observed during August, mainly due to vacation traffic. This suggests that traffic volumes taken either during the winter months of January and February or during the summer months of July and August cannot be representative of the average annual traffic. If this information is presented for a number of consecutive years, the repetitive nature of the variation will be observed since the pattern of the variation will be similar for all years, although the actual volumes may not necessarily be the same.

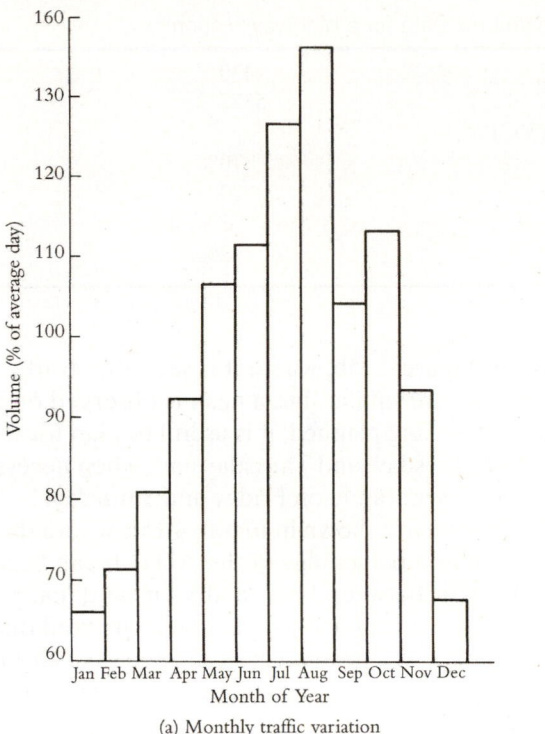

(a) Monthly traffic variation

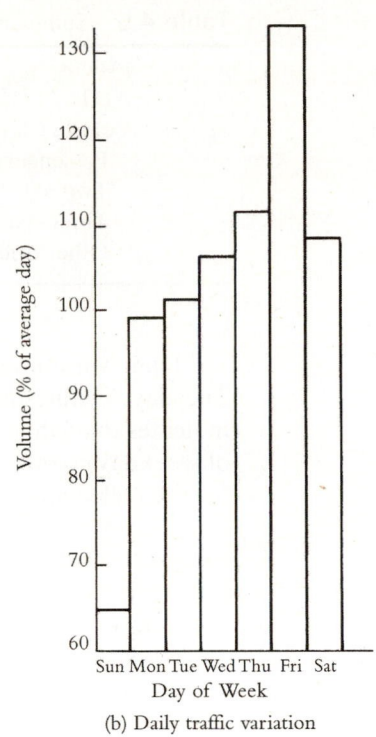

(b) Daily traffic variation

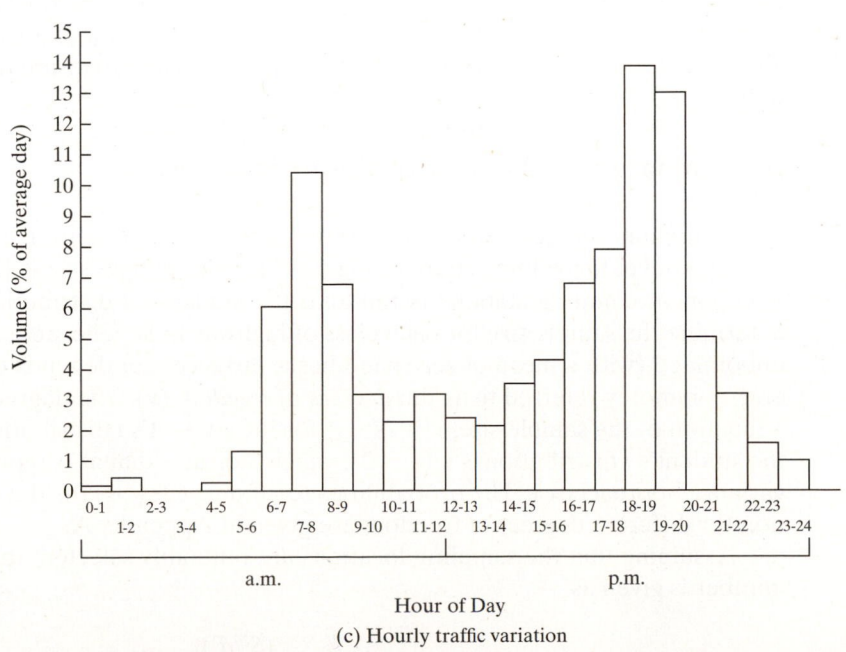

(c) Hourly traffic variation

Figure 4.15   Traffic Volumes on an Urban Highway

Table 4.5   Summary of Traffic Volume Data for a Highway Section

PHV	430
ADT	5375
Vehicle Classification (VC)	
Passenger cars	70%
Two-axle trucks	20%
Three-axle trucks	8%
Other trucks	2%

Daily variations are shown in Figure 4.15b, where it is seen that traffic volumes on Tuesday, Wednesday, and Thursday are similar, but a peak is observed on Friday. This indicates that when short counts are being planned, it is useful to plan for the collection of weekday counts on Tuesday, Wednesday, and Thursday and, when necessary, to plan for the collection of weekend counts separately on Friday and Saturday.

Hourly variations in traffic volume are shown in Figure 4.15c, where the volume for each hour of the day is represented as a percentage of the ADT. It can be seen that for this roadway, there is hardly any traffic between 1 a.m. and 5 a.m. and that peak volumes occur between 7 a.m. and 8 a.m., and 4 p.m. and 6 p.m. It can be inferred that work trips are primarily responsible for the peaks. If such data are collected on every weekday for one week, the hourly variations will be similar to each other, although the actual volumes may not be the same from day to day.

4.5.8 Sample Size and Adjustment of Periodic Counts

The impracticality of collecting data continuously every day of the year at all counting stations makes it necessary to collect sample data from each class of highway and to estimate annual traffic volumes from periodic counts. This involves the determination of the minimum sample size (number of count stations) for a required level of accuracy and the determination of daily, monthly, and/or seasonal expansion factors for each class of highway.

Determination of Number of Count Stations

The minimum sample size depends on the precision level desired. The commonly used precision level for volume counts is 95-5. When the sample size is less than 30 and the selection of counting stations is random, the student's t distribution may be used to determine the sample size for each class of highway links. The student's t distribution is unbounded (with a mean of zero) and has a variance that depends on the scale parameter, commonly referred to as the *degrees of freedom* (v). The degrees of freedom (v) is a function of the sample size; $v = N - 1$ for the student's t distribution. The variance of the student's t distribution is $v/(v - 2)$, which indicates that as v approaches infinity, the variance approaches 1. The probabilities (confidence levels) for the student's t distribution for different degrees of freedom are given in Appendix A.

Assuming that the sampling locations are randomly selected, the minimum sample number is given as

$$n = \frac{t_{\alpha/2,N-1}^2 (S^2/d^2)}{1 + (1/N)(t_{\alpha/2,N-1}^2)(S^2/d^2)}$$

(4.9)

where

n = minimum number of count locations required
t = value of the student's t distribution with $(1 - \alpha/2)$ confidence level $(N - 1$
degrees of freedom)
N = total number of links (population) from which a sample is to be selected
α = significance level
S = estimate of the spatial standard deviation of the link volumes
d = allowable range of error

To use Eq. 4.9, estimates of the mean and standard deviation of the link volumes are required. These estimates can be obtained by taking volume counts at a few links or by using known values for other, similar highways.

Example 4.5 Minimum Number of Count Stations

To determine a representative value for the ADT on 100 highway links that have similar volume characteristics, it was decided to collect 24-hour volume counts on a random sample of these links. Estimates of mean and standard deviation of the link volumes for the type of highways in which these links are located are 32,500 and 5500, respectively. Determine the minimum number of stations at which volume counts should be taken if a 95-5 precision level is required with a 10 percent allowable error.

Solution:

- Establish the data.

$$\alpha = (100 - 95) = 5 \text{ percent}$$
$$S = 5500$$
$$m = 32,500$$
$$d = 0.1 \times 32,500 = 3250 \text{ (allowable range of error)}$$
$$v = 100 - 1 = 99$$
$$t_{\alpha/2,99} \approx 1.984 \text{ (from Appendix A)}$$

- Use Eq. 4.7 to solve for n.

$$n = \frac{t_{\alpha/2,N-1}^2 (S^2/d^2)}{1 + (1/N)(t_{\alpha/2,N-1}^2)(S^2/d^2)}$$

$$= \frac{(1.984^2 \times 5500^2)/3250^2}{1 + (1/100)(1.984^2 \times 5500^2)/3250^2} = \frac{11.27}{1.11} = 10.1$$

Counts should be taken at a minimum of 11 stations. When sample sizes are greater than 30, the normal distribution is used instead of the student's t distribution.

However, the Federal Highway Administration (FHWA) has suggested that although it is feasible to develop a valid statistical sample for statewide traffic counts independent of the Highway Performance Monitoring System (HPMS) sample design, it is more realistic to use the HPMS sample design. This results in much less effort, because it is available, is clearly defined, and has been implemented. The HPMS sample has been implemented in each state, the District of Columbia, and Puerto Rico; it provides a statistically valid, reliable, and consistent database for analysis within states, between states, and for any aggregation of states up to the national level.

The HPMS sample design is a stratified simple random sample based on AADT, although about 100 data items are collected. The population from which the sample is obtained includes all public highways or roads within a state but excludes local roads. The sampling element is defined as a road section that includes all travel lanes and the volumes in both directions. The data are stratified by (1) type of area (rural, small urban, and individual or collective urbanized areas), and (2) functional class, which in rural areas includes interstate highways, other principal arterials, minor arterials, major collectors, and minor collectors and in urban areas includes interstate highways, other freeways or expressways, other principal arterials, minor arterials, and collectors.

Adjustment of Periodic Counts

Expansion factors, used to adjust periodic counts, are determined either from continuous count stations or from control count stations.

Expansion Factors from Continuous Count Stations. Hourly, daily, and monthly expansion factors can be determined using data obtained at continuous count stations. Hourly expansion factors (HEFs) are determined by the formula

$$HEF = \frac{\text{total volume for 24-hr period}}{\text{volume for particular hour}}$$

These factors are used to expand counts of durations shorter than 24-hour to 24-hour volumes by multiplying the hourly volume for each hour during the count period by the HEF for that hour and finding the mean of these products. Daily expansion factors (DEFs) are computed as

$$DEF = \frac{\text{average total volume for week}}{\text{average volume for particular day}}$$

These factors are used to determine weekly volumes from counts of 24-hour duration by multiplying the 24-hour volume by the DEF. Monthly expansion factors (MEFs) are computed as

$$MEF = \frac{AADT}{\text{ADT for particular month}}$$

The AADT for a given year may be obtained from the ADT for a given month by multiplying this volume by the MEF.

Tables 4.6, 4.7, and 4.8 give expansion factors for a particular primary road in Virginia. Such expansion factors should be determined for each class of road in the classification system established for an area.

Table 4.6 Hourly Expansion Factors for a Rural Primary Road

Hour	Volume	HEF	Hour	Volume	HEF
6:00–7:00 a.m.	294	42.00	6:00–7:00 p.m.	743	16.62
7:00–8:00 a.m.	426	29.00	7:00–8:00 p.m.	706	17.49
8:00–9:00 a.m.	560	22.05	8:00–9:00 p.m.	606	20.38
9:00–10:00 a.m.	657	18.80	9:00–10:00 p.m.	489	25.26
10:00–11:00 a.m.	722	17.10	10:00–11:00 p.m.	396	31.19
11:00–12:00 p.m.	667	18.52	11:00–12:00 a.m.	360	34.31
12:00–1:00 p.m.	660	18.71	12:00–1:00 a.m.	241	51.24
1:00–2:00 p.m.	739	16.71	1:00–2:00 a.m.	150	82.33
2:00–3:00 p.m.	832	14.84	2:00–3:00 a.m.	100	123.50
3:00–4:00 p.m.	836	14.77	3:00–4:00 a.m.	90	137.22
4:00–5:00 p.m.	961	12.85	4:00–5:00 a.m.	86	143.60
5:00–6:00 p.m.	892	13.85	5:00–6:00 a.m.	137	90.14

Total daily volume = 12,350.

Table 4.7 Daily Expansion Factors for a Rural Primary Road

Day of Week	Volume	DEF
Sunday	7895	9.515
Monday	10,714	7.012
Tuesday	9722	7.727
Wednesday	11,413	6.582
Thursday	10,714	7.012
Friday	13,125	5.724
Saturday	11,539	6.510

Total weekly volume = 75,122.

Table 4.8 Monthly Expansion Factors for a Rural Primary Road

Month	ADT	MEF
January	1350	1.756
February	1200	1.975
March	1450	1.635
April	1600	1.481
May	1700	1.394
June	2500	0.948
July	4100	0.578
August	4550	0.521
September	3750	0.632
October	2500	0.948
November	2000	1.185
December	1750	1.354

Total yearly volume = 28,450.

Mean average daily volume = 2370.

Example 4.6 Calculating AADT Using Expansion Factors

A traffic engineer urgently needs to determine the AADT on a rural primary road that has the volume distribution characteristics shown in Tables 4.6, 4.7, and 4.8. She collected the data shown below on a Tuesday during the month of May. Determine the AADT of the road.

7:00–8:00 a.m.	400
8:00–9:00 a.m.	535
9:00–10:00 a.m.	650
10:00–11:00 a.m.	710
11:00–12 noon	650

Solution:

- Estimate the 24-hr volume for Tuesday using the factors given in Table 4.6.

$$\frac{(400 \times 29.0 + 535 \times 22.05 + 650 \times 18.80 + 710 \times 17.10 + 650 \times 18.52)}{5} \approx 11{,}959$$

- Adjust the 24-hr volume for Tuesday to an average volume for the week using the factors given in Table 4.7.

$$\text{Total 7-day volume} = 11{,}959 \times 7.727$$

$$\text{Average 24-hr volume} = \frac{11{,}959 \times 7.727}{7} = 13{,}201$$

- Since the data were collected in May, use the factor shown for May in Table 4.8 to obtain the AADT.

$$\text{AADT} = 13{,}201 \times 1.394 = 18{,}402$$

4.6 TRAVEL TIME AND DELAY STUDIES

A travel time study determines the amount of time required to travel from one point to another on a given route. In conducting such a study, information may also be collected on the locations, durations, and causes of delays. When this is done, the study is known as a travel time and delay study. Data obtained from travel time and delay studies give a good indication of the level of service on the study section. These data also aid the traffic engineer in identifying problem locations, which may require special attention in order to improve the overall flow of traffic on the route.

4.6.1 Applications of Travel Time and Delay Data

The data obtained from travel time and delay studies may be used in any one of the following traffic engineering tasks:

- Determination of the efficiency of a route with respect to its ability to carry traffic
- Identification of locations with relatively high delays and the causes for those delays

- Performance of before-and-after studies to evaluate the effectiveness of traffic operation improvements
- Determination of relative efficiency of a route by developing sufficiency ratings or congestion indices
- Determination of travel times on specific links for use in trip assignment models and calibration/validation of simulation models
- Compilation of travel time data that may be used in trend studies to evaluate the changes in efficiency and level of service with time
- Performance of economic studies in the evaluation of traffic operation alternatives that reduce travel time

4.6.2   Definition of Terms Related to Time and Delay Studies

Let us now define certain terms commonly used in travel time and delay studies:

1. **Travel time** is the time it takes a vehicle to traverse a given section of a highway.
2. **Running time** is the time a vehicle is actually in motion while traversing a given section of a highway.
3. **Delay** is the time lost by a vehicle due to causes beyond the control of the driver.
4. **Operational delay** is that part of the delay caused by the impedance of other traffic. This impedance can occur either as side friction, where the stream flow is interfered with by other traffic (for example, parking or unparking vehicles); or as internal friction, where the interference is within the traffic stream (for example, reduction in capacity of the highway).
5. **Stopped-time delay** is that part of the delay during which the vehicle is at rest.
6. **Fixed delay** is that part of the delay caused by control devices such as traffic signals. This delay occurs regardless of the traffic volume or the impedance that may exist.
7. **Travel time delay** is the difference between the actual travel time and the travel time that will be obtained by assuming that a vehicle traverses the study section at an average speed equal to that for an uncongested traffic flow on the section being studied.

4.6.3   Methods for Conducting Travel Time and Delay Studies

Several methods have been used to conduct travel time and delay studies. These methods can be grouped into three general categories: (1) those using a test vehicle, (2) those not requiring a test vehicle and (3) those using Intelligent Transportation Systems Technologies. The particular technique used for any specific study depends on the reason for conducting the study and the available personnel and equipment.

Methods Requiring a Test Vehicle

This category involves three possible techniques: floating-car, average-speed, and moving-vehicle techniques. These methods require drivers to travel in the traffic stream while approximating the average travel speed, recording data either using a stopwatch, computer, or GPS device. GPS devices are most commonly used with test vehicles since the time-stamped location information is highly accurate and can be easily processed using geographic information systems. Test vehicle methods are useful for short-term studies, but it is sometimes difficult to get large sample sizes unless significant time and personnel are invested.

Floating-Car Technique. The floating-car technique is the most commonly used test vehicle method to collect travel time data. In this method, the test car is driven by an observer along the test section so that the test car "floats" with the traffic. The driver of the test vehicle attempts to pass as many vehicles as those that pass his test vehicle. The time taken to traverse the study section is recorded. This is repeated, and the average time is recorded as the travel time. The minimum number of test runs can be determined using an equation similar to Eq. 4.5, using values of the *t* distribution rather than the *z* values. The reason is that the sample size for this type of study is usually less than 30, which makes the *t* distribution more appropriate. The equation is

$$N = \left(\frac{t_\alpha \sigma}{d}\right)^2$$

(4.10)

where

N = sample size (minimum number of test runs)
σ = standard deviation (km/h)
d = limit of acceptable error in the speed estimate (km/h)
t_α = value of the student's *t* distribution with $(1 - \alpha/2)$ confidence level and $(N - 1)$ degrees of freedom
α = significance level

The limit of acceptable error used depends on the purpose of the study. The following limits are commonly used:

* Before-and-after studies: ±1.5 to ±5.0 km/h
* Traffic operation, economic evaluations, and trend analyses: ±3.0 to ±6.0 km/h
* Highway needs and transportation planning studies: ±5.0 to ±8.0 km/h

Average-Speed Technique. This technique involves driving the test car along the length of the test section at a speed that, in the opinion of the driver, is the average speed of the traffic stream. The time required to traverse the test section is noted. The test run is repeated for the minimum number of times, determined from Eq. 4.9, and the average time is recorded as the travel time.

In each of these methods, it is first necessary to clearly identify the test section. The way the travel time is usually obtained is that the observer starts a stopwatch at the beginning point of the test section and stops at the end. Additional data also may be obtained by recording the times at which the test vehicle arrives at specific locations that have been identified before the start of the test runs. A second stopwatch also may be used to determine the time that passes each time the vehicle is stopped. The sum of these times for any test run will give the stopped-time delay for that run. Table 4.9 shows an example of a set of data obtained for such a study.

Alternatively, the driver alone can collect the data by using a laptop computer with internal clock and distance functions. The predetermined locations (control points) are first programmed into the computer. At the start of the run, the driver activates the clock and distance functions; then the driver presses the appropriate computer key for each specified location. The data are then recorded automatically. The causes of delay are then recorded by the driver on a tape recorder.

Moving-Vehicle Technique. In this technique, the observer makes a round trip on a test section like the one shown in Figure 4.16, where it is assumed that the road runs east to west. The observer starts collecting the relevant data at section X-X, drives

Table 4.9 Speed and Delay Information

Street Name: 29 North   Date: July 7, 1994
Weather: Clear   Non-peak   Time: 2:00–3:00 p.m.

Cross Streets	Distance (m)	Travel Time (sec)	Segment Speed (km/h)	Stop Time (sec)	Reason for Stoppage	Speed Limit (km/h)	Ideal Travel Time (sec)	Segment Delay (sec)	Net Speed (km/h)
Ivy Road	0	0	—	0		—	0	0	–
Massie Road	475.2	42.6	40.2	20.1	Signal	64	27	15.6	27.3
Arlington Blvd.	396	27.7	51.5	0		64	22.5	5.2	51.5
Wise Street	237.6	19.7	43.4	8.9	Signal	64	13.5	6.2	29.9
Barracks Road	396	32.1	44.4	15.4	Signal	64	22.5	9.6	30.0
Angus Road	673.2	49.8	48.7	9.2	Signal	64	38.3	11.5	41.1
Hydraulic Road	475.2	24.4	70.1	0		72	24	0.4	70.1
Seminole Court	475.2	42.6	40.2	19.5	Signal	72	24	18.6	27.5
Greenbrier Drive	554.4	41.5	48.1	15.6	Signal	72	28	13.5	35.0
Premier Court	396	37.4	38.1	11.8	Signal	72	20	17.4	29.0
Fashion Square I	475.2	23.6	72.5	4.9	Signal	72	24	−0.4	60.0
Fashion Square II	316.8	19.7	57.9	0		72	16	3.7	57.9
Rio Road	316.8	20.2	56.5	14.1	Signal	72	16.4	2	33.3
Totals	5187.6	381.3	49.0	119.5		0	275.8	105.5	37.3

Note: Segment delay is the difference between observed travel time and calculated ideal travel time.
SOURCE: Study conducted in Charlottesville, VA., by Justin Black and John Ponder.

the car eastward to section Y-Y, then turns the vehicle around and drives westward to section X-X again. The following data are collected as the test vehicle makes the round trip:

- The time it takes to travel east from X-X to Y-Y (T_e), in minutes
- The time it takes to travel west from Y-Y to X-X (T_w), in minutes
- The number of vehicles traveling west in the opposite lane while the test car is traveling east (N_e)
- The number of vehicles that overtake the test car while it is traveling west from Y-Y to X-X; that is, traveling in the westbound direction (O_w)
- The number of vehicles that the test car passes while it is traveling west from Y-Y to X-X; that is, traveling in the westbound direction (P_w)

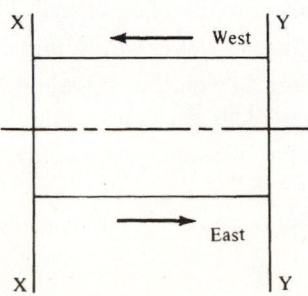

Figure 4.16 Test Site for Moving-Vehicle Method

The volume (V_w) in the westbound direction can then be obtained from the expression

$$V_w = \frac{(N_e + O_w - P_w)60}{T_e + T_w} \tag{4.11}$$

where $(N_e + O_w - P_w)$ is the number of vehicles traveling westward that cross the line X-X during the time $(T_e + T_w)$. Note that when the test vehicle starts at X-X, traveling eastward, all vehicles traveling westward should get to X-X before the test vehicle, except those that are passed by the test vehicle when it is traveling westward. Similarly, all vehicles that pass the test vehicle when it is traveling westward will get to X-X before the test vehicle. The test vehicle will also get to X-X before all vehicles it passes while traveling westward. These vehicles have, however, been counted as part of N_e or O_w and, therefore, should be subtracted from the sum of N_e and O_w to determine the number of westbound vehicles that cross X-X during the time the test vehicle travels from X-X to Y-Y and back to X-X. These considerations lead to Eq. 4.11.

Similarly, the average travel time $\overline{T_w}$ in the westbound direction is obtained from

$$\frac{\overline{T_w}}{60} = \frac{T_w}{60} - \frac{O_w - P_w}{V_w}$$

$$\overline{T_w} = T_w - \frac{60(O_w - P_w)}{V_w} \tag{4.12}$$

If the test car is traveling at the average speed of all vehicles, it will most likely pass the same number of vehicles as the number of vehicles that overtake it. Since it is probable that the test car will not be traveling at the average speed, the second term of Eq. 4.12 corrects for the difference between the number of vehicles that overtake the test car and the number of vehicles that are overtaken by the test car.

Example 4.7 Volume and Travel Time Using Moving-Vehicle Technique

The data in Table 4.10 were obtained in a travel time study on a section of highway using the moving-vehicle technique. Determine the travel time and volume in each direction at this section of the highway.

Mean time it takes to travel eastward (T_e) = 2.85 min

Mean time it takes to travel westbound (T_w) = 3.07 min

Average number of vehicles traveling westward when test vehicle is traveling eastward (N_e) = 79.50

Average number of vehicles traveling eastward when test vehicle is traveling westward (N_w) = 82.25

Average number of vehicles that overtake test vehicle while it is traveling westward (O_w) = 1.25

Table 4.10   Data from Travel Time Study Using the Moving-Vehicle Technique

Run Direction/ Number	Travel Time (min)	No. of Vehicles Traveling in Opposite Direction	No. of Vehicles That Overtook Test Vehicle	No. of Vehicles Overtaken by Test Vehicle
Eastward				
1	2.75	80	1	1
2	2.55	75	2	1
3	2.85	83	0	3
4	3.00	78	0	1
5	3.05	81	1	1
6	2.70	79	3	2
7	2.82	82	1	1
8	3.08	78	0	2
Average	2.85	79.50	1.00	1.50
Westward				
1	2.95	78	2	0
2	3.15	83	1	1
3	3.20	89	1	1
4	2.83	86	1	0
5	3.30	80	2	1
6	3.00	79	1	2
7	3.22	82	2	1
8	2.91	81	0	1
Average	3.07	82.25	1.25	0.875

Average number of vehicles that overtake test vehicle while it is traveling eastward $(O_e) = 1.00$

Average number of vehicles the test vehicle passes while traveling westward $(P_w) = 0.875$

Average number of vehicles the test vehicle passes while traveling eastward $(P_e) = 1.5$

Solution:

- From Eq. 4.11, find the volume in the westbound direction.

$$V_w = \frac{(N_e + O_w - P_w)60}{T_e + T_w}$$

$$= \frac{(79.50 + 1.25 - 0.875)60}{2.85 + 3.07} = 809.5 \quad \text{(or 810 veh/h)}$$

- Similarly, calculate the volume in the eastbound direction.

$$V_e = \frac{(82.25 + 1.00 - 1.50)60}{2.85 + 3.07} = 828.5 \quad \text{(or 829 veh/h)}$$

- Find the average travel time in the westbound direction.

$$\overline{T}_w = 3.07 - \frac{(1.25 - 0.875)}{810}60 = 3.0 \text{ min}$$

- Find the average travel time in the eastbound direction.

$$\overline{T}_e = 2.85 - \frac{(1.00 - 1.50)}{829}60 = 2.9 \text{ min}$$

4.6.4 Methods Not Requiring a Test Vehicle

This category includes the license-plate method.

License-Plate Observations. The license-plate method requires that observers be positioned at the beginning and end of the test section. Observers also can be positioned at other locations if elapsed times to those locations are required. Each observer records the last three or four digits of the license plate of each car that passes, together with the time at which the car passes. The reduction of the data is accomplished in the office by matching the times of arrival at the beginning and end of the test section for each license plate recorded. The difference between these times is the traveling time of each vehicle. The average of these is the average traveling time on the test section. It has been suggested that a sample size of 50 matched license plates will give reasonably accurate results. License plate matching can also be done using automated systems with character recognition software.

4.6.5 Intelligent Transportation Systems (ITS)

ITS, also referred to as Telematics, generally can be described as the process through which data on the movement of people and goods can be collected, stored, and analyzed and related information disseminated. The process has been used in many areas of transportation engineering, including the management of traffic, public transportation, traveler information, electronic toll payment, and safety.

Several ITS approaches are in use to determine travel time and speed data. These methods typically rely on the automated re-identification of a unique attribute of a vehicle at multiple known locations along a road. The elapsed time that occurs between when a vehicle is identified at these known points can be used to determine travel time and speed. One example of an ITS re-identification system uses Automatic Vehicle Identification (AVI). AVI systems wirelessly read a unique identifier from an electronic toll collection tag (like EZ-PASS) at multiple points on a road and uses matched identifiers to determine travel times. Houston, Texas is an example of one jurisdiction that uses this approach. Its AVI system is used to produce a Web site with maps of travel times for Houston's freeway system.

A number of private companies, such as INRIX and NAVTEQ, have also recently begun to sell travel time data on major interstates and arterials. These services derive travel times in part from aggregated GPS data from trucking fleets, taxis, and private vehicles. These data are available for purchase and offer the potential to significantly expand the number of lane-miles of road with travel time data versus field data collection of travel times.

Another recent method for collecting travel time data is Bluetooth-based re-identification. Bluetooth is a wireless communications protocol that is commonly used in consumer electronics. In this case, Bluetooth readers with antennas are mounted on the side of the road. The reader logs the unique media access control (MAC) address of Bluetooth devices (like phones, music players, and gaming systems) in vehicles. When MAC addresses are matched at multiple sites, travel times can be determined. This method can capture at least 3 to 5 percent of all vehicles on a road. This approach is gaining increased popularity in the United States since many more samples can be collected than if a test vehicle approach were used.

An emerging set of ITS technologies that can be used for collecting traffic data is Connected Vehicles, formally referred to as Vehicle Infrastructure Integration (VII). Connected technologies link vehicles on the road with other vehicles and their road environment, such as roadside equipment. These technologies are being developed through the Connected Vehicle initiative, which supports cooperative research effort involving federal and state Departments of Transportation and automobile manufacturers. It is envisioned that Connected Vehicle technology will eventually be widely deployed across the entire vehicle fleet. It can be used to collect, store, and analyze data on the movement of people and goods and disseminate relevant information.

4.7 PARKING STUDIES

Any vehicle traveling on a highway will at one time or another be parked for either a relatively short time or a much longer time, depending on the reason for parking. The provision of parking facilities is therefore an essential element of the highway mode of transportation. The need for parking spaces is usually very great in areas where land uses include business, residential, or commercial activities. The growing use of the automobile as a personal feeder service to transit systems ("park-and-ride") has also increased the demand for parking spaces at transit stations. In areas of high density, where space is very expensive, the space provided for automobiles usually must be divided between that allocated for their movement and that allocated for parking them.

Providing adequate parking space to meet the demand for parking in the Central Business District (CDB) may necessitate the provision of parking bays along curbs, which reduces the capacity of the streets and may affect the level of service. This problem usually confronts a city traffic engineer. The solution is not simple, since the allocation of available space will depend on the goals of the community, which the traffic engineer must take into consideration when trying to solve the problem. Parking studies are therefore used to determine the demand for and the supply of parking facilities in an area, the projection of the demand, and the views of various interest groups on how best to solve the problem. Before we discuss the details of parking studies, it is necessary to discuss the different types of parking facilities.

4.7.1 Types of Parking Facilities

Parking facilities can be divided into two main groups: on-street and off-street.

On-Street Parking Facilities

These are also known as curb facilities. Parking bays are provided alongside the curb on one or both sides of the street. These bays can be unrestricted parking facilities if the duration of parking is unlimited and parking is free, or they can be restricted parking facilities if parking is limited to specific times of the day for a maximum duration. Parking at restricted facilities may or may not be free. Restricted facilities also may be provided for specific purposes, such as to provide handicapped parking or as bus stops or loading bays.

Off-Street Parking Facilities

These facilities may be privately or publicly owned; they include surface lots and garages. Self-parking garages require that drivers park their own automobiles; attendant-parking garages maintain personnel to park the automobiles.

4.7.2 Definitions of Parking Terms

Before discussing the different methods for conducting a parking study, it is necessary to define some terms commonly used in parking studies, including *space-hour, parking volume, parking accumulation, parking load, parking duration,* and *parking turnover.*

1. A **space-hour** is a unit of parking that defines the use of a single parking space for a period of one hour.
2. **Parking volume** is the total number of vehicles that park in a study area during a specific length of time, usually a day.
3. **Parking accumulation** is the number of parked vehicles in a study area at any specified time. These data can be plotted as a curve of parking accumulation against time, which shows the variation of the parking accumulation during the day.
4. The **parking load** is the area under the accumulation curve between two specific times. It is usually given as the number of space-hours used during the specified period of time.
5. **Parking duration** is the length of time a vehicle is parked at a parking bay or stall. When the parking duration is given as an average, it gives an indication of how frequently a parking space becomes available.
6. **Parking turnover** is the rate of use of a parking space. It is obtained by dividing the parking volume for a specified period by the number of parking spaces.

4.7.3 Methodology of Parking Studies

A comprehensive parking study usually involves (1) inventory of existing parking facilities; (2) collection of data on parking accumulation, parking turnover, and parking duration; (3) identification of parking generators; and (4) collection of information on parking demand. Information on related factors, such as financial, legal, and administrative matters, also may be collected.

Inventory of Existing Parking Facilities

An inventory of existing parking facilities is a detailed listing of the location and all other relevant characteristics of each legal parking facility, private and public, in the study area.

The inventory includes both on- and off-street facilities. The relevant characteristics usually listed include the following:

- Type and number of parking spaces at each parking facility
- Times of operation and limit on duration of parking, if any
- Type of ownership (private or public)
- Parking fees, if any, and method of collection
- Restrictions on use (open or closed to the public)
- Other restrictions, if any (such as loading and unloading zones, bus stops, or taxi ranks)
- Probable degree of permanency (can the facility be regarded as permanent or is it just a temporary facility?)

When carrying out an inventory of parking facilities, a coding system should be used to facilitate the location of the different parking facilities. For example, when an inventory is to be undertaken in a section of a city, the area can be divided into identifiable blocks, using letters A, B, C, D, etc., as shown in Figure 4.17a. The curb sides are then identified using numbers from 1 upward to the maximum number of curb faces in any one block, as illustrated in Figure 4.17a. Note that usually this number is 4, except for irregular blocks, as shown in block B of Figure 4.17a. Individual parking facilities within any one block are then coded using numbers that start with the number that is one higher than the maximum used for curb sides. For example, if the maximum number used for curb sides is 5, then 6, 7, 8, etc. are used to identify the individual parking facilities, as shown in Figure 4.17b. Each parking facility is identified by the associated codes, starting with the block number and the facility number. For example, parking facility 8 located in block C is coded as C-8. It is important to obtain all information for each parking facility, including angle of inclination for curb-side parking, number of free parking stalls separately from metered stalls, whether an off-street parking facility is private or public, and any time limits that exist. A summary table for each block is then prepared as shown in Table 4.11. A summary of different types of parking facilities can then be prepared. The extent to which the types are classified usually depends on what the data are to be used for. A typical summary may include the following classes:

1. **Off-Street Parking (Garages)**
   (i)   Number of limited opening-time public garages
   (ii) Number of 24-hr public garages
   (iii) Number of private garages
2. **Off-Street Parking (Parking Lots)**
   (i)   Number of metered stalls for large trucks
   (ii) Number of metered stalls for passenger cars
   (iii) Number of free stalls for large trucks
   (iv) Number of free stalls for passenger cars
3. **Curbside Parking**
   (i)   Number of free-parking stalls
   (ii) Number of metered-parking stalls

In some cases, special studies such as an inventory of large truck parking facilities along an interstate highway could be carried out. This requires segmenting the corridor along the section of the highway for which the parking inventory is required. The width of the corridor will depend on the distance from the interstate highway that truck drivers are willing to search for parking. This information could be obtained from a preliminary survey of truck drivers using the corridor. Figure 4.18 is an example of such a corridor, and Table 4.12 is an example of an inventory summary data form that can be used for such a study.

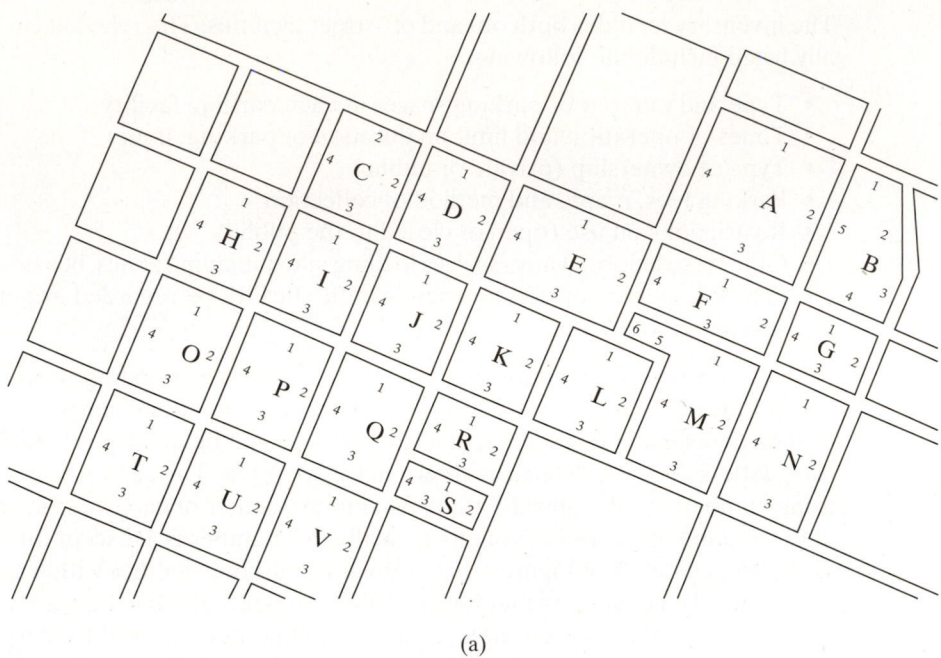

(a)

Example of a coding system for a central business district

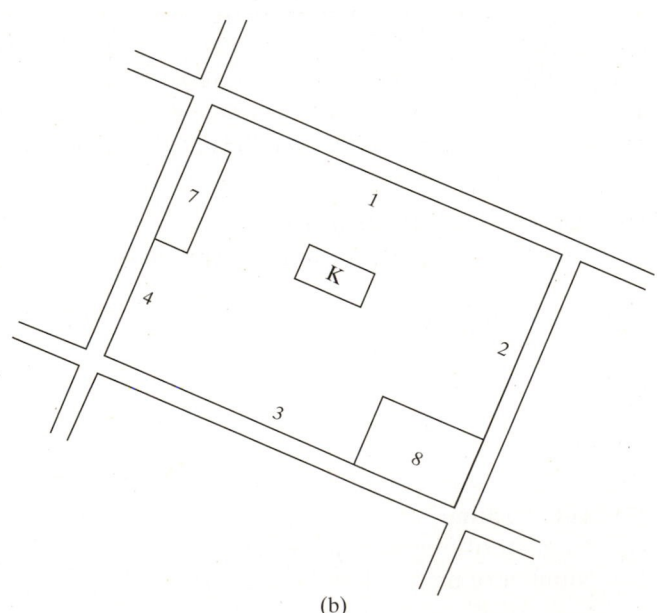

(b)

Example of a coding system for parking inventory in a block

Figure 4.17 An Example of a Coding System for Parking Inventory

Table 4.11   Parking Inventory Summary Form

Block	Facility Identity	Off-Street Parking (Public Garages)		Off-Street Parking (Private Garages), # of Stalls	Off-Street Parking Lots, # of Stalls				Free Curb-Side Parking, # of Stalls				Metered Curb-Side Parking, # of Stalls				Total
		Limited Time Slots (#)	24-Hr Time Slots (#)		Metered Stalls for Large Trucks	Metered Stalls for Passenger Cars	Free Stalls for Large Trucks	Free Stalls for Passenger Cars	Parallel Parking	90°	60°	45°	Parallel	90°	60°	45°	

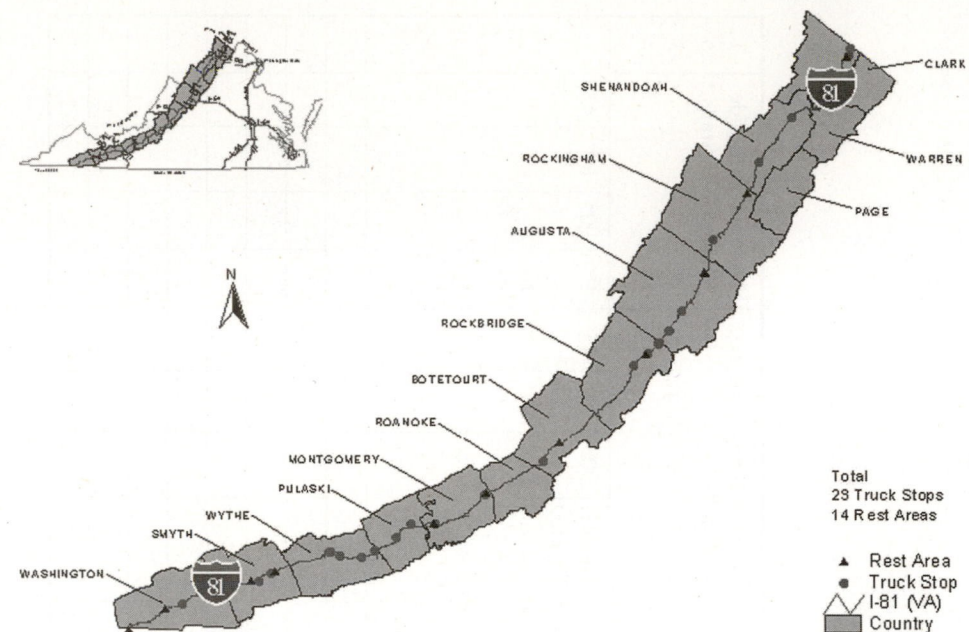

Figure 4.18   Distribution of Commercial Vehicle Parking Spaces along I-81 within Virginia

SOURCE: Figure 2, p. 11 in Garber, N.J., Hua Wang, and Dares Charoenphol, "Estimating the Supply and Demand for Commercial Heavy Truck Parking on Interstate Highways: A Case Study of I-81 in Virginia," Virginia Transportation Research Council, 2002. Copyright 2002 by the Commonwealth of Virginia.

The information obtained from an inventory of parking facilities is useful both to the traffic engineer and to public agencies, such as zoning commissions and planning departments. The inventory should be updated at regular intervals of about four to five years.

Table 4.12   An Example Form Used for a Large Truck Parking Inventory along an Interstate Highway

Section	Begin Milepost	End MilePost	Length	Parking Spaces in Rest Areas	Parking Spaces in Private Truck Stops	Total

4.7.4 Collection of Parking Data

Accumulation. Accumulation data are obtained by checking the amount of parking during regular intervals on different days of the week. The checks are usually carried out on an hourly or 2-hour basis between 6:00 a.m. and 12 midnight. The selection of the times depends on the operation times of land-use activities that act as parking generators. For example, if a commercial zone is included, checks should be made during the times when retail shops are open, which may include periods up to 9:30 p.m. on some days. On the other hand, at truck stops, the highest accumulation may occur around midnight, which requires information to be collected at that time. The information obtained is used to determine hourly variations of parking and peak periods of parking demand. (See Figure 4.19.)

Turnover and Duration. Information on turnover and duration is usually obtained by collecting data on a sample of parking spaces in a given block. This is done by recording the license plate of the vehicle parked on each parking space in the sample at the ends of fixed intervals during the study period. The length of the fixed intervals depends on the maximum permissible duration. For example, if the maximum permissible duration of parking at a curb face is 1 hour, a suitable interval is every 20 minutes. If the permissible duration is 2 hours, checking every 30 minutes would be appropriate. Numbers from 1 upward may be used to identify each parking bay or stall in each parking facility. For example, the first parking bay or stall in block K facility 7 is identified as K-7-1. An example of the form that can be used for a license plate survey is shown in Table 4.13. Turnover is then obtained from the equation

$$T = \frac{\text{Number of different vehicles parked}}{\text{Number of parking spaces}} \tag{4.13}$$

Although the manual collection of parking data is still commonly used, it is now possible for all parking data to be collected electronically. Some of these electronic systems use wireless sensors to detect the arrival and departure of a vehicle at a parking space and the information sent to a central location through the Internet.

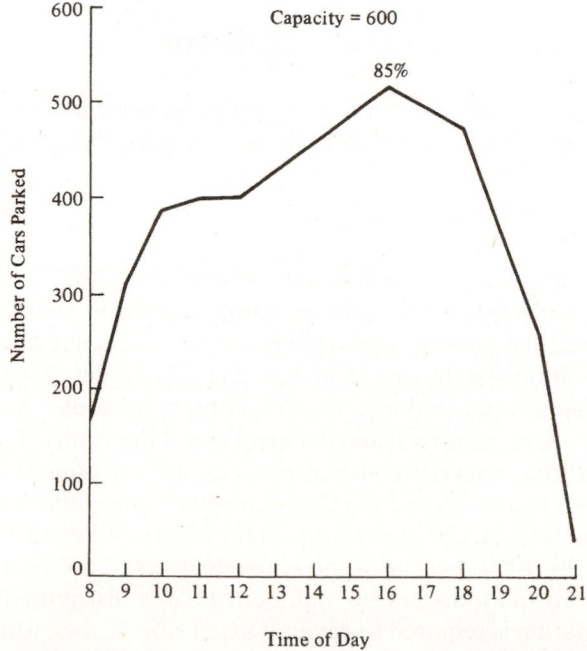

Figure 4.19 Parking Accumulation at a Parking Lot

Table 4.13   Example Form for Collecting Turnover and Duration Data

Parking Stall	Start of Time Interval												
	08:00 AM	08:30 AM	09:00 AM	09:30 AM	10 AM	10:30 AM	11 AM	11:30 AM	12 Noon	12:30 PM	1:00 PM		
K-7-1	623	451	√	√	232	√							
K-7-2	—	293	—	√	√	√							
K-7-3	631	√	√	√	√	—							
K-7-4	951	√	√	663	√	√							
K-7-5	875	930	450	√	212	√							
K-7-6	776	—	341	√	√	223							
K-7-7	441	503	√	831	√	√							
K-7-8	230	333	402	√	673	√							

Location: ———————— Recorder: ———————— Date: ————————

Symbols: 614 (three numbers) indicates last three numbers of vehicle's license plate, √ indicates same vehicle as previous time, — indicates vacant space

An example of this is the Spark Parking Inc. System. In addition to collecting data on parking, the Spark Parking System can be used to collect parking fees. The system enables drivers to occupy a parking space, place a call from their mobile phone, and follow the prompts to record their credit card data and other pertinent personal information. The credit card data are then used for automatic payment of the parking fees. Figure 4.20 illustrates the general principles of the system.

Identification of Parking Generators

This phase involves identifying parking generators (for example, shopping centers or transit terminals) and locating these on a map of the study area.

Parking Demand

Information on parking demand is obtained by interviewing drivers at the various parking facilities listed during the inventory. An effort should be made to interview all drivers using the parking facilities on a typical weekday between 8:00 a.m. and 10:00 p.m. Information sought should include (1) trip origin, (2) purpose of trip, and (3) driver's destination after parking. The interviewer must also note the location of the parking facility, times of arrival and departure, and the vehicle type.

Parking interviews also can be carried out using the postcard technique, in which stamped postcards bearing the appropriate questions and a return address are handed to drivers or placed under windshield wipers. When this technique is used, usually only about 30 to 50 percent of the cards distributed are returned. It is therefore necessary to record the time and the number of cards distributed at each location, because this information is required to develop expansion factors, which are later used to expand the sample. It is also feasible to estimate parking demand from models developed from data obtained during a parking study for different parking variables. For example, a study

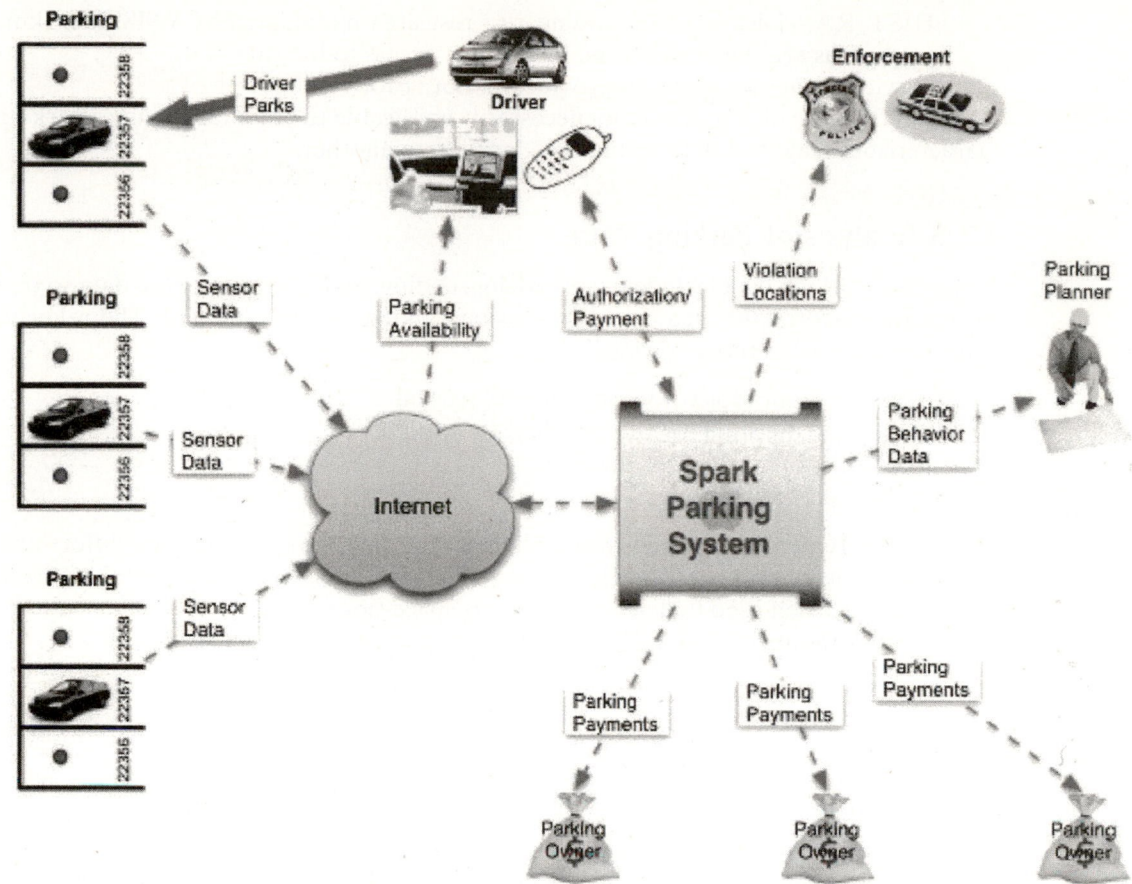

Figure 4.20   The Spark Service System

SOURCE: Courtesy of Spark Parking (www.sparkparking.com).

on a section of I-81 corridor in Virginia showed that large truck parking accumulation, which is also an estimate of demand, can be made from Eq. 4.14:

$$\text{Accumulation} = -1586.89036 + 1.41039 \times \text{percentTRUCK} + 0.1556301 \times \text{Duration}$$

$$+ 0.06955 \times \text{TotalTruck} - 123.29288 \times \text{DIST\_81} + 111.95632 \times \text{DIST\_TS}$$

$$+ 14.22398 \times \text{DIST\_RA} + 988.99725 \times \text{SERVICE}$$

$$(R^2 = 0.9486) \tag{4.14}$$

where
   percentTRUCK—percent of truck in the traffic stream
   Duration—average parking duration of trucks at the parking location (min)
   TotalTRUCK—Total truck volume (veh/h)
   DIST_81—distance from I-81 (km)
   DIST_TS—Distance to the nearest truck stop (km)

DIST_RA—Distance from the nearest rest area maintained by VMS, Inc. (km) (rest areas are state owned and have direct access to the Interstate)

Service—whether service is provided or not (1 for yes, 0 for no)

Note that this type of equation is not necessarily applicable to other locations, as parking characteristics may be different from one location to another.

4.7.5 Analysis of Parking Data

Analysis of parking data includes summarizing, coding, and interpreting the data so that the relevant information required for decision making can be obtained. The relevant information includes the following:

- Number and duration for vehicles legally parked
- Number and duration for vehicles illegally parked
- Space-hours of demand for parking
- Supply of parking facilities

The analysis required to obtain information on the first two items is straightforward; it usually involves simple arithmetical and statistical calculations. Data obtained from these items are then used to determine parking space-hours.

The space-hours of demand for parking are obtained from the expression

$$D = \sum_{i=1}^{N} (n_i t_i) \tag{4.15}$$

where

D = space vehicle-hours demand for a specific period of time
N = number of classes of parking duration ranges
t_i = midparking duration of the ith class
n_i = number of vehicles parked for the ith duration range

The space-hours of supply are obtained from the expression

$$S = f \sum_{i=1}^{N} (t_i) \tag{4.16}$$

where

S = practical number of space-hours of supply for a specific period of time
N = number of parking spaces available
t_i = total length of time in hours when the ith space can be legally parked in during the specific period
f = efficiency factor

The efficiency factor f is used to correct for time lost in each turnover. It is determined on the basis of the best performance a parking facility is expected to produce. Efficiency factors therefore should be determined for different types of parking facilities—for example, surface lots, curb parking, and garages. Efficiency factors for curb parking, during highest demand, vary from 78 percent to 96 percent; for surface lots and garages, from 75 percent to 92 percent. Average values of f are are 90 percent for curb parking, 80 percent for garages, and 85 percent for surface lots.

Example 4.8 Space Requirements for a Parking Garage

The owner of a parking garage located in a CBD has observed that 20 percent of those wishing to park are turned back every day during the open hours of 8 a.m. to 6 p.m. because of a lack of parking spaces. An analysis of data collected at the garage indicates that 60 percent of those who park are commuters, with an average parking duration of 9 hr, and the remaining percentage are shoppers, whose average parking duration is 2 hr. If 20 percent of those who cannot park are commuters and the rest are shoppers, and a total of 200 vehicles currently park daily in the garage, determine the number of additional spaces required to meet the excess demand. Assume parking efficiency is 0.90.

Solution:

- Calculate the space-hours of demand using Eq. 4.15.

$$D = \sum_{i=1}^{N}(n_i t_i)$$

Commuters now being served = 0.6 × 200 × 9 = 1080 space-hr
Shoppers now being served = 0.4 × 200 × 2 = 160 space-hr

Total number of vehicles turned away = $\dfrac{200}{0.8} - 200 = 50$

Commuters not being served = 0.2 × 50 × 9 = 90 space-hr
Shoppers not being served = 0.8 × 50 × 2 = 80 space-hr
Total space-hours of demand = (1080 + 160 + 90 + 80) = 1410
Total space-hours served = 1080 + 160 = 1240
Number of additional space-hours required = 1410 − 1240 = 170

- Determine the number of additional parking spaces required from Eq. 4.16.

$$S = f\sum_{i=1}^{N}t_i = 170 \text{ space-hr}$$

- Use the maximum length of time each space can be legally parked on (8 a.m. through 6 p.m. = 10 hr) to determine the number of additional spaces.

$$0.9 \times 10 \times N = 170$$
$$N = 18.89$$

At least 19 additional spaces will be required, since a fraction of a space cannot be used.

4.8 SUMMARY

Highway transportation has provided considerable opportunities for people, particularly the freedom to move from place to place at one's will and convenience. The positive aspects of the highway mode, however, go hand in hand with numerous negative aspects, which include traffic congestion, crashes, pollution, and parking difficulties. Traffic and transportation engineers are continually involved in determining ways to reduce these negative effects. The effective reduction of the negative impact of the highway mode of transportation at any location can be achieved only after adequate information is obtained to define the problem and the extent to which the problem has a negative impact on the highway system. This information is obtained by conducting studies to collect and analyze the relevant data. These are generally referred to as traffic engineering studies.

This chapter has presented the basic concepts of different traffic engineering studies: speed studies, volume studies, travel time and delay studies, and parking studies. Spot speed studies are conducted to estimate the distribution of speeds of vehicles in a traffic stream at a particular location on a highway. This is done by recording the speeds of a sample of vehicles at the specified location. Speeds of vehicles along sections of highways also can be collected using advanced technologies. These data are used to obtain speed characteristics, such as the average speed, the median speed, the modal speed, the 85th-percentile speed, the pace, and the standard deviation of the speed. Important factors that should be considered in planning a speed study include the location for the study, time of day, duration of the study, and the minimum sample size necessary for the limit of acceptable error. Traffic volume studies entail the collection of data on the number of vehicles and/or pedestrians that pass a point on a highway during a specified time period. The data on vehicular volume can be used to determine the average daily traffic, average peak-hour volume, vehicle classification, and vehicle-miles of travel. Volume data are usually collected manually or by using electronic or mechanical counters; video imaging also can be used. It should be noted, however, that traffic volume varies from hour to hour and from day to day. It is therefore necessary to use expansion factors to adjust periodic counts to obtain representative 24-hour, weekly, monthly, and annual volumes. A travel time study determines the amount of time required to travel from one point to another on a given route. This information is used to determine the delay, which gives a good indication of the level of service on the study section. The methods used to conduct travel time and delay data can be grouped into two general categories: (1) those that require a test vehicle and (2) those that do not. Parking studies are used to determine the demand for and supply of parking facilities in an area. A comprehensive parking study usually involves (1) inventory of existing parking facilities; (2) collection of data on parking accumulation, parking turnover, and parking duration; (3) identification of parking generators; and (4) collection of information on parking demand.

This chapter has not attempted to present an in-depth discussion of any of these studies as such a discussion is beyond the scope of this book. However, enough material has been provided to introduce readers to the subject so that they will be able to understand the more advanced literature on the subject.

PROBLEMS

4-1   What are the advantages and disadvantages of machine vision (video image detection) when compared with other forms of detection?

4-2   Select and describe the method and equipment you will recommend be used for traffic data collection for each of the road sections given below. Give reasons for your recommendations.

(a) A private road leading to an industrial development
(b) A residential street
(c) A rural collector road
(d) A section of an interstate highway

4-3   Speed data collected on an urban roadway yielded a standard deviation in speeds of ±5.6 km/h.

(a) If an engineer wishes to estimate the average speed on the roadway at a 95 percent confidence level so that the estimate is within ±2 km/h of the true average, how many spot speed observations should be collected?

(b) If the estimate of the average must be within ±1 km/h, what should the sample size be?

4-4   An engineer wishing to obtain the speed characteristics on a bypass around her city at a confidence level of 95 percent and an acceptable error limit of ±2.0 km/h collected a total of 104 spot speed observations and determined that the standard deviation is 4.8 km/h. Has the engineer met with all of the requirements of the study?

4-5   A spot speed study was conducted on a freeway on which the mean speed and standard deviation were found to be 72.4 km/h and 5.4 km/h, respectively. What is the range of values that would be expected to contain the middle 95 percent of speeds?

4-6   For a spot speed study in which 100 observations were obtained, the mean speed was 52 km/h and the standard deviation was 4.6 km/h. What is the margin of error on the estimate of true mean speed obtained from this sample? What is a reasonable estimate of the 85th percentile speed based on the information available?

4-7   An engineer wishing to determine whether there is a statistically significant difference between the average speed of passenger cars and that of large trucks on a section of highway collected the data shown below. Determine whether the engineer can conclude that the average speed of large trucks is the same as for passenger cars.

	Trucks	Passenger Cars
Average Speed (km/h)	70	73
Standard deviation of speed ± km/h	5.9	7.3
Sample size	150	300

4-8   Assume that the data shown in Table 4.2 were collected on a rural road in your state and consideration is being made to set the speed limit on the road. Speed limits of 50, 55, 60, and 65 km/h are being considered. Plot the expected noncompliance percentages versus the associated speed limit on a graph and recommend the speed for the road. Give reasons for your selection.

4-9   The accompanying data show spot speeds collected at a section of highway located in a residential area before and after an increase in speed enforcement activities. Using the student's t test, determine whether there was a statistically significant difference in the average speeds at a significance level of $\alpha = 0.05$ (the 95 percent confidence level). Also report, for both the before and after cases, the mean speed, standard deviation, 85th percentile speed, and percentage of traffic exceeding the posted speed limit of 30 km/h.

Before	After	Before	After
40	23	38	25
35	33	35	21
38	25	30	35
37	36	30	30
33	37	38	33
30	34	39	21

Before	After	Before	After
28	23	35	28
35	28	36	23
35	24	34	24
40	31	33	27
33	24	31	20
35	20	36	20
36	21	35	30
36	28	33	32
40	35	39	33

4-10 Using the data furnished in Problem 4-9, draw the histogram frequency distribution and cumulative percentage distribution for each set of data and determine (a) average speed, (b) 85th-percentile speed, (c) 15th-percentile speed, (d) mode, (e) median, and (f) pace.

4-11 Define the following terms and cite examples of how they are used.

> Average annual daily traffic (AADT)
> Average daily traffic (ADT)
> Vehicle-kms of travel (VKT)
> Peak hour volume (PHV)

4-12 Describe the different traffic count programs carried out in your state. What data are collected in each program?

4-13 A traffic engineer, wishing to determine a representative value of the ADT on 250 highway links having similar volume characteristics, conducted a preliminary study from which the following estimates were made: mean volume = 45,750 veh/day, standard deviation = 3750 veh/day. Determine the minimum number of stations for which the engineer should obtain 24-hr volume counts for a 95-5 precision level and an allowable error of ± 1.5 mi/h.

4-14 Describe the following types of traffic counts and when they are used.

(a) Screen line counts
(b) Cordon counts
(c) Intersection counts
(d) Control counts

4-15 A traffic count taken on a rural highway between 7:00 a.m. and 2:00 p.m. on a rural highway found hourly volumes as follows:

Time	Hourly volume
7:00–8:00 a.m.	310
8:00–9:00 a.m.	289
9:00–10:00 a.m.	241
10:00–11:00 a.m.	251
11:00 a.m.–12:00 p.m.	267
12:00–1:00 p.m.	264
1:00–2:00 p.m.	243

If these data were collected on a Tuesday in March, estimate the AADT on this section of highway. Assume that the expansion factors given in Tables 4.6, 4.7, and 4.8 apply.

4-16   How are travel time and delay studies used? Describe one method for collecting travel time and delay data at a section of a highway. Explain how to obtain the following information from the data collected: (a) travel time, (b) operational delay, (c) stopped time delay, (d) fixed delay, and (e) travel time delay.

Table 4.14   Travel Time Data for Problem 4-17

Run Direction/ Number	Travel Time (min)	No. of Vehicles Traveling in Opposite Direction	No. of Vehicles That Overtook Test Vehicle	No. of Vehicles Overtaken by Test Vehicle
Northward				
1	5.25	100	2	2
2	5.08	105	2	1
3	5.30	103	3	1
4	5.15	110	1	0
5	5.00	101	0	0
6	5.51	98	2	2
7	5.38	97	1	1
8	5.41	112	2	3
9	5.12	109	3	1
10	5.31	107	0	0
Southward				
1	4.95	85	1	0
2	4.85	88	0	1
3	5.00	95	0	1
4	4.91	100	2	1
5	4.63	102	1	2
6	5.11	90	1	1
7	4.83	95	2	0
8	4.91	96	3	1
9	4.95	98	1	2
10	4.83	90	0	1

4-17   Table 4.14 shows data obtained in a travel time study on a section of highway using the moving-vehicle technique. Estimate (a) the travel time and (b) the volume in each direction at this section of the highway.

4-18   An engineer, wishing to determine the travel time and average speed along a section of an urban highway as part of an annual trend analysis on traffic operations, conducted a travel time study using the floating-car technique. He carried out 10 runs and obtained a standard deviation of ± 3.4 km/h in the speeds obtained. If a 5 percent significance level is assumed, is the number of test runs adequate?

4-19   Briefly describe the tasks you would include in a comprehensive parking study for your college campus. Indicate how you would perform each task and the way you would present the data collected.

4-20   Select a parking lot on your campus. For several hours, conduct a study of the lot using the methods described in this chapter. From the data collected, determine the turnover and duration. Draw a parking accumulation curve for the lot. Identify the capacity of the parking lot, the peak demand observed, when the peak demand occurred, and occupancy of the lot at that time.

4-21 Data collected at a parking lot indicate that a total of 300 cars park between 8 a.m. and 5 p.m. Five percent of these cars are parked for an average of 2 hr, 20 percent for an average of 3 hr, 10 percent for an average of 4 hr, and the remaining cars are parked for an average of 9 hr. Determine the space-hours of demand at the lot.

4-22 If 10 percent of the parking bays are vacant on average (between 8 a.m. and 5 p.m.) at the parking lot of Problem 4-18, determine the number of parking bays in the parking lot. Assume an efficiency factor of 0.90.

4-23 The owner of the parking lot of Problems 4-21 and 4-22 is planning an expansion of her lot to provide adequate demand for the following 10 years. If she has estimated that parking demand for all categories will increase by 3 percent a year, determine the number of additional parking bays that will be required.

REFERENCES

Highway Performance Monitoring Systems (HPMS) Field Manual. U.S. Department of Transportation, Federal Highway Administration, Washington, D.C., March 2013. IEEE Transactions on Vehicular Technology, vol. 40, no. 1, February 1991.

Manual of Transportation Engineering Studies. 2nd Edition Institute of Transportation Engineers, Washington D.C. November 2010.

Siemens, E. (www.siemens.com.co/SiemensDotNetClient_Andina/medias/PDFS/63.pdf).

Spark Parking System (http://www.sparkparking.com).

Traffic Monitoring Guide. U.S. Department of Transportation, Federal Highway Administration, Washington, D.C., May 2001.

CHAPTER 5

Highway Safety

Traffic statistics in the Highway Performance Monitoring System (HPMS), published by the Federal Highway Administration, and in the Fatal Analysis Reporting System, published by the National Highway Safety Administration (NHTSA), show that while vehicle km of travel remained relatively flat between 2005 and 2011, the number of traffic fatalities decreased. For example, these data show that vehicle km of travel (VKT) in the United States were about 4,765 billion in the year 2005, 4,742 billion in 2009, and 4,759 billion in 2011. The number of traffic fatalities were 43,510 in 2005, 41,259 in 2007, and 32,310 in 2011. Comparing the data for 2005 and 2011, the numbers indicate that the fatality rates dropped from about 1.46 fatalities per 100 million VKT in 2005 to about 1.09 in 2011. This represents a drop of about 25 percent in fatality rates between 2005 and 2011, while the drop in VKT for the same period is about 0.36 percent. Although this is a positive trend, many safety-concerned individuals advocate a "toward zero" fatality rate for the nation, a goal many states have adopted. This goal will not be easy to achieve, which means that for many years to come, traffic and highway engineers must continue to work toward reducing traffic fatalities on our nation's highways and streets. Highway safety is a worldwide problem; with over 500 million cars and trucks in use, more than 500,000 people die each year in motor vehicle crashes, and about 15 million are injured. In the United States, motor vehicle crashes are the leading cause of death for people between the ages of 1 to 34 years and rank third as the most significant cause of years of potential life lost—after cardiac disease and cancer.

Traffic and highway engineers work with law enforcement officials and educators in a team effort to ensure that traffic laws, such as those regarding speed limits and drinking, are enforced and that motorists are educated about their responsibility to drive defensively and to understand and obey traffic regulations. Traffic and highway engineers are also continually engaged in working to ensure that the street and highway system is designed and operated such that highway crash rates can be reduced. This includes identifying effective traffic safety countermeasures, developing methodologies for accurately collecting and analyzing safety data, and determining the effects of safety programs. This chapter briefly describes the main categories of highway crash causal factors, introduces the reader to the recent methodologies traffic engineers use to identify locations that have good potential for safety improvement and effective traffic safety countermeasures, discusses how to collect and analyze traffic crash data, and describes how to conduct before-and-after studies to evaluate the effectiveness of different safety programs.

CHAPTER OBJECTIVES:

- Understand the issues involved in highway safety.
- Become familiar with the causes of highway crashes.
- Understand the procedures for collecting, presenting, and analyzing highway crash data.
- Become familiar with the sources of historic traffic data.
- Understand the procedures of establishing project priorities based on safety.
- Become familiar with the relative safety effectiveness of highway design features.

5.1 ISSUES INVOLVED IN TRANSPORTATION SAFETY

Several issues are involved in transportation safety. These include whether accidents should be referred to as crashes, the causes of transportation crashes, and the factors involved in transportation crashes.

5.1.1 Crashes or Accidents

Accident is the commonly accepted word for an occurrence involving one or more transportation vehicles in a collision that results in property damages, injury, or death. The term *accident* implies a random event that occurs for no apparent reason other than "it just happened."

The National Highway Traffic Safety Administration has suggested replacing the word *accident* with the word *crash* because *crash* implies that the collision could have been prevented or its effect minimized by modifying driver behavior, vehicle design (called "crashworthiness"), roadway geometry, or the traveling environment. The word *crash* is not universally accepted terminology for all transportation modes and is most common in the context of highway and traffic incidents. In this chapter, both terms—*crashes* and *accidents*—are used because while *crash* is the preferred term, in some situations the word *accident* may be more appropriate.

5.1.2 What Causes Transportation Crashes?

The occurrence of a transportation crash presents a challenge to safety investigators. In every instance, the question arises, "What sequence of events or circumstances contributed to the incident that resulted in injury, loss of life, or property damage?" In some cases, the answer may be a simple one. For example, the cause of a single-car crash may be that the driver fell asleep at the wheel, crossed the highway shoulder, and crashed into a tree. In other cases, the answer may be complex, involving many factors that, acting together, caused the crash.

It is therefore possible to construct a general list of the categories of circumstances that could influence the occurrence of transportation crashes. If the factors that have contributed to crash events are identified, it is then possible to modify and improve the transportation system. In the future, with the reduction or elimination of the crash-causing factors, a safer transportation system is likely to result.

5.1.3 Factors Involved in Highway Crashes

While the causes of highway crashes are usually complex and involve several factors, they can be considered in four separate categories: human factors (actions by the driver or operator), factors related to the mechanical condition of the vehicle, factors related to the geometric characteristics of the roadway, and environmental factors (the physical or climatic environment in which the vehicle operates). These different categories of factors will be reviewed in the following sections.

Human Factors (Driver or Operator Action)

The major contributing cause of many traffic crashes is driver error. The human errors leading to highway crashes are related to the complex interaction of the driver's psychological and physiological conditions, the system design, and the existing environmental conditions. Driver actions that lead to errors include driving at an inappropriate speed for the existing physical and/or environmental condition, driver inattentiveness, failure to yield the right-of-way, wrongly negotiating curves, the use of in-vehicle communication tools such as use of cell phones, and other conditions that may lead to driver errors include vision capability and perception response time (both of which were discussed in Chapter 3), roadway distractions, driver weariness, information overload, and driver expectancy. For example, the rate at which an individual can process information is limited. Therefore, although a driver can quickly transfer his/her attention from one information source to another, only a single source can be attended to at a time. In order to compensate for this limitation, drivers tend to subconsciously determine the level of information load they can handle. When conditions are such that this level of load is surpassed, drivers tend to select the information they regard as important and ignore the others. This may result in the driver ignoring an important piece of information and holding on to one that is less critical.

Also, a stretch of road not designed in a consistent way may result in a driver losing the ability to depend on experience gained, which may violate his/her expectancy. For example, if a freeway exit is located on the left-hand side of the road, this causes confusion for drivers who are used to exits being on the right-hand side of the road and may result in unnecessary weaving maneuvers.

The main objective of the traffic engineer regarding human factors in highway safety is to understand how these human factors influence the driver, so that the probability of occurrence of the associated safety consequences of these factors is reduced. For example, rumble strips are placed at the side and medians of high-speed roads in rural areas to alert inattentive or sleepy drivers when they are leaving the travel lane. Another example is the avoidance of information overload. The *Highway Safety Manual* (HSM), which is a publication that gives a quantitative methodology to determine changes in crash frequency as a function of safety treatment and cross-sectional features, gives the following roadway design considerations to reduce information overload:

(a) Giving traffic information in a consistent way. For example, specific colors and shapes are assigned to specific types of roadway signs.
(b) Not giving many pieces of information at the same time, but presenting them one after the other in an orderly way.
(c) Providing clues to help drivers rank the importance of the different information.

When driver expectancy is taken into consideration in the design of roadways, the negative effect of the driver's limitations in processing information is reduced. The HSM gives the following examples of long- and short-term expectancies.

Examples of long-term expectancies:

- Freeway exits are located on the right-hand side of the road.
- At an intersection of a major and minor road controlled by a stop sign, the stop sign is on the approaches of the road that appears to be the minor road.
- At an intersection approach, a driver wishing to turn left will be in the left lane or on a through lane that allows left turns.

Examples of short-term expectancies:

- A long section of roadway with gently winding characteristics is contiguous with a roadway section that has gentle curves.
- A long section of roadway that allows for high-speed driving is contiguous with a roadway sections that also allow for high-speed driving.
- Driving at a consistent speed along a well-coordinated system of traffic signals on an arterial should not suddenly lead to an isolated intersection with a significantly different cycle length.

The Vehicle Condition

The mechanical condition of an automobile can be the cause of highway crashes. Faulty brakes in heavy trucks have caused crashes. Other vehicle conditions include failure of the electrical system, worn tires, and the location of the vehicle's center of gravity. However, many vehicle manufacturers are now installing in-vehicle equipment that helps to reduce the potential of the vehicles being involved in crashes and/or the severity of a crash when the vehicle is involved in one. These include seat belt reminders, antilock braking systems (ABS), reversing warning devices and cameras, speed limiter/intelligent speed adaption, automatic reversing brakes, collision avoidance sensors, and alcohol ignition interlock. Most of these have been shown to have some impact on crash occurrence and/or crash severity. Further discussion of the impact of this equipment is beyond the scope of this book. Interested readers may refer to available literature for additional information.

The Roadway Condition

The condition and quality of the roadway, which includes the pavement, shoulders, intersections, and the traffic-control system, can be a factor in a crash. Highways must be designed to provide adequate stopping sight distance at the design speed, or motorists will be unable to take remedial action to avoid a crash. Traffic signals must provide adequate decision sight distance when the signal goes from green to yellow and then to red (see Chapter 8). Railroad grade crossings must be designed to operate safely and thus minimize crashes between highway traffic and rail cars (see Chapter 7). Highway curves must be carefully designed to accommodate vehicles traveling at or below the design speed of the road (see Chapter 15).

In addition, traffic and highway engineers use a "positive guidance approach," which combines human factors and traffic engineering principles to the design of highway facilities. This approach uses the knowledge gained on human limitations in information processing as well as human dependence on expectation to make up for these limitations. The basic principle that governs this approach is that if a road is designed to incorporate the limitations and expectancies of the driver, the probability of a driver reacting and making the right decision for a given condition is maximized. On the other hand, when the positive guidance approach is not used in the design of the roadway, it is likely that drivers will slowly respond and make wrong decisions when they are overloaded with information or their expectancy is violated. An example of the positive guidance approach is related to the way traffic-control devices are placed. In this case, the traffic-control devices are placed in a way that emphasizes aiding the driver to process information in time and correctly. The HSM gives the following guidelines for achieving this:

- *Primacy*—Signs should be placed based on the importance of the information contained in them. Avoid placing signs at locations where they are not necessary or essential.
- *Spreading*—In situations where the necessary information cannot be placed on one sign or several signs at a single location, the information should be given in portions along the road to reduce the information load.
- *Coding*—Several small pieces of information should be organized into larger groups. Use color and shape coding to accomplish this. For example, warning signs are yellow and regulatory signs are white; also, stop signs are octagonal and yield signs are triangular.
- *Redundancy*—Repeat the same information in different forms. For example, indicate a "no passing zone" by both pavement marking and signing.

The Environment

The physical and climatic environment surrounding an automobile can also be a factor in the occurrence of highway crashes; the most common environmental factor is weather. Weather on roads can contribute to highway crashes; for example, wet pavement reduces stopping friction and can cause vehicles to hydroplane. Many severe crashes have been caused by fog because vehicles traveling at high speeds are unable to see other vehicles ahead that may have stopped or slowed down; the result can be a multivehicle pile-up. The level of lighting also has an effect on the frequency and severity of crashes at some locations, particularly at intersections. It has been shown that 40 percent of intersection fatalities take place during the late-night/early morning hours, and that the probability of a fatal crash occurring is about three times greater during the late-night/early morning hours than during the day. Poor intersection visibility is a major contribution to this as drivers are restricted from having a clear view of conflicting traffic and other users of the intersection. Geography is another environmental cause of automobile crashes. Flooded river plains, swollen rivers, and mud slides on the pavement have caused railroad and highway crashes.

5.2 STRATEGIC HIGHWAY SAFETY PLANS

The Safe Accountable, Flexible, Efficient Transportation Equity Act: A Legacy for Users (SAFETEA-LU) legislation of 2005, which authorized the five-year federal surface transportation program for highways, highway safety, and transit, requires that each state develop a *Strategic Highway Safety Plan (SHSP).* The purpose of this plan is to develop a process through which each state would identify its key safety needs such that investment decisions can be made that will result in significant reductions in highway fatalities and serious injuries on public roads. Suggested activities that could be included in this plan are as follows:

Gain leadership support and initiative.

Identify a champion.

Initiate the development process.

Gather data.

Analyze data.

Establish a working group.

Bring safety partners together.

Adopt a strategic goal.

Identify key emphasis areas.

Form task groups.

Identify key emphasis area performance-based goals.

Identify strategies and countermeasures.

Determine priorities for implementation.

Write a SHSP.

Discussion of each of these activities is beyond the scope of this book. Interested readers may wish to refer to the reference, *Guidance for Implementation of the AASHTO Strategic Safety Plan.* The activities discussed below are those normally included in a roadway safety management process as given in the HSM, which deals with monitoring and analyzing crash data and implementing effective safety countermeasures with the objective of reducing crash frequency. The HSM provides quantitative tools that can be used to make safety decisions based on safety performance. These tools include methodologies to measure, estimate, and evaluate the level of safety at a location in terms of the crash frequency (number of crashes per year) and crash severity (level of crash injuries). Predictive methodologies are provided to improve crash estimation for past and future periods. These methodologies are statistically rigorous, and they alleviate the inherent errors in using historical crash data base and assuming that crashes are related in a one-to-one ratio with traffic volume or amount of travel, as well as assuming the normal distribution for crash occurrence. They instead assume random variation of crash occurrence and provide procedures to estimate crash occurrence based on geometry, operating characteristics, and traffic volumes. An opportunity is therefore provided for an analyst to (i) obtain more reliable results in conducting activities such as screening a number of sites for the reduction of crashes, and (ii) consider new or alternative geometric and operational characteristics in assessing safety impact. The HSM recommends that the roadway safety management process involve the following steps:

- *Network screening*—In this step, sites at which the potential for reducing average crash frequency exists are identified and ranked, through a review of the transportation network.
- *Diagnosis*—In this step, crash patterns are determined through the collection and analysis of historic data and evaluation of site conditions.
- *Select countermeasures*—This step involves determining probable crash-related factors at a site and selecting countermeasures that will be effective in reducing crash frequency.
- *Economic appraisal*—This step involves conducting an appropriate economic analysis to identify specific projects that are economically justifiable.

- *Prioritize projects*—In this step, the projects that are economically justifiable at specific sites and across multiple sites are prioritized with respect to their potential for achieving safety objectives within the available budget. Prioritization factors may include low cost, mobility enhancement potential, or reduced negative environmental impact.
- *Safety effectiveness evaluation*—In this step, the selected countermeasures at a site or multiple sites are evaluated to determine how effective they are or would be in reducing annual crash frequency or severity.

Figure 5.1 gives a schematic presentation of the roadway safety management process and indicates its dynamic nature. Note that an in-depth description of the tasks involved in each of these steps is beyond the scope of this book. However, adequate material is included to give the reader an overview of each step and to provide a full understanding of the whole process. Additional information can be obtained from the HSM.

In order to facilitate comprehension of the steps involved in a roadway safety management process, the definitions of several terms as provided in the HSM are as follows:

Crash: The occurrence or existence of a set of events leading to death, injury, or property damage as a result of a collision involving one or more motorized vehicles.

Crash Frequency: The number of crashes that take place at a site, facility, or network during a period of one year and usually given as the number of crashes per year.

Crash Estimation: A process used to determine the expected crash frequency during a past or future period at an existing or new roadway.

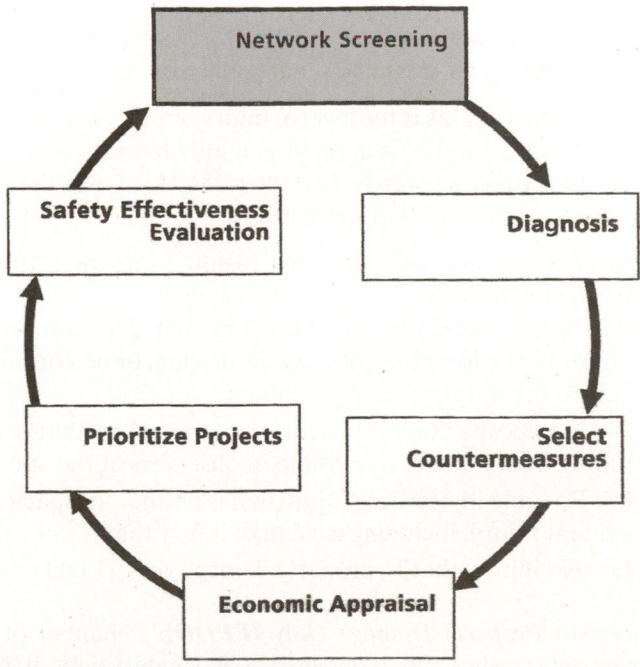

Figure 5.1   Roadway Safety Management Process

SOURCE: From *Highway Safety Manual* 1st Edition, 2010, AASHTO, Washington, D.C. Used by permission.

Crash Rate: The number of crashes per a specified exposure data set (for example, crashes per million vehicles entering an intersection or crashes per 100 million vehicle km traveled along a segment of roadway).

Performance Measure: A procedure or tool used to determine the potential of reducing frequency of crashes at a location.

Safety Performance Function (SPF):   A regression equation that adequately relates the number of crashes per unit of time and road length or intersection by severity with one or more independent variables, such as Average Annual Daily Traffic (AADT) for a given type of highway. These functions are developed through the use of multivariate techniques and assuming a negative binomial distribution rather than the Poisson distribution, which is suitable for data sets with the mean and variance being the same. When a set of data exhibits higher variance than the mean, the data are considered overdispersed. Crash data tend to exhibit overdispersion, which makes the negative binomial distribution more suitable. The extent of overdispersion is reflected in a parameter of the distribution known as the *overdispersion parameter (k),* which is computed as part of the coefficients of the regression equation.

Predictive Method: A procedure or tool that is used to estimate average crash frequency at a location, roadway section, or facility for a specific set of geometric design and traffic volume.

Expected Crash Frequency: The long-term average crash frequency at a site, facility, or network over a period of years for a specific set of geometric design and traffic volume. Note that the expected crash frequency is an estimate, as roadway conditions and traffic volumes do not necessarily remain constant over a period of many years. Note also that because of the random nature of crashes, a crash frequency observed over a short period of time does not necessarily represent the long-term crash frequency at a site, facility, or network even under the same set of geometric design and traffic volume.

Crash Severity: This is the level of injury or property damage that is incurred as a result of a crash, where injury is defined as bodily harm to a person. There are several different methods to rank severity. However, the HSM uses the KABCO scale. There are five levels in this scale and they are given as:

> K—Fatal injury: an injury that results in death within a specified period of time (typically 30 days).
> A—Incapacitating injury: An injury that does not result in death, but causes the injured to be incapable of walking, driving, or performing activities that he/she could before the occurrence of the injury.
> B—Nonincapacitating evident injury: An injury that is neither a fatal nor an incapacitating injury and that is visible to observers at the site at which the crash occurred.
> C—Possible injury: Any injury that is neither incapacitating nor a nonincapacitating evident injury, including a claim of injury that is not evident.
> O—No injury; that is, property damage only (PDO).

Equivalent Property Damage Only (EPDO):   Number of crashes in terms of property damage only crashes. This is obtained by assigning weights to injury and fatal crashes relative to PDO crashes and is intended to address the range of severity levels in a set of crash data.

Crash Evaluation: The process of determining the effectiveness of implementing a specific safety countermeasure or a treatment program in reducing average crash frequency or severity at a site or project.

Countermeasure: A strategy that is expected to decrease crash frequency and/or severity when it is deployed on a roadway.

Regression to the Mean: Crash occurrence naturally fluctuates, often causing average crash frequencies over a short period to be either higher or lower than the mean over a long period. Therefore, if during a couple of years a relatively high crash frequency occurs at a site, it is very likely that this period will be followed by a short period of relatively low crash frequency. This phenomenon is known as *regression to the mean* (RTM).

Let us now discuss the activities that are involved in each of the steps in the roadway safety management process as depicted in Figure 5.1.

5.2.1 Network Screening Process

This is the initial step in the safety management process and it consists of the following five tasks:

- Establish focus.
- Identify network and establish reference population.
- Select performance measures.
- Select screening methods.
- Screen and evaluate results.

Establish Focus

The focus of the network screening is established with regard to the types of sites and crashes to be considered. For example, if a jurisdiction has received funding to implement a program to reduce the relatively high number of rear-end crashes at intersections located on urban arterials, the sites selected for screening will be limited to intersections on arterials in urban areas with rear-end crashes. On the other hand, if a jurisdiction has obtained funding to implement a set of countermeasures on rural two-lane roads to reduce crash frequency due to single-vehicle run-off-the-road crashes, then the sites selected for screening will include curved sections of rural two-lane roads.

Identify Network and Establish Reference Population

In this subtask, the network elements to be screened are identified from the focused sites and crash types established in the previous subtask. These are then organized into reference populations. Roadway elements include roadway segments (not including intersections), intersections, ramps, ramp terminals, facilities (roadway segments including intersections), and at-grade rail crossings. The reference populations are obtained by forming clusters of elements with similar characteristics, such as unsignalized intersections, three-leg signalized intersections, and multilane rural highways. The characteristics that are used to form a reference group may differ from one reference group to another and depend on the extent of information available for each element, the reason for the network screening, the size of the network being screened, and the type or types of performance measures to be used in prioritizing the elements. For example, the HSM gives the following characteristics that can be used:

For Intersections:

- Traffic control (e.g., unsignalized, signalized, two-way or four-way stop control, yield, roundabout)

- Number of approaches (e.g., three-leg or four-leg intersections)
- Cross section (e.g., number of through lanes and turning lanes)
- Functional classification (e.g., arterial, collector, local)
- Area type (e.g., urban, suburban, rural)
- Traffic volume ranges (e.g., total entering volume at an intersection [TEV], peak hour volume, average annual daily traffic [AADT])
- Terrain (e.g., flat, rolling, mountainous)

For Road Segments:

- Number of lanes per direction
- Access density (e.g., driver and intersection spacing)
- Traffic volume ranges (e.g., TEV, peak hour volumes, AADT)
- Median type or width, or both
- Operating speed or posted speed
- Adjacent land use (e.g., urban, suburban, rural)
- Terrain (e.g., flat, rolling, mountainous)
- Functional classification (e.g., arterial, collector, local)

Note that eventual prioritization of elements is carried out within a reference population.

Select Performance Measures

In this subtask, the performance measure/measures to be used in determining the potential to reduce frequency of crashes or crash severity at a site are selected. Note that using more than one performance measure to evaluate each site in network screening tends to increase the level of confidence in the results. The HSM gives the following performance measures from which one or more can be used in the network screening process:

(i) Average Crash Frequency

(ii) Crash Rate

(iii) Equivalent Property Damage Only (EPDO) Average Crash Frequency

(iv) Relative Severity Index—The average cost of a crash based on the number of crashes of each type at a location and the cost for that type of crash).

(v) Critical Rate—The expected crash rate at a site based on the average crash rate at locations on the same type of facility, the traffic volume, and a selected accuracy level.

(vi) Excessive Predicted Average Crash Frequency (Using Method of Moments)—The difference between the adjusted observed crash frequency (based on the variance of crashes) at a site and the average crash frequency in a reference population.

(vii) Level of Service of Safety—Comparison of the observed crash frequency of each site under consideration with the predicted average crash frequency of all the sites being considered, obtained from an appropriate Safety Performance Function (SPF).

(viii) Excess Predicted Average Crash Frequency (Using Safety Performance Functions [SPFs])

(ix) Expected Average Crash Frequency with Empirical Bayes (EB) Adjustments—Empirical Bayes adjustments are made to account for regression to the mean (RTM) bias.

(x) EPDO Average Frequency with EB Adjustment

(xi) Excess Expected Average Crash Frequency with EB Adjustment

The procedures used to compute these performance measures are briefly described in Section 5.3.

The main factors that should be considered in selecting the performance measures are the availability of data, regression to the mean, and the way in which the threshold of the performance measure is determined. Let us first discuss each of these factors.

Availability of Data.   The extent to which data are available affects the number and type of performance measures that can be used. For example, if traffic and detailed crash data are not readily available or are expensive to collect, then a significant decrease in the number of performance measures that can be used in the screening is observed. Table 5.1 shows data requirements for different performance measures. The methods of collecting different types of traffic data were discussed in Chapter 4. The next section describes the different sources of crash data and how they are collected and recorded.

Collecting and Maintaining Data.   Crash data are usually obtained from state and local transportation and police agencies. All relevant information is usually recorded by the police on an accident report form. The type of form used differs from state to state, but a typical completed form will include information on the location, time of occurrence, roadway and environmental conditions, types and number of vehicles involved, a sketch showing the original paths of the maneuver or maneuvers of the vehicles involved, and the severity of the crash (fatal, injury, or property damage only). Figure 5.2 shows the Virginia report form, which is completed by the investigating police officer. Information on minor crashes that do not involve police investigation may be obtained from routine reports given at the police station by the drivers involved, as is required in some states. Sometimes drivers involved in crashes are required to complete accident report forms when the crash is not investigated by the police. Figure 5.3 shows an example of such a form.

Storage and Retrieval of Crash Data.   Two techniques are used in the basic storage of crash data. The first involves the manual filing of each completed accident report form in the offices of the appropriate police agency. These forms are usually filed either by date, by crash report number, by name or number of the routes, or by location, which may be identified by intersection and roadway links. Summary tables, which give the number and percentage of each type of crash occurring during a given year at a given location, are also prepared. The location can be a specific spot on the highway or an identifiable length of the highway. This technique is not commonly used nowadays, but it is suitable for areas where the total number of crashes is less than 500 per year and may be used when the total number is between 500 and 1000 annually. This technique, however, becomes time consuming and inefficient when there are more than 1000 crashes per year.

The second technique involves the use of a computer; each item of information on the report form is coded and stored in a computer file. This technique is suitable for areas where the total number of crashes per year is greater than 500. With this technique, facilities are provided for storing a large amount of data in a small space. The technique also facilitates flexibility in the choice of methods used for data analysis and permits the study of a large number of crash locations in a short time. There are, however, some disadvantages associated with this technique, such as the high cost of equipment and the requirement of trained computer personnel for the operation of the system. The advent of microcomputers has, however, made it feasible for relatively small agencies to purchase individual systems. Several national databanks use computerized systems to store data on national crash statistics. These include the following:

- The ***Highway Performance Monitoring System (HPMS)*** is compiled by the Federal Highway Administration (FHWA) of the U.S. Department of Transportation. The system contains data on the extent, condition, performance, use, and operating

Table 5.1   Summary of Data Needs for Performance Measures

		Data and Inputs			
Performance Measure	*Crash Data*	*Roadway Information for Categorization*	*Traffic Volume[a]*	*Calibrated Safety Performance Function and Overdispersion Parameter*	*Other*
Average Crash Frequency	X	X			
Crash Rate	X	X	X		
Equivalent Property Damage Only (EPDO) Average Crash Frequency	X	X			EPDO Weighting Factors
Relative Severity Index	X	X			Relative Severity Index
Critical Rate	X	X	X		
Excess Predictive Average Crash Frequency Using Method of Moments[b]	X	X	X		
Level of Service of Safety	X	X	X	X	
Excess Predictive Average Crash Frequency Using Safety Performance Functions (SPFs)	X	X	X	X	
Probability of Specific Crash Types Exceeding Threshold Proportion	X	X			
Excess Proportion of Specific Crash Types	X	X			
Expected Average Crash Frequency with EB Adjustment	X	X	X	X	
Equivalent Property Damage Only (EPDO) Average Crash Frequency with EB Adjustment	X	X	X	X	EPDO Weighting Factor
Excess Expected Average Crash Frequency with EB Adjustment	X	X	X	X	

[a]Traffic volume could be *AADT*, ADT, or peak hour volume.

[b]The method of moments consists of adjusting a site's observed crash frequency based on the variance in the crash data and average crash counts for the site's reference population. Traffic volume is needed to apply method of moments to establish the reference populations, based on ranges of traffic volumes as well as site geometric characteristics.

SOURCE: From *Highway Safety Manual* 1st Edition, 2010, AASHTO, Washington, D.C. Used by permission.

characteristics of the nation's highways. Data relating to these characteristics are compiled on each of a representative sample of road sections. The sample is a stratified random sample, based on the geographic location (rural, small urban, and urbanized), the roadway functional system, and traffic volume. Crash data are obtained as part of the operational characteristics and include information on the number of fatal crashes, the number of nonfatal crashes, and the number of fatalities, etc.

Figure 5.2a   Virginia Accident Report Form—Crash Details Sheet

SOURCE: Virginia Department of Motor Vehicles

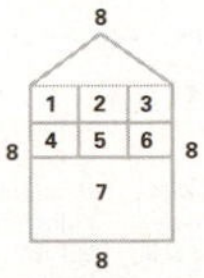

Officer Initials_____ Badge # _____   Commonwealth of Virginia • Department of Motor Vehicles | | | | | | | 07 07D   FR300P (Rev 1/12)

Revised Report ◯   **Police Crash Report**   Page _____ of _____

CRASH

| Crash Date | MM DD YYYY | MILITARY Time (24 hr clock) | County of Crash | ◯ City of ◯ Town of | Local Case Number |

CRASH DIAGRAM

VEHICLE # ☐

Fill In Impact Area(s). Initial Impact. ☐

12
11 ◯   ◯ 1
10 ◯   ◯ 2
9 ◯ 13 ◯ 3
8 ◯   ◯ 4
7 ◯   ◯ 5
6

Veh Dir of Travel–N/S/E/W

VEHICLE # ☐

Fill In Impact Area(s). Initial Impact. ☐

12
11 ◯   ◯ 1
10 ◯   ◯ 2
9 ◯ 13 ◯ 3
8 ◯   ◯ 4
7 ◯   ◯ 5
6

Veh Dir of Travel–N/S/E/W

VEHICLE # ☐

Fill In Impact Area(s). Initial Impact. ☐

12
11 ◯   ◯ 1
10 ◯   ◯ 2
9 ◯ 13 ◯ 3
8 ◯   ◯ 4
7 ◯   ◯ 5
6

Veh Dir of Travel–N/S/E/W

VEHICLE # ☐

Fill In Impact Area(s). Initial Impact. ☐

12
11 ◯   ◯ 1
10 ◯   ◯ 2
9 ◯ 13 ◯ 3
8 ◯   ◯ 4
7 ◯   ◯ 5
6

Indicate North by Arrow

Veh Dir of Travel–N/S/E/W

DAMAGE TO PROPERTY OTHER THAN VEHICLES

| Approx. Repair Cost | Object Struck (Tree, Fence, etc.) | Property Owners Name (Last, First, Middle) | Address (Street and Number) | VDOT Property ◯ ◯ Yes No |

CRASH DESCRIPTION

CRASH EVENTS

Vehicle #	First Event	Second Event	Third Event	Fourth Event	Most Harmful Event

Vehicle #	First Event	Second Event	Third Event	Fourth Event	Most Harmful Event

Vehicle #	First Event	Second Event	Third Event	Fourth Event	Most Harmful Event

Vehicle #	First Event	Second Event	Third Event	Fourth Event	Most Harmful Event

First Harmful Event of Entire Crash that Results in First Injury or Damage.

COLLISION WITH FIXED OBJECT
1. Bank Or Ledge
2. Trees
3. Utility Pole
4. Fence Or Post
5. Guard Rail
6. Parked Vehicle
7. Tunnel, Bridge, Underpass, Culvert, etc.
8. Sign, Traffic Signal
9. Impact Cushioning Device
10. Other
11. Jersey Wall
12. Building/Structure
13. Curb
14. Ditch
15. Other Fixed Object
16. Other Traffic Barrier
17. Traffic Sign Support
18. Mailbox

COLLISION WITH PERSON, MOTOR VEHICLE OR NON-FIXED OBJECT
19. Pedestrian
20. Motor Vehicle In Transport
21. Train
22. Bicycle
23. Animal
24. Work Zone Maintenance Equipment
25. Other Movable Object
26. Unknown Movable Object
27. Other

NON-COLLISION
28. Ran Off Road
29. Jack Knife
30. Overturn (Rollover)
31. Downhill Runaway
32. Cargo Loss or Shift
33. Explosion or Fire
34. Separation of Units
35. Cross Median
36. Cross Centerline
37. Equipment Failure (Tire, etc)
38. Immersion
39. Fell/Jumped From Vehicle
40. Thrown or Falling Object
41. Non-Collision Unknown
42. Other Non-Collision

Figure 5.2b Virginia Accident Report Form—Crash Diagram Sheet

SOURCE: Virginia Department of Motor Vehicles

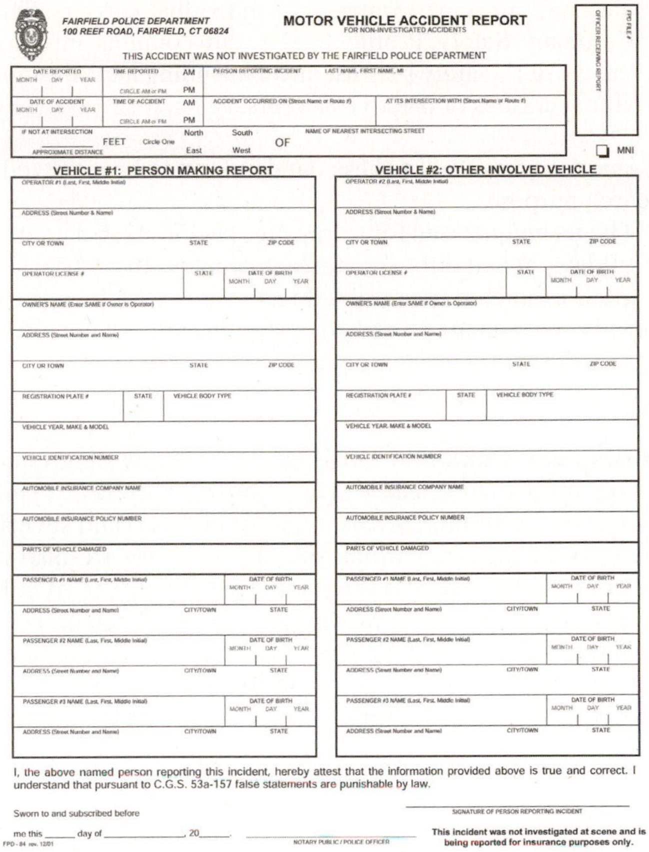

FAIRFIELD POLICE DEPARTMENT
100 REEF ROAD, FAIRFIELD, CT 06824

MOTOR VEHICLE ACCIDENT REPORT
FOR NON-INVESTIGATED ACCIDENTS

OFFICER RECEIVING REPORT | FPD FILE #

THIS ACCIDENT WAS NOT INVESTIGATED BY THE FAIRFIELD POLICE DEPARTMENT

DATE REPORTED MONTH DAY YEAR	TIME REPORTED AM PM CIRCLE AM or PM	PERSON REPORTING INCIDENT LAST NAME, FIRST NAME, MI
DATE OF ACCIDENT MONTH DAY YEAR	TIME OF ACCIDENT AM PM CIRCLE AM or PM	ACCIDENT OCCURRED ON (Street Name or Route #) / AT ITS INTERSECTION WITH (Street Name or Route #)
IF NOT AT INTERSECTION ____ FEET APPROXIMATE DISTANCE	Circle One North South East West OF	NAME OF NEAREST INTERSECTING STREET

MNI

VEHICLE #1: PERSON MAKING REPORT	VEHICLE #2: OTHER INVOLVED VEHICLE
OPERATOR #1 (Last, First, Middle Initial)	OPERATOR #2 (Last, First, Middle Initial)
ADDRESS (Street Number & Name)	ADDRESS (Street Number & Name)
CITY OR TOWN / STATE / ZIP CODE	CITY OR TOWN / STATE / ZIP CODE
OPERATOR LICENSE # / STATE / DATE OF BIRTH MONTH DAY YEAR	OPERATOR LICENSE # / STATE / DATE OF BIRTH MONTH DAY YEAR
OWNER'S NAME (Enter SAME if Owner is Operator)	OWNER'S NAME (Enter SAME if Owner is Operator)
ADDRESS (Street Number and Name)	ADDRESS (Street Number and Name)
CITY OR TOWN / STATE / ZIP CODE	CITY OR TOWN / STATE / ZIP CODE
REGISTRATION PLATE # / STATE / VEHICLE BODY TYPE	REGISTRATION PLATE # / STATE / VEHICLE BODY TYPE
VEHICLE YEAR, MAKE & MODEL	VEHICLE YEAR, MAKE & MODEL
VEHICLE IDENTIFICATION NUMBER	VEHICLE IDENTIFICATION NUMBER
AUTOMOBILE INSURANCE COMPANY NAME	AUTOMOBILE INSURANCE COMPANY NAME
AUTOMOBILE INSURANCE POLICY NUMBER	AUTOMOBILE INSURANCE POLICY NUMBER
PARTS OF VEHICLE DAMAGED	PARTS OF VEHICLE DAMAGED
PASSENGER #1 NAME (Last, First, Middle Initial) / DATE OF BIRTH MONTH DAY YEAR	PASSENGER #1 NAME (Last, First, Middle Initial) / DATE OF BIRTH MONTH DAY YEAR
ADDRESS (Street Number and Name) / CITY/TOWN / STATE	ADDRESS (Street Number and Name) / CITY/TOWN / STATE
PASSENGER #2 NAME (Last, First, Middle Initial) / DATE OF BIRTH MONTH DAY YEAR	PASSENGER #2 NAME (Last, First, Middle Initial) / DATE OF BIRTH MONTH DAY YEAR
ADDRESS (Street Number and Name) / CITY/TOWN / STATE	ADDRESS (Street Number and Name) / CITY/TOWN / STATE
PASSENGER #3 NAME (Last, First, Middle Initial) / DATE OF BIRTH MONTH DAY YEAR	PASSENGER #3 NAME (Last, First, Middle Initial) / DATE OF BIRTH MONTH DAY YEAR
ADDRESS (Street Number and Name) / CITY/TOWN / STATE	ADDRESS (Street Number and Name) / CITY/TOWN / STATE

I, the above named person reporting this incident, hereby attest that the information provided above is true and correct. I understand that pursuant to C.G.S. 53a-157 false statements are punishable by law.

Sworn to and subscribed before

me this ______ day of ______________, 20______.

SIGNATURE OF PERSON REPORTING INCIDENT

NOTARY PUBLIC / POLICE OFFICER

This incident was not investigated at scene and is being reported for insurance purposes only.

FPD - 84 rev. 12/01

Figure 5.3 Example of a Driver Accident Report Form

SOURCE: Courtesy of Fairfield Police Department, Fairfield, CT.

- The ***Fatality Analysis Reporting System (FARS)*** is compiled by the National Highway Traffic Safety Administration of the U.S. Department of Transportation. The system contains data on all fatal traffic crashes occurring within the 50 states, the District of Columbia, and Puerto Rico. The information on each of these fatal crashes is recorded and coded in a standard format by trained personnel of the different states' Departments of Transportation. The criterion for including a crash in the database is that the crash must involve a motor vehicle and result in at least one fatality within 30 days of the crash.
- The ***National Electronic Injury Surveillance System (NEISS)*** is compiled by the Consumer Product Safety Commission (CPSC) and consists of data obtained from emergency departments of 100 hospitals, representing a sample of over 5300 U.S. hospitals. The data on the injury of each patient brought into the emergency room of each of the selected hospitals are collected by a staff member of the hospital's emergency department, who obtains information on how the injury occurred. With this information, traffic-related injuries can be identified.
- The ***Motor Carrier Management Information System (MCMIS)*** is compiled by the Federal Motor Carrier Safety Administration and contains summaries on the

national safety performance of individual carriers. This summary is known as the Company Safety Profile (CSP) and contains information on several aspects of the carrier's safety performance, including a crash summary of four years and individual crashes of one to two years.

Information provided by these databanks may be retrieved by computer techniques for research purposes.

The technique used for retrieving specific crash data depends on the method of storage of data. When data are stored manually, the retrieval is also manual. In this case, the file is examined by a trained technician, who then retrieves the appropriate report forms. When data are stored on computer, retrieval requires only the input of appropriate commands into the computer for any specific data required, and those data are given immediately as output.

Collision and Condition Diagrams.   Collision diagrams present pictorial information on individual crashes at a location. Different symbols are used to represent different types of maneuvers, types of crashes, and severity of crashes. The date and time (day or night) at which the crash occurs are also indicated. Figure 5.4 shows an example of a collision diagram. One advantage of collision diagrams is that they give information on the location of the crash, which is not available with statistical summaries. Collision diagrams may be prepared manually either by retrieving data filed manually or by computer when data are stored in a computer file. A condition diagram gives a pictorial general survey of a site and can be used to relate the

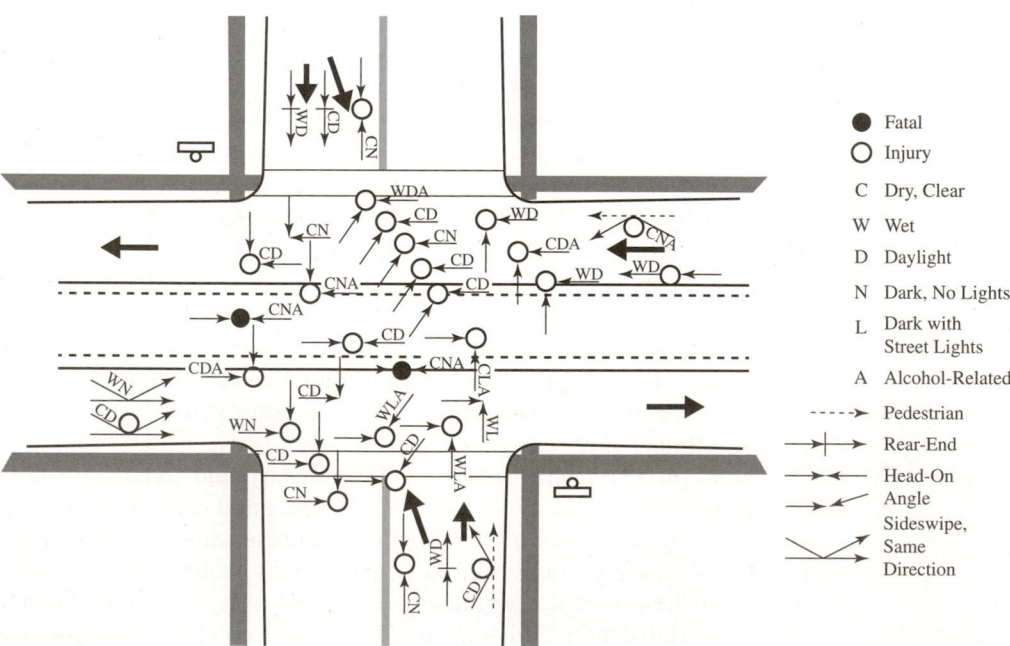

Figure 5.4   Example of a Collision Diagram

SOURCE: From *Highway Safety Manual* 1st Edition, 2010, AASHTO, Washington, D.C. Used by permission.

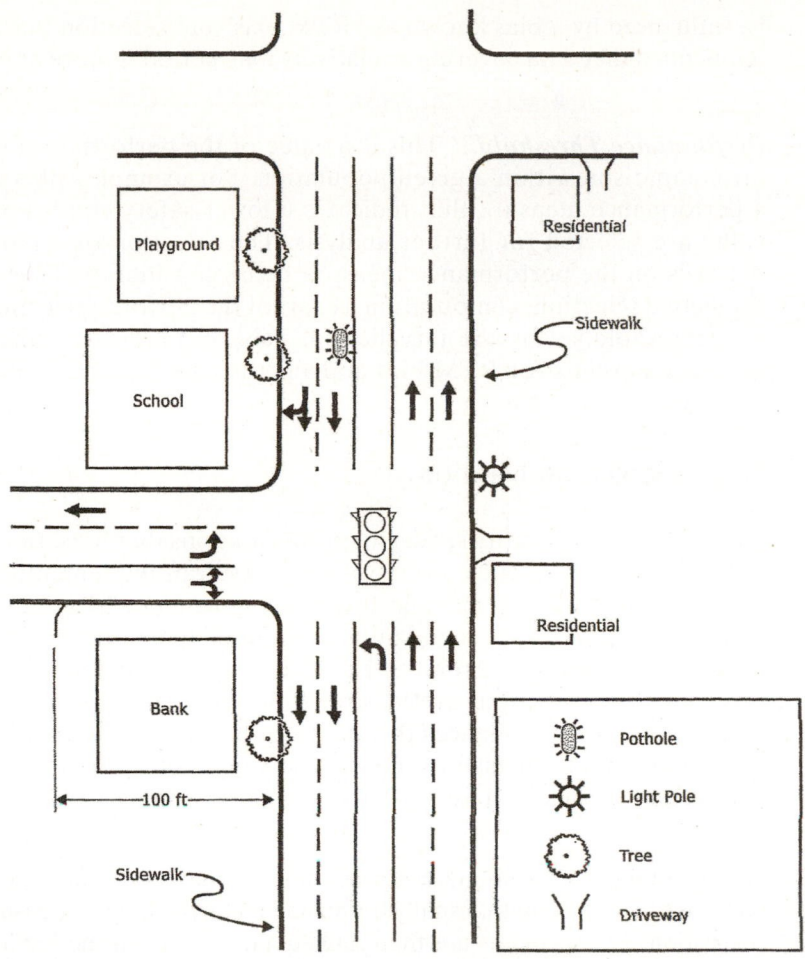

Figure 5.5   Example of a Condition Diagram

SOURCE: From *Highway Safety Manual* 1st Edition, 2010, AASHTO, Washington, D.C. Used by permission.

physical perspective of a site with crashes that occurred at that site. Figure 5.5 shows an example of a condition diagram.

Regression to the Mean (RTM).   The occurrence of a crash is a random event which results in the fluctuation of crash frequencies. This natural fluctuation often causes average crash frequencies over a short period to be either higher or lower than the mean over a long period. Therefore, if during a couple of years a relatively high crash frequency occurs at a site, it is very likely that this period will be followed by a short period of relatively low crash frequency. This phenomenon is known as *regression to the mean* (RTM). Data collected over a short period may therefore give a very different short-time average crash frequency than that for a long-term period. It is therefore difficult to ascertain how short-term crash data reflects the actual mean over a long period. If an adjustment is not made for this phenomenon, it is likely that the data will be influenced by a bias known as "RTM bias" or "selection bias." It is therefore recommended that data covering a relatively long period is more appropriate for network screening.

Performance Threshold. This is a value of the performance measure that is used for comparison within a given population. For example, sites that have a value of a performance measure that indicates a lower safety condition than the threshold value are selected for further analysis. The selection of a performance threshold depends on the performance measure being considered. These methods include a subjective selection, computation as part of the performance measure methodology, or a threshold set by the jurisdiction. Table 5.2 indicates whether a performance measure accounts for RTM bias and whether the method estimates a performance threshold.

Select Screening Method

It is necessary to identify subsegments or locations that have the highest potential for reducing crash frequency or severity as a result of implementing a countermeasure. The procedure for determining these subsegments or locations is known as *screening*. These identified subsegments are then used to determine critical crash frequency for the whole stretch of road to be selected for additional study. Knowledge of the subsegments or locations on the stretch of road that significantly influence the critical crash frequency enhances the identification of the causal factors and the selection of effective countermeasures. One of three methods can be used to screen a stretch or segment of road. These are simple ranking, sliding window, and peak searching methods.

Simple Ranking. In this method, the whole section or segment of road is divided into subsections of a given length, usually 0.1 mi and the performance measure computed for each subsection. The segments are then ranked and those with the highest performance measures are selected for further study to identify causal factors and suitable countermeasures.

Sliding Window Method. In this method, a short length (0.1 to 0.5 km) of a road segment is established at the starting end of the road segment, and this short length (usually referred to as a window) is conceptually moved each time by a specific distance (0.05 to 0.25 km) along the entire stretch of the road segment while keeping its length constant as the original length. Note that a road section being studied can be divided into several road segments based on traffic, geometric factors, or topographical conditions. The performance measure selected for screening is then computed for each window of a segment and noted. The window which exhibits the highest potential for crash frequency reduction is selected and its potential for crash frequency reduction is used to represent that for the entire segment. A window is considered part of a segment if a portion of the window exists in the segment. All segments on the road section under investigation are then ranked using the highest subsegment value. Further studies are conducted on those segments with the highest potential for reduction in crash frequency or severity to determine causal factors and selection of appropriate countermeasures. It is likely that windows will overlap two or more contiguous roadway segments, but the length of the window should be kept constant and the incremental length adjusted so that the last window ends at the end of the roadway segment. Also, in some cases the lengths of roadway segments that are not contiguous to other segments are less than the window length. When this occurs, the length of the window is considered as the length of the segment.

Table 5.2 Stability of Performance Measures

Performance Measure	Accounts for RTM Bias	Method Estimates a Performance Threshold
Average Crash Frequency	No	No
Crash Rate	No	No
Equivalent Property Damage Only (EPDO) Average Crash Frequency	No	No
Relative Severity Index	No	Yes
Critical Rate	No	Yes
Excess Predictive Average Crash Frequency Using Method of Moments	Consider data variance but does not account for RTM bias	Yes
Level of Service of Safety	Consider data variance but does not account for RTM bias	Expected average crash frequency plus/minus 1.5 standard deviations
Excess Predictive Average Crash Frequency Using Safety Performance Functions (SPFs)	No	Predicted average crash at the site
Probability of Specific Crash Types Exceeding Threshold Proportion	Consider data variance; not affected by RTM bias	Yes
Excess Proportion of Specific Crash Types	Consider data variance; not affected by RTM bias	Yes
Expected Average Crash Frequency with EB Adjustment	Yes	Expected average crash frequency at the site
Equivalent Property Damage Only (EPDO) Average Crash Frequency with EB Adjustment	Yes	Expected average crash frequency at the site
Excess Expected Average Crash Frequency with EB Adjustment		

SOURCE: From *Highway Safety Manual* 1st Edition, 2010, AASHTO, Washington, D.C. Used by permission.

Example 5.1 Use of Sliding Window Method

Segment D on a rural two-lane highway undergoing a safety study is to be screened using the sliding window method and the average crash frequency as the performance measure. Based on information given below, determine:

(i) the position of each window
(ii) the window that should be selected to represent the segment
(iii) the value of the performance measure that should be selected to represent the segment.

1. Start of segment is at km 100.00
2. End of segment is at km 101.00
3. Window length = 0.2 km
4. Increment = 0.1 km

5. Average annual crash frequency for each segment:

First segment = 1.45 crashes/yr

Second segment = 1.20 crashes/yr

Third segment = 1.85 crashes/yr

Fourth segment = 1.65 crashes/yr

Fifth segment = 2.05 crashes/yr

Sixth segment = 1.50 crashes/yr

Seventh segment = 1.45 crashes/yr

Eighth segment = 1.20 crashes/yr

Ninth segment = 1.32 crashes/yr

Solution:

The length of the segment is 1.0 km, starting at km 100.0 and ending at km 101.0.

Table 5.3 shows the location of the segments and their respective average crash frequencies. The table shows that subsegment D5, starting at km 100.4 and ending at km 100.6, has the highest average crash frequency of 2.05 crashes/100 million vehicle-km. Therefore, the window that should be selected to represent the segment is window D5, starting at km 100.4 and ending at km 100.6. The value of the performance measure to be selected for ranking of segment D among all segments in the study road section is 2.05 crashes/yr.

Table 5.3 Data and Solution for Example 5.1

Window #	Start km	End km	Crashes/yr
D1	100.0	100.2	1.45
D2	100.1	100.3	1.20
D3	100.2	100.4	1.85
D4	100.3	100.5	1.65
D5	100.4	100.6	2.05
D6	100.5	100.7	1.50
D7	100.6	100.8	1.45
D8	100.7	100.9	1.20
D9	100.8	101.0	1.32

Peak Searching Method. In this method, the procedure starts with establishing subsegments (windows) usually 0.1 km long, covering the entire road segment under consideration. If the length of the road segment is less than 0.1 km, then the window length is the segment length. There is no overlapping of the windows, although it is possible that for certain facilities, the last window may overlap with the first. The selected performance measure is then computed for each window and the coefficient of variation

(*CV*) of the values obtained is determined. This gives an indication of the precision of the values of the determined performance measures. A high *CV* implies a low level of precision, and a low *CV* implies a high level of precision. If the coefficient of variation obtained for any subsegment is lower than a pre-established limit (satisfies the precision requirement), then the best performance measure (e.g., largest excess predictive average crash frequency using SPFs) among those for all subsegments (windows) within the road segment under consideration that satisfy the precision level is selected for ranking the road segment. If, however, no subsegment satisfies the precision level, the length of each segment is increased and the procedure repeated. If the precision level is still not obtained for any subsegment, the length of the subsegments is increased by 0.1 km until the required precision level is achieved by at least one subsegment or until the length of the subsegment is the same as the length of the entire road segment being considered. The coefficient of variation of the performance measure for a given subsegment i (CV_i) is given as:

$$CV_i = \frac{\sqrt{\text{Var(Performance Measure)}}}{\text{Performance Measure}} \tag{5.1}$$

where
Var(Performance Measure) = the variance of the performance measures for all subsegments on the segment under consideration
Performance Measure = the computed performance measure of subsection i

No specific precision level is appropriate for all performance measures used in screening, but different levels of precision levels will result in varying number of identified sites that can be used for additional studies. The HSM suggests a *CV* of 0.5 as an initial or default value.

Example 5.2   Use of Peak Searching Method

The first three columns of the Table 5.4 show the particulars of subsegments of roadway segment D and their respective excess expected average crash frequencies. Using the peak searching method, determine the expected average crash frequency that should be used to rank segment D in comparison to other roadway segments of the road section being studied. Take 0.2 as the limiting value for CV_i.

Solution

Table 5.4 shows the computed value of the CV_i for each subsegment. Since the limiting value of CV_i is 0.20, the results indicate that all subsegments satisfy the precision requirement. Among these, D3 has the highest performance measure value of 5.6. The value of the performance measure to be used for segment D ranking is therefore 5.6 excess expected average crash frequency. Note that in this problem, subsegments satisfy the precision level requirement. If no subsegment had satisfied the precision requirement, the procedure would be repeated after increasing the lengths of the subsegments to 0.2 km.

Table 5.4   Data and solution for Example 5.2

Subsegment	Window Position	Excess Expected Average Crash Frequency (crashes/yr)	Coefficient of Variation
D1	km15.50 to km15.60	4.8	0.098
D2	km15.60 to km15.70	4.5	0.104
D3	km15.70 to km15.80	5.6	0.084
D4	km15.80 to km15.90	5.1	0.092
D5	km15.90 to km16.00	5.5	0.085
Average		5.1	

$$VAR_{segmentD} = \frac{(4.8 - 5.1)^2 + (4.5 - 5.1)^2 + (5.6 - 5.1)^2 + (5.1 - 5.1)^2 + (5.5 - 5.1)^2}{5 - 1}$$

$$= 0.22$$

The coefficient of variation for subsegment D1 is

$$CV_i = (\sqrt{(0.22)})/(4.8) = 0.098$$

The CV_i for each other segment is computed and shown in Table 5.4.

Note that screening can be carried out on nodes, segments, or facilities. Nodes are typically intersections, ramp terminal intersections, and at-grade railway crossings. Segments are typically stretches of roadways between intersections, and facilities are sections of highways consisting of connected roadway segments and intersections. Note that not all screening methods are suitable for all combinations of performance measures and segments, nodes and facilities. Table 5.5 shows which screening method is appropriate for each combination.

Screen and Evaluate Results

After the screening has been applied to one or more of the segments, nodes, or facilities in a given study, a list is prepared with the ordered ranking of all the segments, nodes or facilities based on the performance measure used. The sites that are in the upper level of the list are regarded as those that have the highest potential to benefit from the implementation of appropriate countermeasures. These sites will be studied further in Section 5.3 to identify the most effective countermeasures. The HSM suggests that more than a single performance measure should be used for each data set, as this will show sites that are consistently at the higher level of the ranking. Sites that are consistently in the higher level of the ranking list are selected for further study while those that are consistently in the lower level are usually ruled out for further investigation. Sites that are located at the middle level of the list will tend to show more variation in their ranks for different performance measures.

5.2.2 Diagnosis

After network screening, the next step in the roadway safety management process is diagnosis. In this step, a detailed study of the crashes on the sites identified in network screening is carried out to determine crash patterns and to identify related causal factors

Table 5.5   Performance Measure Consistency with Screening Methods

Performance Measure	Segments			Nodes	Facilities
	Simple Ranking	Sliding Window	Peak Searching	Simple Ranking	Simple Ranking
Average Crash Frequency	Yes	Yes	No	Yes	Yes
Crash Rate	Yes	Yes	No	Yes	Yes
Equivalent Property Damage Only (EPDO) Average Crash Frequency	Yes	Yes	No	Yes	Yes
Relative Severity Index	Yes	Yes	No	Yes	Yes
Critical Rate	Yes	Yes	No	Yes	Yes
Excess Predictive Average Crash Frequency Using Method of Moments	Yes	Yes	No	Yes	Yes
Level of Service of Safety	Yes	Yes	No	Yes	Yes
Excess Predictive Average Crash Frequency Using Safety Performance Functions (SPFs)	Yes	Yes	No	Yes	Yes
Probability of Specific Crash Types Exceeding Threshold Proportion	Yes	Yes	No	Yes	Yes
Excess Proportion of Specific Crash Types	Yes	Yes	No	Yes	Yes
Expected Average Crash Frequency with EB Adjustment	Yes	Yes	Yes	Yes	No
Equivalent Property Damage Only (EPDO) Average Crash Frequency with EB Adjustment	Yes	Yes	Yes	Yes	No
Excess Expected Average Crash Frequency with EB Adjustment	Yes	Yes	Yes	Yes	No

SOURCE: From *Highway Safety Manual* 1st Edition, 2010, AASHTO, Washington, D.C. Used by permission.

and other potential safety concerns that will aid in the selection of potential countermeasures. Diagnosis is carried out by conducting the following tasks:

- Review safety data.
- Assess supporting documentation.
- Assess field conditions.

A brief discussion of each of these tasks is given in this section.

Review Safety Data

This task involves a detailed review of the crash data to identify the type or types of crashes that occurred on the selected sites during the screening process, the patterns in crash types, severity of these crashes, roadway and environmental conditions that are associated with these crashes, and contributing circumstances. Crash types include rear-end, sideswipe, angle, turning, head-on, run-off-the-road, fixed object, animal, and out-of-control. Patterns relate to time of day that a crash occurred and the direction of travel prior to the crash. The severity of the crash is usually selected based on the KABCO scale. Roadway conditions include geometric characteristics such as curvature, and environmental conditions include weather and lighting conditions that existed at the time of a crash. Other contributing factors include vehicle or vehicles involved and impairments due to alcohol, drugs, or fatigue. These crash data are usually retrieved from computerized crash archival files existing in state reports or individual police crash reports. The information obtained is usually summarized in one or more forms, including bar charts, tables, and collision and condition diagrams.

Assess Supporting Documentation

In this task, available documented information or personal testimony of local transportation professionals is reviewed to obtain additional insight to the data summarized in the previous task. Documented information can give historical data on physical changes that were made during the period for which the crash data were obtained; for example, widening of a road segment resulting in higher operational speeds that led to an increase in rear-end crashes at intersections. Supporting documentation that can be used in this task include as-built construction plans, recent traffic operation or transportation studies, historic patterns of adverse weather, and relevant photo or video logs. Interested readers may refer to the HSM for additional information on the questions and data that should be reviewed in this task.

Assess Field Conditions

In this task, field observations are conducted by traveling along the site from as many directions and in many modes as possible. The objective here is to replicate the experience of a person traveling to and through the site, and to obtain first-hand information that will help an analyst understand the circumstances that prevail as one traverses the site by either a motorized or nonmotorized mode. It is useful to make multiple visits to the sites on different days and under different weather and lighting conditions as this may give additional information to the analyst. Examples of factors that should be considered during a field review can be categorized as follows: roadway and roadside characteristics (e.g., posted speed limits, pavement condition, geometric design), traffic conditions (e.g., types of facility, free-flow or congested condition), travel behavior (e.g., aggressive driving, ignoring traffic control, insufficient pedestrian crossing space), roadway consistency, land uses, weather conditions and evidence of problems (e.g., skid marks, damaged road furniture). A prompt list that reminds the observer of different assessments and considerations to be made in the field is a useful tool that can be used when conducting assessing field conditions. The HSM provides prompt lists for different types of roadway environments.

The information gathered from conducting the diagnostic activities can be used to see whether there are any crash patterns that can be addressed by a countermeasure. By comparing the information obtained from each of the three tasks in this step with each other, certain findings can be extracted that otherwise might not have been observed. For example, if the crash data show a high fatal crash frequency at a curve and the assessment of field conditions indicates that there are no guard rails at the location, it is likely that these two pieces of information are related.

5.2.3 Select Countermeasures

After diagnosis, the next step in the roadway safety management process is to select countermeasures. This step first involves identifying associated contributing factors for the different types of crashes and their patterns that evolved from the screening and diagnostic steps, and then selecting appropriate countermeasures to eliminate these contributing factors or alleviate their negative safety effects. It is necessary to first identify a broad range of probable contributing factors for each crash pattern observed, as often more than one contributing factor is associated with a particular crash pattern. Reviewing a range of factors reduces the probability of overlooking an important contributing factor. This broad range of countermeasures is then refined by using information obtained during the field review to confirm the physical deficiencies at the study site. The refined list is used to determine what data will be required to identify the safety deficiencies at the study site. The report, *Guidance for the Implementation of the AASHTO Strategic Highway Safety Plan, NCHRP Report 500,* discusses many contributing factors for different crash types and patterns. Tables 5.6 through 5.8 give examples of associated contributing factors for different types of crashes on roadway segments, signalized intersections, and unsignalized intersections, respectively. Tables 5.9 through 5.11 give contributing associated factors for crashes involving highway–rail grade crossings, pedestrians, and bicyclists, respectively. Table 5.12 gives data needs and procedures to be performed for different possible causes of crashes, and Table 5.13 gives countermeasures for safety deficiencies.

Table 5.6   Associated Contributing Factors for Crashes on Roadway Segments

Crash Type	Associated Contributing Factors
Vehicle rollover	Roadside design (e.g., nontraversable slide slopes, pavement edge drop-off); inadequate shoulder width; excessive speed; pavement design
Fixed object	Obstruction in or near roadway; inadequate lighting; inadequate pavement markings; slippery pavement; roadside design (e.g., inadequate clear distance); inadequate roadway geometry; excessive speed
Nighttime	Poor nighttime visibility or lighting; poor sign visibility; inadequate channelization or delineation; excessive speed; inadequate sight distance
Wet pavement	Pavement design (e.g., drainage, permeability (see Chapter 16)); inadequate pavement markings; inadequate maintenance; excessive speed
Opposite-direction sideswipe or head-on	Inadequate roadway geometry; inadequate shoulders; excessive speed; inadequate pavement markings; inadequate signing
Run-off-the-road	Inadequate lane width; slippery pavement; inadequate roadway shoulders; poor delineation; poor visibility; excessive speed
Bridges	Alignment; narrow roadway; visibility; vertical clearance; slippery pavement; rough surface; inadequate barrier system

SOURCE: Based on *Highway Safety Manual*, 1st Edition, 2010, AASHTO, Washington, D.C.

Table 5.7   Associated Contributing Factors for Crashes at Signalized Intersections

Crash Type	Associated Contributing Factors
Right angle	Poor visibility of signals; inadequate signal timing; excessive speed; slippery pavement; inadequate sight distance; drivers running red light
Rear-end or sideswipe	Inappropriate approach speeds; poor visibility of signals; unexpected lane changes on approach; unexpected stops on approach; slippery pavement; excessive speed
Left or right-turn movement	Misjudged speed of oncoming traffic; pedestrian or bicycle conflicts; inadequate signal timing; inadequate sight distance; conflict with right-turn-on-red vehicles
Nighttime	Poor nighttime visibility or lighting; poor sign visibility; inadequate channelization or delineation; inadequate maintenance; excessive speed; inadequate sight distance
Wet pavement	Slippery pavement; inadequate pavement markings; inadquate maintenance; excessive speed

SOURCE: Based on *Highway Safety Manual*, 1st Edition, 2010, AASHTO, Washington, D.C.

Table 5.8   Associated Contributing Factors for Different Types of Crashes at Unsignalized Intersections

Crash Type	Associated Contributing Factors
Angle	Restricted sight distance; high traffic volume; high approach speed; unexpected crossing traffic; drivers running stop sign; slippery pavement
Rear-end	Pedestrian crossing; driver inattention; slippery pavement; large number of turning vehicles; unexpected lane change; narrow lanes; restricted sight distance; inadequate gaps in traffic; excessive speed
Collision at driveways	Left-turning vehicles; improperly located driveway; right-turning vehicles; large volume of through vehicles; large volume of driveway traffic; restricted sight distance; excessive speed
Head-on or sideswipe	Inadequate pavement marking; narrow lane
Left or right-turn	Inadequate gaps in traffic; restricted sight distance
Nighttime	Poor nighttime visibility or lighting; poor sign visibility; inadequate channelization or delineation; excessive speed; inadequate sight diatance
Wet pavement	Slippery pavement; inadequate pavement marking; excessive speed

SOURCE: Based on *Highway Safety Manual*, 1st Edition, 2010, AASHTO, Washington, D.C.

Table 5.9   Associated Contributing Factors for Different Types of Crashes on Highway–Rail Grade Crossings

Crash Type	Associated Contributing Factors
All	Restricted sight distance; poor visibility of traffic control devices; inadequate pavement markings; rough or wet crossing surface; sharp crossing angle; improper pre-emption timing (see Chapter 8); excessive speed; drivers performing impatient maneuvers

SOURCE: Based on *Highway Safety Manual*, 1st Edition, 2010, AASHTO, Washington, D.C.

Table 5.10   Associated Contributing Factors for Crashes Involving Pedestrians

Crash Type	*Associated Contributing Factors*
Crashes involving pedestrians	Limited sight distance; inadequate barrier between pedestrian and vehicle facilities; inadequate signals/signs; inadequate signal phasing; inadequate pavement markings; inadequate lighting; driver has inadequate warning of mid-block crossings; lack of crossing opportunity; excessive speed; pedestrians on roadway; long distance to nearest crosswalk; sidewalk too close to travel way; school crossing area

SOURCE: Based on *Highway Safety Manual*, 1st Edition, 2010, AASHTO, Washington, D.C.

Table 5.11   Associated Contributing Factors for Different Crashes Involving Bicyclists

Crash Type	*Associated Contributing Factors*
All	Limited sight distance; inadequate signs; inadequate pavement markings; inadequate lighting; excessive speed; bicycles on roadway; bicycle path too close to roadway; narrow lanes for bicyclists

SOURCE: Based on *Highway Safety Manual*, 1st Edition, 2010, AASHTO, Washington, D.C.

Table 5.12   Data Needs for Different Possible Causes of Crashes

Possible Causes	*Data Needs*	*Procedures to Be Performed*
	Left-Turn Head-On Collisions	
Large volume of left turns	• Volume data • Vehicle conflicts • Roadway inventory • Signal timing and phasing • Travel time and delay data	• Volume Study • Traffic Conflict Study • Roadway Inventory Study • Capacity Study • Travel Time and Delay Study
Restricted sight distance	• Roadway inventory • Sight-distance characteristics • Speed characteristics	• Roadway Inventory Study • Sight-Distance Study • Spot Speed Study
Too short amber phase	• Speed characteristics • Volume data • Roadway inventory • Signal timing and phasing	• Spot Speed Study • Volume Study • Roadway Inventory Study • Capacity Study

(Continued)

Table 5.12   Data Needs for Different Possible Causes of Crashes (*continued*)

Absence of special left-turning phase	• Volume data • Roadway inventory • Signal timing and phasing • Delay data	• Volume Study • Roadway Inventory Study • Capacity Study • Travel Time and Delay Study
Excessive speed on approaches	• Speed characteristics	• Spot Speed Study

Rear-End Collisions at Unsignalized Intersections

Driver not aware of intersection	• Roadway inventory • Sight-distance characteristics • Speed characteristics	• Roadway Inventory Study • Sight-Distance Study • Spot Speed Study
Slippery surface	• Pavement skid resistance characteristics • Conflicts resulting from slippery surface	• Skid Resistance Study • Weather-Related Study • Traffic Conflict Study
Large number of turning vehicles	• Volume data • Roadway inventory • Conflict data	• Volume Study • Roadway Inventory Study • Traffic Conflict Study
Inadequate roadway lighting	• Roadway inventory • Volume data • Data on existing lighting	• Roadway Inventory Study • Volume Study • Highway Lighting Study
Excessive speed on approaches	• Speed characteristics	• Spot Speed Study
Lack of adequate gaps	• Roadway inventory • Volume data • Gap data	• Roadway Inventory Study • Volume Study • Gap Study
Crossing pedestrians	• Pedestrian volumes • Pedestrian/vehicle conflicts • Signal inventory	• Volume Study • Pedestrian Study • Roadway Inventory Study

Rear-End Collisions at Signalized Intersections

Slippery surfaces	• Pavement skid resistance characteristics • Conflicts resulting from slippery surface	• Skid Resistance Study • Weather Related Study • Traffic Conflict Study
Large number of turning vehicles	• Volume data • Roadway inventory • Conflict data • Travel time and delay data	• Volume Study • Roadway Inventory Study • Traffic-Conflict Study • Delay Study
Poor visibility of signals	• Roadway inventory • Signal review • Traffic conflicts	• Roadway Inventory Study • Traffic-Control Device Study • Traffic Conflict Study

SOURCE: Adapted from *Highway Safety Engineering Studies Procedural Guide*, U.S. Department of Transportation, Washington, D.C., June 1981

Table 5.13   General Countermeasures for Different Safety Deficiencies

Probable Cause	General Countermeasure
Left-Turn Head-On Collisions	
Large volume of left turns	• Create one-way street • Widen road • Provide left-turn signal phases • Prohibit left turns • Reroute left-turn traffic • Channelize intersection • Install stop signs (see MUTCD)* • Revise signal sequence • Provide turning guidelines (if there is a dual left-turn lane) • Provide traffic signal if warranted by MUTCD*
Restricted sight distance	• Retime signals • Remove obstacles • Provide adequate channelization • Provide special phase for left-turning traffic • Provide left-turn slots • Install warning signs • Reduce speed limit on approaches
Too short amber phase	• Increase amber phase • Provide all-red phase
Absence of special left-turning phase	• Provide special phase for left-turning traffic
Excessive speed on approaches	• Reduce speed limit on approaches
Rear-End Collisions at Unsignalized Intersections	
Driver not aware of intersection	• Install/improve warning signs
Slippery surface	• Overlay pavement • Provide adequate drainage • Groove pavement • Reduce speed limit on approaches • Provide "slippery when wet" signs
Large number of turning vehicles	• Create left- or right-turn lanes • Prohibit turns • Increase curb radii
Inadequate roadway lighting	• Improve roadway lighting
Excessive speed on approach	• Reduce speed limit on approaches
Lack of adequate gaps	• Provide traffic signal if warranted (see MUTCD)* • Provide stop signs
Crossing pedestrians	• Install/improve signing or marking of pedestrian crosswalks
Rear-End Collision at Signalized Intersections	
Slippery surface	• Overlay pavement • Provide adequate drainage • Groove pavement • Reduce speed limit on approaches • Provide "slippery when wet" signs

(Continued)

Table 5.13   General Countermeasures for Different Safety Deficiencies (*continued*)

Large number of turning vehicles	• Create left- or right-turn lanes • Prohibit turns • Increase curb radii • Provide special phase for left-turning traffic
Poor visibility of signals	• Install/improve advance warning devices • Install overhead signals • Install 12-in. signal lenses (see MUTCD)* • Install visors • Install back plates • Relocate signals • Add additional signal heads • Remove obstacles • Reduce speed limit on approaches

Rear-End Collision at Signalized Intersections

Inadequate signal timing	• Adjust amber phase • Provide progression through a set of signalized intersections • Add all-red clearance
Unwarranted signals	• Remove signals (see MUTCD)*
Inadequate roadway lighting	• Improve roadway lighting

Manual on Uniform Traffic Control Devices, published by FHWA
SOURCE: Adapted from *Highway Safety Engineering Studies Procedural Guide*, U.S. Department of Transportation, Washington, D.C., June 1981.

Crash Reduction Capabilities

Crash reduction capabilities are used to estimate the expected reduction that will occur during a given period as a result of implementing a proposed countermeasure. Crash reduction capabilities usually are expressed as crash reduction factors (CRFs) or crash modification factors (*CMFs*). A CRF gives an estimate of the percent reduction in the number of crashes due to the implementation of a countermeasure, while a *CMF* gives the percentage of crash reduction as $100(1 - \text{CRF})$. Therefore, *CMFs* that have values less than one are effective in reducing the predicted average crash frequency at a site when properly applied, while a *CMF* with a value higher than one will increase the predicted average frequency. Some states have developed their own CRFs or *CMFs*, while others have adopted those developed by other states. The problem with adopting other states' factors is that roadway, traffic, weather, and driver characteristics may vary significantly among states. Factors developed by a state usually are based on the evaluation of data obtained from safety projects conducted in the state and can be obtained from state agencies involved in crash analysis. The HSM, however, provides an extensive list of *CMFs* for many different countermeasures when applied to varying facilities. Tables 5.14 and 5.15 give a selected number of *CMFs* for rural roadway segments and intersections, respectively. These tables show that *CMFs* can indicate whether a treatment will improve or degrade safety. For example, Table 5.14 shows that lane widths narrower than 12 ft will increase crashes (*CMFs* > 1.0) while Table 5.15 shows that traffic signals will decrease some crash types and increase rear-end crashes.

Table 5.14   Crash Modification Factors[a] (*CMFs*) for Lane Width on Roadway Segments

Roadway Type	Lane Width	Average Annual Daily Traffic (AADT)(veh/day)			Equation Number
		< 400	*400 to 2000*	*>2000*	
Rural Two-Lane[a]	2.7 m or less	1.05	$1.05 + 2.81 \times 10^{-4}(AADT - 400)$	1.50	5.2
	3 m	1.02	$1.02 + 1.75 \times 10^{-4}(AADT - 400)$	1.30	5.3
	3.3 m	1.01	$1.01 + 2.5 \times 10^{-5}(AADT - 400)$	1.05	5.4
	3.6 m or more	1.00	1.00	1.00	5.5
Undivided Rural Multilane[b]	2.7 m or less	1.04	$1.04 + 2.13 \times 10^{-4}(AADT - 400)$	1.38	5.6
	3 m	1.02	$1.02 + 1.31 \times 10^{-4}(AADT - 400)$	1.23	5.7
	3.3 m	1.01	$1.01 + 1.88 \times 10^{-5}(AADT - 400)$	1.04	5.8
	3.6 m or more	1.00	1.00	1.00	5.9
Divided Rural Multilane[c]	2.7 m or less	1.03	$1.03 + 1.38 \times 10^{-4}(AADT - 400)$	1.25	5.10
	3 m	1.01	$1.01 + 8.75 \times 10^{-5}(AADT - 400)$	1.15	5.11
	3.3 m	1.01	$1.01 + 1.25 \times 10^{-5}(AADT - 400)$	1.03	5.12
	3.6 m or more	1.00	1.00	1.00	5.13

[a]Note: The collision types related to lane width to which these *CMFs* apply are single-vehicle run-off-the-road and multiple-vehicle head-on, opposite direction sideswipe, and same direction sideswipe crashes.
[b]Note: Standard error of *CMF* is unknown.
[c]Note: To determine the *CMF* for changing lane width and/or *AADT*, divide the "new" condition *CMF* by the "existing" condition *CMF*.
SOURCE: Based on *Highway Safety Manual*, 1st Edition, 2010, AASHTO, Washington, D.C.

Table 5.15   Potential Crash Effects of Converting from Stop Control to Signal Control[a]

Treatment	Setting (Intersection Type)	Traffic Volume AADT (veh/day)	Crash Type (severity)	CMF	Std. Error
Install a traffic signal	Urban (major road speed limit at least 64 km/h; four-leg)	Unspecified	All types (All severities)	**0.95**[*]	**0.09**
			Right-angle (All severities)	**0.33**	**0.06**
			Rear-end (All severities)	*2.43*	*0.4*
	Rural (three-leg and four-leg)	Major road (3,261 to 29,926)	All types (All severities)	**0.56**	**0.03**
		Minor road (101 to 10,300)	Right-angle (All severities)	**0.23**	**0.02**
			Left-turn (All severities)	**0.40**	**0.06**
			Rear-end (All severities)	*1.58*	*0.2*

[a]Base Condition: Minor-road, stop-controlled intersection
Note: **Bold** text is used for more reliable *CMFs*. These *CMFs* have a standard error of 0.1 or less.
Italic text is used for less reliable *CMFs*. These *CMFs* have standard errors of 0.2 or higher.
[*]Observed variability suggests that this treatment could result in an increase, decrease or no changes in crashes
SOURCE: Based on *Highway Safety Manual*, 1st Edition, 2010, AASHTO, Washington, D.C.

Note that the *CMFs* given in Table 5.14 for the effect of lane width are applicable only to a particular set of crash types (e.g., single-vehicle run-off-the-road, opposite direction sideswipe, same direction sideswipe) that is affected by the specific associated countermeasure, and other crashes are not affected. The HSM suggests that a *CMF* for a particular set of crash types can be adjusted to that for total crashes. For example, a *CMF* for the effect of lane width on single-vehicle run-off-the-road crashes on two-lane roads can be converted to that for total crashes on two-lane roads by using Eq. 5.14.

$$CMF_{lr} = (CMF_{ra} - 1)p_{ra} + 1.0 \qquad (5.14)$$

where

CMF_{lr} = crash modification for the effect of lane width on total crashes.
CMF_{ra} = crash modification for the effect of lane width on single-vehicle run-off-the-road and/or related crashes on two-lane roads.
p_{ra} = proportion of total crashes constituted by single-vehicle run-off-the-road and/or related crashes on two-lane roads. HSM default value is 0.574, but a value based on actual local data is preferable when available.

Example 5.3 Use of Crash Modification Factor

An undivided rural multilane highway segment is to be widened so as to increase its existing lane widths from 3 to 3.6 m. Compute the expected average total crash frequency after the widening for the following conditions.

Expected average crash frequency before widening = 15 total crashes/yr
AADT on highway segment = 1500 veh/day
Proportion of total crashes constituted by CMF-related crashes = 0.55

Solution:
• Determine crash modification factor (*CMF*).

The *CMF* is obtained from Eq. 5.7 in Table 5.14, given as

$$CMF = 1.02 + 1.31 \times 10^{-4}(AADT - 400)$$
$$= 1.02 + 1.31 \times 10^{-4}(1500 - 400)$$
$$= 1.02 + 0.1441 = 1.164$$

• Compute crash modification factor for total crashes.

Since the crash modification factor given in Table 5.14 is for related crashes (single-vehicle run-off-the-road and multiple-vehicle head-on, opposition direction sideswipe, and same direction sideswipe crashes), Eq. 5.14 is used to compute the *CMF* for total crashes.

$$CMF_{lr} = (CMF_{ra} - 1)p_{ra} + 1.0$$
$$CMF_{lr} = (1.164 - 1)0.55 + 1.0$$
$$= 1.09$$

- Compute the expected average total crash frequency after the widening.

  Since the *CMF*, when the width = 3.6 m, is 1.0, the expected crash frequency after widening is equal to the current expected crash frequency for 3 m lanes divided by the *CMF* for 3 m lanes.

  Expected frequency after widening = 15/1.09 = 13.76 crashes/yr

5.2.4 Economic Appraisal

After appropriate countermeasures have been selected, the next step in the roadway safety management process is to conduct an economic analysis. This involves the determination of the efficacy of implementing a countermeasure at a site in terms of a criterion based on economics. There are two general methodologies that are used: *benefit–cost analysis* and *cost-effectiveness*. In benefit–cost analysis, the number or severity of crashes reduced as a consequence of implementing one or more countermeasures is converted to monetary value and economically compared with the cost of implementing and maintaining the countermeasure. In the cost-effectiveness procedure, the reduction in crash frequency and the associated cost of applying the countermeasure are directly compared. Note that in a comprehensive economic study of a project, the benefits accrued from other factors such as travel time, congestion relief, and environmental impacts are included as discussed in Chapter 13. However, in this chapter, only safety-related benefits are considered. The HSM gives a predictive method that can be used to estimate changes in average crash frequency for different types of facilities. An overview of this method is given in Section 5.5. The benefit–cost ratio and the cost-effective analyses are discussed in Chapter 13.

5.2.5 Prioritize Projects

In this step, the countermeasures that have been proved economically viable from the economic appraisal are prioritized, as funds are usually not available to carry out all projects at the same time. This prioritization is achieved by using one of several ranking procedures to develop a list of ordered projects based on their respective benefits, or by optimization, by which the selection of a project is based on maximizing benefits with respect to budget or other constraints. Only the ranking procedures are discussed here. The ranking procedures can be grouped into two general groups: (1) ranking by economic effectiveness measures and (2) ranking by incremental benefit–cost analysis.

Ranking by Economic Effectiveness Measures

Methods that can be used to rank by economic effectiveness include:

- Monetary value of project benefits
- Number of total crashes reduced
- Number of fatal and incapacitating injury crashes reduced
- Number of fatal and injury crashes reduced
- Cost-effectiveness index
- Net present value (NPV)

For each of the first four methods, the respective number or monetary value is computed and a list is made up with the project having the highest value ranked first, followed by other projects in a descending order based on the value of each project. The cost-effectiveness index and net present value methods are discussed in Chapter 13. Reviewing the rankings obtained by two or more of these methods can be used to make decisions when considering a few sites or a limited number of alternatives at each site. However, since they do not take into consideration factors such as budget constraints or other project impacts, they are not sophisticated enough for use when multiple competing priorities are being considered.

Ranking by Incremental Benefit–Cost Analysis

The incremental benefit–cost analysis compares the benefit–cost ratio determined for each of the alternatives under consideration incrementally to rank the projects. The steps involved are:

- The benefit–cost ratio (B/C) for each project under consideration is determined.
- All projects with (B/C) values higher than 1.0 are ranked with the project having the least cost ranked first.
- The difference between the benefits and costs of the first and second projects is determined. The ratio of these differences (difference of benefits/difference of costs) is determined. This is the incremental benefit–cost ratio between the first and second projects. If this value is greater than 1.0, then the second project (project with higher cost) is tentatively retained and the first project (project with lower cost) is set aside. If this value is less than 1.0, then the first project (project with lower cost) is tentatively retained and the second project (project with higher cost) is set aside.
- The selected project is then compared with the third initially ranked project (i.e., project with the third lowest cost) by determining the incremental benefit–cost ratio between the selected project above and the third project.
- This procedure is repeated until the last pair, and the selected project is considered as the most economically feasible.

The ranking of several projects can be obtained by removing each selected project from the list of projects and repeating the procedure. An example of the procedure is given in Chapter 13.

5.2.6 Safety Effectiveness Evaluation

This step involves the assessment of how the implementation of a particular safety countermeasure or a set of treatments or a project has affected crash frequency or severity. Results obtained from this assessment when a single countermeasure is implemented can also be used to determine a *CMF* for that treatment. Safety effectiveness evaluation can be conducted by using one of three study design methods: (1) observational before/after studies, (2) observational cross-sectional studies, and (3) experimental before/after studies. Observational studies are used to make conclusions on the impact of treatments that have been applied as part of a regular continuing effort to improve the road network, by analyzing and comparing the available crash data at sites with and without the treatments. Experimental studies, on the other hand, are conducted to determine the safety impact of treatments that are implemented to specifically study their effectiveness in reducing average crash frequency or severity. Experimental studies therefore require the selection of sites that have the potential for improvement and are randomly assigned for the treatment to be implemented (treatment group) or for no implementation of the treatment (comparison group). A detailed analysis of crash data collected over several years at the treatment and comparison sites before and after the implementation of the treatment is then conducted. The results of this analysis are used to determine whether there is any significant difference in the average crash frequency or severity between the treated and comparison sites. Although experimental studies are sometimes used in highway safety, they are used less often than observational studies. A detailed discussion of the computational procedures for these studies is beyond the scope of this book. However, two of these procedures are briefly discussed in Section 5.4.

5.3 PERFORMANCE MEASURES

As noted in Section 5.2.1, the HSM gives the following performance measures from which one or more can be used in the network screening process:

   (i) Average Crash Frequency
  (ii) Crash Rate
 (iii) Equivalent Property Damage Only (EPDO) Average Crash Frequency
  (iv) Relative Severity Index
   (v) Critical Rate
  (vi) Excessive Predicted Average Crash Frequency (Using Method of Moments)
 (vii) Level of Service of Safety
(viii) Excess Predicted Average Crash Frequency (Using Safety Performance Functions [SPFs])
  (ix) Expected Average Crash Frequency with Empirical Bayes (EB) Adjustment
   (x) EPDO Average Frequency with EB Adjustment
  (xi) Excess Expected Average Crash Frequency with EB Adjustment

5.3.1 Average Crash Frequency

A typical average crash frequency at a segment of road or facility is often computed over a period of three recent years. Unfortunately, this procedure does not account for any changes in geometric, traffic, or weather conditions that might have affected the rate of crash occurrence within the three-year period. A major error that is inherent in this performance measure is that average crash frequency obtained over a period of n years may not necessarily be the same for another period of n years. In order to account for this anomaly, one of the two following methods—(1) estimating crash frequency without assuming similar crash frequencies in all periods and (2) evaluating safety performance functions—can be used.

Estimating Crash Frequency without Assuming Similar Crash Frequency in All Periods

In this method, crash counts from other periods are incorporated in the computation of the average crash frequency using Eq. 5.15:

$$\hat{\mu}_Y = \sum_{y=1}^{Y} X_y \bigg/ \sum_{y=1}^{Y} d_y \tag{5.15}$$

and the estimate of the variance of $\hat{\mu}_Y$ is given as:

$$\hat{V}(\hat{\mu}_Y) = \sum_{y=1}^{Y} X_y \Big/ \left(\sum_{y=1}^{Y} d_y \right)^2 \tag{5.16}$$

where

$\hat{\mu}_Y$ = most likely estimate for u_Y (last period or year)

$\hat{\mu}_y = u_y \times d_y$, where y denotes a period or a year ($y = 1, 2, ..., Y$) and Y denotes the last period or last year; e.g., $d_1 = \dfrac{\mu_1}{\mu_Y}$

X_y = the number of crashes for each period or year y

Evaluating Safety Performance Functions

Safety performance functions (SPFs) are regression equations that relate average crash frequency for the selected year and specific base conditions at a site or road-way segment with one or more traffic and/or geometric independent variables. For example, the independent variables for a road segment are usually the *AADT* and the length of the segment, while those for an intersection are usually the *AADT*s on the minor and major approaches. SPFs are developed from available crash data covering many years at sites with similar roadway characteristics. The distribution of the crashes is usually assumed to be the negative binomial distribution as experience has shown that this distribution describes crash occurrence better than the Poisson distribution, which is appropriate when the mean and variance of the data are the same. When a set of data exhibits higher variance than the mean, the data are considered to be overdispersed. Crash data tend to exhibit overdispersion, which makes the negative binomial distribution more suitable. The extent of overdispersion is reflected in a parameter of the distribution known as the *overdispersion parameter (k)*, which is computed as part of the computation of the coefficients of the regression equation. The higher the *k* value, the more the crash data vary in comparison with the Poisson distribution with the same mean. The HSM gives SPFs for three-leg signalized inter-sections on urban and suburban arterial highways; for segments on rural two-lane highways, rural multilane highways, and urban and suburban arterial highways; and for four-leg signalized intersections and three- and four-leg intersections with stop control on minor legs. Procedures are provided in a predictive method that allows for categorizing the predicted crashes into different crash severity levels and colli-sion types when necessary. Tables 5.16 and 5.17 give the HSM SPFs for the different types of segments and intersections, respectively, and Tables 5.18 through 5.21 give the regression parameters (*a, b, c, d*) for intersection SPFs. It is, however, useful for states to develop their own SPFs as these may better reflect the conditions in an indi-vidual state. For example, experience has shown that even within a state, it may be necessary to develop more than one SPF for a given facility type because of the varied topographical conditions that exist in different parts of the state. However, the HSM provides a procedure where the given SPFs can be calibrated to local conditions by using calibration factors (see Section 5.4).

When average crash frequency is used for ranking, the site with the highest crash frequency is ranked as the first site, followed by the other sites in descending order based on their crash frequencies. Note that this procedure does not account for the effect of RTM or variations in geometric conditions.

Table 5.16 Safety Performance Functions for Rural Roadway Segments

Roadway Segment Type	SPF[c]	Overdispersion Parameter (k)	Equation #
Rural two-lane, two-way roadways[1,4]	$N_{spfrs} = (AADT) \times (L) \times (365) \times (10^{-6}) \times (e^{-0.312})$	0.236/L	**5.17**
Undivided rural multilane roadways[2,5,7]	$N_{spfru} = e^{(a+(b)\times(\ln AADT) + \ln L)}$	$k = \dfrac{1}{e^{(c+\ln(L))}}$	5.18
Divided rural multilane roadway segments[3,6,8]	$N_{spfru} = e^{(a+(b)\times(\ln AADT) + \ln L)}$	$k = \dfrac{1}{e^{(c+\ln(L))}}$	5.19

Note: SPF[c] gives the value of $N_{sp,i}$, which is the predicted total crash frequency for road segment i
  L = length of segment, km
  a, b, c = regression parameters (see Tables 5.18 and 5.19)
  [1]Base conditions for roadway segments on rural two-lane, two-way roads are
    Lane width (LW)   = 3.6 m
    Shoulder width (SW)   = 1.8 m
    Shoulder type   paved
  [2]Base conditions for undivided rural four-lane roadway segments are
    Lane width (LW)   = 3.6 m
    Shoulder width (SW)   = 1.8 m
    Shoulder type   paved
    Sideslopes   = 1V:7H or flatter
    Lighting   None
    Automated speed enforcement   None
  [3]Base conditions for divided rural multilane roadway segments are
    Lane width (LW)   = 3.6 m
    Shoulder width (SW)   = 2.4 m
    Median width   = 9 m
    Lighting   None
    Automated speed enforcement   None
  [4]*AADT* range for rural two-lane roadways is from zero to 17,800 veh/day.
  [5]*AADT* range for undivided rural four-lane roadways is from zero to 33,200 veh/day.
  [6]*AADT* range for divided rural four-lane roadways is from zero to 89,300 veh/day.
  [7]See Table 5.18 for values of $a, b,$ and c in Eq. 5.18 for undivided rural multilane segments.
  [8]See Table 5.19 for values of $a, b,$ and c in Eq. 5.19 for undivided rural multilane segments.
  SOURCE: Based on *Highway Safety Manual*, 1st Edition, 2010, AASHTO, Washington, D.C.

Table 5.17 Safety Performance Functions for Intersections

Intersection Type	SPF[10] (Total Predicted Average Crash Frequency)	Overdispersion Parameter (k)	Equation #
Three-leg stop-controlled intersections on rural two-lane two-way roads[1]	$N_{spf3sT} = e^{-9.86 + (0.79)\times \ln AADT_{maj} + (0.49)\times \ln AADT_{min}}$	0.54	5.20
Four-leg stop-controlled intersections on rural two-lane two-way roads[2]	$N_{spf4sT} = e^{-8.56 + (0.60)\times \ln AADT_{maj} + (0.61)\times \ln AADT_{min}}$	0.24	5.21
Four-leg signalized intersections on rural two-lane two-way roads[3]	$N_{spf4sG} = e^{-5.13 + (0.60)\times \ln AADT_{maj} + (0.20)\times \ln AADT_{min}}$	0.11	5.22

(Continued)

Table 5.17   Safety Performance Functions for Intersections (*continued*)

Three-leg stop-controlled intersections on rural multilane roads[4,7] and four-leg stop-controlled intersections on rural multilane roads[5,7]	$N_{spf3or4T} = e^{(a+b\times \ln AADT_{maj} + c\times \ln AADT_{min})}$	see Table 5.20	5.23
Four-leg signalized intersections on rural multilane highways[6,8]	$N_{spf3sT} = e^{(a+b\times \ln AADT_{maj} + c\times \ln AADT_{min})}$ or $N_{spf3sT} = e^{(a+d\times \ln AADT_{min})}$	see Table 5.21	5.24 5.25

Note: Equation 5.24 for four-leg signalized intersections on rural multilane highways is suitable for only generalized predictions as there are no associated base conditions for it. Also, Eq. 5.25 is suitable for only fatal and injury crashes (excluding possible injuries) on four-leg signalized intersections.

[1]$AADT_{maj}$ range is from zero to 19,500 veh/day, $AADT_{min}$ range from zero to 4300 veh/day.
[2]$AADT_{maj}$ range is from zero to 14,700 veh/day, $AADT_{min}$ range is from zero to 3500 veh/day.
[3]$AADT_{maj}$ range is from zero to 25,200 veh/day, $AADT_{min}$ range is from zero to 3500 veh/day.
[4]$AADT_{maj}$ range is from zero to 78,300 veh/day, $AADT_{min}$ range is from zero to 23,000 veh/day.
[5]$AADT_{maj}$ range is from zero to 78,300 veh/day, $AADT_{min}$ range is from zero to 7400 veh/day.
[6]$AADT_{maj}$ range is from zero to 43,300 veh/day, $AADT_{min}$ range is from zero to 18,500 veh/day.
[7]See Table 5. 20 for regression parameter values for *a*, *b*, and *c* in Eq. 5.23.
[8]See Table 5. 21 for regression parameter values for *a*, *b*, *c*, and *d* in Eqs. 5.24 and 5.25.
[10] Base conditions for three- and four-leg stop controlled intersections are
    Intersection skew angle = 0°
    Intersection left-turn lanes 0, except for stop controlled approaches
    Intersection right-turn lanes 0, except for stop controlled approaches
    Lighting   None
SOURCE: Based on *Highway Safety Manual*, 1st Edition, 2010, AASHTO, Washington, D.C.

Table 5.18   SPF Coefficients for Total and Fatal-and-Injury Crashes on Undivided Roadway Segments (for Use in Equation 5.18)

Crash Severity Level	a	b	c
4-lane total	−9.653	1.176	1.675
4-lane fatal and injury	−9.410	1.094	1.796
4-lane fatal and injury[a]	−8.577	0.938	2.003

[a] Using KABCO scale, these include only KAB crashes. Crashes with severity level C (possible injury) are not included.
SOURCE: Based on *Highway Safety Manual*, 1st Edition, 2010, AASHTO, Washington, D.C.

Table 5.19   SPF Coefficients for Total and Fatal-and-Injury Crashes on Divided Roadway Segments (for Use in Equation 5.19)

Crash Severity Level	a	b	c
4-lane total	−9.025	1.049	1.549
4-lane fatal and injury	−8.837	0.958	1.687
4-lane fatal and injury[a]	−8.505	0.874	1.740

[a]Using KABCO scale, these include only KAB crashes. Crashes with severity level C (possible injury) are not included.
SOURCE: Based on *Highway Safety Manual*, 1st Edition, 2010, AASHTO, Washington, D.C.

Table 5.20   SPF Coefficients for Three- and Four-Leg Intersections with Minor Road Stop Control for Total and Fatal-and-Injury Crashes (for Use in Eq. 5.23)

Intersection Type/severity Level	a	b	c	Overdispersion Parameter (Fixed k)[a]
Four-leg stop control (4ST) Total	−10.008	0.848	0.448	0.494
Four-leg stop control (4ST) Fatal and Injury	−11.554	0.888	0.525	0.742
Four-leg stop control (4ST) Fatal and Injury[b]	−10.734	0.828	0.412	0.655
Three-leg stop control (3ST) Total	−12.526	1.204	0.236	0.460
Three-leg stop control (3ST) Fatal and Injury	−12.664	1.107	0.272	0.569
Three-leg stop control (3ST) Fatal and Injury[b]	−11.989	1.013	0.228	0.566

[a]This value should be used directly as the overdispersion parameter; no further computation is required.
[b]Using KABCO scale, these include only KAB crashes. Crashes with severity level C (possible injury) are not included.
SOURCE: Based on *Highway Safety Manual*, 1st Edition, 2010, AASHTO, Washington, D.C.

Table 5.21   SPF Coefficients for Four-Leg Signalized Intersections for Total and Fatal and Injury Crashes (for Use in Equations 5.24 and 5.25

Intersection Type/severity Level	a	b	c	d	Overdispersion Parameter (Fixed k)[a]
Four-leg signalized intersections on rural multilane highways (Total Crashes)	−7.182	0.722	0.337		0.277
Four-leg signalized intersections on rural multilane highways (Fatal and Injury Crashes)	−6.393	0.638	0.232		0.218
Four-leg signalized intersections on rural multilane highways (Fatal and Injury Crashes)[b]	−12.011			1.279	0.566

[a]This value should be used directly as the overdispersion parameter; no further computation is required.
[b]Using KABCO scale, these include only KAB crashes. Crashes with severity level C (possible injury) are not included.
SOURCE: Based on *Highway Safety Manual*, 1st Edition, 2010, AASHTO, Washington, D.C.

Note that the SPFs that are computed by using the information given in Tables 5.16 through 5.21 are applicable only to the base conditions shown below Tables 5.16 and 5.17. When conditions at a particular site are different from the base conditions, the average predicted crash frequency should be corrected by applying an appropriate *CMF*. The HSM gives *CMFs* for a wide range of different features.

Example 5.4 Use of Safety Performance Functions

A four-lane undivided rural roadway with the same conditions as the base conditions for its SPF is 3 mi (4.8 km) long and carries an $AADT$ of 15,000 vehicles. Determine:

  (i) the total predicted average crash frequency on the segment
 (ii) the fatal and injury (KABC) predicted average crash frequency on the segment
(iii) the overdispersion parameters for (i) and (ii)

Solution:

Determine correct SPFs (see Eq. 5.18 and Table 5.17); the appropriate SPFs are given as:

$$N_{spfru} = e^{(-9.653 + 1.176 \times \ln(AADT) + \ln(L))} \quad \text{for total crashes and}$$

$$N_{spfru} = e^{(-9.410 + 1.094 \times \ln(AADT) + \ln(L))} \quad \text{for fatal and injury crashes}$$

and the associated expression for the overdispersion factor is:

$$k = \frac{1}{e^{(1.675 + \ln(L))}} \quad \text{for total crashes}$$

$$k = \frac{1}{e^{(1.796 + \ln(L))}} \quad \text{for fatal and injury crashes}$$

where

$$N_{spfru} = \text{base total expected average crash frequency for a roadway segment}$$
$$AADT = \text{annual average daily traffic}$$
$$L = \text{length (mi)}$$
$$k = \text{overdispersion parameter}$$

Total predicted average crash frequency is:

$$N_{spfru} = e^{(-9.653 + 1.176 \times \ln(AADT) + \ln(L))}$$

$$N_{spfru} = e^{(-9.653 + 1.176 \times \ln(15,000) + \ln(3))}$$

$$N_{spfru} = e^{(-9.653 + 1.176 \times 9.62 + 1.099)}$$

$$N_{spfru} = e^{(-9.653 + 11.31 + 1.099)}$$

$$= e^{2.756} = 15.739,$$

Determine fatal and injury predicted average crash frequency:

$$N_{spfru} = e^{(-9.410 + 1.094 \times \ln(AADT) + \ln(3))}$$

$$N_{spfru} = e^{(-9.410 + 1.094 \times \ln(15,000) + \ln(3))}$$

$$N_{spfru} = e^{(-9.410 + 1.094 \times (9.62) + (1.099))}$$

$$N_{spfru} = e^{2.213} = 9.14$$

Determine overdispersion parameter:

$$k = \frac{1}{e^{(1.675 + \ln (L))}}$$

$$k = \frac{1}{e^{(2.774)}} = 0.062 \text{ for total crashes}$$

$$k = \frac{1}{e^{(1.796 + 1.099)}} = 0.055 \text{ for fatal and injury crashes}$$

Note that the expected crash frequency determined does not have to be adjusted as the problem assumes that the road segment has conditions that are the base condition. Otherwise, the value obtained from the base condition will have to be adjusted by incorporating the appropriate *CMFs*. See Example 5.5.

Example 5.5 Use of SPF to Determine Predicted Crash Frequency on a Two-Lane, Two-Way Roadway Segment

Determine the predicted average crash frequency on a two-lane, two-way rural segment 3 mi (4.8 km) long with the following characteristics:

$$AADT = 1800 \text{ veh/day}$$

$$\text{Lane width} = 3.3 \text{ m}$$

$$\text{Shoulder width} = 1.8 \text{ m}$$

$$\text{Shoulder type}\quad \text{Paved}$$

Solution:

Step 1. Compute the expected total crash frequency. Use the appropriate SPF, which is Eq. 5.17 and given as

$$N_{spfrs} = (AADT) \times (L) \times (365) \times (10^{-6}) \times (e^{-0.312})$$

$$N_{spfrs} = (1800) \times (3) \times (365) \times (10^{-6}) \times (e^{-0.312})$$

$$= 1800 \times 3 \times 365 \times 10^{-6} \times 0.732 = 1.44 \text{ crashes/yr}$$

Step 2. Correct for nonbase conditions. The lane width of the segment is 11 ft and not 12 ft. Use Eq. 5.4 in Table 5.14.

$$CMF_{ra} = 1.01 + 2.5 \times 10^{-5}(AADT - 400)$$

$$= 1.01 + 2.5 \times 10^{-5}(1800 - 400)$$

$$= 1.045$$

Step 3. Correct CMF_{ra} for total crashes. Use Eq. 5.14.

$$CMF_{lr} = (CMF_{ra} - 1)p_{ra} + 1.0$$

where

CMF_{lr} = crash modification for the effect of lane width on total crashes

CMF_{ra} = crash modification for the effect of lane width on single-vehicle run-off-the-road and/or related crashes on two-lane roads.

p_{ra} = proportion of total crashes constituted by single-vehicle run-off-the-road and/or related crashes on two-lane roads. HSM default value is 0.574, but a value based on actual local data is preferable when available.

Since the actual proportion of total crashes constituted by related crashes is not known, use the default value of 0.574.

$$CMF_{lr} = (CMF_{ra} - 1)p_{ra} + 1.0 = (1.045 - 1) \times 0.574 + 1 = 1.026$$

Step 4. Compute predicted average crash frequency.

Predicted average total crash frequency = $1.44 \times 1.026 = 1.477$

5.3.2 Crash Rates

Crash rates are determined on the basis of exposure data, such as traffic volume and the length of road section being considered. Commonly used rates are rate per million of entering vehicles (RMEVs) for intersections and rate per 100 million vehicle miles (RMVK) for road segments.

The **rate per million of entering vehicles** (RMEVs) is the number of crashes per million vehicles entering the intersection during the study period. It is expressed as

$$RMEV = \frac{A \times 1,000,000}{V} \tag{5.26}$$

where

$RMEV$ = crash rate per million entering vehicles

A = number of crashes, total or by type occurring in a single year at the location

V = average daily traffic $(ADT) \times 365$

This rate is often used as a measure of crash rates at intersections.

Example 5.6 Computing Crash Rates at Intersections

The number of all crashes recorded at an intersection in a year was 23, and the average 24-hr volume entering from all approaches was 6500. Determine the crash rate per million entering vehicles (RMEV).

Solution:

$$RMEV = \frac{23 \times 1,000,000}{6500 \times 365} = 9.69 \, \text{crashes/million entering vehicles}$$

The **rate per 100 million vehicle km** (RMVK) is the number of crashes per 100 million vehicle km of travel. It is obtained from the expression:

$$RMVK = \frac{A \times 100{,}000{,}000}{VKT} \tag{5.27}$$

where

A = number of crashes, total or by type at the study location, during a given period

VKT = vehicle km of travel during the given period

= $ADT \times$ (number of days in study period) $\times$ (length of road)

This rate is often used as a measure of crash rates on a stretch of highway with similar traffic and geometric characteristics.

Example 5.7 Computing Crash Rates on Roadway Sections

It is observed that 40 traffic crashes occurred on a 17.5-km-long section of highway in one year. The ADT on the section was 5000 vehicles.

(a) Determine the rate of total crashes per 100 million vehicle-km.

(b) Determine the rate of fatal crashes per 100 million vehicle-km if 5% of the crashes involved fatalities.

Solution:

(a) $RMVK_T = \dfrac{40 \times 100{,}000{,}000}{17.5 \times 5000 \times 365} = 125.24$ crashes/100 million veh-km

(b) $RMVK_F = 125.24 \times 0.05 = 6.26$ fatal crashes/100 million veh-km

---

Note that any crash rate may be given in terms of the total number of crashes occurring or in terms of a specific type of crash. Therefore, it is important that the basis on which crash rates are determined is clearly stated. A major deficiency in using crash rates is the inherent assumption of a linear relationship between crashes and the exposure data used, which is often not the case. Also, although the use of crash rates considers the effect of an exposure, it does not take into consideration other factors, usually referred to as confounding factors, that may affect the occurrence of crashes. Research has also shown that it tends to be biased toward low volume sites. Care should therefore be taken in making conclusions by simply comparing crash rates.

When crash rates are used for ranking, the site with the highest crash rate is ranked as the first site, followed by the other sites in descending order based on their crash rates.

5.3.3 Equivalent Property Damage Only (EPDO) Average Crash Frequency

A single value of crash frequency is determined by assigning weighting factors to the different crash severity levels relative to property damage only (PDO) crashes. A common basis for determining the weights is by determining the ratio of the societal cost of each crash to that of a PDO crash. The Federal Highway Administration has suggested the following societal costs:

(i)   Fatal (K)   = \$4,008,900
(ii) Injury crashes (A/B/C)   = \$82,600
(iii) PDO (O)   = \$7,400

Based on these costs, the weighting factors are

(i)   Fatal = 4,008,900/7400   = 541.74, say 542
(ii) Injury = 82,600/7400   = 11.162, say 11
(iii) PDO   = 7400/7400   = 1

Where the state or local jurisdiction has its own societal costs, it is preferable for these to be used. It can be seen from the weights computed that a single fatal crash at a location will surpass many injury crashes at another location. A combined weight for fatal and injury crashes is therefore sometimes used to avoid overemphasizing fatal crashes. When equivalent property damage only (EPDO) average crash frequency is used for ranking, the site with the highest EPDO score is ranked as the first site, followed by the other sites in descending order based on their EPDO scores.

Example 5.8 Computing the EPDO Score at a Site.

The severity and number of associated crashes at an intersection during a year are:

$$\text{Fatal crashes} = 1$$

$$\text{Injury crashes} = 4$$

$$\text{PDO crashes} = 6$$

Determine the EPDO at the site, using the weights given above.

Solution:

Determine EPDO score:

$$\text{EPDO score} = 1 \times 542 + 4 \times 11 + 6 \times 1 = 592$$

5.3.4 Relative Severity Index (RSI)

The relative severity index (RSI) at a site is based on specific social costs that have been developed by a jurisdiction for each crash type at a location. The RSI cost at a site is obtained by first multiplying the number of crashes in each type at the location by the social cost for that type of crash. The sum of these costs is then divided by the number of crashes at the site to determine the RSI for the site. Table 5.22 gives suggested crash costs for different types of crashes. The limitations of the RSI include: (1) does not account for regression to the mean bias; (2) does not consider the effect of traffic volume; (3) there is a possibility of overemphasizing sites with a small number of severe crashes, which depends on the social costs assigned to the different crash types; and (4) there is a possibility of wrongly assigning high priority to sites with low traffic volume, but at which a high number of noninjury crashes occur.

Table 5.22   Crash Cost Estimates by Crash Type

Crash Type	Crash Cost (2001 Dollars)
Rear-End, Signalized Intersection	$26,700
Rear-End, Unsignalized Intersection	$13,200
Sideswipe/Overtaking	$34,000
Angle, Signalized Intersection	$47,300
Angle, Unsignalized Intersection	$61,100
Pedestrian/Bike at an Intersection	$158,900
Head-On, Signalized Intersection	$24,100
Head-On, Unsignalized Intersection	$47,500
Fixed Object	$94,700
Other/Undefined	$55,100

SOURCE: *Crash Cost Estimates by Maximum Police-Reported Injury Severity within Selected Crash Geometrics*, FHWA-HRT-05-051, October 2005. United States Department of Transportation - Federal Highway Administration.

When the average RSI is used for ranking, the average RSI cost for all sites within the population (e.g., rural two-lane roads, rural four-leg intersections, undivided multilane highways) is first determined by dividing the sum of RSIs for all sites within the population by the total number of crashes that occurred at all sites within the group. Sites with RSIs higher than the average RSI are considered for further study and are ranked in order, starting with the highest RSI, followed by the other sites in descending order based on their RSIs.

Example 5.9 Computing the RSI for a Site

The number of observed crashes by type at a signalized intersection during a period of three years is shown below. Determine the RSI for the site using the appropriate costs from Table 5.22.

Rear-end crashes—15

Sideswipe crashes—10

Angle crashes—7

Solution:

- Identify appropriate crash costs.

Rear-end crash at a signalized intersection = $26,700

Sideswipe crash = $34,000

Angle crash at a signalized intersection = $47,300

Determine total cost of all crashes.

Total cost = $(15 \times 26,700 + 10 \times 34,000 + 7 \times 47,300)$

= $1,071,600

Determine RSI for site.

$$RSI = 1,071,600/(15 + 10 + 7)$$

$$= \$33,487.50$$

Example 5.10 Ranking by Average Relative Severity Index

The number of observed crashes by type at three rural four-leg signalized intersections during a period of three years is shown below. Using the RSI and the appropriate associated costs given in Table 5.22, determine the ranking order for the intersections.

Intersection A	Intersection B	Intersection C
Rear-end crashes—8	Rear-end crashes—5	Rear-end crashes—10
Angle—12	Angle crashes—15	Angle crashes—5
Head-on—3	Pedestrian—4	Sideswipe—9

Solution:

Compute RSI for each intersection.

$$\text{Intersection A, RSI} = \$(8 \times 26{,}700 + 12 \times 47{,}300 + 3 \times 24{,}100)/23$$

$$= \$37{,}108.70$$

$$\text{Intersection B, RSI} = \$(5 \times 26{,}700 + 15 \times 47{,}300 + 4 \times 158{,}900)/24$$

$$= \$61{,}608.33$$

$$\text{Intersection C, RSI} = \$(10 \times 26{,}700 + 5 \times 47{,}300 + 9 \times 34{,}000)/24$$

$$= \$33{,}729.17$$

Rank intersections based on the RSIs.

Intersection B is ranked first.
Intersection A is ranked second.
Intersection C is ranked third.

Note that although the number of crashes that occurred over the three years at each intersection is about the same as the others, the RSIs are significantly different because of the severities of the crashes at the different intersections.

5.3.5 Critical Crash Rate Factor

The critical rate method incorporates the traffic volume to determine if the crash rate at a particular location is significantly higher than its critical crash rate which is based on the average for the type of facility. Statistics are typically maintained by facility type, which is determined by factors such as traffic volume, traffic control, number of lanes, land-use density, and functional classification (see Chapter 15). The procedure involves selecting a number of sites similar to the site under consideration in terms of type, traffic volume, and geometric characteristics. The average crash rate (AVR_a) for all the selected sites is determined and the critical crash rate (CR_i) for the site is determined from Eq. 5.28:

$$CR_i = AVR_a + \frac{0.5}{MEV_i} + P\sqrt{\frac{AVR_a}{MEV_i}} \tag{5.28}$$

where

CR_i = critical crash rate for site i, per 100 million vehicle-km for segments or per million entering vehicles for intersections

MEV_i = traffic base, per 100 million vehicle-km for segments or million entering vehicles for intersections

$$= \frac{(n) \times (AADT_i) \times (L_i) \times 365}{100,000,000} \text{ (for roadway segments)} \qquad (5.29)$$

$$= \frac{(n) \times (TEV_i) \times 365}{1,000,000} \text{ (for intersections)} \qquad (5.30)$$

n = number of years for which crash data were obtained
TEV_i = total number of vehicles entering the intersection i per day
$AADT_i$ = average annual daily traffic on a roadway segment
AVR_a = weighted average crash rate for the facility type

$$= \frac{\sum_{i=1}(AADT_i) \times (L_i) \times (R_i)}{\sum_{i=1}(AADT_i) \times (L_i)} \text{ (for roadway segments)} \qquad (5.31)$$

$$= \frac{\sum_{i=1}TEV_i \times (R_i)}{\sum_{i=1}TEV_i} \text{ (for intersections)} \qquad (5.32)$$

L_i = segment length (km)
R_i = number of crashes observed at site i
P = p value for the corresponding confidence level
  = 1.036 for 85% confidence level
  = 1.282 for 90% confidence level
  = 1.645 for 95% confidence level
  = 2.326 for 99% confidence level
  = 2.576 for 99.5% confidence level

The ratio of the crash rate at the site i to the critical crash rate for the road or facility type (CCR_i) is then determined. If this ratio is greater than 1.00, the site is selected for further review. When the critical crash rate is used for ranking, the site with the highest CCR among those selected for further review is ranked as the first site, followed by the other sites in descending order based on their CCR values. The method of critical crash rate also has the inherent associated deficiency in that only the number of crashes, or crash rates, is used to make conclusions on safety, as discussed in the previous section.

Example 5.11 Using the Critical Crash Rate Procedure to Determine Whether a Site Should be Selected for Further Review

An urban arterial street segment 0.2 km long has an average annual daily traffic ($AADT$) of 15,400 veh/day. In a three-year period, there have been 8 crashes resulting in death and/or injuries and 15 involving property damage only. The statewide average crash experience for similar types of roadway is 375 per 100 million vehicle km (mvk) for a three-year period, of which 120 involved death and/or injury and 255 caused property damage only. Should the 0.2-km-long street be considered for further review? Assume that a single death/injury crash is equivalent to three property damage crashes. Use a 95 percent confidence level.

Solution:

Step 1. Calculate the traffic base, MEV_i. Use Eq. 5.8.

$$MEV_i = \frac{(n) \times (AADT_i) \times (L_i) \times 365}{100,000,000}$$

$$= \frac{(3) \times (15,400) \times (0.2) \times 365}{100,000,000}$$

$$= 0.0337/100 \text{ mvk}$$

Step 2. Calculate the 3-year average crash rate (AVR_a) for this type of facility.

$$AVR_a = 3 \times 120 + 255 = 615 \text{ equivalent PDO crashes per 100 mvm per year}$$

Step 3. Select a p value based on confidence level. Since a confidence level of 95 percent is specified, the test factor is 1.645.

Step 4. Compute the critical rate.

$$CR = AVR_a + \frac{0.5}{MEV_i} + TF\sqrt{\frac{AVR_a}{MEV_i}}$$

$$= 615 + \frac{0.5}{0.0337} + 1.645\sqrt{\frac{615}{0.0337}}$$

$$= 852.1 \text{ EPDO crashes per 100 mvk}$$

Step 5. Determine the actual crash rate for the segment.

$$\text{Crash rate for the segment} = \frac{3 \times 8 + 15}{0.0337}$$

$$= 1157 \text{ equivalent PDO crashes/100 mvk}$$

Step 6. Determine critical crash ratio.

$$\text{Critical crash ratio} = 1157/895 = 1.36.$$

Since this ratio is greater than 1.00, the segment should be selected for review.

5.3.6 Excess Predicted Average Crash Frequency (Using Method of Moments)

This procedure includes an attempt to account for regression to the mean (RTM). In conducting this procedure all sites under consideration are first formed into groups or clusters based on the facility type, location, or other defining characteristics (e.g., intersection types, suburban/urban, four-lane divided segments, four-lane undivided segments). Then the number of crashes that occurred at each site over a selected time period is recorded. The mean crash frequency for each cluster or group of sites is then determined as:

$$N_{observed,rp} = \frac{\sum_{i=1}^{n} N_{observed,i}}{n_{sites}}$$

(5.33)

where

$N_{observed,\,rp}$ = average crash frequency, per reference population
$N_{observed,\,i}$ = observed crash frequency at site i
$n_{(sites)}$ = number of sites in the reference cluster or group population

The variance of the crashes within each cluster or group is then determined as:

$$Var(N) = \frac{\sum \left(N_{observed,\,i} - N_{observed,\,rp} \right)^2}{n_{sites} - 1}$$

(5.34)

where

$Var(N)$ = variance of crash frequencies within each cluster
$N_{observed,\,rp}$ = average crash frequency, per reference population
$N_{observed,\,i}$ = observed crash frequency at site i
$n_{(sites)}$ = number of sites in the reference cluster or group population

Next, an adjusted crash frequency for each site in a group or cluster is determined using the variance of crash frequencies in the group, and the difference between the mean of the observed frequencies in a group and the observed frequency at the site. Using Eq. 5.35,

$$N_{observed,i,(adj)} = N_{observed,i} + \frac{N_{observed,rp}}{Var(N)} \times (N_{observed,rp} - N_{observed,i})$$

(5.35)

where

$N_{observed,\,i,(adj)}$ = adjusted observed crash frequency at site i
$Var(N_{observed,\,i})$ = variance of crash frequencies in a reference site
$N_{observed,\,rp}$ = average crash frequency, per reference population
$N_{observed,\,i}$ = observed crash frequency at site i

The potential for improvement at a site (PI_i) is then determined as the difference between the adjusted crash frequency at the site and the mean of crash frequencies in the group or cluster, and given as

$$PI_i = N_{observed,i,\,(adj)} - N_{observed,\,rp}$$

(5.36)

where

PI_i = potential of improvement at site i
$N_{observed,\,rp}$ = average crash frequency, per reference population
$N_{observed,\,i\,(adj)}$ = adjusted observed crash frequency at site i

When the excessive predicted average crash frequency (using the method of moments) is used for ranking, the site with highest PI is ranked as the first site, followed by the other sites in descending order of their *PI* values. Note that although the variance of crash frequency in each group is incorporated in the procedure, there is no guarantee that the effect of RTM is fully accounted for.

Example 5.12 Using Potential for Improvement Computed from Excess Predicted Average Crash Frequency (Using Method of Moments) to Rank Sites.

Column 3 of Table 5.23 gives two sets of data on total crash frequencies over a period of three years on undivided urban four-lane segments and divided urban four-lane segments. Determine the potential for improvement *(PI)* at each site and rank the sites based on their *PI* values. Use the potential for improvement computed from excess predicted average crash frequency procedure. Note that a negative value could be obtained for a *PI*. This indicates a low potential for crash reduction.

Table 5.23   Data and Solution for Example 5.12

Segment ID (1)	Segment Type (2)	Observed Crash Frequency (3)	Variance of Crash Frequencies in a Group (4)	Adjusted Observed Crash Frequency (5)	Potential for Improvement (6)
1	undivided urban 4-leg segment	10		12.39	1.56
2		12		8.64	−2.19
3		9	3.767	14.26	3.43
4		9		14.26	3.43
5		14		4.89	−5.94
6		11		10.51	−0.32
7	divided urban 4-leg segment	7		9.32	0.49
8		8		9.06	0.22
9		11	6.967	8.25	−0.58
10		10		8.52	−0.31
11		5		9.86	1.03
12		12		7.99	−0.84

Solution:

Determine mean crash frequency for each group of data.

$$\text{For undivided urban 4-leg segment} = \frac{10 + 12 + 9 + 9 + 14 + 11}{6}$$

$$= 10.83$$

$$\text{For divided urban 4-leg segment} = \frac{7 + 8 + 11 + 10 + 5 + 12}{6}$$

$$= 8.83$$

Determine crash variance for each group.

For undivided urban 4-leg segment:

$$\frac{(10 - 10.83)^2 + (12 - 10.83)^2 + (9 - 10.83)^2 + (9 - 10.83)^2 + (14 - 10.83)^2 + (11 - 10.83)^2}{6 - 1}$$

$$= \frac{0.689 + 1.369 + 3.349 + 3.349 + 10.049 + 0.029}{5}$$

$$= 18.834/5 = 3.767$$

For divided urban 4-leg segment:

$$= \frac{(7 - 8.83)^2 + (8 - 8.83)^2 + (11 - 8.83)^2 + (10 - 8.83)^2 + (5 - 8.83)^2 + (12 - 8.83)^2}{6 - 1}$$

$$= \frac{3.349 + 0.689 + 4.709 + 1.369 + 14.669 + 10.049}{6 - 1}$$

$$= 34.83/5 = 6.967$$

Determine adjusted observed crash frequency for each site.
   Use Eq. 10.35.

$$N_{observed,\,i,\,(adj)} = N_{observed,\,i} + \frac{N_{observed,\,rp}}{\text{Var(N)}} \times (N_{observed,\,rp} - N_{observed,\,i})$$

For example, for site 5:

$$N_{observed,\,i,\,(adj)} = 14 + \frac{10.83}{3.767} \times (10.83 - 14)$$

$$= 14 - 9.11 = 4.89$$

See column 5 of Table 5.23 for adjusted crash frequency for other sites.
Determine PI for each site.
   Use Eq. 5.36.

$$PI_i = N_{observed,\,i,\,(adj)} - N_{observed,\,rp}$$

For example, for site 5:

$$PI_5 = 4.89 - 10.83 = -5.94$$

See column 6 of Table 5.23 for *PI* values for other sites.

Based on the *PI* values obtained, the ranking is as given in the following table:

Rank	Segment Number	PI	Rank	Segment Number	PI
1	3	3.43	7	10	−0.31
1	4	3.43	8	6	−0.32
3	1	1.56	9	9	−0.58
4	11	1.03	10	12	−0.84
5	7	0.49	11	2	−2.19
6	8	0.22	12	5	−5.94

5.3.7 Level of Service of Safety (LOSS)

This procedure given in the HSM involves the comparison of the observed crash frequency of each site under consideration with the predicted average crash frequency of all the sites being considered, obtained from an appropriate SPF. This comparison is carried out by establishing four levels of service of safety that are based on deviations from the predicted average crash frequency using the standard deviation of the predicted crashes. The standard deviation is based on the overdispersion coefficient and the predicted average crash frequency. Each site is then assigned to one of these levels based on its observed crash frequency. The levels of safety are given as follows:

$$\sigma < N_{observed} < (N_{predicted} - 1.5 \times (\sigma)) \text{ indicates a low potential for crash reduction}$$

$$(N_{predicted} - 1.5 \times (\sigma)) \leq N_{observed} < N \text{ indicates low to moderate potential for crash reduction}$$

$$N_{predicted} \leq N_{observed} < (N_{predicted} + 1.5 \times (\sigma)) \text{ indicates moderate to high potential for crash reduction}$$

$$N_{observed} \geq (N_{predicted} \times 1.5(\sigma) \text{ indicates a high potential for crash reduction}$$

where

σ = standard deviation

$\sigma = \sqrt{(k)N^2_{predicted}}$ This form for standard deviation is valid as the SPF assumes a Negative binomial distribution.

k = overdispersion parameter of the SPF

$N_{predicted}$ = predicted average crash frequency from SPF

When LOSS is used for ranking, the number of crashes observed at each site is compared with the limits of each LOSS category to determine in which category the sites fall. Sites that fall within category IV are then ranked first followed with sites ranked in category III and so on. Note that if an SPF in the HSM is to be used to obtain the predicted average crash frequency, the procedure given in Section 5.4 should be incorporated to account for the specific site conditions.

Example 5.13 Ranking of Intersections by Level of Service of Safety

The observed average crash frequency and associated AADTs on a four-leg stop-controlled intersection on a rural multilane road are shown below. Assuming that the conditions at the intersection are the same as the base conditions for the appropriate SPF, determine the LOSS category for the intersection.

Observed average crash frequency (total crashes) = 13.5 crashes/yr

$$AADT_{major} = 20,000 \text{ veh/day}$$

$$AADT_{minor} = 5500 \text{ veh/day}$$

Solution

Estimate predicted average crash frequency from SPF.
The appropriate SPF is Eq. 5.23, given as

$$N_{spf3or4T} = e^{(a+b\times \ln AADT_{maj}+c\times \ln AADT_{min})}$$

Values of $a, b,$ and c are obtained from Table 5.20 as

$$a = -10.008$$
$$b = 0.848$$
$$c = 0.448$$

$$N_{spf4T} = e^{(-10.008+0.848\times \ln AADT_{maj}+0.448\times \ln AADT_{min})}$$

$$N_{spf4T} = e^{(-10.008+8.398+3.86)} = 9.48 \text{ crashes/yr}$$

Compute standard deviation using

$$\sigma = \sqrt{(k)N^2_{predicted}} = \sigma = \sqrt{(0.494)9.48^2}$$
$$= 6.66$$

Compute limits for LOSS categories.
  LOSS I limit is 0 to $9.48 - 1.5 \times 6.66 = -0.51$, say 0
  LOSS II limit is 0 to 9.48
  LOSS III limit = 9.48 to $9.48 + 1.5 \times 6.66$; i.e., 9.48 to 19.47
  LOSS IV limit = > 19.47
Since the observed average crash frequency is 13.5, the intersection is in LOSS III category and therefore has a moderate to high potential for crash reduction.

5.3.8 Excess Predicted Average Crash Frequency Using Safety Performance Functions (SPFs)

This procedure given in the HSM involves determining the difference between the observed average crash frequency at a site and the predicted crash frequency obtained from an appropriate SPF for the site. First, the sites under consideration are categorized with respect to their characteristics (e.g., four-leg signalized intersections, unsignalized intersections, two-way stop-controlled intersections, etc.); then a summary table is prepared with the number of crashes at each site indicated. The average observed crash frequency is then determined for each referenced category using all years with available data. An appropriate SPF is then used to compute the predicted crash frequency for each category of site. The difference between the observed crash frequency at each site and the predicted crash frequency for the site's category is determined. This difference is referred to as the excess predicted average crash frequency. When the excess predicted average crash frequency is used for ranking, sites in each reference population are ranked with the site having the largest excess predicted average crash frequency listed first, followed by the others in decreasing values of the excess predicted average crash frequencies. Note that this process does not guarantee the elimination of RTM effects.

5.3.9 Expected Average Crash Frequency with Empirical Bayes (EB) Adjustment

The procedure described here is similar to that given in the HSM but assumes that the conditions at the site are the same as those for the SPF used or that all necessary adjustments have been made as described in Section 5.4 under predictive methods. However, the procedure incorporates yearly correction factors to be consistent with the procedure for network screening in the safety analysis software tools. The procedure consists of the following steps:

   (i)   Compute the predicted average crash frequency from an SPF.
  (ii)   Compute the annual correction factor.
 (iii)   Compute the empirical Bayes weight.
  (iv)   Compute the first-year EB-adjusted expected average crash frequency.
   (v)   Compute the final-year EB-adjusted expected average crash frequency.
  (vi)   Compute the variance of the EB-adjusted average crash frequency.

Each step is briefly described in the following sections.

Step (i): Compute the Predicted Average Crash Frequency from an SPF

In this step, an appropriate SPF for the site is selected and the average crash frequency $N_{predicted}$ for each year n ($n = 1, 2, 3, \ldots Y$) computed. Note that an appropriate SPF can be one given in the HSM for which the prediction adjustments will have to be carried out as discussed in Section 5.4 or one developed using locally available data. For example, the SPF given in the HSM for undivided rural four-lane roadway segments is

$$N_{spfru} = e^{(a + b \times \ln(AADT) + \ln(L))} \quad \text{(see Table 5.16)}$$

This SPF is applicable only to undivided roadway segments on rural multilane highways with the following base conditions:

Lane width	= 3.6 m
Shoulder width	= 1.8 m
Shoulder type	= paved
Sideslopes	= 1V:7H or flatter
Lighting	None
Automated speed enforcement	None

When this SPF is used to predict average crash frequency at a rural multilane segment with one or more different characteristics than those shown above, the prediction should be adjusted as shown in Eq. 5.53 of Section 5.4.

Step (ii): Compute the Annual Correction Factor (C_n)

The annual correction factor is used to adjust for the effect on crash occurrence of variation in traffic, weather, and vehicle mix. It is obtained as

$$C_{n(total)} = \frac{N_{predicted,n(total)}}{N_{predicted,n1(total)}} \qquad (5.37)$$

$$C_{n(FI)} = \frac{N_{predicted,n(FI)}}{N_{predicted,n1(FI)}} \qquad (5.38)$$

where

$C_{n(total)}$ = annual correction factor for total crashes
$N_{predicted,\,n(total)}$ = predicted number of total crashes for year n
$N_{predictedn\,1(total)}$ = predicted number of total crashes for year 1
$C_{n(FI)}$ = annual correction factor for fatal and injury crashes
$N_{predicted,\,n(FI)}$ = predicted number of total fatal and injury crashes for year n
$N_{predictedn\,1(FI)}$ = predicted number of total fatal and injury crashes for year 1

Step (iii): Compute the Empirical Bayes Weights

In this step, the empirical Bayes weight for each crash type and/or severity is computed using the overdispersion of the SPF and the number of predicted crashes for year n. These weights are given as:

$$w_{total} = \frac{1}{1 + k_{total} \times \sum_{n=1}^{Y} N_{predicted,n(total)}} \quad \text{or} \qquad (5.39)$$

$$w_{FI} = \frac{1}{1 + k_{FI} \times \sum_{n=1}^{Y} N_{predicted,n(FI)}} \qquad (5.40)$$

where

w_{total} = empirical Bayes weight for total crashes
w_{FI} = empirical Bayes weight for fatal and injury crashes
k_{FI} = overdispersion parameter for fatal and injury crashes
k_{total} = overdispersion parameter for total crashes
$N_{predicted,n(FI)}$ = predicted average fatal and injury crash frequency in year n from SPF
$N_{predicted,n(total)}$ = predicted average total crash frequency in year n from SPF

Step (iv): Compute the First-Year EB-Adjusted Expected Average Crash Frequency

In this step, the observed crash frequency and the computed predicted crash frequency are integrated to determine a long-term (expected) crash frequency for year 1 of the study period using the following expressions.

$$N_{expected,1(total)} = (w_{total}) \times N_{predicted,1(total)} + (1 - w_{total}) \times \left[\frac{\sum\limits_{n=1}^{Y} N_{observed,y(total)}}{\sum\limits_{n=1}^{Y} C_{n(total)}} \right] \qquad (5.41)$$

$$N_{expected,1(FI)} = (w_{total}) \times N_{predicted,1(FI)} + (1 - w_{FI}) \times \left[\frac{\sum\limits_{n=1}^{Y} N_{observed,y(FI)}}{\sum\limits_{n=1}^{Y} C_{n(FI)}} \right] \qquad (5.42)$$

where

$$
\begin{aligned}
N_{expected1,(FI)} &= \text{EB-adjusted estimated average crash frequency for year} \\
N_{expected1,(total)} &= \text{EB-adjusted estimated average crash frequency for year} \\
w_{total} &= \text{empirical Bayes weight for total crashes} \\
w_{FI} &= \text{empirical Bayes weight for fatal and injury crashes} \\
N_{observed,n(FI)} &= \text{observed fatal and injury crash frequency in year } n \\
N_{observed,n(total)} &= \text{observed average total crash frequency in year } n \\
N_{predicted,n(FI)} &= \text{predicted average fatal and injury crash frequency in year } n \text{ from} \\
&\quad \text{SPF} \\
N_{predicted,n(total)} &= \text{predicted average total crash frequency in year } n \text{ from SPF} \\
C_{n(total)} &= \text{annual correction factor for total crashes in year } n \\
C_{n(FI)} &= \text{annual correction factor for fatal crashes in year } n \\
n &= \text{year} \\
Y &= \text{final year}
\end{aligned}
$$

Step (v): Compute the EB-Adjusted Expected Average Frequency for the Final Year (Y) of the Study Period

In this step, the final year EB-adjusted average frequency is computed from the following equations:

$$N_{expected,Y(total)} = N_{expected,1(total)} \times C_{Y(total)} \qquad (5.43)$$

$$N_{expected,Y(FI)} = N_{expected,1(FI)} \times C_{Y(FI)} \qquad (5.44)$$

where

$$
\begin{aligned}
N_{expected,Y(total)} &= \text{EB-adjusted average total crash frequency for final year } (Y) \\
N_{expected,Y(FI)} &= \text{EB-adjusted average FI crash frequency for final year } (Y) \\
C_{Y(total)} &= \text{annual correction factor for total crashes in year final year } (Y) \\
C_{Y(FI)} &= \text{annual correction factor for fatal crashes in final year } (Y)
\end{aligned}
$$

Step (vi): Compute the Variance of the EB–Adjusted Average Crash Frequency

This step is suggested in the HSM although it is optional as it is not required for ranking. The HSM gives the following expressions for computing the variances:

$$Var_{(expected,n)roadways} = (N_{expected,n}) \times \left(\frac{1-w}{L}\right) \times \left(\frac{C_n}{\sum_{n=1}^{Y} C_n}\right) \text{ (for roadway segments)} \quad (5.45)$$

$$Var_{(expected,n)inter sections} = (N_{expected,n}) \times (1-w) \times \left(\frac{C_n}{\sum_{n=1}^{Y} C_n}\right) \text{ (for intersections)} \quad (5.46)$$

When the expected average crash frequency with EB adjustments procedure is used for ranking, sites under consideration are ranked with the site having the largest expected average crash frequency for the final year (Y) listed first, followed by the others in decreasing values of their average expected crash frequencies for the final year. The main advantage of this procedure is that it accounts for the effect of RTM, but it has the disadvantage of requiring an appropriate SPF for the type of sites under consideration. Because of the high level of computational effort involved in this method, a computer program is commonly used.

5.3.10 EPDO Average Crash Frequency with EB Adjustment

This procedure given in the HSM and described here is similar to that described in Section 5.3.9, but with the main parameter being the equivalent average property damage only (EPDO) crash frequency and not the average crash frequency. The procedure consists of the following steps:

(i)   Compute the weighting factor crash severity (see Section 5.3.3).
(ii) Compute the predicted crash frequency from an appropriate SPF (see step (i) of Section 5.3.9).
(iii) Compute the annual correction factor (see step (ii) of Section 5.3.9).
(iv) Compute the empirical Bayes weights (see step (iii) of Section 5.3.9).
(v)   Compute the first-year EB-adjusted expected EPDO average crash frequency with EB adjustment (see step (iv) of Section 5.3.9).
(vi) Compute the final-year EPDO average crash frequency with EB adjustment (see step (v) of Section 5.3.9).
(vii) Compute the proportion of fatal and injury crashes P_F and P_I as:

$$P_F = \frac{\sum N_{observed,(F)}}{\sum N_{observed,(FI)}} \quad (5.47)$$

$$P_I = \frac{\sum N_{observed,(I)}}{\sum N_{observed,(FI)}} \quad (5.48)$$

where

P_F = proportion of observed number of fatal crashes to total of fatal and injury crashes in the reference population

P_I = proportion of observed number of injury crashes to total of fatal and injury crashes in the reference population

$N_{observed,(F)}$ = observed number of fatal crashes in reference population

$N_{observed,(I)}$ = observed number of injury crashes in reference population

$N_{observed,(FI)}$ = observed number of fatal and injury crashes in the reference population

(viii) Compute the relative EPDO weight of fatal and injury crashes as:

$$w_{EPDO,FI} = (P_F) \times f_{F,(weight)} + (P_I) \times f_{I,(weight)} \tag{5.49}$$

where

$w_{EPDO,FI}$ = relative EPDO weight of fatal and injury crashes

P_F = proportion of observed number of fatal crashes to total of fatal and injury crashes in the reference population

P_I = proportion of observed number of injury crashes to total of fatal and injury crashes in the reference population

$f_{F,(weight)}$ = EPDO fatal weighting factor

$f_{I,(weight)}$ = EPDO injury weighting factor

(xi) Compute the final-year (Y) EPDO expected crash frequency with EB adjustment as:

$$N_{expected,Y(EPDO)} = N_{expected,Y(PDO)} + (W_{EPDO,FI}) \times N_{expected,Y(FI)} \tag{5.50}$$

where

$N_{expected,Y(EPDO)}$ = final-year (Y) equivalent expected PDO (EPDO) crash frequency with EB adjustment

$N_{expected,Y(PDO)}$ = final-year (Y) expected PDO crash frequency with EB adjustment

$W_{EPDO,FI}$ = relative EPDO weight of fatal and injury crashes (see Eq. 5.49)

$N_{expected,Y(FI)}$ = final-year (Y) expected fatal and injury (FI) crash frequency with EB adjustment

When the EPDO average crash frequency with EB adjustment procedure is used for ranking, sites under consideration are ranked with the site having the highest average EPDO crash frequency for the final year (Y) listed first, followed by the others in decreasing values of their average EPDO expected crash frequencies for the final year. This method accounts for RTM bias and considers the proportion of different crash severities observed. However, the weighting factors selected may result in emphasis being placed on sites with few fatal crashes. The computational effort involved in this method requires the use of a computer program.

5.3.11 Excess Expected Average Crash Frequency with EB Adjustments

The procedure given in the HSM and described here is similar to that described in Section 5.3.9, but with the main parameter being excess expected average crash frequency and not the average crash frequency. The procedure consists of the following steps.

Step (i): Prepare a summary table showing for each site within a reference population the following:

- Observed average crash frequency for each year by severity (*PDO, FI*)
- Predicted average crash frequency by severity (*PDO, FI*) for each year computed from an appropriate SPF for the reference population
- EB-adjusted expected average crash frequency by severity (*PDO, FI*) for each year using the procedure described in Section 5.3.9

Step (ii): Compute the excess expected average crash frequency with EB adjustments as

$$Excess_n = (N_{expected,n(PDO)} - N_{predicted,n(PDO)}) + (N_{expected,n(FI)} - N_{predicted,n(FI)}) \qquad (5.51)$$

where

$$
\begin{aligned}
Excess_n &= \text{excess expected crashes for year } n(N_{expected,n(PDO)}) \\
N_{expected,n(PDO)} &= \text{EB-adjusted expected average PDO crash frequency for year } n \\
N_{predicted,n(PDO)} &= \text{SPF predicted average PDO crash frequency for year } n \\
N_{expected,n(FI)} &= \text{EB-adjusted expected average FI crash frequency for year } n \\
N_{predicted,n(FI)} &= \text{SPF predicted average FI crash frequency for year } n
\end{aligned}
$$

Step (iii): Compute the severity weighted excess as

$$
\begin{aligned}
Excess_{,(sw)} &= (N_{expected,n(PDO)} - N_{predicted,n(PDO)}) \times CC_{(PDO)} \\
&+ (N_{expected,n(FI)} - N_{predicted,n(FI)}) \times CC_{(FI)}
\end{aligned} \qquad (5.52)
$$

where

$$
\begin{aligned}
Excess_{(sw)} &= \text{severity weighted EB-adjusted expected excess crash value for year } n \\
N_{expected,n(PDO)} &= \text{EB-adjusted expected average PDO crash frequency for year } n \\
N_{predicted,n(PDO)} &= \text{SPF predicted average PDO crash frequency for year } n \\
N_{expected,n(FI)} &= \text{EB-adjusted expected average FI crash frequency for year } n \\
N_{predicted,n(FI)} &= \text{SPF predicted average FI crash frequency for year } n \\
CC_j &= \text{crash cost for crash severity } j
\end{aligned}
$$

When the excess expected average crash frequency with EB adjustments procedure is used for ranking, sites under consideration are ranked using either the excess expected average crash frequency with EB adjustments or the severity weighted EB-adjusted expected excess crash value. The site having the highest excess expected average crash frequency with EB adjustments or the highest severity weighted excess is listed first, followed by the others in decreasing values. This method also accounts for RTM bias and considers sites that exhibit higher than expected crashes for its reference population. The computational effort involved in this method requires the use of a computer program.

5.4 COMPUTATIONAL PROCEDURES FOR SAFETY EFFECTIVENESS EVALUATION METHODS

The evaluation procedures described in this section incorporate some aspects of the predictive method given in the HSM, as it is beyond the scope of this book to give a comprehensive and detailed description of the method given in the HSM. However, the method is illustrated by describing the computational procedures of two methods that can be used to evaluate the effectiveness of implementing a countermeasure, a treatment at a site or facility that has been ranked for safety improvement. The procedures are the Empirical Bayes (EB) Before and After, and the Comparison-Group methods as given in the HSM.

It is also beyond the scope of this book to give a detailed description of each of these procedures, but enough material is presented to give the reader the necessary basic understanding for applying them.

5.4.1 Empirical EB Before and After Effectiveness Evaluation Method

This procedure involves the computation of the difference between the average crash frequencies for the before and after periods. It is given as a percentage change for all sites under consideration. The procedure consists of the following steps:

- Step 1. Compute the expected average crash frequency for site type x for each year of the before period using an appropriate SPF.
- Step 2. Compute and sum up the expected average crash frequency for each site i over the before period.
- Step 3. Compute the predicted average crash frequency for each site i and for each year y ($N_{predicted,i,y,A}$) during the after period, assuming that the treatment was not implemented.
- Step 4. Compute an adjustment factor (r_i).
- Step 5. Compute the expected average crash frequency $N_{expected}$ for each site i and for each year in the after period, assuming the treatment was not implemented.
- Step 6. Compute the ratio (OR_i) of $N_{observed,i,A}$ to $N_{expected,i,A}$. This ratio is an indication of the effectiveness of the treatment at each site.
- Step 7. Compute the safety effectiveness at each site.
- Step 8. Compute the overall treatment effectiveness (OR') at all sites combined.
- Step 9. Compute an unbiased (OR).
- Step 10. Compute the overall unbiased safety effectiveness.
- Step 11. Compute the variance of the unbiased estimated safety effectiveness.
- Step 12. Compute the standard error of the odds ratio (OR).
- Step 13. Determine the standard error of the safety effectiveness.
- Step 14. Evaluate the statistical significance of the computed safety effectiveness.

Step 1. Compute the expected average crash frequency for site type x for each year of the before period using an appropriate SPF.

In this step, an SPF that is applicable to type x sites is used to compute the predicted average crash frequency $N_{predicted}$ for each site of type x. This is computed by first determining the predicted average crash frequency ($N_{spf,x}$) at a site within type x, using an SPF that is applicable to type x sites. This computed predicted average crash frequency is then calibrated for differences between the site characteristics and the base characteristics for the SPF using applicable *CMF* values and for local conditions using a calibration factor (C_x). The predicted average crash frequency is given as:

$$N_{predicted} = (N_{spf,x}) \times (CMF_{1x}) \times (CMF_{2x}) \times \ldots \times (CMF_{yx}) \times C_x \qquad (5.53)$$

where

$N_{predicted,i,}$ = expected average crash frequency at a site within type x for a specific year i

$N_{spf,x}$ = predicted average crash frequency based on the applicable SPF for type x sites

$CMF_{y,x}$ = crash modification factor for a site within type x having specific geometric design and traffic conditions y

C_x = calibration factor to adjust SPF for a specific year for site type x (C_r for roadway segments and C_i for intersections)

C_x = total observed crash frequencies sites type x divided by the total expected average crash frequencies for sites type x computed from an appropriate SPF. Note that both the observed and expected crashes should be for the same selected sites and during the same time period.

Note that when the SPF used was developed using available data for a given jurisdiction, the calibration is not necessary.

Step 2. Compute and sum up the expected average crash frequency for each site i over the before period.

The expected average crash frequency ($N_{expected,i,B}$) is obtained from the predicted crash frequency and the weight, the latter of which is based on the overdispersion parameter and is given as

$$N_{expected,i,B} = w_{i,B} N_{predicted} + (1 - w_{i,B}) N_{observed,i,B} \qquad (5.54)$$

where

$w_{i,B}$ = weight for each site i

$$w_{i,B} = \frac{1}{1 + k \sum\limits_{\substack{Before \\ years}} N_{predicted,i,B}} \qquad (5.55)$$

$N_{expected,i,B}$ = expected average crash frequency at site i within type x for the entire before period

$N_{observed,i,B}$ = observed crash frequency at site i for the entire before period

k = overdispersion parameter for the applicable SPF

Step 3. Compute the predicted average crash frequency for each site *i* and for each year *y* ($N_{predicted,i,y,A}$) during the after period, assuming that the treatment was not implemented.

This task involves repeating task 1, but for the after period.

Step 4. Compute an adjustment factor (r_i).

This factor will be used to correct for variations in before and after period characteristics such as duration and traffic volume at each site. It is given as

$$r_i = \frac{\sum\limits_{\substack{After \\ years}} N_{predicted,i,A}}{\sum\limits_{\substack{Before \\ years}} N_{predicted,i,B}} \tag{5.56}$$

where

$N_{predicted,i,A}$ = predicted average crash frequency at site *i*, for the after period, assuming the treatment was not implemented

$N_{predicted,i,B}$ = predicted average crash frequency for site *i*, for the before period, assuming the treatment was not implemented

Step 5. Compute the expected average crash frequency $N_{expected}$ for each site *i* and for each year in the after period, assuming the treatment was not implemented.

$$N_{expected,i,y,A} = N_{expected,i,y,B} \times r_i$$

where

$N_{expected,i,y,A}$ = expected average crash frequency for site *i*, and for each year *y*, in the after period assuming the treatment was not implemented for the after period

$N_{expected,i,y,B}$ = expected average crash frequency for site *i*, and for each year *y*, in the after before assuming the treatment was not implemented for the after period

r_i = adjustment factor, see Eq. 5.56

Step 6. Compute the ratio (OR_i) of $N_{observed,i,A}$ to $N_{expected, i,A}$. This ratio is an indication of the effectiveness of the treatment at each site.

$$(OR_i) = \frac{N_{observed,i,A}}{N_{expected,i,A}} \tag{5.57}$$

where

(OR_i) = odds ratio at site

$N_{observed,i,A}$ = observed crash frequency at site *i* for the extent of the after period

$N_{expected,i,A}$ = expected crash frequency at site *i* for the extent of the after period

Step 7.   Compute the safety effectiveness at each site.

In this task, the safety effectiveness at each site i is computed as:

$$\text{Safety effectiveness}_i = 100(1 - OR_i) \tag{5.58}$$

Step 8.   Compute the overall treatment effectiveness (OR') at all sites combined.

The overall effectiveness (OR') is given as

$$OR' = \frac{\displaystyle\sum_{All\ sites} N_{observed,i,A}}{\displaystyle\sum_{All\ sites} N_{expected,i,A}} \tag{5.59}$$

Step 9.   Compute an unbiased (OR).

The overall treatment effectiveness (OR') obtained from Eq. 5.59 is adjusted for the potential bias that it may contain. This adjustment gives an overall unbiased treatment effectiveness (OR). The adjustment is made by using the variance of the expected crashes at all sites for the after period and the weight factor, and it is given as

$$OR = \frac{OR'}{1 + \dfrac{Var\left(\displaystyle\sum_{All\ sites} N_{expected,A}\right)}{\left(\displaystyle\sum_{All\ sites} N_{expected,A}\right)^2}} \tag{5.60}$$

where

$$Var\left(\sum_{All\ sites} N_{expected,A}\right) = \sum_{All\ sites} ((r_i)^2 \times (N_{expected,i,B}) \times (1 - w_{i,B})) \tag{5.61}$$

$N_{expected,i,B}$ = expected crash frequency at site i for the extent of the before period
$w_{i,B}$ = weight for site i for the before period (see Eq. 5.55)

Step 10.   Compute the overall unbiased safety effectiveness.

In this step, the overall unbiased safety effectiveness is computed as:

$$\text{Overall unbiased safety effectiveness}_i = 100(1 - OR) \tag{5.62}$$

Step 11. Compute the variance of the unbiased estimated safety effectiveness.

In this step, the variance of the unbiased estimated safety effectiveness is computed as an odds ratio and it is given as

$$Var(OR) = \frac{(OR')^2 \left[\dfrac{1}{N_{observed,A}} + \dfrac{Var\left(\sum\limits_{All\ sites} N_{expected,A} \right)}{\left(\sum N_{expected,A} \right)^2} \right]}{1 + \dfrac{Var\left[\sum\limits_{All\ sites} N_{expected,A} \right]}{\left(\sum\limits_{All\ sites} N_{expected,A} \right)^2}} \tag{5.63}$$

Step 12. Compute the standard error of the odds ratio (OR).

The standard error of OR is given as

$$SE(OR) = \sqrt{Var(OR)} \tag{5.64}$$

Step 13. Determine the standard error of the safety effectiveness.

The relationship between the the unbiased odds ratio and the safety effectiveness as shown in Eq. 5.40 is used to determine the standard error of the safety effectiveness as

$$SE(Safety\ Effectiveness) = 100 \times SE(OR) \tag{5.65}$$

Step 14. Evaluate the statistical significance of the computed safety effectiveness.

The statistical significance of the safety effectiveness is determined by first computing the absolute ratio of the safety effectiveness and the standard error of the safety effectiveness as

$$Abs[Safety\,Effectiveness/SE(Safety\,Effectiveness)]$$

If this ratio < 1.7, the treatment effect is not significant at approximately the 90 percent confidence level.
If this ratio ≥ 1.7, the treatment effect is significant at approximately the 90 percent confidence level.
If this ratio ≥ 2.0, the treatment effect is significant at approximately the 95 percent confidence level.

This procedure is usually carried out with a computer program.

5.4.2 Comparison–Group Methods

In this method, two sets of sites are a selected for the analysis. The first set consists of one or more sites on which the treatment or countermeasure has been implemented, usually referred to as the treatment sites, and the other is a set of similar sites without the treatment, usually known as the comparison sites. The crash frequencies on both the treatment and comparison sites are recorded for a number of years (usually at least 3 years) before the application of the treatment at the treatment sites, and after the application of the treatment. The procedure consists of the following steps:

Step 1. Compute the predicted before and after crashes on the treatment and comparison sites.

In this task the appropriate SPF together with the site specific *AADTs* are used to determine the before and after crashes at each site, and for each of the before and after years. Then determine the following:

(i) sum of the predicted average crash frequencies at each treatment site i and for each year of the before period as $(\Sigma N_{predicted,i,T,B}) = N_{(predicted,T,B)i}$

(ii) sum of the predicted average crash frequencies at each treatment site i for each year of the after period $(\Sigma N_{predicted,i,T,A}) = N_{(predicted,T,A)i}$

(iii) sum of the predicted average crash frequencies at each comparison site and for each year of the before period $(\Sigma N_{predicted,j,C,B}) = N_{(predicted,C,B)j}$

(iv) sum of the predicted average crash frequencies at each comparison site for each of the after period $(\Sigma N_{predicted,j,C,A}) = N_{(predicted,C,A)j}$

Step 2. Compute adjustment factors to correct for differences between the treatment and comparison sites in their traffic volumes and number of years.

In some cases the number of years of available data are not the same for the before and after periods for either and/or the comparison and treatment sites. This difference is corrected for by computing an appropriate adjustment factor for each treatment site and comparison site j combination given as:

$$Adj_{i,j,B} = \left(\frac{N_{(predicted,T,B)_i}}{N_{(predicted,C,B)_j}} \right) \times \left(\frac{Y_{B,T}}{Y_{B,C}} \right) \tag{5.66}$$

$$Adj_{i,j,A} = \left(\frac{N_{(predicted,T,A)i}}{N_{(predicted,C,A)_j}} \right) \times \left(\frac{Y_{A,T}}{Y_{A,C}} \right) \tag{5.67}$$

where

$Adj_{i,j,B}$ = adjusted factor for predicted crashes during the before period

$Adj_{i,j,A}$ = adjusted factor for predicted crashes during the after period

$N_{(predicted,T,B)i}$ = sum of predicted average crash frequencies at treatment site i in the before period

$N_{(predicted,C,B)j}$ = sum of predicted average crash frequencies at comparison site j in the before period

$N_{(predicted,T,A)i}$ = sum of predicted average crash frequencies at treatment site i in the after period

$N_{(predicted,T,A)j}$ = sum of predicted average crash frequencies at comparison site j in the after period

$Y_{i,B,T}$ = duration of the before period for treatment site i (years)

$Y_{j,B,C}$ = duration of the before period for comparison site j (years)
$Y_{i,A,T}$ = duration of the after period for treatment site i (years)
$Y_{j,A,T}$ = duration of the after period for comparison site j (years)

Step 3. Compute the expected average crash frequencies for each treatment site i and for each comparison site j for the before and after periods.

In this task the expected average crash frequency is obtained as the product of the respective observed crash frequency and the adjustment factor, and is given as:

$$N_{(expected,C,B)_j} = \sum_{All\ sites,j} (N_{observed,j,C,B}) \times Adj_{i,j,B} \tag{5.68}$$

$$N_{(expected,C,A)_j} = \sum_{All\ sites,j} (N_{observed,C,B}) \times Adj_{i,j,A} \tag{5.69}$$

where

$N_{(expected,C,B)j}$ = expected average crash frequency for comparison site j for the before period

$N_{(expected,C,A)j}$ = expected average crash frequency for comparison site j for the after period

$\sum (N_{observed,C,B})_j$ = sum of observed crash frequencies at comparison sites j in the before period

$Adj_{i,j,B}$ = adjusted factor for predicted crashes during the before period (see Eq. 5.66)

$Adj_{i,j,A}$ = adjusted factor for predicted crashes during the after period (see Eq. 5.67)

Step 4. Compute the total comparison group expected average crash frequency for the before period.

In this task, the total comparison-group expected average crash frequency for the before ($N_{expected,C,B,total}$) is computed by summing up the expected crash frequencies for all comparison sites for the before period. It is obtained as:

$$N_{expected,C,B,total} = \sum_{\substack{All\ comparison \\ sites}} N_{(expected,C,B)_j} \tag{5.70}$$

Step 5. Compute the total comparison-group expected average crash frequency for the after period.

In this task, the total comparison-group expected average crash frequency for the after period ($N_{expected,C,A,total}$) is computed by summing up the expected crash frequencies for all comparison sites for the after period. It is obtained as:

$$N_{expected,C,A,total} = \sum_{\substack{All\ comparison \\ sites}} N_{(expected,C,A)_j} \tag{5.71}$$

Step 6. Compute the comparison ratio ($r_{i,c}$) for each treatment site i.

In this task, the comparison ratio of each treatment site i is computed as the quotient of the comparison group total expected crash frequency for the after period and the total expected crash frequency for the before period. It is given as:

$$r_{i,c} = \frac{N_{expected,C,A,total}}{N_{expected,C,B,total}} \tag{5.72}$$

Step 7. Compute the expected average crash frequency for each treatment site i for the after period without the application of the treatment.

This task involves the use of the comparison ratio obtained in Eq. 5.72 to compute the expected average crash frequency at each treatment site i for the after period and summing these up for all treatment sites to obtain the total expected crash frequencies for all treatment sites without the implementation of the treatment. It is given as:

$$N_{expected,T,A} = \sum_{All\ sites} (N_{observed,T,B})_i \times r_{i,c} \tag{5.73}$$

Step 8. Compute the safety effectiveness of the treatment.

In this step, the safety effectiveness of the treatment at each treatment site is computed as an odds ratio. It is computed as the ratio of the expected average crash frequency with the treatment to the expected average crash frequency without the treatment, and it is given as:

$$OR_i = \sum_{All\ sites} \left(\frac{N_{observed,T,A}}{N_{expected,T,A}} \right) \tag{5.74}$$

or, alternatively,

$$OR_i = \frac{N_{observed,T,A,total}}{N_{observed,T,B,total}} \times \frac{N_{expected,C,B,total}}{N_{expected,C,A,total}} \tag{5.75}$$

Step 9. Compute the log odds ratio R_i for each treatment site.

This task involves the computation of the log of the odds ratio for each treatment site as:

$$R_i = ln(OR_i) \tag{5.76}$$

Step 10. Compute the weight (w_i) for each treatment site i.

The weight at each site i is given as

$$w_i = \frac{1}{R^2_{i(se)}} \tag{5.77}$$

where

$$R^2_{i(SE)} = \frac{1}{N_{observed,T,B,total}} + \frac{1}{N_{observed,T,A,total}} + \frac{1}{N_{observed,C,B,total}} + \frac{1}{N_{observed,C,A,total}} \tag{5.78}$$

Step 11. Compute the weighted average log odds ratio (R) across all n treatment sites.

The weighted average log odds is given as:

$$R = \frac{\sum_n w_i R_i}{\sum_n w_i} \tag{5.79}$$

Step 12   Compute the overall safety effectiveness of the treatment expressed as a percentage.

The overall safety effectiveness of the treatment is given as

$$\text{Safety effectiveness} = 100 \times (1 - \text{OR})$$

Step 13   Compute the standard error of the precision $SE(Safety\ Effectiveness)$.

In this step, the standard error of the precision is computed and used as the measure of the precision. The standard error is given as:

$$SE(Safety\,Effectiveness) = 100\,\frac{OR}{\sqrt{\sum_n w_i}} \tag{5.80}$$

where

$OR = e^R$ (overall effectiveness of the treatment)
R = the weighted average log odds (see Eq. 5.79)

Step 14   Evaluate the statistical significance of the computed safety effectiveness.

The statistical significance of the safety effectiveness is determined by first computing the absolute ratio of the safety effectiveness and the standard error of the safety effectiveness as

$$Abs\ [Safety\,Effectiveness/SE(Safety\,Effectiveness)]$$

If this ratio < 1.7, the treatment effect is not significant at approximately the 90 percent confidence level

If this ratio ≥ 1.7, the treatment effect is significant at approximately the 90 percent confidence level.

If this ratio ≥ 2.0, the treatment effect is significant at approximately the 95 percent confidence level.

This procedure is usually carried out with a computer program.

5.5 CRASH PATTERNS

Two commonly used techniques to determine crash patterns are (1) expected value analysis and (2) cluster analysis. A suitable summary of crash data also can be used to determine crash patterns

5.5.1 Expected Value Analysis

This is a mathematical method that can be used to identify locations with abnormal crash characteristics. It should be used only to compare sites with similar characteristics (for example, geometrics, volume, or traffic control), since the analysis does not consider exposure levels. The analysis is carried out by determining the average number of a specific type of crash occurring at several locations with similar geometric and traffic characteristics. This average is adjusted by incorporating a weight factor which depends on the variance that is based on data for similar roadways or facilities. The best estimate of the expected number of crashes at a given roadway or facility can be obtained as

$$n = \omega \times n_s + (1 - \omega) \times n_a \tag{5.81}$$

where

n = the best estimate of the expected average number of crashes at given roadway or facility

n_s = the estimate of the number of crashes based on the data of a group of similar roadways or facilities

n_a = estimate of the number of crashes based on crash counts on the given roadway or facility

$V\{n_s\}$ = variance of the estimated number of crashes based on data for similar roadways or facilities

$E\{n_s\}$ = the estimate of the expected average number of crashes

ω = weight generated by the variance of the estimate of expected average crash frequency

$$\omega = \frac{1}{\left(1 + \dfrac{V\{n_s\}}{E\{n_s\}}\right)} \tag{5.82}$$

and the variance of the best estimate is given as

$$V(n) = \omega \times V\{n_s\} = (1 - \omega) \times E\{n_s\} \tag{5.83}$$

Example 5.14 Determining Overrepresentation of Different Crash Types

Data collected for three consecutive years at an intersection study site show that 14 rear-end collision and 10 left-turn collisions occurred during a 3-year period. Data collected at 10 other intersections with similar geometric and traffic characteristics give the information shown in Table 5.24. Determine whether rear-end crashes are overrepresented at the study site.

Solution:

- Determine that the estimated average rear-end crashes based on data for similar intersections = 7.40. See Table 5.24.
- Determine variance based on data for similar 10 intersections = 2.25 (see Table 5.24).

$$\omega = \frac{1}{\left(1 + \dfrac{2.25}{7.4}\right)} = 0.767$$

- Determine the best estimate of rear-end crashes on the study intersection n. Use Eq. 5.83.

$$n = \omega \times n_s + (1 - \omega) \times n_a = 0.767 \times 7.4 + (1 - 0.767) \times 14 = 8.9$$

Actual number of rear-end crashes > than the best estimate of rear-end crashes. Results suggest that rear-end crashes are overrepresented at the study site.

Table 5.24   Total Number of Rear-End Collisions at 10 Control Stations for Three Consecutive Years for Example 5.14

Control Site	Rear-End Collisions
1	8.00
2	5.00
3	7.00
4	8.00
5	6.00
6	8.00
7	9.00
8	10.00
9	6.00
10	7.00
Average	7.40
Standard deviation	1.50

5.5.2 Cluster Analysis

Cluster analysis involves the identification of a particular characteristic from the crash data obtained at a site. It identifies any abnormal occurrence of a specific crash type in comparison with other types of crashes at the site. For example, if there are two rear-end, one right-angle, and six left-turn collisions at an intersection during a given year, the left-turn collisions could be defined as a *cluster* or grouping, with abnormal occurrence at the site.

However, it is very difficult to assign discrete values that can be used to identify crash patterns. This is because crash frequencies, which are the basis for determining patterns, differ considerably from site to site. It is sometimes useful to use exposure data, such as traffic volumes, to define patterns of crash rates. Care must be taken, however, to use correct exposure data. For example, if total intersection volume is used to determine left-turn crash rates at different sites, these rates are not directly comparable because the percentages of left-turn vehicles at these sites may be significantly different. Because of these difficulties, it is desirable to use good engineering judgment when this approach is being used.

5.5.3 Statistical Comparison

There are several statistical tests that do not require knowledge of the distribution of crashes and can be used to compare the level of safety between two or more sites. One of such tests is the Kruskal-Wallis H test.

Kruskal–Wallis H Test

This is a nonparametric technique that does not require the assumption of the distribution of the populations being tested. It can be used to test the null hypothesis that the distributions associated with two populations are not significantly different. Consider two sets of data for large truck crashes: one set is from interstate highways with different speed limits (DSLs) for trucks and passenger cars, and the other is from interstate highways with the same (uniform) speed limit (USL) for trucks and passenger cars. This test can be used to determine whether the distribution of truck crashes for the DSL and USL highways are similar. The procedure involves the following steps.

Step 1.   Combine the two data sets as though they are one and rank according to the relative magnitude of the measurements. For example, in this case, the lowest number of crashes is ranked as 1, regardless of whether it occurs on a DSL or USL highway. The other values are then ranked increasingly to the highest value that will have the rank of the total number of data points in the combined data. When there is a tie between two or more observations, the mean of the ranks is assigned to each of the tied observations. The null hypothesis in this case is that the distribution of large-truck crashes on DSL highways is similar to that for USL highways and is given as:

H_0: The probability distributions for truck crashes on DSL and USL interstate highways are the same

H_1: The probability distributions are not the same

Step 2.   Determine the test statistic (H) using Eq. 5.7.

$$H = \frac{12}{n(n+1)} \sum \frac{R_j^2}{n_j} - 3(n+1) \qquad (5.84)$$

where

n_j = number of measurements in the jth sample
n = the total sample size ($n_1 + n_2 + n_3 + n_4 + \ldots + n_k$)
R_j = rank sum for sample j

It has been shown that if the null hypothesis is true, the distribution of the test statistic H for repeated samples is approximately the same as the χ^2 distribution when each of the sample sizes is five or higher. Values of the χ^2 distribution are given in Appendix A. The theoretical values of the χ^2 distribution depend on the degree of freedom, which depends on the specific distribution being considered. However, the degree of freedom for the approximate sampling distribution H is always ($k - 1$), where k is the number of distributions being considered. For the example being discussed, there are two distributions being tested: the distribution of truck crashes on DSL highways and the distribution of truck crashes on USL highways. The degree of freedom is 1. The rejection region for this test is located in the upper tail of the χ^2 distribution for the level of significance (α) used.

Example 5.15

Table 5.25 gives the number of large-truck crashes that occur on 20-km segments of highways with DSL and USL with the same $AADT$ over a two-year period. Using the Kruskal-Wallis H test, determine whether it can be concluded that the distribution of large-truck crashes are similar at the 5% significance level.

Solution:
- Rank the number of crashes as shown in Table 5.26.
- Determine the statistic H using Eq. 5.7.

$$H = \frac{12}{n(n+1)} \sum \frac{R_j^2}{n_j} - 3(n+1)$$

Table 5.25   Data for Example 5.15

Number of Large-Truck Crashes on DSL Segments		Number of Large-Truck Crashes on USL Segments	
Site Number	Number of Crashes	Site Number	Number of Crashes
1	6	1	5
2	10	2	6
3	8	3	10
4	12	4	7
5	15	5	9
6	2	6	4
7	11	7	13
8	9	8	14
9	18	9	16
10	25	10	24
11	17	11	27
12	28	12	30
13	33	13	31
14	29	14	34
15	35	15	33

Table 5.26   Ranking for Example 5.15

DSL Segments		USL Segments	
Number of Crashes	Rank	Number of Crashes	Rank
6	5 (4.5)	5	3
10	10 (10.5)	6	4 (4.5)
8	7	10	11 (10.5)
12	13	7	6
15	16	9	8
2	1	4	2
11	12	13	14
9	9	14	15
18	19	16	17
25	21	24	20
17	18	27	22
28	23	30	25
33	27 (27.5)	31	26
29	24	34	29
35	30	33	28 (27.5)
	Σ 235.5		Σ 229.5

$$H = \frac{12}{30(30+1)} \sum \left(\frac{235.5^2}{15} + \frac{229.5^2}{15} \right) - 3(30+1)$$

$$= 0.013(3697.35 + 3511.35) - 93$$

$$= 0.71$$

- Determine whether we should accept or reject the null hypothesis.

Degrees of freedom $= 1$

$\alpha = 0.05$

From the χ^2 table in Appendix A, we obtain $\chi^2_{0.05,1} = 3.84146$, $H < \chi^2_{0.05,1}$. We therefore accept the null hypothesis and conclude that the distributions of the crashes on DSL and USL segments are the same.

5.6 EFFECTIVENESS OF SAFETY DESIGN FEATURES

The document, *Guidance for Implementation of the AASHTO Strategic Highway Safety Plan,* published by the Transportation Research Board (TRB), consists of several guides, each of which gives a set of objectives and strategies to improve safety at specific locations. Table 5.27 gives some of these objectives and strategies for a few locations.

Note that Table 5.27 gives only a representative sample of objectives and strategies for selected types of crashes and is not comprehensive. Interested readers should refer to the reference given.

Table 5.27 Objectives and Strategies for Different Crash Types

Objectives	Strategies
Crash Types	
Collisions with Trees in Hazardous Locations	
Prevent trees from growing in hazardous locations	Develop, revise, and implement guidelines to prevent placing trees in hazardous locations (T) Develop mowing and vegetation control guidelines (P)
Eliminate hazardous condition and/or reduce severity of the crash	Remove trees in hazardous locations (P) Shield motorists from striking trees (P) Modify roadside clear zone in vicinity of trees (P) Delineate trees in hazardous locations (E)
Head-On Collisions	
Keep vehicles from encroaching into opposite lanes	Install centerline rumble strips for two-lane roads (T) Install profiled thermo-plastic strips for centerlines (T) Provide wider cross-sections for two-lane roads (T) Provide center two-way, left-turn lanes for four- and two-lane roads (T) Reallocate total two-lane roadway width (lane and shoulder) to include a narrow buffer median

(Continued)

Table 5.27 Objectives and Strategies for Different Crash Types (*continued*)

<table>
<tr><td colspan="2" align="center">*Head-On Collisions*</td></tr>
<tr><td>Minimize the likelihood of crashing into an oncoming vehicle</td><td>Use alternating passing lanes or four-lane sections at key locations (T)
Install median barriers for narrow-width medians on multilane roads (T)</td></tr>
<tr><td colspan="2" align="center">*Unsignalized Intersection Collisions*</td></tr>
<tr><td>Improve management of access near unsignalized intersections</td><td>Implement driveway closures/relocations (T)
Implement driveway turn restrictions (T)</td></tr>
<tr><td>Reduce frequency and severity of intersection conflicts through geometric design</td><td>Provide left-turn lanes at intersections (T)
Provide offset left-turn lanes at intersections (T)
Provide right- and left-turn acceleration lanes at divided highway intersections (T)
Restrict or eliminate turning maneuvers by signing or channelization or closing median openings</td></tr>
<tr><td>Improve sight distance at signalized intersections</td><td>Clear sight triangles on stop or yield intersections (T)
Clear sight triangles in the median of divided highways (T)
Change horizontal and/or vertical alignment of approaches to provide more sight distance (T)</td></tr>
<tr><td colspan="2" align="center">*Collisions on Horizontal Curves*</td></tr>
<tr><td>Reduce likelihood of a vehicle leaving its lane and either crossing the roadway centerline or leaving the roadway at a horizontal curve</td><td>Provide advance warning of unexpected changes in horizontal curves (T)
Enhance delineation along the curve
Install shoulder and centerline rumble strips (T)
Provide skid-resistant pavement surfaces (T)
Improve or restore superelevation (P)
Widen roadway (T)
Modify horizontal alignment</td></tr>
<tr><td>Minimize adverse consequences of leaving roadway at the horizontal curve</td><td>Design safer slopes and ditches to prevent rollovers (P)
Remove/relocate objects in hazardous locations
Delineate roadside objects (E)
Improve design and application of barrier and attenuation systems (T)</td></tr>
<tr><td colspan="2" align="center">*Collisions Involving Pedestrians*</td></tr>
<tr><td>Reduce pedestrian exposure to vehicular traffic</td><td>Provide sidewalks/walkways and curb ramps (P)
Install or upgrade traffic and pedestrian refuge islands and raised medians (P)
Install or upgrade traffic and pedestrian signals (P, T, E)</td></tr>
<tr><td>Improve sight distance and/or visibility between motor vehicles and pedestrians</td><td>Provide crosswalk enhancements (T, E)
Signals to alert motorists that pedestrians are crossing (T)
Eliminate screening by physical objects (T)</td></tr>
<tr><td>Reduce vehicle speeds</td><td>Implement road narrowing measures (T)
Install traffic calming road sections (P, T)
Install traffic calming intersections (P, T)</td></tr>
</table>

Table 5.27   Objectives and Strategies for Different Crash Types (*continued*)

Collisions at Signalized Intersections	
Reduce frequency and severity of intersection conflicts through traffic control and operational improvements	Employ multi-phase signal operation (P,T) Optimize clearance intervals (P) Employ signal coordination (P) Employ emergency vehicle pre-emption (P)
Reduce frequency and severity of intersection conflicts through geometric improvements	Provide/improve left-turn and right-turn channelization (P) Improve geometry of pedestrian and bicycle facilities (P,T) Revise geometry of complex intersections (P, T)
Improve sight distance at signalized intersections	Clear sight triangles (T) Redesign intersection approaches (P)

(P) Proven, strategy has been used and proven to be effective.
(T) Strategy has been implemented and accepted as a standard or standard approach, but valid evaluation is not available.
(E) Experimental, strategy has been suggested and considered sufficiently promising by at least one agency.
SOURCE: Based on data from from *Guidelines for the Implementation of the AASHTO Strategic Highway Safety Plan*, National Cooperative Highway Research Program, NCHRP, Transportation Research Board, Washington, D.C., 2004.

5.7 SAFETY EFFECTIVENESS OF SOME COMMONLY USED HIGHWAY DESIGN FEATURES

Although it is not feasible to include all the *CMFs* provided in the HSM in this chapter, this section gives a brief summary of the safety effectiveness of certain highway design features commonly used, so that the reader will have a fundamental knowledge of how these features could impact safety. The features considered in this chapter are related to (1) access control, (2) alignment, and (3) cross sections.

5.7.1 Access Control

The effects of geometrics on traffic crashes have produced a variety of findings which are not always definitive because often more than one factor may have caused the crash to occur. Furthermore, it is difficult to conduct studies in a controlled environment, and often researchers must rely on data collected by others under a variety of circumstances. Despite these difficulties, research findings over an extended period have confirmed a strong relationship between access control and safety.

Access control is defined as some combination of at-grade intersections, business and private driveways, and median crossovers. For any given highway, access control can range from full control (such as an interstate highway) to no access control (common on most urban highways). The reason why access control improves safety is because there are fewer unexpected events caused by vehicles entering and leaving the traffic stream at slower speeds, resulting in less interference with through traffic. Also, the increase in roadside development, which creates an increased number of at-grade intersections and businesses with direct access to the highway, will also significantly increase crashes.

The effect of control of access is illustrated in the *CMF* given in Eq. 5.85 for the effect of driveway density on total crashes at a roadway segment. This *CMF* indicates that the impact of access control is dependent on the density of driveways on the roadway segment in terms of number of driveways per mile and the *AADT*.

$$CMF_{dr} = \frac{0.322 + 0.625 \times (DD) \times [0.05 - 0.005 \times \ln(AADT)]}{0.322 + [0.05 - 0.005 \times \ln(AADT)]} \tag{5.85}$$

where

CMF_{dr} = crash modification factor for the effect of driveway density on total crashes

$AADT$ = average annual daily traffic volume of the roadway being evaluated

DD = driveway density considering driveways on both sides of the highways (driveways/km).

Note: When the driveway density is less than 5 per mile, the *CMF* is taken as 1.
Also the *CMF* can be used for roadway crashes of all levels of severity.
A driveway is considered when traffic uses it to enter or leave the roadway daily.

There are several mechanisms for reducing crashes due to access, all of which require the elimination of access points from through traffic. Examples include (1) removal of the access points by closing median openings, (2) frontage road access for business driveways, (3) special turning lanes to separate through vehicles from those vehicles using the access point, and (4) proper signing and pavement markings to warn motorists of changing conditions along the roadway.

5.7.2 Alignment

The geometric design of highways, discussed in Chapter 15, involves three elements: (1) vertical alignment, (2) horizontal alignment, and (3) cross section. Design speed is the determining factor in the selection of the alignment needed for the motorist to have sufficient sight distance to safely stop or reduce speed as required by changing traffic and environmental conditions. A safe design ensures that traffic can flow at a uniform speed while traveling on a roadway that changes in a horizontal or vertical direction.

The design of the vertical alignment (which includes tangent grades and sag or crest vertical curves) is influenced by consideration of terrain, cost, and safety. Generally, crash rates for downgrades are higher than for upgrades. One study reported that only 34.6 percent of crashes occurred on level grade, whereas 65.4 percent occurred on a grade or at the location where grades change. The HSM gives the following *CMFs* for grade of two-lane, two-way roadway segments:

Level grades ($< 3\%$), $CMF_{gr} = 1.00$

Moderate terrain ($3\% < $ grade $< 6\%$), $CMF_{gr} = 1.10$

Steep terrain ($> 6\%$), $CMF_{gr} = 1.16$

The design of the horizontal alignment (which consists of level tangents connected by circular curves) is influenced by design speed and superelevation (see Chapter 15 for definition) of the curve itself. Crash rates for horizontal curves are higher than on tangent sections, with rates ranging between 1.5 and 4 times greater than on straight sections. The main factors that influence the safety performance of horizontal curves are the radius of curvature, the length of the horizontal curve, and whether the curve has a spiral transition curve (see Chapter 15 for the definition of *spiral curve*). The HSM gives the crash modification factor for horizontal curves (CMF_{hcr}) on two-lane, two-way roadway segments as:

$$CMF_{hcr} = \frac{(0.97 \times L_c) + \left(\dfrac{24}{R}\right) - (0.012 \times S)}{(0.97 \times L_c)} \tag{5.86}$$

where

$\quad CMF_{hcr}$ = crash modification factor for horizontal curves on two-way, two-lane
$\qquad\qquad$ segment total crashes
$\qquad R$ = radius of curve (m)
$\qquad L_c$ = length of horizontal curve including length of spiral (km)
$\qquad S$ = 1 if spiral transition curve exists at both ends of the curve
$\qquad\quad$ = 0.5 if spiral transition curve exists only at one end of the curve
$\qquad\quad$ = 0 if spiral transition curve does not exist

5.7.3 Cross Sections

One of the most important roadway features affecting safety is the highway cross section. The road may be constructed on an embankment (fill) section or depressed below the natural grade (cut). Cross-section elements (including through and passing lanes, medians, and left-turn lanes) may be added when a two-lane road is inadequate, possibly improving both traffic operations and safety. Safety improvements in the highway cross section are often more focused on two-lane roads, with the exception of clear zone treatments and median design for multilane highways.

Lane Widths

The *CMFs* given in Table 5.14 indicate that, in general, roadway segments with lane widths less than 12 ft are more susceptible to certain crashes (single-vehicle run-off-the-road and multiple-vehicle head-on, opposite direction sideswipe, and same direction sideswipe crashes) than segments with 12-ft lanes. The reason is that a wider distance is provided between vehicles on adjacent lanes, which provides additional buffer for vehicles that deviate from their lanes. However, it has been shown that travel speeds tend to increase on segments with wider lanes and may also encourage drivers to drive with lesser headways (distance between front of a vehicle and the front of the following vehicle). Also, wider lanes result in a longer distance for crossing the street, which may increase the potential of conflict between pedestrians and vehicles on the roadway.

Medians

A median is installed on a multilane road to separate the lanes serving vehicles in opposite directions. The distance between the outer edges of the outside lanes is the width of the median. Chapter 15 discusses the functions of a median. Because there are several interrelated factors (e.g., width, shape, the necessity to provide one) that influence the impact of the existence of a median on a roadway segment, it has not been easy to extract the influence of changing an undivided road to a divided road on crashes.

Shoulders

The shoulder is a contiguous section of the pavement that can be paved or unpaved. It primarily serves as an area along the highway that can be used for vehicles to be stopped when necessary, and also provides a margin of safety for vehicles that stray from the travel lane. When paved it also serves as a protection of the structure of the travel way from water damage and from erosion. Its width ranges from 0.6 m on minor roads to 3.6 m on major arterials. Although a few studies have indicated that wider shoulders may improve safety on two lane roads, wider shoulders tend to encourage drivers to voluntarily stop on the shoulder, which may result in a crash. In fact, it has been shown that stopped-on-shoulder vehicles and maneuvering onto or from the shoulder are associated with 10 percent of fatal crashes on freeways. Table 5.28 gives *CMFs* for different shoulder widths.

Passing and Climbing Lanes

A passing lane is an added lane to a two-lane, two-way roadway segment to provide motorists additional opportunities to overtake slower vehicles (see Chapter 9). A climbing lane is one in an upgrade direction for use by heavy vehicles whose speeds are significantly reduced by the grade (see Chapter 15). The HSM considers the basic condition for passing lanes as a normal two-lane cross section without a passing lane. The HSM suggests a value of 0.75 for the crash modification factor (CMF_{pl}) for total crashes in both directions of travel over the entire length (start of upstream lane addition taper to end of downstream lane drop taper) of the climbing or passing lane. This value is applicable when the passing or climbing lane is warranted and its length satisfies the operational characteristics of the roadway segment.

In some cases, a two-lane roadway segment is widened to a four-lane segment for a short length (a passing lane in each direction) to facilitate passing in both directions. The HSM suggests a value of 0.65 for the *CMF*.

Table 5.28   *CMFs for Collision Types Related to Shoulder Width (CMF_{WRA})*

Shoulder Width	Average Annual Daily Traffic (AADT) Vehicles/Day)		
	<400	400 to 2000	>2000
0 m	1.10	$1.10 + 2.5 \times 10^{-4}(AADT - 400)$	1.50
0.6 m	1.07	$1.07 + 1.43 \times 10^{-4}(AADT - 400)$	1.30
1.2 m	1.02	$1.02 + 8.125 \times 10^{-5}(AADT - 400)$	1.15
1.8 m	1.00	1.00	1.00
2.4 m or more	0.98	$0.98 + 6.875 \times 10^{-5}(AADT - 400)$	0.87

Note: The collision types related to shoulder width to which the *CMF* applies include single-vehicle run-off-the-road and multiple-vehicle head-on, opposite direction sideswipe, and same direction sideswipe crashes.
SOURCE: Based on *Highway Safety Manual*, 1st Edition, 2010, AASHTO, Washington, D.C.

Two-Way, Left-Turn Lanes (TWLTLs)

Rural two-lane, two-way highways are sometimes converted to three-lane cross sections by installing a center lane that provides for left turns in both directions. These are known as two-way, left-turn lanes (TWLTLs). The HSM gives the crash modification factor for TWLTL (CMF_{TWLTL}) as:

$$CMF_{TWLTL} = 1.0 - (0.7) \times (P_{dwy}) \times (P_{LT/D}) \tag{5.87}$$

where

CMF_{TWLTL} = crash modification factor for the effect of two-way left-turn lanes on total crashes

P_{dwy} = driveway-related crashes as a proportion of total crashes

$P_{LT/D}$ = left-turn crashes susceptible to correction by a TWLTL as a proportion of driveway-related crashes (estimated as 0.5)

$$P_{dwy} = \frac{(0.00752 \times DD) + (0.006144 \times DD^2)}{1.199 + (0.00752 \times DD) + (0.006144 \times DD^2)} \quad \text{and} \tag{5.88}$$

DD = driveway density considering driveways on both sides of the highway (driveways/km)

Note that Eq. 5.88 is applicable only when the driveway density (DD) is 5 or higher. Otherwise, $CMF_{TWLTL} = 1$.

Rumble Strips

Rumble strips are safety features installed at the shoulder near the travel lane along a paved roadway segment or at the centerline of a two-lane, two-way highway. When used on divided highways, they are usually installed on both the median and right shoulders. They are usually constructed of a series of elements that are indented, milled, or raised in the transverse direction and installed along the traveling direction. They are mainly used to alert inattentive drivers of the existence of a potential danger (leaving the roadway or travel lane) through the audible rumbling noise and vibration caused when the vehicle's wheels are driven over the strips.

The HSM suggests a value of 0.94 for the *CMF* for centerline rumble strips for total crashes on two-lane, two-way roadway segments. This value is applicable only to two-lane undivided highways that have only the centerline marking between the opposing travel lanes. The base condition for this *CMF* is a two-lane, two-way segment without a rumble strip. The *CMFs* for continuous shoulder rumble strips on rural multilane highways and freeways are given in Tables 5.29 and 5.30, respectively.

Roadside Recovery (Clear Zone) Distance

The distance available for a motorist to recover and either stop or return safely to the paved surface is referred to as the "roadside recovery distance" (also called the "clear zone" distance) and is a factor in crash reduction. Roadside recovery distance is measured from the edge of pavement to the nearest rigid obstacle, steep slope, nontraversable ditch, cliff, or body of water. Recovery distances are determined by averaging the clear zone distances measured at 3 to 5 locations for each mile. Although the effect of

Table 5.29 Potential Crash Effects for Installing Continuous Shoulder Rumble Strips on Multilane Highways

Treatment	Setting (Road Type)	Traffic Volume (AADT)	Crash Type (severity)	CMF	Std. Error
Install continuous milled-in shoulder rumble strips	Rural (multilane-divided)	2,000 to 50,000	All Types (All severities)	**0.84**	**0.1**
			All Types (Injury)	0.83	0.2
			SVROR (All severities)	*0.90**	0.3
			SVROR (Injury)	*0.78**	0.3

Base condition: Absence of shoulder rumble strip
Note: **Bold** text is used for the most reliable *CMFs*. These *CMFs* have a standard error of 0.1 or less.
Italic text is used for less reliable *CMFs*. These *CMFs* have a standard errors of 0.2 to 0.3.
SVOR = single-vehicle run-off-the-road crashes
*Observed variability suggests that this treatment could result in an increase, decrease, or no change in crashes.
SOURCE: Based on *Highway Safety Manual*, 1st Edition, 2010, AASHTO, Washington, D.C.

increasing clear zone distance is not quantitatively definitive, there is strong indication that the longer the clear zone distance, the less the average crash frequency for related crashes (run-off-the-road, head-on, and sideswipe). For example, Table 5.31 shows the percent reduction in related crashes as a function of recovery distance. It indicates that if roadside recovery is increased by 2.4 m, from 2.1 to 4.5 m, a 21 percent reduction in related crash types can be expected. Note that the percentage reductions are not *CMFs*, but results obtained from studies. Among the means to increase the roadside recovery distance are (1) relocating utility poles, (2) removing trees, (3) flattening sideslopes to a

Table 5.30 Potential Crash Effects for Installing Continuous Shoulder Rumble Strips on Freeways

Treatment	Setting (Road Type)	Traffic Volume	Crash Type (severity)	CMF	Std. Error
Install continuous, milled-in shoulder rumble strips	Urban/Rural		Specific SVROR (All severities)	**0.21**	**0.07**
Install continuous, milled-in shoulder rumble strips	Urban/Rural	unspecified	SVROR (All severities)	**0.82**	**0.1**
			SVROR (Injury)	*0.87*	*0.2*
	Rural (Freeway)		SVROR (All severities)	*0.79*	*0.2*
			SVROR (Injury)	*0.93**	*0.3*

Base condition : Absence of shoulder rumble strip
Note: **Bold** text is used for the most reliable *CMFs*. These *CMFs* have a standard error of 0.1 or less.
Italic text is used for less reliable *CMFs*. These *CMFs* have a standard errors of 0.2 to 0.3.
Specific SVOR = single-vehicle run-off-the-road crashes that have certain causes, including alcohol, drugs, inattention, inexperience, fatigue, illness, distraction, and glare.
*Observed variability suggests that this treatment could result in an increase, decrease, or no change in crashes.
SOURCE: Based on *Highway Safety Manual*, 1st Edition, 2010, AASHTO, Washington, D.C.

Table 5.31   Effect of Roadside Recovery Distance for Related Crashes

Amount of Increased Roadside Recovery Distance (m)	Reduction in Related Crash Types (percent)
1.5	13
2.4	21
3.0	25
3.6	29
4.5	35
6.0	44

SOURCE: *Safety Effectiveness of Highway Design Features*, Volume III, U.S. Department of Transportation, Federal Highway Administration, Washington, D.C., November 1992.

maximum 4:1 ratio, and (4) removing other obstacles, such as bridge abutments, fences, mailboxes, and guardrails. When highway signs or obstacles such as mailboxes cannot be relocated, they should be mounted so as to break away when struck by a moving vehicle, thus minimizing crash severity.

Intersection Sight Distance

Crash rates also are affected by the sight distance available to motorists as they approach an intersection. Stopping sight distance is affected by horizontal and vertical alignment. Vertical curve lengths and horizontal curve radii should be selected to conform to design speeds; when this is not feasible, advisory speed limit signs should be posted. The ability to see traffic that approaches from a cross street is dependent on obtaining a clear diagonal line of sight (see Chapter 7). When blocked by foliage, buildings, or other obstructions, the sight line may be insufficient to permit a vehicle from stopping in time to avoid colliding with side street traffic. Figure 5.6 illustrates how sight distance is improved

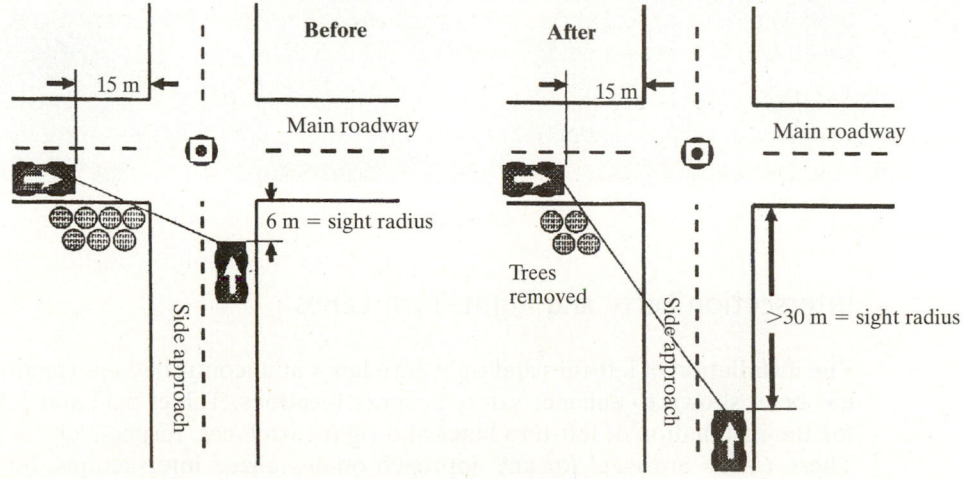

Figure 5.6   Increased Sight Radius by Removal of Obstacles

SOURCE: *Safety Effectiveness of Highway Design Features*, Volume V, U.S. Department of Transportation, Federal Highway Administration, Washington, D.C., November 1992

Table 5.32   Crash Reduction, Per Year, Due to Increased Intersection Sight Distance

ADT[2]	Increased Sight Radius[1]		
	6–14.7 m	15–29.7 m	>30 m
<5000	0.18	0.20	0.30
5000–10,000	1.00	1.3	1.40
10000–15,000	0.87	2.26	3.46
>15,000	5.25	7.41	11.26

[1]At 15 m from intersection, increasing obstruction on approaching leg from initial < 6 m from intersection.
[2]Average Daily Traffic
SOURCE: *Safety Effectiveness of Highway Design Features*, Volume V, U.S. Department of Transportation, Federal Highway Administration, Washington, D.C., November 1992

when trees are removed near an intersection, and Table 5.32 indicates the expected reduction in number of crashes per year as a function of ADT and increased sight radius. Note that the values shown in Table 5.32 are not *CMFs* given in the HSM, but are indications of results from studies. However, in the design of an intersection, in order to account for adequate sight distance, the approach velocity must be taken into account, as explained in Chapter 7.

Example 5.16 Calculating the Crash Reduction Due to Increased Sight Radius

A motorist is 15 m from an intersection and sees a vehicle approaching from the right when it is 6 m from the intersection. After removal of the foliage that has been blocking the sight line, it is now possible to see the same vehicle when it is 22.5 m from the intersection. Average daily traffic volumes on the main roadway are 12,000 veh/day. Prior to removal of the obstructing foliage, the average number of crashes per year was 8.6. Determine the expected number of crashes per year after the foliage has been removed based on the research data provided in Table 5.32.

Solution: From Table 5.32, the crash reduction (CR) is 2.26 crashes/year. The expected average number of crashes/year = 8.60 − 2.26 = 6.34 crashes/year.

Intersection Left- and Right-Turn Lanes

The installation of left-turn and right-turn lanes at uncontrolled intersection approaches has been shown to enhance safety at these locations. Tables 5.33 and 5.34 give *CMFs* for the installation of left-turn lanes and right-turn lanes, respectively, at intersections. These *CMFs* are valid for any approach on signalized intersections, but only for the uncontrolled major road approaches at unsignalized intersections. Also, there is an indication that the safety effect of installing a left-turn lane or a right-turn lane on an approach that is controlled by a stop sign is negligible and is therefore not considered in using the *CMFs* given in Tables 5.33 and 5.34.

Table 5.33   Crash Modification Factors (*CMFs*) for Installing Left-Turn Lanes on Intersection Approaches

Intersection Type	Intersection Traffic Control	Number of Approaches with Left-Turn Lanes[a]			
		One Approach	Two Approaches	Three Approaches	Four Approaches
Three-Leg	Minor road stop[b] control	0.86	0.74	—	—
Four-Leg	Minor road stop[b] control	0.86	0.74	—	—
	Traffic signal	0.96	0.92	0.88	0.85

[a]Stop controlled approaches are not considered in determining the number of approaches with left-turn lanes.
[b]Stop signs present on minor road approaches only.
Note: Base condition is no left-turn lane at the intersection approaches.
SOURCE: Based on *Highway Safety Manual*, 1st Edition, 2010, AASHTO, Washington, D.C.

Table 5.34   Crash Modification Factors (*CMFs*) for Installing Right-Turn Lanes on Intersection Approaches

Intersection Type	Intersection Traffic Control	Number of Approaches with Left-Turn Lanes[a]			
		One Approach	Two Approaches	Three Approaches	Four Approaches
Three-Leg	Minor road stop[b] control	0.56	0.31	—	—
Four-Leg	Minor road stop[b] control	0.72	0.52	—	—
	Traffic signal	0.82	0.67	0.55	0.45

[a]Stop controlled approaches are not considered in determining the number of approaches with left-turn lanes.
[b]Stop signs present on minor road approaches only.
Note: Base condition is no right-turn lane at the intersection approaches.
SOURCE: Based on *Highway Safety Manual*, 1st Edition, 2010, AASHTO, Washington, D.C.

5.8 SAFETY EFFECTS OF PEDESTRIAN FACILITIES

The safety of pedestrians is of great concern to traffic and highway engineers. Efforts to reduce pedestrian and bicycle crashes involve education, enforcement, and engineering measures, as is the case for motor vehicle crashes. In addition, characteristics of pedestrian crashes indicate that factors related to occurrence include age, sex, alcohol use, time of day, urban or rural area type, and intersection or midblock crossing locations. For example, it is known that fatality rates increase sharply for pedestrians over 70 years of age, and that the highest crash rates occur for males 15 to 19 years old. Peak crash periods occur in the afternoon and evening hours, and over 85 percent of all nonfatal crashes occur in urban areas. Approximately 65 percent of all pedestrian crashes occur at locations other than intersections, and many of these involve younger children who dart out into the street. The various types of pedestrian crashes and percentage occurrence are listed in Table 5.35. Note that dart-out at locations account for over one-third of the 14 crash types listed. The most common types of pedestrian crashes are illustrated in Figure 5.7. The principal geometric design elements that are used to improve pedestrian safety are (1) sidewalks, (2) overpasses or tunnels, (3) raised islands, (4) auto-free shopping streets, (5) neighborhood traffic

Table 5.35   Pedestrian Crash Types and Frequency

DART-OUT (FIRST HALF) (24%)
  Midblock (not at intersection)
  Pedestrian sudden appearance and short time exposure (driver does not have time to react)
  Pedestrian crossed less than halfway

DART-OUT (SECOND HALF) (10%)
  Same as above, except pedestrian gets at least halfway across before being struck

MIDBLOCK DASH (8%)
  Midblock (not at intersection)
  Pedestrian is running but *not* sudden appearance or short time exposure as above

INTERSECTION DASH (13%)
  Intersection
  Same as dart-out (short time exposure or running) except it occurs at an intersection

VEHICLE TURN-MERGE WITH ATTENTION CONFLICT (4%)
  Vehicle is turning or merging into traffic
  Driver is attending to traffic in one direction and hits pedestrian from a different direction

TURNING VEHICLE (5%)
  Vehicle is turning or merging into traffic
  Driver attention is *not* documented
  Pedestrian is *not* running

MULTIPLE THREAT (3%)
  Pedestrian is hit as he steps into the next traffic lane by a vehicle moving in the same direction as vehicle(s)
that stopped for the pedestrian, because driver's vision of pedestrian obstructed by the stopped vehicle(s)

BUS STOP–RELATED (2%)
  Pedestrian steps out from in front of bus at a bus stop and is struck by vehicle moving in same direction as
bus while passing bus

VENDOR–ICE CREAM TRUCK (2%)
  Pedestrian struck while going to or from a vendor in a vehicle on the street

DISABLED VEHICLE–RELATED (1%)
  Pedestrian struck while working on or next to a disabled vehicle

RESULT OF VEHICLE-VEHICLE CRASH (3%)
  Pedestrian hit by vehicle(s) as a result of a vehicle-vehicle collision

TRAPPED (1%)
  Pedestrian hit when traffic light turned red (for pedestrian) and vehicles started moving

WALKING ALONG ROADWAY (1%)
  Pedestrian struck while walking along the edge of the highway or on the shoulder

OTHER (23%)
  Unusual circumstances, not countermeasure corrective

SOURCE: *Safety Effectiveness of Highway Design Features*, Volume VI, U.S. Department of Transportation, Federal Highway Administration, Washington, D.C., November 1992

control to limit speeding and through traffic (see Section 5.9, Safety Effect of Traffic Calming), (6) curb cuts that assist wheelchair users and pedestrians with baby carriages, and (7) shoulders that are paved and widened. Other traffic-control measures that may assist pedestrians include crosswalks, traffic signs and signals, parking regulations, and lighting. Sidewalks and pedestrian paths can significantly improve safety

Figure 5.7 Common Types of Pedestrian Crashes

SOURCE: Adapted from *Safety Effectiveness of Highway Design Features*, Volume VI, U.S. Department of Transportation, Federal Highway Administration, Washington, D.C., November 1992

in areas where the volumes of automobile and pedestrian traffic are high. The sidewalk provides a safe and separated lane intended for the exclusive use of pedestrians. However, sidewalks should not be used by higher-speed nonmotorized vehicles, such as bicycles. Guidelines for the minimum width and location of sidewalks are given in

Safety Effectiveness of Highway Design Features, published by the Federal Highway Administration. Sidewalks are often provided on grade-separated structures, such as overpasses or subways, when crossings involve freeways or expressways that carry high-speed and high-volume traffic. They are most effective when pedestrian demand is high (for example, as a connector between a residential area and destinations such as schools, hospitals, and shopping areas). Pedestrians will use these facilities if they are convenient and do not require circuitous travel, but they will select the alternative unsafe path if they are required to walk significantly farther on the overcrossing or through the tunnel.

Traffic safety in residential neighborhoods is a major concern, especially in suburban areas where through traffic uses residential streets as a shortcut, thus bypassing congested arterials and expressways. Typically, citizens protest when they perceive that their neighborhoods are becoming more dangerous for children and others who are required to walk along the same roadways as moving traffic. Several geometric designs have been developed in the United States and in Europe to create more pedestrian-friendly environments (see Section 5.9).

5.9 SAFETY EFFECTS OF TRAFFIC CALMING STRATEGIES

Traffic calming can be described as the implementation of strategies such as changes in street alignment, installation of barriers, and other physical measures to change driver behavior in order to reduce traffic speeds and or cut-through volumes. Successful implementation of one or more appropriate calming strategies for a given roadway environment will usually result in one or more safety benefits that include the reduction of conflicts between local traffic and through traffic, reduction of conflicts between traffic and pedestrians, reduction in the crash risk involving pedestrians crossing the roadway, and reduction in traffic noise. Traffic calming strategies can be generally grouped into two categories: (i) those that reduce traffic speeds and (2) those that reduce traffic volume. Note that although each of these strategies is primarily used to either reduce speeds or volumes, some can also have significant impact on both speeds and volumes. Although the HSM does not give *CMFs* for all of these strategies, a brief description of each is given followed by a summary of their respective effectiveness.

5.9.1 Traffic Calming Strategies that Reduce Traffic Speeds

The traffic calming strategies primarily used to reduce traffic speeds include:

- Speed humps
- Speed tables
- Raised crosswalks
- Raised intersections
- Textured pavements
- Neighborhood traffic circles
- Roundabouts
- Chicanes
- Neckdowns
- Center-island narrowings
- Chokers

Speed Humps

These are rounded raised areas across the roadway pavement, usually about 10 to 14 ft long in the direction of travel and about 3 to 4 inches high. The rounded area is usually of a circular, parabolic, or sinusoidal shape, and in order to facilitate drainage, the ends are tapered toward the curb. They are used mainly in neighborhoods where traffic speeds are required to be very low. The associated advantages are that (1) they can be installed at a relatively low cost, (2) bicycles can easily transverse it when properly designed and constructed, and (3) they are effective in reducing traffic speeds (see Table 5.36). The disadvantages associated with speed humps are that (1) emergency vehicles such as ambulances and fire trucks have to travel at a much lower speed, (2) riding over these may be rough and result in pain for some people, and (3) noise and air pollution may increase.

Speed Tables

Speed tables are similar to speed humps, but with the top being flat instead of being rounded. Brick is often used as the construction material for the flat section, which is usually of a length that allows a passenger car to rest on it. Also, because of the flat top

Table 5.36   Impact of Traffic Calming Measures on Speed

Calming Measure	Sample Size	85th Percentile Speed Afterward* (mi/h)	Average Change in 85th Percentile Speed*(mi/h)	Average Percentage Change
12′ Speed Hump	179	27.4 (4.0)	−7.6 (3.5)	−22 (9)
14′ Speed Hump	15	25.6 (2.1)	−7.7 (2.1)	−23 (6)
22′ Speed Table	58	30.1 (2.7)	−6.6 (3.2)	−18 (8)
Longer Speed Table (> 22′)	10	31.6 (2.8)	−3. (2.4)	−9 (7)
Raised Intersection	3	34.3 (6.0)	−0.3 (3.8)	−1 (10)
Traffic Circle	45	30.3 (4.3)	−3.9 (3.2)	−11 (10)
Narrowing	7	32.3 (2.8)	−2.6 (5.5)	−7 (22)
Choker	5	28.6 (3.1)	−2.6 (1.3)	−14 (4)
Half Closure	16	26.3 (5.2)	−6.0 (3.6)	−19 (11)
Diagonal Diverter	7	27.9 (5.2)	−1.4 (4.7)	0 (17)

*(Standard deviations in parentheses)
Note: Speeds are measured at midpoints between measures.
SOURCE: Adapted from Effectiveness Speed Impacts of Traffic Calming Measures, Fehr & Peers, http://www.TrafficCalming.org, 2012 Used by permission

section, speeds could be higher than for those on speed humps. The main advantage associated with speed tables is that they are effective in lowering speed (see Table 5.36) while providing a smoother ride for large vehicles such as fire trucks, and the main disadvantage is that noise and air pollution may increase when used.

Raised Crosswalks

Raised crosswalks are speed tables with crosswalk markings and appropriate signs that provide a crossing facility for pedestrians. They are commonly used at locations where traffic speeds are high and pedestrians tend to cross randomly. The main associated advantages are that (1) speeds tend to be reduced when used, although not to the same extent as speed humps, and (2) they tend to improve pedestrian and vehicle safety. The main associated disadvantages are that (1) noise and air pollution may increase, and (2) they may have a negative impact on drainage if not properly constructed.

Raised Intersections

A raised intersection has its whole area constructed as a flat raised pavement that is connected to the approaches by ramps. Raised intersections are often raised to the sidewalk level, although they are sometimes raised just below the side walk level to facilitate their detection by the vision-impaired pedestrian. The flat raised area is usually constructed with brick or other textured materials. Raised intersections are used mainly at locations where pedestrian volume is high and where space restriction precludes the use of other traffic calming features. The main associated advantages are as follows: (1) pedestrian and vehicle safety is improved, (2) the calming effect is on two streets at the same time, and (3) motorists are influenced to perceive the crosswalks as areas for pedestrians. The associated disadvantages are that (1) they can be expensive, depending on the type of material used to construct the flat area, (2) they may negatively impact the drainage at the intersection, and (3) they are not as effective in reducing speeds when compared to other traffic calming features, such as speed humps or raised crosswalks.

Textured Pavements

These are pavements that are constructed of paving materials that give an uneven texture to the surface that vehicles travel on. They are often used where pedestrian volume is high and noise is not considered a problem. They can be used to highlight either a specific location such as an intersection, or pedestrian crossing, or can be used on a street segment that is several blocks long. The main associated advantages are that (1) speeds on extended lengths of streets can be reduced, and (2) the calming effect is on two streets at the same time. The associated disadvantages are that (1) they can be expensive, depending on the type of material used for the construction, and (2) when implemented, wheelchair users and vision-impaired pedestrians find it difficult to cross the street.

Neighborhood Traffic Circles

These are raised circular islands or circular pavement markings at intersections that require vehicles to travel around them rather than moving directly across the intersections. They are mainly used in neighborhoods to reduce speeds and to reduce through vehicular volumes on the streets. The main associated advantages are that (1) speeds are effectively reduced, and (2) the calming effect is on two streets at the same time. The main associated disadvantages are that (1) they create difficulties for large vehicles such as fire trucks to transverse the intersection, and (2) they may encroach on the crosswalk if not properly designed.

Roundabouts

These are usually raised circular islands located at intersections with diameters much larger than those for traffic circles (see Chapter 7 for design characteristics). They are mainly used at high vehicular volume locations, and they provide for an orderly allocation of right-of-way among different conflicting traffic movements. They have been used to reduce crashes at intersections with high crash history, to reduce queue lengths, and to effectively cope with a high proportion of U-turns. The main advantages associated with roundabouts are that they (1) may improve safety at some intersections when compared with traffic signals, and (2) tend to reduce traffic speeds on arterials. The main disadvantage associated with roundabouts is that they may require a much larger right-of-way when compared with traffic signals.

Chicanes

These are S-shaped curves formed on the roadway by extending the curb on one side of the road to the other, alternatively. Alternate on-street parking between one side of the street and the other can also be used to form chicanes. They are mainly used in neighborhoods where it is necessary to reduce speeds but noise will not be tolerated. The main advantage of chicanes is that they effectively reduce vehicular speed, and the main disadvantage is that curb realignment is required, which may be costly.

Neckdowns

A neckdown is formed by extending the curbs at an intersection and thereby reducing the roadway width between curbs at the intersection approaches. This facilitates pedestrian crossing as the crossing distance is reduced. The curb radii at the corners are also sometimes reduced so that the speeds of turning vehicles have to be reduced. Neckdowns are mainly used in neighborhoods where intersections have significant pedestrian volumes but increase in noise level is not acceptable. The main advantages of neckdowns are that (1) pedestrian space and circulation are improved, (2) large vehicles can easily make left and through movements, (3) protected on-street parking can be provided, and (4) speeds are reduced, particularly for vehicles making right turns. The main disadvantage is that speeds of emergency vehicles may be reduced.

Center-Island Narrowings

A center-island narrowing is formed by constructing a raised island along the centerline of a street, so as to reduce the width of the travel lanes at the location of the narrowing. Center-island narrowings are usually installed at the entrance of a neighborhood so as to influence drivers to reduce their speeds as they enter the neighborhood and are therefore alternatively called "gateway islands." In order to capture the attention of drivers, they are usually combined with textured pavement and landscaped to provide an attractive feature. They are also sometimes referred to as "pedestrian refuges" when a gap is fitted within the island to form a walkway for pedestrian crossing. They are mainly used at the entrances of neighborhoods with wide streets that have to be crossed by pedestrians. The main associated advantages are that (1) pedestrian safety is increased, and (2) they may reduce traffic volumes. The main disadvantage is that they are not as effective to reduce speeds when compared with other traffic calming features because of the lack of vertical or horizontal deflection.

Chokers

Chokers effectively reduce the width of the street at mid-blocks by increasing the width of the sidewalk. Chokers can also be marked as a crosswalks, and they are then known as safe crosses. There are two types of chokers: (1) two-lane chokers, which provide for narrower but adequate width for vehicular travel in both directions, and (2) one-lane chokers, which provide adequate width for vehicular travel in only one direction. In this case, the section operates like a one-lane bridge. Chokers are used mainly at neighborhood locations where substantial speed reduction is required and there is adequate on-street parking. The main advantages associated with the use of chokers are that (1) large trucks such as fire trucks can be easily maneuvered at chokers, and (2) speeds and volumes are reduced. The main disadvantage is that bicyclists may have to merge with vehicular traffic during short periods.

Figure 5.8 gives examples of some traffic calming features. Table 5.36 gives some study results on the speed impacts of traffic calming features, and Table 5.37 gives some study results on the safety impacts of some traffic calming features.

5.9.2 Traffic Calming Strategies That Reduce Volumes

The traffic calming strategies primarily used to reduce volumes include:

- Full street closure
- Half street closure
- Diagonal diverters
- Median barriers

Table 5.37   Impact of Traffic Calming Measures on Crashes

Calming Measure	Number of Observations	Average Number of Collisions Without Measure	Average Number of Collisions With Measure	Percentage Change in Number of Collisions
12' Speed Hump	49	2.7	2.4	−11
14' Speed Hump	5	4.4	2.6	−41
22' Speed Table	8	6.7	3.7	−45
Traffic Circle Without Seattle[1]	17	5.9	4.2	−29
Traffic Circle With Seattle[2]	130	2.2	0.6	−73

[1]excluding Seattle data
[2]including Seattle data
SOURCE: Adapted from *Effectiveness, Speed Impacts on Traffic Calming Measures*, Fehr & Peers, http://www.TrafficCalming.org, 2012. Used by permission.

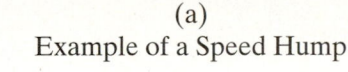

(a)
Example of a Speed Hump

(b)
Example of a Speed Table

SOURCE: Tomas Skopal/Shutterstock.com

SOURCE: Courtesy of Traffic Logix

Figure 5.8   Examples of Traffic Calming Features (*continued*)

SOURCE: Photo by Andy Hamilton, WalkSanDiego

(c)
Raised Crosswalk

Figure 5.8 Examples of Traffic Calming Features (*continued*)

SOURCE: Photo by Andy Hamilton, WalkSanDiego

(d)
Example of a Neighborhood Traffic Circle

Figure 5.8 Examples of Traffic Calming Features (*continued*)

SOURCE: Jorg Hackemann/Shutterstock.com

(e)
Example of Chicanes

Figure 5.8   Examples of Traffic Calming Features (*continued*)

SOURCE: Richard Drdul

(f)
Example of Textured Pavement

Figure 5.8   Examples of Traffic Calming Features

SOURCE: littleny / Shutterstock.com

Full Street Closure

This involves the complete closure of the street by placing barriers across the street. No through traffic is allowed on the street, but the sidewalks are open. Full street closure is used where traffic-volume-related problems have been very serious and other measures have not been successful. The main associated advantages are that (1) both pedestrian access and bicycle access are maintained and (2) full street closure is extremely effective in reducing traffic volume. The main associated disadvantages are that (1) implementation may result in local residents and emergency vehicles having to take circuitous routes to enter or leave the area, (2) full street closure may have significant negative effect on businesses, and (3) there may be legal barriers to overcome before implementing full street closure.

Half Closures

Half closures involve the closing of the road by barriers that allow traffic flow in only one direction on streets that are normally two-way streets. They are also used where traffic-volume-related problems have been very serious and other measures have not been successful. The main associated advantage of half closure are that (1) bicycles can be ridden in both directions, and (2) they are effective in reducing traffic volume. The main associated disadvantages are that (1) implementation may result in local residents and emergency vehicles having to take circuitous routes to enter or leave the area, and (2) they may restrict access to local businesses.

Diagonal Diverters

The main objective of diagonal diverters is to discourage through-vehicles in neighborhoods. This is achieved by placing diagonal barriers at intersections to form two separate L-shaped streets at each intersection, thereby blocking any through-traffic movement. These diagonal barriers are placed in a staggered manner so that through-traffic will have to use a circuitous route, thereby discouraging nonresident drivers from using the neighborhood streets. Diagonal diverters are mainly used at inner neighborhoods that have through-traffic volume problems. The main associated advantages are that (1) streets are not necessarily closed, but traffic is redirected; (2) access is provided to bicycle and pedestrian movements, and (3) they are effective in reducing traffic volume.

Median Barriers

A median barrier is formed by installing islands along the centerline of a street across an intersection, so that traffic from the cross street cannot go through the intersection. Traffic on the cross street will therefore have to make a right turn, with left turns permitted or not permitted, depending on the design of the barrier. Median barriers are mainly used at locations where the cross street through-traffic has historically experienced problems, and where left-turn traffic to and/or from the side street is not safe. The main advantage associated with median barriers is that it improves safety by disallowing unsafe turning movements. The main associated disadvantage is that local traffic and emergency traffic are restricted from making turns freely from and onto the side streets.

Table 5.38 shows some study results on the safety impacts of some traffic calming features on volume.

Table 5.38   Impact of Traffic Calming Measures on Volume

Calming Measure	Sample Size	Average Change in Volume* (veh/day)	Average Percentage Change*
Choker	5	−392 (384)	−20 (19)
Half Closure	19	−671 (786)	−44 (36)
Half Closure	53	−1611 (2444)	−42 (41)
Diagonal Diverter	27	−501 (622)	−35 (46)

(Standard deviations in parentheses)
Note: Speeds are measured at midpoints between measures.
SOURCE: Adapted from *Effectiveness, Speed Impacts on Traffic Calming Measures*, Fehr & Peers, http://www.TrafficCalming.org, 2012. Used by permission.

5.10 SAFETY IMPACT OF INTELLIGENT TRANSPORTATION SYSTEMS (ITS)

ITS, which is also referred to as telematics, generally can be described as the process through which data on the movement of people and goods can be collected, stored, analyzed, and disseminated. The process has been used in many areas of transportation engineering, including the management of traffic, public transportation, traveler information, electronic toll payment, and safety. ITS as related to highways can be described as control and electronic technologies that incorporate communication-based information to manage flow, reduce congestion, and improve safety. These technologies consist of two main categories: (1) intelligent vehicles, and (2) intelligent infrastructure. Both categories of technologies have the potential to reduce the probability of a crash occurring on a highway and/or reduce the severity of a crash when it does occur. This section briefly discusses those technologies that are in the category of intelligent infrastructure and are highway safety related. Many of the intelligent infrastructure technologies that are related to the collection of data were discussed in Chapter 4. The ITS technologies that are in the category of intelligent infrastructure and are related to highway safety include:

- Arterial and freeway management systems
- Incident management and emergency response systems
- Crash prevention and safety strategies
- Roadway operations and maintenance systems

5.10.1 Arterial and Freeway Management Systems

These systems generally employ traffic detectors, dynamic message signs, highway advisory radio, and traffic signals to give safety-related information to drivers. For example, in Colorado, truck crashes were reduced by 13 percent and runaway ramp usage reduced by 24 percent when drivers were warned of a steep downhill and to reduce their speeds. Also, in Minneapolis–St. Paul, crashes were reduced by 22 percent as a result of ramp metering (see Chapter 8). Emerging applications of active traffic management (ATM) on freeways have also been shown to improve safety. These techniques include queue warning systems, which alert drivers to slow or stopped traffic ahead, and variable speed limit systems, which provide speed limits that are based on current observed travel conditions on a road.

5.10.2 Incident Management and Emergency Response Systems

These systems are used to give real-time information to drivers and incident responders, which often results in reduction of response time and the time taken to achieve normal traffic conditions after a crash, and in turn, results in fewer related crashes. For example, the Coordinated Highways Action Response Team program in Maryland resulted in 55 percent reduction in incidents, and incidents were reduced in Pennsylvania by about 40 percent with the implementation of traffic and incident management systems.

5.10.3 Crash Prevention and Safety Strategies

These are also used to warn drivers of upcoming dangerous locations, such as dangerous ramps, sharp curves on gradients, and highway-rail crossings, by providing advanced warning signs. For example, drivers are immediately warned of pedestrians entering a crosswalk by the activation of an automatic in-pavement lighting system or by a bicycle warning system that alerts drivers of the presence of a bicyclist on an approaching segment of road. In areas where collisions with animals are prevalent, the presence of animals on the road can be detected by using radar.

5.10.4 Roadway Operations and Maintenance Systems

These systems can be used to optimize traffic flow while improving safety, particularly at work zones (for example, the proper use of the early or late merging systems for vehicles approaching a work zone with a lane reduction). Using appropriate signing, the early merge system encourages drivers to merge onto the open lane at a significant distance before the beginning of the lane drop, while the late merge system encourages drivers to use all open lanes until they get to the start of the lane drop, where drivers should alternatively yield to each other. There are several safety benefits associated with both the early merge and late merge systems depending on the prevailing traffic conditions.

5.11 SUMMARY

Safety is an important consideration in the design and operation of streets and highways. Traffic and highway engineers working with law enforcement officials are constantly seeking better ways to ensure safety for motorists and pedestrians. A highway safety improvement process involves planning, implementation, and evaluation. The planning process requires that engineers collect and maintain traffic safety data, identify hazardous locations, conduct engineering studies, and establish project priorities. This chapter presents the methods of recording and retrieving crash data. Current methods to adjust crash frequency to account for regression to the mean (RTM) bias are discussed and methods for predicting average crash frequency at a site, including the use of safety performance functions given in the *Highway Safety Manual*, are presented. Different methods for screening and prioritizing sites for detailed safety studies are also discussed. The effects of physical and geometric characteristics on safety are also discussed, including the quantification of these effects by the use of appropriate crash modification factors. The quality of geometric design will also influence the safety of streets and highways.

Roadways should be designed to provide adequate sight distance, to separate through-traffic from local traffic, to avoid speed variations, and to ensure that the driver is aware of changes occurring on the roadway and has adequate time and distance to change speed or direction without becoming involved in a collision with another vehicle, a fixed object, or a pedestrian. In this chapter, we have described a variety of actions that can be taken to improve the highway itself. However, safety on the highway is also dependent on a well-educated population that drives courteously and defensively; in addition, highway safety requires enforcement of the rules of the road.

PROBLEMS

5-1   Explain the difference between short-term and long-term driver expectancy, and provide an example of each.

5-2   Briefly explain the steps in the roadway safety management process as outlined in the *Highway Safety Manual*.

5-3   Describe the type of information on a collision diagram.

5-4   Identify and explain the system for identifying levels of crash severity as used in the *Highway Safety Manual*.

5-5   Determine the crash modification factor (using Table 5.14) that can be used to estimate the change in frequency of the crash type targeted by a proposed improvement that will widen a two-lane rural highway from 5.4 to 6.6 m (18 to 22 ft) if the highway has an *AADT* of 1250 veh/day.

5-6   In Problem 5-5, if there are an average of 25 crashes per year on the segment proposed for improvement, and 62 percent of the crashes are of the type targeted by the improvement, estimate the expected number of crashes (per year) after the improvement is constructed.

5-7   Determine the predicted average crash frequency on a 8.32-km (5.2-mile) segment of a rural divided multilane highway with 3.3-m (11-ft) lanes and 2.4-m (8-ft) paved shoulders that carries an *AADT* of 11,200 vehicles per day.

5-8   A local jurisdiction has determined that for a given set of geometric conditions, a maximum rate of 8 crashes/million entering vehicles can be tolerated. At an intersection of two roadways with ADTs of 10,000 and 7500, how many crashes can occur before corrective action must be sought?

5-9   Studies were conducted at two sites on rural roads with similar characteristics. The first site was 5.1 mi (8.16 km) in length with an ADT of 6700. Over the year-long study period, 28 crashes occurred on this portion of roadway, five of them resulting in fatalities. The second site was a 11.2 mi (17.92 km) section with an ADT of 5300. There were 32 crashes in this section with four crashes resulting in fatalities. Determine the total crash rate (all types of crashes) and the fatal crash rates for both locations, and discuss the implications.

5-10   A 5.4-km segment of highway has the following traffic and safety data:

	2005	2006	2007
Annual average daily traffic (*AADT*)	15,200	16,300	17,400
Number of fatal crashes	3	2	3
Number of crashes	29	31	27

Determine the fatal crash and total crash rates (per 100 million VKT) for each year and for the entire three-year period. Comment on and qualify the results as necessary.

5-11   Describe the categories used to summarize crash data. Give an example of how each category would be used to evaluate a given location.

5-12   A review of crash records shows that a signalized intersection is a hazardous location because of an abnormally high number of rear-end collisions. What are the possible causes of these crashes, and what data should be collected to determine the actual causes?

5-13   A particular section of highway has experienced 3 fatal crashes, 29 injury crashes, and 32 PDO crashes during the study period. Determine the equivalent number of PDO crashes.

5-14   At an unsignalized intersection, 8 rear-end crashes, 21 angle crashes, and 4 head-on crashes were observed during a three-year period. Using the cost data in Table 5.22, determine the relative severity index at this site.

5-15   The crash rate on a heavily-traveled two-lane rural highway is abnormally high. The corridor is 14 km long with an ADT of 34,000. An investigation has determined that head-on collisions are most common with an $RMVM_T$ of 4.5, and are caused by vehicles attempting to pass. Determine an appropriate countermeasure and calculate the estimated yearly reduction in total crashes.

5-16   The numbers of crashes occurring for over a period of three years for different levels (high, medium, low) of $AADT$ at unsignalized rural intersections are given in the table below. Using the Kruskal-Wallis H test determine whether it can be concluded that the distribution of crashes at unsignalized rural intersections is the same for all $AADT$ levels. Use a significance level of 5 percent.

Number of Crashes

Low AADT	Medium AADT	High AADT
6	34	13
35	28	35
3	42	19
17	13	4
11	40	29
30	31	0
15	9	7
16	32	33
25	39	18
5	27	24

5-17   A rural primary road segment of 2 km long has an average annual daily traffic ($AADT$) of 11,350. The number of crashes that have occurred over the past 5 years are 5 fatal, 55 injury crashes, and 100 property damage only crashes. Statewide average crashes for this type of road are 2 fatal, 130 injury, and 300 property damage only crashes per 100 MVMT. The weight factors for fatal, injury, and property damage only crashes are 8, 3, and 1 respectively. Using the critical crash ratio methodology, determine whether this site can be labeled a hazardous site, using a 95 percent confidence limit.

5-18   The applicable safety performance functions for injury, fatal, and property damage only crashes are:

$$SP_{inj} = 1.602 \times 10^{-6} (length\ in\ m)(AADT)^{0.491}$$

$$SP_{PDO} = 3.811 \times 10^{-6} (length\ in\ m)(AADT)^{0.491}$$

For the data given below, determine:

  i.   The long term average number of injury crashes per year at the site
  ii. The long term average number of fatal crashes per year at the site
  iii. The long term average number of property damage only crashes per year at the site

Data:

Length of road segment = 7585 m

$AADT$ = 5250 veh/day

Total number of injury crashes for 10 years = 12

Total number of fatal crashes for 10 years = 3

Total number of property only crashes for 10 years = 48

Assume k = 2.51

5-19   A state transportation agency wants to evaluate engineering countermeasures applied to a two-lane, two-way rural highway segment. This 4-mile segment has been identified as having 14, 19, and 23 total crashes in the last three years, before the implementation of the countermeasures. The AADT at the segment for the last three years is 5,630 vehicles per day, and the AADT for after implementation is 6,240 vehicles per day. Compute the expected average crash frequency for the after period, assuming the treatment was not implemented. The segment has the same conditions as the base conditions established for the SPF.

5-20   Determine the crash modification factor for a horizontal curve on a two-lane rural highway if the curve has a radius of 430 m, a length of 188 m, and no spiral transitions.

5-21   Determine the crash modification factor for access control applied to a highway with an $AADT$ of 32,100 vehicles per day and a driveway density of 18 driveways per km.

5-22   Residents of a local neighborhood have been complaining to city officials that vehicles are using their side streets as shortcuts to avoid rush hour traffic. Discuss the options available to the city transportation officials to address the residents' concerns.

5-23   A two-lane rural highway, four-leg intersection will have its stop control replaced by a signal control. Compute the expected average total crash frequency (all types, all severities) after the change for the following conditions:

Expected average crash frequency before change: 23 total crashes/yr

$AADT$ on minor road: 1500 veh/day

$AADT$ on major road: 5000 veh/day

(a) Determine crash modification factor (CMF). Use Table 5.15:

   $CMF = 0.56$

(b) The expected crash frequency after replacing the stop sign by a signal, will be:

   Expected frequency after change = 23 × 0.56 = 12.88 crashes/year

5-24   Survey your local college campus. What pedestrian facilities are provided? How might pedestrian safety be improved?

REFERENCES

Ezra Hauer, *Observational Before-After Studies in Road Safety, Estimating the Effect of Highway and Traffic Engineering Measures on Road Safety*, Pergamon, 1997.

Highway Safety Manual, 1st edition, American Association of State Highway and Transportation Officials, Washington, D.C., 2010.

Guidance for Implementation of the AASHTO Strategic Highway Safety Plan, National Cooperative Highway Research Program, NCHRP, Transportation Research Board, Washington, D.C., 2003.

Safety Effectiveness of Highway Design Features, U.S. Department of Transportation, Federal Highway Administration, Washington, D.C., November 1992. Volume I, *Access Control*; Volume II, *Alignment*; Volume III, *Cross-Sections*; Volume IV, *Interchanges*; Volume V, *Intersections*; Volume VI, *Pedestrians and Bicyclists*.

Safety Management System, U.S. Department of Transportation, Federal Highway Administration, FHWA-HI-95-012, February 1995.

Status Report: Effective Highway Accident Countermeasures, U.S. Department of Transportation, Federal Highway Administration, Washington, D.C., June 1993.

Symposium on Effective Highway Accident Countermeasures, U.S. Department of Transportation, Federal Highway Administration, and National Highway Traffic Safety Administration, Washington, D.C., August 1990.

CHAPTER 6

Fundamental Principles of Traffic Flow

Traffic flow theory involves the development of mathematical relationships among the primary elements of a traffic stream: flow, density, and speed. These relationships help the traffic engineer in planning, designing, and evaluating the effectiveness of implementing traffic engineering measures on a highway system. Traffic flow theory is used in many aspects of design; for example, to determine adequate lane lengths for storing left-turn vehicles on separate left-turn lanes, the average delay at intersections and freeway ramp merging areas, and changes in the level of freeway performance due to the installation of improved vehicular control devices on ramps. Another important application of traffic flow theory is simulation, where mathematical algorithms are used to study the complex interrelationships that exist among the elements of a traffic stream or network and to estimate the effect of changes in traffic flow on factors such as crashes, travel time, air pollution, and gasoline consumption.

Methods ranging from physical to empirical have been used in studies related to the description and quantification of traffic flow. This chapter, however, will introduce only those aspects of traffic flow theory that can be used in the planning, design, and operation of highway systems.

CHAPTER OBJECTIVES:

- Become familiar with the different elements of traffic flow.
- Understand the relationships among the different elements of traffic flow.
- Become familiar with the fundamental diagram of traffic flow.
- Understand the difference between macroscopic and microscopic traffic models.
- Understand how the different types of traffic shock waves are formed.
- Learn the procedures for determining queue lengths due to different types of shock waves.
- Understand the fundamental principles of gap and gap acceptance.

6.1 TRAFFIC FLOW ELEMENTS

Let us first define the elements of traffic flow before discussing the relationships among them. However, before we do that, we will describe the time-space diagram that serves as a useful device for defining the elements of traffic flow.

6.1.1 Time–Space Diagram

The time-space diagram is a graph that describes the relationship between the location of vehicles in a traffic stream and, time as the vehicles progress along the highway. Figure 6.1 shows a time-space diagram for six vehicles with distance plotted on the vertical axis and time on the horizontal axis. At time zero, vehicles 1, 2, 3, and 4 are at respective distances d_1, d_2, d_3, and d_4 from a reference point whereas vehicles 5 and 6 cross the reference point later at times t_5 and t_6, respectively.

6.1.2 Primary Elements of Traffic Flow

The primary elements of traffic flow are flow, density, and speed. Another element, associated with density, is the gap or headway between two vehicles in a traffic stream. The definitions of these elements follow.

Flow

Flow (q) is the equivalent hourly rate at which vehicles pass a point on a highway during a time period less than 1 hour. It can be determined by:

$$q = \frac{n \times 3600}{T} \text{ veh/h} \tag{6.1}$$

where
　　n = the number of vehicles passing a point in the roadway in T sec
　　q = the equivalent hourly flow

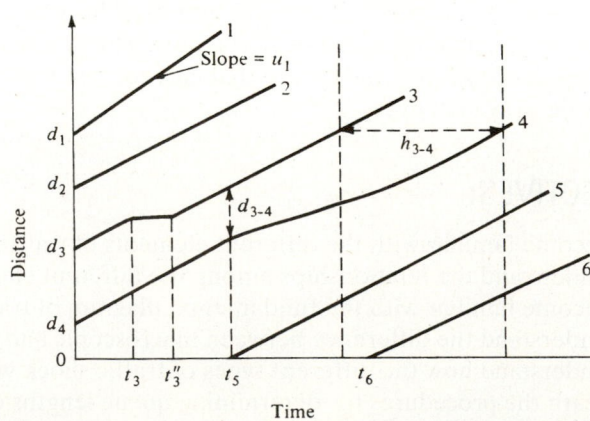

Figure 6.1 Time-Space Diagram

Density

Density (k), sometimes referred to as *concentration,* is the number of vehicles traveling over a unit length of highway at an instant in time. The unit length is usually 1 kilometer (km), thereby making vehicles per km (veh/km) the unit of density.

Speed

[handwritten: Space-time speed]

Speed (u) is the distance traveled by a vehicle during a unit of time. It can be expressed in kms per hour (km/h), kilometers per hour (km/h), or meter per second (m/sec). The speed of a vehicle at any time *t* is the slope of the time-space diagram for that vehicle at time *t*. Vehicles 1 and 2 in Figure 6.1, for example, are moving at constant speeds because the slopes of the associated graphs are constant. Vehicle 3 moves at a constant speed between time zero and time t_3, then stops for the period t_3 to t''_3 (the slope of graph equals 0), and then accelerates and eventually moves at a constant speed. There are two types of mean speeds: time mean speed and space mean speed.

Time mean speed $(\bar{u}_t)$ is the arithmetic mean of the speeds of vehicles passing a point on a highway during an interval of time. The time mean speed is found by:

$$\bar{u}_t = \frac{1}{n} \sum_{t=1}^{n} u_t \qquad (6.2)$$

where

n = number of vehicles passing a point on the highway
u_i = speed of the ith vehicle (m/sec)

Space mean speed $(\bar{u}_s)$ is the harmonic mean of the speeds of vehicles passing a point on a highway during an interval of time. It is obtained by dividing the total distance traveled by two or more vehicles on a section of highway by the total time required by these vehicles to travel that distance. This is the speed that is used in flow-density relationships. The space mean speed is found by

$$\bar{u}_s = \frac{n}{\sum_{i=1}^{n} (1/u_i)}$$

$$= \frac{nL}{\sum_{i=1}^{n} t_i} \qquad (6.3)$$

where

$\bar{u}_s$ = space mean speed (m/sec)
n = number of vehicles
t_i = the time it takes the *i*th vehicle to travel across a section of highway (sec)
u_i = speed of the *i*th vehicle (m/sec)
L = length of section of highway (m)

The time mean speed is always higher than the space mean speed. The difference between these speeds tends to decrease as the absolute values of speeds increase. It has been shown from field data that the relationship between time mean speed and space mean speed can be given as

$$\bar{u}_t = \bar{u}_s + \frac{\sigma^2}{u_s} \tag{6.4}$$

Equation 6.5 shows a more direct relationship developed by Garber and Sankar using data collected at several sites on freeways. Figure 6.2 also shows a plot of time mean speeds against space mean speeds using the same data.

$$\bar{u}_t = 0.966\bar{u}_s + 3.541 \tag{6.5}$$

where
$\bar{u}_t$ = time mean speed, km/h
$\bar{u}_s$ = space mean speed, km/h

Note that the values of the coefficients given in Eq. 6.5 are unique to the sites at which data were collected and may slightly differ from site to site.

Time Headway

Time headway (h) is the difference between the time the front of a vehicle arrives at a point on the highway and the time the front of the next vehicle arrives at that same point. Time headway is usually expressed in seconds. For example, in the time-space diagram (Figure 6.1), the time headway between vehicles 3 and 4 at d_1 is h_{3-4}.

Space Headway

Space headway (d) is the distance between the front of a vehicle and the front of the following vehicle and is usually expressed in meters. The space headway between vehicles 3 and 4 at time t_5 is d_{3-4} (see Figure 6.1).

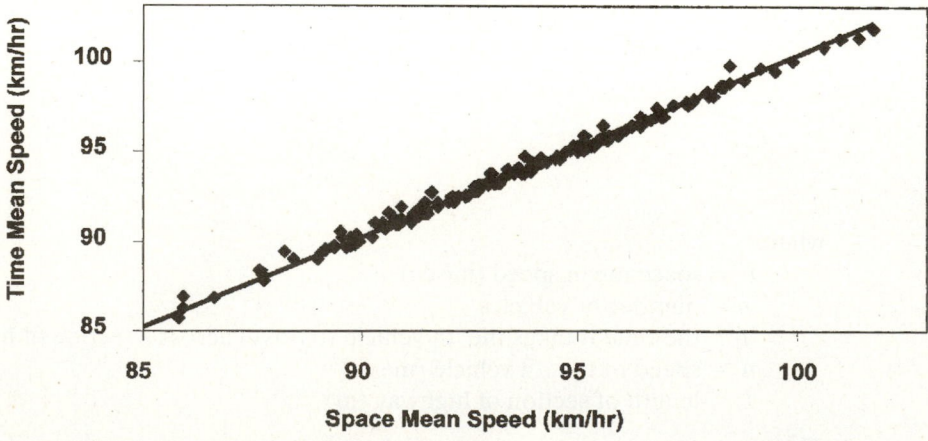

Figure 6.2   Space Mean Speed versus Time Mean Speed

Example 6.1 Determining Flow, Density, Time Mean Speed, and Space Mean Speed

Figure 6.3 shows vehicles traveling at constant speeds on a two-lane highway between sections X and Y with their positions and speeds obtained at an instant of time by photography. An observer located at point X observes the four vehicles passing point X during a period of T sec. The velocities of the vehicles are measured as 70, 70, 65, and 50 (km/h), respectively. Calculate the flow, density, time mean speed, and space mean speed.

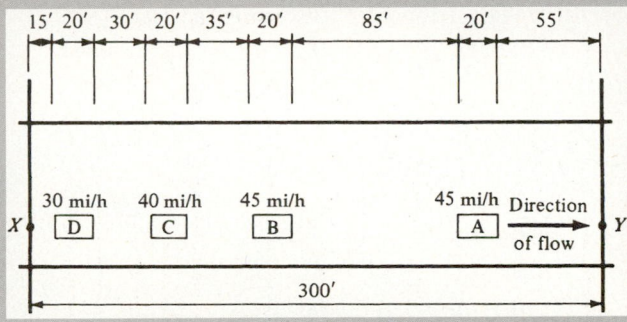

Figure 6.3 Locations and Speeds of Four Vehicles on a Two-Lane Highway at an Instant of Time

Solution: The flow is calculated by

$$q = \frac{n \times 3600}{T} \tag{6.6}$$

$$= \frac{4 \times 3600}{T} = \frac{14{,}400}{T} \text{ veh/h}$$

With L equal to the distance between X and Y (m), density is obtained by

$$k = \frac{n}{L}$$

$$= \frac{4}{90} \times 1000 = 44.4 \text{ veh/km}$$

converting m to km

The time mean speed is found by

$$u_t = \frac{1}{n} \sum_{i=1}^{n} u_i$$

$$= \frac{50 + 65 + 70 + 70}{4} = 64 \text{ km/h}$$

The space mean speed is found by

$$\bar{u}_s = \frac{n}{\sum_{i=1}^{n} (1/u_i)}$$

$$= \frac{Ln}{\sum_{i=1}^{n} t_i}$$

$$= \frac{300n}{\sum_{t=1}^{n} t_i}$$

where t_i is the time it takes the ith vehicle to travel from X to Y at speed u_i, and L (m) is the distance between X and Y.

$$t_i = \frac{L}{0.278u_i} \text{ sec}$$

$$t_A = \frac{90}{0.278 \times 70} = 4.63 \text{ sec}$$

$$t_B = \frac{90}{0.278 \times 70} = 4.63 \text{ sec}$$

$$t_C = \frac{90}{0.278 \times 65} = 4.98 \text{ sec}$$

$$t_D = \frac{90}{0.278 \times 50} = 6.48 \text{ sec}$$

$$\bar{u}_s = \frac{4 \times 90}{4.62 + 4.62 + 4.98 + 6.48} = 17.4 \text{ m/sec}$$

$$= 62.64 \text{ km/h}$$

6.2 FLOW–DENSITY RELATIONSHIPS

The general equation relating flow, density, and space mean speed is given as

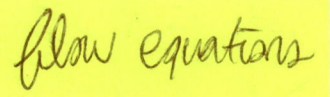

$$\text{Flow} = \text{density} \times \text{space mean speed} \tag{6.7}$$

$$q = k\bar{u}_s$$

Each of the variables in Eq. 6.7 also depends on several other factors, including the characteristics of the roadway, characteristics of the vehicle, characteristics of the driver, and environmental factors such as the weather.

Other relationships that exist among the traffic flow variables are given here.

$$\text{Space mean speed} = (\text{flow}) \times (\text{space headway}) \tag{6.8}$$

$$\bar{u}_s = q\bar{d}$$

where

$$\bar{d} = (1/k) = \text{average space headway} \tag{6.9}$$

$$\text{Density} = (\text{flow}) \times (\text{travel time for unit distance}) \tag{6.10}$$

$$k = q\bar{t}$$

where $\bar{t}$ is the average time for unit distance.

Average space headway = (space mean speed) × (average time headway)

$$\bar{d} = \bar{u}_s \bar{h} \tag{6.11}$$

Average time headway = (average travel time for unit distance)

× (average space headway) (6.12)

$$\bar{h} = \bar{t}\bar{d}$$

Note that it is not easy to directly obtain density from speed data collected in the field using the conventional methods discussed in Chapter 4. A flow characteristic that is related to density and can be obtained from speed data is *occupancy*. Occupancy is defined as the proportion of a specific data collection time interval that vehicles are present in the detection zone. It is given as:

$$\text{occupancy} = \frac{\sum\limits_{i}(L_i + d)/u_i}{T}, \text{ which can be written as}$$

$$\text{occupancy} = \frac{1}{T}\sum \frac{L_i}{u_i} + \frac{d}{T}\sum \frac{1}{u_i} \tag{6.13}$$

where

L_i = length of ith vehicle
d = length of detector
T = specific data collection time interval

Note that the numerator of Eq. 6.13 indicates the amount of time each vehicle affects the detector. By multiplying the second term of Eq. 6.13 by $\frac{N}{N}$, it can be written as:

$$\text{occupancy} = \frac{1}{T}\sum \frac{L_i}{u_i} + \frac{dN}{NT}\sum \frac{1}{u_i} \tag{6.14}$$

Substituting $\frac{N}{T}$ for q and $\frac{1}{N}\sum\limits_{i} \frac{1}{u_i}$ for $\frac{1}{\bar{u}_s}$ (see Eq. 6.3), we obtain

$$\text{occupancy} = \frac{1}{T}\sum \frac{L_i}{u_i} + d\frac{q}{\bar{u}_s} \tag{6.15}$$

Substituting k for $\frac{q}{\bar{u}_s}$ in Eq. 6.15, we obtain

$$\text{occupancy} = \frac{1}{T}\sum_1 \frac{L_i}{u_i} + dk \tag{6.16}$$

Note that T is the sum of the headways of the individual vehicles $= \Sigma h_i$. Substituting Σh_i for T and multiplying the first term of Eq. 6.16 by $\frac{N}{N}$, we obtain

$$\text{occupancy} = \frac{\frac{1}{N}\sum\limits_{i} \frac{L_i}{u_i}}{\bar{h}} + dk \tag{6.17}$$

Assuming that the lengths of the vehicles are approximately the same, we obtain

$$\text{occupancy} = \frac{1}{h}L\frac{1}{N}\sum_i \frac{1}{u_i} + dk = L\frac{q}{u_s} + dk \tag{6.18}$$

$$\text{occupancy} = (L + d)k = c_k k \tag{6.19}$$

where c_k is a constant as d at a given detector is constant and L is assumed to be constant. It should be emphasized, however, that Eq. 6.19 is applicable only when the variation in vehicle lengths is not significant. When the lengths of the vehicles vary significantly (for example, when a significant number of different truck types is in the traffic stream), Eq. 6.19 is not applicable, and Eq. 6.16 or 6.17 should be used.

Example 6.2 Determining Density for Different Assumptions of Vehicles Lengths

Columns 1, 2, and 3 of Table 6.1 give data obtained on vehicles on a lane traversing a detection zone as recorded by a presence detector. If the length of the detector zone is 1.8 m:

(i) Determine the density on the lane without assuming that the lengths of the vehicles are approximately the same.
(ii) Determine the density on the lane assuming that the lengths of the vehicles are approximately the same.
(iii) Comment on your results.

Table 6.1 Data and Solution for Example 6.2

Vehicle # (1)	Length of Vehicle, L_i (m) (2)	Time Headway between Consecutive Vehicles (sec) (3)	Speed of Vehicle (km/h) (4)	Time Vehicle Spends in Detector Zone (sec) (6)	L_i/u_i (7)
1	5.7		88	0.307	0.233
2	5.7	5.5	88	0.307	0.233
3	5.7	4.5	80	0.338	0.257
4	9	4	72	0.540	0.450
5	9	5.5	76.8	0.506	0.422
6	9	5	72	0.540	0.450
7	5.7	5.5	96	0.281	0.214
8	5.7	5.5	96	0.281	0.214
9	9	3.5	72	0.540	0.450
10	5.7	4.5	88	0.307	0.233
11	5.7	5	80	0.338	0.257
12	5.7	5.5	96	0.281	0.214
13	11.85	6	96	0.512	0.444
	$\Sigma = 93.45$	$\Sigma = 60.0$ sec		$\Sigma = 5.077$	$\Sigma = 4.070$

Solution:

- Determine occupancy without assuming similar lengths for vehicles.

  Determine specific data collection time interval. This is given as the sum of the headways = 60 sec (see col. 3).
  Determine time spent by each vehicle over detector zone $(L_i + d)/u_i$.
  For example, time spent over detector zone by vehicle 5 is given as

  $(9 + 1.8)/(76.8 \times (5/18)) = 0.506$ sec (See col. 6 for the other vehicles.)

  Determine total time spent over detector by all vehicles. This is obtained by summing the values in col. 6 as 5.114 sec.

  $$\text{occupancy} = (\text{total time over detector/data collection} \\ \text{time interval})$$

  $$= 5.077/60 = 0.085, \text{ i.e., } 8.5\%$$

- Determine density, *not* assuming that the lengths of vehicles are approximately the same. Use Eq. 6.16.

  $$\text{occupancy} = \frac{1}{T} \sum_1 \frac{L_i}{u_i} + dk$$

  $$0.085 = (1/60) \times 4.070 + 1.8k$$

  $$0.085 = 0.0678 + 1.8k$$

  Giving $k = 0.00954$ veh/m

  $$= 9.54 \text{ veh/km}$$

- Determine density, assuming similar lengths for vehicles. Use Eq. 6.19.

  $$\text{Occupancy} = (L + d)k$$

  where L is the mean length

  $$L = 93.45/13 = 7.2 \text{ m}$$

  $$0.085 = (7.2 + 1.8)k$$

  Giving $k = 0.00944$ veh/m

  $$= 9.44 \text{ veh/km}$$

- Comment: The results indicate that assuming that the lengths of the vehicles are approximately the same and using the mean length gives a different density for the same set of data. In this problem, the difference is about 0.1 veh/km, which could be considered as not significant. It should be noted that the difference for other sets of data will depend on the variability of the vehicle lengths in the data set. The use of either one method or the other will therefore depend on the extent of the variability of the vehicle lengths in the data set and the accuracy that the analyst requires.

6.2.1 Fundamental Diagram of Traffic Flow

The relationship between the density (veh/km) and the corresponding flow of traffic on a highway is generally referred to as the fundamental diagram of traffic flow. The following theory has been postulated with respect to the shape of the curve depicting this relationship:

1. When the density on the highway is 0, the flow is also 0 because there are no vehicles on the highway.
2. As the density increases, the flow also increases.
3. However, when the density reaches its maximum, generally referred to as the *jam density* (k_j), the flow must be 0 because vehicles will tend to line up end to end.
4. It follows that as density increases from 0, the flow will also initially increase from 0 to a maximum value. Further continuous increase in density will then result in continuous reduction of the flow, which will eventually be 0 when the density is equal to the jam density. The shape of the curve therefore takes the form in Figure 6.4a.

Data have been collected that tend to confirm the argument postulated above, but there is some controversy regarding the exact shape of the curve. A similar argument can be postulated for the general relationship between the space mean speed and the flow.

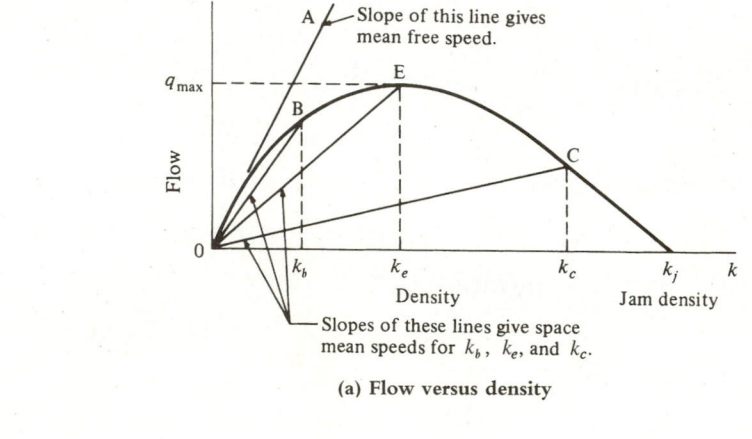

(a) Flow versus density

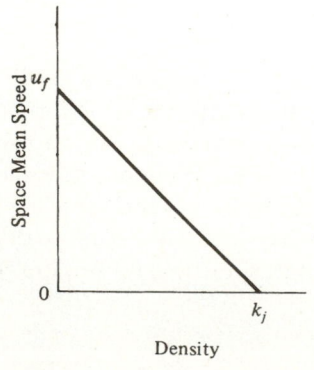

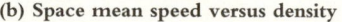

(b) Space mean speed versus density

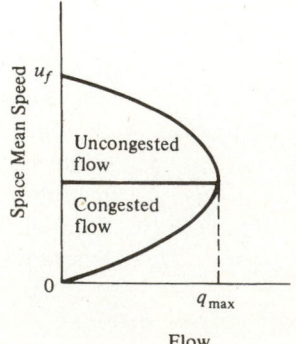

(c) Space mean speed versus volume

Figure 6.4 Fundamental Diagrams of Traffic Flow

When the flow is very low, there is little interaction between individual vehicles. Drivers are therefore free to travel at the maximum possible speed. The absolute maximum speed is obtained as the flow tends to 0, and it is known as the *mean free-flow speed* (u_f). The magnitude of the mean free speed depends on the physical characteristics of the highway. Continuous increase in flow will result in a continuous decrease in speed. A point will be reached, however, when the further addition of vehicles will result in the reduction of the actual number of vehicles that pass a point on the highway (that is, reduction of flow). This results in congestion, and eventually both the speed and the flow become 0. Figure 6.4c shows this general relationship. Figure 6.4b shows the direct relationship between speed and density.

From Eq. 6.7, we know that space mean speed is flow divided by density, which makes the slopes of lines 0B, 0C, and 0E in Figure 6.4a represent the space mean speeds at densities k_b, k_c, and k_e, respectively. The slope of line 0A is the speed as the density tends to 0 and little interaction exists between vehicles. The slope of this line is therefore the mean free speed (u_f); it is the maximum speed that can be attained on the highway. The slope of line 0E is the space mean speed for maximum flow. This maximum flow is the capacity of the highway. Thus, it can be seen that it is desirable for highways to operate at densities not greater than that required for maximum flow.

6.2.2 Mathematical Relationships Describing Traffic Flow

Mathematical relationships describing traffic flow can be classified into two general categories—macroscopic and microscopic—depending on the approach used in the development of these relationships. The macroscopic approach considers flow-density relationships, whereas the microscopic approach considers the spacing between two consecutive vehicles and the speeds of individual vehicles.

Macroscopic Approach

The macroscopic approach considers traffic streams and develops algorithms that relate the flow to the density and space mean speeds. The two most commonly used macroscopic models are the Greenshields and Greenberg models.

Greenshields Model.   Greenshields carried out one of the earliest recorded works in which he studied the relationship between speed and density. He hypothesized that a linear relationship existed between speed and density, which he expressed as

$$\bar{u}_s = u_f - \frac{u_f}{k_j}k \tag{6.20}$$

Corresponding relationships for flow and density and for flow and speed can be developed. Since $q = \bar{u}_s k$, substituting $q/\bar{u}_s$ for k in Eq. 6.20 gives

$$\bar{u}_s^2 = u_f\bar{u}_s - \frac{u_f}{k_j}q \tag{6.21}$$

Also, substituting q/k for $\bar{u}_s$ in Eq. 6.20 gives

$$q = u_f k - \frac{u_f}{k_j}k^2 \tag{6.22}$$

Equations 6.21 and 6.22 indicate that if a linear relationship in the form of Eq. 6.20 is assumed for speed and density, then parabolic relationships are obtained between flow and density and between flow and speed. The shape of the curve shown in Figure 6.4a will therefore be a parabola. Also, Eqs. 6.21 and 6.22 can be used to determine the corresponding speed and the corresponding density for maximum flow. Consider Eq. 6.21:

$$\bar{u}_s^2 = u_f \bar{u}_s - \frac{u_f}{k_j} q$$

Differentiating q with respect to $\bar{u}_s$, we obtain

$$2\bar{u}_s = u_f - \frac{u_f}{k_j} \frac{dq}{du_s}$$

that is,

$$\frac{dq}{du_s} = u_f \frac{k_j}{u_f} - 2\bar{u}_s \frac{k_j}{u_f} = k_j - 2\bar{u}_s \frac{k_j}{u_f}$$

For maximum flow,

$$\frac{dq}{du_s} = 0 \qquad k_j = 2\bar{u}_s \frac{k_j}{u_f} \qquad u_o = \frac{u_f}{2} \tag{6.23}$$

Thus, the space mean speed u_o at which the volume is maximum is equal to half the mean free-flow speed.

Consider Eq. 6.22:

$$q = u_f k - \frac{u_f}{k_j} k^2$$

Differentiating q with respect to k, we obtain:

$$\frac{dq}{dk} = u_f - 2k \frac{u_f}{k_j}$$

For maximum flow,

$$\frac{dq}{dk} = 0 \tag{6.24}$$

$$u_f = 2k \frac{u_f}{k_j}$$

$$\frac{k_j}{2} = k_o$$

Thus, at the maximum flow, the density k_o is half the jam density. The maximum flow for the Greenshields relationship can therefore be obtained from Eqs. 6.7, 6.23, and 6.24, as shown in Eq. 6.25:

$$q_{max} = \frac{k_j u_f}{4} \tag{6.25}$$

Greenberg Model. Several researchers have used the analogy of fluid flow to develop macroscopic relationships for traffic flow. One of the major contributions using the fluid-flow analogy was developed by Greenberg in the form

$$\bar{u}_s = c \ln \frac{k_j}{k} \tag{6.26}$$

Multiplying each side of Eq. 6.26 by k, we obtain

$$\bar{u}_s k = q = ck \ln \frac{k_j}{k}$$

Differentiating q with respect to k, we obtain

$$\frac{dq}{dk} = c \ln \frac{k_j}{k} - c$$

For maximum flow,

$$\frac{dq}{dk} = 0$$

giving

$$\ln \frac{k_f}{k_o} = 1 \tag{6.27}$$

Substituting 1 for (k_j/k_o) in Eq. 6.26 gives

$$u_o = c$$

Thus, the value of c is the speed at maximum flow.

Model Application

Use of these macroscopic models depends on whether they satisfy the boundary criteria of the fundamental diagram of traffic flow at the region that describes the traffic conditions. For example, the Greenshields model satisfies the boundary conditions when the density k is approaching zero as well as when the density is approaching the jam density k_j. The Greenshields model can therefore be used for light or dense traffic. The Greenberg model, on the other hand, satisfies the boundary conditions when the density is approaching the jam density but it does not satisfy the boundary conditions when k is approaching zero. The Greenberg model is therefore useful only for dense traffic conditions.

Calibration of Macroscopic Traffic Flow Models. The traffic models discussed thus far can be used to determine specific characteristics, such as the speed and density at which maximum flow occurs, and the jam density of a facility. This usually involves collecting appropriate data on the particular facility of interest and fitting the data points obtained to a suitable model. The most common method of approach is *regression analysis*. This is done by minimizing the squares of the differences between the observed and expected values of a dependent variable. When the dependent variable is linearly related to the independent variable, the process is known as *linear regression analysis*. When the relationship is with two or more independent variables, the process is known as *multiple linear regression analysis*.

If a dependent variable y and an independent variable x are related by an estimated regression function, then

$$y = a + bx \tag{6.28}$$

The constants a and b could be determined from Eqs. 6.29 and 6.30. (For development of these equations, see Appendix B.)

$$a = \frac{1}{n}\sum_{i=1}^{n} y_i - \frac{b}{n}\sum_{i=1}^{n} x_i = \bar{y} - b\bar{x} \tag{6.29}$$

and

$$b = \frac{\sum_{i=1}^{n} x_i y_i - \frac{1}{n}\left(\sum_{i=1}^{n} x_i\right)\left(\sum_{i=1}^{n} y_i\right)}{\sum_{i=1}^{n} x_i^2 - \frac{1}{n}\left(\sum_{i=1}^{n} x_i\right)^2} \tag{6.30}$$

where

n = number of sets of observations
x_i = ith observation for x
y_i = ith observation for y

A measure commonly used to determine the suitability of an estimated regression function is the coefficient of determination (or square of the estimated correlation coefficient) R^2, which is given by

$$R^2 = \frac{\sum_{i=1}^{n}(Y_i - \bar{y})^2}{\sum_{i=1}^{n}(y_i - \bar{y})^2} \tag{6.31}$$

where Y_i is the value of the dependent variable as computed from the regression equations and $\bar{y}_i$ is the mean of the y_i values. The closer R^2 is to 1, the better the regression fits.

Example 6.3 Fitting Speed and Density Data to the Greenshields Model

Let us now use the data shown in Table 6.2 (columns 1 and 2) to demonstrate the use of the method of regression analysis in fitting speed and density data to the macroscopic models discussed earlier.

Solution: Let us first consider the Greenshields expression

$$\bar{u}_s = u_f - \frac{u_f}{k_j} k$$

Table 6.2 Speed and Density Observations at a Rural Road

(a) Computations for Example 6.3

Speed, u_s (km/h) y_i	Density, k (veh/km) x_i	$x_i y_i$	x_i^2
53.2	20	1064.0	400
48.1	27	1298.7	729
44.8	35	1568.0	1,225
40.1	44	1764.4	1,936
37.3	52	1939.6	2,704
35.2	58	2041.6	3,364
34.1	60	2046.0	3,600
27.2	64	1740.8	4,096
20.4	70	1428.0	4,900
17.5	75	1312.5	5,625
14.6	82	1197.2	6,724
13.1	90	1179.0	8,100
11.2	100	1120.0	10,000
8.0	115	920.0	13,225
$\Sigma = 404.8$	$\Sigma = 892$	$\Sigma = 20{,}619.8$	$\Sigma = 66{,}628.0$
$\bar{y} = 28.91$	$\bar{x} = 63.71$		

(b) Computations for Example 6.4

Speed, u_s (km/h) y_i	Density, k (veh/km)	$Ln\ k_i\ x_i$	$x_i y_i$	x_i^2
53.2	20	2.995732	159.3730	8.974412
48.1	27	3.295837	158.5298	10.86254
44.8	35	3.555348	159.2796	12.64050
40.1	44	3.784190	151.746	14.32009
37.3	52	3.951244	147.3814	15.61233
35.2	58	4.060443	142.9276	16.48720
34.1	60	4.094344	139.6171	16.76365
27.2	64	4.158883	113.1216	17.29631
20.4	70	4.248495	86.66929	18.04971
17.5	75	4.317488	75.55605	18.64071
14.6	82	4.406719	64.33811	19.41917
13.1	90	4.499810	58.94750	20.24828
11.2	100	4.605170	51.57791	21.20759
8.0	115	4.744932	37.95946	22.51438
$\Sigma = 404.8$		$\Sigma = 56.71864$	$\Sigma = 1547.024$	$\Sigma = 233.0369$
$\bar{y} = 28.91$		$\bar{x} = 4.05$		

Comparing this expression with our estimated regression function, Eq. 6.28, we see that the speed $\bar{u}_s$ in the Greenshields expression is represented by y in the estimated regression function, the mean free speed u_f is represented by a, and the value of the mean free speed u_f divided by the jam density k_j is represented by $-b$. We therefore obtain

$$\sum y_i = 404.8 \qquad \sum x_i = 892 \qquad \bar{y} = 28.91$$

$$\sum x_i y_i = 20{,}619.8 \qquad \sum x_i^2 = 66{,}628 \qquad \bar{x} = 63.71$$

- Using Eqs. 6.29 and 6.30, we obtain

$$a = 28.91 - 63.71b$$

$$b = \frac{20{,}619.8 - \dfrac{(892)(404.8)}{14}}{66{,}628 - \dfrac{(892)^2}{14}} = -0.53$$

or

$$a = 28.91 - 63.71(-0.53) = 62.68$$

Since $a = 62.68$ and $b = -0.53$, then $u_f = 62.68$ km/h, $u_f/k_j = 0.53$, and so $k_j = 118$ veh/km, and $\bar{u}_s = 62.68 - 0.53k$.

- Using Eq. 6.31 to determine the value of R^2, we obtain $R^2 = 0.95$.
- Using the above estimated values for u_f and k_j, we can determine the maximum flow from Eq. 6.25 as

$$q_{max} = \frac{k_j u_f}{4} = \frac{118 \times 62.68}{4}$$

$$= 1849 \text{ veh/h}$$

- Using Eq. 6.23, we also obtain the velocity at which flow is maximum, that is, $(62.68/2) = 50.4$ km/h, and Eq. 6.24, the density at which flow is maximum, or $(118/2) = 59$ veh/h.

Example 6.4 Fitting Speed and Density Data to the Greenberg Model

The data in Table 6.2b can also be fitted into the Greenberg model shown in Eq. 6.26:

$$\bar{u}_s = c \ln \frac{k_j}{k}$$

which can be written as

$$\bar{u}_s = c \ln k_j - c \ln k \qquad (6.32)$$

Solution: Comparing Eq. 6.32 and the estimated regression function Eq. 6.28, we see that $\bar{u}_s$ in the Greenberg expression is represented by y in the estimated regression function, $c \ln k_j$ is represented by a, c is represented by $-b$, and $\ln k$ is represented by x. Table 6.1b shows values for x_i, $x_i y_i$, and x_i^2. (Note that these values are computed to a higher degree of accuracy since they involve logarithmic values.) We therefore obtain

$$\sum y_i = 404.8 \qquad \sum x_i = 56.72 \qquad \bar{y} = 28.91$$

$$\sum x_i y_i = 1547.02 \qquad \sum x_i^2 = 233.04 \qquad \bar{x} = 4.05$$

Using Eqs. 6.29 and 6.30, we obtain

$$a = 28.91 - 4.05b$$

$$b = \frac{1547.02 - \dfrac{(56.72)(404.8)}{14}}{233.04 - \dfrac{56.72^2}{14}} = -28.68$$

or

$$a = 28.91 - 4.05(-28.68) = 145.06$$

Since $a = 145.06$ and $b = -28.68$, the speed for maximum flow is $c = 46.16$ km/h. Finally, since

$$c \ln k_j = 145.06$$

$$\ln k_j = \frac{145.06}{28.68} = 5.06$$

$$k_j = 157 \text{ veh/km}$$

then

$$\bar{u}_s = 28.68 \ln \frac{157}{k}$$

Obtaining k_o, the density for maximum flow from Eq. 6.27, we then use Eq. 6.7 to determine the value of the maximum flow.

$$\ln k_j = 1 + \ln k_o$$

$$\ln 157 = 1 + \ln k_o$$

$$5.06 = 1 + \ln k_o$$

$$58.0 = k_o$$

$$q_{max} = 58.0 \times 28.68 \text{ veh/h}$$

$$q_{max} = 1663 \text{ veh/h}$$

The R^2 based on the Greenberg expression is 0.93, which indicates that the Greenshields expression is a better fit for the data in Table 6.2. Figure 6.5 shows plots of speed

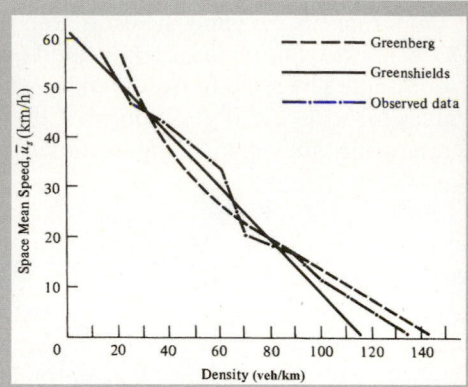

Figure 6.5   Speed versus Density

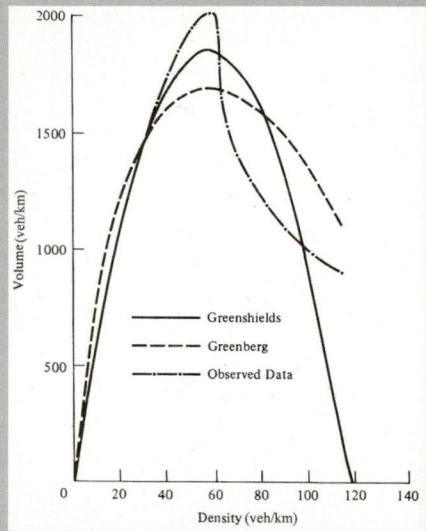

Figure 6.6   Volume versus Density

versus density for the two estimated regression functions obtained and also for the actual data points. Figure 6.6 shows similar plots for the volume against speed.

Software Packages for Linear Regression Analysis.   Several software packages are available that can be used to solve the linear regression problem. These include Excel, MiniTab, SAS, and SPSS. Appendix C illustrates the use of the Excel spreadsheet to solve Example 6.3.

Microscopic Approach

The microscopic approach, which is sometimes referred to as the car-following theory or the follow-the-leader theory, considers spacings between consecutive vehicles and speeds of individual vehicles. Consider two consecutive vehicles, A and B, on a single lane of a highway, as shown in Figure 6.7. If the leading vehicle (A) is considered to be the nth vehicle and the following vehicle (B) is considered the $(n + 1)$th vehicle, then the distances of these vehicles from a fixed section at any time t can be taken as x_n and x_{n+1}, respectively.

If the driver of vehicle B maintains an additional separation distance P above the separation distance at rest S such that P is proportional to the speed of vehicle B, then

$$P = \rho \dot{x}_{n+1} \tag{6.33}$$

where

ρ = factor of proportionality with units of time
$\dot{x}_{n+1}$ = speed of the $(n + 1)$th vehicle

We can write

$$x_n - x_{n+1} = \rho \dot{x}_{n+1} + S \tag{6.34}$$

where S is the distance between front bumpers of vehicles at rest.

Differentiating Eq. 6.34 gives

$$\ddot{x}_{n+1} = \frac{1}{\rho}[\dot{x}_n - \dot{x}_{n+1}] \tag{6.35}$$

Equation 6.35 is the basic equation of microscopic models and it describes the stimulus response of the models. Researchers have shown that a time lag exists for a driver to respond to any stimulus that is induced by the vehicle just ahead and Eq. 6.35 can therefore be written as

$$\ddot{x}_{n+1}(t + T) = \lambda[\dot{x}_n(t) - \dot{x}_{n+1}(t)] \tag{6.36}$$

where

T = time lag of response to the stimulus
$\lambda = (1/p)$ (sometimes called the sensitivity)

A general expression for λ is given in the form

$$\lambda = a \frac{\dot{x}_{n+1}^m(t + T)}{[x_n(t) - x_{n+1}(t)]^\ell} \tag{6.37}$$

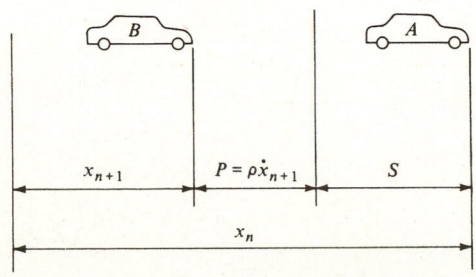

Figure 6.7   Basic Assumptions in the Follow-the-Leader Theory

The general expression for the microscopic models can then be written as

$$\ddot{x}_{n+1}(t + T) = a\frac{\dot{x}_{n+1}^{m}(t + T)}{[x_n(t) - x_{n+1}(t)]^{\ell}}[\dot{x}_n(t) - \dot{x}_{n+1}(t)] \tag{6.38}$$

where a, ℓ, and m are constants.

The microscopic model (Eq. 6.31) can be used to determine the velocity, flow, and density of a traffic stream when the traffic stream is moving in a steady state. The direct analytical solution of either Eq. 6.36 or Eq. 6.38 is not easy. It can be shown, however, that the macroscopic models discussed earlier can all be obtained from Eq. 6.38.

For example, if $m = 0$ and $\ell = 1$, the acceleration of the (n + 1)th vehicle is given as

$$\ddot{x}_{n+1}(t + T) = a\frac{\dot{x}_n(t) - \dot{x}_{n+1}(t)}{x_n(t) - x_{n+1}(t)}$$

Integrating the above expression, we find that the velocity of the $(n + 1)$th vehicle is

$$\dot{x}_{n+1}(t + T) = a \ln [x_n(t) - x_{n+1}(t)] + C$$

Since we are considering the steady-state condition,

$$\dot{x}_n(t + T) = \dot{x}(t) = u$$

and

$$u = a \ln [x_n - x_{n+1}] + C$$

Also,

$$x_n - x_{n+1} = \text{average space headway} = \frac{1}{k}$$

and

$$u = a \ln \left(\frac{1}{k}\right) + C$$

Using the boundary condition,

$$u = 0$$

when

$$k = k_j$$

and

$$u = 0 = a \ln \left(\frac{1}{k_j}\right) + C$$

$$C = -a \ln \left(\frac{1}{k_j}\right)$$

Substituting for C in the equation for u, we obtain

$$u = a \ln\left(\frac{1}{k}\right) - a \ln\left(\frac{1}{k_j}\right)$$

$$= a \ln\left(\frac{k_j}{k}\right)$$

which is the Greenberg model given in Eq. 6.26. Similarly, if m is allowed to be 0 and $\ell = 2$, we obtain the Greenshields model.

6.3 SHOCK WAVES IN TRAFFIC STREAMS

The fundamental diagram of traffic flow for two adjacent sections of a highway with different capacities (maximum flows) is shown in Figure 6.8. This figure describes the phenomenon of backups and queuing on a highway due to a sudden reduction of the capacity of the highway (known as a *bottleneck condition*). The sudden reduction in capacity could be due to a crash, reduction in the number of lanes, restricted bridge sizes, work zones, a signal turning red, and so forth, creating a situation where the capacity on the highway suddenly changes from C_1 to a lower value of C_2, with a corresponding change in optimum density from k_o^a to a value of k_o^b.

When such a condition exists and the normal flow and density on the highway are relatively large, the speeds of the vehicles will have to be reduced while passing the bottleneck. The point at which the speed reduction takes place can be approximately noted by the turning on of the brake lights of the vehicles. An observer will see that this point moves upstream as traffic continues to approach the vicinity of the bottleneck, indicating an upstream movement of the point at which flow and density change. This phenomenon is usually referred to as a *shock wave* in the traffic stream. The phenomenon also exists when the capacity suddenly increases, but in this case, the speeds of the vehicles tend to increase as the vehicles pass the section of the road where the capacity increases.

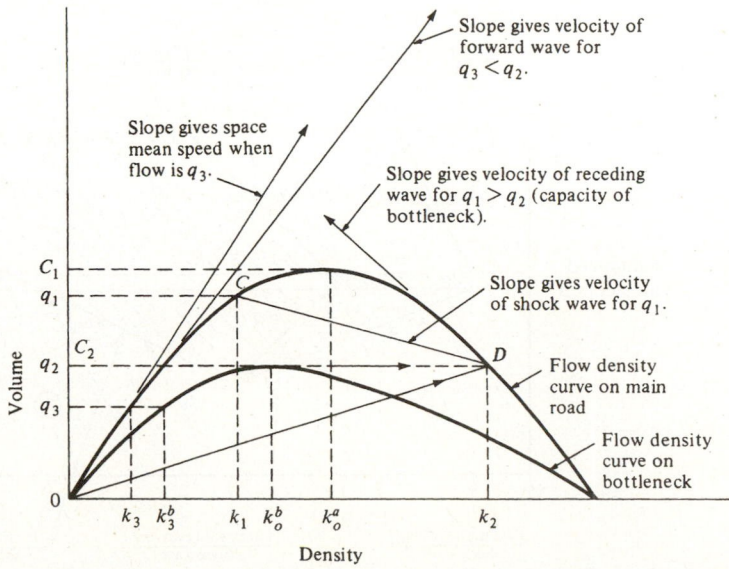

Figure 6.8   Kinematic and Shock Wave Measurements Related to Flow-Density Curve

6.3.1 Types of Shock Waves

Several types of shock waves can be formed, depending on the traffic conditions that lead to their formation. These include *frontal stationary, backward forming, backward recovery, rear stationary,* and *forward recovery* shock waves.

Frontal stationary shock waves are formed when the capacity suddenly reduces to zero at an approach (for example, when a set of lanes has the red indication at a signalized intersection or when a highway is completely closed because of a serious incident). In this case, a frontal stationary shock wave is formed when traffic comes to a stop and the capacity is reduced to zero. This type occurs at the location where the capacity is reduced to zero. For example, at a signalized intersection, the red signal indicates that traffic on the approach or set of lanes cannot move across the intersection, which implies that the capacity is temporarily reduced to zero resulting in the formation of a frontal stationary shock wave, as shown in Figure 6.9.

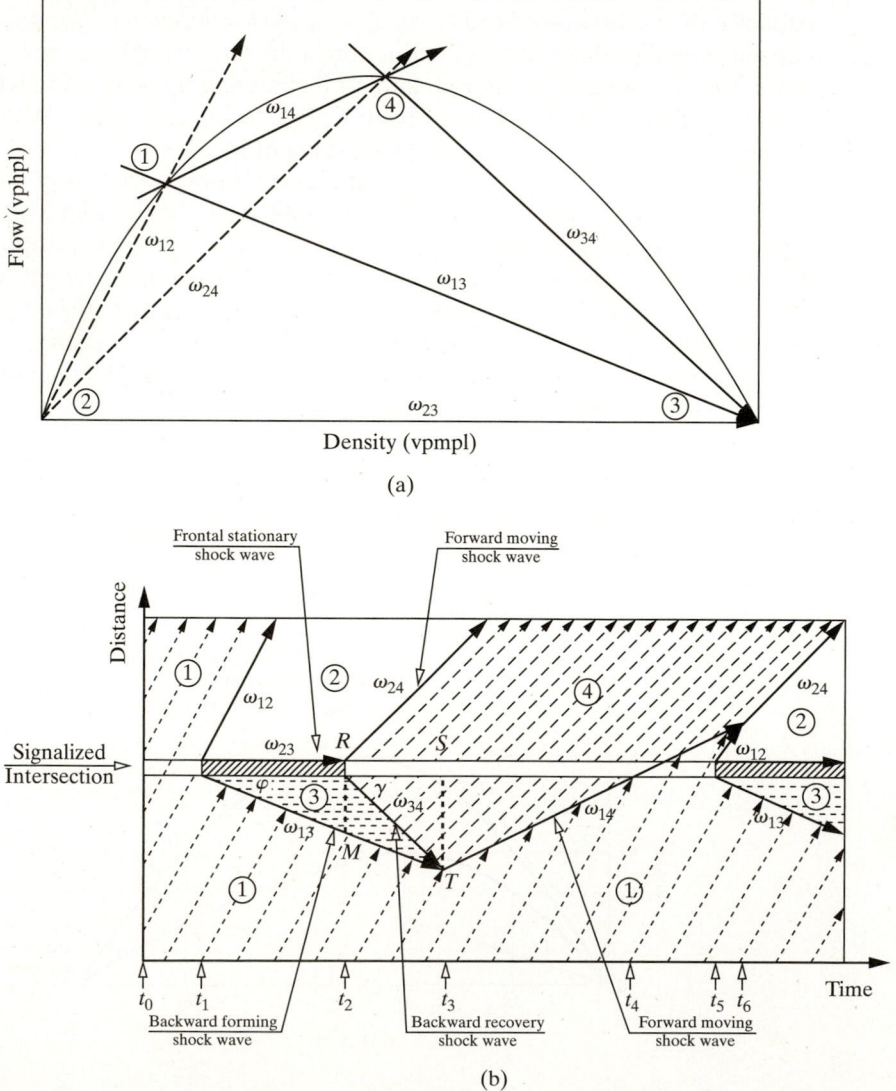

Figure 6.9   Shock Wave at Signalized Intersection

Backward forming shock waves are formed when the capacity is reduced below the demand flow rate resulting in the formation of a queue upstream of the bottleneck. The shock wave moves upstream, with its location at any time indicating the end of the queue at that time. This may occur at the approach of a signalized intersection when the signal indication is red, as shown in Figure 6.9, or at a location on a highway where the number of lanes is reduced.

Backward recovery shock waves are formed when the demand flow rate becomes less than the capacity of the bottleneck or when the restriction causing the capacity reduction at the bottleneck is removed. For example, when the signal indication for a set of lanes on a signalized intersection changes from red to green, the traffic flow restriction is removed, and traffic on that approach or set of lanes is free to move across the intersection, causing a backward recovery shock wave, as shown in Figure 6.9. The intersection of the backward forming shock wave and the backward recovery shock wave indicates the end of the queue shown as point *T* in Figure 6.9.

Rear stationary and forward recovery shock waves are formed when demand flow rate upstream of a bottleneck is first higher than the capacity of the bottleneck and then the demand flow rate reduces to the capacity of the bottleneck. For example, consider a four-lane (one direction) highway that leads to a two-lane tunnel in an urban area as shown in Figure 6.10. During the off-peak period when the demand capacity is less than the tunnel capacity, no shock wave is formed. However, when the demand capacity becomes higher than the tunnel capacity during the peak hour, a backward forming shock wave is formed. This shock wave continues to move upstream of the bottleneck as long as the demand flow is higher than the tunnel capacity, as shown in Figure 6.10. However, as the end of the peak period approaches, the demand flow rate tends to decrease until it is the same as the tunnel capacity. At this point, a rear stationary shock wave is formed until the demand flow becomes less than the tunnel capacity, resulting in the formation of a forward recovery shock wave, as shown in Figure 6.10.

6.3.2 Velocity of Shock Waves

Let us consider two different densities of traffic, k_1 and k_2, along a straight highway as shown in Figure 6.11, where $k_1 > k_2$. Let us also assume that these densities are separated by the line w representing the shock wave moving at a speed u_w. If the line w moves in the direction of the arrow (that is, in the direction of the traffic flow), u_w is positive.

With u_1 equal to the space mean speed of vehicles in the area with density k_1 (section P), the speed of the vehicle in this area relative to line w is

$$u_{r_1} = (u_1 - u_w)$$

The number of vehicles crossing line w from area P during a time period t is

$$N_1 = u_{r_1} k_1 t$$

Similarly, the speed of vehicles in the area with density k_2 (section Q) relative to line w is

$$u_{r_2} = (u_2 - u_w)$$

and the number of vehicles crossing line w during a time period t is

$$N_2 = u_{r_2} k_2 t$$

274

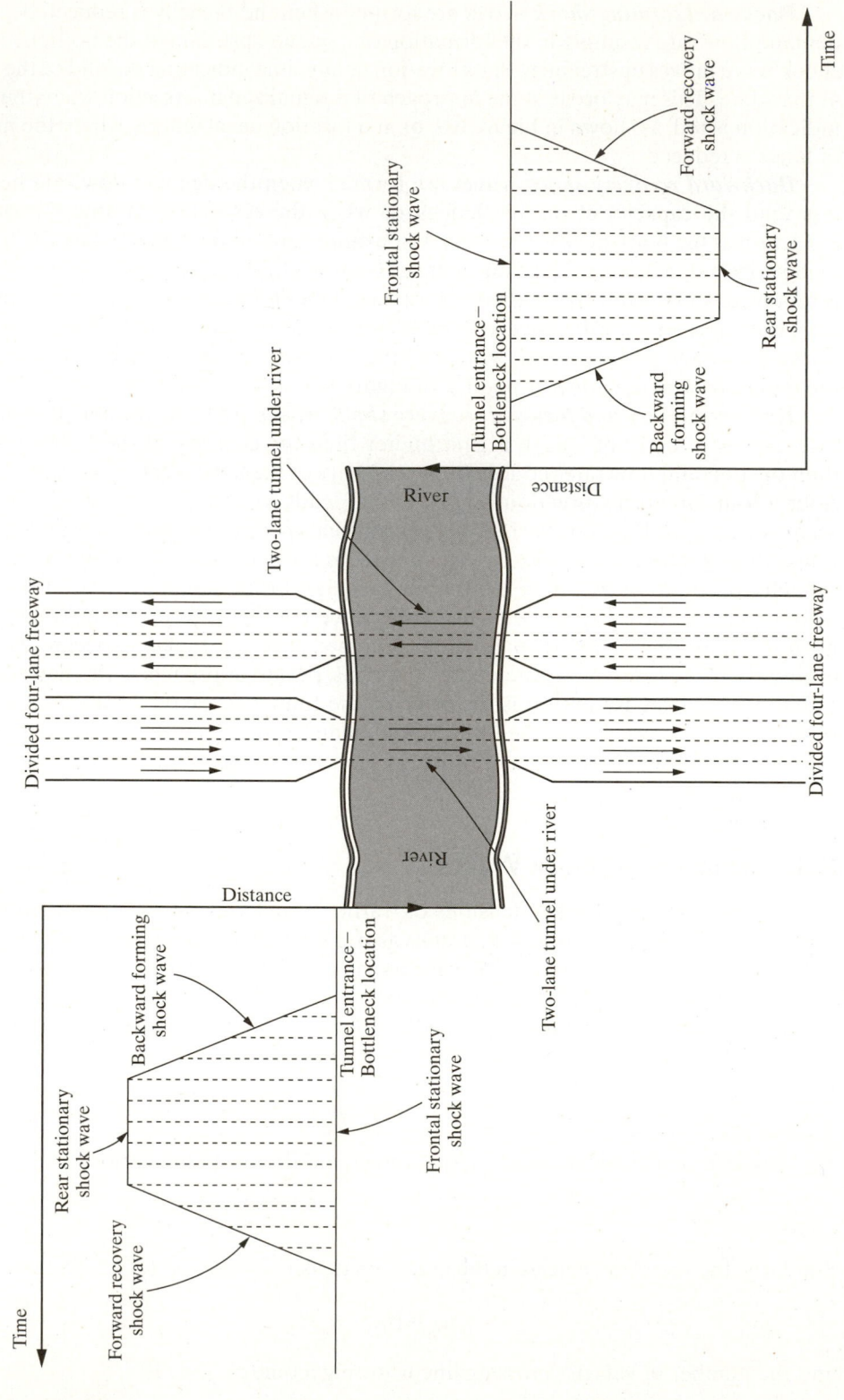

Figure 6.10 Shock Waves Due to a Bottleneck

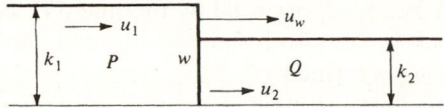

Figure 6.11   Movement of Shock Wave Due to Change in Densities

Since the net change is zero—that is, $N_1 = N_2$ and $(u_1 - u_w)k_1 = (u_2 - u_w)k_2$—we have

$$u_2 k_2 - u_1 k_1 = u_w (k_2 - k_1) \tag{6.39}$$

If the flow rates in sections P and Q are q_1 and q_2, respectively, then

$$q_1 = k_1 u_1 \qquad q_2 = k_2 u_2$$

Substituting q_1 and q_2 for $k_1 u_1$ and $k_2 u_2$ in Eq. 6.39 gives

$$q_2 - q_1 = u_w (k_2 - k_1)$$

That is,

$$u_w = \frac{q_2 - q_1}{k_2 - k_1} \tag{6.40}$$

which is also the slope of the line CD shown in Figure 6.8. This indicates that the velocity of the shock wave created by a sudden change of density from k_1 to k_2 on a traffic stream is the slope of the chord joining the points associated with k_1 and k_2 on the volume density curve for that traffic stream.

6.3.3 Shock Waves and Queue Lengths Due to a Red Phase at a Signalized Intersection

Figure 6.9b also shows the traffic conditions that exist at an approach of a signalized intersection when the signal indication is green then changes to red at the end of the green phase (start of the red phase) and changes to green again at the end of the red phase (start of the green phase). When the signal indication is green, the flow is normal, as shown in section 1. When the signals change to red at time t_1, two new conditions are formed immediately. Flow from this approach is stopped creating section 2, immediately downstream of the stop line with a density of zero and flow of zero. At the same time, all vehicles immediately upstream of the stop line are stationary, forming section 3, where the flow is zero and the density is the jam density. This results in the formation of the frontal stationary shock wave with velocity ω_{23} and the backward forming shock wave with velocity ω_{13}.

At the end of the red phase at time t_2 when the signal indication changes to green again, the flow rate at the stop line changes from zero to the saturation flow rate (see Chapter 8 for definition), as shown in section 4. This results in the forward moving shock wave ω_{24}. The queue length at this time—that is, at the end of the red phase—is represented by the line RM. Also at this time, the backward recovery shock wave with velocity of ω_{34} is formed that releases the queue as it moves upstream of the stop line. The intersection of the backward forming and backward recovery shock waves at point T and time t_3 indicates the position where the queue is completely dissipated, with the maximum queue length being represented by the line ST. The backward forming and backward recovery shock waves also terminate at time t_3, and a new forward moving shock wave with velocity ω_{14} is formed.

When the forward moving shock wave crosses the stop line, at time t_4, the flow changes at the stop line from the saturated flow rate to the original flow rate in section 1, and this continues until time t_5, when the signals change again to red.

Using Eq. 6.33, we can determine expressions for the velocities of the different shock waves and the queue lengths:

$$\text{The shock wave velocity } \omega_{12} = \frac{q_2 - q_1}{k_2 - k_1} = \frac{q_1 - q_2}{k_1 - k_2} = \frac{q_1 - 0}{k_1 - 0} = u_1 \qquad (6.41)$$

$$\text{The shock wave velocity } \omega_{13} = \frac{q_1 - q_3}{k_1 - k_3} = \frac{q_1 - 0}{k_1 - k_j} = \frac{q_1}{k_1 - k_j} \qquad (6.42)$$

$$\text{The shock wave velocity } \omega_{23} = \frac{q_2 - q_3}{k_2 - k_3} = \frac{0 - 0}{0 - k_j} = \frac{0}{k_j} = 0 \qquad (6.43)$$

This confirms that this wave is a stationary wave.

$$\text{The shock wave velocity } \omega_{24} = \frac{q_2 - q_4}{k_2 - k_4} = \frac{0 - q_4}{0 - k_4} = u_4 \qquad (6.44)$$

$$\text{The shock wave velocity } \omega_{34} = \frac{q_3 - q_4}{k_3 - k_4} = \frac{0 - q_4}{k_j - k_4} = \frac{-q_4}{k_j - k_4} \qquad (6.45)$$

$$\text{The length of the queue at the end of the red signal} = r \times \omega_{13} = \frac{rq_1}{k_1 - k_j} \qquad (6.46)$$

where r = the length of the red signal indication. Note that consistent units should be used for all variables in Eq. 6.46.

The maximum queue length $\overline{ST}$ can be determined from Figure 6.9, from where it can be seen that $\omega_{34} = \tan \gamma = \dfrac{\overline{ST}}{\overline{RS}}$, which gives $\overline{RS} = \dfrac{\overline{ST}}{\tan \gamma}$. Also, ω_{13} is $\tan \varphi$:

$$\tan \varphi = \frac{\overline{ST}}{r + \overline{RS}}$$

$$\overline{ST} = \tan \varphi (r + \overline{RS})$$

$$\overline{ST} = \tan \varphi \left(r + \frac{\overline{ST}}{\tan \gamma} \right)$$

$$r = \frac{\overline{ST}}{\tan \varphi} - \frac{\overline{ST}}{\tan \gamma}$$

$$\overline{ST} = \frac{r}{\dfrac{1}{\tan \varphi} - \dfrac{1}{\tan \gamma}}$$

$$\overline{ST} = \frac{r \tan \varphi \tan \gamma}{\tan \gamma - \tan \varphi}$$

$$\overline{ST} = \frac{r\omega_{13}\omega_{34}}{\omega_{34} - \omega_{13}} \tag{6.47}$$

The additional time $\overline{RS}$ (i.e., $t_3 - t_2$) after the end of the red signal it takes for the maximum queue to be formed can be obtained from the expression $\tan \varphi = \dfrac{\overline{ST}}{r + \overline{RS}}$, which gives

$$\overline{RS} = \frac{\overline{ST}}{\tan \varphi} - r \tag{6.48}$$

$$= \frac{r\omega_{13}\omega_{34}}{\omega_{13}(\omega_{34} - \omega_{13})} - r$$

$$= \frac{r\omega_{13}}{\omega_{13} - \omega_{34}}$$

Note that consistent units should be used for all variables in Eq. 6.48.

Example 6.5 Queue Lengths at a Signalized Intersection

The southbound approach of a signalized intersection carries a flow of 1000 veh/h/ln at a velocity of 50 km/h. The duration of the red signal indication for this approach is 15 sec. If the saturation flow is 2000 veh/h/ln with a density of 75 veh/ln, the jam density is 150 veh/km, determine the following:

    a. The length of the queue at the end of the red phase
    b. Speed of backward recovery wave velocity
    c. The maximum queue length

Solution:

    a. Determine speed of backward forming shock wave ω_{13} when signals turn to red. Use Eq. 6.40.

$$\omega_w = \frac{q_2 - q_1}{k_2 - k_1}$$

$$\omega_{13} = \frac{q_1 - q_3}{k_1 - k_3}$$

$$q_1 = 1000 \text{ veh/h/ln}$$

$$q_3 = 0 \text{ veh/h/ln}$$

$$k_1 = \frac{1000}{50} = 20 \text{ veh/km (see Eq. 6.7.)}$$

$$\omega_{13} = \frac{1000 - 0}{20 - 150} \text{ km/h} = -7.69 \text{ km/h}$$

$$= -7.69 \times 0.28 \text{ m/sec} = -2.14 \text{ m/sec}$$

Length of queue at end of red phase = $15 \times 11.31 = 169.65$ m

b. Determine speed of backward recovery wave velocity. Use Eq. 6.40.

$$\omega_{34} = \frac{q_3 - q_4}{k_3 - k_4} = \frac{0 - 2000}{150 - 75} = -26.67 \text{ km/h} = -26.677 \times 0.28 \text{ m/sec} = -7.41 \text{ m/sec}$$

c. Determine the maximum queue length. Use Eq. 6.40.

$$\text{Maximum queue length} = \frac{r\omega_{13}\omega_{34}}{\omega_{34} - \omega_{13}} = \frac{15 \times 2.14 \times 7.41}{7.41 - 2.14} = 45 \text{ m}$$

6.3.4 Shock Waves and Queue Lengths Due to Temporary Speed Reduction at a Section of Highway

Let us now consider the situation where the normal speed on a highway is temporarily reduced at a section of a highway where the flow is relatively high but lower than its capacity. For example, consider a truck that enters a two-lane highway at time t_1 and traveling at a much lower speed than the speed of the vehicles driving behind it. The truck travels for some time on the highway and eventually leaves the highway at time t_2. If the traffic condition is such that the vehicles cannot pass the truck, the shock waves that will be formed are shown in Figure 6.12. The traffic condition prior to the truck entering the highway at time t_1 is depicted as section 1.

At time t_1, vehicles immediately behind the truck will reduce their speed to that of the truck. This results in an increased density immediately behind the truck, resulting in traffic condition 2. The moving shock wave with a velocity of ω_{12} is formed. Also, because vehicles ahead of the truck will continue to travel at their original speed, a section on the highway just downstream of the truck will have no vehicles, thereby creating traffic condition 3. This also results in the formation of the forward moving shock waves with velocities of ω_{13} and ω_{32}. At time t_2, when the truck leaves the highway, the flow will be increased to the capacity of the highway with traffic condition 4. This results in the formation of a backward moving shock wave velocity ω_{24} and a forward moving shock wave with velocity ω_{34}. At time t_3, shock waves with velocities ω_{12} and ω_{24} coincide, resulting in a new forward moving shock wave with a velocity ω_{41}. It should be noted that the actual traffic conditions 2 and 4 depend on the original traffic condition 1 and the speed of the truck.

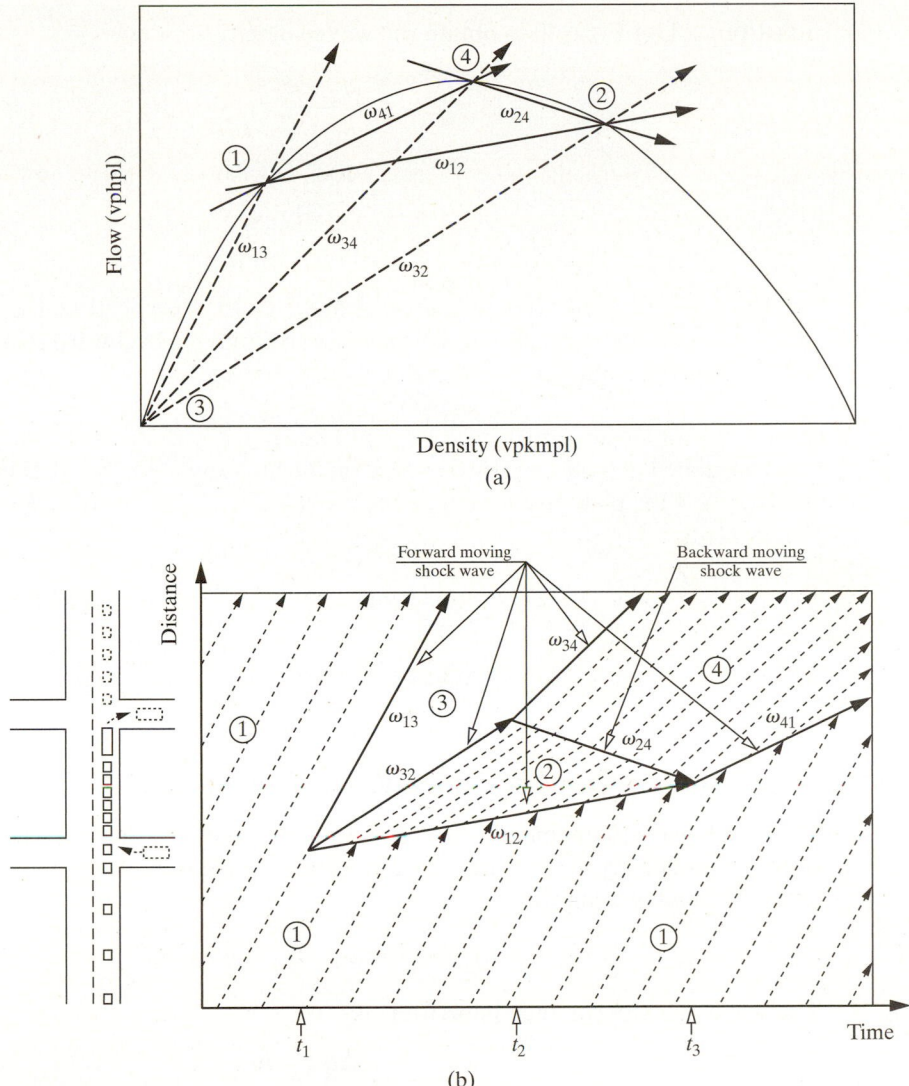

Figure 6.12   Shock Wave Created by Slow Traffic

Example 6.6 Length of Queue Due to a Speed Reduction

The volume at a section of a two-lane highway is 1500 veh/h in each direction and the density is about 25 veh/km. A large dump truck loaded with soil from an adjacent construction site joins the traffic stream and travels at a speed of 10 km/h for a length of 2.5 km along the upgrade before turning off onto a dump site. Due to the relatively high flow in the opposite direction, it is impossible for any car to pass the truck. Vehicles just behind the truck therefore have to travel at the speed of the truck, which results in the formation of a platoon having a density of 100 veh/km and a flow of 1000 veh/h. Determine how many vehicles will be in the platoon by the time the truck leaves the highway.

Solution: Use Eq. 6.40 to obtain the wave velocity.

$$u_w = \frac{q_2 - q_1}{k_2 - q_1}$$

$$u_w = \frac{1000 - 1500}{100 - 25}$$

$$= -6.7 \text{ km/h}$$

Knowing that the truck is traveling at 10 km/h, the speed of the vehicles in the platoon is also 10 km/h and that the shock wave is moving backward relative to the road at 6.7 km/h, determine the growth rate of the platoon.

$$10 - (-6.7) = 16.7 \text{ km/h}$$

Calculate the time spent by the truck on the highway—2.5/10 = 0.25 h—to determine the length of the platoon by the time the truck leaves the highway.

$$0.25 \times 16.7 = 4.2 \text{ km}$$

Use the density of 100 veh/km to calculate the number of vehicles in the platoon.

$$100 \times 4.2 = 420 \text{ vehicles}$$

6.3.5 Special Cases of Shock Wave Propagation

The shock wave phenomenon can also be explained by considering a continuous change of flow and density in the traffic stream. If the change in flow and the change in density are very small, we can write

$$(q_2 - q_1) = \Delta q \quad (k_2 - k_1) = \Delta k$$

The wave velocity can then be written as

$$u_w = \frac{\Delta q}{\Delta k} = \frac{dq}{dk} \tag{6.49}$$

Since $q = k\bar{u}_s$, substituting $k\bar{u}_s$ for q in Eq. 6.49 gives

$$u_w = \frac{d(k\bar{u}_s)}{dk} \tag{6.50}$$

$$= \bar{u}_s + k\frac{d\bar{u}_s}{dk} \tag{6.51}$$

When such a continuous change of volume occurs in a vehicular flow, a phenomenon similar to that of fluid flow exists in which the waves created in the traffic stream transport the continuous changes of flow and density. The speed of these waves is dq/dk and is given by Eq. 6.51.

We have already seen that as density increases, the space mean speed decreases (see Eq. 6.7), giving a negative value for $d\bar{u}_s/dk$. This shows that at any point on the fundamental diagram, the speed of the wave is theoretically less than the space mean speed of the traffic stream. Thus, the wave moves in the opposite direction relative to that of the traffic stream. The actual direction and speed of the wave will depend on the point at which we are on the curve (that is, the flow and density on the highway), and the resultant effect on the traffic downstream will depend on the capacity of the restricted area (bottleneck).

When both the flow and the density of the traffic stream are very low, that is, approaching zero, the flow is much lower than the capacity of the restricted area and there is very little interaction between the vehicles. The differential of $\bar{u}_s$ with respect to k ($d\bar{u}_s/dk$) then tends to zero, and the wave velocity approximately equals the space mean speed. The wave therefore moves forward with respect to the road, and no backups result.

As the flow of the traffic stream increases to a value much higher than zero but still lower than the capacity of the restricted area (say, q_3 in Figure 6.8), the wave velocity is still less than the space mean speed of the traffic stream, and the wave moves forward relative to the road. This results in a reduction in speed and an increase in the density from k_3 to k_3^b as vehicles enter the bottleneck but no backups occur. When the volume on the highway is equal to the capacity of the restricted area (C_2 in Figure 6.8), the speed of the wave is zero and the wave does not move. This results in a much slower speed and a greater increase in the density to k_o^b as the vehicles enter the restricted area. Again, delay occurs but there are no backups.

However, when the flow on the highway is greater than the capacity of the restricted area, not only is the speed of the wave less than the space mean speed of the vehicle stream, but it moves backward relative to the road. As vehicles enter the restricted area, a complex queuing condition arises, resulting in an immediate increase in the density from k_1 to k_2 in the upstream section of the road and a considerable decrease in speed. The movement of the wave toward the upstream section of the traffic stream creates a shock wave in the traffic stream, eventually resulting in backups, which gradually move upstream of the traffic stream.

The expressions developed for the speed of the shock wave, Eqs. 6.40 and 6.51, can be applied to any of the specific models described earlier. For example, the Greenshields model can be written as

$$\bar{u}_{si} = u_f\left(1 - \frac{k_i}{k_j}\right) \quad \bar{u}_{si} = u_f(1 - \eta_i) \tag{6.52}$$

where $\eta_i = (k_i/k_j)$ (normalized density).

where u_f = mean free speed

If the Greenshields model fits the flow density relationship for a particular traffic stream, Eq. 6.40 can be used to determine the speed of a shock wave as

$$u_w = \frac{\left[k_2 u_f\left(1 - \dfrac{k_2}{k_j}\right)\right] - \left[k_1 u_f\left(1 - \dfrac{k_1}{k_j}\right)\right]}{k_2 - k_1}$$

$$= \frac{k_2 u_f(1 - \eta_2) - k_1 u_f(1 - \eta_1)}{k_2 - k_1}$$

$$= \frac{u_f(k_2 - k_1) - k_2 u_f \eta_2 + k_1 u_f \eta_1}{k_2 - k_1}$$

$$= \frac{u_f(k_2 - k_1) - \dfrac{u_f}{k_j}(k_2^2 - k_1^2)}{k_2 - k_1}$$

$$= \frac{u_f(k_2 - k_1) - \dfrac{u_f}{k_j}(k_2 - k_1)(k_2 + k_1)}{(k_2 - k_1)}$$

$$= u_f[1 - (\eta_1 + \eta_2)]$$

The speed of a shock wave for the Greenshields model is therefore given as

$$u_w = u_f[1 - (\eta_1 + \eta_2)] \tag{6.53}$$

Density Nearly Equal

When there is only a small difference between k_1 and k_2 (that is, $\eta_1 \approx \eta_2$),

$$u_w = u_f[1 - \eta_1 + \eta_2] \quad \text{(neglecting the small change in } \eta_1)$$
$$= u_f[1 - 2\eta_1]$$

where u_f = mean free speed

Stopping Waves

Equation 6.53 can also be used to determine the velocity of the shock wave due to the change from green to red of a signal at an intersection approach if the Greenshields model is applicable. During the green phase, the normalized density is η_1. When the traffic signal changes to red, the traffic at the stop line of the approach comes to a halt, which results in a density equal to the jam density. The value of η_2 is then equal to 1.

The speed of the shock wave, which in this case is a stopping wave, can be obtained by

$$u_w = u_f[1 - (\eta_1 + 1)] = -u_f\eta_1 \tag{6.54}$$

where u_f = mean free speed

Equation 6.54 indicates that in this case the shock wave travels upstream of the traffic with a velocity of $u_f\eta_1$. If the length of the red phase is t sec, then the length of the line of cars upstream of the stopline at the end of the red interval is $u_f\eta_1 t$.

Starting Waves

At the instant when the signal again changes from red to green, η_1 equals 1. Vehicles will then move forward at a speed of $\bar{u}_{s2}$, resulting in a density of η_2. The speed of the shock wave, which in this case is a starting wave, is obtained by

$$u_w = u_f[1 - (1 + \eta_2)] = -u_f\eta_2 \tag{6.55}$$

Equation 6.52, $\bar{u}_{s2} = u_f(1 - \eta_2)$, gives

$$\eta_2 = 1 - \frac{\bar{u}_{s2}}{u_f}$$

The velocity of the shock wave is then obtained as

$$u_w = -u_f + \bar{u}_{s2}$$

Since the starting velocity $\bar{u}_{s2}$ just after the signal changes to green is usually small, the velocity of the starting shock wave approximately equals $-u_f$.

Example 6.7 Length of Queue Due to a Stopping Shock Wave

Studies have shown that the traffic flow on a single-lane approach to a signalized intersection can be described by the Greenshields model. If the jam density on the approach is 130 veh/km, determine the velocity of the stopping wave when the approach signal changes to red if the density on the approach is 45 veh/km and the space mean speed is 40 km/h. At the end of the red interval, what length of the approach upstream from the stop line will vehicles be affected if the red interval is 35 sec?

Solution:

- Use the Greenshields model.

$$\bar{u}_s = u_f - \frac{u_f}{k_j} k$$

$$40 = u_f - \frac{u_f}{130} 45$$

$$5200 = 130 u_f - 45 u_f$$

$$u_f = 61.2 \text{ km/h}$$

- Use Eq. 6.54 for a stopping wave.

$$u_w = -u_f \eta_1$$

$$= -61.2 \times \frac{45}{130}$$

$$= -21.2 \text{ km/h}$$

Since u_w is negative, the wave moves upstream.

- Determine the approach length that will be affected in 35 sec.

$$21.2 \times 0.28 \times 35 = 207.8 \text{ m}$$

6.4 GAP AND GAP ACCEPTANCE

Thus far, we have been considering the theory of traffic flow as it relates to the flow of vehicles in a single stream. Another important aspect of traffic flow is the interaction of vehicles as they join, leave, or cross a traffic stream. Examples of these include ramp vehicles merging onto an expressway stream, freeway vehicles leaving the freeway onto frontage roads, the changing of lanes by vehicles on a multilane highway and vehicles turning left or right at a stop sign. The most important factor a driver considers in making any one of these maneuvers is the availability of a gap between two vehicles that, in the driver's judgment, is adequate for him or her to complete the maneuver. The evaluation of available gaps and the decision to carry out a specific maneuver within a particular gap are inherent in the concept of gap acceptance.

Following are the important measures that involve the concept of gap acceptance:

1. **Merging** is the process by which a vehicle in one traffic stream joins another traffic stream moving in the same direction, such as a ramp vehicle joining a freeway stream.

2. **Diverging** is the process by which a vehicle in a traffic stream leaves that traffic stream, such as a vehicle leaving the outside lane of an expressway.

3. **Weaving** is the process by which a vehicle first merges into a stream of traffic, obliquely crosses that stream, and then merges into a second stream moving in the same direction; for example, the maneuver required for a ramp vehicle to join the far side stream of flow on an expressway.

4. **Gap** is the the distance between the rear bumper of a vehicle and the front bumper of the following vehicle. It is evaluated by a vehicle driver in a minor stream who wishes to merge into the major stream. It is expressed either in units of time (time gap) or in units of distance (space gap).

5. **Time lag** is the difference between the time a vehicle that merges into a main traffic stream reaches a point on the highway in the area of merge and the time a vehicle in the main stream reaches the same point.

6. **Space lag** is the difference, at an instant of time, between the distance a merging vehicle is away from a reference point in the area of merge and the distance a vehicle in the main stream is away from the same point.

Figure 6.13 depicts the time-distance relationships for a vehicle at a stop sign waiting to merge and for vehicles on the near lane of the main traffic stream.

A driver who intends to merge must first evaluate the gaps that become available to determine which gap (if any) is large enough to accept the vehicle, in his or her opinion. In accepting that gap, the driver feels that he or she will be able to complete the merging maneuver and safely join the main stream within the length of the gap. This phenomenon is generally referred to as *gap acceptance.* It is of importance when engineers are considering the delay of vehicles on minor roads wishing to join a major-road traffic stream at unsignalized intersections, and also the delay of ramp vehicles wishing to join expressways. It can also be used in timing the release of vehicles at an on-ramp of an expressway, such that the probability of the released vehicle finding an acceptable gap in arriving at the freeway shoulder lane is maximized.

To use the phenomenon of gap acceptance in evaluating delays, waiting times, queue lengths, and so forth, at unsignalized intersections and at on-ramps, the average minimum

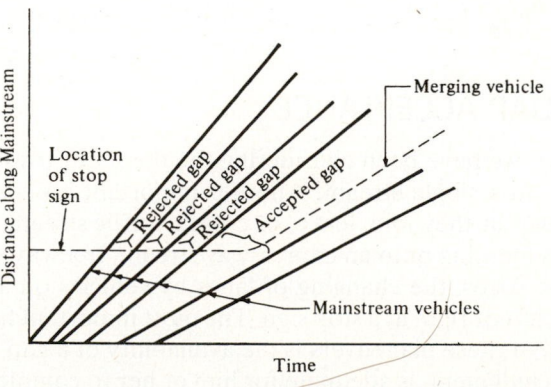

Figure 6.13   Time-Space Diagrams for Vehicles in the Vicinity of a Stop Sign

gap length that will be accepted by drivers should be determined first. Several definitions have been given to this "critical" value. Greenshields referred to it as the "acceptable average minimum time gap" and defined it as the gap accepted by 50 percent of the drivers. The concept of "critical gap" was used by Raff, who defined it as the gap for which the number of accepted gaps shorter than it is equal to the number of rejected gaps longer than it. The data in Table 6.3 are used to demonstrate the determination of the critical gap using Raff's definition. Either a graphical or an algebraic method can be used.

In using the graphical method, two cumulative distribution curves are drawn as shown in Figure 6.14. One relates gap lengths t with the number of accepted gaps less than t, and the other relates t with the number of rejected gaps greater than t. The intersection of these two curves gives the value of t for the critical gap.

In using the algebraic method, it is necessary to first identify the gap lengths between where the critical gap lies. This is done by comparing the change in number of accepted gaps less than t sec (column 2 of Table 6.3b) for two consecutive gap lengths, with the change in number of rejected gaps greater than t sec (column 3 of Table 6.3b) for the same two consecutive gap lengths. The critical gap length lies between the two consecutive gap lengths where the difference between the two changes is minimal. Table 6.3b shows the computation and indicates that the critical gap for this case lies between 3 and 4 seconds.

For example, in Figure 6.14, with Δt equal to the time increment used for gap analysis, the critical gap lies between t_1 and $t_2 = t_1 + \Delta t$,

Table 6.3   Computation of Critical Gap (t_c)

(a) Gaps Accepted and Rejected		
1	*2*	*3*
Length of Gap (t sec)	*Number of Accepted Gaps (less than t sec)*	*Number of Rejected Gaps (greater than t sec)*
0.0	0	116
1.0	2	103
2.0	12	66
3.0	$m = 32$	$r = 38$
4.0	$n = 57$	$p = 19$
5.0	84	6
6.0	116	0

(b) Difference in Gaps Accepted and Rejected			
1	*2*	*3*	*4*
Consecutive Gap Lengths (t sec)	*Change in Number of Accepted Gaps (less than t sec)*	*Change in Number of Rejected Gaps (greater than t sec)*	*Difference Between Columns 2 and 3*
0.0–1.0	2	13	11
1.0–2.0	10	37	27
2.0–3.0	20	28	8
3.0–4.0	25	19	6
4.0–5.0	27	13	14
5.0–6.0	32	6	26

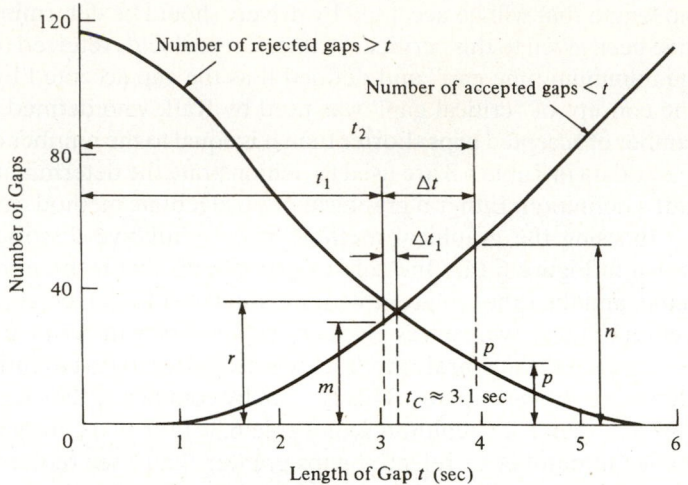

Figure 6.14   Cumulative Distribution Curves for Accepted and Rejected Gaps

where

m = number of accepted gaps less than t_1
r = number of rejected gaps greater than t_1
n = number of accepted gaps less than t_2
p = number of rejected gaps greater than t_2

Assuming that the curves are linear between t_1 and t_2, the point of intersection of these two lines represents the critical gap. From Figure 6.14, the critical gap expression can be written as

$$t_c = t_1 + \Delta t_1$$

Using the properties of similar triangles,

$$\frac{\Delta t_1}{r - m} = \frac{\Delta t - \Delta t_1}{n - p}$$

$$\Delta t_1 = \frac{\Delta t(r - m)}{(n - p) + (r - m)}$$

we obtain

$$t_c = t_1 + \frac{\Delta t(r - m)}{(n - p) + (r - m)} \qquad (6.56)$$

For the data given in Table 6.3, we thus have

$$t_c = 3 + \frac{1(38 - 32)}{(57 - 19) + (38 - 32)} = 3 + \frac{6}{38 + 6}$$

$$\approx 3.16 \text{ sec}$$

6.4.1 Stochastic Approach to Gap and Gap Acceptance Problems

The use of gap acceptance to determine the delay of vehicles in minor streams wishing to merge onto major streams requires a knowledge of the frequency of arrivals of gaps that are at least equal to the critical gap. This in turn depends on the distribution of arrivals of mainstream vehicles at the area of merge. It is generally accepted that for light to medium traffic flow on a highway, the arrival of vehicles is randomly distributed. It is therefore important that the probabilistic approach to the subject be discussed. It is usually assumed that for light-to-medium traffic the distribution is Poisson, although assumptions of gamma and exponential distributions have also been made.

Assuming that the distribution of mainstream arrival is Poisson, then the probability of x arrivals in any interval of time t sec can be obtained from the expression

$$P(x) = \frac{\mu^x e^{-\mu}}{x!} \qquad \text{(for } x = 0, 1, 2 \cdots, \infty) \tag{6.57}$$

where

$P(x)$ = the probability of x vehicles arriving in time t sec
μ = average number of vehicles arriving in time t

If V represents the total number of vehicles arriving in time T sec, then the average number of vehicles arriving per second is

$$\lambda = \frac{V}{T} \qquad \mu = \lambda t$$

We can therefore write Eq. 6.57 as

$$P(x) = \frac{(\lambda t)^x e^{-\lambda t}}{x!} \tag{6.58}$$

Now consider a vehicle at an unsignalized intersection or at a ramp waiting to merge into the mainstream flow, arrivals of which can be described by Eq. 6.58. The minor stream vehicle will merge only if there is a gap of t sec equal to or greater than its critical gap. This will occur when no vehicles arrive during a period t sec long. The probability of this is the probability of zero cars arriving (that is, when x in Eq. 6.58 is zero). Substituting zero for x in Eq. 6.58 will therefore give a probability of a gap $(h \geq t)$ occurring. Thus,

$$P(0) = P(h \geq t) = e^{-\lambda t} \qquad \text{for } t \geq 0 \tag{6.59}$$

$$P(h < t) = 1 - e^{-\lambda t} \qquad \text{for } t \geq 0 \tag{6.60}$$

Since

$$P(h < t) + P(h \geq t) = 1$$

it can be seen that t can take all values from 0 to ∞, which therefore makes Eqs. 6.59 and 6.60 continuous functions. The probability function described by Eq. 6.59 is known as the *exponential distribution*.

Equation 6.59 can be used to determine the expected number of acceptable gaps that will occur at an unsignalized intersection or at the merging area of an expressway on-ramp during a period T, if the Poisson distribution is assumed for the mainstream flow and the volume V is also known. Let us assume that T is equal to 1 hr and that V is the volume in veh/h on the mainstream flow. Since $(V - 1)$ gaps occur between V successive vehicles in a stream of vehicles, then the expected number of gaps greater or equal to t is given as

$$\text{Frequency } (h \geq t) = (V - 1)e^{-\lambda t} \tag{6.61}$$

and the expected number of gaps less than t is given as

$$\text{Frequency } (h < t) = (V - 1)(1 - e^{-\lambda t}) \tag{6.62}$$

Example 6.8 Number of Acceptable Gaps for Vehicles on an Expressway Ramp

The peak hour volume on an expressway at the vicinity of the merging area of an on-ramp was determined to be 1800 veh/h. If it is assumed that the arrival of expressway vehicles can be described by a Poisson distribution, and the critical gap for merging vehicles is 3.5 sec, determine the expected number of acceptable gaps for ramp vehicles that will occur on the expressway during the peak hour.

Solution: List the data.

$$V = 1800$$

$$T = 3600 \text{ sec}$$

$$\lambda = (1800/3600) = 0.5 \text{ veh/sec}$$

Calculate the expected number of acceptable gaps in 1 hr using Eq. 6.61.

$$(h \geq t) = (1800 - 1)e^{(-0.5 \times 3.5)} = 1799e^{-1.75} = 312$$

The expected number of occurrences of different gaps t for the previous example have been calculated and are shown in Table 6.4.

Table 6.4 Number of Different Lengths of Gaps Occurring During a Period of 1 hr for $V = 1800$ veh/h and an Assumed Distribution of Poisson for Arrivals

Gap (t sec)	Probability P(h ≥ t)	P (h < t)	No. of Gaps h ≥ t	h ≤ t
0	1.0000	0.0000	1799	0
0.5	0.7788	0.2212	1401	398
1.0	0.6065	0.3935	1091	708
1.5	0.4724	0.5276	849	950
2.0	0.3679	0.6321	661	1138
2.5	0.2865	0.7135	515	1284
3.0	0.2231	0.7769	401	1398
3.5	0.1738	0.8262	312	1487
4.0	0.1353	0.8647	243	1556
4.5	0.1054	0.8946	189	1610
5.0	0.0821	0.9179	147	1652

The basic assumption made in this analysis is that the arrival of mainstream vehicles can be described by a Poisson distribution. This assumption is reasonable for light-to-medium traffic but may not be acceptable for conditions of heavy traffic. Analyses of the occurrence of different gap sizes when traffic volume is heavy have shown that the main discrepancies occur at gaps of short lengths (that is, less than 1 sec). The reason for this is that although theoretically there are definite probabilities for the occurrence of gaps between 0 and 1 sec, in reality these gaps very rarely occur, since a driver will tend to keep a safe distance between his or her vehicle and the vehicle immediately in front. One alternative used to deal with this situation is to restrict the range of headways by introducing a minimum gap. Equations 6.61 and 6.62 can then be written as

$$P(h \geq t) = e^{-\lambda(t-\tau)} \qquad \text{(for } t \geq 0) \qquad (6.63)$$

$$P(h < t) = 1 - e^{-\lambda(t-\tau)} \qquad \text{(for } t \leq 0) \qquad (6.64)$$

where τ is the minimum gap introduced.

Example 6.9 Number of Acceptable Gaps with a Restrictive Range, for Vehicles on an Expressway Ramp

Repeat Example 6.7 using a minimum gap in the expressway traffic stream of 1.0 sec and the data:

$$V = 1800$$

$$T = 3600$$

$$\lambda = (1800/3600) = 0.5 \text{ veh/sec}$$

$$t = 3.5 \text{ sec}$$

Solution: Calculate the expected number of acceptable gaps in 1 hr.

$$(h \geq t) = (1800 - 1)e^{-0.5(3.5-1.0)} = 1799e^{-0.5 \times 2.5}$$

$$= 515$$

6.5 INTRODUCTION TO QUEUING THEORY

One of the greatest concerns of traffic engineers is the serious congestion that exists on urban highways, especially during peak hours. This congestion results in the formation of queues on expressway on-ramps and off-ramps, at signalized and unsignalized intersections, and on arterials, where moving queues may occur. An understanding of the processes that lead to the occurrence of queues and the subsequent delays on highways is essential for the proper analysis of the effects of queuing. The theory of queuing therefore concerns the use of mathematical algorithms to describe the processes that result in the formation of queues, so that a detailed analysis of the effects of queues can be undertaken. The analysis of queues can be undertaken by assuming either deterministic or stochastic queue characteristics.

6.5.1 Deterministic Analysis of Queues

The deterministic analysis assumes that all the traffic characteristics of the queue are deterministic and demand volumes and capacities are known. There are two common traffic conditions for which the deterministic approach has been used. The first is when an incident occurs on a highway resulting in a significant reduction on the capacity of the highway. This can be described as a varying service rate and constant demand condition. The second is significant increase in demand flow exceeding the capacity of a section of highway which can be described as a varying demand and constant service rate condition.

Varying Service Rate and Constant Demand

Consider a section of three-lane (one direction) highway with a capacity of c veh/h; that is, it can serve a maximum volume of c veh/h. (See Chapters 9 and 10 for discussion on capacity.) An incident occurs which resulted in the closure of one lane, thereby reducing its capacity to c_R for a period of t hr, which is the time it takes to clear the incident. The demand volume continues to be V veh/h throughout the period of the incident as shown in Figure 6.15a. The demand volume is less than the capacity of the highway section but greater than the reduced capacity. Before the incident, there is no queue as the demand volume is less than the capacity of the highway. However, during the incident, the demand volume is higher than the reduced capacity, resulting in the formation of a queue, as shown in Figure 6.15b. Several important parameters can be determined to describe the effect of this reduction in the highway capacity. These include the maximum queue length, duration of the queue, average queue length, maximum individual delay, time a driver spends in the queue, average queue length while the queue exists, maximum individual delay, and the total delay.

The maximum queue length (q_{max}) is the excess demand rate multiplied by the duration of the incident (t_{inc}) and is given as

$$q_{max} = (V - c_R)t_{inc} \text{ vehicles} \qquad (6.65)$$

The time duration of the queue (t_q) is the queue length divided by the difference between the capacity and the demand rate and is given as

$$t_q = \frac{(V - c_R)t_{inc}}{(c - V)} \text{ hr} \qquad (6.66)$$

The average queue length (q_{av}) is

$$q_{av} = \frac{(V - c_R)t_{inc}}{2} \text{ veh} \qquad (6.67)$$

The total delay (d_T) is the time duration of the queue multiplied by the average queue length and is given as

$$d_T = \frac{(V - c_R)t_{inc}}{2} \frac{(V - c_R)t_{inc}}{(c - V)} = \frac{t^2(V - c_R)(c - c_R)}{2(c - V)} \text{ hr} \qquad (6.68)$$

When using Eqs. 6.65 to 6.68, care should be taken to ensure that the same unit is used for all variables.

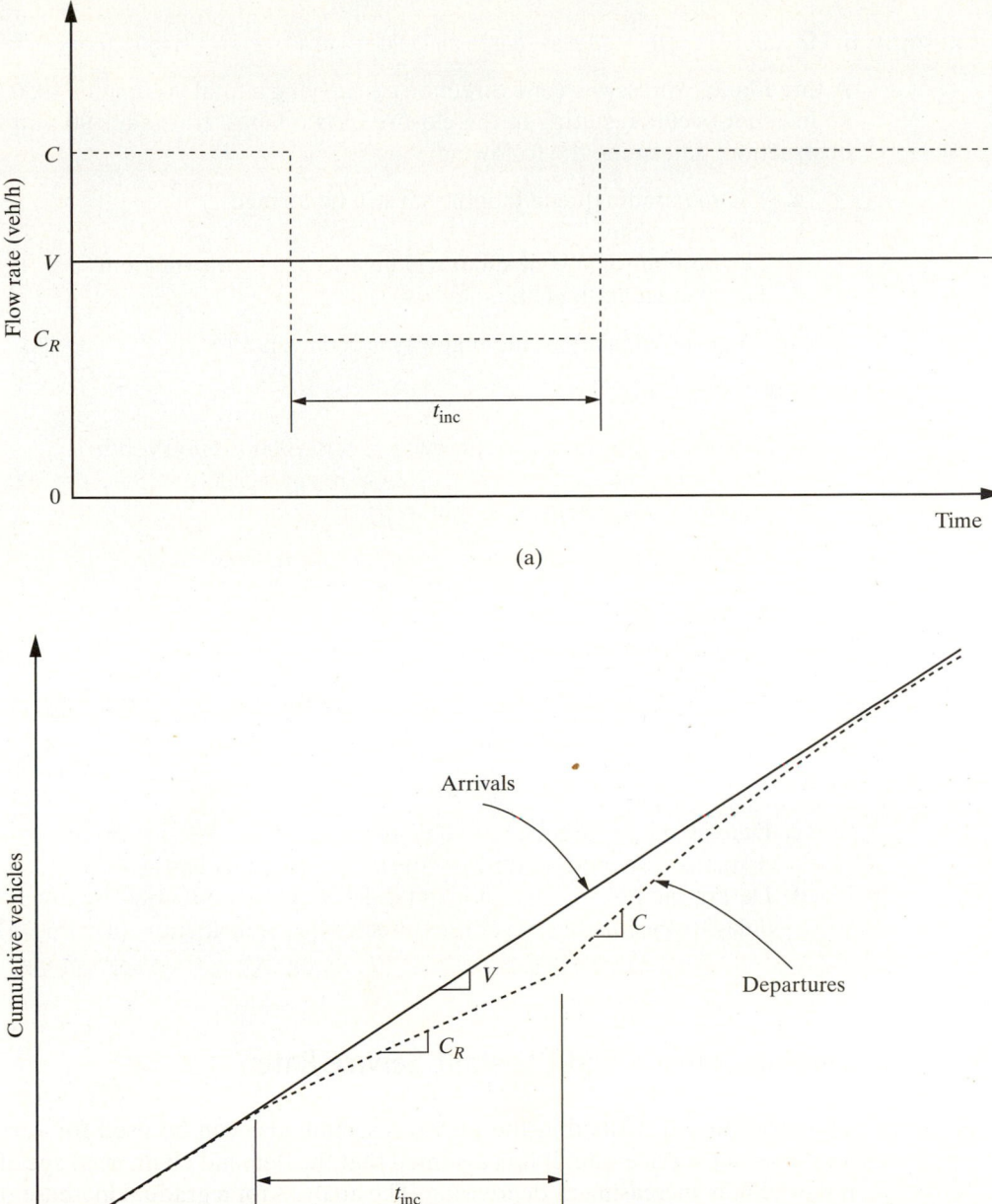

(a)

(b)

Figure 6.15 Queuing Diagram for Incident Situation

Example 6.10 Queue Length and Delay Due to an Incident on a Freeway Using Deterministic Analysis

A three-lane expressway (one direction) is carrying a total volume of 4050 veh/h when an incident occurs resulting in the closure of two lanes. If it takes 90 min to clear the obstruction, determine the following:

  a. The maximum queue length that will be formed
  b. The total delay
  c. The number of vehicles that will be affected by the incident
  d. The average individual delay

Assume that the capacity of the highway is 2000 veh/h/ln.

Solution:

- Determine capacity, c, of highway $= 3 \times 2000 = 6000$ veh/h
- Determine reduced capacity, c_R, of highway $= 2000 \times (3 - 2) = 2000$ veh/h
- Duration of incident $= 90$ min $= 1.5$ h

a. Determine maximum queue length. Use Eq. 6.65.

$$q_{max} = (V - c_R)t_{inc} \text{ vehicles} = (4050 - 2000) \times 1.5 \text{ veh} = 3075 \text{ veh}$$

b. Determine the total delay—use Eq. 6.61.

$$d_T = \frac{t_{inc}^2(V - c_R)(c - c_R)}{2(c - V)} = \frac{1.5^2(4050 - 2000)(6000 - 2000)}{2(6000 - 4050)}$$

$$= 4731 \text{ hr}$$

c. Determine the number of vehicles that will be affected by the incident = the demand rate multiplied by the duration of the incident $= 4050 \times 1.5 = 6075$ veh

d. Determine the average individual delay. This is obtained by dividing the total delay by the number of vehicles affected by the incident $= (4731/6075) = 0.779$ hr

Varying Demand and Constant Service Rate

The procedure described in the previous section also can be used for varying demand and constant service rate, if it is assumed that the demand changes at specific times and not gradually increasing or decreasing. The analysis for a gradual increase or decrease is beyond the scope of this book. Interested readers may refer to any book on traffic flow theory for additional information on this topic.

6.5.2 Stochastic Analyses of Queues

Using a stochastic approach to analyze queues considers the fact that certain traffic characteristics such as arrival rates are not always deterministic. In fact, arrivals at an intersection, for example, are deterministic or regular only when approach volumes are high. Arrival rates tend to be random for light to medium traffic. The stochastic approach is used to determine the probability that an arrival will be delayed, the expected waiting time for all arrivals, the expected waiting time of an arrival that waits, and so forth.

Several models have been developed that can be applied to traffic situations such as the merging of ramp traffic to freeway traffic, interactions at pedestrian crossings, and sudden reduction of capacity on freeways. This section will give only the elementary queuing theory relationships for a specific type of queue; that is, the single-channel queue. The theoretical development of these relationships is not included here. Interested readers may refer to any traffic flow theory book for a more detailed treatment of the topic.

A queue is formed when arrivals wait for a service or an opportunity, such as the arrival of an accepted gap in a main traffic stream, the collection of tolls at a tollbooth or of parking fees at a parking garage, and so forth. The service can be provided in a single channel or in several channels. Proper analysis of the effects of such a queue can be carried out only if the queue is fully specified. This requires that the following characteristics of the queue be given: (1) the characteristic distribution of arrivals, such as uniform, Poisson, and so on; (2) the method of service, such as first come–first served, random, and priority; (3) the characteristic of the queue length, that is, whether it is finite or infinite; (4) the distribution of service times; and (5) the channel layout, that is, whether there are single or multiple channels and, in the case of multiple channels, whether they are in series or parallel. Several methods for the classification of queues based on the above characteristics have been used—some of which are discussed below.

Arrival Distribution

The arrivals can be described as either a deterministic distribution or a random distribution. Light-to-medium traffic is usually described by a Poisson distribution, and this is generally used in queuing theories related to traffic flow.

Service Method

Queues also can be classified by the method used in servicing the arrivals. These include first come–first served where units are served in order of their arrivals, and last in–first served, where the service is reversed to the order of arrival. The service method can also be based on priority, where arrivals are directed to specific queues of appropriate priority levels—for example, giving priority to buses. Queues are then serviced in order of their priority level.

Characteristics of the Queue Length

The maximum length of the queue, that is, the maximum number of units in the queue, is specified, in which case the queue is a finite or truncated queue, or else there may be no restriction on the length of the queue. Finite queues are sometimes necessary when the waiting area is limited.

Service Distribution

The Poisson and negative exponential distributions have been used as the random distributions.

Number of Channels

The number of channels usually corresponds to the number of waiting lines and is therefore used to classify queues, for example, as a single-channel or multichannel queue.

Oversaturated and Undersaturated Queues

Oversaturated queues are those in which the arrival rate is greater than the service rate, and undersaturated queues are those in which the arrival rate is less than the service rate. The length of an undersaturated queue may vary but will reach a steady state with the arrival of units. The length of an oversaturated queue, however, will never reach a steady state but will continue to increase with the arrival of units.

Single-Channel, Undersaturated, Infinite Queues

Figure 6.16 is a schematic of a single-channel queue in which the rate of arrival is q veh/h and the service rate is Q veh/h. For an undersaturated queue, $Q > q$, assuming that both the rate of arrivals and the rate of service are random, the following relationships can be developed:

1. Probability of n units in the system, $P(n)$:

$$P(n) = \left(\frac{q}{Q}\right)^n \left(1 - \frac{q}{Q}\right) \tag{6.69}$$

where n is the number of units in the system, including the unit being serviced.

2. The expected number of units in the system, $E(n)$:

$$E(n) = \frac{q}{Q - q} \tag{6.70}$$

3. The expected number of units waiting to be served (that is, the mean queue length) in the system, $E(m)$:

$$E(m) = \frac{q^2}{Q(Q - q)} \tag{6.71}$$

Note that $E(m)$ is not exactly equal to $E(n) - 1$, the reason being that there is a definite probability of zero units being in the system, $P(0)$.

4. Average waiting time in the queue, $E(w)$:

$$E(w) = \frac{q}{Q(Q - q)} \tag{6.72}$$

5. Average waiting time of an arrival, including queue and service, $E(v)$:

$$E(v) = \frac{1}{Q - q} \tag{6.73}$$

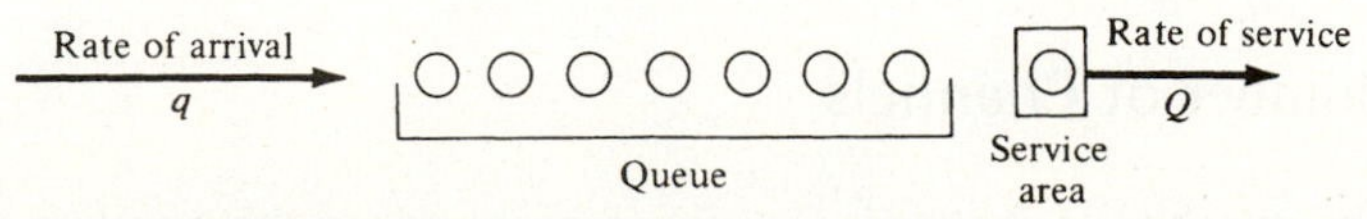

Figure 6.16 A Single-Channel Queue

6. Probability of spending time t or less in the system:

$$P(v \leq t) = 1 - e^{-(1-\frac{q}{Q})qt}$$ (6.74)

7. Probability of waiting for time t or less in the queue:

$$P(w \leq t) = 1 - \frac{q}{Q} e^{-(1-\frac{q}{Q})qt}$$ (6.75)

8. Probability of more than N vehicles being in the system, that is, $P(n > N)$:

$$P(n > N) = \left(\frac{q}{Q}\right)^{N+1}$$ (6.76)

Equation 6.70 can be used to produce a graph of the relationship between the expected number of units in the system, $E(n)$, and the ratio of the rate of arrival to the rate of service, $p = q/Q$. Figure 6.17 is such a representation for different values of p. It should be noted that as this ratio tends to 1 (that is, approaching saturation), the expected number of vehicles in the system tends to infinity. This shows that q/Q, which is usually referred to as the *traffic intensity,* is an important factor in the queuing process. The figure also indicates that queuing is of no significance when p is less than 0.5, but at values of 0.75 and above, the average queue lengths tend to increase rapidly. Figure 6.18 is also a graph of the probability of n units being in the system versus q/Q.

Single–Channel, Undersaturated, Finite Queues

In the case of a finite queue, the maximum number of units in the system is specified. Let this number be N. Let the rate of arrival be q and the service rate be Q. If it is also assumed that both the rate of arrival and the rate of service are random, the following relationships can be developed for the finite queue.

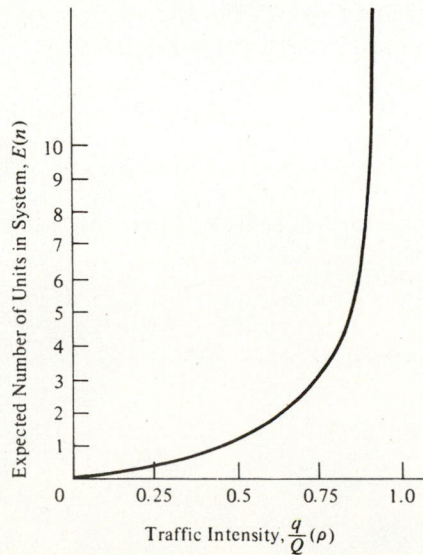

Figure 6.17   Expected Number of Vehicles in the System $E(n)$ versus Traffic Intensity (ρ)

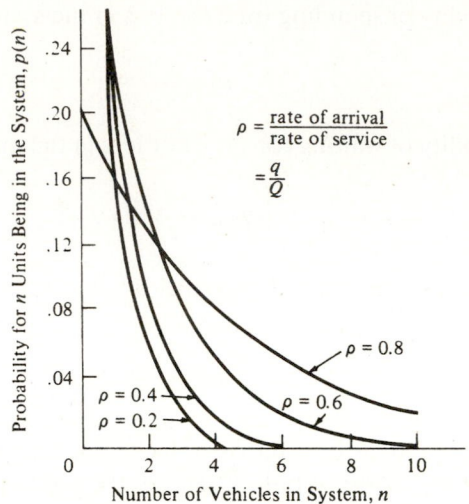

Figure 6.18   Probability of *n* Vehicles Being in the System for Different Traffic Intensities (*ρ*)

Example 6.11   Application of the Single-Channel, Undersaturated, Infinite Queue Theory to a Tollbooth Operation

On a given day, 425 veh/h arrive at a tollbooth located at the end of an off-ramp of a rural expressway. If the vehicles can be serviced by only a single channel at the service rate of 625 veh/h, determine (a) the percentage of time the operator of the tollbooth will be free, (b) the average number of vehicles in the system, and (c) the average waiting time for the vehicles that wait. (Assume Poisson arrival and negative exponential service time.)

Solution:

    a. $q = 425$ and $Q = 625$. For the operator to be free, the number of vehicles in the system must be zero. From Eq. 6.69,

$$P(0) = 1 - \frac{q}{Q} = 1 - \frac{425}{625}$$

$$= 0.32$$

    The operator will be free 32 percent of the time.

    b. From Eq. 6.70,

$$E(n) = \frac{425}{625 - 425}$$

$$= 2$$

    c. From Eq. 6.73,

$$E(v) = \frac{1}{625 - 425} = 0.005 \text{ hr}$$

$$= 18.0 \text{ sec}$$

1. Probability of n units in the system:

$$P(n) = \frac{1-\rho}{1-\rho^{N+1}}\rho^n \tag{6.77}$$

where $p = q/Q$.

2. The expected number of units in the system:

$$E(n) = \frac{\rho}{1-\rho}\frac{1-(N+1)\rho^N + N\rho^{N+1}}{1-\rho^{N+1}} \tag{6.78}$$

Example 6.12 Application of the Single-Channel, Undersaturated, Finite Queue Theory to an Expressway Ramp

The number of vehicles that can enter the on ramp of an expressway is controlled by a metering system which allows a maximum of 10 vehicles to be on the ramp at any one time. If the vehicles can enter the expressway at a rate of 500 veh/h and the rate of arrival of vehicles at the on-ramp is 400 veh/h during the peak hour, determine (a) the probability of 5 cars being on the on-ramp, (b) the percent of time the ramp is full, and (c) the expected number of vehicles on the ramp during the peak hour.

Solution:

a. Probability of 5 cars being on the on-ramp: $q = 400$, $Q = 500$, and $\rho = (400/500) = 0.8$. From Eq. 6.77,

$$P(5) = \frac{1-0.8}{1-(0.8)^{11}}(0.8)^5$$

$$= 0.072$$

b. From Eq. 6.77, the probability of 10 cars being on the ramp is

$$P(10) = \frac{1-0.8}{1-(0.8)^{11}}(0.8)^{10}$$

$$= 0.023$$

That is, the ramp is full only 2.3 percent of the time.

c. The expected number of vehicles on the ramp is obtained from Eq. 6.78:

$$E(n) = \frac{0.8}{1-0.8}\frac{1-(11)(0.8)^{10}+10(0.8)^{11}}{1-(0.8)^{11}} = 2.97$$

The expected number of vehicles on the ramp is 3.

6.6 SUMMARY

One of the most important functions of a traffic engineer is to implement traffic control measures that will facilitate the efficient use of existing highway facilities, since extensive highway construction is no longer taking place at the rate it once was. Efficient use of any highway system entails the flow of the maximum volume of traffic without causing excessive delay to the traffic and inconvenience to the motorist. It is therefore essential that the traffic engineer understands the basic characteristics of the elements of a traffic stream, since these characteristics play an important role in the success or failure of any traffic engineering action to achieve an efficient use of the existing highway system.

This chapter has furnished the fundamental theories that are used to determine the effect of these characteristics. The definitions of the different elements have been presented, together with mathematical relationships of these elements. These relationships are given in the form of macroscopic models, which consider the traffic stream as a whole, and microscopic models, which deal with individual vehicles in the traffic stream. Using the appropriate model for a traffic flow will facilitate the computation of any change in one or more elements due to a change in another element. An introduction to queuing theory is also presented to provide the reader with simple equations that can be used to determine delay and queue lengths in simple traffic queuing systems.

PROBLEMS

6-1 Observers stationed at two sections XX and YY, 152 m apart on a highway, record the arrival times of four vehicles as shown in the accompanying table. If the total time of observation at XX was 15 sec, determine (a) the time mean speed, (b) the space mean speed, and (c) the flow at section XX.

	Time of Arrival	
Vehicle	*Section XX*	*Section YY*
A	T_0	$T_0 + 7.58$ sec
B	$T_0 + 3$ sec	$T_0 + 9.18$ sec
C	$T_0 + 6$ sec	$T_0 + 12.36$ sec
D	$T_0 + 12$ sec	$T_0 + 21.74$ sec

6-2 Data obtained from aerial photography showed six vehicles on a 210 m-long section of road. Traffic data collected at the same time indicated an average time headway of 3.7 sec. Determine (a) the density on the highway, (b) the flow on the road, and (c) the space mean speed.

6-3 Two sets of students are collecting traffic data at two sections, *xx* and *yy,* of a highway 450 m apart. Observations at *xx* show that five vehicles passed that section at intervals of 3, 4, 3, and 5 sec, respectively. If the speeds of the vehicles were 80, 70, 65, 55, and 50 km/h respectively, draw a schematic showing the locations of the vehicles 20 sec after the first vehicle passed section *xx*. Also, determine (a) the time mean speed, (b) the space mean speed, and (c) the density on the highway.

6-4 Determine the space mean speed for the data given in Problem 6-3 using Equation 6.5. Compare your answer with that obtained in Problem 6-3 for the space mean speed and discuss the results.

6-5 The following dataset consists of 30 observations of vehicle speed and length taken from a 1.8 m by 1.8 m inductive loop detector during a 60-sec time period. Determine the occupancy, density, and flow rate.

Vehicle	Speed (km/h)	Length (m)
1	97.6	5.4
2	105.6	5.1
3	99.2	5.7
4	112	6.3
5	104	4.8
6	110.4	7.8
7	115.2	6.3
8	105.6	5.7
9	104	6
10	102.4	6
11	107.2	7.5
12	108.8	21
13	104	10.5
14	105.6	6
15	113.6	19.5
16	102.4	7.2
17	94.4	6.9
18	92.8	6.6
19	102.4	19.5
20	102.4	9
21	108.8	7.2
22	92.8	6.3
23	105.6	16.8
24	91.2	6.3
25	102.4	6
26	97.6	15
27	110.4	5.7
28	100.8	6.9
29	100.8	5.1
30	105.6	5.4

6-6   Data from a 1.8 m by 1.8 m inductive loop detector collected during a 30-sec time period indicate that the mean speed of traffic is 80 km/h among the 16 vehicles counted. Assume an average vehicle length of 5.8 m. Determine the density and occupancy.

6-7   The data shown below were obtained by time-lapse photography on a highway. Use regression analysis to fit these data to the Greenshields model and determine (a) the mean free speed, (b) the jam density, (c) the capacity, and (d) the speed at maximum flow.

Speed (km/h)	Density (veh/km)
14.2	85
24.1	70
30.3	55
40.1	41
50.6	20
55.0	15

6-8   Under what traffic conditions will you be able to use the Greenshields model but not the Greenberg model? Give the reason for your answer.

6-9   In a freeway traffic stream, the capacity flow was observed to be 2200 veh/h/ln, and the jam density at this location had been observed to be 125 veh/ln/mi. If the traffic stream is modeled using Greenberg's model, determine the optimum speed and optimum density. If the traffic stream is modeled using Greenshields' model, determine the free flow speed, optimum density, and optimum speed.

6-10   The table below shows data on speeds and corresponding densities on a section of a rural collector road. If it can be assumed that the traffic flow characteristics can be described by the Greenberg model, develop an appropriate relationship between the flow and density. Also determine the capacity of this section of the road.

Speed (km/h)	Density (veh/km)	Speed (km/h)	Density (veh/km)
60.0	20	32.6	50
46.0	32	30.8	53
40.8	38	28.4	57
39.3	40	24.7	65
35.7	45	18.5	80

6-11   Researchers have used analogies between the flow of fluids and the movement of vehicular traffic to develop mathematical algorithms describing the relationship among traffic flow elements. Discuss in one or two paragraphs the main deficiencies in this approach.

6-12   Assuming that the expression:

$$\bar{u}_s = u_f e^{-k/k_j}$$

can be used to describe the speed-density relationship of a highway, determine the capacity of the highway from the data below using regression analysis.

k (veh/km)	$\bar{u}_s$ (km/h)
43	38.4
50	33.8
8	53.2
31	42.3

Under what flow conditions is the above model valid?

6-13   Results of traffic flow studies on a highway indicate that the flow-density relationship can be described by the expression:

$$q = u_f k - \frac{u_f}{k_j} k^2$$

If speed and density observations give the data shown below, develop an appropriate expression for speed versus density for this highway, and determine the density at which the maximum volume will occur as well as the value of the maximum volume. Also, plot speed versus density and volume versus speed for both the expression developed and the data shown. Comment on the differences between the two sets of curves.

Speed (km/h)	Density (veh/km)
50	18
45	25
40	41
34	58
22	71
13	88
12	99

6-14   Traffic on the eastbound approach of a signalized intersection is traveling at 56 km/h, with a density of 46 veh/km/ln. The duration of the red signal indication for this approach is 30 sec. If the saturation flow is 1900 veh/h/ln with a density of 52 veh/mi/ln, and the jam density is 125 veh/mi/ln, determine the following:

    (i) The length of the queue at the end of the red phase
    (ii) The maximum queue length
    (iii) The time it takes for the queue to dissipate after the end of the red indication.

6-15   A developer wants to provide access to a new building from a driveway placed 305 m upstream of a busy intersection. He is concerned that queues developing during the red phase of the signal at the intersection will block access. If the speed on the approach averages 56 km/h, the density is 50 veh/km, and the red phase is 20 sec, determine if the driveway will be affected. Assume that the traffic flow has a jam density of 110 veh/km and can be described by the Greenshields model.

6-16   Studies have shown that the traffic flow on a two-lane road adjacent to a school can be described by the Greenshields model. A length of 0.8 km adjacent to a school is described as a school zone (see Figure 6.19) and operates for a period of 30 min just before the start of school and just after the close of school. The posted speed limit for the school zone during its operation is 40 kmph. Data collected at the site when the school zone is not in operation show that the jam density and mean free speed for each lane are 125 veh/km and 92 kmph. If the demand flow on the highway at the times of operation of the school zone is 90% of the capacity of the highway, determine:

    (i) The speeds of the shock waves created by the operation of the school zone
    (ii) The number of vehicles affected by the school zone during this 30-minute operation

6-17   Briefly describe the different shock waves that can be formed and the traffic conditions that will result in each of these shock waves.

6-18   Traffic flow on a three-lane (one direction) freeway can be described by the Greenshields model. One lane of the three lanes on a section of this freeway will have to be closed to undertake an emergency bridge repair that is expected to take 2 hours. It is estimated that the capacity at the work zone will be reduced by 30 percent of that of the section just upstream stream of the work zone. The mean free flow speed of the highway is 89 km/h and the jam density is 135 veh/mi/ln. If it is estimated that the demand flow on the highway during the emergency repairs is 90 percent of the capacity, using the deterministic approach, determine:

    (i) The maximum queue length that will be formed
    (ii) The total delay
    (iii) The number of vehicles that will be affected by the incident
    (iv) The average individual delay

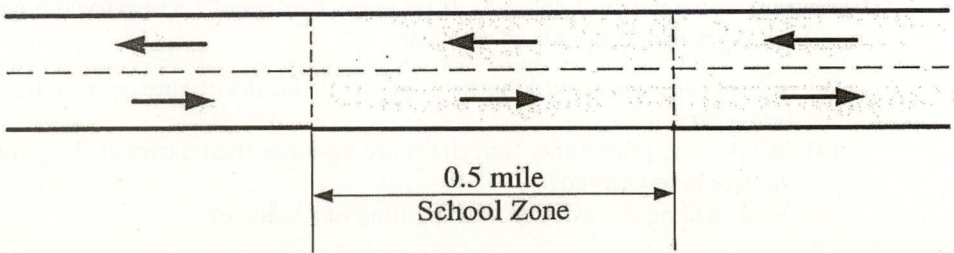

Figure 6.19   Layout of School Zone for Problem 6-16

6-19   Repeat Problem 6-18 for the expected repair periods of 1 h, 1.5 h, 2.5 h, 2.75 h, and 3 h. Plot a graph of average individual delay versus the repair period and use this graph to discuss the effect of the expected repair time on the average delay.

6-20   Repeat Problem 6-18 for the expected demand flows of 70 percent, 75 percent, 80 percent, and 85 percent of the capacity of the highway. Plot a graph of average individual delay vs the expected demand flow and use this graph to discuss the effect of the expected demand flow on the average delay.

6-21   Traffic flow on a section of a two-lane highway can be described by the Greenshields model, with a mean free speed of 90 kmph and a jam density of 145 veh/mi/ln. At the time when the flow was 90 percent of the capacity of the highway, a large dump truck loaded with heavy industrial machinery from an adjacent construction site joins the traffic stream and travels at a speed of 24 km/h for a length of 5.6 km along the upgrade before turning off onto a dump site. Due to the relatively high flow in the opposite direction, it is impossible for any car to pass the truck. Determine how many vehicles will be in the platoon behind the truck by the time the truck leaves the highway.

6-22   Briefly discuss the phenomenon of gap acceptance with respect to merging and weaving maneuvers in traffic streams.

6-23   The table below gives data on accepted and rejected gaps of vehicles on the minor road of an unsignalized intersection. If the arrival of major road vehicles can be described by the Poisson distribution, and the peak hour volume is 1100 veh/h, determine the expected number of accepted gaps that will be available for minor road vehicles during the peak hour.

Gap (t) (s)	Number of Rejected Gaps > t	Number of Accepted Gaps < t
1.5	92	3
2.5	52	18
3.5	30	35
4.5	10	62
5.5	2	100

6-24   Using appropriate diagrams, describe the resultant effect of a sudden reduction of the capacity (bottleneck) on a highway both upstream and downstream of the bottleneck.

6-25   The capacity of a highway is suddenly reduced to 50 percent of its normal capacity due to closure of certain lanes in a work zone. If the Greenshields model describes the relationship between speed and density on the highway, the jam density of the highway is 112 veh/km, and the mean free speed is 109 km/h, determine by what percentage the space mean speed at the vicinity of the work zone will be reduced if the flow upstream is 70 percent of the capacity of the highway.

6-26   The arrival times of vehicles at the ticket gate of a sports stadium may be assumed to be Poisson with a mean of 48 km/h. It takes an average of 1.5 min for the necessary tickets to be bought for occupants of each car.

(a) What is the expected length of queue at the ticket gate, not including the vehicle being served?

(b) What is the probability that there are no more than 5 cars at the gate, including the vehicle being served?

(c) What will be the average waiting time of a vehicle?

6-27   An expressway off-ramp consisting of a single lane leads directly to a tollbooth. The rate of arrival of vehicles at the expressway can be considered to be Poisson with a mean of

45 veh/h, and the rate of service to vehicles can be assumed to be exponentially distributed with a mean of 1 min.

(a) What is the average number of vehicles waiting to be served at the booth (that is, the number of vehicles in queue, not including the vehicle being served)?

(b) What is the length of the ramp required to provide storage for all exiting vehicles 90 percent of the time? Assume the average length of a vehicle is 5 m and that there is an average space of 3 m between consecutive vehicles waiting to be served.

(c) What is the average waiting time a driver waits before being served at the tollbooth (that is, the average waiting time in the queue)?

CHAPTER 7

Intersection Design

An intersection is an area, shared by two or more roads, whose main function is to provide for the change of route directions. Intersections vary in complexity from simple intersections, which have only two roads crossing at a right angle to each other, to more complex intersections, at which three or more roads cross within the same area. Drivers therefore have to make a decision at an intersection concerning which of the alternative routes they wish to take. This effort, which is not required at non-intersection areas of the highway, is part of the reason why intersections tend to have a high potential for crashes. The overall traffic flow on any highway depends to a great extent on the performance of the intersections, since intersections usually operate at a lower capacity than through sections of the road.

Intersections are classified into three general categories: grade-separated without ramps, grade-separated with ramps (commonly known as interchanges), and at-grade. Grade-separated intersections without ramps usually consist of structures that provide for traffic to cross at different levels (vertical distances) without interruption. Provision is not provided for one of the intersecting roads to connect with the other intersecting road, thereby eliminating all potential conflicts between intersecting streams of traffic. Grade separated with ramps intersections also consist of structures that provide for traffic to cross at different levels (vertical distances) without interruption, but ramps are provided that allow traffic from one intersecting road to connect with another intersecting road. The potential for crashes at grade-separated intersections with ramps is reduced because many potential conflicts between intersecting streams of traffic are eliminated. At-grade intersections do not provide for the flow of traffic at different levels and therefore there exist conflicts between intersecting streams of traffic. Figure 7.1 shows examples of some types of grade-separated intersections, and Figures 7.2 and 7.3 show examples of some types of at-grade intersections.

Interchange design is beyond the scope of this book, but this chapter presents the basic principles of the design of at-grade intersections.

CHAPTER OBJECTIVES:

- Become familiar with the different at-grade and grade-separated intersections.
- Become familiar with the design principles for at-grade intersections.
- Learn the procedures for determining sight distance requirements at at-grade intersections.
- Become familiar with the procedure for the design of railroad grade crossings.

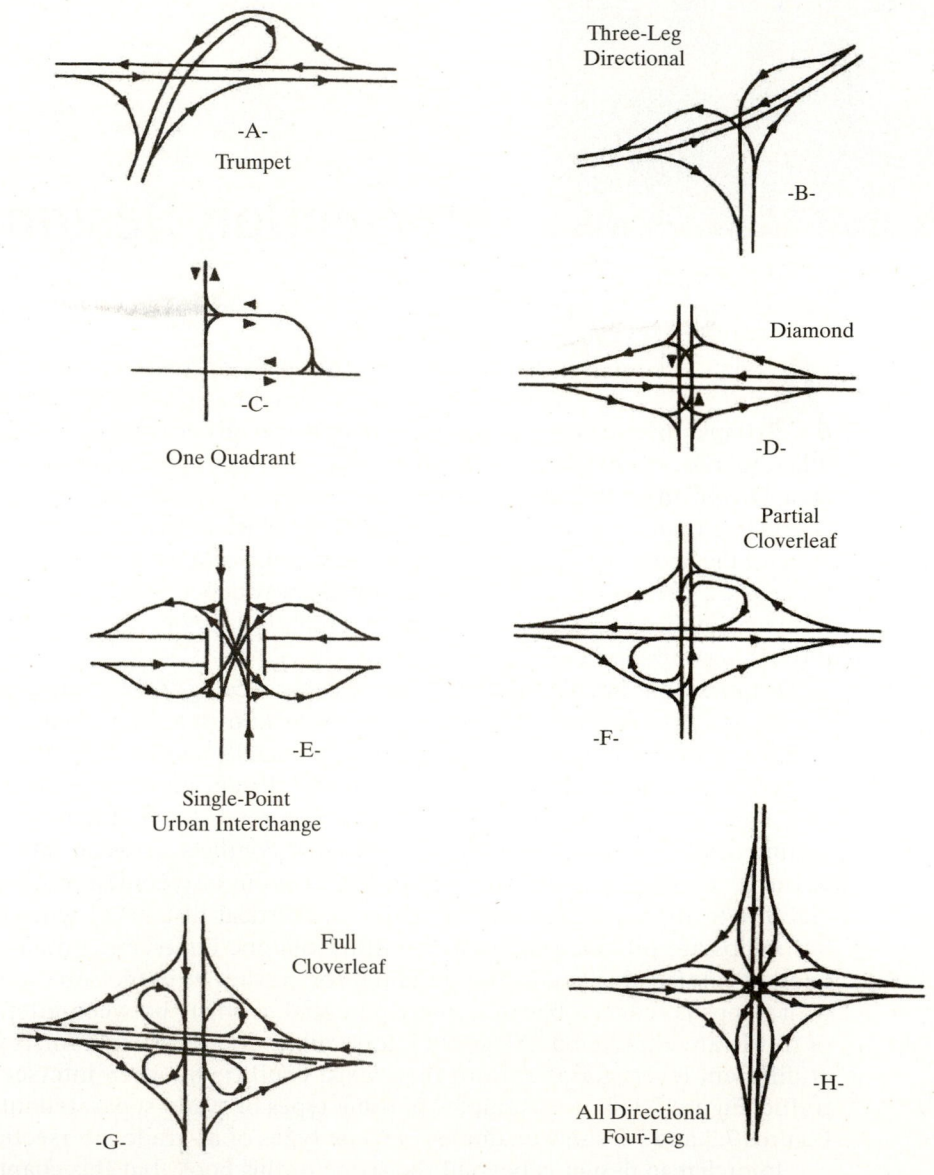

Figure 7.1 Examples of Grade-Separated Interchanges

SOURCE: From *A Policy on Geometric Design of Highways and Streets*, 2011, AASHTO, Washington, D.C. Used by permission.

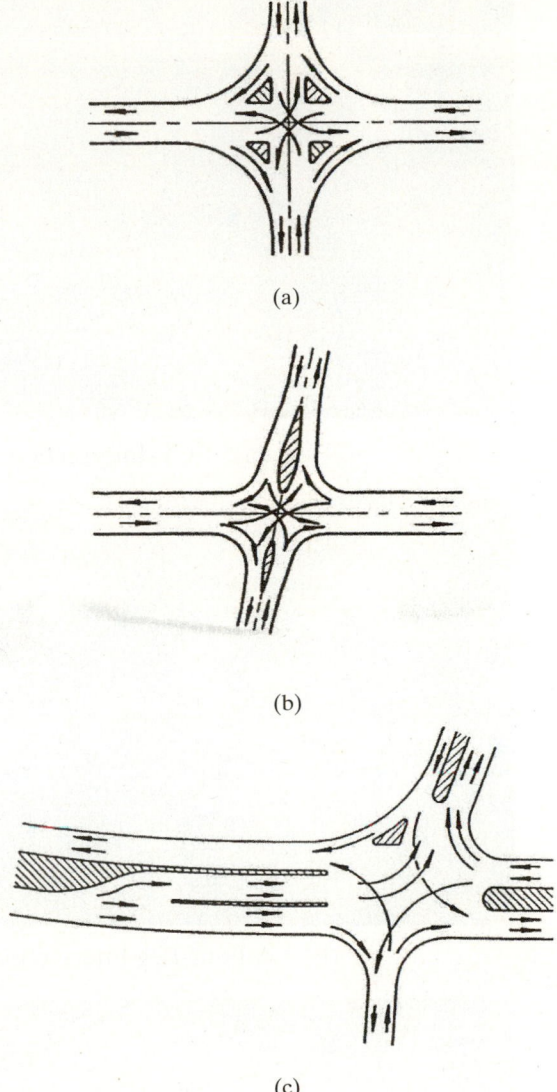

(a)

(b)

(c)

Figure 7.2   Examples of At-Grade Intersections

SOURCE: From *A Policy on Geometric Design of Highways and Streets*, 2011, AASHTO, Washington, D.C. Used by permission.

7.1 TYPES OF AT-GRADE INTERSECTIONS

The basic types of at-grade intersections are T or three-leg intersections which consist of three approaches; four-leg or cross intersections, which consist of four approaches; and multileg intersections, which consist of five or more approaches.

7.1.1 T Intersections

Figure 7.4 shows examples of different types of T intersections ranging from the simplest shown in Figure 7.4a to a channelized one with divisional islands and turning roadways shown in Figure 7.4d. Channelization involves the provision of facilities such as

(a)   A Y-Intersection

(b)   A Four-Leg Intersection Leg

(c)   A T-Intersection with Raised Islands

Figure 7.3   Examples of At-Grade Intersections in Urban Areas

SOURCE: (a) Photographs by Winston Lung, Used with permission. (b) Photographs by Winston Lung, Used with Permission Intersection, (c) Photograph by Winston Lung. Used by permission

pavement markings and traffic islands to regulate and direct conflicting traffic streams into specific travel paths. The intersection shown in Figure 7.4a is suitable for minor or local roads and may be used when minor roads intersect important highways with an intersection angle less than 30 degrees from the normal. This type of intersection is also

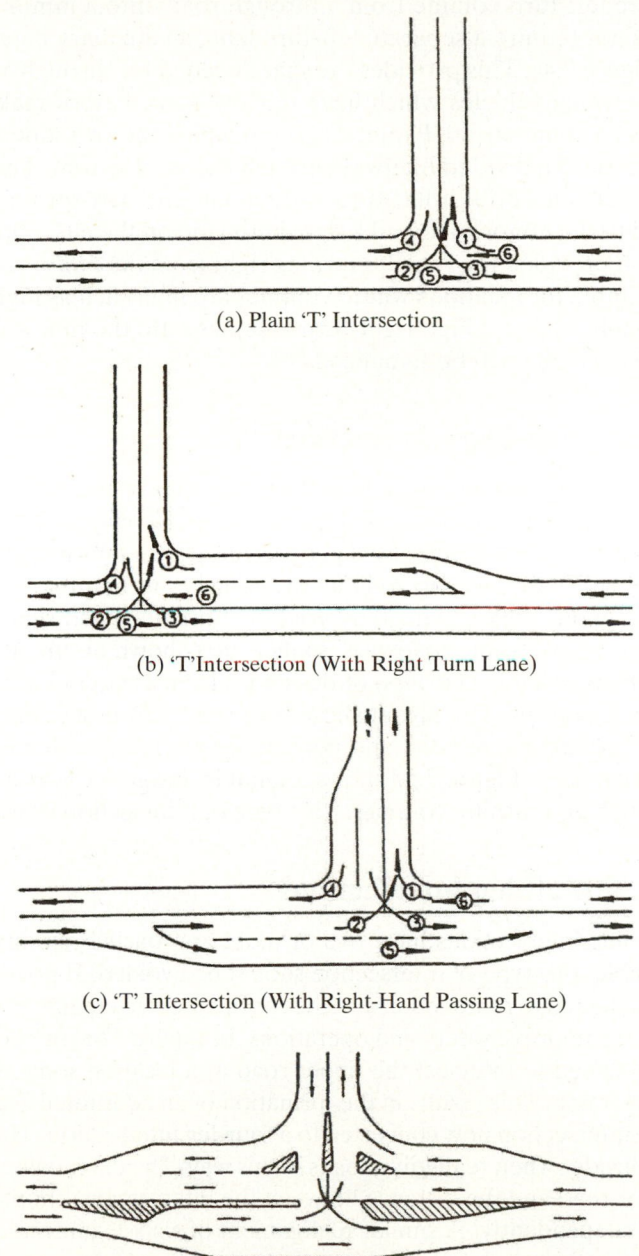

(a) Plain 'T' Intersection

(b) 'T' Intersection (With Right Turn Lane)

(c) 'T' Intersection (With Right-Hand Passing Lane)

(d) 'T' Intersection (With Divisional Island and Turning Roadways)

Figure 7.4 Examples of T Intersections

SOURCE: From *A Policy on Geometric Design of Highways and Streets*, 2011, AASHTO, Washington, D.C. Used by permission.

suitable for use on rural two-lane highways that carry light traffic. At locations with higher speeds and turning volumes, which increase the potential for rear-end collisions between through vehicles and turning vehicles, usually an additional area of surfacing or flaring is provided, as shown in Figure 7.4b. In this case, the flare is provided to separate right-turning vehicles from through vehicles approaching from the east. In cases where left-turn volume from a through road onto a minor road is sufficiently high but does not require a separate left-turn lane, an auxiliary lane may be provided, as shown in Figure 7.4c. This provides the space needed for through-vehicles to maneuver around left-turning vehicles which have to slow down before making their turns. Figure 7.4d shows a channelized T intersection in which the two-lane through road has been converted into a divided highway through the intersection. The channelized T intersection also provides both a left-turn storage lane for left-turning vehicles from the through road to the minor road and a right-turn lane on the east approach. An acceleration lane is also provided for vehicles turning right from the minor road. This type of intersection is suitable for locations where volumes are high such as high left-turn volumes from the through road and high right-turn volumes onto the minor road. An intersection of this type probably will be signalized.

7.1.2 Four–Leg Intersections

Figure 7.5 shows varying levels of channelization at a four-leg intersection. The unchannelized intersection shown in Figure 7.5a is used mainly at locations where minor or local roads cross, although it also can be used where a minor road crosses a major highway. In these cases, the turning volumes are usually low and the roads intersect at an angle that is not greater than 30 degrees from the normal. When right-turning movements are frequent, right-turning roadways, such as that shown on the north approach of Figure 7.5c, can be provided. This type of design is also common in suburban areas where pedestrians are present. The layout shown in Figure 7.5c is suitable for a two-lane highway that is not a minor crossroad and that carries moderate volumes at high speeds or operates near capacity. Figure 7.5d shows a suitable design for four-lane approaches carrying high through and turning volumes. This type of intersection is usually signalized.

7.1.3 Multileg Intersections

Multileg intersections have five or more approaches, as shown in Figure 7.6. Whenever possible, this type of intersection should be avoided. If possible, one or more legs should be realigned to remove some of the conflicting movements from the major intersection and thereby improve safety and operations. In Figure 7.6a, the diagonal leg of the intersection is realigned to intersect the upper road at a location some distance away from the main intersection. This results in the formation of an additional T intersection but with the multileg intersection now converted to a four-leg intersection. There are two important factors to consider when realigning roads in this way: The diagonal road should be realigned to the minor road and the distance between the intersections should be such that they can operate independently. A similar realignment of a six-leg intersection is shown in Figure 7.6b, resulting in two four-leg intersections. In this case, it is also necessary for a realignment to be made to the minor road. For example, if the road in the right-to-left direction is the major road, it may be better to realign each diagonal road to the road in the top-to-bottom direction, thereby forming two additional T intersections and resulting in a total of three intersections. Again, the distances between these intersections should be great enough to allow for the independent operation of each intersection.

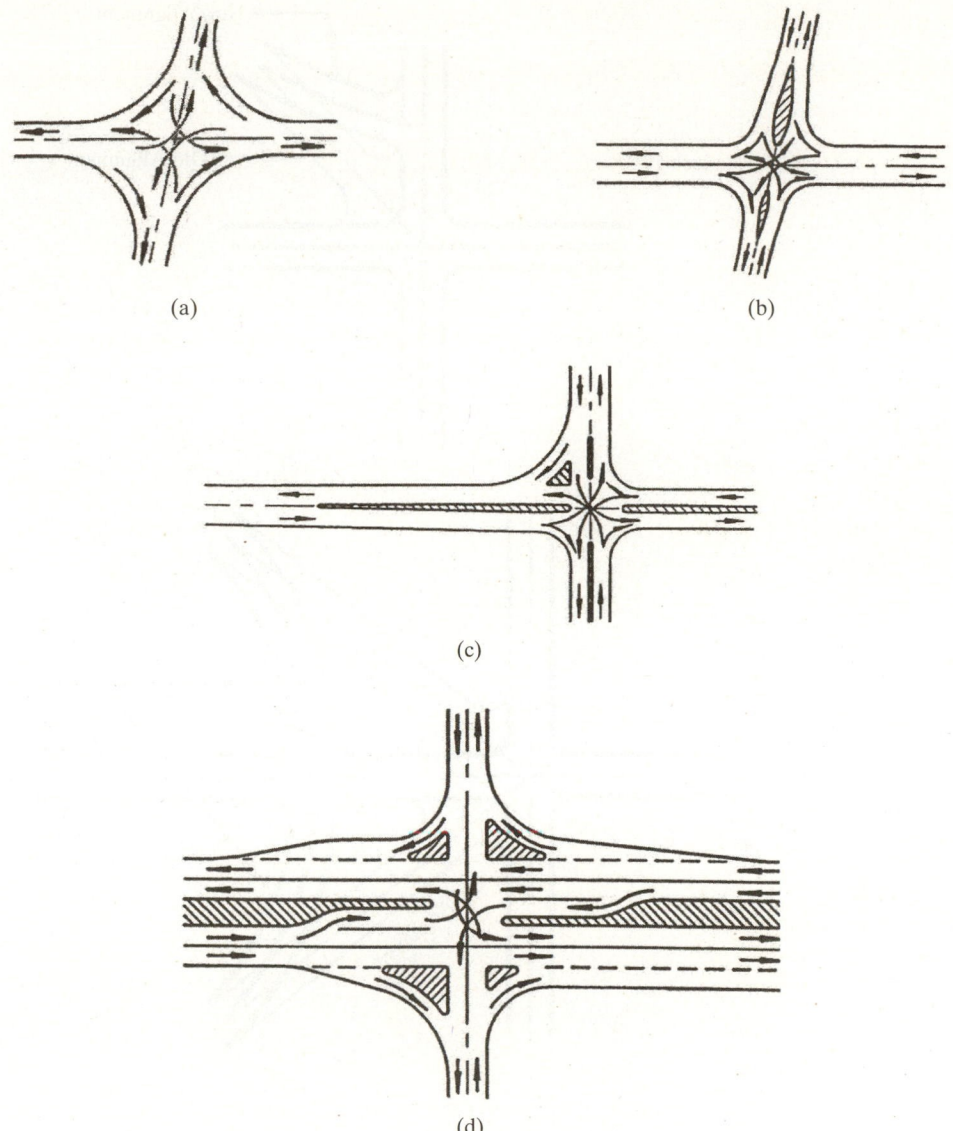

Figure 7.5   Examples of Four-Leg Intersections

SOURCE: From *A Policy on Geometric Design of Highways and Streets*, 2011, AASHTO, Washington, D.C. Used by permission.

7.1.4 Traffic Circles

A traffic circle is a circular intersection that provides a circular traffic pattern with significant reduction in the crossing conflict points. The Transportation Research Board publication, *Roundabouts: An Informational Guide,* NCHRP No. 672, describes three types of traffic circles: rotaries, neighborhood traffic circles, and roundabouts.

Rotaries have large diameters that are usually greater than 90 m, thereby allowing speeds exceeding 48 km/h, with a minimum horizontal deflection of the path of the through traffic.

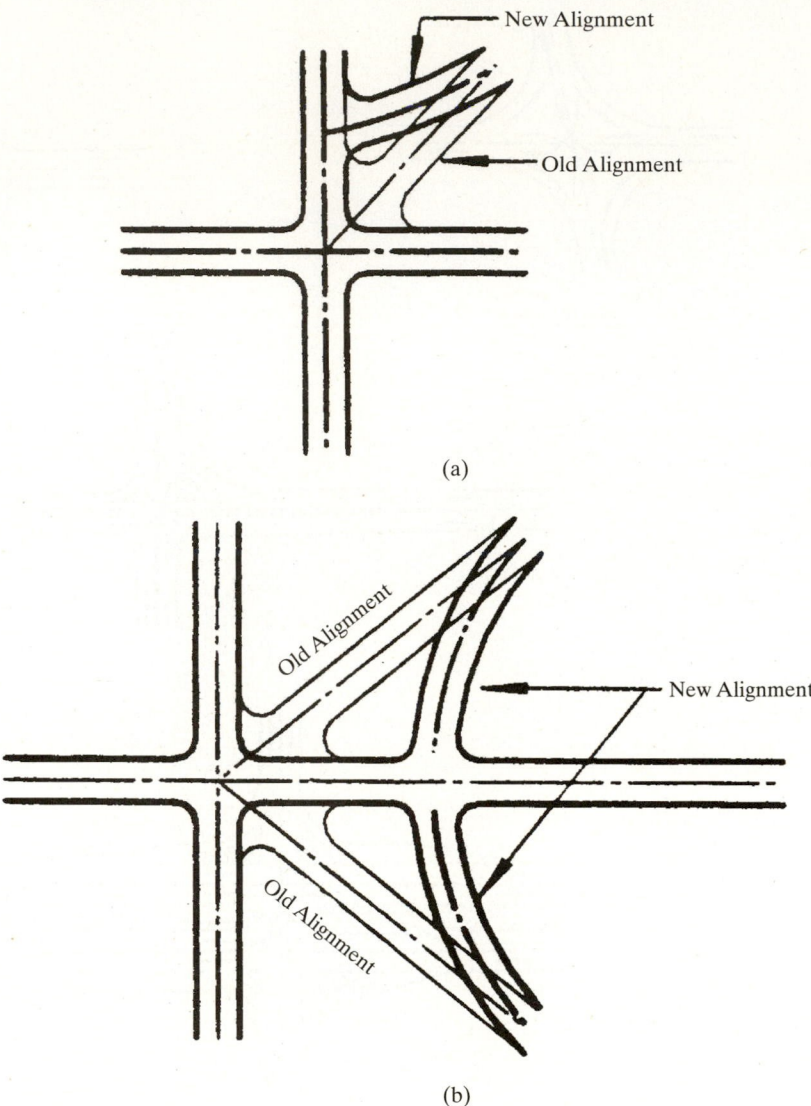

(a)

(b)

Figure 7.6   Examples of Multileg Intersections

SOURCE: From *A Policy on Geometric Design of Highways and Streets*, 2011, AASHTO, Washington, D.C. Used by permission.

Neighborhood traffic circles have diameters that are much smaller than rotaries and therefore allow much lower speeds. Consequently, they are used mainly at the intersections of local streets, as a means of traffic calming and/or as an aesthetic device. As a rule, they consist of pavement markings and do not always employ raised islands. Neighborhood traffic circles may use stop control or no control at the approaches and may or may not allow pedestrian access to the central circle. Parking also may be allowed within the circulatory roadway.

Roundabouts have specific defining characteristics that separate them from other circular intersections. These include:

Yield control at each approach

Separation of conflicting traffic movements through channelization, by pavement markings or raised islands

Geometric characteristics of the central island that typically allow travel speeds of less than 48 km/h

Parking not usually allowed within the circulating roadway

Figure 7.7a shows the geometric elements of a single-lane modern roundabout, while Figure 7.7b shows a photograph of an existing roundabout. Roundabouts can be further categorized into six classes based on the size and environment in which they are located. These are:

Mini roundabouts

Urban compact roundabouts

Urban single-lane roundabouts

Urban double-lane roundabouts

Rural single-lane roundabouts

Rural double-lane roundabouts

The characteristics of each of these categories are shown in Table 7.1.

7.2 DESIGN PRINCIPLES FOR AT-GRADE INTERSECTIONS

The fundamental objective in the design of at-grade intersections is to minimize the severity of potential conflicts among different streams of traffic and between pedestrians and turning vehicles. At the same time, it is necessary to provide for the smooth flow of traffic across the intersection. The design should therefore incorporate the operating characteristics of both vehicles and pedestrians using the intersection. For example, the corner radius of an intersection pavement or surfacing should not be less than either the turning radius of the design vehicle or the radius required for design velocity of the turning roadway under consideration. The design should also ensure adequate pavement widths of turning roadways and approach sight distances. This suggests that at-grade intersections should not be located at or just beyond sharp crest vertical curves or at sharp horizontal curves.

The design of an at-grade intersection involves the design of the alignment, the design of a suitable channelizing system for the traffic pattern, the determination of the minimum required widths of turning roadways when traffic is expected to make turns at speeds higher than 24 km/h, and the assurance that the sight distances are adequate for the type of control at the intersection. The methodology presented later in this chapter for determining minimum sight distances should be used to ensure that the minimum required sight distance is available on each approach. The sight distance at an approach of an at-grade intersection can be improved by flattening cut slopes, removing vegetation, and lengthening vertical and horizontal curves. Approaches of the intersection should preferably intersect at angles which are not greater than 30 degrees from the normal.

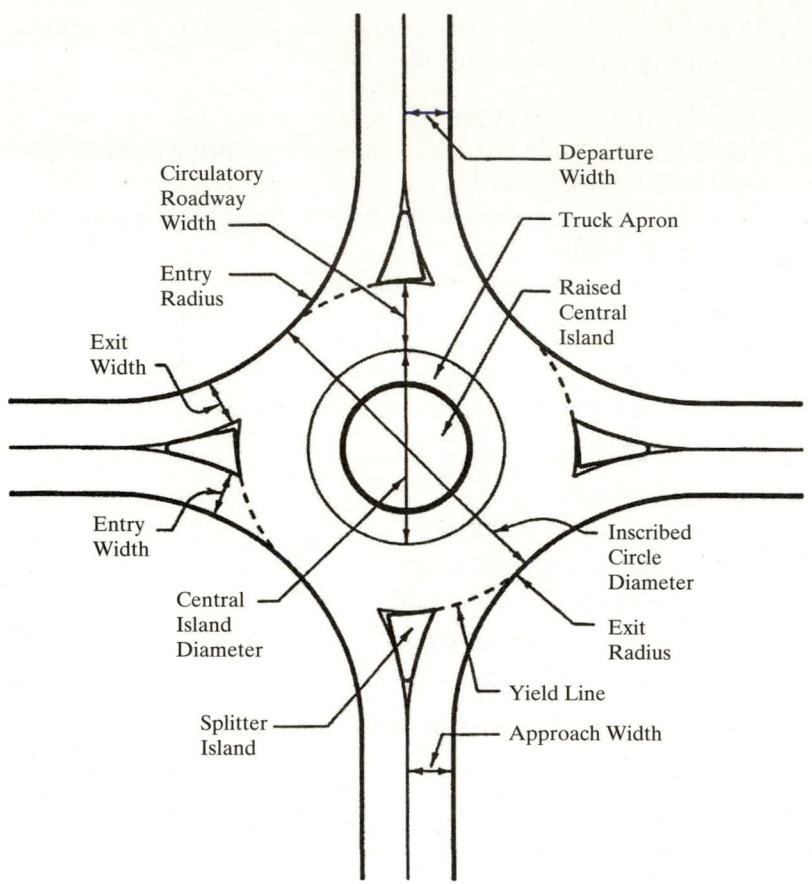

(a) Geometric Elements of a Single-Lane Modern Roundabout

(b) An Example of a Roundabout

Figure 7.7   Geometric Elements and Example of Roundabout

SOURCE: (a) From *A Policy on Geometric Design of Highways and Streets*, 2011, AASHTO, Washington, D.C. Used by permission. (b) Photograph by Winston Lung, Used with permission.

Table 7.1   Characteristics of Roundabout Categories

Design Element	Mini Roundabout	Urban Compact	Urban Single-Lane	Urban Double-Lane	Rural Single-Lane	Rural Double-Lane
Recommended maximum entry design speed	25 km/h (15 mi/h)	25 km/h (15 mi/h)	35 km/h (20 mi/h)	40 km/h (25 mi/h)	40 km/h (25 mi/h)	50 km/h (30 mi/h)
Maximum number of entering lanes per approach	1	1	1	2	1	2
Typical inscribed circle diameter[1]	13 to 25 m (45 ft to 80 ft)	25 to 30 m (80 to 100 ft)	30 to 40 m (100 to 130 ft)	45 to 55 m (150 to 180 ft)	35 to 40 m (115 to 130 ft)	55 to 60 m (180 to 200 ft)
Splitter island treatment	Raised if possible, crosswalk cut if raised	Raised, with crosswalk cut	Raised, with crosswalk cut	Raised, with crosswalk cut	Raised and extended, with crosswalk cut	Raised and extended, with crosswalk cut
Typical daily service volumes on four-leg roundabout (veh/day)	10,000	15,000	20,000	Refer to the source	20,000	Refer to the source

[1]Assumes 90° entries and no more than four legs.
SOURCE: *Roundabouts: An Informational Guide.* U.S. Department of Transportation, Federal Highway Administration, Publication No. FHWA-RD-00-067, Washington, D.C., 2011.

7.2.1 Alignment of At–Grade Intersections

The best alignment for an at-grade intersection is when the intersecting roads meet at right or nearly right angles. This alignment is superior to acute-angle alignments. Much less road area is required for turning at the intersection, there is a lower exposure time for vehicles crossing the main traffic flow, and visibility limitations (particularly for trucks) are not as serious as those at acute-angle intersections. Figure 7.8 shows alternative methods for realigning roads intersecting at acute angles to obtain a nearly right-angle intersection. The dashed lines in this figure represent the original minor road as it intersected the major road at an acute angle. The solid lines that connect both ends of the dashed lines represent the realignment of the minor road across the major road. The methods illustrated in Figures 7.8a and 7.8b have been used successfully, but care must be taken to ensure that the realignment provides for a safe operating speed, which, to avoid hazardous situations, should not be much less than the speeds on the approaches.

The methods illustrated in Figures 7.8c and 7.8d involve the creation of a staggered intersection, in that a single curve is placed at each crossroad leg. This requires a vehicle on the minor road crossing the intersection to turn first onto the major highway and then back onto the minor highway. The realignment illustrated in Figure 7.8d is preferable because the minor-road vehicle crossing the intersection is required to make a right turn rather than a left turn from the major road to reenter the minor road. Therefore, the method illustrated in Figure 7.8c should be used only when traffic on the minor road is light and when most of this traffic is turning onto and continuing on the major road

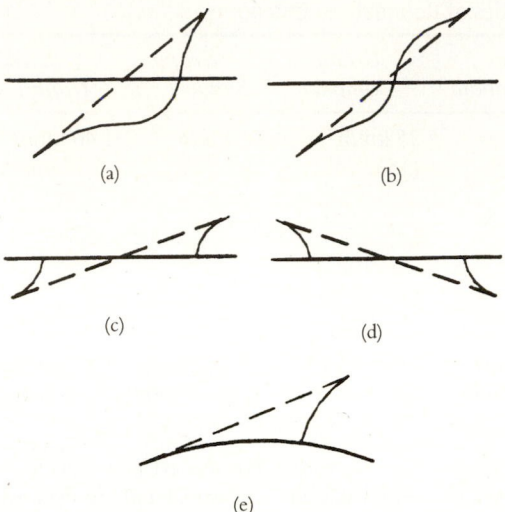

Figure 7.8   Alternative Methods of Realigning Skewed Intersections

SOURCE: From *A Policy on Geometric Design of Highways and Streets*, 2011, AASHTO, Washington, D.C. Used by permission.

rather than crossing the intersection. A major consideration in roadway realignment at intersections is that every effort should be made to avoid creating short-radii horizontal curves, since such curves result in the encroachment of drivers on sections of the opposite lanes.

7.2.2 Profile of At-Grade Intersections

In designing the profile (vertical alignment) at the intersection, a combination of grade lines should be provided to facilitate the driver's control of the vehicle. For example, wherever possible, large changes in grade should be avoided; preferably, grades should not be greater than 3 percent approaching the intersection. The stopping and accelerating distances for passenger cars on grades of 3 percent or less are not much different from those of cars on flat grades; however, significant differences start to occur at grades higher than 3 percent. When it is unavoidable to use grades of 3 percent or more, design factors such as stopping distances and acceleration distances should be adjusted so that conditions equivalent to those on level ground exist. In any case, it is not advisable to use grades higher than 6 percent at intersections.

When it is necessary to adjust the grade lines of the approaches at an intersection, it is preferable that the grade line of the major highway be continued across the intersection and that of the minor road be altered to obtain the desired result. However, any adjustment to a grade line of an approach should be made at a suitable distance from the intersection in order to provide a smooth junction and proper drainage. It always should be remembered that the combination of alignment and grades at an intersection should produce traffic lanes that are clearly seen by motorists at all times, without the sudden appearance of potential hazards. Also, motorists should be able to easily understand the path they should take for any desired direction.

7.2.3 Curves at At-Grade Intersections

The angle of turn, turning speed, design vehicle, and traffic volume are the main factors governing the design of curves at at-grade intersections. When the turning speed at an intersection is assumed to be 24 km/h or less, the curves for the pavement edges are designed to conform to at least the minimum turning path of the design vehicle. When the turning speed is expected to be greater than 24 km/h, the design speed is also considered.

The three types of design commonly used when turning speeds are 24 km/h or less are the simple curve (an arc of a circular curve), the simple curve with taper, and the three-centered compound curve (three simple curves joined together and turning in the same direction). (Simple curves are discussed in Chapter 15.) Figure 7.9 shows the minimum designs necessary for a passenger car making a 90-degree right turn. Figure 7.9a shows the minimum design using a simple curve. The radius of the inner edge pavement, shown as a solid line, should not be less than 7.5 m, since this is the sharpest simple curvature that provides adequate space for the path of a passenger car's inner wheels to clear the pavement's edge. This design will provide for a clearance of about 8 in. near the end of the arc. Increasing the radius of the inner pavement edge to 9 m, as shown by the dotted line, will provide clearances of 16 in. at the end of the curve and 1.62 m at the middle of the curve. The design shown in Figure 7.9b is a simple curve with tapers of 1:10 at each end and an offset of 0.75 m. In this case, it is feasible to use the lower radius of 6 m. The layout of a three-centered compound curve is shown in Figure 7.9c. This type of curve is composed of three circular curves of radii of 30, 6 and 30 m, with the center of the middle curve located at a distance of 6.75 m, including the 0.75 m offset, from the tangent edges. This design is preferable to the simple curve because it provides for a smoother transition and because the resulting edge of the pavement fits the design vehicle path more closely. In fact, in comparison with the 9.0 m radius simple curve, this design results in little additional pavement. The simple curve with a taper shown in Figure 7.9b closely approximates the three-centered curve in the field. Similar designs for single-unit (SU) trucks are shown in Figure 7.10, where the minimum radii are 15 m for the simple curve, 12 m for the simple curve with taper, and 36, 12, and 36 m for the three-centered curve.

The minimum design for passenger cars shown in Figure 7.9 is used only at locations where the absolute minimum turns will occur, such as the intersections of local roads with major highways where only occasional turns are made and at intersections of two minor highways carrying low volumes. It is recommended, when conditions permit, that the minimum design for the SU truck shown in Figure 7.10 be used. The minimum design layouts for larger design vehicles turning at 90 degrees are given in *A Policy on Geometric Design of Highways and Streets*. Minimum edge-of-pavement designs for different angles of turn and design vehicles are given in Table 7.2 for simple curves and simple curves with taper and in Table 7.3 for symmetric and asymmetric three-centered curves. Table 7.2 indicates that it is not feasible to have simple curves for large trucks such as WB-40, WB-50 and WB-62 when the angle of turn is 75 degrees or greater. When the turning speed at an intersection is greater than 24 km/h, the expected turning speed is used to determine the minimum radius required using the procedure presented in Chapter 3.

7.2.4 Channelization of At-Grade Intersections

AASHTO defines *channelization* as the separation of conflicting traffic movements into definite paths of travel by traffic islands or pavement markings to facilitate the safe and orderly movements of both vehicles and pedestrians. A *traffic island* is a defined area between traffic lanes that is used to regulate the movement of vehicles or to serve as a

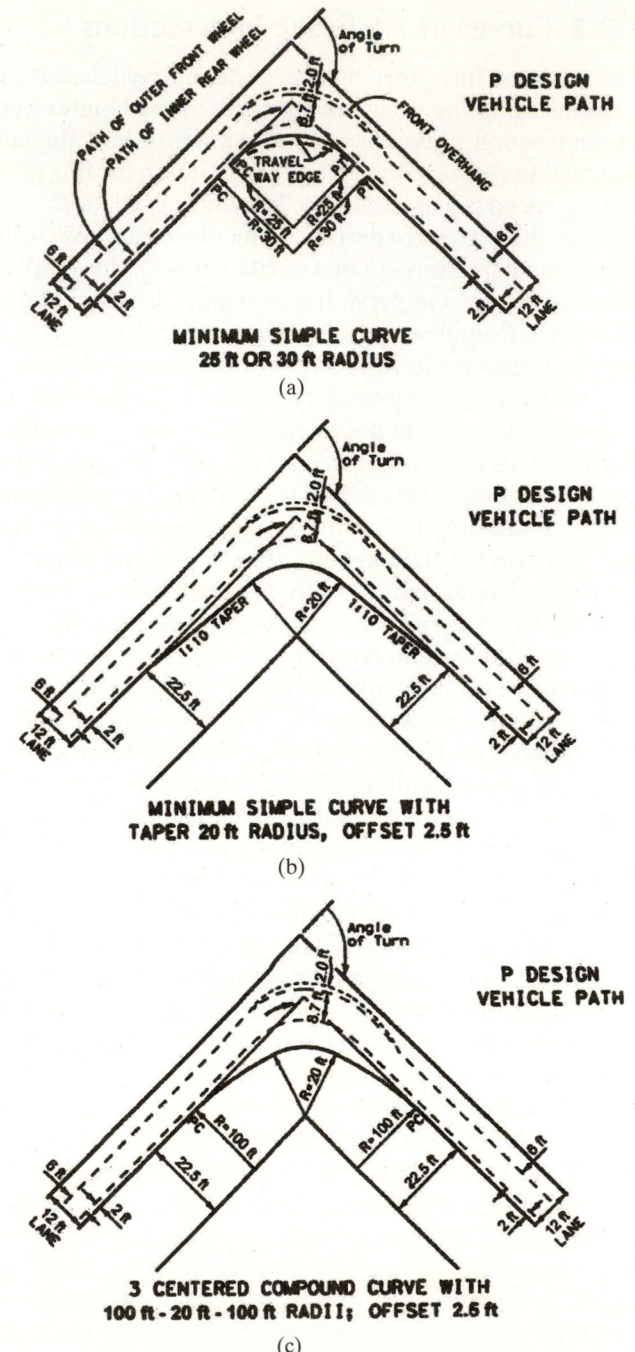

Figure 7.9   Minimum Edge-of-Traveled-Way Designs (Passenger Vehicles)

SOURCE: From *A Policy on Geometric Design of Highways and Streets*, 2011, AASHTO, Washington, D.C. Used by permission.

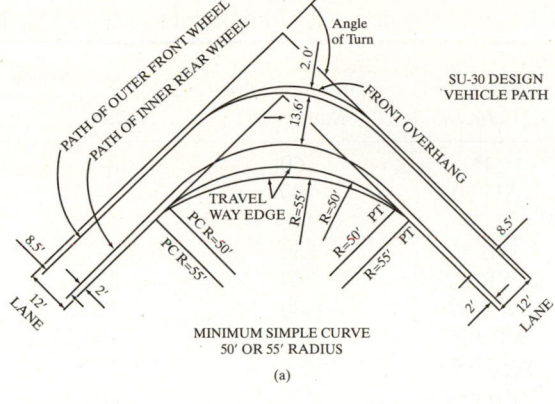

MINIMUM SIMPLE CURVE
50' OR 55' RADIUS

(a)

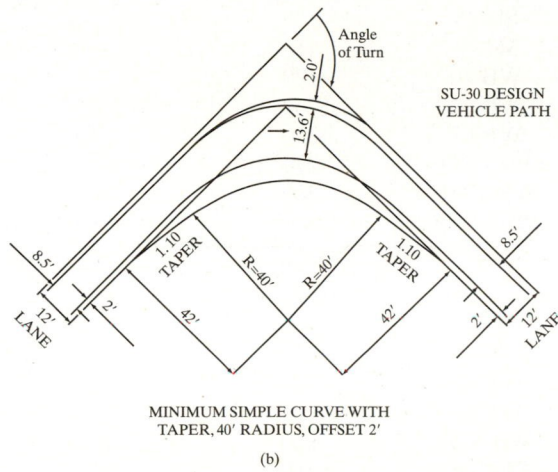

MINIMUM SIMPLE CURVE WITH
TAPER, 40' RADIUS, OFFSET 2'

(b)

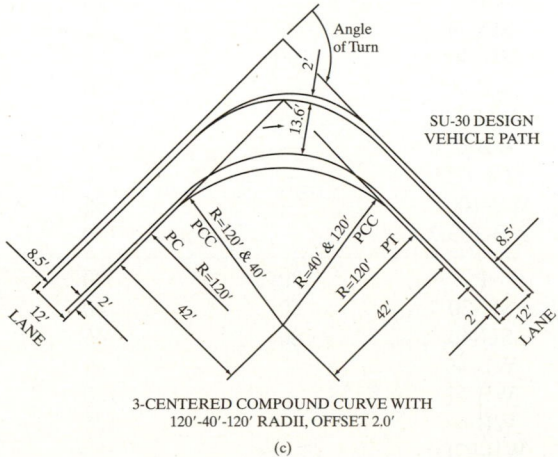

3-CENTERED COMPOUND CURVE WITH
120'-40'-120' RADII, OFFSET 2.0'

(c)

Figure 7.10   Minimum Edge-of-Traveled-Way Designs (Su-9 [Su-30]) Trucks and City Transit Buses

SOURCE: From *A Policy on Geometric Design of Highways and Streets*, 2011, AASHTO, Washington, D.C. Used by permission.

Table 7.2 Edge-of-Traveled-Way design for Turns at Intersections - Simple Curve Radius with Taper

Angle of Turn (°)	Design Vehicle	Simple Curve Radius (ft)	Simple Curve Radius with Taper		
			Radius (ft)	Offset (ft)	Taper L:T
30	P	60	—	—	—
	SU-30	100	—	—	—
	SU-40	140	—	—	—
	WB-40	150	—	—	—
	WB-62	360	220	3.0	15:1
	WB-67	380	220	3.0	15:1
	WB-92D	365	190	3.0	15:1
	WB-100T	260	125	3.0	15:1
	WB-109D	475	260	3.5	20:1
45	P	50	—	—	—
	SU-30	75	—	—	—
	SU-40	115	—	—	—
	WB-40	120	—	—	—
	WB-62	230	145	4.0	15:1
	WB-67	250	145	4.5	15:1
	WB-92D	270	145	4.0	15:1
	WB-100T	200	115	2.5	15:1
	WB-109D	—	200	4.5	20:1
60	P	40	—	—	—
	SU-30	60	—	—	—
	SU-40	100	—	—	—
	WB-40	90	—	—	—
	WB-62	170	140	4.0	15:1
	WB-67	200	140	4.5	15:1
	WB-92D	230	120	5.0	15:1
	WB-100T	150	95	2.5	15:1
	WB-109D	—	180	4.5	20:1
75	P	35	25	2.0	10:1
	SU-30	55	45	2.0	10:1
	SU-40	90	60	2.0	10:1
	WB-40	—	60	2.0	15:1
	WB-62	—	145	4.0	20:1
	WB-67	—	145	4.5	20:1
	WB-92D	—	110	5.0	15:1
	WB-100T	—	85	3.0	15:1
	WB-109D	—	140	5.5	20:1
90	P	30	20	2.5	10:1
	SU-30	50	40	2.0	10:1
	SU-40	80	45	4.0	10:1
	WB-40	—	45	4.0	10:1
	WB-62	—	120	4.5	30:1
	WB-67	—	125	4.5	30:1
	WB-92D	—	95	6.0	10:1
	WB-100T	—	85	2.5	15:1
	WB-109D	—	115	2.9	15:1
105	P	—	20	2.5	8:1
	SU-30	—	35	3.0	10:1

Table 7.2 Edge-of-Traveled-Way design for Turns at Intersections – Simple Curve Radius with Taper (*continued*)

Angle of Turn (°)	Design Vehicle	Simple Curve Radius (ft)	Simple Curve Radius with Taper		
			Radius (ft)	Offset (ft)	Taper L:T
	SU-40	—	45	4.0	10:1
	WB-40	—	40	4.0	10:1
	WB-62	—	115	3.0	15:1
	WB-67	—	115	3.0	15:1
	WB-92D	—	80	8.0	10:1
	WB-100T	—	75	3.0	15:1
	WB-109D	—	90	9.2	20:1
120	P	—	20	2.0	10:1
	SU-30	—	30	3.0	10:1
	SU-40	—	35	6.0	8:1
	WB-40	—	35	5.0	8:1
	WB-62	—	100	5.0	15:1
	WB-67	—	105	5.2	15:1
	WB-92D	—	80	7.0	10:1
	WB-100T	—	65	3.5	15:1
	WB-109D	—	85	9.2	20:1
135	P	—	20	1.5	10:1
	SU-30	—	30	4.0	10:1
	SU-40	—	40	4.0	8:1
	WB-40	—	30	8.0	15:1
	WB-62	—	80	5.0	20:1
	WB-67	—	85	5.2	20:1
	WB-92D	—	75	7.3	10:1
	WB-100T	—	65	5.5	15:1
	WB-109D	—	85	8.5	20:1
150	P	—	18	2.0	10:1
	SU-30	—	30	4.0	8:1
	SU-40	—	35	7.0	8:1
	WB-40	—	30	6.0	8:1
	WB-62	—	60	10.0	10:1
	WB-67	—	65	10.2	10:1
	WB-92D	—	65	11.0	10:1
	WB-100T	—	65	7.3	10:1
	WB-109D	—	65	15.1	10:1
180	P	—	15	0.5	20:1
	SU-30	—	30	1.5	10:1
	SU-40	—	35	6.4	10:1
	WB-40	—	20	9.5	5:1
	WB-62	—	55	10.0	15:1
	WB-67	—	55	13.8	10:1
	WB-92D	—	55	16.8	10:1
	WB-100T	—	55	10.2	10:1
	WB-109D	—	55	20.0	10:1

SOURCE: From *A Policy on Geometric Design of Highways and Streets*, 2011, AASHTO, Washington, D.C. Used by permission.

Table 7.3 Edge-of-Traveled-Way Design for Turns at Intersections – Three-Centered Curves

Angle of Turn (°)	Design Vehicle	Three-Centered Compound Curve Radii (ft)	Symmetric Offset (ft)	Three-Centered Compound Curve Radii (ft)	Asymmetric Offset (ft)
30	P	—	—	—	—
	SU-30	—	—	—	—
	SU-40	—	—	—	—
	WB-40	—	—	—	—
	WB-62	—	—	—	—
	WB-67	460-175-460	4.0	300-175-550	2.0-4.5
	WB-92D	550-155-550	4.0	200-150-500	2.0-6.0
	WB-100T	220-80-220	4.5	200-80-300	2.5-5.0
	WB-109D	550-250-550	5.0	250-200-650	1.5-7.0
45	P	—	—	—	—
	SU-30	—	—	—	—
	SU-40	—	—	—	—
	WB-40	—	—	—	—
	WB-62	460-240-460	2.0	120-140-500	3.0-8.5
	WB-67	460-175-460	4.0	250-125-600	1.0-6.0
	WB-92D	525-155-525	5.0	200-140-500	1.5-6.0
	WB-100T	250-80-250	4.5	200-80-300	2.5-5.5
	WB-109D	550-200-550	5.0	200-170-650	1.5-7.0
60	P	—	—	—	—
	SU-30	—	—	—	—
	SU-40	—	—	—	—
	WB-40	—	—	—	—
	WB-62	400-100-400	15.0	110-100-220	10.0-12.5
	WB-67	400-100-400	8.0	250-125-600	1.0-6.0
	WB-92D	480-110-480	6.0	150-110-500	3.0-5.0
	WB-100T	250-80-250	4.5	200-80-300	2.0-5.5
	WB-109D	650-150-650	5.5	200-140-600	1.5-8.0

Angle of Turn (°)	Design Vehicle	Three-Centered Compound Curve Radii (ft)	Symmetric Offset (ft)	Three-Centered Compound Curve Radii (ft)	Asymmetric Offset (ft)
75	P	100-25-100	2.0	—	—
	SU-30	120-45-120	2.0	—	—
	SU-40	200-35-200	5.0	60-45-200	1.0-4.5
	WB-40	120-45-120	5.0	120-45-195	2.0-6.5
	WB-62	440-75-440	15.0	140-100-540	5.0-12.0
	WB-67	420-75-420	10.0	200-80-600	1.0-10.0
	WB-92D	500-95-500	7.0	150-100-500	1.0-8.0
	WB-100T	250-80-250	4.5	100-80-300	1.5-5.0
	WB-109D	700-125-700	6.5	150-110-550	1.5-11.5
90	P	100-20-100	2.5	—	—
	SU-30	120-40-120	2.0	—	—
	SU-40	200-30-200	7.0	60-45-200	1.0-4.5
	WB-40	120-40-120	5.0	120-40-200	2.0-6.5
	WB-62	400-70-400	10.0	160-70-360	6.0-10.0
	WB-67	440-65-440	10.0	200-70-600	1.0-11.0
	WB-92D	470-75-470	10.0	150-90-500	1.5-8.5
	WB-100T	250-70-250	4.5	200-70-300	1.0-5.0
	WB-109D	700-110-700	6.5	100-95-550	2.0-11.5
105	P	100-20-100	2.5	—	—
	SU-30	100-35-100	3.0	—	—
	SU-40	200-35-200	6.0	60-40-190	1.5-6.0
	WB-40	100-35-100	5.0	100-55-200	2.0-8.0
	WB-62	520-50-520	15.0	360-75-600	4.0-10.5
	WB-67	500-50-500	13.0	200-65-600	1.0-11.0
	WB-92D	500-80-500	8.0	150-80-500	2.0-10.0
	WB-100T	250-60-250	5.0	100-60-300	1.5-6.0
	WB-109D	700-95-700	8.0	150-80-500	3.0-15.0

Table 7.3 Edge-of-Traveled-Way Design for Turns at Intersections – Three-Centered Curves (*continued*)

Angle of Turn (°)	Design Vehicle	Three-Centered Compound Curve Radii (ft)	Symmetric Offset (ft)	Three-Centered Compound Curve Radii (ft)	Asymmetric Offset (ft)
120	P	100-20-100	2.0	—	—
	SU-30	100-30-100	3.0	—	—
	SU-40	200-35-200	6.0	60-40-190	1.5-5.0
	WB-40	120-30-120	6.0	100-30-180	2.0-9.0
	WB-62	520-70-520	10.0	80-55-520	24.0-17.0
	WB-67	550-45-550	15.0	200-60-600	2.0-12.5
	WB-92D	500-70-500	10.0	150-70-450	3.0-10.5
	WB-100T	250-60-250	5.0	100-60-300	1.5-6.0
	WB-109D	700-85-700	9.0	150-70-500	7.0-17.4
135	P	100-20-100	1.5	—	—
	SU-30	100-30-100	4.0	—	—
	SU-40	200-40-200	4.0	60-40-180	1.5-5.0
	WB-40	120-30-120	6.5	100-25-180	3.0-13.0
	WB-62	600-60-600	12.0	100-60-640	14.0-7.0
	WB-67	550-45-550	16.0	200-60-600	2.0-12.5
	WB-92D	450-70-450	9.0	150-65-450	7.0-13.5
	WB-100T	250-60-250	5.5	100-60-300	2.5-7.0
	WB-109D	700-70-700	12.5	150-65-500	14.0-18.4
150	P	75-20-75	2.0	—	—
	SU-30	100-30-100	4.0	—	—
	SU-40	200-35-200	6.5	60-40-200	1.0-4.5
	WB-40	100-30-100	6.0	90-25-160	1.0-12.0
	WB-62	480-55-480	15.0	140-60-560	8.0-10.0
	WB-67	550-45-550	19.0	200-55-600	7.0-16.4
	WB-92D	350-60-350	15.0	120-65-450	6.0-13.0
	WB-100T	250-60-250	7.0	100-60-300	5.0-8.0
	WB-109D	700-65-700	15.0	200-65-500	9.0-18.4
180	P	50-15-50	0.5	—	—
	SU-30	100-30-100	1.5	—	—
	SU-40	150-35-150	6.2	50-35-130	5.5-7.0
	WB-40	100-20-100	9.5	85-20-150	6.0-13.0
	WB-62	800-45-800	20.0	100-55-900	15.0-15.0
	WB-67	600-45-600	20.5	100-55-400	6.0-15.0
	WB-92D	400-55-400	16.8	120-60-400	9.0-14.5
	WB-100T	250-55-250	9.5	100-55-300	8.5-10.5
	WB-109D	700-55-700	20.0	200-60-500	10.0-21.0

SOURCE: From *A Policy on Geometric Design of Highways and Streets*, 2011, AASHTO, Washington, D.C. Used by permission.

pedestrian refuge. Vehicular traffic is excluded from the island area. A properly channelized intersection will result in increased capacity, enhanced safety, and increased driver confidence. On the other hand, an intersection that is not properly channelized may have the opposite effect. Care should always be taken to avoid overchannelization since this frequently creates confusion for the motorist and may even result in a lower operating level than that for an intersection without any channelization. When islands are used for channelization, they should be designed and located at the intersection so that they do not create an undue hazard to vehicles; at the same time, they should be commanding enough to prevent motorists from driving over them.

Channelization at an intersection is normally used to achieve one or more of the following objectives:

1. Direct the paths of vehicles so that not more than two paths cross at any one point.
2. Control the merging, diverging, or crossing angle of vehicles.
3. Decrease vehicle wander and the area of conflict among vehicles by reducing the amount of paved area.
4. Provide a clear indication of the proper path for different movements.
5. Give priority to the predominant movements.
6. Provide pedestrian refuge.
7. Provide separate storage lanes for turning vehicles, thereby creating space away from the path of through vehicles for turning vehicles to wait.
8. Provide space for traffic control devices so that they can be readily seen.
9. Control prohibited turns.
10. Separate different traffic movements at signalized intersections with multiple-phase signals.
11. Restrict the speeds of vehicles.

The factors that influence the design of a channelized intersection are availability of right-of-way, terrain, type of design vehicle, expected vehicular and pedestrian volumes, cross sections of crossing roads, approach speeds, bus stop requirements, and the location and type of traffic-control device. For example, factors such as available right-of-way, terrain, bus stop requirements, and vehicular and pedestrian volumes influence the extent to which channelization can be undertaken at a given location while factors such as type of design vehicle and approach speeds influence the design of the edge of pavement.

The design of a channelized intersection also always should be governed by the following principles:

1. Motorists should not be required to make more than one decision at a time.
2. Sharp reverse curves and turning paths greater than 90 degrees should be avoided.
3. Merging and weaving areas should be as long as possible, but other areas of conflict between vehicles should be reduced to a minimum.
4. Crossing traffic streams that do not weave or merge should intersect at 90 degrees, although a range of 60 to 120 degrees is acceptable.
5. The intersecting angle of merging streams should be such that adequate sight distance is provided.
6. Refuge areas for turning vehicles should not interfere with the movement of through vehicles.
7. Prohibited turns should be physically blocked wherever possible.
8. Decisions on the location of essential traffic control devices should be a component of the design process.

General Characteristics of Traffic Islands

The definition given for traffic islands in the previous section clearly indicates that they are not all of one physical type. These islands can be formed by using raised curbs, pavement markings, or the pavement edges as shown in Figure 7.11.

Curbed Traffic Islands.   A curbed island is usually formed by the construction of a concrete curb that delineates the area of the island, as shown in Figure 7.11a. Curbs are generally classified as mountable or barrier. *Mountable curbs* are constructed with their

(a)   Curbed Island at an Intersection

(b)   Island Formed by Pavement Markings (Flush Island)

Figure 7.11   Examples of Traffic Islands

SOURCE: Photographs by Winston Lung. Used by permission.

faces inclined at an angle of 45 degrees or less so that vehicles may mount them without difficulty if necessary. The faces of *barrier curbs* are usually vertical. Specific designs for different types of curbs are presented in Chapter 15. It should be noted, however, that because of glare, curbed islands may be difficult to see at night which makes it necessary that intersections with curbed islands have fixed-source lighting. Curbed islands are used mainly in urban highways where approach speed is not excessively high and pedestrian volume is relatively high.

Traffic Islands Formed by Pavement Markings.   This type of island is sometimes referred to as a *flush island* because it is flushed with the pavement, as shown in Figure 7.11b. Flush islands are formed by pavement markings that delineate the area of the island. Markers include paint, thermoplastic striping, and raised retroreflective markers. Flushed islands are preferred over curbed islands at intersections where approach speeds are relatively high, pedestrian traffic is low, and signals or sign mountings are not located on the island.

Islands Formed by Pavement Edges.   These islands are usually unpaved and are mainly used at rural intersections where there is space for large intersection curves.

Functions of Traffic Islands

Traffic islands also can be classified into three categories based on their functions: channelized, divisional, and refuge. *Channelized islands* are mainly used to control and direct traffic. *Divisional islands* are mainly used to divide opposing or same-directional traffic streams. *Refuge islands* are used primarily to provide refuge for pedestrians. In most cases, however, traffic islands perform two or more of these functions rather than a single function, although each island may have a primary function.

Channelized Islands.   The objective of channelized islands is to eliminate confusion for motorists at intersections with different traffic movements by guiding them into the correct lane for their intended movement. This is achieved by converting excess space at the intersection into islands in a manner that leaves very little to the discretion of the motorist. A channelized island may take one of many shapes (d, e, f of Figure 7.12b), depending on its specific purpose. For example, a triangularly shaped channelized island is often used to separate right-turning traffic from through traffic (see Figure 7.12a) whereas a curved, central island is frequently used to guide turning vehicles (see Figure 7.12b). In any case, the outlines of a channelized island should be nearly parallel to the lines of traffic it is channeling. Where the island is used to separate turning traffic from through traffic, the radii of the curved sections must be equal to or greater than the minimum radius required for the expected turning speed.

The number of islands used for channelization at an intersection should be kept to a practical minimum, since the presence of several islands may cause confusion to the motorist. For example, the use of a set of islands to delineate several one-way lanes may cause unfamiliar drivers to enter the intersection in the wrong lane.

Divisional Islands.   These are frequently used at intersections of undivided highways to alert drivers that they are approaching an intersection and to control traffic at the intersection. They also can be used effectively to control left turns at skewed intersections. Examples of divisional islands are shown in Figure 7.13. When it is necessary to widen a road at an intersection so that a divisional island can be included, every effort should be made to ensure that the path a driver is expected to take is made quite clear.

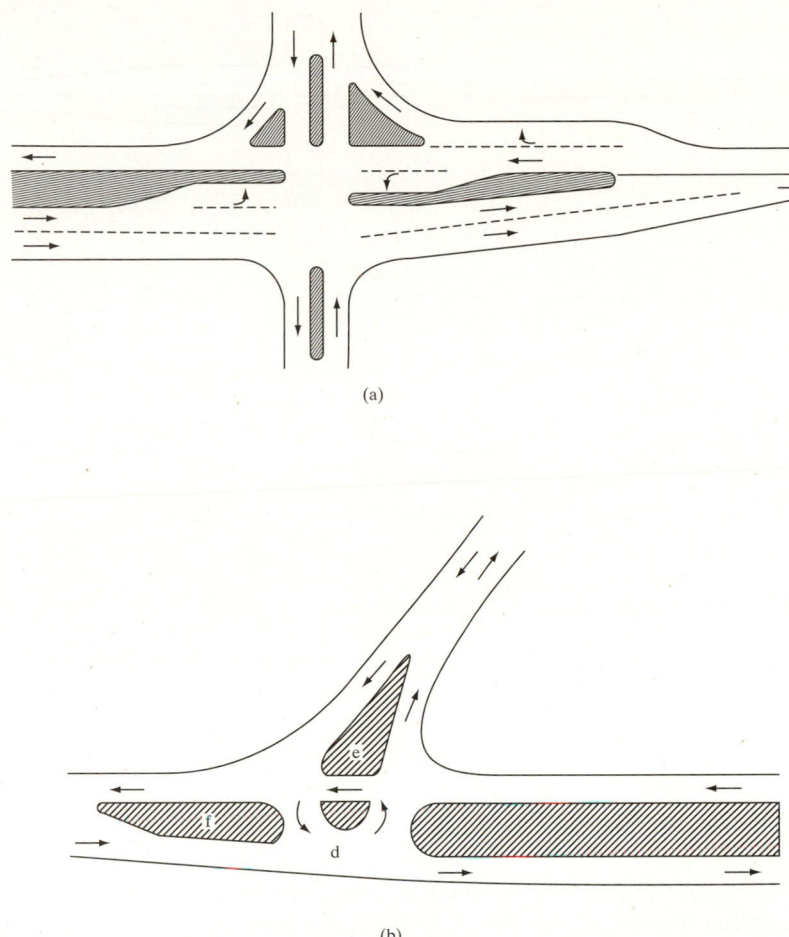

(a)

(b)

Figure 7.12   Examples of Channelized Islands

SOURCE: From *A Policy on Geometric Design of Highways and Streets*, 2011, AASHTO, Washington, D.C. Used by permission.

The alignment also should be designed so that the driver can traverse the intersection easily without any excessive steering.

It is sometimes necessary to use reverse curves (two simple curves with opposite curvatures, forming a compound curve) when divisional islands are introduced, particularly when the location is at a tangent. At locations where speeds tend to be high, particularly in rural areas, it is recommended that the reversal in curvature be no greater than 1 degree (see Chapter 15). Sharper curves can be used when speeds are relatively low, but a maximum of 2 degrees is recommended.

Refuge Islands.   Refuge islands, sometimes referred to as *pedestrian islands,* are used mainly at urban intersections to serve as refuge areas for wheelchairs and pedestrians crossing wide intersections. They also may be used for loading and unloading transit passengers. Figure 7.14 shows examples of islands that provide refuge as well as function as channelized islands.

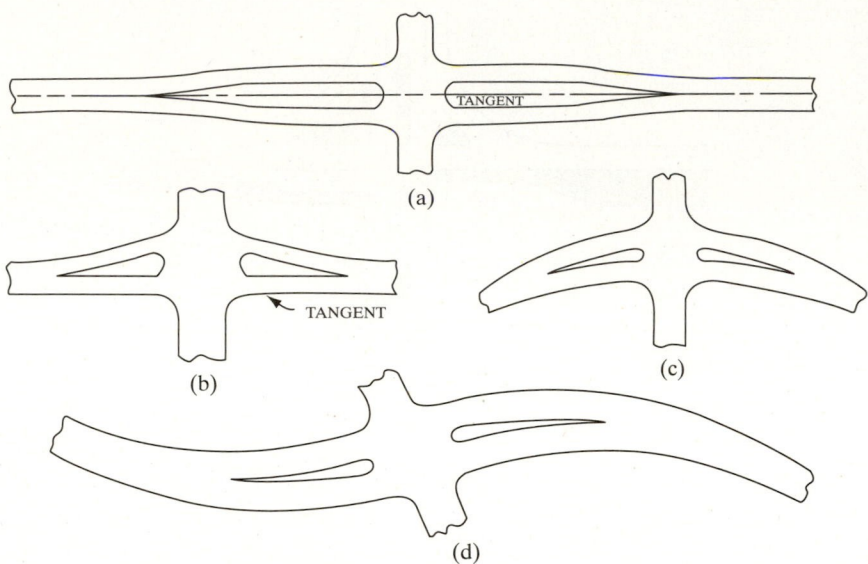

Figure 7.13   Examples of Divisional Islands

SOURCE: From *A Policy on Geometric Design of Highways and Streets*, 2004, AASHTO, Washington, D.C. Used by permission.

Minimum Sizes of Islands

It is essential that islands be large enough to command the necessary attention by drivers. In order to achieve this, AASHTO recommends that curbed islands have a minimum area of approximately 4.5 sq m for urban intersections and 6.75 sq m for rural intersections, although 9 sq m is preferable for both. The minimum side lengths recommended are 1.2 m (but preferably 4.5 m) for triangular islands after the rounding of corners, 6.0 to 7.5 m for elongated or divisional islands, and 30 m (but preferably few hundred meters) for curbed divisional islands that are located at isolated intersections on high-speed highways. It is not advisable to introduce curbed divisional islands at isolated intersections on high-speed roads, since this may create a hazardous situation unless the island is made visible enough to attract the attention of the driver.

Islands having side lengths near the minimum are considered to be small islands whereas those with side lengths of 30 m or greater are considered large. Those with side lengths less than those for large islands but greater than the minimum are considered to be intermediate islands.

In general, the width of elongated islands should not be less than 1.2 m, although this dimension can be reduced to an absolute minimum of 0.6 m in special cases when space is limited. Also, the Institute of Transportation Engineers, in its publication, *Context Sensitive Solutions in Designing Major Urban Thoroughfares for Walkable Communities*, recommended that for medians serving pedestrians, a minimum width of 1.8 m should be used but 2.4 m is preferable. In cases where signs are located on the island, the width of the sign must be considered in selecting the width of the island to ensure that the sign does not extend beyond the limits of the island.

(a)

(b)

Figure 7.14   Examples of Refuge Islands at Wide Intersections

SOURCE: (a) Photographs by Lewis Woodson, Virginia Transportation Research Council, Charlottesville, VA. Used with permission. (b) Photographs by Lewis Woodson, Virginia Transportation Research Council, Charlottesville, VA. Used with permission.

Location and Treatment of Approach Ends of Curbed Islands

The location of a curbed island at an intersection is dictated by the edge of the through traffic lanes and the turning roadways. Figures 7.15 and 7.16 show design details of curbed islands at urban and rural intersections without and with shoulders, respectively. Figures 7.15 and 7.16, respectively, illustrate the condition where the curbed island edge is located on one approach by providing an offset to the through traffic lane, and where it is located outside a shoulder that is carried through the intersection. The offset from the through traffic lane should be 0.6 to 0.9 m, depending on factors such as the type of edge treatment, island contrast, length of taper or auxiliary pavement preceding the curbed island, and traffic speed. This offset is required for the sides of curbed islands adjacent to the through traffic lanes for both barrier and mountable curbs, and for the sides of barrier curbs adjacent to turning roadways. However, it is not necessary to offset the side of a mountable curb adjacent to a turning roadway except as needed to provide additional protection for the island. In cases where there are no shoulders (Figure 7.15), an offset of 0.6 to 0.9 m should be maintained when there are no curbs on the approach pavement. When there is a mountable curb on the approach pavement, a similar curb can be used on the curbed island at the edge of the through lane. However, this requires that the length of the curbed island be adequate to obtain a gradual taper from the nose offset. AASHTO recommends that the offset of the approach nose of a curbed island from the travel lane should normally be about 0.6 m greater than that of the side of the island from the travel lane. However, for median curbed islands, an offset of at least 0.6 m but preferably 1.2 m of the approach nose from the normal median edge of pavement is recommended. It should be emphasized that in order to prevent the perception of lateral constraint by drivers, the required offset from the edge of through pavement lanes should always be used when barrier curbed islands are introduced. When uncurbed large or intermediate islands are used, the required offsets may be eliminated, although it is still preferable to provide the offsets.

At intersections with approach shoulders but without deceleration or turn lanes, the offset of curbed islands from the through travel lane should be equal to the width of the shoulder. When a deceleration lane precedes the curbed island, or when a gradually widened auxiliary pavement exists and speeds are within the intermediate-to-high range, it is desirable to increase the offset of the nose by an additional 0.6 to 1.2 m.

The end treatments for curbed islands are also shown in Figures 7.15 and 7.16. These figures show that the approach noses for intermediate and large curbed islands are rounded using curves of radii 0.6 to 0.9 m, while the merging ends are rounded with curves of radii 0.3 to 0.45 m; the other corners are rounded with curves of radii of 0.6 to 1.5 m.

7.2.5 Minimum Pavement Widths of Turning Roadways at At-Grade Intersections

In cases where vehicle speeds are expected to be greater than 24 km/h, such as at channelized intersections and where ramps intersect with local roads, it is necessary to increase the pavement widths of the turning roadways. Three classifications of pavement widths are used:

Case I: one-lane, one-way operation with no provision for passing a stalled vehicle
Case II: one-lane, one-way operation with provision for passing a stalled vehicle
Case III: two-lane operation, either one-way or two-way

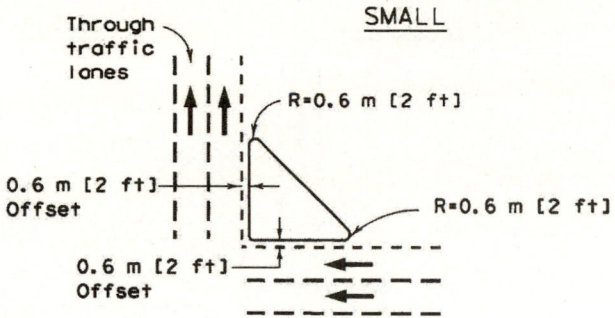

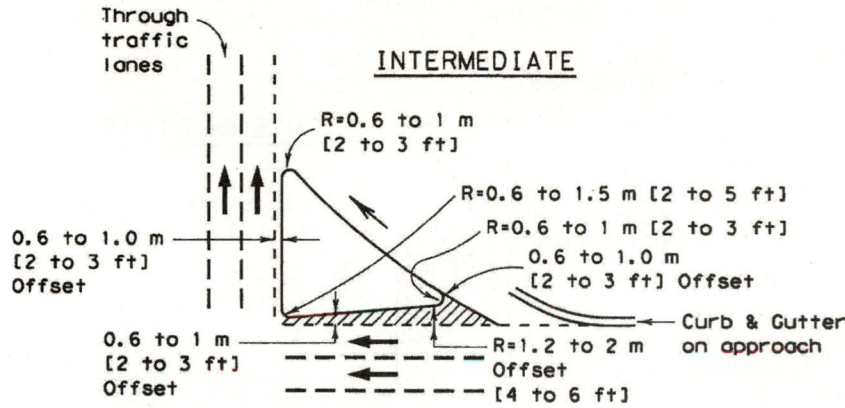

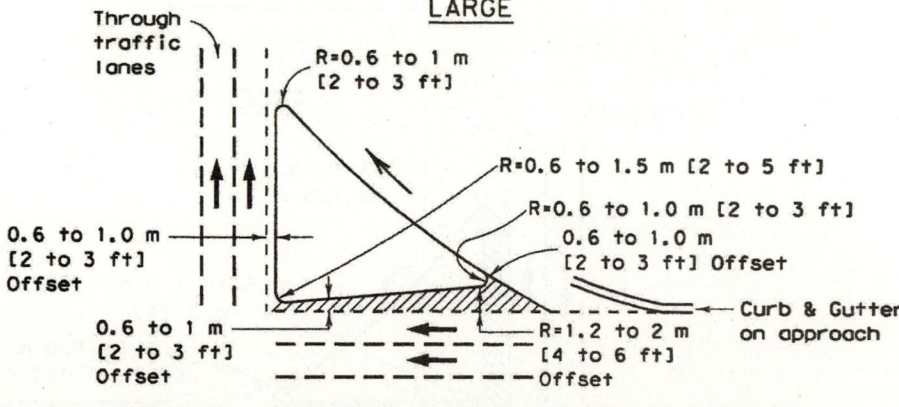

TRIANGULAR CURBED ISLAND ON URBAN STREETS

Figure 7.15 Layouts of Curbed Islands without Shoulders

SOURCE: From *A Policy on Geometric Design of Highways and Streets*, 2011, AASHTO, Washington, D.C. Used by permission.

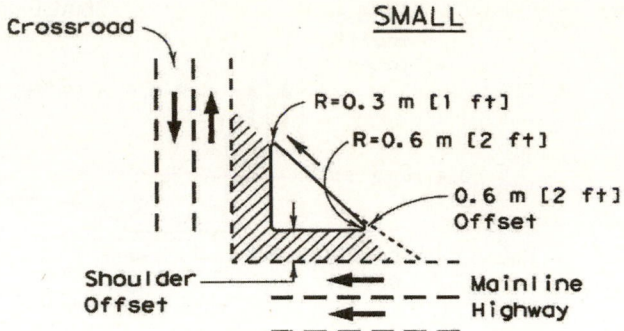

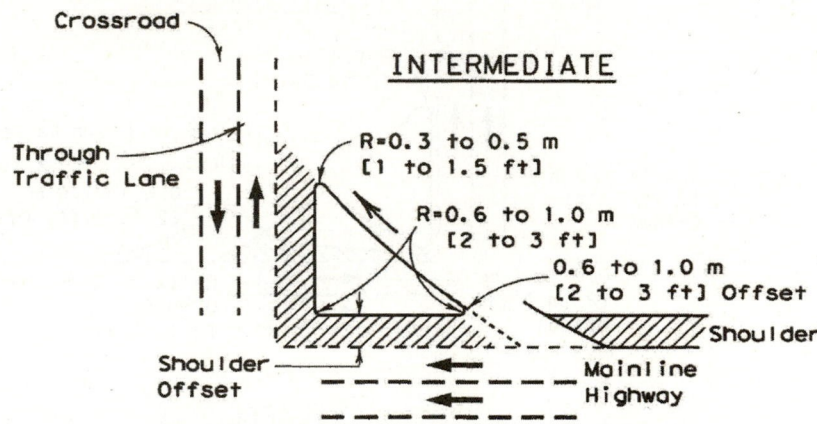

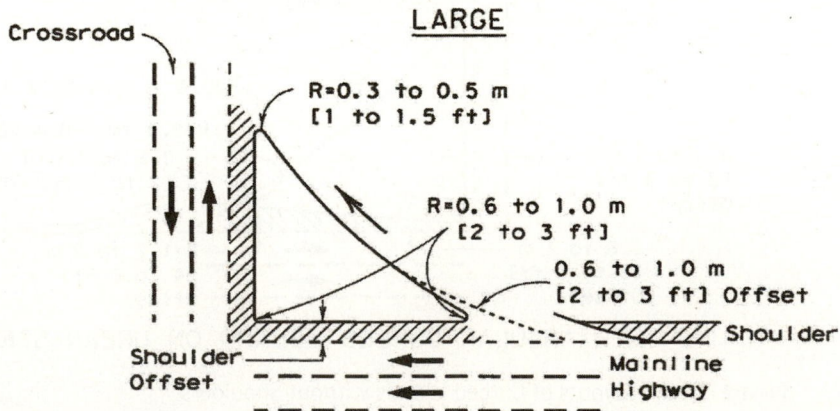

Figure 7.16   Layouts of Curbed Islands with Shoulders

SOURCE: From *A Policy on Geometric Design of Highways and Streets*, 2011, AASHTO, Washington, D.C. Used by permission.

Case I is used mainly at relatively short connecting roads with moderate turning volumes. Case II, which provides for the passing of a stalled vehicle, is commonly used at locations where ramps intersect with local roads and at channelized intersections.

Case III is used at one-way, high-volume locations that require two lanes or at two-way locations. The pavement width for each case depends on the radius of the turning roadway and the characteristics of the design vehicle. Figure 7.17 shows the basis suggested by AASHTO for deriving the appropriate pavement width for any design vehicle. The pavement width depends on the widths of the front and rear overhangs F_A and F_B of the design vehicle, the total clearance per vehicle C, an extra width allowance due to difficulty of driving on curves Z, and the track width U of the vehicle as it moves around the curve. Values for the front overhang F_A for different vehicle types can be obtained directly from Figure 7.18. Note, however, that F_A can be ignored in Case I as no passing maneuver is involved. The width of the rear overhang F_B is usually taken as 0.15 m for passenger cars, since the width of the body of a typical passenger car is 0.3 m greater than the out-to-out width of the rear wheels. For truck vehicles, the rear overhang is 0 m, since the width of truck bodies is usually the same as the out-to-out width of the rear wheels.

The pavement width provides for some lateral clearance both between the edge of the pavement and the nearest wheel path and between the sides of the vehicles passing or meeting. AASHTO suggests that the total clearance per vehicle be taken as 1.2 m for Case I, 0.6 m for the stopped vehicle and 0.6 m for the passing vehicle in Case II, and 1.2 m for Case III. An extra width allowance Z is provided to compensate for the difficulty of maneuvering on a curve and for variation in driver operation. It is obtained from the empirical expression:

$$Z = \frac{v}{\sqrt{R}} \tag{7.1}$$

where

Z = extra width allowance to compensate for the difficulty in maneuvering (m)
v = design speed (km/h)
R = radius of curve (m)

Because of the relationship between speed and velocity at intersections, it has been found that Z is usually about 0.6 m for radii between 15 to 150 m.

The track width U for passenger cars and single-unit trucks is given as:

$$U = u + R - \sqrt{\left(R^2 - \sum_i L_i^2\right)} \tag{7.2}$$

where

U = track width on curve (m)
u = track width on tangent (out-to-out of tires) (m)
R = radius of curve or turn (m)
L_i = wheelbase of design vehicle between consecutive axles (or sets of tandem axles) and articulation points (m)

These different width elements are summed to determine the total traveled way width W at the intersection, as shown in Figure 7.17. Table 7.4 gives values for the required

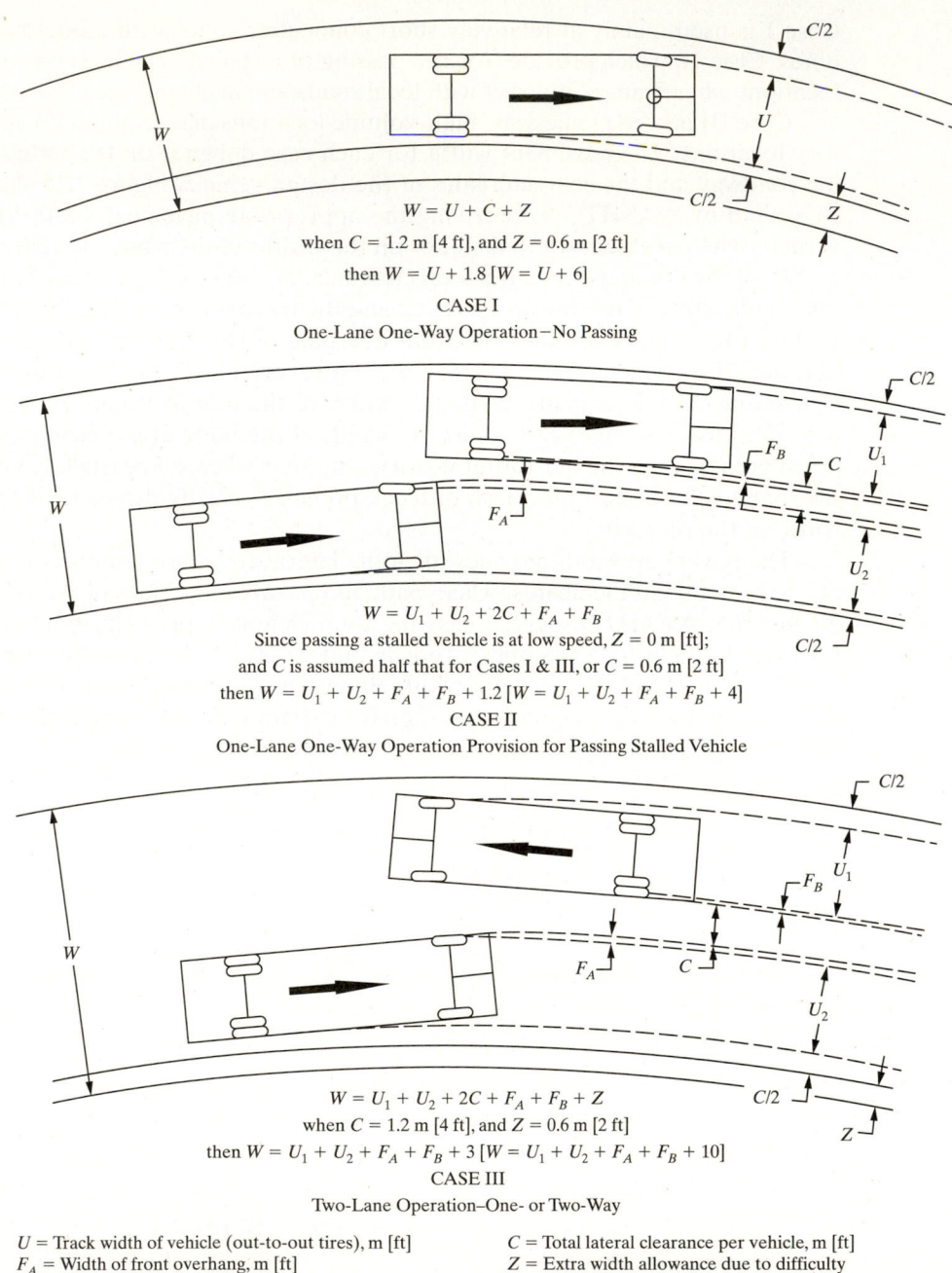

$$W = U + C + Z$$

when $C = 1.2$ m [4 ft], and $Z = 0.6$ m [2 ft]

then $W = U + 1.8$ [$W = U + 6$]

CASE I
One-Lane One-Way Operation—No Passing

$$W = U_1 + U_2 + 2C + F_A + F_B$$

Since passing a stalled vehicle is at low speed, $Z = 0$ m [ft];

and C is assumed half that for Cases I & III, or $C = 0.6$ m [2 ft]

then $W = U_1 + U_2 + F_A + F_B + 1.2$ [$W = U_1 + U_2 + F_A + F_B + 4$]

CASE II
One-Lane One-Way Operation Provision for Passing Stalled Vehicle

$$W = U_1 + U_2 + 2C + F_A + F_B + Z$$

when $C = 1.2$ m [4 ft], and $Z = 0.6$ m [2 ft]

then $W = U_1 + U_2 + F_A + F_B + 3$ [$W = U_1 + U_2 + F_A + F_B + 10$]

CASE III
Two-Lane Operation–One- or Two-Way

U = Track width of vehicle (out-to-out tires), m [ft]
F_A = Width of front overhang, m [ft]
F_B = Width of rear overhang, m [ft]

C = Total lateral clearance per vehicle, m [ft]
Z = Extra width allowance due to difficulty of driving on curves, m [ft]

Figure 7.17 Basis for Deriving the Appropriate Pavement Width at Turning Roadways for Any Design Vehicle

SOURCE: From *A Policy on Geometric Design of Highways and Streets*, 2011, AASHTO, Washington, D.C. Used by permission.

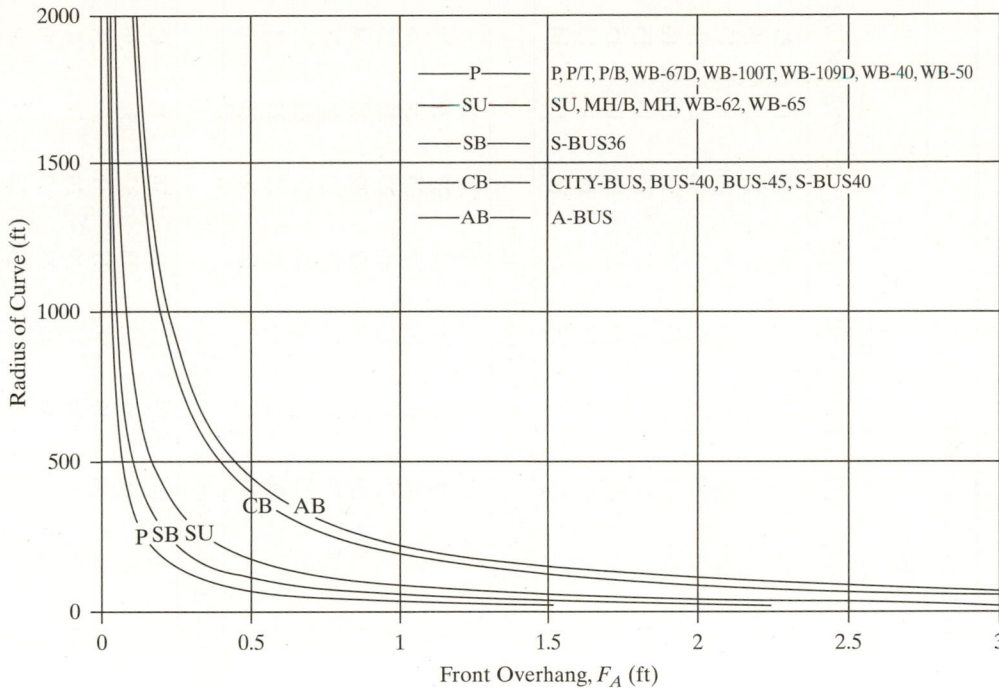

Figure 7.18   Front Overhang for Widening of Traveled Way on Curves

SOURCE: From *A Policy on Geometric Design of Highways and Streets*, 2011, AASHTO, Washington, D.C. Used by permission.

pavement widths for different design vehicles. In practice, however, pavement widths are seldom designed for a single vehicle type, since the highway should be capable of accommodating different types of vehicles, especially when provision is made for passing. The design of the pavement width is therefore based on three types of traffic conditions, listed below, each concerning a specific mix of vehicle types:

- *Traffic Condition A:* Passenger vehicles are predominant, but this traffic condition also provides some consideration for the occasional SU truck.
- *Traffic Condition B:* Proportion of SU vehicles warrants this vehicle type to be the design vehicle but it allows for the accommodation of some tractor-trailer combination trucks (5 to 10 percent).
- *Traffic Condition C:* Proportion of tractor-semitrailer combination trucks WB-12 or WB-15 (WB-40 or WB-50) vehicles in the traffic stream warrants one of these vehicle types to be the design vehicle.

Table 7.5 gives pavement design widths for the different operational cases and traffic conditions. The pavement widths given in Table 7.5 are further modified with respect to the treatment at the edge of the pavement. For example, when the pavement is designed for small trucks but there are space and stability outside the pavement and no barrier preventing its use, a large truck can occasionally pass another. The pavement width can therefore be a little narrower than the tabulated value. On the other hand, the existence of a barrier curb along the edge of the pavement creates a sense of restriction to the

Table 7.4 Derived Pavement Widths for Turning Roadways for Different Design Vehicles

Case I, One-Lane Operation, No Provision for Passing a Stalled Vehicle

Radius on inner Edge of Pavement, R(ft)	P	SU-30	SU-40	BUS-40	BUS-45	CITY-BUS	S-BUS-36	S-BUS-40	A-BUS-11	WB-40	WB-62	WB-67	WB-67D	WB-92D	WB-100T	WB-109D	MH	P/T	P/B	MH/B
50	13	18	21	22	23	21	19	18	22	23	44	57	29	—	37	—	18	19	18	21
75	13	17	18	19	20	19	17	17	19	20	30	33	23	34	27	43	17	17	17	19
100	13	16	17	18	19	18	16	16	18	18	25	28	21	28	24	34	16	16	16	17
150	12	15	16	17	17	17	16	15	17	17	22	23	19	23	21	27	15	16	15	16
200	12	15	16	16	17	16	15	15	16	16	20	21	18	21	19	23	15	15	15	16
300	12	15	15	16	16	16	15	15	16	15	18	19	17	19	17	20	15	15	15	15
400	12	15	15	15	15	15	15	15	15	15	17	18	16	18	17	19	15	15	14	15
500	12	14	15	15	15	15	14	14	15	15	17	17	16	17	16	18	14	14	14	15
Tangent	12	14	14	15	15	15	14	14	15	14	15	15	15	15	15	15	14	14	14	14

Case II, One-Lane, One-Way Operation with Provision for Passing a Stalled Vehicle by Another of the Same Type

Radius on inner Edge of Pavement, R(ft)	P	SU-30	SU-40	BUS-40	BUS-45	CITY-BUS	S-BUS-36	S-BUS-40	A-BUS-11	WB-40	WB-62	WB-67	WB-67D	WB-92D	WB-100T	WB-109D	MH	P/T	P/B	MH/B
50	20	30	36	39	42	38	31	32	40	39	81	109	50	—	67	—	30	30	28	36
75	19	27	30	32	35	32	27	28	34	32	53	59	39	60	47	79	27	27	26	30
100	18	25	27	30	31	29	25	26	30	29	44	48	34	48	40	60	25	25	24	28
150	18	23	25	27	28	27	23	24	27	26	36	38	29	39	33	45	23	23	23	25
200	17	22	24	25	26	25	23	23	26	24	32	34	27	34	30	39	22	22	22	24
300	17	22	22	24	24	24	22	22	24	23	28	30	25	30	27	33	22	22	21	23
400	17	21	22	23	24	23	21	21	23	22	26	27	24	27	25	30	21	21	21	22
500	17	21	21	23	23	23	21	21	23	22	25	26	23	26	25	28	21	21	21	21
Tangent	17	20	20	21	21	21	20	20	21	20	21	21	21	21	21	21	20	20	20	20

Case III, Two-Lane Operation, Either One- or Two-Way (Same Type Vehicle in Both Lanes)

Radius on inner Edge of Pavement, R(ft)	P	SU-30	SU-40	BUS-40	BUS-45	CITY-BUS	S-BUS-36	S-BUS-40	A-BUS-11	WB-40	WB-62	WB-67	WB-67D	WB-92D	WB-100T	WB-109D	MH	P/T	P/B	MH/B
50	26	36	42	45	48	44	37	38	46	45	87	115	56	—	73	—	36	36	34	42
75	25	33	36	38	41	38	33	34	40	38	59	65	45	66	53	85	33	33	32	36
100	24	31	33	36	37	35	31	32	36	35	50	54	40	54	46	66	31	31	30	34
150	24	29	31	33	34	33	29	30	33	32	42	44	35	45	39	51	29	29	29	31
200	23	28	30	31	32	31	29	29	32	30	38	40	33	40	36	45	28	28	28	30
300	23	28	28	30	30	30	28	28	30	29	34	36	31	36	33	39	28	28	27	29
400	23	27	28	29	30	29	27	27	29	28	32	33	30	33	31	36	27	27	27	28
500	23	27	27	29	29	29	27	27	29	28	31	32	29	32	31	34	27	27	27	27
Tangent	23	26	26	27	27	27	26	26	27	26	27	27	27	27	27	27	26	26	26	26

SOURCE: From *A Policy on Geometric Design of Highways and Streets*, 2004, AASHTO, Washington, D.C. Used by permission.

Table 7.5 Design Widths of Pavements for Turning Roadways

U.S. CUSTOMARY

Radius on inner edge of pavement R(ft)	Case I One-Lane, One-Way Operation—No Provision for Passing a Stalled Vehicle			Case II One-Lane, One-Way Operation—with Provision for Passing a Stalled Vehicle			Case III Two-Lane Operation— Either One-Way or Two-Way		
	Design Traffic Conditions								
	A	B	C	A	B	C	A	B	C
50	18	18	23	20	26	30	31	36	45
75	16	17	20	19	23	27	29	33	38
100	15	16	18	18	22	25	28	31	35
150	14	15	17	18	21	23	26	29	32
200	13	15	16	17	20	22	26	28	30
300	13	15	15	17	20	22	25	28	29
400	13	15	15	17	19	21	25	27	28
500	12	15	15	17	19	21	25	27	28
Tangent	12	14	14	17	18	20	24	26	26

Width Modification Regarding Edge Treatment

No stabilized shoulder	None	None	None
Sloping curb	None	None	None
Vertical curb: One side / Two sides	Add 1 ft / Add 2 ft	None / Add 1 ft	Add 1 ft / Add 2 ft
Stabilized shoulder, one or both sides	Lane width for conditions B&C on tangent may be reduced to 12 ft where shoulder is 4 ft or wider	Deduct shoulder width; minimum pavement width as under Case I	Deduct 2 ft where shoulder is 4 ft or wider

SOURCE: From *A Policy on Geometric Design of Highways and Streets*, 2011, AASHTO, Washington, D.C. Used by permission.

driver, and so the occasional truck has no additional space to maneuver. This requires additional space to the widths shown in the table. Consideration of these factors results in the modifications given at the bottom of Table 7.5.

The values given in Table 7.5 are based on the assumption that passing is rather infrequent in Case II and full offtracking does not necessarily occur for both the stalled and passing vehicles, since the stalled vehicle can be placed closer to the inner edge of the pavement and thus provide additional clearance for the passing vehicle. Therefore, smaller vehicles in combination are used in deriving the values for Case II rather than for Case III. The design vehicles in combination that were used to determine the values given in Table 7.5 are given in Table 7.6a, where the vehicle designations indicate the

Table 7.6   Design Vehicles in Combination

(a) Design Vehicles in Combination that Provide the Full Clear Width (*C*)

	Design Traffic Condition		
Case	A	B	C
I	P	SU-30	WB-40
II	P-P	P-SU-30	SU-30-SU-30
III	P-SU-30	SU-30-SU-30	WB-40-WB-40

(b) Larger Vehicles that Can Be Operated

	Design Traffic Condition		
Case	A	B	C
I	WB-40	WB-40	WB-62
II	P-SU-30	P-WB-40	SU-30-WB-40
III	SU-30-WB-40	WB-40-WB-40	WB-62-WB-82

SOURCE: From *A Policy on Geometric Design of Highways and Streets*, 2011, AASHTO, Washington, D.C. Used by permission.

types of vehicles that can pass each other. For example, the designation "P-SU" indicates that the design width will allow a passenger car to pass a stalled single-unit truck, or vice versa. Although it is feasible for larger vehicles to pass each other on turning roadways with the widths shown in the table for Cases II and III, this will occur at lower speeds and requires more caution and skill by drivers. The larger vehicles that can be operated under these conditions are shown in Table 7.6b. This will result in reduced clearance between vehicles that will vary from about one-half the value of *C* for sharper curves to nearly the full value of *C* for flatter curves.

When larger design vehicles, such as WB-62 or WB-114, are expected to be using the turning roadway, the width of the pavement should not be less than the minimum width required for these vehicles for the Case I classification in Table 7.4. In such cases, values obtained from Table 7.5 must be compared with those given in Table 7.4 for Case I, and the higher values should be used.

Example 7.1   Determining the Width of a Turning Roadway at an Intersection

A ramp from an urban expressway with a design speed of 48 km/h connects with a local road forming a T intersection. An additional lane is provided on the local road to allow vehicles on the ramp to turn right onto the local road without stopping. The turning roadway has a mountable curb on one side and will provide for a one-lane, one-way operation with provision for passing a stalled vehicle. Determine the width of the turning roadway if the predominant vehicles on the ramp are single-unit trucks but give some consideration to semitrailer vehicles. Use 0.08 for the superelevation.

Solution:

- Determine the minimum radius of the curve.
  Because the speed is greater than 24 km/h, use Eq. 3.34 of Chapter 3 and a value of 0.20 for f_s (from Table 3.3):

$$R = \frac{u^2}{127(e + f_s)}$$

$$= \frac{48^2}{127(0.08 + 0.2)}$$

$$= 64.79 \, \text{m}$$

Use 65 m.

- Determine the type of operation and traffic condition.
  The operational requirements, one-lane, one-way with provision for passing a stalled vehicle require Case II operation.
  The traffic condition, single-unit trucks but with some consideration given to semitrailer vehicles, requires Traffic Condition B.
- Determine the turning roadway width.
  Use Table 7.5, with $R = 63.6$ m, Traffic Condition B, and Case II.

  Pavement width = 6 m

Since the turning roadway has a mountable curb only on one side, no modification to the width is required.

7.2.6 Sight Distance at Intersections

The high crash potential at an intersection can be reduced by providing sight distances that allow drivers to have an unobstructed view of the entire intersection at a distance great enough to permit good decision-making and proper control of the vehicle. At signalized intersections, the unobstructed view may be limited to the area where the signals are located, but for unsignalized intersections, it is necessary to provide an adequate view of the crossroads or intersecting highways to reduce the potential of collision with crossing vehicles. This requires an unobstructed triangular area (sight triangle) that allows a clear view for drivers on the minor and major roads to see an approaching vehicle on the crossing road in time to avoid a potential conflict, as shown in Figure 7.19. There are two types of sight triangles, *approach sight* triangles and *departure sight* triangles. The *approach sight* triangle allows for the drivers on both the major roads and minor roads to see approaching intersecting vehicles in sufficient time to avoid a potential collision by reducing the vehicle's speed or by stopping. The decision point on a minor road of an uncontrolled or yield control intersection is the location where the minor road driver should start his/her braking or deceleration maneuver to avoid a potential conflict with an approaching major road vehicle. The *departure sight* triangle allows for the driver of a stopped vehicle on the minor road to enter or cross the major road without conflicting with an approaching vehicle from either direction of the major road, as shown in Figure 7.19. A sight obstruction within a sight triangle is considered as an object having a minimum height of 1.30 m that can be seen by a driver with an eye height of 1.05 m

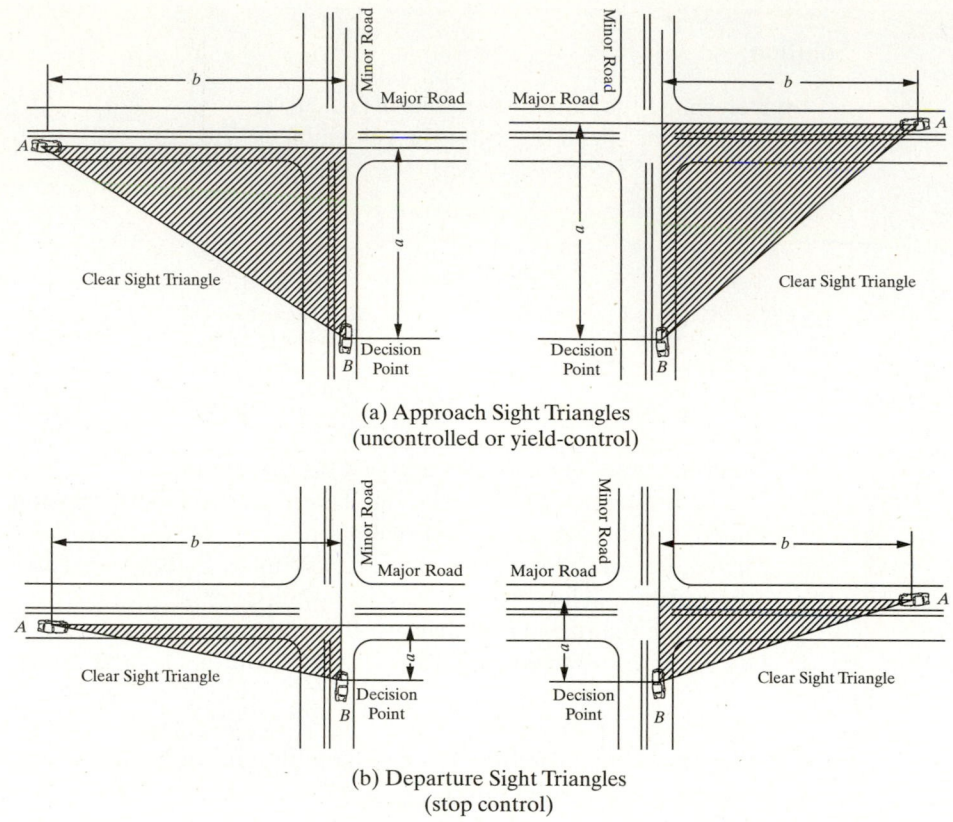

Figure 7.19   Sight Triangles at Intersections

above the surface for a passenger car design vehicle. When the design vehicle is a truck, AASHTO recommends a driver's eye height of 2.28 m. The lengths of the legs of a sight triangle depends on the speed of the minor road traffic, but for some types of maneuvers, the length of the major road leg also depends on the speed of the major road vehicle. In addition to the speeds of the approaching vehicles, the lengths of the legs of the sight triangle also depend on the type of control at the intersection (see Chapter 8). At-grade intersections either have no control (Case A) or are controlled by one of the following methods: stop control on the minor road (Case B), yield control on the minor road (Case C), traffic signal control (Case D), or all-way stop control (Case E). Consideration should also be given to the maneuver of left turns from the major road (Case F).

Sight Distance Requirements for No–Control Intersections—Case A

In this situation, the intersection is not controlled by a yield sign, stop sign, or traffic signal, but sufficient sight distance is provided for the operator of a vehicle approaching the intersection to see a crossing vehicle and, if necessary, to adjust the vehicle's speed so as to avoid a collision. This distance must include the distance traveled by the vehicle both during the driver's perception reaction time and during brake actuation or the acceleration to regulate speed. AASHTO has suggested that a driver may take up to 2.5 sec to detect and recognize a vehicle at intersections that is near the limits of his/her peripheral vision. Also, AASHTO has noted that field observations have indicated drivers tend to decrease their speeds to about 50 percent of their mid-block speed as they approach intersections that have no control. Based on this information, AASHTO has suggested the distances shown in Table 7.7 for different approach speeds. Figure 7.20 shows a schematic of the sight triangle required for the location of an obstruction that will allow for the provision of the minimum distances d_a and d_b. These minimum distances depend on the approaching speed, as shown in Table 7.7. For example, if a road with a speed limit of 64 km/h intersects with a road with a speed limit of 40 km/h, the distances d_a and d_b are 58.5 m and 34.5 m, respectively. It should be noted that the distances shown in Table 7.7 allow time for drivers to come to a stop before reaching the intersection if he or she decides that this is necessary. However, it should be noted that these distances tend to be lower than those given in Chapter 3 for the corresponding speeds because of the phenomenon of drivers reducing their speeds as they approach intersections with no controls.

It can be seen from Figure 7.20 that triangles ABC and ADE are similar, which gives

$$\frac{CB}{AB} = \frac{ED}{AD} \tag{7.3}$$

and

$$\frac{d_b}{d_a} = \frac{a}{d_a - b} \tag{7.4}$$

If any three of the variables d_a, d_b, a, and b are known, the fourth can be determined using Eq. 7.4.

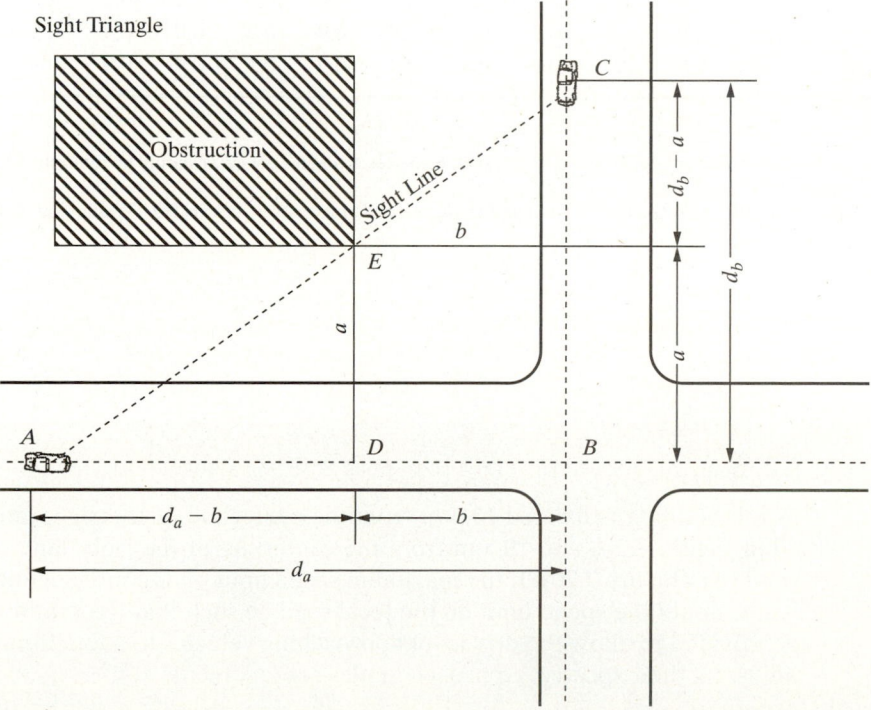

Figure 7.20 Minimum Sight Triangle at a No-Control or Yield-Control Intersection—Cases A and C

Table 7.7 Suggested Lengths and Adjustments of Sight-Triangle Leg Case A—No Traffic Control

Design Speed (mi/h)	Length of Leg (ft)
15	70
20	90
25	115
30	140
35	165
40	195
45	220
50	245
55	285
60	325
65	365
70	405
75	445
80	485

(a)

Approach Grade (%)	Design Speed (mi/h)													
	15	20	25	30	35	40	45	50	55	60	65	70	75	80
−6	1.1	1.1	1.1	1.1	1.1	1.1	1.1	1.2	1.2	1.2	1.2	1.2	1.2	1.2
−5	1.0	1.0	1.1	1.1	1.1	1.1	1.1	1.1	1.1	1.1	1.2	1.2	1.2	1.2
−4	1.0	1.0	1.0	1.1	1.1	1.1	1.1	1.1	1.1	1.1	1.1	1.1	1.1	1.1
−3 to 3	1.0	1.0	1.0	1.0	1.0	1.0	1.0	1.0	1.0	1.0	1.0	1.0	1.0	1.0
+4	1.0	1.0	1.0	1.0	1.0	0.9	0.9	0.9	0.9	0.9	0.9	0.9	0.9	0.9
+5	1.0	1.0	1.0	0.9	0.9	0.9	0.9	0.9	0.9	0.9	0.9	0.9	0.9	0.9
+6	1.0	1.0	0.9	0.9	0.9	0.9	0.9	0.9	0.9	0.9	0.9	0.9	0.9	0.9

(b)

Note: For approach grades greater than 3 percent, sight distances given in Table 7.7a should be multiplied by the appropriate values given in Table 7.7b.
SOURCE: From *A Policy on Geometric Design of Highways and Streets*, 2011, AASHTO, Washington, D.C. Used by permission.

Example 7.2 Computing Speed Limit on a Local Road Based on Sight Distance

A tall building is located 13.5 m from the centerline of the right lane of a local road (*b* in Figure 7.20) and 19.5 m from the centerline of the right lane of an intersecting road (*a* in Figure 7.20). If the maximum speed limit on the intersecting road is 56 km/h, what should the speed limit on the local road be such that the minimum sight distance is provided to allow the drivers of approaching vehicles to avoid imminent collision by adjusting their speeds? Approach grades are 2 percent.

Solution:

- Determine the distance on the local road at which the driver first sees traffic on the intersecting road.

  Speed limit on intersecting road = 56 km/h

  Distance required on intersecting road (d_a) = 49.5 m (from Table 7.7)

  Calculate the distance available on local road by using Eq. 7.4.

  $$d_b = a \frac{d_a}{d_a - b}$$

  $$= 19.5 \frac{49.5}{49.5 - 13.5}$$

  $$= 26.81\,\text{m}$$

- Determine the maximum speed allowable on the local road.

  The maximum speed allowable on the local road is 32 km/h (from Table 7.7). No correction is required for the approach grade as it is less than 3 percent.

Sight Distance Requirement for Stop Control Intersections on Minor Roads—Case B

When vehicles are required to stop at an intersection, the drivers of such vehicles should be provided sufficient sight distance to allow for a safe departure from the stopped position for the three basic maneuvers that occur at an average intersection. These maneuvers are:

1. Turning left onto the major road, which requires clearing the traffic approaching from the left and then joining the traffic stream on the major road with vehicles approaching from the right (see Figure 7.21a), Case B1
2. Turning right onto the major road by joining the traffic on the major road with vehicles approaching from the left (see Figure 7.21b), Case B2
3. Crossing the intersection, thereby clearing traffic approaching from both sides of the intersection (see Figure 7.21c), Case B3.

Case B1: Sight Distance Requirement for Left Turns at Stop Control Intersections on Minor Roads.   Adequate sight triangles for traffic approaching both from the right and the left should be provided for this condition as shown in Figure 7.19. AASHTO assumes that the decision point or vertex of the sight triangle is 4.35 m from the edge of the major-road traveled way, as this is a good representation of the location of the driver's eye of a vehicle on the minor road that is stopped relatively close to the major road. However, AASHTO also suggests that this distance be increased to 5.4 m when practicable. The sight distance required for this maneuver depends on the time the vehicle will take to cross the intersection and the distance an approaching vehicle on the crossroad (usually the major highway) will travel during that time. Equation 7.5 can be used to determine this distance:

$$d_{\text{ISD}} = 0.278v_{\text{major}}t_g \tag{7.5}$$

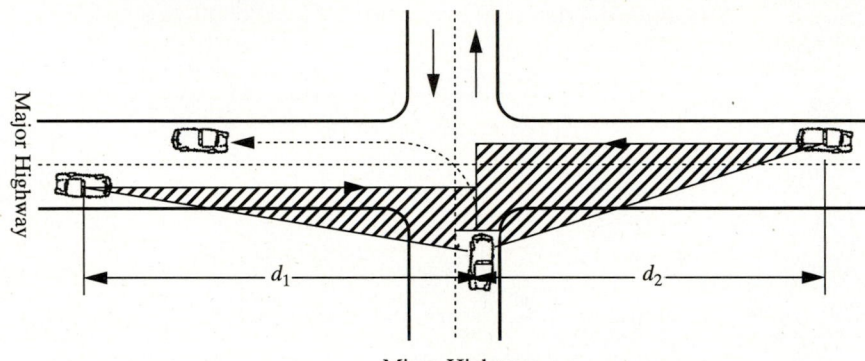

(a) Case B1—Stopped Vehicle Turning Left into Two-Lane Major Highway

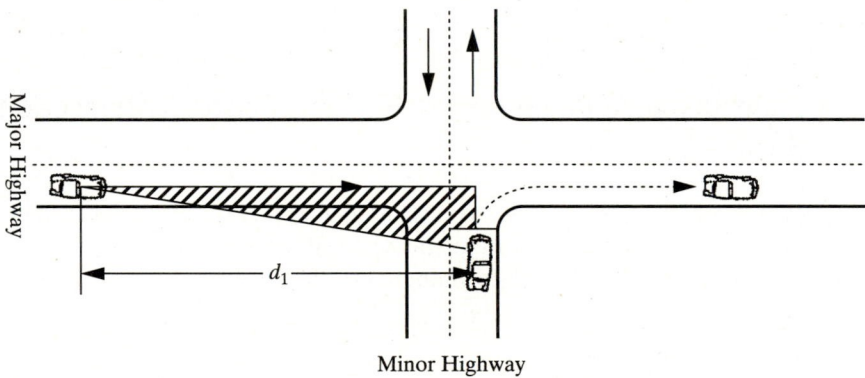

(b) Case B2—Stopped Vehicle Turning Right into
Two-Lane Major Highway or Right Turn on a Red Signal

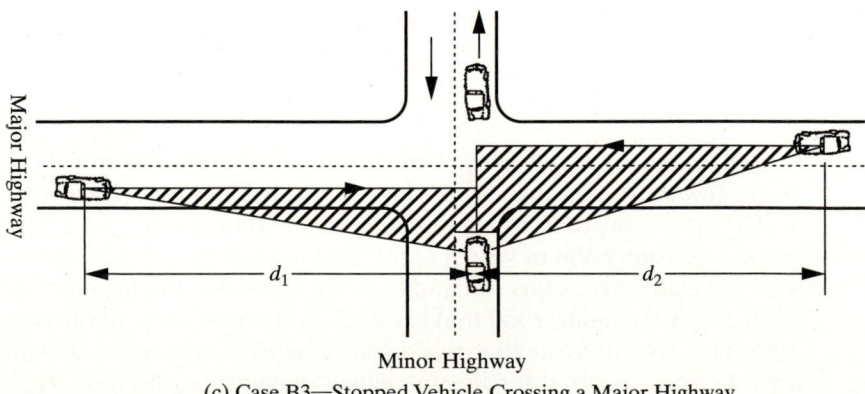

(c) Case B3—Stopped Vehicle Crossing a Major Highway

Figure 7.21 Maneuvers at a Stop Control Intersection

Table 7.8   Time Gap for Case B1—Left Turn from Stop

Design Vehicle	Time Gap (t_g) (sec) at Design Speed of Major Road
Passenger Car	7.5
Single-Unit Truck	9.5
Combination Truck	11.5

Note: Time gaps are for a stopped vehicle to turn left onto a two-lane highway with no median and grade 3 percent or less. The table values require adjustment as follows:

For multilane highways:

   For left turns onto two-way highways with more than two lanes, add 0.5 sec for passenger cars or 0.7 sec for trucks for each additional lane, from the left, in excess of one, to be crossed by the turning vehicle.

For minor road approach grades:

   If the approach grade is an upgrade that exceeds 3 percent, add 0.2 sec for each percent grade for left turns.

SOURCE: *A Policy on Geometric Design of Highways and Streets,* American Association of State Highway and Transportation Officials, Washington, D.C., 2011, p. 9-37. Used with permission.

where

$$d_{\text{ISD}} = \text{sight distance (length of the leg of sight triangle) along the major road from the intersection}$$

$$v_{\text{major road}} = \text{design speed on the major highway from the intersection (km/h)}$$

$$t_g = \text{time gap for the minor road vehicle to enter the major road (sec)}$$

   Table 7.8 gives suggested values for the time gap for the minor road vehicle to enter the major road. Note that these values do not depend on the approach speeds on the major road as studies have shown that a value based on the type of vehicle is appropriate for determining sight distances at intersections. Also, it has been observed that major road drivers tend to reduce their speeds when they observe a minor-road vehicle turning onto the major road. AASHTO suggests that when values given in Table 7.8 are used to determine the length of the leg of the sight triangle along the major road, it is not necessary for drivers to reduce their speeds to less than 70 percent of their original speeds. It should be noted that these values should be corrected for multilane highways and minor approach grades, as shown in the table. When the major road is a divided highway and the design vehicle is larger than a passenger car, it is necessary to check the required sight distance for both the design vehicle and for smaller design vehicles. Also, if the median is wide enough to accommodate the design vehicle with clearances to the through lanes of at least 0.9 m, it is not necessary to carry out a separate computation for the departure sight triangle for left turns from the minor road. This is because values obtained for the right-turn departure triangle will usually satisfy the requirement for crossing the near roadway to the median. However, in this case, it is necessary to carry out the analysis for the right sight triangle for left turns from the median.

Example 7.3   Computing Sight Distance Requirement for a Vehicle Turning Left from a Minor Road with a Stop Sign Control

A minor road intersects a major four-lane undivided road with a design speed of 104 km/h. The intersection is controlled with a stop sign on the minor road. If the design vehicle is a single-unit truck, determine the minimum sight distance required on the major road that will allow a stopped vehicle on the minor road to safely turn left if the approach grade on the minor road is 2 percent.

Solution:

- Use Eq. 7.5.

$$d_{\text{ISD}} = 0.278 v_{\text{major}} t_g$$

- Determine t_g.
  From Table 7.8, $t_g = 9.5$ sec (for a single unit truck).
  Correct for number of lanes:

$$t_g = (9.5 + 0.7) \text{ sec} = 10.2 \text{ sec}$$

(Note no adjustment is necessary for approach grade as it is not higher than 3 percent.)
- Determine minimum sight distance $= 0.278 \times 104 \times 10.2 = 298.9$ m.

Example 7.4 Computing Sight Distance Requirement for a Vehicle Turning Left from a Minor Road with a Stop Sign Control

A minor road intersects a major four-lane divided road with a design speed of 104 km/h and a median width of 1.8 m. The intersection is controlled with a stop sign on the minor road. If the design vehicle is a passenger car, determine the minimum sight distance required on the major road for the stopped vehicle to turn left onto the major road if the approach grade on the minor road is 4 percent.

Solution:

- Use Eq. 7.5.

$$d_{\text{ISD}} = 0.278 v_{\text{major}} t_g$$

- Determine t_g.
  From Table 7.8, $t_g = 7.5$ sec.
  Correct for number of lanes:

$$t_g = (7.5 + 0.5 \times 1.5) \text{ sec} = 8.25 \text{ sec}$$

(This assumes that the 6-ft median is equivalent to half a lane.)
  Correct for approach grade:

$$t_g = (8.25 + 0.2 \times 4) \text{ sec} = 9.05 \text{ sec}$$

(Grade is 1 percent higher than 3 percent.)
- Determine minimum sight distance $= 0.278 \times 104 \times 9.05 = 261.65$ m.

Case B2: Sight Distance Requirement for Right Turns on Stop Control Intersections on Minor Roads. The computational procedure used for this case is similar to that for left turns discussed for Case B1, but the values of the time gap for the minor road vehicle to enter the major road (t_g) are adjusted in consideration of the fact that drivers tend to accept gaps that are slightly lower than those for left turns. AASHTO suggests that values shown on Table 7.8 should be decreased by 1 sec, and the correction for grades greater than 3 percent is 0.1 sec for each percent grade. In cases where this sight distance is not available, reducing the speed limit on the major road could be considered.

Case B3: Sight Distance Requirement for Crossing Major Roads from Stop-Controlled Intersections on Minor Roads. Minimum requirements determined for right and left turns as presented for Cases B1 and B2 will usually satisfy the requirements for the crossing maneuver. AASHTO, however, recommends that the available sight distance for crossing be checked when the following conditions exist:

- When only a crossing maneuver is allowed at the intersection,
- When the crossing maneuver will involve an equivalent width of more than six lanes, or
- Where the vehicle mix of the crossing traffic includes a substantial number of heavy vehicles and the existence of steep grades that may slow down these heavy vehicles while their back portion is still in the intersection.

Example 7.5 Determining whether a Construction Obstructs the Required Sight Triangle

A developer has applied for a construction permit to build a supermarket at the southeast corner of an intersection that is formed by a four-lane divided highway in the east–west direction and a two-lane road in the north–south direction. Traffic on the two-lane road is controlled by a stop sign at the intersection. The speed limit on the four-lane highway is 88 km/h. Using the sight triangle requirements at the intersection, determine whether permission should be given for the construction if the following conditions exist.

Major road lane width = 3.6 m

Minor road lane width = 3.3 m

Median width = 1.8 m

Design vehicle on minor road is a passenger car. Length = 6.6 m

Approach grade = 2%

Maximum feasible distance of building from the center of the nearest lane of the four-lane road = 5.4 m (due to available land restriction)

Maximum feasible distance of building from the center of the nearest lane of the two-lane road = 6.15 m (due to available land restriction)

Solution: Adequacy of sight triangle for left turns (Case B1) should be checked. Note that checking for right turns crossing (Case B2) is not necessary because the building is expected to be located in the south-east corner, and therefore does not interfere with the sight distance triangle for right turns. Also, checking for crossing (Case B3) is not necessary as minimum requirements for Case B1 and Case B2 usually satisfy the requirements for Case B3.

(A) Check the sight triangle for left turns (Case B1). Note that this condition has to be checked as the median is not large enough to accommodate the design vehicle.
  • Use Table 7.8 to determine t_g (time gap for the minor road to enter the major road).

$$t_g = 7.5 \text{ sec (for passenger cars)}$$

Correct for number of lanes crossed.
$t_g = 7.5 + 0.5(1.5 + 0.75)$ ($1\frac{1}{2}$ half lanes and a median equivalent to $\frac{3}{4}$ lane)

$$t_g = 7.5 + 1.125 = 8.625 \text{ sec}$$

  • Use Eq. 7.5 to determine the length of the sight triangle on the major road.

$$d_{ISD} = 0.278 v_{major} t_g$$

$$d_{ISD} = 0.278 \times 88 \times 8.625 = 211.0 \text{ m}$$

  • Sight triangle is therefore set by d_{ISD} and the distance from the driver's eye location to the center line of the inner lane of the major road in the westbound direction, which is $(4.35 + 2 \times 3.6 + 2.7 + 1.8)$ ft = 16.05 m (AB in Figure 7.22)

(B) Check whether corner of proposed building is within the sight triangle. This can be done graphically or by computation. The graphical method involves drawing

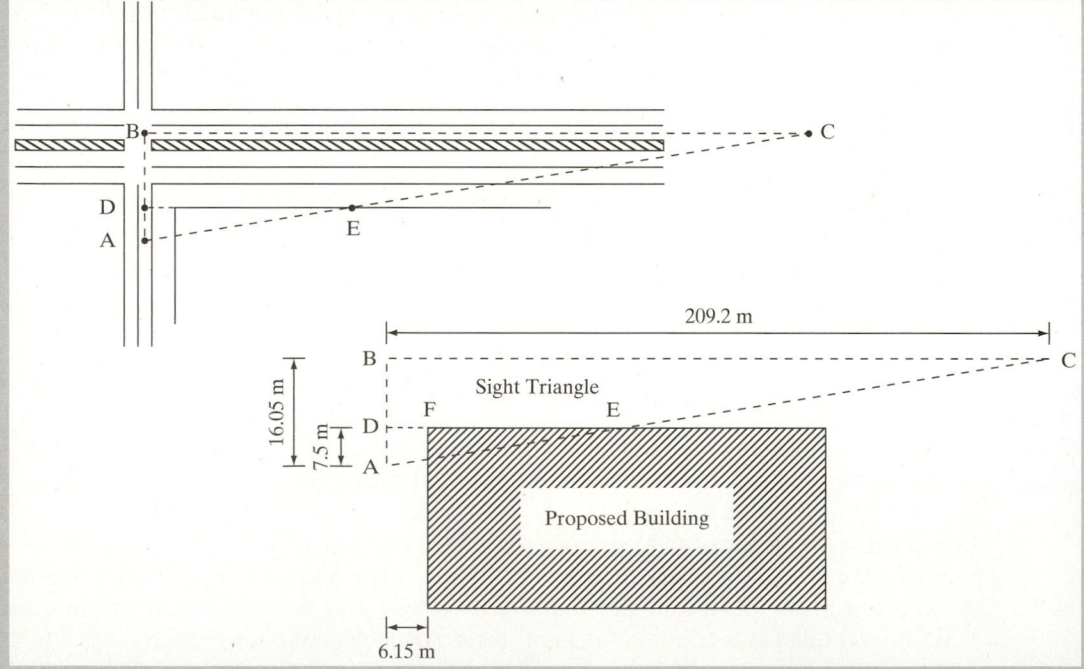

Figure 7.22 Sight Triangle for Example 7.5

the intersection to scale and, using the same scale, drawing the departure sight triangle starting from the driver's eye location, which is 4.35 m from the edge of the minor road, and using the appropriate dimensions. Then the corner of the proposed building is located to determine whether it is within the sight triangle. If it is within the sight triangle, permission should be denied.

Alternately, the position of the corner of the building is computed using similar triangles as shown below.

Using similar triangle ABC and ADE (see Figure 7.22):

$$AB = 16.05 \text{ m}$$

$$BC = d_{ISD} = 209.2 \text{ m}$$

$$AD = 16.05 - (5.4 + 1.8 + 3.6 + 2.7 + 1.8) = 0.75 \text{ m}$$

$$\frac{16.05}{0.75} = \frac{209.2 \text{ m}}{DE}$$

$$DE = 9.78 \text{ m}$$

$$DF = 6.15 \text{ ft} < 9.78 \text{ m}$$

Therefore, the building is located within the sight triangle and the permit should be refused.

Sight Distance Requirement for Yield Control on the Minor Road (Case C)

Drivers on a minor road approaching a yield-controlled intersection with a major road can enter or cross the intersection without stopping if the driver does not perceive any conflict with oncoming major road traffic. Adequate sight distance on the major road therefore should be provided for crossing the intersection (Case C1) and for making right and left turns (Case C2). For three-leg intersections, only case C2 exists as no through movement can occur.

Case C1: Sight Distance Requirement for Crossing a Yield-Controlled Intersection from a Minor Road.   The assumption made to determine the minimum sight distance for this maneuver is similar to that used for the no-control maneuver in Case A, but with the following modifications:

- Drivers on minor roads approaching a yield sign tend to decelerate to 60 percent of the minor road design speed and not 50 percent, as assumed for the no-control condition. This assumption is based on field observation.
  The rate of deceleration is 1.5 m/sec^2.
- The time t_g to cross the intersection should include the time taken for the vehicle to travel from the decision point where the deceleration begins to where the speed is reduced to 60 percent of the minor road design speed.
- The vehicle then travels at the reduced speed (60 percent of the minor road design speed) until it crosses and clears the intersection.

Based on these assumptions, the length of the sight distance (d_{ISD}) on the major road can be obtained from the following equations:

$$t_g = t_a + (w + L_a)/0.167v_{minor} \tag{7.6}$$

$$d_{ISD} = 0.278v_{major}t_g \tag{7.7}$$

where

t_g = travel time to reach and clear the intersection (sec)
t_a = travel time to reach the major road from the decision point for a vehicle that does not stop
w = width of the intersection to be crossed (m)
L_a = length of design vehicle (m)
v_{minor} = design speed of minor road (km/h)
v_{major} = design speed of major road (km/h)

Table 7.9 gives the AASHTO suggested t_a and t_g values with the corresponding lengths for the minor leg sight triangle for crossing a two-lane highway at a yield-controlled intersection for different design speeds. Also, computed values for the length of the major leg of the sight triangle are given in Table 7.10 for different two-lane major-road design speeds. It should be noted that these values are for a passenger car as the design vehicle

Table 7.9   Case C1—Crossing Maneuvers from Yield-Controlled Approaches—Length of Minor Road Leg and Travel Times

Design Speed (mi/h)	Minor-Road Approach			Travel Time t_g (seconds)	
	Length of Leg[1] (ft)	Travel Time $t_a^{1,2}$ (seconds)		Calculated Value	Design Value[3,4]
15	75	3.4		6.7	6.7
20	100	3.7		6.1	6.5
25	130	4.0		6.0	6.5
30	160	4.3		5.9	6.5
35	195	4.6		6.0	6.5
40	235	4.9		6.1	6.5
45	275	5.2		6.3	6.5
50	320	5.5		6.5	6.5
55	370	5.8		6.7	6.7
60	420	6.1		6.9	6.9
65	470	6.4		7.2	7.2
70	530	6.7		7.4	7.4
75	590	7.0		7.7	7.7
80	660	7.3		7.9	7.9

[1]For minor-road approach grades that exceed 3 percent, multiply the distance or the time in this table by the appropriate adjustment factor from Table 7.7.
[2]Travel time applies to a vehicle that slows before crossing the intersection but does not stop.
[3]The value of t_g should equal or exceed the appropriate time gap for crossing the major road from a stop-controlled approach.
[4]Values shown are for a passenger car crossing a two-lane highway with no median and grades 3 percent or less.
SOURCE: From *A Policy on Geometric Design of Highways and Streets*, 2011, AASHTO, Washington, D.C. Used by permission.

Table 7.10 Length of Sight Triangle Leg along Major Road—Case C1—Crossing Maneuver at Yield-Controlled Intersection

Major Road Design Speed (mi/h)	Stopping Sight Distance (ft)	Minor Road Design Speed (mi/h)							
		15	20–50	55	60	65	70	75	80
15	80	150	145	150	155	160	165	170	175
20	115	200	195	200	205	215	220	230	235
25	155	250	240	250	255	265	275	285	295
30	200	300	290	300	305	320	330	340	350
35	250	345	335	345	360	375	385	400	410
40	305	395	385	395	410	425	440	455	465
45	360	445	430	445	460	480	490	510	525
50	425	495	480	495	510	530	545	570	585
55	495	545	530	545	560	585	600	625	640
60	570	595	575	595	610	640	655	680	700
65	645	645	625	645	660	690	710	740	755
70	730	690	670	690	715	745	765	795	815
75	820	740	720	740	765	795	795	850	875
80	910	790	765	790	815	850	850	910	930

Note: Values in the table are for passenger cars and are based on the unadjusted distance and times in Table 7.9. The distance and times in Table 7.9 need to be adjusted using the factors in Table 7.7b.
SOURCE: From *A Policy on Geometric Design of Highways and Streets*, 2011, AASHTO, Washington, D.C. Used by permission.

and approach grades of 3 percent or less. Appropriate corrections should be made to t_g for cases where the design vehicle is a truck and/or when the grade is higher than 3 percent using correction factors shown in Table 7.7.

Example 7.6 Determining the Minimum Sight Distance for Crossing at an Intersection with Yield Control

An urban two-lane minor road crosses a four-lane divided highway with a speed limit of 88 km/h. If the minor road has a speed limit of 72 km/h and the intersection is controlled by a yield sign on the minor road, determine the sight distance from the intersection that is required along the major road such that the driver of a vehicle on the minor road can safely cross the intersection. The following conditions exist at the intersection.

Major road lane width = 3.3 m

Median width = 2.4 m

Design vehicle on minor road is a passenger car length = 6.6 m

Approach grade on minor road = 3%

Solution: Use Eq. 7.6 to determine time (t_a) as the travel time to reach and clear the intersection.

$$t_g = t_a + (w + L_a)/0.167v_{min}$$

$t_a = 5.2$ sec for passenger vehicles from Table 7.9

$w = (4 \times 3.3 + 2.4) = 15.6$ m

$t_g = 5.2 + (15.6 + 6.6)/(0.167 \times 72)$

$\quad = 7.06$ sec

Use Eq. 7.7 to determine the length of the sight on the major road.

$$d_{ISD} = 0.278v_{major}t_g$$

$$= 0.278 \times 88 \times 7.06 = 172.71 \text{ m}$$

Case C2: Sight Distance Requirements for Turning Right or Left from a Minor Road at a Yield-Controlled Intersection. For this maneuver, it is assumed that a driver will reduce his/her speed to about 16 km/h. Based on this assumption, the length of the minor road leg of the sight triangle is taken as 24.6 m. The length of the major road leg is computed using the same principles for the stop control of Case B1 and B2. However, the t_g values used are 0.5 sec higher than those shown in Table 7.8. Also, adjustment should be made for major highways with more than two lanes, although it is not necessary to make this adjustment for right turns. It should be noted that sight distance requirements for yield-controlled intersections are usually larger than those for stop-controlled intersections, which makes it unnecessary to check for the stopped condition to accommodate those vehicles that are stopped to avoid approaching vehicles on the major road.

Sight Distance Requirements at Signalized Intersections—Case D

The two main requirements at signalized intersections are as follows: (1) The first vehicle stopped at the stop line of each approach should be visible to the driver of the first vehicle stopped on all other approaches, and (2) adequate sight distance should be provided for left-turning vehicles to enable drivers of these vehicles to select adequate gaps. However, when the signals are to be placed in a flashing operation for all approaches during off-peak periods, then the sight distance requirements for the appropriate condition of Case B should be provided. Similarly, if right turn on red is permitted, then the appropriate sight requirement for right turns of Case B should be provided.

Sight Distance Requirements at All-Way Stop-Controlled Intersections—Case E

The only sight distance required in this case is that the first vehicle stopped at the stop line of each approach should be visible to the driver of the first vehicle stopped on all other approaches.

Sight Distance Requirements for Turning Left from a Major Road—Case F

In this case, the turning vehicle is assumed to start its turning movement from a stopped position, as vehicles that do not stop will require less sight distance. Adequate sight distance along the major road is the distance traveled by an oncoming vehicle along the major road during the time it takes the stopped vehicle to cross the major road. Tables 7.11 and 7.12 give AASHTO-suggested values for t_g for two-lane highways and a passenger car design vehicle. The necessary adjustment for t_g should be made for a different design vehicle, higher number of lanes, median widths, and the sight distance computed.

Table 7.11   Time Gap for Case F—Left Turns from the Major Road

Design Vehicle	Time Gap t_g (sec) at Design Speed of Major Road
Passenger Car	5.5
Single-Unit Truck	6.5
Combination Truck	7.5

Adjustment for multilane highways: For left-turning vehicles that cross more than one opposing lane, add 0.5 sec for passenger cars and 0.7 sec for trucks for each additional lane to be crossed.
SOURCE: From *A Policy on Geometric Design of Highways and Streets*, 2011, AASHTO, Washington, D.C. Used by permission.

Table 7.12   Intersection Sight Distance—Case F—Left Turn from Major Road

Design Speed (mi/h)	Stopping Sight Distance (ft)	Intersection Sight Distance Passenger Cars Calculated (ft)	Design (ft)
15	80	121.3	125
20	115	161.7	165
25	155	202.1	205
30	200	242.6	245
35	250	283.0	285
40	305	323.4	325
45	360	363.8	365
50	425	404.3	405
55	495	444.7	445
60	570	485.1	490
65	645	525.5	530
70	730	566.0	570
75	820	606.4	610
80	910	646.8	650

Note: Intersection sight distance shown is for a passenger car making a left turn from an undivided highway. For other conditions and design vehicles, the time gap should be adjusted and the sight distance recalculated.
SOURCE: From *A Policy on Geometric Design of Highways and Streets*, 2011, AASHTO, Washington, D.C. Used by permission.

Effect of Skew on Intersection Sight Distance

As noted earlier, it is desirable for roads to intersect at or nearly 90 degrees. When it is not feasible to obtain a normal or near-normal intersection, it is necessary to consider the way in which sight distances are computed. The legs of the sight triangles should be parallel to the intersection approaches, as shown in Figure 7.23. The actual distance traveled by a vehicle crossing the intersection is obtained by dividing the width of the major road by the sine of the intersection angle, as shown in Figure 7.23. This results in a longer path than the width of the intersection. When the computed path length is longer than the total widths of the lane by more than 3.6 m, the appropriate number of lanes should be determined and used for adjusting t_g for number of lanes. Drivers in the obtuse-angle quadrant of a skewed intersection can easily see the full sight triangle by only slightly turning their heads as the angle between the approach leg and the sight line is usually very small. However, in the acute-angle quadrant, drivers often have to significantly turn their heads to see the full sight triangle. AASHTO therefore recommends that the requirements for Case A not be used for skewed intersections, but the requirements for Case B can be applied if this is practical.

7.3 DESIGN OF RAILROAD GRADE CROSSINGS

Railroad crossings are similar to four-leg intersections and can also be either at-grade or grade-separated. When a railroad crosses a highway at grade, it is essential that the highway approaches be designed to provide the safest condition so that the likelihood of a collision between a vehicle crossing the rail tracks and an oncoming train is reduced to the absolute minimum. This involves both the proper selection of the traffic control systems, the appropriate design of the horizontal and vertical alignments of the highway approaches, and assuring the availability of adequate sight distance triangles.

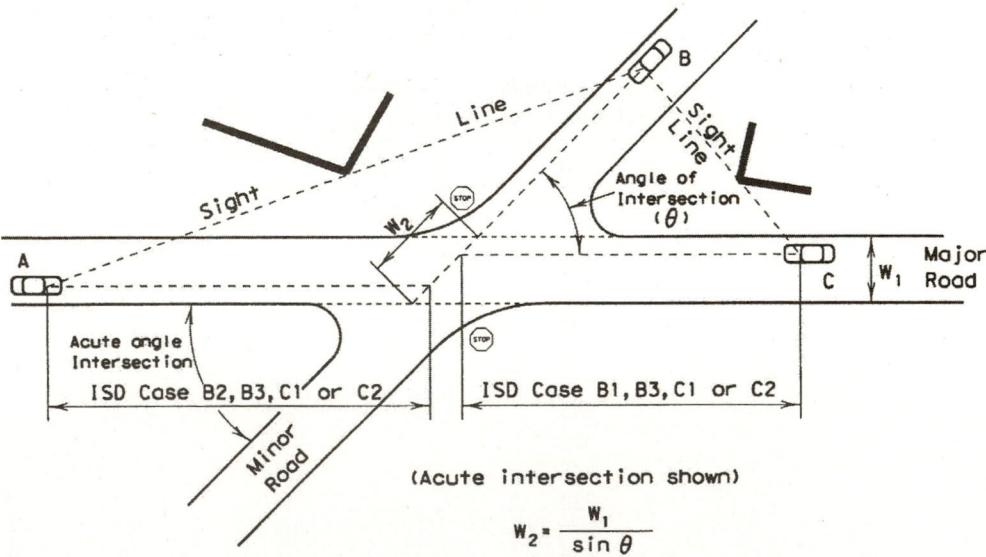

Figure 7.23 Sight Triangles at Skewed Intersections

SOURCE: From *A Policy on Geometric Design of Highways and Streets*, 2011, AASHTO, Washington, D.C. Used by permission.

7.3.1 Selection of Traffic-Control Devices

The selection of the appropriate traffic-control systems used at the railroad crossing should be made at the same time as the design of the vertical and horizontal alignments. These include both passive and active warning devices. Passive devices include signs, pavement markings, and grade-crossing illumination that warn an approaching driver of the crossing location. This type of warning device allows the driver of an approaching vehicle to determine whether a danger exists because of an approaching train; in fact, in this case, the decision on whether to stop or proceed belongs entirely to the driver. Several factors influence the designer's decision to use passive or active control devices. Passive warning devices are used mainly at low-volume, low-speed roadways with adequate sight distance that cross minor spurs or tracks seldom used by trains. Active warning devices include flashing light signals and automatic gates that give the driver a positive indication of an approaching train at the intersection. When it is necessary to use an active control device at a railroad grade crossing, the basic device used is the flashing light signal. An automatic gate is usually added after an evaluation of the conditions at the location. The factors considered in this evaluation include the highway type; the volumes of vehicular, railroad, and pedestrian traffic; the expected speeds of both the vehicular and the railroad traffic; accident history at the site; and the geometric characteristics of the approaches, including available sight distances. For example, when multiple main-line tracks exist or when moderate volumes of railroad and vehicular traffic exist with high speeds, the automatic gate is frequently used. AASHTO, however, suggests that the potential of eliminating grade crossings with only passive control devices should be given serious consideration. This can be achieved by closing crossings that are infrequently used and installing active controls at others.

7.3.2 Design of the Horizontal Alignment

Guidelines similar to those for the design of intersections formed by two highways are used in the design of a railroad grade crossing. It is desirable that the highway intersects the rail tracks at a right angle and that other intersections or driveways be located at distances far enough away from the crossing to ensure that traffic operation at the crossing is not affected by vehicular movements at these other intersections and driveways. In order to enhance the sight distance, railroad grade crossings should not be placed at curves.

7.3.3 Design of the Vertical Alignment

The basic requirements for the vertical alignment at a railway grade crossing are the provision of suitable grades and adequate sight distances. Grades at the approaches should be as flat as possible to enhance the view across the crossing. When vertical curves are used, their lengths should be adequate to ensure that the driver on the highway clearly sees the crossing. In order to eliminate the possibility of low-clearance vehicles being trapped on the tracks, the surface of the crossing roadway should be at the same level as the top of the tracks for a distance of 0.6 m from the outside of the rails. Also, at a point on the roadway 9 m from the nearest track, the elevation of the road should not be higher or lower than 0.075 m of the elevation of the tracks unless this cannot be achieved because of the superelevation of the track.

Sight Distance Requirements at Railroad Crossings

The provision of adequate sight distances at the crossing is very important, especially at crossings where train-activated warning devices are not installed. The two principal actions the driver of a vehicle on the road can take when approaching a railroad crossing are:

1. The driver, having seen the train, brings the vehicle to a stop at the stop line on the road.
2. The driver, having seen the train, continues safely across the crossing before the train arrives.

The minimum sight distances required to satisfy these conditions should be provided at the crossing, as illustrated in Figure 7.24.

Driver Stops at the Stop Line.   The total distance required for the driver to stop a vehicle at the stop line consists of the following:

The distance traveled during the driver's perception/reaction time, or $0.278v_vt$, where v_v is the velocity of the approaching vehicle in mi/h, and t is the perception reaction time, usually taken as 2.5 sec

The actual distance traveled during braking, or $v_v^2/\left(254\dfrac{a}{g}\right)$ (assuming a flat grade), as discussed in Chapter 3

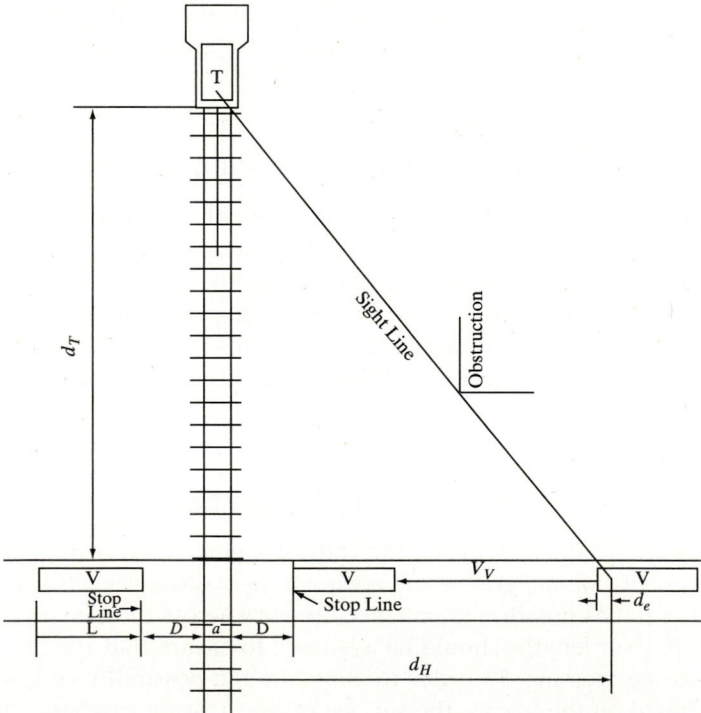

Figure 7.24   Layout of Sight Distance Requirements for a Moving Vehicle to Cross or Stop at a Railroad Crossing Safely—Case I

SOURCE: From *A Policy on Geometric Design of Highways and Streets*, 2011, AASHTO, Washington, D.C. Used by permission.

This gives

$$\text{Total distance traveled during braking} = 0.278v_v t + \frac{v_v^2}{254\dfrac{a}{g}}$$

Therefore, the distance d_H (sight-distance leg along the highway) the driver's eye should be away from the track to safely complete a stopping maneuver is the sum of the total distance traveled during the braking maneuver, the distance between the stop line and the nearest rail, and the distance between the driver's eyes and the front of the vehicle as shown in Eq. 7.8:

$$d_H = 0.278v_v t + \frac{v_v^2}{254\left(\dfrac{a}{g}\right)} + D + d_e \qquad (7.8)$$

where

  d_H = sight-distance leg along the highway that allows a vehicle approaching the crossing at a speed v_v to stop at the stop line at a distance D from the nearest rail
  D = distance from the stop line to the nearest rail (assumed to be 4.5 m)
  d_e = distance from the driver's eye to the front of the vehicle (assumed to be 2.4 m)
  v_v = velocity of the vehicle

Driver Continues Safely across Crossing.   In this case, the sight-distance leg d_T along the railroad depends on both the time it takes the vehicle on the highway to cross the tracks from the time the driver first sees the train and the speed at which the train is approaching the crossing. The time at which the driver first sees the train must, however, provide a distance that will allow the driver either to cross the tracks safely or to stop the vehicle safely without encroachment of the crossing area. This distance therefore must be at least the sum of the distance required to stop the vehicle at the stop line and that between the two stop lines. It can be seen from Figure 7.24 that this total distance is

$$0.278v_v t + \frac{v_v^2}{254\left(\dfrac{a}{g}\right)} + 2D + L + W$$

where

  L = length of the vehicle, assumed as 19.5 m
  W = the distance between the nearest and farthest rails (i.e., 1.5 m)

The time taken for the vehicle to traverse this total distance is obtained by dividing this distance by the speed of the vehicle (v_v). The distance traveled by the train during this time is then obtained by multiplying the time by the speed of the train. This gives the sight distance leg d_T along the railroad track as:

$$d_T = \frac{v_T}{v_v}\left(0.278v_v t + \frac{v_v^2}{254\left(\dfrac{a}{g}\right)} + 2D + L + W\right) \qquad (7.9)$$

where

> D = distance from the stop line to the nearest rail (assumed to be 4.5 m)
> v_v = velocity of the vehicle
> v_T = velocity of train (km/h)

The sight distance legs d_H along the highway and d_T along the track and the sight line connecting the driver's eye and the train form the sides of the sight triangle, which should be kept clear of all obstructions to the driver's view. Table 7.13 gives AASHTO-computed values for these sight-distance legs for different approach speeds of the train and the vehicle and are designated as Case A.

Driver Departs from Stopped Position and Safely Clears Crossing.   In this case, an adequate sight distance should be provided along the track to allow enough time for a driver to accelerate the vehicle and safely cross the tracks before the arrival of a train that comes into view just as this maneuver begins. It is assumed that the driver will go through the crossing in first gear. The time required to go through the crossing, t_s, is the sum of (a) the time it takes to accelerate the vehicle to its maximum speed in the first gear, or v_g/a; and (b) the time it takes to travel the distance from the point at which the vehicle attains maximum speed in first gear to the position where the rear of the vehicle is at a distance D from the farthest tracks (i.e., the time it takes to travel the distance from the stopped position to the position where the rear of the vehicle is at D from the farthest tracks minus the distance traveled

Table 7.13   Required Design Sight Distance for Combination of Highway and Train Vehicle Speeds; 73.5-ft Truck Crossing a Single Set of Tracks at 90 Degrees

Train Speed (mi/h)	Case B Departure from Stop	Case A Moving Vehicle							
		Vehicle Speed (mi/h)							
	0	10	20	30	40	50	60	70	80
			Distance Along Railroad from Crossing, d_H (ft)						
10	255	155	110	102	102	106	112	119	127
20	509	310	220	203	205	213	225	239	254
30	794	465	331	305	307	319	337	358	381
40	1019	619	441	407	409	426	450	478	508
50	1273	774	551	509	511	532	562	597	635
60	1528	929	661	610	614	639	675	717	763
70	1783	1084	771	712	716	745	787	836	890
80	2037	1239	882	814	818	852	899	956	1017
90	2292	1394	992	915	920	958	1012	1075	1144
			Distance Along Highway from Crossing, d_H (ft)						
		69	135	220	324	447	589	751	931

SOURCE: From *A Policy on Geometric Design of Highways and Streets*, 2011, AASHTO, Washington, D.C. Used by permission.

when the vehicle is being accelerated to its maximum speed in first gear); and (c) the perception-reaction time *J*.

$$t_s = \frac{v_g}{a_1} + (L + 2D + W - d_a)\frac{1}{v_g} + J$$

where

v_g = maximum velocity of vehicle in first gear, assumed to be 8.8 ft/sec
a_1 = acceleration of vehicle in first gear, assumed to be 0.44 m/sec²
L = length of vehicle, assumed to be 23.05 m
D = distance of stop line to nearest rail, assumed to be 4.5 m
J = perception-reaction time, assumed to be 2.0 sec
W = distance between outer rails (for a single track this is equal to 5 ft)
d_a = distance vehicle travels while accelerating to maximum speed in the first gear,

$$d_a = \frac{v_g^2}{2a_1}$$

The distance d_T traveled by the train along the track during this time is obtained by multiplying this time by the speed of the train:

$$d_T = 0.44v_T\left[\frac{v_g}{a_1} + (L + 2D + W - d_a)\frac{1}{v_g} + J\right] \tag{7.10}$$

where

d_T = sight distance along railroad tracks to allow a stopped vehicle to depart and safely cross the railroad tracks
v_T = velocity of train
v_G = maximum speed of vehicle in first gear (assumed 2.64 m/sec)
a_1 = acceleration of vehicle in first gear (assumed 0.44 m/sec²)
$d_a = \dfrac{v_G^2}{2a_1}$ = or distance vehicle travels while accelerating to maximum speed
D = distance from stop line to near rail (assumed 4.5 m)
W = distance between outer rails (single track W = 1.5 m)
L = length of vehicle (assumed 23.05 m)
J = perception/reaction time (assumed 2.0 sec)

Adjustments must be made for skewed crossings.

Assume flat highway grades adjacent to and at crossings.

The sight triangle is formed by the lengths 2D, d_T, and the sight line between the eyes of the driver of the stopped vehicle and the approaching train, as shown in Figure 7.25, should therefore be kept clear of any obstruction to the vehicle driver's view. Values for d_T for the stopped condition (Case B) are also given in Table 7.13.

Equations 7.8, 7.9, and 7.10 have been developed with the assumptions that the approach grades are level and that the road crosses the tracks at 90 degrees. When one or both assumptions do not apply, the necessary corrections should be made (as discussed earlier) for highway–highway at-grade intersections. Also, when it is not feasible to provide the required sight distances for the expected speed on the highway, the appropriate speed for which the sight distances are available should be determined, and

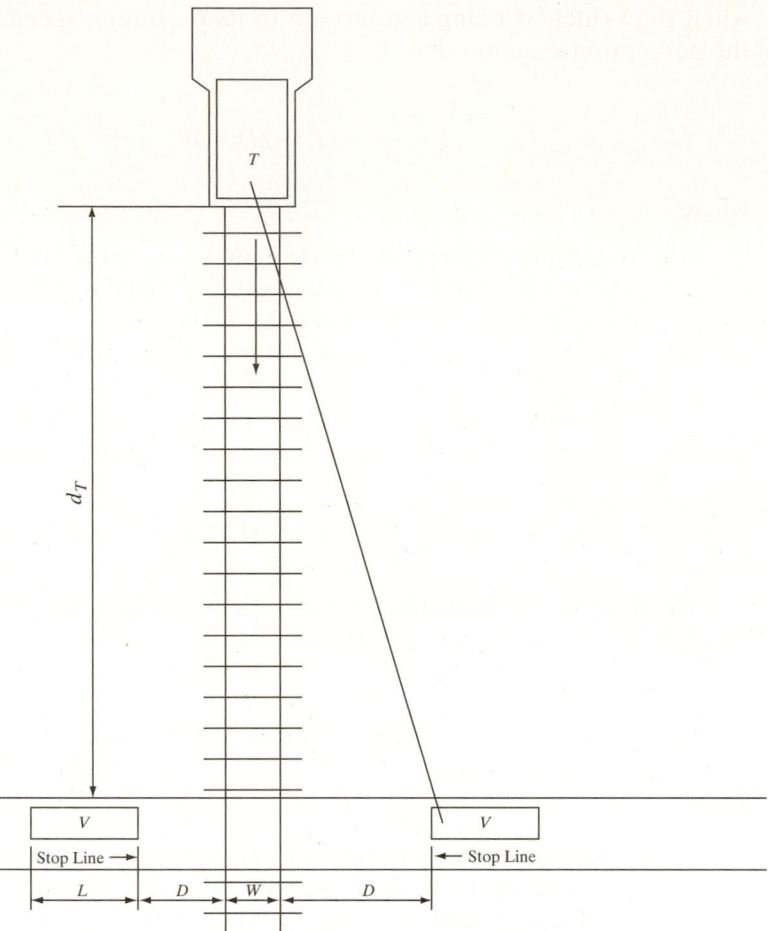

Figure 7.25   Layout of Sight Distance Requirements for Departure of a Vehicle at a Stopped Position to Cross a Single Railroad Track

SOURCE: From *A Policy on Geometric Design of Highways and Streets*, 2011, AASHTO, Washington, D.C. Used by permission.

speed-control signs and devices should be used to advise drivers of the maximum speed allowable at the approach.

7.4 SUMMARY

The basic principles involved in the design of at-grade intersections are presented in this chapter. The fundamental objective in the design of at-grade intersections is to minimize the severity of potential conflicts among different streams of traffic and between pedestrians and turning movements. This design process involves the selection of an appropriate layout of the intersection and the appropriate design of the horizontal and vertical alignments. The selection of the appropriate layout depends on the turning volumes, the available space, and whether the intersection is in an urban or rural area. An important factor in designing the layout is whether the intersection should be channelized. When traffic volumes warrant channelization, care should be taken not to overchannelize, since this may confuse drivers. Channelization usually involves the use of islands which may or may not be raised. Raised islands may be either of the mountable or barrier type. It is advisable not to use raised barriers at intersections on high-speed roads. Care also should be taken to ensure that the minimum sizes for the type of island selected be provided.

In designing horizontal alignments, sharp horizontal curves should be avoided, and every effort should be made to ensure that the intersecting roads cross at 90 degrees (or as near to 90 degrees as possible).

The most important factor in the design of the vertical alignment is to provide sight distances that allow drivers to have an unobstructed view of the whole intersection. The minimum sight distance required at any approach of an intersection depends on the design vehicle on the approach, the expected velocity on the approach, and the type of control at the approach.

An at-grade railroad crossing is similar to an at-grade highway–highway intersection in that similar principles are used in their design. For example, in the design of the vertical alignment of a railroad crossing, suitable grades and sight distances must be provided at the approaches. In the design of the railroad crossing, however, additional factors come into play. For example, at a railroad crossing, the elevation of the surface of the crossing roadway should be the same as the top of the tracks for a distance of 0.6 m from the outside rails, and at a point on the roadway 9 m from the nearest track, the elevation of the roadway should not be more than 0.075 m higher or lower than that of the track.

In general, all intersections are points of conflicts between vehicles and vehicles, and vehicles and pedestrians. The basic objective of their design should be to eliminate these conflict points as much as possible.

PROBLEMS

7-1   Briefly describe the different principles involved in the design of at-grade intersections.

7-2   Describe the different types of at-grade intersections. Also, give an example of an appropriate location for the use of each type.

7-3   Describe the different types of traffic circles, indicating under what conditions you would recommend the use of each.

7-4   What are the key defining characteristics of roundabouts that distinguish them from other traffic circles?

7-5   What are the main functions of channelization at an at-grade intersection?

7-6   Discuss the fundamental general principles that should be used in designing a channelized at-grade intersection.

7-7   Describe the different types of islands used in channelizing at-grade intersections, indicating the principal function of each type.

7-8   Figure 7.26a illustrates a three-leg intersection of State Route 150 and State Route 30. Both roads carry relatively low traffic with most of the traffic oriented along State Route 150. The layout of the intersection, coupled with the high-speed traffic on State Route 150, have made this intersection a hazardous location. Drivers on State Route 30 tend to violate the stop sign at the intersection because of the mild turn onto westbound State Route 150, and they also experience difficulty in seeing the high-speed vehicles approaching from the left on State Route 150. Design a new layout for the intersection to eliminate these difficulties for the volumes shown in Figure 7.26b. Design vehicle is a passenger car.

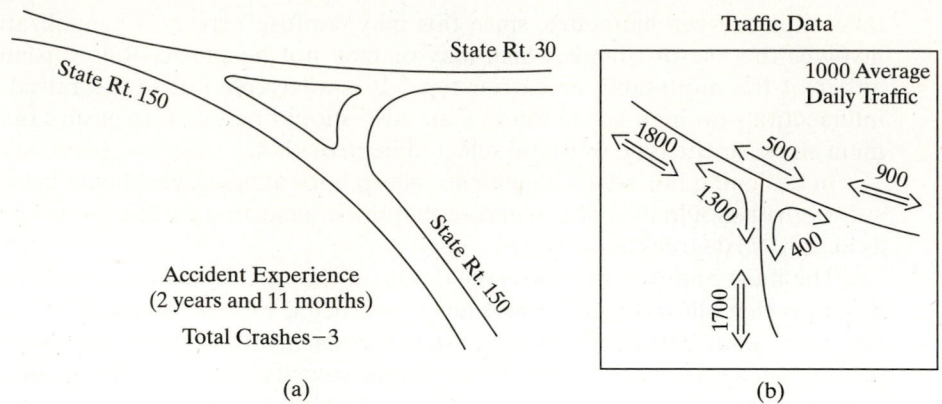

Figure 7.26   Intersection Layout for Problem 7-8

7-9   Figure 7.27a shows the staggered unsignalized intersection of Patton Avenue and Goree Street. The distance between the T intersections is about 48 m. The general layout and striping of the lanes at this intersection result in confusion to drivers and create multiple conflicts. Design an improved layout for the intersection for the traffic volumes shown in Figure 7.27b. Design vehicle is a passenger car.

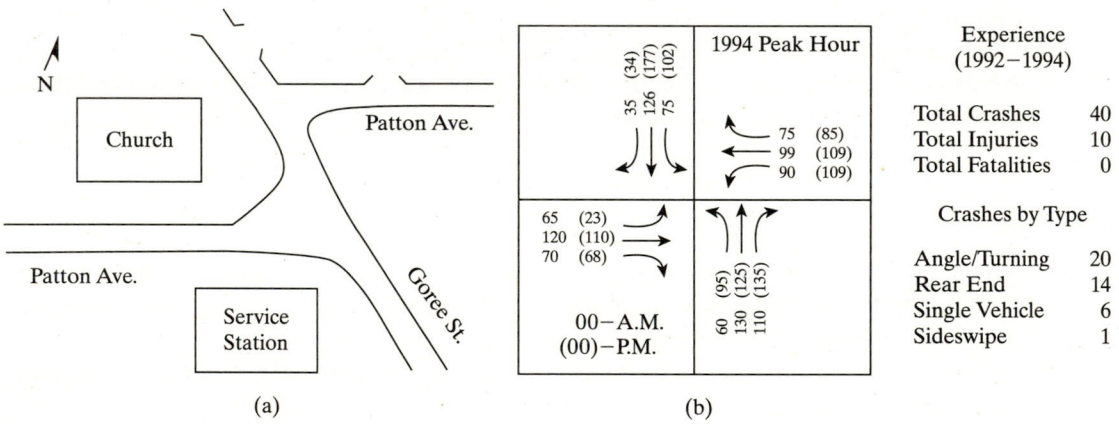

Figure 7.27   Intersection Layout for Problem 7-9

7-10   An existing intersection consists of a turning movement of 75 degrees, and the radius of the edge of pavement is 18 m. The turning radius has no entry taper. What is the largest design vehicle that can be accommodated at the intersection? What improvements would you recommend be made so if this movement could accommodate a WB-67 design vehicle?

7-11   An intersection is being designed for two roads that intersect at approximately 60 degrees. For an SU-30 design vehicle, recommend a design for pavement edge for the right turn movement for the 60-degree turns and the 120-degree turns.

7-12   A ramp from an expressway with a design speed of 48 km/h connects with a local road, forming a T intersection. An additional lane is provided on the local road to allow vehicles from the ramp to turn right onto the local road without stopping. The turning roadway has stabilized shoulders on both sides and will provide for a one-lane, one-way operation with no provision for passing a stalled vehicle. Determine the width of the turning roadway if the design vehicle is a single-unit truck. Use 0.08 for superelevation.

7-13   Determine the required width of the turning roadway in Problem 7-12 for a two-lane operation with barrier curbs on both sides.

7-14   Repeat Problem 7-12 for a one-lane, one-way operation with provision for passing a stalled vehicle.

7-15   A four-leg intersection with no traffic control is formed by two two-lane roads with the speed limits on the minor and major roads being 40 to 64 km/h, respectively. If the roads cross at 90° and a building is to be located at a distance of 15 m from the centerline of the nearest lane on the minor road, determine the minimum distance at which the building should be located from the centerline of the outside lane of the major road so that adequate sight distances are provided. Grades are approximately level.

7-16   A major roadway with a speed limit of 72 km/h has an intersection that has no intersection control with a minor roadway at a right angle. A building is located next to the intersection at a distance of 12 m from the center of the near lane on the major roadway and 12 m from the center of the near lane on the minor roadway. At what speed for the minor roadway is the intersection safe? Grades are approximately level.

7-17   What are the main deficiencies of multi-leg intersections? Using a suitable diagram, show how you would correct for these deficiencies.

7-18   A two-lane minor road intersects a two-lane major road at 90°, forming a four-leg intersection with traffic on the minor road controlled by a yield sign. A building is located 33 m from the centerline of the outside lane of the major road and 12 m from the centerline of the nearest lane of the minor road. Determine the maximum speed that can be allowed on the minor road if the speed limit on the major road is 64 km/h. Grades are approximately level.

7-19   If the speed limit were 72 km/h on the major road, and 40 km/h on the minor road in Problem 7-18, determine the minimum distance that the building can be located from the centerline of the outside lane of the major street.

7-20   A developer has requested permission to build a large retail store at a location adjacent to the intersection of an undivided four-lane major road and a two-lane minor road. Traffic on the minor road is controlled by a stop sign. The speed limits are 56 to 80 km/h on the minor and major roads, respectively. The building is to be located at a distance of 65 ft from the near lane of one of the approaches of the minor road. Determine where the building should be located relative to the centerline of the outside lane of the major road in order to provide adequate sight distance for a driver on the minor road to turn right onto the major road after stopping at the stop line. Design vehicle is a single-unit truck. Lanes on the both roadways are 3.6 m wide. Grades are approximately level.

7-21   Repeat Problem 7-20 for an intersection of a divided four-lane major road and a two-lane minor road if the median width on the major road is 3.6 m and the approach grade on the minor road is 3 percent.

7-22   A minor road intersects a four-lane divided highway at 90° forming a T intersection. The median width on the major road is 2.4 m. The speed limits on the major and minor roads are 88 and 56 km/h, respectively. Determine the minimum sight distance required for a single-unit truck on the minor road to depart from a stopped position and turn left onto the major road.

7-23   What additional consideration should be given to the sight distances computed in Problems 7-20 and 7-22 for the design vehicle crossing the intersection?

7-24   For the information given in Problem 7-22, determine the sight distance required by the minor-road vehicle to safely complete a right turn onto the major road.

7-25   Repeat Problem 7-22 for an oblique intersection with an acute angle of 35°.

7-26   Briefly discuss what factors are considered in the design of an at-grade railway crossing in addition to those for an at-grade crossing formed by two highways.

7-27   A two-lane road crosses an at-grade railroad track at 90°. If the design speed of the two-lane road is 72 km/h and the velocity of the train when crossing the highway is 129 km/h, determine the sight distance leg along the railroad tracks to permit a vehicle traveling at the design speed to safely cross the tracks when a train is observed at a distance equal to the sight distance leg.

7-28   A stop sign controls all vehicles on the highway at a railroad crossing. Determine the minimum distance a building should be placed from the centerline of the tracks to allow a stopped vehicle to safely clear the intersection. Assume that the building is located 10.5 m from the centerline of the near lane. The velocity of trains approaching the crossing is 112 km/h.

REFERENCES

A Policy on Geometric Design of Highways and Streets. American Association of State Highway and Transportation Officials, Washington, D.C., 2004.

A Policy on Geometric Design of Highways and Streets. American Association of State Highway and Transportation Officials, Washington, D.C., 2011.

Manual on Uniform Traffic Control Devices. Federal Highway Administration, Washington, D.C., 2009, Revision 1.

Roundabouts: An Informational Guide. U.S. Department of Transportation, National Cooperative Highway Research Program (NCHRP), Report No. 672, Transportation Research Board, National Research Council, The National Academies, Washington, D.C., Nov. 2011.

Traffic-Control Devices for Passive Railroad-Highway Grade Crossings. Transportation Research Board, National Research Council, National Cooperative Highway Research Program, NCHRP Report 470, Transportation Research Board, Washington, D.C., 2002.

Traffic Control Devices and Rail-Highway Grade Crossings. Transportation Research Board, National Research Council, Transportation Research Record 1114, Washington, D.C., 1987.

CHAPTER 8

Intersection Control

An intersection is an area shared by two or more roads. Its main function is to allow the change of route directions. While Chapter 7 describes the different types of intersections, this chapter also gives a brief description to facilitate a clear understanding of the elements of intersection control. A simple intersection consists of two intersecting roads; a complex intersection serves several intersecting roads within the same area. The intersection is therefore an area of decision for all drivers; each must select one of the available choices to proceed. This requires an additional effort by the driver that is not necessary in non-intersection areas of a highway. The flow of traffic on any street or highway is greatly affected by the flow of traffic through the intersection points on that street or highway because the intersection usually performs at a level below that of any other section of the road.

Intersections can be classified as grade-separated without ramps, grade-separated with ramps (commonly known as interchanges), or at-grade. Interchanges consist of structures that provide for the cross flow of traffic at different levels without interruption, thus reducing delay, particularly when volumes are high. Figure 8.1 shows some types of at-grade intersections; Figure 8.2 shows some innovative types of interchanges. Several types of traffic-control systems are used to reduce traffic delays and crashes on at-grade intersections and to increase the capacity of highways and streets. However, appropriate regulations must be enforced if these systems are to be effective. This chapter describes the different methods of controlling traffic on at-grade intersections and freeway ramps.

CHAPTER OBJECTIVES:

- Understand the basic principles in intersection control.
- Become familiar with the different types of intersection control signs.
- Become familiar with the warrants for installing traffic signals at intersections.
- Understand the procedures for signal timing for different color indications.
- Become familiar with the principles of signal preemption and priority.
- Become familiar with the principles of freeway ramp control.

(a)

(b)

(c)

Figure 8.1   Examples of At-Grade Intersections

SOURCE: Photograph by Winston Lung. University of Virginia. Used with permission.

Figure 8.2a   An Aerial View of a Single-Point Interchange

SOURCE: Connecticut Department of Transportation. Used with permission.

Figure 8.2b   Aerial View of a Diverging Diamond Interchange

SOURCE: Missouri Department of Transportation (MoDOT). Used with permission.

8.1 GENERAL CONCEPTS OF TRAFFIC CONTROL

The purpose of traffic control is to assign the right-of-way to drivers and thus to facilitate highway safety and efficiency by ensuring the orderly and predictable movement of all users of roadway systems, including highways, streets (including private roads open to the public), and bikeways. Control may be achieved by using traffic signals, signs, or markings that regulate, guide, warn, and/or channel traffic. The more complex the maneuvering area is, the greater the need for a properly designed traffic-control system. Many intersections are complex maneuvering areas and therefore require properly designed traffic-control systems. Guidelines for determining whether a particular control type is suitable for a given intersection are provided in the *Manual on Uniform Traffic Control Devices* (MUTCD).

To be effective, a traffic-control device must:

- Fulfill a need
- Command attention
- Convey a clear, simple meaning
- Command the respect of road users
- Give adequate time for proper response

To ensure that a traffic-control device possesses these five properties, the MUTCD recommends that engineers consider the following five factors:

Design. The device should be designed with a combination of size, color, shape, and a simple message that will convey a message and command the respect and attention of the roadway users.

Placement. The device should be located so that it is within the cone of vision of the road user in order to provide adequate visibility and response time for him/her to respond properly and adequately both during the day and night.

Operation. The device should be used in a manner that ensures the fulfillment of traffic requirements in a consistent and uniform way.

Maintenance. Maintenance includes both functional and physical. Functional maintenance involves determining whether a device meets the current traffic conditions, while physical maintenance involves assuring that the legibility and visibility of the device are maintained and the device is functioning properly.

Uniformity. To facilitate the recognition and understanding of these devices by drivers, similar devices should be used at locations with similar traffic and geometric characteristics. It should be noted that merely using a standard device does not imply that uniformity is attained. This is because, a standard device at an inappropriate location may result in a lack of respect by users at locations where the sign is appropriately located.

In addition to these considerations, it is essential that engineers avoid using control devices that conflict with one another at the same location. It is imperative that control devices aid each other in transmitting the required message to the driver.

8.2 CONFLICT POINTS AT INTERSECTIONS

Conflicts occur when traffic streams moving in different directions interfere with each other. The three types of conflicts are merging, diverging, and crossing. Figure 8.3 shows the different conflict points that exist at a four-approach unsignalized intersection.

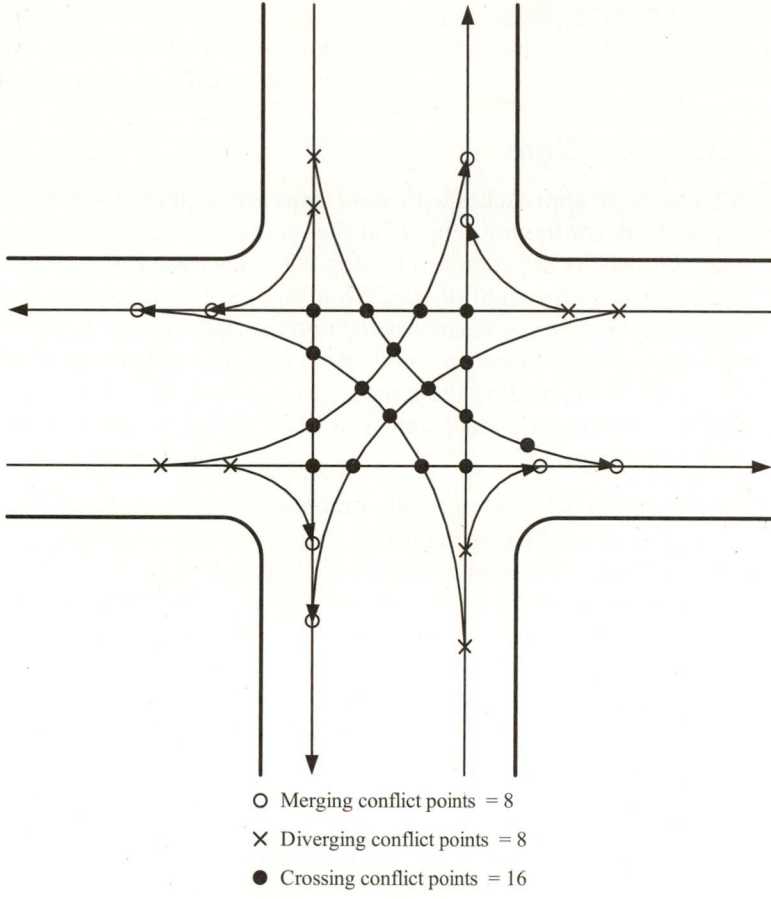

○ Merging conflict points = 8
✕ Diverging conflict points = 8
● Crossing conflict points = 16

Figure 8.3   Conflict Points at a Four-Approach Unsignalized Intersection

There are 32 conflict points in this case. The number of possible conflict points at any intersection depends on the number of approaches, the turning movements, and the type of traffic control at the intersection.

   The primary objective in the design of a traffic-control system at an intersection is to reduce the number of significant conflict points. In designing such a system, it is first necessary to undertake an analysis of the turning movements at the intersection which will indicate the significant types of conflicts. Factors that influence the significance of a conflict include the type of conflict, the number of vehicles in each conflicting stream, and the speeds of the vehicles in those streams. Crossing conflicts, however, tend to have the most severe effect on traffic flow and safety and should be reduced to a minimum whenever possible.

8.3 TYPES OF INTERSECTION CONTROL

Several methods of controlling conflicting streams of vehicles at intersections are in use. The choice of one of these methods depends on the type of intersection and the volume of traffic in each of the conflicting streams. Guidelines for determining whether a particular control type is suitable for a given intersection have been developed and are given in the MUTCD. These guidelines are presented in the form of warrants, which have to be compared with the traffic and geometric characteristics at the intersection being considered. The different types of intersection control are described here.

8.3.1 Yield Signs

All drivers on approaches with yield signs are required to slow down and yield the right-of-way to all conflicting vehicles at the intersection. Stopping at yield signs is not mandatory, but drivers are required to stop when necessary to avoid interfering with a traffic stream that has the right-of-way. Yield signs are therefore usually placed on minor-road approaches where it is necessary to yield the right-of-way to the major-road traffic. Yield signs shall also be placed at roundabout approaches to control traffic on the approaches, but shall not be used on the circulatory roadway. Figure 8.4 shows the regulated shape and dimensions for a yield sign. The MUTCD suggests the following guidelines for the use of yield signs:

A. At approaches to through streets or highways where a stop sign is not required because of the existing traffic and geometric conditions.
B. At intersections with medians of 9 m or higher. The yield sign may be installed at the entrances of both roadways of the divided highway, although the yield sign at the first crossroad may be replaced with a stop sign.
C. At a turn lane that is separated from adjacent travel lanes by a channelized island, even if the intersection is controlled by traffic signals or a stop sign.
D. At intersections where engineering judgment indicates that a specific problem can be solved by the use of a yield sign.
E. At merging lanes where engineering judgment indicates that merging maneuvers may be hampered due to existing conditions such as inadequate acceleration lane length and/or inadequate sight distance.

8.3.2 Stop Signs

A stop sign is used where an approaching vehicle is required to stop before entering the intersection. Stop signs should be used only when they are warranted, since the use of these signs results in considerable inconvenience to motorists. At intersections where a full stop is not necessary at all times, the less restrictive yield sign should be

R1-1

R1-2

Stop Sign Sizes	
Conventional Roads (single lane)	762 mm × 762 mm
Conventional Roads (multilane)	914 mm × 914 mm
Expressways	914 mm × 914 mm
Minimum	762 mm × 762 mm
Oversized	1219 mm × 1219 mm

Yield Sign Sizes	
Conventional Roads (single lane)	914 mm × 914 mm × 914 mm
Conventional Roads (multilane)	1218 mm × 1219 mm × 1219 mm
Expressways	1218 mm × 1219 mm × 1219 mm
Freeways	1524 mm × 1524 mm × 1524 mm
Minimum	762 mm × 762 mm × 762 mm

Figure 8.4   Stop Sign and Yield Sign

SOURCE: Page 5B-2, Manual on Uniform Traffic Control Devices for Streets and Highways - 2003 Edition

first considered. Stop signs should not be used at signalized intersections or on through roadways of expressways.

Figure 8.4 shows the regulated shape and dimensions of a stop sign. The MUTCD suggests that a stop sign should be considered for placement on minor street approaches if engineering judgment indicates that it is required because one or more of the following conditions exist:

A. At intersections where the traffic volumes on the through street or highway is higher than 6000 vehicles per day.
B. At intersections where adequate observation of conflicting traffic on the through street or highway requires drivers to stop.
C. At intersections where three or more crashes that can be avoided by a stop sign have occurred within a period of 12 months, or five or more of these crashes have occurred within a period of 2 years. These crashes include right-angle crashes, in which minor street drivers fail to yield the right-of-way to through traffic.

These guidelines suggest that stop signs may be used on a minor road when it intersects a major road, at an unsignalized intersection, and where a combination of high speed, restricted view, and serious crashes indicates the necessity for such a control.

8.3.3 Multiway Stop Signs

Multiway stop signs require that all vehicles approaching the intersection stop before entering it. They are used as a safety measure at some intersections and normally are used when the traffic volumes on all of the approaches are approximately equal. When traffic volumes are high, however, the use of signals is recommended. The MUTCD suggests consideration should be given to the following guidelines for the placement of multiway stop signs:

A. Multiway stop signs may be used as an interim measure to control traffic at intersections where a signalized system is justified until the signalized system is in place.
B. They should also be considered when five or more crashes occur at an intersection in a 12-month period, and these crashes (e.g., right-turn, left-turn, and right-angle crashes) could be avoided with a multiway stop control.
C. Multiway stop signs may also be used when the following traffic conditions exist:
   a. The volume warrants for this control specify that the average of the total volume on both major street approaches should not be less than 300 veh/h for any eight hours of an average day nor should the combined volume of vehicles, pedestrians, and bicycles from both minor approaches be less than 200 units/h for the same eight hours.
   b. The average delay of the vehicles on the minor street should also be not less than 30 sec/veh during the hour with the maximum volume, when the combined average volume of vehicles, pedestrians and bicycles entering the intersection (total of both approaches) is not less than 200 units per hour for the same 8 hours.
   c. The minimum requirement for vehicular volume can be reduced by 30 percent if the 85th-percentile approach speed on the major street is greater than 65 km/h.
D. If none of the safety or volume criteria (excluding that for the 30 percent reduction for speeds of 65 km/h or more) is fully met, but each satisfies 80 percent of the minimum requirement, the installation of a multiway stop sign is justified.

Example 8.1   Evaluating the Need for a Multiway Stop Sign at an Intersection

A minor road carrying 75 veh/h on each approach for eight hours of an average day crosses a major road carrying 145 veh/h on each approach for the same eight hours, forming a four-leg intersection. Determine whether a multiway stop sign is justified at this location if the following conditions exist:

1. Total pedestrian volume from the approaches of the minor street for the same eight hours as the traffic volumes is 40 ped/h.
2. The average delay to minor-street vehicular traffic during the maximum hours is 27 sec/veh.
3. There are an average of four crashes per year that may be corrected by a multiway stop control.

85th percentile approach speed on the major road = 35 km/h.

Solution:

- Determine whether traffic volume on the major street satisfies the warrant. Total vehicular volume entering the intersection from the major approaches is 145 + 145 = 290 veh/h. The major-road traffic volume criterion is not satisfied.
- Determine whether total minor-road traffic and pedestrian volume satisfies the warrant.
  Total minor-road traffic and pedestrian volume = 2 × 75 + 40 = 190.
  Total minor-road and pedestrian volume is not satisfied.
- Determine whether crash criterion is satisfied.
  Total number of crashes per year = 4.
  Crash criterion is not satisfied.
  Average delay to minor street vehicles = 27 sec < 30 sec (Delay criterion is not satisfied.)
  85th percentile speed on on major street = 35 km/h (70 percent criterion does not apply.)

Check whether 80 percent crash and volume criterion (Guideline D) is met.
Minimum number of crashes = 0.8 × 5 = 4 (Criterion is satisfied.)
Minimum volume requirement on both major street approaches
= 0.8 × 300 = 240 veh/h (Criterion is satisfied.)
Minimum volume of vehicles, pedestrians and bicycles on minor-street approaches requirement = 0.8 × 200 = 160 (Criterion is satisfied.)
Minimum delay = 0.8 × 30 = 24 sec (Criterion is satisfied.)
Each crash and volume criterion is satisfied up to 80 percent of the minimum required. The installation of a multiway stop control is justified.

8.3.4 Intersection Channelization

Intersection channelization is used mainly to separate turn lanes from through lanes. A channelized intersection consists of solid white lines or raised islands that guide traffic within a lane so that vehicles can safely negotiate a complex intersection. When raised islands are used, they can also provide a refuge for pedestrians.

Channelization design criteria have been developed by many individual states to provide guidelines for the cost-effective design of channelized intersections. A detailed description of the techniques that have proven effective for both simple and complicated intersections is given in Chapter 7 and in the *Intersection Channelization Design Guide.* Guidelines for the use of channels at intersections include:

- Laying out islands or channel lines to allow a natural, convenient flow of traffic
- Avoiding confusion by using a few well-located islands
- Providing adequate radii of curves and widths of lanes for the prevailing type of vehicle

8.3.5 Traffic Signals

One of the most effective ways of controlling traffic at an intersection is the use of traffic signals. Traffic signals can be used to eliminate many conflicts because different traffic streams can be assigned the use of the intersection at different times. Since this results in a delay to vehicles in all streams, it is important that traffic signals be used only when necessary. The most important factor that determines the need for traffic signals at a particular intersection is the intersection's approach traffic volume, although other factors such as pedestrian volume and crash experience may also play a significant role. The *Manual on Traffic Signal Design* gives the fundamental concepts and standard practices used in the design of traffic signals. In addition, the MUTCD describes nine warrants in detail, at least one of which should be satisfied for an intersection to be signalized. However, satisfying a warrant does not necessarily mean that a signal should be installed, only that it meets the minimum threshold for considering installing a signal. Therefore, these warrants should be considered only as a guide, and before installing signals at an intersection, an engineering study of traffic conditions, pedestrian characteristics, and physical characteristics should be conducted to ascertain that the signal would provide operational and/or safety benefits. At locations under development or construction, it may not be possible to conduct traffic volume studies that will yield results that are representative of future traffic. If projected volumes are used to justify the installation of signals, a traffic study should be conducted within one year of the start of operation of the signals to ascertain that the signals are justified. If the signals are not justified, they should be taken out of operation. Also, several traffic and/or geometric conditions may exist at an intersection that will require engineering judgment on whether a signal control should or should not be installed. These include:

- Disruption of progressive traffic flow. Installing a signal control at an intersection may result in a disruption to an existing progressive flow. Engineering judgment should be used to evaluate the seriousness of the disruption and then to decide whether installing the signal control is justified.
- Effect of right turns from the minor street approaches. A high number of right turns from the minor street approaches could have an impact on the actual volume that should be considered in evaluating the minor street vehicular volume. Engineering judgment is therefore needed to determine whether the minor street traffic count should be reduced by any portion of the right-turning traffic when the traffic count is being evaluated with respect to the requirements of the signal warrants.
- Approaches consisting of one lane plus one left-turn or right-turn lane. An approach consisting of one lane plus one left-turn or right-turn lane could be considered as a one-lane or two-lane approach, depending on the traffic characteristics at the intersection. Engineering judgment is used to make this decision. For example, engineering judgment could be used to consider an approach consisting of one through and right-turn lane plus a left-turn lane to be a single lane approach if the traffic on the left-turn lane is minor. The total traffic volume is then applied to a single-lane approach in evaluating the traffic count with respect to the requirements of the signal warrants. However, if the left-turn traffic is about 50 percent of the approach traffic and the left-turn lane is long enough to accommodate the left-turn traffic, then the approach is considered a two-lane approach.

- Approaches consisting of one through/left lane plus a right-turn lane. A major influencing factor is the extent to which minor-street right-turn traffic conflicts with the traffic on the major street. Engineering judgment is used to evaluate this degree of conflict, and if it is concluded that this conflict is minimal, then the right-turn traffic volume is not included as part of the minor-street traffic volume and the approach is evaluated as a single-lane approach carrying only the through- and left-turn traffic.

The factors considered in the warrants are:

- Warrant 1. Eight-hour vehicular volume
- Warrant 2. Four-hour vehicular volume
- Warrant 3. Peak hour
- Warrant 4. Pedestrian volume
- Warrant 5. School crossing
- Warrant 6. Coordinated signal system
- Warrant 7. Crash experience
- Warrant 8. Roadway network
- Warrant 9. Intersection near a grade crossing

A brief discussion of each of these is given next. Interested readers are referred to the MUTCD for details. Also, note that a location with a wide median is analyzed as a single intersection if the width of the median is greater than 9 m.

Warrant 1. Eight-Hour Vehicular Volume

This warrant is applied when the principal factor for considering signalization is the high intersection traffic volume. Tables 8.1 and 8.2 give minimum volumes that should exist at the intersection for consideration of traffic signals under this warrant. Either Condition A or B should be satisfied. Condition A considers minimum vehicular volumes on the major and higher-volume minor streets, while Condition B can be used for locations where Condition A is not satisfied but the high volume on the major street causes the traffic on the minor street to experience excessive delay or conflict with major-street traffic while crossing or turning onto the major street.

Condition A (Minimum Vehicular Volume). The warrant is satisfied when traffic volumes on the major street and the higher-volume minor-street approaches for each of any eight hours of an average day are at least equal to the volumes specified in the 100 percent columns of Table 8.1. An "average" day is a weekday whose traffic volumes are normally and repeatedly observed at the location.

Condition B (Interruption of Continuous Flow). The warrant is satisfied when traffic volumes on the major street and on the higher-volume minor-street approaches for each of any eight hours of an average day are at least equal to the volumes specified in the 100 percent columns of Table 8.2.

Also, if both Conditions A and B are not satisfied, a combination of Conditions A and B can be utilized. The criteria for satisfying this combination are:

1. Traffic volume on the major street and the higher-volume minor-street traffic volume are at least equal to the 80 percent values given in Table 8.1 (minimum vehicular volume), and
2. Traffic volume on the major street and the higher-volume minor street traffic volume are at least equal to the 80 percent values given in Table 8.2 (interruption of continuous traffic)

Also, in cases where the 85th-percentile speed on the major street, or the posted or statuary speed limit on the major street is higher than 40 mph, or the intersection is located

Table 8.1 Volume Requirements for Warrant 1, Condition A, Eight-Hour Vehicular Volumes

Number of Lanes for Moving Traffic on Each Approach		Vehicles Per Hour on Major Street (Total of Both Approach)				Vehicles Per Hour on Higher Volume Minor-Street Approach (One Direction Only)			
Major Street	Minor Street	100%[a]	80%[b]	70%[c]	56%[d]	100%[a]	80%[b]	70%[c]	56%[d]
1	1	500	400	350	280	150	120	105	84
2 or more	1	600	480	420	336	150	120	105	84
2 or more	2 or more	600	480	420	336	200	160	140	112
1	2 or more	500	400	350	280	200	160	140	112

Condition A—Minimum Vehicular Volume

[a]Basic minimum hourly volume.
[b]Used for combination of Condition A and B (see Table 8.2) after adequate trial of other remedial measures.
[c]May be used when the major-street speed exceeds 40 mi/h or in an isolated community with a population of less than 10,000.
[d]May be used for combination of Conditions A and B after adequate trial of other remedial measures when the major-street speed exceeds 40 mi/h or in an isolated community with a population of less than 10,000.

SOURCE: Adapted from *Manual on Uniform Traffic Control Devices*, U.S. Department of Transportation, Federal Highway Administration, Washington, DC., 2009, http://mutcd.fhwa.dot.gov/pdfs/2009/part4.pdf

Table 8.2 Volume Requirements for Warrant 1, Condition B, Interruption of Continuous Traffic

Condition B—Interruption of Continuous Traffic

Number of Lanes for Moving Traffic on Each Approach		Vehicles Per Hour on Major Street (Total of Both Approach)				Vehicles Per Hour on Higher Volume Minor-Street Approach (One Direction Only)			
Major Street	Minor Street	100%[a]	80%[b]	70%[c]	56%[d]	100%[a]	80%[b]	70%[c]	56%[d]
1	1	750	600	525	420	75	60	53	42
2 or more	1	900	720	630	504	75	60	53	42
2 or more	2 or more	900	720	630	504	100	80	70	56
1	2 or more	750	600	525	420	100	80	70	56

[a]Basic minimum hourly volume.
[b]Used for combination of Condition A (see Table 8.1) and B after adequate trial of other remedial measures.
[c]May be used when the major-street speed exceeds 40 mi/h or in an isolated community with a population of less than 10,000.
[d]May be used for combination of Conditions A and B after adequate trial of other remedial measures when the major-street speed exceeds 40 mi/h or in an isolated community with a population of less than 10,000.

SOURCE: Adapted from *Manual on Uniform Traffic Control Devices*, U.S. Department of Transportation, Federal Highway Administration, Washington, DC., 2009, http://mutcd.fhwa.dot.gov/pdfs/2009/part4.pdf

in an isolated community with a population of less than 10,000, the 56 percent column values can be used instead of the 80 percent column values when the combination of Conditions A and B is utilized.

Note that for these conditions, the traffic volumes for the major-street and the high-volume minor-street approaches should be for the same eight hours, but the higher volume on the minor street does not have to be on the same approach during each of the eight hours being considered.

Warrant 2. Four-Hour Vehicular Volume

This warrant is considered at locations where the main reason for installing a signal is the high intersecting volume. It is based on the comparison of standard graphs given in

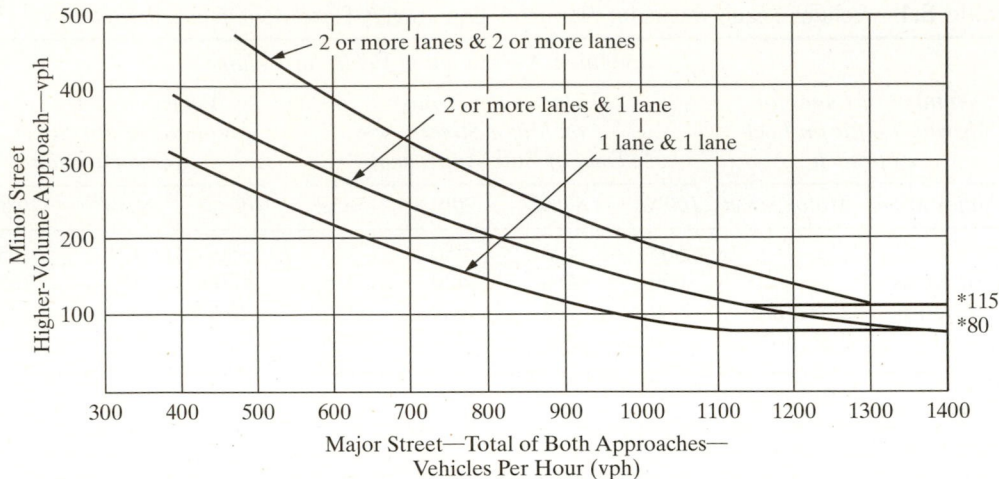

Figure 8.5   Graphs for Four-Hour Vehicular Volume Warrant

Note: 115 vph applies as the lower threshold volume for a minor-street approach with two or more lanes and 80 vph applies as the lower threshold volume for a minor-street approach with one lane.

SOURCE: Redrawn from *Mannual on Uniform Traffic Control Devices*, U.S. Department of Transportation, Federal Highway Administration, Washington D.C., 2009

the MUTCD, one of which is shown in Figure 8.5. This figure shows a plot of the total volume in veh/h on the approach with the higher volume on the minor street (vertical scale) against the total volume in veh/h on both approaches of the major street (horizontal scale) at the intersection. When the plot for each of any four hours of an average day falls above the appropriate standard graph, this warrant is satisfied. Standard plots are given for different types of lane configurations at the intersection. The MUTCD also gives a different set of standard plots for intersections where the 85th-percentile speed of the major-street traffic is higher than 40 mi/h or where the intersection is located in a built-up area of an isolated community whose population is less than 10,000.

Warrant 3. Peak Hour

This warrant is used to justify the installation of traffic signals at intersections where traffic conditions during one hour or longer of an average day result in undue delay to traffic on the minor street entering or crossing the intersection. One of two conditions (A or B) should be satisfied for the warrant to be satisfied. The conditions are:

A. The three following criteria should be satisfied for the same hour (any four consecutive 15-minute periods) of an average day:

1. The total stopped delay during any four consecutive 15-minute periods on one of the minor-street approaches (one direction only) controlled by a stop sign is equal to or greater than 4 veh-h for a one-lane approach and 5 veh-h for a two-lane approach.
2. The same minor-street approach (one direction only) volume should be equal to or exceed 100 and 150 vehicles per hour for one moving lane of traffic and two moving lanes of traffic, respectively.
3. The total intersection entering volume is equal to or greater than 650 vehicles per hour for three-leg intersections and 800 vehicles per hour for four-leg and multileg intersections.

B. Figure 8.6 illustrates Condition B. This condition is satisfied when the plot of the vehicles per hour on the major street (total of both approaches) and the corresponding vehicles per hour on the higher-volume minor-street approach (one direction only) is above the appropriate curve in Figure 8.6. At locations where the speed on the major street is higher than 40 mi/h, or where the intersection is within an isolated built-up area with a population less than 10,000, Figure 8.7 may be consulted.

This warrant is only applicable at locations where large number of vehicles are attracted or discharged during a short time period. These include office complexes, manufacturing plants, industrial complexes or high-occupancy vehicle facilities.

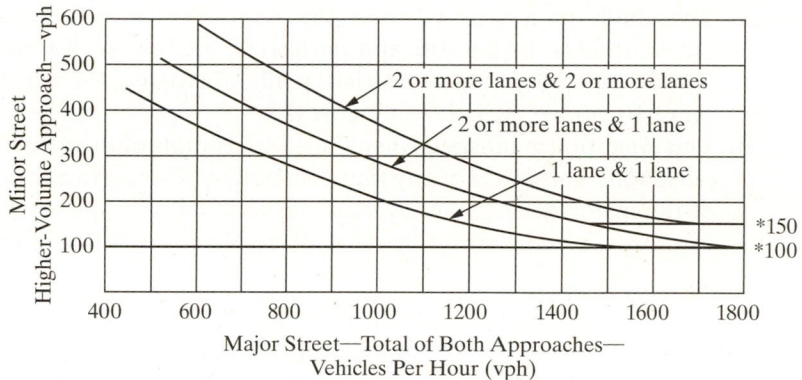

Figure 8.6   Graphs for Peak Hour Volume Warrant

Note: 150 vph applies as the lower threshold volume for a minor-street approach with two or more lanes and 100 vph applies as the lower threshold volume for a minor-street approach with one lane.

SOURCE: Redrawn from *Manual on Uniform Traffic Control Devices*, U.S. Department of Transportation, Federal Highway Administration, Washington, D.C., 2009

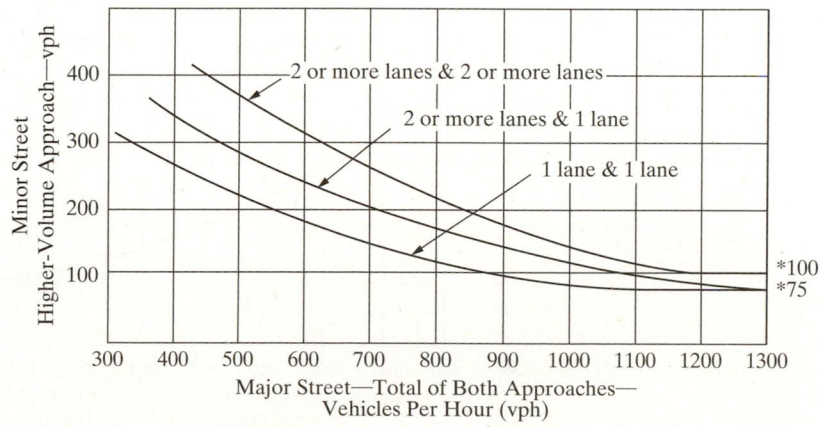

Figure 8.7   Graphs for Peak Hour Volume Counts (70% Condition)

Note: 100 vph applies as the lower threshold volume for a minor-street approach with two or more lanes and 75 vph applies as the lower threshold volume for a minor-street approach with one lane.

SOURCE: Redrawn from *Manual on Uniform Traffic Control Devices*, U.S. Department of Transportation, Federal Highway Administration, Washington, D.C., 2009

Warrant 4. Minimum Pedestrian Volume

This warrant is used to justify the installation of traffic signals at intersections or mid-block crossings where traffic conditions result in excessive delay to pedestrians wishing to cross the major street. One of two conditions (A or B) should be satisfied for this warrant to be met. Figures 8.8 and 8.9 are two of four charts that can be used. Figure 8.8 shows a plot of the total of all pedestrians crossing the major street (pedestrians per hour, vertical scale) over a four-hour period versus the corresponding total vehicles per hour on both approaches of the major street (vehicles per hour horizontal scale). This figure represents the lowest threshold of 107 ped/h. Figure 8.9 is similar to Figure 8.8, but is based on the peak hour only and has a lower threshold volume of 133 ped/h.

A. This condition is satisfied when the plotted points of the vehicles per hour on the major street (total for both approaches) for each of any four hours of an average day and the corresponding pedestrians per hour crossing the major street (one direction only) are above the curve in Figure 8.8.

B. This condition is satisfied when the plotted points of the vehicles per hour during the peak hour, consisting of any four consecutive 15-minute periods on both approaches of the major street on an average day and the corresponding pedestrians per hour crossing the major street (total of all crossings), are above the curve in Figure 8.9.

In cases where the 85th-percentile speed on the major street, or the posted or statuatory speed limit on the major street is higher than 55 km/h, or the intersection is located in an isolated community with a population of less than 10,000, alternative figures with lower threshold values than those for Figures 8.8 and 8.9 are provided for use in the MUTCD. The MUTCD also suggests that when the 15th-percentile crossing speed of pedestrians is less than 1.05 m/sec, the pedestrian volume criterion may be reduced by up to 50 percent.

Also, note that the nearest traffic signal or stop sign along the major street should be at least 90 m away from the proposed intersection for the installation of a signal system, unless the installation of the proposed traffic signal will not negatively impact the progression of traffic. When this warrant is used, the signal should be of the traffic-actuated type with pushbuttons for pedestrian crossing.

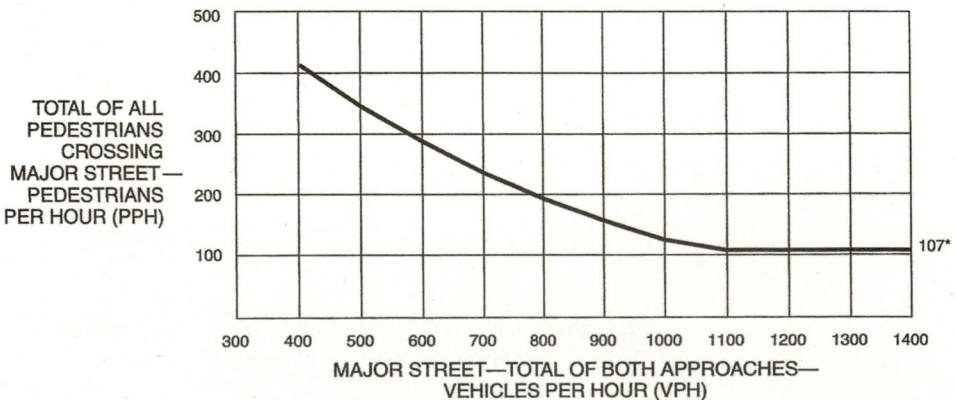

Figure 8.8   Graph for Pedestrian Four-Hour Volume

Note: 107 pph applies as the lower threshold volume.
SOURCE: Adapted from *Manual on Uniform Traffic Control Devices*, U.S. Department of Transportation, Federal Highway Administration, Washington, D.C., 2009, http://fmutcd.fhwa.dot.gov/pdfs/2009rlr2/part4.pdf

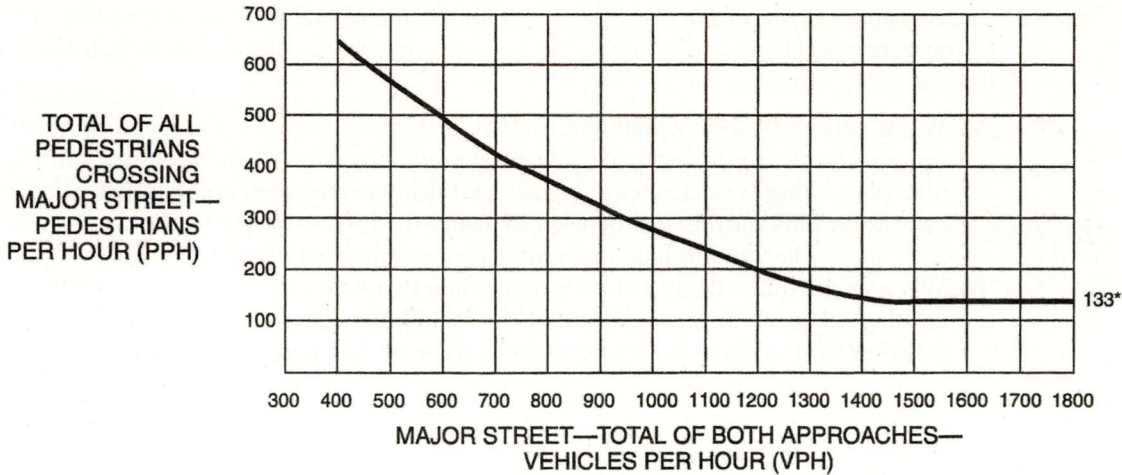

Figure 8.9   Graph for Pedestrian Peak-Hour Volume

Note: 133 pph applies as the lowest threshold volume.
SOURCE: Adapted from *Manual on Uniform Traffic Control Devices*, U.S. Department of Transportation, Federal Highway Administration, Washington, D.C., 2009, http://fmutcd.fhwa.dot.gov/pdfs/2009rlr2/part4.pdf

Warrant 5. School Crossing

This warrant is used when the main reason for installing a traffic signal control is to accommodate the crossing of the major street by schoolchildren (elementary through high school students). When an analysis of gap data at an established school zone shows that the frequency of occurrence of gaps and the lengths of gaps are inadequate for safe crossing of the street by schoolchildren, this warrant is applied. It stipulates that if during the period when schoolchildren are using the crossing, the number of acceptable gaps is less than the number of minutes in that period and there are at least 20 students during the highest crossing hour, the use of traffic signals is warranted. The signal in this case should be pedestrian actuated, and all obstructions to view (such as parked vehicles) should be prohibited for at least 30 m before and 6.1 m after the crosswalk. Also, this warrant is not applicable at locations that are less than 90 m from the nearest traffic control signal along the major road, unless the progression of traffic will not be negatively impacted by the installation of the proposed traffic control signal. When a proposed signal control is being considered at a non-intersection crossing based on this warrant, it should be located at least 100 ft from side streets or driveways with stop or yield controls. The MUTCD recommends that prior to the installation of traffic signals justified by this warrant, other remedial measures (e.g., warning signs and flashers, school speed zones, school crossing guards, or grade separated crossing) shall be considered.

Warrant 6. Coordinated Signal System

This warrant may be used to justify the installation of traffic signal control at an intersection where otherwise it would not have been installed, but the installation will enhance the progressive movement of traffic along a highway segment with a coordinated traffic signal system. This warrant, however, is not applicable when the resultant spacing of the traffic signals will be less than 300 m. One of two criteria (A or B) should be satisfied for this warrant to apply.

A. When adjacent traffic signal controls are at long distances from each other, on a one-way street, or on a street with traffic predominantly in one direction, unfavorable platooning (vehicles traveling at fixed distance between consecutive vehicles) can result. This warrant can be used to justify the installation of traffic signals when such an installation will help maintain a proper grouping of vehicles.

B. When an adequate degree of platooning on a two-way street is not provided by the existing adjacent traffic signals, and the installation of the proposed traffic signal together with the existing traffic signals will result in a better progression of traffic.

Warrant 7. Crash Experience

This warrant is used when the purpose of installing a traffic signal control is to reduce the number and severity of crashes at the intersection. The following three criteria should be achieved for this warrant to be justified.

A. Crash frequency has not been reduced by an adequate trial of less restrictive measures.

B. Five or more injury or reportable property-damage-only crashes have occurred within a 12-month period, and that signal control is a suitable countermeasure for these crashes.

C. The traffic volume should not be less than those given in both of the 80 percent columns of the requirements specified in the minimum vehicular volume warrant (see Table 8.1), or the 80 percent columns of the interruption of continuous traffic warrant (see Table 8.2), or the minimum pedestrian volume is at least equal to 80 percent requirements for the pedestrian volume warrant. Note that the volumes for the major street and the high-volume minor street should be for the same eight hours, but the higher volume on the minor street does not have to be on the same approach during each of the eight hours being considered.

In cases where the 85th-percentile speed on the major street is higher than 65 km/h or the intersection is located in an isolated community with a population of less than 10,000, the volumes shown under the 56 percent columns of Tables 8.1 and 8.2 may be used instead of the 80 percent column.

Warrant 8. Roadway Network

This warrant justifies the installation of signals at some intersections when such an installation will help to encourage concentration and organization of traffic flow on the networks. The warrant is justified when one of two criteria (A or B) is satisfied.

A. The intersection of two or more major roads has a total existing or immediately projected entering volume of at least 1000 vehicles during the peak hour of a typical weekday and five-year projected traffic volumes, based on an engineering study, satisfy one or more of the requirements of the following warrants: eight-hour vehicular volume, four-hour vehicular volume, and peak hour volume during an average weekday.

B. The total existing volume or immediately projected entering volume is 1000 veh/h or greater for each of any five hours of a non-normal business day such as Saturday or Sunday.

A major street considered for this warrant shall possess at least one of the following characteristics:

- It is a component of a street or highway system that serves as the principal roadway network for through traffic flow.
- It is a component of a street or highway system that includes rural or suburban highways outside, entering, or traversing a city.
- It is designated as a major route on an official transportation plan or study.

Warrant 9. Intersection Near a Grade Crossing

This warrant is considered at stop or yield control intersections, near rail road grade crossings where the criteria for installing a traffic signal control for all of the eight warrants discussed above are not met. Also, a grade crossing is located relatively close to an intersection, such that consideration should be given to the installation of a traffic-control signal. However, installation of a traffic signal should only be considered if alternatives like extending the storage length or changing which approach must stop are not feasible. The warrant is justified if the following two criteria are met.

A. The center of the rail track closest to the stop or yield controlled intersection is located at a distance of 40 m or less from the stop or yield line of the approach.

B. The plot of the highest number of vehicles per hour on the major street (total of both approaches) when rail traffic is using the crossing and the corresponding minor street volume in vehicles per hour that crosses the track (one direction only, approaching the rail track) is located above the appropriate chart shown in Figures 8.10 or 8.11 for the appropriate clear storage distance (D), and there is a one-approach lane at the track (Figure 8.10) or two or more approach lanes (Figure 8.11). The MUTCD defines the clear storage distance (D) as the distance that can be used for the storage of vehicles, and it is the distance between 1.8 meter from the rail nearest the intersection to the intersection stop line or

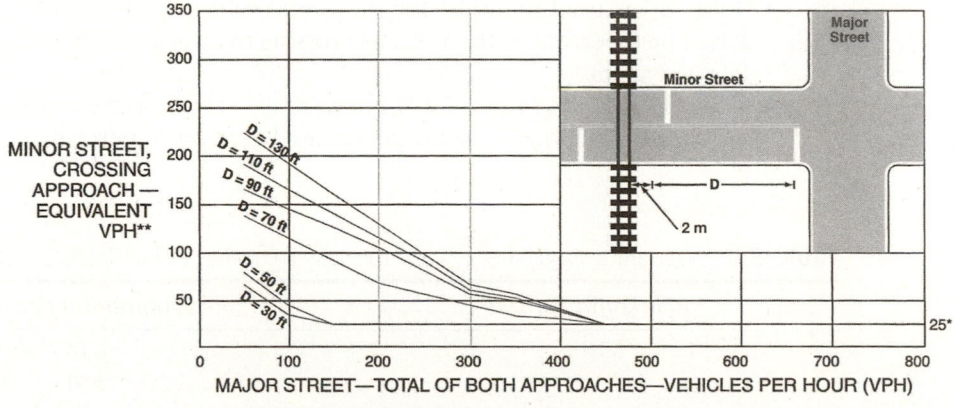

Figure 8.10   Chart for Intersection Near a Grade Crossing (One Approach Lane at the Track)

SOURCE: Adapted from *Manual on Uniform Traffic Control Devices*, U.S. Department of Transportation, Federal Highway Administration, Washington, D.C., 2009, http://fmutcd.fhwa.dot.gov/pdfs/2009rlr2/part4.pdf

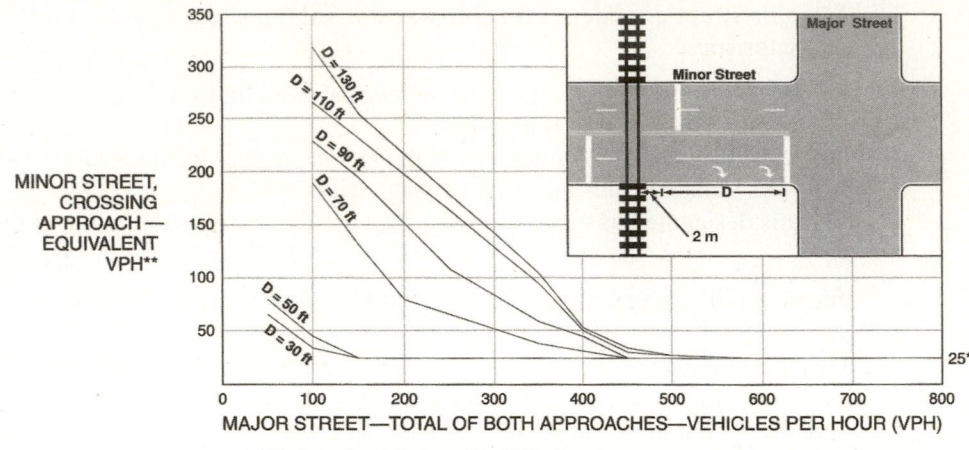

* 25 vph applies as the lower threshold volume
** VPH after applying the adjustment factors in Tables 4C-2, 4C-3, and/or 4C-4, if appropriate

Figure 8.11 Chart for intersection Near a Grade Crossing (Two or More Approach Lanes at the Track)

SOURCE: Adapted from *Manual on Uniform Traffic Control Devices*, U.S. Department of Transportation, Federal Highway Administration, Washington, D.C., 2009, http://fmutcd.fhwa.dot.gov/pdfs/2009rlr2/part4.pdf

the normal stopping point of the highway. Note that when using Figure 8.10 or 8.11, the curve used for distance D is that nearest to the actual value of D determined. For example, if the actual distance is 35 m, the curve for D = 30 m should be used. Also, the minor street approach volumes should be multiplied by adjusted factors given in Tables 8.3, 8.4, and 8.5 to account for the basic traffic characteristics used to develop the curves in Figures 8.10 and 8.11, as indicated below:

- Table 8.3 is used to adjust for the number of rail traffic per day, as the curves were developed for an average of four per day.
- Table 8.4 is used to adjust for the actual vehicle occupancy, as the curves are based on 2 percent of the vehicles crossing the tracks are buses carrying at least 20 passengers.
- Table 8.5 is used to adjust for the percentage of tractor-trailer trucks in the traffic crossing the rail tracks, as the curves are based on 10 percent.

Table 8.3 Warrant 9, Adjustment Factor for Daily Frequency of Rail Traffic

Rail Traffic per Day	Adjustment Factor
1	0.67
2	0.91
3 to 5	1.00
6 to 8	1.18
9 to 11	1.25
12 or more	1.33

SOURCE: *Manual on Uniform Traffic Control Devices*, U.S. Department of Transportation, Federal Highway Administration, Washington, D.C., 2009, http://mutcd.fhwa.dot.gov/HTM/2009rl/part4/part4c.htm

Table 8.4   Warrant 9, Adjustment Factor for Percentage of High-Occupancy Buses

Percent of High-Occupancy Buses* on Minor-Street Adjustment Factor	Approach
0%	1.00
2%	1.09
4%	1.19
6% or more	1.32

*A high-occupancy bus is defined as a bus occupied by at least 20 people.

SOURCE: *Manual on Uniform Traffic Control Devices*, U.S. Department of Transportation, Federal Highway Administration, Washington, D.C., 2009, http://mutcd.fhwa.dot.gov/HTM/2009rl/part4/part4c.htm

Table 8.5   Warrant 9, Adjustment Factor for Percentage of Tractor-Trailer Trucks

Percent of Tractor-Trailer Trucks on Minor-Street Approach	Adjustment Factor	
	D less than 20 meter	D of 20 meter or more
0% to 2.5%	0.50	0.50
2.6% to 7.5%	0.75	0.75
7.6% to 12.5%	1.00	1.00
12.6% to 17.5%	2.30	1.15
17.6% to 22.5%	2.70	1.35
22.6% to 27.5%	3.28	1.64
More than 27.5%	4.18	2.09

SOURCE: *Manual on Uniform Traffic Control Devices*, U.S. Department of Transportation, Federal Highway Administration, Washington, D.C., 2009, http://mutcd.fhwa.dot.gov/HTM/2009rl/part4/part4c.htm

Example 8.2 Determining Whether the Conditions at an Intersection Warrant Installing Traffic Signal Controls

A two-lane minor street crosses a four-lane major street. If the traffic conditions are as given, determine whether installing a traffic signal at this intersection is warranted.

1. The traffic volumes for each eight hours of an average day (both directions on major street) total 400 veh/h. For the higher volume minor-street approach (one direction only), the total is 100 for each of the eight hours.
2. The 85th-percentile speed of major-street traffic is 33 mi/h.
3. The pedestrian volume crossing the major street during each of any four hours of an average day is 450 ped/h. The nearest traffic signal is located 137 m from this location.

Solution: Since the eight-hour vehicular and pedestrian volumes are given, these warrants are checked to determine whether the conditions of either are satisfied.

• Check minimum vehicular volume condition (Warrant 1, Condition A). Minimum volume requirements for this intersection (two lanes for moving traffic on major street and one lane on minor street) are 600 veh/h on the major street (total of both approaches) and 150 veh/h on the minor street (one direction only) (see Table 8.1). Volumes at the intersection are less than the minimum volumes; that is, $400 < 600$ and $100 < 150$. Note that the 85th-percentile speed is 53 km/h < 65 km/h, and the 70 percent minimum volumes cannot be used.

- Check minimum vehicular volume condition (Condition B).
  Note that it is not necessary to check for Condition B, as the required major-street volume is higher than those for Condition A.
- Check minimum vehicular volume condition (combination of Conditions A and B).
  In this case, the 80 percent volumes can be used if adequate trial of other remedial measures has been made. (See Tables 8.1 and 8.2 footnotes.)
  Minimum volume for major road:
  Condition A = 480 $(400 < 480)$ condition is not met (see Table 8.1)
  Condition B = 720 $(400 < 720)$ condition is not met (see Table 8.2)
  The minimum vehicular volumes warrant therefore is not satisfied.
- Check minimum pedestrian volume (Warrant 4).
  Using Figure 8.8, the minimum pedestrian volume required for major-street total of both approaches is approximately 410, which is less than 450. The minimum requirement for pedestrian volumes is satisfied.
- Check location of nearest traffic signal.
  For the location of the nearest traffic signal along the major street, the minimum distance required is 90 m. The nearest traffic signal to the intersection is 137 m away; $137 > 90$. Thus, the location of the nearest traffic signal condition is satisfied.

A traffic signal therefore is justified under the minimum pedestrian volume warrant. The signal should be of the traffic-actuated type with pushbuttons for pedestrians who are crossing.

Example 8.3 Determining the Necessity of Installing a Traffic Signal at an Intersection Near a Grade Crossing

A large residential development is being proposed for a site that is east of an existing four-lane highway that runs in the north-south direction. A two-lane access road is planned to serve traffic leaving and entering the proposed development and provide access to a shopping center located about a mile west of the existing road. This will require the proposed access road to intersect with the existing road and then cross a railroad track downstream of the west-bound traffic. Also, the construction schedule for the proposed development indicates that full usage of the facilities at the proposed development will occur in three years' time. Determine whether a traffic signal system should be used at the intersection of the proposed access road and the existing road if the following conditions exist.

- Current vehicles per hour for each hour of an eight-hour duration on an average day on each approach of the existing four-lane highway = 100 veh/h
- Current vehicles per hour for each hour of a four-hour duration on an average day on each approach of the existing four-lane highway = 130 veh/h
- Estimated vehicles per hour for each hour of an eight-hour duration on an average day on the higher-volume approach of the access (minor) road at opening of the road = 50 veh/h

- Estimated vehicles per hour for each hour of a four-hour duration on an average day on the higher-volume approach access (minor) road at opening of the road = 55 veh/h
- 85th percentile speed on the existing four-lane road = 65 km/h
- Distance from the edge of the rail track closest to the proposed intersection = 23.8 m
- Estimated traffic growth rate on the existing road = 3.0%
- Clearance storage distance (D) = 22 m = (23.8 − 1.8) m (see Figure 8.10)
- Number of rail traffic/day = 7
- Percentage of high-occupancy buses on the access road approach = 0%
- Percent of tractor-trailer trucks on the access road approach = 2%

Solution: Since only volume data are given, Warrant 1 (eight-hour vehicular volume) and Warrant 2 (four-hour vehicular volume) should first be checked.

- Check eight-hour vehicular volume (Warrant 1) requirement.
  Current volume on existing road = 100 veh/h
  Traffic growth rate = 3.0%
  Traffic volume at opening of access road = $100(1.03)^3 = 109$
  Total volume on both approaches of the existing road = $109 \times 2 = 218$ veh/h
  Estimated volume on proposed access (minor) road = 50 veh/h
  The hourly vehicular volumes do not satisfy any of the requirements for conditions A or B. (See Tables 8.1 and 8.2.) Warrant 1 is therefore not satisfied.
- Check four-hour vehicular volume (Warrant 2) requirement.
  Current volume on existing road = 130 veh/h
  Traffic growth rate = 3.0%
  Traffic volume at opening of access road = $130(1.03)^3 = 142$ veh/h
  Total volume on both approaches of the existing road = $142 \times 2 = 284$ veh/h
  Estimated higher approach volume on proposed access (minor) road = 55 veh/h
  The plot of these volumes on Figure 8.5 is below the appropriate graph (2 or more lanes and 1 lane). Warrant 2 is therefore not satisfied.
- Since the distance of the intersection is less than 140 ft (80.5 ft) and neither Warrant 1 nor Warrant 2 is satisfied, then Warrant 9 (intersection near a grade crossing) is checked:
  - Major street volume − total volume of both approaches = 284 veh/h
    Minor-street volume = 55 veh/h
    Adjustment for frequency of rail traffic = 1.18 (7/day; see Table 8.3)
    Adjustment for percentage of high-occupancy buses = 1.0 (see Table 8.4)
    Adjustment for percent of tractor-trailer trucks = 0.5
    (D = 72 ft, 2 percentage tractor-trailer trucks; see Table 8.5)
    Adjusted minor-street crossing approach equivalent volume
    $= 55 \times 1.18 \times 1 \times 0.5 = 32.45$
    Plot of minor-street crossing approach volume versus major-street total of both approaches falls below appropriate graph for D = 70 ft (see Figure 8.10). Warrant 9 is not satisfied.

The traffic conditions do not satisfy any of the relevant warrants for a signalized intersection. A stop sign or yield sign should be considered.

8.4 SIGNAL TIMING FOR DIFFERENT COLOR INDICATIONS

The warrants described earlier will help the engineer only in deciding whether a traffic signal should be used at an intersection. The efficient operation of the signal also requires proper timing of the different color indications, which is obtained by implementing the necessary design. Before presenting the different methods of signal timing design, however, first it is necessary to define a number of terms commonly used in the design of traffic signal times.

1. *Controller.* A device in a traffic signal installation that changes the colors indicated by the signal lamps according to a fixed or variable plan. It assigns the right-of-way to different approaches at appropriate times. The National Electrical Manufacturers' Association (NEMA) TS2 standard, which is an updated version of the NEMA TS1 standard, is incorporated in many of the controllers currently in use. Basic specifications for interval (pretimed) and phase (actuated) controllers are included in this standard. It includes definitions for the functionality of actuated signal controllers in the areas of the arrangement of phases, overlapping, the selection of phases, and the option of permitting pedestrians to start walking other than at the start of the green phase. It does not, however, include standards for the physical size, shape, or appearance of most components, except in cases where whole components from different manufacturers have to be changed. The controllers that meet the physical requirements specified in NEMA TS2 are usually referred to as NEMA controllers. Another set of standards known as the Advanced Transportation Controller (ATC) family of standards is maintained by a consortium of NEMA, the Institute of Transportation Engineers (ITE), and the American Association of State Highway and Transportation Officials (AASHTO). The ATC standard was developed as a result of a new direction in traffic signals initiated by the passage of the Intermodal Surface Transportation Act (ISTEA). The ATC 2070 standard, which is also commonly used, is based on the specifications of the 2070 controller developed by the California Department of Transportation (Caltrans). A significant difference between the ATC 2070 standard and the NEMA TS2 standard is that the ATC 2070 includes specifications for every detail of the controller hardware and internal subcomponents, and it also does not give specifications for the functionality of any software that is used. Caltrans and the New York Department of Transportation (NYDOT) developed the Model 170 specifications, which also do not include specifications for the functionality of any software used, but include specifications for cabinet hardware and all components, including the controller. The main disadvantage of the Model 170 is its limited ability to support advanced software applications.
2. *Cycle (cycle length).* The time in seconds required for one complete color sequence of signal indication. Figure 8.12 is a schematic of a cycle. In Figure 8.12, for example, the cycle length is the time that elapses from the start of the green indication to the end of the red indication.
3. *Phase (signal phase).* That part of a cycle allocated to a stream of traffic or a combination of two or more streams of traffic having the right-of-way simultaneously during one or more intervals (see Figure 8.12).
4. *Interval.* Any part of the cycle length during which signal indications do not change.
5. *Change interval.* The total length of time in seconds of the yellow change and red clearance intervals that occur between phases. This time is provided for vehicles to clear the intersection after the green interval before conflicting movements are released.

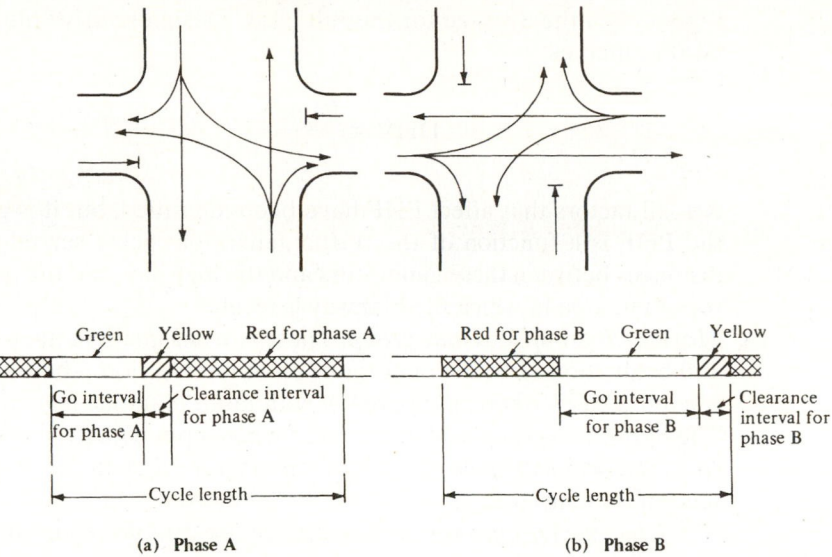

Figure 8.12   Two-Phase Signal System

6. **Yellow change interval.** This is typically taken as the sum of the driver's perception-reaction time and the time required for a driver traveling at the speed limit of the road to safely stop or safely go through the intersection. The equation for determining the yellow change interval is given in Section 8.4.4.

7. **Red clearance interval.** This is a display time of a red indication that appears after the yellow change interval. It is optionally provided to give additional time before conflicting movements are given the green indication. This is sometimes referred to as the "all red interval."

8. **Red time.** The time within the cycle during which the signal indication is red for a given phase.

9. **Split phase.** The part of a phase that is set apart from the primary movement, thus forming a special phase that relates to the parent phase.

10. **Effective green time.** The time during which a combination of traffic streams could effectively move at the saturation flow rate.

11. **Effective red time.** This is the difference between the cycle length and the effective green time. It is the time period during which a combination of traffic movements do not effectively move at the saturation flow.

12. **Extension of effective green.** The portion of time at the beginning of the yellow change interval during which a combination of traffic streams effectively move at the saturation flow rate.

13. **Peak hour factor (PHF).** A measure of the variability of demand during the peak hour. It is the ratio of the volume during the peak hour to the maximum rate of flow during the peak 15-minute period within the peak hour. The PHF is given as

$$\text{PHF} = \frac{\text{Volume during peak hour}}{4 \times \text{volume during peak 15 min within peak hour}} \tag{8.1}$$

The PHF may be used in signal timing design to compensate for the possibility that peak arrival rates for short periods during the peak hour may be much higher than the average for the full hour. Design hourly volume (DHV) can then be obtained as

$$DHV = \frac{\text{Peak hour volume}}{\text{PHF}} \qquad (8.2)$$

Not all factors that affect PHF have been identified, but it is generally known that the PHF is a function of the traffic generators being served by the highway, the distances between these generators and the highway, and the population of the metropolitan area in which the highway is located.

14. ***Movement group and lane group.*** These two designations are used in the description and evaluation of traffic operations at an intersection. Their definitions are similar except in cases where an approach with more than one lane also has a shared lane. The *movement group* is of importance when input data for the intersection are being specified, while the *lane group* is of importance when the level of service at the intersection is being computed.

    Movement groups for each approach are established using the following guidelines:

    • A turning movement on one or more exclusive lanes with no shared lanes should be designated a movement group.
    • All lanes that cannot be designated within a movement group based on the definition above should be classified into one movement group.

    Lane groups for each approach are established using the following guidelines:

    • Separate lane groups should be established for exclusive left-turn lane(s).
    • Separate lane groups should be established for exclusive right-turn lane(s).
    • A separate lane group should be established for each shared lane.
    • A single lane group should be established for all lanes that are neither exclusive turn lanes or shared lanes.
    • When exclusive left-turn lane(s) and/or exclusive right-turn lane(s) are provided on an approach, all other lanes are generally established as a single lane group.
    • When an approach with more than one lane also has a shared left-turn lane, the operation of the shared left-turn lane should be evaluated as shown in Section 10.3 of Chapter 10 to determine whether it is effectively operating as an exclusive left-turn lane because of the high volume of left-turn vehicles on it.

    Figure 8.13 shows typical movement and lane groups. Note that when two or more lanes have been established as a single lane group for analysis, all subsequent computations must consider these lanes as a single entity.

15. ***Critical lane group.*** The lane group that requires the longest green time in a phase; that is, the lane group with the highest flow ratio in a phase. This lane group, therefore, determines the green time that is allocated to that phase.

16. ***Adjusted saturation flow rate.*** The flow rate in veh/h that the lane group can carry if it has the green indication continuously (i.e., if $g/C = 1$) and no lost time. The adjusted saturation flow rate depends on an ideal saturation flow (S_0), which is usually taken as 1900 veh/h of green time per lane. The ideal saturation flow is then adjusted for the prevailing conditions to obtain the saturation flow for the lane group being considered. The adjustment is made by introducing factors that adjust for the number of lanes, lane width, the percent of heavy vehicles in the traffic, approach grade, parking activity, local buses stopping within the intersection, area

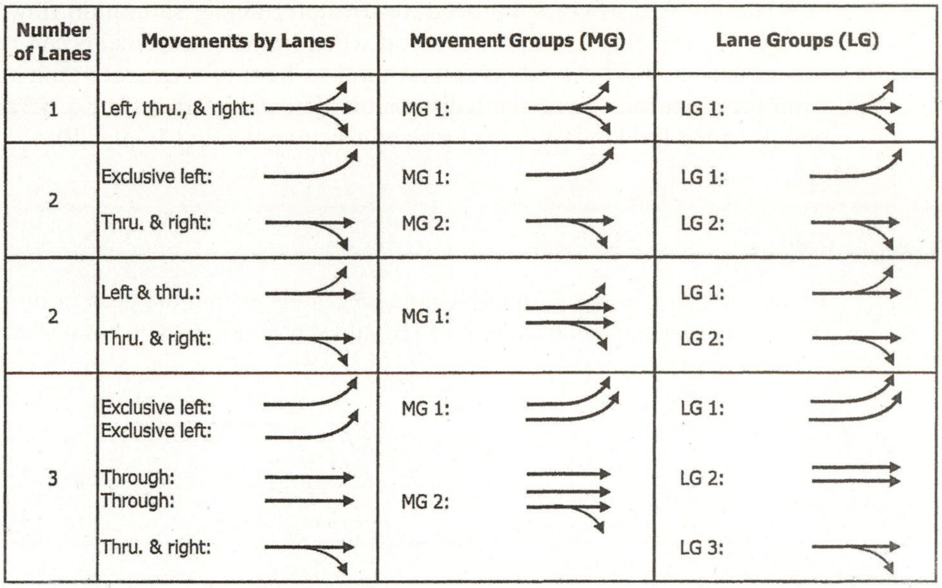

Number of Lanes	Movements by Lanes	Movement Groups (MG)	Lane Groups (LG)
1	Left, thru., & right:	MG 1:	LG 1:
2	Exclusive left:	MG 1:	LG 1:
	Thru. & right:	MG 2:	LG 2:
2	Left & thru.:	MG 1:	LG 1:
	Thru. & right:		LG 2:
3	Exclusive left: Exclusive left:	MG 1:	LG 1:
	Through: Through:	MG 2:	LG 2:
	Thru. & right:		LG 3:

Figure 8.13   Typical Movement and Lane Groups for Analysis

SOURCE: From *Highway Capacity Manual 2010.* Copyright, National Academy of Sciences, Washington, D.C. Reproduced with permission of the Transportation Research Board.

type, lane utilization, right and left turns, pedestrians, and bicycles. An equation given in the *Highway Capacity Manual* (HCM) and shown as Eq. 8.3 can be used to compute the saturation flow rate.

$$s = s_0 N f_w f_{HV} f_g f_p f_{bb} f_a f_{LU} f_{LT} f_{RT} f_{Lpb} f_{Rpb} \tag{8.3}$$

where

s = saturation flow rate for the subject lane group expressed as a total for all lanes in the lane group (veh/h)

s_0 = base saturation flow rate per lane (pc/h/ln)

N = number of lanes in the group

f_w = adjustment factor for lane width

f_{HV} = adjustment factor for heavy vehicles in the traffic stream

f_g = adjustment factor for approach grade

f_p = adjustment factor for the existence of a parking lane and parking activity adjacent to the lane group

f_{bb} = adjustment factor for the blocking effect of local buses that stop within the intersection area

f_a = adjustment factor for area type

f_{LU} = adjustment factor for lane utilization

f_{LT} = adjustment factor for left turns in the lane group

f_{RT} = adjustment factor for right turns in the lane group

f_{Lpb} = pedestrian adjustment factor for left-turn groups

f_{Rpb} = pedestrian/bicycle adjustment factor for right-turn group movements

The HCM also gives a procedure for determining saturation flow rate using field measurements. It is recommended that when field measurements can be obtained, that procedure should be used, as results tend to be more accurate than values computed from the equation. An in-depth discussion of how the factors in Eq. 8.3 are obtained and details of the field measurement procedure are given in Chapter 10.

Example 8.4 Determining the Peak Hour Factor and the Design Hourly Volume at an Intersection

The table below shows 15-minute volume counts during the peak hour on an approach of an intersection. Determine the PHF and the design hourly volume of the approach.

Time	Volume
6:00–6:15 p.m.	375
6:15–6:30 p.m.	380
6:30–6:45 p.m.	412
6:45–7:00 p.m.	390

Total volume during peak hour = (375 + 380 + 412 + 390) = 1557
Volume during peak 15 min = 412

Solution:

Obtain peak hour factor from Eq. 8.1.

$$PHF = \frac{\text{Volume during peak hour}}{4 \times \text{volume during peak 15 min within peak hour}}$$

$$= \frac{1557}{4 \times 412}$$

$$= 0.945$$

Obtain design hourly volume from Eq. 8.2.

$$DHV = \frac{\text{Peak hour volume}}{PHF}$$

$$= \frac{1557}{0.945}$$

$$= 1648$$

8.4.1 Vehicle and Pedestrian Movements and Phase Numbering

Intersection flows typically consist of several pedestrian and vehicle movements. Figure 8.14 shows the pedestrian and vehicular movements at a typical four-leg intersection. There are three vehicular movements and one pedestrian movement at each approach of the intersection. Each movement is assigned a number using the numbering system adopted in the *Highway Capacity Manual*. The pedestrian movements are designated by a number and the letter P, while the vehicular movements are designated by numbers only. These movements are used to select an appropriate phasing system at an intersection. Most current

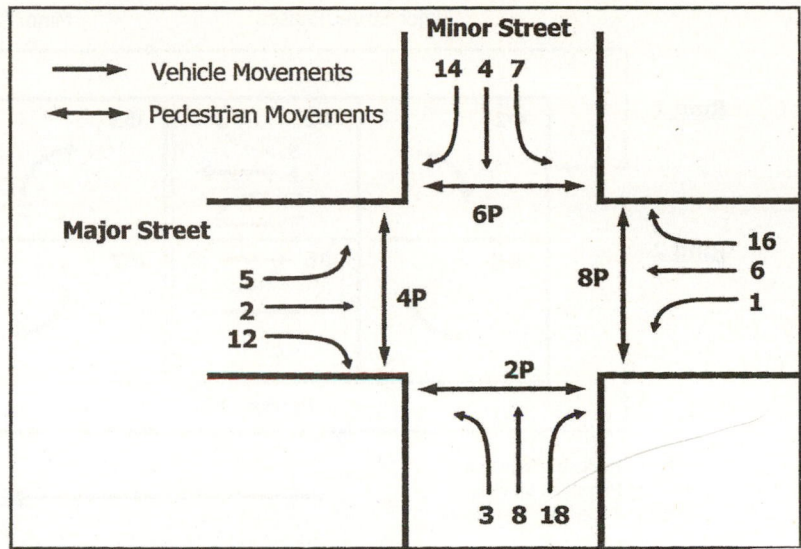

Figure 8.14   Typical Vehicle and Pedestrian Movements and Numbering Scheme at a Four-Leg Intersection

SOURCE: From *Highway Capacity Manual 2010*. Copyright, National Academy of Sciences, Washington, D.C. Reproduced with permission of the Transportation Research Board.

actuated controllers use rings and barriers to structure the signal phases. Rings group phases into different continuous loops, where conflicting and crossing movements are separated by barriers. The rings are used to develop a structure that makes it possible for two different phases that do not serve conflicting movements have the green indications concurrently (concurrent phases). The barriers are used to separate conflicting movements, such as the west bound through movement and the southbound left-turn movement. Figure 8.15 shows an example of a ring and barrier structure for an eight-phase system at a four-leg intersection. In this case, the left turns are each assigned a protected phase and are given the green arrow, while the dashed lines indicate permitted turns that should yield the right-of-way to conflicting movements. Note that this is not the only ring and barrier structure that is feasible at a four-leg intersection. However, it is usual to assign phases 2 and 6 to movements on the major street while the movements on the other side of the barrier are assigned to the minor street. The reader may refer to the *Signal Timing Manual* for other ring and barrier structures.

8.4.2 Signal Timing Policies and Process

The quality of traffic flow at intersections significantly affects the overall traffic flow and safety of an arterial or a network of roads within which the intersections are located. Signal timing is therefore a major factor that impacts the overall travel time, safety, and the resultant environmental impact of a highway transportation system. It is therefore essential that signal timing be based on a regional transportation policy that reflects the regional values of its transportation network. These values are usually incorporated in a Transportation Improvement Program (TIP), a Capital Improvement Program (CIP), a Long-Range Transportation Plan (LRTP), or other similar plans. Signal timing policies that are based on these values then can be developed to incorporate relevant issues,

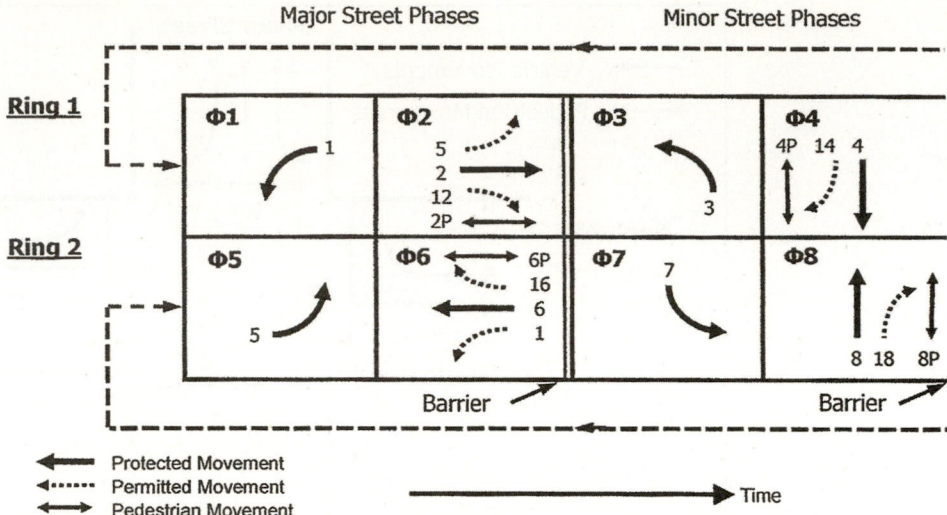

Figure 8.15 Dual-Ring Structure with Illustrative Assignments

SOURCE: From *Highway Capacity Manual* 2010. Copyright, National Academy of Sciences, Washington, D.C. Reproduced with permission of the Transportation Research Board.

including the user needs and specific objectives that should be optimized, the amount and extent of data that should be utilized to develop the signal timing, and the performance measures that should be used to evaluate the signal system. This process will lead to answers for specific questions, such as: Should priority be given to certain types of users of the highway system? What measures of effectiveness, such as vehicle stops, network delay, fuel consumption, personal delay, etc., should be used to evaluate the system? Should approaches on different road classifications be treated the same or differently?

8.4.3 Objectives of Signal Timing

Two primary objectives of signal timing at an intersection are to reduce the average delay of all vehicles and the likelihood of crashes. These objectives are achieved by minimizing the possible conflict points when assigning the right of way to different traffic streams at different times. The objective of reducing delay, however, sometimes conflicts with that of crash reduction. This is because the number of distinct phases should be kept to a minimum to reduce average delay whereas many more distinct phases may be required to separate all traffic streams from each other. When this situation exists, it is essential that engineering judgment be used to determine a compromise solution. In general, however, it is usual to adopt a two-phase system whenever possible, using the shortest practical cycle length that is consistent with the demand. At a complex intersection, though, it may be necessary to use a multiphase (three or more phases) system to achieve the main design objectives.

8.4.4 Signal Timing at Isolated Intersections

An isolated intersection is one in which the signal timing is not coordinated with that of any other intersection and therefore operates independently. The cycle length for an intersection of this type should be short, preferably between 35 and 60 sec, although it may be necessary to use longer cycles when approach volumes are very high. Several methods have been developed for determining the optimal cycle length at an intersection, and in most cases the yellow interval is considered as a component of the green time. Before discussing two of these methods, we will discuss the basis for selecting the yellow interval at an intersection.

Yellow Change Interval

The main purpose of the yellow indication after the green is to alert motorists to the fact that the green light is about to change to red and to allow vehicles already in the intersection to cross it. An inappropriate choice of yellow interval may lead to the creation of a *dilemma zone,* an area close to an intersection in which a vehicle can neither stop safely before the intersection nor clear the intersection without speeding before the red signal comes on. The required yellow interval is the time period that guarantees that an approaching vehicle can either stop safely or proceed through the intersection without speeding.

Figure 8.16 is a schematic of a dilemma zone. For the dilemma zone to be eliminated, the distance X_o should be equal to the distance X_c. Let Y_{min} be the yellow interval (sec) and let the distance traveled during the change interval without accelerating be $u_0 \times (Y_{min})$ with u_0 = speed limit on approach (m/sec). If the vehicle just clears the intersection, then

$$X_c = u_0(Y_{min}) - (W + L)$$

where X_c is the distance within which a vehicle traveling at the speed limit (u_0) during the yellow interval Y_{min} cannot stop before encroaching on the intersection. Vehicles within this distance at the start of the yellow interval will therefore have to go through the intersection.

W = width of intersection (m)
L = length of vehicle (m)
u_0 = traveling speed of vehicle m/sec (usually assumed to be the posted speed limit)

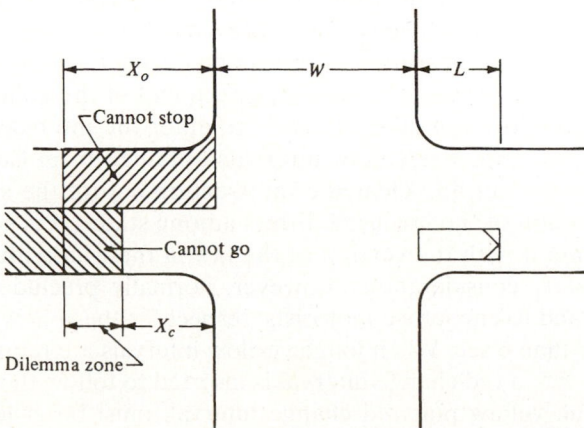

Figure 8.16   Schematic of a Dilemma Zone at an Intersection

For vehicles to be able to stop, however,

$$X_0 = u_0 \delta + \frac{u_0^2}{2a}$$

where

X_0 = the minimum distance from the intersection for which a vehicle traveling at the speed limit u_0 during the yellow interval Y_{min} cannot go through the intersection without accelerating; any vehicle at this distance or at a distance greater than this has to stop

δ = perception-reaction time (sec), assumed to be 1 sec

a = constant rate of braking deceleration (m/sec²)

For the dilemma zones to be eliminated, X_0 must be equal to X_c. Accordingly,

$$u_0(Y_{min}) - (W + L) = u_0 \delta + \frac{u_0^2}{2a}$$

$$Y_{min} = \delta + \frac{W + L}{u_0} + \frac{u_0}{2a} \tag{8.4}$$

If the effect of grade is added, Eq. 8.4 becomes

$$Y_{min} = \delta + \frac{W + L}{u_0} + \frac{u_0}{2(a + Gg)} \tag{8.5}$$

where G is the grade of the approach and g is the acceleration due to gravity (9.81 m/sec²). Note that the grade is in decimal. For example, 4 percent grade is 0.04 in Eq. 8.5. Note that there are two laws relating to a yellow change interval: (1) permissive yellow law and (2) restrictive yellow law. The permissive yellow law allows drivers to enter the intersection any time during the full time period of the yellow interval and be in the intersection during the red indication, as long as they enter during the yellow change interval. When the permissive law applies, an all-red interval must be included. The restrictive law has two variations. The first variation does not allow a driver to enter the intersection during the yellow indication unless the vehicle can go through the intersection by the end of the yellow change interval. The other variation does not allow a driver to enter the intersection unless it is impossible or unsafe to stop. The yellow intervals obtained from Eqs. 8.4 and 8.5 provide for the driver to enter and clear the intersection before the start of the red interval. Since the version of the law used differs among states, it is advisable that traffic engineers be familiar with the version of the law in their individual states.

Safety considerations, however, normally preclude yellow intervals of less than 3 sec, and to encourage motorists' respect for the yellow interval, it is usually not made longer than 6 sec. When longer yellow intervals are required as computed from Eq. 8.4 or Eq. 8.5, a red change interval is inserted to follow the yellow indication. The change interval, yellow plus red change interval, must be at least the value computed from Eq. 8.4 or 8.5 as appropriate.

Example 8.5 Determining the Minimum Yellow Interval at an Intersection

Determine the minimum yellow interval at an intersection whose width is 12 m if the maximum allowable speed on the approach roads is 48 km/h. Assume average length of vehicle is 6 m.

Solution: We must first decide on a deceleration rate. AASHTO recommends a deceleration rate of 3.4 m/sec². Assuming this value for a and taking δ as 1.0 sec, we obtain

$$Y_{min} = 1.0 + \frac{12 + 6}{48 \times 0.278} + \frac{48 \times 0.278}{2 \times 3.4}$$

$$= 4.3 \text{ sec}$$

In this case, a yellow period of 4.5 sec will be needed.

Cycle Lengths of Fixed (Pretimed) Signals

The signals at isolated intersections can be pretimed (fixed), semiactuated, or fully actuated. Pretimed signals assign the right-of-way to different traffic streams in accordance with a preset timing program. Each signal has a preset cycle length that remains fixed for a specific period of the day or for the entire day and do not change based on changes in traffic volume. Several design strategies exist for determining the optimum cycle length. Three strategies frequently used for pretimed signals are:

- Minimizing the total delay at the intersection. In some cases, other performance measures such as number of stops and fuel consumption are incorporated in the optimization factor. An example methodology of this strategy is the Webster method.
- A second strategy is to ensure that the volume-to-capacity ratios for the critical lane groups are the same by assigning the green times among the different phases in the same proportion as the flow ratio of their respective critical lane groups.
- A third strategy is to ensure that the levels of service for all critical lane groups are the same. Although this strategy is an improvement on the two described above, as it leads to obtaining a level of service at each approach that is similar to the overall level of service of the intersection, it tends to give the movements on the minor approaches a higher delay per vehicle.

The Webster and the HCM methods are presented here.

Webster Method.   Webster has shown that for a wide range of practical conditions minimum intersection delay is obtained when the cycle length is obtained by the equation

$$C_0 = \frac{1.5L + 5}{1 - \sum_{i=1}^{\phi} Y_i} \tag{8.6}$$

where
C_0 = optimum cycle length
L = total lost time per cycle (sec)
Y_i = maximum value of the ratios of approach flows to saturation flows for all lane groups using phase i (i.e., q_{ij}/S_j)

φ = numbers of phases
q_{ij} = flow on lane groups having the right of way during phase i
S_j = saturation flow on lane group j

This equation assumes that arrivals are random and traffic conditions at the intersection are undersaturated (i.e., demand flow is less than capacity).

Total Lost Time. Figure 8.17 shows a graph of rate of discharge of vehicles at various times during a green phase of a signal cycle at an intersection. At the beginning of the green phase, the drivers of the first few vehicles in the queue take some additional time to react and accelerate and achieve their running speeds. This results in these vehicles initially traveling with headways that are higher than the saturation flow headway, thereby losing some time before the vehicles move with saturation headways. This lost time is the *start-up lost time*, designated as l_1. At the end of the green indication, there is a portion of the yellow indication that is always used by drivers to go through the intersection. This initial part is known as the *extension of the effective green (e)*. However, sometime toward the end of the change period (yellow change interval and red clearance interval) is not used by vehicles to go through the intersection, because drivers are concerned about not complying with the yellow indication. The difference between the change period (yellow change interval and red clearance interval) and the extension of the effective green time (e) is designated as the *clearance lost time* (l_2). Suggested values for l_1 and l_2 are 2 seconds each. The total lost time (l_t) for each phase is the sum of l_1 and l_2 and can be expressed as:

$$l_t = l_1 + l_2 = l_1 + Y + R_c - e \qquad (8.7)$$

where

l_t = phase lost time
l_1 = start-up lost time = 2.0 sec
l_2 = clearance lost time = $Y + R_c - e$ (sec)
e = extension of effective green time = 2.0 sec
Y = yellow change interval (sec) and
R_c = red clearance interval (sec)

Total lost time for a cycle is given as

$$L = \sum_1^\varphi l_i \qquad (8.8)$$

where φ is the number of phases in a cycle.

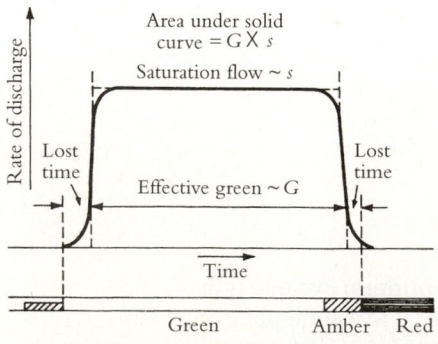

Figure 8.17   Discharge of Vehicles at Various Times during a Green Phase

Allocation of Green Times. In general, the total effective green time available per cycle is given by

$$G_{te} = C - L = C - \left(\sum_{i=1}^{\phi} l_i \right) \tag{8.9}$$

where

C = actual cycle length used (usually obtained by rounding off C_0 to the nearest 5 sec)

G_{te} = total effective green time per cycle

To obtain minimum overall delay, the total effective green time should be distributed among the different phases in proportion to their Y values to obtain the effective green time for each phase.

$$G_{ei} = \frac{Y_i}{Y_1 + Y_2 + \cdots Y_{\phi}} G_{te} \tag{8.10}$$

and the actual green time for each phase is obtained as

$$G_{a1} = G_{e1} + l_1 - Y_1 \tag{8.11}$$

$$G_{a2} = G_{e2} + l_2 - Y_2$$

$$G_{a3} = G_{e3} + l_3 - Y_3$$

$$G_{a\phi} = G_{e\phi} + l_{\phi} - Y_{\phi}$$

Minimum Green Time for Pedestrian Movement. At an intersection where a significant number of pedestrians cross, it is necessary to provide a minimum green time that will allow the pedestrians to safely cross the intersection. The length of this minimum green time may be higher than that needed for vehicular traffic to go through the intersection. The green time allocated to the traffic moving in the north-south direction should, therefore, not be less than the green time required for pedestrians to cross the east-west approaches at the intersection. Similarly, the green time allocated to the traffic moving in the east-west direction cannot be less than that required for pedestrians to cross the north-south approaches. The HCM gives three equations that should be considered. The first shown as Eq. 8.12 considers the time taken by the pedestrian to perceive the signal indication and the time taken by the pedestrian to cross the approach along the crosswalk. The second and third shown as Eqs. 8.13 and 8.14 take into consideration the time that is required for cyclic pedestrian demand.

$$G_{p,\min} = t_{pr} + \frac{L_{cc}}{S_p} - y - R_c \tag{8.12}$$

$$t_{ps} = 3.2 + \frac{L_{cc}}{S_p} + 9 \frac{N_{ped}}{W_E} \qquad \text{for } W_E > 3 \, \text{m} \tag{8.13}$$

$$t_{ps} = 3.2 + \frac{L}{S_p} + (0.27 N_{ped}) \qquad \text{for } W_E \leq 3 \, \text{m} \tag{8.14}$$

where

$G_{p,min}$ = minimum green time based on pedestrian crossing time (sec)

t_{pr} = time taken by the pedestrian to perceive the signal indication and depart the curb (HCM suggested value = 7 sec)

L_{cc} = curb to curb crossing distance (m) or crosswalk length (m)

S_p = average speed of pedestrians, usually taken as 1.0 m/sec (however, HCM 2010 suggests 1.2 m/sec when the percentage of elderly [65 yrs or older] pedestrians is 20 percent or less and 1.0 m/sec when the percentage of elderly pedestrians is higher than 20 percent; also, an upgrade of 10 percent or more results in reduction of pedestrian speeds by 0.1 m/sec)

Y = yellow change interval (sec)

R_c = red clearance interval (sec)

W_E = effective crosswalk width (m)

N_{ped} = number pedestrians crossing during an interval = $N_{ped} = \dfrac{v_{ped,i}}{3600} C$

$v_{ped,i}$ = pedestrian flow rate in the subject crossing for travel direction i (p/h)

Note that in some cases when left turns are permitted or protected-permitted, the reduction of the minimum green time by the yellow change interval (Y) and the red change interval (R_c) as shown in Eq. 8.12 result in pedestrians conflicting with left-turning vehicles. When it is feasible, the reduction of the minimum green by Y and R_c should be discarded and the minimum green is given as $G_{p,min} = t_{pr} + (L_{cc}/S_p)$.

Example 8.6 Signal Timing Using the Webster Method

Figure 8.18a shows peak hour volumes for a four-leg intersection on a highway. Using the Webster method, determine a suitable signal timing for the intersection using the four-phase system shown in the following table. Use a yellow change + red change interval of 3 sec and the saturation flows given in the table.

Note: The influences of heavy vehicles and turning movements and all other factors that affect the saturation flow have already been considered.

Phase	Lane Group	Saturation Flow (Veh/h/ln)
A	1 Exclusive left-turn movements	1615
	2 Through movements	1800
	3 Shared through & right-turn movements	1725
B	1 Exclusive left-turn movements	1615
	2 Through movements	1800
	3 Shared through & right-turn movements	1725
C	1 Exclusive left-turn movements	1615
	2 Through movements	1800
	3 Shared through & right-turn movements	1725
D	1 Exclusive left-turn movement	1615
	2 Through movements	1800
	3 Shared through & right-turn movements	1725

Solution:

- Determine equivalent hourly flows by dividing the peak hour volumes by the PHF peak hour factor (PHF) (e.g., for left-turn lane group of phase A, equivalent hourly flow = 222/0.95 = 234). See Figure 8.18b for all equivalent hourly flows.

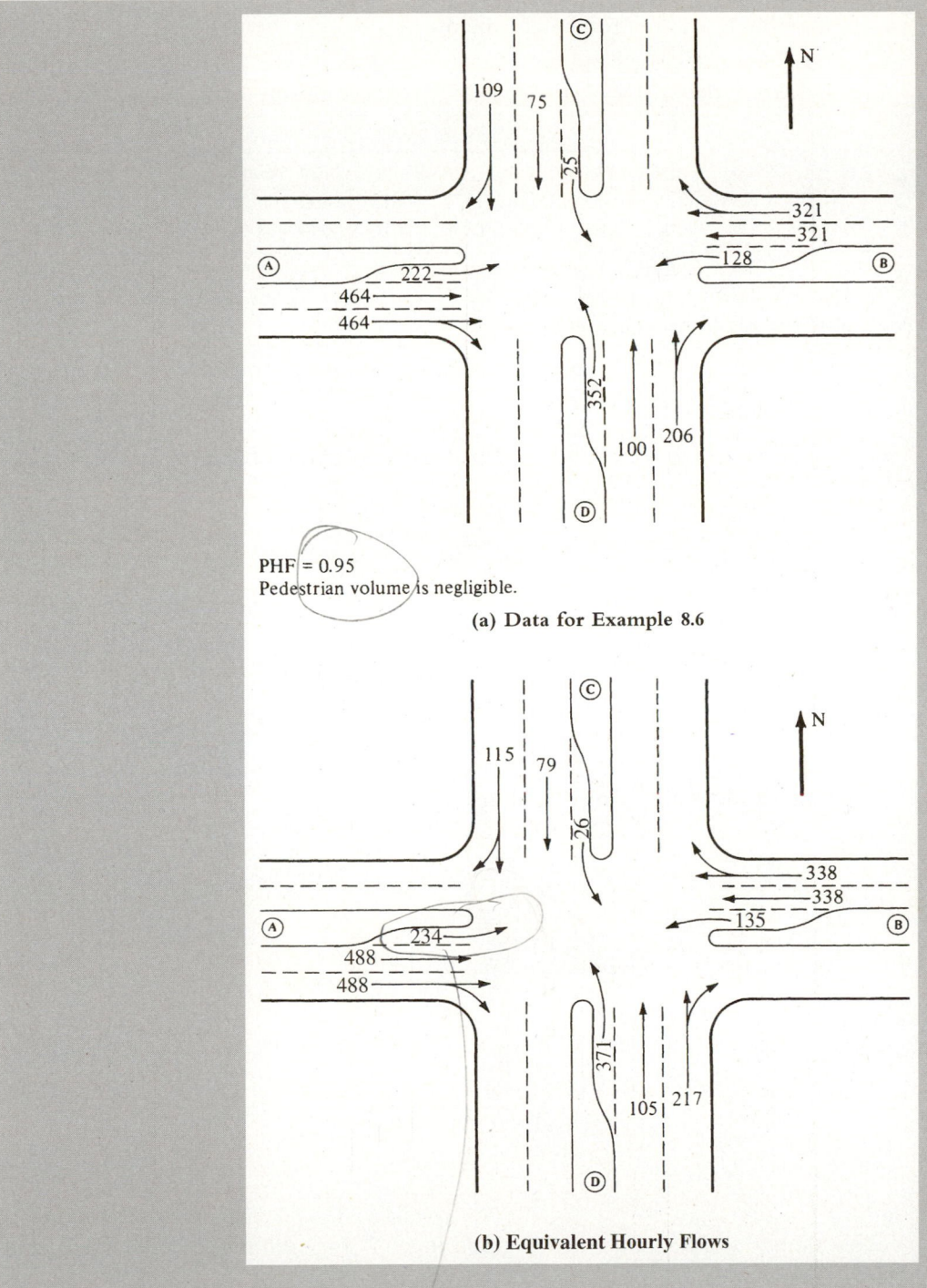

PHF = 0.95
Pedestrian volume is negligible.

(a) Data for Example 8.6

(b) Equivalent Hourly Flows

Figure 8.18   Data for Example 8.6

- Determine Y_i, as shown in table below.

	Phase A (EB) Lane Groups			Phase B (WB) Lane Group			Phase C (SB) Lane Groups			Phase D (NB) Lane Groups		
	1	*2*	*3*	*1*	*2*	*3*	*1*	*2*	*3*	*1*	*2*	*3*
q_{ij}	234	488	488	135	338	338	26	79	115	371	105	217
S_j	1615	1800	1725	1615	1800	1725	1615	1800	1725	1615	1800	1725
q_{ij}/S_j	0.145	0.271	0.283	0.084	0.188	0.196	0.016	0.044	0.067	0.230	0.058	0.126
Y_i		0.283			0.196			0.067			0.230	

Compute the total lost time using Eq. 8.7.

$$L = \sum \ell_i = 4 \times 3.0 = 12 \text{ sec (assuming lost time per phase is 3.0 sec)}$$

- Determine ΣY_i.

$$\sum Y_i = (0.283 + 0.196 + 0.067 + 0.230) = 0.776$$

- Determine the optimum cycle length using Eq. 8.6.

$\Sigma y_i \rightarrow$ choose the highest.

$$C_0 = \frac{1.5L + 5}{1 - \sum_{i=1}^{\phi} Y_i}$$

$$= \frac{(1.5 \times 12) + 5}{1 - 0.776}$$

$$= 102.68 \text{ sec}$$

Use 105 sec, as cycle lengths are usually multiples of 5 or 10 sec.
- Find the total effective green time.

$$G_{te} = C - L$$
$$= (105 - 12)$$
$$= 93 \text{ sec}$$

Effective time for phase i is obtained from Eq. 8.10.

$$G_{ei} = \frac{Y_i}{Y_1 + \cdots + Y_n} G_{te}$$

$$= \frac{Y_i}{0.283 + 0.196 + 0.067 + 0.230} 93$$

$$= \frac{Y_i}{0.776} 93$$

Yellow time $Y = 3.0$ sec; the actual green time G_{ai} for each phase is obtained from Eq. 8.11 as

$$G_{ai} = G_{ei} + \ell_i - 3.0$$

Actual green time for Phase A:

$$(G_{aA}) = \frac{0.283}{0.776} \times 93 + 3.0 - 3.0$$

$$\approx 34 \text{ sec}$$

Actual green time for Phase B:

$$(G_{aB}) = \frac{0.196}{0.776} \times 93 + 3.0 - 3.0$$
$$\approx 23.5 \text{ sec}$$

Actual green time for Phase C:

$$(G_{aC}) = \frac{0.067}{0.776} \times 93 + 3.0 - 3.0$$
$$\approx 8 \text{ sec}$$

Actual green time for Phase D:

$$(G_{aD}) = \frac{0.230}{0.776} \times 93 + 3.0 - 3.0$$
$$\approx 27.5 \text{ sec}$$

The Highway Capacity Method.   This method is used to determine the cycle length, and it is based on the capacity (the maximum flow based on the available effective green time) of a lane group. Since the saturation flow rate is the maximum flow rate on an approach or lane group when 100 percent effective green time is available, the capacity of an approach or lane group depends on the percentage of the cycle length that is given to that approach or lane group.

The capacity of an approach or lane group is given as

$$c_i = s_i(g_i/C) \tag{8.15}$$

where

c_i = capacity of lane group i (veh/h)
s_i = saturation flow rate for lane group or approach i (veh/h of green, or veh/h/g)
(g_i/C) = green ratio for lane group or approach
g_i = effective green for lane group i or approach i
C = cycle length

The ratio of flow to capacity (v/c) is usually referred to as the *degree of saturation* and can be expressed as

$$(v/c)_i = X_i = \frac{v_i}{s_i(g_i/C)} \tag{8.16}$$

where

X_i = (v/c) ratio for lane group or approach i
v_i = actual flow rate for lane group or approach i (veh/h)
s_i = saturation flow for lane group or approach i (veh/h)
g_i = effective green time for lane group i or approach i (sec)

It can be seen that when the flow rate equals capacity, X_i equals 1.00; when flow rate equals zero, X_i equals zero.

When the overall intersection is to be evaluated with respect to its geometry and the total cycle time, the concept of critical volume-to-capacity ratio (X_c) is used. The critical (v/c) ratio is usually obtained for the overall intersection but considers only the critical lane groups or approaches, which are those lane groups or approaches that have the maximum flow ratio (v/s) for each signal phase. For example, in a two-phase signalized intersection, if the north approach has a higher (v/s) ratio than the south approach, more time will be required for vehicles on the north approach to go through the intersection during the north-south green phase, and the phase length will be based on the green time requirements for the north approach. The north approach will therefore be the critical approach for the north-south phase. The critical v/c ratio for the whole intersection is given as

$$X_c = \sum_i (v/s)_{ci} \frac{C}{C - L}$$

and the cycle length is given as

$$C = \frac{L X_c}{X_c - \sum_{i=1}^{\phi}\left(\frac{v}{s}\right)_{ci}} \tag{8.17}$$

and the effective green time for the lane group is given as:

$$g_i = \frac{v_i C}{N_i s_i X_i} = \left(\frac{v}{Ns}\right)_i \left(\frac{C}{X_i}\right) \tag{8.18}$$

where

X_c = critical v/c ratio for the intersection
$\Sigma_i (v/s)_{ci}$ = summation of the ratios of actual flows to saturation flow for all critical lanes, groups, or approaches
C = cycle length (sec)
L = total lost time per cycle computed as the sum of the lost time (l_i) for each critical signal phase, $L = \Sigma l_i$
X_i = volume-to-capacity ratio for lane group i
N_i = number of lanes in lane group i

Equation 8.17 can be used to estimate the cycle length for the intersection if it is unknown and a critical (v/c) ratio is specified for the intersection. Alternatively, this equation can be used to obtain a broader indicator of the overall sufficiency of the intersection by substituting the maximum permitted cycle length for the jurisdiction and determining the resultant critical (v/c) ratio for the intersection. When the critical (v/c) ratio is less than 1.00, the cycle length provided is adequate for all critical movements to go through the intersection if the green time is proportionately distributed to the different phases, as shown in Eq. 8.18. That is, for the assumed phase sequence, all movements in the intersection will be provided with adequate green times if the total green time is proportionately divided among all phases. If the total green time is not properly allocated to the different phases, it is possible to have a critical (v/c) ratio of less than 1.00, but with one or more individual oversaturated movements within a cycle.

Example 8.7 Determining Cycle Lengths from *v*/*c* Criteria

A four-phase signal system is to be designed for a major intersection in an urban area. The flow ratios are

$$Phase\ A\ (v/s)_A = 0.25$$
$$Phase\ B\ (v/s)_B = 0.25$$
$$Phase\ C\ (v/s)_C = 0.20$$
$$Phase\ D\ (v/s)_D = 0.15$$

If the total lost time (L) is 14 secs/cycle, determine

1. The shortest cycle length that will avoid oversaturation
2. The cycle length if the desired critical *v*/*c* ratio (X_c) is 0.95
3. The critical *v*/*c* ratio (X_c) if a cycle length of 90 seconds is used

Solution: The shortest cycle length that will avoid oversaturation is the cycle length corresponding to the critical (v/c) ratio $X_c = 1$. Determine C from the expression.

$$X_c = \sum \left(\frac{v}{s}\right)_{ci} \frac{C}{C - L}$$

$$1 = (0.25 + 0.25 + 0.20 + 0.15) \frac{C}{C - 14} \quad (for\ X_c = 1)$$

$$1 = 0.85 \frac{C}{C - 14}$$

$$C - 14 = 0.85C$$

$$0.15C = 14$$

$$C = 93.3\ sec,\ say\ 95\ sec$$

- Determine C if the desired critical (v/c) ratio X_c is 0.95.

$$0.95 = (0.25 + 0.25 + 0.20 + 0.15) \frac{C}{C - 14}$$

$$0.95 = 0.85 \frac{C}{C - 14}$$

$$0.95C - 13.3 = 0.85C$$

$$0.95C - 0.85C = 13.3$$

$$0.1C = 13.3$$

$$C = 133\ sec,\ say\ 135\ sec$$

- Determine X_c for a cycle length of 90 seconds. Use the expression.

$$X_c = \sum \left(\frac{v}{s}\right)_{ci} \frac{C}{C - L}$$

$$= 0.85 \frac{90}{90 - 14}$$

$$= \frac{0.85 \times 90}{76}$$

$$= 1.01 \text{ (this will result in oversaturation)}$$

Example 8.8 Determining Cycle Length Using the HCM Method and Pedestrian Criteria

Repeat Example 8.6 using the HCM method with the following additional information:

- The desired critical v/c ratio is 0.90.
- Number of pedestrians crossing the east approach = 35 per interval in each direction.
- Number of pedestrians crossing the west approach = 25 per interval in each direction.
- Number of pedestrians crossing the north approach = 30 per interval in each direction.
- Number of pedestrians crossing the south approach = 30 per interval in each direction.
- Effective crosswalk width for each crosswalk = 2.7 m.
- Crosswalk length in E-W direction = 12 m.
- Crosswalk length in the N-S direction = 12 m.
- Pedestrian speed = 1.2 m/sec.
- Speed limit on each approach = 48 km/h.

Solution:

- Determine $\Sigma(v/s)$ for the critical lane groups.

  Phase A (EB), $(v/s)_c = 0.283$
  Phase B (WB), $(v/s)_c = 0.196$
  Phase C (SB), $(v/s)_c = 0.067$
  Phase D (NB), $(v/s)_c = 0.230$

  $$\Sigma (v/s)_{ci} = 0.776$$

- Determine cycle length for $X_c = 0.90$. Use the expression.

  $$X_c = 0.90 = \Sigma \left(\frac{v}{s}\right) \frac{C}{C - L}$$

  $$= 0.776(C/(C - 12))$$

  $$0.90(C - 12) = 0.776C$$

  $$0.90C - 0.776C = 0.90 \times 12$$

  $$C = 87.1 \text{ sec}$$

  Say $C = 90$ sec

- Find the minimum yellow interval by using Eq. 8.5 for the N-S and E-W movement.

$$Y_{\min} = 1.0 + \frac{40 + 20}{30 \times 1.47} + \frac{30 \times 1.47}{2 \times 11.2}$$

$$= 1.0 + 1.36 + 1.97$$

$$= 4.33 \text{ sec, say 5 sec}$$

- Determine actual green for each phase. Allow a yellow interval of 5 sec for each phase and assume a lost time of 3.0 sec/phase. In this case, the total effective green time $(G_{te}) = C{-}L{-}R$. In this case, $R = 0$.

$$G_{te} = 90 - 4 \times 3.0 = 78 \text{ sec}$$

Actual green time for (EB) Phase A $(G_{ta}) = (0.283/0.776) \times 78 + 3.0 - 5 = 26.4$ sec, say 26 sec.
Actual green time for (WB) Phase B $(G_{tb}) = (0.196/0.776) \times 78 + 3.0 - 5 = 17.7$ sec, say 18 sec.
Actual green time for (SB) Phase C $(G_{tc}) = (0.067/0.776) \times 78 + 3.0 - 5 = 4.7$ sec, say 5 sec.
Actual green time for (NB) Phase D $(G_{tD}) = (0.230/0.776) \times 78 + 3.0 - 5 = 21.1$ sec, say 21 sec.

- Check for minimum green times required for pedestrian crossing.
  Since W_E ,<3 m, use Eq. 8.14 and assume speed of pedestrian is 1.2 m/sec.

$$G_p = 3.2 + \frac{L_{cc}}{S_p} + (0.27 N_{\text{ped}})$$

Minimum time required for east approach $= 3.2 + 12/1.2 + 0.27 \times 35 = 22.65$ sec
Minimum time required for west approach $= 3.2 + 12/1.2 + 0.27 \times 25 = 19.95$ sec
Minimum time required for south approach $= 3.2 + 12/1.2 + 0.27 \times 30 = 21.30$ sec
Minimum time required for north approach $= 3.2 + 12/1.2 + 0.27 \times 30 = 21.30$ sec

Because of the phasing system used, the total time available to cross each approach is

East-west approaches $= (18 + 26)$ sec $= 44$ sec

North-south approaches $= (5 + 21)$ sec $= 26$ sec

The minimum time requirements are therefore satisfied.

Determination of Left-Turn Treatment

Left-turn vehicles at signalized intersections can proceed under one of three signal conditions: permitted, protected, and protected-permissive turning movements.

Permitted turning movements are those made within gaps of an opposing traffic stream or through a conflicting pedestrian flow and should yield to conflicting traffic and pedestrian movements. For example, when a right turn is made while pedestrians are crossing a conflicting crosswalk, the right turn is a permitted turning movement. Similarly, when a left turn is made between two consecutive vehicles of the opposing traffic stream, the left turn is a permitted turn. The suitability of permitted turns at a given intersection depends on the geometric characteristics of the intersection, the turning volume, and the opposing volume.

Protected turns are those turns protected from any conflicts with vehicles in an opposing stream or pedestrians on a conflicting crosswalk. All conflicting movements must therefore yield to protected turns. A permitted turn takes more time than a similar protected turn and will use more of the available green time.

Protected-Permited is a combination of the protected and permissive conditions, in which vehicles are first allowed to make left turns under the protected condition and then allowed to make left turns under the permissive condition.

Left–Turn Phase Treatment

When only permitted left turns are allowed at an approach, a left-turn phase is not provided for that approach. Figure 8.19 shows an example of a phasing system in which only permitted left turns are allowed on the minor street (north-south) approaches. However, when a protected or a protected-permitted left-turn mode is provided, a *leading, lagging,* or split phase system is used to accommodate the left turns. A leading left-turn phase is when the protected green phase is provided prior to releasing the conflicting through movements, and a lagging left-turn phase is when the protected green phase is provided after the conflicting through movement phase is completed. Figure 8.19 shows an example of a lag-lag protected phasing system for the major street, while Figure 8.20 shows an example of a lead-lead phasing system for both the major and minor streets.

Split phasing is when a single phase is provided for all movements on an approach and another phase is provided for all movements on the opposing approach. Figure 8.21 shows an example of a split phasing system for the minor street. Note that protected left

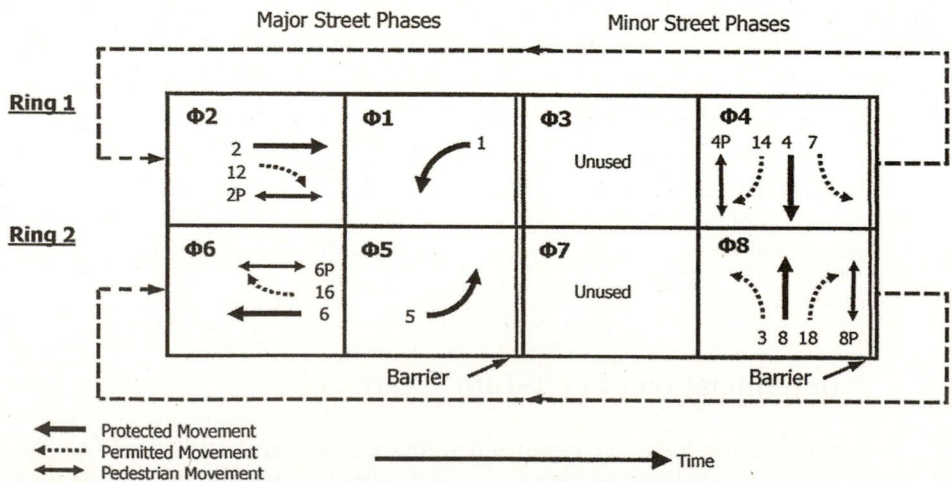

Figure 8.19 Illustrative Protected Lag-Lag and Permitted-Only Phasing

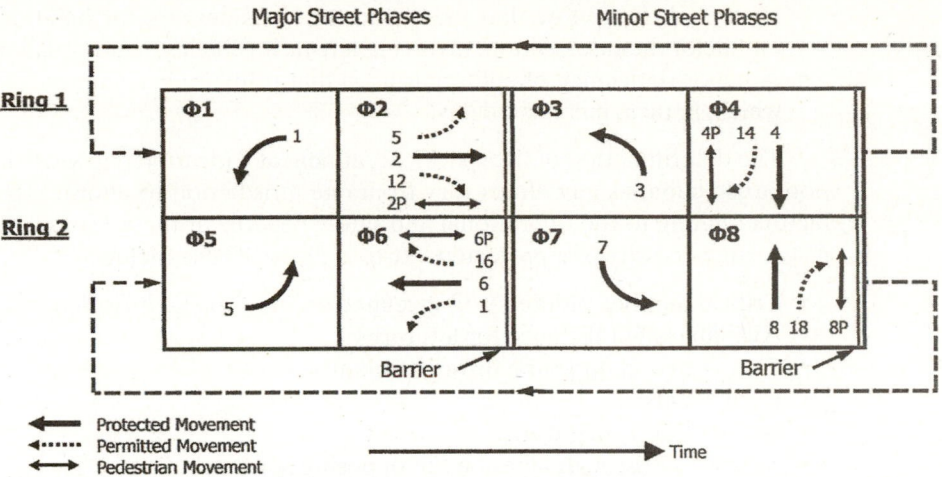

Figure 8.20   Illustrative Protected Lead-Lead and Permitted-Only Phasing

SOURCE: From *Highway Capacity Manual 2010*. Copyright, National Academy of Sciences, Washington, D.C. Reproduced with permission of the Transportation Research Board.

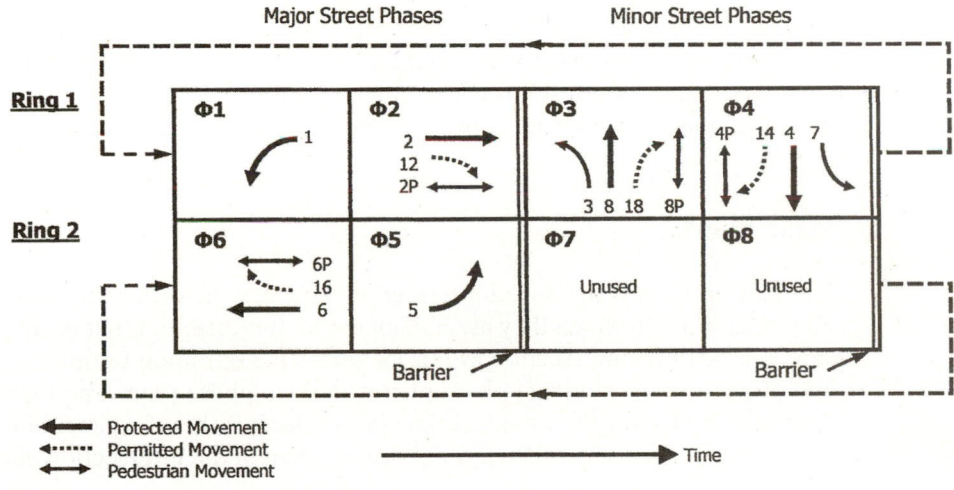

Figure 8.21   Illustrative Protected Lead-Lag and Split Phasing

SOURCE: From *Highway Capacity Manual 2010*. Copyright, National Academy of Sciences, Washington, D.C. Reproduced with permission of the Transportation Research Board.

turns are usually provided in split phasing except when pedestrian conflicts exist. Conditions under which split phasing may be appropriate include:

- It is necessary to provide one or more left-turn lanes on each approach, but sufficient width is not available in the middle of the intersection for this to be done. This is sometimes caused by a large intersection skew angle.
- The higher left-turn volume and opposing through-lane volume (volume by lane) are about the same throughout most of the hours of the day.

- Historical data show that the occurrence of sideswipe or head-on crashes in the middle of the intersection, involving left-turn vehicles, is unusually high.
- A major deficiency of split phasing is that it invariably increases the cycle length, which, in turn, increases delay.

The determination of the specific treatment at a location depends on the transportation jurisdiction, as guidelines vary from one jurisdiction to another. However, several factors relating to the operational and safety benefits at the intersection can be used to justify the necessity of a protected left-turn phase. These include:

- Critical number of intersection-related crashes for the jurisdiction
- Available sight distance for left turns
- Left-turn and opposing through volumes
- Cycle length
- Speed of opposing traffic
- Number of through lanes on the opposing approach

Figure 8.22 shows a chart proposed in the *Signal Timing Manual* that can be used to justify the necessity of a protected left-turn phase.

When a protected left-turn phase is provided, an exclusive left-turn lane must also be provided. The length of this storage lane should be adequate for the turning volume so that the safety or capacity of the approach is not affected negatively. It is a function of several traffic parameters, including the cycle length, signal phasing, arrival rate of left-turn vehicles, and the vehicle mix. It is suggested in the FHWA publication, *Signalized Intersection Informational Guide*, that a rule of thumb for determining the length is that the length should be adequate to store $1\frac{1}{2}$ to 2 times the average number of vehicle queued per cycle. A procedure for determining the queue length is given in Chapter 10.

Phase Plans

The phase plan at a signalized intersection indicates the different phases used and the sequential order in which they are implemented. It is essential that an appropriate phase plan be used at an intersection as this facilitates the optimum use of the effective green time provided. The simplest phase plan is the two-phase plan. The higher the number of phases, the higher the total lost time in a cycle. It is therefore recommended that the two-phase plan be used unless the traffic conditions at the intersection dictate otherwise.

8.4.5 Concepts of Actuated Traffic Signals

A major disadvantage of fixed or pretimed signals is that they cannot adjust themselves to handle fluctuating volumes. When the fluctuation of traffic volumes warrants it, a vehicle-actuated signal is used. When such a signal is used, vehicles arriving at the intersection are registered by detectors that transmit this information to a controller. The controller then adjusts the phase lengths to meet the requirements of the prevailing traffic condition.

The following terms are associated with actuated signals.

Recall. This is a request to the controller from a vehicle or pedestrian for the right-of-way (service) for a particular phase. This is sometimes referred to as a call. The different types of recalls are minimum or vehicle, maximum, pedestrian, and soft recalls.

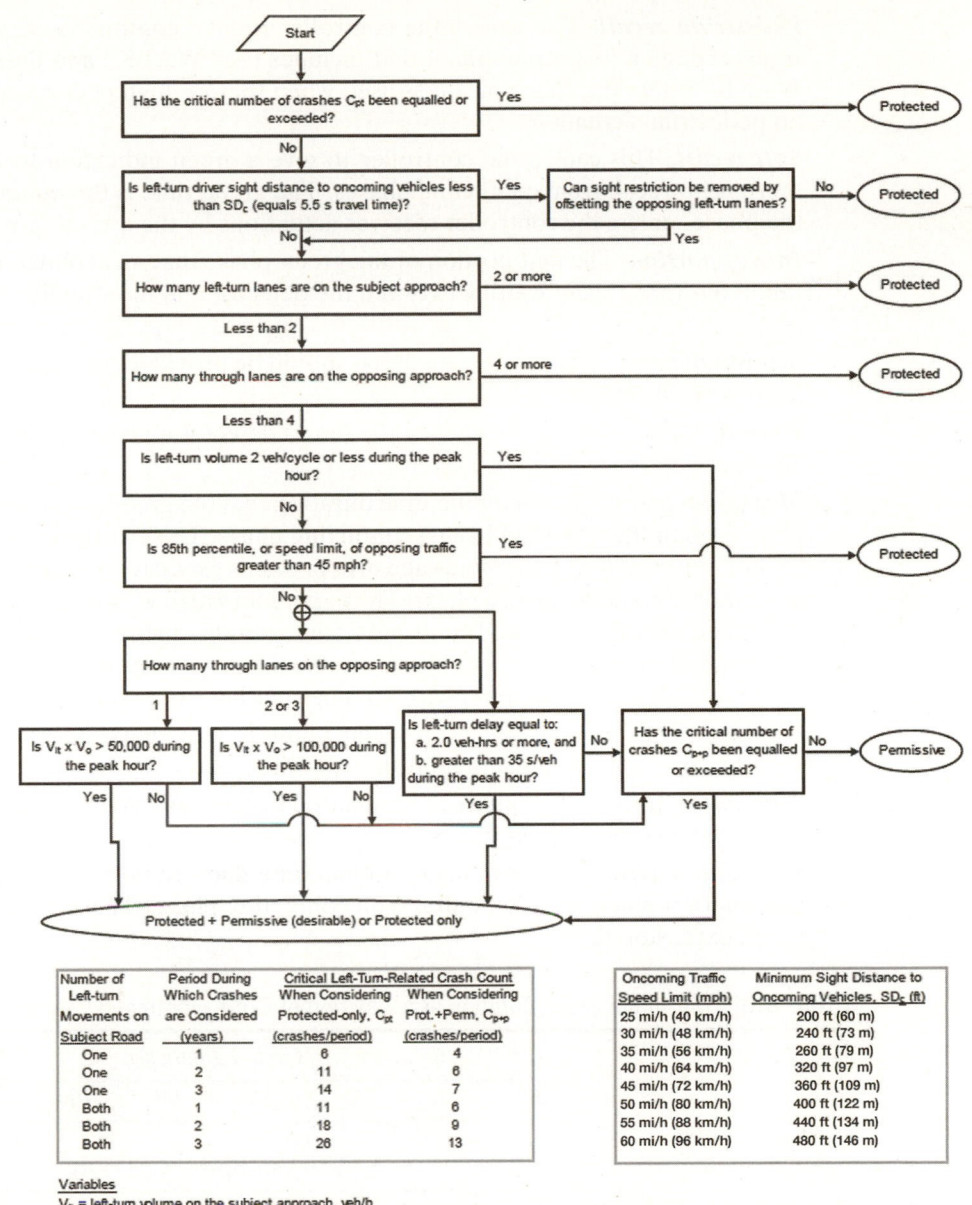

Figure 8.22   Guidelines for Determining the Potential Need for a Left-Turn Phase

SOURCE: *Traffic Signal Timing Manual*, Figure 4-11, pg. 4-13, Publication Number: FHWA-HOP-08-024 US Department of Transportation, Federal Highway Administration, Washington D.C., June 2008.

Minimum recall (vehicle recall). This parameter makes the controller place a continuous call for service on a given phase. When service is provided, the phase is timed for at least the minimum green, regardless of whether there is demand for vehicle movement.

Maximum recall. This causes the controller to give a green indication for the maximum green time assigned to a phase regardless of any recall from conflicting phases. If all phases are set to maximum recall, the signal operates like a pretimed signal.

Pedestrian recall. This causes the controller to give continuous service to pedestrian through a pedestrian phase that includes the "WALK" and flashing "DON'T WALK" intervals. This typically is used when there is high pedestrian demand and no pedestrian actuation.

Soft recall. This causes the controller to give a green indication to a phase when there is no recall from conflicting phases. The green time is the *minimum green* for the phase, unless the controller receives actuations by the arrival of vehicles.

Initial portion. The first portion of the green phase that an actuated controller has timed out for vehicles waiting between the detector and the stop line during the red phase.

Minimum green. The shortest time that should be provided for a green interval during any traffic phase.

Extend. A parameter that increases the time a detector is actuated by a defined fixed amount.

Maximum green. The maximum time duration that the green signal indication is displayed when there is a call from a conflicting phase. This parameter limits the delay to other intersection movements and also prevents excessive green times when there is a continuous demand on a phase. The maximum green value should be higher than the green duration required to disperse the average queue. This will facilitate the servicing of varying cyclical demands. Table 8.6 gives example values of maximum green durations for different phase volumes per lane and cycle lengths.

Gap out. This occurs when a particular phase is terminated because there are no vehicle extensions within a predetermined time period.

Max out. This is when a phase is terminated because the green indication has been on for its maximum green time.

Extension limit. The maximum additional time that can be given to the extendable portion of a phase after actuation on a conflicting phase. This is also known as the green extension time.

Table 8.6   Maximum Green Duration as a Function of Cycle Length and Volume

Phase Volume per Lane, veh/h/ln	Cycle Length, sec							
	50	60	70	80	90	100	110	120
	Maximum Green (G_{max})[1], sec							
100	15	15	15	15	15	15	15	15
200	15	15	15	15	16	18	19	21
300	15	16	19	21	24	26	29	31
400	18	21	24	28	31	34	38	41
500	22	26	30	34	39	43	47	51
600	26	31	36	41	46	51	56	61
700	30	36	42	48	54	59	65	71
800	34	41	48	54	61	68	74	81

[1]Values listed are computed as: $Gmax = (VC)/(1200n) + 1$ where V = design hourly volume served by subject phase (in vehicles per hour); n = number of lanes served by subject phase; and C = cycle length (in seconds). A 15-sec minimum duration is imposed on the computed values.

SOURCE: *Traffic Signal Timing Manual*, Publication Number: FHWA-HOP-08-024 US Department of Transportation, Federal Highway Administration, Washington D.C., June 2008.

Passage time. The minimum time by which a green phase could be increased during the extendable portion, due to the actuation by a vehicle on a phase. However, the extension period is terminated when the maximum green time for the phase is achieved. This action is known as a "force off." Passage time is also known as *unit extension, passage gap,* or *vehicle extension.* The three objectives that should be considered in setting the passage time are: ensuring "queue clearance," satisfying "driver expectancy," and reducing "max-out" frequency. It is essential for the duration of the passage time to be high enough to ensure that the green indication does not end before adequately serving the vehicular movement. This phenomenon is known as *premature gap out.* A shorter duration of the passage time will lead to increased delays and, in some cases, queue spillback. It is also essential that the passage time should not be so long as to create a long period of green after the queue is cleared. This will result in drivers at conflicting phases becoming impatient, which may result in them disrespecting the signal indication. The duration of the passage time should not be so large to cause frequent *max outs*, as this may lead to unfairly increased delay to drivers on higher-volume conflicting phases. The *Signal Timing Manual* suggests that passage time for efficient intersection operations range from 1 to 4 sec for presence mode detections, with the lower values more suitable for higher volume conditions. Delay tends to increase when passage time is outside this duration range.

Maximum allowable headway. This is the maximum time between the actuation of two consecutive vehicles without gapping out the phase. Note that when there is only one traffic lane being served by a phase, maximum allowable headway is the same as passage time, but they are different when several lanes are served by the same phase. However, the term *maximum allowable headway* is still used, as the time interval is that between consecutive actuations and not necessarily between two vehicles on the same lane.

Simultaneous gap out. This determines how a barrier is crossed when a conflicting phase is actuated. When simultaneous gap out is enabled, all phases that are concurrently timed reach a point together when they are committed to terminate either by a gap out, max out, or a force off, before they are jointly allowed to terminate. When it is not enabled, each of the concurrent phases can reach a committed terminating point separately, and then stay in that state until all the other concurrent phases attain a terminating status and then be allowed to terminate.

There are other terms that are used in advanced actuated signal design which are beyond the scope of this text. Interested readers may refer to the *Signal Timing Manual* or the *Highway Capacity Manual* for definitions of these terms.

Semiactuated Signals

Actuated signals can be either semiactuated or fully actuated. A semiactuated signal uses detectors only in the minor stream flow, while the through movements on the major stream flow operate in a "non-actuated" mode. Such a system can be installed even when the minor-stream volume does not satisfy the volume requirements for signalization. The operation of the semiactuated signal is based on the ability of the controllers to vary the lengths of the different phases to meet the demand on the minor approach. The signals are set as follows.

1. The green signal on the major approach is preset for a minimum period (minimum green), and it will stay on until the signal is actuated by a minor-stream vehicle.

2. If the green signal on the major approach has been on for a period equal to or greater than the preset minimum, the signal will change to red in response to the actuation of the minor-street vehicle.
3. The green signal on the minor stream will then come on for at least a period equal to the preset minimum for this stream. This minimum is given an extendable green for each vehicle arriving, up to a preset extension limit.
4. The signal on the minor stream then changes to red, and that on the major stream changes to green.

Note that when the volume is high on the minor stream, the signal acts as a pretimed one.

Several advantages are associated with the semiactuated signal controls. These include:

- They are effective in a coordinated-actuated signal system.
- They reduce the delay of the major road through volume when traffic is light, in comparison with that of a pretimed control signal system.
- The major road through-movements are not affected by failure of detectors, as these do not exist on the major-road approaches.

The main disadvantage of the semiactuated signal system is that an inappropriate maximum green time for the minor movements may lead to significant delay for the major through movements, especially when there is continuous demand from the minor movements.

The operation of a semiactuated signal requires certain times to be set for both the minor and major streams. For the minor streams, times should be set for the initial portion, unit extension, maximum green (sum of initial portion and extension limit), and change intervals. For the major streams, times should be set for the minimum green and change intervals. When pedestrian actuators are installed, it is also necessary to set a time for pedestrian clearance. Several factors should be taken into consideration when setting these times. The major factor, however, is that a semiactuated signal works as a pretimed signal during peak periods. It is therefore important that the time set for the maximum green in the minor stream be adequate to meet the demand during the peak period. Similarly, the time set for the minimum green on the major approach should be adequate to provide for the movement through the intersection of the expected number of vehicles waiting between the stop line and the detector whenever the signals change to green during the peak period. However, these settings should not be so large that the resulting cycle length becomes undesirable. In general, the procedures described in Section 8.4.5 can be used to obtain some indication of the required lengths of the different set times.

Fully Actuated Signals

Fully actuated signals are suitable for intersections at which large fluctuations of traffic volumes occur on all approaches during the day. Maximum and minimum green times are set for each approach. The basic operation of the fully actuated signal can be described using an intersection with four approaches and a two-phase signal system. Let phase A be assigned to the north-south direction and phase B to the east-west direction. If phase A is given the right-of-way, the green signal will stay on until the phase at least reaches the end of the minimum green if an approaching vehicle in the east-west direction actuates one of the detectors for phase B. Next, a demand for the right-of-way in the east-west direction is registered. If the north-south direction has reached its maximum green or has gapped out, the red signal will come on for phase A and the right-of-way will be given to vehicles in the east-west direction; that is, the green indicator will come on for phase B. This right-of-way will be held by phase B until at least the minimum green time expires. At the expiration of the minimum green time, the right-of-way will be given to phase A—that is, the north-south direction—only if during the period of the minimum green, a demand is registered by an approaching vehicle in this direction. If no demand is registered in the north-south direction and vehicles continue to arrive in the east-west direction, the right-of-way will continue to be given to phase B until the maximum green is reached. At this time, the right-of-way is given to phase A, and so on.

Fully actuated control has the following advantages:

- It responds to traffic demand and changes in traffic pattern, and therefore produces a relatively lower delay than that for a pretimed control.
- It has an efficient cycle time as it is dependent on the detection information, which may result in different durations from one cycle to the other.
- Phases with no demand can be skipped, thereby providing extra time to phases with demand.

The main disadvantage of the fully actuated controller is that it is more expensive to install and operate than the other control types as more detection is needed.

8.4.6. Average Phase Duration for Actuated Traffic Signals (HCM Procedure)

The HCM procedure for computing the average phase duration is iterative, as duration of the green interval is required for the computation of several of the intermediate equations. The green interval duration for each phase is therefore estimated initially, and the procedure carried out. A comparison is made between the computed green interval and the initially assumed value. If they are not effectively equal, the process is repeated using the computed value as the new initial value. This process is repeated until the initial and computed values are essentially equal.

The procedure for isolated intersections consists of the following tasks.

(i) Determine Effective Change Period for Each Phase

This involves computing the yellow change (Y_i) interval using Eq. 8.4 or 8.5 and selecting an appropriate red change interval (R_c) for each group. The sum of these two $(Y_c + R_c)$ gives the change period for each phase.

(ii) Estimate Green Interval

The green interval for each phase is first estimated and used as the initial green interval. Any of the methods discussed above, such as the Webster method or the HCM method, can be used to compute the initial green time. Note that for a fully actuated control, the maximum green time selected for the phase is used as the initial green time.

(iii) Compute Queue Accumulation Polygon

The HCM uses queue accumulation polygons depicting the arrival rate during the red time period, the queue discharge rate during the green phase, and the different time intervals

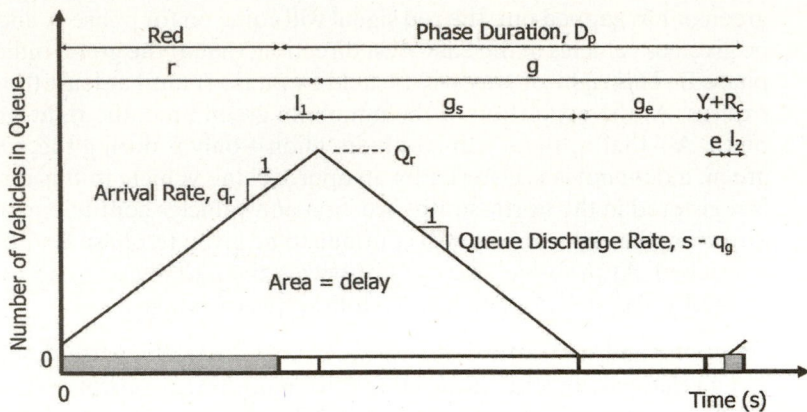

Figure 8.23   Time Elements Influencing Actuated Phase Duration

SOURCE: From *Highway Capacity Manual 2010*. Copyright, National Academy of Sciences, Washington, D.C. Reproduced with permission of the Transportation Research Board.

during the phases to determine the relationship between the phase duration (D_p) and queue size for the average signal length. The flow rates are given in a common unit of vehicles per second per lane (veh/sec/ln). Figure 8.23 shows a polygon for a through movement in an exclusive lane. Although this polygon can be used for several types of lane groups, other polygon shapes are possible. The HCM gives detailed procedures for constructing other polygons. In this text, the polygon shown in Figure 8.23 is used to illustrate the development of the relationship between phase duration and the associated time elements.

(iv) Compute Queue Service Time (g_s)

Figure 8.23 shows that a queue is formed at a rate of q_r by vehicles arriving during the red time. The maximum queue size in number of vehicles (Q_r) is formed at the end of the initial time lost, (i.e., l_1 seconds after the start of the green indication), and the queue starts discharging at a rate that is less than the saturation flow rate (s) by the arrival flow rate (q_g) during the effective green time. The queue is cleared at the end of the queue service time (g_s), followed by the random arrivals of vehicles, which cause the green interval to be extended by the extension time (g_e). This extension is then terminated at the end of the maximum green time (g_{max}) or when a headway that is larger than the maximum allowable occurs. The end of the extension time defines the end of the green interval. Note that the queue service time includes the time taken to clear the queue formed up to the end of the effective red and that taken to clear any additional vehicles that arrive during the time period the queue is being cleared. Based on this sequence of events, the HCM has shown that the queue service time (g_s) can be obtained from Eq. 8.19.

$$g_s = \frac{qC(1-P)}{s/3600 - qC(P/g)}$$

(8.19)

where

g_s = queue service time (sec)
q = arrival flow rate for the subject phase (veh/sec)

C = cycle length (sec)
P = proportion of vehicles arriving during the green indication
s = adjusted saturation flow rate (veh/h)
g = effective green time for the subject phase (sec)

(v) Compute Call Rate to Extend Green

The flow rate parameter (λ) represents the call rate to extend the green. It is an adjusted flow rate that takes into account the natural tendency for vehicles to form random platoons (bunches). This task involves computing λ_i for each lane group of an actuated phase. These are then summed up to determine a representative value (λ^*) for the phase. The HCM gives the flow rate parameter as:

$$\lambda^* = \sum_{i=1}^{m} \lambda_i \tag{8.20}$$

$$\lambda_i = \frac{\varphi_i q_i}{1 - \Delta_i q_i} \tag{8.21}$$

where

λ_i = flow parameter for lane group i ($i = 1, 2, \ldots m$)
φ_i = proportion of free (unbunched) vehicles in lane group i (decimal)

$$\varphi_1 = e^{-b_i \Delta_i q_i} \tag{8.22}$$

q_i = arrival rate for lane group (i) = $v_i/3600$ veh/sec
Δ_i = headway of unbunched vehicle stream in lane group i = 1.5 sec for single-lane group, 0.5 sec otherwise
m = number of lane groups served during the phase
b_i = bunching factor for lane group i (0.6, 0.5, and 0.8 for lane groups with 1, 2, and 3 or more lanes, respectively)

(vi) Compute a Maximum Allowable Headway (MAH)

The maximum allowable headway is related to the passage time, as shown in Figure 8.24. The two vehicles are traveling in the westbound direction with the maximum allowable headway. The second vehicle therefore will enter the detection zone at the same time the passage time is set to time out. Equation 8.23, which gives the relationship between the

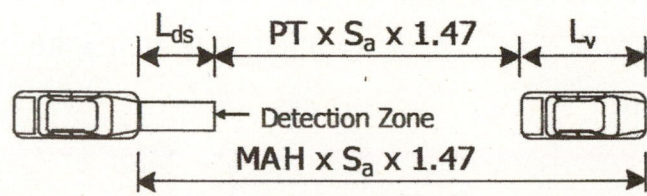

Figure 8.24 Relationship between Passage Time and Maximum Allowable Headway

maximum allowable headway and the passage time for presence mode detectors, can be derived from Figure 8.24 as:

$$MAH = PT + \frac{L_{ds} + L_v}{0.278u_a} \tag{8.23}$$

where

$$L_v = \mathrm{L}_{pc}(1 - 0.01P_{HV}) + 0.01L_{HV}P_{HV} - D_{sv}$$

MAH = maximum allowable headway (sec)

PT = passage time (sec)

L_v = detected length of the vehicle

L_{ds} = length of stop-line detection zone (m)

u_a = average approach speed (km/h) = $0.90(25.6 + 0.47u_{pl})$ (as given in the HCM)

u_{pl} = posted speed limit (km/h)

L_{pc} = stored passenger car lane length = 7.6 m

P_{HV} = percent of heavy vehicles in the corresponding movement group (%)

L_{HV} = stored heavy-vehicle lane length = 13.7 m

D_{sv} = distance between stored vehicles = 2 m

Note that if the detector is of the pulse-mode type, the values for L_{pc} and L_d are zero, which makes $MAH=PT$. The signal timing manual suggests that for average speeds (computed as 88 percent of 85th percentile approach speed) varying from 25 km/h to 45 km/h, PT varies from about 0.6 to 3.6 sec, for detector zone lengths varying from 1.8 m to 22 m, with PT increasing as speed increases but decreasing as the detection zone length increases.

(vii) Compute Equivalent Maximum Allowable Headway (MAH*)

When a phase serves more than one lane group, it is necessary to determine an equivalent maximum allowable headway (MAH*) for the phase. However, the MAH* for a phase that serves only one lane group without simultaneous gap out or that does not end at the barrier is the MAH for associated lane group. Note that different MAHs can be developed for different lane groups by modifying Eq. 8.23. The detailed development and discussion of these MAHs are beyond the scope of this text. Interested readers may refer to the HCM. However, Tables 8.7 and 8.8 give the HCM equations for the MAHs and equivalent MAHs, respectively.

(viii) Compute Number of Extensions (*n*) before Max Out

This task involves the computation of the number of extensions before max out using the HCM expression given in Eq. 8.38:

$$n = q^*[G_{max} - (g_s + l_1)] \geq 0.0 \tag{8.38}$$

where

$$q^* = \sum_{i=1}^{m} q_i$$

G_{max} = maximum green setting (sec)

g_s = queue service time

l_1 = time lost at the beginning of the green time

Table 8.7   Maximum Allowable Headway (MAH) Equations for Different Lane Groups

Lane Group Type	MAH Equation	Equation Number
Lane groups serving through vehicle	$MAH_{th} = PT_{th} + \dfrac{L_{ds,th} + L_v}{1.47u_a}$	8.24
Lane groups serving left-turn movements on exclusive lanes with the protected mode or (protected-permitted) mode	$MAH_{lt,e,p} = PT_{lt} + \dfrac{L_{ds,lt} + L_v}{1.47u_a} + \dfrac{E_L - 1}{s_o/3600}$	8.25
Lane groups serving left-turning vehicles in a shared lane with the protected-permitted mode	$MAH_{lt,s,p} = MAH_{th} + \dfrac{E_L - 1}{s_o/3600}$	8.26
Lane groups serving left-turning vehicles in an exclusive lane with the permitted mode	$MAH_{lt,e} = PT_{th} + \dfrac{L_{ds,lt} + L_v}{1.47u_a} + \dfrac{3600}{S_l} - t_{f_{th}}$	8.27
Lane groups serving right-turning vehicles in an exclusive lane with the protected mode	$MAH_{rt,e,p} = PT_{rt} + \dfrac{L_{ds,rt} + L_v}{1.47u_a} + \dfrac{E_R - 1}{s_o/3600}$	8.28
Lane groups consisting of permitted right-turning vehicles in an exclusive lane	$MAH_{rt,e} = PT_{th} + \dfrac{L_{ds,rt} + L_v}{1.47u_a} + \dfrac{\left(E_R/f_{Rpb}\right) - 1}{s_o/3600}$	8.29
Lane groups consisting of left turning vehicles in a shared lane with the permitted mode	$MAH_{lt,s} = MAH_{th} + \dfrac{3600}{S_l} - t_{f_{th}}$	8.30
Lane groups consisting of right-turning vehicles in a shared lane with the permitted mode	$MAH_{rt,s} = MAH_{th} + \dfrac{\left(E_R/f_{Rpb}\right) - 1}{s_o/3600}$	8.31

MAH_i = maximum allowable headway for lane group i (sec)
PT_i = passage time for phase serving through and turning vehicles (sec) (th =through, rt = right turning, lt = left-turning)
L_v = detected length of the vehicle
    = $L_{pc}(1 - 0.01P_{HV}) + 0.01 L_{HV}P_{HV} - D_{sv}$
L_{ds} = length of stop-line detection zone (ft)
u_a = average approach speed (mi/h) = $0.90(25.6 + 0.47u_{pl})$ (as given in the HCM)
u_{pl} = posted speed limit (mi/h)
L_{pc} = stored passenger car lane length = 25 ft
P_{HV} = percent of heavy vehicles in the corresponding movement group (%)
L_{HV} = stored heavy-vehicle lane length = 45 ft
D_{sv} = distance between stored vehicles = 8 ft
P_{HV} = percent of heavy vehicles in the corresponding movement group (%)
E_L = equivalent number of through cars for a protected left turn = 1.05
E_R = equivalent number of through cars for a protected right-turning vehicles = 1.18
s_o = base saturation flow rate (pc/h/ln)
S_l = saturation flow rate in exclusive left-turn lane with permitted mode (pc/h/ln)
$t_{f_{th}}$ = follow-up headway (4.5 if the subject left turn is served in a shared lane, 2.5 if the subject left turn is served in an exclusive lane [sec])
$L_{ds,rt}$ = length of the stop-line detection zone in the right-turning lanes
$L_{ds,lt}$ = length of the stop-line detection zone in the left-turn lanes (ft)
f_{Rpb} = pedestrian-bicycle saturation flow rate adjustment factor

SOURCE: From *Highway Capacity Manual 2010*. Copyright, National Academy of Sciences, Washington, D.C. Reproduced with permission of the Transportation Research Board.

Table 8.8   Equivalent Maximum Allowable Headway Equations for Different Lane Groups

Lane Group Type	MAH* Equation	Equation Number
No left-turn lane or right-turn lane group (approach contains only shared lanes)	$MAH^* = \dfrac{P_L\lambda_{sl}MAH_{lt,s} + [(1-P_L)\lambda_{sl} + \lambda_t + (1-P_R)\lambda_{sr}]MAH_{th} + P_R\lambda_{sr}MAH_{rt,s}}{\lambda_{sl} + \lambda_t + \lambda_{sr}}$	8.32
A right-turn lane group but no left-turn lane group	$MAH^* = \dfrac{P_L\lambda_{sl}MAH_{lt,s} + [(1-P_L)\lambda_{sl} + \lambda_t]MAH_{th} + \lambda_r MAH_{rt,e}}{\lambda_{sl} + \lambda_t + \lambda_r}$	8.33
A left-turn lane group but no right-turn lane group	$MAH^* = \dfrac{\lambda_l MAH_{lt,e} + [\lambda_t + (1-P_R)\lambda_{sr}]MAH_{th} + P_R\lambda_{sr}MAH_{rt,s}}{\lambda_l + \lambda_t + \lambda_{sr}}$	8.34
A left-turn lane group and a right-turn lane group	$MAH^* = \dfrac{\lambda_l MAH_{lt,e} + \lambda_t MAH_{th} + \lambda_r MAH_{rt,e}}{\lambda_l + \lambda_t + \lambda_r}$	8.35
A right-turn lane group and a through-lane group	$MAH^* = \dfrac{\lambda_t MAH_{th} + \lambda_r MAH_{rt,e}}{\lambda_t + \lambda_r}$	8.36
A shared right-turn and through-lane group	$MAH^* = \dfrac{[\lambda_t + (1-P_R)\lambda_{sr}]MAH_{th} + P_R\lambda_{sr}MAH_{rt,s}}{\lambda_t + \lambda_{sr}}$	8.37

MAH^* = equivalent MAH for condition stated
MAH_{th} = MAH for through vehicles
λ_{sl} = flow rate parameter for shard left-turn and through-lane group (veh/sec)
λ_t = flow rate parameter for exclusive through-lane group (veh/sec)
λ_{sr} = flow rate parameter for shared right-turn and through-lane group (veh/sec)
λ_r = flow rate parameter for exclusive right-lane group (veh/sec)
P_L = proportion of left-turning vehicles in shared lane (decimal)
P_R = proportion of right-turning vehicles in shared lane (decimal)

(ix) Determine the Probability (p) of a Green Interval Being Extended by Randomly Arriving Vehicles

This is the probability that the headway of a call is less than the MAH and is computed using the expression given in the HCM and shown as Eq. 8.39:

$$p = 1 - \varphi^* e^{-\lambda^*(MAH^* - \Delta^*)} \tag{8.39}$$

where

φ^* = combined proportion of free (unbunched) vehicles for the phase (decimal)
= $e^{-\sum_{i=1}^{m} b_i\Delta_i q_i}$ (see Eq. 8.22)
m = number of lane groups served during the phase

$\Delta^* =$ equivalent headway of bunched vehicle stream served by the phase (sec/veh)

$$= \frac{\sum_{i=1}^{m} \lambda_i \Delta_i}{\lambda^*} \tag{8.40}$$

$\Delta_i =$ headway of bunched vehicle stream in lane group i
$\quad = 1.5$ sec for single lane groups, and 0.5 sec otherwise

$$\lambda^* = \sum_{i=1}^{m} \lambda_i \tag{see Eq. 8.20}$$

$$\lambda_i = \frac{\varphi_i q_i}{1 - \Delta_i q_i} \tag{see Eq. 8.21}$$

$\varphi_i =$ proportion of free (unbunched) vehicles in lane group i (decimal) (see Eq. 8.22)

$q_i =$ arrival rate for lane group $i = \dfrac{v_i}{3600}$ veh/sec

(x) Determine the Green Extension Time (g_e)

The *green extension time* depends on the maximum allowable headway and the number of extensions before the termination of the phase. In this task, the green extension time for each phase is computed using the HCM expression given in Eq. 8.41.

The average green extension time (g_e) is given as:

$$g_e = \frac{p^2(1 - p^n)}{q^*(1 - p)} \tag{8.41}$$

where

$g_e =$ green extension time (sec)
$q^* =$ arrival rate for the phase $= \sum_{i=1}^{m} q_i$
$q_i =$ arrival rate for lane group $i = v_i/3600$ (veh/sec)
$p =$ the probability of a call headway being less than the equivalent maximum allowable headway (i.e., the probability of a green interval being extended by randomly arriving vehicles)

$$= 1 - \varphi^* e^{-\lambda^*(MAH - \Delta^*)} \tag{see Eq. 8.39}$$

(xi) Compute Activating Call Rate

This is the rate at which an actuated phase is activated, and it is used to compute the probability that in the following cycle, the phase will be activated. Separate call rates for vehicles and pedestrians are computed. These flow rates are based on the rate at which the different traffic movements for a given phase arrive at the approach. The HCM gives the following rules for determining phase vehicular flow rates:

- For a phase exclusively serving a left-turn movement, the left-turn movement flow rate is the vehicular flow rate.
- For a phase serving a through or right-turn movement whose adjacent left-turn movement is not served by an exclusive left-turn phase, the approach flow rate is the vehicular flow rate.
- For a phase serving a through or right-turn movement whose adjacent left-turn movement is served by an exclusive left-turn phase, then:
  (i) When a left-turn bay exists, the vehicular flow rate = (through movement flow rate + right-turn movement flow rate).

(ii) When a left-turn bay does not exist, the vehicular flow rate is the approach flow rate.

(iii) When a split phasing system is used, the vehicular flow rate is the approach flow rate.

The *activating vehicular call rate* (q_v^*) is then computed in units of vehicles per second by dividing the vehicular flow rate by 3600. Note that additional modification is required when dual entry is activated for a phase. Interested readers should refer to the HCM for details of this modification.

The *activating pedestrian* call rate (q_p^*) is obtained by dividing the associated phase pedestrian flow rate by 3600. Note that additional modification is required when dual entry is activated for a phase. Interested readers should refer to the HCM for details of this modification.

(xii) Compute Probability of Phase Call (p_c)

This probability is based on whether the actuated phase is set on recall in the controller. The probability of phase call is 1.0 when the actuated phase is set on recall. When the actuated phase is not set on recall, the probability of phase call is computed using the HCM equation shown as Eq. 8.42.

$$p_c = p_v(1 - p_p) + p_p(1 - p_v) + p_v p_p \qquad (8.42)$$

where

p_v = the probability that the subject phase is called by a vehicle detector

$$= 1 - e^{-q_v^* c} \qquad (8.43)$$

p_p = the probability that the subject phase is called by a pedestrian detector

$$= 1 - e^{-q_p^* P_p C} \qquad (8.44)$$

(xiii) Compute Unbalanced Green Duration

The green interval for the phase is then computed, taking into consideration the probabilities of the phase being called by a vehicle detector and the phase being called by a pedestrian detector. However, this green interval does not take into consideration the constraints that are placed by the structure of the controller structure and the associated barriers that reflect whether the controller is fully actuated or is coordinated actuated, as discussed in tasks (xv) and (xvi). It is therefore referred to as the "unbalanced" green interval. It is given as:

$$G_u = G_{veh,call} p_v(1 - p_p) + G_{ped,call} p_p(1 - p_v) + \max \left| G_{veh,call}, G_{ped,call} \right| p_v p_p \leq G_{max} \quad (8.45)$$

where

G_u = unbalanced green interval duration for a phase (sec)

$G_{veh, call}$ = average green interval given that the phase is called by a vehicle detection

$$= \max \left[\begin{array}{c} l_1 + g_s + g_e \\ G_{min} \end{array} \right] \qquad (8.46)$$

G_{min} = minimum green setting

$G_{ped, call}$ = average green interval given that the phase is called by a pedestrian detector

$$= Walk + PC \qquad (8.47)$$

$Walk$ = pedestrian walk setting time (sec) (usually between 4 and 7 sec)

PC = pedestrian clear setting (sec)

G_{max} = maximum green time (sec)

p_v = the probability that the subject phase is called by a vehicle detector

$= 1 - e^{-q_v^* C}$ (see Eq. 8.43)

q_v^* = activating vehicle call rate for the phase (see task xi)

p_p = the probability that the subject phase is called by a pedestrian detector

$= 1 - e^{-q_p^* P_p C}$ (see Eq. 8.44)

q_p^* = activating pedestrian call rate for the phase (see task xi)

P_p = probability of a pedestrian pressing the detector button = 0.51

Note the following:

- the value of G_u is G_{max} when a maximum recall is set for the phase.
- the probability of a pedestrian pressing the detector is usually taken as 0.51, as research has shown that about 51% of crossing pedestrians will initiate a call by pushing the pedestrian call button.
- the probability that the subject phase is called by a pedestrian detector is 0.0 when a protected left turn movement is served by the phase.
- when a protected permitted left-turn movement is served by the phase, and the left-turn movement shares a lane with through vehicles, the unbalanced green interval duration is the minimum green setting of the phase.

(xiv) Compute Unbalanced Phase Duration (D_{up})

The unbalanced duration is then obtained as:

$$D_{up} = G_u + Y + R_c \qquad (8.48)$$

where

D_{up} = unbalanced phase duration (sec)

G_u = unbalanced green duration for the phase (sec)

Y = yellow clearance interval (sec)

R_c = red clearance interval (sec)

The *yellow change* interval can be obtained by using Eq. 8.4 or 8.5 as appropriate. Also, the MUTCD suggests that the yellow change interval should be between 3 and 6 sec.

The *red clearance interval* starts at the end of the yellow change interval during which time the phase has a red indication, before the green indication of the next phase. It allows for vehicles entering the intersection during the yellow change interval to clear the intersection before the next phase. It is therefore optional and may not be required if the yellow change interval is obtained from Eq. 8.4 or 8.5. However, a range of 0.5 to 2 sec has been used by traffic engineers.

Note that the green extension time computed in task (x) is too long when simultaneous gap is enabled and the green signal of the current phase has reached its maximum duration. This increase in the green extension is the result of the green intervals of both the subject and current phases being extended only because of the green interval extension of the subject phase. This extension is therefore based only on the call flow rate of the subject phase. To adjust for this phenomenon in this task, the green extension obtained in task (x) is multiplied by an adjusted flow rate ratio, which is given as:

$$\lambda^{**} = \frac{\displaystyle\sum_{i=1}^{m_{isp}} \lambda_{isp}}{\displaystyle\sum_{i=1}^{m_{isp}} \lambda_{isp} + \sum_{i=1}^{m_{icp}} \lambda_{icp}} \tag{8.49}$$

where

λ^{**} = adjusted flow ratio
λ_{isp} = flow ratio parameter for lane group i of subject phase
λ_{icp} = flow ratio for lane group i of concurrent phase
$\lambda_i = \dfrac{\varphi_i q_i}{1 - \Delta_i q_i}$ (see Eq. 8.21)
m_{isp} = number of lane groups served during the subject phase
m_{icp} = number of lane groups served during the concurrent phase

(xv) Compute Average Phase Duration—Fully Actuated Control

Since the computed unbalanced phase duration does not reflect the structure of the controller at the intersection, the average phase duration is obtained by taking into consideration the type and sequence of the phases at the intersection. The procedure is illustrated for a fully actuated system with the phasing system shown in Figure 8.20 and shown again as Figure 8.25. Let us assume that on the major road, movements 2 and 6 are served by phases 2 and 6, respectively. Then, phases 1 and 5 will serve movements 1 and 5, respectively, if the left turns are protected or protected-permitted. Similarly, if movements 4 and 8 on the minor street are served by phases 4 and 8, respectively, and the left-turn movements 3 and 7 on the minor streets are permitted or protected-permitted, then they are served by phases 3 and 7, respectively.

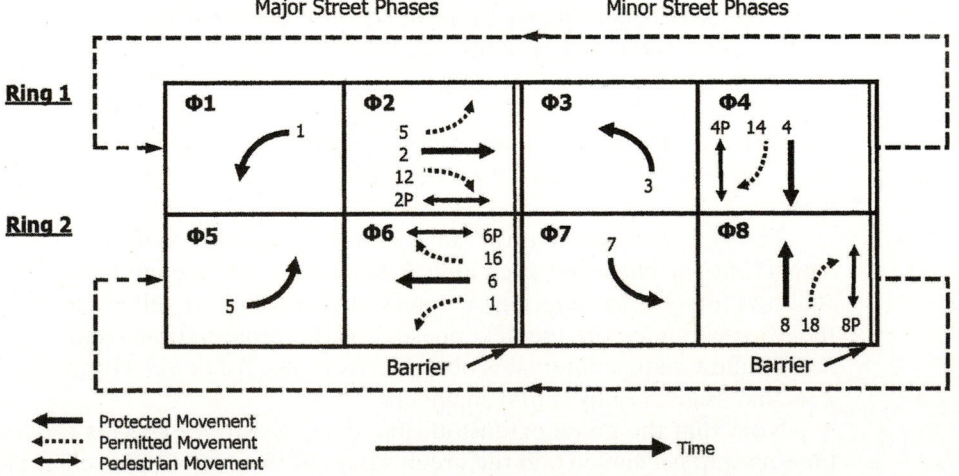

Figure 8.25   Illustrative Protected Lead-Lead and Permitted-Only Phasing

SOURCE: From *Highway Capacity Manual 2010*. Copyright, National Academy of Sciences, Washington, D.C. Reproduced with permission of the Transportation Research Board.

Also, if the through movement phase in a phase pair occurs before the opposing left-turn movement, then it is a lagging left-turn phase.

The average duration of each phase is computed using the following rules:

1. When two phases occur in sequence between barriers, (for example, Phase B follows A), the duration for phase B for the major street is given as:

$$D_{p,b} = \max[D_{up,1} + D_{up,2} + D_{up,5} + D_{up,6}] - D_{p,a} \tag{8.50}$$

and the average duration for the minor approach is given as

$$D_{p,b} = \max[D_{up,3} + D_{up,4} + D_{up,7} + D_{up,8}] - D_{p,a} \tag{8.51}$$

where

$D_{p,b}$ = phase duration of phase B, which occurs just after phase A (sec)
$D_{p,a}$ = phase duration of phase A, which occurs just before phase B (sec) = $D_{up,a}$
$D_{up,i}$ = unbalanced phase duration for phase i = 1, 2, 5, and 6 for major road approaches and i = 3, 4, 7, and 8 for minor road approaches.

2. When one permitted phase not in a split phasing system serves an approach, $D_{p,a}$ is zero and the Eqs. 8.50 are 8.51 are used to estimate $D_{p,b}$.
3. When split phasing is used, $D_{p,a}$ is the same as the unbalanced phase duration for one approach and $D_{p,b}$ is the unbalanced phase duration for the other approach.

With all the five time periods given in Eq. 8.45 determined, the phase duration ($D_{p,i}$) for each phase can then be obtained.

(xvi) Computed Green Interval Duration (G)

The computed green interval duration for the phase is then obtained from Eq. 8.52.

$$G = D_p - Y - R_c \tag{8.52}$$

(xvii) Compare Computed and Estimated Green Interval Durations

The estimated green interval used in task (ii) is then compared with the computed green interval obtained in task (xviii). If the difference between the two green intervals is greater than 0.1 sec, the computed green interval is then used as the new initial estimate and the procedure repeated starting with task (iii). This procedure is repeated until the difference between the two green intervals is less than 0.1 sec.

(xviii) Determine Equilibrium Cycle Length (C_e)

The intersection equilibrium cycle (C_e) for semiactuated or fully actuated systems is given as:

$$C_e = \sum_{i=1}^{\Phi} D_{p,i} \tag{8.53}$$

where

Φ = number of phases in the cycle (4 for the illustration above)
C_e = equilibrium cycle length (sec)
$D_{p,i}$ = average phase duration for phase i

The extensive computational effort involved in the procedure essentially precludes a manual computation to determine the green phase using this HCM procedure. Available computer software packages are therefore commonly used. These are discussed later in this chapter and in Chapter 10. Examples 8.9 and 8.10 are, however, presented to illustrate some of the computational tasks that are involved in this procedure.

Example 8.9 Determining the Queue Service Time for Each Phase at an Intersection

The intersection given in Example 8.6 is being considered for a fully actuated signal system. Take the computed cycle length and phase lengths as the initial trial and determine the queue service length for each phase for the following additional data:

(i) The proportion of vehicles arriving during the green interval = 0.8 for the major streets and 0.9 for the minor streets.

Solution: Use Eq. 8.19 to determine the queue service time.

Note that this problem illustrates the computations for steps (i) through (iv) of the procedure.

- Steps (i) and (ii) have been carried out as the cycle length = 105 sec (from Example 8.6).
- Step (iii): Assume Figure 8.23 for the queue accumulation polygon.
- Step (iv): Determine the queue service time: With the assumption of Figure 8.23 as the queue accumulation polygon, the queue service time is given as

$$g_s = \frac{qC(1 - P)}{s/3600 - qC(P/g)} \qquad \text{(see Eq. 8.19)}$$

- Determine arrival rate for each phase. Note that in Example 8.6, the split phasing system was used to compute the cycle and phase lengths and the effective green time for each phase was determined using the lane group with the highest v/s ratio for the phase. The arrival rate in veh/sec/ln for each phase is therefore that for the lane group with the highest v/s ratio for the phase.
  For example, for the eastbound approach phase,

$$q = 488/3600 = 0.136 \text{ veh/sec/ln}$$

- Determine the effective green time. Note that the total effective green time was distributed in the same proportion as the proportion of the v/s ratio for each phase to the sum of all v/s ratios.
  For example, the effective green time (g) for the eastbound phase is

$$(0.283/0.776)93 = 33.92 \text{ sec}$$

Table 8.9 gives the effective green times for the three other phases.

- Equation 8.19 is then used to determine the queue service time.
  For the eastbound approach phase,

$$g_s = \frac{0.136 \times 105(1 - 0.8)}{(1725/3600) - 0.136 \times 105 (0.8/33.92)}$$

$$= \frac{2.856}{0.479 - 0.337} = 20.1 \text{ sec}$$

Table 8.9   Data and Solutions for Example 8.9

Parameter	Phase 2 Eastbound Major Street	Phase 6 Westbound Major Street	Phase 4 Southbound Minor Street	Phase 8 Northbound Minor Street
Arrival rate (q) veh/sec/ln	488/3600 = 0.136	338/3600 = 0.094	115/3600 = 0.032	371/3600 = 0.103
Proportion of vehicles arriving during the green interval	0.8	0.8	0.9	0.9
Adjusted saturation flow (s) veh/l/h	1725	1725	1725	1615
Effective green time (sec)	(0.283/0.776)93 = 33.92	(0.196/0.776)93 = 23.49	(0.067/0.776)93 = 8.03	(0.23/0.776)93 = 27.56

For the westbound approach phase,

$$g_s = \frac{0.094 \times 105(1 - 0.8)}{1725/3600 - 0.094 \times 105\left(0.8/23.49\right)}$$

$$= \frac{1.974}{0.479 - 0.336} = 13.8 \text{ sec}$$

For the southbound approach phase,

$$g_s = \frac{0.032 \times 105(1 - 0.9)}{1725/3600 - 0.032 \times 105\left(0.9/8.03\right)}$$

$$= \frac{0.336}{0.479 - 0.377} = 3.29 \text{ sec}$$

For the northbound approach phase,

$$g_s = \frac{0.103 \times 105(1 - 0.9)}{1615/3600 - 0.103 \times 105\left(0.9/27.56\right)}$$

$$= \frac{1.082}{0.449 - 0.353} = 11.27 \text{ sec}$$

Example 8.10   Determination of Average Green Extension Time

Using the additional information given below and all necessary information that can be obtained from Examples 8.8 and 8.9, determine the average green extension time for the eastbound (major street) phase of the intersection in Example 8.8.

Maximum green setting = 35 sec

Minimum green setting = 15 sec

Passage time for through vehicles $(P_{th}) = 1.1$ sec

Passage time for left turning vehicles $(P_{lt}) = 1.5$ sec

Passage time for right turning vehicles $(P_{rt}) = 1.3$ sec

Queue service time = 20.1 sec (see Example 8.9)

Base saturation flow rate = 1900 veh/h

Length of stop line detection zone for left turns = 75 ft

Length of stop line detection zone for right turns = 100 ft

Posted speed limit = 30 mi/h

Percentage of heavy vehicles = 3%

Solution:

Note that this problem illustrates the computations for steps (i) to (x). Steps (i) to (iv) have been illustrated in Example 8.9. The computations in steps (v) to (x) have to be carried out.

Step (v):

- Step (v): Compute Call Rate to Extend Green
  Call rate for a phase is obtained from the flow rate parameters given as

$$\lambda^* = \sum_{i=1}^{m} \lambda_i \qquad \text{(see Eq. 8.20)}$$

$$\lambda_i = \frac{\varphi_i q_i}{1 - \Delta_i q_i} \qquad \text{(see Eq. 8.21)}$$

$$\varphi_i = e^{-b_i \Delta_i q_i} \qquad \text{(see Eq. 8.22)}$$

Δ_i = headway of unbunched vehicle stream in lane group i = 1.5 sec for single–lane group, 0.5 sec otherwise

m = number of lane groups served during the phase

b_i = bunching factor for lane group i (0.6, 0.5, and 0.8 for lane groups with 1, 2, and 3 or more lanes, respectively)

$$\varphi_i = e^{-b_i \Delta_i q_i} = \varphi_{lft} = e^{-0.6 \times 1.5 \times \frac{234}{3600} = 0.0585} = 0.9432$$

$$\varphi_i = e^{-b_i \Delta_i q_i} = \varphi_{th} = e^{-0.6 \times 1.5 \times \frac{488}{3600} = 0.122} = 0.8851$$

$$\varphi_i = e^{-b_i \Delta_i q_i} = \varphi_{th,rt} = e^{-0.6 \times 1.5 \times \frac{488}{3600} = 0.122} = 0.8851$$

$$\lambda = \frac{\varphi_i q_i}{1 - \Delta_i q_i} = \lambda_i = \frac{0.9432 \times \dfrac{234}{3600}}{1 - 1.5 \times \dfrac{234}{3600}} = 0.0679$$

$$\lambda = \frac{\varphi_i q_i}{1 - \Delta_i q_i} = \lambda_{th} = \frac{0.8851 \times \dfrac{488}{3600}}{1 - 1.5 \times \dfrac{488}{3600}} = 0.1506$$

$$\lambda = \frac{\varphi_i q_i}{1 - \Delta_i q_i} = \lambda_{th,rt} = \frac{0.8851 \times \dfrac{488}{3600}}{1 - 1.5 \times \dfrac{488}{3600}} = 0.1506$$

$$\lambda^* = \sum_{i=1}^{m} \lambda_i = (0.0679 + 0.1506 + 0.1506) = 0.369$$

φ^* = combined proportion of free (unbunched) vehicles for the
phase (decimal) = $e^{-\sum_{i=1}^{i=m} b_i \Delta_i q_i}$ \hfill (see Eq. 8.22)

$$\varphi^*_{lft,th\&rt} = e^{-((0.6 \times 1.5 \times 234/3600) + (0.6 \times 1.5 \times 488/3600) + (0.61 \times 1.5 \times 488/3600))}$$
$$= e^{-(0.059 + 0.122 + 0.122)} = 0.739$$

Note that values for b_i and Δ_i are given in the definition of the flow parameter λ_i in Eqs. 8.20 and 8.21.

Step (vi) Compute Equivalent Maximum Allowable Headway (MAH*):

- Step (vi): Compute Maximum Allowable Headway (MAH*):
  For the left-turn lane group, use Eq. 8.25.

$$MAH_{lt,e,p} = PT_{lt} + \frac{L_{ds,lt} + L_v}{1.47 u_a} + \frac{E_L - 1}{s_o / 3600}$$

$$L_v = L_{pc}(1 - 0.01 P_{HV}) + 0.01 L_{HV} P_{HV} - D_{sv}$$
$$= 25(1 - 0.01 \times 3) + 0.01 \times 45 \times 3 - 8$$
$$= 24.25 + 1.35 - 8 = 17.6$$

u_a = average approach speed (mi/h) = $0.90(25.6 + 0.47 u_{pl})$
(as given in the HCM)

$u_a = 0.90(25.6 + 0.47 \times 30)$

$= 35.73$ mi/h

$$MAH_{lt,e,p} = 1.5 + \frac{75 + 17.6}{1.47 \times 35.73} + \frac{1.05 - 1}{1900 / 3600}$$
$$= 1.5 + 1.763 + 0.095 = 3.358 \text{ sec}$$

For the through-lane group, use Eq. 8.24.

$$MAH_{th} = PT_{th} + \frac{L_{ds,th} + L_v}{1.47 u_a}$$

$$MAH_{th} = 1.1 + \frac{100 + 17.6}{1.47 \times 35.73} = 3.34 \text{ sec}$$

For the right-turning shared lane group, use Eq. 8.31.

$$MAH_{rt,s} = MAH_{th} + \frac{\left(E_R / f_{Rpb}\right) - 1}{s_o / 3600}$$

Take f_{Rpb} as 0.99; the computation of this factor will be discussed in Chapter 10.

$$E_R = 1.18 \text{ (see Table 8.7)}$$

$$MAH_{rt,s} = 3.34 + \frac{\left(1.18 / 0.99\right) - 1}{1900 / 3600} = 3.34 + 0.364 = 3.7 \text{ sec}$$

Step (vii) Compute Equivalent Maximum Allowable Headway.
Equivalent maximum allowable headway for the approach is given as:

$$MAH^* = \frac{\lambda_l MAH_{lt,e} + \lambda_t MAH_{th} + \lambda_r MAH_{rt,s}}{\lambda_l + \lambda_t + \lambda_r} \qquad \text{(see Eq. 8.35)}$$

Note that Eq. 8.35 is applicable as the split phasing system is used and the $MAH_{rt,s}$ is used, as the right-lane group is a shared right-lane group.

$$MAH^* = \frac{0.0679 \times 3.358 + 0.1506 \times 3.34 + 0.1506 \times 3.70}{0.0679 + 0.1506 + 0.1506} = \frac{1.288}{0.3691} = 3.49$$

Step (viii): Compute number of extensions (n) before max out.
Use Eq. 8.38.

$$n = q^*[G_{max} - (g_s + l_1)] \geq 0.0$$

$$q^* = \sum_{i=1}^{m} q_i = (234 + 488 + 488)/3600 = 0.336$$

$$G_{max} = \text{maximum green setting (sec)} = 35 \text{ sec}$$

$$g_s = \text{queue service time} = 20.1 \text{ sec}$$

$$l_1 = \text{time lost at the beginning of the green time} = 3.0 \text{ sec}$$

$$g_s = 20.1 \text{ sec}$$

$$l_1 = 2.0 \text{ sec}$$

$$n = 0.336(35 - (20.1 + 3.0))$$

$$= 4.0$$

- Step (ix): Determine the probability (p) of a green interval being extended by randomly arriving vehicles. Use Eq. 8.39

$$p = 1 - \varphi^* e^{-\lambda^*(MAH^* - \Delta^*)}$$

φ^* = combined proportion of free (unbunched) vehicles for the
phase (decimal) = $e^{-\sum_{i=1}^{i=m} b_i \Delta_i q_i}$ (see Eq. 8.22)

$$\varphi^*_{lt,th\&rt} = e^{-\left((0.6 \times 1.5 \times \frac{234}{3600}) + (0.61 \times 1.5 \times \frac{488}{3600}) + (0.61 \times 1.5 \times \frac{488}{3600})\right)}$$

$$= e^{-(0.060 + 0.124 + 0.124)} = 0.739$$

$$\Delta^* = \frac{\sum_{i=1}^{m} \lambda_i \Delta_i}{\lambda^*} \qquad \text{(see Eq. 8.40)}$$

$$\Delta^* = \frac{0.0679 \times 1.5 + 0.1506 \times 1.5 + 0.1506 \times 1.5}{0.3691} = 1.5$$

Note that values for b_i and Δ_i are given in the definition of the flow parameter λ_i in Eqs. 8.20 and 8.21.

$$P = 1 - \varphi^* e^{-\lambda^*(MAH^* - \Delta^*)}$$

$$= 1 - 0.739 e^{-0.3691(3.49 - 1.5)} = 0.3545$$

Step (x): Determine green extension. Use Eq. 8.41.

$$g_e = \frac{p^2(1 - p^n)}{q^*(1 - p)}$$

$$g_e = \frac{0.3545^2(1 - 0.3545^4)}{0.336(1 - 0.3545)} = g_e = \frac{0.1257(1 - 0.0158)}{0.336(1 - 0.3545)} = 0.57 \text{ s}$$

8.4.7 Signal Timing of Arterial Routes

In urban areas where two or more intersections are adjacent to each other, the signals should be timed so that when a queue of vehicles is released by receiving the right-of-way at an intersection, these vehicles also will have the right-of-way at the adjacent intersections. This is known as a coordinated signal system. The FHWA report, *Signal Timing on a Shoestring*, suggests that coordinated signal systems are advantageous when intersections are within 1.2 km of each other. However, when this distance is greater, traffic conditions such as volumes and the potential for platoons to form should be evaluated to determine the efficacy of using a coordinated system. A properly coordinated traffic control system will reduce the delay experienced by vehicles on the arterial. To obtain this coordination, all intersections in the system should have the same cycle length. In some instances, however, some intersections in the system may have cycle lengths equal to half or twice the common cycle length. The definitions of some basic terminologies that are used in coordinated systems are:

Coordination. A phenomenon that enhances the operation of a set of adjacent intersections in one or more directions

Double cycle. A cycle length that provides for phases at an intersection to have the green indication at a rate that is twice that of the other intersections. It is also known as a "half cycle."

Early return to green. A coordinated phase is given the green indication prior to its programmed start time because of unused time from noncoordinated phases.

Force off. A point within the cycle where a phase ends even if there is continued demand for it.

Fixed force off. This does not allow force-off points to be moved and provides for noncoordinated phases to use unused time from previous phases.

Floating force off. This allows for force-off points to be moved as a result of the demand of earlier phases. It provides for the maximum green duration for an uncoordinated phase to be the split time assigned to that phase and any unused time is given to the coordinated phase.

Master clock. In a coordinated operation, it is necessary for the controller at each intersection be referenced to a timing mechanism within the controller logic. This mechanism is known as the master clock.

Offset. This is the time lapse in seconds or the percentage of the cycle length between the beginning of a green phase at a coordinated intersection and the beginning of a corresponding green phase at the next coordinated intersection. It is the time base of the system controller.

Splits. This is the proportion of the cycle duration that is assigned to each of the phases at an intersection. It is given as a percentage of the cycle or in seconds.

Offset reference point (coordination point). This provides an association between the intersection's local clock and the master clock.

Yield point. This stipulates where the controller terminates the coordinated phase.

Permissive period. This is a time period after the yield point during which a noncoordinated phase can have the green indication without any delay to the start of the green indication to the coordinated phase

Progression Speed. Speed at which a platoon of vehicles released at an intersection will proceed along the arterial

Fundamentals of Traffic Signal Coordination

The basic difference between the operation of an isolated intersection discussed above and that for a coordinated system is the additional time constraint that is required for a coordinated system. In the operation of an isolated intersection, the controller settings are the maximum green, minimum green, pedestrian time, etc. In signal coordination, an additional set of time constraints is prescribed among the coordinated intersection signals by introducing a background cycle length on each ring of the phase diagram. This cycle length consists of a number of phases with timers for each, with one of these phases specified as the coordination phase and the phase that will be first (or last) to receive its green indication. This coordination phase is different from other actuated phases in that it is always given a prescribed minimum green time. This provides for the coordinated phase to have a nonactuated guaranteed time interval in each cycle that is used for coordination purposes. The major street through phase is usually the designated coordinated phase as it usually services highest volumes, which results in better performance of the signalized system.

The primary parameters that characterize a coordinated traffic signal system are the cycle length, offset, and split.

Selection of Cycle Length: It is usual for a common cycle length to be set with an offset (time lapse between the start of the green phases of two adjacent intersections on an arterial) that is suitable for the major street. Traffic conditions at a given intersection are used to determine the appropriate phases of green, red, and yellow times for that intersection. The minimum cycle length of each intersection is first estimated, which often results in different "optimal cycle lengths" for these intersections. This makes it necessary to seriously consider the intersections that should be included in a coordinated system. For example, if the inclusion of an intersection in a coordinated system results in higher cycle lengths for other intersections, which may lead to higher delays, the exclusion of that intersection from the coordinated system should be evaluated, although capacity may be marginally increased. Another factor that influences the selection of an appropriate cycle length for coordinated systems is the progression speed for the road.

The main objective that should influence the decision to select a cycle length for a coordinated system is that the cycle length selected should be the smallest time period that is adequate for vehicular demand at the intersection, while the needs of other users such as pedestrians are also met. Any of the methods presented for determining the cycle lengths for isolated intersections can be used to estimate the cycle length for the intersections on a coordinated system. For example, the HCM procedure for actuated intersections may be used with the following illustrated example.

Using Figure 8.25, assuming that phases 2 and 6 are the coordinated phases and they serve movements 2 and 6, respectively, and the left turn movements are protected or protected-permitted, then phases 1 and 5 serve the left-turn movements. For a phase that serves a coordinated movement and for which there is a left-turn phase for the associated approach, then:

$$D_{p,t} = C - \max \left[D_{up,3} + D_{up,4}, D_{up,7} + D_{up,8} \right] - D_{p,l} \tag{8.54}$$

where

$D_{p,t}$ = duration for coordinated phase t (t = 2 or 6), sec
$D_{p,l}$ = phase duration for left-turn phase l (l = 1 or 5), sec

For a phase that serves a noncoordinated movement, then Eqs. 49 and 50 are used.

Offset: The offset is influenced by the offset reference point because it serves as a means of measuring the local controller's offset relative to the master clock. The offset point in each controller is related to the master clock, which provides for offsets for the different intersections to be related to each other. This provides for the alignment of the coordinated phase between intersections, which establishes a relationship for the synchronized movements. The offset selected for a coordinated signal system should ideally provide for a platoon of vehicles leaving an upstream intersection at the start of the green interval to arrive at the adjacent downstream intersection near the start of its green interval. This requires that in selecting an optimal offset, consideration should be given to factors such as the desired or actual speeds of the vehicles between intersections, the distance between intersections, and the traffic volumes. This ideal offset applies only when the time-distance relationship is observable and supports progression. However, since the controller's actuated logic can provide for an early return to green, due to the variable demand on the noncoordinated phases, which in turn results in varying green time allocated to the side streets from one cycle to another, this ideal offset is not always observed by the user. Field observation can be used to determine how effectively traffic progresses between intersections with different offsets. The HCM also suggests that the time-space diagrams should be reviewed to evaluate the arterial progression and how effective the offsets are for a set of signal plans.

Split Distribution: The split distribution essentially determines the portion of cycle length that is given to each phase. The split for a given phase determines the maximum green duration that is given to the phase before it must end and service is provided for the next phase. In a coordinated system, if a noncoordinated phase does not use its entire split duration, the remaining available time is given to the coordinated phase. Split durations should not be too long to avoid possible excessive delays at other approaches, nor should it be too short, as this may result in demand not being met. The basic objective in determining split times is that adequate time should be provided for the phase so that oversaturated conditions are avoided. Several policies have been used to determine the split time for a phase. These include obtaining a *v/s* ratio that meets the operating agency's design standards, obtaining volume-to-capacity ratio for the critical movements of the intersection, and enhancing progression opportunities by allocating the minimum amount of time to the minor streets and allocating the rest to the coordinated phases.

8.4.8 Time–Space Diagram

A time-space diagram is a plot of vehicle trajectories along an arterial with intersections that have a coordinated signal system. It is used by engineers to evaluate a coordination strategy and make changes to timing plans. The main parameters used to construct a time-space diagram are the locations of the individual intersections, the cycle length, the offset, and the splits. A typical time-space diagram is shown in Figure 8.26. It is plotted with distance from a reference point on the vertical axis and time on the horizontal axis. The trajectories are shown from left to right as the vehicles traverse through the intersections during the green phases of the respective coordinated intersection. The distance traveled in opposite directions can be depicted (for example, northbound and southbound, with the northbound trajectory plotted from bottom to top of the figure and the southbound trajectory plotted from the top to the bottom). The speed of the vehicles is represented by the slope of the trajectory lines. The progression speed could be a desired speed, the 85th-percentile speed on the arterial, or the posted speed limit. The trajectories representing the first and last vehicles that can go through the intersections without being stopped by a red signal indication are shown as the two parallel progressive speed lines. The space between the two progressive speed lines is the *bandwith,* which delineates the amount of time provided for the vehicles to travel through the intersections at

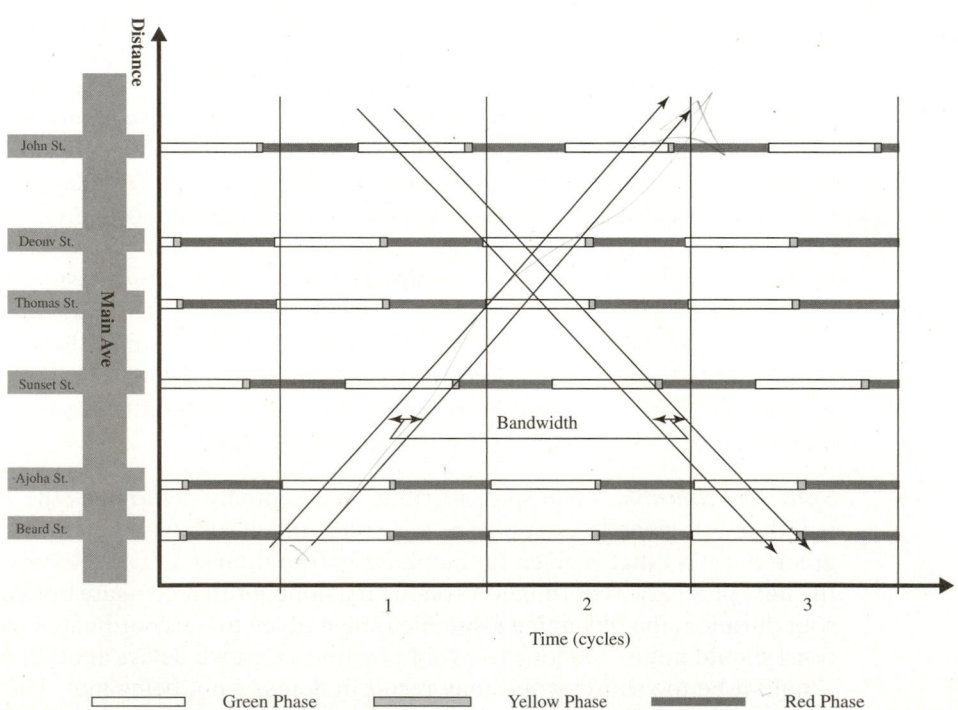

Figure 8.26 A Typical Time-Space Diagram along an Arterial with Intersections Controlled by a Coordinated Signal System

the progression speed without stopping. The bandwidth is dependent on the green interval for the coordinated phase at each intersection, the assumed progression speed, and the offsets between intersections. The maximization of bandwith is sometimes used as the optimization criterion for developing a signal timing plan. It should be noted that this may result in increased network delay and fuel consumption, and in poor performance with oversaturated conditions.

8.4.9 Signal Preemption and/or Priority

Preemption is the transfer of the normal operation of the signals to a special mode that allows for trains crossing at-grade railroad intersections with streets, allowing emergency vehicles and mass transit vehicles to easily cross an intersection and for other special tasks. This involves terminating the normal operation of the signal traffic control. Priority is when preferred treatment is given to a particular class of vehicles, such as a transit vehicle or an emergency vehicle, without the traffic signal controllers dropping from the coordinated mode. It can be obtained by changing the start and end times of a green phase, changing the phase sequence, or the addition of special phases, such that the continuity of the general timing relationship between green indications at adjacent intersections is not broken.

Vehicle Preemption

Preemption is mainly used for trains, light rail transit, and emergency vehicles, such as an ambulance, police car, or fire truck. In this case, the emergency vehicle is allowed to interrupt the normal signal cycle so that it can quickly cross the intersection safely. This may involve extending the green phase as the emergency vehicle approaches or changing the phasing and timings for the whole cycle. Preemption can also be used to give the right-of-way to specific types of vehicles when approaching certain types of non-intersection locations such as tunnels, one-lane bridges and tunnels, moveable bridges, work zones, and metered freeway ramps. The required time for a signal to transition into preemption is based on the distance the preempting vehicle is away from the signals when it is detected.

The MUTCD gives the following examples of preemption control.

- Promptly displaying the green signal indication at the approach of an emergency vehicle
- Using a special signal timing and phasing plan to provide additional clearance time for vehicles to clear tracks prior to the arrival of a train
- Prohibiting left-turning movements toward the tracks prior to the passage of a train

The MUTCD also gives standards that should be adhered to when preemption is used. These can be grouped into two classes—those that apply during the transition into preemption control and those that apply during preemption and transition out of preemption. During preemption control, the following standards apply.

1. No omission of or reduction to the length of the yellow change interval and any red clearance interval that follows should be made.
2. Shortening or omission of any pedestrian walk interval and/or pedestrian change interval is permitted.
3. It is permitted for the signal indication to be returned to the previous steady green indication after a yellow signal indication in the same signal phase, omitting any existing red clearance interval.

During preemption control and during the transition out of preemption, the following standards apply.

1. The shortening or omission of any yellow change interval and of any clearance interval that follows is not allowed.
2. It is not allowable for signal indication to be changed from a steady yellow to a steady green.

Several types of technologies exist that can be used to identify vehicles that request preemption. These include light (strobe), pavement loops, sound, radio transmission, and push buttons. Figure 8.27 shows an example of signal preemption for an emergency vehicle using optical detection. Signal timing is affected by preemption, in that the preemption signal timing and logic take over the signal operation from the normal signal timing and logic to provide the green indication for a specific type of vehicle. Preemptive operation can therefore extend the green time on an approach to a maximum preemptive green time that is not related to the stipulated maximum green or the settings required for coordination. Another impact of preemptive service is that it takes some time for coordinated systems to recover after preemption. Experience has shown that this recovery time ranges from 30 sec to 7 min, although values of up to 30 min have been reported. When more than one signal preemption can be requested at an intersection, it is necessary that these requests be prioritized by the importance of the vehicle

Figure 8.27   Example of Using Optical Detection for Preemption

SOURCE: *Traffic Signal Timing Manual*, Figure 9-1, pg. 9-1. Publication Number: FHWA-HOP-08-024 US Department of Transportation, Federal Highway Administration, Washington D.C. 2008. http://ops.fhwa.dot.gov/publications/fhwahop08024/fhwa_hop_08_024.pdf

right-of-way and/or how difficult it is to stop the vehicle type. However, preemption can result in the following benefits.

- Reduction in response times for emergency vehicles
- Increased safety and reliability for preempting vehicles
- Increased safety for other users of the roadway

Priority to Particular Class of Vehicles

Priority may be given to some nonemergency vehicles (such as buses and light rail vehicles) by modifying the signal timing and/or phase plan so that the green phase can be assigned to the vehicle with the preferred treatment. This is accomplished by the vehicle communicating with the traffic signal, and thereby influencing the signal to change its timing and assign priority to the transit vehicle. This priority is achieved by either extending the green times on specified phases, making changes to the sequence of the phases, or including distinctive phases while maintaining the green coordination between consecutive intersections. Priority operation is therefore different from preemption, in that preemption interferes with the normal cycle, while priority operation does not. The MUTCD gives the following standards that should be adhered to during priority control and during the transition to priority control.

1. The shortening or omission of any yellow change interval and of any clearance interval that follows is not allowed.
2. The reduction of any pedestrian walk interval below that determined as discussed in Section 8.4.4 is not allowed.
3. The omission of a pedestrian walk interval and its associated change interval is not allowed unless the associated vehicular phase is also omitted or the pedestrian phase is exclusive.
4. The shortening or omission of any pedestrian change interval is not allowed.
5. It is not allowed for a signal indication to be changed from a steady yellow to a steady green.

These different phenomena in intersection signals timing have led to the establishment of a set of communication standards known as the National Transportation Communications for ITS Protocol (NTCIP) by the Signal Control and Protection working group to facilitate information among signal control and prioritization systems.

Adaptive Traffic Signal Control

This is a system in which the arrival of traffic at a point within a network is determined by an algorithm, after detecting vehicular traffic upstream or downstream of the point. These detected traffic volumes are used to determine optimal signal timings that are implemented simultaneously in real time within the network. Thus, the timings are reflective of the real-time variation in traffic volumes, thereby reducing vehicular delay, queue lengths, and travel times. Adaptive traffic signal control systems are effective where:

- Random fluctuation of traffic occurs on a day-to-day basis
- Frequent changes in traffic conditions result from land-use activities
- Unexpected variation in traffic demand occurs due to incidents
- Response to disruptive events is implemented (e.g., preemption)

However, when traffic is very low, fixed-time and fixed-parameter systems often function as well as adaptive systems but at a lower cost.

The basic process used in an adaptive control system consists of four tasks: (1) use sensors to collect real-time traffic data, (2) develop and evaluate alternative signal timing strategies, (3) use a performance metric such as total intersection delay to identify the most appropriate strategy, and (4) implement the most appropriate strategy identified. Note that this is a difficult process that requires a traffic model that is capable of predicting the movement of vehicles within the network and estimating platoons of vehicles. Although several adaptive traffic control systems have been developed, usage of the system is less than 1 percent of all traffic signals in the nation. Historically, cost and required technical expertise have acted as barriers to deployment of these systems. Several vendors have recently developed cost-effective systems that can be applied with limited specialized expertise, so the number of deployments has begun to increase.

8.5 FREEWAY RAMPS

Ramps are usually part of grade-separated intersections, where they serve as interconnecting roadways for traffic streams at different levels. They are also sometimes constructed between two parallel highways to allow vehicles to change from one highway to the other. Freeway ramps can be divided into two groups: (1) entrance ramps to allow the merging of vehicles into the freeway stream, and (2) exit ramps to allow vehicles to leave the freeway stream. When it becomes necessary to control the number of vehicles entering or leaving a freeway at a particular location, access to the entrance or exit ramp is controlled in one of several ways.

8.5.1 Freeway Entrance Ramp Control

The control of entrance ramps is essential to the efficient operation of freeways, particularly when volumes are high. The main objectives in controlling entrance ramps are to regulate the number of vehicles entering the freeway so that the volume is kept lower than the capacity of the freeway (reduction of freeway congestion) and to reduce freeway crashes (improvement of freeway safety). The reduction of congestion ensures that freeway traffic moves at a speed approximately equal to the optimum speed (which will result in maximum flow rates). The control of entrance ramps also allows a better level of service on the freeway and safer overall operation of both the freeway and the ramp. On the other hand, entrance-ramp control may result in long queues on the ramps formed by vehicles waiting to join the freeway traffic stream or the diversion of traffic to local roads, which may result in serious congestion on those roads. It is therefore essential that the control of freeway entrance ramps be undertaken only when certain conditions are satisfied. The MUTCD provides general guidelines for the successful application of ramp control. The guidelines are mainly qualitative because there are too many variables that affect the flow of traffic on freeways. It is therefore extremely difficult to develop numerical volume warrants that will be applicable to the wide variety of conditions found in practice. The MUTCD suggests the following:

Installation of entrance-ramp control signals may be justified when it will result in a reduction of the total expected delay to traffic in the freeway corridor, including freeway ramps and local streets, and when at least one of the following conditions exists.

1. There is recurring congestion on the freeway due to traffic demand in excess of the capacity, or there is recurring congestion or a severe accident hazard at the freeway entrance because of an inadequate ramp merging area. An indication of recurring freeway congestion is when an operating speed of less than 50 mi/h occurs regularly for at least a half-hour period.

2. The signals are needed to accomplish transportation system management objectives identified locally for freeway traffic flow. This includes maintaining the level of service of a freeway at a specific level, providing priority treatment for mass transit and carpools, and redistributing access demand from a ramp to other ramps.

3. The signals are needed to reduce (predictable) sporadic congestion on isolated sections of freeway caused by short-period peak traffic loads from special events or from severe peak loads of recreational traffic.

Methods for Controlling Freeway Entrance Ramps

The common methods used in controlling freeway entrance ramps are

- Closure
- Ramp Metering
  - Local (isolated) metering control
  - System-wide (coordinated) metering control

Closure.   Closure entails the physical closure of the ramp by using "Do Not Enter" signs or by placing barriers at the entrance to the ramp. This is the simplest form of ramp control, but unfortunately, it is the most restrictive. The types of ramp closures used can be classified into three general categories: (1) permanent, (2) temporary, and (3) time-of-day or scheduled.

Permanent ramp closures do not provide any future opportunity for use of the ramp and therefore are the most restrictive of the three types of ramp closures. Implementing the permanent closure of a ramp will often result in permanent alteration of motorists' travel patterns, have impact on surrounding land value, and affect access to and from nearby businesses. Permanent closure should therefore be used only when absolutely necessary; for example, at locations where severe safety problems exist that cannot be resolved by other efforts.

Temporary closures are often used during maintenance or construction activities at the ramp in order to eliminate the potential conflict between ramp and construction vehicles, and between construction workers and ramp vehicles. This also may result in increased productivity and reduction of the associated cost for the project.

Temporary closure also may be used to assist in the management of traffic on a freeway near the location of a special event. Bad weather conditions may also necessitate the temporary closure of a ramp, particularly when a large amount of ice, snow, or water is on the roadway.

Time-of-day or scheduled closure is mainly used during the morning and peak hours at locations when recurring traffic congestion may lead to a severe safety problem. When this type of closure is used, it is recommended that adequate signing be provided to avoid any confusion by motorists unfamiliar with the area.

Barriers used for ramp closure can be either manually placed or automated.

Manual barricades can be either portable or fixed. Portable barricades can be moved from place to place and include barricades and cones. These are therefore mainly used for temporary closures. Fixed barricades are usually permanently installed and provide the flexibility of opening and closing of the ramp in response to traffic conditions. These include vertical and horizontal swing gates that are installed alongside the ramp. Automated barricades are fixed barricades that can be operated from a Traffic Management Center (TMC) or from another remote location.

The use of *enforcement personnel* is another method of ramp closure. This method is normally used on a temporary basis when automated systems are either not installed or when maintenance personnel are unavailable to install temporary barricades. It is used to prevent traffic from entering the ramp until maintenance personnel are available to install the temporary barricade.

Ramp metering.   Ramp metering involves the use of traffic signals to control the rate at which vehicles enter a freeway. This facilitates a more harmonious flow of traffic on the freeway, which in turn leads to an efficient use of the available freeway capacity. Ramps may be metered so that only one vehicle or a platoon consisting of two vehicles enter the freeway at a time. The metering could also be coordinated among several ramps along a freeway to facilitate a smooth flow along a segment of the freeway. Ramp metering is mainly deployed on ramps that connect local or arterial streets to freeways, although it has been used to connect freeway to freeway. Ramp metering can be deployed in a local or isolated mode, or in a system-wide mode.

Local or Isolated Metering Control. This method uses conditions at a specific ramp to determine the metering rate. It is therefore appropriate only for ramps with isolated problems and when no effort is made to consider conditions at adjacent ramps. The primary objective is to improve safety or alleviate congestion at or near the ramp. Although this metering method can be used at two or more ramps on a freeway, the effect of metering at one ramp is not coordinated with the effects at another. It is therefore possible for a desired solution to be obtained at a given metered ramp, but a negative effect may occur at another ramp as a result of the metering at the given ramp. When this condition exists it is not recommended to use local control.

System-Wide or Coordinated Metering Control. Here conditions are used that include those at other adjacent ramps to determine the metering rate. This method can therefore be used for ramps along a freeway segment, an entire corridor, or several corridors where problems at one ramp may affect the other adjacent ramps. It is therefore more effective than local control in dealing with capacity reduction that occurs as a result of road blockages and collisions. However, detection requirements for this approach are also more extensive.

Methods of Controlling Ramp Meters

Pretimed metering and traffic responsive metering are the two types of control used for ramp metering.

Pretimed Metering.   This is the simpler of the two methods, as the metering rate is based on historical conditions and there is no requirement for communication with a Traffic Management Center (TMC). This form of metering consists of setting up a pretimed signal with extremely short cycles at the ramp entrance. The time settings are usually made for different times of the day and/or days of the week. Pretimed metering can be used to reduce the flow of traffic on the ramp from about 1200 veh/h, which is the normal capacity of a properly designed ramp, to about 250 veh/h. Figure 8.28 shows the layout of a typical simple metering system for an entrance ramp, including some optional features that can be added to the basic system. The basic system consists of a traffic signal with a two-section (green-red) or three-section (green-yellow-red) indicator located on the ramp; a warning sign, which informs motorists that the ramp is being metered; and a controller, which is actuated by a time clock. The detectors shown are optional, but when

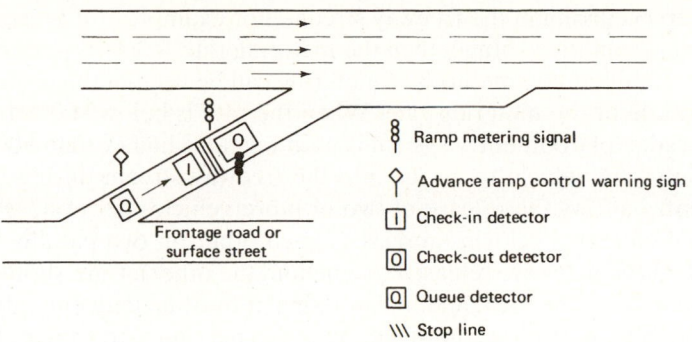

Figure 8.28   Layout of Pretimed Entrance Ramp Metering System

SOURCE: Adapted from *Ramp Management and Control Handbook*, Report No. FHWA-HOP-06-001, Figure 10-2, pg 10-12, Federal Highway Administration, Department of Transportation, Washington, D.C., 2003

used will enhance the efficient operation of the system. For example, the check-in detector allows the signal to change to green only when a vehicle is waiting, which means that the signal will stay red until a vehicle is detected by the check-in detector and the minimum red time has elapsed. The check-out detector is useful when a single-entry system is desired. The green interval is terminated immediately when a vehicle is detected by the check-out detector.

The calculation of the metering rate depends on the primary objective of the control. When this objective is to reduce congestion on the highway, the difference between the upstream volume and the downstream capacity (maximum flow that can occur) is used as the metering rate. This is demonstrated in Figure 8.29, where the metering rate is 400 veh/h. It must be remembered, however, that the guidelines given by the MUTCD must be taken into consideration. For example, if the storage space on the ramp is not adequate to accommodate this volume, the signal should be pretimed for a metering rate that can be accommodated on the ramp. When the objective is to enhance safety in the merging area of the ramp and the freeway, the metering rate will be such that only one vehicle at a time is within the merging area. This will allow an individual vehicle to merge into the freeway traffic stream before the next vehicle reaches the merging area. The metering rate in this case will depend on the average time it takes a stopped vehicle to merge, which in turn depends on the ramp geometry and the probability of an acceptable

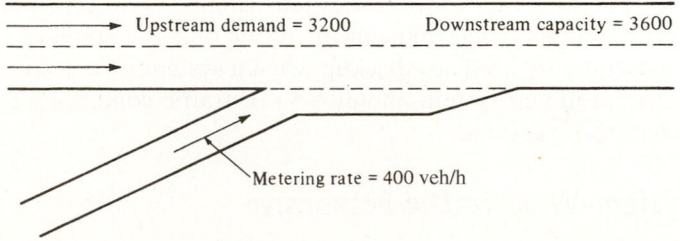

Figure 8.29   Relations among Metering Rate, Upstream Demand, and Downstream Capacity

SOURCE: Adapted from *Ramp Management and Control Handbook*, Report No. FHWA-HOP-06-001, Figure 10-2, pg 10-12, Federal Highway Administration, Department of Transportation, Washington, D.C., 2003

gap occurring in the freeway stream. For example, if it is estimated that it takes 9 sec to merge on the average, then the metering rate is 3600/9—that is, 400 veh/h.

One of two methods of metering can be used in this system, depending on the magnitude of the metering rate. When the rate is below 900 veh/h, a single-entry system is used; a platoon-entry system is used for rates higher than 900 veh/h. Single entry allows only one vehicle to merge into the freeway stream during the green interval; platoon entry allows the release of two or more vehicles per cycle. Platoon entry can be either parallel (two vehicles abreast of each other on two parallel lanes are released) or tandem (vehicles are released one behind the other). Care should be taken in designing the green interval for tandem platooning; it must be long enough to allow all the vehicles in the platoon to pass the signal. It is also necessary to frequently collect data at locations with pretimed metering systems so that metering rates can be adjusted to meet changes in the traffic conditions.

Traffic Responsive Metering.   This control system is based on the same principles as the pretimed metering system, but the traffic response system uses actual current information on traffic conditions obtained from freeway loop detectors or other methods to determine the metering rates, whereas historical data on traffic volumes are used to determine metering rates in the pretimed metering system. The traffic responsive system therefore has the advantage of being capable of responding to short-term changes in traffic conditions. The system layout for traffic responsive metering is similar to that for the pretimed ramp metering but with additional features, including some that are optional. The basic requirements include a traffic signal, detectors, a ramp-control sign, and a controller that can monitor the variation of traffic conditions. The optional features include the queue detector, which when continuously actuated indicates that the vehicles queued on the ramp may interfere with traffic on the local road; a merging detector, which indicates whether a merging vehicle is still in the merging area; and a check-out detector, which indicates whether a vehicle uses the green interval to proceed to the merging area. A detailed description of the design of a traffic response metering system is beyond the scope of this book, but interested readers may consult the *Ramp Management and Control Handbook* for further discussion of the subject. However, a brief discussion of the methods used to determine the metering rates follows.

Two methods are used to determine the metering rate: (1) local traffic responsive metering and (2) system-wide traffic responsive.

Local Traffic Responsive

This method uses traffic conditions immediately upstream and downstream of the metered ramp to determine the metering rate. It is similar to the pretimed system in that it is frequently used as a backup when a system-wide traffic algorithm fails. In contrast to the pretimed system, monitoring of traffic conditions along the freeway with loop detectors is required.

System–Wide Traffic Responsive

These systems are used to optimize the flow of traffic along the freeway rather than at a single location. Therefore, the selection of a metering rate for a location will be influenced by the traffic conditions at other ramps within the system or along the corridor to be metered. It is therefore necessary for data to be collected from many upstream and downstream ramps. It may also be necessary to obtain data from cross-street signal controllers and the central computer. Linear programming can be used to determine a set of integrated metering rates for each ramp, based on the expected range of capacity and demand conditions. The proper metering rate at each intersection is then selected from the appropriate set of metering rates, based on current traffic conditions on the freeway.

8.5.2 Benefits of Ramp Metering

The Federal Highway Administration has associated several benefits with the deployment of ramp metering systems on freeways. These benefits can be classified into four main groups; (1) improved system operation, (2) improved safety, (3) reduced environmental effects, and (4) promotion of multimodal operation. Improved system operation includes increasing vehicle throughput, vehicle speeds, and more efficient use of existing capacity. Improved safety is reflected in the reduction of the number of crashes and crash rates at locations upstream of the freeway ramp or within the freeway merge zone. Within the environmental category is the reduction of vehicle emissions and fuel consumption. The category of multimodal promotion involves the ease at which different modes of transportation can be accommodated.

8.6 EVALUATION AND OPTIMIZATION OF INTERSECTION TIMING PLANS

As noted previously, the procedure for computing the phase durations in an actuated signal timing system requires some knowledge of the cycle length. This requires an assumption of an initial cycle length, followed by an iterative procedure to obtain an optimal cycle length. The iteration usually involves the use of one or more performance measures to evaluate the effectiveness of the signal times. Performance measures that are commonly used include vehicle delay, person delay, travel time, queue lengths, number of stops, and air quality measures. The use of computers to reduce the computational effort and increase analysis flexibility has made the design of signalized systems less taxing. Several computer models are available for evaluating and/or optimizing signal timings. These include TRANSYT-7F, Highway Capacity Software (HCS), CORSIM, and Synchro. One of the more commonly used software packages is Synchro 8, which supports the HCM 2010 methodology for computing capacity at signalized intersections and roundabouts. It is a macroscopic capacity analysis and optimization software application that provides for data at many intersections to be entered using a single file. This eliminates the cumbersome data management problem that usually exists when the data for each signalized intersection is in a separate file. Synchro optimizes signal systems by attempting to minimize delays and stops while considering all intersections in the file. It allows the user to obtain measures of delays, queues, etc. that are based on specific equations. TRANSYT-7F uses a generic algorithm to optimize cycle length, phasing, and offsets. It can be used for several intersections along a street or a single intersection with complex or simple characteristics. It provides a time-space diagram screen that gives a graphical representation of signal progression along the major street. The user can also customize the screen to show progression bands in the forward, reverse, or both directions. The program has been integrated with the Highway Capacity Software (HCS) and CORSIM, so that candidates of timing plans obtained by TRANSYT-7F can be evaluated. This version of TRANSYT-7F also can be used for oversaturated conditions.

8.7 SUMMARY

Control of traffic at highway intersections is of fundamental importance to traffic engineers seeking ways to achieve efficient operation of any highway system. Several methods of controlling traffic at intersections are presented in this chapter. These include yield signs, stop signs, multiway stop signs, channelization, and traffic signals. When traffic conditions warrant signalization, it may be an effective means of traffic control at intersections, since the number of conflict points at an intersection could be significantly reduced by signalization. Note, however, that an attempt to significantly reduce the number of conflict points will increase the number of phases required, which may result in increased delay. The two methods—Webster and HCM—that are only used to determine the cycle and phase lengths of pretimed signal systems are discussed, together with the methods for determining the important parameters for actuated and coordinated signals using the 2010 HCM procedure. Any of these methods could be used to obtain reasonable cycle lengths for different traffic conditions at intersections.

Ramp control can be used to limit the number of vehicles entering or leaving an expressway at an off- or on-ramp. However, care should be taken in using ramp control because this may result in longer travel times or in congestion on local streets.

Although mathematical algorithms are presented for computing the important parameters required for signalization and ramp control, traffic engineers always must be aware that a good design is based both on correct mathematical computation and on engineering judgment.

PROBLEMS

8-1   Using an appropriate diagram, identify all the possible conflict points at an unsignalized T intersection.

8-2   Explain the considerations in deciding between stop control and yield control on a minor road approach at an intersection with a major road.

8-3   Determine whether the criteria for installation of multi-way stop control are satisfied for an intersection with a total entering volume of 230 veh/h on the major street approaches, 220 veh/h on the minor street approaches, and 6 crashes within the last 12 months that could be avoided had multi-way stop control been in place.

8-4   A two-phase signal system is installed at the intersection described in Problem 8-1, with channelized left-turn lanes and shared through and right-turn lanes. Using a suitable diagram, determine the possible conflict points. Indicate the phasing system used.

8-5   Using appropriate diagrams, determine the possible conflict points on a four-leg signalized intersection for a two-phase system. Assume no turn on red.

8-6   Repeat Problem 8-5 for the following phasing systems:

(a) Four-phase with separate phases for left turns
(b) Four-phase with separate phase for each approach

8-7   Determine whether this intersection for which data are provided below meets *MUTCD* Warrants 1, 2, and/or 3 for signalization. These data were collected using pneumatic tubes and the data were binned in 60-minute increments (15-minute data are not available). All approaches have one lane entering the intersection. For any warrants that are met, explain how the warrant was met (rules applied, hours in which the volume criteria were met, etc.) The major street speed limit is 72 kmph, and the minor street speed limit is 40 kmph. The minor street has only one approach (i.e. it is a "T" intersection.)

Hour Beginning	Major Street Volume	Minor Street Volume
0	25	3
1	28	2
2	31	1
3	59	2
4	270	27
5	714	115
6	774	159
7	460	151
8	479	123
9	430	93
10	508	98
11	455	85
12	537	112
13	558	82
14	736	91
15	785	99
16	689	81
17	546	71
18	396	60
19	264	42
20	193	29
21	149	21
22	84	15
23	36	8

8-8 A railroad track is crossed by a minor street that intersects a major street that runs parallel to the railroad track. There is a concern about queuing of minor street traffic, controlled by a STOP sign at the intersection with the major street, backward onto the track. Determine whether this intersection meets *MUTCD* Warrant 9, given the following data:

- Major street peak hour volume (two-way): 320 veh/h
- Minor street peak hour volume (on the one-lane approach that crosses the tracks): 120 veh/h; 6% of which are tractor-trailer trucks; bus traffic is negligible.
- Rail traffic is an average of 8 trains per day distributed in a scattered manner throughout the day
- The distance on the minor street between the nearest rail of the track and the STOP line at the intersection with the major street is 29 m.

8-9 Under what conditions would you recommend the use of each of the following intersection control devices at urban intersections:

(a) yield sign
(b) stop sign
(c) multiway stop sign

8-10 Both crash rates and traffic volumes at an unsignalized urban intersection have steadily increased during the past few years. Briefly describe the types of data you will collect and how you will use those data to justify the installation of a traffic signal at the intersection.

8-11   A traffic signal control is being designed for a four-leg intersection on a divided highway with the characteristics shown in the table below. Determine an appropriate length of the yellow interval for each approach and how you will provide it.

	N–S Approaches	E–W Approaches
Median width, (m)	4.8 m	3.6 m
Number of 3.6 m lanes on each approach	3	2
Design speed, km/h	50	40
Grade	0%	3%

8-12   Explain the conventional numbering of phases and the relationship among the phase numbers at an eight-phase signal in which left-turn phasing is provided on all 4 approaches.

8-13   Explain the implications of increasing the number of phases at a signalized intersection on safety and delay.

8-14   Determine an appropriate yellow interval for a signal phase under the following conditions:

- Approach speed limit: 45 km/h
- Approach grade: 2% downgrade
- Assumed perception-reaction time: 1.0 sec
- Assumed deceleration rate: 3.36 km/sec^2
- Assumed average vehicle length: 6 m
- Width of intersection to be crossed: 18 m

8-15   Determine the minimum green times for each approach in Problem 8-8 if the effective crosswalk width in each direction is 8 ft and the number of pedestrians crossing during an interval is 30 in the E-W direction and 25 in the N-S direction.

8-16   For the geometric and traffic characteristics shown below determine a suitable signal phasing system and phase lengths for the intersection using the Webster method. Show a detailed layout of the phasing system and the intersection geometry used.

Approach (Width)	North (16.8 m)	South (16.8 m)	East (20.4 m)	West (20.4 m)
Peakhour approach volumes				
Left turn	133	73	168	134
Through movement	420	373	563	516
Right turn	140	135	169	178
Conflicting pedestrian volumes	900	1200	1200	900
PHF	0.95	0.95	0.95	0.95

Assume the following saturation flows:

Through lanes:	1600 veh/h/ln
Through-right lanes:	1400 veh/h/ln
Left lanes:	1000 veh/h/ln
Left-through lanes:	1200 veh/h/ln
Left-through-right lanes:	1100 veh/h/ln

8-17   Repeat Problem 8-16 using saturation flow rates that are 10% higher. What effect does this have on cycle length?

8-18   Repeat Problem 8-16 using pedestrian flow rates that are 20% higher. What effect does this have on cycle length?

8-19   Repeat Problem 8-16 using the HCM method and a critical v/c of 0.9.

8-20   Using the results for Problems 8-16 and 8-19, compare the two different approaches used for computing cycle length.

8-21   Briefly describe the different ways the traffic signals at the intersection of an arterial route could be coordinated, stating under what conditions you would use each of them.

8-22   You have been asked to design a simultaneous traffic signal system for six intersections on a suburban arterial. The distances between consecutive intersections are:

Intersection A to Intersection B - 1140 m
Intersection B to Intersection C - 1200 m
Intersection C to Intersection D - 1188 m
Intersection D to Intersection E - 1155 m
Intersection E to Intersection F - 1185 m

Suitable cycle lengths for the intersections are:

Intersection A - 60 sec
Intersection B - 55 sec
Intersection C - 65 sec
Intersection D - 60 sec
Intersection E - 55 sec
Intersection F - 60 sec

If an appropriate progression speed for the arterial is 80 kmph, what cycle length would you use if the signals are not actuated? Give a reason for your choice.

8-23   In Problem 8-22, if conditions at intersection C require that a cycle length of 65 sec be maintained through the corridor, what will be a suitable progression speed?

8-24   In Problem 8-22, an intersection located midway between intersections B and C is to be signalized. Recommend a cycle length and progression speed for the corridor that includes the new signal if the signals are not actuated.

8-25   Describe the effects of the three left-turn treatments at signalized intersections on saturation flow rate and geometric design.

8-26   Using the results of Problem 8-16, determine the queue service length for each phase assuming the proportion of vehicles arriving during the green interval is 0.8 for the north and south approaches and 0.85 for the east and west approaches.

8-27   Briefly discuss the different methods by which freeway entrance ramps can be controlled. Clearly indicate the advantages and disadvantages of each method, and give the conditions under which each of them can be used.

8-28   Compare and contrast the different metering systems that are used in traffic signal ramp control indicating under what conditions you will use each.

REFERENCES

Highway Capacity Manual. HCM 2010, Transportation Research Board, National Research Council, Washington, D.C., 2010.

Intersection Channelization Design Guide. National Cooperative Highway Research Program Report 279, National Research Council, Transportation Research Board, Washington, D.C., November 1985.

Manual of Traffic Signal Design, 2nd Edition, Institute of Transportation Engineers Washington D.C. 1998.

Manual on Uniform Traffic Control Devices. U.S. Department of Transportation, Federal Highway Administration, Washington, D.C., 2009.

Ramp Management and Control Handbook, Report No. FHWA-HOP-06-001, Federal Highway Administration, U.S. Department of Transportation, Washington, D.C., 2006.

Signalized Intersection, Information Guide. U.S. Department of Transportation, Federal Highway Administration, Report No. FHWA-HRT-04-091, Washington, D.C., August 2004. *www.tfhrc.gov/safety/pubs/04091/*

Traffic Analysis Toolbox: Guidelines for Applying CORSIM Microsimulation Modeling Software. Publication No. FHWA-HOP-07-079, U.S. Department of Transportation, Federal Highway Administration, Washington, D.C., 2007. *www.ops.fhwa.dot.gov/trafficanalysistools/toolbox.htm*

Traffic Engineering Handbook, 6th edition, Institute of Transportation Engineering, Washington, D.C., 2009.

Capacity and Level of Service for Highway Segments

The fundamental diagram of traffic flow was used in Chapter 6 to illustrate the relationship between flow and density. It was shown that traffic flows reasonably well when the flow rate is less than at capacity, but excessive delay and congestion can occur when the flow rate is at or near capacity. This phenomenon is a primary consideration in the planning and design of highway facilities, because a main objective is to design or plan facilities that will operate at flow rates below their maximum flow rate, known as capacity. This objective can be achieved when an estimate of the capacity of a highway facility can be made. Capacity analysis therefore involves the quantitative evaluation of the capability of a road section to carry traffic, based upon procedures to determine the maximum flow of traffic that a given section of highway will carry under prevailing roadway, environmental, traffic, and control conditions.

It was also shown in Chapter 6 that the maximum speed that can be achieved on a uniform section of highway is the mean free speed, which depends solely on the physical characteristics of the highway. This speed can be achieved when traffic demand volumes are low (< 1000 pc/h/ln) and there is little interaction between vehicles on the highway segment. Under these conditions, motorists are able to drive at a desired speed up to the mean free speed and, accordingly, they perceive a high level of service in the quality of traffic flow on the highway. As the demand volume increases, vehicle interaction and density increase, resulting in the gradual lowering of mean speeds and the quality of traffic flow and level of service

As the interaction among vehicles increases, motorists are increasingly influenced by the actions of others. Individual drivers find it more difficult to achieve their desired speeds, and there is a perceived deterioration in the quality of flow as density (pc/mi) increases. As discussed in Chapter 6, the level of operating performance changes with traffic density. The measure of quality of flow is the level of service (LOS), a qualitative measure, ranging from A to F, that characterizes both operational conditions within a traffic stream and highway users' perception.

This chapter presents the procedures used to determine the level of service and other performance measures of uniform road segments on freeways, multilane highways, and two-lane roads. The procedures described in this chapter are based on the *Highway Capacity Manual*, HCM 2010, published by the Transportation Research Board of the National Academies.

CHAPTER OBJECTIVES:

- Learn how to use the HCM 2010 procedures to determine the level of service and other performance measures of uniform road segments.
- Become familiar with highway capacity terminology and definitions.
- Apply the operational analysis procedure for estimating speed (S) and density (D) to determine LOS (A–F) for basic freeway segments.
- Apply the equations and procedures developed for basic freeway sections to analyze the capacity and LOS for multilane highways.
- Become familiar with the procedures developed to evaluate level of service and capacity for two-lane highway segments.

9.1 FREEWAYS

A freeway is a divided highway with full access control and two or more lanes in each direction for the exclusive use of moving traffic. Signalized or stop controlled, at-grade intersections and direct access to adjacent land use are not permitted in order to ensure the uninterrupted flow of vehicles. A raised barrier, an at-grade median, or a raised traffic island separates opposing traffic. A freeway is composed of three elements: basic freeway sections, weaving areas, and ramp junctions.

Basic freeway sections are segments of the freeway that are outside of the influence area of ramps or weaving areas. Merging or diverging occurs where on – or off-ramps join the basic freeway section. Weaving occurs when vehicles cross each others' path while traveling on freeway lanes.

Figure 9.1a illustrates basic freeway segments for an urban freeway and depicts the influence areas of three segment types; (1) merge, 457 m downstream of the merge point; (2) diverge, 457 m upstream from the point where the diverging lane exits the freeway; and (3) weaving, 150 m both upstream and downstream of the entry and exit point. Figure 9.1b schematically depicts a 2.7 km (2770 m) section of an urban freeway that includes a weaving section and two merge sections. These segments decrease the length of the basic segment by 1700 m, creating two basic segments of 300 m and one basic segment of 457 m.

The exact point at which a basic freeway segment begins or ends—that is, where the influence of weaving areas and ramp junctions has dissipated—depends on local conditions, particularly the level of service operating at the time. If traffic flow is light, the influence may be negligible, whereas under congested conditions, queues may be extensive. In the absence of other data, the points where the edges of the travel lanes meet are used and are typically defined by pavement markings.

The *capacity* of a freeway is the maximum sustainable hourly flow rate in one direction, based on the peak 15-minute rate of flow, expressed in passenger cars per hour. Capacity flow can be accommodated through a uniform freeway segment under prevailing traffic and roadway conditions. The roadway conditions are the geometric characteristics, which include the number and width of lanes, right-shoulder lateral clearance, ramp density, and grade. The traffic conditions are flow characteristics, which include

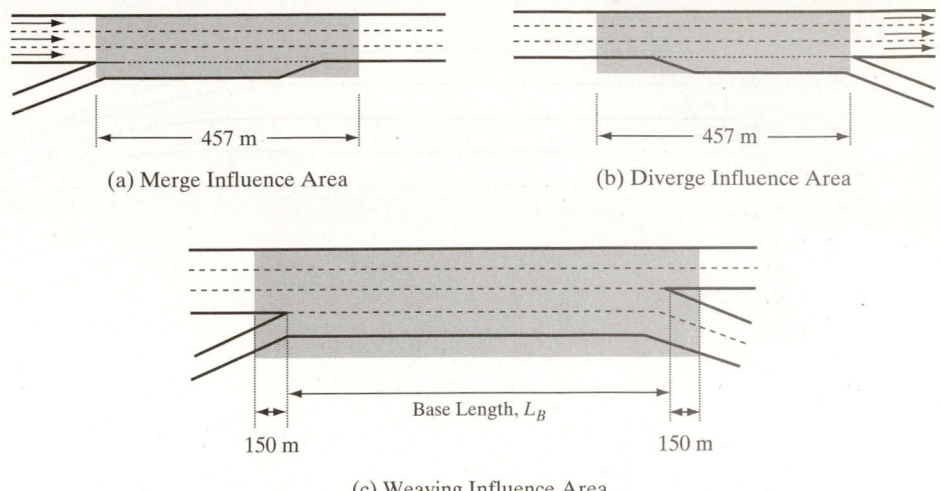

(a) Merge Influence Area   (b) Diverge Influence Area

(c) Weaving Influence Area

Figure 9.1(a)   Influence Areas of Merge, Diverge, and Weaving Segments

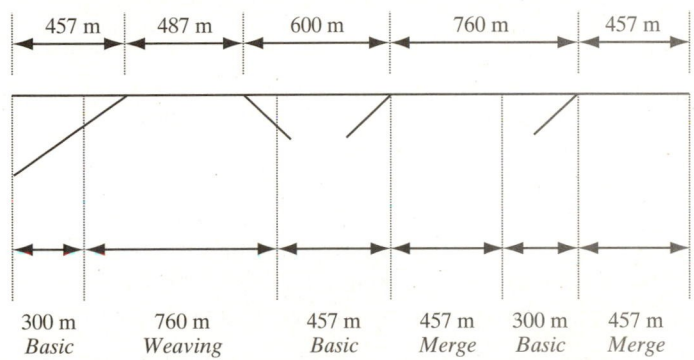

Figure 9.1(b)   Basic Freeway Segments on an Urban Freeway

SOURCE: From *Highway Capacity Manual 2010*. Copyright, National Academy of Sciences, Washington, D.C. Reproduced with permission of the Transportation Research Board.

the percentage composition of passenger cars and heavy vehicles and the extent to which drivers are familiar with the freeway segment.

The speed-flow relationship for basic freeway segments operating under base free-flow conditions is illustrated in Figure 9.2. The five curves are for free-flow speeds between 55 and 75 mi/h (88 and 120 km/h). Capacity is reached at a density of 45 pc/mi/ln and capacity values range from 2250 to 2500 vph. The flow rate at the inflection point of each curve (called the breakpoint) is where speed begins to diminish.

9.1.1. Highway Capacity Terminology and Definitions

The variables used in Figure 9.2, "Speed-Flow Curves for Basic Freeway Segments under Base Conditions," are defined as follows. An understanding of the meaning these terms is essential if the HCM 2010 procedures are to be followed correctly.

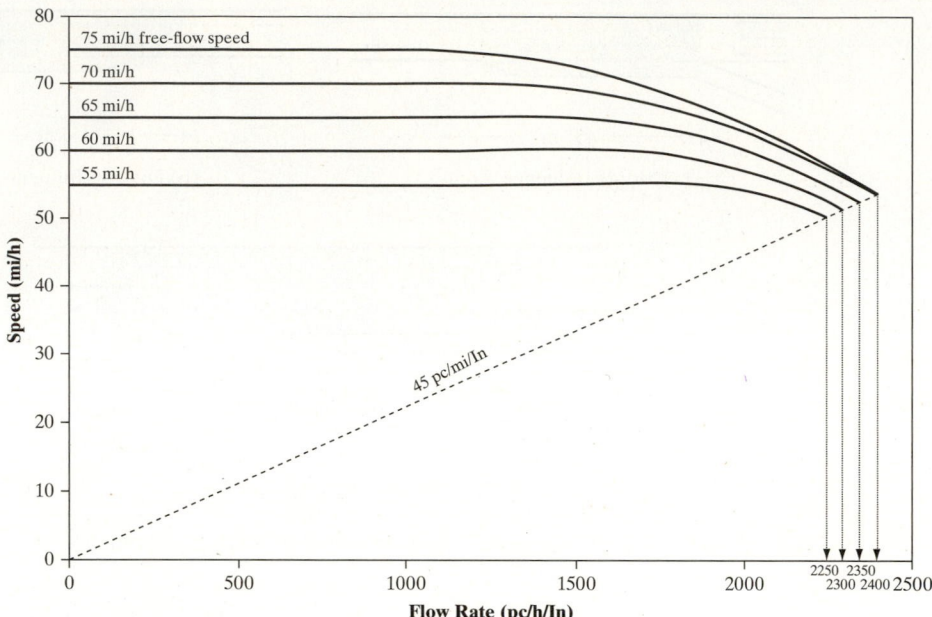

Figure 9.2   Speed-Flow Curves for Basic Freeway Segments under Base Conditions

SOURCE: From *Highway Capacity Manual 2010*. Copyright, National Academy of Sciences, Washington, D.C. Reproduced with permission of the Transportation Research Board.

Base Conditions

These are the criteria that must be satisfied if a basic freeway segment is to operate at the maximum capacities shown in Figure 9.2. If one or more of the following conditions is not met, the capacity of the segment will be reduced.

1. Weather and driver visibility are good and do not affect traffic flow. There are no obstructions that could impede traffic flow, such as incidents, accidents, or roadwork. The HCM 2010 capacity procedure does not offer a computational compensation factor for these conditions. Accordingly, the procedure is based on the assumption that each condition is met.
2. The traffic stream is composed solely of passenger cars and there are no heavy vehicles such as trucks, buses, and recreational vehicles (RVs).
3. Traffic lanes are at least 3.5 m wide, and a right-side clearance exists that is at least 1.8 m wide.
4. All drivers in the traffic stream are regular users and are familiar with the freeway segment.
5. The total ramp density is zero. Ramp density is the total number of on + off ramps per mile in one direction within the freeway segment under study. For the ramp density to be zero, there are no ramps within the segment. The HCM 2010 capacity procedure for basic freeway segments applies to segments no greater than 6 mi (9.6 km) long.
6. A state of unsaturated flow must exist for the HCM 2010 procedures to apply. This state occurs when the traffic stream is unaffected by upstream or downstream bottlenecks.

Free-Flow Speed

As shown in Figure 9.2, free-flow speed (FFS) is a constant value beginning at the *y*-intercept of each curve and extending to the breakpoint, where the mean speed of the traffic stream diminishes. The volume range where FFS is constant varies as a function of the value of FFS as follows.

FFS (km/h)	Range of Flow Rate (pc/h/ln)
120	0–1000
112	0–1200
104	0–1400
96	0–1600
88	0–1800

Flow Rates beyond the Breakpoint

For demand volumes beyond the breakpoint, the flow rate declines in value until it reaches the capacity of the freeway segment. Beyond this point the HCM 2010 procedures are no longer valid. Equations 9.1a through 9.1e are used to compute free-flow speeds beyond the breakpoint.

$$S_{BP-75} = 75 - 0.00001107(v_p - 1000)^2 \tag{9.1a}$$

$$S_{BP-70} = 70 - 0.00001160(v_p - 1200)^2 \tag{9.1b}$$

$$S_{BP-65} = 65 - 0.00001418(v_p - 1400)^2 \tag{9.1c}$$

$$S_{BP-60} = 60 - 0.00001816(v_p - 1600)^2 \tag{9.1d}$$

$$S_{BP-55} = 55 - 0.00002469(v_p - 1800)^2 \tag{9.1e}$$

where

S_{BP-75} = flow rate beyond the breakpoint at FFS of 75 mi/h (120 km/h) (similar notation is used for all other values of FFS, 70–55 mi/h (112–88 km/h)

v_p = demand flow rate (pc/h/ln) under equivalent base conditions

Computing Free-Flow Speed

Free-flow speed is determined in two ways: (1) by field measurements or (2) by formula estimation.

1. *Field measurement.* Free-flow speed is the mean speed of passenger cars during a period of low to moderate flow no greater that 1000 pc/h/ln. Methods for conducting speed studies are described in Chapter 4. Select all passenger cars or a systematic sample and use at least 100 passenger-car speeds.
2. *Estimating FFS.* If measurement of FFS is not practical, then the value may be estimated by using Eq. 9.2.

$$FFS = 75.4 - f_{LW} - f_{LC} - 3.22 TRD^{0.84} \tag{9.2}$$

where

FFS = free-flow speed
f_{LW} = adjustment for lane width (mi/h(km/h)) (Table 9.1)
f_{LC} = adjustment for right side lateral clearance (mi/h(km/h)) (Table 9.2)
TRD = total ramp density (ramps per mile)

9.1.2 Level of Service for Freeway Segments

Level of service (LOS) is a qualitative measure that describes the operating conditions within a given freeway segment using density (pc/mi) as the quantitative variable and an alphabetic grading system, ranging from A to F, to convey how the density values are perceived by drivers and passengers when traveling in the traffic stream at various demand flow rates (pc/h/ln).

The relationship among speed, flow rate, and density is shown in Eq. 9.3.

$$D = \frac{y_p}{S} \tag{9.3}$$

where

v_p = flow rate (pc/h/ln)
S = Average passenger car speed mi/h (km/h)
D = Density (pc/mi/ln (pc/km/ln))

Density describes the proximity to other vehicles and thus is a measure of the freedom to maneuver. Level of service describes how the motorist would evaluate the quality of the driving experience by using a letter scale very similar to that used to grade the performance of a student. Figure 9.3a depicts the five speed-flow curves (75–55 mi/h(120–88 km/h)) together with dotted lines showing the demarcation in density for level of service A–F. Figure 9.3b illustrates examples of each level of service as perceived by motorists. The density ranges and a description of each LOS are as follows.

LOS	Density (pc/mi/ln)
A	≤11
B	>11–18
C	>18–26
D	>26–35
E	>35–45
F	Demand exceeds capacity >45

Level of Service A

Free-flow operations exist in which vehicles are almost completely unimpeded in their ability to maneuver within the traffic stream. FFS prevails and the effects of incidents or point breakdowns are easily absorbed.

Level of Service B

Traffic is moving under reasonably free-flow conditions, and FFS is maintained. The ability to maneuver within the traffic stream is only slightly restricted and the general level of physical and psychological comfort provided to drivers is still high. The effects of minor incidents and point breakdowns are still easily absorbed.

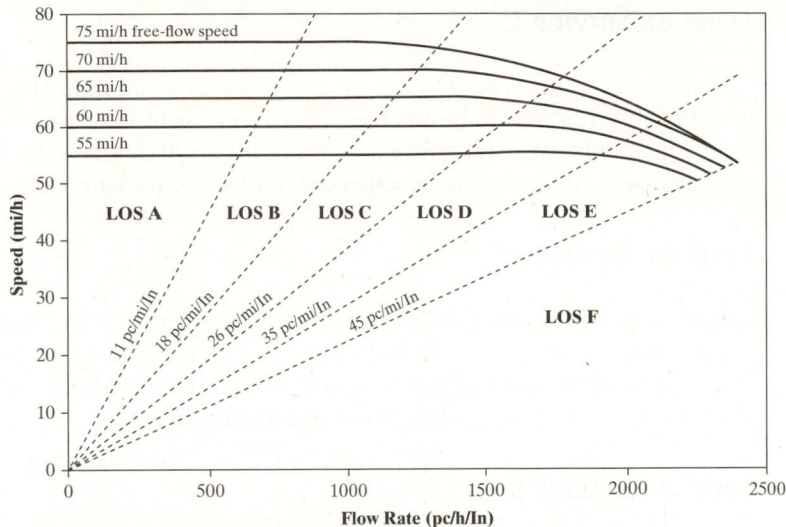

Figure 9.3(a)   LOS for Basic Freeway Segments

Figure 9.3(b)   LOS Examples

SOURCE: From *Highway Capacity Manual 2010*. Copyright, National Academy of Sciences, Washington, D.C. Reproduced with permission of the Transportation Research Board.

Level of Service C

Speeds are near the free-flow speed, and freedom to maneuver within the traffic stream is noticeably restricted. Lane changes require more care and vigilance by the driver. Minor incidents may still be absorbed, but the local deterioration in service quality will be significant. Queues may be expected to form behind any significant blockage.

Level of Service D

Speeds begin to decline with increasing flows, with density increasing more quickly. Freedom to maneuver within the traffic stream is seriously limited and drivers experience reduced physical and psychological comfort levels. Minor incidents can be expected to create queuing because the traffic stream has little space to absorb disruptions.

Level of Service E

Operation is at or near capacity. Freeway operations are highly volatile because there are virtually no useable gaps within the traffic stream, leaving little room to maneuver. Any disruption within the traffic stream, such as vehicles entering from a ramp or a vehicle changing lanes, can establish a disruption wave that propagates throughout the upstream traffic flow. At capacity, the traffic stream has no ability to dissipate even the most minor disruption, and any incident can be expected to produce a serious breakdown and substantial queuing. Maneuvers such as lane changes or merging of traffic from entrance ramps will result in a disturbance of the traffic stream. Minor incidents result in immediate and extensive queuing. The physical and psychological comfort afforded to drivers is poor.

Level of Service F

Operation is at breakdown conditions or unstable flow. These conditions exist within queues forming behind bottlenecks. Breakdowns occur for the following reasons: (1) Traffic incidents that temporarily reduce the capacity of a short segment such that the number of vehicles arriving at a point is greater than the number of vehicles that can move through it. (2) Points of recurring congestion such as merge or weaving segments and lane drops. Freeway sections experience very high demand such that the number of vehicles arriving is greater than the number discharged. (3) In analysis using forecast volumes, the projected flow rate can exceed the estimated capacity at the segment being analyzed. Breakdown occurs when the ratio of existing demand to actual capacity or of forecasted demand to estimated capacity exceeds 1.00.

9.1.3 Calculating the Flow Rate for a Basic Freeway Section

The speed-flow curves for basic freeway segments shown in Figure 9.2 are based on traffic streams with only passenger cars (no heavy vehicles) and a driver population consisting of regular users. If actual demand volumes include heavy vehicles and a different driver population mix, Eq. 9.4 is used to convert the actual demand volume, V, to a demand flow rate, v_p, under equivalent base conditions.

$$v_p = \frac{V}{(PHF)(N)(f_p)(f_{HV})} \qquad (9.4)$$

where

v_p = demand flow rate under equivalent base conditions (pc/h/ln)
V = demand volume under prevailing conditions (veh/h)
PHF = peak hour factor
N = number of lanes in the analysis direction
f_p = adjustment factor for unfamiliar driver populations. Range is 0.85–1.00 (In general, use 1.00 for commuters or other accustomed drivers unless evidence exists to the contrary.)
f_{HV} = adjustment factor for presence of heavy vehicles in traffic stream (Eq. 9.5)

$$f_{HV} = \frac{1}{1 + P_T(E_T - 1) + P_R(E_R - 1)} \tag{9.5}$$

where

P_T = proportion of trucks in the traffic stream
P_R = proportion of RVs in the traffic stream
E_T = passenger car equivalent (PCE) of one truck or bus in traffic stream
E_R = PCE of one RV in the traffic stream

The passenger car equivalent (PCE) of each truck/bus (E_T) or recreational vehicle (E_R) is determined for the prevailing traffic and roadway conditions. These tabular values represent the number of passenger cars that would use up the same space on the highway as one truck, bus, or recreational vehicle under prevailing conditions. Trucks and buses are treated identically because it has been determined that their traffic flow characteristics are similar. The proportion of each type of vehicle, P_T and P_R, completes the necessary adjustment factors to compute f_{HV} using Eq. 9.5.

The extent to which the presence of a truck affects the traffic stream also depends on the grade of the segment being considered. PCEs for trucks can be determined for three grade conditions: (1) extended segments in general terrain, (2) specific upgrades, or (3) specific downgrades. A procedure for including the effect of grades on the 15-minute passenger car equivalent flow rate is discussed in the following sections.

Equivalents for General Terrain Segments

General terrain refers to a series of single grades that are not too long or too steep and as such do not have a significant impact on the operation of the overall segment. Such upgrades, downgrades, and level sections are considered as extended general freeway segments. They encompass grades of ≤ 2 percent and segment length ≤ 0.40 km long or grades between 2 percent and 3 percent and segment lengths ≤ 0.80 km long. The PCEs for these conditions are given in Table 9.3. The type of terrain, which is classified as level, rolling, or mountainous, affects PCE value. Each category is described as follows.

Freeway segments are considered to be on *level terrain* if the combination of grades and horizontal alignment permits heavy vehicles to maintain the same speed as passenger cars. Grades are generally short and not greater than 2 percent.

Freeway segments are considered to be on *rolling terrain* if the combination of grades and horizontal alignment causes heavy vehicles to reduce their speeds to values substantially below those of passenger cars but not to travel at crawl speeds for any significant length of time or at frequent intervals. Crawl speed is the maximum sustained speed that trucks can maintain on an extended upgrade of a given percent.

Freeway segments are considered to be in *mountainous terrain* if the combination of grades and horizontal alignment causes heavy vehicles to operate at crawl speeds for a significant distance or at frequent intervals.

Equivalents for Specific Upgrades

Any freeway grade of 3 percent or greater and longer than 0.40 km, or a grade between 2 percent and 3 percent and longer than 0.80 km, should be considered as a separate segment. Specific grades are analyzed individually for downgrade and upgrade conditions. The segment is considered either as a single isolated grade of constant percentage or as a series forming a composite grade, which is discussed later in this chapter.

The variety of trucks and recreational vehicles with varying characteristics results in a wide range of performance capabilities on specific grades. The truck population on freeways has an average weight-to-horsepower ratio of 57 to 68 kg/hp, and this range is used in determining PCEs for trucks and buses on specific upgrades. RVs vary considerably in both type and characteristics and range from cars pulling trailers to large, self-contained mobile campers. Further, unlike trucks, professionals typically do not drive RVs, and the skill levels of recreational vehicle drivers greatly vary. Typical weight-to-horsepower ratios for RVs are between 13 to 26 kg/hp.

Tables 9.4 and 9.5 list PCE values for trucks/buses (E_T) and for recreational vehicles (E_R), respectively, traveling on specific upgrades for different grades and different percentages of heavy vehicles within the traffic stream. For example, the effect on other traffic of a truck or bus traveling up a grade is magnified with increasing segment length and grade because the vehicle slows down. The PCEs selected should be associated with the point on the freeway where the effect is greatest, which is usually at the end of the grade, although a ramp junction at midgrade could be a critical point. The grade length should be obtained from a profile of the road, including the tangent portion, plus one-fourth of the length of the vertical curves at the beginning and end of the grade. In situations where two consecutive upgrades are connected by a vertical curve, half the vertical curve length is added to each portion of the grade. However, this guideline may not apply for some specific conditions. For example, to determine the effect of an on-ramp at an upgrade section of a freeway, the length used is that up to the ramp junction.

Equivalents for Specific Downgrades

If the downgrade is not so severe as to cause trucks to shift into low gear, they may be treated as if they were on level segments. In cases where grades are severe and require that trucks downshift, the effect on car equivalents (E_T) is greater, as shown in Table 9.6. For recreational vehicles, a downgrade is treated as if it were level.

Equivalents for Composite Grades

When a segment of freeway consists of two or more consecutive upgrades with different slopes, the PCE of heavy vehicles is determined by using one of two techniques. One technique is to determine the average grade of the segment by dividing the total rise in elevation by the total horizontal distance of the segment. The computed average grade is then used with Tables 9.4 and 9.5 to determine the PCE values for trucks and RVs. The average grade technique is valid when grades in all subsections are less than 4 percent or the total length of the composite grade is less than 1200 m. The average grade method is illustrated in Example 9.1.

Example 9.1 Computation of PCE for Consecutive Upgrades Using Average Grades

A segment of freeway consists of two consecutive upgrades of 3 percent, 600 m long and 2 percent, 450 m long. Determine the PCE of trucks/buses and recreational vehicles on this composite upgrade if 6 percent of the vehicles are trucks and buses and 10 percent are recreational vehicles.

Solution: The average grade technique can be used since subsection grades are less than 4 percent and the total length is less than 1200 m.

$$\text{Total rise} = (0.03 \times 600) + (0.02 \times 450) = 27 \text{ m}$$

$$\text{Average grade} = 27/(600 + 450) = 0.026, \text{ or } 2.6\%$$

$$\text{Total length} = 1050/100 = 1.05 \text{ km } (0.66 \text{ mi})$$

Enter Table 9.4, with >2 to 3 percent grade and length >0.5–0.75 (0.8–1.2 km) and 6 percent trucks to obtain E_T for trucks/buses of 1.5.

Enter Table 9.5, with >2 to 3 percent grade and length > 0.5 mi (0.8 km), and 10 percent RVs to obtain E, for recreational vehicles of 1.5.

The second method for determining the PCE of heavy vehicles on consecutive upgrades applies to more severe composite grades. The procedure involves estimating the value of an equivalent continuous grade G_E that would result in a truck speed at the end of the segment upgrade that is the same as the actual exit speed from the segment when traversing the composite grades. This technique should always be used if any single portion of the consecutive grade exceeds 4 percent or if the total length of grade (measured horizontally) exceeds 1200 m.

Truck acceleration/deceleration curves are used based on a vehicle with an average weight-to-horsepower ratio of 900 N/hp, which represents a somewhat heavier vehicle than the "typical" truck of 550 to 700 N/hp used to determine PCE values. This is a conservative approach that accounts for the greater influence heavier vehicles have over light vehicles in operation. Figure 9.4 is a performance curve for a standard 90 kg/hp truck, depicting the relationship between speed (km/h) and length of grade for upgrades (deceleration—solid lines) ranging from 1 to 8 percent and downgrades (acceleration—dashed lines) ranging from 0 to 5 percent. The curves are conservative as a maximum truck speed of 88 km/h for trucks entering a grade and 95 km/h for truck acceleration on the grade is assumed. For example, at a 5 percent upgrade, a truck will slow to a crawl speed of 43 km/h in a distance of 1500 m.

Figure 9.4 can be used to determine the single constant grade G_E for a series of consecutive grades of length L_{CG}. The procedure, illustrated in Example 9.2, requires that the speed V_{EG} of a truck at the end of the consecutive grades be determined. Then, given the length of the composite grade, L_{CG}, and knowing the truck speed at the end of the grade, V_{EG}, using Figure 9.4, an equivalent grade G_E is determined.

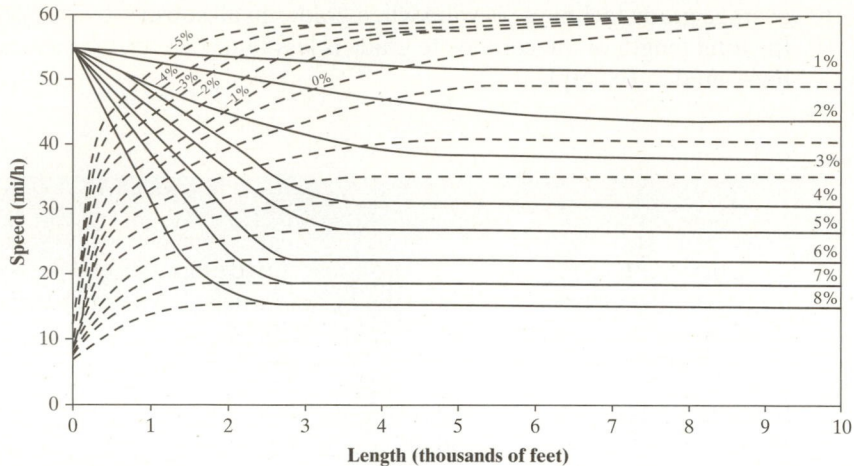

Figure 9.4   Performance Curves for 200-lb/hp Truck

SOURCE: From *Highway Capacity Manual 2010*. Copyright, National Academy of Sciences, Washington, D.C. Reproduced with permission of the Transportation Research Board.

Example 9.2 Computation of PCE for a Consecutive Upgrade Using Performance Curves

A consecutive upgrade consists of two sections, the first of 2 percent grade and 1500 m long, and the second of 6 percent grade and 1500 m long. Trucks comprise 10 percent of traffic, and recreational vehicles comprise 6 percent. Determine (a) the equivalent grade and (b) the PCEs. Entry speed is 88 km/h (55 mi/h).

Solution:

(a) Since the lengths of both grades exceed 1200 m, the average grade technique does not apply. Instead, the performance curve for standard trucks must be used, as illustrated in Figure 9.5.

1. Find the point on the 2 percent curve that intersects with length = 1500 m. In this case (point 1), that point signifies that the speed of a truck at the end of the first grade (point 2) is 75 km/h.
2. Since 75 km/h is also the speed of the truck at the beginning of the 6 percent grade, draw a line horizontally, at 75 km/h, intersecting the 6 percent grade curve (point 3). The reference distance is located on the horizontal axis (point 4).
3. Find the point on the horizontal axis that has a value 1500 km greater than that at point 4. This point (point 5) represents the distance traveled by a truck on a 6 percent grade whose initial speed is 75 km/h. The final speed for this truck is found to be 37 km/h (point 7), which is also the crawl speed, since the 6 percent curve is horizontal at that point.
4. Finally, determine the intersection of the lines L_{CG} = 3000 m and V_{EG} = 37 km/h (point 8). Since this point lies on the 6 percent grade curve, the equivalent grade, G_E, is 6 percent.

(b) PCEs for typical trucks on specific freeway upgrades:

$$L_{CG} = 10{,}000/5280 = 1.89 \text{ mi } (3000/1584 = 3.02 \text{ km})$$
$$G_E = 6\%$$

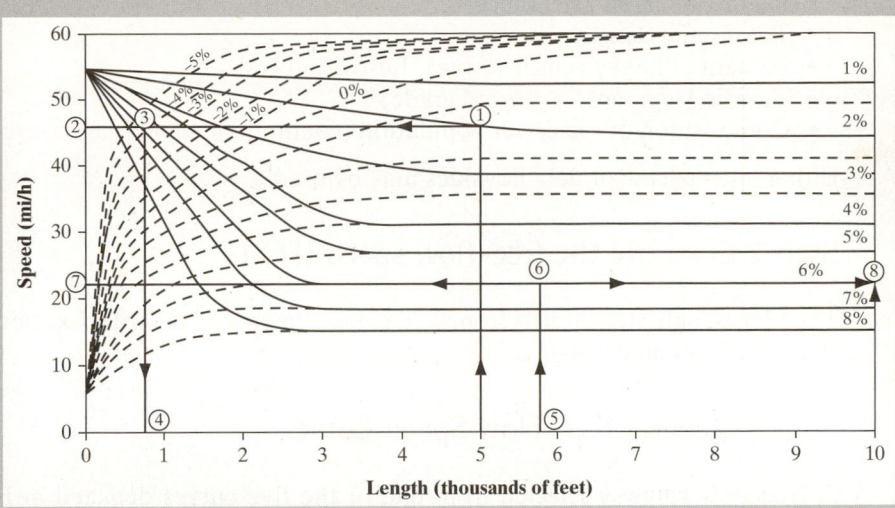

Figure 9.5   Solution Using Composite Grade Procedure

SOURCE: From *Highway Capacity Manual 2010*. Copyright, National Academy of Sciences, Washington, D.C. Reproduced with permission of the Transportation Research Board.

$$P_T = 10\%$$
$$P_R = 6\%$$
$$E_T = 3.5\,(\text{from Table 9.4})$$
$$E_R = 4.0\,(\text{from Table 9.5})$$

Note: If the average grade technique had been used, G_A would have been calculated as $(100 + 300)/10{,}000$, or 4%, resulting in values for E_T of 3.0 and E_R of 2.5, both lower than the correct values of 3.5 and 4.0.

9.1.4 Operational Analysis Procedure for Estimating Speed (*S*) and Density (*D*), to Determine LOS (A–F) for Basic Freeway Segments

An operational analysis of a basic freeway segment requires specific geometric data, traffic characteristics of vehicles in the travel stream, and all relevant demand flow rates. Where field data are unavailable, default values are used.

The output of an operational analysis for a basic freeway segment may include: (1) average speed (km/h), (2) average density (pc/mi/ln), (3) vehicle-miles of travel (VMT), (4) vehicle hours of travel (VHT), and (5) travel time (h/veh). The operational analysis procedure involves six steps that are described as follows.

Step 1 Specify Input Data

The following data must be provided in order to conduct an operational analysis of a basic freeway section.

- Demand volume
- Number and width of lanes

- Right-side lateral clearance
- Total ramp density
- Percent of heavy vehicles (truck, buses, and RVs)
- Terrain (segment length and grade)
- Composition of the driver population (commuters, regular users, etc.)

Either site-specific or default values may be used.

Step 2 Compute the free-flow speed (FFS)

The FFS is computed directly from field measurements, as described earlier in this chapter, or can be estimated using Eq. 9.2.

Step 3 Select a Free-Flow Speed Curve

A free-flow value is selected from one of the five curves depicted in Figure 9.2. The curve selected is based on the value of FFS computed in Step 2. HCM 2010 recommends that the analyst not interpolate between curves but select one of the five FFS values, 55-75 (mi/h) based on the following limits:

$$\geq 72.5\,\text{mi/h} < 77.5\,\text{mi/h: use FFS} = 75\,\text{mi/h}$$

$$\geq 67.5\,\text{mi/h} < 72.5\,\text{mi/h: use FFS} = 70\,\text{mi/h}$$

$$\geq 62.5\,\text{mi/h} < 67.5\,\text{mi/h: use FFS} = 65\,\text{mi/h}$$

$$\geq 57.5\,\text{mi/h} < 62.5\,\text{mi/h: use FFS} = 60\,\text{mi/h}$$

$$\geq 52.5\,\text{mi/h} < 57.5\,\text{mi/h: use FFS} = 55\,\text{mi/h}$$

Step 4 Adjust Demand Volume

If the traffic stream includes heavy vehicles and/or a noncommuter driver population, the demand volume must be converted to equivalent base conditions by using Eq. 9.4 and 9.5 to determine v_p, demand flow rate under prevailing conditions (pc/h/ln), and the heavy vehicle factor, f_{HV}.

The heavy vehicle factor is a function of the proportion of heavy vehicles and RVs in the traffic stream and the PCE equivalents for each vehicle category. Furthermore, the PCE values are a function of the type of terrain categorized as general: level, rolling and mountainous (Table 9.3), specific upgrades (Tables 9.4 and 9.5), and specific downgrades (Table 9.6). A special procedure is followed to determine a single equivalent value for a composite grade, which consists of two or more contiguous grades in the segment.

The adjustment factor for driver population ranges from 0.85 to 1.00. In most cases, a value of 1.00 is used unless there is evidence that a less familiar driver population, such as recreation travel, is prevalent.

Step 5 Estimate Speed and Density

The preceding steps have produced the value for v_p, the equivalent passenger car demand flow rate under prevailing conditions (pc/h/ln), and the FFS. The values can be used to calculate the mean speed of the traffic stream using Eq. 9.2 or Eqs. 9.1(a)–9.1(f) if the demand flow rate under base conditions exceeds the breakpoint. The density is computed using Eq. 9.3.

If the density exceeds 45 pc/mi/ln as shown in Figure 9.3, the segment is operating at LOS = F. This condition reflects traffic flow at capacity, and as such, the operational analysis procedure described in this chapter is not valid. The corresponding capacity value for each FFS is shown as follows. If the computed value for demand flow rate is less than capacity, the analysis proceeds to Step 6.

FFS (mi/h)	Capacity (pc/h/ln)
75	2500
70	2400
65	2350
60	2300
55	2250

Step 6 Determine LOS

Using the computed density value, refer to the definitions of LOS (A–D) and select the LOS corresponding to this value or use Figure 9.3. The density values are as follows.

LOS	Density
A	≤ 11 pc/mi/ln
B	>11–18 pc/mi/ln
C	>18–26 pc/mi/ln
D	>26–35 pc/mi/ln
E	>35–45 pc/mi/ln

Example 9.3 Operational Analysis to Determine the Free-Flow Speed (FFS), Density (*D*), and Level of Service (LOS) for a Basic Freeway Segment

Determine the free-flow speed, density, and level of service for an urban freeway segment.

Solution:

Follow the procedure for an operational analysis.

Step 1.   Input data.

- Demand volume: 4000 veh/h
- *PHF* = 0.92
- Terrain: Level
- Number of lanes: 3 in each direction
- Lane width: 11ft (3.3 m)

- Lateral clearance: 3 ft
- Ramp density: 4 Diamond interchanges spaced 1.5 mi apart
- Heavy vehicles in traffic stream during the peak hour: None
- Driver population: Commuters and familiar drivers

Step 2. Compute FFS.

Use Eq. 9.2.

$FFS = 75.4 - f_{LW} - f_{LC} - 3.22TRD^{0.84}$

$fLW = 1.9$ (Table 9.1)

$fLC = 1.2$ (Table 9.2)

Ramps/interchange $= 2$, number of interchanges in 6 mi $= 4$

$$TRD = \frac{(2 \times 4)}{6} = 1.33$$

$3.22TRD^{0.84} = 3.22(1.33)^{0.84} = 4.09$ (use 4.1)

$FFS = 75.4 - 1.9 - 1.2 - 4.1 = 68.2$ mi/h

Step 3. Select FFS curve.

Refer to Figure 9.2. The FFS curve for 70 mi/h is the one nearest to the computed FFS of 68.2 mi/h. The breakpoint is 1200 pc/h/ln.

Step 4. Adjust demand volume.

Refer to Eq. 9.4.

$$v_p = \frac{V}{(PHF)(N)(f_p)(f_{HV})} = \frac{4000}{(0.92)(3)(1.00)(1.00)} = 1450 \text{ pc/h/ln}$$

Step 5. Estimate speed and density.

Refer to Eq. 9.1(b) as v_p exceeds the breakpoint of 1200 pc/h/ln at a FFS of 70 mi/h.

$$S_{BP\text{-}70} = 70 - 0.00001160(v_p - 1200)^2 = 70 - 0.00001160(1450 - 1200)^2$$
$$= 70 - 0.73 = 69.3 \text{ mi/h}$$

Refer to Eq. 9.3 to compute density.

$$D = \frac{v_p}{S} = \frac{1450}{69.3} = 20.9 \text{ pc/mi/ln } (13.06 \text{ pc/km/In})$$

Step 6. Determine LOS.

Refer to Figure 9.3 or Table 9.10.

Since the density falls in the range of 18–26 pc/mi/ln, the LOS is C.

Example 9.4 Computation of Speed, Density, and Level of Service for an Extended Freeway Segment with Mixed Traffic in General Terrain

Determine the LOS on a regular weekday on a 0.40-mi section of a six-lane freeway with a grade of 2 percent, using the data shown in Step 1.

Solution:

Step 1. Input data.

Hourly volume, V = 3000 veh/h

FFS = 70 mi/h (112 km/h)(measured)

PHF = 0.85

Traffic composition:
  Trucks = 12%
  RVs = 2%

Lane width = 11 ft (3.3 m)

Number of lanes = 3

Terrain category: Level

Shoulder width = 6 ft (1.8 m)

Driver population adjustment factor f_p = 0.9 (unfamiliar drivers)

Step 2. Compute FFS.

It is not necessary to compute the FFS, as the given value of 70 mi/h was measured in the field.

Step 3. Select FFS curve.

Refer to Figure 9.2. The FFS curve is for 70 mi/h (the measured FFS). The breakpoint is 1200 pc/h/ln.

Step 4. Adjust demand volume.

Refer to Eq. 9.4.

$$v_p = \frac{V}{(PHF)(N)(f_p)(f_{HV})} = \frac{3000}{(0.85)(3)(0.9)(0.94)} = 1391 \text{ pc/h/ln}$$

$N = 3$

$PHF = 0.85$

$f_p = 0.9$

Compute the heavy-vehicle factor using Eq. 9.5.

PCEs:

E_T = 1.5 (from Table 9.3)

E_R = 1.2 (from Table 9.3)

$$f_{HV} = \frac{1}{1 + P_T(E_T - 1) + P_R(E_R - 1)}$$

$$= \frac{1}{1 + 0.12(1.5 - 1) + 0.02(1.2 - 1)}$$

$$= \frac{1}{1 + 0.06 + 0.004}$$

$$= 0.94$$

Step 5. Estimate speed and density.

Refer to Eq.9.1(b) as v_p exceeds the breakpoint of 1200 pc/h/ln at FFS = 70 mi/h.

$S_{\text{BP-70}} = 70 - 0.00001160 (v_p - 1200)^2 = 70 - 0.00001160 (1391 - 1200)^2$

$= 70 - 0.42 = 69.86 \text{ mi/h} (111 \text{ km/h})$

Refer to Eq. 9.3 to compute density.

> **Step 6.** Determine LOS.
> Refer to Figure 9.3a.
> Since the density falls in the range: 18–26 pc/mi/ln, the LOS is C.

Example 9.5 Required Number of Lanes for a Given Level of Service

Determine the number of lanes required for a freeway section if the section is to operate at LOS C.

Solution:

Step 1. Input data.
Number of lanes: Must be 2 or greater per direction
Lane width = 12 ft (3.6 m)
Right-side lateral clearance = 6 ft (1.8 m)
Analysis segment is (0.35 mi (0.56 km)) long with a 4.5% grade
V = 3000 veh/h (primarily commuter traffic and familiar drivers)
PHF = 0.95
Traffic composition: 10% Trucks, 2% RVs
One cloverleaf interchange per mile (ramp density = 4 ramps/mi (2.5 ramps (km))

Step 2. Compute FFS.
Use Eq.9.2

$$FFS = 75.4 - f_{LW} - f_{LC} - 3.22 TRD^{0.84}$$

$$= 75.4 - 0 - 0 - 3.22(4)^{0.84} = 75.4 - 10.3 = 65.1 \text{ mi/h}$$

Step 3. Select FFS curve.
The FFS calculated in Step 2 is ≥ 62.5 mi/h and < 67.5 mi/h. Thus FFS = 65 mi/h.
The breakpoint is 1400 pc/h/ln.

Step 4. Adjust demand volume.
PCE equivalents:

$$E_T = 2.0 \text{ (Table 9.4)}$$

$$E_R = 4.0 \text{ (Table 9.5)}$$

Heavy-vehicle adjustment factor

$$f_{HV} = \frac{1}{1 + P_T(E_T - 1) + P_R(E_R - 1)}$$

$$f_{HV} = \frac{1}{1 + 0.1(2 - 1) + 0.02(4 - 1)} = 0.86$$

Demand volume:
Convert vehicle/hour to peak 15-minute passenger-car equivalent flow rate for two, three, and four lanes.

For $N = 2$,

$$v_P = \frac{V}{PHF \times N \times f_p \times f_{HV}}$$

$$= \frac{3000}{0.95 \times 2 \times 1.00 \times 0.86}$$

$$= 1834 \text{ pc/h/ln}$$

For $N = 3$, $v_p = 1223$ pc/h/ln

For $N = 4$, $v_p = 917$ pc/h/ln

Step 5. Estimate speed and density.
$S_{BP\text{-}65} = 65 - 0.00001418(v_p - 1400)^2$
$N = 2$
$S_{BP\text{-}65} = 65 - 0.00001418(1834 - 1400)^2 = 65 - 2.67 = 62.3$ mi/h
$FFS = 62.3$ mi/h
$N = 3$ and 4
$FFS = 65$ mi/h since v_p is less than the breakpoint.

$$D_{N=4} = \frac{y_p}{S} = \frac{917}{65.0} = 14.1 \text{ pc/mi/ln}$$

Step 6. Determine LOS.
Refer to Figure 9.3.

$D_{N=2} = 29.4$ LOS = D
$D_{N=3} = 18.8$ LOS = C
$D_{N=4} = 14.1$ LOS = B

To achieve a LOS = C, use three lanes in the upgrade direction.
Repeat the process for the downgrade section to show that two lanes are required to maintain a LOS of C. Note that for downgrades, RVs are treated as if on level terrain.

9.2 MULTILANE HIGHWAYS

The procedures developed for basic freeway sections can be applied to an analysis of the capacity and LOS for multilane highways. The results obtained are used in either the planning or design phase. Lane requirements necessary to achieve a given LOS are determined and the impacts of traffic and design features in rural and suburban environments are evaluated.

Multilane highways differ from two-lane highways since the cross section consists of at least two or more lanes. Unlike freeways, they span a spectrum between freeway-like conditions of limited access to urban street conditions with frequent traffic-controlled intersections. Illustrations of the variety of multilane highway configurations are provided in Figure 9.6.

(a) Divided suburban multilane highway   (b) Undivided suburban multilane highway

(c) Suburban multilane highway with TWLTL   (d) Undivided rural multilane highway

Figure 9.6   Multilane Highways

SOURCE: From *Highway Capacity Manual 2010*. Copyright, National Academy of Sciences, Washington, D.C. Reproduced with permission of the Transportation Research Board.

Multilane highways may exhibit the following characteristics:

- Posted speed limits are usually between 40 and 55 mi/h (64 and 88 km).
- They may be undivided or include medians.
- They are typically located in suburban areas or in high-volume rural corridors.
- They may include a two-way, left-turn median lane (TWLTL).
- Traffic volumes range from 15,000 to 40,000/day.
- Traffic volumes may be as great as 100,000/day with grade separations and no cross-median access.
- Traffic signals at major crossing points are possible.
- There is partial control of access.

9.2.1 Base Conditions for Multilane Highways

Base conditions include the following elements (other characteristics such as total lateral clearance [TLC], median type, and access-point density will also impact the FFS of multilane highways).

1. Weather and driver visibility are good and do not affect traffic flow. There are no obstructions that could impede traffic flow, such as incidents, accidents, or roadwork. The HCM procedures are based on the assumption that each condition is met. If this is not the case, the speed, LOS, and capacity are likely to be lower than estimated.

2. The traffic stream is composed solely of passenger cars, and there are no heavy vehicles such as trucks, buses, and RVs.
3. All drivers in the traffic stream are regular users and are familiar with the freeway segment.

Free–Flow Speed

As shown in Figure 9.7, free-flow speed (FFS) is a constant value beginning at the *y*-intercept of each curve and extending to the breakpoint, at which point the value of speed in the traffic stream diminishes. The breakpoint value is the same value for all free-flow speeds and occurs at a flow rate of 1400 pc/h/ln. The curves are between 60 and 45 mi/h in 5-mi/h (96 and 72 km/h) 8 km increments. Interpolation is not advised, and the range of calculated values are as follows:

$$\geq 42.5 \text{ mi/h} < 47.5 \text{ mi/h: use FFS} = 45 \text{ mi/h}$$

$$\geq 47.5 \text{ mi/h} < 52.5 \text{ mi/h: use FFS} = 50 \text{ mi/h}$$

$$\geq 52.5 \text{ mi/h} < 57.5 \text{ mi/h: use FFS} = 55 \text{ mi/h}$$

$$\geq 57.5 \text{ mi/h} < 62.5 \text{ mi/h: use FFS} = 60 \text{ mi/h}$$

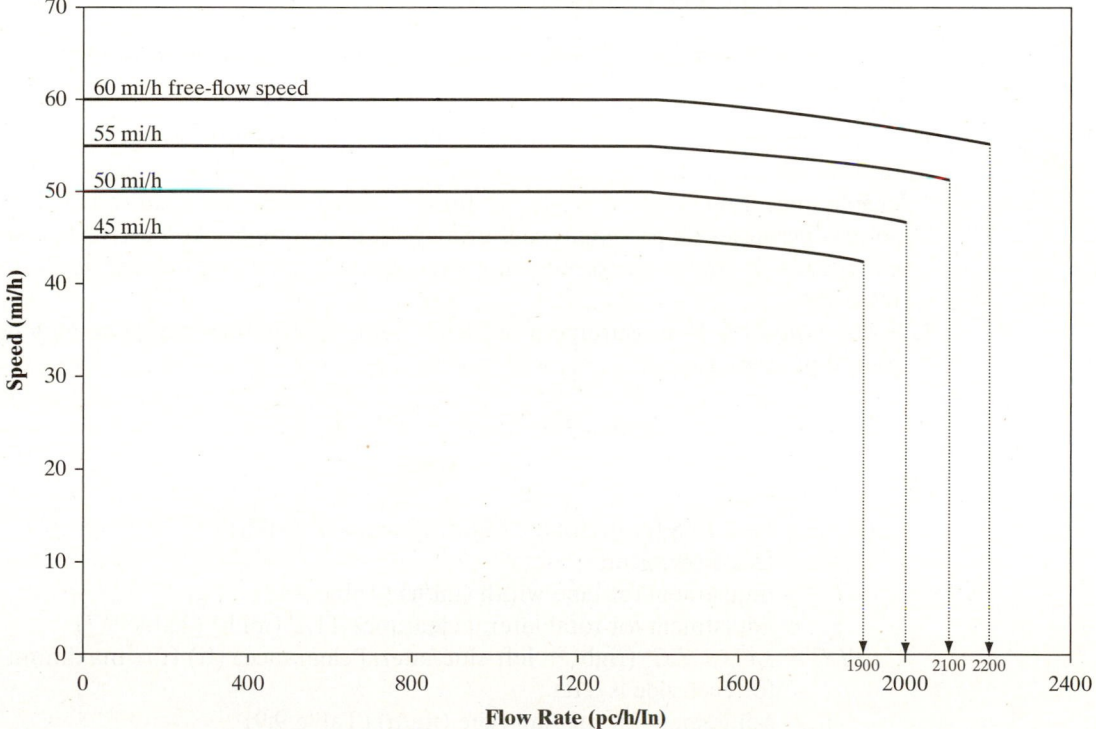

Figure 9.7   Speed-Flow Curves for Multilane Highways under Base Conditions

Note: Maximum densities for LOS E occur at a *v/c* ratio of 1.00. These are 40, 41, 43, and 45 pc/mi/in for FFSs of 60, 55, 50, and 45 mi/h, respectively.

Flow Rates beyond the Breakpoint

For demand volumes beyond the breakpoint, the flow rate declines in value until it reaches the capacity of the freeway segment. Beyond this point the HCM 2010 procedures are no longer valid. Equations 9.6(a) through 9.6(d) are used to compute free-flow speeds beyond the breakpoint.

$$S_{BP\text{-}60} = 60 - [5.00 \times (v_p - 1400)/800)^{1.31}] \tag{9.6a}$$

$$S_{BP\text{-}55} = 55 - [3.78 \times (v_p - 1400)/700)^{1.31}] \tag{9.6b}$$

$$S_{BP\text{-}50} = 50 - [3.49 \times (v_p - 1400)/600)^{1.31}] \tag{9.6c}$$

$$S_{BP\text{-}45} = 45 - [2.78 \times (v_p - 1400)/500)^{1.31}] \tag{9.6d}$$

where

$S_{BP\text{-}60}$ = flow rate beyond the breakpoint of 1400 pc/h/ln. Similar notation is used for values of FFS = 55, 50, and 45 mi/h.

v_p = demand flow rate (pc/h/ln) under equivalent base conditions

Computing Free–Flow Speed

Free-flow speed is determined in two ways: (1) by field measurements or (2) by formula estimation.

1. *Field measurement.* Free-flow speed is the mean speed of passenger cars during a period of low to moderate flow up to 1400 pc/h/ln. Field determination of FFS is preferable and, if measured directly, no further adjustments are required. The sample should be either all passenger cars or a systematic sample. At least 100 passenger-car speeds should be measured using a recognized traffic engineering measurement technique.
2. *Estimating FFS.* If measurement of FFS is not practical, then the value may be estimated by using Eq. 9.7:

$$FFS = BFFS - f_{LW} - f_{LC} - f_M - f_A \tag{9.7}$$

where

$BFFS$ = base FFS for multilane highway segment (mi/h)
FFS = free-flow speed
f_{LW} = adjustment for lane width (mi/h) (Table 9.1)
f_{LC} = adjustment for total lateral clearance TLC (mi/h) (Table 9.7)
TLC = $LC_R + LC_L$ (right + left-side lateral clearance) (ft) (the maximum value for each side is 6 ft)
f_M = adjustment for median type (mi/h) (Table 9.9)
f_A = adjustment for access-point density (mi/h) (Table 9.9)

Capacity of Multilane Highway Segments

The capacity of a multilane highway under base conditions ranges from 2400 pc/h/ln at 60 mi/h and decreases in 1000 pc/h/ln increments to 1900 pc/h/ln at a FFS of 45 mi/h.

These are national averages and may vary at different locations. They are also averages across all lanes, which may actually be distributed in a nonuniform manner such that a single lane could have stable flows at volumes greater than 2200 pc/h/ln.

9.2.2 LOS for Multilane Highway Segments

The description for LOS of freeway segments also applies to multilane highways. The density criteria are also similar with the exception of LOS F, which begins at 40 pc/mi/ln for 60 mi/h and increases to a density of > 45 pc/mi/ln at 45 mi/h. Figure 9.8 illustrates the relationship among base free-flow speed, density, and level of service.

Table 9.10 lists the level of service criteria for multilane highways. Any two of the following three performance characteristics can describe the LOS for a multilane highway:

v_p = flow rate (pc/h/ln)

S = average passenger car speed (mi/h)

D = density defined as number of cars per mi (pc/mi/ln)

The relationship among the three performance characteristics can be computed using Eq. 9.3:

$$D = \frac{v_p}{S} \tag{9.3}$$

9.2.3 Calculating the Flow Rate for a Multilane Highway

The flow rate in pc/h/ln for a multilane highway is computed using Eq. 9.4:

$$v_p = \frac{V}{(PHF)(N)(f_p)(f_{HV})} \tag{9.4}$$

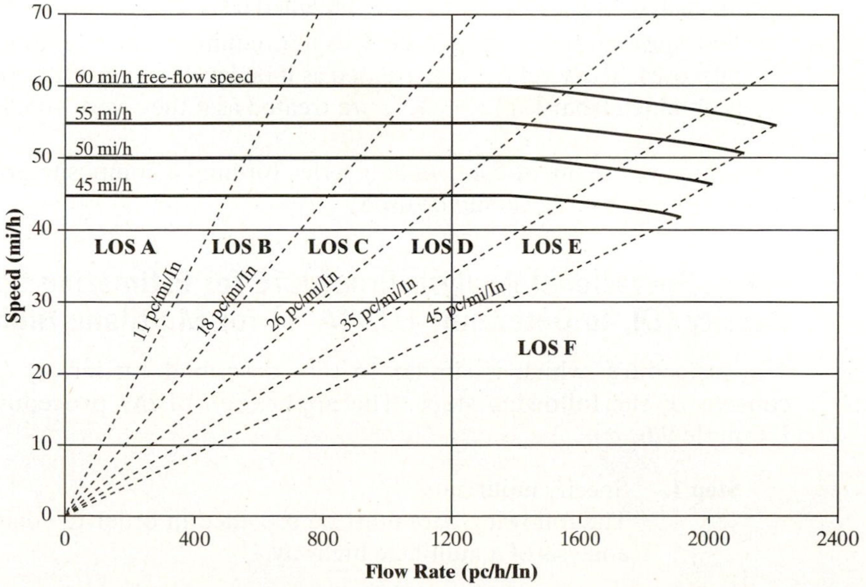

Figure 9.8 LOS on Base Speed-Flow Curves

SOURCE: From *Highway Capacity Manual 2010*. Copyright, National Academy of Sciences, Washington, D.C. Reproduced with permission of the Transportation Research Board.

where

v_p = 15-minute passenger car equivalent flow rate (pc/h/ln)
V = hourly peak vehicle volume (veh/h) in one direction
N = number of travel lanes in one direction (2 or 3)
f_p = driver population factor with a range of 0.85 to 1.00. Use 1.00 for commuter traffic.

If there is significant recreational or weekend traffic, the value is reduced. f_{HV} = heavy-vehicle adjustment factor (Eq. 9.5)

$$ f_{HV} = \frac{1}{1 + P_T(E_T - 1) + P_R(E_R - 1)} \tag{9.5} $$

where

P_T and P_R = decimal portion of trucks/buses and recreational vehicles in the traffic stream
E_T and E_R = passenger car equivalents. Number of cars using the same space as a single truck/bus or a recreational vehicle.

There are three situations that must be considered:

Extended general segments, in which the terrain is level, rolling, or mountainous. No single grade of 3 percent or more is longer than 0.25 mi and no single grade between 2 percent and 3 percent is longer than 0.50 mi. The values of E_T and E_R are obtained from Table 9.3.

Upgrades: Segments that are between 2 percent and 3 percent and longer than 0.25 mi, or 3 percent or greater and longer than 0.25 mi should be considered as a separate segment. The values of E_T and E_R are obtained from Tables 9.4 and 9.5.

Downgrades: If the downgrade does not require a shift to a lower gear for breaking purposes, heavy vehicles are treated as if on level terrain. Otherwise, the value of E_T is obtained from Table 9.6. RVs are treated as if they were on level terrain.

If the highway segment consists of a series forming a composite grade, the procedure shown in Example 9.2 should be used.

9.2.4 Operational Analysis Procedure for Estimating Speed (*S*) and Density (*D*), to Determine LOS (A–F) for Multilane Highways

The procedure, which is similar to that described earlier for freeway segments, consists of the following steps. The application of the procedure is illustrated in Example 9.6.

Step 1.   Specify input data.
The following data must be provided in order to conduct an operational analysis of a multilane highway.

- Demand volume
- Number and width of lanes
- Right-side and median lateral clearance

- Roadside access points per mile
- Percent of heavy vehicles (trucks, buses, and RVs), terrain
- Driver population

Either site-specific or default values may be used.

Step 2. Compute the FFS.
The FFS is computed directly from field measurements, as described earlier in this chapter, or can be estimated using Eq. 9.7.

Step 3. Select a FFS curve.
A free-flow value is selected from one of the five curves depicted in Figure 9.7. The curve selected is based on the value of FFS computed in Step 2. Do not interpolate between curves, but select one of the four FFS values 45–60 (mi/h) based on the ranges specified in Figure 9.7.

Step 4. Adjust demand volume.
If the traffic stream includes heavy vehicles and/or a noncommuter driver population, the demand volume must be converted to equivalent base conditions by using Eq. 9.4 and 9.5 to determine v_p, demand flow rate under prevailing conditions (pc/h/ln), and f_{HV}, heavy-vehicle factor.

Step 5. Estimate speed and density.
The preceding steps have produced the value for v_p, the equivalent passenger car demand flow rate under prevailing conditions (pc/h/ln), and the FFS. The values can be used to calculate the mean speed of the traffic stream using Eq. 9.6(a)–9.6(d) for FFS = 45–60 mi/h. The density is then computed using Eq. 9.3.

Step 6. Determine LOS.
Using the computed density value, refer to the definitions of LOS (A–F) and select the LOS corresponding to this value, or use Figure 9.8 or Table 9.10.

Example 9.6 Determining the LOS of a Multilane Highway Segment of Uniform Grade Solution

Step 1. Input data.
Length of highway: 3.25 mi (5.2 km)
Length of segment: 3200 ft (960 m)
Terrain: Level
Lane width: 11 ft (3.3 m)
Number of lanes: 2 (each direction)
Location: Suburban area
Free-flow speed: 46 mi/h (74 km/h) (measured)
Demand volume: 1900 veh/h
PHF: 0.90
Driver population: Commuters and residents
Trucks: 13%
RVs: 2%

Step 2. Compute FFS.
Since the FFS is measured, no computations are required.
FFS = 46.0 mi/h (74 km/h)

Step 3. Select FFS curve.
Since the FFS is 46 mi/h, the nearest FFS curve is 45 mi/h (72.4 km/h).

Step 4.   Adjust demand volume.
Compute v_p using Eqs. 9.4 and 9.5.

$$v_p = \frac{v}{(PHF)(N)(f_p)(f_{HV})}$$

where

$PHF = 0.90$

$N = 2$ lanes in the analysis direction

$f_p = 1.00$

$f_{HV} = 0.94$ (computed from Eq. 9.5)

$$f_{HV} = \frac{1}{1 + P_T(E_T - 1) + P_R(E_R - 1)}$$

$P_T = 0.13$

$P_R = 0.02$

$E_T = 1.5$ (Table 9.3, level terrain)

$E_T = 1.2$ (Table 9.3, level terrain)

$$f_{HV} = \frac{1}{1 + 0.13(1.5 - 1) + 0.2(1.2 - 1)} = 0.935$$

$$v_p = \frac{1900}{(0.90)(2)(1.00)(0.935)} = 1129 \text{ pc/h/ln}$$

Step 5.   Estimate speed and density.

$$S = FFS = 46 \text{ mi/h} (73.6 \text{ km/h}) (\text{since } v_p < 1400 \text{ pc/h/ln})$$

$$D = \frac{1129}{46} = 24.5 \text{ pc/h/ln}$$

Step 6.   Determine LOS.

Refer to Table 9.10.   LOS = C

9.3 TWO-LANE HIGHWAYS

The procedures developed for two-lane highway segments provide the basis to evaluate level of service and capacity. Three classes of two-lane highways are analyzed. They are defined according to their function in the following manner and are illustrated in Figure 9.9.

Class I. Two-lane highways whose function it is to serve as primary arterials, daily commuter routes, and links to other state or national highway networks. Motorists' expectations are that travel will be at relatively high speeds.

(a) Examples of Class I Two-Lane Highways

(b) Examples of Class II Two-Lane Highways

(c) Examples of Class III Two-Lane Highways

Figure 9.9   Two-Lane Highway Classification Illustrated

SOURCE: From *Highway Capacity Manual 2010*. Copyright, National Academy of Sciences, Washington, D.C. Reproduced with permission of the Transportation Research Board.

Class II. Two-lane highways whose function it is to serve as access to Class I highways. They also serve as scenic byways and can be used by motorists for sightseeing; some are located in rugged terrain. Average trip lengths are shorter than on Class I highways. The expectation of motorists is that travel speeds will be lower than for Class I roads.

Class III. Two-lane highways that serve moderately developed areas. They may be a portion of a Class I or Class II highway passing through a small town or a recreational area. These segments may be used by local traffic, and the number of unsignalized access points is greater than in rural areas. They may contain longer segments that pass through spread-out recreational areas with increased roadside activity and reduced speed limits.

There are three measures used to describe the service quality of a two-lane highway. These are (1) percent time following another vehicle (PTSF), (2) average travel speed (ATS), (3) and percent of free-flow speed (PFFS). They are defined as follows:

1. **Percent time-spent-following another vehicle (PTSF)** is the average percentage of time that vehicles are traveling behind slower vehicles. When the headway or time between consecutive vehicles is less than 3 sec, the trailing vehicle is considered to be following the lead vehicle.
2. **Average travel speed (ATS)** is the space mean speed of vehicles in the traffic stream. Space mean speed is the segment length divided by average time for all vehicles to traverse the segment in both directions during a designated time interval.
3. **Percent of free-flow speed (PFFS)** is a measure of the ability of vehicles to travel at the posted speed limit.

Figure 9.10a depicts the relationship between flow rate and average travel speed (ATS) for directional segments of a two-lane highway under base conditions. Figure 9.10b depicts the relationship between flow rate and percent time spent following (PTSF) for directional segments of a two-lane highway for base conditions. It demonstrates that low directional volumes result in high values of PTSF, due to the reduced opportunities for passing slower vehicles. The HCM 2010 methodology analyzes each direction separately.

9.3.1 Base Conditions for Two-Lane Highways

Base conditions for two-lane highways are the absence of restrictive conditions, including geometric elements, traffic control, and the environment. Base conditions represent the best conditions that can be expected. Where these conditions are not met, procedures such as were explained for freeways and multilane highways are used that reflect the reality of the design situation. The base conditions for two-lane highways are:

- Level terrain
- Lane widths 12 ft (3.6 m) or greater
- Clear shoulders 6 ft (1.8 m) wide or greater
- Passing permitted throughout with passing sight distances > 1000 ft (300 m)
- No impediments to through traffic due to traffic control or turning vehicles
- Passenger cars only in the traffic stream

9.3.2 Capacity and Level of Service for Two-Lane Highways

Capacity of a two-lane highway under base conditions is 1700 passenger cars per hour (pc/h) for each direction of travel and is nearly independent of the directional distribution of traffic. For extended segments, the capacity of a two-lane highway will not exceed a combined total of 3200 pc/h. When the capacity of 1700 pc/h is reached in one direction, the flow in the opposite direction is limited to 1500 pc/h. Short sections of two-lane highway, such as a tunnel or bridge, may reach a capacity of 3200 to 3400 pc/h.

LOS expresses the performance of a highway at traffic volumes less than capacity. The criterion for LOS depends on the class of highway.

LOS for **Class I** highways is based on two measures: PTSF and ATS. Speed and delay due to passing restrictions are important factors to motorists on these highways.

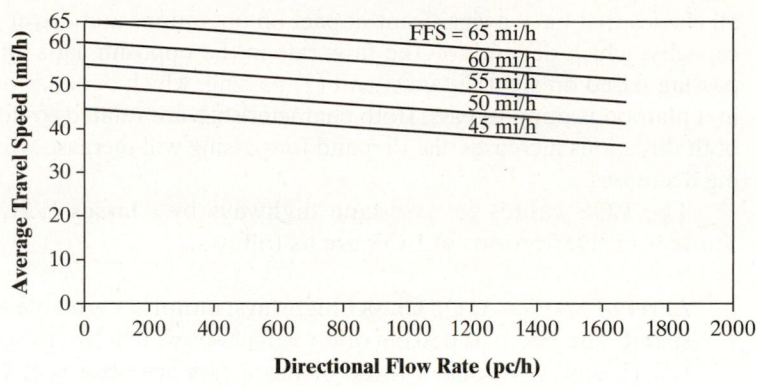

(a) ATS versus Directional Flow Rate

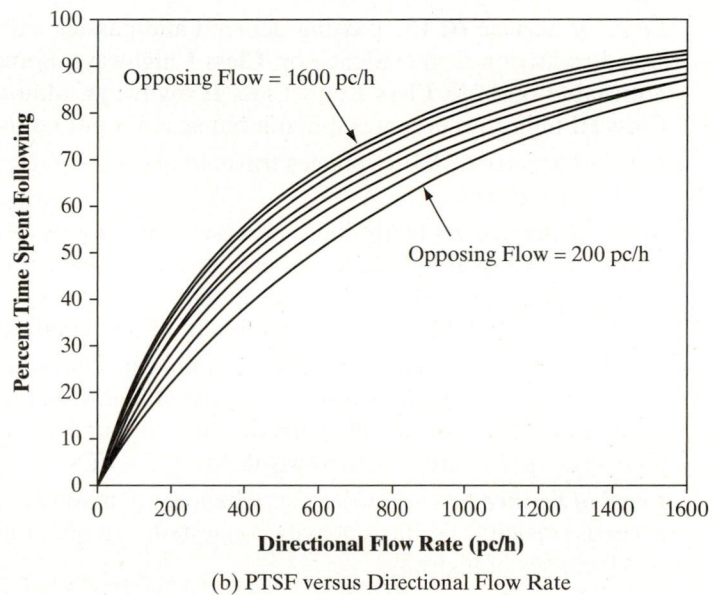

(b) PTSF versus Directional Flow Rate

Figure 9.10   Speed-Flow and PTSF Relationships for Directional Segments with Base Conditions

SOURCE: From *Highway Capacity Manual 2010*. Copyright, National Academy of Sciences, Washington, D.C. Reproduced with permission of the Transportation Research Board.

LOS for **Class II** highways is based on a single measure: PTSF. Travel speed is not a key factor, but delay measures the motorist's sense of the quality of the trip.

LOS for **Class III** highways is based on a single measure: PFFS. In this situation, where the length of the Class III segment is limited, neither high speeds nor passing restrictions are major concerns. Motorists in a Class III situation are most concerned with the ability to maintain steady progress at or near the speed limit.

Level of service criteria are applied to travel during the peak 15 min of travel and on highway segments of significant length. Level of service designations are from A (highest) to F (lowest). A single definition of LOS is not possible, as driver expectations vary depending on the class of two-lane highway. However, there are two characteristics common t all classes that have a significant impact on operations and driver perceptions: (1) passing capacity, which depends on the flow rate in the opposing lane and the gaps available for passing based on sight distance; and (2) passing, which depends on the number of drivers in a platoon waiting to pass. Both characteristics are related to flow rates. As flow rate in both directions increases, the demand for passing will increase while the capacity for passing decreases.

The LOS values for two-lane highways by **Classes I, II,** and **III** are shown in Table 9.11. Definitions of LOS are as follows.

Level of Service A: On **Class I** highways, motorists are able to travel at their desired speed. The need for passing other vehicles is well below the capacity for passing, and few (if any) platoons of three or more cars are observed. **Class II** highway speeds are controlled by roadway conditions and a small amount of platooning is likely. **Class III** highway users should be able to maintain speeds close or equal to FFS.

Level of Service B: The passing demand and passing capacity are balanced. Some speed reduction is in evidence on **Class I** highways. Some degree of platooning is noticeable on both **Class I** and **Class II** highways. Maintaining FFS operation on **Class III** highways becomes difficult but speed is not noticeably reduced.

Level of Service C: Most vehicles travel in platoons and speeds decrease in all three classes of highway.

Level of Service D: Platooning increases significantly. Passing demand is high on **Class I** and **Class II** highways, but passing maneuvers are difficult, if not impossible, to complete. Most vehicles are travelling in platoons with a substantial increase in PTSF. On Class III highways, reduction in FFS is significant.

Level of Service E: Demand is approaching capacity, which is reached at the lower limit of LOS. Passing on **Class I** and **Class II** highways has become virtually impossible, and PTSF exceeds 80 percent. Speeds diminish considerably. On **Class III** highways, speeds are less than two-thirds of the FFS.

Level of Service F: This LOS exists whenever demand flow in one or both directions exceeds capacity. Traffic is heavily congested, and operating conditions are unstable on all classes of highway.

Example 9.7 Determine the Level of Service of Class I, II, and III Two-Lane Roads If PTSF, ATS, and FFS Are Known

The following values of PTSF, ATS, and FFS have been determined based on the analysis of four roadway segments. (Methods for performing the analysis are described later in this chapter.) Determine the LOS if the roadway segments are: (a) Class I, (b) Class II, or (c) Class III.

Segment	PTSF (%)	ATS (mi/h)	FFS (mi/h)
1	36	54	55
2	54	47	55
3	72	42	55
4	90	30	55

Solution:

Use Table 9.11 to determine LOS for each class. If values of LOS differ for PTSF and ATS, the lower LOS value is used.

(a) Class I Highways
Segment 1: PTSF: LOS B ATS: LOS B
Segment 2: PTSF: LOS C ATS: LOS C
Segment 3: PTSF: LOS D ATS: LOS D
Segment 4: PTSF: LOS E ATS: LOS E

(b) Class II Highways
Segment 1: PTSF: LOS A
Segment 2: PTSF: LOS B
Segment 3: PTSF: LOS D
Segment 4: PTSF: LOS E

(c) Class III
Segment 1 PFFS = ATS/FFS = 98% LOS A
Segment 2 PFFS = ATS/FFS = 85% LOS B
Segment 3 PFFS = ATS/FFS = 76% LOS C
Segment 4 PFFS = ATS/FFS = 54% LOS E

9.3.3 Operational Analysis to Determine LOS (A–F) for Two-Lane Highway Segments

The procedures for determining density, speed, and LOS are followed separately for each of the three classes of two-lane highways. Class I highways require seven steps. Class II highways require five steps, and Class III highways require six steps. However, since the procedure applies to three classes of two-lane highway, not all steps are followed by each facility class. The procedural steps to compute ATS apply only to Class I highways. The procedural steps to compute PTSF apply only to Class I and Class II highways, and the procedural steps to compute PFFS apply only to Class III highways.

Three segment types are analyzed:

1. Direction segments in general terrain (level or rolling)
2. Direction segments on specific grades (mountainous terrain or grades ≥ 3 percent and length ≥ 0.6 mi (0.96 km))
3. Direction segments that include passing and truck climbing lanes

The procedure applies only to directional segments (one lane) of a two-lane highway. Accordingly, each direction is analyzed separately, even though there is interaction between directions when passing occurs. The procedural steps that are required for each of the three highway classes are as follows:

Develop Input Data: Classes I, II, and III
Geometric characteristics
Demand volume

Highway class

Field-measured speed (S_{FM}), or

Base free-flow speed (BFFS)

Estimate Free-Flow Speed: Classes I, II, and III

Field-Measured Speed Adjustments:

Flow rate (Eq. 9.8) and heavy vehicles (Eq. 9.11) or

BFFS adjustments (Eq.9.9)

Lane width and shoulder width (Table 9.12)

Access-point density (Table 9.13)

Compute Demand Adjustments for Average Travel Speed: Classes I and III

(Eqs. 9.10, 9.11, 9.12)

Peak hour factor

Heavy-vehicle adjustment:

General terrain (Table 9.16)

Specific grade (Tables 9.17, 9.18, 9.19)

Grade adjustment:

General terrain (Table 9.14)

Specific upgrade (Table 9.16)

Estimate Average Travel Speed: Classes I and III

(Eq. 9.13)

No-passing-zone adjustment (Table 9.20)

Compute Demand Adjustment for Percent Time Spent Following: Classes I and II

(Eqs. 9.14 and 9.15)

Peak hour factor

Heavy-vehicle adjustment:

General terrain (Table 9.23)

Specific grade (Table 9.24)

Grade adjustment:

General terrain (Table 9.21)

Specific grade (Table 9.22)

Estimate Percent Time Spent Following: Classes I and II

(Eqs. 9.16 and 9.17, Table 9.26)

No-passing-zone adjustment (Table 9.25)

Estimate Percent of Free-Flow Speed: Class III

(Eq. 9.18)

Determine Level of Service and Capacity: Classes I, II, and III

(Table 9.11)

9.3.4 Estimating the Input Values to Determine the LOS of Two–Lane Highways: For Directional Segments without Passing Lanes

This section describes the computational procedures to obtain estimates of the following variables: (1) FFS, (2) ATS, (3) PTSF, and (4) PFFS for directional segments in general terrain (level or rolling) and directional segments on specific grades.

Estimating the FFS

The estimation of FFS can be accomplished in the following three ways:

Direct Field Measurement.   This applies only for two-way volumes < 200 veh/h. These are taken only in the direction of analysis with a random sample of at least 100 vehicle speeds. Where conditions do not permit direct measurement, another two-lane highway with similar characteristics may be used.

Field Measurements at Higher Flow Rates.   When volumes are > 200 veh/h, it is often difficult to observe flow rates that are less than 200 veh/h. In such cases, the mean speed of the sample S_{FM} is measured based on at least 100 observations. The measured speed is then adjusted using Eq. 9.8.

$$FFS = S_{FM} + 0.00776\left(\frac{v}{f_{HV,ATS}}\right) \tag{9.8}$$

where

FFS = free-flow speed
S_{SM} = mean speed of sample ($v > 200$ veh/h) (mi/h)
V = total demand flow rate, both directions, during period of speed measurements (veh/h)
$f_{HV,ATS}$ = heavy-vehicle adjustment factor for ATS using either Eq. 9.11 or 9.12

Estimating FFS.   When field data are unavailable, FFS is computed using Eq. 9.9.

$$FFS = BFFS - f_{LS} - f_A \tag{9.9}$$

where

FFS = estimated free-flow speed (mi/h)
$BFFS$ = base free-flow speed (mi/h)
f_{LS} = adjustment for lane and shoulder width (Table 9.12)
f_A = adjustment for number of access points per mile (Table 9.13)

The base free-flow speed (BFFS) depends upon local conditions regarding the desired speeds of drivers. The transportation engineer estimates BFFS based on knowledge of the area and the speeds on similar facilities. The range of BFFS is 45 to 65 mi/h (72 to 104 km/h). Posted speed limits or design speeds may serve as surrogates for BFFS.

Demand Adjustment for ATS

The demand volume for each direction (analysis and opposing) is converted to flow rates under equivalent base (passenger car) conditions using Eq. 9.10.

$$v_{i,ATS} = \frac{V_i}{PHF \times f_{g,ATS} \times f_{HV,ATS}} \tag{9.10}$$

where

$v_{i,ATS}$ = demand flow rate i for ATS estimation (pc/h)
$i = d$ (analysis direction) or o (opposing direction)

V_i = demand volume for direction i (veh/h)

$f_{g,ATS}$ = grade adjustment factor, from Table 9.14 or Table 9.15

$f_{HV,ATS}$ = heavy-vehicle adjustment factor, from Eq. 9.11 or 9.12

The grade adjustment factor is determined as follows.

- For level or rolling terrain ≥2 mi (3.2 km) (extended segments) and specific downgrades not traveling at crawl speeds, Table 9.14 is used.
- For specific downgrades ≥3 percent and ≥0.6 mi (0.96 km), the grade adjustment factors listed in Table 9.14 are used.
- For specific upgrades ≥3 percent and ≥0.6 mi (0.96 km), the values in Table 9.15 are used.

The heavy-vehicle adjustment factor is determined as follows.

- For extended sections of general level or rolling terrain and specific downgrades, Eq. 9.11 and Table 9.16 are used.
- For specific upgrades, Eq. 9.11 and Table 9.17 are used.

$$f_{HV,ATS} = \frac{1}{1 + P_T(E_T - 1) + P_R(E_R - 1)} \tag{9.11}$$

where

$f_{HV,ATS}$ = heavy-vehicle adjustment factor for ATS estimation

P_T = the decimal portion of trucks in the traffic stream

P_R = the decimal portion of RVs in the traffic stream

E_T = the passenger car equivalent of trucks in the traffic stream (Table 9.16 or 9.17)

E_R = the passenger car equivalent of RVs in the traffic stream (Table 9.16 or 9.18)

- For specific downgrades where trucks travel at crawl speed, Eq. 9.12 and Table 9.19 are used.

$$f_{HV} = \frac{1}{1 + P_{TC} \times P_T(E_{TC} - 1) + (1 - P_{TC}) \times P_T(E_T - 1) + P_R(E_R - 1)} \tag{9.12}$$

where

P_{TC} = the decimal portion of trucks operating at crawl speed

E_T = the passenger car equivalent of trucks operating at crawl speed (Table 9.19)

Estimate the ATS

The ATS is estimated by use of Eq. 9.13.

$$ATS_d = FFS - 0.00776(v_{d,ATS} + v_{vo,ATS}) - f_{np,ATS} \tag{9.13}$$

where

ATS_d = average travel speed in the analysis direction (mi/h)

FFS = free-flow speed (mi/h)

$v_{d,ATS}$ = the demand flow rate for ATS determination analysis direction (pc/h)

$v_{o,ATS}$ = the demand flow rate for ATS determination opposite direction (pc/h)
$f_{dnp,ATS}$ = adjustment factor for ATS determination for the percentage of no-passing zones in the analysis direction (Table 9.20)

Note: This table is entered with $v_{o,ATS}$ in pc/h, not in veh/h.

Demand Adjustment for PTSF

The procedure followed to determine the demand flow rate and estimate the PTSF is similar to that for the ATS. Accordingly, a further discussion of the process itself is not needed. Equations 9.14 and 9.15 contain the same variables as in Eqs. 9.10 and 9.12, with ATS replaced by PTSF. The equations are as follows:

$$v_{i,PTSF} = \frac{V_i}{PHF \times f_{g,PTSF} \times f_{HV,PTSF}} \tag{9.14}$$

$$f_{HV,PTSF} = \frac{1}{1 + P_T(E_T - 1) + P_R(E_R - 1)} \tag{9.15}$$

where

$v_{i,PTSF}$ = demand flow rate i for determination of PTSF (pc/h)
i = "d (analysis direction) or o" (opposing direction)
V_i = demand volume for direction i (veh/h)
$f_{g,PTSF}$ = grade adjustment factor for PTSF determination; Table 9.21 or Table 9.22
$f_{HV,PTSF}$ = heavy-vehicle adjustment factor for PTSF determination
E_R and E_T = passenger car equivalent of RVs and trucks in the traffic stream

For level and rolling terrain and specific downgrades, use Table 9.23 or 9.24.
For specific upgrades, use Table 9.22.

Enter demand flow rates in pc/h: $v_i = \dfrac{v}{PHF}$.

Estimate the PTSF

The PTSF is estimated using Eq. 9.16 as follows:

$$PTSF_d = BPTSf_d + f_{np,PTSF}\left\{\frac{v_{d,PTSF}}{v_{d,PTSF} + v_{o,PTSF}}\right] \tag{9.16}$$

where

$PTSF_d$ = the percent time spent following in the analysis direction (decimal)
$BPTSF_d$ = base percent time spent following in the analysis direction, Eq. 9.17
$f_{np,PTSF}$ = adjustment to PTSF for the percentage of no-passing zones in the analysis segment, Table 9.25
$v_{d,PTSF}$ = demand flow rate in the analysis direction for estimation of PTSF (pc/h)
$v_{o,PTSF}$ = demand flow rate in the opposing direction for estimation of PTSF (pc/h)

$$BPTSF_d = 100\{1 - \exp(av_d^b)\} \tag{9.17}$$

where

a and b are constants from Table 9.26 and other term are as defined previously

Estimate the PFFS

The PFFS is determined using Eq. 9.18.

$$PFFS = \frac{ATS_d}{FFS} \tag{9.18}$$

Determine LOS and Capacity

The LOS is determined from Table 9.11 based on the measures developed in the previous sections for a given class of the two-lane highway. The measures are as follows:

Class I ATS and PTSF (Two values of LOS are obtained. The poorest LOS is selected.)
Class II PTSF
Class III PFFS

The *capacity* of a two-lane highway (which exists at the boundary of LOS E and F) is determined as follows. Under base conditions the capacity of a two-lane highway in one direction is 1700 pc/h. The capacity under prevailing conditions will be lower and requires that relevant adjustment factors be applied to Eq. 9.10, which is the ATS demand flow rate, and Eq. 9.14, which is the PTSF demand flow rate. (All adjustment factors used for capacity determination should be based on a flow rate > 900 veh/h). In this situation the demand flow rate under base conditions is known (1700 pc/h) whereas the demand flow rate under prevailing conditions is unknown. Since capacity is defined as a flow rate, the PHF = 1.0 in Eq. 9.10 and Eq. 9.14.

Use Eq. 9.19 to compute the capacity based on ATS (pc/h) and Eq. 9.20 to compute the capacity based on PTSF (pc/h).

$$C_{dATS} = 1700 f_{g,ATS} f_{HV,ATS} \tag{9.19}$$

$$C_{dPTSF} = 1700 f_{g,PTSF} f_{HV,PTSF} \tag{9.20}$$

where

C_{dATS} = the capacity in the analysis direction under prevailing conditions based on ATS (pc/h)
C_{dPTSF} = the capacity in the analysis direction under prevailing conditions based on PTSF (pc/h)

Class I Highways: Compute both capacities (ATS and PTSF). Select the lowest value.
Class II Highways: Compute the capacity based on PTSF.
Class III Highways: Compute the capacity based on PFFS.

When the directional distribution in level and rolling terrain is not 50/50, the base capacity is restricted to 1700 pc/h in the direction of heaviest flow and the capacity in the opposing direction is based on the lower proportion of flow, with a maximum value of 1500 pc/h.

Example 9.8 LOS and Capacity for a Class I, Class II, and Class III Two-Lane Highway

Determine the LOS and capacity for the following classes of two-lane highways.

(a) Class I

(b) Class II

(c) Class III

Input data is as follows:

Volume = 1600 veh/h (two-way)

Percent trucks = 14

Percent RVs = 4

Peak hour factor = 0.9

Rolling terrain

Percent directional split = 50/50

50% no-passing zones in the analysis segment (both directions)

Access points: 20 per mi (12.5 per km)

BFFS = 60 mi/h (96 km/h)

Segment length = 10 mi (16 km)

Lane width = 11 ft (3.3 m)

Shoulder width = 4 ft (1.3 m)

Solution:

The steps to complete an operational analysis procedure for two-lane highways as described in Section 9.3.3 are used.

Step 1. Specify Input Data.

Input data are as previously specified.

Step 2. Estimate Free-Flow Speed, FFS–Classes I, II, and III.

Compute FFS (Eq. 9.9).

$$FFS = BFFS - f_{LS} - f_A$$

$$FFS = 60.0 - 1.7 - 5.0 = 53.3 \text{ mi/h } (85.2 \text{ km/h})$$

$f_{LS} = 1.7$ (Table 9.12), $f_A = 5.0$ (Table 9.13)

Step 3. Compute Demand Adjustments for Average Travel Speed, ATS—Classes I and III.

Compute demand-adjusted ATS volume (Eq. 9.10).

$$v_{i,ATS} = \frac{V_i}{PHF \times f_{g,ATS} \times f_{HV,ATS}}$$

$V_i = 800$ veh/h since directional split = 50/50

$PHF = 0.95$

$f_{g,ATS} = 0.99$ (Table 9.14 by interpolation between 800 and 900)

Compute heavy-vehicle adjustment factor (Eq. 9.11).

$$f_{HV,ATS} = \frac{1}{1 + P_T(E_T - 1) + P_R(E_R - 1)}$$

$E_T = 1.4$, $E_r = 1.1$ (Table 9.16)

$$f_{HV,ATS} = \frac{1}{1 + 0.14(1.4 - 1) + 0.04(1.1 - 1)} = 0.943$$

$$v_{i,ATS} = \frac{800}{0.95 \times 0.99 \times 0.943} = 902 \text{ pc/h}$$

Step 4. Estimate Average Travel Speed (ATS)—Classes I and III. Compute ATS (Eq. 9.13).

$$ATS_d = FFS - 0.00776(v_{d,ATS} + V_{vo,ATS}) - f_{np,ATS}$$

No-passing-zone adjustment, $f_{np,ATS} = 0.7$ (Table 9.20 by interpolation)

$$ATS_d = 53.3 - 0.00776(902 + 902) - 0.7 = 38.6 \text{ mi/h } (61.76 \text{ km/h})$$

Step 5. Compute Demand Adjustments for Percent Time Spent Following, PTSF—Classes I and II. Use Eq. 9.14.

$$v_{i,PTSF} = \frac{V_i}{PHF \times f_{g,PTSF} \times f_{HV,PTSF}}$$

$f_{g,PTSF} = 1.00$ (Table 9.21 by interpolation between 800 and 900)
$E_T = 1.0$, $E_r = 1.0$ (Table 9.23)
Use Eq. 9.15.

$$f_{HV,PTSF} = \frac{1}{1 + P_T(E_T - 1) + P_R(E_R - 1)}$$

$$f_{HV,PTSF} = \frac{1}{1 + 0.14(1 - 1) + 0.04(1 - 1)} = 1.00$$

$$v_{i,PTSF} = \frac{800}{0.95 \times 1.00 \times 1.00} = 842 \text{ pc/h}$$

Step 6. Estimate Percent Time Spent Following, PTSF—Classes I and II. Use Eq. 9.17.

$$BPTSF_d = 100\{1 - \exp(av_d^b)\}$$

Use Table 9.26 (by interpolation between 800–1000 pc/h).
$a = -0.0046$
$b = 0.832$

$$BPTSF_d = 100\left\{1 - e^{-0.0046 * 842^{0.832}}\right\}$$
$$BPTSF_d = 71.3\%$$

Use Eq. 9.16.

$$PTSF_d = BPTSf_d + f_{np,PTSF}\left\{\frac{v_{d,PTSF}}{v_{d,PTSF} + v_{o,PTSF}}\right\}$$

$f_{np,PTSF} = 21$ (by interpolation in Table 9.25 using total flow rate = 1684 pc/h)

$$PTSF_d = 713 + 21\left\{\frac{842}{842 + 842}\right\} = 81.8\%$$

Step 7.   Estimate Percent of Free-Flow Speed, PFFS—Class III.
Use Eq. 9.18.

$$PFFS = \frac{ATS_d}{FFS} = \frac{38.6}{53.3} = 72.4\%$$

Step 8.   Determine Level of Service and Capacity.
The following results were obtained.

$$ATS = 38.6 \text{ mi/h (61.76 km/h)}$$

$$PTSF = 81.8\%$$

$$PFFS = 72.4\%$$

(a) Determine LOS.
Use Table 9.11.
Class I

$$LOS_{ATS} = E$$

$$LOS_{PTSF} = E$$

Thus: $LOS_{Class\ I} = E$
Class II
$LOS_{PTSF} = D$
Thus: $LOS_{Class\ II} = D$
Class III
$LOS_{PFFS} = D$
Thus: $LOS_{Class\ III} = D$
(b) Determine Capacity.
Use Eqs. 9.19 and 9.20.

$$C_{d,ATS} = 1700 f_{g,ATS} f_{HV,ATS}$$

$$C_{d,ATS} = 1700 \times 0.99 \times 0.943 = 1587 \text{ veh/h}$$

$$C_{d,PTSF} = 1700 \times 1.00 \times 1.00 = 1700 \text{ veh/h}$$

$C_{d,ATS}$ = capacity in the analysis direction under prevailing conditions based on ATS (pc/h)

$C_{d,PTSF}$ = capacity in the analysis direction under prevailing conditions based on PTSF (pc/h)

$f_{g,ATS} = 0.99$
$f_{hv,ATS} = 0.943$

$$f_{g,PTSF} = 1.00$$
$$f_{hv,PTSF} = 1.00$$

Class I Highways: Compute both capacities (ATS and PTSF). Select the lower value.

Capacity = 1587 veh/h

Class: II Highways: Compute the capacity based on PTSF.

Capacity = 1700 veh/h

Class: III Highways: Compute the capacity based on ATS.

Capacity = 1587 veh/h

9.3.5 Estimating the Input Values to Determine the LOS of Two-Lane Highways: For Directional Segments with Passing Lanes

A passing lane is added to a two-lane highway to provide the motorist with additional opportunities to overtake slower vehicles. By adding a lane in one direction of travel, the percentage of time spent following can be reduced in the widened section and in a portion of the section that follows. The net effect of a passing lane is to reduce the overall percent time spent following for the segment being analyzed. The greatest effect would result if the passing lane were as long as the entire length of the segment. The effect decreases as the length of the passing lane is reduced.

A typical passing lane includes tapers on both ends. Passing lanes are provided in a variety of formats. They may be exclusively for a single direction of traffic, or opposing traffic may be permitted to use them. Passing lanes may be provided intermittently or at fixed intervals for each direction of travel. They also may be provided for both directions of travel at the same location, resulting in a short section of four-lane undivided highway.

Figure 9.11 illustrates how a passing lane through the analysis segment influences the PTSF. Note that the length of the passing lane is considerably shorter than its length of influence on the value of PTSF, which decreases significantly at the beginning of the passing lane but remains at a lower value than for a normal two-lane highway for a considerable distance beyond the end of the passing lane itself. Accordingly, the effective length of a passing lane is significantly greater than its actual length.

Table 9.27 provides the values of downstream length of two-lane highways for PTSF and ATS (which is a constant value of 1.7 mi (2.72 km)). PTSF values are affected by the directional demand flow rate (pc/h). At 200 pc/h, the length of roadway affected is 13 mi (20.8 km) and diminishes to 3.6 mi (5.76 km) for flow rates ≥ 1000 pc/h.

The LOS on a Class I or Class II two-lane highway is determined by the PTSF and as such is significantly improved with the addition of a passing lane. However, the impact on service volumes or service flow rates is dependent on the segment used in the analysis. If the segment includes only the passing lane and the length of its downstream effect on PTSF, the service rate at LOS (A–D) can appear to be significantly improved. However, if the analysis segment extends beyond the point of PTSF influence, the improvement obtained from the passing lane is reduced. Thus, an apparent increase in service flow rate should be evaluated with respect to the overall roadway conditions.

The procedure to estimate the effect of a passing lane on a two-lane highway (PTSF and ATS) is valid only in level and rolling terrain and is described in the following sections.

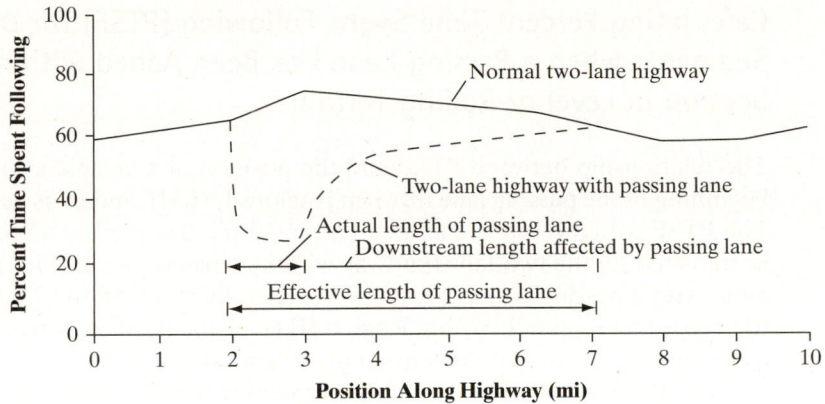

Figure 9.11 Operational Effect of a Passing Lane on PTSF

Passing lanes on specific grades are considered as *climbing* lanes and are described separately. For both situations an analysis of the two-lane highway should be completed without passing lanes (as described in previous sections) in order to provide a basis to compare the added value provided by the passing lane at the segment selected for analysis.

As illustrated in Figure 9.11, there are four regions where changes occur to the value of PTSF. The sum of the lengths of the four regions adds up to the total length of the analysis. These total lengths differ when estimating PTSF and ATS as shown in Table 9.27 because downstream lengths of roadway affected, L_{de} (Region III), are not the same value for these two measures of effectiveness.

Region I. Length upstream of the passing lane, L_u

The PTSF in the immediate region preceding the passing lane (upstream of the passing lane) of length L_u will be the PTSF value computed as described earlier for a directional segment. The length of Region I is determined by deciding where the planned or actual passing lane should be placed relative to the beginning point of the analysis segment.

Region II. Length of the passing lane, L_{pl}

This is the constructed or planned length including tapers. The procedure is calibrated for an optimal length of Region II shown in Table 9.28 and ranges from 0.50 to 2.0 mi, depending on the directional flow rate. Passing lanes that are shorter or longer than optimal will usually provide a lower benefit then the lengths used in the analysis procedure.

Region III. Length downstream of the passing lane, L_{de}

This region is considered to be within the effective length of the passing lane. The length of Region III, L_{de}, is determined from Table 9.26 as a function of the directional flow rate.

Region IV. Length downstream of the passing lane beyond its effective length, L_d

This is the remaining downstream region within the segment. The length of this region is the difference between the total segment length L_t and the sum of the lengths of Regions I, II, and III as shown in Eq. 9.21. Thus,

$$L_d = L_t - (L_u + L_{pl} + L_{de}) \qquad (9.21)$$

Calculating Percent Time Spent Following (PTSF) for Directional Segments when a Passing Lane Has Been Added within an Analysis Section in Level or Rolling Terrain

The relationship between PTSF and the position of a vehicle as it progresses from the beginning of the passing lane through Regions I, II, III, and IV is depicted in Figure 9.12. The PTSF within Regions I and IV (L_u and L_d) is assumed to be equal to the PTSF that is predicted for the two-lane roadway without a passing lane. With Region II for the segment with a passing lane (L_{pl}), PTSF is generally about 58 to 62 percent of the PTSF in the preceding Region I. Within Region III (L_{de}), the PTSF is assumed to increase linearly from the passing lane to the normal upstream value.

The PTSF for the entire analysis segment with a passing lane in place is $PTSF_{pl}$. The value is determined as the weighted average of the PTSF values in each region weighted by the region length and computed by using Eq. 9.22.

$$PTSF_{pl} = \frac{PTSF_d\left[L_u + L_d + f_{pl,PTSF}L_{pl} + \left(\dfrac{1 + f_{pl,PTSF}}{2}\right)L_{de}\right]}{L_t} \qquad (9.22)$$

where

$PTSF_{pl}$ = percent time spent following for the segment as affected by the passing lane

$PTSF_d$ = the percent time spent following in the analysis direction (decimal) for the segment without the passing lane from Eq. 9.12

f_{pl} = adjustment factor for the effect of a passing lane on percent time spent following from Table 9.29

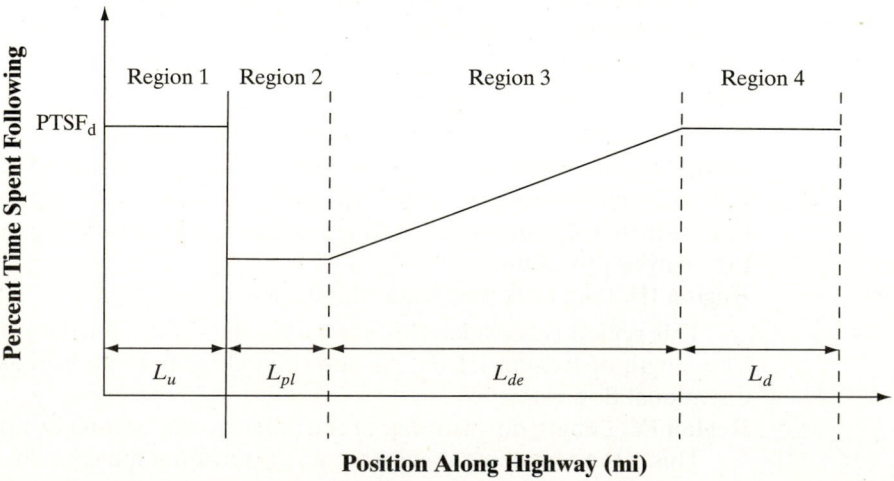

Figure 9.12   Effect of a Passing Lane on PTSF

If the full downstream length is not reached because a traffic signal or a town interrupts the analysis section, then L_{de} is replaced by a shorter length L'_{de}. In this instance, Eq. 9.22 is modified as Eq. 9.23.

$$PTSF_{pl} = \frac{PTSF_d \left[L_u + f_{pl,PTSF} L_{pl} + f_{pl,PTSF} L'_{de} + \left(\frac{1 - f_{pl,PTSF}}{2} \right) \left(\frac{L'^2_{de}}{L_{de}} \right) \right]}{L_t} \qquad (9.23)$$

where L'_{de} = actual distance from end of passing lane to end of analysis segment in miles. *Note:* L'_{de} must be less than or equal to the value of L_{de} (obtained from Table 9.27).

Generally, the effective downstream distance of the passing lane is not truncated and should only be at the point where one of the following situations exists.

- The environment of the highway changes radically. (Small town, developed area)
- A major unsignalized intersection
- A proximate signalized intersection
- The terrain changes significantly
- Lane or shoulder widths change significantly

Calculating Average Travel Speed (ATS) for Directional Segments when a Passing Lane Has Been Added within an Analysis Section in Level or Rolling Terrain

Figure 9.13 illustrates the changes in the value of average travel speed (ATS) that occur within each of the four regions when a passing lane has been added. The values are:

Region I. Average travel speed in the upstream region is ATS_d for length L_u.
Region II. Average travel speed in the passing lane increases by a factor L_{pl} provided in Table 9.20. Values range from 1.08 to 1.11, depending on the directional flow rate v_d.

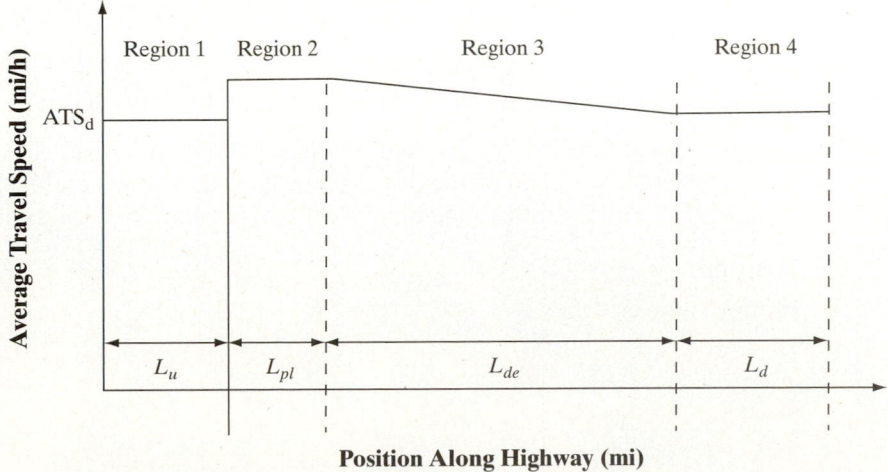

Figure 9.13   Impact of a Passing Lane on ATS

Region III. Average travel speed within the downstream length decreases linearly from the value in Region II to ATS_d. $L_{de} = 1.7$ mi, as shown in Table 9.30.

Region IV. Average travel speed in the downstream region is ATS_d for length L_d. To compute the average speed for the analysis segment, it is first necessary to compute the individual travel time within regions t_I, t_{II}, t_{III}, and t_{IV}. The average travel speed is the total analysis length L_t, divided by the sum of the four travel times. Equation 9.24, for ATS_{pl}, is the average travel speed in the analysis section as the result of adding a passing lane:

$$ATS_{pl} = \frac{ATS_d L_t}{L_u + L_d + \left[\dfrac{Lpl}{f_{pl,ATS}}\right\} + \left[\dfrac{2L_{de}}{1 + f_{pl,ATS}}\right]} \tag{9.24}$$

where

ATS_{pl} = average travel speed in the analysis segment as affected by a passing lane (mi/h)

$f_{p,lATS}$ = adjustment factor for the effect of a passing lane on ATS from Table 9.30

In the case where the analysis segment cannot include all of the effective downstream distance, L_{de}, because a town or major intersections cause the segment to be truncated, distance L'_{de} is less than the value of L_{de}. In this case, Eq. 9.25 is used instead of Eq. 9.24 to compute ATS.

$$ATS_{pl} = \frac{ATS_d L_t}{L_u + \dfrac{L_{pl}}{f_{pl,ATS}} + \dfrac{2L'_{de}}{\left[1 + f_{pl,ATS} + (f_{pl,ATS} - 1)\left(\dfrac{L_{de} - L'_{de}}{L_{de}}\right)\right]}} \tag{9.25}$$

Example 9.9 Determine the Level of Service for a Class I Two-Lane Directional Highway with a Passing Lane

Use the data and results for Example 9.8 to determine the level of service if a passing lane is added. The length of the passing lane is 2 mi (3.2 km) and is designed to begin 1 mi downstream from the beginning of a 10-mi (16-km) analysis section.

Solution:

From Examples 9.8:

$PTSF = 81.8\%$

$ATS = 38.6$ mi/h

$PTSF$ demand flow rate (each direction) = 842 pc/h

ATS demand flow rate (each direction) = 902 pc/h

$$FFS = 53.3 \, \text{mi/h}$$

$$PHF = 0.95$$

Rolling terrain

Step 1. Determine the length of each region.

Region I: $L_u = 1$ mi

Region II: $L_{pl} = 2$ mi

Region III:

For $PTSF$, $L_{de} = 4.7$ mi (Table 9.22) $f_{pl} = 0.62$ (Table 9.29)

For ATS, $L_{de} = 1.7$ mi (Table 9.22) $f_{pl} = 1.11$ (Table 9.30)

Region IV:

For $PTSF$, $L_d = 10 - 1 - 2 - 4.7 = 2.3$ mi

For ATS, $L_d = 10 - 1 - 2 - 1.7 = 5.3$ mi

Step 2. Compute $PTSF_{pl}$ using Eq. 9.22.

$$PTSF_{pl} = \frac{PTSF_d\left[L_u + L_d + f_{pl,PTSF}L_{pl} + \left(\frac{1 + f_{pl,PTSF}}{2} \right)L_{de} \right]}{L_t}$$

$$PTSF_{pl} = \frac{81.8[1.0 + 2.3 + (0.62 \times 2.00) + \left[\frac{1 + 0.62}{2} \right]4.7}{10} = 68.3\%$$

Step 3. Compute ATS_{pl} using Eq. 9.24.

$$ATS_{pl} = \frac{ATS_d L_t}{L_u + L_d + \left[\dfrac{L_{pl}}{f_{pl,ATS}} \right] + \left[\dfrac{2L_{de}}{1 + f_{pl,ATS}} \right]}$$

$$ATS_{pl} = \frac{38.6 \times 10}{1.00 + 5.3 + \left[\dfrac{2.00}{1.11} \right] + \left[\dfrac{2 \times 11.7}{1 + 1.11} \right]} = 39.6 \, \text{mi/h} \, (63.36 \, \text{km/h})$$

Step 4. Determine level of service.

Use Table 9.11.

$$LOS(PTSF) = D \, (\text{with a passing lane})$$

$$LOS(ATS) = E \, (\text{with a passing lane})$$

Since the lower value is selected, the LOS overall was not improved significantly by adding the passing lane.

Calculating Percent Time Spent Following (PTSF) and Average Travel Speed (ATS) for Directional Segments when a Climbing Lane Has Been Added within an Analysis Section on an Upgrade

Climbing lanes are provided on long upgrade sections to provide an opportunity for faster moving vehicles to pass heavy trucks whose speed has been reduced. Climbing lanes are warranted on two-lane highways when one of the following conditions is met:

- Directional flow rate on the upgrade >200 veh/h
- Directional flow rate for trucks on the upgrade >20 veh/h
- When any of the following conditions apply:
    - Speed reduction is $\geq$ 10 mi/h for a typical truck.
    - LOS E or F exists on the grade without a climbing lane.
    - LOS on the upgrade without a climbing lane is two or more levels of service values higher than the approach segment to the grade.

Operational analysis of climbing lanes follows the same procedural steps as described earlier for computing PTSF and ATS in level and rolling terrain with the following differences.

1. Table 9.31 is used to obtain adjustment factors for the existence of the climbing lane.
2. The analysis without a climbing lane is completed using the specific grade procedure.
3. L_u and L_d are zero because climbing lanes are analyzed as part of a specific upgrade. L_{de} is usually zero unless the climbing lane ends before the top of the grade.

9.3.6 LOS Assessment for Contiguous Segments of Two-Lane Highways

Several contiguous two-lane highway segments (in the same directions) may be combined to look at a longer section (with varying characteristics) as a facility. Weighted-average values of PTSF and ATS may be estimated for the facility. The weighting is on the basis of total travel time within the 15-min analysis period. The total travel time of all vehicles within the 15-min analysis period is estimated with Eq. 9.26 and Eq. 9.27.

$$VMT_{i15} = 0.25\left(\frac{V_i}{PHF}\right)L_t \tag{9.26}$$

$$TT_{i15} = \frac{VMT_{i15}}{ATS_i} \tag{9.27}$$

where

VMT_i = total vehicle miles traveled by all vehicles in directional segment i during the 15-min analysis period (veh/mi),

V_i = demand volume in directional segment i (veh/h)

PHF = peak hour factor

L_t = total length of directional segment i

TT_{i15} = total travel time consumed by all vehicles traversing directional segment i during the 15-min analysis period (veh/h)

A combined value for PTSF and ATS can be used to determine the overall LOS of several contiguous two-lane highway segments by using Eqs. 9.28 and 9.29 to obtain a weighted-average value based on the peak 15-min analysis period.

$$PTSF_F = \sum_{i=1}^{n} \frac{(TT_i)(PTSF_i)}{\sum_{i=1}^{n}(TT_i)} \qquad (9.28)$$

$$ATS_F = \sum_{i=1}^{n} \frac{(VMT_i)}{\sum_{i=1}^{n}(TT_i)} \qquad (9.29)$$

where

$PTSF_t$ = percent time spent following for segment i
$PTSF_F$ = percent time spent following for the facility
ATS_i = average travel speed for directional segment i (mi/h)
ATS_F = average travel speed for t
TT_i = total travel time for all vehicles on analysis segment (veh/h)

9.4 SUMMARY

Highway and traffic engineers are frequently engaged in evaluating the performance of different facilities of the highway system. These facilities include freeway segments, intersections, ramps, and other highway elements. The level of service is a measure of how well the facility is operating. It is both a qualitative measure of motorists' perceptions of the operational conditions existing on the facility, as well as a measure of the density of vehicular travel.

A primary objective in highway and traffic engineering is to provide highway facilities that operate at levels of service acceptable to the users of those facilities. Regular evaluation of the level of service at the facilities will help the engineer to determine whether acceptable conditions exist and to identify those locations where improvements may be necessary. Levels of different operating conditions are assigned to varying levels of service, ranging from level of service A to level of service F for each facility. The different levels of operating conditions are also related to the volume of traffic that can be accommodated by the specific component. This amount of traffic is also related to the capacity of the facility.

This chapter has described the procedures for determining the level of service on freeway sections as presented in the *Highway Capacity Manual 2010*. The procedures are based on the results of several major studies, from which several empirical expressions have been developed to determine the maximum flow rate at a specific facility under prevailing conditions. The chapter has focused on two-lane highways, freeways, and multilane highways. In the next chapter, we discuss capacity and level of service for signalized intersections.

PROBLEMS

9-1 Briefly describe the traffic characteristics associated with each of the levels of service for basic freeway sections.

9-2 Describe the factors that affect the level of service of a freeway section and the impact each has on flow.

9-3   Describe the possible situations that would require adjustments from the base conditions for freeway capacity analysis.

9-4   A freeway is to be designed to provide LOS C for the following conditions: design hourly volume of 5600 veh/h; PHF: 0.92; trucks: 6%; free flow speed: 70 mi/h (112 km/h); no lateral obstructions; rolling terrain; total ramp density of 0.75 ramps per mile.
Determine number of 12-ft (3.6 m) lanes required in each direction.

9-5   An existing rural freeway in rolling terrain is to be analyzed to determine LOS using the following information:

> Number of lanes in each direction: 2
> Peak hour volume: 2640 veh/h (in the peak direction)
> 18% trucks
> 2% recreational vehicles
> PHF = 0.91
> Lane width: 12 ft (3.6 m)
> Lateral clearance: 10 ft (3 m)
> Average interchange spacing: 3 mi (4.8 km); all interchanges are diamond interchanges.

9-6   An existing urban freeway with 4 lanes in each direction has the following characteristics:

> Traffic data:
>   Peak hour volume (in the peak direction): 7070 veh/h
>   Trucks: 10% of peak hour volume
>   PHF = 0.94
> Geometric data:
>   Lane width: 11 ft (3.3 m)
>   Shoulder width: 6 ft (1.8 m)
>   Total ramp density: 1.8 ramps per mile
>   Terrain: rolling

Determine the LOS in the peak hour. Clearly state assumptions used for any values not given. Show the demand flow rate, mean speed, and density for the given conditions.

9-7   An urban freeway is to be designed using the following information:

> AADT = 52,000 veh/day
> K (proportion of AADT occurring during the peak hour): 0.11
> D (proportion of peak hour traffic traveling in the peak direction): 0.65
> Trucks: 8% of peak hour volume
> PHF = 0.95
> Lane width: 12 ft (3.6 m)
> Shoulder width: 10 ft (3.0 m)
> Total ramp density: 0.5 interchange/mile; all interchanges are to be cloverleaf interchanges.
> Terrain: rolling

Determine the number of lanes required to provide LOS C. Clearly state assumptions used for any values not given, and show all calculations required.

9-8   An existing 4-lane freeway (2 lanes in each direction) is to be expanded. The segment length is 2 mi (3.2 km); sustained grade: 4%; design volume of 3000 veh/h; trucks: 10%; buses: 2%; RVs: 3%; PHF: 0.95; free-flow speed: 70 mi/h (112 km/h); right side lateral obstruction: 5 ft (1.5 m); design LOS: B.
Determine number of additional lanes required in each direction.

9-9   Given: Roadway segment with 6000 ft (1800 m) of 3% upgrade, followed by 5000 ft (1500 m) of 5% upgrade; trucks: 8%; RVs: 4%.
Determine number of PCEs.

9-10   Describe the situations that require adjustments from the base conditions for multilane highway capacity analysis.

9-11   A new section of Richmond Highway is being designed as a six-lane facility (three in each direction) with a two-way left-turn lane. Determine the peak-hour LOS.
Traffic data:

- Directional design hourly volume = 3600 veh/h
- PHF = 0.94
- Assumed base free-flow speed = 55 mi/h (88 km/h)

Geometric data:

- Urban setting
- Rolling terrain
- Lane width = 11 ft (3.3 m)
- Shoulder widths = 4 ft (1.2 m) (right side) and 1 ft (0.3 m) (left side)
- Average access point spacing = 12 points per mile on each side

9-12   Define the elements of a Class I, Class II, and Class III two-lane highway.

9-13   What are the three measures used to describe service quality for a two-lane highway? Which of the measures are used to describe level-of-service for Class I Class II, and Class III highways?

9-14   Describe the traffic characteristics associated with each of the six levels of service for two-lane highways.

9-15   The following values of PTSF, FFS, and ATS have been determined for three separate two-lane segments. Determine LOS if the segments are: (a) Class I, (b) Class II, and (c) Class III.

Segment	PTSF (%)	ATS (mi/h)	FFS (mi/h)
10	25	52	60
11	46	49	55
12	67	39	60

9-16   Determine the PTSF in each direction for a 4.5 mi (7.2 km) two-lane highway segment in level terrain. Traffic volumes (two-way) are 1100 veh/h. Trucks: 10%; RVs: 7%; PHF: 0.97; directional split: 60/40; no passing zones: 40%.

9-17   Use the data provided in Problem 9-16 to estimate the ATS in each direction. Base free-flow speed: 55 mi/h (88 km/h); lane width: 11 ft (3.3 m); shoulder width: 3 ft (0.9 m); access points per mile: 15.

9-18   Use the results of Problems 9-16 and 9-17 to compute LOS, v/c, and veh-mi in the peak 15 min and peak hr, and total travel time in the peak 15 min.

Segment	PTSF (%)	ATS (mi/h)
10	25	52
11	46	39
12	67	39

9-19   Use the data and results obtained in Problem 9-18 to determine the level of service of a two-lane section if a passing lane 1.5 mi (2.4 km) long is added. The passing lane begins 0.75 mi (1.2 km) from the starting point of the analysis segment.

9-20   An existing *Class I* two-lane highway is to be analyzed to determine the level of service in each direction, given the following information:

> Traffic data:
> PHV = 600 veh/h
> 60% in the peak direction
> 8% trucks
> 2% recreational vehicles
> PHF = 0.86
> No passing zones: 40%
> Geometric data:
> Rolling terrain
> BFFS = 55 mi/h (88 km/h)
> Lane width = 11 ft (3.3 m)
> Shoulder width = 2 ft (0.6 m)
> 8 access points per mile

9-21   An existing *Class II* two-lane highway is to be analyzed to determine the level of service in each direction given the following information:

> Peak hourly volume in the analysis direction: 900 veh/h
> Peak hourly volume in the opposing direction: 400 veh/h
> Trucks: 12% of total volume
> Recreational vehicles: 2% of total volume
> PHF: 0.95
> Lane width: 12 ft (3.6 m)
> Shoulder width: 10 ft (3.0 m)
> Access points per mile: 20
> Terrain: rolling
> Base free-flow speed: 60 mi/h (96 km/h)
> No passing zones: 40% of analysis segment length

9-22   An existing *Class III* two-lane highway is to be analyzed to determine the level of service in each direction given the following information:

> Peak hourly volume in the analysis direction: 900 veh/h
> Peak hourly volume in the opposing direction: 720 veh/h
> Trucks: 10% of total volume
> Recreational vehicles: 2% of total volume
> PHF: 0.94
> Lane width: 12 ft (3.6 m)
> Shoulder width: 2 ft (0.6 m)
> Access points per mile: 30
> Terrain: Level
> Base free flow speed: 45 mi/h (72 km/h)
> No passing zones: 60% of analysis segment length

REFERENCE

Highway Capacity Manual. HCM 2010, Transportation Research Board, National Research Council, Washington, D.C., 2010.

CHAPTER 9 APPENDIX: TABLES (These Tables are for U.S. only. Specific tables by country may be required).

Table 9.1   Adjustments to FFS for Average Lane Width

Average Lane Width (ft)	Reduction in FFS, f_{LW} (mi/h)
≥12	0.0
≥11–12	1.9
≥10–11	6.6

SOURCE: From *Highway Capacity Manual 2010.* Copyright, National Academy of Sciences, Washington, D.C. Reproduced with permission of the Transportation Research Board.

Table 9.2   Adjustments to FFS for Right-Side Lateral Clearance, f_{LC} (mi/h)

Right-Side Lateral Clearance (ft)	Lanes in One Direction			
	2	3	4	≥5
≥6	0.0	0.0	0.0	0.0
5	0.6	0.4	0.2	0.1
4	1.2	0.8	0.4	0.2
3	1.8	1.2	0.6	0.3
2	2.4	1.6	0.8	0.4
1	3.0	2.0	1.0	0.5
0	3.6	2.4	1.2	0.6

SOURCE: From *Highway Capacity Manual 2010.* Copyright, National Academy of Sciences, Washington, D.C. Reproduced with permission of the Transportation Research Board.

Table 9.3   PCEs for Heavy Vehicles in General Terrain Segments

Vehicle	PCE by Type of Terrain		
	Level	Rolling	Mountainous
Trucks and buses, E_T	1.5	2.5	4.5
RVs, E_R	1.2	2.0	4.0

SOURCE: From *Highway Capacity Manual 2010.* Copyright, National Academy of Sciences, Washington, D.C. Reproduced with permission of the Transportation Research Board.

Table 9.4   PCEs for Trucks and Buses (E_T) on Upgrades

Upgrade (%)	Length (mi)	Proportion of Trucks and Buses								
		2%	4%	5%	6%	8%	10%	15%	20%	≥25%
≤2	All	1.5	1.5	1.5	1.5	1.5	1.5	1.5	1.5	1.5
>2–3	0.00–0.25	1.5	1.5	1.5	1.5	1.5	1.5	1.5	1.5	1.5
	>0.25–0.50	1.5	1.5	1.5	1.5	1.5	1.5	1.5	1.5	1.5
	>0.50–0.75	1.5	1.5	1.5	1.5	1.5	1.5	1.5	1.5	1.5
	>0.75–1.00	2.0	2.0	2.0	2.0	1.5	1.5	1.5	1.5	1.5
	>1.00–1.50	2.5	2.5	2.5	2.5	2.0	2.0	2.0	2.0	2.0
	>1.50	3.0	3.0	2.5	2.5	2.0	2.0	2.0	2.0	2.0
>3–4	0.00–0.25	1.5	1.5	1.5	1.5	1.5	1.5	1.5	1.5	1.5
	>0.25–0.50	2.0	2.0	2.0	2.0	2.0	2.0	1.5	1.5	1.5
	>0.50–0.75	2.5	2.5	2.0	2.0	2.0	2.0	2.0	2.0	2.0
	>0.75–1.00	3.0	3.0	2.5	2.5	2.5	2.5	2.0	2.0	2.0
	>1.00–1.50	3.5	3.5	3.0	3.0	3.0	3.0	2.5	2.5	2.5
	>1.50	4.0	3.5	3.0	3.0	3.0	3.0	2.5	2.5	2.5
>4–5	0.00–0.25	1.5	1.5	1.5	1.5	1.5	1.5	1.5	1.5	1.5
	>0.25–0.50	3.0	2.5	2.5	2.5	2.0	2.0	2.0	2.0	2.0
	>0.50–0.75	3.5	3.0	3.0	3.0	2.5	2.5	2.5	2.5	2.5
	>0.75–1.00	4.0	3.5	3.5	3.5	3.0	3.0	3.0	3.0	3.0
	>1.00	5.0	4.0	4.0	4.0	3.5	3.5	3.0	3.0	3.0
>5–6	0.00–0.25	2.0	2.0	1.5	1.5	1.5	1.5	1.5	1.5	1.5
	>0.25–0.30	4.0	3.0	2.5	2.5	2.0	2.0	2.0	2.0	2.0
	>0.30–0.50	4.5	4.0	3.5	3.0	2.5	2.5	2.5	2.5	2.5
	>0.50–0.75	5.0	4.5	4.0	3.5	3.0	3.0	3.0	3.0	3.0
	>0.75–1.00	5.5	5.0	4.5	4.0	3.0	3.0	3.0	3.0	3.0
	>1.00	6.0	5.0	5.0	4.5	3.5	3.5	3.5	3.5	3.5
>6	0.00–0.25	4.0	3.0	2.5	2.5	2.5	2.5	2.0	2.0	1.0
	>0.25–0.30	4.5	4.0	3.5	3.5	3.5	3.0	2.5	2.5	2.5
	>0.30–0.50	5.0	4.5	4.0	4.0	3.5	3.0	2.5	2.5	2.5
	>0.50–0.75	5.5	5.0	4.5	4.5	4.0	3.5	3.0	3.0	3.0
	>0.75–1.00	6.0	5.5	5.0	5.0	4.5	4.0	3.5	3.5	3.5
	>1.00	7.0	6.0	5.5	5.5	5.0	4.5	4.0	4.0	4.0

Note: Interpolation for percentage of trucks and buses is recommended to the nearest 0.1.

Table 9.5 PCEs for RVs (E_R) on Upgrades

Upgrade (%)	Length (mi)	Proportion of RVs								
		2%	4%	5%	6%	8%	10%	15%	20%	≥25%
≤2	All	1.2	1.2	1.2	1.2	1.2	1.2	1.2	1.2	1.2
>2–3	0.00–0.50	1.2	1.2	1.2	1.2	1.2	1.2	1.2	1.2	1.2
	>0.50	3.0	1.5	1.5	1.5	1.5	1.5	1.2	1.2	1.2
>3–4	0.00–0.25	1.2	1.2	1.2	1.2	1.2	1.2	1.2	1.2	1.2
	>0.25–0.50	2.5	2.5	2.0	2.0	2.0	2.0	1.5	1.5	1.5
	>0.50	3.0	2.5	2.5	2.5	2.0	2.0	2.0	1.5	1.5
>4–5	0.00–0.25	2.5	2.0	2.0	2.0	1.5	1.5	1.5	1.5	1.5
	>0.25–0.50	4.0	3.0	3.0	3.0	2.5	2.5	2.0	2.0	2.0
	>0.50	4.5	3.5	3.0	3.0	3.0	2.5	2.5	2.0	2.0
>5	0.00–0.25	4.0	3.0	2.5	2.5	2.5	2.0	2.0	2.0	1.5
	>0.25–0.50	6.0	4.0	4.0	3.5	3.0	3.0	2.5	2.5	2.0
	>0.50	6.0	4.5	4.0	4.0	3.5	3.0	3.0	2.5	2.0

Note: Interpolation for percentage of RVs is recommended to the nearest 0.1.

SOURCE: From *Highway Capacity Manual 2010*. Copyright, National Academy of Sciences, Washington, D.C. Reproduced with permission of the Transportation Research Board.

Table 9.6 PCEs for Trucks and Buses (E_T) on Specific Downgrades

Downgrade (%)	Length of Grade (mi)	Proportion of Trucks and Buses			
		5%	10%	15%	≥20%
<4	All	1.5	1.5	1.5	1.5
4–5	≤4	1.5	1.5	1.5	1.5
	>4	2.0	2.0	2.0	1.5
>5–6	≤4	1.5	1.5	1.5	1.5
	>4	5.5	4.0	4.0	3.0
>6	≤4	1.5	1.5	1.5	1.5
	>4	7.5	6.0	5.5	4.5

SOURCE: From *Highway Capacity Manual 2010*. Copyright, National Academy of Sciences, Washington, D.C. Reproduced with permission of the Transportation Research Board.

Table 9.7 Adjustment to FFS for Lateral Clearances

Four-Lane Highways		Six-Lane Highways	
TLC (ft)	Reduction in FFS (mi/h)	TLC (ft)	Reduction in FFS (mi/h)
12	0.0	12	0.0
10	0.4	10	0.4
8	0.9	8	0.9
6	1.3	6	1.3
4	1.8	4	1.7
2	3.6	2	2.8
0	5.4	0	3.9

Note: Interpolation to the nearest 0.1 is recommended.

SOURCE: From *Highway Capacity Manual 2010*. Copyright, National Academy of Sciences, Washington, D.C. Reproduced with permission of the Transportation Research Board.

Table 9.8 Adjustment to FFS for Median Type

Median Type	Reduction in FFS, f_M (mi/h)
Undivided	1.6
TWLTL	0.0
Divided	0.0

Table 9.9 Adjustment to FFS for Access-Point Density

Access-Point Density (access points/mi)	Reduction in FFS, f_A (mi/h)
0	0.0
10	2.5
20	5.0
30	7.5
≥40	10.0

Note: Interpolation to the nearest 0.1 is recommended.

Table 9.10 Automobile LOS for Multilane Highway Segments

LOS	FFS (mi/h)	Density (pc/mi/ln)
A	All	>0–11
B	All	>11–18
C	All	>18–26
D	All	>26–35
E	60	>35–40
	55	>35–41
	50	>35–43
	45	>35–45
	Demand Exceeds Capacity	
F	60	>40
	55	>41
	50	>43
	45	>45

Table 9.11 Automobile LOS for Two-Lane Highways

	Class I Highways		Class II Highways	Class III Highways
LOS	ATS (mi/h)	PTSF (%)	PTSF (%)	PFFS (%)
A	>55	≤35	≤40	>91.7
B	>50–55	>35–50	>40–55	>83.3–91.7
C	>45–50	>50–65	>55–70	>75.0–83.3
D	>40–45	>65–80	>70–85	>66.7–75.0
E	≤40	>80	>85	≤66.7

SOURCE: From *Highway Capacity Manual 2010*. Copyright, National Academy of Sciences, Washington, D.C. Reproduced with permission of the Transportation Research Board.

Table 9.12 Adjustment Factor for Lane and Shoulder Width (f_{LS})

Lane Width (ft)	Shoulder Width (ft)			
	≥0 <2	≥2 <4	≥4 <6	≥6
≥9 <10	6.4	4.8	3.5	2.2
≥10 <11	5.3	3.7	2.4	1.1
≥11 <12	4.7	3.0	1.7	0.4
≥12	4.2	2.6	1.3	0.0

SOURCE: From *Highway Capacity Manual 2010*. Copyright, National Academy of Sciences, Washington, D.C. Reproduced with permission of the Transportation Research Board.

Table 9.13 Adjustment Factor for Access-Point Density (f_A)

Access Points per Mile (Two Directions)	Reduction in FFS (mi/h)
0	0.0
10	2.5
20	5.0
30	7.5
40	10.0

Note: Interpolation to the nearest 0.1 is recommended.

SOURCE: From *Highway Capacity Manual 2010*. Copyright, National Academy of Sciences, Washington, D.C. Reproduced with permission of the Transportation Research Board.

Table 9.14   ATS Grade Adjustment Factor ($f_{g,ATS}$) for Level Terrain, Rolling Terrain, and Specific Downgrades

One-Direction Demand Flow Rate, V_{vph} (veh/h)	Adjustment Factor	
	Level Terrain and Specific Downgrades	Rolling Terrain
≤100	1.00	0.67
200	1.00	0.75
300	1.00	0.83
400	1.00	0.90
500	1.00	0.95
600	1.00	0.97
700	1.00	0.98
800	1.00	0.99
≥900	1.00	1.00

Note: Interpolation to the nearest 0.01 is recommended.

SOURCE: From *Highway Capacity Manual 2010*. Copyright, National Academy of Sciences, Washington, D.C. Reproduced with permission of the Transportation Research Board.

Table 9.15   ATS Grade Adjustment Factor ($f_{g,ATS}$) for Specific Upgrades

Grade (%)	Grade Length (mi)	Directional Demand Flow Rate, v_{vph} (veh/h)								
		≤100	200	300	400	500	600	700	800	≥900
≥3 <3.5	0.25	0.78	0.84	0.87	0.91	1.00	1.00	1.00	1.00	1.00
	0.50	0.75	0.83	0.86	0.90	1.00	1.00	1.00	1.00	1.00
	0.75	0.73	0.81	0.85	0.89	1.00	1.00	1.00	1.00	1.00
	1.00	0.73	0.79	0.83	0.88	1.00	1.00	1.00	1.00	1.00
	1.50	0.73	0.79	0.83	0.87	0.99	0.99	1.00	1.00	1.00
	2.00	0.73	0.79	0.82	0.86	0.98	0.98	0.99	1.00	1.00
	3.00	0.73	0.78	0.82	0.85	0.95	0.96	0.96	0.97	0.98
	≥4.00	0.73	0.78	0.81	0.85	0.94	0.94	0.95	0.95	0.96
≥3.5 <4.5	0.25	0.75	0.83	0.86	0.90	1.00	1.00	1.00	1.00	1.00
	0.50	0.72	0.80	0.84	0.88	1.00	1.00	1.00	1.00	1.00
	0.75	0.67	0.77	0.81	0.86	1.00	1.00	1.00	1.00	1.00
	1.00	0.65	0.73	0.77	0.81	0.94	0.95	0.97	1.00	1.00
	1.50	0.63	0.72	0.76	0.80	0.93	0.95	0.96	1.00	1.00
	2.00	0.62	0.70	0.74	0.79	0.93	0.94	0.96	1.00	1.00
	3.00	0.61	0.69	0.74	0.78	0.92	0.93	0.94	0.98	1.00
	≥4.00	0.61	0.69	0.73	0.78	0.91	0.91	0.92	0.96	1.00
≥4.5 <5.5	0.25	0.71	0.79	0.83	0.88	1.00	1.00	1.00	1.00	1.00
	0.50	0.60	0.70	0.74	0.79	0.94	0.95	0.97	1.00	1.00
	0.75	0.55	0.65	0.70	0.75	0.91	0.93	0.95	1.00	1.00
	1.00	0.54	0.64	0.69	0.74	0.91	0.93	0.95	1.00	1.00
	1.50	0.52	0.62	0.67	0.72	0.88	0.90	0.93	1.00	1.00
	2.00	0.51	0.61	0.66	0.71	0.87	0.89	0.92	0.99	1.00
	3.00	0.51	0.61	0.65	0.70	0.86	0.88	0.91	0.98	0.99
	≥4.00	0.51	0.60	0.65	0.69	0.84	0.86	0.88	0.95	0.97

(Continued)

Table 9.15 ATS Grade Adjustment Factor ($f_{g,ATS}$) for Specific Upgrades (*continued*)

	0.25	0.57	0.68	0.72	0.77	0.93	0.94	0.96	1.00	1.00
	0.50	0.52	0.62	0.66	0.71	0.87	0.90	0.92	1.00	1.00
	0.75	0.49	0.57	0.62	0.68	0.85	0.88	0.90	1.00	1.00
$\geq 5.5 <6.5$	1.00	0.46	0.56	0.60	0.65	0.82	0.85	0.88	1.00	1.00
	1.50	0.44	0.54	0.59	0.64	0.81	0.84	0.87	0.98	1.00
	2.00	0.43	0.53	0.58	0.63	0.81	0.83	0.86	0.97	0.99
	3.00	0.41	0.51	0.56	0.61	0.79	0.82	0.85	0.97	0.99
	≥ 4.00	0.40	0.50	0.55	0.61	0.79	0.82	0.85	0.97	0.99
	0.25	0.54	0.64	0.68	0.73	0.88	0.90	0.92	1.00	1.00
	0.50	0.43	0.53	0.57	0.62	0.79	0.82	0.85	0.98	1.00
	0.75	0.39	0.49	0.54	0.59	0.77	0.80	0.83	0.96	1.00
≥ 6.5	1.00	0.37	0.45	0.50	0.54	0.74	0.77	0.81	0.96	1.00
	1.50	0.35	0.45	0.49	0.54	0.71	0.75	0.79	0.96	1.00
	2.00	0.34	0.44	0.48	0.53	0.71	0.74	0.78	0.94	0.99
	3.00	0.34	0.44	0.48	0.53	0.70	0.73	0.77	0.93	0.98
	≥ 4.00	0.33	0.43	0.47	0.52	0.70	0.73	0.77	0.91	0.95

Note: Straight-line interpolation of $f_{g,ATS}$ for length of grade and demand flow permitted to the nearest 0.01.

SOURCE: From *Highway Capacity Manual 2010*. Copyright, National Academy of Sciences, Washington, D.C. Reproduced with permission of the Transportation Research Board.

Table 9.16 ATS Passenger Car Equivalents for Trucks (E_T) and RVs (E_R) for Level Terrain, Rolling Terrain, and Specific Downgrades

Vehicle Type	Directional Demand Flow Rate, v_{vph} (veh/h)	Level Terrain and Specific Downgrades	Rolling Terrain
Trucks, E_T	≤ 100	1.9	2.7
	200	1.5	2.3
	300	1.4	2.1
	400	1.3	2.0
	500	1.2	1.8
	600	1.1	1.7
	700	1.1	1.6
	800	1.1	1.4
	≥ 900	1.0	1.3
RVs, E_R	All flows	1.0	1.1

Note: Interpolation to the nearest 0.1 is recommended.

SOURCE: From *Highway Capacity Manual 2010*. Copyright, National Academy of Sciences, Washington, D.C. Reproduced with permission of the Transportation Research Board.

Table 9.17   ATS Passenger Car Equivalents for Trucks (E_T) on Specific Upgrades

Grade (%)	Grade Length (mi)	Directional Demand Flow Rate, v_{vph} (veh/h)								
		≤100	200	300	400	500	600	700	800	≥900
≥3 <3.5	0.25	2.6	2.4	2.3	2.2	1.8	1.8	1.7	1.3	1.1
	0.50	3.7	3.4	3.3	3.2	2.7	2.6	2.6	2.3	2.0
	0.75	4.6	4.4	4.3	4.2	3.7	3.6	3.4	2.4	1.9
	1.00	5.2	5.0	4.9	4.9	4.4	4.2	4.1	3.0	1.6
	1.50	6.2	6.0	5.9	5.8	5.3	5.0	4.8	3.6	2.9
	2.00	7.3	6.9	6.7	6.5	5.7	5.5	5.3	4.1	3.5
	3.00	8.4	8.0	7.7	7.5	6.5	6.2	6.0	4.6	3.9
	≥4.00	9.4	8.8	8.6	8.3	7.2	6.9	6.6	4.8	3.7
≥3.5 <4.5	0.25	3.8	3.4	3.2	3.0	2.3	2.2	2.2	1.7	1.5
	0.50	5.5	5.3	5.1	5.0	4.4	4.2	4.0	2.8	2.2
	0.75	6.5	6.4	6.5	6.5	6.3	5.9	5.6	3.6	2.6
	1.00	7.9	7.6	7.4	7.3	6.7	6.6	6.4	5.3	4.7
	1.50	9.6	9.2	9.0	8.9	8.1	7.9	7.7	6.5	5.9
	2.00	10.3	10.1	10.0	9.9	9.4	9.1	8.9	7.4	6.7
	3.00	11.4	11.3	11.2	11.2	10.7	10.3	10.0	8.0	7.0
	≥4.00	12.4	12.2	12.2	12.1	11.5	11.2	10.8	8.6	7.5
≥4.5 <5.5	0.25	4.4	4.0	3.7	3.5	2.7	2.7	2.7	2.6	2.5
	0.50	6.0	6.0	6.0	6.0	5.9	5.7	5.6	4.6	4.2
	0.75	7.5	7.5	7.5	7.5	7.5	7.5	7.5	7.5	7.5
	1.00	9.2	9.2	9.1	9.1	9.0	9.0	9.0	8.9	8.8
	1.50	10.6	10.6	10.6	10.6	10.5	10.4	10.4	10.2	10.1
	2.00	11.8	11.8	11.8	11.8	11.6	11.6	11.5	11.1	10.9
	3.00	13.7	13.7	13.6	13.6	13.3	13.1	13.0	11.9	11.3
	≥4.00	15.3	15.3	15.2	15.2	14.6	14.2	13.8	11.3	10.0
≥5.5 <6.5	0.25	4.8	4.6	4.5	4.4	4.0	3.9	3.8	3.2	2.9
	0.50	7.2	7.2	7.2	7.2	7.2	7.2	7.2	7.2	7.2
	0.75	9.1	9.1	9.1	9.1	9.1	9.1	9.1	9.1	9.1
	1.00	10.3	10.3	10.3	10.3	10.3	10.3	10.3	10.2	10.1
	1.50	11.9	11.9	11.9	11.9	11.8	11.8	11.8	11.7	11.6
	2.00	12.8	12.8	12.8	12.8	12.7	12.7	12.7	12.6	12.5
	3.00	14.4	14.4	14.4	14.4	14.3	14.3	14.3	14.2	14.1
	≥4.00	15.4	15.4	15.3	15.3	15.2	15.1	15.1	14.9	14.8
≥6.5	0.25	5.1	5.1	5.0	5.0	4.8	4.7	4.7	4.5	4.4
	0.50	7.8	7.8	7.8	7.8	7.8	7.8	7.8	7.8	7.8
	0.75	9.8	9.8	9.8	9.8	9.8	9.8	9.8	9.8	9.8
	1.00	10.4	10.4	10.4	10.4	10.4	10.4	10.4	10.3	10.2
	1.50	12.0	12.0	12.0	12.0	11.9	11.9	11.9	11.8	11.7
	2.00	12.9	12.9	12.9	12.9	12.8	12.8	12.8	12.7	12.6
	3.00	14.5	14.5	14.5	14.5	14.4	14.4	14.4	14.3	14.2
	≥4.00	15.4	15.4	15.4	15.4	15.3	15.3	15.3	15.2	15.1

Note: Interpolation for length of grade and demand flow rate to the nearest 0.1 is recommended.

SOURCE: From *Highway Capacity Manual 2010*. Copyright, National Academy of Sciences, Washington, D.C. Reproduced with permission of the Transportation Research Board.

Table 9.18   ATS Passenger Car Equivalents for RVs (E_R) on Specific Upgrades

Grade (%)	Grade Length (mi)	Directional Demand Flow Rate, v_{vph} (veh/h)								
		≤100	200	300	400	500	600	700	800	≥900
≥3 <3.5	≤0.25	1.1	1.1	1.1	1.0	1.0	1.0	1.0	1.0	1.0
	>0.25 ≤0.75	1.2	1.2	1.1	1.1	1.0	1.0	1.0	1.0	1.0
	>0.75 ≤1.25	1.3	1.2	1.2	1.1	1.0	1.0	1.0	1.0	1.0
	>1.25 ≤2.25	1.4	1.3	1.2	1.1	1.0	1.0	1.0	1.0	1.0
	>2.25	1.5	1.4	1.3	1.2	1.0	1.0	1.0	1.0	1.0
≥3.5 <4.5	≤0.75	1.3	1.2	1.2	1.1	1.0	1.0	1.0	1.0	1.0
	>0.75 ≤3.50	1.4	1.3	1.2	1.1	1.0	1.0	1.0	1.0	1.0
	>3.50	1.5	1.4	1.3	1.2	1.0	1.0	1.0	1.0	1.0
≥4.5 <5.5	≤2.50	1.5	1.4	1.3	1.2	1.0	1.0	1.0	1.0	1.0
	>2.50	1.6	1.5	1.4	1.2	1.0	1.0	1.0	1.0	1.0
≥5.5 <6.5	≤0.75	1.5	1.4	1.3	1.1	1.0	1.0	1.0	1.0	1.0
	>0.75 ≤2.50	1.6	1.5	1.4	1.2	1.0	1.0	1.0	1.0	1.0
	>2.50 ≤3.50	1.6	1.5	1.4	1.3	1.2	1.1	1.0	1.0	1.0
	>3.50	1.6	1.6	1.6	1.5	1.5	1.4	1.3	1.2	1.1
≥6.5	≤2.50	1.6	1.5	1.4	1.2	1.0	1.0	1.0	1.0	1.0
	>2.50 ≤3.50	1.6	1.5	1.4	1.2	1.3	1.3	1.3	1.3	1.3
	>3.50	1.6	1.6	1.6	1.5	1.5	1.5	1.4	1.4	1.4

Note: Interpolation in this exhibit is not recommended.

SOURCE: From *Highway Capacity Manual 2010*. Copyright, National Academy of Sciences, Washington, D.C. Reproduced with permission of the Transportation Research Board.

Table 9.19   ATS Passenger Car Equivalents (E_{TC}) for Trucks on Downgrades Traveling at Crawl Speed

Difference Between FFS and Truck Crawl Speed (mi/h)	Directional Demand Flow Rate, V_{vph} (veh/h)								
	≤100	200	300	400	500	600	700	800	≥900
≤15	4.7	4.1	3.6	3.1	2.6	2.1	1.6	1.0	1.0
20	9.9	8.7	7.8	6.7	5.8	4.9	4.0	2.7	1.0
25	15.1	13.5	12.0	10.4	9.0	7.7	6.4	5.1	3.8
30	22.0	19.8	17.5	15.6	13.1	11.6	9.2	6.1	4.1
35	29.0	26.0	23.1	20.1	17.3	14.6	11.9	9.2	6.5
≥40	35.9	32.3	28.6	24.9	21.4	18.1	14.7	11.3	7.9

Note: Interpolation against both speed difference and demand flow rate to the nearest 0.1 is recommended.

SOURCE: From *Highway Capacity Manual 2010*. Copyright, National Academy of Sciences, Washington, D.C. Reproduced with permission of the Transportation Research Board.

Table 9.20   ATS Adjustment Factor for No-Passing Zones ($f_{np,ATS}$)

Opposing Demand Flow Rate, v_o (pc/h)	Percent No-Passing Zones				
	≤20	40	60	80	100
FFS ≥ 65 mi/h					
≤100	1.1	2.2	2.8	3.0	3.1
200	2.2	3.3	3.9	4.0	4.2
400	1.6	2.3	2.7	2.8	2.9
600	1.4	1.5	1.7	1.9	2.0
800	0.7	1.0	1.2	1.4	1.5
1000	0.6	0.8	1.1	1.1	1.2
1200	0.6	0.8	0.9	1.0	1.1
1400	0.6	0.7	0.9	0.9	0.9
≥1600	0.6	0.7	0.7	0.7	0.8
FFS = 60 mi/h					
≤100	0.7	1.7	2.5	2.8	2.9
200	1.9	2.9	3.7	4.0	4.2
400	1.4	2.0	2.5	2.7	3.9
600	1.1	1.3	1.6	1.9	2.0
800	0.6	0.9	1.1	1.3	1.4
1000	0.6	0.7	0.9	1.1	1.2
1200	0.5	0.7	0.9	0.9	1.1
1400	0.5	0.6	0.8	0.8	0.9
≥1600	0.5	0.6	0.7	0.7	0.7
FFS = 55 mi/h					
≤100	0.5	1.2	2.2	2.6	2.7
200	1.5	2.4	3.5	3.9	4.1
400	1.3	1.9	2.4	2.7	2.8
600	0.9	1.1	1.6	1.8	1.9
800	0.5	0.7	1.1	1.2	1.4
1000	0.5	0.6	0.8	0.9	1.1
1200	0.5	0.6	0.7	0.9	1.0
1400	0.5	0.6	0.7	0.7	0.9
≥1600	0.5	0.6	0.6	0.6	0.7
FFS = 50 mi/h					
≤100	0.2	0.7	1.9	2.4	2.5
200	1.2	2.0	3.3	3.9	4.0
400	1.1	1.6	2.2	2.6	2.7
600	0.6	0.9	1.4	1.7	1.9
800	0.4	0.6	0.9	1.2	1.3
1000	0.4	0.4	0.7	0.9	1.1
1200	0.4	0.4	0.7	0.8	1.0
1400	0.4	0.4	0.6	0.7	0.8
≥1600	0.4	0.4	0.5	0.5	0.5

(Continued)

Table 9.20   ATS Adjustment Factor for No-Passing Zones ($f_{np,ATS}$) (continued)

FFS ≤ 45 mi/h					
≤100	0.1	0.4	1.7	2.2	2.4
200	0.9	1.6	3.1	3.8	4.0
400	0.9	0.5	2.0	2.5	2.7
600	0.4	0.3	1.3	1.7	1.8
800	0.3	0.3	0.8	1.1	1.2
1000	0.3	0.3	0.6	0.8	1.1
1200	0.3	0.3	0.6	0.7	1.0
1400	0.3	0.3	0.6	0.6	0.7
≥1600	0.3	0.3	0.4	0.4	0.6

Note: Interpolation of $f_{np,ATS}$ for percent no-passing zones, demand flow rate, and FFS to the nearest 0.1 is recommended.

SOURCE: From *Highway Capacity Manual 2010*. Copyright, National Academy of Sciences, Washington, D.C. Reproduced with permission of the Transportation Research Board.

Table 9.21   PTSF Grade Adjustment Factor ($f_{g,PTSF}$) for Level Terrain, Rolling Terrain, and Specific Downgrades

Directional Demand Flow Rate, v_{vph} (veh/h)	Level Terrain and Specific Downgrades	Rolling Terrain
≤100	1.00	0.73
200	1.00	0.80
300	1.00	0.85
400	1.00	0.90
500	1.00	0.96
600	1.00	0.97
700	1.00	0.99
800	1.00	1.00
≥900	1.00	1.00

Note: Interpolation to the nearest 0.01 is recommended.

SOURCE: From *Highway Capacity Manual 2010*. Copyright, National Academy of Sciences, Washington, D.C. Reproduced with permission of the Transportation Research Board.

Table 9.22   PTSF Grade Adjustment Factor $(f_{g,PTSF})$ for Specific Upgrades

Grade (%)	Grade Length (mi)	Directional Demand Flow Rate, v_{vph} (veh/h)								
		≤100	200	300	400	500	600	700	800	≥900
≥3 <3.5	0.25	1.00	0.99	0.97	0.96	0.92	0.92	0.92	0.92	0.92
	0.50	1.00	0.99	0.98	0.97	0.93	0.93	0.93	0.93	0.93
	0.75	1.00	0.99	0.98	0.97	0.93	0.93	0.93	0.93	0.93
	1.00	1.00	0.99	0.98	0.97	0.93	0.93	0.93	0.93	0.93
	1.50	1.00	0.99	0.98	0.97	0.94	0.94	0.94	0.94	0.94
	2.00	1.00	0.99	0.98	0.98	0.95	0.95	0.95	0.95	0.95
	3.00	1.00	1.00	0.99	0.99	0.97	0.97	0.97	0.96	0.96
	≥4.00	1.00	1.00	1.00	1.00	1.00	0.99	0.99	0.97	0.97
≥3.5 <4.5	0.25	1.00	0.99	0.98	0.97	0.94	0.93	0.93	0.92	0.92
	0.50	1.00	1.00	0.99	0.99	0.97	0.97	0.97	0.96	0.95
	0.75	1.00	1.00	0.99	0.99	0.97	0.97	0.97	0.96	0.96
	1.00	1.00	1.00	0.99	0.99	0.97	0.97	0.97	0.97	0.97
	1.50	1.00	1.00	0.99	0.99	0.97	0.97	0.97	0.97	0.97
	2.00	1.00	1.00	0.99	0.99	0.98	0.98	0.98	0.98	0.98
	3.00	1.00	1.00	1.00	1.00	1.00	1.00	1.00	1.00	1.00
	≥4.00	1.00	1.00	1.00	1.00	1.00	1.00	1.00	1.00	1.00
≥4.5 <5.5	0.25	1.00	1.00	1.00	1.00	1.00	0.99	0.99	0.97	0.97
	≥0.50	1.00	1.00	1.00	1.00	1.00	1.00	1.00	1.00	1.00
≥5.5	All	1.00	1.00	1.00	1.00	1.00	1.00	1.00	1.00	1.00

Note: Interpolation for length of grade and demand flow rate to the nearest 0.01 is recommended.

SOURCE: From *Highway Capacity Manual 2010.* Copyright, National Academy of Sciences, Washington, D.C. Reproduced with permission of the Transportation Research Board.

Table 9.23   PTSF Passenger Car Equivalents for Trucks (E_T) and RVs (E_R) for Level Terrain, Rolling Terrain, and Specific Downgrades

Vehicle Type	Directional Demand Flow Rate, v_{vph} (veh/h)	Level and Specific Downgrade	Rolling
Trucks, E_T	≤100	1.1	1.9
	200	1.1	1.8
	300	1.1	1.7
	400	1.1	1.6
	500	1.0	1.4
	600	1.0	1.2
	700	1.0	1.0
	800	1.0	1.0
	≥900	1.0	1.0
RVs, E_R	All	1.0	1.0

Note: Interpolation in this exhibit is not recommended.

SOURCE: From *Highway Capacity Manual 2010.* Copyright, National Academy of Sciences, Washington, D.C. Reproduced with permission of the Transportation Research Board.

Table 9.24   PTSF Passenger Car Equivalents for Trucks (E_T) and RVs (E_R) on Specific Upgrades

Grade (%)	Grade Length (mi)	Directional Demand Flow Rate, V_{vph} (veh/h)								
		≤100	200	300	400	500	600	700	800	≥900
		Passenger Car Equivalents for Trucks (E_T)								
≥3 <3.5	≤2.00	1.0	1.0	1.0	1.0	1.0	1.0	1.0	1.0	1.0
	3.00	1.5	1.3	1.3	1.2	1.0	1.0	1.0	1.0	1.0
	≥4.00	1.6	1.4	1.3	1.3	1.0	1.0	1.0	1.0	1.0
≥3.5 <4.5	≤1.00	1.0	1.0	1.0	1.0	1.0	1.0	1.0	1.0	1.0
	1.50	1.1	1.1	1.0	1.0	1.0	1.0	1.0	1.0	1.0
	2.00	1.6	1.3	1.0	1.0	1.0	1.0	1.0	1.0	1.0
	3.00	1.8	1.4	1.1	1.2	1.2	1.2	1.2	1.2	1.2
	≥4.00	2.1	1.9	1.8	1.7	1.4	1.4	1.4	1.4	1.4
≥4.5 <5.5	≤1.00	1.0	1.0	1.0	1.0	1.0	1.0	1.0	1.0	1.0
	1.50	1.1	1.1	1.1	1.2	1.2	1.2	1.2	1.2	1.2
	2.00	1.7	1.6	1.6	1.6	1.5	1.4	1.4	1.3	1.3
	3.00	2.4	2.2	2.2	2.1	1.9	1.8	1.8	1.7	1.7
	≥4.00	3.5	3.1	2.9	2.7	2.1	2.0	2.0	1.8	1.8
≥5.5 <6.5	≤0.75	1.0	1.0	1.0	1.0	1.0	1.0	1.0	1.0	1.0
	1.00	1.0	1.0	1.1	1.1	1.2	1.2	1.2	1.2	1.2
	1.50	1.5	1.5	1.5	1.6	1.6	1.6	1.6	1.6	1.6
	2.00	1.9	1.9	1.9	1.9	1.9	1.9	1.9	1.8	1.8
	3.00	3.4	3.2	3.0	2.9	2.4	2.3	2.3	1.9	1.9
	≥4.00	4.5	4.1	3.9	3.7	2.9	2.7	2.6	2.0	2.0
≥6.5	≤0.50	1.0	1.0	1.0	1.0	1.0	1.0	1.0	1.0	1.0
	0.75	1.0	1.0	1.0	1.0	1.1	1.1	1.1	1.0	1.0
	1.00	1.3	1.3	1.3	1.4	1.4	1.5	1.5	1.4	1.4
	1.50	2.1	2.1	2.1	2.1	2.0	2.0	2.0	2.0	2.0
	2.00	2.9	2.8	2.7	2.7	2.4	2.4	2.3	2.3	2.3
	3.00	4.2	3.9	3.7	3.6	3.0	2.8	2.7	2.2	2.2
	≥4.00	5.0	4.6	4.4	4.2	3.3	3.1	2.9	2.7	2.5
		Passenger Car Equivalents for RVs (E_R)								
All	All	1.0	1.0	1.0	1.0	1.0	1.0	1.0	1.0	1.0

Note: Interpolation for length of grade and demand flow rate to the nearest 0.1 is recommended.

SOURCE: From *Highway Capacity Manual 2010*. Copyright, National Academy of Sciences, Washington, D.C. Reproduced with permission of the Transportation Research Board.

Table 9.25   No-Passing-Zone Adjustment Factor $(f_{np,PTSF})$ for Determination of PTSF

Total Two-Way Flow Rate, $v = v_d + v_o$ (pc/h)	Percent No-Passing Zones					
	0	20	40	60	80	100
Directional Split = 50/50						
≤200	9.0	29.2	43.4	49.4	51.0	52.6
400	16.2	41.0	54.2	61.6	63.8	65.8
600	15.8	38.2	47.8	53.2	55.2	56.8
800	15.8	33.8	40.4	44.0	44.8	46.6
1400	12.8	20.0	23.8	26.2	27.4	28.6
2000	10.0	13.6	15.8	17.4	18.2	18.8
2600	5.5	7.7	8.7	9.5	10.1	10.3
3200	3.3	4.7	5.1	5.5	5.7	6.1
Directional Split = 60/40						
≤200	11.0	30.6	41.0	51.2	52.3	53.5
400	14.6	36.1	44.8	53.4	55.0	56.3
600	14.8	36.9	44.0	51.1	52.8	54.6
800	13.6	28.2	33.4	38.6	39.9	41.3
1400	11.8	18.9	22.1	25.4	26.4	27.3
2000	9.1	13.5	15.6	16.0	16.8	17.3
2600	5.9	7.7	8.6	9.6	10.0	10.2
Directional Split = 70/30						
≤200	9.9	28.1	38.0	47.8	48.5	49.0
400	10.6	30.3	38.6	46.7	47.7	48.8
600	10.9	30.9	37.5	43.9	45.4	47.0
800	10.3	23.6	28.4	33.3	34.5	35.5
1400	8.0	14.6	17.7	20.8	21.6	22.3
2000	7.3	9.7	11.7	13.3	14.0	14.5
Directional Split = 80/20						
≤200	8.9	27.1	37.1	47.0	47.4	47.9
400	6.6	26.1	34.5	42.7	43.5	44.1
600	4.0	24.5	31.3	38.1	39.1	40.0
800	3.8	18.5	23.5	28.4	29.1	29.9
1400	3.5	10.3	13.3	16.3	16.9	32.2
2000	3.5	7.0	8.5	10.1	10.4	10.7
Directional Split = 90/10						
≤200	4.6	24.1	33.6	43.1	43.4	43.6
400	0.0	20.2	28.3	36.3	36.7	37.0
600	−3.1	16.8	23.5	30.1	30.6	31.1
800	−2.8	10.5	15.2	19.9	20.3	20.8
1400	−1.2	5.5	8.3	11.0	11.5	11.9

Note: Straight-line interpolation of $f_{np,PTSF}$ for percent no-passing zones, demand flow rate, and directional split is recommended to the nearest 0.1.

SOURCE: From *Highway Capacity Manual 2010*. Copyright, National Academy of Sciences, Washington, D.C. Reproduced with permission of the Transportation Research Board.

Table 9.26   PTSF Coefficients for Use in Equation 9.17 for Estimating BPTSF

Opposing Demand Flow Rate, v_o (pc/h)	Coefficient a	Coefficient b
≤200	−0.0014	0.973
400	−0.0022	0.923
600	−0.0033	0.870
800	−0.0045	0.833
1000	−0.0049	0.829
1200	−0.0054	0.825
1400	−0.0058	0.821
≥1600	−0.0062	0.817

Note: Straight-line interpolation of *a* to the nearest 0.0001 and *b* to the nearest 0.001 is recommended.

SOURCE: From *Highway Capacity Manual 2010*. Copyright, National Academy of Sciences, Washington, D.C. Reproduced with permission of the Transportation Research Board.

Table 9.27   Downstream Length of Roadway Affected by Passing Lanes on Directional Segments in Level and Rolling Terrain

Directional Demand Flow Rate, v_d (pc/h)	Downstream Length of Roadway Affected, L_{de} (mi)	
	PTSF	ATS
≤200	13.0	1.7
300	11.6	1.7
400	8.1	1.7
500	7.3	1.7
600	6.5	1.7
700	5.7	1.7
800	5.0	1.7
900	4.3	1.7
≥1,000	3.6	1.7

Note: Interpolation to the nearest 0.1 is recommended.

SOURCE: From *Highway Capacity Manual 2010*. Copyright, National Academy of Sciences, Washington, D.C. Reproduced with permission of the Transportation Research Board.

Table 9.28   Optimal Lengths of Passing Lanes on Two-Lane Highways

Directional Demand Flow Rate, v_d (pc/h)	Optimal Passing Lane Length (mi)
≤100	≤0.50
>100 ≤400	>0.50 ≤0.75
>400 ≤700	>0.75 ≤1.00
≥700	>1.00 ≤2.00

SOURCE: From *Highway Capacity Manual 2010*. Copyright, National Academy of Sciences, Washington, D.C. Reproduced with permission of the Transportation Research Board.

Table 9.29   Adjustment Factor for the Impact of a Passing Lane on PTSF ($f_{pl,PTSF}$)

Directional Demand Flow Rate, v_d (pc/h)	$f_{pl,PTSF}$
≤100	0.58
200	0.59
300	0.60
400	0.61
500	0.61
600	0.61
700	0.62
800	0.62
≥900	0.62

Note: Interpolation is not recommended; use closest value.

SOURCE: From *Highway Capacity Manual 2010.* Copyright, National Academy of Sciences, Washington, D.C. Reproduced with permission of the Transportation Research Board.

Table 9.30   Adjustment Factor for Estimating the Impact of a Passing Lane on ATS ($f_{pl,ATS}$)

Directional Demand Flow Rate, v_d (pc/h)	$f_{pl,ATS}$
≤100	1.08
200	1.09
300	1.10
400	1.10
500	1.10
600	1.11
700	1.11
800	1.11
≥900	1.11

Note: Interpolation is not recommended; use closest value.

SOURCE: From *Highway Capacity Manual 2010.* Copyright, National Academy of Sciences, Washington, D.C. Reproduced with permission of the Transportation Research Board.

Table 9.31   Adjustment Factors (f_{pl}) for Estimating ATS and PTSF within a Climbing Lane

Directional Demand Flow Rate, v_d (pc/h)	f_{pl}	
	ATS	PTSF
0–300	1.02	0.20
>300–600	1.07	0.21
>600	1.14	0.23

SOURCE: From *Highway Capacity Manual 2010.* Copyright, National Academy of Sciences, Washington, D.C. Reproduced with permission of the Transportation Research Board.

CHAPTER 10

Capacity and Level of Service at Signalized Intersections

The level of service at any intersection on a highway has a significant effect on the overall operating performance of that highway. Thus, improvement of the level of service at each intersection usually results in an improvement of the overall operating performance of the highway. An analysis procedure that provides for the determination of capacity or level of service at intersections is therefore an important tool for designers, operation personnel, and policy makers. Factors that affect the level of service at intersections include the flow and distribution of traffic, the geometric characteristics, and the signal system.

A major difference between consideration of level of service on highway segments and level of service at intersections is that only through flows are used in computing the levels of service at highway segments (see Chapter 9), whereas turning flows are significant when computing the levels of service at signalized intersections. The signalization system (which includes the allocation of time among the conflicting movements of traffic and pedestrians at the intersection) is also an important factor. For example, the distribution of green times among these conflicting flows significantly affects both capacity and operation of the intersection. Other factors such as lane widths, traffic composition, grade, and speed also affect the level of service at intersections in a similar manner as for highway segments.

This chapter describes the methodology given in the 2010 edition of the *Highway Capacity Manual* (HCM). The HCM procedure can be used to evaluate the performance of a signalized intersection with respect to the quality of service provided to the road users at the intersection. It also provides performance measures for different travel modes that can be used to identify problem sources and help the analyst to develop appropriate strategies for development. Although the procedure is primarily for an isolated intersection, the impact of a preceding signalized intersection is considered by introducing input variables that reflect the arrival time destribution of vehicles from an upstream signalized intersection. Thus, the procedure considers the extent to which vehicles arrive in platoons at the subject intersection. The procedure can only be used for three- or four-leg intersections formed by two highways or streets.

CHAPTER OBJECTIVES:

- Become familiar with the definitions of terms commonly used in capacity and level of service.
- Understand the procedure for determining the critical v/c ratio at an intersection.
- Become familiar with the different analysis levels for level of service at an intersection.
- Become familiar with the *Highway Capacity Manual* (HCM) procedures for determining the operational and planning levels of service at an intersection.

10.1 DEFINITIONS OF SOME COMMON TERMS

Most of the terms commonly used in capacity and level-of-service analyses of signalized intersections were defined in Chapter 8. However, some additional terms need to be defined and others need to be redefined to understand their use in this chapter.

1. **Permitted turning movements** are those made within gaps of an opposing traffic stream or through a conflicting pedestrian flow. For example, when a right turn is made while pedestrians are crossing a conflicting crosswalk, the right turn is a permitted turning movement. Similarly, when a left turn is made between two consecutive vehicles of the opposing traffic stream, the left turn is a permitted turn. The suitability of permitted turns at a given intersection depends on the geometric characteristics of the intersection, the turning volume, and the opposing volume. These movements would be shown a green ball on the signal.

2. **Protected turns** are those turns protected from any conflicts with vehicles in an opposing stream or pedestrians on a conflicting crosswalk. A permitted turn takes more time than a similar protected turn and will use more of the available green time. Protected turns would be shown a green arrow on the relevant signal head.

3. **Yellow change and red clearance interval** is the sum of the "yellow" and "red" intervals (given in seconds) that are provided between phases to allow vehicular and pedestrian traffic to clear the intersection before conflicting movements are released.

4. **Geometric conditions** is a term used to describe the roadway characteristics of the approach. They include the number and width of lanes, grades, and the allocation of the lanes for different uses, including the designation of a parking lane.

5. **Signalization conditions** is a term used to describe the details of the signal operation. They include the type of signal control, phasing sequence, timing, and an evaluation of signal progression on each approach.

6. **Flow ratio** v/s is the ratio of the actual flow rate or projected demand v on an approach or lane group to the saturation flow rate s.

7. **Lane group** consists of one or more lanes that have a common stop line, carry a set of traffic streams, and whose capacity is shared by all vehicles in the group.

8. An **analysis period** (T) is the time interval selected to evaluate an intersection by applying the procedure once. The analysis period varies between 0.25 h and 1 h as it is assumed that traffic conditions do not change significantly during this period. One of three approaches can be used to select an analysis period: (i) An analysis period of 0.25 h (i.e., $T = 0.25$ h) within the peak hour, and the equivalent hourly flow in veh/h, determined by multiplying the peak 15-min volume by 4 or dividing the hourly volume by the peak hour factor. (ii) An analysis period of 1 h (i.e., $T = 1$ h), with the equivalent flow rate being the demand flow during the hour and the peak hour factor not used. This approach is appropriate when the traffic condition is stable during the hour. The effect of large variations of traffic volumes during the hour is not

captured when this approach is used. (iii) Four consecutive 0.25 h analysis periods covering duration of 1 h. In this case, the procedure is carried out once for each of four consecutive 0.25-h analysis periods. This approach gives a more representative value for delay when there is a large variation of traffic volumes within the hour, and when queues extend from one analysis period to another. Although analysis periods can be longer than 1 h, the HCM cautions against using analysis periods greater than 1 h, as traffic conditions do not usually stay stable for long durations, and the effect of traffic peaking may not be reflected in the analysis.

9. The **study period** is the time period for which the results of the analysis are applicable and may consist of one or more consecutive analysis periods.

10. When the **rest in walk** mode is enabled for a phase, this mode will dwell in walk except when there are conflicting calls. A conflicting call will result in the pedestrian phase ending after duration of the pedestrian clear interval setting.

11. **Pedestrian clear interval** (PC), sometimes known as the *pedestrian change interval*, is provided to allow a pedestrian who leaves the curb during the walk interval to arrive at the opposite curb or median. Some agencies set the PC as the time for the pedestrian to cross from curb to curb using a walking speed of 3.5 ft/sec (consistent with requirements in the *Manual on Uniform Traffic Control Devices* [MUTCD], while others use the difference between the walking time and the vehicle change period (yellow change + red clearance).

10.1.1 Capacity at Signalized Intersections

The capacity at a signalized intersection is given for each lane group and is defined as the maximum rate of flow for the subject lane group that can go through the intersection under prevailing traffic, roadway, and signalized conditions. Capacity is given in vehicles per hour (veh/h) but is based on the flow during a peak 15-min period. The capacity of the intersection as a whole is not considered; rather, emphasis is placed on providing suitable facilities for the major movements of the intersections. Capacity therefore is applied meaningfully only to major movements or approaches of the intersection. Note also that in comparison with other locations such as freeway segments, the capacity of an intersection approach is not as strongly correlated with the level of service. It is therefore necessary that both the level of service and capacity be analyzed separately when signalized intersections are being evaluated.

10.1.2 Saturation Flow or Saturation Flow Rate

The concept of a saturation flow or saturation flow rate (s) is used to determine the capacity of a lane group. The saturation flow rate is the maximum flow rate on the approach or lane group that can go through the intersection under prevailing traffic and roadway conditions when 100 percent of the green time is available. The saturation flow rate is given in units of veh/h of effective green time.

The capacity of an approach or lane group is given as

$$c_i = s_i(g_i/C) \tag{10.1}$$

where

$$
\begin{aligned}
c_i &= \text{capacity of lane group } i \text{ (veh/h)} \\
s_i &= \text{saturation flow rate for lane group or approach } i \\
(g_i/C) &= \text{green ratio for lane group or approach } i \\
g_i &= \text{effective green for lane group } i \text{ or approach } i \\
C &= \text{cycle length}
\end{aligned}
$$

The ratio of flow to capacity v/c is usually referred to as the *degree of saturation* and can be expressed as

$$(v/c)_i = X_i = \frac{v_i}{s_i(g_i/C)} \tag{10.2}$$

where

v_i = flow rate for lane i
X_i = (v/c) ratio for lane group or approach i
v_i = actual flow rate or projected demand for lane group or approach i veh/h)
s_i = saturation flow for lane group or approach i (veh/h/g)
g_i = effective green time for lane group i or approach i (sec)
C = cycle length (sec)

It can be seen that when the flow rate equals capacity, X_i equals 1.00; when flow rate equals zero, X_i equals zero.

When the overall intersection is to be evaluated with respect to its geometry and total cycle time, the concept of critical volume-to-capacity ratio (X_c) is used. The critical v/c ratio is usually obtained for the overall intersection but considers only the critical lane groups or approaches, which are those lane groups or approaches that have the maximum flow ratio v/s for each signal phase. For example, in a two-phase signalized intersection, if the north approach has a higher v/s ratio than the south approach, more time will be required for vehicles on the north approach to go through the intersection during the north-south green phase, and the phase length will be based on the green time requirements for the north approach. The north approach will therefore be the critical approach for the north-south phase. As shown in Chapter 8, the critical v/c ratio for the whole intersection is given as:

$$X_c = \sum_i (v/s)_{ci} \frac{C}{C-L} \tag{10.3}$$

where

X_c = critical v/c ratio for the intersection
$\sum_i (v/s)_{ci}$ = summation of the ratios of actual flows to saturation flow (flow ratios) for all critical lanes, groups, or approaches
C = cycle length (sec)
L = total lost time per cycle computed as the sum of the lost time, l_i for each critical signal phase, $L = \sum_{i}^{j} l_{t,i}$

Equation 10.3 can be used to estimate the signal timing for the intersection if this is unknown and a critical v/c ratio is specified for the intersection. Alternatively, this equation can be used to obtain a broader indicator of the overall sufficiency of the intersection by substituting the maximum permitted cycle length for the jurisdiction and determining the resultant critical v/c ratio for the intersection. When the critical v/c ratio is less than 1.00, the cycle length provided is adequate for all critical movements to go through the intersection if the green time is proportionately distributed to the different phases; that is, for the assumed phase sequence, all movements in the intersection will be provided with adequate green times if the total green time is proportionately divided among all phases. If the total green time is not properly allocated to the different phases, it is possible to have a critical v/c ratio of less than 1.00 but with one or more individual oversaturated movements within a cycle.

10.2 ANALYSIS LEVELS AND PERFORMANCE MEASURES FOR LEVEL OF SERVICE AT SIGNALIZED INTERSECTIONS

The procedures for determining the level of service (LOS) at an intersection can be used for three analysis levels. These are:

- Operational
- Design and
- Planning and preliminary engineering

The *operational level* uses detailed information on traffic, geometric, and signalization characteristics for a comprehensive analysis of the intersection. The *design analysis level* is used when an intersection is being designed for a given LOS. Therefore, in addition to the available information on traffic, geometric, and signalized characteristics, this level of analysis also requires the desired LOS. It is then used to determine values for unknown characteristics (e.g., number of lanes) that will reasonably achieve the desired LOS. It is also possible at this level of analysis to determine the effect of changing signal timing. The **planning and preliminary engineering** analysis level uses basic and default values to evaluate a given intersection or a general planning estimate of the overall performance of an existing or planned signalized intersection. This level of analysis requires the least amount of site-specific data.

The procedure evaluates several performance measures that can be used to describe the quality of operation at the intersection as perceived by users of different travel modes. These modes are the automobile, pedestrian, and the bicycle. The performance measures include demand volume-to-capacity ratio v/c, queue storage ratio, and automobile delay for automobiles; pedestrian circulation area, pedestrian perception score, and pedestrian delay for walking; and bicycle delay and bicycle perception score for bicycles.

The LOS is also considered as a performance measure and it is computed separately for each of the modes considered, using one or more of the related performance measures listed above. The influence area that should be considered in the analysis of the LOS at an intersection should be at least 250 ft backward from the stop line of each leg of the intersection. This is not a fixed distance for all intersections, but depends on the traffic conditions on each leg. The basic guideline is that the influence area should cover the expected length of the queue on each leg of the intersection due to the operation of the signals during the analysis period. Thus, the distances backward from the stop lines on all legs of the intersection may be different, depending on the length of the queue.

10.3 LEVEL OF SERVICE CRITERIA AT SIGNALIZED INTERSECTIONS

The computational procedures presented here are for the operational level of analysis as given in the 2010 edition of the HCM. The HCM procedure provides different criteria for the LOS for the different travel modes. For example, while the criteria for the automobile mode are based on performance measures that can be measured in the field and can also be perceived by the automobile user, the criteria for the nonautomobile modes are based on scores computed from factors that influence pedestrians and bicyclists in their evaluation of the quality of service provided to them at an intersection. These factors are discussed below under the LOS criteria discussion for the different modes.

10.3.1 LOS Criteria for the Automobile Mode

These procedures deal with the computation of the level of service at each lane group, intersection approach, and at the intersection as a whole. The LOS at each intersection approach and the entire intersection is represented by only the control delay, while the LOS for a lane group is represented by both the control delay and the volume-to-capacity ratio. Control delay does not only indicate the amount of lost travel time and fuel consumption, but it is also a measure of the frustration and discomfort of motorists. Control delay, which is that portion of total delay that is attributed to the control facility, includes the delay due to the initial deceleration, queue move-up time, stopped time, and final acceleration. Delay, however, depends on the red time, which in turn depends on the length of the cycle. Reasonable levels of service can therefore be obtained for short cycle lengths, even though the v/c ratio is as high as 0.9. To the extent that signal coordination reduces delay, different levels of service may also be obtained for the same v/c ratio when the effect of signal coordination changes. The LOS criteria for the automobile mode are given in terms of the average control delay per vehicle. Six levels of service are prescribed. The criteria for each are described below and are shown in Table 10.1.

Level of Service A describes that level of operation at which the average delay per vehicle is 10.0 seconds or less and the volume-to-capacity ratio is 1.0 or less. At LOS A, vehicles arrive mainly during the green phase, resulting in only a few vehicles stopping at the intersection. Short cycle lengths may help in obtaining low delays.

Level of Service B describes that level of operation at which delay per vehicle is greater than 10 sec but not greater than 20 sec and the volume-to-capacity ratio is 1.0 or less. At LOS B, the number of vehicles stopped at the intersection is greater than that for LOS A, but progression is still good, and cycle length also may be short.

Level of Service C describes that level of operation at which delay per vehicle is greater than 20 sec but not greater than 35 sec and the volume-to-capacity ratio is 1.0 or less. At LOS C, many vehicles go through the intersection without stopping, but a significant number of vehicles are stopped. In addition, some vehicles at an approach may not clear the intersection during the first cycle (cycle failure). The higher delay may be due to the significant number of vehicles arriving during the red phase (fair progression) and/or moderate cycle lengths.

Table 10.1   LOS Criteria for the Automobile Mode

Control Delay (sec/veh)	LOS by Volume-to-Capacity Ratio[a]	
	≤1.0	>1.0
≤10	A	F
>10–20	B	F
>20–35	C	F
>35–55	D	F
>55–80	E	F
>80	F	F

[a]For approach-based and intersection-wide assessments, LOS is defined solely by control delay.
SOURCE: From *Highway Capacity Manual 2010*. Copyright, National Academy of Sciences, Washington, D.C. Reproduced with permission of the Transportation Research Board.

Level of Service D describes that level of operation at which the delay per vehicle is greater than 35 sec but not greater than 55 sec and the volume-to-capacity ratio is 1.0 or less. At LOS D, more vehicles are stopped at the intersection, resulting in a longer delay. The number of individual cycles failing is now noticeable. The longer delay at this level of service is due to a combination of two or more of several factors that include long cycle lengths, high v/c ratios, and unfavorable progression.

Level of Service E describes that level of operation at which the delay per vehicle is greater than 55 sec but not greater than 80 sec and the volume-to-capacity ratio is 1.0 or less. At LOS E, individual cycles frequently fail. This long delay, which is usually taken as the limit of acceptable delay by many agencies, generally includes high v/c ratios, long cycle lengths, and poor progression.

Level of Service F describes that level of operation at which the delay per vehicle is greater than 80 sec or the volume-to-capacity ratio is greater than 1.0. This long delay is usually unacceptable to most motorists. At LOS F, *oversaturation* usually occurs—that is, arrival flow rates are greater than the capacity at the intersection. Long delay can also occur as a result of poor progression and long cycle lengths. Note that this level of service can occur when approaches have high v/c ratios that are less than 1.00 but also have many individual cycles failing.

It should be noted that the average delay for a lane group can be less than 80 sec, while the v/c ratio is greater than 1.0. The conditions that facilitate this are short cycle lengths and/or favorable signal progression. A v/c ratio greater than 1.0 for a lane group indicates that the capacity is less than the lane group's demand flow and therefore capacity failure occurs. It is therefore necessary to consider both the v/c ratio and the delay when the LOS for a lane group is being evaluated.

10.3.2 LOS Criteria for Nonautomobile (Pedestrian and Bicycle) Modes

The criteria for the nonautomobile modes are based on scores that are computed from factors that can be described as either performance measures, such as pedestrian delay, or are indicators of the intersection characteristics, such as the pedestrian corner circulation area. The relationship between the computed scores and the LOS was obtained from the results of a survey in which travelers rated their perceived quality of service for a specific crossing of a signalized intersection. LOS of F was assigned the worst quality of service and LOS of A the best. Table 10.2 gives the relationship between the scores and the LOS for the nonautomobile mode.

Table 10.2   Level of Service Criteria for Pedestrian and Bicycle Modes

LOS	LOS Score
A	≤2.0
B	>2.0–2.75
C	>2.75–3.5
D	>3.5–4.25
E	>4.25–5.00
F	>5.0

SOURCE: From *Highway Capacity Manual 2010*. Copyright, National Academy of Sciences, Washington, D.C. Reproduced with permission of the Transportation Research Board.

10.3.3 Required Input Data for the Automobile Mode

The data that are required to carry out the LOS analysis for the automobile mode can be categorized into the following four groups:

- Traffic characteristics
- Geometric design
- Signal control
- Other

Table 10.3 gives details of the elements of each category for a pretimed, fully actuated, or semiactuated signal control, and Table 10.4 gives details of additional data required for a coordinated-actuated signal control.

Many of the elements in Tables 10.3 and 10.4 have been defined in Chapter 8, and the reader should refer to Chapter 8 for the definitions of those elements. The definitions of some of the elements in Chapter 8 are repeated here for easy comprehension, and those elements that are not in Chapter 8 are also defined in this section.

Traffic Characteristics

The elements within this category provide the input data for the traffic that goes through the intersection during the study period and are described below.

The *demand flow rate* is the number of vehicles that arrive at the intersection divided by the analysis period. It is given in terms of an hourly flow rate, which requires the use of a peak hour factor when a 15-min analysis period is used. Note that the flow rate is required for each analysis period in an evaluation using multiple sequential analysis periods.

The *right-turn-on-red* (**RTOR**) flow rate is the quotient of the number of vehicles that turn right when the through signal indication is red by the duration of the analysis period and is expressed in terms of an hourly flow. Many factors influence the RTOR flow rate, which makes it difficult to predict. These include whether right turns are on a shared or exclusive lane, the right-turn flow rate, v/c ratio for conflicting movements, volume of conflicting pedestrians, etc. The HCM suggests that actual data should be obtained in the field for this traffic element. The HCM also recommends the conservative approach of using a value of 0 veh/h when field data are not available, or when future conditions are being considered.

Platoon ratio represents how well a movement group progresses through the intersection and depends on the percentage of vehicles that arrive during the green indication. It is correlated with the arrival type, which determines the level of coordination between the signals at the intersection being studied and those at adjacent intersections.

The HCM gives the platoon ratio as

$$R_p = P(C/g) \tag{10.4}$$

where

R_p = platoon ratio
P = proportion of all vehicles in the movement arriving during the green indication (decimal)
C = cycle length (sec)
g = effective green time for the movement (sec)

Table 10.3 Input Data Requirements: Automobile Mode with Pretimed, Fully Actuated, or Semiactuated Signal Control

Data Category	Input Data Element	Basis
Traffic characteristics	Demand flow rate	Movement
	Right-turn-on-red flow rate	Approach
	Percent heavy vehicles	Movement group
	Intersection pack hour factor	Intersection
	Platoon ratio	Movement group
	Upstream filtering adjustment factor	Movement group
	Initial queue	Movement group
	Base saturation flow rate	Movement group
	Lane utilization adjustment factor	Movement group
	Pedestrian flow rate	Approach
	Bicycle flow rate	Approach
	On-street parking maneuver rate	Movement group
	Local bus stopping rate	Approach
Geometric design	Number of lanes	Movement group
	Average lane width	Movement group
	Number of receiving lanes	Approach
	Turn bay length	Movement group
	Presence of on-street parking	Movement group
	Approach grade	Approach
Signal control	Type of signal control	Intersection
	Phase sequence	Intersection
	Left-turn operational mode	Approach
	Dallas left-turn phasing option	Approach
	Passage time (if actuated)	Phase
	Maximum green (or green duration if pretimed)	Phase
	Minimum green	Phase
	Yellow change	Phase
	Red clearance	Phase
	Walk	Phase
	Pedestrian clear	Phase
	Phase recall	Phase
	Dual entry (if actuated)	Phase
	Simultaneous gap out (if actuated)	Approach
Other	Analysis period duration	Intersection
	Speed limit	Approach
	Stop-line detector length and detection mode	Movement group
	Area type	Intersection

Notes: Movement = one value for each left-turn, through, and right-turn movement.
Movement group = one value for each turn movement with exclusive turn lanes and one value for the through movement (inclusive of any turn movements in a shared lane).
Approach = one value or condition for the intersection approach.
Intersection = one value or condition for the intersection.
Phase = one value or condition for each signal phase.
SOURCE: From *Highway Capacity Manual 2010*. Copyright, National Academy of Sciences, Washington, D.C. Reproduced with permission of the Transportation Research Board.

Table 10.4   Input Data Requirements: Automobile Mode with Coordinated-Actuated Signal Control

Data Category	Input Data Element	Basis
Signal Control	Cycle length	Intersection
	Phase splits	Phase
	Offset	Intersection
	Offset reference point	Intersection
	Force mode	Intersection

SOURCE: From *Highway Capacity Manual 2010*. Copyright, National Academy of Sciences, Washington, D.C. Reproduced with permission of the Transportation Research Board.

The following rules apply regarding the applicable flow rate for different traffic operational conditions.

- When a left-turn lane group in an exclusive lane operates with a protected or protected-permitted mode, the platoon ratio should be computed using the flow rate that occurs during the green interval of the left-turn phase.
- When a left-turn lane group in an exclusive lane operates with a permitted mode, the platoon ratio should be computed using the flow rate that occurs during the green interval of the phase that provides the permitted operation.
- When a right-turn lane group in an exclusive lane operates with either a permitted or a protected-permitted mode, the platoon ratio should be computed using the through flow rate that occurs during the green interval of the phase that provides the permitted operation.
- For a through-lane group in an exclusive lane, the platoon ratio should be computed using the through flow rate that occurs during the green interval of the associated phase.
- When a split phasing system (see Chapter 8) is used on an approach, the platoon ratio should be computed using the flow rate that occurs during the green interval of the common phase.
- When there is one or more shared lanes on an approach, one platoon ratio is computed for the shared movement group using the flow rate of all shared lanes, including any exclusive lanes that are also served during the green interval of the common phase.

The arrival types and their corresponding platoon ratios and quality of progression are shown in Table 10.5.

Table 10.5   Relationship between Arrival Type and Progression Quality

Platoon Ratio	Arrival Type	Progression Quality
0.33	1	Very Poor
0.67	2	Unfavorable
1.00	3	Random Arrivals
1.33	4	Favorable
1.67	5	Highly Favorable
2.00	6	Exceptionally Favorable

SOURCE: From *Highway Capacity Manual 2010*. Copyright, National Academy of Sciences, Washington, D.C. Reproduced with permission of the Transportation Research Board.

The level of coordination between the signals at the intersection being studied and those at adjacent intersections is a critical characteristic and is determined in terms of the type of vehicle arrival at the intersection. Six arrival types (AT) have been identified:

- **Arrival Type 1**, which represents the worst condition of arrival, is a dense platoon containing over 80 percent of the lane group volume arriving at the beginning of the red interval.
- **Arrival Type 2**, which, while better than Type 1, is still considered unfavorable, is either a dense platoon arriving in the middle of the red interval or a dispersed platoon containing 40 to 80 percent of the lane group volume arriving throughout the red phase.
- **Arrival Type 3** occurs for one of two conditions. First, for coordinated signals, when the platoon partly arrives during the red interval and partly during the green interval and consists of less than 40 percent of the movement group volume. Second, for noncoordinated signals, when vehicles arrive randomly at the intersection. This often occurs at isolated intersections.
- **Arrival Type 4**, which is usually considered a favorable platoon condition, is either a moderately dense platoon arriving in the middle of a green interval or a dispersed platoon containing 40 to 80 percent of the lane group volume arriving throughout the green interval. It often occurs when progression is favorable in the travel direction and segment lengths are average.
- **Arrival Type 5**, which represents the best condition of arrival that usually occurs, is a dense platoon containing over 80 percent of the lane group volume arriving at the start of the green phase.
- **Arrival Type 6**, which represents exceptional progression quality, is a very dense platoon containing more than 80 percent of the movement group volume arriving at the beginning of the green interval. It occurs when the platoon progresses through several closely spaced intersections with very low traffic from the side streets.

It is necessary to determine, as accurately as possible, the type of arrival for the intersection being considered, since both the estimate of delay and the determination of the level of service will be significantly affected by the arrival type used in the analysis. Field observation is the best way to determine the arrival type, although time-space diagrams for the street being considered could be used for an approximate estimation. In using field observations, the percentage of vehicles arriving during the green phase is determined and the arrival type is then obtained for the platoon ratio for the approach.

The **upstream filtering adjustment factor (I)** accounts for the effect of filtered arrivals from upstream signals, which reflects how the variance of the number of vehicles arriving at the movement group under consideration is altered by an upstream signal. The value of this factor ranges from 0.09 to 1.0, where a value of 1.0 can be used for an isolated intersection that is located at least 0.6 mi from the nearest upstream signalized intersection. The value of I for a nonisolated intersection can be obtained from the HCM expression given as

$$I = 1.0 - 0.91X_u^{2.68} \geq 0.090 \qquad (10.5)$$

where

I = upstream filtering adjustment factor
X_u = weighted v/c ratio for all upstream movements contributing to the volume in the movement group under consideration

The ***initial queue*** is the number of vehicles that are at the intersection at the start of the analysis period of the movement group under consideration. The initial queue should represent the queue that is created from oversaturation for an extended period of time, and should not include vehicles that join the queue as a result of the random variations of the cycle intervals. It is computed from field data by obtaining information on queue lengths continuously for each of the three cycle intervals that just precede the start of the analysis period and the average of the three smallest counts determined as the initial queue.

The ***base saturation flow rate*** (s_o) is the maximum rate of flow across the stop line that can occur in a traffic lane. This flow rate is associated with the following conditions:

- Traffic lanes are at least 12 ft wide
- No heavy vehicles in the traffic stream
- A flat grade
- No parking
- No buses that stop at the intersection
- Even lane utilization
- No turning vehicles

Default values given by the HCM are1900 pc/ln/h for metropolitan areas having a population of 250,000 and 1700 pc/ln/h for all other areas.

The ***lane utilization adjustment factor*** (f_{Lu}) is used to adjust the ideal saturation flow rate to account for the unequal utilization of the lanes in a lane group. When a movement group consists of only one lane or it can be assumed that the distribution of traffic is uniform across all lanes in a movement group, the value of f_{Lu} can be taken as 1.0. However, the value of f_{Lu} is less than 1.0 when traffic volume is not uniform across lanes or when the demand flow is less than the capacity of the movement group.

This factor is computed using Eq. 10.6.

$$f_{Lu} = \frac{v_g}{v_{gl}N} \tag{10.6}$$

where

v_g = unadjusted demand flow rate for lane group (veh/h)
v_{gl} = unadjusted demand flow rate on the single lane in the lane group with the highest volume
N = number of lanes in the lane group

The ***pedestrian flow rate*** is computed from the number of pedestrians in the crosswalk that interrupt the flow of right-turning vehicles from the approach under consideration, during the analysis period. For example, the pedestrian flow rate on the east approach is associated with vehicles turning right from the south approach. Pedestrian counts should be taken for each walking direction of the pedestrians. Pedestrian flow rate is given as pedestrians per hour.

The ***bicycle flow rate*** is computed from the number of bicycles with a travel path that results in an interruption of the flow of right-turning vehicles from the approach under consideration during the analysis period. The bicycle count should include bicycles traveling on the shoulder or in a bike lane, but should not include those traveling on the right lane with automobile traffic. Bicycle flow rate is given as bicycles per hour.

The *on-street parking maneuver rate* is the number of parking maneuvers per hour during the analysis period that have an impact on an intersection leg. The extent upstream from the stop line in which parking influences the movement group is usually taken as 250 ft, and the maneuvers should occur directly adjacent to the movement group. The HCM suggests that a practical limit of 180 maneuvers/h should be used. It should be measured on one side of the road on two-way legs and on both sides on a one-way leg.

The *local bus stopping rate* is given as the number of buses per hour during the analysis period that block traffic by stopping for embarking or disembarking passengers within 250 ft upstream of the stop line. The HCM suggests that a practical upper limit of 250 buses/h should be used.

Geometric Design

The elements within this category provide the input data relating to the geometric elements of the intersection that have some impact on the flow of traffic. These elements are the number of lanes, lane width, grade, presence or not of on-street parking, lengths of turn bays, etc. In cases where the physical configuration of the intersection is unknown, the quick estimation method of analysis (see Section 10.12) may be used to determine a suitable configuration, or state and local policies and/or guidelines can be used. If no guidelines are available, guidelines given in Chapter 8 may be used.

The *number of lanes* gives the count of lanes on which each intersection movement occurs. For turning movements, only the exclusive turn lanes are considered, as lanes that serve through and turn movements are counted as part of the through-lane counts and are known as *shared lanes*. The turn movement lanes include turn bays and lanes that extend backward from the intersection.

The *average lane width* is the mean of the widths of all lanes serving a movement group. An average lane width of 8 ft is considered the minimum and the standard lane width is 12 ft. Although lane widths can be greater than 16 ft, it is necessary to evaluate the usage of such a lane to ascertain whether it is being used as two narrow lanes.

The *number of receiving lanes* is the count of lanes on which a movement departs from the intersection. It should be noted for each of the left- and right-turn movements, as turning traffic can be significantly affected by vehicles double parking on the receiving approach. The HCM recommends that the number of receiving lanes be determined from field studies.

The *turn bay length* gives the length of the bay having a full width that can store vehicles that are queued. When the bay consists of more than one lane, the average length of all lanes is used to represent the turn bay length.

Presence of on-street parking is an indication of parking being allowed at the approach within 250 ft of the stop line.

Approach grade is the average grade along a length on the approach, starting from the stop line to a distance of 100 ft upstream, measured on a line that is parallel to the direction of travel.

Signal Control

The elements within this category provide the input data for the signal control data and are given for an actuated signal controller with a pretimed, semiactuated, fully actuated, or coordinated-actuated mode. Nearly all of these elements have been defined and discussed in Chapter 8. The reader may refer to Chapter 8 for details. One element that has not been defined is the ***Dallas-left-turn*** phasing system. This phasing system uses a flashing arrow left-turn signal display and operates with the protected-permitted mode, in which the permitted period of the left turn is effectively tied with the signal indication of the opposing through movement in a way that avoids the safety concern known as a "yellow trap." The yellow trap occurs when the permitted left turns have the circular yellow indication, which is followed by a red indication while at the same time a circular green is displayed for the opposing through traffic. Some drivers in the permissive left turn movement assume that the opposing traffic has the same yellow indication and will tend to take the left turn, assuming that the opposing through traffic will stop. This may lead to serious crashes at the intersection.

Other

The ***area type*** describes the activities in the area at which the intersection is located. These activities have a significant effect on speed and therefore on saturation volume at an approach. For example, because of the complexity of intersections located in areas with typical central business district (CBD) characteristics, such as narrow sidewalks, frequent parking maneuvers, vehicle blockades, narrow streets, and high-pedestrian activities, these intersections operate less efficiently than intersections in other areas.

All of the elements in this category have been defined above or in Chapter 8.

10.3.4 Required Input Data for the Nonautomobile (Pedestrian and Bicycle) Modes

The data that are required to carry out the LOS analysis for the nonautomobile mode can be categorized into the following four groups:

- Traffic characteristics
- Geometric design
- Signal control
- Other

Table 10.6 gives details of the elements for each category. The table also shows those elements that are required for the pedestrian and bicycle (nonautomobile) modes by indicating this in the appropriate column. Most of the elements in the table have been defined either in Chapter 8 or in Section 10.3.3. The few remaining elements will be defined when they are first used in the procedure, as this will facilitate comprehension of the material. Note that a blank cell under a mode column indicates that the element is not required for that mode.

10.4 METHODOLOGY OF OPERATIONAL ANALYSIS FOR THE AUTOMOBILE MODE

The intense computational effort of the procedure precludes a manual process. The methodology is therefore usually carried out by using an available software package. Therefore, only a brief description of each step is given to enable the reader to a have a fundamental knowledge of the process involved. Interested readers may refer to the HCM for further details of the procedure. The computational steps involved in the operational analysis for

Table 10.6   Input Data Requirements for Nonautomobile Mode

Data Category	Input Data Element	Pedestrian Mode[a]	Bicycle Mode[a]
Traffic characteristics	Demand flow rate of motorized vehicles	Movement	Approach
	Right-turn-on-red flow rate	Approach	
	Permitted left-turn flow rate	Movement	
	Midsegment 85th percentile speed	Approach	
	Pedestrian flow rate	Movement	
	Bicycle flow rate		Approach
	Proportion of on-street parking occupied		Approach
Geometric design	Street width		Approach
	Number of lanes	Leg	Approach
	Number of right-turn islands	Leg	
	Width of outside through lane		Approach
	Width of bicycle lane		Approach
	Width of paved outside shoulder (or parking lane)		Approach
	Total walkway width	Approach	
	Crosswalk width	Leg	
	Crosswalk length	Leg	
	Corner radius	Approach	
Signal control	Walk	Phase	
	Pedestrian clear	Phase	
	Rest in walk	Phase	
	Cycle length	Intersection	Intersection
	Yellow change	Phase	Phase
	Red clearance	Phase	Phase
	Duration of phase serving pedestrians and bicycles	Phase	Phase
	Pedestrian signal head presence	Phase	
Other	Analysis period duration[b]	Intersection	Intersection

Notes: [a]Movement = one value for each left-turn, through, and right-turn movement.
Approach = one value for the intersection approach.
Leg = one value for the intersection leg (approach plus departure sides).
Intersection = one value for the intersection.
Phase = one value or condition for each signal phase.
[b]Analysis period duration is as defined for Table 10.3.
SOURCE: From *Highway Capacity Manual 2010*. Copyright, National Academy of Sciences, Washington, D.C. Reproduced with permission of the Transportation Research Board.

the automobile mode are presented in the flow chart shown in Figure 10.1 and consists of the following:

Step 1.   Determine Movement Groups and Lane Groups
Step 2.   Determine Movement Group Flow Rate
Step 3.   Determine Lane Group Flow Rate
Step 4.   Determine Adjusted Saturation Flow Rate
Step 5.   Determine Proportion Arriving during Green
Step 6.   Determine Signal Phase Duration
Step 7.   Determine Capacity and Volume-to-Capacity Ratio
Step 8.   Determine Delay
Step 9.   Determine LOS

The input variables used in the lane group procedure are shown in Figure 10.2.

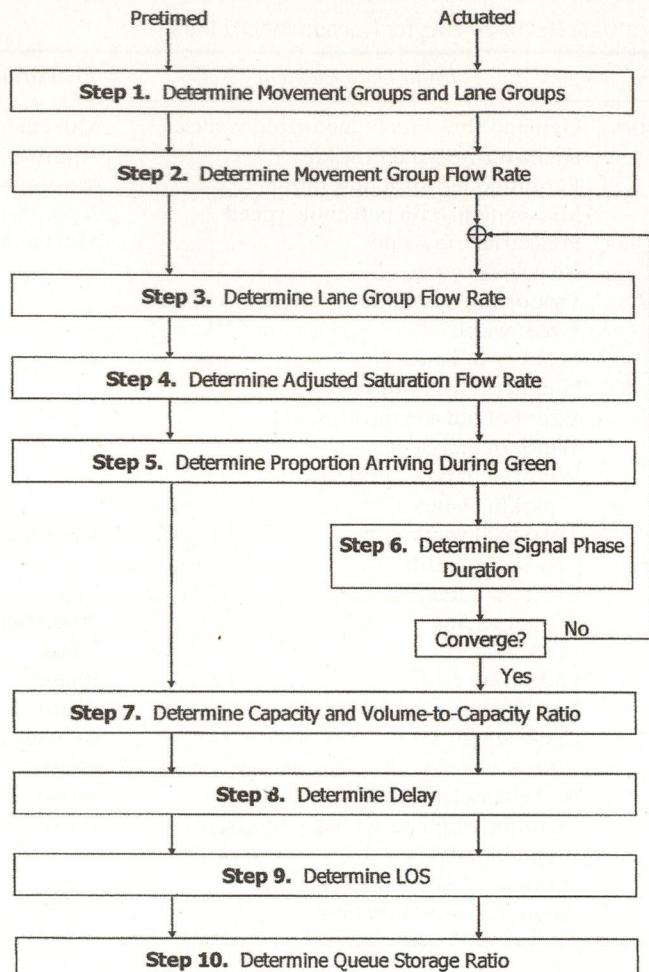

Figure 10.1   Framework for Automobile Methodology for Signalized Intersections

SOURCE: From *Highway Capacity Manual 2010*. Copyright, National Academy of Sciences, Washington, D.C. Reproduced with permission of the Transportation Research Board.

10.4.1 Step 1. Determine Movement Groups and Lane Groups

The traffic at each approach of the intersection is represented by movement and lane groups. These have been defined in Chapter 8, and guidelines on how to identify them are given in Section 8.4. The different lane groups and movement groups are given in Figure 8.14 and are presented again as Figure 10.3 to facilitate comprehension of this step.

10.4.2 Step 2. Determine Movement Group Flow Rate

This step involves the computation of the flow rate for each of the different movement groups. Note that the right-turn-on-red flow rate is deducted from the right-turn flow rate whether or not the right turns are on exclusive lanes.

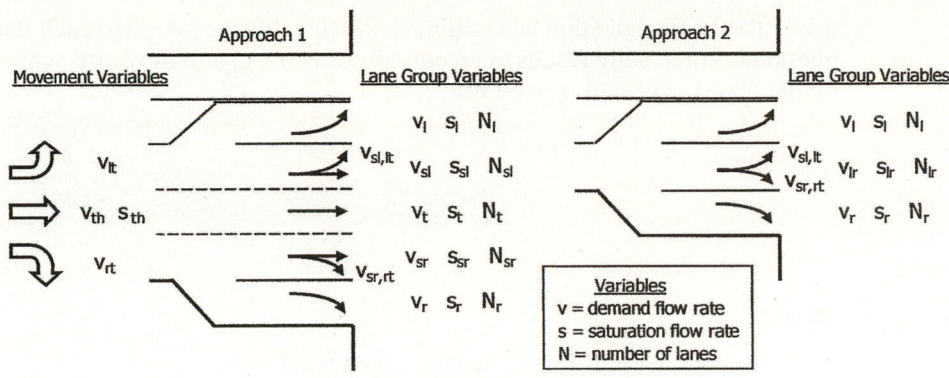

Figure 10.2   Input Variables for Lane Group Flow Rate Procedure

SOURCE: From *Highway Capacity Manual 2010*. Copyright, National Academy of Sciences, Washington, D.C. Reproduced with permission of the Transportation Research Board.

Number of Lanes	Movements by Lanes	Movement Groups (MG)	Lane Groups (LG)
1	Left, thru., & right:	MG 1:	LG 1:
2	Exclusive left:	MG 1:	LG 1:
	Thru. & right:	MG 2:	LG 2:
2	Left & thru.:	MG 1:	LG 1:
	Thru. & right:		LG 2:
3	Exclusive left: Exclusive left:	MG 1:	LG 1:
	Through: Through:	MG 2:	LG 2:
	Thru. & right:		LG 3:

Figure 10.3   Typical Movement and Lane Groups

SOURCE: From *Highway Capacity Manual 2010*. Copyright, National Academy of Sciences, Washington, D.C. Reproduced with permission of the Transportation Research Board.

10.4.3 Step 3. Determine Lane Group Flow Rate

This step involves computation of the lane group flow rates. Note that on an approach where only one lane exists or no shared lanes exist, then the lane group corresponds to the movement group, and the flow rates for the movement group and lane groups are the same. However, when there are two or more lanes at an approach with at least a shared lane, the lane group flow rate is determined by considering the inclination of the driver to use the lane that will minimize the time he/she takes to reach the stop line. This phenomenon usually results in an equilibrium lane group flow rate, which is determined by the HCM expression given in Eq. 10.7.

$$\frac{v_i}{s_i} = \frac{\displaystyle\sum_{i=1}^{i=N_{th}} v_i}{\displaystyle\sum_{i=1}^{N_{th}} s_i}$$

(10.7)

where

v_i = demand flow rate in lane i (veh/h/ln)
s_i = saturation flow rate in lane i (veh/h/ln)
N_{th} = number of through lanes (shared or exclusive) (ln)

10.4.4 Step 4. Determine Adjusted Saturation Flow Rate

This module provides for the computation of an adjusted saturation flow rate for each lane group. The adjusted saturation flow rate is defined as the flow rate in veh/h that the lane group can carry if it has the green indication continuously; that is, if $g/C = 1$.

Equation for Adjusted Saturation Flow Rate

The adjusted saturation flow rate (s) depends on a base saturation flow (s_o) rate, which is usually taken as 1900 passenger cars/h of green time per lane. This ideal saturation flow is then adjusted for the prevailing conditions to obtain the saturation flow for the lane group being considered. The computed saturation flow rate is known as the "adjusted" saturation flow rate. The adjustment is made by introducing factors that correct for the number of lanes, lane width, the percent of heavy vehicles in the traffic stream, approach grade, parking activity, local buses stopping within the intersection, area type, lane utilization factor, and right and left turns. The HCM gives the saturation flow as

$$s = (s_o)(f_w)(f_{HV})(f_g)(f_p)(f_a)(f_{bb})(f_{Lu})(f_{RT})(f_{LT})(f_{LPb})(f_{RPb})$$

(10.8)

where

s = saturation flow rate for the subject lane group, expressed as a total for all lanes in the lane group under prevailing conditions (veh/h/ln)
s_o = base saturation flow rate per lane, usually taken as 1900 veh/h/ln
N = number of lanes in lane group
f_w = adjustment factor for lane width
f_{HV} = adjustment factor for heavy vehicles in the traffic stream
f_g = adjustment factor for approach grade
f_p = adjustment factor for the existence of a parking lane adjacent to the lane group and the parking activity on that lane
f_a = adjustment factor for area type (for CBD, 0.90; for all other areas, 1.00)
f_{bb} = adjustment factor for the blocking effect of local buses stopping within the intersection area
f_{Lu} = adjustment factor for lane utilization

f_{RT} = adjustment factor for right turns in the lane group
f_{LT} = adjustment factor for left turns in the lane group
f_{Lpb} = pedestrian adjustment factor for left-turn movements
f_{Rpb} = pedestrian adjustment factor for right-turn movements

Adjustment Factors

Although the necessity for using some of these adjustment factors was presented in Chapter 8, the basis for using each of them is given again here to facilitate comprehension of the material.

Lane Width Adjustment Factor, f_w. This factor depends on the average width of the lanes in a lane group. It is used to account for both the reduction in saturation flow rates when lane widths are less than 12 ft and the increase in saturation flow rates when lane widths are greater than 12 ft. When lane widths are 16 ft or greater, such lanes may be divided into two narrow lanes of 8 ft each. Lane width factors should not be computed for lanes less than 8 ft wide. See Table 10.7 for recommended lane width adjustment factors.

Heavy-Vehicle Adjustment Factor, f_{HV}. The heavy-vehicle adjustment factor is related to the percentage of heavy vehicles in the lane group. This factor corrects for the additional delay and reduction in saturation flow due to the presence of heavy vehicles in the traffic stream. The additional delay and reduction in saturation flow are due mainly to the difference between the operational capabilities of heavy vehicles and passenger cars and the additional space taken up by heavy vehicles. In this procedure, heavy vehicles are defined as any vehicle that has more than four tires touching the pavement. A passenger-car equivalent (E_T) of two is used for each heavy vehicle. This factor is computed by using the HCM expression given in Eq. 10.9.

$$f_{HV} = \frac{100}{100 + P_{HV}(E_T - 1)} \tag{10.9}$$

where

P_{HV} = percent of heavy vehicles in the subject movement group (%).

Grade Adjustment Factor, f_g. This factor is related to the slope of the approach being considered. It is used to correct for the effect of slopes on the speed of vehicles, including both passenger cars and heavy vehicles, since passenger cars are also affected by grade. This effect is different for up-slope and down-slope conditions; therefore, the direction

Table 10.7   Lane Width Adjustment Factor

Average lane width (ft)	Adjustment factor (l_w)
<10*	0.96
≥10–12.9	1.00
>12.9	1.04

*Factors apply to average lane width of 8 ft or more.
SOURCE: From *Highway Capacity Manual 2010*. Copyright, National Academy of Sciences, Washington, D.C. Reproduced with permission of the Transportation Research Board.

of the slope should be taken into consideration. This factor is computed by using the HCM expression given as

$$f_g = 1 - \frac{P_g}{200} \qquad (10.10)$$

where

P_g = approach grade for the subject movement group (%)

Note that Eq. 10.10 is only applicable for grades ranging from −6 percent to +10 percent.

Parking Adjustment Factor, f_p. On-street parking within 250 ft upstream of the stop line of an intersection causes friction between parking and nonparking vehicles, which results in a reduction of the maximum flow rate that the adjacent lane group can handle. This effect is corrected for by using a parking adjustment factor on the base saturation flow. This factor depends on the number of lanes in the lane group and the number of parking maneuvers/h. The equation given below for the parking adjustment factor indicates that the higher the number of lanes in a given lane group, the less effect parking has on the saturation flow; the higher the number of parking maneuvers, the greater the effect. In determining these factors, it is assumed that each parking maneuver (either in or out) blocks traffic on the adjacent lane group for an average duration of 18 sec. It should be noted that when the number of parking maneuvers/h is greater than 180 (equivalent to more than 54 min), a practical limit of 180 should be used. This adjustment factor should be applied only to the lane group immediately adjacent to the parking lane. When parking occurs on both sides of a single lane group, the sum of the number of parking maneuvers on both sides should be used.

$$f_p = \frac{N - 0.1 - \dfrac{18N_m}{3600}}{N} \geq 0.050 \qquad (10.11)$$

where

f_p = parking adjustment factor

N_m = parking maneuver rate adjacent to lane group (maneuvers/h)

N = number of lanes in lane group

Area Type Adjustment Factor, f_a. The general types of activities in the area at which the intersection is located have a significant effect on speed and therefore on saturation volume at an approach. For example, because of the complexity of intersections located in areas with typical central business district characteristics, such as narrow sidewalks, frequent parking maneuvers, vehicle blockades, narrow streets, and high-pedestrian activities, these intersections operate less efficiently than intersections at other areas. This is corrected for by using the area type adjustment factor f_a, which is 0.90 for a CBD and 1.0 for all other areas. It should be noted, however, that 0.90 is not automatically used for all areas designated as CBDs, nor should it be limited only to CBDs. It should be used for locations that exhibit the characteristics referred to earlier that result in a significant impact on the intersection capacity.

Bus Blockage Adjustment Factor, f_{bb}. When buses have to stop on a travel lane to discharge or pick up passengers, some of the vehicles immediately behind the bus will also have to stop. This results in a decrease in the maximum volume that can be handled by that lane. This effect is corrected for by using the bus blockage adjustment factor, which is related to the number of buses in an hour that stop on the travel lane within 250 ft upstream or downstream from the stop line to pick up or discharge passengers, as well as the number of lanes in the lane group. This factor only applies to the lane group blocked by buses, and a practical upper limit of 250 bus/h should be used for the bus stopping rate. This factor is computed using the HCM expression given in Eq. 10.12.

$$f_{bb} = \frac{N - \dfrac{14.4N_b}{3600}}{N} \geq 0.050 \tag{10.12}$$

where

N = number of lanes in the subject lane group (ln)

N_b = bus stopping rate at the approach buses/h

Lane Utilization Adjustment Factor, f_{Lu}.   The lane utilization factor is used to adjust the ideal saturation flow rate to account for the unequal utilization of the lane in a lane group. It is used for lane groups that have more than one exclusive lane. The lane utilization factor is 1.0 when the lane group consists of one shared lane or one exclusive lane. This factor is also computed using the HCM expression given as

$$f_{Lu} = \frac{v_g}{v_{gi}N} \tag{10.13}$$

where

v_g = unadjusted demand flow rate for lane group (veh/h)

v_{gi} = unadjusted demand flow rate on the single lane in the lane group with the highest volume

N = number of lanes in the lane group

It is recommended that actual field data be used for computing f_{Lui}.

Right-Turn Adjustment Factor for a Protected Movement on an Exclusive Lane, f_{RT}.   This factor reflects the extent to which the right-turn geometry affects the saturation flow. It is obtained for exclusive right-turn lane groups operating in a protected mode as

$$f_{RT} = 1/E_R \tag{10.14}$$

where

E_R = equivalent number of through cars for a protected right-turning vehicle

     = 1.18

If the right-turn movement is on a shared lane or operating in a permitted mode, then the procedure given in Section 10.7 should be used to compute the adjusted saturation flow rate.

Left-Turn Adjustment Factor for a Protected Movement on an Exclusive Lane, f_{LT}.   This adjustment factor is used to account for the fact that left-turn movements take more time than through movements. It is obtained for exclusive left-turn lane groups operating in a protected mode as

$$f_{LT} = 1/E_L \tag{10.15}$$

where

E_L = equivalent number of through cars for a protected left-turning vehicle
= 1.05

The procedures for determining the adjusted saturation flow rates for shared lanes with a permitted mode are given in Section 10.7.

Adjustment for Pedestrians and Bicycles (f_{Lpb}, f_{Rpb}). The left-turn pedestrian bicycle adjustment factor (f_{Lpb}) and right-turn pedestrian-bicycle adjustment factor reflect the conflict among the different modes (turning vehicles, pedestrians, and bicycles) at the intersection. The basis for developing these factors is the concept of *relevant conflict zone occupancy*, which also reflects the impact of the conflict caused by the opposing vehicle and the left-turning movement. The relevant occupancy and the number of receiving lanes for the turning vehicles are used to determine the proportion of green time during which the conflict zone is occupied.

Guidelines for determining the pedestrian-bicycle adjustment factors for left-turn lane groups (f_{Lpb}) are:

- f_{Lpb} = 1.0 when the conflicting pedestrian flow = 0.0
- f_{Lpb} = 1.0 when the phase operates in a protected or split mode and the lane group is on a two-way street
- f_{Lpb} is computed using the appropriate procedure described in Section 10.5, when the phase operates in a permitted or protected-permitted mode and the lane group is on a two-way street.

Guidelines for determining the pedestrian-bicycle adjustment factor (f_{Rpb}) for lane groups that include right-turn movement are:

- f_{Rpb} = 1.0 when the conflicting pedestrian flow = 0.0
- f_{Rpb} = 1.0 when the lane group operates in a protected mode
- f_{Rpb} is computed using the appropriate procedure described in Section 10.5, when the lane group operates in a permitted or protected-permitted mode.

10.4.5 Step 5. Determine Proportion Arriving during Green

This step involves computation of the proportion of vehicles arriving during the green interval, as discussed in Section 10.3.3 under Platoon Ratio.

10.4.6 Step 6. Determine Signal Phase Duration

This step involves the computation of the duration of the different phases in a cycle, as discussed in Chapter 8.

10.4.7 Step 7. Determine Capacity and Volume-to-Capacity Ratio

The capacity of a given lane group for which left turns are not permitted and there is a single traffic movement (as discussed in Chapter 8) is given as

$$c = Ns\frac{g}{C}$$ (10.16)

where

c = capacity (veh/h)
N = number of lanes in the lane group
s = the adjusted saturated flow rate (veh/h/ln)
g = effective green time for the lane group (sec)
C = cycle length

As discussed in Chapter 8, the volume-to-capacity ratio (X) is given as v/c and the critical-volume-to-capacity ratio is given as

$$X_c = \left(\frac{C}{C-L}\right)\sum_{i\varepsilon c_i} y_{c,i} \tag{10.17}$$

$$L = \sum_{i\varepsilon ci} l_{t,i}$$

where

X_c = critical intersection volume-to-capacity ratio
C = cycle length (sec)
y_{ci} = critical flow ratio for phase $i = v_i/(Ns_i)$ (see Chapter 8)
l_{ti} = phase i lost time = $l_{1i} + l_{2i}$ (sec)
c_i = set of critical phases on the critical path (see Chapter 8)
L = cycle lost time (sec) (see Chapter 8)

10.4.8 Step 8. Determine Delay

This step involves computation of the average control delay of vehicles arriving during the analysis period and includes the delay of vehicles that do not clear the queue during the analysis period. The control delay (d) for a lane group is the sum of three delay components: (i) uniform delay (d_1), (ii) incremental delay (d_2), and (iii) initial queue delay (d_3).

Uniform Delay

The uniform delay is that which will occur in a lane group if vehicles arrive uniformly at the intersection and saturation does not occur during any cycle. Also, only one effective green time for the cycle and one saturation flow rate for the analysis period are assumed. Uniform delay can be determined as

$$d_{li} = 0.50C \frac{(1 - g_i/C)^2}{1 - (g_i/C)\left[\min\left(X_i,1.0\right)\right]} \tag{10.18}$$

where

d_{li} = uniform delay (sec/vehicle) for lane group i
C = cycle length (sec)
g_i = effective green time for lane group i (sec)
X_i = v/c ratio for lane group i

Note that an alternate procedure to determine delay for protected-plus-permitted left-turn operation is given in the HCM.

Incremental Delay

Incremental delay takes into consideration that the arrivals are not uniform but random, with the number of arrivals varying from cycle to cycle. It also takes into consideration the delay that occurs when demand is higher than the capacity during the analysis period. However, in developing the expression given in Eq. 10.19 for incremental delay, it was assumed that there is no unmet demand during the period preceding the analysis period and therefore no initial queue at the start of the analysis period. A detailed description of the development of this delay is given in the HCM. It is given as

$$d_{2i} = 900T \left[(X_{Ai} - 1) + \sqrt{(X_{Ai} - 1)^2 + \frac{8k_i I_i X_{Ai}}{c_{ai}T}} \right] \qquad (10.19)$$

where

d_{2i} = incremental delay (sec/veh) for lane group i
X_{Ai} = average volume-to-capacity ratio for group $i = v_i/c_{Ai}$
c_i = capacity of lane group i (veh/h) (see Eq. 10.16 of Section 10.4.7)
T = duration of analysis period (h)
k_i = incremental delay factor that is dependent on controller settings

$$k_i = (1 - 2k_{min})(v_i/c_{ai} - 0.5) + k_{min} \leq 0.50 \qquad (10.20)$$

$$k_{min} = -0.375 + 0.354PT - 0.0910PT^2 + 0.0089PT^3 \geq 0.04 \qquad (10.21)$$

$$c_{ai} = 3600\frac{g_{ai}sN}{C} \text{ veh/h} \qquad (10.22)$$

g_{ai} = available green time = $G_{max} + Y + R_c - l_1 - l_2$
c_{ai} = available capacity for a lane group served by an actuated phase (veh/h)
k_{min} = minimum incremental delay factor
I_i = upstream filtering metering adjustment factor accounts for the effect of filtered arrivals from upstream signals (for isolated intersections, $I = 1$; for nonisolated intersections, see Eq. 10.5)
X_i = v/c ratio for lane group i
PT = passage time; the minimum time by which a green phase could be increased during the extendable portion due to the actuation by a vehicle on a phase (range: from 1 to 4 sec) (see Section 8.4.5)

Note that the incremental delay factor (k_i) used in Eq. 10.19 varies from a minimum value of 0.04 to a maximum value of 0.50. The HCM recommends that the maximum value of 0.50 be used for pretimed phases, for coordinated phases, and for those phases that are set to "recall-to-maximum."

Initial Queue Delay

This delay occurs as a result of an initial unmet demand (O_b) by vehicles at the start of the analysis period T. That is, a residual event of length Q_b exists at the start of the analysis period. This does not include vehicles in the queue due to the random variation in demand from cycle to cycle. The initial queue delay for each lane group is given as

$$d_{3i} = \frac{3600}{v_i T}\left(t_{Ai}\frac{Q_{bi} + Q_{ei} - Q_{eoi}}{2} + \frac{Q_{ei}^2 - Q_{eoi}^2}{2c_{Ai}} - \frac{Q_{bi}^2}{2c_{Ai}}\right) \tag{10.23}$$

$$Q_{ei} = Q_{bi} + t_A(v_i - c_{Ai}) \tag{10.24}$$

If $v \geq c_{Ai}$, then

$$Q_{ei} = T(v_i - c_{Ai})$$

$$t_{Ai} = T \tag{10.25}$$

If $v < c_{Ai}$, then

$$Q_{eo} = 0.0 \text{ veh}$$

$$t_{Ai} = Q_{bi}/(c_{Ai} - v_i) \leq T \tag{10.26}$$

where

v_i = demand flow rate (veh/h)
t_{Ai} = adjusted duration of unmet demand in the analysis period (h)
Q_{bi} = initial queue in lane group i at the start of the analysis period
Q_{ei} = queue in lane group i at end of analysis period
Q_{eoi} = queue in lane group i at the end of the analysis period when $v_i \geq c_{Ai}$ and Q_{bi} = 0.0 veh
T = duration of analysis period (h)
c_A = average lane group capacity
c_A = c (if no lane group has an initial queue)

$$c_{A,i} = c_{s,i}\frac{t_a}{T} + c_i\frac{T - t_i}{T} \quad \text{Initial queue exists in lane group } i. \tag{10.27}$$

$$c_{Ai} = c_{si}\frac{t_a}{T} + c_i\frac{(T - t_a)}{T} \quad \text{Initial queue does not exist in lane group } i. \tag{10.28}$$

t_i = duration of unmet demand for lane group i in the analysis period (h)
t_i = T when $v_i \geq c_{si}$
t_i = $Q_{bi}/(c_{si} - v_i) \leq T$, when $v_i < c_{si}$
c_{si} = saturated capacity (veh/h) for lane group i.
t_a = adjusted duration of unmet demand in the analysis period for lane group i (h)

$$t_a = \frac{1}{N_g}\sum_{i\varepsilon N_g}t_i \tag{10.29}$$

N_g = number of lane groups for which $t > 0.0$ h

The time the last vehicle arrives during the analysis period when a queue exists at the start of the analysis period is known as the queue clearing time (t_c) and is given as

$$t_c = t_A + Q_e/c_A \tag{10.30}$$

Total Control Delay

The sum of the uniform, incremental, and initial queue delays for each lane group on each approach is then determined to obtain the average delay for each approach as shown in Eq. 10.32. The total control delay for lane group i is given as

$$d_i = d_{1i}PF + d_{2i} + d_{3i} \tag{10.31}$$

where

d_i = the average control delay per vehicle for a given lane group
PF = uniform delay adjustment factor for quality of progression (See Table 10.8)
d_{1i} = uniform control delay component assuming uniform arrival
d_{2i} = incremental delay component for lane group i, no residual demand at the start of the analysis period
d_{3i} = initial queue delay for lane group i

Note that when Eq. 10.18 is used to determine the average control delay (d_{1i}), it is multiplied by the progression adjustment factor (PF) obtained from Table 10.8 to compute the total control delay. This factor accounts for the quality of progression.

It should be noted that while the adjustment factor for controller type (k) accounts for the ability of actuated controllers to adjust timing from cycle to cycle, the delay adjustment factor (PF) accounts for the effect of quality of signal progression at the intersection. PF accounts for the positive effect that good signal progression has on the flow of traffic through the intersection and depends on the arrival type. It has a value of one for isolated intersections (arrival type 3). The six different arrival types were defined earlier in Table 10.5.

Table 10.8   Progression Adjustment Factors (PF)

Arrival Type	Progression Adjustment Factor (PF) as a Function of Green Ratio					
	0.2	0.3	0.4	0.5	0.6	0.7
Uncoordinated	1.00	1.00	1.00	1.00	1.00	1.00
Coordinated[a]	0.92	0.86	0.78	0.67	0.50	0.22

Note: [a]$PF = (1 - [1.33g/C])/(1 - g/C)$ $\tag{10.32}$

SOURCE: From *Highway Capacity Manual 2010*. Copyright, National Academy of Sciences, Washington, D.C. Reproduced with permission of the Transportation Research Board.

Approach Delay

Having determined the average stopped delay for each lane group, we can now determine the average stopped delay for any approach as the weighted average of the stopped delays of all lane groups on that approach. The approach delay is given as

$$d_A = \frac{\sum_{i=1}^{n_A}(d_{ia}v_i)}{\sum_{i=1}^{n_A}v_i} \tag{10.33}$$

where

d_A = delay for approach A (sec/veh)
d_{ia} = adjusted delay for lane group i on approach A (sec/veh)
v_i = adjusted flow rate for lane group i (veh/h)
n_A = number of lane groups on approach A

The level of service of approach A then can be determined from Table 10.1.

Intersection Delay

The average intersection control delay is found in a manner similar to the approach delay. In this case the weighted average of the delays at all approaches is the average stopped delay at the intersection. The average intersection delay is therefore given as

$$d_I = \sum_{A=1}^{A_n} \frac{d_A v_A}{\sum_A^{A_n} v_A} \tag{10.34}$$

where

d_I = average stopped delay for the intersection (sec/veh)
d_A = adjusted delay for approach A (sec/veh)
v_A = adjusted flow rate for approach A (veh/h)
A_n = number of approaches at the intersection

Note that the HCM gives an alternative procedure for computing control delay known as the *incremental queue procedure,* which does not incorporate the assumptions made in the example presented above. For example, rather than assuming one saturation flow rate for the analysis period, the procedure uses multiple green periods and movements with multiple saturation flow rates. Further discussion of the procedure is beyond the scope of this text, and interested readers should refer to the HCM for additional details.

Note that it is also feasible for delay to be directly measured in the field at an existing intersection as an alternative to using the computational method described herein. One such method is the test-car survey described in Chapter 4. The HCM recommends the queue-count technique for measuring the control delay at an intersection. It entails the direct counting of the vehicles in the queue for each lane group under observation. The technique can be used for all undersaturated lane groups, but it is not practical for oversaturated lane groups with significant queue buildup or when the storage length of a lane group is limited. Interested readers should refer to the HCM, which gives a detailed description of the technique.

10.4.9 Step 9. Determine LOS for the Automobile Mode

The level of service for each lane group, each approach, and the whole intersection is then determined from Table 10.1 using the appropriate average control delay. It should be emphasized again that short or acceptable delays do not automatically indicate adequate capacity. Both the capacity and the delay should be considered in the evaluation of any intersection. Where long and unacceptable delays are determined, it is necessary to find the cause of the delay. For example, if v/c ratios are low and the delay is long, the most probable cause for the long delay is that the cycle length is too long and/or the progression (arrival type) is unfavorable. Delay therefore can be reduced by improving the arrival type, by coordinating the intersection signal with the signals at adjacent intersections, and/or by reducing the cycle length at the entire intersection. The procedure is illustrated in Appendix D, using the Highway Capacity Software (HCS).

10.5 COMPUTATION OF PEDESTRIAN AND BICYCLE FACTORS (f_{Lpb}, f_{Rpb}) FOR RIGHT- AND LEFT-TURN MOVEMENTS FROM ONE-WAY STREETS

When the users of the intersection compete for space, the saturation flow rate of vehicles making turns is reduced. These factors are included in the saturation flow equation to account for the reduction in the saturation flow rate resulting from the conflicts among automobiles, pedestrians, and bicycles. The specific zones within the intersection where these conflicts occur are shown in Figure 10.4. Recall that the values for f_{Lpb} and f_{Rpb} are given in section 10.4.4 for specific conditions; when there are no opposing bicycles and pedestrians, or for protected operations on exclusive lanes, or split mode phasing on a two-way highway. The procedures described in this section are for shared lanes and permitted modes. The procedure can be divided into the following five main tasks:

1. Determine average pedestrian occupancy, OCC_{pedg}.
2. Determine average bicycle occupancy (OCC_{bicg}).
3. Determine relevant conflict zone occupancy, OCC_r.
4. Determine unoccupied time, A_{pBT}.
5. Determine saturation flow adjustment factors for right-turn movements f_{Rpb} and for left-turn movements (f_{Lpb}).

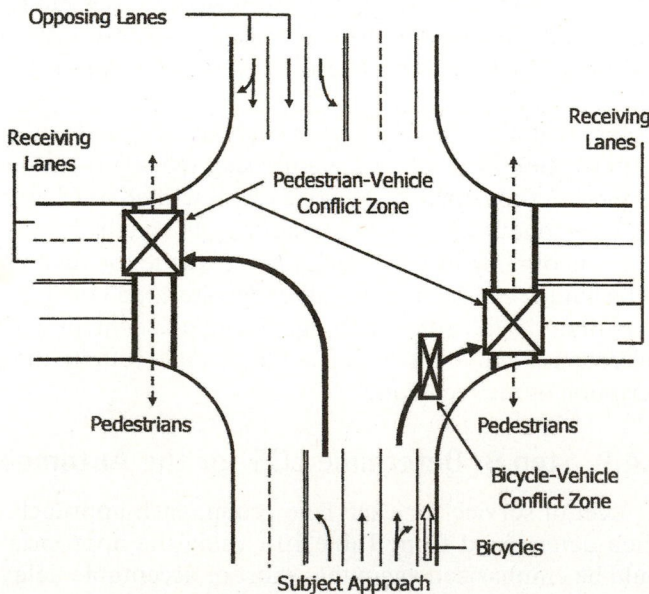

Figure 10.4   Conflict Zone Locations

10.5.1 Step 1. Determine Average Pedestrian Occupancy (OCC_{pedg})

In this task, the average pedestrian flow rate (v_{pedg}) during the pedestrian service time is first computed from the pedestrian volume using Eq. 10.35, and the average pedestrian occupancy is then computed from the average pedestrian flow rate using Eqs. 10.36 or 10.37.

$$v_{pedg} = v_{ped} \frac{C}{g_{ped}} \le 5000 \tag{10.35}$$

$$OCC_{pedg} = \frac{v_{pedg}}{2000} \text{ for pedestrian flow rate} \le 1000 \text{ p/h} \tag{10.36}$$

$$OCC_{pedg} = 0.4 + \frac{v_{pedg}}{10{,}000} \le 0.90 \text{ for pedestrian flow rate} > 1000 \tag{10.37}$$

where

v_{pedg} = pedestrian flow rate during the pedestrian service time (ped/h)
v_{ped} = pedestrian flow rate in the subject crossing (walking in both directions) (ped/h)
C = cycle length (sec)
g_{ped} = pedestrian service time (sec)

For a pedestrian-actuated phase with a pedestrian signal head and rest-in-walk not enabled, g_{ped} is the smaller of (a) and (b), where

(a) g_{ped} = g (effective green time for the phase) (sec)
(b) g_{ped} = min(g, *Walk* + PC)
  PC = pedestrian clear interval (sec)

For all other conditions,

$$g_{ped} = g \text{ (effective green time for the phase) (sec)}$$

Note that the HCM stipulates that the practical maximum pedestrian flow rate during the pedestrian service time (v_{pedg}) is 5000 ped/h and should apply when Eq. 10.35 is used. Also note that in using Eqs. 10.36 and 10.37, it is not necessary to compute v_{pedg} from Eq. 10.35 if pedestrian flow rate is collected directly from the field.

10.5.2 Step 2. Determine Average Bicycle Occupancy (OCC_{bicg})

In this task, the bicycle flow rate (v_{bicg}) during the green is first determined using Eq. 10.38, and the average bicycle occupancy (OCC_{bicg}) is then computed from the average bicycle flow rate using Eq. 10.39.

$$v_{bicg} = v_{bic} \frac{C}{g} \le 1900 \tag{10.38}$$

$$OCC_{bicg} = 0.02 + \frac{v_{bicg}}{2700} \tag{10.39}$$

where

v_{bicg} = bicycle flow rate during the green indication (bicycles/h)
v_{bic} = bicycle flow rate (bicycles/h)
g = effective green time (sec)
C = cycle length (sec)

10.5.3 Step 3. Determine Relevant Conflict Zone Occupancy (OCC_r)

Two conditions influence the computation of this factor. These are (1) when there is no bicycle interference for right-turn movements or for left-turn movements when the approach is a one-way street. Under these conditions, Eq. 10.40 is applicable, and (2) when there are pedestrian and bicycle interferences for right-turn movements. Under this condition, Eq. 10.41 is used.

$$OCC_r = \frac{g_{ped}}{g} OCC_{pedg} \tag{10.40}$$

Eq. 10.40 is associated with two assumptions. First is that pedestrians are crossing during the pedestrian service time (g_{ped}), and second, there is no pedestrian crossing during the green time period of ($g - g_{ped}$).

$$OCC_r = \left(\frac{g_{ped}}{g} OCC_{pedg}\right) + OCC_{bicg} - \left(\frac{g_{ped}}{g} OCC_{pedg} OCC_{bicg}\right) \tag{10.41}$$

where

OCC_{pedg} = average pedestrian occupancy
OCC_{bicg} = average bicycle occupancy
g_{ped} = pedestrian service time (sec)
g = effective green time of the subject phase (sec)

10.5.4 Step 4. Determine Unoccupied Time (A_{pbT})

Two conditions are considered in the determination of the A_{pbT}. These are: (1) number of turning lanes (N_{turn}) is the same as the number of the cross-street receiving lanes (N_{rec}), and (2) number of turning lanes is less than the number of the cross-street receiving lanes.

When N_{turn} is equal to N_{rec}, it is unlikely that the turning vehicles will be able to move around pedestrians or bicycles. The time the conflict zone is unoccupied (A_{pbt}) is therefore obtained from Eq. 10.42:

$$A_{pbT} = 1 - OCC_r \quad (N_{turn} = N_{rec}) \tag{10.42}$$

When N_{turn} is less than the N_{rec}, the impact of pedestrians and bicycles on the saturation flow is reduced, as it is more likely that the turning vehicles will be able to move around pedestrians and bicycles. The A_{pbT} in this case is obtained from Eq. 10.43.

$$A_{pbT} = 1 - 0.6(OCC_r) \quad (N_{turn} < N_{rec}) \tag{10.43}$$

where

OCC_r = relevant conflict zone occupancy (see Eq. 10.39 or 10.40)
N_{turn} = the number of turning lanes
N_{rec} = the number of receiving lanes

It is recommended that actual field observation be carried out to determine the number of turning lanes (N_{turn}) and the number of receiving lanes (N_{rec}). The reason for this is that at some intersections, left turns are illegally made deliberately from an outer lane or the receiving lane is blocked by vehicles that are double parked, making it difficult for the turning vehicles to make a proper turn. Simply reviewing the intersection plans and noting the striping cannot identify these conditions.

10.5.5 Step 5. Determine Saturation Flow Adjustment Factors for Right–Turn Movements (f_{Rpb}) and for Left–Turn Movements from One–Way Streets (f_{Lpb})

These factors depend on A_{pbT}, and the following rules apply.
For permitted right turns in an exclusive lane,

$$f_{Rpb} = A_{pbT} \text{ (see Eq. 10.42 or 10.43)}$$

For protected-permitted right turns in an exclusive lane,

$$f_{Rpb} = 1 \text{ for the protected period}$$

$$f_{Rpb} = A_{pbT} \text{ (see Eq. 10.42 or 10.43)}$$

For left-turn movements from a one-way street,

$$f_{Lpb} = A_{pbT} \text{ (see Eq. 10.42 or 10.43)}$$

10.6 COMPUTATION OF PEDESTRIANS AND BICYCLES FACTOR (f_{Lpb}), FOR PROTECTED OR PROTECTED-PERMITTED LEFT–TURN MOVEMENTS ON TWO-WAY STREETS

In this case, the procedure can be divided into four steps as follows:

1. Compute pedestrian occupancy after queue clears
2. Determine relevant conflict zone occupancy
3. Determine unoccupied time
4. Determine saturation flow rate adjustment factor

10.6.1 Step 1. Compute Pedestrian Occupancy after Queue Clears (OCC_{pedu})

This factor depends on the opposing queue service time, which is determined as the difference between the effective permitted green time (g_p) and the duration of the permitted green time that is not blocked by an opposing queue (g_u). It is given for two conditions as:

$$OCC_{pedu} = OCC_{pedg}(1 - (0.5g_q/g_{ped})) \quad (g_q < g_{ped}) \tag{10.44}$$

$$OCC_{pedu} = 0.0 \quad (g_q \geq g_{ped}) \tag{10.45}$$

where

g_{ped} = pedestrian service time (sec)

$g_q = g_p - g_u$ (sec)

g_p = effective permitted green time = g_s for the opposing movement

g_u = duration of permitted green time that is not blocked by an opposing queue (sec)

10.6.2 Step 2. Determine Relevant Conflict Zone Occupancy (OCC_r)

This factor depends on the pedestrian occupancy and the probability of acceptable gaps occurring for left turns to be made after the opposing queue clears. It is given as

$$OCC_r = \frac{g_{ped} - g_q}{g_p - g_q}(OCC_{pedu})e^{-5.00v_o/3600} \qquad (10.46)$$

where

OCC_r = relevant conflict zone occupancy

OCC_{pedu} = pedestrian occupancy after queue clears (see Eq. 10.44 or 10.45)

g_{ped} = pedestrian service time (sec)

$g_q = g_p - g_u$

g_p = effective permitted green time

g_u = duration of permitted green time that is not blocked by an opposing queue

v_o = opposing demand flow

10.6.3 Step 3. Determine Unoccupied Time

The unoccupied time is determined from the equations given in Section 10.5 and repeated here as

$$A_{pbT} = 1 - OCC_r \quad (N_{turn} = N_{rec}) \qquad (10.47)$$

$$A_{pbT} = 1 - 0.6(OCC_r) \quad (N_{turn} < N_{rec}) \qquad (10.48)$$

where

OCC_r = relevant conflict zone occupancy

N_{turn} = the number of turning lanes

N_{rec} = the number of receiving lanes

10.6.4 Step 4. Determine Saturation Flow Rate Adjustment Factor

The saturation flow rate adjustment factor for permitted and protected permitted left turns is given as

$$f_{Lpb} = A_{pbt} \qquad (10.49)$$

10.7 DETERMINATION OF LANE GROUP ADJUSTED SATURATION FLOW RATE

The HCM gives the following procedures for computing the saturation flow rates for different lane groups operating with different modes. For an exclusive right-turn or left-turn lane group with a protected mode, Eq. 10.8 is used, with the computed values for the appropriate adjusted factors as given in Eq. 10.14 or Eq. 10.15. The computational procedures for permitted, protected-permitted, and shared lane groups are given in the following sections.

10.7.1 Shared Right–Turn and Through–Lane Groups with Permitted Operation

The adjusted saturation flow rate for a shared right-turn and through lane group with permitted operation is given as

$$s_{sr} = \frac{s_{th}}{1 + P_R\left(\dfrac{E_R}{f_{Rpb}} - 1\right)} \tag{10.50}$$

where

s_{sr} = saturation flow rate of a shared right-turn and through lane group with permitted operation (veh/h/ln)

s_{th} = saturation flow rate of an exclusive through lane, which is the base saturation flow for a through lane adjusted for lane width, heavy vehicles, grade, parking, buses, and area type

P_R = percentage of right-turning vehicles in the shared lane

E_R = equivalent number of through cars for a protected right-turning vehicle = 1.18

f_{Rpb} = pedestrian-bicycle adjustment factor for right-turn groups (see Section 10.5.5)

10.7.2 Adjusted Saturation Flow Rate for Shared Right–Turn and Through–Lane Group with Protected Operation

The adjusted saturation flow rate for a shared right-turn and through-lane group with protected operation is obtained from Eq. 10.50, with f_{Rp} being 1.0, and it is given as

$$s_{sr,per} = \frac{s_{th}}{1 + P_R(E_R - 1)} \tag{10.51}$$

where

$s_{sr,per}$ = saturation flow rate of a shared right-turn and through lane group with permitted operation (veh/h/ln)

s_{th} = saturation flow rate of an exclusive through, which is the base saturation flow for a through lane adjusted for lane width, heavy vehicles, grade, parking, buses, and area type (veh/h/ln)

P_R = percentage of right-turning vehicles in the shared lane

E_R = equivalent number of through cars for a protected right-turning vehicle = 1.18

10.7.3 Adjusted Saturation Flow Rate for Shared Right–Turn and Through–Lane Group with Protected–Permitted Operation $(s_{sr,pro-per})$

The adjusted saturation flow rate for shared right-turn and through lane group with a protected-permitted operation consists of two parts. The first part is the adjusted saturation flow for the protected period and is obtained from Eq. 10.51, and the second part is that for the permitted period obtained from Eq. 10.50 and is given as

$$s_{sr,\,pro-per} = \frac{s_{th}}{g_{pro-per}} \left(\frac{g_{pro}}{1 + P_R(E_R - 1)} + \frac{g_{per}}{1 + P_R\left(\dfrac{E_R}{f_{Rpb}} - 1\right)} \right) \tag{10.52}$$

$s_{sr,\,pro-per}$ = adjusted saturation flow for shared right-turn and through-lane group with a protected-permitted operation

s_{th} = saturation flow rate of an exclusive through lane, which is the base saturation flow for a through lane adjusted for lane width, heavy vehicles, grade, parking, buses, and area type

P_R = percentage of right-turning vehicles in the shared lane

E_R = equivalent number of through cars for a protected right-turning vehicle = 1.18

f_{Rpb} = pedestrian-bicycle adjustment factor for right-turn groups (see Section 10.5)

g_{pro} = effective protected green time for the shared-right and through-lane group

g_{per} = effective permitted green time for the shared-right and through-lane group

$g_{pro-per}$ = effective protected plus permitted green time for the shared-right and through-lane group

10.7.4 Adjusted Saturation Flow Rate for an Exclusive Left–Turn Lane Group with Permitted Operation

Equation 10.53 gives the expression for the adjusted saturation flow for a permitted left-turn operation in an exclusive lane.

$$s_{l,per,ex} = s_{p,lf} f_w f_g f_p f_{bb} f_a f_{Lu} f_{Lpb} f_{HV} \tag{10.53}$$

where

$s_{l,per,ex}$ = saturation flow rate of a permitted left-turn movement in an exclusive lane (veh/h/ln)

$$s_{p,lf} = \frac{v_o e^{-v_o t_{cg}/3600}}{1 - e^{-v_o t_{fh}/3600}} \tag{10.54}$$

v_o = opposing demand flow rate (veh/h)

t_{cg} = critical headway = 4.5 sec

t_{fh} = 2.5 sec (subject left lane is an exclusive lane)

f_w = adjustment factor for lane width

f_{HV} = adjustment factor for heavy vehicles in the traffic stream

f_g = adjustment factor for approach grade

f_p = adjustment factor for the existence of a parking lane adjacent to the lane group and the parking activity on that lane

f_a = adjustment factor for area type (for CBD, 0.90; for all other areas, 1.00)

f_{bb} = adjustment factor for the blocking effect of local buses stopping within the intersection area

f_{Lu} = adjustment factor for lane utilization

f_{Lpb} = pedestrian adjustment factor for left-turn movements (see Section 10.5.5)

Note that the the HCM stipulates that the minimum value of the opposing demand flow that should be used in Eq. 10.54 is 0.1. Also, no adjustment is made to the opposing demand flow (v_o) for unequal lane use, as doing this will result in an inaccurate representation of the frequency and size of headways in the opposing flow. This, in turn, will result in an underestimation of the left-turn saturation flow rate.

10.7.5 Adjusted Saturation Flow Rate for a Shared Left–Turn and Through–Lane Group with Permitted Operation

The operation of a shared left-turn and through-lane group with a permitted operation involves movements at three different time periods. The first is from the start of the green interval (g_p) to the time the first left-turn vehicle arrives on the shared lane and blocks the shared lane. The blockage will continue until the opposing queue clears and there is an acceptable gap in the opposing traffic stream. This period continues for a length of time designated as g_f, with a saturation flow of s_{th}. The second period starts at the end of g_f and ends when the opposing queue is cleared. This time period is designated g_{diff} and is given as

$$g_{diff} = g_p - g_u - g_f \geq 0.0 \tag{10.55}$$

where

g_u = duration of permitted left-turn green time that is not blocked by an opposing queue (sec)

Note that as shown in Eq. 10.55, this period may not exist. Also, when there is more than one opposing traffic lane, the saturation flow during this second period is zero.

The third period starts after the opposing queue is cleared or the arrival of the first left-turn vehicle that blocks the shared lane.

The saturation flow rate for this second time period is given as

$$s_{sl2, per} = \frac{s_{th}}{1 + P_L\left(\dfrac{E_{L2}}{f_{Lpb}} - 1\right)} \tag{10.56}$$

and the saturation flow rate for the third period $(s_{sl3, per})$ is

$$s_{sl3, per} = \frac{s_{th}}{1 + P_L\left(\dfrac{E_{L1}}{f_{Lpb}} - 1\right)} \tag{10.57}$$

The saturation flow rate for a shared left turn and through movement with a permitted operation is therefore given as

$$s_{sl, per} = \frac{s_{th}}{g_p}\left(g_f + \frac{g_{diff}}{1 + P_L\left(\dfrac{E_{L2}}{f_{Lpb}} - 1\right)} + \frac{\min(g_p - g_f, g_u)}{1 + P_L\left(\dfrac{E_{L1}}{f_{Lpb}} - 1\right)}\right) \tag{10.58}$$

where

$s_{sl2,per}$ = saturation flow rate in shared left-turn and through-lane group in a permitted mode during period 2 (veh/h/ln)

$s_{sl3,per}$ = saturation flow rate in shared left-turn and through-lane group in a permitted mode during period 3 (veh/h/ln)

s_{th} = saturation flow rate of an exclusive through lane, which is the base saturation flow for a through lane adjusted for lane width, heavy vehicles, grade, parking, buses, and area type

P_L = percentage of left-turning vehicles in the shared lane

f_{Lpb} = pedestrian-bicycle adjustment factor for left-turn groups (see Section 10.5.5)

E_{L2} = equivalent number of through cars for a permitted left-turning vehicle when opposed by a queue on a single-lane approach (i.e., during period 2)

$$E_{L2} = \frac{1 - (1 - P_{lto})^{n_q}}{P_{lto}} \geq 1.0 \tag{10.59}$$

with

$$n_q = 0.5(g_p - g_u - g_f) \geq 0.0 \tag{10.60}$$

P_{lto} = proportion of left-turning vehicles in the opposing traffic stream

n_q = maximum number of opposing vehicles that could arrive after g_f and before g_u

E_{L1} = equivalent number of through cars for a permitted left-turning vehicle

$= s_o/s_p$

s_o = base saturation flow rate (pc/h/ln)

Maximum Duration of Time before the First Left–Turn Vehicle Arrives $(g_{f,max})$

The time period between the start of the permitted green indication to the time the first left-turn vehicle arrives has been designated g_f. This parameter should have a maximum value because consideration should be given to vehicle distribution. The HCM gives the following expression for this maximum value.

$$g_{f, max} = \frac{(1 - P_L)}{0.5P_L}\left(1 - [1 - P_L]^{0.5g_p}\right) - l_{l, p} \geq 0.0 \tag{10.61}$$

The value of g_f to be used is given as

$$g_f = \max\left(G_p e^{-0.860 LTC^{0.629}} - l_{l,p}, 0.0\right) \le g_{f,\max} \quad \text{(for single-lane approaches)} \quad (10.62)$$

or

$$g_f = \max\left(G_p e^{-0.882 LTC^{0.717}} - l_{l,p}, 0.0\right) \le g_{f,\max} \quad \text{(for multiple-lane approaches)} \quad (10.63)$$

where

$$LTC = \text{left-turn flow rate per cycle (veh/cycle)} = \frac{v_{lt} C}{3600} \quad (10.64)$$

$g_{f,\max}$ = maximum value of g_f
P_L = percentage of left-turning vehicles in the shared lane
$l_{l,p}$ = permitted start-up lost time
G_p = displayed green interval corresponding to g_p (sec)
g_p = effective green time for permitted left-turn operation (sec)
v_{lt} = left-turn demand flow rate (veh/h)

10.7.6 Adjusted Saturation Flow Rate for a Protected–Permitted Left Turn in an Exclusive Lane

The adjusted saturation flow rate for a protected-permitted left turn in an exclusive lane consists of two parts. The first part is the adjusted saturation flow obtained for the protected period by using the f_{lt} obtained from Eq. 10.15 and the second part is that obtained for the permitted period obtained from Eq. 10.53 with Eq. 10.54 in Section 10.7.4.

10.7.7 Adjusted Saturation Flow Rate for a Protected–Permitted Left Turn in a Shared Lane

This mode of operation stipulates certain conditions to ensure the safe and efficient operation of the signal. These conditions are:

- The left-turn phase should be set to minimum recall.
- The left-turn phase maximum green should be set at a value that is less than or equal to the minimum green setting for the adjacent through phase.
- The lead-lag phasing sequence should be used when the protected-permitted mode is used for both opposing approaches.
- The left-turn phase should not be assigned a vehicle detector.
- The vehicle detector in the shared lane should be assigned to the adjacent through movement phase.

Four possible saturation flow rates for four different time periods exist for this mode of operation during the effective green time. The saturation rates for the first three time periods are the same as those discussed in Section 10.7.5 for a shared left-turn and through-lane group with permitted operation. The fourth period is the protected period for the left turns, which is the duration of the effective green time for left-turn phase (g_l). It is obtained as $s_{sl4,pro-per}$ from Eq. 10.65.

$$s_{sl4,pro-per} = \frac{s_{th}}{1 + P_L(E_L - 1)} \quad (10.65)$$

where

$s_{sl4,pro-per}$ = saturation flow rate of a shared left-turn and through-lane group during period 4 (veh/h/ln)

s_{th} = saturation flow rate of an exclusive through lane, which is the base saturation flow for a through lane adjusted for lane width, heavy vehicles, grade, parking, buses, and area type

P_L = percentage of left-turning vehicles in the shared lane

The shared lane in an approach with multiple lanes also has an impact on the adjacent through lanes. This impact is taken into consideration by reducing the saturation flow rate for the adjacent through lanes by 9.0 percent.

10.7.8 Adjusted Saturation Flow Rate for a Protected Left- and Right-Turn Lane Group in a Shared Lane ($s_{lr,pro}$)

The saturation flow rate for a shared left- and right-turn lane group with a protected mode is obtained from Eq. 10.66.

$$s_{lr,pro} = \frac{s_{th}}{1 + P_L(E_L - 1) + P_R(E_R - 1)} \tag{10.66}$$

where

$s_{lr,pro}$ = saturation flow rate for a shared left- and right-turn lane group with a protected mode

P_L = percentage of left-turning vehicles in the shared lane

P_R = percentage of right-turning vehicles in the shared lane

E_R = equivalent number of through cars for a protected right-turning vehicle = 1.18

E_L = equivalent number of through cars for a protected left-turning vehicle = 1.05

10.8 LANE GROUP CAPACITY

The capacity of each lane group is required to further evaluate the quality of service for the intersections. The equations were developed by taking into consideration all the available opportunities for service during the cycle.

10.8.1 Capacity of an Exclusive Left-Turn Lane Group with a Protected Operation

$$c_{lt,ex,pro} = \frac{gs_{lt,ex,pro}}{C}N_L \tag{10.67}$$

where

$c_{lt,ex,pro}$ = capacity of a protected left-turn lane group with operation in an exclusive lane

$s_{lt,ex,pro}$ = saturation flow rate for an exclusive left-turn lane group with a protected operation

g = effective green time for the phase

C = cycle length (sec)

N_L = number of lanes in the lane group

The capacity for an exclusive right-turn lane group with a protected operation can also be determined from Eq. 10.67, using the appropriate saturation flow, the number of lanes and the effective green time.

10.8.2 Capacity of a Permitted Left–Turn Lane Group in Exclusive Lanes

$$c_{lt,ex,per} = \frac{\min(g_p - g_f, g_u)s_{lt,per,ex} + 3600n_s}{C}N_L \tag{10.68}$$

where

$c_{lt,ex,per}$ = capacity of a permitted left-turn lane operation on exclusive lanes
g_p = effective green time for permitted left-turn operation
g_f = duration of blockage until opposing queue clears
g_u = duration of permitted left-turn green time that is not blocked by an opposing queue.
$s_{lt,per,ex}$ = saturation flow rate of the permitted left-turn lane group (see section 10.7.4)
n_s = number of sneakers per cycle (usually taken as =2)
N_L = number of lanes in the lane group
C = cycle length (sec)

10.8.3 Capacity of a Shared Left–Turn Lane Group with Permitted Operation

The capacity for a shared left- and through-lane group is given as:

$$c_{sl,per} = \frac{g_{per}s_{sl,per} + 3600(1 + P_L)}{C} \tag{10.69}$$

where

$c_{sl,per}$ = capacity of shared lane group with permitted left-turn operation (veh/h/ln)
$s_{sl,per}$ = saturation flow rate for a shared left-turn and through-lane group with permitted operation (see Eq. 10.58)
g_{per} = effective green time for the permitted phase
C = cycle length (sec)
P_L = proportion of left turns in the shared lane

10.8.4 Capacity of an Exclusive Left–Turn Lane Group with a Protected–Permitted Operation

$$c_{lt,ex,pro+per} = \left(\frac{g_{lt}s_{lt,ex,pro+per}}{C} + \frac{\min(g_p - g_f, g_u)s_{lt} + 3600n_s}{C}\right)N_l \tag{10.70}$$

where

$c_{lt,ex,pro+per}$ = capacity of a protected + permitted left-turn operation on exclusive lanes
$s_{lt,ex,pro+per}$ = saturation flow rate of the protected + permitted left-turn lane group (see section 10.7.6)
g_p = effective green time for protected left-turn operation
g_f = duration of blockage until opposing queue clears
g_u = duration of permitted left-turn green time that is not blocked by an opposing queue.
s_{lt} = saturation flow rate for an exclusive lane group with a protected left-turn

n_s = number of sneakers per cycle (usually taken as =2)
N_L = number of lanes in the lane group
C = cycle length (sec)

10.8.5 Capacity of a Shared Through- and Left-Turn Lane Group with a Protected-Permitted Operation

$$c_{sl,pro+per} = \frac{g_l s_{sl4}}{C} + \frac{g_p s_{slt} + 3600(1 + P_L)}{C} \qquad (10.71)$$

where

$C_{sl,pro+per}$ = capacity of a protected + permitted left-turn operation on shared lanes
s_{sl4} = saturation flow rate of the protected + permitted left-turn lane group (see section 10.7.7)
g_p = effective green time for protected left-turn operation
g_l = effective green time of the left-turn phase
C = cycle length(sec)
P_L = percentage of left-turning vehicles in the shared lane
f_{slt} = saturated flow rate for a shared left-turn lane group with a protected operation

10.9 LEVEL OF SERVICE COMPUTATION FOR PEDESTRIAN MODE

The procedure for computing the LOS for pedestrian mode has certain limitations and is not applicable when any of the following conditions exist:

- Approach grades greater than 2 percent
- Existence of railroad crossings
- Unpaved sidewalks
- Uncontrolled channelized right turns on multiple lanes or high-speed operation

The procedure evaluates the performance of only a signalized intersection with respect to the pedestrian mode. The reader is referred to the *Highway Capacity Manual* for other intersection controls. The procedure evaluates each crosswalk and intersection corner and should be carried out on all crosswalks and intersection corners for a full evaluation of the intersection. Therefore, the variables used are specific to each intersection corner. It is assumed that each leg of the intersection has a crosswalk, unless pedestrian crossing is prohibited by signing.
The *performance measures* used in the procedure are:

- Corner circulation area
- Crosswalk circulation area
- Pedestrian delay
- Pedestrian LOS score

Corner circulation area reflects the area available to the pedestrians while waiting for the opportunity to cross, while *crosswalk circulation area* reflects the area available to the pedestrian while crossing the road on the crosswalk. These are given in terms of average space available to a pedestrian. Table 10.9 gives qualitative descriptions of different areas per pedestrian, showing that the higher the area, the better the perception by the pedestrian.
Pedestrian delay reflects the average time a pedestrian has to wait before a legal opportunity to cross the intersection occurs. The *LOS score* reflects the overall service

Table 10.9   Qualitative Description of Pedestrian Space

Pedestrian area (ft²/p)	Description
>60	Ability to move in desired path, no need to alter movements
>40–60	Occasional need to adjust path to avoid conflicts
>24–40	Frequent need to adjust path to avoid conflicts
>15–24	Speed and ability to pass slower pedestrians restricted
>8–15	Speed restricted, very limited ability to pass slower pedestrians
≤8	Speed severely restricted, frequent contact with other users

SOURCE: From *Highway Capacity Manual 2010*. Copyright, National Academy of Sciences, Washington, D.C. Reproduced with permission of the Transportation Research Board.

quality as perceived by the average pedestrian wishing to cross the approach. Table 10.2 shows the scores for different levels of service.

The computational process to obtain the LOS for the pedestrian mode consists of the following five steps:

Step 1:   Determine street corner circulation area
Step 2:   Determine crosswalk circulation area
Step 3:   Determine pedestrian delay
Step 4:   Determine pedestrian LOS score for intersection
Step 5:   Determine LOS

10.9.1 Step 1: Determine Street Corner Circulation Area

The computation involved in this step is for one corner at a time and should be repeated for all the corners involved in the analysis. It evaluates the time and space available with respect to the pedestrian demand at each corner of the intersection. It combines the limitations of the physical design in terms of the space available for pedestrians, and the limitations of the signal design in terms of the time available for pedestrians to cross the intersection. The combined effect of these is obtained by multiplying time and space to obtain a parameter known as the **time-space**. The associated crosswalks and walkways for crossing the minor and major streets are identified as A or B, as shown in Figures 10.5 and 10.6.

This step consists of the following three substeps:

- Compute the available time-space.
- Compute the holding-area waiting time.
- Compute the circulation time-space.

The **available time-space** is the product of the net corner area and the cycle length and obtained from the equation given in the HCM and shown as Eq. 10.72.

$$TS_{corner} = C(W_a W_b - 0.215R^2) \tag{10.72}$$

where

$$
\begin{aligned}
TS_{corner} &= \text{available corner time-space (ft}^2\text{-sec)} \\
C &= \text{cycle length (sec)} \\
W_a &= \text{total walkway width of Sidewalk A (ft)} \\
W_b &= \text{total walkway width of Sidewalk B (ft)} \\
R &= \text{radius of corner curb (ft)}
\end{aligned}
$$

Note: When $R > W_a$ or W_b, then R is taken as the smaller of W_a and W_b.

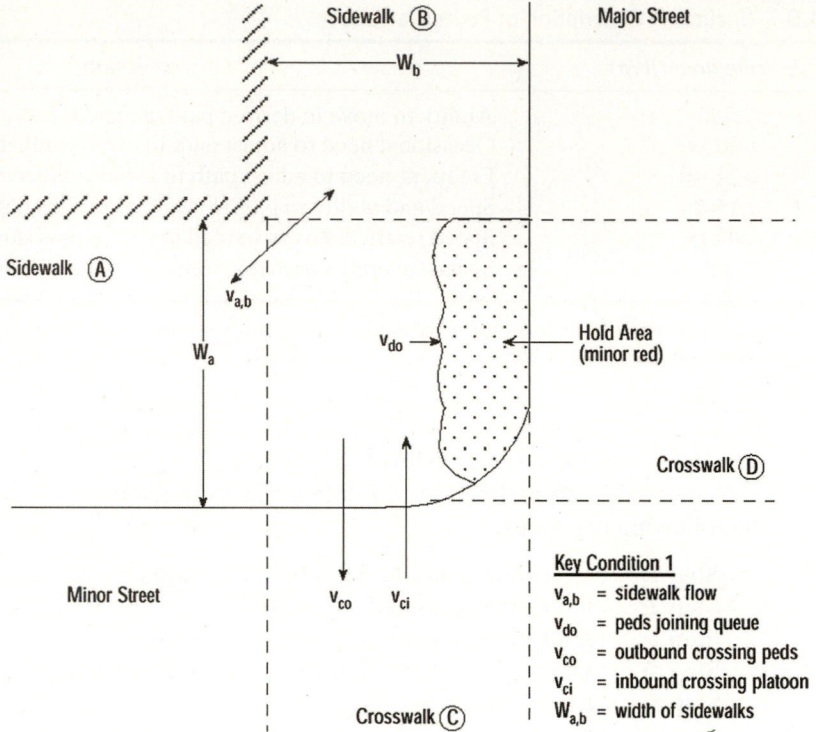

Figure 10.5   Condition 1: Minor-Street Crossing

SOURCE: From *Highway Capacity Manual 2010*. Copyright, National Academy of Sciences, Washington, D.C. Reproduced with permission of the Transportation Research Board.

The ***holding-area waiting time*** represents the average time a pedestrian waits at the corner before crossing the intersection. Pedestrian arrivals are assumed to be uniformly distributed. The two crossing conditions for minor-street and major-street crossings are shown in Figures 10.5 and 10.6, respectively. The HCM gives the following equations for computing the holding-area waiting time for pedestrians crossing the major street (condition 1):

$$Q_{tdo} = \frac{N_{do}(C - g_{walk,mi})^2}{2C} \qquad (10.73)$$

$$N_{do} = \frac{v_{do}}{3600} C \qquad (10.74)$$

where

Q_{tdo} = total time spent by pedestrians waiting to cross the major street during one cycle (p-sec)

N_{do} = number of pedestrians arriving at the corner each cycle to cross the major street (p)

$g_{walk,mi}$ = effective walk time for the phase serving the minor street through movement (sec)

= $Walk_{mi}$ + 4.0 (when pedestrian phase is either (a) actuated with a pedestrian signal head with rest-in-walk not enabled, or (b) pretimed with a pedestrian signal head

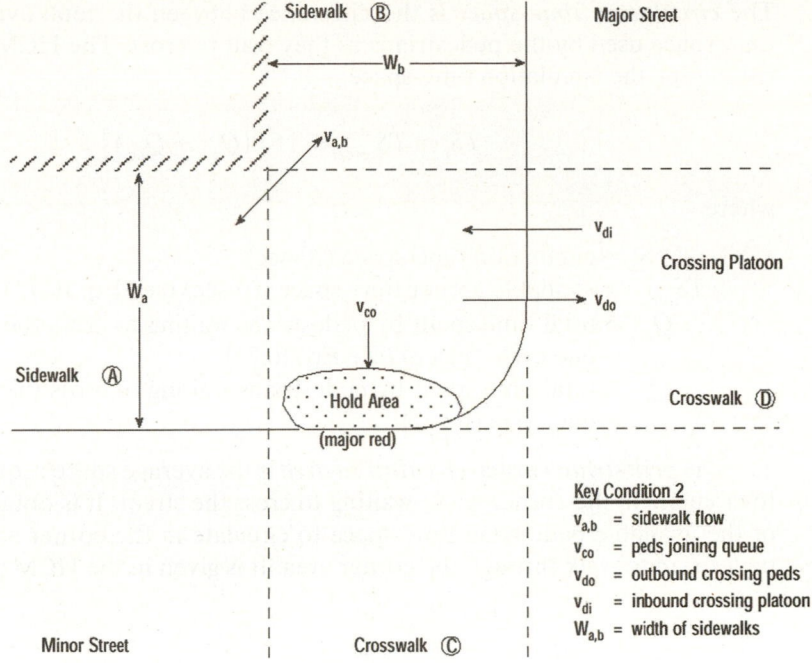

Figure 10.6   Condition 2: Major-Street Crossing

SOURCE: From *Highway Capacity Manual 2010*. Copyright, National Academy of Sciences, Washington, D.C. Reproduced with permission of the Transportation Research Board.

$$g_{walk,mi} = D_{p,mi} - Y_{mi} - R_{c,mi} \text{ (when pedestrian phase is actuated with a pedestrian signal}$$
   head and rest-in-walk enabled

$$g_{walk,mi} = D_{p,mi} - Y_{mi} - R_{c,mi} - PC_{mi} + 4 \text{ (when no pedestrian signal head exists)}$$

   C = cycle length (sec)

   v_{do} = flow rate of pedestrians arriving at the corner to cross the major street (p/h)

where

   $g_{walk,mi}$ = effective walk time for the phase serving the minor street through movement (sec)

   $walk,mi$ = pedestrian walk setting for the phase serving the minor street through movement (sec)

   PC_{mi} = pedestrian clear setting for the phase serving the minor street (sec)

   $Dp_{,mi}$ = duration of the phase serving the minor street through movement (sec)

   Y_{mi} = yellow change interval of the phase serving the minor street through movement (sec)

   Rc,mi = red clearance interval of the phase serving the minor street through movement (sec)

Note that Eqs. 10.73 and 10.74 are also used to compute the holding-area waiting time for pedestrians crossing the major street. For this case, *"co"* replaces *"do"* and *"mj"* replaces *"mi"*.

The *circulation time-space* is the difference between the total available time and the time-space used by the pedestrians as they wait to cross. The HCM gives Eq. 10.75 for computing the circulation time-space.

$$TS_c = TS_{corner} - [5.0(Q_{tdo} + Q_{tco})]$$ (10.75)

where

TS_c = circulation time-space (ft²-sec)
TS_{corner} = available corner time-space (ft²-sec) (see Eq. 10.72)
Q_{tdo} = total time spent by pedestrians waiting to cross the major street during one cycle (p-sec) (see Eq. 10.73)
Q_{tco} = total time spent by pedestrians waiting to cross the minor street during one cycle (p-sec)

The *pedestrian corner-circulation area* is the average space required for pedestrians to circulate in the corner while waiting to cross the street. It is obtained as the quotient of the available pedestrian time-space to circulate in the corner and the time used by pedestrian to walk through the corner area. It is given in the HCM as:

$$M_{corner} = \frac{TS_c}{4.0N_{tot}}$$ (10.76)

where

$$N_{total} = \frac{v_{ci} + v_{co} + v_{di} + v_{do} + v_{a,b}}{3600} C$$ (10.77)

M_{corner} = pedestrian corner-circulation area (ft²/p)
v_{ci} = flow rate of pedestrians arriving at the corner after crossing the minor street (p/hr)
v_{co} = flow rate of pedestrians arriving at the corner to cross the minor street (p/hr)
v_{di} = flow rate of pedestrians arriving at the corner after crossing the minor street (p/hr)
v_{do} = flow rate of pedestrians arriving at the corner to cross the major street (p/hr)
$v_{a,b}$ = flow rate of pedestrians traveling through corner from Sidewalk A or Sidewalk B or vice versa (p/hr)

10.9.2 Step 2: Determine Crosswalk Circulation Area

This step evaluates how effectively a crosswalk performs and should be carried out for all crosswalks at the intersection. It consists of the following six substeps:

- Establish pedestrian walking speed.
- Compute available time-space.
- Compute effective available time-space.
- Compute pedestrian service time.
- Compute crosswalk occupancy time.
- Compute pedestrian crosswalk circulation area.

The steps are described for crosswalk D, shown in Figure 10.6, for crossing the major street. Note that the appropriate subscript should be used for the other crosswalks.

The ***pedestrian walking speed*** is used in the evaluation of the corner and sidewalk performances. The HCM recommends an average speed of 4.0ft/sec^2 when 20 percent or less of the pedestrians are aged 65 years or older (elderly pedestrians), and 3.3 ft/sec when elderly pedestrians exceed 20 percent. The HCM also recommends that the average walking speed should be decreased by 0.3 ft/sec when the approach grade is 10 percent or greater.

The ***available time-space*** is computed using the HCM expression given as

$$TS_{cw} = L_d W_d g_{walk,mi} \tag{10.78}$$

where

TS_{cw} = available time-space in the crosswalk (ft^2/sec)
L_d = length of crosswalk D (ft)
W_d = effective width of crosswalk D (ft)
$g_{walk,mi}$ = effective walk time for the phase serving the minor-street through movement

The ***effective available time-space*** is obtained from the available time-space by adjusting the available time-space to account for the impact of turning vehicles on pedestrians. This impact depends on an assumed occupancy of a turning vehicle in the crosswalk by a vehicle, which is determined as the product of the vehicle-swept path (8 ft for most vehicles), the width of the crosswalk, and the time the vehicle occupies the crosswalk space (5 sec). The HCM gives the following equation for computing it:

$$TS^*_{cw} = TS_{cw} - TS_{tv} \tag{10.79}$$

where

TS^*_{cw} = effective available crosswalk time-space in the crosswalk (ft^2/sec)
TS_{cw} = available crosswalk time-space in the crosswalk (see Eq. 10.77) (ft^2/sec)

$TS_{tv} = 40N_{tv}W$ (Note that 40 is obtained from 8 ft × 5 sec) $\tag{10.80}$

N_{tv} = number of turning vehicles during the walk and pedestrian clear intervals

$$N_{tv} = \frac{v_{lt,per} + v_{rt} - v_{rtor}}{3600} C \tag{10.81}$$

C = cycle length (sec)
$v_{lt,per}$ = permitted left-turn demand flow rate onto the crosswalk under consideration
v_{rt} = right-turn demand flow rate onto the crosswalk under consideration
v_{rtor} = right-turn-on-red demand flow rate onto the crosswalk under consideration

Figure 10.7 illustrates the different vehicular movement volumes associated with the pedestrians crossing crosswalk D.

The ***pedestrian service time*** is the duration that starts at the time when the first pedestrian leaves the corner to the time the last pedestrian reaches the far side of the crosswalk. It is obtained from either Eq. 10.82 or Eq. 10.83, depending on the width of the crosswalk.

$$t_{ps,do} = 3.2 + \frac{L_d}{S_p} + 2.7 \frac{N_{ped,do}}{W_d} \quad \text{for } W_d > 10 \text{ ft} \tag{10.82}$$

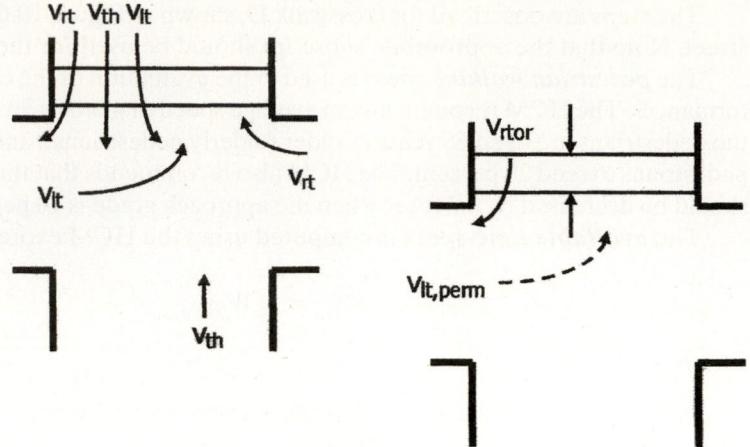

Figure 10.7   Movements Crossing Crosswalk D.

SOURCE: From *Highway Capacity Manual 2010*. Copyright, National Academy of Sciences, Washington, D.C. Reproduced with permission of the Transportation Research Board.

$$t_{ps,do} = 3.2 + \frac{L_d}{S_p} + (0.27N_{ped,do}) \quad \text{for } W_d \leq 10 \text{ ft} \tag{10.83}$$

$$N_{ped,do} = \frac{C - g_{Walk,mi}}{C} \tag{10.84}$$

where

$t_{ps,do}$ = service time for pedestrians that arrive at the corner to cross the major street (sec)

$N_{ped,do}$ = number of pedestrians waiting at the corner to cross the major street (p)

C = cycle length (sec)

S_p = pedestrian walking speed (ft/sec)

W_d = width of crosswalk

$g_{Walk,mi}$ = effective walk time for the phase serving the minor-street through movement

L_d = length of crosswalk (ft)

The ***crosswalk occupancy time*** is obtained by multiplying the number of pedestrians using the crosswalk by the pedestrian service time. It is given as

$$T_{occ} = t_{ps,do}N_{do} + t_{ps,di}N_{di} \tag{10.85}$$

where

T_{occ} = crosswalk occupancy time (p-sec)

$t_{ps,do}$ = service time for pedestrians that arrive at the corner to cross the major street (sec) (see Eq. 10.83)

$N_{ped,do}$ = number of pedestrians waiting at the corner to cross the major street (p) (see Eq. 10.84)

N_{di} = number of pedestrians that during each cycle cross the major street and get to the corner

$= \frac{v_{di}}{3600} C$, v_{di} is the flow rate of pedestrians arriving at the corner after crossing the major street (p/h)

The *pedestrian crosswalk circulation area* is obtained as the quotient of the time-space available for pedestrians to cross and the total occupancy time. It is given as

$$M_{cw} = \frac{TS^*_{cw}}{T_{occ}}$$

(10.86)

where

M_{cw} = crosswalk circulation area per pedestrian (ft²/p)
TS^*_{cw} = effective available crosswalk time-space in the crosswalk (ft²/sec) (see Eq. 10.79)
T_{occ} = total occupancy time (see Eq. 10.85)

The analyst can evaluate the performance of the crosswalk in the direction of travel under consideration by comparing the value obtained for the pedestrian crosswalk circulation area (M_{cw}) using Eq. 10.86 with those given in Table 10.9. The HCM suggests that for the analyst to have a complete indication of the performance of the crosswalk under consideration, the procedure should be carried out for the other direction along the crosswalk using the opposite corner.

10.9.3 Step 3: Determine Pedestrian Delay (d_p)

This step describes the procedure for determining the delay of the pedestrian while he/she is waiting to cross the major street. It is described for crosswalk D in Figure 10.5 and should be repeated for crosswalk C in Figure 10.6. with subscripts of "*mi*" changed to "*mj*". The delay is given by the HCM as

$$d_p = \frac{(C - g_{Walk,mi})^2}{2C}$$

(10.87)

where

d_p = pedestrian delay (sec/p)
C = cycle length (sec)
$g_{Walk,mi}$ = effective walk time for the phase serving the minor-street through movement

The delay obtained from Eq. 10.87 is applicable to both directions of crosswalk crossing. Note that oversaturation is not considered in this expression, as it has been shown that pedestrian delay while crossing a signalized intersection is not restricted by capacity even with pedestrian flow rates as high as 5000 p/h. The pedestrian delay when the crosswalk is closed is obtained as the sum of the delay obtained from Eq. 10.87 and twice the delay for the perpendicular crosswalk, also determined from Eq. 10.87. The HCM suggests that an evaluation of how the pedestrian complies with the signals at the crosswalk can be done by considering that pedestrians will very likely adhere to the pedestrian signal indication when the expected delay is 10 sec/p or less, but will most likely ignore the signals when the expected delay is higher than 30 sec/p. This step is repeated for each crosswalk.

10.9.4 Step 4: Determine Pedestrian LOS Score for Intersection

The LOS score for pedestrians is determined by using the HCM expression given as

$$I_{p,int} = 0.5997 + F_w + F_v + F_s + F_{delay}$$

(10.88)

$$F_w = 0.681(N_d)^{0.514} \tag{10.89}$$

$$F_v = 0.00569\left(\frac{v_{rtor,c} + v_{lt,per}}{4}\right) - N_{rtci,d}(0.0027n_{15,mj} - 0.1946) \tag{10.90}$$

$$F_s = 0.00013n_{15,mj}S_{85,mj} \tag{10.91}$$

$$F_{delay} = 0.0401Ln(d_{p,d}) \tag{10.92}$$

$$n_{15,mj} = \frac{0.25}{N_d}\sum_{i\in m_d}v_i \tag{10.93}$$

where

$I_{p,int}$ = pedestrian LOS score for intersection
F_w = cross-section adjustment factor
F_v = motorized vehicle volume adjusted factor
F_s = motorized vehicle speed adjustment factor
F_{delay} = pedestrian delay adjusted factor
N_d = number of traffic lanes crossed when traversing Crosswalk D
$S_{85,mj}$ = 85th-percentile speed at midsegment location on the major street (veh/h)
$d_{p,d}$ = pedestrian delay when traversing Crosswalk D (sec/p)
v_i = demand flow rate for movement i (veh/h) (see Figure 10.7 for the direction of movement i crossing Crosswalk D)
m_d = set of all automobile movements that cross Crosswalk D
$N_{rtci,d}$ = number of right-turn channelizing island along crosswalk D
$v_{rtor,c}$ = right-turn on red that is associated with the approach being crossed and that turns across the subject crosswalk
$n_{15,mj}$ = count of vehicles traveling on major street during a 15 min period.
$v_{lt,per}$ = left-turn flow rate that is associated with the left-turn movement that receives a green concurrently with the subject pedestrian crossing and turns across the subject crosswalk.

10.9.5 Step 5: Determine Pedestrian LOS

This step involves determination of the LOS of the pedestrian by comparing the LOS score computed in Section 10.9.4 with the threshold values given in Table 10.2. The procedure is computed for each crosswalk of interest at the intersection.

10.10 LEVEL OF SERVICE FOR BICYCLE MODE

This procedure provides for the determination of the bicycle LOS at each approach of the intersection. It is assumed that the bicycle is traveling in the same direction as the adjacent traffic. The bicycle delay is computed only for an approach that has an on-street bicycle lane, or a wide shoulder that is available for bicyclists to use as a bicycle lane. The procedure consists of the following three steps:

- Determine bicycle delay.
- Determine bicycle LOS score for the intersection.
- Determine LOS for bicycles.

10.10.1 Step 1: Determine Bicycle Delay

This step consists of following two subsets:

- Compute bicycle lane capacity.
- Compute bicycle delay.

Compute Bicycle Lane Capacity

Several values have been suggested for the capacity and the base saturation flow rate for bicycle lanes, but unfortunately, adequate data are not available for the base conditions for bicycle lanes at intersections. Therefore, the information necessary to calibrate appropriate adjustment factors is not available. Although a base saturation flow rate of 2600 bicycles/h has been reported, the HCM suggests a value of 2000 bicycles/h until the appropriate adjustment factors can be developed. This value assumes that automobiles turning right will yield the right-of-way to through bicyclists. The HCM also recommends that local observation be conducted, as in some instances, when drivers of right-turning vehicles are aggressive, the saturation flow rate of 2000 bicycles/hr may not be achieved. The HCM gives the capacity of a bicycle lane as

$$c_b = s_b \frac{g_b}{C} \tag{10.94}$$

where

c_b = capacity of bicycle lane (bicycles/h)
s_b = saturation flow rate of the bicycle lane
C = cycle length (sec)
g_b = effective green time for the bicycle lane (sec)

The effective green time for the bicycle lane can be taken as the same as the effective green time for the motor vehicle traffic that is adjacent to the bicycle lane. It can be computed as

$$g_p = D_p - l_1 - l_2$$

where

D_p = green phase duration for the adjacent vehicle stream
l_1 and l_2 = lost times for the adjacent vehicle stream

Compute Bicycle Delay

Because bicyclists under normal conditions will avoid routes where they will experience oversaturated conditions, the components of delay associated with oversaturation (incremental delay and initial queue delay) are assumed to be negligible. The bicycle delay is therefore given as

$$d_b = \frac{0.5C(1 - g_b/C)^2}{1 - \min\left[\dfrac{v_{bic}}{c_b}, 1.0\right]\dfrac{g_b}{C}} \tag{10.95}$$

where

d_b = bicycle delay (sec/bicycle)
v_{bic} = bicycle flow rate (bicycles/h)

g_b = effective green time for the bicycle lane (sec)

c_b = capacity of bicycle lane (bicycles/h) (see Eq. 10.94)

C = cycle length (sec)

10.10.2 Step 2: Determine Bicycle LOS Score for Intersection

The score for bicycles at an intersection is obtained from a series of equations given in the HCM. These are:

$$I_{b,int} = 4.1324 + F_w + F_v \tag{10.96}$$

$$F_w = 0.0153Wcd - 0.2144W_t \tag{10.97}$$

$$F_v = 0.0066\frac{v_{lt} + v_{th} + v_{rt}}{4N_{th}} \tag{10.98}$$

$$W_t = W_{ol} + W_{bl} + I_{pk}W_{os}^* \tag{10.99}$$

where

$I_{b,int}$ = bicycle LOS score for intersection

W_{cd} = curb-to-curb width of the cross street (ft)

W_t = total width of outside through lane, bicycle lane, and paved shoulder (ft)

v_{lt} = left-turn demand flow rate (veh/h)

v_{th} = through demand flow rate (veh/h)

v_{rt} = right-turn demand flow rate (veh/h)

W_{ol} = width of the outside through lane (ft)

W_{bl} = width of the bicycle lane (ft) = 0.0 if bicycle lane not provided

I_{pk} = indicator variable for on-street parking occupancy

     = 0 if $p_{pk} > 0.0$, otherwise 1

p_{pk} = proportion of on-street parking occupied (decimal)

W_{os} = width of paved outside shoulder lane (ft)

W_{os}^* = adjusted width of paved outside shoulder

     = $W_{os} - 1.5 \geq 0.0$, if curb is present

     = W_{os} if curb is not present

Note that the effect of off-street parking and activity on the intersection approach and departure legs on which the bicycle under consideration travels is taken into consideration by the inclusion of the variable "proportion of on-street parking occupied."

10.10.3 Step 3: Determine LOS

The LOS for the bicycle mode on the approach under consideration is then determined by comparing the LOS factor obtained in Step 2 with the threshold values given in Table 10.2.

10.11 QUICK ESTIMATION METHOD (QEM)

The HCM gives a method that can be used to determine the LOS at an intersection when only minimal data are available. This is known as the *quick estimation method (QEM)* and it can also be used to determine the critical volume-to-capacity ratio (X_{cr}) and signal timing. The data requirements for this method are shown in Table 10.10. When some

Table 10.10   Input Data Requirements for Quick Estimation Method

Data Item	Comments
Volumes	By movement as projected
Lanes	Left, through, or right: exclusive or shared
Adjusted saturation flow rate	Includes all adjustments for PHF, CBD, grades, etc.
Left-turn phasing treatment (phasing plan)	Use actual treatment, if known. See discussion of phasing plan development
Cycle length (minimum and maximum)	Use actual value if known. May be estimated by using control delay and LOS worksheet
Lost time	May be estimated by using control delay and LOS worksheet
Green times	Use actual values. May be estimated by using control delay and LOS worksheet
Coordination	Isolated intersection versus intersection influenced by upstream signals
Peak hour factor	Use default value of 0.9 if not known
Parking	On-street parking is or is not present
Area type	Signal is or is not in CBD

SOURCE: From *Highway Capacity Manual 2010*. Copyright, National Academy of Sciences, Washington, D.C. Reproduced with permission of the Transportation Research Board.

of these data elements are not available, default values given in Chapter 8 may be used. When default values for specific data elements are not available, these most likely are site-specific and should be obtained by field studies. Figure 10.8 is an input worksheet that can be used for the QEM.

The QEM method consists of the following steps:

Step 1.   Determine left-turn treatment
Step 2.   Determine lane volume
Step 3.   Determine signal timing
Step 4.   Determine critical intersection volume-to-capacity ratio
Step 5.   Determine control delay

10.11.1 Determine Left–Turn Treatment

Since permitted left-turn movements are not included in the traffic analysis procedure of this method, the HCM suggests that if protected left-turn treatments are not assumed, this may result in a predicted higher quality of the intersection operation and optimistic assessment of the volume-to-capacity ratio. The alternative left-turn phase treatments used in the procedure are (i) no left-turn (i.e., permitted only), (2) left-turn phase (i.e., protected), and (3) split phasing (i.e., not opposed). Figure 10.9 is a worksheet that can be used to determine the left-turn treatment for each approach at the intersection. Four guidelines that are used to determine the left-turn treatment are given as

(i)   Use left-turn phase when the number of left-turn lanes on the approach > 1.
(ii)   Use left-turn phase when the unadjusted left-turn volume > 240 veh/h.
(iii) Use left-turn phase when the cross product of the unadjusted left-turn and opposing mainline volumes (opposing through plus right-turn volumes) > minimum values given in Figure 10.9. Note that the opposing right-turn volumes may be excluded when the drivers of the left turns under consideration can safely make the turn without any interference from the opposing right turns. Such a condition

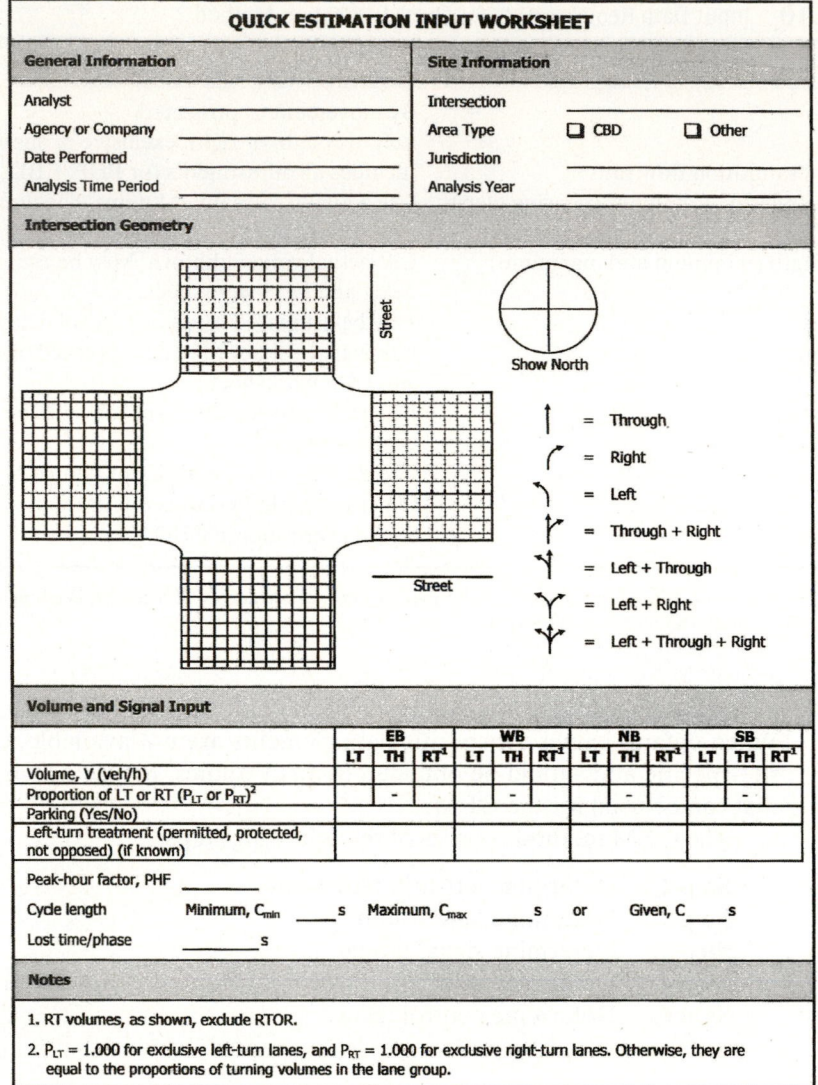

Figure 10.8   Quick Estimation Input Worksheet

SOURCE: From *Highway Capacity Manual 2010*. Copyright, National Academy of Sciences, Washington, D.C. Reproduced with permission of the Transportation Research Board.

may exist at an intersection where the opposing right-turn volume is on an exclusive right lane (i.e., when the number of receiving lanes on the cross street is more than one).

(iv) Use left-turn phase when unadjusted left-turn volume > sneaker capacity and equivalency factor > 3.5 (see Step 2).

10.11.2 Determine Lane Volume

Figure 10.10 is used to specify the flow rates in veh/h/ln of individual lanes at each approach of the intersection. The lane volumes are computed differently based on the specific left-turn treatment that is assigned to the left turn (i.e., permitted, protected, and

QUICK ESTIMATION LEFT-TURN TREATMENT WORKSHEET

General Information

Description _____

Check # 1. Left-Turn Lane Check

Approach	EB	WB	NB	SB
Number of left-turn lanes				
Protect left turn (Y or N)?				

If the number of left-turn lanes on any approach exceeds 1, then it is recommended that the left turns on that approach be protected. Those approaches with protected left turns need not be evaluated in subsequent checks.

Check # 2. Minimum Volume Check

Approach	EB	WB	NB	SB
Left-turn volume				
Protect left turn (Y or N)?				

If left-turn volume on any approach exceeds 240 veh/h, then it is recommended that the left turns on that approach be protected. Those approaches with protected left turns need not be evaluated in subsequent checks.

Check # 3. Minimum Cross-Product Check

Approach	EB	WB	NB	SB
Left-turn volume, V_L (veh/h)				
Opposing mainline volume, V_o (veh/h)				
Cross-product ($V_L * V_o$)				
Opposing through lanes				
Protected left turn (Y or N)?				

Minimum Cross-Product Values for Recommending Left-Turn Protection

Number of Through Lanes	Minimum Cross-Product
1	50,000
2	90,000
3	110,000

If the cross-product on any approach exceeds the above values, then it is recommended that the left turns on that approach be protected. Those approaches with protected left turns need not be evaluated in subsequent checks.

Check # 4. Sneaker Check

Approach	EB	WB	NB	SB
Left-turn volume, V_L (veh/h)				
Sneaker capacity, c_s (veh/h) $c_s = 7200/C$				
Equivalence factor, E_{L1}				
Protected left turn (Y or N)?				

If the equivalence factor is 3.5 or higher (computed in the Quick Estimation Lane Volume Worksheet) and the unadjusted left turn is greater than the sneaker capacity, then it is recommended that the left turns on that approach be protected.

Notes

1. If any approach is recommended for left-turn protection but the analyst evaluates it as having permitted operation, then this quick estimation method may give overly optimistic results. The analyst should instead use the methodology described in   Section 10.4
2. All volumes used in this worksheet are unadjusted hourly volumes.

Figure 10.9   Quick Estimation Left-Turn Treatment Worksheet

not opposed). One of the three columns is completed for each approach depending on the left-turn treatment assigned to the approach. The procedure under this step consists of the following substeps:

(a) Compute lane volume for right-turn movement
(b) Compute volume for left-turn movement
(c) Compute through-movement volume
(d) Compute lane volume for through movement with exclusive turn lane
(e) Compute lane volume for through movement with shared lane

Compute Lane Volume Right-Turn Movement

In this substep the right-turn volume (V_R) is first determined by subtracting the right-turn-on-red volume from the right-turn traffic. The equivalent through movement per lane (V_{RT}) is then obtained as the quotient of V_{RT} and the product of the right-turn adjustment factor (f_{RT}) and the number of exclusive right-turn lanes, as shown in Figure 10.10. Note that the value of the right-turn adjustment factor f_{RT} is 0.85 for a single lane or 0.75 for two lanes.

QUICK ESTIMATION LANE VOLUME WORKSHEET		
General Information		
Description/Approach		
Right-Turn Movement		
	Exclusive RT Lane	**Shared RT Lane**
RT volume, V_R (veh/h)		
Number of exclusive RT lanes, N_{RT}		use 1
RT adjustment factor,[1] f_{RT}		
RT volume per lane, V_{RT} (veh/h/ln) $V_{RT} = \dfrac{V_R}{(N_{RT} \times f_{RT})}$		
Left-Turn Movement		
LT volume, V_L (veh/h)		
Opposing mainline volume, V_o (veh/h)		
Number of exclusive LT lanes, N_{LT}		
LT adjustment factor,[2] f_{LT}		
LT volume per lane,[3] V_{LT} (veh/h/ln) $V_{LT} = \dfrac{V_L}{(N_{LT} \times f_{LT})}$	Permitted LT, use 0 ___ Protected LT ___	Not Opposed LT
Through Movement		
	Permitted LT	**Protected LT**	**Not Opposed LT**
Through volume, V_T (veh/h)			
Parking adjustment factor, f_p			
Number of through lanes, N_{TH}			
Total approach volume,[4] V_{tot} (veh/h) $V_{tot} = \dfrac{V_{RT}(\text{shared}) + V_T + V_{LT(\text{not.opp})}}{f_p}$			
Through Movement with Exclusive LT Lane			
Through volume per lane, V_{TH} (veh/h/ln) $V_{TH} = \dfrac{V_{tot}}{N_{TH}}$			
Critical lane volume,[5] V_{CL} (veh/h) Max[V_{LT}, V_{RT} (exclusive), V_{TH}]			
Through Movement with Shared LT Lane			
Proportion of left turns, P_{LT}		Does not apply	Does not apply
Equivalence factor, E_{L1}		Does not apply	Does not apply
Shared lane LT adjustment factor, f_{DL}			use 1.0
Through volume per lane, V_{TH} (veh/h/ln) $V_{TH} = \dfrac{V_{tot}}{(N_{TH} \times f_{DL})}$			
Critical lane volume,[5] V_{CL} (veh/h) Max[V_{RT} (exclusive), V_{TH}]			

Notes

1. For RT shared or single lanes, use 0.85. For RT double lanes, use 0.75.
2. For LT single lanes, use 0.95. For LT double lanes, use 0.92. For a one-way street or T-intersection, use 0.85 for one lane and 0.75 for two lanes.
3. For unopposed LT shared lanes, $N_{LT} = 1$.
4. For exclusive RT lanes, V_{RT} (shared) = 0. If not opposed, add V_{LT} to V_T and set $V_{LT(\text{not opp})} = 0$.
5. V_{LT} is included only if LT is unopposed. V_{RT} (exclusive) is included only if RT is exclusive.

Figure 10.10 Quick Estimation Lane Volume Worksheet

Compute Lane Volume for Left–Turn Movement

The left-turn volume (V_{LT}) is determined in a similar way as the right-turn volume. In this case, the left-turn volume (V_L) replaces the right-turn volume and the left-turn adjustment factor (f_{LT}) replaces the right-turn adjustment factor, as shown in Figure 10.10. The left-turn volume (V_L) is also reduced by two vehicles per cycle to account for the impact of sneakers when the mode of operation of the left turns is protected-permitted on an exclusive left-turn lane. The left-turn adjustment factor is used for either when the left turn operates in an exclusive mode or when unopposed by a conflicting movement. When the left turn is in a single lane, then f_{LT} is 0.95, and 0.92 when it is in two lanes. The interference of pedestrians should be considered when the left-turn volume is unopposed because of a one-way street or a T-intersection. Under these geometric conditions the left-turn adjustment factor is reduced to 0.85 for one lane and 0.75 for two lanes. Note also that the opposing mainline volume V_o is obtained as the difference between the total approach volume and the volume from exclusive lanes or from a single-lane approach. The HCM suggests that in this step, a left-turn volume that is less than four vehicles per cycle should be avoided, as this may result in unreasonable cycle lengths. Note, however, that when the left turns operate with a permitted mode, this volume is zero as the procedure does not consider the needs of permitted left-turn movements.

Compute Through–Movement Volume

The through movement (V_{th}) of the approach is recorded in the appropriate cell in Figure 10.10, based on the mode of operation of the left turns (permitted, protected, or not opposed). Also, the value of V_{th} depends on the lane configuration of the approach as shown in Figure 10.10. The parking adjustment factor (f_p) is determined from Eq. 10.5, which is repeated here as Eq. 10.100.

$$f_p = \frac{N_{th} - 0.1 - \dfrac{18N_m}{3600}}{N_{th}} \geq 0.050 \tag{10.100}$$

where

N_m = parking maneuver rate adjacent to the lane group (maneuvers/h)
N_{th} = number of through lanes (shared or exclusive)

Note that all lanes that serve the through movement should be included as part of the through lanes, but all exclusive turn lanes should not count as part of the through lanes.

Compute Lane Volume for Through Movement with Exclusive Turn Lane (V_{th})

When the through volume is on an exclusive lane or lanes, the through-lane volume is obtained as the quotient of the total through volume and the number of through lanes. This is typically the critical lane volume (V_{CL}), except in cases where there is no opposing traffic to the left turns or the right-turn movement is in an exclusive lane and one of these movements has a higher *v/s* ratio than the through-lane movement. When this condition exists, then the critical lane volume is the highest of all lane volumes (through-lane volume, left-lane volume, and exclusive right-lane volume).

Compute Lane Volume for Shared Lane through Left–Turn Movement

In the case of shared left-turn lanes, it is first necessary to determine the left-turn equivalence factor (E_{L1}) by either using Table 10.11 or computing it from the HCM expression given in Eq. 10.101. When the opposing flow is greater than 1200 veh/h

$$E_{L1} = \frac{s_o}{s_{p,lf}} - I_{sh} \tag{10.101}$$

$$s_{p,lf} = \frac{v_o e^{-v_o t_{cg}/3600}}{1 - e^{-v_o t_{fh}/3600}} \tag{10.102}$$

where

E_{L1} = equivalent number of through cars for a permitted left-turn vehicle
v_o = opposing demand flow rate (veh/h)
s_o = base saturation flow rate (pc/h/ln)
t_{cg} = critical headway = 4.5 sec
I_{sh} = indicator variable for shared lane = 1 for a shared lane
= 0 for an exclusive lane
t_{flt} = follow-up headway (sec)
= 4.5 (left turn in a shared lane)
= 2.5 (left turn in an exclusive lane)

An equivalence factor greater than 3.5 suggests that the capacity of the left turn is obtained predominantly from sneakers. When this occurs and the left-turn volume per cycle is also greater than two, there is a reasonable probability that adequate capacity will not be available if either a protected- or protected-permitted left-turn phase is not provided for the movement. The left-turn adjustment factor for the QEM is then obtained by using the appropriate HCM expression from the set of equations given in Eqs. 10.102 through 10.106.

Table 10.11   Through-Car Equivalent (E_{L1}) for Shared Through- and Left-Turn Movement

Type of Left Turn	Through-Car Equivalent (E_{L1}) as a Function of Opposing Flow Rate (veh/h) V_o						
	1	200	400	600	800	1000	1200[a]
Shared	1.4	1.7	2.1	2.5	3.1	3.7	4.5
Exclusive	1.3	1.6	1.9	2.3	2.8	3.3	4.0

[a]Use Eq. 10.101 with Eq. 10.102 for opposing flow in excess of 1200 veh/h.
Note: V_o must be > 0.1 veh/h.
SOURCE: From *Highway Capacity Manual 2010*. Copyright, National Academy of Sciences, Washington, D.C. Reproduced with permission of the Transportation Research Board.

For Permitted Left-Turn
Lane groups with two or more Lanes

$$f_{DL} = \frac{(N_{th} - 1) + e^{-(N_{th}V_L E_{L1})/3600}}{N_{th}} \qquad (10.103)$$

Note that the left-turn adjustment factor for the QEM is subjected to a minimum value obtained from Eq. 10.104. This minimum value is applicable when the left-turn volumes are low, with no left turns arriving during some cycles.

$$f_{DL(\min)} = \frac{(N_{th} - 1) + e^{-(V_L C_{\max})/3,600}}{N_{th}} \qquad (10.104)$$

Lane groups with only one lane for all movements

$$f_{DL} = e^{-0.02(E_{L1} + 10P_{LT})V_L C_{\max}/3600} \qquad (10.105)$$

For Protected-Permitted Left Turn (one direction only)

$$f_{DL} = \frac{1}{1 + \left(\dfrac{P_{LT}(235 + 0.435V_o)}{1400 - V_o} \right)} \quad (V_o) < 1220 \text{ veh/h} \qquad (10.106)$$

$$f_{DL} = \frac{1}{1 + 4.525P_{LT}} \quad (V_o) \geq 1220 \text{ veh/h} \qquad (10.107)$$

The equivalent through-lane volume (V_{th}) is then obtained as the quotient of the total approach volume and the product of the total number of through lanes and the shared left-turn adjustment factor; that is,

$$V_{th} = (\text{total approach volume})/(N_{th} \times f_{DL})$$

The higher of the computed through-lane volume and the right-turn volume is the critical lane volume (V_{CL}). It is necessary to investigate the necessity of providing a left-turn phase at this juncture if one or more left-turn movements operate in the permissive mode.

10.11.3 Determine Signal Timing

This step involves the determination of a feasible signal timing plan, which may not necessarily be the optimal timing plan but which can be used to estimate the delay at the intersection. The procedure consists of the following steps.

- Develop phasing plan
- Compute critical phase volume and lost time
- Compute critical sum and cycle lost time
- Compute cycle length
- Compute green time

The results obtained from executing these steps are recorded in the control delay and level of service worksheet shown in Figure 10.11.

10.11.4 Develop Phasing Plan

Table 10.12 gives alternative phase plans from which the analyst can select an appropriate plan. The plan should be based on the left-turn treatment specified by the analyst and the left-turn treatments that have been identified in the treatment worksheet as being dominant. The appropriate movement codes are then recorded in the first two sections of the control-delay and level-of-service worksheet shown in Figure 10.11.

QUICK ESTIMATION CONTROL DELAY AND LOS WORKSHEET

General Information

Description _____

East-West Phasing Plan

Selected plan _____	Phase No. 1	Phase No. 2	Phase No. 3
Movement codes			
Critical phase volume, CV (veh/h)			
Lost time/phase, t_L (s)			

North-South Phasing Plan

Selected plan _____	Phase No. 1	Phase No. 2	Phase No. 3
Movement codes			
Critical phase volume, CV (veh/h)			
Lost time/phase, t_L (s)			

Intersection Status Computation

Critical sum, CS (veh/h) $CS = \Sigma CV$	
Lost time/cycle, L (s) $L = \Sigma t_L$	
Reference sum flow rate RS (veh/h)[1]	
Cycle length, C (s) $C_{min} \le C \le C_{max}$ $C = C_{max}$ when $CS \ge RS$ $C = \dfrac{L}{1 - \dfrac{CS}{RS}}$	
Critical v/c ratio, X_c $X_c = \dfrac{CS}{1700\, PHF\, f_a\left(1-\dfrac{L}{C}\right)}$	
Intersection status (relationship to capacity)	Under ___ Near ___ At ___ Over ___

Green Time Calculation

East-West Phasing	Phase No. 1	Phase No. 2	Phase No. 3
Green time, g (s) $g = \left[(C-L)\left(\dfrac{CV}{CS}\right)+t_L\right]$			
North-South Phasing	Phase No. 1	Phase No. 2	Phase No. 3
Green time, g (s) $g = \left[(C-L)\left(\dfrac{CV}{CS}\right)+t_L\right]$			

Control Delay and LOS

	EB	WB	NB	SB
Lane group				
Lane group adjusted volume from Lane Volume worksheet, V (veh/h)				
Green ratio, g/C				
Lane group saturation flow rate, s (veh/h) s = RS * number of lanes in lane group				
v/c ratio, X $X = \dfrac{V/s}{g/C}$				
Lane group capacity, c (veh/h) $c = \dfrac{V}{X}$				
Progression adjustment factor, PF				
Uniform delay, d_1 (s/veh)				
Incremental delay, d_2 (s/veh)				
Control delay, $d = d_1(PF) + d_2$ (s/veh)				
Delay by approach, d_A (s/veh) $\dfrac{\Sigma(d)(V)}{\Sigma V}$				
Approach flow rate, V_A (veh/h)				
Intersection delay, d_I (s/veh) $d_I = \dfrac{\Sigma(d_A)(V_A)}{\Sigma V_A}$		Intersection LOS (Exhibit 18-4)		

Notes

1. RS = 1530 x PHFx f_a, where f_a is area adjustment factor (= 0.90 for CBD and 1.0 for all other area types).

Figure 10.11   Quick Estimation Control-Delay and Level-of-Service Worksheet

SOURCE: From *Highway Capacity Manual 2010*. Copyright, National Academy of Sciences, Washington, D.C. Reproduced with permission of the Transportation Research Board.

Table 10.12   Phase Plan for Quick Estimation Method

Phase Plan	Eastbound	Westbound	Northbound	Southbound
1a	Permitted	Permitted	Permitted	Permitted
1b	Permitted	Not opposed	Permitted	Not opposed
1c	Not opposed	Permitted	Not opposed	Permitted
2a	Permitted	Protected	Permitted	Protected
2b	Protected	Permitted	Protected	Permitted
3a	Protected[a]	Protected	Protected[a]	Protected
3b	Protected	Protected[a]	Protected	Protected[a]
4	Not opposed	Not opposed	Not opposed	Not opposed

[a]Dominant left turn of each opposing movement.
SOURCE: From *Highway Capacity Manual 2010*. Copyright, National Academy of Sciences, Washington, D.C. Reproduced with permission of the Transportation Research Board.

10.11.5 Compute Critical Phase Volume and Lost Time

As discussed in Chapter 8, the critical phase volume for a given phase is that which requires the longest green time during the phase. Table 10.13 shows how the critical lane volume can be selected based on the movement code. The movement code indicates the movements that have the right-of-way during the given phase. For example, the code EBWBTH shows that the eastbound and westbound through movements have the right-of-way during the subject phase. The lost time (l_t) for each phase is also given in Table 10.13. The appropriate lost time is selected and recorded in the first two sections of the control-delay and level-of-service worksheet (Figure 10.11). Note that whenever a movement starts and stops during a given phase, a lost time is experienced during that phase.

10.11.6 Compute Critical Sum and Cycle Lost Time

At the completion of computing the critical phase volume and lost times for all phases, the critical sum (CS) of the critical phase volumes is then recorded in the third section of worksheet as shown in Figure 10.11. The cycle lost time (L) is also determined as the sum of the lost times of each of the critical phases and recorded in the worksheet as shown in Figure 10.11.

10.11.7 Compute Cycle Length

If a specific cycle length has been selected for the analysis, then it is not necessary to perform this step. If a cycle length has not been selected, the cycle length can be computed by using the HCM expression given in Eq. 10.108.

$$C = \frac{L}{1 - \left(\dfrac{CS}{RS}\right)} \tag{10.108}$$

Table 10.13 Phase Plan Summary for Quick Estimation

Phase Plan	Phase No.	Lost Time (s)	East–West Movement Code	East–West Critical Volume	North–South Movement Code	North–South Critical Volume
1a, 1b, 1c	1	4	EBWBTH	Max(EBTH, WBTH, WBLT)	NBSBTH	Max(NBTH, NBLT, SBTH, SBLT)
2a	1	4	WBTHLT	WBLT	SBTHLT	SBLT
	2	4	EBWBTH	Max(WBTH-WBLT, EBTH)	NBSBTH	Max(SBTH-SBLT, NBTH)
2b	1	4	EBTHLT	EBLT	NBTHLT	NBLT
	2	4	EBWBTH	Max(EBTH-EBLT, WBTH)	NBSBTH	Max(NBTH-NBLT, SBTH)
3a	1	4	EBWBLT	WBLT	NBSBLT	SBLT
	2	0	EBTHLT	EBLT-WBLT	NBTHLT	NBLT-SBLT
	3	4	EBWBTH	Max(WBTH, EBTH-(EBLT-WBLT))	NBSBTH	Max(SBTH, NBTH-(NBLT-SBLT))
3b	1	4	EBWBLT	EBLT	NBSBLT	NBLT
	2	0	WBTHLT	WBLT-EBLT	SBTHLT	SBLT-NBLT
	3	4	EBWBTH	Max(EBTH, WBTH-(WBLT-EBLT))	NBSBTH	Max(NBTH, SBTH-(SBLT-NBLT))
4	1	4	EBTHLT	Max(EBTH, EBLT)	NBTHLT	Max(NBTH, NBLT)
	2	4	WBTHLT	Max(WBTH, WBLT)	SBTHLT	Max(SBTH, SBLT)

SOURCE: From *Highway Capacity Manual 2010*. Copyright, National Academy of Sciences, Washington, D.C. Reproduced with permission of the Transportation Research Board.

where

$C_{min} \leq C \leq C_{max}$

$C = C_{max}$ when CS $\geq$ RS

C = cycle length (sec)

C_{min} = minimum cycle length; at least equal to time for pedestrian crossing; use 60 sec if no data available (sec)

C_{max} = maximum cycle length as set by local jurisdiction (sec)

L = cycle lost time

CS = critical sum (veh/h)

RS = reference sum flow rate = $1530 \times PHF \times f_a$ (veh/h)

PHF = peak hour factor

f_a = adjustment factor for area type
= 0.9 for CBD
= 1.00 for non-CBD

Note that the reference sum (RS) is estimated to be the theoretical maximum volume that can go through the intersection with a continuous green. It is determined with the objective of achieving a volume-to-capacity ratio of 0.9 for all critical movements. Using a theoretical saturation flow rate of 1700 pc/h/ln and a *v/c* ratio of 0.9 gives a value that is 1530 pc/h/ln.

10.11.8 Compute Effective Green Time (g_e)

The basic principle used in determining the effective green time (g_e) in the QEM is that the critical movements will all have the same volume-to-capacity ratio v/c. The effective green time for each phase is computed as

$$g_{e,i} = (C - L)\frac{CV_i}{CS} \tag{10.109}$$

where

$g_{e,i}$ = effective green time for critical phase i (sec)
C = cycle length (sec)
L = cycle lost time
CV_i = flow rate for critical phase i (veh/h)
CS = critical sum (veh/h)

Note that the effective green times computed for the critical phase using this procedure may not result in the minimum delay for the intersection.

10.11.9 Determine Critical Intersection Volume–to–Capacity Ratio (X_c)

The intersection volume-to-capacity ratio is dependent on the critical sum, the total lost time per cycle, the cycle length and the area type adjustment factor. It is obtained by using the HCM expression given as

$$X_c = \frac{CS}{1700PHFf_a\left(1 - \dfrac{L}{C}\right)} \tag{10.110}$$

where

(X_c) = critical intersection volume-to-capacity ratio
CS = critical sum (veh/h)
L = cycle lost time
PHF = peak hour factor
C = cycle length (sec)
f_a = adjustment factor for area type
= 0.9 for CBD
= 1.00 for non-CBD

Note that in addition to considering the critical volume-to-capacity ratio when determining the delay (as will be shown in the next section), critical volume-to-capacity ratio is also used to evaluate the operational status of the whole intersection. Although specific levels of service are not assigned to different values of X_c, qualitative evaluations are given for different ranges, as shown in Table 10.14.

10.11.10 Determine Control Delay

In this process, the control delay consists of two components: the uniform delay and the incremental delay. The uniform delay is given as

$$d_{1i} = \frac{0.5C(1 - g_i/C)^2}{1 - [\min(1, X_i)g_i/C]} \tag{10.111}$$

Table 10.14   Intersection Status Criteria for Quick Estimation Method

Critical Volume-to-Capacity Ratio (X_c)	Relationship to Capacity
<0.85	Under capacity
>0.85–0.95	Near capacity
>0.95–1.00	At capacity
>1.00	Over capacity

SOURCE: From *Highway Capacity Manual 2010*. Copyright, National Academy of Sciences, Washington, D.C. Reproduced with permission of the Transportation Research Board.

and the incremental delay is given as

$$d_{2i} = 900T\left[(X_i - 1) + \sqrt{(X_i - 1)^2 + \frac{4X_i}{c_i T}} \right]$$ (10.112)

where

d_{1i} = uniform delay (sec/vehicle) for lane group i
C = cycle length (sec)
g_i = effective green time for lane group i (sec)
X_i = v/c ratio for lane group i
d_{2i} = incremental delay for lane group i (sec/vehicle)
c_i = capacity of lane group i (veh/h) (see Section 10.1.2)
T = duration of analysis period (h)

The control delay (d) is given as

$$d = d_1(PF) + d_2$$

where

PF = progression adjustment factor, as discussed in Section 10.4.8 and given in Table 10.8

As previously indicated in Section 10.1, when a peak hour factor is used, the analysis period is 0.25 h, and when it is not used, the input volumes are the hourly volumes and the analysis period is 1h. Also, as discussed in Section 10.4.8, the approach delay is a weighted average of the delays of the critical lane groups on the approach, and the intersection delay is the weighted average value of the approach delays. The weights are based on the volumes recorded on the input worksheet and not the adjusted volumes that are used to determine the capacity. The estimated LOS for the intersection is then obtained from Table 10.5.

Note that in the QEM, the delay for a permitted left-turn movement from an exclusive turn lane is not determined as adequate information is not provided. Also, the delay associated with sneakers is not included as the sneaker volume is deducted from left turns operating with a permitted or a protected-permitted mode.

10.12 FIELD DETERMINATION OF SATURATION FLOW

An alternative to the use of adjustment factors is to determine directly the saturation flow in the field. It was shown in Chapter 8 that the saturation flow rate is the maximum discharge flow rate during the green time. This flow rate is usually achieved 10 to

14 sec after the start of the green phase, which is usually the time the fourth, fifth, or sixth passenger car crosses the stop line. Therefore, saturation flow rates are computed starting with the headway after the fourth vehicle in the queue.

Two people are needed to carry out the procedure, with one being the timer equipped with a stopwatch, and the other the recorder equipped with a push-button event recorder or a notebook computer with appropriate software. The form shown in Figure 10.12 is used to record the data. It is suggested that the general information section of the form be completed and other details such as area type, width, and grade of the lane being evaluated be measured and recorded. An observation point is selected at the intersection such that a clear view of the traffic signals and the stop line is maintained. A reference point is selected to indicate when a vehicle has entered the intersection. This reference point is usually the stop line such that all vehicles that cross the stop line are considered as having entered the intersection. When the stop line cannot be clearly seen, or when vehicles often tend to stop beyond the stop line, an alternate reference point is used to indicate when a vehicle enters the intersection. The collection of data starts at the beginning of the green or when the front axle of the first vehicle crosses the stop line or the alternate reference point. The following steps are then carried out for each cycle and for each lane.

Step 1. The timer starts the stopwatch at the beginning of the green phase and notifies the recorder.

Step 2. The recorder immediately notes the last vehicle in the stopped queue and describes it to the timer and also notes which vehicles are heavy vehicles and which vehicles turn left or right.

Step 3. The timer then counts aloud each vehicle in the queue as its rear axle crosses the reference point (that is, "one," "two," "three," and so on). Note that right- or left-turning vehicles that are yielding to either pedestrians or opposing vehicles are not counted until they have gone through the opposing traffic.

Step 4. The timer calls out the times that the fourth, tenth, and last vehicles in the queue cross the stop line or reference point, and these are noted by the recorder.

Step 5. In cases where queued vehicles are still entering the intersection at the end of the green phase, the number of the last vehicle at the end of the green phase is identified by the timer and told to the recorder so that that number can be recorded. Any event that is unusual and that will affect the saturation flow should be recorded. These events include buses, stalled vehicles, and unloading trucks.

Step 6. The width of the lane and the slope of the approach are then measured and recorded together with any unusual occurrences that might have affected the saturation flow.

Step 7. Since the flow just after the start of the green phase is less than the saturation flow, the time considered for calculating the saturation flow is that between the time the rear axle of the fourth car crosses the reference point (t_4) and the time the rear axle of the last vehicle queued at the beginning of the green crosses the same reference point (t_n). The saturation flow is then determined from Eq. 10.113.

$$\text{Saturation flow} = \frac{3600}{(t_4 - t_n)/(n - 4)} \text{ veh/h/ln} \tag{10.113}$$

FIELD SATURATION FLOW RATE STUDY WORKSHEET

General Information

Analyst _____
Agency or Company _____
Date Performed _____
Analysis Time Period _____

Site Information

Intersection
Area Type   ☐ CBD   ☐ Other
Jurisdiction
Analysis Year _____

Lane Movement Input

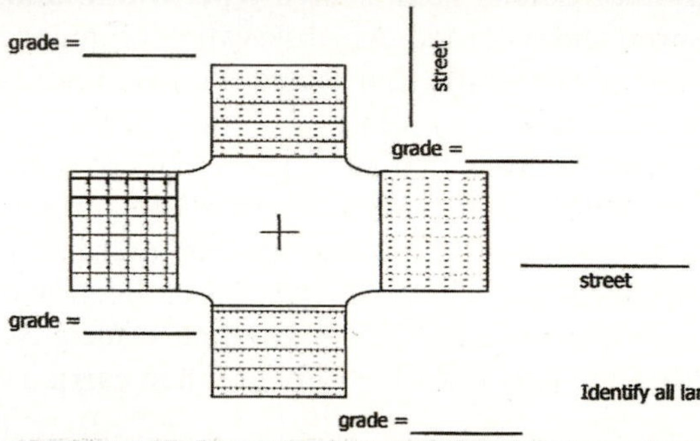

grade = _____

grade = _____

street

grade = _____

grade = _____

street

Movements Allowed
☐ Through
☐ Right turn
☐ Left turn

Identify all lane movements and the lane studied

Input Field Measurement

Veh. in queue	Cycle 1			Cycle 2			Cycle 3			Cycle 4			Cycle 5			Cycle 6		
	Time	HV	T	Time	HV	T	Time	HV	T	Time	HV	T	Time	HV	T	Time	HV	T
1																		
2																		
3																		
4																		
5																		
6																		
7																		
8																		
9																		
10																		
11																		
12																		
13																		
14																		
15																		
16																		
17																		
18																		
19																		
20																		
End of saturation																		
End of green																		
No. veh. > 20																		
No. veh. on yellow																		

Glossary and Notes

HV   = Heavy vehicles (vehicles with more than 4 tires on pavement)
T   = Turning vehicles (L = Left, R = Right)
Pedestrians and buses that block vehicles should be noted with the time that they block traffic, for example,
P12   = Pedestrians blocked traffic for 12 s
B15   = Bus blocked for 15 s

Figure 10.12   Field Saturation Flow Rate Study Worksheet

SOURCE: From *Highway Capacity Manual 2010*. Copyright, National Academy of Sciences, Washington, D.C. Reproduced with permission of the Transportation Research Board.

where *n* is the number of the last vehicle surveyed. The data record on heavy vehicles, turning vehicles, and approach geometrics can be used in the future if adjustment factors are to be applied.

10.3 SUMMARY

The complexity of a signalized intersection requires that several factors be considered when its quality of operation is being evaluated. In particular, the geometric characteristics, prevailing traffic conditions, and signal characteristics must be used in determining the level of service at the intersection. This requirement makes the determination and improvement of level of service at a signalized intersection much more complex than at other locations of a highway. For example, in Chapter 9, the primary characteristics used in determining capacity and level of service at segments of highways are those of geometry of the highway and traffic composition. Since the geometry is usually fixed, the capacity of a highway segment can be enhanced by improving the highway geometry if consideration is given to some variations over time in traffic composition. However, this cannot be done easily at signalized intersections because of the added factor of the green time allocation to different traffic streams, which has a significant impact on the operation of the intersection.

The procedures presented in this chapter for the determination of delay and level of service at a signalized intersection take into consideration the most recent results of research in this area. In particular, the procedures not only take into consideration such factors as traffic mix, lane width, and grades but also factors inherent in the signal system itself—such as whether the signal is pretimed, semiactuated, or fully actuated—and the impact of pedestrians and bicycles.

In using the procedures for design, remember that the cycle length has a significant effect on the delay at an intersection and therefore on the level of service. Thus, it is useful to start with an initial cycle and suitable phase lengths, using one of the methods discussed in Chapter 8, and then to use this information to determine the level of service.

PROBLEMS

10-1 Explain fundamental differences between how LOS is determined for the automobile mode compared with human-powered modes.

10-2 What factors influence the volume of vehicles turning right on a red indication and why is this important? What does the *Highway Capacity Manual* recommend if field data on right-turn-on-reds cannot be collected?

10-3 Name and briefly describe the five flow or movement rates that are inputs to the automobile LOS procedure for signalized intersections.

10-4 Determine the adjusted saturation flow rate for a lane group comprised of one lane for the through movement, under the following conditions:

> Base rate = 1900 pc/h/ln
> Lane width = 11 ft
> Heavy vehicles = 4% of the traffic stream
> Approach grade = +3%
> No on-street parking
> No bus stops
> Bicycle and pedestrian traffic conflicting with this lane group is negligible
> Intersection is in a central business district

10-5 Determine the capacity of the lane group described in Problem 10-4 if the effective green time for this movement is 35 sec and the total cycle length is 60 sec.

10-6 Describe the three components of total control delay for a given lane group.

10-7   Explain how the total control delay for each lane group at an intersection is used to determine the total control delay for the entire intersection.

10-8   Determine the saturation flow rate for a left-turn lane that operates in a permitted-only mode, under the following conditions:

> Opposing demand flow rate = 1100 veh/h
> Lane width = 11 ft
> Heavy vehicles = 3% of the traffic stream
> Approach grade = −2%
> No on-street parking
> No bus stops
> Bicycle and pedestrian traffic conflicting with this lane group is negligible Intersection is not in a central business district

10-9   Determine the pedestrian LOS for pedestrians using a crosswalk under the following peak hour conditions:

> Number of traffic lanes crossed = 3
> Conflicting right-turn-on-red flow rate = 25 veh/h
> Conflicting right turns are not channelized
> Conflicting permitted left turn flow rate = 35 veh/h
> Total vehicular movements conflicting with the subject crosswalk = 960 veh/h
> 85th-percentile speed of street being crossed = 40 mi/h
> Additionally, a previous analysis found pedestrian delay to be 24 seconds per pedestrian.

10-10   Determine the bicycle LOS for bicyclists under the following peak hour conditions:

> The street consists of three 12-ft travel lanes (one in each direction and a two-way left-turn lane) and a 5-ft bicycle lane on each side (for a total pavement width of 46 ft)
> The street has a total approach volume of 650 veh/h

REFERENCES

Highway Capacity Manual. HCM 2010, Transportation Research Board, National Research Council, Washington, D.C., 2010.

Highway Capacity Software. McTrans, *http://mctrans.ce.ufl.edu/hcs/hcsplus/*

The Traffic Software Integrated System (TSIS). McTrans, *http://mctrans.ce.ufl.edu/hcs/hcsplus/*

PART 3

Transportation Planning

The process of transportation planning involves the elements of situation and problem definition, search for solutions and performance analysis, as well as evaluation and choice of project. The process is useful for describing the effects of a proposed transportation alternative and for explaining the benefits to the traveler of a new transportation system and its impacts on the community. The highway and traffic engineer is responsible for developing forecasts of travel demand, conducting evaluations based on economic and noneconomic factors, and identifying alternatives for short-, medium-, and long-range purposes.

CHAPTER 11

The Transportation Planning Process

This chapter explains how decisions to build transportation facilities are made and highlights the major elements of the process. Transportation planning has become institutionalized; federal guidelines, regulations, and requirements for local planning are often driving forces behind existing planning methods.

The formation of the nation's transportation system has been evolutionary, not the result of a grand plan. The system now in place is the product of many individual decisions to select projects for construction or improvement, such as bridges, highways, tunnels, harbors, railway stations, and airport runways. These transportation projects were selected because a conclusion was reached that the project would result in overall improvements to the system.

Among the factors that may justify a transportation project are improvements in traffic flow and safety, energy consumption, travel time, economic growth, air quality, and accessibility. Some transportation projects may have been selected for reasons unrelated to specific benefits; for example, to stimulate employment in a particular region, to compete with other cities or states for prestige, to attract industry, to respond to pressures from a political constituency, or to gain personal benefit from a particular route location or construction project. In some instances, transportation projects are not selected because of opposition from those who would be adversely affected. For example, a new highway may require the taking of community property, or the construction of an airport may introduce undesirable noise due to low-flying planes or take residential or wetland acreage to accommodate runway expansion. Whatever the reason for selecting or rejecting a transportation project, a specific process led to the conclusion to build or not to build.

The process for planning transportation systems should be a rational one that serves to furnish unbiased information about the effects that the proposed transportation project will have on the affected community and on users. For example, if noise or air pollution is a concern, the process will examine and estimate how much additional noise or air pollution will occur if the transportation facility is built. Usually, cost is a major factor, and so the process will include estimates of the construction, maintenance, and operating costs.

The process must be flexible enough to be applicable to any transportation project or system, because the kinds of problems that transportation engineers work on will vary over time. Transportation has undergone considerable change in emphasis over a 200-year period; such modes as canals, railroads, highways, air, and public transit have each been dominant at one time or another. Thus, the activities of transportation engineers have varied considerably during this period, depending on society's needs and concerns. Examples of changing societal concerns include energy conservation, traffic congestion, environmental impacts, safety, security, efficiency, productivity, land consumption, and community preservation.

The transportation planning process is not intended to furnish a decision or to give a single result that must be followed, although it can do so in relatively simple situations. Rather, the process is intended to provide the appropriate information to those who will be affected and those responsible for deciding whether the transportation project should go forward.

CHAPTER OBJECTIVES:

- Understand the major elements of the transportation planning process and how the process influences decision making.
- Explain the seven basic elements of the transportation planning process.
- Describe the five relevant processes for implementing transportation planning recommendations.
- Become familiar with the elements of urban transportation planning and travel forecasting.

11.1 BASIC ELEMENTS OF TRANSPORTATION PLANNING

The transportation planning process comprises seven basic elements, which are interrelated and not necessarily carried out sequentially. The information acquired in one phase of the process may be helpful in some earlier or later phase, so there is a continuity of effort that should eventually result in a decision. The elements in the process are:

- Situation definition
- Problem definition
- Search for solutions
- Analysis of performance
- Evaluation of alternatives
- Choice of project
- Specification and construction

These elements are described and illustrated in Figure 11.1, using a scenario involving the feasibility of constructing a new bridge.

11.1.1 Situation Definition

The first step in the planning process is *situation definition,* which involves all of the activities required to understand the situation that gave rise to the perceived need for a transportation improvement. In this phase, the basic factors that created the present situation are described, and the scope of the system to be studied is delineated. The present system is analyzed and its characteristics are described. Information about the surrounding area, its people, and their travel habits may be obtained. Previous reports

The Process   Application to Bridge Study

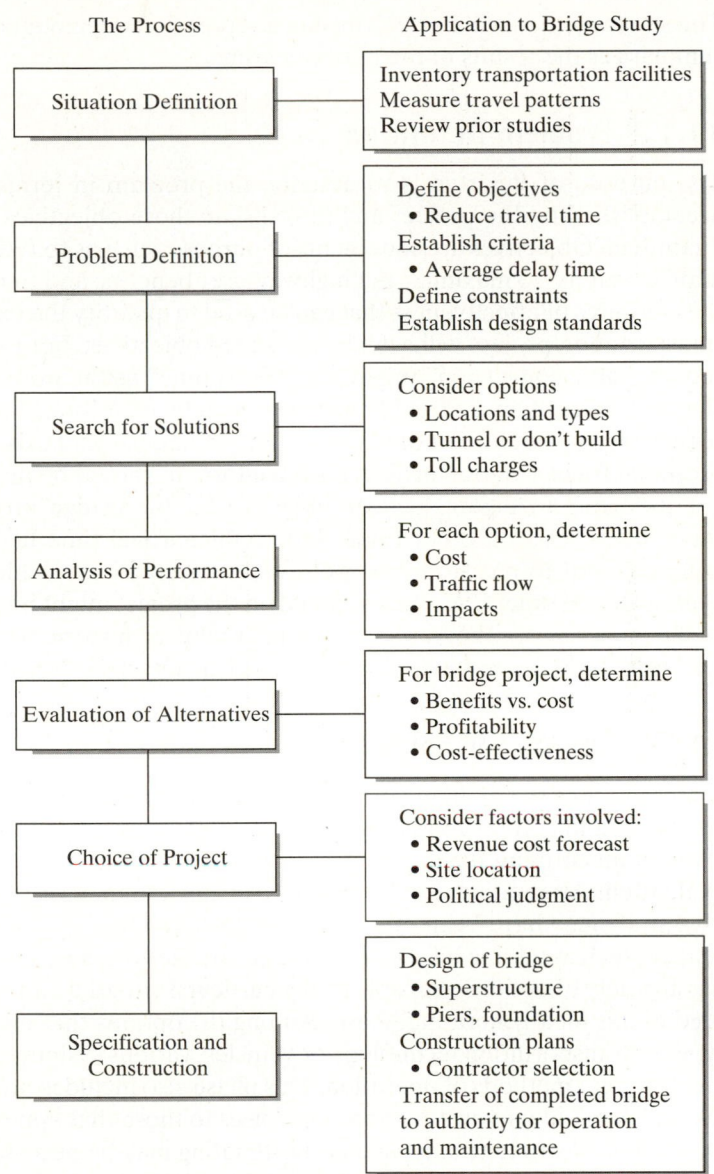

Figure 11.1   Basic Elements in the Transportation Planning Process Applied to Consider the Feasibility of a New Bridge

and studies that may be relevant to the present situation are reviewed and summarized. Both the scope of the study and the domain of the system to be investigated are delineated.

In the example described in Figure 11.1, a new bridge is being considered. Situation definition involves developing a description of the present highway and transportation services in the region; measuring present travel patterns and highway traffic volumes; reviewing prior studies, geological maps, and soil conditions; and delineating the scope of the study and the area affected. A public hearing is often held to obtain citizen input and in some instances, such as to comply with federal aid requirements, may be required.

The situation then will be described in a report that documents the overall situation and summarizes the results of the public hearing.

11.1.2 Problem Definition

The purpose of this step is to describe the problem in terms of the objectives to be accomplished by the project and to translate those objectives into criteria that can be quantified. Objectives are statements of purpose, such as to reduce traffic congestion; to improve safety; to maximize net highway-user benefits; and to reduce noise. Criteria are the measures of effectiveness that can be used to quantify the extent to which a proposed transportation project will achieve the stated objectives. For example, the objective "to reduce traffic congestion" might use "travel time" as the measure of effectiveness. The characteristics of an acceptable system should be identified, and specific limitations and requirements should be noted. Also, any pertinent standards and restrictions that the proposed transportation project must conform to should be understood.

Referring to Figure 11.1, an objective for the bridge project might be to reduce travel congestion on other roads or to reduce travel time between certain areas. The criterion used to measure how well these objectives are achieved is average delay or average travel time. Constraints placed on the project might be physical limitations, such as the presence of other structures, topography, or historic buildings. Design standards for bridge width, clearances, loadings, and capacity also should be noted.

11.1.3 Search for Solutions

In this phase of the planning process, consideration is given to a variety of ideas, designs, locations, and system configurations that might provide solutions to the problem. This is the brainstorming stage, in which many options may be proposed for later testing and evaluation. Alternatives can be proposed by any group or organization. In fact, the planning study may have been originated to determine the feasibility of a particular project or idea, such as adding bike lanes to reduce traffic volumes. The transportation engineer has a variety of options available in any particular situation, and any or all may be considered in this idea-generating phase. Among the options that might be used are different types of transportation technology or vehicles, various system or network arrangements, and different methods of operation. This phase also includes preliminary feasibility studies, which might narrow the range of choices to those that appear most promising. Some data gathering, field testing, and cost estimating may be necessary at this stage to determine the practicality and financial feasibility of the alternatives being proposed.

In the case of the bridge project, a variety of options may be considered, including different locations and bridge types. The study should also include the option of not building the bridge and might also consider what other alternatives are available, such as a tunnel or an alternate route. Operating policies should be considered, including various toll charges and methods of collection.

11.1.4 Analysis of Performance

The purpose of performance analysis is to estimate how each of the proposed alternatives would perform under present and future conditions. The criteria identified in the previous steps are calculated for each transportation option. Included in this step is a determination of the investment cost of building the transportation project, as well as annual costs for maintenance and operation. This element also involves the use of mathematical models for estimating travel demand. The number of persons or vehicles that will use the system is determined, and these results, expressed in vehicles or persons/hour, serve as the basis for project design. Other information about the use of the system (such as trip length, travel by time of day, and vehicle occupancy) is also determined and used in calculating user benefits for various criteria or measures of effectiveness. Environmental effects of the transportation project (such as noise and air pollution levels and hectares of land required) are estimated. These nonuser impacts are calculated in situations where the transportation project could have significant impacts on the community or as required by law.

This task is sometimes referred to as the transportation planning process, but it is really a systems analysis process that integrates system supply on a network with travel demand forecasts to show equilibrium travel flows. The forecasting-model system and related network simulation are discussed in Chapter 12.

To analyze the performance of the new bridge project, preliminary cost estimates must be prepared for each location being considered. Then estimates of the traffic that would use the bridge are developed, given various toll levels and bridge widths. The average trip length and average travel time for bridge users would be determined and compared with existing or no-build conditions. Other impacts (such as land required, visual effects, noise levels, and air or water quality changes) also would be computed.

11.1.5 Evaluation of Alternatives

The purpose of the evaluation phase is to determine how well each alternative will achieve the objectives of the project as defined by the criteria. The performance data produced in the analysis phase are used to compute the benefits and costs that will result if the project is selected. In cases where the results cannot be reduced to a single monetary value, a weighted ranking for each alternative might be produced and compared with other proposed projects. For those effects that can be described in monetary terms, the benefit–cost ratio (described in Chapter 13) for each project is calculated to show the extent to which the project would be a sound investment. Other economic tests might also be applied, including the net present worth of benefits and costs.

In situations where there are many criteria, particularly in an environmental analysis, the results can be shown in a cost-effectiveness matrix (for example, project cost versus number of homes displaced) that will furnish a better understanding as to how each alternative performs for each of the criteria and at what cost. The results can be plotted to provide a visual comparison of each alternative and its performance.

In the evaluation of the bridge project, first determine the benefits and costs and compute the benefit–cost ratio. If the result is greater than one, the evaluation of alternative sites requires additional comparison of factors, both for engineering and economic feasibility and for environmental impact. A cost-effectiveness matrix that compares the cost of each alternative with its effectiveness in achieving certain goals will further assist in the evaluation.

11.1.6 Choice of Project

Project selection is made after considering all the factors involved. In a simple situation—for example, where the project has been authorized and is in the design phase—a single criterion (such as cost) might be used and the chosen project would be the one with the lowest cost. With a more complex project, however, many factors have to be considered, and selection is based on how the results are perceived by those involved in decision making. If the project involves the community, it may be necessary to hold additional public hearings. A bond issue or referendum may be required. It is possible that none of the alternatives will meet the criteria or standards, and additional investigations will be necessary. The transportation engineer, who participates in the planning process, may have developed a strong opinion as to which alternative to select. Such bias could result in the early elimination of promising alternatives or the presentation to decision makers of inferior projects. If the engineer is acting professionally and ethically, he or she will perform the task such that the appropriate information is provided to make an informed choice and that every feasible alternative has been considered.

Before deciding whether or not to build the proposed bridge, decision makers look carefully at the results of both revenue and cost estimates and would likely consider projects that appear to be financially sound. The site location is selected based on a careful study of the factors involved. The information gathered in the earlier phases would be used, together with engineering judgment and political considerations, to arrive at a final project selection.

11.1.7 Specification and Construction

Once the transportation project has been selected, the project moves into a detailed design phase in which each of the components of the facility is specified. For a transportation facility, this involves its physical location, geometric dimensions, and structural configuration. Design plans are produced that can be used by contractors to estimate the construction cost of building the project. When a construction firm is selected, these plans will be the basis on which the project will be built.

For the bridge project, once a decision to proceed has been made, a design is produced that includes the type of superstructure, piers and foundations, roadway widths and approach treatment, as well as appurtenances such as tollbooths, traffic signals, and lighting. These plans are made available to contractors, who submit bids for the construction of the bridge. If a bid does not exceed the amount of funds available and the contractor is deemed qualified to do the work, the project proceeds to the construction phase. Upon completion, the new bridge is turned over to the local transportation authority for operation and maintenance.

Example 11.1 Planning the Relocation of a Rural Road

To illustrate the transportation planning process, a situation that involves a rural road relocation project is described. Each of the activities that are part of the project is discussed in terms of the seven-step planning process previously described. This project includes both a traffic analysis and an environmental assessment and is typical of those conducted by transportation consultants or metropolitan transportation organizations. (This example is based on a study completed by the engineering firm Edwards and Kelsey.)

   Step 1.   **Situation definition.** The project is a proposed relocation or reconstruction of 5.3 km of U.S. 1A located in the coastal town of Harrington, Maine. The town center, a focal point of the project, is located near the intersection of highways U.S. 1 and U.S. 1A on the banks of the Harrington River, an estuary of the Gulf of Maine. (See Figure 11.2.) The town of Harrington has 553 residents, of whom 420 live within the study

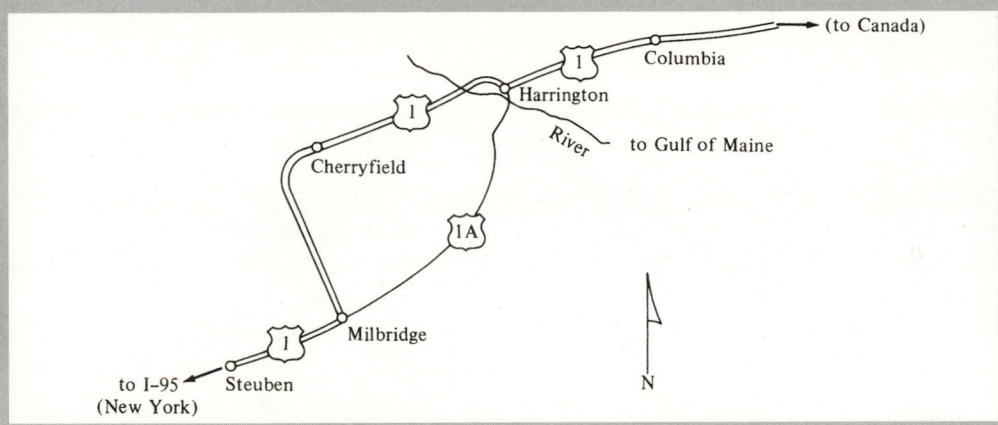

Figure 11.2   Location Map for Highways U.S. 1 and U.S. 1A

area and 350 live in the town center. The population has been declining in recent years; many young people have left because of the lack of employment opportunities. Most of the town's industry consists of agriculture or fishing, so a realignment of the road that damages the environment would also affect the town's livelihood. There are 10 business establishments within the study area; 20 percent of the town's retail sales are tourism related. The average daily traffic is 2620 vehicles/day, of which 69 percent represent through traffic and 31 percent represent local traffic.

Step 2.   **Problem definition.** The Maine Department of Transportation wishes to improve U.S. 1A, primarily to reduce the high accident rate on this road in the vicinity of the town center. The problem is caused by a narrow bridge that carries the traffic on U.S. 1A into the town center, the poor horizontal and vertical alignment of the road within the town center, and a dangerous intersection where U.S. 1A and U.S. 1 meet. The accident rate on U.S. 1A in the vicinity of the town center is four times the statewide average. A secondary purpose of the proposed relocation is to improve the level of service for through traffic by increasing the average speed on the relocated highway.

   The measures of effectiveness for the project will be the accident rate, travel time, and construction cost. Other aspects that will be considered are the effects that each alternative would have on a number of businesses and residences that would be displaced, the changes in noise levels and air quality, and the changes in natural ecology. The criteria that will be used to measure these effects will be the number of businesses and homes displaced, noise levels and air quality, and the acreage of salt marsh and trees affected.

Step 3.   **Search for solutions.** The Department of Transportation has identified four alternative routes, as illustrated in Figure 11.3, in addition to the present route—Alternative 0—referred to as the null or "do-nothing" alternative. All routes begin at the same location—4.8 km southwest of

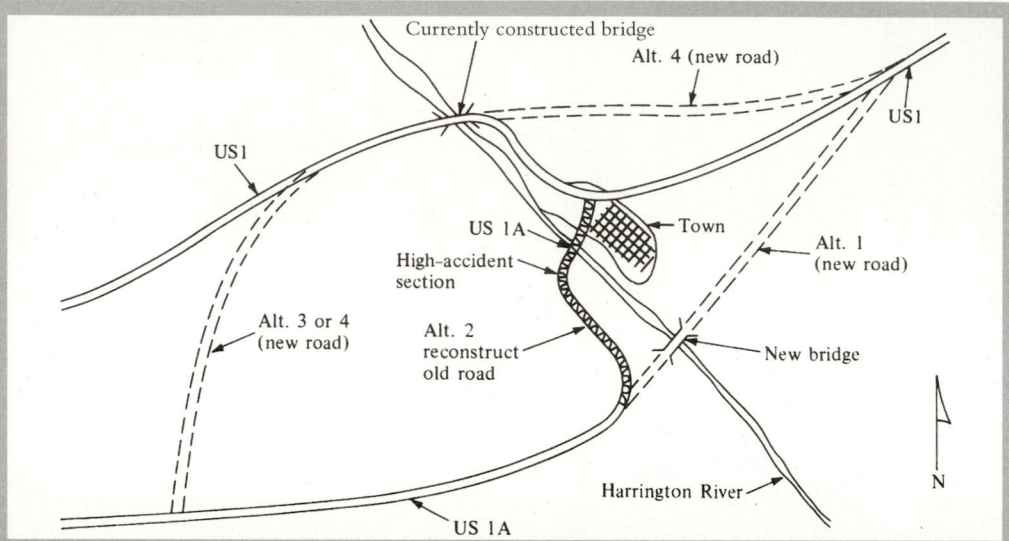

Figure 11.3   Alternative Routes for Highway Relocation

the center of Harrington—and end at a common point northeast of the town center. The alternatives are as follows:

- *Alternative 1:*   This road bypasses the town to the south on a new location across the Harrington River. The road would have two lanes, each 3.7 m wide with 2.4 m shoulders. A new bridge would be constructed about one-half mi downstream from the old bridge.
- *Alternative 2:*   This alternative would use the existing U.S. 1A into town, but with improvements to the horizontal and vertical alignment throughout its length and the construction of a new bridge. The geometric specifications would be the same as for Alternative 1.
- *Alternative 3:*   This new road would merge with U.S. 1 west of Harrington, and then continue through town. It would use the Route 1 Bridge, which was recently constructed. Geometric specifications are the same as those for the other alternatives.
- *Alternative 4:*   This road would merge with U.S. 1 and use the Route 1 Bridge, as in Alternative 3. However, it would bypass the town center on a new alignment.

Step 4.   **Analysis of performance.** The measures of effectiveness are calculated for each alternative. The results of these calculations are shown in Table 11.1 for Alternatives 1 through 4 and for the null alternative. The relative ranking of each alternative is presented in Table 11.2. For example, the average speed on the existing road is 40 km/h, whereas for Alternatives 1 and 4, the speed is 88.5 km/h, and for Alternatives 2 and 3, the speed is 48 km/h. Similarly, the accident factor, which is now four times the statewide average, would be reduced to 0.6 for Alternative 4 and 1.2 for Alternative 1. The project cost ranges from $1.18 million for Alternative 3 to $1.58 million for Alternative 2. Other items that are calculated include the number of

Table 11.1   Measures of Effectiveness for Rural Road

| | Alternatives | | | | |
Criteria	0	1	2	3	4
Speed (km/h)	40	88.5	48	48	88.5
Distance (km)	6	5	6.1	6.1	6
Travel time (min)	8.9	3.5	7.6	7.6	4.0
Accident factor	4	1.2	3.5	2.5	0.6
(Relative to statewide average)					
Construction cost ($ million)	0	1.50	1.58	1.18	1.54
Residences displaced	0	0	7	3	0
City traffic					
Present	2620	1400	2620	2520	1250
Future (20 years)	4350	2325	4350	4180	2075
Air quality (μg/m^3 CO)	825	306	825	536	386
Noise (dBA)	73	70	73	73	70
Tax loss	None	Slight	High	Moderate	Slight
Trees removed (hectares)	None	Slight	Slight	10.1	11.3
Runoff	None	Some	Some	Much	Much

residences displaced, the volume of traffic within the town both now and in the future, air quality, noise, lost taxes, and hectares of trees removed.

Step 5.   Evaluation of alternatives. Each of the alternatives is compared with the others to assess the improvements that would occur based on a given criterion. In this example, we consider the following measures of effectiveness and their relationship to project cost.

Comparison of Each Criterion

- *Travel time:* Every alternative improves the travel time. As shown in Figure 11.4, the best is Alternative 1, followed by Alternative 4. Alternatives 2 and 3 are equal, but neither reduces travel time significantly.

Table 11.2   Ranking of Alternatives

| | Alternatives | | | | |
Criterion/Alternative	0	1	2	3	4
Travel time	4	1	3	3	2
Accident factor	5	2	4	3	1
Cost	1	3	5	2	4
Residences displaced	1	1	3	2	1
Air quality	4	1	4	3	2
Noise	2	1	2	2	1
Tax loss	1	2	4	3	2
Trees removed	1	2	2	3	4
Runoff	1	2	2	3	3

Note: 1 = highest; 5 = lowest

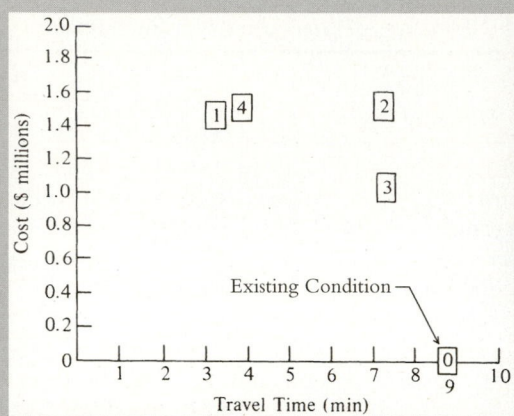

Figure 11.4 Travel Time between West Harrington and U.S. 1 versus Cost

- *Accident factor:* Figure 11.5 shows that the best accident record will occur with Alternative 4, followed by Alternatives 1, 3, and 2.
- *Cost:* The least costly alternative is simply to do nothing, but the dramatic potential improvements in travel time and safety would indicate that the proposed project should probably be undertaken. Alternative 3 is lowest in cost at $1.18 million. Alternative 2 is highest in cost, would not be as safe as Alternative 3, and would produce the same travel time. Thus, Alternative 2 would be eliminated. Alternative 1 would cost $0.32 million more than Alternative 3, but would reduce the accident factor by 1.3 and travel time by 4.1 minutes. Alternative 4 would cost $0.04 million more than Alternative 1 and would increase travel time, but would decrease the accident factor. These cost-effectiveness values are shown in Figures 11.4 and 11.5. They indicate that Alternatives 1 and 4 are both more attractive than Alternatives 2 and 3 because the former would produce significant improvements in travel time and accidents.
- *Residences:* Three residences would be displaced if Alternative 3 were selected; seven residences would be displaced if Alternative 2 were selected. No residences would have to be removed if Alternatives 1 or 4 were selected.

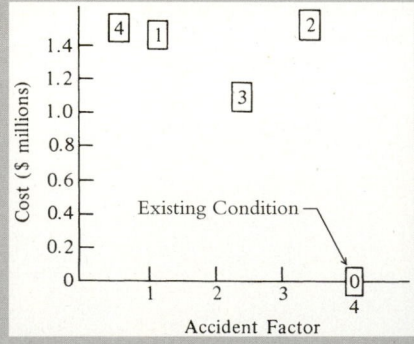

Figure 11.5 Accident Factor (relative to statewide average) versus Cost

- *Air quality:* Alternative 1 would produce the highest air quality, followed by Alternatives 4, 3, and 2. The air quality improvement would result from removing a significant amount of the slow-moving through traffic from the center of the city to a high-speed road where most of the pollution would be dispersed.
- *Noise:* Noise levels are lower for Alternatives 1 and 4.
- *Tax loss:* Tax losses would be slight for Alternatives 1 and 4, moderate for Alternative 3, and high for Alternative 2.
- *Trees removed:* Alternatives 3 and 4 would eliminate 10.1 and 11.3 hectares of trees, respectively. Alternative 1 would result in slight losses; Alternative 2, no loss.
- *Runoff:* There would be no runoff for Alternative 0, some for Alternatives 1 and 2, and a considerable amount for Alternatives 3 and 4.

Step 6. **Choice of project.** From a cost point of view, the Department of Transportation would select Alternative 3, since it results in travel time and safety improvements at the lowest cost. However, if additional funds are available, then Alterative 1 or 4 would be considered. Since Alternative 1 is lower in cost than Alternative 4 and is equal or better than Alternative 3 for each criterion related to community impacts, this alternative would be the one most likely to be selected. In the selection process, each alternative would be reviewed. Also, comments would be received from citizens and elected officials to assist in the design process so that environmental and community effects would be minimized.

Step 7. **Specifications and construction.** The choice has been made, and Alternative 1, a bypass south of Harrington, has been ranked of sufficiently high priority so that it will be constructed. This alternative involves building both a new bridge across the Harrington River and a new road connecting U.S. 1A with U.S. 1. The designs for the bridge and road will be prepared. Detailed estimates of the cost to construct will be made, and the project will be announced for bid. The construction company that produces the lowest bid and can meet other qualifications will be awarded the contract, and the road will be built. Upon completion, the road will be turned over to the Department of Transportation, who will be responsible for its maintenance and operation. Follow-up studies will be conducted to determine how successful the road was in meeting its objectives; where necessary, modifications will be made to improve its performance.

11.2 TRANSPORTATION PLANNING INSTITUTIONS

This transportation planning process is based on a systems approach to problem solving and is quite general in its structure. The process is not confined to highways but can be applied to many other situations, such as intercity high-speed rail feasibility studies, airport location, port and harbor development, and urban transportation systems. The most common application is in urban areas, where it has been mandated by law since 1962, when the Federal Aid Highway Act required that all transportation projects in urbanized areas with populations of 50,000 or more be based on a transportation planning process that was continuing, comprehensive, and cooperative, sometimes referred to as the "3C" process. The term *continuing* implies that the process be revisited frequently and viewed as an ongoing concern. *Comprehensive* in this context ensures that all transportation modes are addressed. A *cooperative* process indicates that the state (or states) and all municipalities in an urbanized area work together.

Because the urban transportation planning process provides an institutionalized and formalized planning structure, it is important to identify the environment in which the transportation planner works. The forecasting modeling process that has evolved is presented in Chapter 12 to provide an illustration of the methodology. The planning process used at other problem levels is a variation of this basic approach.

11.2.1 Transportation Planning Organization

In carrying out the urban transportation planning process, several committees represent various community interests and viewpoints. These committees are the policy committee, the technical committee, and the citizens' advisory committee. They also interact with permanent planning entities, such as the regional metropolitan planning organization (MPO).

Policy Committee

The policy committee is composed of elected or appointed officials, such as the mayor and director of public works, who represent the governing bodies or agencies that will be affected by the results. This committee makes the basic policy decisions and acts as a board of directors for the study. Committee members will decide on management aspects of the study as well as key issues of a financial or political nature.

Technical Committee

The technical committee is composed of the engineering and planning staffs that are responsible for carrying out the work or evaluating the technical aspects of the project prepared by consultants. This group will assure that the necessary evaluations and cost comparisons for each project alternative are complete and will supervise the technical details of the entire process. Typically, the technical committee will include highway, transit, and traffic engineers, as well as other specialists in land-use planning, economics, and environmental engineering.

Citizens' Advisory Committee

The citizens' advisory committee is composed of a cross section of the community and may include representatives from labor and business, interested citizens, and members of community interest groups. The committee's function is to express community goals and objectives, to suggest alternatives, and to react to proposed alternatives. Through this committee structure, an open dialogue is facilitated among the policy makers, technical staff, and the community. When a selection is made and recommendations are produced by the study, they should be based on consensus of all interested parties. Although agreement is not always achieved, the citizens' advisory committee serves as a means to increase communication to assure that the final plan which results from the process reflects community interests.

Metropolitan Planning Organization (MPO)

A metropolitan planning organization (MPO) is a transportation policy-making organization made up of representatives from local government and transportation authorities. For example, an MPO representing a city and a county might have a policy board with five voting members: two from the city council, two from the county Board of Supervisors, and one from the state transportation department. The MPO policy board might also include nonvoting members representing various local transit providers, local planning commissions, and other transportation agencies, such as state public transportation departments and the Federal Highway Administration.

11.2.2 Implementation of Transportation Planning Recommendations

There is no single model that represents how each state or region implements projects recommended as a result of the transportation planning process, as implementation is governed by state laws and processes (in addition to federal requirements). Figure 11.6 illustrates how recommendations from the planning process might be implemented in a typical state. There are five relevant processes:

- The *transportation planning process* entails the generation of plans for various types of transportation facilities and programs. Such studies may include a 20-year comprehensive plan created by a regional planning body or specific jurisdiction, project-specific studies of a particular corridor or location, and statewide plans. The creation of these plans typically includes some degree of public involvement, such as public hearings, citizen surveys, or meetings with citizen committees. This process is also guided by federal requirements for state and MPO planning processes. For example, the metropolitan area's long-range plan must be financially constrained such that the projects recommended by the plan do not exceed the forecast for revenue that will be available.

- The *transportation programming process* is the act of reconciling recommended projects from the planning process with the amount of funds available over an expected period of time, usually between three and six years. The transportation program is thus a list of projects and costs that can be supported by the expected revenues.

- The *preliminary engineering* and *right-of-way process* occurs after a project has been selected. Within this process, variations of the alignment may be considered and environmental studies may be conducted, including solicitation of public input through project-specific hearings. The outcome of these environmental studies may result in no change to the project, a decision to implement mitigation measures such as the creation of wetlands to compensate for wetlands destroyed, a modification to the project such as a new alignment, or a termination of the project.

- The *construction process* entails advertising the project, soliciting bids, and constructing the project as well as specifications, construction inspection, and acceptance by the owner.

- The *operations and maintenance* process entails steps to ensure that the investment remains effective, such as pavement management and signal retiming. This phase may also include monitoring of system performance, such as changes in travel time, reliability, or use, which would subsequently influence transportation planning process.

Figure 11.6 illustrates the complexity of the relationship between the transportation planning process and these other project development processes. Clearly, stakeholder involvement does not stop with the completion of a region's transportation

VDOT Project Development Process

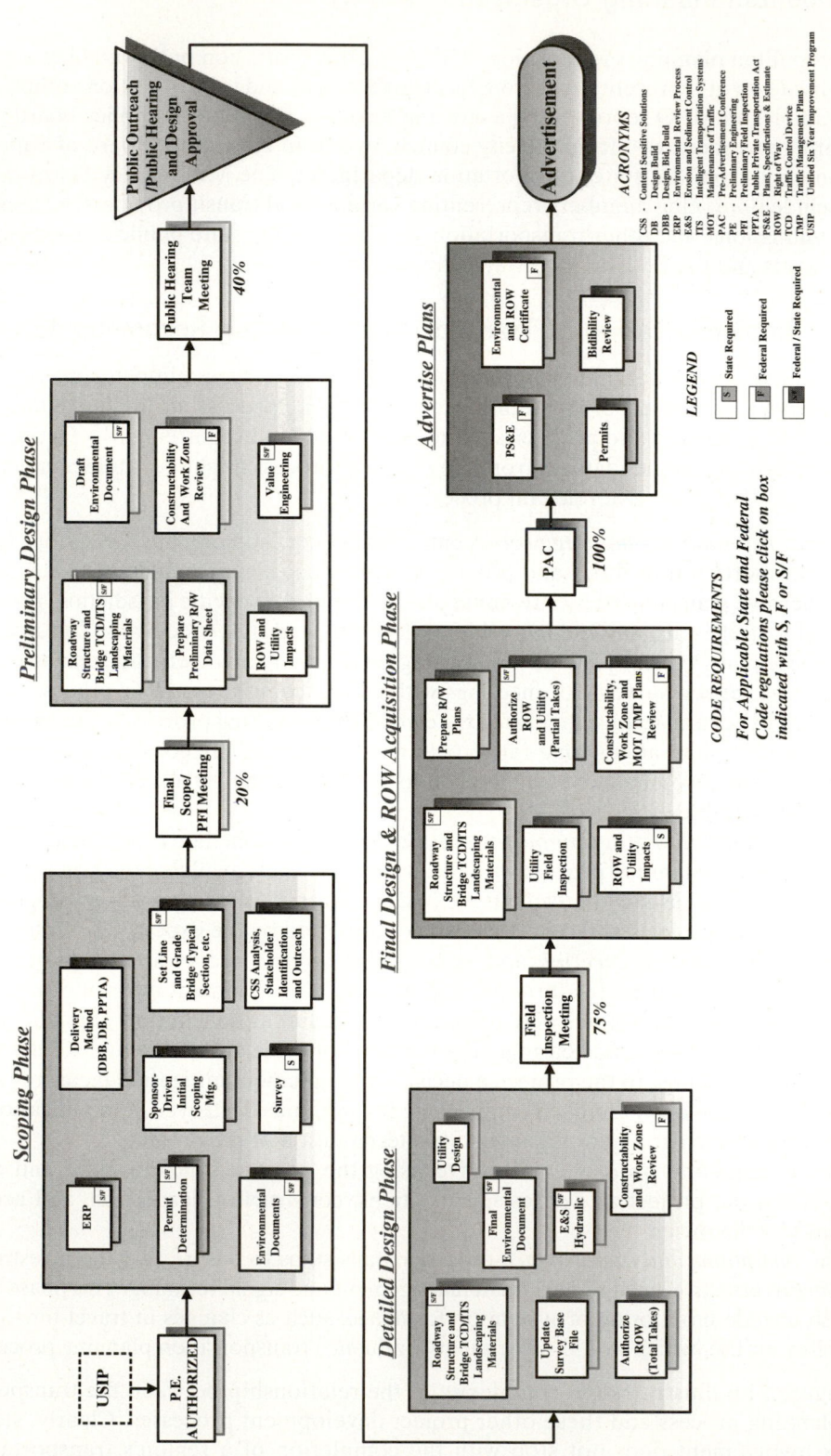

Figure 11.6   Project Development Process

SOURCE: © 2012 Commonwealth of Virginia

plan, meaning there are a variety of opportunities to change the outcome of the plan's recommendations once they have been approved by the MPO and other governing bodies. This figure also illustrates that the practice of transportation planning is incremental, with changes being made until the construction process is underway. There are also a variety of requirements that require coordination of individual bodies in order for a recommendation to be implemented. For example, Figure 11.6 shows that for projects in the MPO area to receive federal funds, the MPO must include the project in its transportation improvement program (TIP), which is a list of projects that the MPO wishes to be undertaken within the next three years (or more). The TIP must be financially constrained; that is, only projects for which funds are available may be placed in the TIP. The TIP in turn is incorporated into the state transportation improvement program (STIP), which is then submitted by the state to FHWA (or FTA) for approval. Finally, the "decision makers" refer to individuals with specific authority at various points throughout the planning, programming, environmental, and construction processes. Decision makers include elected officials in the legislative branches who establish total funds available for the transportation program, special boards that approve the projects that comprise the transportation program, citizens and advocacy groups who influence the outcome of project and program hearings, local and regional officials who make recommendations from their transportation plans, and federal officials who approve projects.

As will be discussed in Section 11.3, the transportation planning process includes operational strategies that can complement or replace traditional capacity expansions. In addition to the TIP and STIP, such operational strategies may be considered through the constrained long-range plan (CLRP) and the region's congestion management process (CMP). Examples of ways to incorporate operations-related initiatives within the transportation planning process include:

- the inclusion of operations-oriented strategies, such as signal retiming, in the TIP
- quantification of the delay reduction impacts of a traffic management system (TMS), thereby enabling it to be compared to delay reduction benefits of a highway investment
- identification of operations-related projects, such as ramp metering, that can delay the need for a more expensive capacity enhancement

11.3 URBAN TRANSPORTATION PLANNING

Urban transportation planning involves the evaluation and selection of highway or transit facilities to serve present and future land uses. For example, the construction of a new shopping center, airport, or convention center will require additional transportation services. Also, new residential development, office space, and industrial parks will generate additional traffic, requiring the creation or expansion of roads and transit services.

The process must also consider other proposed developments and improvements that will occur within the planning period. The urban transportation planning process has been enhanced through the efforts of the Federal Highway Administration and the Federal Transit Administration of the U.S. Department of Transportation by the preparation of manuals and computer programs that assist in organizing data and forecasting travel flows.

Urban transportation planning is concerned with two separate time horizons. The first is a short-term emphasis intended to select projects that can be implemented within a one-to four-year period. These projects are designed to provide better management of existing facilities by making them as efficient as possible. The second time horizon deals with the long-range transportation needs of an area and identifies the projects to be constructed over a 20-year period.

Short-term projects involve programs such as traffic signal timing to improve flow, car and van pooling to reduce congestion, park-and-ride fringe parking lots to increase transit ridership, and transit improvements.

Long-term projects involve programs such as adding new highway elements, additional bus lines or freeway lanes, rapid transit systems and extensions, or access roads to airports or shopping malls.

The urban transportation planning process can be carried out in terms of the procedures outlined previously and is usually described as follows. Figure 11.7 illustrates the comprehensive urban area transportation planning process.

11.3.1 Inventory of Existing Travel and Facilities

This is the data-gathering activity in which urban travel characteristics are described for each defined geographic unit or traffic zone within the study area. Inventories and surveys are made to determine traffic volumes, land uses, origins and destinations of travelers, population, employment, and economic activity. Inventories are made of existing transportation facilities, both highway and transit. Capacity, speed, travel time, and traffic volume are determined. The information gathered is summarized by geographic areas called transportation analysis zones (TAZ).

The size of the TAZ will depend on the nature of the transportation study, and it is important that the number of zones be adequate for the type of problem being investigated. Often, census tracts or census enumeration districts are used as TAZs because population data are easily available by this geographic designation.

11.3.2 Establishment of Goals and Objectives

The urban transportation study is carried out to develop a program of highway and transit projects that should be completed in the future. Thus, a statement of goals, objectives, and standards is prepared that identifies deficiencies in the existing system, desired improvements, and what is to be achieved by the transportation improvements.

For example, if a transit authority is considering the possibility of extending an existing rail line into a newly developed area of the city, its objectives for the new service might be to maximize its revenue from operations, maximize ridership, promote development, and attract the largest number of auto users so as to relieve traffic congestion.

11.3.3 Generation of Alternatives

In this phase of the urban transportation planning process, the alternatives to be analyzed will be identified. It also may be necessary to analyze the travel effects of different land-use plans and to consider various lifestyle scenarios. The options available to the urban transportation planner include various technologies, network configurations, vehicles, operating policies, and organizational arrangements.

In the case of a transit line extension, the technologies could be rail rapid transit or bus. The network configuration could be defined by a single line, two branches, or a geometric configuration such as a radial or grid pattern. The guideway, which represents a homogeneous section of the transportation system, could be varied in length, speed, waiting time, capacity, and direction. The intersections, which represent the end points of the guideway, could be a transit station or the line terminus. The vehicles could

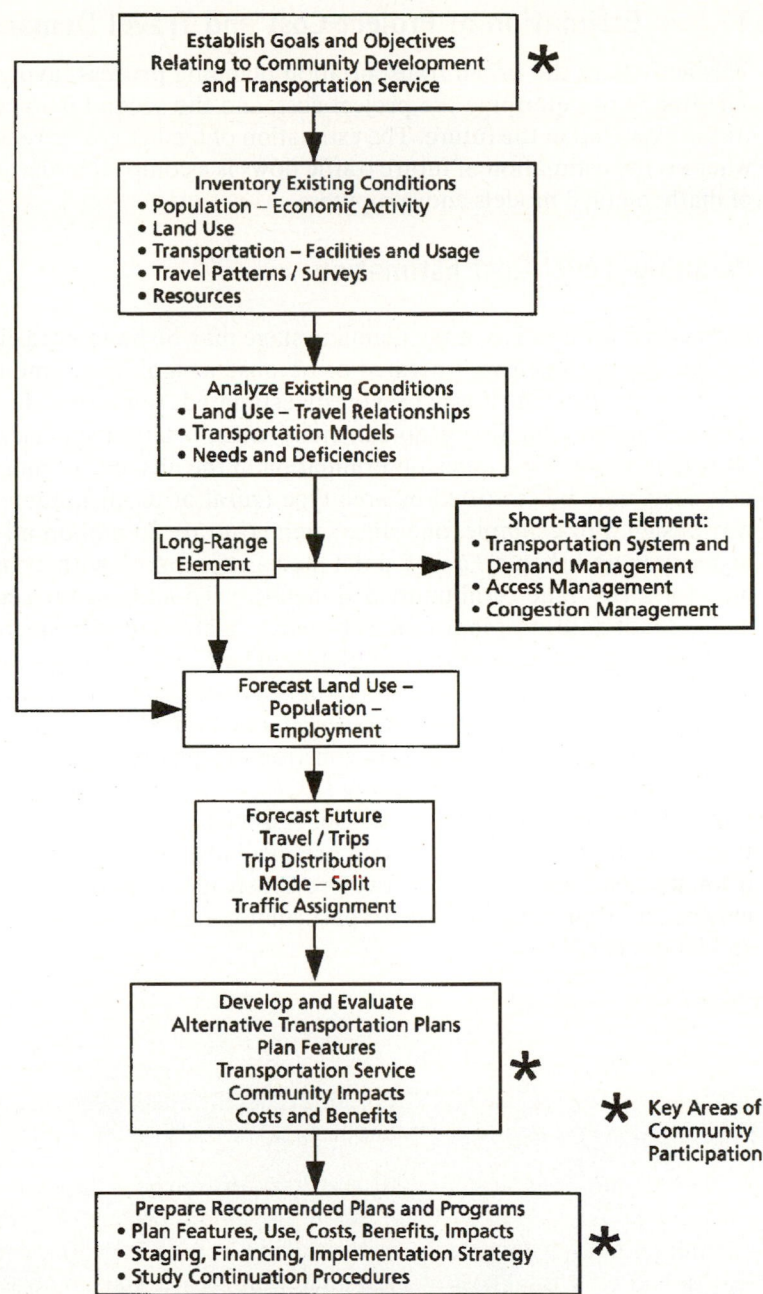

Figure 11.7   Comprehensive Urban Area Transportation Planning Process

SOURCE: Redrawn from *Transportation Planning Handbook*, Institute of Transportation Engineers, 2nd Edition, Institute of Transportation Engineers, 1999. www.ite.org.

be singly driven buses or multicar trains. The operating policy could involve 10-minute headways during peak hours and 30-minute headways during off-peak hours, or other combinations. The organizational arrangements could be private or public. These and other alternatives would be considered in this phase of the planning process.

11.3.4 Estimation of Project Cost and Travel Demand

This activity in the urban transportation planning process involves two separate tasks. The first is to determine the project cost, and the second is to estimate the amount of traffic expected in the future. The estimation of facility cost is relatively straightforward, whereas the estimation of future traffic flows is a complex undertaking requiring the use of mathematical models and computers.

Planning-Level Cost Estimation

Project cost estimation at the planning stage may be hampered either because the project has not yet been well defined or because a significant amount of time has passed since the project's cost was previously estimated, rendering the older estimate out of date. To address the first problem, many transportation agencies maintain a set of *unit costs* that allows for a quick determination in the absence of more detailed data. These unit costs may be stratified by area type (rural or urban), number of lanes, and roadway design. For example, one state's unit costs are $5 million/mi of a four-lane divided highway in an urban area with a flat median compared with a cost of about $1 million/mile for a two-lane rural undivided highway. To address the second problem of costs being out of date, cost indices may be used which convert costs from a historical year to a current year by accounting for inflation. The *Consumer Price Index* (CPI) provides an average rate of inflation for all goods and services. Indices specific to the transportation field are the *National Highway Construction Cost Index* (compiled by the Federal Highway Administration for highway construction projects) and various railroad indices, such as the *Railroad Cost Recovery (RCR) Index* or the *All–Inclusive Index–Less Fuel (AII-LF)* (compiled by the American Association of Railroads for railroad projects). If more detailed project cost data are available (e.g., highway labor costs separated from materials costs), then the two previously mentioned sources may be consulted for indices that represent changes for particular goods and services, such as concrete, steel, and labor costs.

Example 11.2 Updating Costs for a Rail Feasibility Study

Table 11.3 shows indices for 2001 and 2005 for railroads, highways, and the *Consumer Price Index*. A study of a freight rail improvement project was completed in 2001 that recommended improvements such as siding, track extension, and track maintenance and estimated a total cost of $120 million in 2001 dollars. The study cost $250,000 to perform, and the state agency would like to convert this cost estimate to 2005 dollars without redoing the entire study. How much should the improvements cost in 2005 dollars?

Solution: Because these are all rail items, the *Railroad Cost Index* is appropriate. This index may be applied as follows:

$$\text{Estimate in 2005 dollars} = (\text{Estimate in 2001 dollars})\frac{2005 \text{ index}}{2001 \text{ index}} \quad (11.1)$$

Table 11.3   Indices for Railroad Projects, Highway Projects, and Consumer Prices

Index	Applies to	Year 2001	Year 2005
Railroad Index[a]	Rail construction	315.7	356.8
Highway Index[b]	At-grade rail highway crossings	144.8	183.6
Consumer Price Index[c]	All goods and services	177.1	195.3

[a]*American Association of Railroads* (2006) (materials prices, wage rates, and supplements combined, excluding fuel).
[b]*Federal Highway Administration* (2006a) (Federal-Aid Highway Construction Composite Index).
[c]*Bureau of Labor Statistics* (2006) (Consumer Price Index for each year).

$$\text{Estimate in 2005 dollars} = (\$120 \text{ million}) \frac{356.8}{315.7} = \$135.6 \text{ million}$$

Thus, the improvements will cost $136 million in 2005 dollars.

Planning–Level Demand Estimation

Future travel is determined by forecasting future land use in terms of the economic activity and population that the land use in each TAZ will produce. With the land-use forecasts established in terms of number of jobs, residents, auto ownership, income, and so forth, the traffic that this land use will add to the highway and transit facility can be determined. This is carried out in a four-step process that includes the determination of the number of trips generated, the origin and destination of trips, the mode of transportation used by each trip (for example, auto, bus, rail), and the route taken by each trip. The urban travel demand forecasting process thus involves four distinct activities: *trip generation, trip distribution, modal split,* and *network assignment,* as illustrated in Figure 11.8. This forecasting process is described in Chapter 12.

When the travel forecasting process is completed, the highway and transit volumes on each link of the system will have been estimated. The actual amount of traffic, however, is not known until it occurs. The results of the travel demand forecast can be compared with the present capacity of the system to determine the operating level of service.

11.3.5 Evaluation of Alternatives

This phase of the process is similar in concept to what was described earlier but can be complex in practice because of the conflicting objectives and diverse groups that will be affected by an urban transportation project.

Among the groups that could be affected are the traveling public (user), the highway or transit agencies (operator), and the nontraveling public (community). Each of these groups will have different objectives and viewpoints concerning how well the system performs. The traveling public wants to improve speed, safety, and comfort; the transportation agency wishes to minimize cost; and the community wants to preserve its quality of life and improve or minimize environmental impacts.

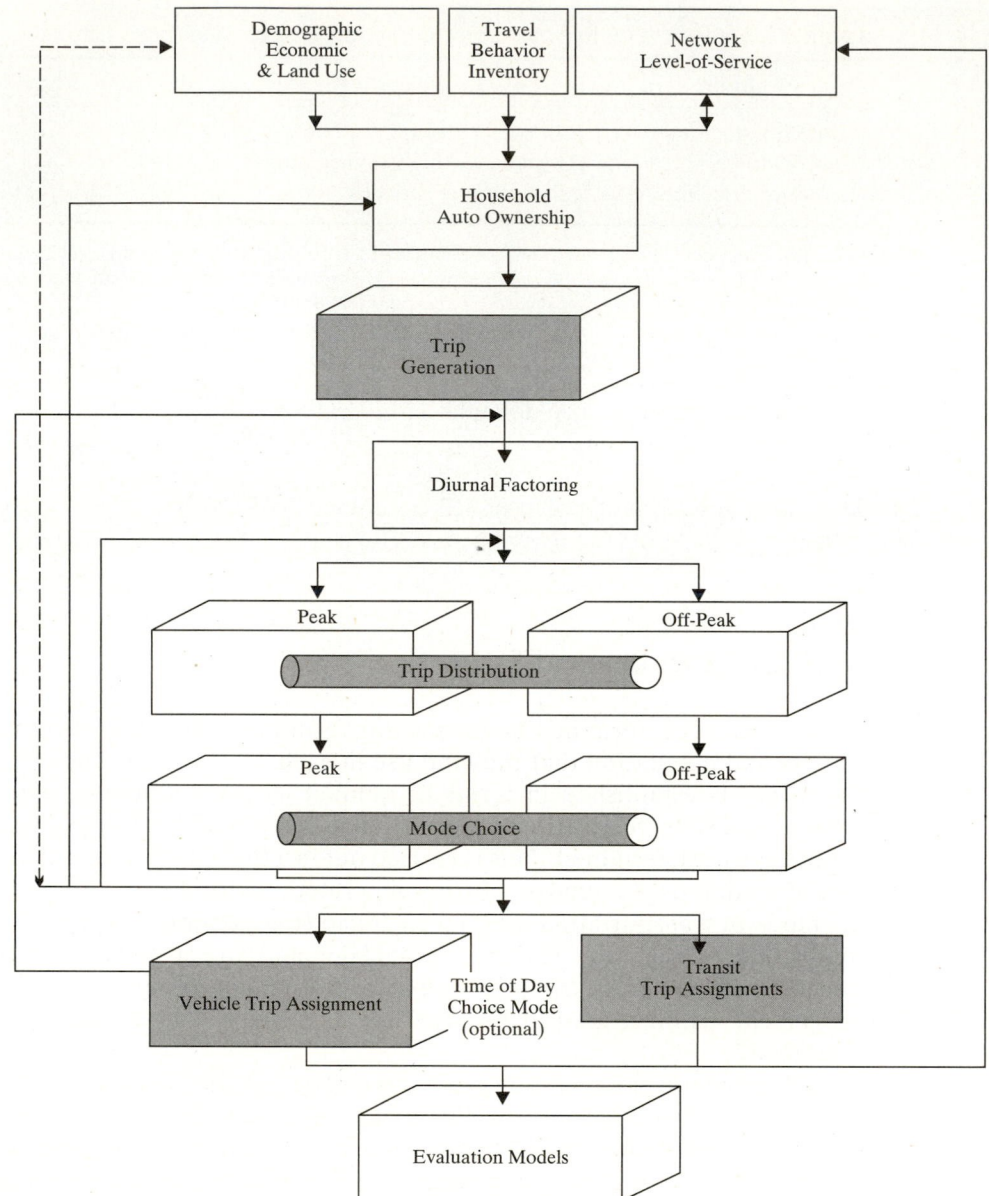

Figure 11.8 Travel Demand Model Flow Chart

SOURCE: *Transportation Planning Handbook*, Third Edition, Institute of Transportation Engineers, 2004.
www.ite.org. Used with permission.
ITE SOURCE: Atlanta Regional Commission, January 2004.

The purpose of the evaluation process is to identify feasible alternatives in terms of cost and traffic capacity, to estimate the effects of each alternative in terms of the objectives expressed, and to assist in identifying those alternatives that will serve the traveling public and be acceptable to the community. Of particular importance are the environmental assessments mandated in most urban transportation studies.

Environmental Impact Statements

Federal and/or state regulations may require that the environmental impacts of proposed projects be assessed. These impacts may include effects on air quality, noise levels, water quality, wetlands, and the preservation of historic sites of interest.

The analytical process through which these effects are identified can take one of three forms, depending on the scope of the proposed project: a full *environmental impact statement* (EIS), a simpler *environmental assessment* (EA), or a cursory checklist of requirements known as a *categorical exclusion* (CE). Examples of projects that might receive the CE designation are utility installations along an existing transportation facility, bicycle lanes, noise barriers, and improvements to rest areas: Generally, projects that receive the CE designation do not require further detailed analysis. Projects for which an EIS is required might include the construction of a new limited-access freeway or a fixed guideway transit line. For some projects, it might not initially be clear whether the project merits a full environmental study. For those projects, an environmental assessment may be conducted, and the EA will either result in a finding of no significant impact (FONSI) or in a finding that a full EIS is warranted. For some large-scale projects that require an EIS, projects may be "tiered." The first tier addresses macroscopic issues, such as whether a proposed facility should be a highway bypass or a light rail line and whether it should follow an eastern or western alignment. The second tier addresses microscopic issues, such as the number of parcels that might be taken once a decision has been made that the facility will be a rail line following a western alignment.

The purpose of this environmental review process is to assure that all potential effects (positive as well as adverse) are addressed in a complete manner so that decision makers can understand the consequences of the proposed project. Once an agency, such as a state department of transportation, completes a draft EIS, the public is given an opportunity to comment. The project cannot proceed until the revised, final EIS has been accepted by the appropriate federal and state regulatory agencies. Generally, these federal regulatory agencies will include FHWA and FTA.

Elements of an EIS

Although the entire environmental review process is beyond the scope of this text, examination of some of the common elements of an EIS illustrate its role in transportation planning.

The project's *purpose and need* section articulates why the project is being undertaken: Is it to improve safety, to increase capacity in response to expected future traffic growth, or is it to upgrade a deficient link in a region's comprehensive transportation network? The purpose and need section should include relevant AADT projections, crash rates, and a description of existing geometric conditions.

The *alternatives* to the proposed project, such as the do-nothing case, should be described, as well as any criteria that have been used to eliminate alternatives from further consideration. For example, if a second bridge crossing over a body of water is being considered, then alternatives could be to change the location, to widen an existing bridge, to improve ferry service, or to do nothing. Criteria that prevent further consideration, such as the presence of an endangered species at what would have been another potential location or the permanent loss of several hectares of wetlands, are given in this section.

The *environmental effects* of the proposed project, such as water quality (during construction and once construction is complete), soil, wetlands, and impacts on plant and animal life, especially endangered species, should be analyzed. Note that environmental effects also include the impact on communities, such as air quality, land use, cultural resources, and noise.

Computing Environmental Impacts for Emissions and Noise

The level of detail in a full EIS can be staggering given the amount of analysis required to answer some seemingly simple questions: What is the noise level? How much will automobile emissions increase?

A number of tools are provided by regulatory agencies that can answer some of these questions. For example, the Environmental Protection Agency (EPA) has developed the Motor Vehicle Emissions Simulator (MOVES) model, which estimates emissions for mobile sources such as cars and trucks. The results of this model, compared with observed carbon monoxide concentrations, can be used to determine the relative impact of different project alternatives on the level of carbon monoxide.

One approach for quantifying noise impacts is to use the Federal Highway Administration's Transportation Noise Model (TNM), a computer program that forecasts noise levels as a function of traffic volumes and other factors. The method by which noise impacts are assessed can vary by regulatory agency. For example, the FHWA will permit one to use the L_{10} descriptor, which is "the percentile noise level that is exceeded for ten percent of the time." A more common noise descriptor is L_{eq}, which is the average noise intensity over time. A variant of this descriptor may be used by the U.S. Department of Housing and Urban Development (HUD), where the L_{eq} for each hour is determined but then a 10-dB "penalty" is added to the values from 10 p.m. to 7 a.m. Since noise is proportional to traffic speed, the impact of this last type of descriptor is to favor projects that would not necessarily result in high speeds in close proximity to populated areas during the evening hours.

An EIS utilizes current and forecasted volume counts (i.e., automobiles, medium trucks, and heavy trucks), speeds, and directional split in order to compare environmental effects for the current conditions, future conditions with the proposed project, and future conditions assuming any other alternatives, which at a minimum should include the no-build option.

11.3.6 Choice of Project

Selection of a project will be based on a process that will ultimately involve elected officials and the public. Quite often, funds to build an urban transportation project (such as a subway system) may involve a public referendum. In other cases, a vote by a state legislature may be required before funds are committed. A multiyear program then will be produced that outlines the projects to be carried out over the next 20 years. With approval in hand, the project can proceed to the specification and construction phase.

11.4 FORECASTING TRAVEL

To accomplish the objectives and tasks of the urban transportation planning process, a technical effort referred to as the *urban transportation forecasting process* is carried out to analyze the performance of various alternatives. There are four basic elements and related tasks in the process: (1) data collection (or inventories), (2) analysis of existing conditions and calibration of forecasting techniques, (3) forecast of future travel demand, and (4) analysis of the results. These elements and related tasks are described in the following sections.

11.4.1 Defining the Study Area

Prior to collecting and summarizing the data, it is usually necessary to delineate the study area boundaries and to further subdivide the area into transportation analysis zones (TAZ) for data tabulation. An illustration of traffic analysis zones for a transportation study is shown in Figure 11.9. The selection of these zones is based on the following criteria:

1. Socioeconomic characteristics should be homogeneous.
2. Intrazonal trips should be minimized.
3. Physical, political, and historical boundaries should be utilized where possible.
4. Zones should not be created within other zones.

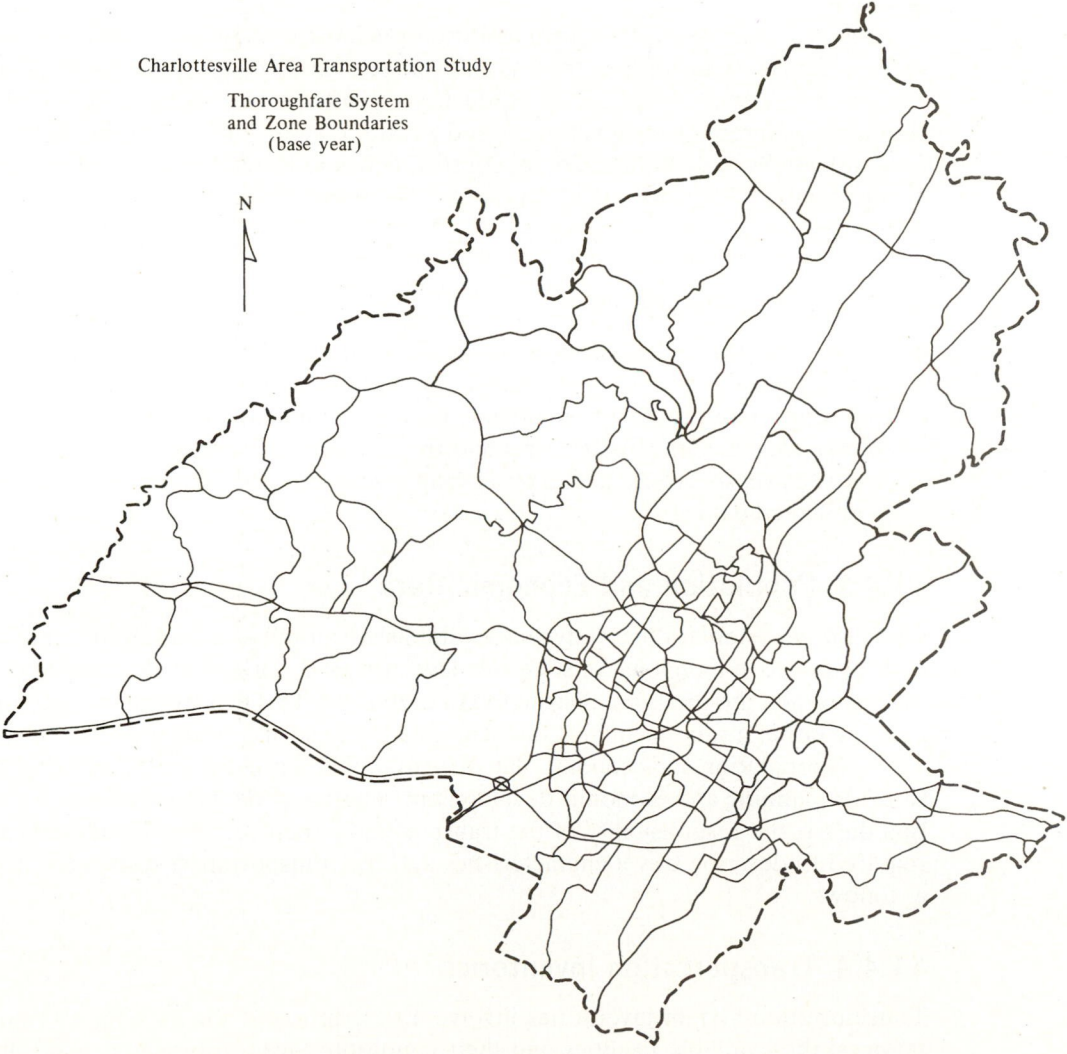

Charlottesville Area Transportation Study

Thoroughfare System
and Zone Boundaries
(base year)

N

Figure 11.9   Traffic Analysis Zones for Transportation Study

SOURCE: © 1985 Commonwealth of Virginia

5. The zone system should generate and attract approximately equal trips, households, population, or area. For example, labor force and employment should be similar.
6. Zones should use census tract boundaries where possible.

It may be necessary to exercise some judgment in determining the total number of zones. For example, one guideline for establishing the total number within a study area is that there should be, on average, one zone per 1000 people (such that an area with 500,000 people would have 500 total zones). The internal trip table for such a study area will thus have $500 \times 500 = 250,000$ cells (e.g., trips from zone 1 to zone 1, trips from zone 1 to zone 2, trips from zone 1 to zone 3, and so forth). By comparison, suppose the study area has a population of 2 million. Application of the 1000 people/zone guideline would yield 2000 total zones and thus $2000 \times 2000 = 4$ million cells for the internal trip table, which requires substantially more computing power and time to process.

Agencies may provide some guidance for achieving these six criteria. Examples of such guidance are as follows: there is an average of 1000 people/zone for smaller areas, a ratio of between 0.9 and 1.1 for productions to attractions, no more than 10,000 trips should be generated for a given zone, and a ratio of labor force to employment must be at least 0.80. Such guidelines do not constitute absolute standards but rather represent a compromise between an ideal data set and available resources for data collection and processing.

11.4.2 Data Collection

The data collection phase provides information about the city and its people that will serve as the basis for developing travel demand estimates. The data include information about economic activity (employment, sales volume, income, etc.), land use (type, intensity), travel characteristics (trip and traveler profile), and transportation facilities (capacity, travel speed, etc.). This phase may involve surveys and can be based on previously collected data.

11.4.3 Population and Economic Data

Once a zone system for the study area is established, population and socioeconomic forecasts prepared at a regional or statewide level are used. These are allocated to the study area, and then the totals are distributed to each zone. This process can be accomplished by using either a ratio technique or small-area land-use allocation models.

The population and economic data usually will be furnished by the agencies responsible for planning and economic development, whereas providing travel and transportation data is the responsibility of the transportation engineer. For this reason, the data required to describe travel characteristics and the transportation system are described as follows.

11.4.4 Transportation Inventories

Transportation system inventories involve a description of the existing transportation services; the available facilities and their condition; location of routes and schedules; maintenance and operating costs; system capacity and existing traffic; volumes, speed, and delay; and property and equipment. The types of data collected about the current system will depend on the specifics of the problem.

For a highway planning study, the system would be classified functionally into categories that reflect their principal use. These are the major arterial system, minor arterials, collector roads, and local service (see Chapter 15). Physical features of the road system would include number of lanes, pavement and approach width, traffic signals, and traffic-control devices. Street and highway capacity would be determined, including capacity of intersections. Traffic volume data would be determined for intersections and highway links. Travel times along the arterial highway system would also be determined.

A computer model of the existing street and highway system is produced. The network consists of a series of links, nodes, and centroids (as illustrated in Figure 11.10). A *link* is a portion of the highway system that can be described by its capacity, lane width, and speed. A *node* is the end point of a link and represents an intersection or location where a link changes direction, capacity, width, or speed. A *centroid* is the location within a zone where trips are considered to begin and end. Coding of the network requires information from the highway inventory in terms of link speeds, length, and capacities. The network is then coded to locate zone centroids, nodes, and the street system.

Figure 11.11 shows *external stations*, which are established at the study area boundary. External stations are those roadways where traffic is likely to enter or exit the study area, such as primary and interstate facilities, and are used to account for the impact of changes outside the study area on the travel network within the study area.

For a transit planning study, the inventory includes present routes and schedules, including headways, transfer points, location of bus stops, terminals, and parking facilities. Information about the bus fleet, such as its number, size, and age, would be identified. Maintenance facilities and maintenance schedules would be determined, as would the organization and financial condition of the transit companies furnishing service in the area. Other data would include revenue and operating expenses.

The transportation facility inventories provide the basis for establishing the networks that will be studied to determine present and future traffic flows. Data needs can include the following items:

- Public streets and highways
  - Rights-of-way
  - Roadway and shoulder widths
  - Locations of curbed sections
  - Locations of structures such as bridges, overpasses, underpasses, and major culverts
  - Overhead structure clearances
  - Railroad crossings
  - Location of critical curves or grades
  - Identification of routes by governmental unit having maintenance jurisdiction
  - Functional classification
  - Street lighting
- Land-use and zoning controls
- Traffic generators
  - Schools
  - Parks
  - Stadiums
  - Shopping centers
  - Office complexes
- Laws, ordinances, and regulations
- Traffic-control devices

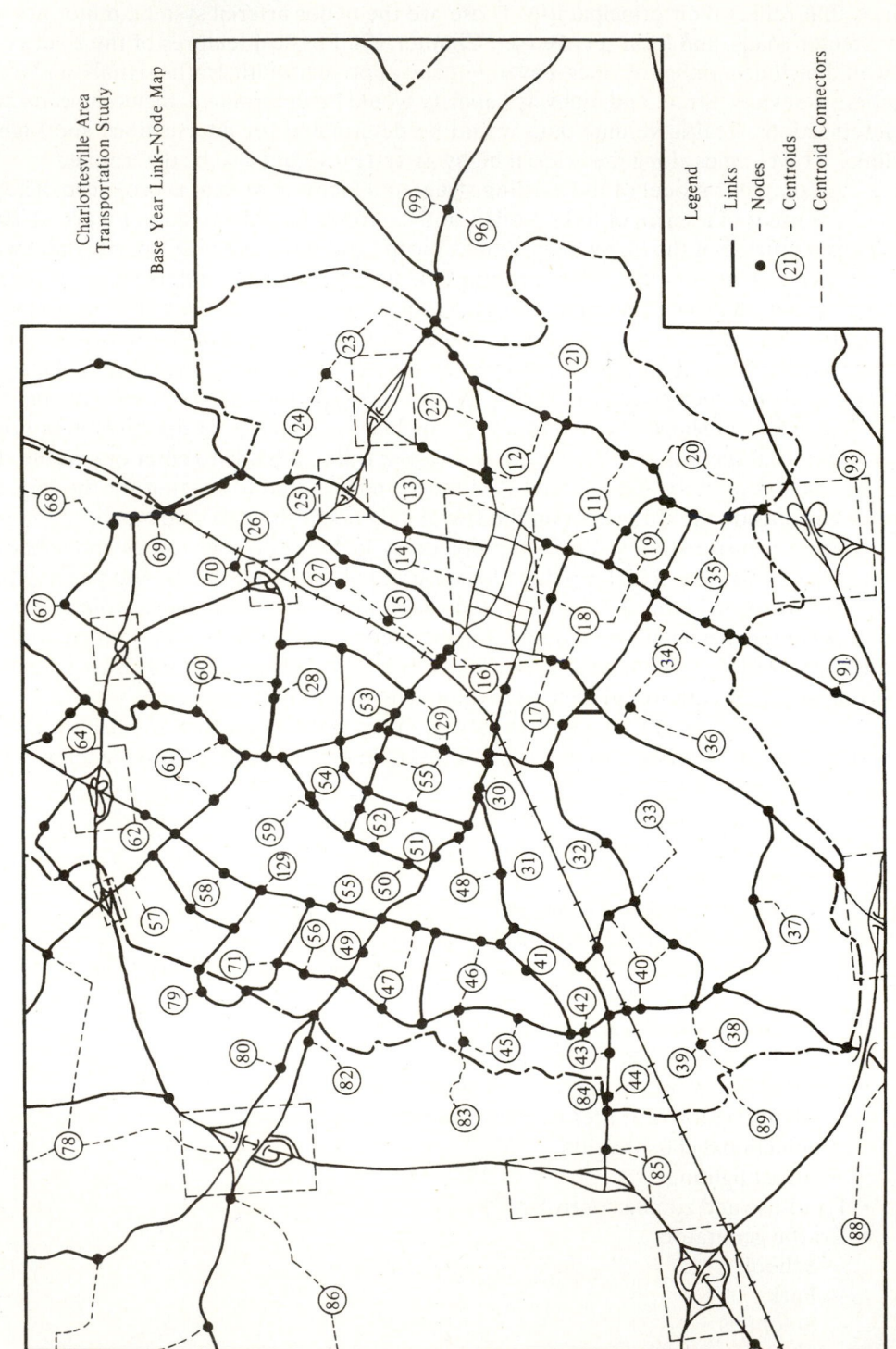

Figure 11.10 Link-Node Map for Highway System

SOURCE: © 1985 Commonwealth of Virginia

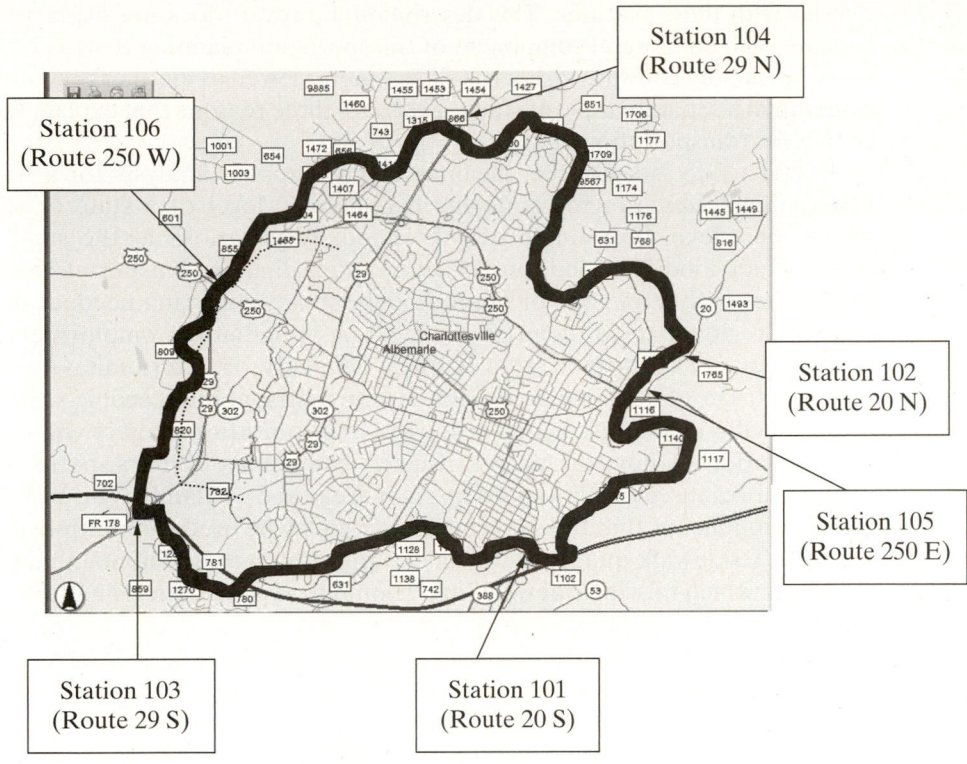

Figure 11.11 External Stations for a Study Area Boundary

SOURCE: Based on Virginia Department of Transportation GIS Integrator

- – Traffic signs
- – Signals
- – Pavement markings
- Transit system
  - – Routes by street
  - – Locations and lengths of stops and bus layover spaces
  - – Location of off-street terminals
  - – Change of mode facilities
- Parking facilities
- Traffic volumes
- Travel times
- Intersection and roadway capacities

In many instances, the data will already have been collected and are available in the files of city, county, or state offices. In other instances, some data may be more essential than others. A careful evaluation of the data needs should be undertaken prior to the study.

11.4.5 Information Systems

Almost all network data are organized within some type of Geographic Information System (GIS). A GIS is a spatially oriented database management system containing location and attribute information for synthetic and natural features and supporting related queries with these features. This description, however, does not explain why GIS has become such an integral component of transportation planning that, as is the case with word processors and spreadsheets, GIS is simply viewed as another practical instrument rather than a separate topic of study. There are three reasons that explain the popularity of GIS for transportation planning.

First, a GIS is scaleable, meaning it may support analysis for a wide range of geographic scales, ranging from the macroscopic level of a state or region to the microscopic level of a single neighborhood. Applications at the state or regional level include modifying the zone structure (e.g., aligning transportation analysis zone boundaries with manmade or natural boundaries), obtaining needed socioeconomic data for regional travel demand models (e.g., population, employment, household size, income, or other indicators of travel demand), or determining the rail facilities within 160 km of a freight generator. By comparison, microscopic-scale queries are often specific to a single neighborhood or transportation project and may include a determination of how the project impacts community centers, parks, schools, and ecological resources such as conservation areas and wetlands. The scalability of this information means that the same information may be used for multiple purposes. For example, a sidewalk quality survey may support both a current maintenance program (dictating which missing sidewalk links should be repaired first) and future pedestrian travel demand estimation.

Second, a GIS contains information used by other professions, enabling planners to access data that already have been collected for other purposes. For example, a county assessor's office may have tax records for various residential parcels, including the square footage of each structure, its appraised value, its street address, and the residential density (e.g., number of dwellings per hectare). Although these tax maps may have been created primarily for the purpose of tax assessment, knowledge of the housing density and price may enable the transportation planner to more accurately estimate the number of automobile trips that will be generated by a neighborhood (since such trips are affected by factors that include housing density and personal wealth). GIS data are thus available from a variety of public and private sources. For example, the Census Bureau provides Topologically Integrated Geographic Encoding and Referencing (TIGER) line files that include transportation facilities (streets, highways, and rail lines), community landmarks (schools and hospitals), environmental features (water, streams, and wetlands), and jurisdictional information (e.g., boundaries and census tracts).

Third, a GIS offers strong spatial analytical capabilities that make use of the point, line, and area features contained within the GIS. For example, to investigate the possibility of runoff from transportation facilities affecting wetlands, an analyst may create a 200-meter (656-ft) buffer around a region's roads and identify any wetlands within that buffer. Such a query is known as a line/area query, since the highways are line features (defined by a series of segments) and the wetlands are an area feature (defined by a polygon). In addition to line features (e.g., streams, rivers, highways, or rail lines) and area features (e.g., cities, counties, lakes, or forests), a GIS may also contain point features (e.g., historic churches, community centers, or wildlife crossings). These features support other types of queries besides those shown in Figure 11.12, such as tabulating the number of persons within 0.8 km of a heavy rail stop (a point/area query), identifying unforested lands that are not within a county's development plan (an area/area query), or determining the number of alternative routes between two points (a network analysis query).

The versatility of GIS to incorporate data from a variety of sources means that substantial data cleansing may be necessary for some applications when GIS data are

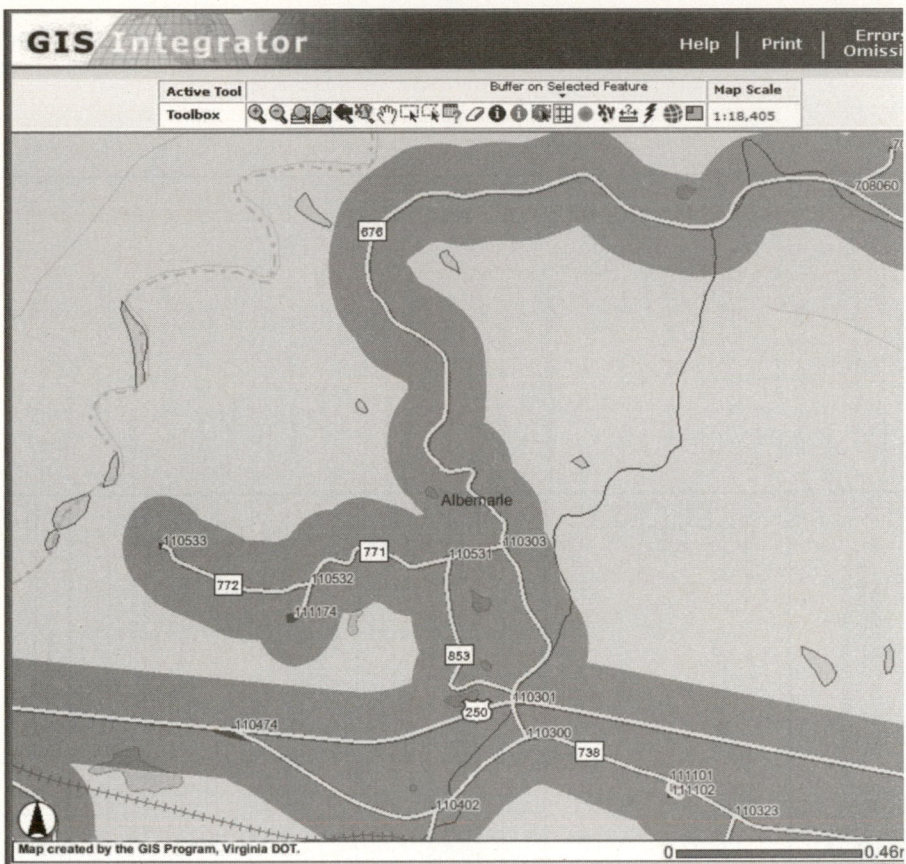

Figure 11.12   GIS Buffer Analysis to Identify Wetlands Impacted by Roadways

SOURCE: Created by using GIS tools within the Virginia Department of Transportation GIS Integrator

imported from other sources. For example, one state's GIS had a grade-separated overpass initially represented as an at-grade intersection (because the GIS had initially not been developed for the purposes of network analysis). Thus, each intersection was reviewed to ensure that representation within the GIS matched intersections in the field.

11.4.6 Travel Surveys

Travel surveys are conducted to establish a complete understanding of the travel patterns within the study area. For single projects (such as a highway project), it may be sufficient to use traffic counts on existing roads or (for transit) counts of passengers riding the present system. However, to understand why people travel and where they wish to go, origin-destination (O-D) survey data can be useful. The O-D survey asks questions about each trip that is made on a specific day—such as where the trip begins and ends, the purpose of the trip, the time of day, and the vehicle involved (auto or transit)—and about the person making the trip—age, sex, income, vehicle owner, and so on. Figure 11.13 illustrates a home interview origin-destination survey form. The O-D survey may be completed as a home interview, or people may be asked questions

Travel Day _____ and Date

Sample Number

NORTH CENTRAL TEXAS
COUNCIL OF GOVERNMENTS
HOME INTERVIEW SURVEY

Section I: Household Data

A. Sample Address _____ House Number, Street Name, Apt. No. _____ City/Town _____ County _____ Zip Code

B. Structure Type _____

C. Number of People Living at this Address _____

D. Number of People Age 5 and Over Living at this Address _____

E. Number of Out-of-Area Visitors Staying at this Address _____

F. Number of Passenger Cars, Vans, and Pickups Available for Use _____

G. Household Income: (Do Not Ask Until Interview Is Complete)

Section II: Data on Persons Age 5 and Over

A	B	C	D	E	F	G	H	I	J	K
Person Number	✓ If Interviewed	Relation To Head	Age	Sex	Licensed to Drive?	Occupation	Industry	Worked on Travel Day?	Made Trips While at Work?	Made Other Trips on Travel Day?
01		Head		1 M / 2 F	1 YES / 2 NO			1 YES 2 NO 3 Worked at Home	1 YES 2 NO	1 YES 2 NO
02				1 M / 2 F	1 YES / 2 NO			1 YES 2 NO 3 Worked at Home	1 YES 2 NO	1 YES 2 NO
03				1 M / 2 F	1 YES / 2 NO			1 YES 2 NO 3 Worked at Home	1 YES 2 NO	1 YES 2 NO
04				1 M / 2 F	1 YES / 2 NO			1 YES 2 NO 3 Worked at Home	1 YES 2 NO	1 YES 2 NO
05				1 M / 2 F	1 YES / 2 NO			1 YES 2 NO 3 Worked at Home	1 YES 2 NO	1 YES 2 NO
06				1 M / 2 F	1 YES / 2 NO			1 YES 2 NO 3 Worked at Home	1 YES 2 NO	1 YES 2 NO
07				1 M / 2 F	1 YES / 2 NO			1 YES 2 NO 3 Worked at Home	1 YES 2 NO	1 YES 2 NO
08				1 M / 2 F	1 YES / 2 NO			1 YES 2 NO 3 Worked at Home	1 YES 2 NO	1 YES 2 NO
09				1 M / 2 F	1 YES / 2 NO			1 YES 2 NO 3 Worked at Home	1 YES 2 NO	1 YES 2 NO
10				1 M / 2 F	1 YES / 2 NO			1 YES 2 NO 3 Worked at Home	1 YES 2 NO	1 YES 2 NO

Age Codes

1	5 - 10	6	36 - 45
2	11 - 15	7	46 - 55
3	16 - 20	8	56 - 65
4	21 - 25	9	65 - OVER
5	26 - 35	0	UNKNOWN

Relation Codes

1	HEAD	6	GRANDCHILD
2	SPOUSE	7	OTHER RELATIVE
3	SON	8	UNRELATED
4	DAUGHTER	9	OUT-OF-AREA VISITORS
5	GRANDPARENT	0	UNKNOWN

Section III: Trip Summary

A. Total Vehicular Trips Reported _____
B. Persons Age 5 and Over Making Trips _____
C. Persons Age 5 and Over Not Making Trips _____
D. Complete or Incomplete Interview Code _____

Section IV: Administrative

A. Household Telephone Number _____

B. Interviewer _____

C. Telephone Contacts (If Any) :

Date _____ Time _____ Purpose/Outcome _____

D. Personal Contacts In Household:

Date _____ Time _____ Talked To/Comments _____

E. Completed Interview Submitted:

Date: _____ By: _____

I Certify That All Information
On This Form Is Correct And True.

_____
Signature of Interviewer

F. If Interview Submitted Incomplete

Interviewer's Reason: _____

_____ Initials
Supervisor's Comments

G. First Edit: Fail Pass

_____ _____
Date   Initials

H. Final Edit: Fail Pass

_____ _____
Date   Initials

I. Coding Complete

_____ _____
Date   Initials

Figure 11.13 Travel Behavior Inventory: Home Interview Survey

SOURCE: North Central Texas Council of Governments

while riding the bus or when stopped at a roadside interview station. Sometimes, the information is requested by telephone, by return postcard, by e-mail, and by the Internet. O-D surveys are rarely completed in communities where these data have been previously collected. Due to their high cost, O-D surveys are being replaced by using U.S. Census travel-to-work data.

O-D data are compared with other sources to ensure the accuracy and consistency of the results. Among the comparisons used are crosschecks between the number of dwelling units or the trips per dwelling unit observed in the survey with published data. Screenline checks (see Chapter 4) can be made to compare the number of reported trips that cross a defined boundary, such as a bridge or two parts of a city, with the number actually observed. For example, the number of cars observed crossing one or more bridges might be compared with the number estimated from the surveys. It is also possible to assign trips to the existing network to compare how well the data replicate actual travel. If the screenline crossings are significantly different from those produced by the data, it is possible to make adjustments in the O-D results so that conformance with the actual conditions is assured.

Following the O-D checking procedure, a set of trip tables is prepared that shows the number of trips between each zone in the study area. These tables can be subdivided, for example, by trip purpose, truck trips, and taxi trips. Tables are also prepared that list the socioeconomic characteristics for each zone and the travel time between zones. Examples of trip tables are shown in Chapter 12.

11.4.7 Calibration

Calibration is concerned with establishing mathematical relationships that can be used to estimate future travel demand. Usually, analysis of the data will reveal the effect on travel demand of factors such as land use, socioeconomic characteristics, or transportation system factors.

Example 11.3 Estimating Trips per Day Using Multiple Regression

A multiple regression analysis shows the following relationship for the number of trips per household.

$$T = 0.82 + 1.3P + 2.1A$$

where

T = number of trips per household per day
P = number of persons per household
A = number of autos per household

If a particular TAZ contains 250 households with an average of 4 persons and 2 autos for each household, determine the average number of trips per day in that zone.

Solution:

Step 1. Calculate the number of trips per household.

$$T = 0.82 + 1.3P + 2.1A$$

$$= 0.82 + (1.3 \times 4) + (2.1 + 2)$$

$$= 10.22 \text{ trips/household/day}$$

Step 2. Determine the number of trips in the entire zone.

$$\text{Total trips in TAZ} = 250(10.22) = 2555 \text{ trips/day}$$

Other mathematical formulas establish the relationships for trip length, percentage of trips by auto or transit, or the particular travel route selected.

Travel forecasts are made by applying the relationships developed in the calibration process. These formulas rely upon estimates of future land use, socioeconomic characteristics, and transportation conditions.

11.4.8 Steps in the Travel Forecasting Process

Forecasting can be summarized in a simplified way by indicating the task that each step in the process is intended to perform. These tasks are as follows (and are described more fully in Chapter 12).

Step 1. **Population and economic analysis** determines the magnitude and extent of activity in the urban area.

Step 2. **Land-use analysis** determines where the activities will be located.

Step 3. **Trip generation** determines how many trips each activity will produce or attract.

Step 4. **Trip distribution** determines the origin or destination of trips that are generated at a given activity.

Step 5. **Modal split** determines which mode of transportation will be used to make the trip.

Step 6. **Traffic assignment** determines which route on the transportation network will be used when making the trip where each user seeks to minimize their travel time on the network.

Computers are used extensively in the urban transportation planning process. A package of programs was developed by the Federal Highway Administration (FHWA) and the Federal Transit Administration (FTA), called the Urban Transportation Planning System (UTPS). The techniques have been computerized, and various versions of the original UTPS program are now available from private vendors. Because the data requirements for applying travel-demand software are substantial, the ability to use a package to import data from other sources is an important consideration in choosing such software. A separate class of computer software packages has been developed to give rough estimates of travel demand and the resultant impact on the transportation network. These packages are known as sketch planning methods and are designed to work as a supplement to the travel-demand packages. One example is the Surface

Transportation Efficiency Analysis Module (STEAM), which provides estimates of the impact various alternatives will have on delay, emissions, accidents, and user costs such as fuel consumption and parking.

11.4.9 Transportation and Land-Use Coordination

Because transportation is a desired demand, decision makers may be interested in evaluating how alternative land-use policies may influence this demand. Such policies may take the form of regulation, incentives, or both.

- *Concurrency requirements* are a regulatory technique, where a state or locale requires that sufficient transportation infrastructure be present to accommodate anticipated growth. One mechanism through which a state may implement concurrency is to require that localities adopt a set of performance standards for various transportation facilities in the county. For roadways, these standards may be based on level of service, functional classification, and location; for example, a locale may set a performance standard of LOS B for principal arterials in rural areas and a standard of E for minor collectors in urban areas. Standards may also reflect transit or bicycle performance through the use of service frequency or breadth of coverage, respectively. When new developments are proposed, a *traffic impact assessment* is undertaken to determine whether the new development will still permit the performance standard to be obtained. If not, improvements may be required, such as a roadway widening, transit service expansion, signal coordination, the construction of bicycle lanes, or other steps that will mitigate the travel impacts of the new development.
- *Grants for public and private entities* are examples of incentives. In contrast with residential development, commercial sites tend to generate more tax revenue than what they require in the form of services, which may encourage localities to have zoning that favors commercial over residential development. It is possible that such a situation may result in longer commuting distances if workers are unable to find affordable housing in proximity to their employment. To encourage an increase in housing stock in areas that have relatively high employment, grants may be provided by a state to localities which (1) have a relatively high ratio of jobs to housing and (2) issue a certain number of residential building permits. Grants may also be provided to the private sector; for example, a locale may permit developers to build extra office space for each increment of housing they provide.
- *Priority funding areas* are a mix of regulation and incentives. For example, in an effort to limit or manage the geographical growth of an urban area, a state may require that certain types of transportation investments, such as state-funded infrastructure capacity expansions, only be undertaken in regions that meet criteria in terms of residential density, water and sewer availability, or other indicators of land development. An illustration is Maryland's 1997 Smart Growth Areas Act, which required that major capacity-expansion construction projects, excluding toll facilities, be restricted to priority funding areas (PFAs). One criterion that counties could use to designate a PFA located outside an existing community was that the area must have densities equal to or greater than 3.5 dwelling units per acre.

A common element of these policies is that they provide guidance for public and private sector entities to plan for transportation systems that achieve one or more policy objectives. This guidance may also pertain to how highways are designed and managed. For example, consider the importance of access control on arterial multi-lane highways. The extent to which this control is managed clearly influences safety, where facilities with full access control have a crash rate that is approximately one-third that of non-access controlled facilities. State and local agencies have an interest in minimizing the need for new access points along an arterial facility and the volume of vehicles using these new access points. An agency may establish a policy that when new residential developments are constructed along an arterial facility, these new developments should have multiple connections between developments rather than having each development be accessible only from the arterial facility. Such a gridded network of streets may reduce the number of local trips on the arterial facility.

Connectivity requirements are one tool an agency may use to implement this policy of a gridded street network. When residential subdivisions and streets are constructed, a landowner may petition the state or county to assume responsibility for maintaining these streets. The public agency may then use connectivity requirements as a factor in deciding whether or not to accept maintenance responsibilities for such streets. Such requirements may concern the extent to which the streets are aligned in a grid pattern as opposed to a dendritic pattern, pedestrian considerations, and interconnections between properties (e.g., a series of highly interconnected streets within a subdivision that do not provide access to any other properties except via a single connection to an arterial highway are not likely to meet the requirements). For a privately built residential development, ways to measure connectivity include, but are not limited to,

- the number of intersections with an external roadway network
- the availability of sidewalks, trails, or other nonmotorized facilities between the development and other developments, businesses, or cultural sites
- an ability to provide future connections to the external roadway network, such as preservation of street end points that can be extended when new properties are developed

---

Example 11.4   Computing a Connectivity Index

A proposed residential development is shown in Figure 11.14a. The jurisdiction in which this parcel is located requires that new residential streets be privately maintained unless the proposed development will have a connectivity index of 1.5 or higher, where the connectivity index is defined by Eq. 11.2. Compute the connectivity index of the proposed development, indicate how the index can be increased to the required threshold of 1.5, and give one dimension of connectivity not quantified by the index.

$$CI = \frac{\text{Segments}}{\text{Nodes}} \qquad (11.2)$$

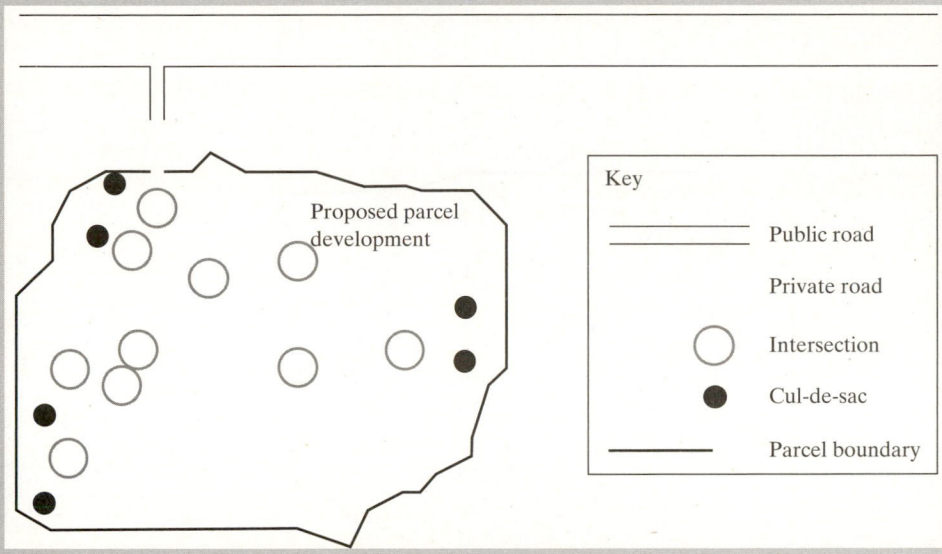

Figure 11.14a   Proposed Parcel Development (Original Proposal)

where

$$CI = \text{connectivity index}$$

Node = intersection with another private street or cul-de-sac

Segments = section of street between two nodes or between one node and the boundary of the private development

Solution:

Figure 11.14a shows 10 intersections and 6 cul-de-sacs, so there are 16 nodes
Figure 11.14a shows 20 segments. The connectivity index is therefore

$$CI = \frac{20 \text{ Segments}}{16 \text{ Nodes}} = 1.25$$

To raise the connectivity index to the required threshold of 1.5, one solution is to convert three of the cul-de-sacs to through streets as shown in Figure 11.14b, thereby reducing the number of nodes from 16 to 13. By extending segments 1, 13, and 17 to the boundary of the development for connection to future (as of yet unbuilt) street networks, the connectivity index will rise to 20/13 = 1.54.

This particular connectivity index does not measure the extent to which sidewalks, trails, and other amenities that would facilitate nonmotorized travel between this and adjacent developments. Further, this index does not differentiate between connections to existing and proposed developments. Such differentiation might affect residents if, for example, segment 1 connects to an existing business area whereas the area to the

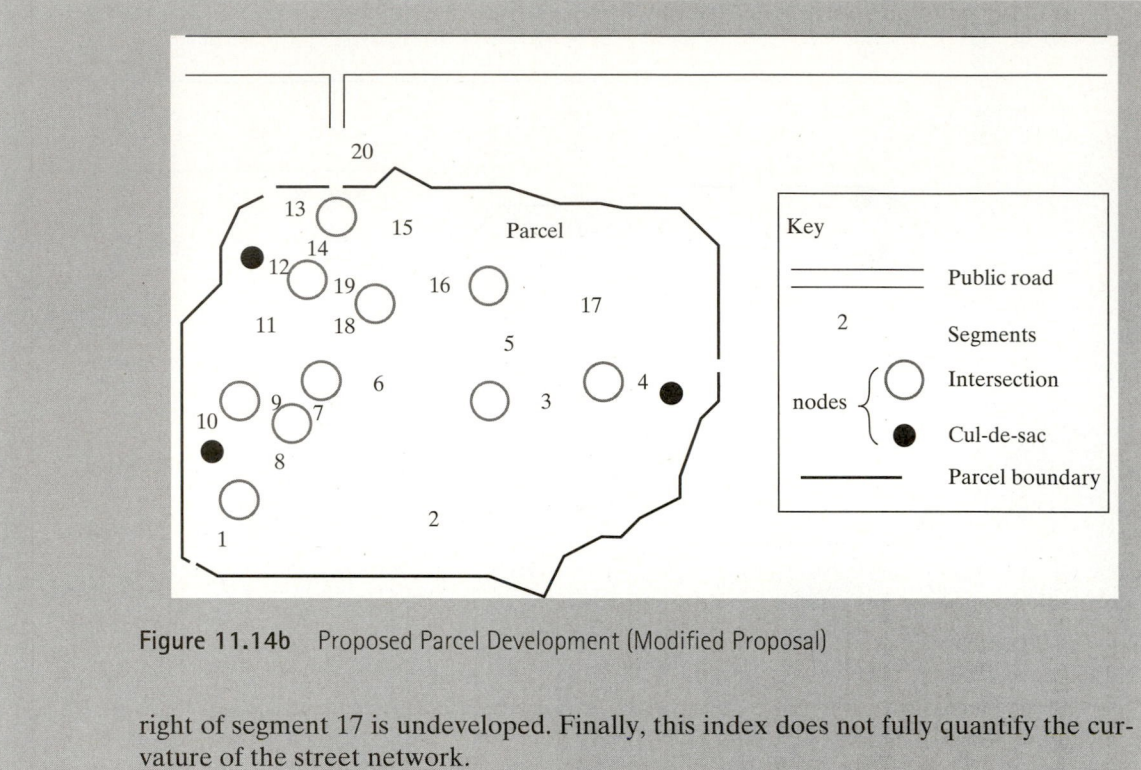

Figure 11.14b   Proposed Parcel Development (Modified Proposal)

right of segment 17 is undeveloped. Finally, this index does not fully quantify the curvature of the street network.

An active field of inquiry is quantifying the extent to which various techniques for coordinating transportation and land use influence travel and land development decisions. Because the impact of such techniques may depend on local conditions, it is often appropriate to assess their impact through the collection of local data that consider activity generators in the area, localized travel patterns, and supporting policies. Table 11.4 shows how such a local study may examine the impact of four potential techniques. Since such studies require that one control for confounding factors, the interested reader should consult the literature for examples of how such factors may be addressed.

11.4.10 Freight Planning

Freight planning is similar to passenger planning in that both may be used to satisfy one or more stated policy goals. Such goals may include providing adequate facilities to meet forecasted demand, mitigating the adverse consequences of such facilities, increasing economic competitiveness, or encouraging a shift between modes. For example, an expected increase in goods movement along a north-south corridor coupled with high congestion levels on the highway facility may lead to a decision to invest in improvements in an adjacent rail line to encourage freight shipments by rail. Alternatively, the feasibility of an inland port may be assessed with respect to the proposed port's ability to increase market share for a state's seaport at the expense of ports in other states. Steps to mitigate adverse impacts of freight facilities might include the relocation of a

Table 11.4 Approaches for Quantifying the Travel Impacts of Transportation/Land-Use Coordination Techniques

Technique	*Potential Method to Determine the Technique's Impact on Travel Behavior*
Concurrency requirement	Compare transportation system performance for areas with a concurrency requirement to areas without such a requirement
Grants to encourage housing in jobs-rich areas	Compare the distances commuters travel when living in (a) areas with a mix of jobs and housing and (b) areas that are primarily residential
Priority funding areas	Compare the rate of land development in areas designated as a PFA with the rate of land development in areas not designated as a PFA
Connectivity index	Compare the number of vehicle trips entering an arterial facility from otherwise similar developments with dissimilar connectivity indices

rail yard from a residential area, reduced idling of trucks and locomotives, the construction of additional truck rest areas (to accommodate truck shipments that arrive prior to the scheduled delivery time), and noise mitigation measures—such as the construction of sound walls (to reduce truck noise) or the use of continuous welded rail to reduce train noise.

Freight planning also may be performed at both the systems level of analysis and at the project level of analysis. At the *systems level,* freight-related projects will fall within the general urban travel demand analysis process. For example, a roadway widening that will support operations at a port in an MPO area will typically be identified within the urban travel demand forecast and then be placed in the Transportation Improvement Program (TIP) for inclusion within the state highway program. At the *project level,* short-term freight projects that are of direct interest to the business community may be considered, such as railroad siding improvements, expansion of turning radii to accommodate large trucks, and traffic signal timings.

There are several possible modes for freight transportation, each with its own advantages and disadvantages. Mode categories include truck (tractor trailers, drayage trips between seaports and rail facilities, and local deliveries), rail (tank cars, containerized shipments, or boxcars), maritime (auto carriers or large vessels), and air cargo. As shown in Figure 11.15, the market share for each mode may depend on whether tonnage or value is used. For the state of Virginia, the amount of freight shipped by rail is much larger than that shipped by air if the unit of measurement is tonnage. Because commodities shipped by air have a much higher value than commodities shipped by rail, air freight has a larger market share than rail freight, as shown in Figure 11.15, if the unit of measurement is value (billions $).

The analysis underpinning freight planning considers three factors that differ somewhat from that of transportation passenger planning. First, unlike passenger freight, some commodities are shipments that are not time sensitive, thereby allowing the shipper's choice of mode for those commodities to be made solely on the basis of cost and convenience. Second, freight-movement data are generally studied at a larger geographic scale than passenger movements, with county-to-county or

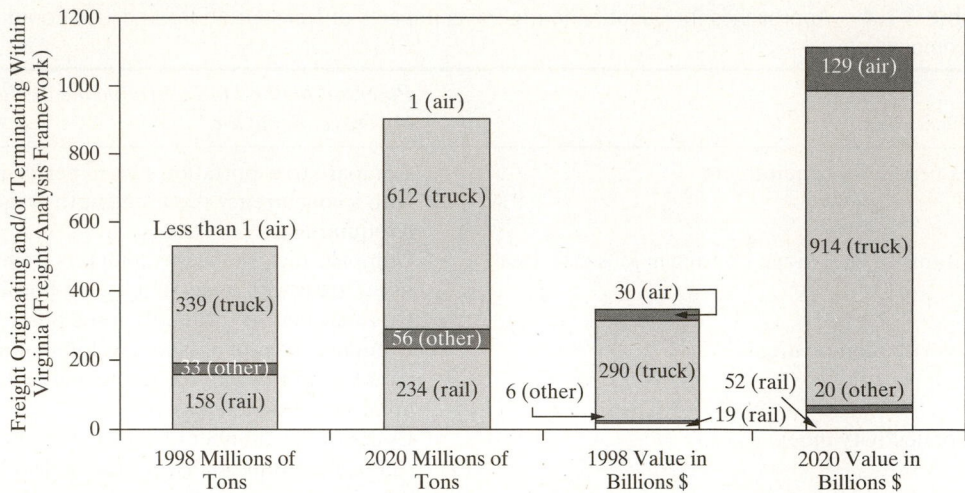

Figure 11.15   Virginia Forecasts for Freight Movements by Mode (U.S. DOT Freight Analysis Framework)

Note: Does not include freight that moves entirely through Virginia without an origin or destination therein. "Other" includes water, pipeline, and shipments that moved by an unspecified mode.
SOURCE: John S. Miller, *Expected Changes in Transportation Demand in Virginia by 2025*, Virginia Transportation Research Council, (VTRC) June 2003, VTRC 03-TAR5, Fig. 26, p. 51; http://www.virginiadot.org/vtrc/main/online_reports/pdf/03-tar5.pdf

state-to-state flows commonly analyzed. Third, whereas each passenger chooses his or her own mode and travel path, a single decision maker makes this choice for a large number of freight parcels. While an equilibrium network assignment for passenger travel may be defined as the assignment where no passenger can reduce his/her travel time (which is not necessarily the lowest system cost), a freight logistics provider may be able to consider the total system cost, where some parcels may have a longer delivery time than others.

11.5 SUMMARY

Transportation projects are selected based on a variety of factors and considerations. The transportation planning process is useful when it can assist decision makers and others in the community to select a course of action for improving transportation services.

The seven-step planning process is a useful guide for organizing the work necessary to develop a plan. The seven steps are (1) situation definition, (2) problem definition, (3) search for solutions, (4) analysis of performance, (5) evaluation of alternatives, (6) choice of project, and (7) specification and construction. Although the process does not produce a single answer, it assists the transportation planner or engineer in carrying out a logical procedure that will result in a solution to the problem. The process is also valuable as a means of describing the effects of each course of action and for explaining to those involved how the new transportation system will benefit the traveler and what its impacts will be on the community.

The elements of the urban transportation planning process are (1) inventory of existing travel and facilities, (2) establishment of goals and objectives, (3) generation of alternatives, (4) estimation of project costs and travel demand, (5) evaluation of alternatives, and (6) choice of project. An understanding of the elements of urban transportation planning is essential to place in perspective the analytical processes for estimating travel demand. Other elements of the transportation planning process are environmental impact statements, geographic information systems, land development policies, and freight planning.

PROBLEMS

11-1   Explain why the transportation planning process is not intended to furnish a decision or give a single result.

11-2   Describe the steps that an engineer must follow if asked to determine the need for a grade-separated railroad grade crossing that would replace an at-grade crossing of a two-lane highway with a rail line.

11-3   Describe the basic steps in the transportation planning process.

11-4   Select a current transportation problem in your community or state. Briefly describe the situation and the problem. Indicate options available and the major impacts of each option on the community.

11-5   Evaluate a proposal to increase tolls on existing roads and bridges. Describe the general planning analysis used.

11-6   Prepare a study to consider improvements to transportation between an airport and the city it serves.

11-7   What caused transportation planning to become institutionalized in urban areas, and on what does the process need to be based?

11-8   Explain the three "C"s concept in the transportation planning process, as mandated in the Federal Aid Highway Act of 1962.

11-9   Explain the difference between transportation planning and transportation programming, including differences between the key documents in each process and their respective time horizons.

11-10   Urban transportation is concerned with two separate time horizons. Briefly describe each and provide examples of the types of projects that can be categorized in each horizon.

11-11   In gathering data to support urban transportation planning, name the unit of geographic area at which data supporting an inventory of existing conditions are summarized. Note the factors that are considered in delineating these geographic units.

11-12   An existing highway-rail at-grade crossing is to be upgraded. Plans were developed in 2001; the cost estimate for that improvement was $570,000 at that time. Due to funding constraints, construction of the improvement was delayed until 2005. Using the data given in Table 11.3, estimate the construction cost in 2005 dollars.

11-13   Given the information in Problem 11-12, assume that construction was delayed until 2014. Using the compound growth rate (see Chapter 13) that can be derived from the data given in Table 11.3, estimate the construction cost in 2014 dollars.

11-14   Describe the three forms of environmental impact analysis documentation.

11-15   Identify the types of impacts and effects that might be addressed in an analysis of environmental impacts.

11-16   What is the purpose of performing inventories and surveys for each defined geographic unit or transportation analysis zone within a study area?

11-17   What are the four basic elements that make up the urban transportation forecasting process?

11-18   In the data collection phase of the urban transportation forecasting process, what type of information should the data reveal for a transportation analysis zone?

11-19   Define the following terms: (a) link, (b) node, (c) centroid, and (d) network.

11-20   Draw a link-node diagram of the streets and highway within your neighborhood or campus. For each link, show travel times and distances (to the nearest 0.16 km).

11-21   Explain the role that geographic information systems have come to play in transportation planning and key reasons why this has happened.

11-22   The initial zone structure for a regional travel demand model has been proposed as summarized below. Name three potential problems with this structure.

Zones:	50
Employment:	100,000
Labor force:	60,000
Population:	200,000
Trips produced by the study area:	100,000
Trips attracted by the study area:	150,000

11-23   Explain the flaw in each of these traffic analysis zones (TAZs):

(a) TAZ 1 contains dormitories, a research park, and 300 single-family detached dwelling units.

(b) TAZ 2 straddles an interstate highway, with half of the zone east of the highway and half to the west of the highway.

(c) TAZ 3 contains 5000 people.

11-24   Define these four acronyms and explain how they affect the transportation planning process: MPO, CLRP, TIP, and STIP. Describe the composition of one of them.

11-25   Name the steps in the four-step travel demand forecasting process where feedback can occur. Which of these feedback loops directly affects land use? (*Hint*: See Figure 11.7.)

11-26   Name and briefly explain the tools available to local governments that may affect how land-use decisions influence travel demand.

11-27   Briefly explain the differences in the factors considered in freight transportation planning versus passenger transportation planning.

REFERENCES

American Association of Railroads, *AAR Railroad Cost Indexes*. Washington, D.C. 2006.

Benz, G. P., Environmental Considerations in the Transportation Planning Process, Chapter 5, Institute of Transportation Engineers, *Transportation Planning Handbook*. 3rd ed., Washington, D.C., 2009.

Bureau of Labor Statistics, *Consumer Price Index—All Urban Consumers*. Washington, D.C., 2006. http://data.bls.gov/cgi-bin/surveymost?cu

Cambridge Systematics, Inc., Transmanagement, Inc., TransTech Management, Inc., and Heanue, K., *NCHRP Report 570: Guidebook for Freight Policy, Planning, and Programming in Small- and Medium-Sized Metropolitan Areas*. TRB, National Research Council, Washington, D.C., 2007.

Center for Microcomputers in Transportation, University of Florida, Transportation Research Center, 2000.

California Department of Housing and Community Development. California's Jobs-Housing Balance Incentive Grant Program: Final Report to the Legislature. Sacramento, 2007. http://www.hcd.ca.gov/fa/jhbig/jhb_rept_legis1007.pdf

Cervero, R., "Jobs-Housing Balancing and Regional Mobility," *Journal of the American Planning Association*, 55(2): 136–150, 1989.

Dahlgren, J. and D. B. Lee, Jr., *Integrating ITS Alternatives into Investment Decisions in California,* Report UCB-ITS-PRR-2004-40, University of California, Berkeley, 2004.

Federal Highway Administration, *Computer Programs for Urban Transportation.* Washington, D.C., April 1977.

Federal Highway Administration, *National Highway Construction Cost Index.* Washington, D.C., undated. http://www.fhwa.dot.gov/policyinformation/nhcci.cfm

Federal Highway Administration, *Price Trends for Federal-Aid Highway Construction.* Washington, D.C., 2006. http://www.fhwa.dot.gov/programadmin/pricetrends.htm

Federal Highway Administration, "Tiering Can Work: Missouri's I-70 Project," *Success in Stewardship,* December 2002. http://environment.fhwa.dot.gov/strmlng/newsletters/dec02nl.asp

Fischer, M. J. and M. Han, *NCHRP Synthesis 298: Truck Trip Generation Data.* TRB, National Research Council, Washington, D.C., 2001.

Forecasting Inputs to Transportation Planning. NCHRP Report 266, Transportation Research Board, National Research Council, Washington, D.C., 1983.

Horowitz, A., *NCHRP Synthesis 358: Statewide Travel Forecasting Models.* Washington, D.C., 2006.

Institute of Transportation Engineers, *Transportation Planning Handbook,* 3rd ed., Washington, D.C., 2009.

Lewis, R., G.-J. Knaap, and J. Sohn. "Managing Growth with Priority Funding Areas: A Good Idea Whose Time Has Yet to Come." *Journal of the American Planning Association,* 75(4): 457–478, 2009. http://www.informaworld.com/smpp/content~content=a915602672

Martin, W. A., and N. A. McGuckin, *NCHRP Report 365: Travel Estimation Techniques for Urban Planning.* TRB, National Research Council, Washington, D.C., 1998.

Office of Freight Management and Operations, *Freight Transportation Profile—Virginia Freight Analysis Framework.* Federal Highway Administration, Washington, D.C., 2002, FHWA-OP-03-036. http://www.ops.fhwa.dot.gov/freight/ state_profiles.htm

Seggerman, K. E., K. M. Williams, and P. S. Lin, *Transportation Concurrency: Best Practices Guide*, University of South Florida and the Florida Department of Community Affairs, Tampa, 2007. http://www.cutr.usf.edu/pdf/DCA_TCBP%20Guide.pdf

Southeastern Michigan Council of Governments. Regional Concept for Transportation Operations (RCTO) Final Report: Southeast Michigan Experience, Detroit, 2007.

Strauss-Wieder, A., *NCHRP Synthesis 320: Integrating Freight Facilities and Operations with Community Goals.* TRB, Washington, D.C., 2003.

U.S. Environmental Protection Agency, MOVES (Motor Vehicle Emission Simulator) Washington, D.C., 2011. http://www.epa.gov/otaq/models/moves/index.htm

Virginia Department of Transportation, *2025 State Highway Plan Technical Report.* Richmond, 2005. http://www.vdot.virginia.gov/projects/resources/Virginia2025StateHighwayPlanTechReport.pdf

Virginia Department of Transportation. Secondary Street Acceptance Requirements, VDOT Internal Training, Richmond, June 2009. http://www.virginiadot.org/projects/resources/SSAR/070609_June_Training_June_10.pdf

Virginia Department of Transportation, *Virginia Travel Demand Modeling Policies and Procedures Manual (version 1.21).* Richmond, VA, October 2006 (version 1.21).

Wasatch Front Regional Council, Wasatch Front Urban Area Regional Transportation Plan: 2007-2030 Technical Report 46, Salt Lake City, 2007. http://www.wfrc.org/cms/publications/Adopted_2007-2030_RTP/Chapter%208%20-%20Recommended%20Improvements.pdf

CHAPTER 12

Forecasting Travel
Demand

Travel demand is expressed as the number of persons or vehicles per unit time that can be expected to travel on a given segment of a transportation system under a set of given land-use, socioeconomic, and environmental conditions. Forecasts of travel demand are used to establish the vehicular volume on future or modified transportation system alternatives. The methods for forecasting travel demand can range from a simple extrapolation of observed trends to a sophisticated computerized process involving extensive data gathering and mathematical modeling. The travel demand forecasting process is as much an art as it is a science. Judgments are required concerning the various parameters—that is, population, car ownership, and so forth—that provide the basis for a travel forecast. The methods used in forecasting demand will depend on the availability of data and on specific constraints on the project, such as availability of funds and project schedules.

CHAPTER OBJECTIVES:

- Understand the difference between urban and intercity travel demand forecasting.
- Calculate trip generation forecasts using the cross-classification and activity unit method.
- Apply the gravity model and growth factor equations to compute trip distributions between zones.
- Learn how to use the three types of transit estimating procedures to compute modal choice.
- Learn how to predict the number of autos and transit trips on roadway segments using three basic procedures of traffic assignment.
- Describe forecasting models other than the 4-step process, including trend analysis and demand elasticity.

12.1 DEMAND FORECASTING APPROACHES

There are two basic demand-forecasting situations in transportation planning. The first involves travel demand studies for urban areas, and the second deals with intercity travel demand. Urban travel demand forecasts, when first developed in the 1950s and 1960s, required that extensive databases be prepared using home interview and/or roadside interview surveys. The information gathered provided useful insight concerning the characteristics of the trip maker, such as age, sex, income, auto ownership, and so forth; the land use at each end of the trip; and the mode of travel. Travel data then could be aggregated by zone to formulate relationships between variables and to calibrate models.

In the intercity case, data are generally aggregated to a greater extent than for urban travel forecasting, such as city population, average city income, and travel time or travel cost between city pairs. The availability of travel data improved considerably with the formation of the Bureau of Transportation Statistics, now within the Research and Innovative Technology Administration (RITA) of the U.S. DOT. The availability of data from the Census Bureau's American Community Survey is another positive development. This chapter describes the urban travel forecasting process. The underlying concepts may also be applied to intercity travel demand.

The databases that were established in many urban transportation studies have been used for the calibration and testing of models for trip generation, distribution, mode choice, and traffic assignment. These data collection and calibration efforts involved a significant investment of money and personnel resources, and consequent studies are based on updating the existing database and using models that had been previously developed.

12.1.1 Factors Influencing Travel Demand

The three factors that influence the demand for urban travel are: (1) the location and intensity of land use; (2) the socioeconomic characteristics of people living in the area; and (3) the extent, cost, and quality of available transportation services. These factors are incorporated in most travel forecasting procedures.

Land-use characteristics are a primary determinant of travel demand. The amount of traffic generated by a parcel of land depends on how the land is used. For example, shopping centers, residential complexes, and office buildings produce different traffic generation patterns.

Socioeconomic characteristics of the people living within the city also influence the demand for transportation. Lifestyles and values affect how people decide to use their resources for transportation. For example, a residential area consisting primarily of high-income workers will generate more trips by automobile per person than a residential area populated primarily by retirees.

The availability of transportation facilities and services, referred to as the *supply,* also affects the demand for travel. Travelers are sensitive to the level of service provided by alternative transportation modes. When deciding whether to travel at all or which mode to use, they consider attributes such as travel time, cost, convenience, comfort, and safety.

12.1.2 Sequential Steps for Travel Forecasting

Prior to the technical task of travel forecasting, the study area must be delineated into a set of traffic analysis zones (TAZ) (also called transportation analysis zones) that form the basis for analysis of travel movements within, into, and out of the urban area, as discussed in Chapter 11. The set of zones can be aggregated into larger units, called

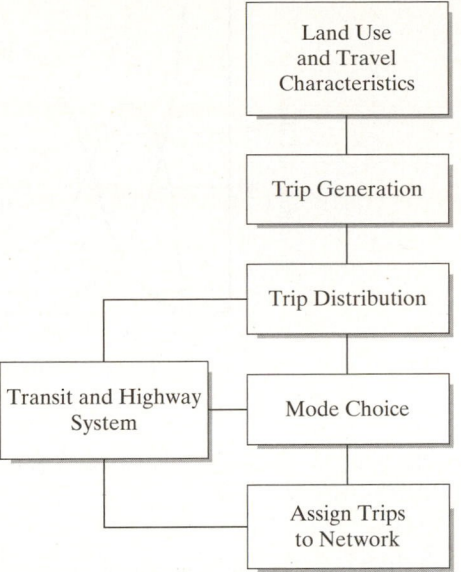

Figure 12.1 Travel Forecasting Process

districts, for certain analytical techniques or analyses that work at such levels. Land-use estimates are also developed.

Travel forecasting is solely within the domain of the transportation planner and is an integral part of site development and traffic engineering studies as well as area-wide transportation planning. Techniques that represent the state-of-the-practice of each task are described in this chapter to introduce the topic and to illustrate how demand forecasts are determined. Variations between forecasting techniques are also described in the literature.

The approach most commonly used to forecast travel demand is based on land-use and travel characteristics that provide the basis for the "four-step process" of trip generation, trip distribution, mode choice, and traffic assignment illustrated in Figure 12.1. Simultaneous model structures have also been used in practice, particularly to forecast intercity travel.

12.2 TRIP GENERATION

Trip generation is the process of determining the number of trips that will begin or end in each traffic analysis zone within a study area. Since the trips are determined without regard to destination, they are referred to as *trip ends.* Each trip has two ends, and these are described in terms of trip purpose, or whether the trips are either produced by a traffic zone or attracted to a traffic zone.

For example, a home-to-work trip would be considered to have a trip end produced in the home zone and attracted to the work zone. Trip generation analysis has two functions: (1) to develop a relationship between trip end production or attraction and land use, and (2) to use the relationship to estimate the number of trips generated at some future date under a new set of land-use conditions. To illustrate the process, two methods are considered: cross-classification and rates based on activity units.

Another commonly used method is regression analysis, which has been applied to estimate both productions and attractions. This method is used infrequently because it relies on zonal aggregated data. Trip generation methods that use a disaggregated analy-

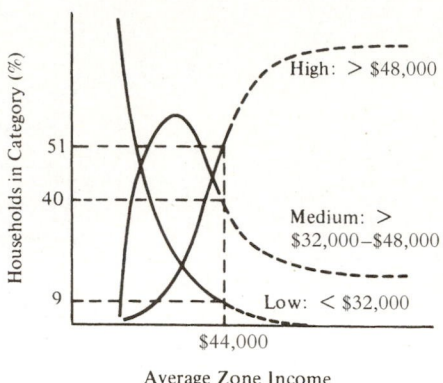

Figure 12.2   Average Zonal Income versus Households in Income Category

SOURCE: Modified from *Computer Programs for Urban Transportation Planning*, U.S. Department of Transportation, Washington, D.C., April 1977

sis, based on individual sample units such as persons, households, income, and vehicle units, are preferred.

12.2.1 Cross–Classification

Cross-classification is a technique developed by the Federal Highway Administration (FHWA) to determine the number of trips that begin or end at the home. Home-based trip generation is useful because it can represent a significant proportion of all trips. A relationship between socioeconomic measures and trip production is developed. The variables most commonly used are average income and auto ownership. Figure 12.2 illustrates the variation in average income within a zone. Other variables that could be considered are household size and stage in the household life cycle. The relationships are developed based on income data and results of O-D surveys.

Example 12.1   Developing Trip Generation Curves from Household Data

A travel survey produced the data shown in Table 12.1. Twenty households were interviewed. The table shows the number of trips produced per day for each of the households (numbered 1 through 20), as well as the corresponding annual household income and the number of automobiles owned. Based on the data provided, develop a set of curves showing the number of trips per household versus income and auto ownership.

Solution:

Step 1.   From the information in Table 12.1, produce a matrix that shows the number and percentage of households as a function of auto ownership and income grouping (see Table 12.2). The numerical values in each cell represent the number of households observed in each combination of income–auto ownership category. The value in parentheses is the percentage observed at each income level. In actual practice, the sample size would be at least 25 data points per cell to ensure statistical accuracy.

Table 12.1   Survey Data Showing Trips per Household, Income, and Auto Ownership

Household Number	Trips Produced per Household	Household Income ($1000s)	Autos per Household
1	2	16	0
2	4	24	0
3	10	68	2
4	5	44	0
5	5	18	1
6	15	68	3
7	7	38	1
8	4	36	0
9	6	28	1
10	13	76	3
11	8	72	1
12	6	32	1
13	9	28	2
14	11	44	2
15	10	44	2
16	11	52	2
17	12	60	2
18	8	44	1
19	8	52	1
20	6	28	1

Figure 12.3 illustrates how the data shown in Table 12.2 are used to develop relationships between the percent of households in each auto ownership category by household income.

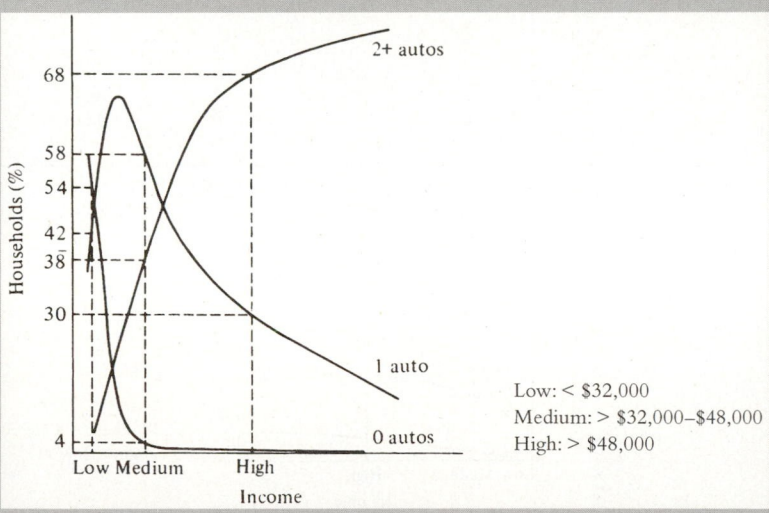

Figure 12.3   Households by Automobile Ownership and Income Category

SOURCE: Modified from *Computer Programs for Urban Transportation Planning*, U.S. Department of Transportation, Washington, D.C., April 1977

Table 12.2   Number and Percent of Household in Each Income Category versus Car Ownership

Income ($1000s)	Autos Owned			
	0	*1*	*2 +*	*Total*
24	2(67)	1(33)	0(0)	3(100)
24–36	1(25)	3(50)	1(25)	5(100)
36–48	1(20)	2(40)	2(40)	5(100)
48–60	—	1(33)	2(67)	3(100)
>60	—	1(25)	3(75)	4(100)
Total	4	8	8	20

Note: Values in parentheses are percent of automobiles owned at each income range.

Step 2.   A second table produced from the data in Table 12.1 shows the average number of trips per household versus income and cars owned. The results shown in Table 12.3 are illustrated in Figure 12.4, which depicts the relationship between trips per household per day by income and auto ownership. The table indicates that for a given income, trip generation increases with the number of cars owned. Similarly, for a given car ownership, trip generation increases with the rise in income.

Table 12.3   Average Trips per Household versus Income and Car Ownership

Income ($1000s)	Autos Owned		
	0	*1*	*2+*
≤24	3	5	—
24–36	4	6	9
36–48	5	7.5	10.5
48–60	—	8.5	11.5
>60	—	8.5	12.7

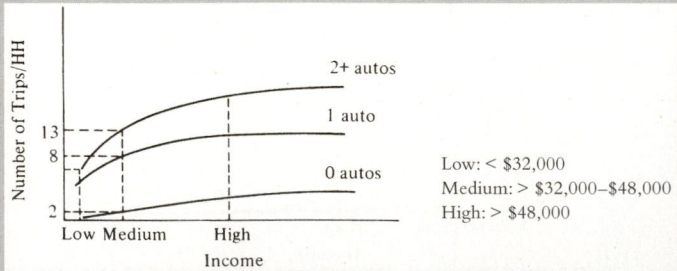

Figure 12.4   Trips per Household per Day by Auto Ownership and Income Category

SOURCE: Modified from *Computer Programs for Urban Transportation Planning*, U.S. Department of Transportation, Washington, D.C., April 1977

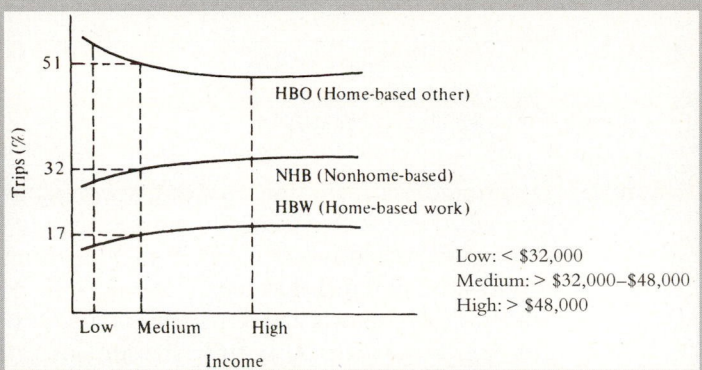

Figure 12.5   Trips by Purpose and Income Category

SOURCE: Modified from *Computer Programs for Urban Transportation Planning*, U.S. Department of Transportation, Washington, D.C., April 1977

Step 3.   As a further refinement, additional O-D data (not shown in Table 12.1) can be used to determine the percentage of trips by each trip purpose for each income category. These results are shown in Figure 12.5, wherein three trip purposes are used: home-based work (HBW), home-based other (HBO), and non-home-based (NHB). The terminology refers to the origination of a trip as either at the home or not at the home.

The trip generation model that has been developed based on survey data can now be used to estimate the number of home- and non-home-based trips for each trip purpose.

Example.12.2   Computing Trips Generated in a Suburban Zone

Consider a zone that is located in a suburban area of a city. The population and income data for the zone are as follows.

Number of dwelling units: 60

Average income per dwelling unit: $44,000

Determine the number of trips per day generated in this zone for each trip purpose, assuming that the characteristics depicted in Figures 12.2 through 12.5 apply in this situation. The problem is solved in four basic steps.

Solution:

Step 1.   Determine the percentage of households in each economic category. These results can be obtained by analysis of census data for the area. A typical plot of average zonal income versus income distribution is shown in Figure 12.2. For an average zonal income of $44,000, the following distribution is observed.

Income ($)	Households (%)
Low (under 32,000)	9
Medium (32,000–48,000)	40
High (over 48,000)	51

Step 2.   Determine the distribution of auto ownership per household for each income category. A typical curve showing percent of households, at each income level, that own 0, 1, or 2+ autos is shown in Figure 12.3, and the results are listed in Table 12.4.

   Table 12.4 shows that 58 percent of medium-income families own one auto per household. Also, from the previous step, we know that a zone, with an average income of $44,000, contains 40 pecent of households in the medium-income category. Thus, we can calculate that of the 60 households in that zone, there will be $60 \times 0.40 \times 0.58 = 14$ medium-income households that own one auto.

Step 3.   Determine the number of trips per household per day for each income–auto ownership category. A typical curve showing the relationship between trips per household, household income, and auto ownership is shown in Figure 12.4. The results are listed in Table 12.5. The table shows that a medium-income household owning one auto will generate eight trips per day.

Table 12.4   Percentage of Households in Each Income Category versus Auto Ownership

Income	Autos/Household		
	0	*1*	*2+*
Low	54	42	4
Medium	4	58	38
High	2	30	68

Table 12.5   Number of Trips per Household per Day

Income	Autos/Household		
	0	*1*	*2+*
Low	1	6	7
Medium	2	8	13
High	3	11	15

Step 4.   Calculate the total number of trips per day generated in the zone. This is done by computing the number of households in each income–auto ownership category, multiplying this result by the number of trips per household, as determined in Step 3, and summing the result. Thus,

$$P_{gh} = HH \times I_g \times A_{gh} \times (P_H)_{gh} \tag{12.1}$$

$$P_T = \sum_{g}^{3} \sum_{h}^{3} P_{gh} \tag{12.2}$$

where

HH = number of households in the zone

I_g = percentage of households (decimal) in zone with income level g (low, medium, or high)

A_{gh} = percentage of households (decimal) in income level g with h autos per household ($h = 0, 1,$ or $2+$)

P_{gh} = number of trips per day generated in the zone by householders with income level g and auto ownership h

$(P_H)_{gh}$ = number of trips per day produced in a household at income level g and auto ownership h

P_T = total number of trips generated in the zone

The calculations are shown in Table 12.6. For a zone with 60 households and an average income of \$44,000, the number of trips generated is 666 auto trips/day

Table 12.6   Number of Trips per Day Generated by Sixty Households

	Income, Auto Ownership	Total Trips by Income Group
$60 \times 0.09 \times 0.54 \times 1 = 3$ trips	L, 0+	
$60 \times 0.09 \times 0.42 \times 6 = 14$ trips	L, 1+	
$60 \times 0.09 \times 0.04 \times 7 = 2$ trips	L, 2+	19
$60 \times 0.40 \times 0.04 \times 2 = 2$ trips	M, 0+	
$60 \times 0.40 \times 0.58 \times 8 = 111$ trips	M, 1+	
$60 \times 0.40 \times 0.38 \times 13 = 119$ trips	M, 2+	232
$60 \times 0.51 \times 0.02 \times 3 = 2$ trips	H, 0+	
$60 \times 0.51 \times 0.30 \times 11 = 101$ trips	H, 1+	
$60 \times 0.51 \times 0.68 \times 15 = 312$ trips	H, 2+	415
Total $= 666$ trips		666

Step 5. Determine the percentage of trips by trip purpose. As a final step, we can calculate the number of trips that are HBW, HBO, and NHB. If these percentages are 17, 51, and 32, respectively (see Figure 12.5), for the medium-income category, then the number of trips from the zone for the three trip purposes are $232 \times 0.17 = 40$ HBW, $232 \times 0.51 = 118$ HBO, and $232 \times 0.32 = 74$ NHB. (Similar calculations would be made for other income groups.) The final result, which is left for the reader to verify, is obtained by using the following percentages: low income at 15, 55, and 30, and high income at 18, 48, and 34. These yield 118 HBW, 327 HBO, and 221 NHB trips.

Trip generation values as determined in the preceding examples are used to calculate the number of trips in each zone. Values for each income or auto ownership category can be developed using survey data or published statistics compiled for other cities. Figure 12.6 illustrates the trip generation rate per household for single-family detached housing. The average rate is 9.57, and the range of rates is 4.31 to 21.85. This table is one of over 1000 included in the Institute of Transportation Engineers publication, *Trip*

Single-Family Detached Housing

Average Vehicle Trip Ends vs: Dwelling Units
On a: Weekday

Number of Studies: 350
Avg. Number of Dwelling Units: 197
Directional Distribution: 50% entering, 50% exiting

Trip Generation per Dwelling Unit

Average Rate	Range of Rates	Standard Deviation
9.57	4.31–21.85	3.69

Data Plot and Equation

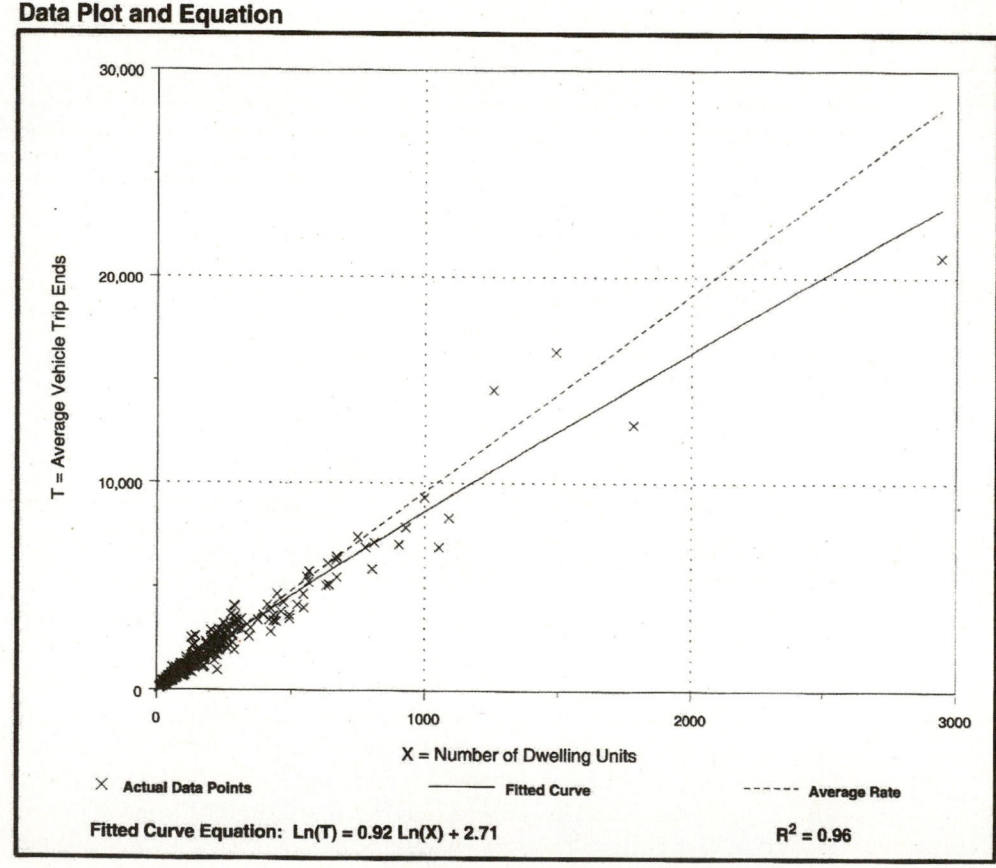

Fitted Curve Equation: $\ln(T) = 0.92 \ln(X) + 2.71$ $R^2 = 0.96$

Figure 12.6   Trip Generation Characteristics

SOURCE: *Trip Generation*, 7th ed., Institute of Transportation Engineers, Washington, D.C., 2003. www.ite.org.
Used by permission.

Generation for 10 Different Land Uses, including port and terminal, industrial, agricultural, residential, lodging, institutional, medical, office, retail, and services. An extensive amount of useful trip generation data are also available in *Quick Response Urban Travel Estimation Techniques and Transferable Parameters*. The ITE *Transportation Planning Handbook* also provides trip generation rates.

12.2.2 Rates Based on Activity Units

The preceding section illustrated how trip generation is determined for residential zones where the basic unit is the household. Trips generated at the household end are referred to as *productions,* and they are *attracted* to zones for purposes such as work, shopping, visiting friends, and medical trips. Thus, an activity unit can be described by measures such as square feet of floor space or number of employees. Trip generation rates for attraction zones can be determined from survey data or are tabulated in some of the reference sources listed at the end of this chapter. Trip attraction rates are illustrated in Table 12.7.

Table 12.7   Trip Generation Rates by Trip Purpose and Employee Category

	Attractions per Household	Attractions per Nonretail Employee	Attractions per Downtown Retail Employee	Attractions per Other Retail Employee
HBW	—	1.7	1.7	1.7
HBO	1.0	2.0	5.0	10.0
NHB	1.0	1.0	3.0	5.0

Example 12.3 Computing Trips Generated in an Activity Zone

A commercial center in the downtown contains several retail establishments and light industries. Employed at the center are 220 retail and 650 nonretail workers. Determine the number of trips per day attracted to this zone.

Solution: Use the trip generation rates listed in Table 12.7.

$$\text{HBW: } (220 \times 1.7) + (650 \times 1.7) = 1479$$

$$\text{HBO: } (220 \times 5.0) + (650 \times 2.0) = 2400$$

$$\text{NHB: } (220 \times 3.0) + (650 \times 1.0) = 1310$$

$$\text{Total} = 5189 \text{ trips/day}$$

Note that three trip purposes are given in Table 12.7: home-based work (HBW), home-based other (HBO), and non-home-based (NHB). For example, for HBO trips, there are 5.0 attractions per downtown retail employee (in trips/day) and 2.0 attractions per nonretail employee.

12.2.3 Balancing Trip Productions and Attractions

A likely result of the trip generation process is that the number of trip productions may not be equal to the number of trip attractions. Trip productions, which are based on census data, are considered to be more accurate than trip attractions. Accordingly, trip attractions are usually modified so that they are equal to trip productions.

Table 12.8a illustrates how adjustments are made. The trip generation process has produced 600 home-based work productions for zones 1 through 3. However, the same process has produced 800 home-based work attractions. To rectify this imbalance, each attraction value for zones 1 through 3 is reduced by a factor equal to 600/800, or 0.75. The result is shown in Table 12.8a in the column "Balanced HBW Trips." Now both productions and attractions are equal. A similar procedure is used for HBO trips.

An extra step is required for balancing NHB trips. This extra step is that after total productions and total attractions are equal, the productions for each zone are set equal to the attractions for each zone. For example, in Table 12.8b, since there are 180 NHB attractions for zone 1 after balancing productions and attractions, then the number of NHB productions for zone 1 is also changed from 100 to 180. The rationale behind this extra step is that the true origin of non-home-based trips is not provided by survey or census data, and thus the best estimate of the number of NHB trips produced in each zone is the number of NHB trips attracted to each zone.

Table 12.8a   Balancing Home-Based Work Trips

	Unbalanced HBW Trips		Balanced HBW Trips	
Zone	Productions	Attractions	Productions	Attractions
1	100	240	100	180
2	200	400	200	300
3	300	160	300	120
Total	600	800	600	600

Table 12.8b   Balancing Non-Home-Based Trips

	Unbalanced NHB Trips		Balanced NHB Trips	
Zone	NHB Productions	NHB Attractions	NHB Productions	NHB Attractions
1	100	240	180	180
2	200	400	300	300
3	300	160	120	120
Total	600	800	600	600

12.3 TRIP DISTRIBUTION

Trip distribution is a process by which the trips generated in one zone are allocated to other zones in the study area. These trips may be within the study area (internal-internal) or between the study area and areas outside the study area (internal-external).

For example, if the trip generation analysis results in an estimate of 200 HBW trips in zone 10, then the trip distribution analysis would determine how many of these trips would be made between zone 10 and each of the other internal zones.

In addition, the trip distribution process considers internal-external trips (or vice versa) where one end of the trip is within the study area and the other end is outside the study area. Chapter 11, Figure 11.11, illustrates external stations for a study area boundary. If a trip begins somewhere south of the study area and ends in the center of the study area using Route 29, then an external-internal trip is defined that begins at external station 103 and ends in a zone located in the center of the study area.

Several basic methods are used for trip distribution. Among these are the gravity model, growth factor models, and intervening opportunities. The gravity model is preferred because it uses the attributes of the transportation system and land-use characteristics and has been calibrated extensively for many urban areas. The gravity model has achieved virtually universal use because of its simplicity, its accuracy, and its support from the U.S. Department of Transportation. Growth factor models, which were used more widely in the 1950s and 1960s, require that the origin-destination matrix be known for the base (or current) year, as well as an estimate of the number of future trip ends in each zone. The intervening opportunities model and other models are available but not widely used in practice.

12.3.1 The Gravity Model

The most widely used and documented trip distribution model is the *gravity model*, which states that the number of trips between two zones is directly proportional to the number of trip attractions generated by the zone of destination and inversely proportional to a function of travel time between the two zones. Mathematically, the gravity model is expressed as

$$T_{ij} = P_i \left[\frac{A_j F_{ij} K_{ij}}{\sum_j A_j F_{ij} K_{ij}} \right] \qquad (12.3)$$

where

T_{ij} = number of trips that are produced in zone i and attracted to zone j
P_i = total number of trips produced in zone i
A_j = number of trips attracted to zone j
F_{ij} = a value which is an inverse function of travel time
K_{ij} = socioeconomic adjustment factor for interchange ij

The values of P_i and A_j have been determined in the trip generation process. The sum of P_i for all zones must equal the sum of A_j for all zones. K_{ij} values are used when the estimated trip interchange must be adjusted to ensure that it agrees with the observed trip interchange.

A calibrating process in which trip generation values as measured in the O-D survey are distributed using the gravity model determines the values for F_{ij}. After each distribution process is completed, the percentage of trips in each trip length category produced by the gravity model is compared with the percentage of trips recorded in the O-D survey. If the percentages do not agree, then the F_{ij} factors that were used in the distribution process are adjusted and another gravity model trip distribution is performed.

Figure 12.7 illustrates F values for calibrations of a gravity model. (Normally this curve is a semilog plot.) F values can also be determined using travel time values and an inverse relationship between F and t. For example, the relationship for F might be in the form t^{-1}, t^{-2}, e^{-t}, and so forth, since F values decrease as travel time increases. The friction factor can be expressed as $F = ab^t e^{-ct}$, where parameters a, b, and c are based

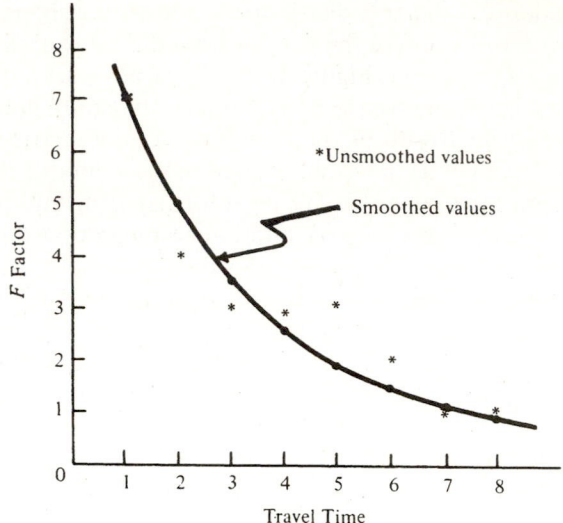

Figure 12.7   Calibration of *F* Factors

SOURCE: Modified from *Computer Programs for Urban Transportation Planning*, U.S. Department of Transportation, Washington, D.C., April 1977

on national data sources, such as NCHRP Report 365, or the formula may be calibrated using local data.

The socioeconomic factor is used to make adjustments of trip distribution T_{ij} values between zones where differences between estimated and actual values are significant. The *K* value is referred to as the "socioeconomic factor" since it accounts for variables other than travel time. The values for *K* are determined in the calibration process, but it is used judiciously when a zone is considered to possess unique characteristics.

Example 12.4   Use of Calibrated *F* Values and Iteration

To illustrate the application of the gravity model, consider a study area consisting of three zones. The data have been determined as follows: the number of productions and attractions has been computed for each zone by methods described in the section on trip generation, and the average travel times between each zone have been determined. Both are shown in Tables 12.9 and 12.10. Assume K_{ij} is the same unit value for all zones. Finally, the *F* values have been calibrated as previously described and are shown in Table 12.11 for each travel time increment. Note that the intra-zonal travel time for zone 1 is larger than those of most other inter-zone times because of the geographical

Table 12.9   Trip Productions and Attractions for a Three-Zone Study Area

Zone	1	2	3	Total
Trip productions	140	330	280	750
Trip attractions	300	270	180	750

Table 12.10   Travel Time between Zones (min)

Zone	1	2	3
1	5	2	3
2	2	6	6
3	3	6	5

Table 12.11   Travel Time versus Friction Factor

Time (min)	F
1	82
2	52
3	50
4	41
5	39
6	26
7	20
8	13

Note: *F* values were obtained from the calibration process.

characteristics of the zone and lack of access within the area. This zone could represent conditions in a congested downtown area.

Determine the number of zone-to-zone trips through two iterations.

Solution: The number of trips between each zone is computed using the gravity model and the given data. (*Note:* F_{ij} is obtained by using the travel times in Table 12.10 and selecting the correct *F* value from Table 12.11. For example, travel time is 2 min between zones 1 and 2. The corresponding *F* value is 52.)

Use Eq. 12.3.

$$T_{ij} = P_i \left[\frac{A_j F_{ij} K_{ij}}{\sum_{j=1}^{n} A_j F_{ij} K_{ij}} \right] \qquad K_{ij} = 1 \text{ for all zones}$$

$$T_{1-1} = 140 \times \frac{300 \times 39}{(300 \times 39) + (270 \times 52) + (180 \times 50)} = 47$$

$$T_{1-2} = 140 \times \frac{270 \times 52}{(300 \times 39) + (270 \times 52) + (180 \times 50)} = 57$$

$$T_{1-3} = 140 \times \frac{180 \times 50}{(300 \times 39) + (270 \times 52) + (180 \times 50)} = 36$$

$$P_1 = 140$$

Make similar calculations for zones 2 and 3.

$$T_{2-1} = 188 \qquad T_{2-2} = 85 \qquad T_{2-3} = 57 \qquad P_2 = 330$$

$$T_{3-1} = 144 \qquad T_{3-2} = 68 \qquad T_{3-3} = 68 \qquad P_3 = 280$$

Table 12.12 Zone-to-Zone Trips: First Iteration, Singly Constrained

Zone	1	2	3	Computed P	Given P
1	47	57	36	140	140
2	188	85	57	330	330
3	144	68	68	280	280
Computed A	379	210	161	750	750
Given A	300	270	180	750	

The results summarized in Table 12.12 represent a *singly constrained* gravity model. This constraint is that the sum of the productions in each zone is equal to the number of productions given in the problem statement. However, the number of attractions estimated in the trip distribution phase differs from the number of attractions given. For zone 1, the correct number is 300, whereas the computed value is 379. Values for zone 2 are 270 versus 210, and for zone 3, they are 180 versus 161.

To create a doubly constrained gravity model where the computed attractions equal the given attractions, calculate the adjusted attraction factors according to the formula

$$A_{jk} = \frac{A_j}{C_{j(k-1)}} A_{j(k-1)}$$ (12.4)

where

A_{jk} = adjusted attraction factor for attraction zone (column)j, iteration k
$A_{jk} = A_j$ when $k = 1$
C_{jk} = actual attraction (column) total for zone j, iteration k
A_j = desired attraction total for attraction zone (column) j
j = attraction zone number, $j = 1, 2,..., n$
n = number of zones
k = iteration number, $k = 1, 2,..., m$
m = number of iterations

Repeat the trip distribution computations using modified attraction values so that the numbers attracted will be increased or reduced as required. For zone 1, for example, the estimated attractions were too great. Therefore, the new attraction factors are adjusted downward by multiplying the original attraction value by the ratio of the original to estimated attraction values.

$$\text{Zone 1:} A_{12} = 300 \times \frac{300}{379} = 237$$

$$\text{Zone 2:} A_{22} = 270 \times \frac{270}{210} = 347$$

$$\text{Zone 3:} A_{32} = 180 \times \frac{180}{161} = 201$$

Apply the gravity model (Eq. 12.3) for all iterations to calculate zonal trip interchanges using the adjusted attraction factors obtained from the preceding iteration. In practice, the gravity model becomes

Table 12.13   Zone-to-Zone Trips: Second Iteration, Doubly Constrained

Zone	1	2	3	Computed P	Given P
1	34	68	38	140	140
2	153	112	65	330	330
3	116	88	76	280	280
Computed A	303	268	179	750	750
Given A	300	270	180	750	

$$T_{ij} = P_i \left[\frac{A_j F_{ij} K_{ij}}{\sum_j A_j F_{ij} K_{ij}} \right]$$

where T_{ijk} is the trip interchange between i and j for iteration k, and $A_{jk} = A_j$, when $k = 1$. Subscript j goes through one complete cycle every time k changes, and i goes through one complete cycle every time j changes. This formula is enclosed in parentheses and subscripted to indicate that the complete process is performed for each trip purpose.

Perform a second iteration using the adjusted attraction values.

$$T_{1-1} = 140 \times \frac{237 \times 39}{(237 \times 39) + (347 \times 52) + (201 \times 50)} = 34$$

$$T_{1-2} = 140 \times \frac{347 \times 52}{(237 \times 39) + (347 \times 52) + (201 \times 50)} = 68$$

$$T_{1-3} = 140 \times \frac{201 \times 50}{(237 \times 39) + (347 \times 52) + (201 \times 50)} = 37$$

$$P_1 = 140$$

Make similar calculations for zones 2 and 3.

$$T_{2-1} = 153 \qquad T_{2-2} = 112 \qquad T_{2-3} = 65 \qquad P_2 = 330$$

$$T_{3-1} = 116 \qquad T_{3-2} = 88 \qquad T_{3-3} = 76 \qquad P_3 = 280$$

The results are summarized in Table 12.13. Note that, in each case, the sum of the attractions is now much closer to the given value. The process will be continued until there is a reasonable agreement (within 5 percent) between the A that is estimated using the gravity model and the values that are furnished in the trip generation phase.

A singly constrained gravity model requires that computed and actual productions must be equal, whereas a doubly constrained gravity model requires that computed and actual productions and attractions must be equal. The singly constrained gravity model may be preferred if the friction factors are more reliable than the attraction values. The doubly constrained gravity model is appropriate if the attraction values are more reliable than friction factors. To illustrate either choice, consider the following example.

Example 12.5 Selecting Singly or Doubly Constrained Gravity Model Results

A three-zone system with 900 home-based shopping productions is shown in Table 12.14. Zones 1 and 2 each generate 400 productions, while zone 3 generates 100 productions. Each zone contains a shopping mall with 300 attractions. The shopping mall in zone 1 can be easily reached due to the parking availability and transit service. Thus, F_{11}, F_{21}, and $F_{31} = 1.0$. Parking costs at the shopping mall in zone 2 are moderate with some transit service. Thus, F_{12}, F_{22}, and $F_{32} = 0.5$. Parking costs at the mall in zone 3 is high and transit service is unavailable. Thus, F_{13}, F_{23}, and $F_{33} = 0.2$.

Application of the singly constrained gravity model yields the results shown in Table 12.15 and application of the doubly constrained gravity model yields the results shown in Table 12.16.

Which of the results shown for the singly constrained gravity model and for the doubly constrained gravity model are more likely to be the most accurate?

Solution: Table 12.15 is more likely to be accurate if engineering judgment suggests the occurrence of travel impedances and thus the friction factors are more accurate than trip attractions. Table 12.16 is more likely to be accurate if the attractions are more accurate than the friction factors.

In practice, these judgments must be made based on the quality of the data set. For example, if local land-use data had been recently used to develop trip attraction rates whereas friction factors had been borrowed from another area, then the selection of the doubly constrained gravity model results in Table 12.16 is recommended.

Table 12.14   Home-Based Shopping Productions and Attractions

Zone	Productions	Attractions
1	400	300
2	400	300
3	100	300
Total	900	900

Table 12.15   Zone-to-Zone Trips: Singly Constrained Gravity Model

Zone	1	2	3	Computed P	Given P
1	235	118	47	400	400
2	235	118	47	400	400
3	59	29	12	100	100
Computed A	529	265	106	900	900
Given A	300	300	300	900	

Table 12.16   Zone-to-Zone Trips: Doubly Constrained Gravity Model

Zone	1	2	3	Computed P	Given P
1	133	133	133	400	400
2	133	133	133	400	400
3	33	33	33	100	100
Computed A	300	300	300	900	900
Given A	300	300	300	900	

12.3.2 Growth Factor Models

Trip distribution can also be computed when the only data available are the origins and destinations between each zone for the current or base year and the trip generation values for each zone for the future year. This method was widely used when O-D data were available but the gravity model and calibrations for *F* factors had not yet become operational. Growth factor models are used primarily to distribute trips between zones in the study area and zones in cities external to the study area. Since they rely upon an existing O-D matrix, they cannot be used to forecast traffic between zones where no traffic currently exists. Further, the only measure of travel friction is the amount of current travel. Thus, the growth factor method cannot reflect changes in travel time between zones, as does the gravity model.

The most popular growth factor model is the *Fratar method* (named for Thomas J. Fratar, who developed the method), which is a mathematical formula that proportions future trip generation estimates to each zone as a func tion of the product of the current trips between the two zones T_{ij} and the growth factor of the attracting zone G_j. Thus,

$$T_{ij} = (t_i G_i) \frac{t_{ij} G_j}{\sum_x t_{ix} G_x} \tag{12.5}$$

where

T_{ij} = number of trips estimated from zone *i* to zone *j*)
t_t = present trip generation in zone *i*
G_x = growth factor of zone *x*
$T_i = t_i G_i$ = future trip generation in zone *i*
t_{ix} = number of trips between zone *i* and other zones *x*
t_{ij} = present trips between zone *i* and zone *j*
G_j = growth factor of zone *j*

The following example illustrates the application of the growth factor model.

Example 12.6   Forecasting Trips Using the Fratar Model

A study area consists of four zones (A, B, C, and D). An O-D survey indicates that the number of trips between each zone is as shown in Table 12.17. Planning estimates for

Table 12.17   Present Trips between Zones

Zone	A	B	C	D
A	—	400	100	100
B	400	—	300	—
C	100	300	—	300
D	100	—	300	—
Total	600	700	700	400

the area indicate that in five years the number of trips in each zone will increase by the growth factor shown in Table 12.18 and that trip generation will be increased to the amounts shown in the last column of the table.

Table 12.18 Present Trip Generation and Growth Factors

Zone	Present Trip Generation (trips/day)	Growth Factor	Trip Generation in Five Years
A	600	1.2	720
B	700	1.1	770
C	700	1.4	980
D	400	1.3	520

Determine the number of trips between each zone for future conditions.

Solution: Using the Fratar formula (Eq. 12.5), calculate the number of trips between zones A and B, A and C, A and D, and so forth. Note that two values are obtained for each zone pair, (that is, T_{AB} and T_{BA}). These values are averaged, yielding a value for $T_{AB} = (T_{AB} + T_{BA})/2$.
The calculations are as follows.

$$T_{ij} = (t_i G_i) \frac{t_{ij} G_j}{\sum_x t_{ix} G_x}$$

$$T_{AB} = 600 \times 1.2 \frac{400 \times 1.1}{(400 \times 1.1) + (100 \times 1.4) + (100 \times 1.3)} = 446$$

$$T_{BA} = 700 \times 1.1 \frac{400 \times 1.2}{(400 \times 1.2) + (300 \times 1.4)} = 411$$

$$\overline{T}_{AB} = \frac{T_{AB} + T_{BA}}{2} = \frac{446 + 411}{2} = 428$$

Similar calculations yield

$$\overline{T}_{AC} = 141 \qquad \overline{T}_{AD} = 124 \qquad \overline{T}_{BC} = 372 \qquad \overline{T}_{CD} = 430$$

The results of the preceding calculations have produced the first estimate (or iteration) of future trip distribution and are shown in Table 12.19. The totals for each zone do not equal the values of future trip generation. For example, the trip generation in zone A is estimated as 693 trips, whereas the actual value is 720 trips. Similarly, the estimate for zone B is 800 trips, whereas the actual value is 770 trips.

Proceed with a second iteration in which the input data are the numbers of trips between zones as previously calculated. Also, new growth factors are computed as the ratio of the trip generation expected to occur in five years and the

Table 12.19 First Estimate of Trips between Zones

Zone	A	B	C	D	Estimated Total Trip Generation	Actual Trip Generation
A	—	428	141	124	693	720
B	428	—	372	—	800	770
C	141	372	—	430	943	980
D	124	—	430	—	554	520
Totals	693	800	943	554		

trip generation estimated in the preceding calculation. The values are given in Table 12.20.

Table 12.20 Growth Factors for Second Iteration

Zone	Estimated Trip Generation	Actual Trip Generation	Growth Factor
A	693	720	1.04
B	800	770	0.96
C	943	980	1.04
D	554	520	0.94

The calculations for the second iteration are left to the reader to complete and the process can be repeated as many times as needed until the estimate and actual trip generation values are close in agreement.

A more general form of the growth factor model is the *average growth factor model*. Rather than weighting the growth of trips between zones i and j by the growth across all zones, as is done in the Fratar method, the growth rate of trips between any zones i and j is simply the average of the growth rates of these zones.

$$T'_{ij} = T_{ij}\left(\frac{G_i + G_j}{2}\right) \tag{12.5a}$$

Application of the average growth factor method proceeds similarly to that of the Fratar method. As iterations continue, the growth factors converge toward unity. Iterations can cease when an acceptable degree of convergence in the values is reached; one such practice is to continue until all growth factors are within 5 percent of unity (i.e., between 0.95 and 1.05).

Example 12.7 Accounting for Trips

Consider a simple region comprised of two zones. One hundred individuals live in zone 1, walk to work in zone 2 in the morning, and then walk home to zone 1 in the evening. Prepare

(a) A production-attraction matrix
(b) An origin-destination trip matrix

Which of these is similar to Table 12.12? Which is similar to Table 12.17?

Solution:

(a) A production-attraction matrix is similar to Table 12.12.

In the morning the 100 individuals generate 100 productions at the home end in zone 1 and 100 attractions at the work end in zone 2. Then, in the evening, these individuals generate another 100 attractions at the home end in zone 1 and 100 productions at the work end in zone 2. Thus, for a 24-hour period, there are 200 productions in zone 1 and 200 attractions in zone 2. The production attraction matrix is as follows:

Production-Attraction Trip Matrix (Zone-to-Zone Trips)

	Zone 1	*Zone 2*	*Total Productions*
Zone 1	0	200	200
Zone 2	200	0	200
Total attractions	200	200	

(b) The origin–destination trip matrix is similar to Table 12.17.

Origin–Destination Trip Matrix (Trips between Zones)

	Zone 1	*Zone 2*	*Total Origins*
Zone 1	0	100	100
Zone 2	100	0	100
Total destinations	100	100	

12.4 MODE CHOICE

Mode choice is that aspect of the demand analysis process that determines the number (or percentage) of trips between zones that are made by automobile and by transit. The selection of one mode or another is a complex process that depends on factors such as the traveler's income, the availability of transit service or auto ownership, and the relative advantages of each mode in terms of travel time, cost, comfort, convenience, and safety. Mode choice models attempt to replicate the relevant characteristics of the traveler, the transportation system, and the trip itself, such that a realistic estimate of the number of trips by each mode for each zonal pair is obtained. A discussion of the many mode choice models is beyond the scope of this chapter, and the interested reader should refer to sources cited.

12.4.1 Types of Mode Choice Models

Since public transportation is a vital transportation component in urban areas, mode choice calculations typically involve distinguishing trip interchanges as either auto or transit. Depending on the level of detail required, three types of transit estimating procedures are used: (1) direct generation of transit trips, (2) use of trip end models, and (3) trip interchange modal split models.

Direct Generation Models

Transit trips can be generated directly, by estimating either total person trips or auto driver trips. Figure 12.8 is a graph that illustrates the relationship between number of

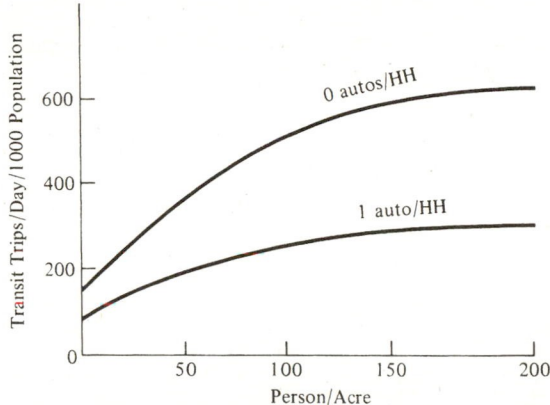

Figure 12.8   Number of Transit Trips by Population Density and Automobile Ownership per Household

Example 12.8   Estimating Mode Choice by Direct Trip Generation

Determine the number of transit trips per day in a zone, which has 5000 people living on 202350 square meter. The auto ownership is 40 percent zero autos per household and 60 percent one auto per household.

Solution: Calculate the number of persons per square meter: 5000/202350 = 0.025. Then determine the number of transit trips per day per 1000 persons (from Figure 12.8) to calculate the total of all transit trips per day for the zone.

Zero autos/HH: 510 trips/day/1000 population

One auto/HH: 250 trips/day/1000 population

Total Transit Trips: (0.40)(510)(5) + (0.60)(250)(5) =

1020 + 750 = 1770 transit trips per day

transit trips per 1000 population and persons per square meter versus automobile ownership. As population density increases, it can be expected that transit ridership will also increase for a given level of auto ownership.

This method assumes that the attributes of the system are not relevant. Factors such as travel time, cost, and convenience are not considered. These so-called "pre-trip" distribution models apply when transit service is poor and riders are "captive," or when transit service is excellent and "choice" clearly favors transit. When highway and transit modes "compete" for auto riders, then system factors are considered.

Trip End Models

To determine the percentage of total person or auto trips that will use transit, estimates are made prior to the trip distribution phase based on land-use or socioeconomic characteristics of the zone. This method does not incorporate the quality of service. The procedure follows:

- Generate total person trip productions and attractions by trip purpose.
- Compute the urban travel factor.
- Determine the percentage of these trips by transit using a mode choice curve.
- Apply auto occupancy factors.
- Distribute transit and auto trips separately.

The mode choice model shown in Figure 12.9 is based on two factors: households per auto and persons per square km. The product of these variables is called the urban travel factor (UTF). Percentage of travel by transit will increase in an "S" curve fashion as the UTF increases.

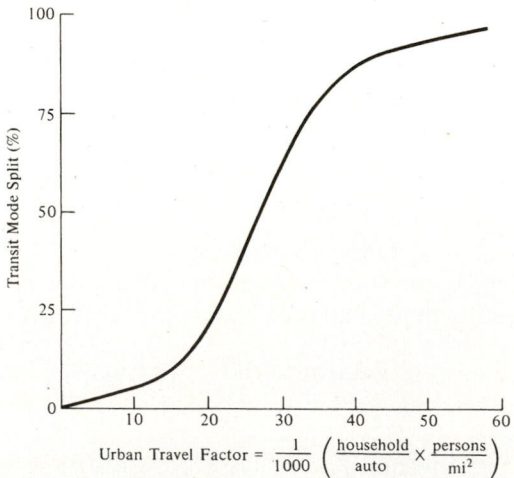

$$\text{Urban Travel Factor} = \frac{1}{1000} \left(\frac{\text{household}}{\text{auto}} \times \frac{\text{persons}}{\text{mi}^2} \right)$$

Figure 12.9 Transit Mode Split versus Urban Travel Factor

Trip Interchange Models

In this method, system level-of-service variables are considered, including relative travel time, relative travel cost, economic status of the trip maker, and relative travel service. An example of this procedure takes account of service parameters in estimating mode choice using the following relationship:

$$MS_a = \frac{I_{ijt}^{-b}}{I_{ija}^{-b} + I_{ija}^{-b}} \times 100 \text{ or } \frac{I_{ija}^{b}}{I_{ijt}^{b} + I_{ija}^{b}} \times 100 \tag{12.6}$$

$$MS_t = (1 - MS_a) \times 100 \tag{12.7}$$

where

MS_t = proportion of trips between zone i and j using transit
MS_a = proportion of trips between zone i and j using auto
I_{ijm} = a value referred to as the *impedance* of travel of mode m, between i and j, which is a measure of the total cost of the trip
Impedance = (in-vehicle time, min) + [(2.5)(excess time, min) + {(3)(trip cost) \$ ÷(income earned min)}]
b = an exponent, which depends on trip purpose
m = t for transit mode; a for auto mode

In-vehicle time is time spent traveling in the vehicle, and excess time is time spent traveling but not in the vehicle, including waiting for the train or bus and walking to the station. The impedance value is determined for each zone pair and represents a measure of the expenditure required to make the trip by either auto or transit. The data required for estimating mode choice include (1) distance between zones by auto and transit, (2) transit fare, (3) out-of-pocket auto cost, (4) parking cost, (5) highway and transit speed, (6) exponent values, *b,* (7) median income, and (8) excess time, which includes the time required to walk to a transit vehicle and time waiting or transferring. Assume that the time worked per year is 120,000 min.

Example 12.10 Computing Mode Choice Using the QRS Model

To illustrate the application of the QRS method, assume that the data shown in Table 12.21 have been developed for travel between a suburban zone S and a downtown zone D. Determine the percent of work trips by auto and transit. An exponent value of 2.0 is used for work travel. Median income is $24,000 per year.

Table 12.21   Travel Data between Two Zones, S and D

	Auto	Transit
Distance	16 km	13 km
Cost per km	$0.09	$=0.06
Excess time	5 min	8 min
Parking cost	$1.50 (or 0.75/trip)	—
Speed	48 km/h	32 km/h

Solution: Use Eq. 12.6.

$$MS_a = \frac{I_{ija}^b}{I_{ijt}^b + I_{ija}^b}$$

$$I_{SDa} = \left(\frac{16}{48} \times 60\right) + (2.5 \times 5) + \left\{\frac{3 \times [(1.50/2) + 0.09 \times 16]}{24,000/120,000}\right\}$$

$$= 20 + 12.5 + 18.45$$

$$= 50.95 \text{ equivalent min}$$

$$I_{SDt} = \left(\frac{13}{22} \times 60\right) + (2.5 \times 8) + \left[\frac{3 \times (13 \times 0.06)}{24,000/120,000}\right] = 24 + 20 + 11.7$$

$$= 55.7 \text{ equivalent min}$$

$$MS_a = \frac{(55.7)^2}{(55.7)^2 + (50.95)^2} \times 100 = 54.4\%$$

$$MS_t = (1 - 0.544) \times 100 = 45.6\%$$

Thus, the mode choice of travel by transit between zones S and D is 68.4 percent, and by highway the value is 41.6 percent. These percentages are applied to the estimated trip distribution values to determine the number of trips by each mode. If, for example, the number of work trips between zones S and D was computed to be 500, then the number by auto would be $500 \times 54.4 = 272$, and by transit, the number of trips would be $500 \times 0.456 = 228$.

12.4.2 Logit Models

An alternative approach used in transportation demand analysis is to consider the relative utility of each mode as a summation of each modal attribute. Then the choice of a mode is expressed as a probability distribution. For example, assume that the utility of each mode is

$$U_x = \sum_{i=1}^{n} a_i X_i \tag{12.8}$$

where

U_x = utility of mode x
n = number of attributes
X_i = attribute value (time, cost, and so forth)
a_i = coefficient value for attributes i (negative, since the values are disutilities)

If two modes, auto *(A)* and transit *(T),* are being considered, the probability of selecting the auto mode A can be written as

$$P(A) = \frac{e^{U_A}}{e^{U_A} + e^{U_T}} \qquad (12.9)$$

This form is called the *logit model,* as illustrated in Figure 12.10, and provides a convenient way to compute mode choice. Choice models are utilized within the urban transportation planning process and in transit marketing studies.

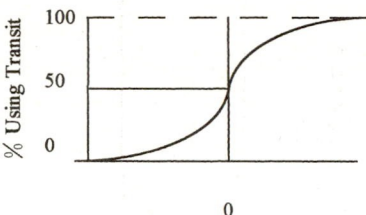

Utility Differences between Modes (U_A - U_T)

Figure 12.10 Modal Choice for Transit versus Automobile

Example 12.11 Use of Logit Model to Compute Mode Choice

The utility functions for auto and transit are as follows.

$$\text{Auto: } U_A = -0.46 - 0.35T_1 - 0.08T_2 - 0.005C$$

$$\text{Transit: } U_T = -0.07 - 0.05T_1 - 0.15T_2 - 0.005C$$

where

T_1 = total travel time (minutes)
T_2 = waiting time (minutes)
C = cost (cents)

The travel characteristics between two zones are as follows:

	Auto	Transit
T_1	20	30
T_2	8	6
C	320	100

Solution: Use the logit model to determine the percent of travel in the zone by auto and transit.

$$U_x = \sum_{i=1}^{n} a_i X_i$$

$$U_A = -0.46 - (0.35 \times 20) - (0.08 \times 8) - (0.005 \times 320) = -9.70$$

$$U_B = -0.07 - (0.35 \times 30) - (0.08 \times 6) - (0.005 \times 100) = -11.55$$

Using Eq.12.9 yields

$$P_A = \frac{e^{U_A}}{e^{U_A} + e^{U_T}} = \frac{e^{-9.70}}{e^{-9.7} + e^{-11.55}} = 0.86$$

$$P_T = \frac{e^{U_T}}{e^{U_A} + e^{U_T}} = \frac{e^{-11.55}}{e^{-9.7} + e^{-11.55}} = 0.14$$

Example 12.12 Role of the Difference in Utilities in the Logit Model

Referring to Example 12.11, suppose rising fuel prices lead to an increase of $1.00 for each mode. How will mode shares be affected?

Solution: An increase of 100 cents will lead to new utilities but not new mode shares.

$$U_A = -9.70 - 0.005(100) = -10.2$$

$$U_T = -11.55 - 0.005(100) = -12.05$$

$$P_A = \frac{e^{U_A}}{e^{U_A} + e^{U_T}} = \frac{e^{-10.2}}{e^{-10.2} + e^{-12.05}} = 0.864$$

$$P_T = \frac{e^{U_T}}{e^{U_A} + e^{U_T}} = \frac{e^{-12.05}}{e^{-10.2} + e^{-12.05}} = 0.136$$

The answer does not change, because the difference between U_A and U_T did not change, and it is this difference, not the utilities themselves, that determines P_A and P_T. This concept is shown in the logit curve of Figure 12.10, where the proportion of individuals using transit is governed by the difference $U_A - U_T$. A change would result only if the increase in fuel price did not have the same impact on costs for transit and auto.

Borrowing Utility Functions from Other Sources

If a utility function such as that shown in Eq. 12.9 is not available, then the coefficients for the function either may be borrowed from another source or derived from survey data. To the extent that the selection of a mode is governed by its in-vehicle travel time, out-of-vehicle travel time, and cost, a utility function may be written as:

$$\text{Utility}_i = b(IVTT) + c(OVTT) + d(COST) \qquad (12.10)$$

where

$$\text{Utility}_i = \text{utility function for mode } i$$
$$IVTT = \text{in-vehicle travel time (min)}$$
$$OVTT = \text{out-of-vehicle travel time (min)}$$
$$COST = \text{out-of-pocket cost (cents)}$$

The following approach for calibrating the coefficients b, c, and d in Eq. 12.10 are based on methods published in NCHRP Report 365.

- In-vehicle travel time ($IVTT$) has a coefficient of $b = -0.025$.
- Out-of-vehicle travel time has a coefficient of $c = -0.050$, which reflects the observation that time waiting for a vehicle is perceived to be twice as great as time spent inside a moving vehicle.
- Cost coefficient d is computed as follows:

$$d = \frac{(b)(1248)}{(TVP)(AI)}$$

where

TVP = the ratio of (value of one hour travel time) ÷ (hourly employment rate). In the absence of other data, $TVP = 0.30$
AI = the average annual regional household income, ($). The number in the numerator is a factor that converts $/yr to cents/min, which is 1248.

Example 12.13 Borrowing Utility Coefficients from Other Sources

A transit authority wishes to determine the number of total travelers in a corridor that will shift from auto to a proposed new bus line. Since local data are unavailable, use of borrowed utility values is the only option. It is believed that the key factors in the decision to use transit will be time and cost. Average annual household income (AI) is $60,000, $TVP = 0.30$, and waiting time is perceived to be twice as long as riding time. System times and cost values are as follows.

Variable	Bus	Auto
IVIT (min)	30	20
OVIT (min)	6	8
Cost (cents)	100	320

Determine the proportion of persons who will use the new bus line.

$$b = -0.025$$
$$c = -0.050$$
$$d = \frac{(b)(1248)}{(TVP)(AI)} = \frac{(-0.025)(1248)}{(0.30)(\$60,000)} = -0.00173$$

$a_t = 0$ since the problem stated *IVTT*, *OVTT*, and *COST* sufficiently explain mode choice.

The utility functions are:

$$U_{auto} = b(IVTT) + c(OVTT) + d(COST)$$

$$= -0.025(20) + -0.050(8) + -0.00173(320) = -1.454$$

$$U_{bus} = b(IVTT) + c(OVTT) + d(COST)$$

$$= -0.025(30) + -0.050(6) + -0.00173(100) = -1.223$$

The proportion of travelers using the bus is computed using Eq. 12.9.

$$P_{bus} = \frac{e^{U_{bus}}}{e^{U_{bus}} + e^{U_{auto}}} = \frac{e^{-1.223}}{e^{-1.223} + e^{-1.454}} = 0.557$$

Thus, this model predicts that 56 percent of travelers will use the new bus line.

Example 12.14 Using Local Data to Improve Utility Coefficients

In Example 12.13, suppose ten residents at the local Department of Motor Vehicles are given a survey where they are asked the following two questions:

- What is your hourly salary?
- What would you be willing to pay to shorten your travel time by one hour?

Data from residents are shown below.

Resident	Hourly Salary	Amount Resident is Willing to Pay to Shorten Travel Time by 1 Hour
1	$8	$3
2	$9	$5
3	$10	$5
4	$15	$8
5	$20	$9
6	$20	$11
7	$24	$12
8	$25	$13
9	$30	$12
10	$40	$24

(a) Use the local data to re-estimate the proportion of travelers using the new bus line.
(b) Interpret the reason for the change in proportion.
(c) Indicate whether the sample of 10 residents likely represents the region as a whole.

Solution:

(a) the local data to re-estimate the proportion of travelers using the new bus line.

The sample data may be used to compute a local TVP. For example, for person 1, the TVP is $3/$8 = 0.375. For person 2, the TVP is $5/$9 = 0.56. The average TVP for all 10 residents is 0.50 rather than the assumed value of 0.30 in the original problem. Accordingly, the cost coefficient d may be recomputed as

$$d = \frac{(b)(1248)}{(TVP)(AI)} = \frac{(-0.025)(1248)}{(0.50)(\$60,000)} = -0.00104$$

Thus the utility functions U_{auto} and U_{bus} and the bus proportion P_{bus} are

$$U_{auto} = b(IVTT) + c(OVTT) + d(COST)$$

$$U_{auto} = -0.025(20) - 0.05(8) - 0.00104(320) = -1.233$$

$$U_{bus} = b(IVTT) + c(OVTT) + d(COST)$$

$$U_{bus} = -0.025(30) - 0.05(6) - 0.00104(320) = -1.154$$

$$P_{bus} = \frac{e^{U_{bus}}}{e^{U_{auto}} + e^{U_{bus}}} = \frac{e^{-1.154}}{e^{-1.233} + e^{-1.154}} = 0.520$$

Thus the proportion using the bus changes from 55.7 percent to 52.0 percent.

(b) Interpret the reason for the change in proportion.

In the original assumption, TVP was assumed to be 30 percent, whereas in part (a), local data was used to calculate a TVP of 50 percent. The higher value of TVP in part (a) means that additional cost has a lesser disutility than what was originally assumed. This change is evident in the new value of the cost coefficient d, which had an original value of -0.00173 and a new value in part (a) of -0.00104. The new value signifies that increased costs will have less of an impact on the utility than was the case in the original problem.

Examination of in-vehicle travel time and cost for each mode shows that the bus generally offers lower cost whereas the auto offers lower travel time. Because the cost parameter d has less of an impact on utility in part (a) than the original problem, the relative importance of lower costs is diminished in part (a). Accordingly, it is not surprising that the mode share of the bus drops in part (a) relative to the original problem.

(c) Indicate whether the sample of 10 residents likely represents the region as a whole.

It is unlikely that these 10 residents represent the region as a whole. The average hourly salary of the 10 residents is $10.20, which, assuming 2080 hours per year, yields an average salary of $41,808, which is considerably smaller than the regional salary of $60,000. A survey with a greater sample size, conducted at multiple locations rather than only the DMV, should attract a more representative population. Such data would need to be examined to determine whether the calculated value of TVP should be modified.

Example 12.15 Adding a Mode-Specific Constant to the Utility Function

Referring to Example 12.13, upon inaugurating bus service, the percentage of travelers that use the new bus service is actually 65 percent. Follow-up surveys confirm that the coefficients *b*, *c*, and *d*, which were used to estimate potential bus service, appear to have been correct. However, the surveys suggest that a further incentive (beyond time and cost) for using the bus is influenced by the availability of laptop outlets at each seat and a complimentary beverage service.

Given this added information, explain how to modify the utility function to reflect the influence of added amenities.

Solution: Because the coefficients *b*, *c*, and *d* do not include the additional features that favor bus usage, a mode-specific coefficient (a_i) should be included in one of the utility functions. This term may either be a positive coefficient that is added to the bus utility function or a negative coefficient that is subtracted from the auto utility function. Using the former approach, simply add a constant value (which in this example is 0.3885) to the bus utility function in order to yield the required 65 percent of travelers using the bus.

$$P_{bus} = \frac{e^{(U_{bus}+0.3885)}}{e^{(U_{bus}+0.3885)} + e^{U_{auto}}} = \frac{e^{(-1.223+0.3885)}}{e^{(-1.223+0.3885)} + e^{-1.454}} = 0.650$$

Thus, the bus utility function is rewritten and the auto utility function is unchanged, as follows.

$$U_{bus} = a_{bus} + b(IVTT) + c(OVTT) + d(COST)$$

$$U_{bus} = 0.3885 + -0.025(IVTT) + -0.050(OVTT) + -0.00173(COST)$$

Modifying a Logit Model for Changes in Service Parameters

If the value of the *IVTT*, *OVTT*, or *COST* parameters has changed, then the new mode share P_i' can be calculated from the original mode share P_i and the change in the utility function value as shown in Eq. 12.11. This property is useful because determination of P_i' does not require knowledge of the mode-specific constant *a*. Since the *a* values cancel when calculating Δu_b, the difference between utility function values, $U_{i\text{-}new}$ and $U_{i\text{-}old}$ in Eq. 12.11 is the incremental logit model and can be applied if the mode is already in service. The incremental logit model cannot be used for new modes where prior data to compute P_t are unavailable.

$$P_i' = \frac{P_i e^{\Delta u_b}}{\sum_i P_i e^{\Delta u_b}} \tag{12.11}$$

where

P_i' = proportion using mode *i* after system changes
P_i = proportion using mode **i** before system changes
Δu_b = difference in utility functions values $U_{i\text{-}new} - U_{i\text{-}old}$

Example 12.16 Applying the Incremental Logit Model

The regional transportation agency in Example 12.15 is considering an investment in signal preemption for transit vehicles, which would reduce the in-vehicle travel time for bus service from 30 to 25 min. All other service amenities will remain.
Determine the percentage of travelers who will use bus service if this investment is made.

Solution: Equation 12.11 may be applied as follows.

$$P_{bus} = 65\%$$

$$P_{auto} = 35\%$$

$$\Delta U_{bus} = U_{busnew} - U_{busold}$$

$$\Delta U_{bus} = [0.3855 + -0.025(IVTT_{new}) + -0.050(OVTT_{new}) + -0.00173(COST_{new})] -$$
$$[0.3855 + -0.025(IVTT_{old}) + -0.050(OVTT_{old}) + -0.00173(COST_{old})]$$

Since bus travel time is the only variable that has been changed, from 30 to 25 min:

$$\Delta U_{bus} = -0.025\,(25 - 30)$$

$$\Delta U_{bus} = 0.125$$

$$\Delta U_{bus} = 0 \text{ (assuming no change in auto travel time or cost)}$$

Using Eq. 12.11:

$$P'_{bus} = \frac{P_{bus}e^{\Delta U_{bus}}}{P_{auto}e^{\Delta U_{bus}} + P_{bus}e^{\Delta U_{bus}}}$$

$$P'_{bus} = \frac{0.65e^{0.125}}{0.35e^0 + 0.65e^{0.125}} = 0.68$$

This answer to Example 12.16 can also be obtained if the logit model was used with all system parameters and the amenity value a, as shown in Example 12.15. The advantage of the incremental logit model applied in Example 12.16 is that knowledge of the mode specific constant a_i is not required.

Calibrating Utility Functions with Survey Data

A second approach to determine utility function coefficients is to calibrate the coefficients based on survey data using the method of maximum likelihood estimation. Software packages such as SAS and ALOGIT are available that support maximum likelihood estimation and replace manual procedures presented here. The utility functions that are best supported by data are determined through a variety of statistical tests that represent a fundamental component of this calibration process. To illustrate this process, a simple calibration of a utility function using survey data is shown in Example 12.17 . For discussion of more complex cases, refer to references at the end of the chapter.

Example 12.17   Calibrating Utility Functions

A regional transportation agency wishes to calibrate a utility function that can be used with the logit model to predict modal choice between bus, auto, and rail. Survey data were obtained by interviewing seven people identified as persons A through G who reported the travel time for three modes they considered (car, bus, and rail) and the mode that was selected. The results of the survey are shown in the following table. The agency has proposed to select a utility function of the form $U = b$ (time).

Use the method of maximum likelihood estimation to calibrate this utility function for the parameter, b.

Sample Interview Survey Data:

Respondent	Auto Time (min)	Bus Time (min)	Rail Time (min)	Mode
A	10	13	15	Auto
B	12	9	8	Auto
C	35	32	20	Rail
D	45	15	44	Bus
E	60	58	64	Bus
F	70	65	60	Auto
G	25	20	15	Rail

Solution: The utility function is

$$U = b(\text{IV}TT)$$

where

> b = a constant to be determined from the calibration process
>
> IVTT = in-vehicle travel time (in minutes)

A maximum likelihood function may be used to derive model coefficients that replicate the observed data. For these data, a "perfect" function would predict that respondents A, B, and F would select auto; C and G would select rail; and D and E would select bus. For respondent A, the utility function is as shown, since A selected auto and not the bus or rail. Thus,

$$L_A = (P_{A-\text{auto}})$$

The probability that A will select a mode is computed using Eqs. 12.8 and 12.9. For example. The probability that respondent A will select auto, rail or bus is:

$$P_{A,\text{auto}} = \frac{e^{U_{1\text{auto}}}}{e^{U_{1\text{auto}}} + e^{U_{1\text{bus}}} + e^{U_{1\text{rail}}}} = \frac{e^{b10}}{e^{b10} + e^{b13} + e^{b15}}$$

$$P_{A,\text{bus}} = \frac{e^{U_{1\text{bus}}}}{e^{U_{1\text{auto}}} + e^{U_{1\text{bus}}} + e^{U_{1\text{rail}}}} = \frac{e^{b13}}{e^{b10} + e^{b13} + e^{b15}}$$

$$P_{A,\text{rail}} = \frac{e^{U_{1\text{rail}}}}{e^{U_{1\text{auto}}} + e^{U_{1\text{bus}}} + e^{U_{1\text{rail}}}} = \frac{e^{b15}}{e^{b10} + e^{b13} + e^{b15}}$$

Substitution of the appropriate equation into the expression for L_A yields the maximum likelihood function for respondent A.

$$L_A = \left(\frac{e^{b10}}{e^{b10} + e^{b13} + e^{b15}} \right)$$

For the entire data set, therefore, the maximum likelihood function may be computed as

$$L = \{L_A)(L_B)(L_C)(L_D)(L_E)(L_P)(L_G)$$

Since b cannot be determined such that L is exactly equal to 1.0, the best possible result is to select a value of b such that L is as close to 1.0 as possible. Theoretically, L could be differentiated with respect to b and equated to zero. However, the nonlinear equations that result usually necessitate the use of specialized software to solve. Plot L versus b is as shown in the figure 12.11. The value of $b = (-0.1504)$ maximizes L. Thus, the utility expression based on the data collected about user behavior is

$$U = (-0.1504)(\text{IVTT})$$

Several tests may be used in logit model calibration to determine which parameters are statistically significant. One example is the likelihood ratio test, defined in Eq. A as

$$-2|L(\mathbf{0}) - L(\mathbf{B})|$$

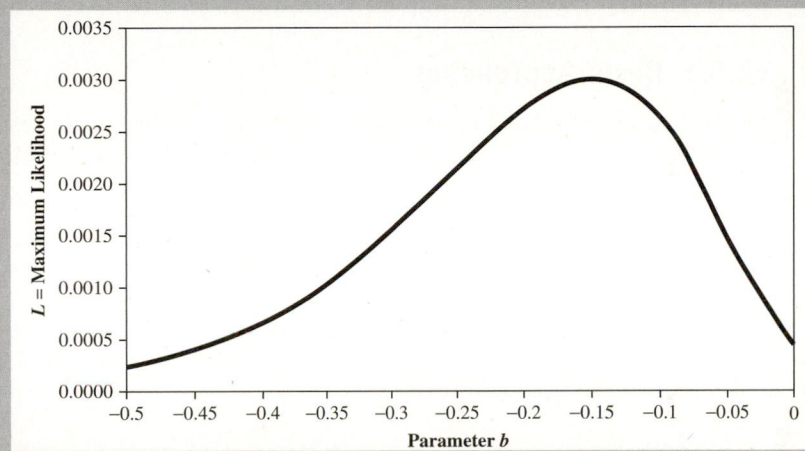

Figure 12.11   Plot of Maximum Likelihood Function versus b.

where

     $L(\mathbf{0})$ is the log-likelihood value when all model parameters are zero

     $L(\mathbf{B})$ is the log-likelihood value of the calibrated model

The quantity is compared to the Chi-square statistic with K degrees of freedom. For this example:

     $K = 1$ since the model has one nonzero parameter.

     $L(\mathbf{0}) = -7.69$, since with no parameters the probability of choosing each mode is
         1/3 and, with seven respondents, $7 \ln(1/3) = -7.69$

     $L(\mathbf{B}) = -5.81$ since $\ln(.003) = -5.81$ (see Figure 12.11)

Thus the likelihood ratio test is:

     $-2|L(\mathbf{0}) - L(\mathbf{B})|$
     $-2|-7.69 - -5.81 = 3.76$

Because 3.76 is less than $\chi_{.025,1} = 5.03$, which can be found from the Excel function CHI-INV (.025,1), the parameter b is not statistically significant. In practice, either more data would be collected or another utility function would be devised.

12.5 TRAFFIC ASSIGNMENT

The final step in the travel demand forecasting process is to determine the street and highway routes that are likely to be used and to estimate the number of automobiles and buses that can be expected on each roadway segment. The procedure used is known as *traffic assignment*. Since the number of trips by transit and auto that will travel between zones are known from the previous steps in the process, each trip O-D can be assigned to a highway or transit route. The sum of the results for each segment of the system is a forecast of the average daily or peak hour traffic volumes that will occur on the urban transportation system that serves the study area.

To carry out a trip assignment, the following data are required: (1) number of trips that will be made from one zone to another (this information was determined in the trip distribution phase), (2) a description of the highway or transit routes between zones, (3) travel time on each route segment, and (4) external trips that were not considered in the previous trip generation and distribution steps. Finally, a decision rule is required (or algorithm) that explains how motorists or transit users select a route.

12.5.1 Basic Approaches

Three basic approaches can be used for traffic assignment purposes: (1) diversion curves, (2) minimum time path (all-or-nothing) assignment, and (3) minimum time path with capacity restraint.

Diversion Curves

This method is similar in approach to a mode choice curve. The traffic between two routes is determined as a function of relative travel time or cost. Figure 12.12 illustrates a diversion curve based on travel time ratio.

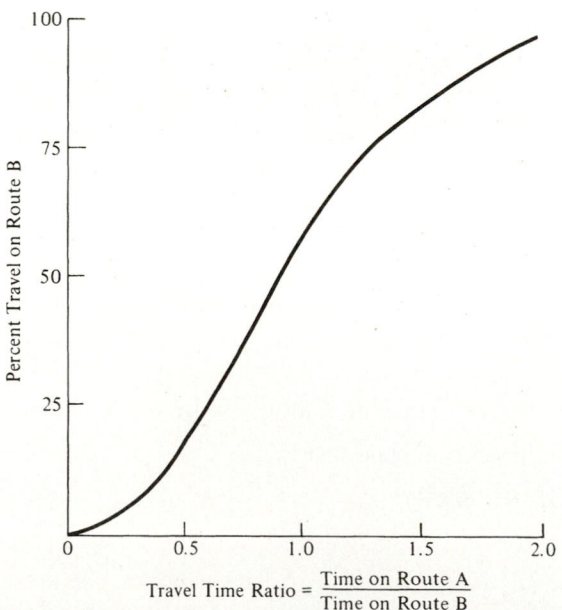

Figure 12.12   Travel Time Ratio versus Percentage of Travel on Route *B*

Minimum Path Algorithm

The traffic assignment process is illustrated using the minimum path algorithm. This method is selected because it is commonly used, generally produces accurate results, and adequately demonstrates the basic principles involved. The *minimum time path* method assigns all trips to those links that comprise the shortest time path between the two zones.

The minimum path assignment is based on the theory that a motorist or transit user will select the quickest route between any O-D pair. In other words, the traveler will always select the route that represents minimum travel time. Thus, to determine which route that will be, it is necessary to find the shortest route from the zone of origin to all other destination zones. The results can be depicted as a tree, referred to as a *skim tree*. All trips from that zone are assigned to links on the skim tree. A node in the area-wide network represents each zone. To determine the minimum path, a procedure is used that finds the shortest path without having to test all possible combinations.

The algorithm that will be used in the next example is to connect all nodes from the home (originating) node and keep all paths as contenders until one path to the same node is a faster route than others, at which juncture those links on the slower path are eliminated.

The general mathematical algorithm that describes the process is to select paths that minimize the expression

$$\sum_{\text{all}\,ij} V_{ij} T_{ij} \tag{12.12}$$

where

$$V_{ij} = \text{volume on link } i,j$$
$$T_{ij} = \text{travel on link } i,j$$
$$i,j = \text{adjacent nodes}$$

Example 12.18 Finding Minimum Paths in a Network

To illustrate the process of path building, consider the following 16-node network with travel times on each link shown for each node (zone) pair.

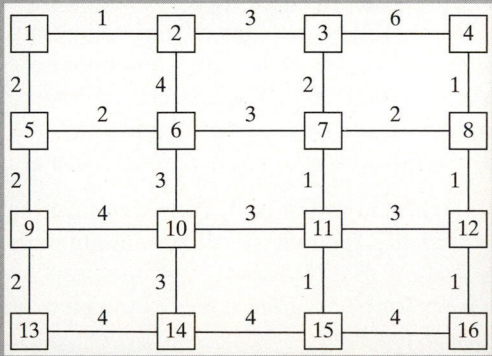

The link and node network is representative of the road and street system. Determine the shortest travel path from node 1 (home node) to all other zones.

Solution: To determine minimum time paths from node 1 to all other nodes, proceed as follows.

Step 1. Determine the time to nodes connected to node 1. Time to node 2 is 1 min. Time to node 5 is 2 min. Times are noted near nodes in diagram.

Step 2. From the node closest to the home node (node 2 is the closest to home node 1), make connections to nearest nodes. These are nodes 3 and 6. Write the cumulative travel times at each node.

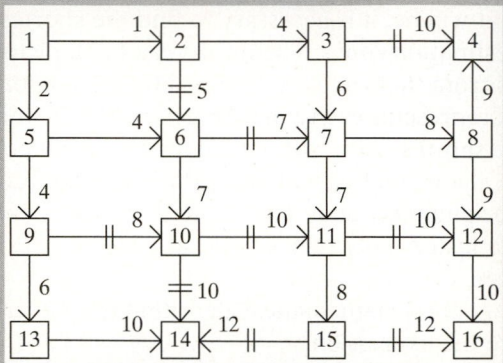

Step 3. From the node that is now closest to the home node (node 5), make connections to the nearest nodes (node 6 and 9). Write the cumulative travel times at each node.

Step 4. Time to node 6 via node 5 is shorter than that via zone 2. Therefore, link 2 to 6 is deleted.

Step 5. Three nodes are equally close to the home node (nodes 3, 6, and 9). Select the lowest-numbered node (3); add corresponding links to nodes 4 and 7.

Step 6. Of the three equally close nodes, node 6 is the next lowest numbered node. Connect to zones 7 and 10. Eliminate link 6 to 7.

Step 7. Building proceeds from node 9 to nodes 10 and 13. Eliminate link 9 to 10.

Step 8. Build from node 7.

Step 9. Build from node 13.

Step 10. Build from node 10, and eliminate link 10 to 11.

Step 11. Build from node 11, and eliminate link 11 to 12.

Step 12. Build from node 8, and eliminate link 3 to 4.

Step 13. Build from node 15, and eliminate link 14 to 15.

Step 14. Build from node 12, and eliminate link 15 to 16.

To find the minimum path from any node to node 1, follow the path backward. Thus, for example, the links on the minimum path from zone 1 to zone 11 are 7 to 11, 3 to 7, 2 to 3, and 1 to 2. This process is then repeated for the other 15 zones to produce the skim trees for each of the zones in the study area. Figure 12.13 illustrates the skim tree produced for zone 1.

Note that link 10 to 14 has been eliminated in the skim tree, although it was not explicitly eliminated in the above analysis. The reason for the elimination of link 10 to 14 is that there was a "tie" between link 10 to 14 and link 13 to 14, where the use of

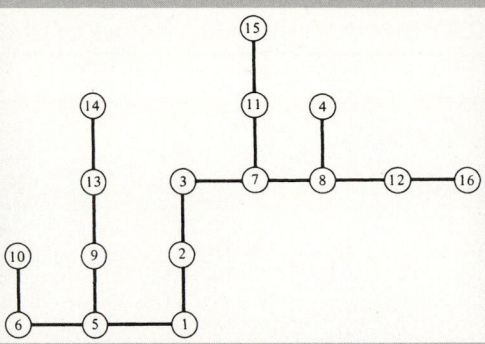

Figure 12.13 Minimum Path Tree for Zone 1

Table 12.22 Dealing with Link Elimination when Travel Times Are Equal

Link 13 to 14 option	Link 10 to 14 option
Link 1–5 (2 units)	Link 1–5 (2 units)
Link 5–9 (2 units)	Link 5–6 (2 units)
Link 9–13 (2 units)	Link 6–10 (3 units)
Link 13–14 (4 units)	Link 10–14 (3 units)
Total to reach node 14 (10 units)	Total to reach node 14 (10 units)
Total to reach the node preceding node 14 (6 units)	Total to reach the node preceding node 14 (7 units)

either link will still result in the same number of minutes (10) to reach node 14. The link is selected by considering how many minutes are required to reach the preceding node (e.g., node 10 for link 10 to 14 or node 13 for link 13 to 14). Table 12.22 shows that link 10 to 14 was eliminated, since 7 min are required to reach node 10 but only six units were required to reach node 13.

Example 12.19 Network Loading Using Minimum Path Method

The links that are on the minimum path for each of the nodes connecting node 1 are shown in Table 12.23. Also shown is the number of auto trips between zone 1 and all other zones. From these results, the number of trips on each link is determined.

To illustrate, link 1 to 2 is used by trips from node 1 to nodes 2, 3, 4, 7, 8, 11, l2, 15, and 16. Thus, the trips between these node pairs are assigned to link 1 to 2 as illustrated in Table 12.23. The volumes are 50, 75, 80, 60, 30, 80, 25, 20, and 85 for a total of 505 trips on link 1 to 2 from node 1.

Solution: Calculate the number of trips that should be assigned to each link of those that have been generated in node 1 and distributed to nodes 2 through 16 (Table 12.24). A similar process of network loading would be completed for all other zone pairs. Calculations for traffic assignment, as well as for other steps in the forecasting model system, can be performed using computer programs for transportation modeling.

Table 12.23   Links on Minimum Path for Trips from Node 1

From	To	Trips	Links on the Minimum Path
1	2	50	1–2
	3	75	1–2, 2–3
	4	80	1–2, 2–3, 3–7, 7–8, 4–8
	5	100	1–5
	6	125	1–5, 5–6
	7	60	1–2, 2–3, 3–7
	8	30	1–2, 2–3, 3–7, 7–8
	9	90	1–5, 5–9
	10	40	1–5, 5–6, 6–10
	11	80	1–2, 2–3, 3–7, 7–11
	12	25	1–2, 2–3, 3–7, 7–8, 8–12
	13	70	1–5, 5–9, 9–13
	14	60	1–5, 5–9, 9–13, 13–14
	15	20	1–2, 2–3, 3–7, 7–11, 11–15
	16	85	1–2, 2–3, 3–7, 7–8, 8–12, 12–16

Table 12.24   Assignment of Trips from Node 1 to Links on Highway Network

Link	Trips on Link
1–2	50, 75, 80, 60, 30, 80, 25, 20, 85 = 505
2–3	75, 80, 60, 30, 80, 25, 20, 85 = 455
3–7	80, 60, 30, 80, 25, 20, 85 = 380
1–5	100, 125, 90, 40, 70, 60 = 485
5–6	125, 40 = 165
7–8	80, 30, 25, 85 = 220
4–8	80 = 80
5–9	90, 70, 60 = 220
6–10	40 = 40
7–11	80, 20 = 100
8–12	25, 85 = 110
9–13	70, 60 = 130
11–15	20 = 20
12–16	85 = 85
13–14	60 = 60

12.5.2 Capacity Restraint

A modification of the process just described is known as *capacity restraint.* The number of trips assigned to each link is compared with the capacity of the link to determine the extent to which link travel times have been increased by the additional volume placed on the formerly empty link. Using relationships between volume and travel time (or speed) similar to those derived in Chapter 6, it is possible to recalculate the new link travel time. A reassignment is then made based on these new values. The iteration process continues until a balance is achieved, such that the link travel time based on the loaded volume does not change with successive assignments.

The speed-volume relationship most commonly used in computer programs was developed by the U.S. Department of Transportation and is depicted in Figure 12.14. It is called a link performance function and expressed in the following formula.

$$t = t_0 \left[1 + 0.15 \left(\frac{V}{C} \right)^4 \right]$$ (12.13)

where

t = travel time on the link
t_0 = free-flow travel time
V = volume on the link
C = capacity of the link

The capacity restraint relationship given in Eq. 12.13 can be generalized by allowing the coefficients to be adjusted to corridor-specific or roadway-type, as follows.

$$t = t_0 \left[1 + \alpha \left(\frac{V}{C} \right)^\beta \right]$$ (12.13a)

where

t = travel time on the link
t_0 = free-flow travel time
V = volume on the link
C = capacity of the link

α and β are link or roadway-type specific parameters.

One study of freeways and multilane highways found the parameters (as a function of free-flow speed) shown in Table 12.25. Alternatively, a traffic engineering study can be conducted for a specific corridor and the model fitted to the collected speed and volume data to determine appropriate values for α and β.

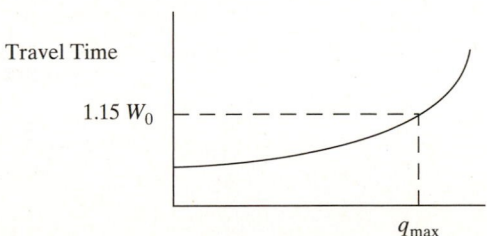

Figure 12.14   Travel Time versus Vehicle Volume

Table 12.25   Roadway-Type Specific Parameters for Capacity Restraint

Coefficient	Freeways			Multilane Highways		
	112 km/h	96 km/h	80 km/h	112 km/h	96 km/h	80 km/h
α	0.88	0.83	0.56	1.00	0.83	0.71
β	9.8	5.5	3.6	5.4	2.7	2.1

Example 12.20 Computing Capacity-Restrained Travel Times

In Example 12.19, the volume on link 1 to 5 was 485, and the travel time was 2 minutes. If the capacity of the link is 500, determine the link travel time that should be used for the next traffic assignment iteration.

Solution:

$$t_1 = t_0\left[1 + 0.15\left(\frac{V}{C}\right)^4\right]$$

$$t_{1-5} = 2\left[1 + 0.15\left(\frac{485}{500}\right)^4\right]$$

$$= 2.27 \text{ min}$$

Total System Cost Assignment

Application of Eq. 12.12 in conjunction with the equation for capacity restraint, Eq. 12.13, will result in an equilibrium assignment, where no single user may reduce their individual travel time by changing travel paths. However, user-equilibrium assignment is not necessarily the method that results in the lowest total travel time for all travelers. Rather, a total system cost assignment may be an option if the lowest total cost (as compared with lowest individual cost) is preferred. If a system cost assignment is used, route selection decisions are no longer made by the motorist but are the responsibility of the transportation agency.

To illustrate the potential benefits of a system cost assignment, consider a simple highway network shown in Figure 12.15. In this situation, there are two origin zones (1 and 2) and one destination zone (3). If 400 travelers desire to travel from zone 2 to zone 3, they have only one option: use $Link_{23}$. If 300 travelers desire to travel from zone 1 to zone 3, they have two options: (1) $Link_{12}$ and $Link_{23}$ in succession, or (2) $Link_{13}$ separately.

$$\text{Travel time}_{12} = \frac{3}{1 - volume_{12}/10,000}$$

$$\text{Travel time}_{23} = \frac{3}{1 - volume_{23}/800}$$

$$\text{Travel time}_{13} = \frac{12}{1 - volume_{13}/100,000}$$

The travel times for each link be given by the following relationships:

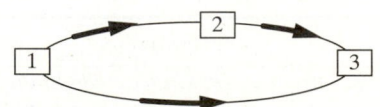

Figure 12.15 Three-Zone Highway System

An equilibrium assignment where each of the travelers from zone 1 to zone 3 individually chooses the fastest route will result in a total system cost of about 7198 min, as shown next. Only those traveling between zone 1 and zone 3 can reduce driving time by changing routes.

Results of an Equilibrium Assignment for a Three-Zone Highway Network

Link	Link Volume	Link Travel Time	Travel Time between Zone 1 and Zone 2
$Link_{12}$	132.73	3.04	12.02
$Link_{23}$	532.73	8.98	
$Link_{13}$	167.27	12.02	12.02
Total system cost			7198

Consider the situation where drivers must use a prescribed route. For example, all 300 motorists traveling from zone 1 to zone 3 could be told to use $Link_{13}$. Under this scenario, the travel time for these motorists will increase slightly—from 12.02 to 12.04 min—and the travel time for the motorists who must travel between zone 2 and travel to zone 3 will experience dramatically lowered travel times. The net result will be a lower system cost, as shown in the following table, where the total travel time has been reduced from 7198 to 6011 minutes.

Lowest System Cost Assignment

Link	Link Volume	Link Travel Time	Travel Time between Zone 1 and Zone 2 (min)
$Link_{12}$	0	3.00	9.00
$Link_{23}$	400	6.00	
$Link_{13}$	300	12.04	12.02
Total system cost			6011

In general, the assumption of user-equilibrium assignment is a more realistic basis for depicting individual decision making, since travelers act on what they consider to be their own best interest. The lowest system cost assignment may be of use for evaluating potential benefits of public interventions, such as traffic management strategies or improvements to infrastructure. For example, a public entity might choose to subsidize rail freight capacity if it found that total system costs for the rail and the adjacent interstate facility could be reduced, or variable message signs may direct motorists away from congested areas.

The process of calculating the travel demand for an urban transportation system is now completed. The results of this work will be used to determine where improvements will be needed in the system, to make economic evaluations of project priority, as well as to assist in the geometric and pavement design phases. In actual practice, computers carry out the calculations because the process becomes computationally more intensive as the number of zones increases.

....

Example 12.21 Capacity Restraint for a More Complex Network

A sample network is shown in Figure 12.16. In this network, there is just one origin zone (zone 1), one destination zone (zone 2), two intersections (shown as 3 and 4), and five unidirectional links designated as L13, L14, L32, L34, and L42.

For the network shown in Figure 12.16,

(a) Write the general capacity restraint relationship if $\alpha = 1$ and $\beta = 2$.
(b) Compute the "cost" of travel for link L13 if $V_{13} = 30$.
(c) Identify all possible travel paths from zone 1 to zone 2.
(d) Perform an equilibrium assignment.
(e) Remove the appropriate constraints to perform a system optimal assignment.

Solution:

(a) A general capacity restraint relationship was given in Eq. 12.13a as

$$\text{Link congested time} = \text{link free-flow time}\left[1 + \alpha\left(\frac{\text{link volume}}{\text{link capacity}}\right)^{\beta}\right] \quad (12.13a)$$

Setting $\alpha = 1$ and $\beta = 2$ yields

$$\text{Link congested time} = \text{link free-flow time}\left[1 + \left(\frac{\text{link volume}}{\text{link capacity}}\right)^{2}\right]$$

(b) Compute the "cost" of travel for link L13 if $V_{13} = 30$.
The cost may be defined as the link travel time multiplied by the link volume, as shown in Eq. 12.12.

$$\text{Link cost} = V_{\text{link}}T_{\text{link}}. \quad (12.12)$$

The free-flow times and capacities for each link are shown in Figure 12.16. For example, consider link L13, with a free-flow time of 5 and a capacity of 20.

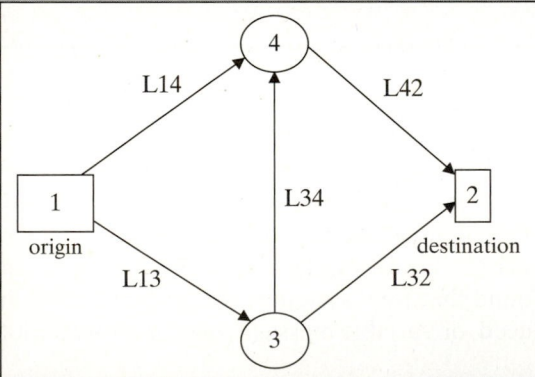

Link	Link Free-Flow Time	Link Capacity
L13	5	20
L14	20	200
L34	3	50
L42	4	20
L32	5	20

Figure 12.16 Sample Network.

With zero volume, the link congested time is simply 5. With a volume of 30, the congested time for L13 is

$$\text{Link L13 congested time} = 5\left[1 + \left(\frac{30}{20}\right)^2\right] = 5[1 + 1.5^2] = 16.25$$

For Link L13, Eq. 12.12 shows that the link cost is thus

$$(30 \text{ vehicles})(16.25 \text{ minutes}) = 487.5 \text{ veh-min}$$

(c) There are three possible paths from zone 1 to zone 2:
- Path 1: Links L14 and L42
- Path 2: Links L13, L34, and L42
- Path 3: Links L13 and L32

(d) An equilibrium assignment may be performed using spreadsheet software, such as Microsoft Excel. Letting i and j be origin and destination zones, respectively, with k travel paths between these zones, deterministic user equilibrium assignment seeks to minimize Eq. 12.12, rewritten as Eq. 12.14, where V_{ijk} denotes volume between zones i and j using path k and T_{ijk} denotes travel time between zones i and j using path k.

$$\overset{\text{origins}}{\underset{i=1}{\sum}} \; \overset{\text{destinations}}{\underset{j=1}{\sum}} \; \overset{\text{paths}}{\underset{k=1}{\sum}} (V_{ijk} \, T_{ijk}) \tag{12.14}$$

subject to the constraint that

$$T_{ij1} = T_{ij2} = \cdots = T_{ijk} \text{ for all trips between zone } i \text{ and zone } j \tag{12.15}$$

Since there is only one origin zone ($i = 1$), one destination zone ($j = 2$), and three paths between these zones ($k = 1, 2,$ or 3), we seek to minimize the system cost, which is

$$\sum_{k=1}^{3} (V_{12k} \, T_{12k})$$

subject to the constraint that

$$T_{121} = T_{122} = T_{123}$$

Thus all three paths (L14–L42, L13–L34–L42, and L13–L32) must have the same travel time. Note that the system cost is the same whether one multiplies the *path* volumes by the *path* travel times or the *link* volumes by the *link* travel times. That is, the minimization function could have been written following the exact formulation of Eq. 12.12 but tailored to this example.

$$\sum_{\text{link}=1}^{5} (V_{\text{link}} \, T_{\text{link}})$$

To obtain an equilibrium assignment such that no traveler can reduce their travel time by changing routes, Excel Solver may be used as shown in Figure 12.17a.

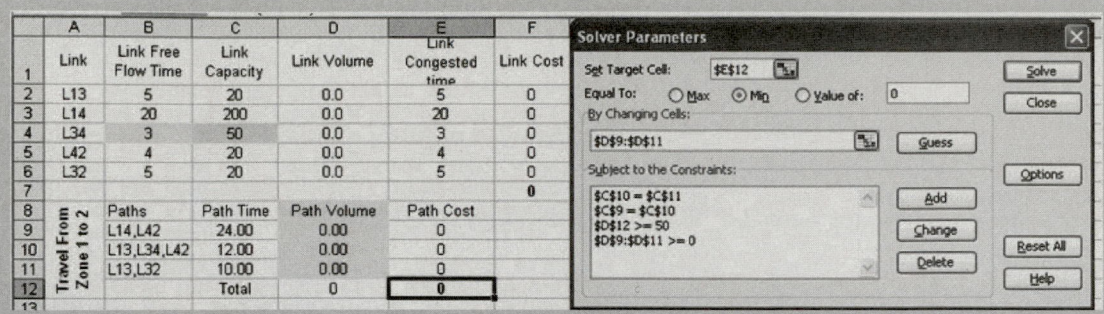

Figure 12.17a Deterministic User Equilibrium Assignment in Excel (Starting Point)

SOURCE: Created using Microsoft Office

To implement the solution shown in Figure 12.17a, notice the following:

Link free-flow times (cells B2–B6) are values.	Example: Cell B2 has "5"
Link capacities (cells C2–C6) are values.	Example: Cell C2 has "20"
Link congested times (cells E2–E6) are formulas (Eq. 12.14).	Example: Cell E2 has "=B2*(1+(D2/C2)^2)"
Link costs (cells F2–F6) are formulas (Eq. 12.15).	Example: Cell F2 has "=D2*E2"
Path times (cells C9–C11) are formulas summing link times.	Example: Cell C9 has "=E3+E5"
Path costs (cells E9–E11) are formulas (Eq. 12.20)	Example: Cell E9 has "=C9*D9"
Path volumes (cells D9–D11) will be computed by Excel	Example: Cell D9 will be found to have 18.8
Total path volume (cell D12) is the sum of all three paths	Example: Cell D12 has "=SUM(D9:D11)"
Link volumes (cells D2–D6) reflect volumes from each path.	Example: Cell D2 has "=D10+D11"

The right side of Eq. 12.17 shows that we seek to minimize Eq. 12.16 (in Cell E12) by changing the volumes using each of the three paths (in cells D9-D11). The constraints are that travel times on each path must be equal as per Eq. 12.17 (hence path travel times in cells C9, C10, and C11 must be equal), 50 vehicles must be assigned to the network (cell D12), and finally Excel Solver must only assign positive values (hence the constraint that cells D9, D10, and D11 must be greater than or equal to zero). The result of applying Excel Solver is shown in Figure 12.17b. Note that the individual path travel times are all 30.22 min and that the total system cost is 1511 veh-min.

(e) Note that the travel time for each path is the same as 30.2 such that $T_{ij1} = T_{ij2} = \cdots = T_{ijk}$. Removal of this constraint from the problem (e.g., eliminating the requirement that cells C9, C10, and C11 have the same value) reduces the system cost to 1388 veh-min. Notice that the paths have unequal travel times but that the total system cost is lower than in part (d).

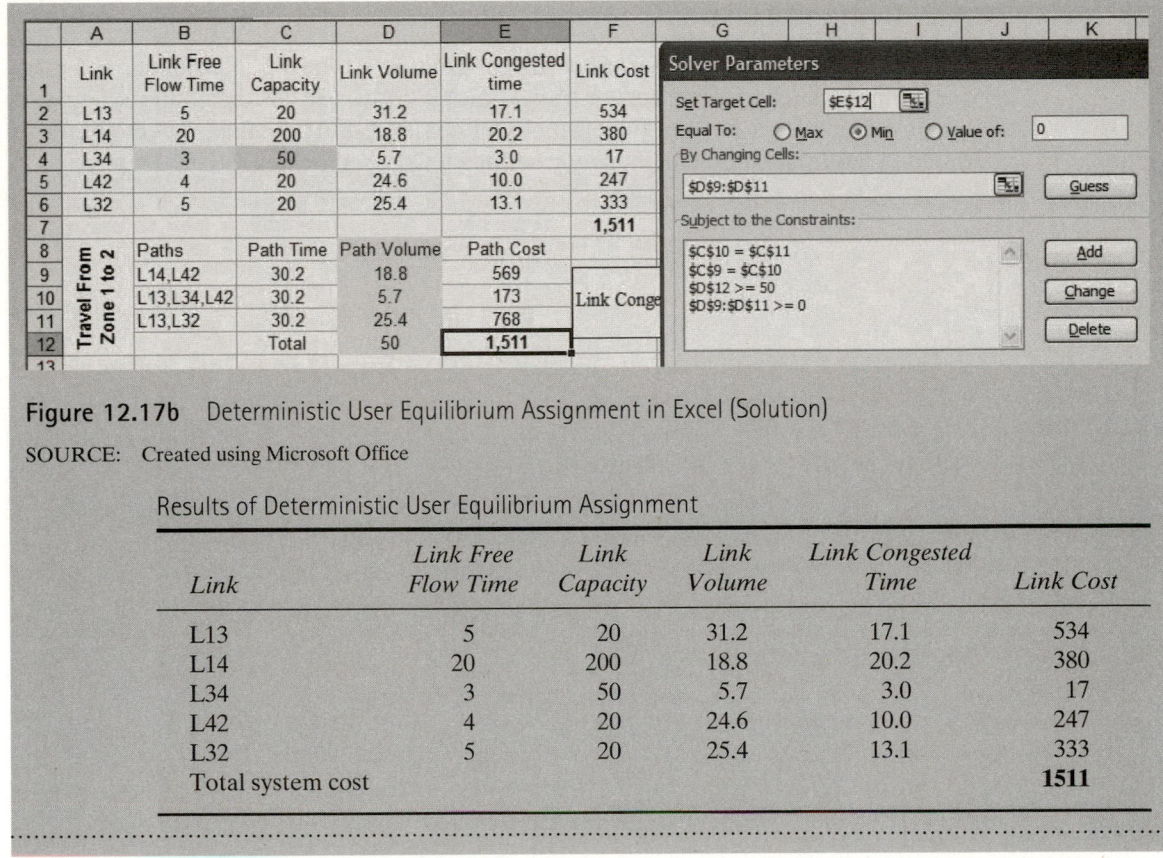

Figure 12.17b   Deterministic User Equilibrium Assignment in Excel (Solution)

SOURCE: Created using Microsoft Office

Results of Deterministic User Equilibrium Assignment

Link	Link Free Flow Time	Link Capacity	Link Volume	Link Congested Time	Link Cost
L13	5	20	31.2	17.1	534
L14	20	200	18.8	20.2	380
L34	3	50	5.7	3.0	17
L42	4	20	24.6	10.0	247
L32	5	20	25.4	13.1	333
Total system cost					**1511**

12.6 OTHER METHODS FOR FORECASTING DEMAND

This chapter has described how travel demand is forecast by using the four-step procedure of trip generation, distribution, mode choice, and traffic assignment. There are many variations within each of these steps, and the interested reader should refer to the references cited for additional details. Furthermore, there are other methods that can be used to forecast demand. Some of these are described in the following section.

12.6.1 Trend Analysis

This approach to demand estimation is based on the extrapolation of past trends. For example, to forecast the amount of traffic on a rural road, traffic count data from previous years are plotted versus time. Then, to compute the volume of traffic at some future date, the *trend line* is extrapolated forward, or else the average growth rate is used. Often a mathematical expression is developed using statistical techniques, and quite often a semi-log relationship is used. Although simple in application, trend line analysis has the disadvantage that future demand estimates are based on extrapolations of the past, and thus no allowance is made for changes that may be time-dependent. For example, rather than presuming that trip generation rate trends will continue to increase as they have in the past, understanding the underlying causes of increased trip generation (e.g., wealth, employment changes, and land-use patterns) may be more productive.

12.6.2 Demand Elasticity

Travel demand can also be determined if the relationship between demand and a key service variable (such as travel cost or travel time) is known. If V is the volume (demand) at a given service level X, then the elasticity of demand, $E(V)$, is the percent change in volume divided by the percent change in service level, as shown in Eq. 12.14.

$$E(V) = \frac{\% \Delta \text{ in } V}{\% \Delta \text{ in } X} \tag{12.14}$$

$$E(V) = \frac{\Delta V/V}{\Delta X/X} = \frac{X}{V} \frac{\Delta V}{\Delta X}.$$

Example 12.22 Forecasting Transit Ridership Reduction Due to Increase in Fares

To illustrate the use of demand elasticity, a rule of thumb in the transit industry states that for each 1 percent increase in fares, there will be one-third of 1 percent reduction in ridership. If current ridership is 2000/day at a fare of 30¢, what will the ridership be if the fare is increased to 40¢?

Solution:

In this case, $E(V) = 1/3$ and, substituting into Eq. 12.16,

$$\frac{1}{3} = \frac{X}{V} \frac{\Delta V}{\Delta X} = \frac{30}{2000} \times \frac{\Delta V}{10} \quad \text{or} \quad \frac{30\Delta V}{(2000)(10)}$$

$$\Delta V = 222.2 \text{ passengers/day}$$

Thus, the ridership will decline to $2000 - 222 = 1778$ passengers/day.

Using Midpoint Arc and Linear Arc Elasticities

The principal advantage of using elasticity methods for demand estimation is that elasticity constants are often available in published sources. There are several methods that can be used to determine the elasticity constant. In addition to the linear formulation illustrated in Example 12.22, two additional methods have been used: (1) *midpoint (for linear) arc elasticity* and (2) *log arc elasticity*. Midpoint arc elasticity is computed as shown in Eq. 12.15. Log arc elasticity is computed as shown in Eq. 12.16.

$$\text{Midpoint arc elasticity, } e_M = \frac{\Delta D/D_{avg}}{\Delta X/X_{avg}} \tag{12.15}$$

$$\text{Log arc elasticity, } e_L = \frac{\log D_o - \log D_n}{\log X_o - \log X_n} \tag{12.16}$$

where

D_0 = the original demand level
D_n = the new demand level
X_o = the original service level

X_n = the new service level

$AD = D_o - D_n$

$D_{avg} = (D_o + D_n)/2$

$AX = X - X_n$

$X_{avg} = (X_o + X_n)/2$

Example 12.23 Computing the Midpoint Arc and Log Arc Elasticity

A transit system has increased its service route mileage by 110 percent. Following this increase in service kilometers, demand increased by 28 percent.

(a) Find the midpoint arc elasticity and the log arc elasticity.
(b) What might be the explanation for the elasticity values obtained?

Solution:

(a) Find the midpoint arc elasticity and the log arc elasticity.

$$D_o = 1.00$$

$$D_n = 1.28$$

$$X_o = 1.00$$

$$X_n = 2.10$$

$$\Delta D = D_o - D_n = 1.0 - 1.28 = -0.28$$

$$\Delta X = X_o - X_n = 1.0 - 2.10 = -1.10$$

$$X_{avg} = (X_o + X_n)/2 = (1.00 + 2.10)/2 = 1.55$$

Midpoint arc elasticity is obtained from Eq. 12.15.

$$\frac{\Delta D/D_{avg}}{\Delta X/X_{avg}} = \frac{-0.28/1.14}{-1.10/1.55} = 0.35$$

Log arc elasticity is obtained from Eq. 12.16.

$$\frac{\log D_o - \log D_n}{\log X_o - \log X_n} = \frac{\log(1.0) - \log(1.28)}{\log(1.0) - \log(2.1)} = 0.33$$

(b) What might be the explanation for the elasticity values obtained?

There are several possible reasons for the relatively low change in demand given a significant increase in system mileage. Among these are:

- Most of the transit patrons had no other modal options available such that the market for transit was fixed.
- In the process of more than doubling system kilometers, the transit routes were restructured although the original routes had greater ridership.
- The buses were unreliable, ranging in age from 6 and 21 years, thus limiting the effect of expanded kilometers.

Geographic Variations in Midpoint Arc Elasticities

Midpoint arc elasticities that show the responsiveness of travelers to fare increases are shown in Table 12.26. The values in the table demonstrate that there is variation in elasticities by geographical area. For example, a fare increase of 10 percent will reduce the ridership by 32 percent in Spokane, Washington; but only by 14 percent in San Francisco, California. Table 12.26 also shows that peak hour travel is less sensitive to fare changes than nonpeak travel, which is due to a larger proportion of work trips in the peak hour.

Table 12.26 Bus Fare Midpoint Arc Elasticities for Urban Areas

Urban Area	Peak Bus Fare Elasticity	Off-Peak Bus Fare Elasticity
Spokane, Washington	−0.32	−0.73
Grand Rapids, Michigan	−0.29	−0.49
Portland, Oregon	−0.20	−0.58
San Francisco, California	−0.14	−0.31
Los Angeles, California	−0.21	−0.29

SOURCE: Linsalata, J., and Pham, L. H., *Fare Elasticity and Its Application to Forecasting Transit Demand*. American Public Transit Association, Washington, D.C., 1991. Copyright © 2003, American Public Transportation Association; All Rights Reserved.

Example 12.24 Using Midpoint Arc Elasticity Data to Forecast Demand

The current peak ridership for urban area X is 1000, the current fare is $1, and a proposed new fare is $2.

Estimate the future peak ridership of this community using published elasticity values for Portland, OR.

Solution: Use Eq. 12.15 to compute the midpoint arc elasticity.

$$e_M = \frac{\Delta D/D_{\text{avg}}}{\Delta X/X_{\text{avg}}}$$

$$e_M = \frac{(D_o - D_n)/[(D_o + D_n)/2]}{(X_o - X_n)/[(X_o + X_n)/2]}$$

Rearrange the expression for e_M to solve for D_n.

$$D_n = \frac{(e_M - 1)(X_o D_o) - (e_M + 1)(X_n D_o)}{(e_M - 1)(X_n) - (e_M + 1)(X_o)}$$

The data for city X are

$$e_M = -0.20 \text{ (from Table 12.26)}$$

$$X_o = \$1$$

$$X_2 = \$2$$

$$D_o = 1000$$

Use Eq. 12.15.

$$D_n = \frac{(-0.20-1)(\$1(1000)) - (-0.20+1)(\$2(1000))}{(-0.20-1)(\$2) - (-0.20+1)(\$1)}$$

$$D_n = \frac{-1200-1600}{-2.4-0.8} = 875$$

Thus 875 riders (a decrease of 125) are estimated if the fare is increased from \$1 to \$2. If the value (-0.20) was a log arc rather than midpoint arc elasticity, the forecasted demand would have been 871 passenger trips. Either result is within estimating accuracy, which was also the case for the results in Example 12.23. Thus, both methods are similar with regard to results obtained. Computations for the log arc method are left for the reader to verify using the expression $D_n = D_o(X_n/X_o)^{-0.20}$.

12.7 ESTIMATING FREIGHT DEMAND

This section discusses procedures available for estimating freight traffic. The first approach is based on observation of vehicle flows. The second method is based on the use of commodity flow data.

12.7.1 Using Trend Analysis of Freight Vehicle Travel

The forecasting methods presented in this chapter for passenger travel are equally suitable for freight travel forecasting. At the state level, freight flows may be computed using

Example 12.25 Estimating Truck Travel Demand Based on Trend Data

It has been observed that interstate truck traffic has been growing at an average rate of 3 percent annually for the past 20 years. Currently, truck ADT volumes are 4000 veh/day.

(a) Provide an estimate of truck traffic 5 years hence if the past 20 years of growth is expected to continue.

(b) Comment on the usefulness of this method of forecasting freight flows.

Solution:

(a) Use the following equation:

$$V_5 = V_o(1+i)^n$$

where

V_o = current truck volume
V_5 = truck volume in 5 years
i = growth rate
n = number of years

$$V_5 = 4000(1+0.03)^5 = 4637 \text{ trucks/day}$$

(b) Comment on the usefulness of this method.

This estimate may be useful for preliminary sketch planning purposes, especially when little or no other information is available.

Table 12.27 Sketch Methods for Estimating Commodity Flows

Method	Use When	Advantages	Disadvantages
Fratar method	An older zone-to-zone commodity flow table is available and simply needs to be updated	Does not require information about impedances between travel zones	Requires a prior commodity flow table Cannot address changes in travel impedance
Gravity model	A new commodity flow table is generated and impedances between zones are available	Does not require an existing commodity flow table	Requires a well-calibrated impedance measure based on time and cost for various commodities
Logit model	Shippers'choices for O–D flows can be represented as a utility function	May allow for representation of specific policy choices within the model	Requires a calibrated utility function that is rare in freight forecasting

the trend analysis of traffic flow that forecasts freight movements by truck and rail or may be determined directly from observations of prior goods movement activity.

12.7.2 Using Commodity Flow Data to Forecast Freight Vehicle Travel

A more productive approach for estimating freight travel is to base forecasts on commodity travel. Once commodity flows are estimated, they can be converted to vehicle flows. This conversion must be based on the payload for each mode (which will vary by commodity), the extent to which there are empty (return) trips, and the number of days per year that the mode will be operable. Table 12.27 compares the advantages and disadvantages of the principal forecasting methods described earlier in this chapter and suggests when their use is appropriate.

To illustrate, one state study found that for lumber, pulp, and paper categories, 5 percent of the goods were moved by light truck (less than 29,000 kg gross vehicle weight [GVW]), 47 percent by medium truck (29,000 to 36,000 kg GVW), and 48 percent by heavy truck (over 36,300 kg GVW). For all three types, it was found that 83 percent of the trucks were loaded. Data for these conversions may be obtained from the literature (e.g., one study suggested a rail car may be assumed to have a 59,000 kg shipment) or from freight-related data sources such as the U.S. Census Bureau's *Vehicle Inventory and Use Survey* (VIUS) (for trucks) and the *Rail Carload Waybill Sample* (for rail).

12.8 TRAFFIC IMPACT STUDIES

The purpose of a *traffic impact study* or a *transportation impact analysis* is to determine the impact on traffic and the need for transportation services in the immediate vicinity of a proposed development or as the result of a change in zoning designation. Impact studies are conducted for a variety of reasons. These include: (1) to determine whether a landowner should be granted permission to develop a given parcel of land that may require rezoning from its current designated use to the use proposed by the developer (e.g., from agricultural to commercial), (2) to permit a landowner to subdivide the property into two or more parcels such that a portion of the subdivision may be used for additional high-density development, (3) to obtain a building permit, (4) to grant direct access from a development to the transportation network, (5) to evaluate the effects of a county-initiated rezoning plan, or (6) to create a special-use tax district.

A traffic impact study provides quantitative information regarding how the construction of a new development will change transportation demand. Consider a proposed 100-dwelling unit subdivision from which each household generates 10 vehicle trips/day. An impact estimate of trip generation from this proposed subdivision is computed as $(10)(100) = 1000$ veh/day that will be added to the current highway network. The study can also quantify the change in demand for other modes of transportation, and it can compute the increase in traffic delay and estimate the cost of necessary improvements to mitigate these effects. This type of data is found in the Institute of Transportation Engineers publication, *Trip Generation*.

The primary difference between a traffic impact study and a regional travel demand forecast is the geographic scope involved. A regional travel demand forecast is suitable for large metropolitan areas containing hundreds or thousands of zones. A traffic impact study is intended to forecast the demand created by a specified neighborhood that may encompass an area smaller than a single traffic analysis zone.

Accordingly, traffic impact studies differentiate trips made solely for the purposes of reaching the new development *(primary trips)* and trips where travel to or through the development is part of an existing trip (either *pass-by trips* or *diverted linked trips)*. For example, if a new restaurant is built adjacent to an arterial highway, restaurant patrons who otherwise would not make the trip to this parcel of land (i.e.g., a family leaves home, visits the restaurant, and returns home) are now making primary trips to the site. Patrons who already use the arterial and choose to stop at the new restaurant, such as a person traveling home from work, are making a pass-by trip. Patrons who make a change in their route, such as a person traveling home from work but who otherwise would use the adjacent interstate, are making a diverted linked trip.

Example 12.26 Impact of Rezoning on Intersection Delay

A county wishes to consider rezoning several land parcels adjacent to reconstructed interchange. The county is interested in knowing how the change in zoning from residential to commercial would increase development and cause increased vehicle delay at a nearby signalized intersection. Figure 12.18 depicts the study area and shows the land parcels A, B, C, D, and E, as well as the critical intersections, I_1, I_2, I_3, I_4, and I_5 that could be affected by developing these parcels. The traffic impact study consists of the following steps:

1. Collect existing traffic counts at each intersection.
2. Determine the development permitted under commercial zoning.
3. Determine the number of trips generated by each type of development.
4. Add new trips generated by the development to existing trips.
5. Determine the vehicle delay at each intersection.

It is expected that an electronics superstore that is 4100 m² in area will be constructed within area E, near intersection I_1 shown in Figure 12.18. A peak hour trip generation rate of 4.5 vehicle trips/100 m² gross floor area is used for this type of development. The calculation below forecasts a total of 198 veh/peak hr, of which 51 percent

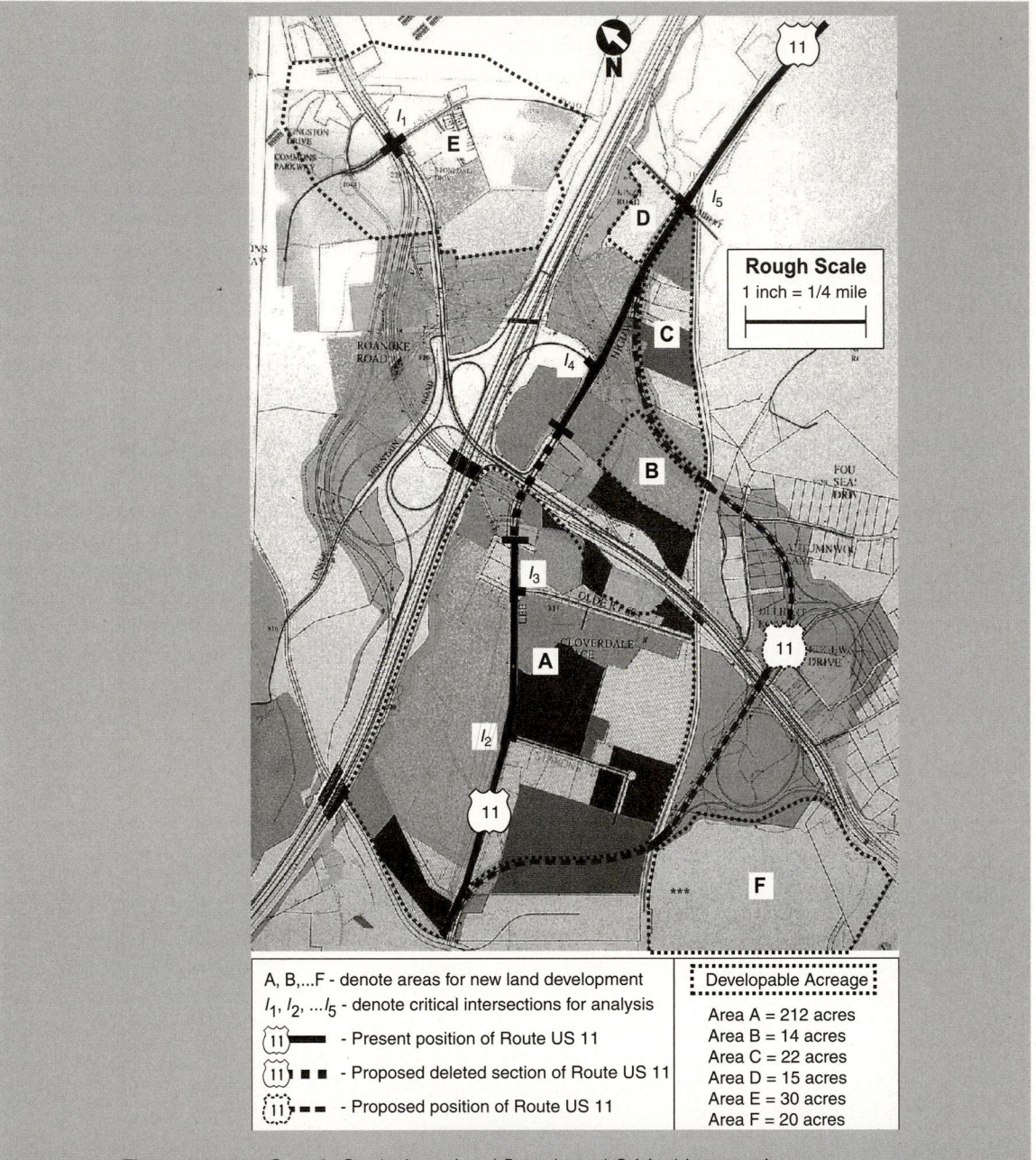

A, B,...F - denote areas for new land development

I_1, I_2, ...I_5 - denote critical intersections for analysis

⬭11 - Present position of Route US 11

⬭11 ▪ ▪ - Proposed deleted section of Route US 11

⬭11 ▬ ▬ - Proposed position of Route US 11

Developable Acreage

Area A = 212 acres
Area B = 14 acres
Area C = 22 acres
Area D = 15 acres
Area E = 30 acres
Area F = 20 acres

Figure 12.18   Sample Study Area: Land Parcels and Critical Intersections

Note: All alignments, zoning, and acreage estimates are tentative and subject to change. Colors denote different zoning at present; however, in the future, all areas will be rezoned B2. The "deleted" section of U.S. 11 refers to the fact that existing U.S. 11 will be split into two dead-end sections terminating on either side of Interstate 81.

SOURCE: J.S. Miller and A.K. Goswami, *Providing Technical Assistance in an Environment of Uncertainty: A Case Study in Coordinating Transportation and Land Use*, Virginia Transportation Research Council, Charlottesville, VA, February 2005.

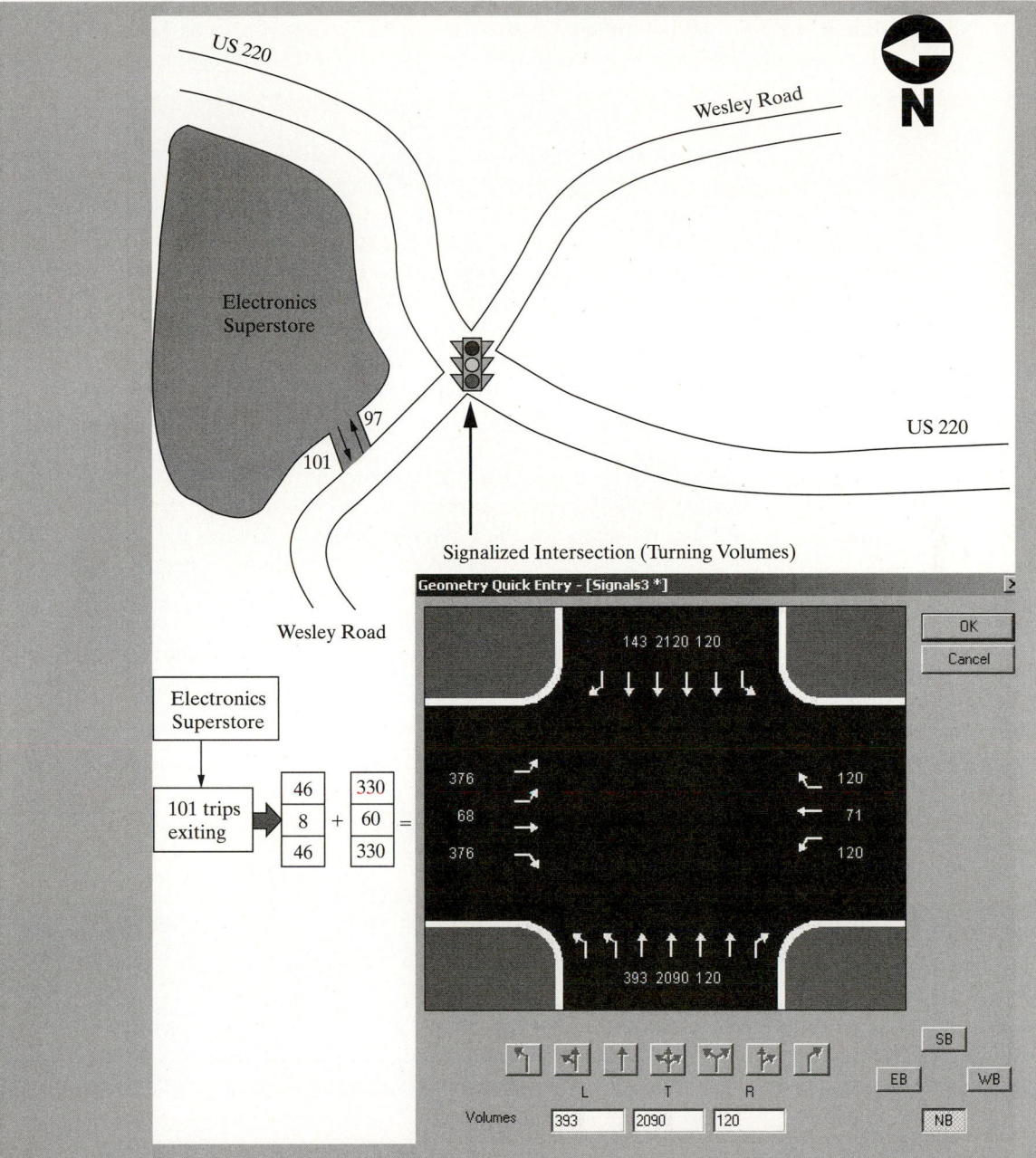

Figure 12.19   Area E, Intersection I_1

SOURCE: J.S. Miller and A.K. Goswami. *Providing Technical Assistance in an Environment of Uncertainty: A Case Study in Coordinating Transportation and Land Use.* Virginia Transportation Research Council, Charlottesville, VA, February 2005.

will exit the store and 49 percent will enter during the evening peak hour (based on ITE recommended values). Accordingly, (0.51)(198) or 101 vehicles exit from the store parking lot during the evening peak hour.

$$\text{Trips generated} = 4100 \text{ m}^2\left(\frac{4.5 \text{ trips}}{100 \text{ m}^2}\right) = 198 \text{ vehicle trips}$$

The addition of primary vehicle trips generated by the superstore to existing vehicles on the network is illustrated with the eastbound approach of the intersection in Figure 12.19. For the 101 exiting vehicles and the eastbound approach for intersection I_1, observations suggest that 8 percent would use the eastbound through movement and 46 percent would use the eastbound left and right movements. Accordingly, (0.08)(101) or 8 vehicles are added to the existing through movement of 60 vehicles, and (0.46)(101) or 46 vehicles are added to the eastbound left and right movements of 330 vehicles. In this analysis, assume that no pass-by or linked trips are destined for the superstore.

Highway Capacity Software (HCS), the computerized version of the *Highway Capacity Manual 2010*, is used to estimate the vehicle delay at the traffic signal. Table 12.28 shows that the delay for the superstore is about 34 sec/veh. Because other land uses will be permitted, and the parcel is configured so the superstore can be developed east or west of the intersection, Table 12.28 also shows the delay values for several additional uses of area E and demonstrates that (even with the same type of zoning) there is variability in delay due to the wide variety of land-use types that are permitted within a given zoning category.

Table 12.28 Intersection Delay as a Result of Single-Parcel Development

	Results for Area E	
Land Development	*Intersection Delay at I_1 If Parcel is West of I_1 (sec/veh)*	*Intersection Delay at I_1 If Parcel is East of I_1 (sec/veh)*
No development	29.7	29.7
Electronics superstore	34.5	33.2
Nursing home	30.2	29.7
Quality restaurant	38.4	37.4
Day care center	51.5	67.3

12.8.1 Data Requirements for a Traffic Impact Study

Tables 12.29 and 12.30 summarize the data elements needed for a traffic impact study. Table 12.29 identifies those data elements that will come from the governmental units that have authority over land development. Examples are the county planning commission and the local land development office. Table 12.30 identifies those elements that are available from the administrative unit responsible for the transportation system; such units may be the city, county, or state traffic engineering department and/or a consultant hired to obtain this information as part of a special study.

Not all data elements will be available, and thus assumptions may be necessary to complete a traffic impact study. Some data are easily obtained, while acquiring other

Table 12.29 Data Available from Governmental Units with Authority over Land Development

Data Element	Example	Effort Level
Zoning according to county ordinance	Commercial, which allows community shopping and service businesses	Low
Maximum floor area ratio (FAR)[a]	0.40	Low
Maximum allowable density	4000 m²/ha	Low
Net acreage land encompassed by parcel	12.14 ha	Medium
Developable acreage as opposed to net acreage	4.86 ha	Medium
Specific land uses to be built on parcel, such that they are compatible with ITE land use codes	Office building, furniture store, and electronic superstore	High
Size of development to be constructed on parcel	2800 m² of office space, 2800 m² for furniture store, and 1900 m² for electronic superstore	High

[a]The FAR is the ratio of square footage of development to square footage of open land. For example, a FAR of 0.40 means that 4047 m² of land (13,277 m²) may have no more than 5211 m² of development since $0.40 \times 13277 = 5,311$.

Table 12.30 Data Available from Sources Responsible for the Transportation System

Data Element	Example	Effort Level
List of affected walkways and bikeways	There is an Appalachian Trail crossing in what is now an urban area	Low
Number of through lanes and posted speed limits	4 total (two in each direction) Speed limit of 72 km/h	Low
24-hour volumes (Average Daily Traffic or ADT)	U.S. 220 has 21,897 vehicles per day between I-81 and Route 779 North of Daleville (both directions)	Low
Traveler characteristics (basic)	*Walking:* Visual observation suggests that most persons in the area are currently drivers, although there is some pedestrian activity *Biking:* None observed *Transit:* No fixed-route public transportation currently serves area	Low
Peak hour directional volumes	U.S. 220 p.m. peak = 1264 veh/h NB and 995 veh/h SB	Medium
Truck volumes	US 220 p.m. peak NB: 2.3% US 220 p.m. peak SB: 4.8%	Medium
Traveler characteristics (advanced)	*Walking:* Pedestrian counts of 20/h on U.S. 220 *Biking:* 5 crossings/h at intersection of U.S. 220 and Wesley Road *Transit:* No fixed route public transportation, but potential for future service exists as result of x industries	High
Peak hour turning movements and cycle lengths of new signals	For Wesley Road (Route 653) and U.S. 220, turning movements are not directly available. However, it is known that only 653 veh/day used Wesley Road between U.S. 220 and Route 1071 (Cedar Ridge Road)	High

data may involve considerable effort in both time and cost. Tables 12.29 and 12.30 show three levels of data collection effort: low, medium, and high, defined as follows.

- *Low-effort level* data are those that can be obtained with a minimal time investment and that yield order of magnitude comparisons only.
- *Medium-effort level* data require more time to obtain but can give more precise predictions.
- *High-effort level* data are quite time-consuming to obtain and are often only available when specific development proposals are being considered.

Table 12.31 illustrates how specific data may increase the precision of the results for a traffic impact study using Figure 12.17 as an illustration. For example, suppose that the county desires to know how a proposed development will increase vehicle delay during the evening peak hour. Table 12.31 shows that with low-effort data, such as knowledge of the type of zoning for the parcel, a delay of between 29 and 128 sec/veh is obtained. The reason for this large range in delay is that commercial zoning permits a wide variety of land developments, all exhibiting different trip generation rates. However, if the county knows the exact size and type of development that will be proposed, this data narrows the range to between 51 and 53 seconds.

In practice, other factors besides those discussed here, such as variation in signal timing and fluctuations in existing traffic, will increase the range of the estimates.

Although Table 12.31 illustrates the general principle that better data improves estimates, it is not always necessary to obtain such data. In any traffic impact study, the level of effort made to obtain data should be consistent with the level of precision needed for a study.

Table 12.31   Impacts of Better Land-Use Data for Peak Hour Intersection Delay Estimates

Data Extent	Low Effort	High Effort
Data elements	Commercial zoning; building size 0.4–1.21 ha	There will be 20,000 m² of office space, 850 m² for quality restaurant, and 3450 m² electronic superstore
Reason for uncertainty	Commercial zoning permits wide range of land uses (e.g., motel, bank, restaurant, office building, school), mean trip generation rates range from 0.5 to 55 trips per 100 m² and size unknown	Exact sizes of buildings known
Range of values for trips generated	The site could have 465 m² furniture store that generates 2.25 trips or a 1200 m² drive-in bank that generates 700 trips	This combination of office, restaurant, and retail uses will generate 518 to 556 trips
Range of delay estimates at intersection	Accordingly, mean delay varies from 29.5 to 128.2 sec/vehicle	Accordingly, mean delay varies from 51.9 to 53.1 sec/vehicle

12.9 SUMMARY

The process of forecasting travel demand is necessary to determine the number of persons or vehicles that will use a new transportation system or component. The methods used to forecast demand include extrapolation of past trends, elasticity of demand, and relating travel demand to socioeconomic variables.

Urban travel demand forecasting is a complex process, because demand for urban travel is influenced by the location and intensity of land use; the socioeconomic characteristics of the population; and the extent, cost, and quality of transportation services.

Forecasting urban travel demand involves a series of tasks. These include population and economic analysis, land-use forecasts, trip generation, trip distribution, mode choice, and traffic assignment. The development of computer programs to calculate the elements within each task has greatly simplified implementation of the demand forecasting process. The inability to foresee unexpected changes in travel trends, of course, remains a part of demand forecasting. Travel demand forecasts are also required for completing an economic evaluation of various system alternatives. This topic is described in the next chapter.

PROBLEMS

12-1   Identify and briefly describe the two basic demand forecasting situations in transportation planning.

12-2   Identify the three factors that affect demand for urban travel.

12-3   Define the following terms:

(a) home-based work (HBW) trips
(b) home-based other (HBO) trips
(c) non-home-based (NHB) trips
(d) production
(e) attraction
(f) origin
(g) destination

12-4   Given cross-classification data for the Jeffersonville Transportation Study Area, develop the family of cross-classification curves.
Determine the number of trips produced (by purpose) for a traffic zone containing 500 houses with an average household income of $35,000. (Use high = $55,000; medium = $25,000; low = $15,000.)

($) Income	HH (%)			Autos/HH (%)				Trip Rate/Auto				Trips (%)		
	High	Med	Low	0	1	2	3	0	1	2	3+	HBW	HBO	NHB
10,000	0	30	70	48	48	4	0	2.0	6.0	11.5	17.0	38	34	28
20,000	0	50	50	4	72	24	0	2.5	7.5	12.5	17.5	38	34	28
30,000	10	70	20	2	53	40	5	4.0	9.0	14.0	19.0	35	34	31
40,000	20	75	5	1	32	52	15	5.5	10.5	15.5	20.5	27	35	38
50,000	50	50	0	0	19	56	25	7.5	12.0	17.0	22.0	20	37	43
60,000	70	30	0	0	10	60	30	8.0	13.5	18.0	23.0	16	40	44

12-5   Given: A person travels to work in the morning and returns home in the evening. Determine productions and attractions generated in the work and residence zones.

12-6   Describe and illustrate cross-classification procedures for (a) trip production, and (b) trip attraction.

12-7   Given socioeconomic data for the Jeffersonville Transportation Study Area as follows:

Population = 72,173
Area = 70 sq km
Registered vehicles = 26,685
Single-family housing units = 15,675
Apartment units = 7567
Retail employment = 5502
Nonretail employment = 27,324
Student attendance = 28,551
Average household income = $17,500
Transportation analysis zones = 129

The results of the cross-classification analysis are as follows:

Total trips produced for study area = 282,150 per day
Home-to-work trips: 13% (36,680)
Home-to-nonwork trips: 62% (174,933)
Non-home-based trips: 25% (70,537)

The attraction rates for the area have been developed using the following assumptions:

100 percent of home-to-work trips go to employment locations.

Home-to-nonwork trips are divided into the following types:

Visit friends: 10%
Shopping: 60%
School: 10%
Nonretail employment: 20%

Non-home-based trips are divided into the following types:

Other employment area (nonretail): 60%
Shopping: 40%

Determine the number of home-to-work, home-to-nonwork, and nonhome-based trips attracted to a zone with the following characteristics: population = 1440; dwelling units = 630; retail employment = 40; nonretail employment = 650; school attendance = 0.

12-8   Given a small town with three transportation analysis zones and origin-destination survey results, provide a trip distribution calculation using the gravity model for two iterations; assume $K_{ij} = 1$. The following table shows the number of productions and attractions in each zone:

Zone	1	2	3	Total
Productions	250	450	300	1000
Attractions	395	180	425	1000

The survey's results for the zones' travel time in minutes were as follows.

Zone	1	2	3
1	6	4	2
2	2	8	3
3	1	3	5

The following table shows travel time versus friction factor.

Time (min)	1	2	3	4	5	6	7	8
Friction Factor	82	52	50	41	39	26	20	13

12-9   Given a study area with four transportation analysis zones, and origin-destination survey results, provide a trip distribution calculation using the gravity model for two iterations; assume $K_{ij} = 1$.

District	Productions	Attractions	Travel Time (min)			
			1	*2*	*3*	*4*
1	3400	2800	4	11	15	10
2	6150	6500	11	6	6	9
3	3900	2550	15	6	6	11
4	2800	4400	10	9	11	4

Friction factors are as follows:

Travel Time (minutes)	4	5	6	7	8	9	10	11	12	13	14	15
F_{ij}	1.51	1.39	1.30	1.22	1.13	1.04	0.97	0.92	0.85	0.80	0.75	0.71

12-10   Given a table with production and attraction data, determine the number of productions and attractions that should be used for each zone in the second iteration.

	1	2	3	4
P	100	200	400	600
A	300	100	200	700
P^1	100	200	400	600
A^1	250	150	300	600

12-11   Given a table with production and attraction data, determine the number of productions and attractions that should be used for each zone in the second iteration.

	1	2	3	4	5
P	100	150	350	500	200
A	250	100	150	500	300
P^1	100	200	350	500	200
A^1	150	100	250	400	350

12-12   The Jeffersonville Transportation Study area has been divided into four traffic zones. The following data have been compiled. Complete the second iteration.

District	Productions	Attractions	Travel Time (min)			
			1	*2*	*3*	*4*
1	1000	1000	5	8	12	15
2	2000	700	8	5	10	8
3	3000	6000	12	10	5	7
4	2200	500	15	8	7	5

Travel Time	1	5	6	7	8	10	12	15
F_{ij}	2.00	1.30	1.10	1.00	0.95	0.85	0.80	0.65

After the first iteration, the trip table was

District	1	2	3	4	P_s
1	183	94	677	46	1000
2	256	244	1372	128	2000
3	250	186	2404	160	3000
4	180	183	1657	180	2200
A_s	869	707	6110	514	8200

12-13 For the travel pattern illustrated in Figure 12.20, develop the Fratar method of trip distribution for two iterations.

12-14 Redo Problem 12-13 using the average growth factor method.

12-15 What data are required in order to use (a) the gravity model, and (b) the Fratar model?

12-16 The amount of lumber produced and consumed by three states is shown in the following table. Intrastate shipment distances are 320 km and interstate distances are 1280 km (between states 1 and 2), 1600 km (between states 1 and 3), and 400 miles (between states 2 and 3). Assuming an impedance function of the form $1/d$, estimate the tonnage of lumber that will travel between the three states:

Tons of Lumber Produced and Consumed Per Year (metric tons)

State	Lumber Produced	Lumber Consumed
1	5334	889
2	2994	9072
3	8890	7257

12-17 Suppose that traffic congestion has rendered the distance-based impedance function unsuitable for Problem 12-16. A detailed survey yields present-day commodity flows shown below. In the future, state #1 lumber production will increase to 18,000 and state #2 lumber consumption will increase to 21,800. Estimate the lumber flows between the three states.

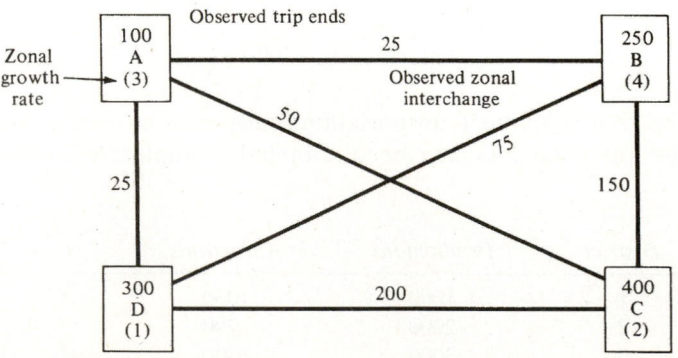

Figure 12.20 Travel Pattern

State	1	2	3	Total Produced
1	200	1800	4000	6000
2	100	200	3000	3300
3	800	7000	1200	9000
Total Consumed	1100	9000	8200	0

12-18   Survey data suggest the trip interchange matrix shown below. Calibrate the friction factors for one iteration, assuming travel times are 2.0 minutes for interzonal trips and 1.0 minute for all intrazonal trips. Assume the friction factor takes the form t^{-b}.

	Given Trips		
Zone	1	2	3
1	10	40	70
2	20	50	80
3	30	60	90

12-19   For Problem 12-18, a field study of travel times was conducted which yielded the travel times of 1.5 minutes for intrazonal trips and 4.0 minutes for interzonal factors. Perform the first iteration of friction factor calibration.

12-20   Determine the share (proportion) of person-trips by each of two modes (private auto and mass transit) using the multinomial logit model and given the following information:

$$\text{Utility function: } U_k = A_k - 0.05T_a - 0.04T_w - 0.03T_r - 0.014C$$

Parameter	Private auto	Mass transit
T_a = access time (min.)	5	10
T_w = waiting time (min.)	0	15
T_r = riding time (min.)	25	40
C = out-of-pocket cost (cents)	150	100
Calibration constant, A_k	−0.01	−0.07

12-21   For Problem 12-20, an increase in the price of fuel has changed the out-of-pocket costs for the auto mode to $1.80. Recalculate the mode shares accordingly.

12-22   A mode choice logit model is to be developed based on the following information. A survey of travelers in an area with bus service found the following data:

Model parameter	Auto	Bus
X_1, waiting time (min.)	0	10
X_2, travel time (min.)	20	35
X_3, parking time (min.)	5	0
X_4, out-of-pocket cost (cents)	225	100
A_k, calibration constant	−0.33	−0.27

The following utility function was calibrated based on an observed mode split of 84.9% private auto use and 15.1% bus use.

$$\text{Utility function: } U_k = A_k - 0.10X_1 - 0.13X_2 - 0.12X_3 - 0.0045X_4$$

After implementing service improvements to the buses, the mode split changed to 81.6% private auto use and 18.4 percent bus use. Determine a value for the calibration constant for the bus mode that reflects this shift in mode split.

12-23   A previously developed elasticity model for improvements to a transit system found a midpoint arc elasticity value of 0.38 to describe the relationship between demand increase and route mileage. The system is now planning a 60 percent increase in route mileage. Estimate the expected demand increase.

12-24   Determine the minimum path for nodes 1, 3, and 9 in Figure 12.21. Sketch the final trees.

12-25   Assign the vehicle trips shown in the O-D trip table to the network in Figure 12.22, using all-or-nothing assignment. Make a list of links in the network and indicate the volume assigned to each. Calculate the total vehicle-minutes of travel. Show the minimum path and assign traffic for each of the five nodes.

12-26   Figure 12.23 represents travel times on links connecting six zonal centroids. Determine the minimum path from zone to zone using all-or-nothing assignment based on the given trip table.

12-27   Given the following information, and using the generalized capacity restraint link performance function, perform two iterations of multipath traffic assignment. A flow of 10,100 vehicles in the peak hour is to be distributed between three routes whose properties are given in the following table.

Link Performance Component	Route 1	Route 2	Route 3
Free-flow travel time (min.)	17.0	15.5	12.5
Capacity (veh/h)	3.8	4.2	6.6
α	0.2	0.3	0.4
β	2.8	3.7	4.3

12-28   Consider the user equilibrium assignment and the lowest system cost assignment discussed in Section 12.5.1.4 (Figure 12.17). Zones 1, 2, and 3 represent a company's warehouses all

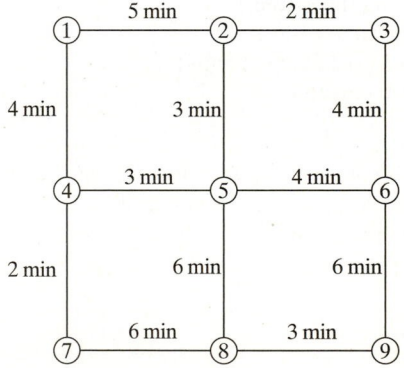

Figure 12.21   Link Node Network

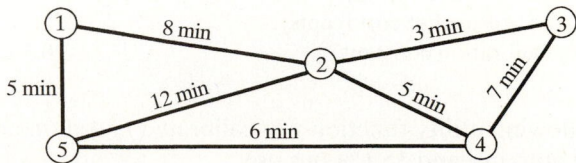

Figure 12.22   Highway Network

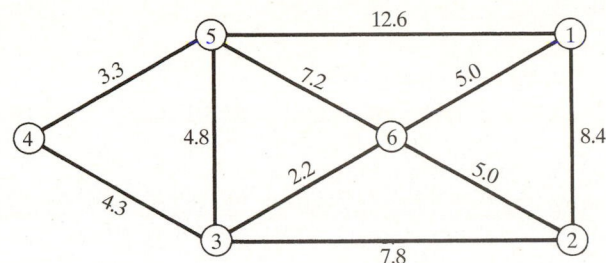

Figure 12.23   Link Travel Times

located on a privately owned parcel of land. The vehicles shown are delivery trucks rather than passenger cars. Which method of traffic assignment is more appropriate?

REFERENCES

Evans, J. E., *Transit Scheduling and Frequency*. TCRP Report 95: Traveler Response to Transit Scheduling and Frequency, TRB, National Research Council, Washington, D.C., 2004.

Hunt, J. D. and B. J. Gregor, *Oregon Generation I Land Use Transport Economic Model Treatment of Commercial Movements*. TRB Conference on Freight Demand Modeling: Tools for Public-Sector Decision Making, TRB, Washington, D.C., 2006.

Institute of Transportation Engineers, *Transportation Planning Handbook,* 3rd ed., Washington, D.C., 2009.

Institute of Transportation Engineers, *Trip Generation,* 7th ed., Washington, D.C., 2003.

Koppelman, F. S. and C. Bhat, *A Self-Taught Course in Mode Choice Modeling: Multinomial and Nested Logit Models,* Federal Transit Administration, 2006.

Martin, W. A. and N. A. McGuckin, *NCHRP Report 365: Travel Estimation Techniques for Urban Planning*. Transportation Research Board of the National Academies, National Research Council, Washington, D.C., 1996.

McCollom, B. E. and R. H. Pratt, *Transit Pricing and Fares*. TCRP Report 95: Traveler Response to Transit Scheduling and Frequency, Transportation Research Board of the National Academies, National Research Council, Washington, D.C., 2004.

Meyer, M. D. and E. J. Miller, *Urban Transportation Planning,* 2nd ed., McGraw-Hill, New York, 2000.

Ortuzar, J. de D. and L. G. Willunsen, *Modeling Transport*. John Wiley and Sons, 2001.

Quick Response Urban Travel Estimation Techniques and Transferable Parameters, Users Guide. NCHRP Report No. 187, National Research Council, Transportation Research Board, Washington, D.C., 1978.

Shunk, G. A., "Urban Transportation Systems." Chapter 4, *Transportation Planning Handbook*. Institute of Transportation Engineers, 1992.

University of Florida, *McTrans—Microcomputers in Transportation*. Transportation Research Center, 2007.

CHAPTER 13

Evaluating Transportation Alternatives

In the previous chapter, methods and techniques were described for establishing the demand for transportation services under a given set of conditions. The results of this process furnish the necessary input data to prepare an evaluation of the relative worth of alternative projects. This chapter describes how transportation project evaluations are conducted and compared based on quantitative information.

CHAPTER OBJECTIVES:

- Understand the basic questions and issues to consider prior to beginning an evaluation analysis.
- Calculate net benefits to road users based on supply-demand relationships.
- Determine project cost including initial investment, vehicle operating, travel time, and accidents.
- Apply the equations used in economic evaluation methods.
- Learn how to use evaluation methods based on multiple criteria.

13.1 BASIC ISSUES IN EVALUATION

The basic concept of an evaluation is simple and straightforward, but the actual process itself can be complex and involved. A transportation project is usually proposed because of a perceived problem or need. For example, a project to improve safety at a railroad grade crossing may be based on citizen complaints about accidents or time delays at the crossing site. In most instances, there are many ways to solve the problem, and each solution or alternative will result in a unique outcome in terms of project cost and results. In the railroad grade crossing example, one solution would be to install gates and flashing lights; another solution would be to construct a grade-separated overpass. These two solutions are quite different in terms of their costs and effectiveness. The first solution will be less costly than the second, but it also will be less effective in reducing accidents and delays.

A transportation improvement can be viewed as a mechanism for producing a result desired by society at a price. The question is, Will the benefits of the project be worth the cost? In some instances, the results may be confined to the users of the system (as in the case of the grade crossing), whereas in other instances, those affected may include persons in the community who do not use the system.

Prior to beginning an analysis to evaluate a transportation alternative, the engineer or planner should consider a number of basic questions and issues. These will assist in determining the proper approach to be taken, what data are needed, and what analytical techniques should be used. These issues are discussed in the following sections.

13.1.1 Objectives of Evaluation

What information is needed for project selection? The objective of an evaluation is to furnish the appropriate information about the outcome of each alternative so that a selection can be made. The evaluation process should be viewed as an activity in which information relevant to the selection is available to the person or group who will make a decision. An essential input in the process is to know what information will be important in making a project selection. In some instances, a single criterion may be paramount (such as cost); in other cases, there may be many objectives to be achieved. The decision maker may wish to have the relative outcome of each alternative expressed as a single number, whereas at other times, it may be more helpful to see the results individually for each criteria and each alternative.

There are many methods and approaches for preparing a transportation project evaluation, and each one can be useful when correctly applied. This chapter describes the considerations in selecting an evaluation method and discusses issues that are raised in the evaluation process. Two classes of evaluation methods are considered that are based on a single measure of effectiveness: The first reduces all outcomes to a monetary value, and the second reduces all outcomes to a numerical relative value. Evaluation can also be viewed as a fact-finding process in which all outcomes are reported separately in a matrix format that provides decision maker with complete information about the project outcome. This information can be used in public forums for citizen input, and the decision process can be extended to include public participation.

Evaluations also can be made after a project is completed to determine if the outcomes for the project are as had been anticipated. *Post facto* evaluation can be very helpful in formulating information useful for evaluating similar projects elsewhere or in making modifications in original designs. Thus, *post facto* evaluations are described and illustrated using the results of evaluations for completed projects.

13.1.2 Identifying Project Stakeholders

Who will use the information, and what are their viewpoints? A transportation project can affect a variety of groups in different ways. In some instances, only one or a few groups are involved; in other cases, many factions have an interest. Examples of groups that could be affected by a transportation project include the system users, transportation management, labor, citizens in the community, business, and local, state, and national governments. Each of these groups may have special concerns. Since these viewpoints may differ from group to group, the elements of the evaluation process itself will be reflective of the viewpoints expressed.

For smaller, self-contained projects, those groups with something to gain or lose by the project—the *stakeholders*—usually will be limited to the system users and transportation management. For larger, regional-scale transportation projects, the number and variety of stakeholders will increase because the project will affect many groups in addition to the users and management. For example, a major project could increase business in the downtown area, or expanded construction activity could trigger an economic boom in the area. The project might also require the taking of land, or it could create other environmental effects. Thus, if the viewpoint is that of an individual traveler or business, the analysis can be made on narrow economic grounds. If the viewpoint is that of the community at large, then the analysis must consider a wider spectrum of concerns.

If the viewpoint were that of a local community, then the transfer of funds by grants from the state or federal government would not be considered to be a cost, whereas increases in land values within the area would be considered a benefit. However, if the viewpoint were expanded to a regional or state level, these grants and land-value increases would be viewed as costs to the region or as transfers of benefits from one area to another. Thus, a clear definition of whose viewpoint is being considered in the evaluation is necessary if proper consideration is to be given to how these groups are either positively or negatively affected by each proposed alternative.

13.1.3 Selecting and Measuring Evaluation Criteria

What are the relevant criteria, and how should these be measured? A transportation project is intended to accomplish one or more goals and objectives, which are made operational as criteria. The numerical or relative results for each criterion are called *measures of effectiveness*. For example, in a railroad grade crossing problem, if the goal is to reduce accidents, the criteria can be measured as the number of accidents expected to occur for each of the alternatives considered. If another goal is to reduce waiting time, the criteria could be the number of minutes per vehicle delayed at the grade crossing. Nonquantifiable criteria also can be used and expressed in a relative scale, such as high, medium, and low.

Criteria selection is a basic element of the evaluation process because the measure used becomes the basis on which each project is compared. Thus, it is important that the criteria be related as closely as possible to the stated objective. To use a nontransportation example for illustration, if the objective of a university course is to learn traffic and highway engineering, then a relevant criterion to measure results is exam grades, whereas a less relevant criterion is the number of class lectures attended. Both are measures of class performance, but the first is more relevant in measuring how well the stated objective was achieved.

Criteria not only must be relevant to the problem but should also have other attributes. They should be easy to measure and sensitive to changes made in each alternative. Also, it is advisable to limit the number of criteria to those that will be most helpful in reaching a decision in order to keep the analysis manageable for both the engineer who is doing the work and the person(s) who will act on the result. Too much information can be confusing and counterproductive and, rather than being helpful, could create uncertainty and encourage a decision on a political or other nonquantitative basis. Some examples of criteria used in transportation evaluation are listed in Table 13.1.

13.1.4 Measures of Effectiveness

How are measures of effectiveness used in the evaluation process itself? One approach is to convert each measure of effectiveness to a common unit, and then, for each alternative, compute the summation for all measures. A common unit is money, and it may be

Table 13.1   Criteria for Evaluating Transportation Alternatives

- Capital costs
  - —Construction
  - —Right-of-way
  - —Vehicles
- Maintenance costs
- Facility operating costs
- Travel time
  - —Total hours and cost of system travel
  - —Average door-to-door speed
  - —Distribution of door-to-door speeds
- Vehicle operating costs
- Safety
- Social and environmental costs
  - —Noise
  - —Visual quality
  - —Community cohesion
  - —Air and water quality

possible to make a transformation of the relevant criteria to equivalent dollars and then compare each alternative from an economic point of view. For example, if the cost of an accident is known and the value of travel time can be determined, then for the railroad grade crossing problem, it would be possible to compute a single number that would represent the total cost involved for each alternative, since construction, maintenance, and operating costs are already known in dollar terms, and the accident and time costs can be computed using conversion rates.

A second approach is to convert each measure of effectiveness to a numerical score. For example, if a project alternative does well in one criterion, it is given a high score; if it does poorly in another criterion, it is given a low score. A single number can be calculated that represents the weighted average score of all the measures of effectiveness that were considered. This approach is similar to calculating grades in a course. The instructor establishes both a set of criteria to measure a student's performance (for example, homework, midterms, finals, class attendance, and a term paper) and weights for each criterion. The overall measure of the student's performance is the weighted sum of the outcome for each measure of effectiveness. Measures of effectiveness should be independent of each other if a summation procedure (such as adding grades) is to be used in the evaluation. If the criteria are correlated, then totaling the weighted scores will bias the outcome. (This would suggest, for example, that homework grades should not be included in a student's final grade since they may correlate with midterm results.)

A third way is to identify the measures of effectiveness for each alternative in a matrix form, with no attempt made to combine them. This approach furnishes the maximum amount of information without prejudging either how the measures of effectiveness should be combined or their relative importance.

13.1.5 Evaluation Procedures and Decision Making

How well will the evaluation process assist in making a decision? The decision maker typically needs to know what the costs of the project will be; in many instances, this alone will determine the outcome. Another question may be, Do the benefits justify the expenditure of funds for transportation, or would the money be better spent elsewhere? The decision maker also will want to know if the proposed project is likely to produce the stated results—that is, How confident can we be of the predicted outcomes?

It may be necessary to carry out a sensitivity analysis that shows a range of values rather than a single number. Also, evaluations of similar projects elsewhere may provide clues to the probable success of the proposed venture. The decision maker also may wish to know if all the alternatives have been considered and how they compare with the one being recommended. Are there other ways to accomplish the objective, such as using management and traffic-control strategies that would eliminate the need for a costly construction project? It may be that providing separate bus and carpool lanes results in significant increases in the passenger-carrying capacity of a freeway, thus eliminating the need to build additional highway lanes.

The decision maker may want to know the cost to highway users as the result of travel delays during construction. Also of interest may be the length of time necessary to finish the project, since public officials are often interested in seeing work completed during their administration. The source of funds for the project and other matters dealing with its implementation will also be of concern. Thus, in addition to the fairly straightforward problem of evaluation based on a selected set of measurable criteria, the transportation engineer must be prepared to answer any and all questions about the project and its implications.

The evaluation process requires that the engineer have all appropriate facts about a proposed project and be able to convey these in a clear and logical manner to facilitate decision making. In addition to the formal numerical summaries of each project, the engineer also must be prepared to answer other questions about the project that relate to its political and financial feasibility. In the final analysis, the selection itself will be based on a variety of factors and considerations that reflect all the inputs that a decision maker receives from the appropriate source.

13.2 EVALUATION BASED ON ECONOMIC CRITERIA

To begin the discussion of economic evaluation, it is helpful to consider the relationship between the supply and demand for transportation services. Consider a particular transportation project, such as a section of roadway or a bridge. Further, assume that we can calculate the cost involved for a motorist to travel on the facility. (These costs would include fuel, tolls, travel time, maintenance, and other actual or perceived out-of-pocket expenses.) Using methods described in Chapter 12, we can calculate the traffic volumes (or demand) for various values of user cost. As explained in Chapter 2, as the cost of using the facility decreases, the number of vehicles per day will increase. This relationship is shown schematically in Figure 13.1, which represents the demand curve for the facility for a particular group of motorists.

A demand curve could shift upward or downward and have a different slope for users with different incomes or for various trip purposes. If the curve moved upward, it would indicate a greater willingness to pay, reflecting perhaps a group with a higher income. If the slope approached horizontal, it would indicate that demand is elastic (i.e., that a small change in price would result in a large change in volume). If the slope approached vertical, it would indicate that the demand is inelastic (i.e., that a large change in price has little effect on demand). As an example, the price of gasoline is said to be inelastic because people seem to drive equally as often after gas prices increase as before the increase.

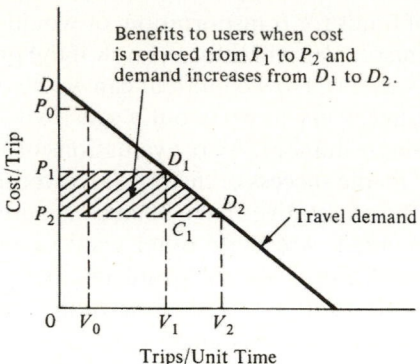

Figure 13.1 Demand Curve for Travel on a Given Facility

Consider the cost to travel on this facility to be P_1. The number of trips per unit time will be V_1 and the total cost for all users over a given period per hour or day will be $(P_1) \times (V_1)$. This amount can be shown graphically as the area of rectangle $0P_1D_1V_1$. As can be seen from the demand curve, all but the last user would have been willing to pay more than the actual price. For example, V_0 users would have been willing to pay P_0 to use the facility whereas they paid the lesser amount P_1. The area under the demand curve $0DD_1V_1$ is the amount that the V_1 users would be willing to pay. If we subtract the area $0P_1D_1V_1$, which is the amount the users actually paid, we are left with the area P_1DD_1. This triangular area is referred to as a *consumer surplus* and represents the value of the economic benefit for the current users of the facility.

Suppose the price for using the facility is reduced to P_2 because of various improvements that have been made to the facility. The total user cost is now equal to $(P_2) \times (V_2)$, and the consumer surplus is the triangular area between the demand curve and P_2D_2 or P_2DD_2. The net benefit of the project is the net increase in consumer surplus or area P_2DD_2 minus area P_1DD_1 and is represented by the shaded trapezoidal area $P_2P_1D_1D_2$. That area is made up of two parts: The first is the reduction in total cost paid by the original travelers, V_1, and is represented by the rectangular area $P_2P_1D_1C_1$. The second is the consumer surplus earned by the new users $V_2 - V_1$ and is represented by the triangular area $C_1D_1D_2$.

Thus, a theoretical basis has been established to calculate the net benefits to users of an improved transportation facility which then can be compared with the improvement cost. For a linear demand curve, the formula for user benefits is

$$B_{2,1} = \frac{1}{2}(P_1 - P_2)(V_1 + V_2) \qquad (13.1)$$

where

$B_{2,1}$ = net benefits to transport users
P_1 = user cost of unimproved facility
P_2 = user cost of improved facility
V_1 = volume of travel on unimproved facility
V_2 = volume of travel on improved facility

It is not practical to develop demand curves, but rather the four-step process described in Chapter 12 is used. In these instances, the value for the volume that is used in economic calculations is taken to be the number of trips that will occur on the improved facility. Equation 13.2, which replaces the term $\frac{1}{2}$ $(V_1 + V_2)$ of Eq. 13.1 with V_2, has been commonly used in highway engineering studies. This formula will overstate benefits unless demand is inelastic; that is, if the demand curve is vertical (i.e., $V_1 = V_2$).

$$B_{2,1} = (P_1 - P_2)(V_2) \tag{13.2}$$

To consider the economic worth of improving this transportation facility, calculate the cost of the improvement and compare it with the cost of maintaining the facility in its present condition (the do-nothing alternative). One approach is to consider the difference in costs, to compare this with the difference in benefits, and then to select the project if the net increase in benefits exceeds the net increase in costs. Another approach is to consider the total costs of each alternative, including user and facility costs, and then to select the project that has the lowest total cost. Thus, to carry out an economic evaluation, it is necessary to develop the elements of cost for both the facility and the users. These include facility costs for construction, maintenance, and operation and user costs for travel time, accidents, and vehicle operations. The elements of cost are discussed in the following section.

13.2.1 Elements of Cost

The cost of a transportation facility improvement includes two components: *first cost* and *continuing costs.* Since an evaluation is concerned with cost differences, those costs that are common to both projects can be excluded. The first cost for a highway or transit project may include engineering design, right-of-way, and construction. Each transportation project is unique, and the specifics of the design will dictate what items will be required and at what cost. Continuing costs include maintenance, operation, and administration. These are recurring costs that will be incurred over the life of the facility and are usually based on historical data for similar projects. For example, if one alternative involves the purchase of buses, then the first (or capital) cost is the price of the bus, and the operating and maintenance cost will be known from manufacturer data or experience.

Expenses for administration or other overhead charges are usually excluded in an economic evaluation, because they will be incurred regardless of whether or not the project is selected. Other excluded costs are those that already have been incurred. These are known as *sunk costs* and as such are not relevant to the decision of what to do in the future since these expenditures have already been made. For most capital projects, a service life must be determined and a salvage value estimated. *Salvage value* is the worth of an asset at the end of its service life. For example, a transit bus costing $150,000 may be considered to have a service life of 12 years and a salvage value of $20,000, and a concrete pavement may have a service life of 15 years and no salvage value. Suggested service lives for various facilities can be obtained from various transportation organizations, such as the American Association of State Highway and Transportation Officials (AASHTO) and the American Public Transportation Association (APTA).

As illustrated in Figure 13.2, three measures of user costs are typically included in a transportation project evaluation: motor vehicle operation, travel time, and traffic accidents. These costs may be referred to as *benefits,* the implication being that the improvements to a transportation facility will reduce the cost for the users—that is, lower the perceived price, as shown on the demand curve—and result in a user benefit. It is more appropriate to consider these in terms of their actual cost, because this format is used in economic evaluations.

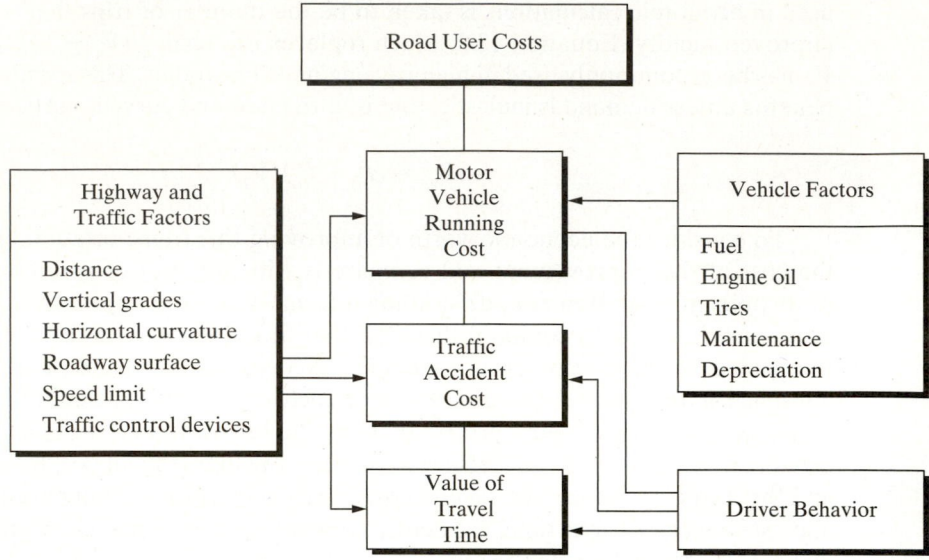

Figure 13.2 Road User Cost Factors

SOURCE: *Highway Engineering Economy,* U.S. Department of Transportation, Federal Highway Administration, April 1983

Vehicle Operating Costs

User costs for motor-vehicle operation are significant items in a highway project evaluation. For example, a road improvement that eliminates grades, curves, and traffic signals (as well as shortening the route) can result in major cost reductions to the motorist. Agencies, such as the U.S. Department of Transportation, The American Association of State and Highway Officials, and various vehicle manufacturers, furnish data useful in cost analysis for highways or at intersections.

Travel Time Costs

One of the most important reasons for making transportation improvements is to increase speed or to reduce travel delay. In the world of trade and commerce, time is equivalent to money. For example, business ventures that furnish overnight delivery of small packages have grown and flourished. Transoceanic airline service replaced steamships because airplanes reduced the time to cross the ocean from 6 to 9 days to 6 to 9 hours. The tunnel between Great Britain and France (known as the Chunnel), which opened in 1994, has shortened trip time by replacing ferries with rail.

The method of handling travel time in an economic analysis has stirred considerable debate. There is general agreement that time savings have an economic value, but the question is, Can and how should these be converted to dollar amounts? One problem is that a typical stream of traffic contains both private and commercial vehicles, each of which values time quite differently. The value of time for a trucking firm can be translated directly into labor cost savings by using an hourly rate for labor and equipment. Personal travel, on the other hand, is made for a variety of reasons; some are work related, but many are not (shopping, school, social, recreation). Time saved in traveling to and from work can be related to wages earned, but time saved in other pursuits may have little, if any, economic basis for conversion.

The value of time saved also depends on the length of trip and family income. If time saved is small—less than 5 min—it will not be perceived as significant and will therefore have little value. If time saved is above a threshold level where it will make a noticeable difference in total travel time—over 15 min—then it could have significant economic value. The apparent monetary savings from even small travel time reductions can be quite large. For example, if a highway project that will carry average daily traffic (ADT) of 50,000 autos saves only 2 min per traveler, and the value of time for the average motorist is estimated conservatively at $5.00/hour, the total minimum annual savings is 50,000 × (2/60) × 365 × 5 = $3,041,667. At 10 percent interest, these savings could justify spending a total of almost $26 million for a 20-year project life. Clearly, this result is an exaggeration of the actual benefits received. Although travel time does represent an economic benefit, the conversion to a dollar value is always open to question. AASHTO has used the average value approach. Others would argue that time savings should be credited only for commercial uses or stated simply in terms of actual value of number of hours saved.

The Federal Highway Administration (FHWA) has published a method for estimating the benefits of highway improvements known as the Highway Economic Requirements System (HERS). Travel time costs are computed using the following equation.

$$TTCST_{vt} = \frac{1000}{AES_{vt}} (TTVAL_{vt}) \qquad (13.3)$$

where

$TTCST_{vt}$ = average travel time cost for vehicles of type vt
AES_{vt} = average effective speed of vehicle type vt (km/h)
$TTVAL_{vt}$ = average value of one hour of travel time for occupants and cargo in vehicle type vt ($/h) in the year 2000 (see Table 13.2)

The value of one hour of travel time, expressed in year 2011 dollars ($), is shown in Table 13.2 for seven classes of vehicles (two auto and five truck types). The values for hourly time cost have been adjusted using an inflation factor based on the cost index. The first row of figures is for business-only travel. The second row of figures is a weighted average of business and nonbusiness travel. For example, for the small auto, 10 percent of trips are for business ($41.52/hr), and 90 percent of trips are for other purposes ($18.95/hr), such that the weighted average is approximately 0.10($41.52) + 0.90($18.95) = $21.22. Because six-tire trucks and larger vehicles are primarily used for business travel, corresponding hourly values for other purposes are not provided in Table 13.2.

Table 13.2 Value of One Hour of Travel Time for Various Vehicle Types (Year 2011 $)

Trip Purpose	Small Auto	Medium Auto	4-Tire Truck	6-Tire Truck	3- or 4-Axle Truck	4-Axle Combo	5-Axle Combo
Business only	$41.52	$41.96	$29.95	$37.02	$40.07	$46.00	$46.83
Other purposes	$18.95	$18.95	$18.95	—	—	—	—
All purposes	$21.22	$21.26	$22.37	$37.02	$40.07	$46.00	$46.83

SOURCE: Based on data from NCHRP Report 456, *Guidebook for Assessing the Social and Economic Effects of Transportation Projects*, Transportation Research Board, National Research Council, Washington, D.C., 2001.

Crash or Accident Costs

Loss of life, injury, and property damage incurred in a transportation crash or accident is a continuing national concern. Following every major air tragedy is an extensive investigation; and following the investigation, expenditures of funds are often authorized to improve the nation's air navigation system. Similarly, the 88.5 km/h highway speed limit imposed by Congress following the oil crisis of 1973 to 1974 was retained long after the crisis was ended because it was credited with saving lives on the nation's highways. In 1995, Congress removed all federal restrictions on speed limits.

It has been well established that the crash rate per million vehicle-km travelled is substantially lower on limited-access highways than on four-lane undivided roads. Reflecting the economic costs of crashes requires both an estimate of the number and type that are likely to occur over the life of the facility and an estimate of the value of each occurrence. Property damage and injury-related crashes can be valued using insurance data. The value of a human life is "priceless," but in economic terms, measures such as future earnings have been used. There is no simple numerical answer to the question What is the value of a human life lost in a highway crash? There is general agreement that economic value does exist and published data vary widely. The most prudent approach, if an economic value is desired, is to select a value that appears most appropriate for the given situation.

13.2.2 Economic Evaluation Methods

An economic evaluation of a transportation project is completed using one of the following methods: present worth (PW), equivalent uniform annual cost (EUAC), benefit–cost ratio (BCR), or internal rate of return (ROR). Each method, when correctly used as shown in Example 13.1, will produce the same results. The reason for selecting one over the other is preference for how the results will be presented. Since transportation projects are usually built to serve traffic over a long period of time, it is necessary to consider the time-dependent value of money over the life of a project.

Present Worth

The most straightforward of the economic evaluation methods is the present worth (PW), since it represents the current value of all the costs that will be incurred over the lifetime of the project. The general expression for present worth of a project is

$$PW = \sum_{n=0}^{N} \frac{C_n}{(1+i)^n} \tag{13.4}$$

where

C_n = facility and user costs incurred in year n
N = service life of the facility (in years)
i = interest rate

Net Present Worth

The present worth of a given cash flow that has both receipts and disbursements is referred to as the net present worth (NPW).

The use of an interest rate in an economic evaluation is common practice because it represents the cost of capital. Money spent on a transportation project is no longer available for other investments. Therefore, a minimal value of interest rate is the rate that would have been earned if the money were invested elsewhere. For example, if $1000 were deposited in a bank at 8 percent interest, its value in five years would be $1000(1 + 08)^5 = \$1469.33$. Thus, the PW of having $1469.33 in five years at 8 percent interest is equivalent to $1000, and the opportunity cost is 8 percent. Discount rates can be higher or lower, depending on risk of investment and economic conditions.

It is helpful to use a cash flow diagram to depict the costs and revenues that will occur over the lifetime of a project. Time is plotted as the horizontal axis and money as the vertical axis, as illustrated in Figure 13.3. Using Eq. 13.5, we can calculate the NPW of the project, which is

$$\text{NPW} = \sum_{n=0}^{N} \frac{R_n}{(1+i)^n} + \frac{S}{(1+i)^n} - \sum_{n=0}^{N} \frac{M_n + O_n + U_n}{(1+i)^n} - C_o \tag{13.5}$$

where

C_o = initial construction cost
n = a specific year
M_n = maintenance cost in year n
O_n = operating cost in year n
U_n = user cost in year n
S = salvage value
R_n = revenues in year n
N = service life, years

In this manner, a time stream of costs and revenues is converted into a single number: the NPW. The term $1/(1 + i)^n$ is known as the present worth factor of a single payment and is written as $P/F - i - N$, or $P/F_{i,N}$, where P is the present value given the future amount F, and N is the years of service life.

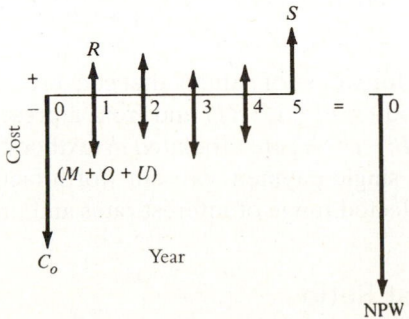

Figure 13.3 Typical Cash Flow Diagram for a Transportation Alternative and Equivalence as Net Present Worth

Equivalent Uniform Annual Worth

The conversion of a given cash flow to a series of equal annual amounts is referred to as the equivalent uniform annual worth (EUAW). If the uniform amounts are considered to occur at the end of the interest period, then the formula is

$$\text{EUAW} = \text{NPW}\left[\frac{i(1+i)^N}{(1+i)^N - 1}\right] = \text{NPW}(A/P - i - N) \tag{13.6}$$

Similarly,

$$\text{NPW} = \text{EUAW}\left[\frac{(1+i)^N - 1}{i(1+i)^N}\right] = \text{EUAW}(P/A - i - N) \tag{13.7}$$

where

$$\begin{aligned}
\text{EUAW} &= \text{equivalent uniform annual worth}\\
\text{NPW} &= \text{net present worth}\\
i &= \text{interest rate, expressed as a decimal}\\
N &= \text{number of years}
\end{aligned}$$

The term in the brackets in Eq. 13.6 is referred to as the capital recovery factor and represents the amount necessary to repay \$1 if N equal payments are made at interest rate i. For example, if a loan of \$5000 to be repaid in equal monthly payments over a five-year period at 1 percent/month, then the amount is

$$\text{EUAW} = 5000\left[\frac{0.01(1+0.01)^{60}}{(1+0.01)^{60} - 1}\right] = 5000(0.02225) = 111.25$$

Thus, 60 payments of \$111.25 would repay a \$5000 debt, including both principal and interest. The NPW of a cash flow is converted to an EUAW by multiplying the NPW by the capital recovery factor.

The inverse of the capital recovery factor is the present worth factor for a uniform series, as stated in Eq. 13.7. Thus, the present value of 60 payments of \$111.25, at 1 percent per month, is

$$\text{NPW} = 111.25\left[\frac{(1+0.01)^{60} - 1}{0.01(1+0.01)^{60}}\right] = 111.25(44.96) = 5000$$

Solutions for values of i and N that convert a monetary value from a future to a present time period ($P/F - i - N$) and from a present time period to equal end-of-period payments ($A/P - i - N$) are tabulated in textbooks on engineering economics. Table 13.3 lists values of single-payment present worth factors (P/F) and capital recovery factors (A/P) for a selected range of interest rates and time periods.

Benefit–Cost Ratio

The ratio of the present worth of net project benefits and net project costs is called the benefit–cost ratio (BCR). This method is used in situations where it is desired to show

Table 13.3 Present Worth and Capital Recovery Factors

N	i = 3 (P/F)	i = 3 (A/P)	i = 5 (P/F)	i = 5 (A/P)	i = 10 (P/F)	i = 10 (A/P)	i = 15 (P/F)	i = 15 (A/P)
1	0.9709	1.0300	0.9524	1.0500	0.9091	1.1000	0.8696	1.1500
2	0.9426	0.5226	0.9070	0.5378	0.8264	0.5762	0.7561	0.6151
3	0.9151	0.3535	0.8638	0.3672	0.7513	0.4021	0.6575	0.4380
4	0.8885	0.2690	0.8227	0.2820	0.6830	0.3155	0.5718	0.3503
5	0.8626	0.2184	0.7835	0.2310	0.6209	0.2638	0.4972	0.2983
10	0.7441	0.1172	0.6139	0.1295	0.3855	0.1627	0.2472	0.1993
15	0.6419	0.0838	0.4810	0.0963	0.2394	0.1315	0.1229	0.1710
20	0.5537	0.0672	0.3769	0.0802	0.1486	0.1175	0.0611	0.1598
25	0.4776	0.0574	0.2953	0.0710	0.0923	0.1102	0.0304	0.1547
30	0.4120	0.0510	0.2314	0.0651	0.0573	0.1061	0.0151	0.1523
35	0.3554	0.0465	0.1813	0.0611	0.0356	0.1037	0.0075	0.1511
40	0.3066	0.0433	0.1420	0.0583	0.0221	0.1023	0.0037	0.1506
45	0.2644	0.0408	0.1113	0.0563	0.0137	0.1014	0.0019	0.1503
50	0.2281	0.0389	0.0872	0.0548	0.0085	0.1009	0.0009	0.1501

the extent to which an investment in a transportation project will result in a benefit to the investor. To do this, it is necessary to make project comparisons to determine how the added investment compares with the added benefits. The formula for BCR is

$$\text{BCR}_{2/1} = \frac{B_{2/1}}{C_{2/1}} \tag{13.8}$$

where

$B_{2/1}$ = reduction in user and operation costs between higher-cost Alternative 2 and lower-cost Alternative 1, expressed as PW or EUAW

$C_{2/1}$ = increase in facility costs, expressed as PW or EUAW

If the BCR is 1 or greater, then the higher cost alternative is economically attractive. If the BCR is less than 1, this alternative is discarded.

Correct application of the BCR method requires that costs for each alternative be converted to PW or EUAW values. The proposals must be ranked in ascending order of capital cost, including the do-nothing alternative, which usually has little, if any, initial cost. The incremental BCR is calculated for pairs of projects, beginning with the lowest cost alternative. If the higher-cost alternative yields a BCR less than 1, it is eliminated and the next-higher-cost alternative is compared with the lower cost alternative. If the higher-cost alternative yields a BCR equal to or greater than 1, it is retained and the lower-cost alternative is eliminated. This process continues until every alternative has been compared. The alternative selected is the one with the highest initial cost and a BCR of 1 or more with respect to lower-cost alternatives and a BCR less than 1 when compared with all higher-cost projects.

Internal Rate of Return

Rate of return (ROR) is the interest rate at which the PW of reductions in user and operation costs $B_{2/1}$ equals the PW of increases in facility costs $C_{2/1}$. If the ROR exceeds the interest rate (referred to as minimum attractive rate of return), the higher-cost project is retained. If the ROR is less than the interest rate, the higher priced project is eliminated. The procedure is similar to that used in the BCR method.

Example 13.1 Illustration of Economic Analysis Methods

The Department of Traffic is considering three improvement plans for a heavily traveled intersection within the city. The intersection improvement is expected to achieve three goals: improve travel speeds, increase safety, and reduce operating expenses for motorists. The annual dollar value of savings compared with existing conditions for each criterion as well as additional construction and maintenance costs is shown in Table 13.4. If the economic life of the road is considered to be 50 years and the discount rate is 3 percent, which alternative should be selected? Solve the problem using the four methods for economic analysis.

Solution:

- Compute the NPW of each project.

$$(P/A - 3 - 50) = \frac{(1+i)^N - 1}{i(1+i)^N} = \frac{(1+0.03)^{50} - 1}{0.03(1+0.03)^{50}} = 25.729$$

$$\text{NPW}_I = -185{,}000 + (-1500 + 5000 + 500)(P/A - 3 - 50)$$

$$= -185{,}000 + (7000)(25.729) = -185{,}000 + 180{,}103$$

$$= -4897$$

$$\text{NPW}_{II} = -220{,}000 + (-2500 + 5000 + 6500 + 500)(P/A - 3 - 50)$$

$$= -220{,}000 + (9500)(25.729) = -220{,}000 + 244{,}425$$

$$= +24{,}465$$

$$\text{NPW}_{III} = -310{,}000 + (-3000 + 7000 + 6000 + 2800)(P/A - 3 - 50)$$

$$= -310{,}000 + (12{,}8000)(25{,}729) = -310{,}000 + 329{,}331$$

$$= +19{,}331$$

Table 13.4 Cost and Benefits for Improvement Plans with Respect to Existing Conditions

Alternative	Construction Cost	Annual Savings in Accidents	Annual Travel Time Benefits	Annual Operating Savings	Annual Additional Maintenance Cost
I	$185,000	$5000	$3000	$ 500	$1500
II	220,000	5000	6500	500	2500
III	310,000	7000	6000	2800	3000

The project with the highest NPW is alternative II.

- Solve by the EUAW method. Note $(A/P - 3 - 50) = 1/25.729 = 0.03887$.

$$\text{EUAW}_I = -185,000(A/P - 3 - 50) - 1500 + 5000 + 3000 + 500$$
$$= -185,000(0.03887) + 7000 = -7190 + 7000$$
$$= -190$$

$$\text{EUAW}_{II} = -220,000(A/P - 3 - 50) - 2500 + 5000 + 6500 + 500$$
$$= -220,000(0.03887) + 9500 = 8551 + 9500$$
$$= +949$$

$$\text{EUAW}_{III} = -310,000(0.03887) - 3000 + 7000 + 6000 + 2800$$
$$= -12,050 + 12,800$$
$$= +750$$

The project with the highest EUAW is Alternative II, which is as expected since $\text{EUAW} = \text{NPW}(0.03887)$.

- Solve by the BCR method.

Step 1. Compare the BCR of Alternative I with respect to do-nothing (DN).

$$\text{BCR}_{I/DN} = \frac{180,103}{185,000} = 0.97$$

Since $\text{BCR}_{I/DN}$ is less than 1, we would not build Alternative I.

Step 2. Compare BCR of Alternative II with respect to DN.

$$\text{BCR}_{II/DN} = \frac{244,425}{220,000} = 1.11$$

Since $\text{BCR} > 1$, we would select Alternative II over DN.

Step 3. Compare BCR of Alternative III with respect to Alternative II.

$$\text{BCR} = \frac{(329,331) - (244,425)}{(310,000) - (220,000)} \quad \frac{84,906}{90,000} = 0.94$$

Since BCR is less than 1, we would not select Alternative III. We reach the same conclusion as previously, which is to select Alternative II.

- Solve by the ROR method. In this situation, we solve for the value of interest rate for which NPW = 0.

Step 1. Compute ROR for Alternative I versus DN. (Recall that all values are with respect to existing conditions.)

$$\text{NPW} = 0 = -185,000 + (-1500 + 5000 + 3000 + 500) \times (P/A - i - 50)$$
$$(P/A - i - 50) = 185,000/7000$$
$$(P/A - i - 50) = 26.428$$
$$i = 2.6\%$$

Since the ROR is lower than 3 percent, we discard Alternative I.

Step 2. Compute ROR for Alternative II versus DN.

$$\text{NPW} = 0 = -220,000 + (-2500 + 5000 + 6500 + 500) \times (P/A - i - 50)$$

$$(P/A - i - 50) = 220,000/9500$$

$$(P/A - i - 50) = 23.16$$

$$i = 3.6\%$$

Since ROR is greater than 3 percent, we select Alternative II over DN.

Step 3. Compute ROR for Alternative III versus Alternative II.

$$\text{NPW} = 0 = -(310,000 - 220,000) + (12,800 - 9500) \times (P/A - i - 50)$$

$$(P/A - i - 50) = 90,000/3300$$

$$(P/A - i - 50) = 27.27$$

$$i = 2.7\%$$

Since the increased investment in Alternative III yields an ROR less than 3 percent, we do not select it but again pick Alternative II.

The preceding example illustrates the basic procedures used in an economic evaluation. Four separate methods were used, each producing the same result. The PW or EUAW method is simplest to understand and apply and is recommended for most purposes when the economic lives of each alternative are equal. The BCR gives less information to the decision maker and must be carefully applied if it is to produce the correct answer. (For example, the alternative with the highest BCR with respect to the do-nothing case is not necessarily the best.) The ROR method requires more calculations but does provide additional information. For example, the highest ROR in the preceding problem was 3.6 percent. This says that if the minimum attractive ROR were greater than 3.6 percent (say 5 percent), none of the projects would be economically attractive.

13.3 EVALUATION BASED ON MULTIPLE CRITERIA

Many problems associated with economic methods limit their usefulness. Among these are

- Converting criteria values directly into dollar amounts
- Choosing the appropriate value of interest rate and service life
- Distinguishing between those groups that benefit from a project and those that pay
- Considering all costs, including external costs

For these reasons, economic evaluation methods should be used primarily for narrowly focused projects or as only one of many inputs for the evaluation of larger projects. The next section discusses evaluation methods that seek to include measurable criteria that are not translated just in monetary terms.

13.3.1 Rating and Ranking

Numerical scores are helpful in comparing the relative worth of alternatives in cases where criteria values cannot be transformed into monetary amounts. The basic equation is as follows.

$$S_i = \sum_{j=1}^{N} K_j V_{ij} \tag{13.9}$$

where

S_i = total value of score of alternative i
K_j = weight placed on criterion j
V_{ij} = relative value achieved by criteria j for alternative i

The application of this method is illustrated by the following example.

Example 13.2 Evaluating Light Rail Transit Alternatives Using the Rating and Ranking Method

A transportation agency is considering the construction of a light rail transit line from the center of town to a growing suburban region. The transit agency wishes to examine five alternative alignments, each of which has advantages and disadvantages in terms of cost, ridership, and service provided. The alternatives differ in length of the line, location, types of vehicles used, seating arrangements, operating speeds, and numbers of stops. The agency wants to evaluate each alternative using a ranking process. Determine which project should be selected.

Solution:

Step 1. Identify the goals and objectives of the project. The transit agency has determined that the new transit line should achieve five major objectives.

1. Net revenue generated by fares should be as large as possible with respect to the capital investment.
2. Ridership on the transit line should be maximized.
3. Service on the system should be comfortable and convenient.
4. The transit line should extend as far as possible to promote development and accessibility.
5. The transit line should divert as many auto users as possible during the peak hour in order to reduce highway congestion.

Step 2. Develop the alternatives that will be tested. In this case five alternatives have been identified as feasible candidates. These vary in length from 8 to 12.8 km. The alignment; the amount of the system below-, at-, and above-grade; vehicle size; headways; number of trains; and other physical and operational features of the line are determined in this step.

Step 3. Define an appropriate measure of effectiveness for each objective. For the objectives listed in Step 1, the following measures of effectiveness are selected.

Objective	Measure of Effectiveness
1	Net annual revenue divided by annual capital cost
2	Total daily ridership
3	Percent of riders seated during the peak hour
4	km of extension into the corridor
5	Number of auto drivers diverted to transit

Step 4. Determine the value of each measure of effectiveness. In this step, the measures of effectiveness are calculated for each alternative. Techniques for demand estimation, as described in Chapter 12, are used to obtain daily and hourly ridership on the line. Cost estimates are developed based on the length of line, number of vehicles and stations, right-of-way costs, electrification, and so forth. Revenues are computed, and ridership volumes during the peak hour are estimated. In some instances, forecasts are difficult to make, so a best or most likely estimate is produced. Since it is the comparative performance of each alternative that is of interest, relative values of effectiveness measures can be used. Estimated values achieved by each criterion for each of the five alternatives are shown in Table 13.5.

Step 5. Determine the relative weight for each objective. This step requires a subjective judgment on the part of the group making the evaluation and will vary among individuals and vested interests. One approach is to allocate the weights on a 100-point scale (just as would be done in developing final grade averages for a course). Another approach is to rank each objective in order of importance and then use a formula of proportionality to obtain relative weights. In this example, the objectives are ranked as shown in Table 13.6 using a 100-point scale. The weighting factor is determined by assigning the value n to the highest ranked alternative, $n-1$ to the next highest (and so forth), and computing a relative weight as

$$K_j = \frac{W_j}{\sum_{j=1}^{n} W_j} \tag{13.10}$$

Table 13.5   Estimated Values for Measures of Effectiveness

		Alternatives				
Number	Measure of Effectiveness	I	II	III	IV	V
1	Annual return on investment(%)	13.0	14.0	11.0	13.5	15.0
2	Daily ridership (1000s)	25	23	20	18	17
3	Passengers seated in peak hour (%)	25	35	40	50	50
4	Length of line (km)	13	11	10	8	8
5	Auto drivers diverted (1000s)	3.5	3.0	2.0	1.5	1.5

Table 13.6   Ranking and Weights for Each Objective

Objective	Ranking	Relative Weight (W_j)	Weighting Factor* ($\times 100$)
1	1	5	30
2	2	4	24
3	3	3	17
4	3	3	17
5	4	2	12
Total		17	100

*Rounded to whole numbers to equal 100.

where

K_j = weighting factor of objective j

W_j = relative weight for objective j

The resulting values for each objective shown in Table 13.6. For example, Objective 1, which is to generate revenue, is ranked 1 and the weighted point value is 30. Objective 5, which is to divert auto drivers, is ranked 4 and the weighted point value is 12. Objective 3 and 4 are weighted equally with a point value of 17.

Other weighting methods (such as by ballot or group consensus) could be used. It is not necessary to use weights that total 100, as any range of values can be selected. The final results are normalized to 100 at the end of the process.

Step 6. Compute a score and ranking for each alternative. The score for each alternative is computed by considering each measure of effectiveness and awarding the maximum score to the alternative with the highest value and a proportionate amount to the other alternatives. Consider the first criterion, return on investment. As shown in Table 13.5, Alternative V achieves the highest value and is awarded 30 points. The value for Alternative I is calculated as $(13/15)(30) = 26$. The results are shown in Table 13.7. (An alternative approach is to award the maximum points to the highest valued alternative and zero points to the lowest.)

Table 13.7   Point Score for Candidate Transit Lines

Measure of Effectiveness	Alternatives I	II	III	IV	V
1	26.0	28.0	22.0	27.0	30.0
2	24.0	22.1	19.2	17.3	16.3
3	8.5	11.9	13.6	17.0	17.0
4	17.0	14.9	12.8	10.6	10.6
5	12.0	10.3	6.9	5.1	5.1
Total	87.5	87.2	74.5	77.0	79.0

The total point score indicates that the ranking of the alternatives in order of preference is I, II, V, IV, and III. Alternatives I and II are clearly superior to the others and are very similar in ranking. These two will bear further investigation prior to making a decision.

Ranking and rating evaluation is an attractive approach because it can accommodate a wide variety of criteria and can incorporate various viewpoints. Reducing all inputs to a single number is a convenient way to rate the alternatives. The principal disadvantage is that the dependence on a numerical outcome masks the major issues underlying the selection and tradeoffs involved.

Another problem with ranking methods is that the mathematical form for the rating value (Eq. 13.9) is a summation of the products of the weight of each criteria and the relative value. For this mathematical operation to be correct, the scale of measurement must be a constant interval. If the relative values are ordinal (listed in a series), the ranking formula cannot be used. Revising the ranking of the objectives and their relative weights also could change results.

There is also the problem of communicating the results to decision makers, since the interpretation is often difficult to visualize. People think in concrete terms and are able to judge alternatives only when they are presented realistically, rather than as numerical values.

The next section describes a more general and comprehensive approach to evaluation that furnishes information for decision making but stops short of computing numerical values for each alternative.

13.3.2 Cost Effectiveness

Cost effectiveness attempts to be comprehensive in its approach while using the best attributes of economic evaluation. In this method, the criteria that reflect the goals of the project are listed separately from project costs. Thus, the project criteria are considered to be measures of its effectiveness, and the costs are considered as the investment required if that effectiveness value is to be achieved. This approach uses data from economic analysis but allows for other intangible effects, such as environmental consequences, which are also measured. The following example illustrates the use of the cost-effectiveness method.

Example 13.3 Evaluating Metropolitan Transportation Plans Using Cost Effectiveness

Five alternative system plans are being considered for a major metropolitan area. They are intended to provide added capacity, improved levels of service, and reductions in travel time during peak hours. Plan A retains the status quo with no major improvements, Plan B is an all-rail system, Plan C is all highways, Plan D is a mix of rail transit and highways, and Plan E is a mix of express buses and highways. An economic evaluation has been completed for the project, with the results shown in Table 13.8.

Plan B, the all-rail system, and Plan D, the combination rail and highway system, have an incremental BCR of less than 1, whereas Plan C, all highways, and Plan E, highways and express buses, have an incremental BCR greater than 1. These results

Table 13.8   Benefit–Cost Comparisons for Highway and Transit Alternatives

Plan Comparisons	Annual Cost Difference ($ million)	Annual Savings ($ million)	BCR
A versus B	28.58	21.26	0.74
A versus C	104.14	116.15	1.12
C versus D	22.66	17.16	0.76
C versus E	16.73	19.75	1.18

SOURCE: Adapted from *Alternative Multimodal Passenger Transportation Systems*, NCHRP Report 146, Transportation Research Board, National Research Council, Washington, D.C., 1973.

would suggest that the highway–bus alternative (Plan E) is preferable to the highway–rail transit alternatives (Plans B and D).

To examine these options more fully, noneconomic impacts have been determined for each and are displayed as an evaluation matrix in Table 13.9. Among the measures of interest are numbers of persons and businesses displaced, number of fatal and personal-injury accidents, emissions of carbon monoxide and hydrocarbons, and average travel speeds by highway and transit.

Solution: An examination of Table 13.9 yields several observations. In terms of number of transit passengers carried, Plan E ranks highest, followed by Plan B. The relationship between annual cost and transit passengers carried is illustrated in Figure 13.4. This

Table 13.9   Measure of Effectiveness Data for Alternative Highway–Transit Plans

Measure of Effectiveness	Plan A Null	Plan B All Rail	Plan C All Highway	Plan D Rail and Highway	Plan E Bus and Highway
Persons displaced	0	660	8000	8000	8000
Businesses displaced	0	15	183	183	183
Annual total fatal accidents	159	158	137	136	134
Annual total personal injuries	6767	6714	5596	5544	5517
Daily emissions of carbon monoxide (tons)	2396	2383	2233	2222	2215
Daily emissions of hydrocarbons (tons)	204	203	190	189	188
Average door-to-door auto trip speed (mi/h)	15.9	16.2	21.0	21.2	21.5
Average door-to-door transit trip speed (mi/h)	6.8	7.6	6.8	7.6	7.8
Annual transit passengers (millions)	154.2	161.7	154.2	161.7	165.2
Total annual cost ($ millions)	2.58	31.16	106.72	129.38	123.44
Interest rate (%)	8.0	8.0	8.0	8.0	8.0

SOURCE: Adapted from *Alternative Multimodal Passenger Transportation Systems,* NCHRP Report 146, Transportation Research Board, National Research Council, Washington, D.C., 1973

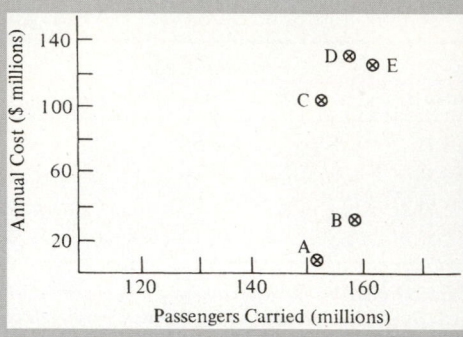

Figure 13.4   Relationship between Annual Cost and Passengers Carried

cost-effectiveness analysis indicates that Plan B produces a significant increase in transit passengers over Plan A. Although Plans C, D, and E are much more costly, they do not produce many more transit riders for the added investment.

Community impacts are reflected in the number of homes and businesses displaced and the extent of environmental pollution. Figure 13.5 illustrates the results for number of businesses displaced, and Figure 13.6 depicts the results for emissions of hydrocarbons.

In terms of businesses displaced versus transit passengers carried, Plans C and D require considerable disruption with very little increase in transit patronage over Plan B,

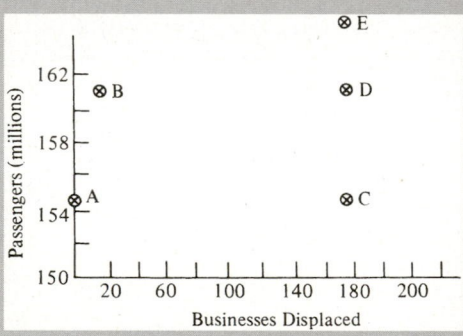

Figure 13.5   Relationship between Passengers Carried and Businesses Displaced

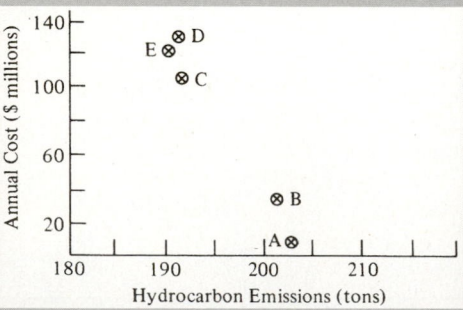

Figure 13.6   Annual Cost versus Hydrocarbon Emissions

which is clearly preferred if the impact on the community is to be minimized. On the other hand, Plan C, which is considerably more costly than Plan B, results in a significant reduction in pollution levels, whereas the other two plans, D and E, although more expensive than C, have little further impact on pollution levels.

The items described are but a few of the many relationships that could be examined. They do, however, illustrate the cost-effectiveness procedure and the various conflicting tradeoffs that can result. One conclusion that seems evident is that, although the BCR for Plans B and D is less than 1, these plans bear further investigation since they produce several environmentally and socially beneficial effects and attract more transit ridership. A sensitivity analysis of the benefit–cost study would show that if the interest rate were reduced to 4 percent or the value of travel time were increased by $0.30 per hour, the rail–transit plan, Plan B, would have a BCR greater than 1.

The cost-effectiveness approach does not yield a recommended result, as do economic methods or ranking schemes. However, it is a valuable tool because it defines more fully the impacts of each course of action and helps to clarify the issues. With more complete information, a better decision should result. Rather than closing out the analysis, the approach opens it up and permits a wide variety of factors to be considered.

13.3.3 Evaluation as a Fact–Finding Process

The preceding discussion of economic and rating methods for evaluation has illustrated the technique and application of these approaches. These so-called rational methods are inadequate when the transportation alternatives create a large number of impacts on a wide variety of individuals and groups. Under these conditions, the evaluation process is primarily one of fact finding to provide the essential information from which a decision can be made. The evaluation procedure for complex projects is illustrated in Figure 13.7 and should include four activities that follow the development and organization of basic data and the identification of the major problems or issues that must be addressed in the evaluation process. The activities are as follows.

Activity 1. View the issues from the perspective of each affected interest group. The thrust of this activity is to view the consequences of proposed alternatives as they affect particular groups and individuals and as those groups perceive them. For each interest group, the information should be examined and a statement prepared that indicates how that group will react to each alternative. This step can be considered as a means of understanding where each of the stakeholders are coming from; what their biases, likes, and dislikes are; what problems they represent; and so forth.

Activity 2. View the issues from the perspective of each action. The purpose of this activity is to describe each alternative in terms of its advantages and disadvantages. Each proposed project is discussed from the point of view of community concern, feasibility, equity, and potential acceptability.

Activity 3. View the issues from the perspective of the process as a whole. In this step, all of the alternatives and each of the issues (criteria) are examined together to see if patterns develop from which general statements can be made. For example, there may be one or more alternatives that prove to have so many disadvantages that they can be eliminated. There may be several groups who share the same viewpoint

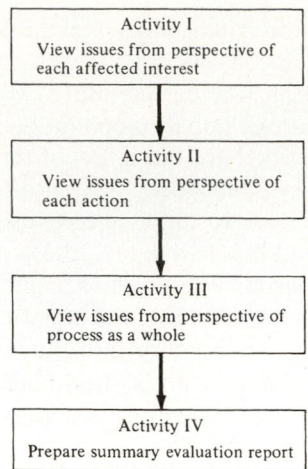

Figure 13.7   Evaluation Procedure for Complex Projects

SOURCE: Redrawn from *Transportation Decision Making,* NCHRP Report 156, Transportation Research Board, National Research Council, Washington, D.C., 1975

or who are in fierce opposition. Compromise solutions may emerge that will reconcile conflicts, or there may be popular alternatives that generate little controversy.

Activity 4. Summarize the results. In this activity, the result of the evaluation is documented and used by decision makers and other interested individuals. The report should include a description of the alternatives, the advantages and disadvantages of each alternative, areas of conflict and of agreement, and identification of the alternatives that have the greatest potential for success in accomplishing major objectives and achieving public acceptance.

This process is comprehensive and goes beyond a simple listing of criteria and alternatives. One essential feature of the approach is that it requires the analyst to furnish values of each measure of effectiveness for each alternative without attempting to reduce the results to a single numerical value. The information is then used to make judgments about the relative merits of each alternative.

13.3.4 Tradeoff and Balance-Sheet Approaches

There are several important conditions that must be met if the difference among alternative projects are to be adequately considered. They are as follows:

1. All alternatives should be evaluated in a framework of common objectives. Measures of effectiveness should be derived from the objectives covering all impact areas.
2. The incidence and timing of impacts on groups and areas should be identified for all impact categories.
3. Standards or accepted impact significance thresholds for measures of effectiveness should be indicated where accepted or required by law.
4. All measures of effectiveness should be treated at an equal level of detail and appropriate scale.
5. Uncertainties or probabilities or both should be expressed for each impact category.
6. A sensitivity analysis should be conducted to describe variations in results for alternatives when values of key parameters are changed.

Balance-sheet or tradeoff approaches satisfy these criteria in contrast to economic or weighting schemes, which provide little in the way of comparative information. These approaches display the impacts of plan alternatives to various groups. The method is based on the viewpoint that individuals or groups that review the data will first introduce their own sets of values and weights, and then reach a judgment based on the merits of each alternative using all the data in a disaggregated fashion. Each impact category or goal may have more than one measure of effectiveness. To determine the cost-effectiveness of each impact category in a balance-sheet framework, it is useful to compare proposals with the do-nothing alternative. In this way, the positive and negative impacts of not constructing a new project are fully understood.

13.3.5 Evaluation of Completed Projects

The material on evaluation discussed thus far has dealt with the evaluation of plans, and the focus has been to answer the "what if" questions in sufficient detail so that a good decision will be made. Another form of evaluation is to examine the results of a project after it has been implemented to determine (1) how effective it has been in accomplishing its objectives, (2) what can be learned that is useful for other project decisions, (3) what changes should be made to improve the current situation, or (4) if the project should be continued or abandoned.

The subject of post-evaluation of transportation projects is closely related to the more general topic of experimental design. If, for example, the effects of a particular medical treatment are to be determined, two population groups with similar characteristics are selected: one serving as the control group and the other as the experimental group. Then the treatment is applied to the experimental group but not to the control group. Differences are measured before and after treatment for both groups, and the net effect is considered to be the result of the treatment itself.

Example 13.4 Evaluating the Effect of Bus Shelters on Transit Ridership

A transit authority wishes to evaluate the effectiveness of new bus shelters on transit ridership as well as acceptance by the community. A series of new shelters was built along one bus route but not on the other lines. Do the shelters affect ridership?

Solution: Bus ridership has been measured before and after the shelters had been installed on the test line and on a control line where nothing new had been added. Both lines serve similar neighborhoods. The ridership results are shown in Table 13.10. The line with new shelters increased ridership by 13.3 percent, whereas the line without shelters increased by only 2.5 percent. It should be stressed that only in the absence of any other factors can we conclude that the effect of the new shelters was to increase ridership by $(13.3 - 2.5) = 10.8\%$.

Table 13.10 Transit Ridership

	Before	*After*	*Change (%)*
Line A: new shelters	1500	1700	13.3
Line B: no shelters	1950	2000	2.5

Another tool for evaluation of completed projects is to conduct a survey of users of the new facility. The questionnaire can probe in greater depth why riders use the facility and to what extent the project improvement influenced their choice. In the bus shelter example, a survey of bus riders would inquire if passengers were longtime bus riders or are new riders. If new riders, the survey would ask the reasons for riding to find out how many of the new riders considered the new shelters a factor. (Other reasons for riding could be that the rider is new in the neighborhood, his or her car is being repaired, gasoline prices have just gone up, and so forth.) The survey would also ask the old riders to comment on the new shelters. Thus, the survey would corroborate the before-and-after ridership data as well as furnish additional information about the riders themselves and how they reacted to the new project. This information would be useful in deciding whether or not to implement a bus shelter program for the entire city.

In the transportation field, it is difficult to achieve an experimental design with a well-defined control and experimental group because (1) transportation projects influence a wide range of outcomes, (2) implementation times are very long and therefore funds may not not available to gather before data, (3) a control group is difficult to identify, and (4) changes that occur over a long time period are difficult to connect with a single event, such as a new transportation system.

An example of a post-evaluation for a major transportation project is the Bay Area Rapid Transit (BART) impact study. The approach to post-evaluation in this instance was to predict the effect on the region of measures such as air pollution, noise, travel time, and so forth without the rail transit system and then measure the actual amounts with the rapid transit system in place. In this approach, a control group was impossible to obtain, but in its place, a forecast was made of conditions in the region if a rapid transit system had not been constructed. The obvious difficulty with this method is that it depends on the accuracy of the forecast and must take into account all the changes that have occurred in the region that might have an effect on the impact measures of interest.

Another type of transportation project post-evaluation is to make comparisons between different systems or technologies that serve similar travel markets. These comparisons can be helpful to decision makers in other localities because they furnish useful information about what happened in an actual situation. A post-evaluation study can be useful because it can consider the actual results for many variables, whereas a pre-project evaluation of a mode or technology tends to focus primarily on cost factors or is based on hypothetical situations that require many questionable assumptions.

Example 13.5 Comparing the Effectiveness of Bus and Rail Transit

Compare the effectiveness of rail and bus based on the experience with a rail transit line serving downtown Philadelphia and a suburb of New Jersey with an express bus line connecting downtown Washington, D.C., with the Virginia suburbs. The rail line, known as the Lindenwold Line, serves 12 stations with 24-hour service per day, whereas the busway, known as the Shirley Highway, extends for 18 km, with no stations along the way and with bus service provided on exclusive lanes only during the peak hour. Both systems serve relatively low-density, auto-oriented residential areas with heavy travel during the peak hours.

Solution: To determine the relative effectiveness, a comparative analysis of each project was made after they had been in operation for several years. Measures of effectiveness were considered from the viewpoint of the passenger, the operator, and the community. Data were collected for each system and for each measure of effectiveness. A detailed evaluation for each parameter was prepared that both described how each system performed and discussed its advantages and disadvantages. To illustrate, consider the evaluation of one service parameter—*reliability*—expressed as schedule adherence. The variance from scheduled travel times may result from traffic delays, vehicle breakdowns, or adverse weather conditions. It depends mostly on the control that the operator has over the entire system. By far, the most significant factor for reliability is availability of exclusive rights-of-way.

- *Lindenwold:* That year, 99.15 percent of all trains ran less than 5 min late, and the following year the figure was 97 percent.
- *Shirley:* Surveys conducted over a 4-day period indicated that 22 percent arrived before the scheduled time, 32 percent were more than 6 min late, and only 46 percent arrived at the scheduled time within a 5-min period.
- *Comparison:* The Lindenwold Line (rail) is superior to the Shirley Highway (bus) with respect to reliability.

A summary of the comparative evaluations of the two systems is shown in Table 13.11.

A detailed analysis of the results would indicate that each system has advantages and disadvantages. The principal reasons why the rail system appears more attractive than the bus are that it provides all-day service, it is simpler to understand and use, and it produces a higher quality of service.

Table 13.11   Comparative Evaluations of Completed Rail and Bus Transit

Measure of Effectiveness	Lindenwold (Rail)	Shirley (Bus)	Higher Rated System
Investment cost	Very poor	Fair	Bus
Operating cost	Good	Fair	Rail
Capacity	Good	Poor	Rail
Passenger attraction	Very good	Good	Rail
System impact	Very good	Good	Rail

SOURCE: Adapted from V. R. Vuchic and R. M. Stanger, "Lindenwold Rail Line and Shirley Busway: A Comparison," *Highway Research Record,* 459, Transportation Research Board, National Research Council, Washington, D.C., 1973 pp 13-28

13.3.6 Evaluating Effects of Transportation on Social and Natural Systems

A comparative evaluation of the impacts of a transportation project involves the following sequence.

Step 1.   Assess the need for the project.
Step 2.   Conduct a feasibility analysis of the alternatives.
Step 3.   Analyze the impact of the project from the following points of view.
   (a) Transportation system effects
   (b) Social and economic effects
   (c) Natural systems effects

Step 4. Organize the results of the evaluation in a manner that is clear and under-standable by the affected parties, such as citizens, stakeholders, and decision makers.

Step 1: Assess the Need for the Project

This step addresses the question: Why do it at all? That is, how does the proposed project advance the stated goals and objectives and does the project represent the best use of funds when compared with other options?

Step 2: Conduct a Feasibility Analysis of the Alternatives

This step addresses the question: Why do it this way? That is, has the project been demonstrated to be a feasible one from an engineering perspective? What are the costs involved in the project? Are there other methods or approaches that could achieve a similar result at a lower cost in time and money? Should the project be included as a budget item for implementation or deferred to a later date?

Step 3: Analyze the Impact of the Project

This step addresses the question: If the project is feasible, what will be its impact on affected groups? These include the users of the transportation improvement, the community, and other stakeholders who will be impacted by the construction of the project. These effects are categorized into three major effects.

- Transportation system effects
- Social and economic effects
- Natural systems effects

A comprehensive treatment of these effects and each element is contained in the textbook, *Transportation Decision Making: Principles of Project Evaluation and Programming*. Each of these is summarized as follows.

Step 3 (a): Transportation System Effects. These are effects experienced by the travelers who use the transportation facility, such as motorists, transit riders, and commercial vehicles. They comprise the following elements.

- Changes in travel time
- Changes in safety
- Changes in vehicle operating costs

To determine system changes, methods of analysis in areas of traffic flow theory, safety engineering, and cost analysis are used. Many of these topics are covered in this textbook and other references.

Step 3 (b): Social and Economic Effects. These are analyzed to determine the impact that a transportation project could have on the community and its residents. These studies are also conducted to meet federal and state requirements regarding environmental impact, civil rights, and environmental justice. They comprise the following elements.

- Accessibility
- Community cohesion
- Economic development

- Traffic noise
- Visual quality
- Property values

To determine social and economic effects of a transportation project, a variety of methods are used. Many of these areas require special expertise in professional fields such as economics, demography, sociology, geography, geographic information systems (GIS), architecture, and urban planning. NCHRP Report 456 can serve as a useful guide for conducting these impact evaluations.

Step 3 (c): Natural Systems Effects.   These refer to those impacts of transportation projects that are related to the environment within which the project will be located. For example, there are short-term effects, such as those caused by displacement of natural soil during construction, and long-term effects related to the sustainability of energy resources and climate change. Among the natural elements that may be affected are

- Air and water quality
- Endangered species
- Wildlife
- Greenhouse gas emissions
- Archeological sites
- Energy conservation
- Areas of cultural or historic significance

To explain natural system effects from transportation projects, some of which are described in NCHRP Report 532, a broad range of expertise is required from many fields of science, medicine, fine arts, and engineering. These include air and water quality modeling, environmental science, chemistry, history, archeology, veterinary medicine, wildlife management, and ecology. Each of the impacts are relevant to categories specified in federal regulations discussed in Chapter 11, such as the National Environmental Policy Act (NEPA) and regulations for Environmental Impact Statements (EIS) as specified by the U.S. Department of Transportation.

13.4 SUMMARY

The evaluation process for selecting a transportation project has been described. Various methods have been presented that, when used in the proper context, can assist a decision maker in making a selection. The most important attribute of an evaluation method is its ability to correctly describe the outcomes of a given alternative. The evaluation process begins with a statement of the goals and objectives of the proposed project, and these are converted into measures of effectiveness. Evaluation methods differ by the way in which measures of effectiveness are considered.

Economic evaluation methods require that each measure of effectiveness be converted into dollar units. Numerical ranking methods require that each measure of effectiveness be translated to an equivalent score. Both methods produce a single number to indicate the total worth of the project. Cost-effectiveness methods require only that each measure of effectiveness be displayed in matrix form, and it is the task of the analysts to develop relationships between various impacts and the costs involved. For projects with many impacts that will influence a wide variety of individuals and groups, the evaluation process is essentially one of fact finding, and the projects must be considered from the viewpoint of the stakeholders and community. The reasons for selecting a project will include many factors in addition to simply how the project performs. A decision maker must consider issues such as implementation, schedules, financing, and legal and political matters.

When a project has been completed and has been in operation for some time, a post-evaluation can be a useful means to examine the effectiveness of the results. To conduct a post-evaluation, it is necessary to separate the effect of the project on each measure of effectiveness from other influencing variables. A standard procedure is the use of a control group for comparative purposes, but this is usually not possible for most transportation projects. A typical procedure is to compare the results with a forecast of the region without the project in place. Post-evaluations also can be used to compare alternative modes and technologies, using a wide range of measures of effectiveness.

The usefulness of an evaluation procedure is measured by its effectiveness in assisting decision makers to arrive at a solution that will best accomplish the intended goals.

PROBLEMS

13-1   What is the main objective of conducting a transportation project evaluation?

13-2   Describe four basic issues that should be considered prior to selection of an evaluation procedure.

13-3   List the basic criteria used for evaluating transportation alternatives. What units are used for measurement?

13-4   Average demand on a rural roadway ranges from zero to 500 veh/day when the cost per trip goes from $1.50 to zero.

   (a) Calculate the net user benefits per year if the cost decreases from $1.00 to $0.75/trip (assume a linear demand function).

   (b) Compare the value calculated in (a) with the benefits as calculated in typical highway studies.

13-5   A ferry is currently transporting 300 veh/day at a cost of $1.50/vehicle. The ferry can attract 600 more veh/day when the cost/veh is $1.00. Calculate the net user benefits/year if the cost/veh decreases from $1.10 to $0.95.

13-6   What are the two components of the cost of a transportation facility improvement? Describe each.

13-7   Estimate the average unit costs for (a) operating a standard vehicle on a level roadway, (b) travel time for a truck company, (c) single-vehicle property damage, (d) personal injury, and (e) fatality.

13-8   An incident that occurs on a particular highway results in traffic incurring an average delay of one hour per vehicle. An estimated 2000 vehicles are impacted by the incident. If the distribution of traffic is comprised of 5 percent five-axle trucks, 2 percent three-axle trucks, 4 percent six-tire trucks, 2 percent four-tire trucks, 4 percent medium autos for business purposes, 3 percent small autos for business purposes, and the remainder automobiles for personal purposes, what is the value of the total delay incurred by the traffic?

13-9   Derive the equation to compute the equivalent annual cost given the capital cost of a highway, such that $A = (A/P) \times P$, where A/P is the capital recovery factor. Compute the equivalent annual cost if the capital cost of a transportation project is $100,000, annual interest = 10% and $n = 15$ years.

13-10   A highway project is expected to cost $1,700,000 initially. The annual operating and maintenance cost after the first year is $5000 and will increase by $500 each year for a project lifespan of 20 years. At the end of the tenth year, the project must be resurfaced at a cost of $300,000.

(a) Calculate the present worth of costs for this project over a 20-year period if the annual interest rate is 5 percent.

(b) Convert the value obtained in (a) to equivalent uniform annual costs.

13-11   The concepts applied in economic evaluation to address the time-dependent value of money can also be applied to other phenomena, such as growth in traffic volume. The traffic volume on a local highway was 3510 vehicles per day on an average day in 2012. Based on historical trends, the traffic volume is expected to increase by 3 percent per year for the foreseeable future. What is the expected traffic volume in 2022?

13-12   Three transportation projects have been proposed to increase the safety in and around a residential neighborhood. Each project consists of upgrading existing street signing to highly retroreflective sheeting to increase visibility. The following table shows the initial construction costs, annual operating costs, useful life of the sheeting, and salvage values for each alternative. Assume that the discount rate is 10 percent. Calculate the present worth for each alternative and determine the preferred project based on the economic criteria.

Alternative	Initial Construction Cost ($)	Annual Operations and Maintenance Costs ($)	Useful Life (years)	Salvage Value ($)
1	19,000	2,500	10	3,000
2	8,000	4,000	5	900
3	20,000	2,500	10	4,600

13-13   Two designs have been proposed for a short-span bridge in a rural area, as shown in the following table. The first proposal is to construct the bridge in two phases (Phase I now and Phase II in 25 years). The second alternative is to construct it in one phase. Assuming that the annual interest rate is 4 percent, use present worth analysis to determine which alternative is preferred.

Alternative	Construction Costs ($)	Annual Maintenance Costs ($)	Service Period (yr)
A (Phase 1)	14,200,000	75,000	1–50
A (Phase 2)	12,600,000	25,000	26–50
B	22,400,000	100,000	1–50

13-14   Three designs have been proposed to improve traffic flow at a major intersection in a heavily traveled suburban area. The first alternative involves improved traffic signaling. The second alternative includes traffic-signal improvements and intersection widening for exclusive left turns. The third alternative includes extensive reconstruction, including a grade separation structure. The construction costs, as well as annual maintenance and user costs, are listed in the following table for each alternative. Determine which alternative is preferred based on economic criteria if the analysis period is 20 years and the annual interest rate is 15 percent. Show that the result is the same using the present worth, equivalent annual cost, benefit–cost ratio, and rate-of-return methods.

Alternative	Capital Costs ($)	Annual Maintenance Costs ($)	Annual User Costs ($)	Salvage Value ($)
Present condition	0	15,000	600,000	0
Traffic signals	340,000	10,000	450,000	25,000
Intersection widening	850,000	5,000	300,000	12,000
Grade separation	2,120,000	5,000	225,000	0

13-15   A road is being proposed to facilitate a housing development on a scenic lake. Two alternatives have been suggested. One of the roadway alignments is to go around the lake and slightly impact a wetland. The second alternative will also go around the lake and will significantly impact two wetlands. The following table shows the anticipated costs for each alternative. Assuming that the annual interest rate is 7 percent, determine which alternative is preferred using equivalent annual cost analysis.

Alternative	First Cost ($)	Annual Maintenance Cost ($)	Service Life (yr)	Salvage Value ($)	Annual Wetland Costs ($)	Annual Lighting Costs ($)
I	75,000	3,000	15	45,000	7,500	1,500
II	125,000	2,000	15	25,000	2,500	2,500

13-16   Two alternatives are under consideration for maintenance of a bridge. Select the most cost-effective alternative using present worth analysis. Assume an interest rate of 10 percent per year and a design life of 50 years for each alternative.

Alternative A consists of annual maintenance costs of $5,000 per year for the design life except for:

Year 20, in which bridge deck repairs will cost $20,000
Year 30, in which a deck overlay and structural repairs will cost $105,000

Alternative B consists of annual maintenance costs of $3,000 per year for the design life except for:

Year 20, in which bridge deck repairs will cost $35,000
Year 30, in which a deck overlay and structural repairs will cost $85,000

13-17   Two pavement maintenance plans have been proposed for a county road. Select the most cost-effective alternative based on equivalent annual cost. Assume an interest rate of 8 percent per year and a design life of 25 years. Show cash flow diagrams for each alternative.

- Alternative A entails expenses of $500 per year for the entire lifespan plus an additional $1000 in year 6, $1500 in year 11, $2000 in year 16, and $2500 in year 21.
- Alternative B entails expenses of $600 per year for the entire lifespan plus an additional $100/yr starting in year 14 and continuing through year 25.

13-18   Two highway capacity improvement plans have been proposed for a congested suburban arterial. Select the most cost-effective alternative using present worth analysis. Assume an interest rate of 5 percent per year and a design life of 20 years for each alternative.

Alternative A entails improvements to traffic signals at an initial cost of $82,000 and a salvage value of $5000. Annual maintenance costs will be $700 per year.

Alternative B entails traffic signal improvements and addition of a left-turn lane at an initial cost of $72,000 and no salvage value. Annual maintenance costs will be $1200 per year, except in year 10, in which a rehabilitation will cost $14,000.

13-19   The light rail transit line described in this chapter is being evaluated by another group of stakeholders. Using the revised information, determine the weighted score for each alternative and comment on your result.

Objective	Ranking
1	5
2	3
3	1
4	2
5	4

13-20   Three alternatives to replace an existing two-lane highway with a four-lane highway that will meet current design standards are proposed. The selected alternative will provide a more direct route between two towns that are 19 km apart along the existing highway. With each alternative operating speeds are expected to be at or near the design speed of 96 km/h. Develop the scores for each alternative and recommend a preferred alternative for development. The following scoring method, developed by the transportation oversight board, is to be used:

Evaluation Criterion	Performance Measure	Weight (%)
Mobility	Travel time of shortest travel time alternative divided by travel time of alternative i	25
Safety	Annual reduction in number of crashes of alternative i divided by highest annual reduction in number of crashes among all alternatives	25
Cost-effectiveness	Project development cost of least expensive alternative (in $ per km) divided by project development cost of alternative i (in $ per km)	20
Environmental impacts	Area of wetlands impacted of least-impacting alternative divided by area of wetlands impacted by alternative i	15
Community impacts	Number of business and residences displaced by least-impacting alternative divided by number of businesses and residences displaced by alternative i	15

The following information has been estimated for each alternative by the planning staff:

Property	Alt. 1	Alt. 2	Alt. 3
Cost of development	$10,900,000	$18,400,000	$16,900,000
Length	18 km	15.8 km	16 km
Annual crash reduction	10	17	19
Business displacements	3	4	5
Residential displacements	4	3	3
Wetlands impacted	6070 m^2	15780 m^2	15780 m^2

13-21   Four alternatives to replace an existing two-lane highway with a four-lane highway that will meet current design standards are proposed. The selected alternative will provide a more direct route between two cities that are 27 km apart along the existing highway. With each alternative, operating speeds are expected to be at or near the design speed of 113 km/h. Develop the scores for each alternative and recommend a preferred alternative for development. The following scoring method, developed by the transportation oversight board, is to be used:

Evaluation Criterion	Performance Measure	Weight (%)
Mobility	Travel time of shortest travel time alternative divided by travel time of alternative i	28
Safety	Annual reduction in number of crashes of alternative i divided by highest annual reduction in number of crashes among all alternatives	30
Cost-effectiveness	Project development cost of least expensive alternative (in $ per km) divided by project development cost of alternative i (in $ per km)	18
Environmental impacts	Area of wetlands impacted of least-impacting alternative divided by area of wetlands impacted by alternative i	14
Community impacts	Number of business and residences displaced by least-impacting alternative divided by number of businesses and residences displaced by alternative i	10

The following information has been estimated for each alternative by the planning staff:

Property	Alt. A	Alt. B	Alt. C	Alt. D
Cost of development	$12,800,000	$12,300,000	$11,900,000	$12,900,000
Length	26 km	25 km	25.9 km	26.5 km
Annual crash reduction	19	24	23	27
Business displacements	2	1	4	4
Residential displacements	10	12	8	6
Wetlands impacted	7690 m^2	16997 m^2	11736 m^2	11736 m^2

13-22   You have been hired as a consultant to a medium-sized city to develop and implement a procedure for evaluating whether or not to build a highway bypass around the CBD. Write a short report describing your proposal and recommendation as to how the city should proceed with this process.

13-23   The following data have been developed for four alternative transportation plans for a high-speed transit line that will connect a major airport with the downtown area of a large city. Prepare an evaluation report for these proposals by considering the cost effectiveness of each attribute. Show your results in graphical form and comment on each proposal.

		Rail Alternatives			
Measure of Effectiveness	Existing Service	Plan A	Plan B	Plan C	Plan D
Persons displaced	0	264	3200	3200	3200
Businesses displaced	0	23	275	275	275
Average door-to-door trip speed (km/h)	16.4	61	72.4	74	77.2
Annual passengers (millions)	118.6	124.4	118.6	124.4	127.0
Annual cost (millions)	—	16.4	20.2	23.8	22.7

13-24   A new carpool lane has replaced one lane of an existing six-lane highway. During peak hours, the lane is restricted to cars carrying three or more passengers. After five months of operation, the carpool lane handles 800 autos/h, whereas the existing lanes are operating at capacity levels of 1500 veh/h/ln at an occupancy rate of 1.2. How would you determine if the new carpool lane is successful or if the lane should be open to all traffic?

REFERENCES

American Association of State Highway and Transportation Officials, *A Manual of User Benefit Analysis for Highways,* 2nd edition, 2003.

American Association of State Highway and Transportation Officials, *User and Non-User Benefit Analysis for Highways,* 3rd edition, 2010.

Bureau of Labor Statistics, *Employment Cost Index.* Washington, D.C., 2011. http://data.bls.gov/pdq/querytool.jsp?survey=ci

Federal Highway Administration, *Highway Economic Requirements System — State Version: Technical Report,* Washington, D.C., updated July 20, 2011. Section 5.5, Travel Time Costs. http://www.fhwa.dot.gov/asset/hersst/pubs/tech/tech05.cfm#table52

Forkenbrock, D. J. and J. Shelley, *Effective Methods for Environmental Justice Assessment.* NCHRP Report 532. Transportation Research Board of the National Academies, National Research Council, 2004.

Forkenbrock, D. J. and G. E. Weisbrod, *Guidebook for Assessing the Social and Economic Effects of Transportation Projects.* NCHRP Report 456. Transportation Research Board of the National Academies, National Research Council, 2001.

Litman, T, *Transportation Cost and Benefit Analysis II—Travel Time Costs,* Victoria Transportation Policy Institute, 2011. http://www.vtpi.org/tca/tca0502.pdf

Park, C. S., *Contemporary Engineering Economics.* 5th edition, Prentice Hall, 2010.

Schrank, D., T. Lomax, and B. Eisele, *Urban Mobility Report.* Texas Transportation Institute, Arlington, 2011. http://tti.tamu.edu/documents/mobility-report-2011.pdf

Sinha, K. C. and S. Labi, *Transportation Decision Making: Principles of Project Evaluation and Programming.* John Wiley & Sons, 2007.

Smalkoski, B. and D. Levinson, "Value of Time for Commercial Vehicle Operators," *Minnesota Journal of the Transportation Research Forum,* Volume 44, No. 1, pp. 89–102, 2005. http://papers.ssrn.com/sol3/papers.cfm?abstract_id=1091828

Sullivan, W. G., E. M. Wicks, and C. P. Koelling, *Engineering Economy,* 15th edition, Prentice Hall, 2011.

P A R T   4

Location, Geometrics, and Drainage

Highway location involves the acquisition of data concerning the terrain upon which the road will traverse and the economical siting of an alignment. To be considered are factors of earthwork, geologic conditions, and land use. Geometric design principles are used to establish the horizontal and vertical alignment, including consideration of the driver, the vehicle, and roadway characteristics. Design of parking and terminal facilities must be considered as they form an integral part of the total system. Since the new highway will alter existing patterns of surface and subsurface flow—and be influenced by it—careful attention to the design of drainage facilities is required.

CHAPTER 14

Highway Surveys and Location

An important initial step in the design of a proposed highway is to determine the location. The basis for selecting the location can be topography, soil characteristics, cost, and environmental factors. The data required are usually obtained from different types of surveys, depending on the factors being considered. Engineering consultants and state agencies that are involved in highway location employ computerized techniques that assemble vast amounts of data compiled using remote sensing, aerial photography and/or computer graphics to produce topographic maps that are used to develop candidate alternatives for highway location.

This chapter presents a brief description of the current techniques used in highway surveys to collect and analyze the required data, and the steps involved in the procedure for locating highways. Also included are earthwork computations using mass diagrams, to estimate of the amount of earthwork associated with any given location. The result is an economic evaluation that aids in the decision to accept, reject, or modify that location.

CHAPTER OBJECTIVES:

- Describe the four phases of the highway location process and related issues of scenic routes, urban areas, and bridges.
- Learn how to apply ground survey and remote sensing methods to determine terrain characteristics.
- Understand how to obtain and interpret aerial photographs.
- Become familiar with advanced scanning techniques.
- Apply the mass diagram to minimize earthwork costs.

14.1 PRINCIPLES OF HIGHWAY LOCATION

The basic principle for locating highways is that roadway elements such as curvature and grade must blend with each other to produce a system that provides for an efficient flow of traffic at the design capacity, while meeting design criteria and safety standards.

The highway should also cause minimal disruption to historic and archeological sites and to other land-use activities. Environmental impact studies are therefore required in most cases before a highway location is selected.

The highway location process involves four phases:

- Office study of existing information
- Reconnaissance survey
- Preliminary location survey
- Final location survey

14.1.1 Office Study of Existing Information

The first phase in any highway location study is the examination of all available data pertaining to the area in which the road is to be constructed. This phase is usually carried out in the office prior to field or photogrammetric investigations. These data can be obtained from existing engineering reports, maps, aerial photographs, and charts, which are usually at one of the state agencies such as departments of transportation, agriculture, geology, hydrology, and mining. The type and amount of data collected and examined depend on the type of highway being considered, but in general, data should be obtained about the following characteristics of the area:

- Engineering, including topography, geology, climate, and traffic volumes
- Social and demographic, including land-use and zoning patterns
- Environmental, including types of wildlife; location of recreational, historic, and archeological sites; and the possible effects of air, noise, and water pollution
- Economic, including unit costs for construction and the trend of agricultural, commercial, and industrial activities

Preliminary analysis of the data will suggest if any of the specific sites should be excluded from further consideration. For example, if it is found that a site of historic and archeological importance is located within an area being considered for the highway, any route that traverses that site could be excluded from further consideration. At the completion of this phase of the study, the engineer will have selected general areas through which the highway can traverse.

14.1.2 Reconnaissance Survey

The object of this phase of the study is to identify several feasible routes, each within a band of a limited width of a few hundred feet. When rural roads are being considered, there is often little information available from maps or photographs, and therefore aerial photography is widely used to obtain the required information. Feasible routes are identified by a stereoscopic examination of the aerial photographs, taking into consideration factors such as:

- Terrain and soil conditions
- Serviceability of route to industrial and population areas
- Crossing of other transportation facilities, such as bodies of water, railroads, and highways
- Directness of route
- Environmentally sensitive areas

Control points between the two endpoints are determined for each feasible route. For example, a unique bridge site with no alternative may be taken as a primary control point. The feasible routes identified are then plotted on photographic base maps.

14.1.3 Preliminary Location Survey

During this phase of the study, the positions of the feasible routes are established as closely as possible by defining the control points and determining preliminary vertical and horizontal alignments. Preliminary alignments are used to evaluate the economic and environmental feasibility of the alternative routes.

Economic Evaluation

An economic evaluation for each alternative route is carried out to determine the resources necessary to construct the highway. The evaluation methods described in Chapter 13 are used for this decision process. Factors taken into consideration include road user costs and benefits, construction costs, maintenance costs, and possible adverse impacts due to dislocation of families, businesses, and so forth.

The results obtained from the economic evaluation of the feasible routes provide valuable information to the decision maker, such as the economic resources that will be gained or lost if a particular location is selected. This information is also used to aid the policy maker in determining whether or not the highway should be built.

Environmental Evaluation

Construction of a highway at any location will have a significant impact on its surroundings because a highway is an integral part of the local environment. The environment includes plant, animal, and human communities and encompasses social, physical, natural, and synthetic variables. These variables are interrelated in a manner that maintains equilibrium and sustains the lifestyle of the different communities. The construction of a highway at a given location may result in significant changes in one or more variables, which can result in significant adverse effects on the environment and lead to a reduction in the quality of life. It is therefore essential to evaluate the environmental impact of all alignments selected.

Federal legislation has been enacted that sets forth the requirements of the environmental evaluation for different types of projects. In general, they call for the preparation of environmental impact statements that include:

- A detailed description of alternatives
- The probable environmental impact, including the assessment of positive and negative effects
- An analysis of both short-term and long-term impacts
- Any secondary effects, which may be in the form of changes in the patterns of social and economic activities
- Adverse environmental effects that cannot be avoided if the project is constructed

In cases where an environmental impact study is required, it is conducted at this stage to determine the environmental impact of each alternative route. Such a study will determine the negative and/or positive effects the highway facility will have on the environment. For example, the construction of a freeway at grade through an urban area may result in an unacceptable noise level for the residents of the area (negative impact), or the highway facility may be located so that it provides better access to jobs and recreation centers (positive impact). Public hearings are also held at this stage to provide an opportunity for constituents to give their views on the positive and negative impacts of the proposed alternatives.

The best alternative, based on all the factors considered, is then selected as the preliminary alignment of the highway.

14.1.4 Final Location Survey

The final location survey is a detailed layout of the selected route. The horizontal and vertical alignments are determined, and the positions of structures and drainage channels are located. The method used is to set out the points of intersections (PI) of the straight portions of the highway and fit a suitable horizontal curve (as described in Chapter 15). This is usually a trial-and-error process until, in the designer's opinion, the best alignment is obtained, taking both engineering and aesthetic factors into consideration.

Splines and curve templates were used in the final location process prior to the availability of computer-aided design. The *spline* is a flexible plastic guide that can be bent into different positions and is used to create different curvilinear alignments, from which the most suitable is selected. *Curve templates* are transparencies giving circular curves, three-center compound curves, and spiral curves of different radii and different standard scales. Figure 14.1 shows circular curve templates, and Figure 14.2 shows three-centered curve templates. The spline is used first to obtain a hand-fitted smooth curve that fits in with the requirements of grade, cross sections, curvature, and drainage. The hand-fitted curve is then changed to a more defined curve by using the standard templates.

The availability of computer-based techniques has significantly enhanced this process since a proposed highway can be displayed on a monitor, enabling the designer to have a driver's eye view of both the horizontal and vertical alignments of the road. The designer can therefore change either or both alignments until the best alignment is achieved.

Detailed design of the vertical and horizontal alignments is then carried out to obtain both the deflection angles for horizontal curves and the cuts or fills for vertical curves and straight sections of the highway. The design of horizontal and vertical curves is presented in Chapter 15.

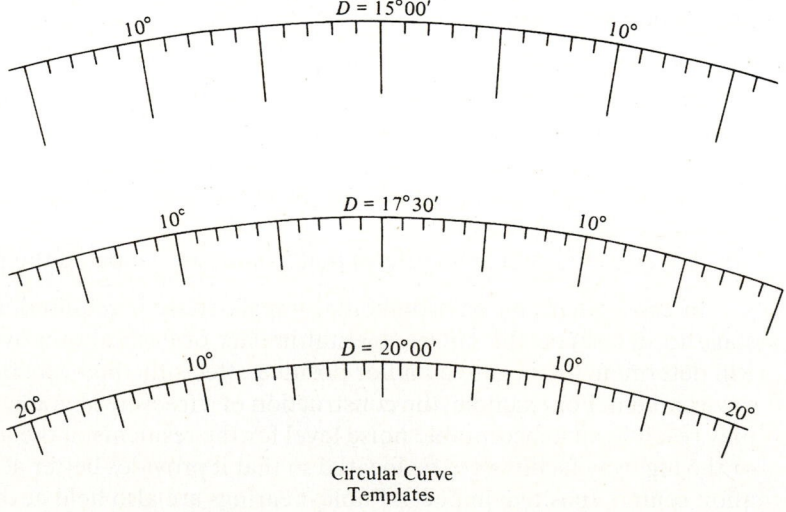

Figure 14.1   Circular Curve Templates

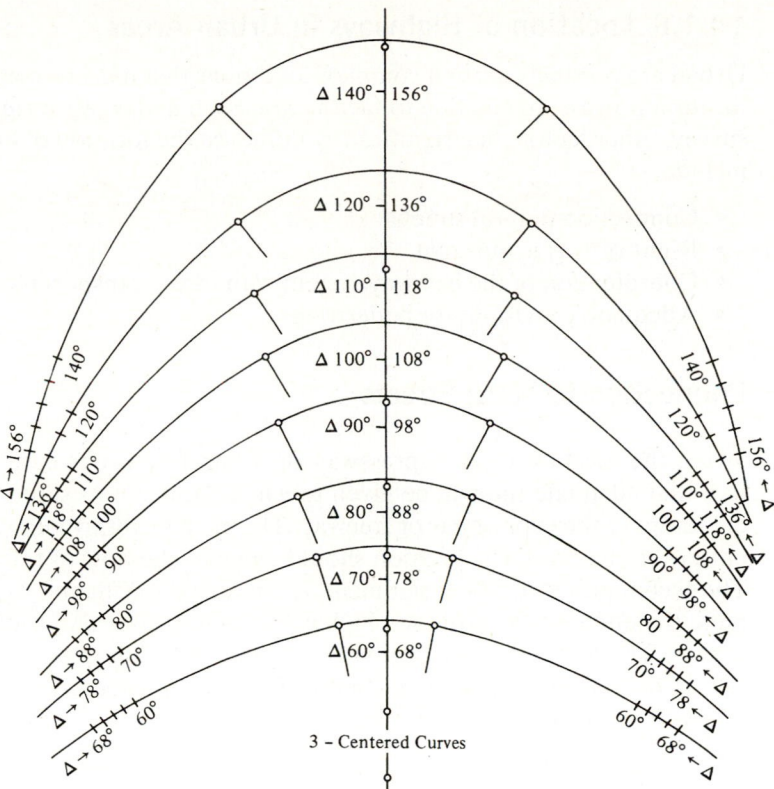

Figure 14.2 Centered Curve Templates

14.1.5 Location of Recreational and Scenic Routes

The location process of recreational and scenic routes follows the same steps as discussed earlier, but the designer of these types of roads must be aware of their primary purpose. For example, although it is essential for freeways and arterial routes to be as direct as possible, a circuitous alignment may be desirable for recreational and scenic routes to provide access to recreational sites (such as lakes or campsites) or to provide special scenic views. The following items are considered in the location of recreational and scenic routes:

- Design speeds are usually low, and therefore special provisions should be made to discourage fast driving (for example, by providing a narrower lane width).
- Location should minimize the conflict between the driver's attention on the road and the need to enjoy the scenic view. This can be achieved by providing turnouts with wide shoulders and adequate turning space at regular intervals, or by providing only straight alignments when the view is spectacular.
- Location should cause minimum disruption to the surrounding area.

Special guides are available for the location and design of recreational and scenic routes. The reader is referred to *A Policy on Geometric Design of Highways and Streets,* published by the American Association of State Highway and Transportation Officials, for more information on this topic.

14.1.6 Location of Highways in Urban Areas

Urban areas usually present complex conditions that must be considered in the highway location process. In addition to factors discussed under office study and reconnaissance survey, other factors that significantly influence the location of highways in urban areas include:

- Connection to local streets
- Right-of-way acquisition
- Coordination of the highway system with other transportation systems
- Adequate provisions for pedestrians

Connection to Local Streets

When the location of an expressway or urban freeway is being planned, it is important that adequate thought be given to which local streets should connect with on- and off-ramps to the expressway or freeway. The main factor to consider is the existing travel pattern in the area. The location should enhance the flow of traffic on the local streets. The techniques of traffic assignment, as discussed in Chapter 12, can be used to determine the effect of the proposed highway on the volume and traffic flow on the existing streets. The location should provide for adequate sight distances at all ramps. Ramps should not be placed at intervals that will cause confusion or increase the crash potential on the freeway or expressway.

Right-of-Way Acquisition

One factor that significantly affects the location of highways in urban areas is the cost of acquiring right-of-way. This cost is largely dependent on the predominant land use in the right-of-way of the proposed highway. Costs tend to be much higher in commercial areas, and landowners in these areas may not be willing to give up their property for highway construction. Thus, freeways and expressways in urban areas are often placed on continuous elevated structures in order to avoid the acquisition of rights of way and the disruption of commercial and residential activities.

Elevated highways have the advantage of minimal interference with existing land-use activities, but they are unpopular with occupants of adjacent land because of noise and aesthetics. Elevated structures are also very expensive to construct and therefore do not completely eliminate the problem of high costs.

Coordination of the Highway System with Other Transportation Systems

Urban planners understand the importance of a balanced transportation system and strive toward providing a fully integrated system of highways and public transportation. This integration should be taken into account during the location process of an urban highway. Several approaches have been considered, but the main objective is to provide new facilities that will increase the overall level of service of the transportation system in the urban area. In Washington, D.C., for example, park-and-ride facilities have been provided at stations to facilitate the use of the heavy rail mass transit system, and exclusive bus lanes have been used to reduce the travel time of express buses during the peak hour.

Another form of transportation system integration is the multiple use of rights-of-way by both highway and transit agencies. In this case, the right-of-way is shared between them, and bus or rail facilities are constructed either in the median or alongside the freeway. Examples include the Washington Metropolitan Area Transit Authority (WMATA) rail system in the median of Interstate I-66 in Northern Virginia; the Congress Street and Dan Ryan Expressway in Chicago, Illinois; sections of the Bay Area Rapid Transit System in San Francisco, California; and freight rail lines in the medians of freeways, as illustrated in Figure 14.3 for the CSX Railroad, on I-95 in Richmond, Virginia.

Adequate Provisions for Bicycles and Pedestrians

Providing adequate facilities for bicycles and pedestrians should be an important factor in deciding the location of highways, particularly for highways in urban areas. Pedestrians are an integral part of any highway system but are more numerous in urban areas than in rural areas. Bicycles are an alternate mode of transportation that can help to reduce energy consumption and traffic congestion.

Therefore, special attention must be given to the provision of adequate bicycle and pedestrian facilities in planning and designing urban highways. Facilities for pedestrians should include sidewalks, crosswalks, traffic-control features, curb cuts, and ramps for the handicapped. Facilities for bicycles should include wide-curb lanes, bicycle paths, and shared-use paths. Design considerations for bicycles are discussed in Chapter 15. In heavily congested urban areas, the need for grade-separated facilities, such as overhead bridges and/or tunnels, may have a significant effect on the final location of the highway. Although vehicular traffic demands in urban areas are of primary concern in deciding the location of highways in these areas, the provision of adequate bicycle and pedestrian facilities must also be of concern because they are an indispensable and vital component of the urban area.

Figure 14.3   CSX Railroad in the median of I-195, Richmond, Virginia

SOURCE: Office of Public Affairs, Virginia Department of Transportation. Used with permission

14.1.7 Principles of Bridge Location

The basic principle for locating highway bridges is that the highway location should determine the bridge location, not the reverse. When the bridge is located first, in most cases the resulting highway alignment is not optimal. The general procedure for most highways, therefore, is to first determine the best highway location and then determine the bridge site. In some cases, this will result in skewed bridges, which are more expensive to construct, or put them in locations where foundation problems exist.

When serious problems of this nature occur, all factors such as highway alignments, construction costs of the bridge deck and its foundation, and construction costs of bridge approaches should be considered in order to determine a compromise route alignment that will give a suitable bridge site. This will include completing the transportation planning process and the economic evaluation of the benefits and costs, as discussed in Chapters 11 and 13.

A detailed report should be prepared for the bridge site selected to determine whether there are any factors that make the site unacceptable. This report should include accurate data on soil stratification, the engineering properties of each soil stratum at the location, the crushing strength of bedrock, and water levels in the channel or waterway.

When the waterway to be crossed requires a major bridge structure, however, it is necessary to first identify a narrow section of the waterway with suitable foundation conditions for the location of the bridge and then determine acceptable highway alignments that cross the waterway at that section. This will significantly reduce the cost of bridge construction in many situations.

14.2 HIGHWAY SURVEY METHODS

Highway surveys usually involve measuring and computing horizontal and vertical angles, vertical heights (elevations), and horizontal distances. The surveys are then used to prepare base maps with contour lines (that is, lines on a map connecting points that have the same elevation) and longitudinal cross sections. Highway surveying techniques have been revolutionized due to the rapid development of electronic equipment and computers. Surveying techniques can be grouped into three general categories:

- Ground surveys
- Remote sensing
- Computer graphics

14.2.1 Ground Surveys

Ground surveys are the basic location technique for highways. The *total station* is used for measuring angles in both vertical and horizontal planes, distances, and changes in elevation. The *level* is used for measuring changes in elevation only. A summary of survey equipment follows. For greater detail, refer to the surveying texts cited at the end of this chapter.

The Total Station

A total station is both an electronic theodolite and electronic distance-measuring device (EDM). The total station enables one to determine angles and distances from the instrument to other points. Angles and distances may be used to calculate the actual positions (coordinates and elevations). An example of a total station is shown in Figure 14.4a.

Figure 14.4a   Topcon Total Station GTS 722

SOURCE: Photograph by Tom Saunders, Virginia Department of Transportation

The standard theodolite consists of a telescope with vertical and horizontal cross hairs, a graduated arc or vernier for reading vertical angles, and a graduated circular plate for reading horizontal angles, whereas the electronic theodolite provides a digital readout of those angles. These readouts are continuous, so angles can be checked at any time. The telescope on both instruments is mounted so that it can rotate vertically about a horizontal axis.

With the standard theodolite, two vertical arms support the telescope on its horizontal axis, with the graduated arc attached to one of the arms. The arms are attached to a circular plate, which can rotate horizontally with reference to the graduated circular plate, thereby providing a means for measuring horizontal angles.

Electronic Distance-Measuring Devices (EDM)

An EDM device consists mainly of a transmitter located at one end of the distance to be measured and a reflector at the other end. The transmitter sends a light beam or a low-power laser pulse, which is reflected back to the transmitter. The difference in phase between the transmitted and reflected pulses is measured electronically and used to determine the distance between the transmitter and the reflector. This equipment can measure distances up to about 1000 m in average atmospheric conditions. Special features permit the operator to change the display from slope to horizontal distance automatically. Units can also be changed from meters to feet. The total station offers these solutions because of internal instantaneous calculations.

Measurement of Horizontal Angles. A *horizontal* angle, such as angle *XOY* in Figure 14.4b, can be measured by setting the theodolite or total station directly over the point *O*. The telescope is then turned toward point *X* and viewed first over its top and then by sighting through the telescope to obtain a sightline of *OX,* using the horizontal tangent screws provided. The graduated horizontal scale is then set to zero and locked into position, using the clamp screws provided. The telescope is rotated in the horizontal

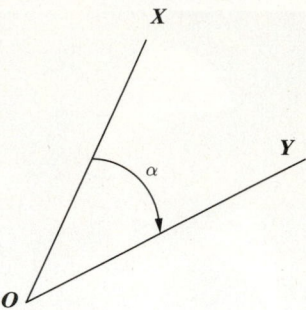

Figure 14.4b   Illustration of Horizontal Angle

plane until the line of sight is OY. The horizontal angle (α) is read from the graduated scale. While similar in method to a theodolite, when using a total station, the angle measured is shown in a digital readout when a *horizontal* readout is selected.

Measurement of Vertical Angles.   The theodolite measures *vertical* angles, such as angle AOB in Figure 14.4c, by using the fixed vertical graduated arc or vernier and a spirit level attached to the telescope. When a theodolite is used to determine the vertical angle to a point, the angle (β) obtained is the angle of elevation or the angle of depression from the horizontal. The theodolite is set up in a way similar to that for measuring horizontal angles, ensuring that the longitudinal axis is in the horizontal position. The telescope is approximately sighted at the point. The tangent screw of the telescope is then used to set the horizontal cross hair at the point. The vertical angle is obtained from the graduated arc or the vertical vernier. Again, while similar in method to a theodolite, when using a total station, the angle measured is shown in a digital readout when a *vertical* readout is selected.

The Level

The essential parts of a level are the telescope, with vertical and horizontal cross hairs, a level bar, a spindle, and a leveling head. The level bar on which the telescope is mounted is rigidly fixed to the spindle. The level tube is attached to the telescope or the level bar so that it is parallel to the telescope. The spindle is fitted into the leveling head in such a way that allows the level to rotate about the spindle as an axis, with the leveling head attached to a tripod. The level also carries a bubble that indicates whether the level is properly centered. Using the leveling screws provided centers the bubble.

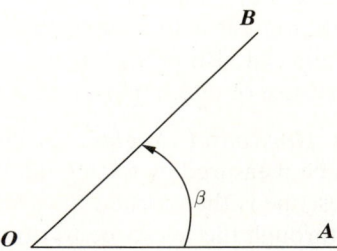

Figure 14.4c   Illustration of Vertical Angle

There are two older types of levels: the "Wye" and the "Dumpy." Today, the more common type is the self-leveling or automatic level, shown in Figure 14.4d. The specific features of the first two types are that the telescope of the Dumpy level is permanently fixed to the level bar; the telescope of the Wye level is supported by a pair of Wye rings that can be opened for the purpose of turning the telescope or rotating it around its horizontal axis; the automatic level has a built-in compensator that automatically adjusts for minor errors in the set-up of the instrument.

A leveling rod, shown opposite the digital level in Figure 14.4d, is required to measure changes in elevation. These are rods of rectangular cross section, graduated in feet, meters, or a barcode for digital systems, so that differences in heights can be measured.

The difference in elevation between two points, *A* and *B,* is obtained by setting the level at a third point. The rod is then placed at the first point, say, point *A,* and the level is sighted at the rod. The reading indicated by the cross hair is recorded. The level rod is transferred to point *B,* the level is again sighted at the rod, and again the reading indicated by the cross hair is recorded. The difference between these readings gives the difference in elevation between points *A* and *B.* If the elevation of one of the points is known, the elevation of the other point can then be determined.

Measuring Tapes

Tapes can be used for direct measurement of horizontal distances. They are available in several materials, but the types used for engineering work are usually made of steel or a woven nonmetallic or metallic material. They are available in both U.S. and metric units.

Figure 14.4d   High-Accuracy Digital Leveling Using a Trimble DiNi 12 Digital Level

SOURCE: Photograph by Tom Saunders, Virginia Department of Transportation

14.2.2 Digital Survey Advancements

Significant advancements in survey technology have been made in recent years that reflect the worldwide revolution in wireless and communications technology. These are summarized in the following sections.

Survey Data Collectors

Survey data collectors are devices that connect to any survey instrument (total station) through a cable or Bluetooth that allows a surveyor to secure information digitally as illustrated in Figure 14.4e. While all modern data collectors use Bluetooth, there are some total stations that predate this era and cables are required. Cable technology will become obsolete when these older total stations units are replaced. The information is stored on the hard drive of the device. The data collector converts the survey instrument's raw data string into coordinates and elevations. The raw data string includes bearing, angle, distance, slope distance, zenith, and more. Each point stored on the data collector has an XYZ value that it calculates from the occupied location of the survey instrument. The data collector can then be attached to a desktop or laptop computer, and the information can then be processed into a computer-aided design drafting (CADD) file. This allows for more accurate three-dimensional surfaces to be created. The triangulated surface created is called a TIN file (Triangulated Irregular Network).

Multilayered Information

With the digital age, many projects are submitted in digital format. Digital formats offer a great deal of flexibility. Prior to digital files, information was often drawn on various-sized paper by hand. The digital files (CADD files) can have certain features turned on

Figure 14.4e   Carlson Explorer 600+ Digital Data Collector

SOURCE: Photograph by Tom Saunders, Virginia Department of Transportation or off. This can be very beneficial in a congested urban area if one wishes to see only selected features. An example would be utility items. If a paper copy is desired, it also can be plotted showing only the desired features.

Global Positioning System Surveys

The Global Positioning System (GPS) is a satellite-based system that uses a constellation of 28 satellites to give a user an accurate position. GPS surveying is an evolving technology that was originally designed for military use at any time anywhere on the surface of the earth. Soon it became clear that civilians could also use GPS, and not only for personal positioning. The first two major civilian applications to emerge were in marine navigation and surveying. Now applications include in-car navigation, truck fleet management, and automation of construction machinery. Accuracy expectations range from sub-centimeter with static surveys to plus or minus 2 cm using real-time kinematic (RTK) methods. The basic GPS survey types are given next.

Static GPS Surveys. Various systematic errors can be resolved when high-accuracy positioning is required. Static procedures are used to produce baselines between stationary GPS units by recording data over an extended period of time during which the satellite geometry changes.

Fast-Static GPS Surveys. These are similar to static GPS surveys, but require shorter observation periods (approximately 5 to 10 min). Fast-static GPS survey procedures require more advanced equipment and data reduction techniques than static GPS methods. Typically, the fast-static GPS method should not be used for corridor control or other surveys requiring horizontal accuracy greater than first order.

Kinematic GPS Surveys. This type makes use of two or more GPS units. At least one unit occupies a known (reference) station and remains stationary, while other (rover) GPS units are moved from point to point. All baselines are produced from the GPS unit occupying a reference station to the rover units. Kinematic GPS surveys can be either continuous or "stop and go," in which observation periods are of short duration, typically under 2 min. Kinematic GPS surveys are employed where third-order or lower accuracy standards are applicable.

Real-Time Kinematic (RTK) GPS Surveys. Figure 14.5 illustrates a recent development in survey technology that uses a method offering positional results that could be as accurate as static surveys, but in real time. RTK requires a communication link between the transmitter at the base station and the receiver at the rover. It is used in engineering, surveying, air navigation, mineral exploration, construction machine control, hydrology, and other applications.

OPUS GPS Surveys. The On-line Positioning User Service (OPUS) allows users to submit individual GPS unit data files directly for automatic processing. Each data file submitted is processed. OPUS solutions are not to be used for producing final coordinates or elevations on any survey; however, OPUS solutions may be used as a verification of other procedures.

Equipment. Post-processed GPS surveying equipment generally consists of two major components, receivers and antennas.

Receiver Requirements. First-order, second-order, and third-order post-processed GPS surveys require GPS receivers that are capable of recording data. When performing specific

Figure 14.5   "Trimble 5800" Setup for RTK (Real-Time Kinematic) Data Collection

SOURCE: Photograph by Tom Saunders, Virginia Department of Transportation

types of GPS surveys (i.e., static, fast-static, and kinematic), receivers and software should be suitable for a specific survey as stated by the manufacturer. Dual frequency receivers are used for observing baselines over 14.5 km in length. During periods of intense solar activity, dual frequency receivers are used for observing baselines over 10 km in length.

Antennas. For vertical control surveys, identical antennas are used unless software is available to accommodate the use of different antennas. For first-order and second-order horizontal surveys, antennas with a ground plane attached are used, and the antennas are mounted on a tripod or a stable supporting tower. When tripods or towers are used, optical plummets or collimators are required to ensure accurate centering over marks. Fixed height tripods are required for third-order or better vertical surveys. Range poles and/or stakeout poles to support GPS antennas are only used for third-order horizontal and general-order surveys.

Miscellaneous Equipment Requirements. All equipment should be properly maintained and regularly checked for accuracy. Errors due to poorly maintained equipment must be eliminated to ensure valid survey results. Level vials, optical plummets, and collimators should be calibrated at the beginning and end of each GPS survey. If the duration of the survey exceeds a week, these calibrations are repeated weekly for the duration of the survey.

Office Procedures

The following office procedures are followed when conducting GPS surveys.

General.   For first-order, second-order, and some third-order post-processed GPS surveys, raw GPS observation (tracking) data are collected and post-processed for results and analysis. Post-processing and analysis are required for first-order and second-order GPS surveys. The primary post-processed results that are analyzed are

- Baseline processing results
- Loop closures
- Repeat baseline differences
- Results from least-squares network adjustments

Post-Processing.   Software should be capable of producing relative-position coordinates and corresponding statistics that can be used in a three-dimensional least squares network adjustment. This software must also allow analysis of loop closures and repeat baseline observations.

Loop Closure and Repeat Baseline Analysis.   Loop closures and differences in repeat baselines are computed to check for errors and to obtain initial estimates of the internal consistency of the GPS network. Loop closures are included and differences in repeat baselines in the project documentation are tabulated. Failure of a baseline in a loop closure does not imply that the baseline in question should be rejected but may be an indication that a portion of the network requires additional analysis.

Least-Squares Network Adjustment.   An unconstrained (free) adjustment is performed, after errors are corrected or removed from the network, to verify the baselines of the network. After a satisfactory standard deviation of unit weight (network reference factor) is achieved using realistic *a priori* error estimates, a constrained adjustment is performed. The constrained network adjustment fixes the coordinates of the known reference stations, thereby adjusting the network to the datum and epoch of the reference stations. A consistent control reference network (datum) and epoch shall be used for the constrained adjustment. The NGS Horizontal Time Dependent Positioning (HTDP) program may be used to translate geodetic positions from one epoch to another.

14.2.3 Remote Sensing

Remote sensing is the measurement of distances and elevations by using devices located above the earth, such as airplanes or orbiting satellites using GPS. The most commonly used remote-sensing method is photogrammetry, which utilizes aerial photography. Photogrammetry is the science of obtaining accurate and reliable information through measurements and interpretation of photographs, and displaying this information in digital form and/or map form. This process is fast and economical for large projects but can be very expensive for small projects. The break-even size for which photogrammetry can be used varies between 30 and 100 acres, depending on the circumstances of the specific project. The successful use of the method depends on the type of terrain. Difficulties will arise when it is used for terrain with the following characteristics.

- Areas of thick forest, such as tropical rain forests, that completely cover the ground surface
- Areas that contain deep canyons or tall buildings, which may conceal the ground surface on the photographs
- Areas that photograph as uniform shades, such as plains and some deserts

The most common uses of photogrammetry in highway engineering are the identification of suitable locations for highways, referred to as corridor study, and the preparation of base maps for design mapping, showing all natural and synthetic features plus contours of 0.6 or 1.5 m intervals. In both of these uses, the first task is to obtain the aerial photographs of the area if none is available. A brief description of the methodology used to obtain and interpret the photographs is presented next. A more detailed treatment of the subject can be obtained from sources listed in the References section at the end of this chapter.

Obtaining and Interpreting Aerial Photographs

Aerial photographs are taken from airplanes with the axis of the camera at a near vertical position. The axis should be exactly vertical, but this position is usually difficult to obtain because the motion of the aircraft may cause some tilting of the camera up to a maximum of about 5 degrees, although on average this value is about 1 degree. Photographs taken this way are defined as vertical aerial photographs and are used for highway mapping. In some cases, however, the axis of the camera may be intentionally tilted so that a single photograph will cover a greater area. Photographs of this type are known as oblique photographs and are not used for mapping.

Vertical aerial photographs are taken in a square format, usually with dimensions 9×9 in., so that the area covered by each photograph is also a square. The airplane flies over the area to be photographed in parallel runs such that any two adjacent photographs overlap both in the direction of flight and in the direction perpendicular to flight. These photographs form a block and are used for corridor studies. The overlap in the direction of flight is the forward or end overlap and provides an overlap of about 60 percent for any two consecutive photographs. The overlap in the direction perpendicular to flight is the side overlap and provides an overlap of about 25 percent between consecutive flight lines, as shown in Figure 14.6. This ensures that each point on the ground is photographed at least twice, which is necessary for obtaining a three-dimensional view of the area.

Figure 14.7 shows a set of consecutive vertical aerial photographs, usually called stereopairs. Certain features of the photographs are as follows. The four marks A, B, C, and D that appear in the middle of each side of the photographs are known as *fiducial marks*. Some aerial cameras provide eight fiducials, with the four additional marks being in the corners. The intersection of the lines joining opposite fiducial marks gives the geometric center of the photograph, which is also called the *principal point*. It is also necessary to select a set of points on the ground that can be easily identified on the photographs as *control points*. These control points are used to bring the photo-coverage area to ground coordinates through the aerotriangulation process. This process involves point transfer from photo to photo, creating a control "mesh" over the area.

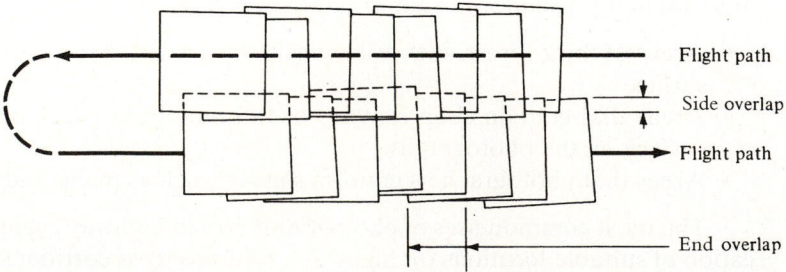

Figure 14.6   End and Side Overlaps in Aerial Photography

Figure 14.7   Set of Stereopairs

SOURCE: © Kristiina Paul

The information on the aerial photographs is then used to convert these photographs into maps. The instruments used for this process are known as *stereoscopes* or *stereoplotters,* and they vary from a simple mirror stereoscope, shown in Figure 14.8a, to more complex types such as the softcopy stereoplotter shown in Figure 14.8b. All of these instruments use the principle of *stereoscopy,* which is the ability to see objects in three dimensions when both eyes view these objects. When a set of stereopairs is properly placed under a stereoscope, so that an object on the left photograph is viewed by the left

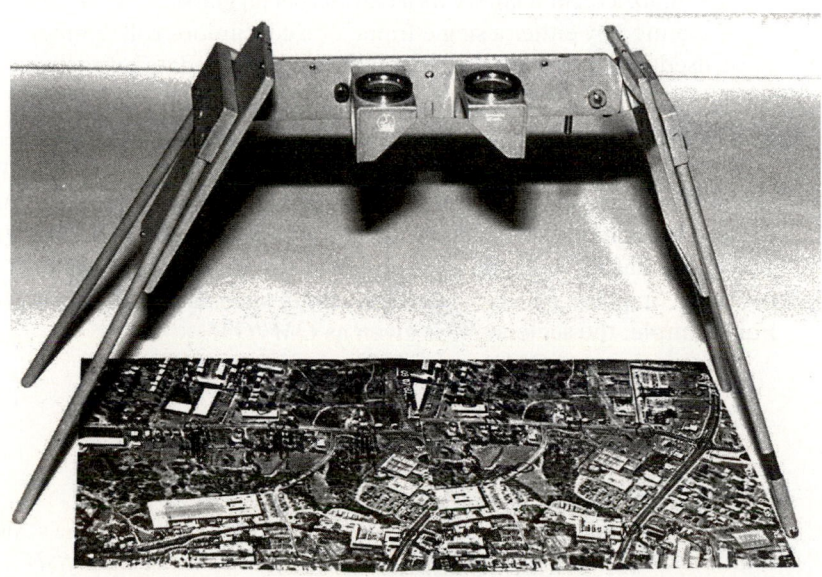

Figure 14.8a   Mirror Stereoscope

SOURCE: Photograph by Nicholas J. Garber

Figure 14.8b   Imagestation Softcopy Workstation

SOURCE: Photograph by Tom Saunders, Virginia Department of Transportation

eye and the same object on the right photograph is viewed by the right eye, the observer perceives the object in three dimensions and therefore sees the area in the photograph as if it were in three dimensions. Using the newer Softcopy Workstations, as illustrated in Figure 14.8b, a technician wears a pair of special glasses to view stereo imagery. This imagery is scanned by either a single frame or a continuous roll scanner, as seen in Figure 14.9, to be used in the aerotriangulation process and the data extraction or mapping phase.

Scale at a Point on a Vertical Photograph

Figure 14.10 shows a schematic of a single photograph taken of points M and N. The camera is located at O, and it is assumed that the axis is vertical. H is the flying height above the datum XX, and h_m and h_n are the elevations of M and N. If the images are recorded on plane FF, the scale at any point such as M'' on the photograph can be found. For example, the scale at M'' is given as OM''/OM.

Since triangles OMK and $OM''N''$ are similar,

$$\frac{OM''}{OM} = \frac{ON''}{OK}$$

but $ON'' = f$, the focal length of camera, and $OK = H - h_m$. The scale at M'' is $f/(H - h_m)$. Similarly, the scale at P'' is $f/(H - h_p)$, and the scale at N'' is $f/(H - h_n)$. In general, the scale at any point is given as

$$s = \frac{f}{H - h} \tag{14.1}$$

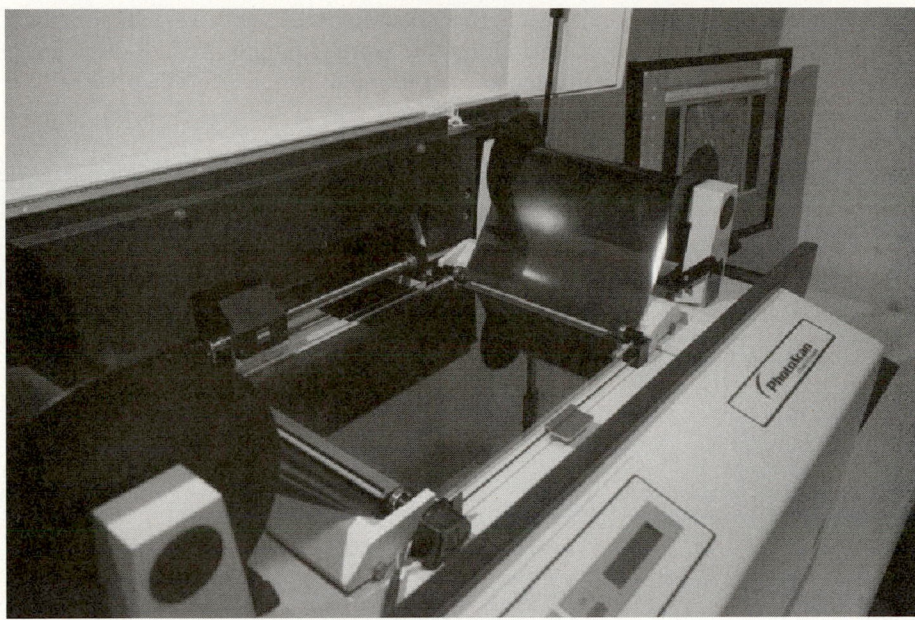

Figure 14.9 Photoscan Roll-Film Photogrammetric Scanning System

SOURCE: Photograph by Tom Saunders, Virginia Department of Transportation

where

S = scale at a given point on the photograph
f = focal length of camera lens
H = flying height (m)
h = elevation of point at which the scale is being determined (m)

Thus, even when the axis of the camera is perfectly vertical, the scale on the photograph varies from point to point and depends on the elevation of the point at which the scale is to be determined. Therefore, an aerial photograph itself does not have a uniform

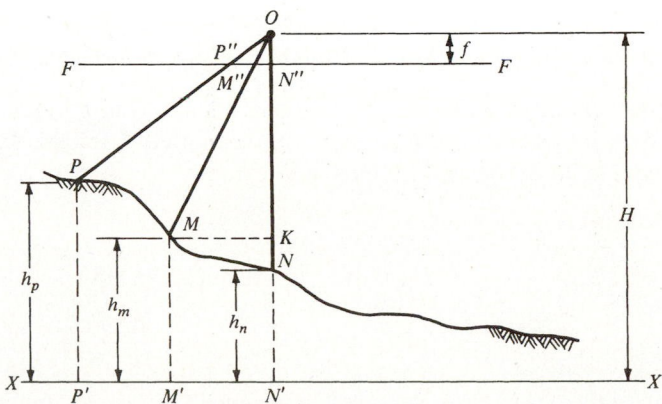

Figure 14.10 Schematic of a Single Aerial Photograph

scale; rather, the scale refers to a specific point. However, an approximate overall scale may be obtained by determining the average height (h_{avg}) of different points of the photograph and by substituting h_{avg} for h in Eq. 14.1.

Example 14.1 Computing Elevation of a Point from an Aerial Photograph

The elevations of two points A and B on an aerial photograph are 150 m and 170 m, respectively. The scales at these points on the photograph are 1:10,000 and 1:9870, respectively. Determine the elevation at point C, if the scale at C is 1:8000.

Solution: Use Eq. 14.1 to solve for focal length, f.

$$s = \frac{f}{H - h}$$

$$\frac{1}{10,000} = \frac{f}{H - 150} \qquad \text{(Point A)}$$

$$\frac{1}{9870} = \frac{f}{H - 170} \qquad \text{(Point B)}$$

$$H - 150 = 10,000\,f$$

$$-H + 170 = -9870\,f$$

$$65 = 130\,f$$

$$f = 0.15 \text{ m}$$

Since $f = 0.15$ m and $H - 500 = 10,000 \times 0.5$, $H = 1650$ m. Elevation at Point C is:

$$\frac{1}{8000} = \frac{0.5}{5500 - h_C}$$

$$h_C = 5500 - 4000 = 450 \text{ m}$$

Example 14.2 Computing Flying Height of an Airplane Used in Aerial Photography

Vertical photographs are to be taken of an area whose mean ground elevation is 600 m. If the scale of the photographs is approximately 1:15,000, determine the flying height. Focal length of the camera lens is 19 cm.

Solution:

$$f = 19 \text{ cm} = 0.19 \text{ m}$$

$$s = \frac{1}{15,000} = \frac{0.19 \text{ m}}{H - 600}$$

$$H = 3450 \text{ m}$$

Distances and Elevations from Aerial Photographs

Figure 14.11 is a schematic of a stereopair of vertical photographs taken on a flight. D is the orthogonal projection to the datum plane of a point A located on the ground. The height AD is therefore the elevation of point A. O and O' are the camera positions for the two photographs, with OO' being the air base (the distance between the centers of two consecutive aerial photographs). Images of A on the photographs are a' and a'', whereas d' and d'' are the images of D. The lines $a'\,d''$ and $a'\,d''$ represent the radial shift, which is caused by the elevation of point A above point D on the datum plane. This radial shift is known as the *parallax* and is an important parameter when elevations and distances are to be obtained from aerial photographs. The parallax at any point of an aerial photograph can be determined by the use of an instrument known as the parallax bar, shown in Figure 14.12. Two identical dots on the bar known as half-marks appear as a single floating dot above the three-dimensional view (terrain model) when stereoscopic viewing is attained.

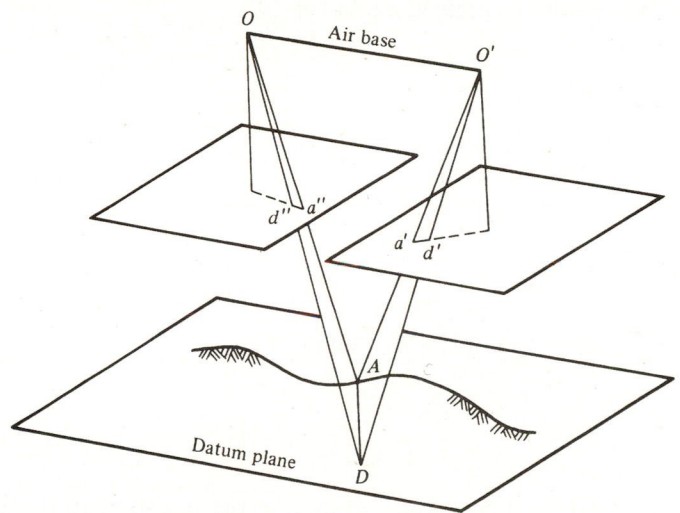

Figure 14.11   Schematic of a Stereopair of Vertical Photographs

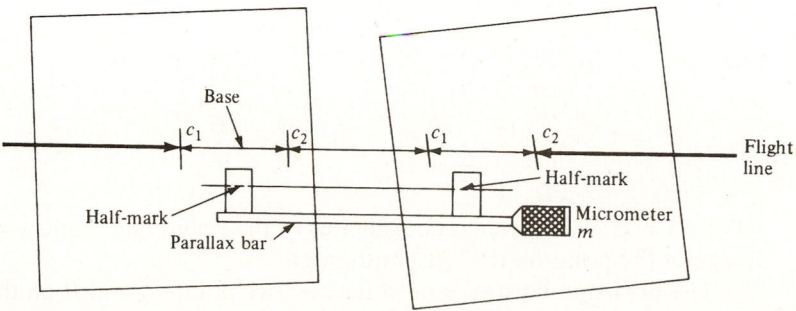

Figure 14.12   Parallax Bar

The distance between the half-marks can be changed by turning the micrometer, which results in the single dot moving up and down over the terrain model. The micrometer is moved until the dot appears to be just touching the point on the model at which the parallax is required. The distance between the half-marks as read by the micrometer is the parallax at that point. It can be shown that

$$\frac{H-h}{f} = \frac{B}{p} \qquad (14.2)$$

where

H = flying height (m)
h = elevation of a given point (m)
p = parallax at the given point (m)
B = air base (m)
f = focal length of the camera (m)

The X and Y coordinates of any point also can be found from the parallax by using the expression for scale given in Eq. 14.1.

$$s = \frac{f}{H-h}$$

From Eq. 14.2,

$$H = \frac{fB}{p} + h$$

Substituting for H in Eq. 14.1 gives

$$s = \frac{p}{B} \qquad (14.3)$$

By establishing a coordinate system in the terrain with the x-axis being parallel to the photographic axis of flight, the coordinates of any point are given as

$$X = \frac{x}{S} = \frac{xB}{p} \qquad (14.4)$$

and

$$Y = \frac{y}{S} = \frac{yB}{p} \qquad (14.5)$$

where X and Y are ground coordinates of the point, and x and y are coordinates of the image of the point on the left photograph.

The horizontal distance between any two points A and B on the ground then can be obtained from Eq. 14.6.

$$D = \sqrt{(X_A - X_B)^2 + (Y_A - Y_B)^2} \qquad (14.6)$$

Example 14.3 Computing the Elevation and Distance between Two Points Using an Aerial Photograph

The scale at the image of a control point on an aerial photograph is 1:15,000, and the elevation of the point is 400 m. The focal length of the camera lens is 15 cm. Determine the elevations of points B and C and the horizontal ground distance between them for the following data. Also, determine the parallax at B if the air base is 60 m.

	B	C
x coordinate on axis of left photograph	15 cm	17.5
y coordinate on axis of left photograph	7.5 cm	10 cm
Scale on photograph	1:17,500	1:17,400

Solution:

For the control point,

$$\frac{1}{15,000} = \frac{0.15}{H - 400} \quad \text{or} \quad H = 2650 \text{ m}$$

Thus, the elevations at B and C are found to be

$$\frac{1}{17,500} = \frac{0.15}{2650 - h_B} \quad \text{or} \quad h_B = 25 \text{ m}$$

$$\frac{1}{17,400} = \frac{0.15}{2650 - h_C} \quad \text{or} \quad h_C = 40 \text{ m}$$

The coordinates for X and Y are determined using equations 14.4 and 14.5, where

$$X_B = 0.15/(1/17500) = 2625 \text{ m (from Eq. 14.4)}$$

$$Y_B = 0.075/(1/17500) = 1313 \text{ m (from Eq. 14.5)}$$

$$X_C = 0.175/(1/17400) = 3045 \text{ m}$$

$$Y_C = 0.1/(1/17400) = 1740 \text{ m}$$

The distance between B and C, using Eq. 14.6, is

$$D = \sqrt{[(2625 - 3045)^2 + (13/3 - 1740)^2]}$$

$$= \sqrt{(176400 + 182329)}$$

$$= 598.94 \text{ m}$$

The parallax at B is

$$p_B = \frac{0.15 \times 60}{2625} = 0.003429 \text{ m (from Eq. 14.4)}$$

$$p_B = \frac{0.075 \times 60}{1313} = 0.003428 \text{ m (from Eq. 14.5)}$$

14.2.4 Computer Graphics

Computer graphics, when used for highway location, is usually the combination of photogrammetry and computer techniques. With the use of mapping software, line styles, and feature tables, objects and photographic features can be recorded digitally and stored in a computer file. This file can then either be plotted out in map form or sent on to the design unit. Figure 14.8b shows the Z/I Imaging Dual-Screen Softcopy Photogrammetric Workstation. A typical workstation is controlled by system software that covers four main areas of design work:

- Preparatory work (project setup)
- Photo orientations and aerotriangulation
- Data transfer
- Plotting and storage

The software for preparatory work is used for the input of control point coordinates, input of camera calibration data, and the selection of the needed image files. The aerotriangulation software is used for automatically locating fiducial marks (interior orientation), removing bad parallax (relative orientation), scaling and leveling the images (absolute orientation), and creating a control mesh. The data transfer programs store and check all data in digital form in a MicroStation file for use by designers. The fourth area is that of file storage and plotting.

Aerial Photography

The most common technology used to obtain aerial photography is the film camera. Modern cameras are very accurate and have excellent resolving power. They function as older cameras did but offer options with major improvements. Every camera is equipped with automatic exposure control, thus eliminating the need for the operator to estimate the proper exposure since the camera adjusts the aperture and shutter speed to achieve the desired exposure. Another advance is the incorporation of forward-motion compensation (FMC), a technique that moves the film at the same apparent speed as the image while the shutter is open. The effect of FMC is to maintain the image at the same location on the film during the exposure, thus eliminating elongation of linear features, such as power lines, striping, and edges of buildings, that are perpendicular to the flight line. FMC also increases accuracy by eliminating image motion. It also facilitates higher aircraft speeds at lower altitudes, which results in greater productivity.

Digital cameras are also used for large-area base mapping at smaller scales. The two dominant technologies in digital cameras are "push broom" scanners and "framing" cameras. Push-broom scanners operate in a manner similar to document scanners by employing a series of linear sensors that continuously collect image data during the sensor operation. The advantage of push-broom scanners is that the sensor heads are oriented at different angles relative to the nadir and can acquire continuous stereo coverage of the image area. Framing digital cameras operate as do film cameras, with the film and lens cone replaced by a series of lens and charge-coupled device (CCD) arrays. The framing camera acquires an image and then overlaps the next image, as would a film camera, to achieve stereo coverage. The advantage of using framing cameras is that accuracy is somewhat better than that of push-broom sensors. Another consideration is that framing cameras can be used at lower flying altitudes. The most significant advantage of using digital cameras rather than film cameras is that the digital camera can acquire images in multiple spectrums (red, blue, green, infrared, and panchromatic) simultaneously, which enables the image analysis to mix the different spectra to produce the desired image.

Airborne GPS

The technology that has been adapted for both film and digital aerial cameras is Airborne GPS (ABGPS) with the kinematic positioning of the airborne sensor and the derivation of the position of the sensor at the time of exposure. These positions are then used in aerial triangulation. Kinematic positioning is achieved when the precise coordinates of the trajectory are derived by post-processing the raw GPS data collected by a GPS receiver coupled to the airborne sensor in contrast to using a fixed receiver receiving the same GPS signal. This post-processing can produce position accuracies of 10 cm or less. An event handler interpolates the photo centers from the sensor trajectory. The advantage of using ABGPS is the reduction in the number of ground control points needed. By having fewer control points, the cost of the field-work is reduced.

LiDAR Scanning

As technology advances and computers become more powerful, so has the technology of highway location. Methods for ground surveys have advanced from the use of labor intensive techniques using transits, levels, and traversing points to information-based technologies that have almost fully automated the process. Now ground survey methods, as described earlier, involve total stations, GPS, and robotics. Another technology is LiDAR (Light Detection and Ranging). There are two forms of LiDAR: one used for airborne applications and the other for terrestrial work.

Airborne LiDAR.   Airborne LiDAR is performed from an aircraft, as illustrated in Figure 14.13. The instrument is attached to a fixed-wing or rotary-wing aircraft and the scanner emits a light laser beam that is reflected from a mirror and directed at the surface of the earth. As the light is transmitted from the surface, the time and reflectance are measured and an elevation is calculated. Surface elevation in airborne LiDAR is calculated through a compilation of different instruments. An inertial measurement unit (IMU) aboard the aircraft calculates the pitch, roll, and heading, while GPS locks onto satellites and registers the platform's spatial position. The accuracy of this technology is within 0.15 m for elevation and 1/1000th of the flight height for horizontal position.

Terrestrial LiDAR.   The terrestrial LiDAR scanner, illustrated in Figure 14.14a, is a specific, localized form of LiDAR. The scanner is mounted on a tripod or stationary vehicle. A light pulse is transmitted, and the time it takes to return is measured, thus providing the software a position on the earth. Using a series of internal mirrors and time-of-flight laser technology, the surveyor can create a point cloud with millions of points. Each point has a position and elevation based on a user-defined coordinate system. The combination of these elements reduces the need for detailed sketches at a project site. This technology expands the options available for survey work; for example, by enabling access to dangerous areas or other locations that are time consuming to measure or

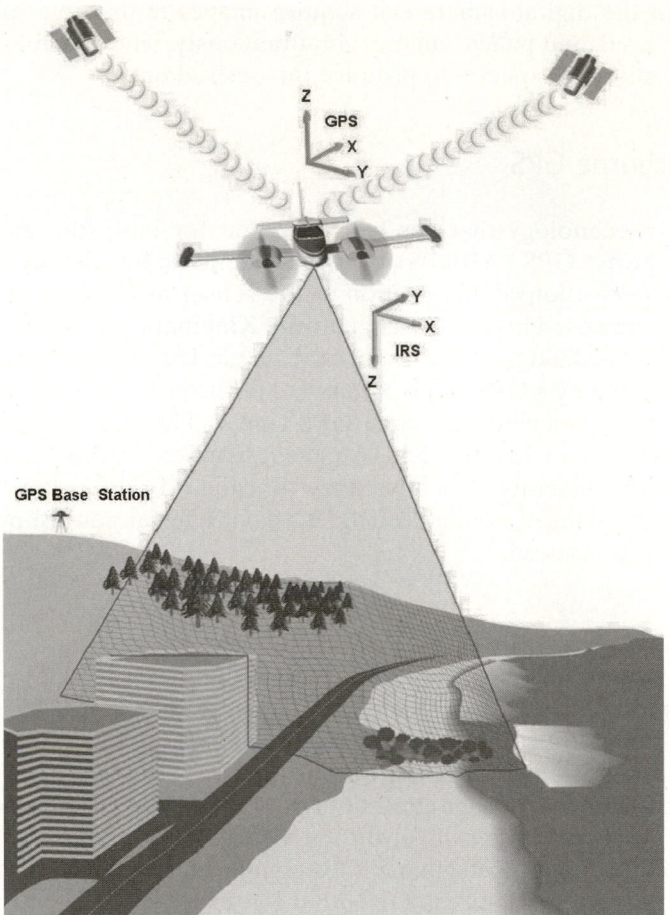

Figure 14.13   Airborne GPS/LiDAR Operation from an Aircraft

SOURCE: Virginia Department of Transportation

where working conditions are hazardous. The laser scanner also lessens the need to revisit a previously worked survey site.

Scanner Advantages.   The scanner is most useful when confidence in the data is important, there are time constraints on a project, fast turn-around is needed, or when an area is difficult or dangerous to access and/or measure. A single setup of the system can cover more area than traditional survey methods that may require many setups, thus reducing field time. The system's ability to capture large amounts of information and gather information from all surrounding elements in a single setup enables the surveyor to secure measurements from the office rather than spending time in the field. Projects such as scanning busy intersections, bridges, or delicate surfaces can capture as-built geometry and associated infrastructure quickly, safely, and accurately. These tasks can be accomplished without disruption to passing traffic or placing individuals in harm's way. A variety of power sources and target choices make scanning versatile and easy.

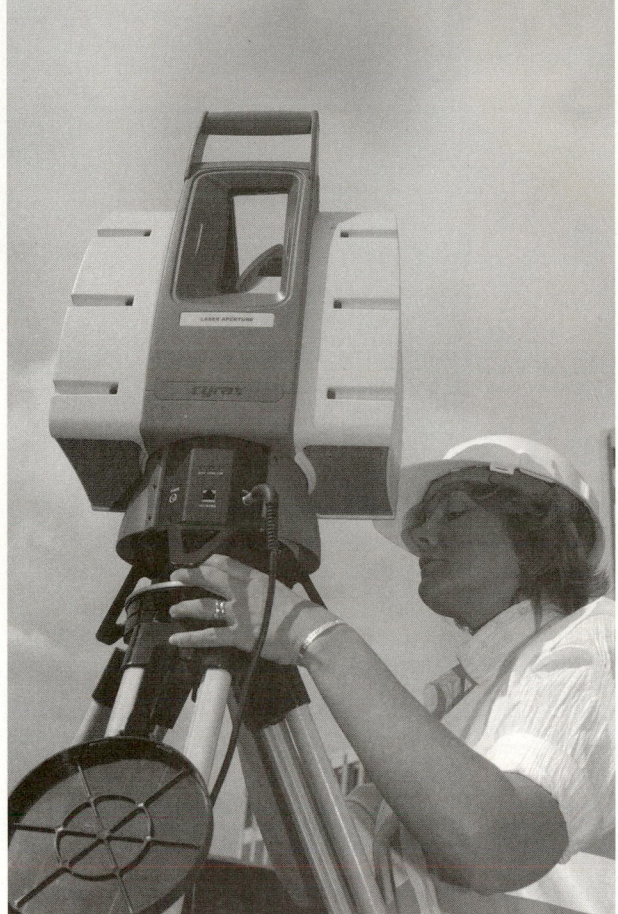

Figure 14.14a   Leica HDS3000 High-Density Scanner

SOURCE: Photograph by Tom Saunders, Virginia Department of Transportation

Terrestrial Mobile LiDAR Scanning (TMLS).   A recent advance in scanning technology collects data in mobile units as illustrated in Figure 14.14b. TMLS acquires data as the vehicle is in motion and can produce 3D images of the terrain. Among the benefits claimed for this device are: safety for roadway surveyors, less traffic disruption, reduction in time devoted to fieldwork, and improved accuracy.

Scanner Software.   Most software programs for scanner technology are user friendly and support easy manipulation of billions of points. Control of the size of point clouds is achieved, and they easily can be converted to conventional MicroStation and AutoCAD files. Software capabilities include referencing to a coordinate system notation on cloud point files, fly-through with clearance measurement, and simple elimination of noise. The software also has the ability to create two-dimensional alignments from three-dimensional scans, complex wire frame, TIN models, and build contours. All of these can be more easily used in design projects without requiring additional person-hours to secure contours in the field. The scan technology provides visualization capabilities,

Figure 14.14b   A Terrestrial Mobile LiDAR Scanning Vehicle

SOURCE: Mike1024

immediate measurements, two-dimensional maps and drawings, cross sectioning, and three-dimension modeling.

Geographic Information Systems

Geographic Information Systems (GIS) capture, store, analyze, and manage data and associated attributes that are spatially referenced to the earth. GIS is a computer system accessing a database of layers of information that is capable of integrating, storing, editing, analyzing, sharing, and displaying geographically referenced information. GIS is a tool that allows users to create interactive queries (user-created searches), analyze the spatial information, edit data, create maps, and present the results of all these operations. GIS software technology is powerful enough to assimilate data in many types of formats (i.e., raster, vector, geocoded, tabular, etc.), provided the information is in a digital format and can be scaled, translated, and "fitted" into the same projected system of the working representation or map. Figure 14.15 illustrates the layered data concept of a GIS and the importance of each layer being in the same coordinate system and projection for spatial integrity.

  Geographic information system technology can be used for scientific investigations, resource management, asset management, environmental impact assessment, urban

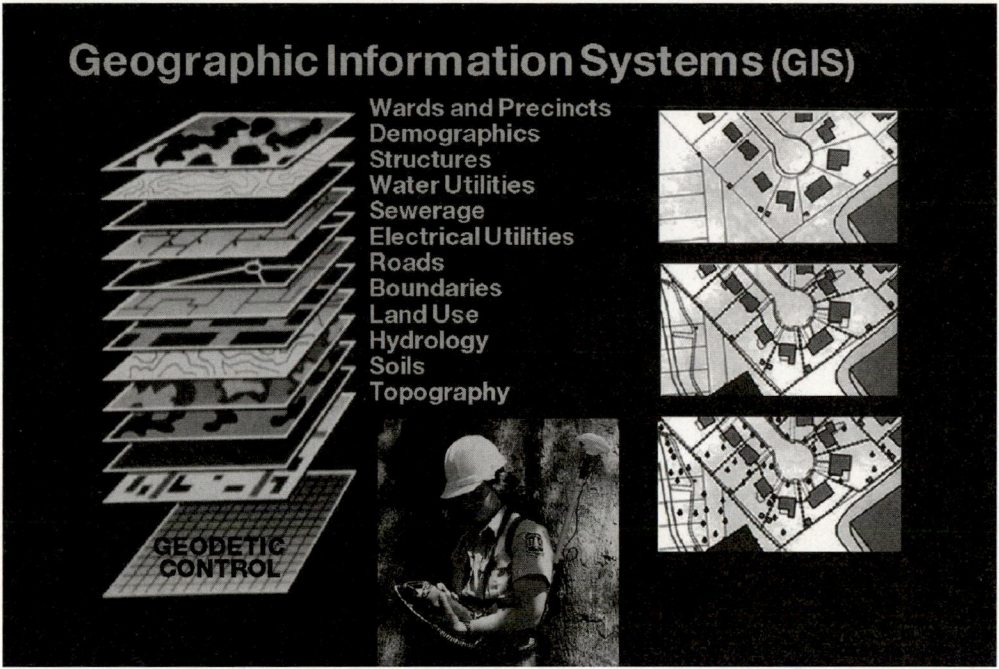

Figure 14.15   Data Layers Found in a Typical GIS

SOURCE: Courtesy of U.S. National Oceanic & Atmospheric Administration (NOAA)

planning, cartography, criminology, history, sales, marketing, and route planning. For example, a GIS might allow planners to calculate emergency response times or evacuation routes in the event of a natural disaster or to identify locations of subsurface utilities in proximity to a highway project. GIS technology can be used to determine which cultural or environmental resources may be impacted by a proposed highway corridor alignment and aid in developing alternative alignments. It is a planning tool that can serve as an excellent aid to engineers by presenting a project's strengths and weaknesses in a wide variety of formats to suit any audience.

14.3 HIGHWAY EARTHWORK AND FINAL PLANS

The final element in the location process is to establish the horizontal and vertical alignments of the highway project and to prepare highway plans and specifications for estimating project costs and preparation of bids by contractors. The following sections explain how the terrain influences the cost to transport earthen materials that will be used to construct the roadbed and how to estimate payment. The final result of the location process is a highway plan used in estimating quantities and computing the overall project cost.

14.3.1 Highway Grades and Terrain

One factor that significantly influences the selection of a highway location is the terrain of the land, which in turn affects the laying of the grade line. The primary factor that the designer considers on laying the grade line is the amount of earthwork that will be necessary for the selected grade line. One method to reduce the amount of earthwork is to set the grade line as closely as possible to the natural ground level. This is not always possible, especially in undulating or hilly terrain. The least overall cost also may be obtained if the grade line is set such that there is a balance between the excavated volume and the volume of embankment. Another factor that should be considered in laying the grade line is the existence of fixed points, such as railway crossings, intersections with other highways, and in some cases, existing bridges, which require that the grade be set to meet them. When the route traverses flat or swampy areas, the grade line must be set high enough above the water level to facilitate proper drainage and to provide adequate cover to the natural soil.

The height of the grade line is usually dictated by the expected floodwater level. Grade lines should also be set such that the minimum sight distance requirements, as discussed in Chapter 3, are obtained. The criteria for selecting maximum and minimum grade lines are presented in Chapter 15. In addition to these guidelines, the amount of earthwork associated with any grade line influences the decision on whether the grade line should be accepted or rejected. The following sections describe how a highway grade is established that minimizes earth moving and maximizes the use of native soil.

Computing Earthwork Volumes

One of the major objectives in selecting a particular location for a highway is to minimize the amount of earthwork required for the project. Therefore, the estimation of the amount of earthwork involved for each alternative location is required at both the preliminary and final stages.

To determine the amount of earthwork involved for a given grade line, cross sections are taken at regular intervals along the grade line. The cross sections are usually spaced 15 m apart, although this distance is sometimes increased for preliminary engineering. These cross sections are obtained by plotting the natural ground levels and proposed grade profile of the highway along a line perpendicular to the grade line to indicate areas of excavation and areas of fill. Figure 14.16 shows three types of cross sections. When the computation is done manually, the cross sections are plotted on standard cross-section paper, usually to a scale of 1 cm to 12 m for both the horizontal and vertical directions. The areas of cuts and fills at each cross section are then determined by the use of a planimeter or by any other suitable method. Surveying books document the different methods for area computation. The volume of earthwork is then computed from the cross-sectional areas and the distances between the cross sections.

A common method of determining the volume is that of average end areas. This procedure is based on the assumption that the volume between two consecutive cross sections is the average of their areas multiplied by the distance between them, computed as follows.

$$V = \frac{L}{2}(A_1 + A_2)$$ (14.7)

where

V = volume (m³)
A_1 and A_2 = end areas (m²)
L = distance between cross sections (m)

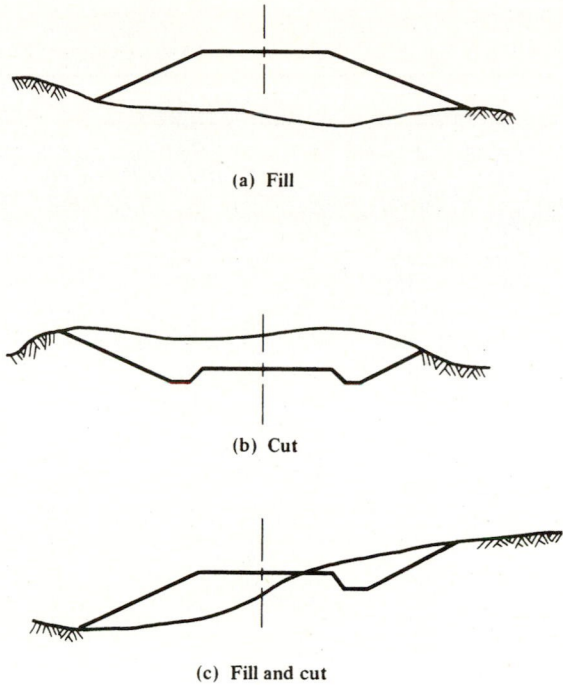

(a) Fill

(b) Cut

(c) Fill and cut

Figure 14.16 Types of Cross Sections

The average end-area method has been found to be sufficiently accurate for most earthwork computations, since cross sections are taken 15 to 30 m apart, and minor irregularities tend to cancel each other. When greater accuracy is required, such as in situations where the grade line moves from a cut to a fill section, the volume may be considered as a pyramid or other geometric shape.

It is common practice in earthwork construction to move suitable materials from cut sections to fill sections to reduce to a minimum the amount of material borrowed from borrow pits. When the materials excavated from cut sections are compacted at the fill sections, they fill less volume than was originally occupied. This phenomenon is referred to as *shrinkage* and should be accounted for when excavated material is to be reused as fill material. The amount of shrinkage depends on the type of material. Shrinkages of up to 50 percent have been observed for some soils. However, shrinkage factors used are generally between 1.10 and 1.25 for high fills and between 1.20 and 1.25 for low fills. These factors are applied to the fill volume in order to determine the required quantity of fill material.

Example 14.4 Computing Fill and Cut Volumes Using the Average End-Area Method

A roadway section is 600 m long (20 stations). The cut and fill volumes are to be computed between each station. Table 14.1 lists the station numbers (column 1) and lists the end area values (m²) between each station that are in cut (column 2) and that are in fill (column 3). Material in a fill section will consolidate (known as shrinkage), and for this road section is 10 percent. [For example, if 100 m³ of net fill is required, the

Table 14.1　Computation of Fill and Cut Volumes and Mass Diagram Ordinate

	End Area (m²)			Volume (m³)				Net Volume (4 to 7)		
1	2	3	4	5	6	7	8	9	10
					Shrinkage	Total			Mass
			Total		10	Fill	Fill	Cut	Diagram
Station	Cut	Fill	Cut	Fill	percent	(5 + 6)	(−)	(+)	Ordinate
0	3	18	75	1020	102	1122	1047	–	0
1	2	50	60	2205	221	2426	2366	–	−1047
2	2	97	90	3405	341	3746	3656	–	−3413
3	4	130	180	2715	272	2987	2807	–	−7069
4	8	51	720	1485	149	1634	914	–	−9876
5	40	45	1275	975	98	1073	–	202	−10790
6	45	20	1875	375	38	413	–	1462	−10588
7	80	5	3030	105	11	116	–	2914	−9126
8	122	2	3780	30	3	33	–	3747	−6212
9	130	0	4050	0	0	0	–	4050	−2465
10	140	0	3600	45	5	50	–	3550	1585
11	100	3	2700	495	50	545	–	2155	5135
12	80	30	2325	750	75	825	–	1500	7290
13	75	20	1875	1050	105	1155	–	720	8790
14	50	50	1050	1950	195	2145	1095	–	9510
15	20	80	450	2700	270	2970	2520	–	8415
16	10	100	150	3300	330	3630	3480	–	5895
17	0	120	45	3600	360	3960	3915	–	2415
18	3	120	645	2550	255	2805	2160	–	−1500
19	40	50	1050	1200	120	1320	270	–	−3660
20	30	30	–	–	–	–	–	–	−3930

total amount of fill material that is supplied by a cut section is 100 + (0.10 × 100) = 100 + 10 = 110 m³.]

Determine the net volume of cut and fill that is required between station 0 and station 1.

Solution:

$$V_{cut} = \frac{30(A_{0C} + A_{1C})}{2} = \frac{30(3 + 2)}{2} = 75 \text{ m}^3$$

$$V_{fill} = \frac{30(A_{0F} + A_{1F})}{2} = \frac{30(18 + 50)}{2} = 1020 \text{ m}^3$$

$$\text{Shrinkage} = 1020(0.010) = 102 \text{ m}^3$$

$$\text{Total fill volume} = 1020 + 102 = 1122 \text{ m}^3$$

The cut and fill volume between station 0 + 00 and 1 + 00 is shown in columns 4 and 7.

Cut: 75 m³ (column 4)

Fill: 1020 m³ (column 5)

Shrinkage: 102 m³ (column 6)

Total fill required: 1122 m³ (column 7)

$$\text{Net volume between stations } 0\text{--}1 = \text{total cut} - \text{total fill} = 75 - 1122$$

$$= -1047 \text{ m}^3 \text{ (column 8)}$$

Note: Net fill volumes are negative (−) (column 8) and net cut volumes are positive (+) (column 9).

Similar calculations are performed between all other stations, from station 1 + 00 to 20 + 00, to obtain the remaining cut or fill values shown in columns 2 through 9.

Computing Ordinates of the Mass Diagram

The mass diagram is a series of connected lines that depicts the *net* accumulation of cut or fill between any two stations. The ordinate of the mass diagram is the net accumulation in cubic yards (yd³) from an arbitrary starting point. Thus, the difference in ordinates between any two stations represents the net accumulation of cut or fill between these stations. If the first station of the roadway is considered to be the starting point, then the net accumulation at this station is zero.

Example 14.5 Computing Mass Diagram Ordinates

Use the data obtained in Example 14.4 to determine the net accumulation of cut or fill beginning with station 0 + 00. Plot the results.

Solution: Columns 8 and 9 show the net cut and fill between each station. To compute the mass diagram ordinate between station X and $X + 1$, add the *net* accumulation from Station X (the first station) to the net cut or fill volume (columns 8 or 9) between stations X and $X + 1$. Enter this value in column 10.

Station 0 + 00 mass diagram ordinate = 0

Station 1 + 00 mass diagram ordinate = 0 − 1047 = −1047 m³

Station 2 + 00 mass diagram ordinate = −1047 − 2366 = −3413 m³

Station 3 + 00 mass diagram ordinate = −3413 − 3656 = −7069 m³

Station 4 + 00 mass diagram ordinate = −7069 − 2807 = −9876 m³

Station 5 + 00 mass diagram ordinate = −9876 − 914 = −10790 m³

Station 6 + 00 mass diagram ordinate = −10790 + 202 = −10588 m³

Station 7 + 00 mass diagram ordinate = −10588 + 1462 = −9126 m³

Continue the calculation process for the remaining 13 stations to obtain the values shown in column 10 of Table 14.1. A plot of the results is shown in Figure 14.17.

Interpretation of the Mass Diagram

Inspection of Figure 14.17 and Table 14.1 reveals the following characteristics.

1. When the mass diagram slopes downward (negative), the preceding section is in fill, and when the slope is upward (positive), the preceding section is in cut.
2. The difference in mass diagram ordinates between any two stations represents the *net accumulation* between the two stations (cut or fill). For example, the net accumulation between station 6 + 00 and 12 + 00 is 10588+7290=17878 m³.
3. A horizontal line on the mass diagram defines the locations where the net accumulation between these two points is zero. These are referred to as "balance points," because there is a balance in cut and fill volumes between these points. In Figure 14.15, the *x* axis represents a balance between points *A'* and *D'* and a balance between points *D'* and *E'*. Beyond point *E'*, the mass diagram indicates a fill condition for which there is no compensating cut. The maximum value is the ordinate at station 20 + 00 of −3930 m³. For this section, imported material (called borrow) will have to be purchased and transported from an off-site location.
4. Other horizontal lines can be drawn connecting portions of the mass diagram. For example, lines *J–K* and *S–T*, which are each five stations long, depict a balance of cut and fill between stations at points *J* and *K* and *S* and *T*.

Computing Overhaul Payments

Contractors are compensated for the cost of earthmoving in the following manner. Typically, the contract price will include a stipulated maximum distance that earth will be moved without the client incurring additional charges. If this distance is exceeded, then the contract stipulates a unit price add-on quoted in additional station m³ of material moved. The maximum distance for which there is no charge is called free haul. The extra distance is called overhaul.

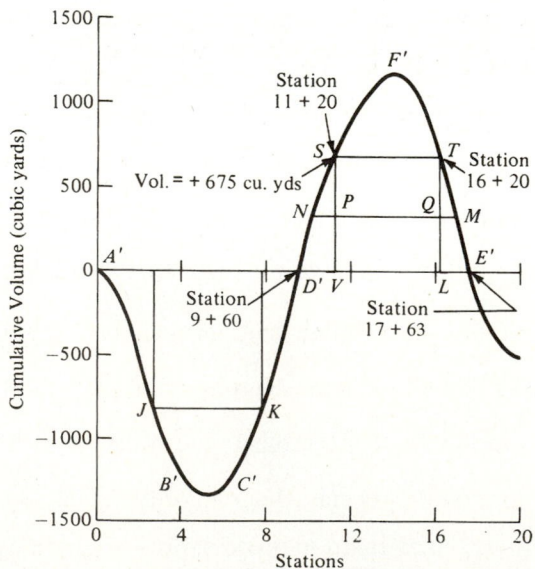

Figure 14.17   Mass Diagram for Computation Shown in Table 14.1

Example 14.6 Computing Balance Point Stations

Compute the value of balance point stations for the mass diagram in Figure 14.17 for the following situations:

- **(a)** The *x*-axis
- **(b)** The horizontal distance *S–T,* which measures 150 m

Solution:

- **(a)** Balance points are computed by interpolation using the even stations where the ordinates change from cut to fill (or vice versa).

  Balance point D' occurs between Station $9 + 00$ and $10 + 00$ (since ordinate values are -2465 and $+1585$).

  Assuming that the mass diagram ordinate changes linearly between stations, by similar triangles, we can write

  Station of the Balance Point $D' = (9 + 00) + [2465/(2465 + 1585)](30) = 9 + 18$

Similarly,

  Station of the Balance Point $E' = (17 + 00) + [2415/(24157 + 1500)](30) = 17 + 19$

- **(b)** To determine the balance point stations for line *S–T,* it is necessary to draw the mass diagram to a larger scale than depicted in the textbook, and to read the station for one of the points directly from the diagram. Using this technique, station $11 + 6$ was measured for point *S,* and from this value the station for point *T* is computed as

$$(11 + 6) + (5 + 00) = \text{Station } 16 + 6$$

Example 14.7 Computing Overhaul Payment

The free-haul distance in a highway construction contract is 150 m and the overhaul price is $11/m³ station. For the mass diagram shown in Figure 14.17, determine the extra compensation that must be paid to a contractor to balance the cut and fill between station $9 + 18$ (*D*) and station $17 + 19$ (*E*).

Solution:

Step 1. Determine the number of cubic yards of overhaul.

The overhaul volume will occur between stations $9 + 18$ and $11 + 6$, and between stations $16 + 6$ and $17 + 19$. The overhaul value is obtained by interpolation between stations $11 + 00$ and $12 + 00$ or by reading the value from the mass diagram.

By interpolation, the value is

Overhaul = Ordinate at station 11 + (difference in ordinates at 12 and 11)(6/30)

$$= 5135 + (7290 - 5135)(0.20) = 5135 + 431 = 5566 \text{ m}^3$$

This overhaul value should equal the value at station 16 + 6. By interpolation, the value is

$$5895 - (5895 - 2415)(0.2) = 5199 \text{ m}^3$$

Since the values are not equal, use the average (5383 m³) or measure the overhaul from a larger scale diagram to obtain a value of 5400 m³. This value is selected for the calculation of contractor compensation.

Step 2.   Determine the overhaul distance.

The method of moments is used to compute the weighted average of the overhaul distances from the balance line to the station where free haul begins.

Beginning with stations 9 + 18 to 10 + 00, the volume moved is 1585 m³, and the average distance to the free-haul station (11 + 6) is (10 + 00 − 9 + 18)/2 + 30 + 6 = 42 m.

From stations 10 + 00 to 11 + 00, the volume moved is (5135 − 1585) = 3550 m³, and the distance moved to the free-haul line is (11 + 00 − 10 + 00)/2 + 6 = 36 m.

From station 11 + 00 to station 11 + 6, the volume moved is 5400 − 5135 = 265 m³, and the average distance is 3 m.

Overhaul distance moved between station 9 + 18 and 11 + 6 is

$$[(1585)(942) + (3550)(21) + (265)(3)]/5400 = 145538/5400 = 26.95 \text{ m}$$

Similarly, compute the overhaul distance between the balance point at station 17 + 19 and the beginning of free haul at station 16 + 6. Beginning with stations 17 + 19 to 17 + 00, the volume moved is 2415 m³, and the average distance to the free-haul station (16 + 6) is (917 + 19 − 17 + 00)/2 + (17 + 00 − 16 + 6) = 33.5 m.

From stations 17 + 00 to 16 + 6, the volume moved is (5400 − 2415) = 2985 m², and the distance moved to the free-haul line is (17 + 00 − 16 + 6)/2 = 12 m.

Overhaul average distance moved between station 16 + 6 and 17 + 19 is

$$[(2415)(33.5) + (2985)(12)]/5400 = 116722.5/5400 = 21.60$$

Total overhaul distance = 26.95 + 21.62 = 48.57 m

Step 3.   Compute overhaul cost due to the contractor.

Overhaul cost = contract price ($/m³ station) × overhaul (m³) × stations

$$= 14 \times 5400 \times (26.25/30 + 21.62/30) = \$122,396$$

Computer programs are now available that can be used to compute cross-sectional areas and volumes directly from the elevations given at cross sections. Some programs will also compute the ordinate values for a mass diagram and determine the overhaul, if necessary.

14.3.2 Preparation of Highway Plans

An example of a highway plan and the proposed vertical alignment are illustrated in Figure 14.18. The solid line is the vertical projection of the centerline of the road profile. The dotted lines represent points along the terrain a distance of 17 m from the centerline. The circles and triangles are points along the terrain that are 26 m from the centerline of the road. This information can be used to plot cross sections that depict the shape of the roadway when completed. The final grade line is adjusted until the amount of excess cut or fill has been minimized. If there is an excess of cut material, then it must be removed and stored at another location. If there is an excess of fill, then material must be purchased and delivered to the site. Thus, an ideal situation occurs when there is a balance between the amount of cut and fill.

Once the final location of the highway system is determined, it is then necessary to provide the plans and specifications for the facility. The plans and specifications of a highway are the instructions under which the highway is constructed. They are also used for the preparation of engineers' estimates and contractors' bids. When a contract is let out for the construction of a highway, the plans and specifications are part of the contract documents and are therefore considered legal documents. The plans are drawings that contain all details necessary for proper construction, whereas the specifications give written instructions on quality and type of materials. Figure 14.18a shows an example of a highway plan and the horizontal alignment. Figure 14.18b shows the highway plan and the vertical alignment. The latter view is sometimes referred to as the *profile,* indicating the natural ground surface and the centerline of the road with details of vertical curves. The horizontal alignment is usually drawn to a scale of 1 cm to 12 m, although in some cases the scale of 1 cm to 6 m is used to provide greater detail. In drawing the vertical alignment, the horizontal scale used is the same as that of the horizontal alignment, but the vertical scale is exaggerated 5 to 10 times. The vertical alignment may also give estimated earthwork quantities at regular intervals, usually at 30 m stations.

Most state agencies require consultants to prepare final design drawings on standard sheets 90 × 55 cm. These drawings are then usually reduced to facilitate easy handling in the field during construction. Other drawings showing typical cross sections and specific features such as pipe culverts and concrete box culverts are also provided. Standard drawings of some of these features that occur frequently in highway construction have been provided by some states and can be obtained directly from the highway agencies. Consultants may not have to produce them as part of their scope of work.

14.4 SUMMARY

The selection of a suitable location for a new highway requires information obtained from highway surveys. These surveys can be carried out by either conventional ground methods or use of electronic equipment and computers. A brief description of some of the more commonly used methods of surveys has been presented to introduce the reader to these techniques.

A detailed discussion of the four phases of the highway location process has been presented to provide the reader with the information required and the tasks involved in selecting the location of a highway. The computation of earthwork volumes is also presented, since the amount of earthwork required for any particular location may significantly influence the decision to either reject or select that location. Note, however, that the final selection of a highway location, particularly in an urban area, is not now purely in the hands of the engineers. The reason is that citizen groups with interest in

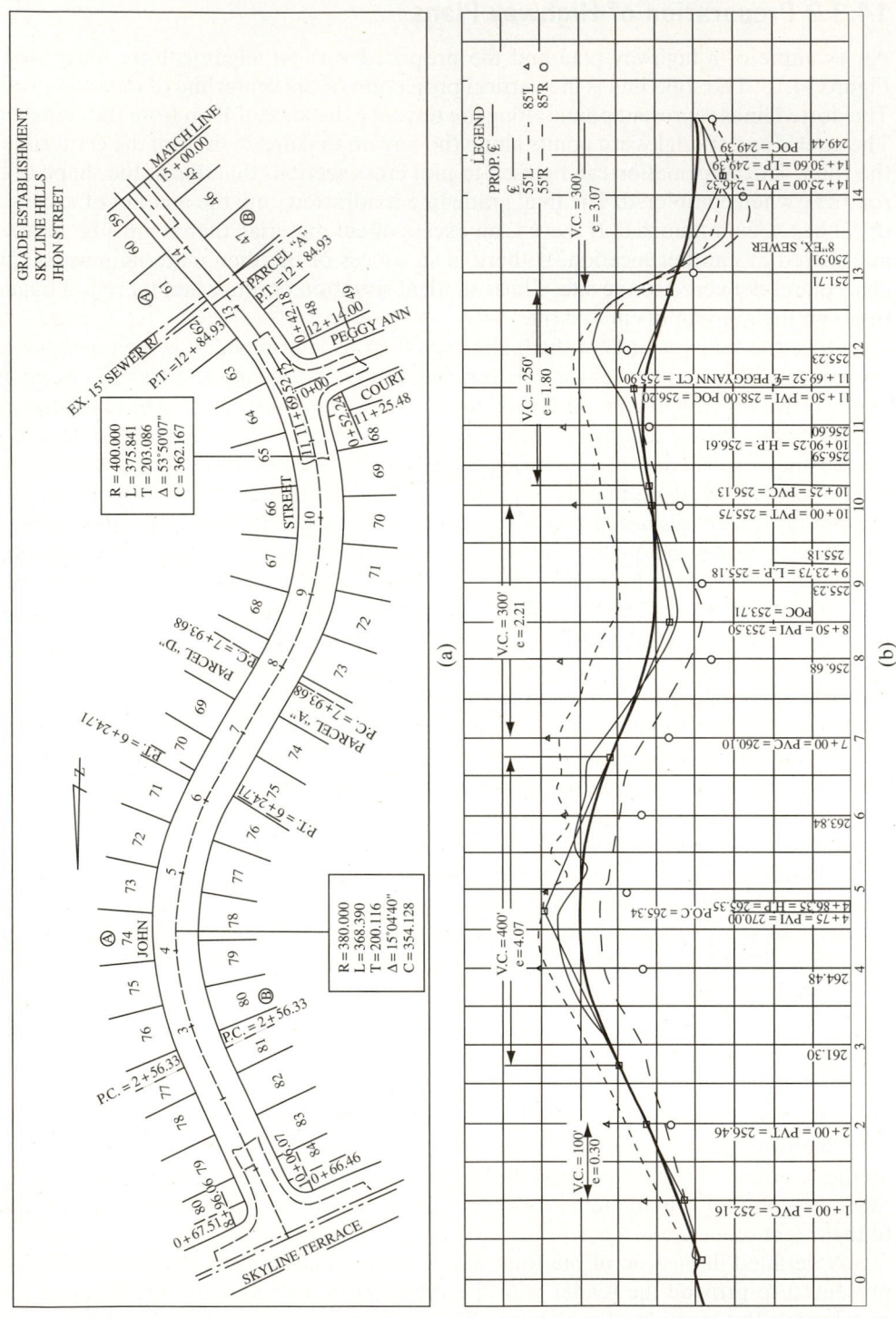

Figure 14.18 Highway Plan (a) Horizontal Alignment; (b) Vertical Alignment

SOURCE: Courtesy of Byrd, Tallamy, McDonald and Lewis, Fairfax, VA

the environment and historical preservation can be extremely vocal in opposing highway locations that, in their opinion, conflict with their objectives. Thus, in selecting a highway location, the engineer must take into consideration the environmental impact of the road on its surroundings. For many types of projects, federal regulations also require environmental impact statements, which require detailed analyses.

PROBLEMS

14-1   Describe the three categories of information gathered, in the office prior to any field survey activity, about the characteristics of the area of a proposed highway location.

14-2   Briefly discuss factors that are of specific importance in the location of scenic routes.

14-3   Describe the factors that significantly influence the location of highways in urban areas.

14-4   What are the three elements that highway surveys usually involve?

14-5   Briefly describe the use of each of the following instruments in conventional ground surveys:

(a) total station
(b) level
(c) measuring tapes
(d) electronic distance measuring devices

14-6   Briefly compare the factors that should be considered in locating an urban freeway with those for a rural highway.

14-7   Describe how each of the following could be used in highway survey location:

(a) aerial photogrammetry
(b) computer graphics
(c) conventional survey techniques

14-8   A photograph is to be obtained at a scale of 1:12,000 by aerial photogrammetry. If the focal length of the camera to be used is 16 cm, determine the height at which the aircraft should be flown if the average elevation of the terrain is 285 m.

14-9   The distance in the x direction between two control points on a vertical aerial photograph is 11.5 cm. If the distance between these same two points is 9.0 cm on another photograph having a scale of 1:24,000, determine the scale of the first vertical aerial photograph. If the focal length of the camera is 15.0 cm and the average elevation at these points is 30 m, determine the flying height from which each photograph was taken.

14-10   The distance in the x direction between two control points on a vertical aerial photograph is 15.4 cm. If the distance between these same two points is 11.8 cm on another photograph having a scale of 1:24,000, determine the scale of the first vertical aerial photograph. If the focal length of the camera is 15 cm and the average elevation at these points is 30 m, determine the flying height from which each photograph was taken.

14-11   The scale at the image of a well-defined object on an aerial photograph is 1:24,000, and the elevation of the object is 450 m. The focal length of the camera lens is 16.5 cm. If the air base (B) is 75 m, determine the elevation of the two points A and C and the distance between them if the coordinates of A and C are as given below.

	A	C
x coordinate	14 cm	16.5 cm
y coordinate	8.9 cm	12.5 cm
Scale of photograph	1:13,000	1:17,400

14-12   A vertical photo has an air base of 720 m. Stereoscopic measurements of parallax at a point representing the top of a 60 m tower is 0.706 cm. The camera focal length is 16.5 cm. Photos were taken at an elevation of 2250 m. Determine the elevation of the *base* of the tower.

14-13   The length of a runway at a national airport is 2250 m long and at elevation 450 m above sea level. The airport was recently expanded to include another runway used primarily for corporate aircraft. It is desired to determine the length of this runway, whose elevation is 540 m. An aerial photograph was taken of the airport. Measurements on the photograph for the national airport runway are 12 cm and for the corporate runway are 8.5 cm. The camera focal length is 15 cm. Determine the length of the corporate runway.

14-14   Using an appropriate diagram, discuss the importance of the side and forward overlaps in aerial photography.

14-15   Under what conditions would the borrowing of new material from a borrow pit for a highway embankment be preferred over using material excavated from an adjacent section of the road?

14-16   Using the data given in Table 14.1, determine the total overhaul cost if the free haul is 213 m and the overhaul cost is $7.50 m^3 station. Stations of the free-haul lines are 1 + 80 and 8 + 80 and 10 + 20 and 17 + 20.

14-17   The following table shows the stations and ordinates for a mass diagram. The free-haul distance is 180 m. Overhaul cost is $ 15.

Station	Ordinate (m^3)
0 + 00	0
1 + 00	50
2 + 00	75
2 + 20	95
4 + 00	135
6 + 00	150
7 + 00	125
8 + 20	95
9 + 00	85
10 + 00	55
10 + 30	0

(a) Use the method of moments to compute the additional cost that must be paid to the contractor.

(b) Sketch the ground profile if the finished grade of this roadway section is level (0 percent).

REFERENCES

A Policy on Geometric Design of Highways and Streets. American Association of State Highway and Transportation Officials, Washington, D.C., 2004.

Anderson, James M. and Edward M. Mikhail, *Surveying: Theory and Practice.* McGraw-Hill, Inc., 1998.

Brinker, R. C. (editor), *The Surveying Handbook.* Chapman and Hall, December 1995.

Glennie, Craig, *A Kinematic Terrestrial LiDAR Scanning System,* presented at the Transportation Research Board Annual Meeting, 2009.

Journal of Surveying Engineering, American Society of Civil Engineering, March 1983–2012, www.asce.org

Mikhail, Edward M., James S. Bethel, and J. Chris McGlone, *Introduction to Modern Photogrammetry.* John Wiley and Sons, 2001.

Van Sickle, Jan, *GPS for Land Surveyors.* Taylor and Francis, New York, 2001.

Wolf, Paul R. and Bon A. DeWitt, *Elements of Photogrammetry with Applications in GIS,* 3rd ed., McGraw-Hill, Inc., 2000.

CHAPTER 15

Geometric Design of Highway Facilities

Geometric design deals with the dimensioning of the elements of highways, such as vertical and horizontal curves, cross sections, truck climbing lanes, bicycle paths, and parking facilities. The characteristics of driver, pedestrian, vehicle, and road, as discussed in Chapter 3, serve as the basis for determining the physical dimensions of these elements. For example, lengths of vertical curves or radii of circular curves are determined to assure that the minimum stopping sight distance is provided to highway users for the design speed of the highway. The fundamental objective of geometric design is to produce a smooth-flowing and safe highway facility, an objective that only can be achieved by providing a consistent design standard that satisfies the characteristics of the driver and the vehicles that use the road.

The American Association of State Highway and Transportation Officials (AASHTO) serves a critical function in developing guidelines and standards used in highway geometric design. The membership of AASHTO includes representatives from every state highway and transportation department in the United States as well as the Federal Highway Administration (FHWA). The association has several technical committees that consider suggested design practices from individual states. When approved, they are incorporated into the AASHTO publication, *A Policy on Geometric Design of Highways and Streets*, which provides a policy or set of guidelines for geometric design of highways. Ultimately, it is the decision of individual state transportation departments to adopt the AASHTO publication guidelines in whole or in part regarding geometric design policy. The most recent edition (2011) does not suggest that existing streets and highways are unsafe if the design was based on prior editions. Also, the latest edition does not represent a mandate for the initiation of improvement projects.

This chapter presents the principles and theories used in the design of horizontal and vertical alignments are the current standards used for geometric design, as recommended by AASHTO. Design principles and standards are also linked to the characteristics of the driver, pedestrian, vehicle, and road, as discussed in Chapter 3.

CHAPTER OBJECTIVES:

- Understand the factors influencing highway design.
- Calculate the elements necessary to design crest or sag vertical curve alignments.
- Calculate the elements necessary to design simple, compound, and reverse horizontal curves.
- Calculate transition spiral curve and superelevation runoff lengths.
- Apply the equations for stopping distance to compute curve radius.
- Apply geometric design principles to climbing lanes, emergency ramps, parking lots, and bicycle facilities.

15.1 FACTORS INFLUENCING HIGHWAY DESIGN

Highway design is based on specified design standards and controls that depend on the following roadway system factors:

- Functional classification
- Design hourly traffic volume and vehicle mix
- Design speed
- Design vehicle
- Cross section of the highway, such as lanes, shoulders, and medians
- Presence of heavy vehicles on steep grades
- Topography of the area that the highway traverses
- Level of service
- Available funds
- Safety
- Social and environmental factors

These factors are often interrelated. For example, design speed depends on functional classification that is usually related to expected traffic volume. The design speed may also depend on the topography, particularly in cases where limited funds are available. In most instances, the principal factors used to determine the standards to which a particular highway will be designed are the level of service to be provided, expected traffic volume, design speed, and the design vehicle. These factors, coupled with the basic characteristics of the driver, vehicle, and road, are used to determine standards for the geometric characteristics of the highway, such as cross sections and horizontal and vertical alignments. For example, appropriate geometric standards should be selected to maintain a desired level of service for a known proportional distribution of different types of vehicles.

15.1.1 Highway Functional Classification

Highways are classified according to their functions in terms of the service they provide. The classification system facilitates a systematic development of highways and the logical assignment of highway responsibilities among different jurisdictions. Highways and streets are categorized as rural or urban roads, depending on the area in which they are located. This initial classification is necessary because urban and rural areas have significantly different characteristics with respect to the type of land use and population density, which in turn influences travel patterns. Within the classification of urban and rural, highways are categorized into the following groups:

- Principal arterials
- Minor arterials

- Major collectors
- Minor collectors
- Local roads and streets

Freeways are not listed as a separate functional class since they are generally classified as part of the principal arterial system. However, they have unique geometric criteria that require special design consideration.

Functional System of Urban Roads

Urban roads comprise highway facilities within urban areas as designated by responsible state and local officials to include communities with a population of at least 5000 people. Some states use other values; for example, the Virginia Department of Transportation uses a population of 3500 to define an urban area. Urban areas are further subdivided into urbanized areas, with populations of 50 000 or more, and small urban areas, with populations between 5 000 and 50 000. Urban roads are functionally classified into principal arterials, minor arterials, collectors, and local roads. A schematic illustrated of a functionally classified suburban street network is illustrated in Figure 15.1.

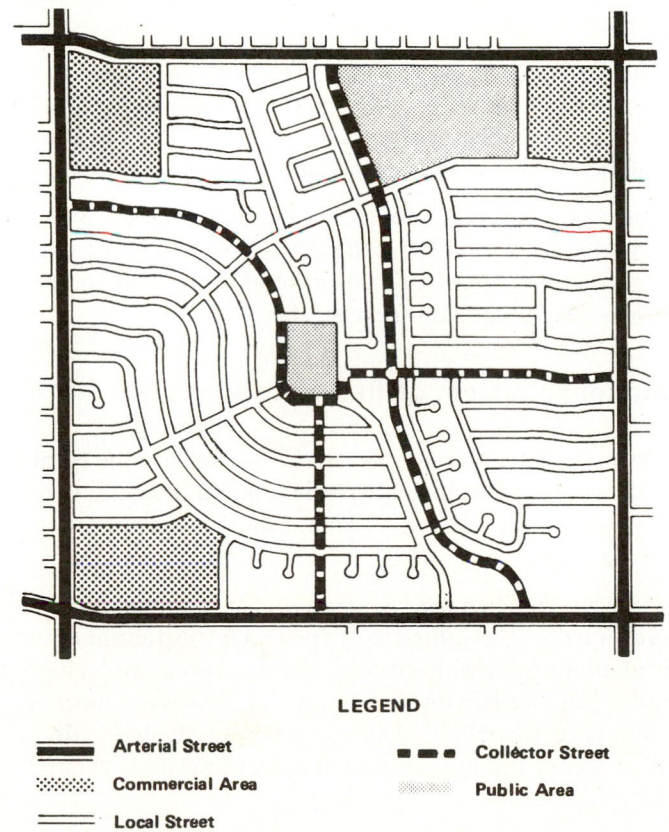

LEGEND

═══ Arterial Street		▪ ▬ ▪ Collector Street	
░░░ Commercial Area		░░░ Public Area	
═══ Local Street			

Figure 15.1   Schematic Illustration of a Portion of a Suburban Street Network

SOURCE: From *A Policy on Geometric Design of Highways and Streets*, 2004, AASHTO, Washington, D.C. Used by permission.

Urban Principal Arterial System.   This system of highways serves the major activity centers of the urban area and consists mainly of the highest-traffic-volume corridors. It carries a high proportion of the total vehicle-miles of travel within the urban area, including most trips with an origin or destination within the urban area. The system also serves trips that bypass the central business districts (CBDs) of urbanized areas. All limited or controlled-access facilities are within this system, although highways without complete access control may be included and classified as an urban principal arterial. Highways within this system are further divided into three subclasses based mainly on the type of access to the facility: (1) interstate, with fully controlled access and grade-separated interchanges; (2) expressways, which have controlled access but may also include at-grade intersections; and (3) other principal arterials (with partial or no controlled access).

Urban Minor Arterial Street System.   Streets and highways that interconnect with and augment the urban primary arterials are classified as urban minor arterials. This system serves trips of moderate length and places more emphasis on land access than the principal arterial system. All arterials not classified as primary are included in this class. Although highways within this system may serve as local bus routes and may connect communities within the urban areas, they do not normally go through identifiable neighborhoods. The spacing between minor arterial streets in fully developed areas is usually not less than 1.6 km, but the spacing can be 3.2 to 4.8 kms in suburban fringes.

Urban Collector Street System.   The main purpose of streets within this system is to collect traffic from local streets in residential areas or in CBDs and convey it to the arterial system. Thus, collector streets usually go through residential areas and facilitate traffic circulation within residential, commercial, and industrial areas.

Urban Local Street System.   This system consists of all other streets within the urban area that are not included in the three systems described earlier. The primary purposes of these streets are to provide access to abutting land and to the collector streets. Through traffic is discouraged on these streets.

Functional System of Rural Roads

Highway facilities outside urban areas comprise the rural road system. These highways are categorized as principal arterials, minor arterials, major collectors, minor collectors, and locals. Figure 15.2 is a schematic illustration of a functionally classified rural highway network.

Rural Principal Arterial System.   This system consists of a network of highways that serves most of the interstate trips and a substantial amount of intrastate trips. Virtually all highway trips between urbanized areas and a high percentage of trips between small urban areas with populations of 25 000 or more are made on this system. The system is further divided into freeways (which are divided highways with fully controlled access and no at-grade intersections) and other principal arterials not classified as freeways.

Rural Minor Arterial System.   This system of roads augments the principal arterial system in the formation of a network of roads that connects cities, large towns, and other traffic generators, such as large resorts. Travel speeds on these roads are relatively high with minimum interference to through movement.

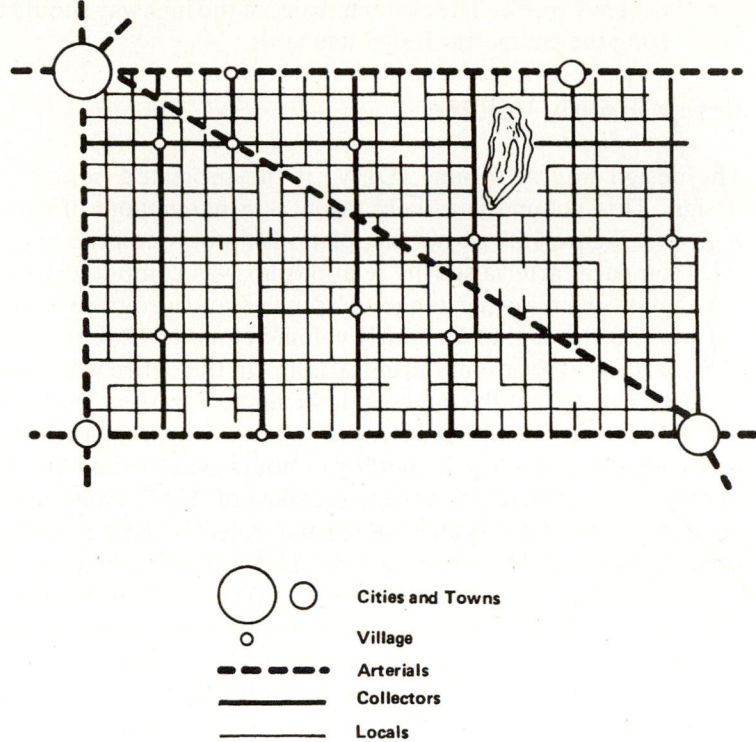

Cities and Towns
Village
Arterials
Collectors
Locals

Figure 15.2   Schematic Illustration of a Functionally Classified Rural Highway Network

SOURCE: From *A Policy on Geometric Design of Highways and Streets*, 2004, AASHTO, Washington, D.C. Used by permission.

Rural Collector System.   Highways within this system carry traffic primarily within individual counties, and trip distances are usually shorter than those on the arterial roads. This system of roads is subdivided into major collector roads and minor collector roads.

- *Major collector roads.* Routes under this system carry traffic within and between county seats and large cities that are not directly served by the arterial system.
- *Minor collector roads.* This system consists of routes that collect traffic from local roads and convey it to other facilities. One important function of minor collector roads is that they provide linkage between rural hinterland and locally important traffic generators such as small communities.

Rural Local Road System   This system consists of all roads within the rural area not classified within the other systems. These roads serve trips of relatively short distances and connect adjacent lands with the collector roads.

15.1.2 Highway Design Standards

Selection of the appropriate set of geometric design standards is the first step in the design of any highway. This is essential because no single set of geometric standards can be used for all highways. For example, geometric standards that may be suitable for a scenic mountain road with low average daily traffic (ADT) are inadequate for a freeway carrying heavy traffic. The characteristics of the highway should therefore be considered in selecting the geometric design standards.

Design Hourly Volume

The design hourly volume (DHV) is the projected hourly volume that is used for design. This volume is usually taken as a percentage of the expected ADT on the highway. Figure 15.3 shows the relationship between the highest hourly volumes and ADT on rural arterials. This relationship was computed from the analysis of traffic count data over a wide range of volumes and geographic conditions. For example, Figure 15.3 shows that an hourly volume equal to 12 percent of the ADT is exceeded at 85 percent of locations during 20 hours in the entire year. This curve also shows that between 0 and 20 of the hours with the highest traffic volumes, a small increase in that number of hours results in a significant reduction in the percentage of ADT, whereas a relatively large increase in number of hours greater than the 30th-highest hour results in only a slight decrease in the percentage of ADT. This peculiar characteristic of the curve suggests that it is uneconomical to select a DHV greater than that which will be exceeded during 29 hours in a year. Thus, the 30th-highest hourly volume is usually selected as the DHV. Experience has also shown that the 30th-highest hourly volume as a percentage of ADT varies only slightly from year to year, even when significant changes of ADT occur. It has also been shown that, excluding rural highways with unusually high or low fluctuation in traffic volume, the 30th-highest hourly volume for rural highways is usually between 12 and 18 percent of the ADT, with the average being 15 percent.

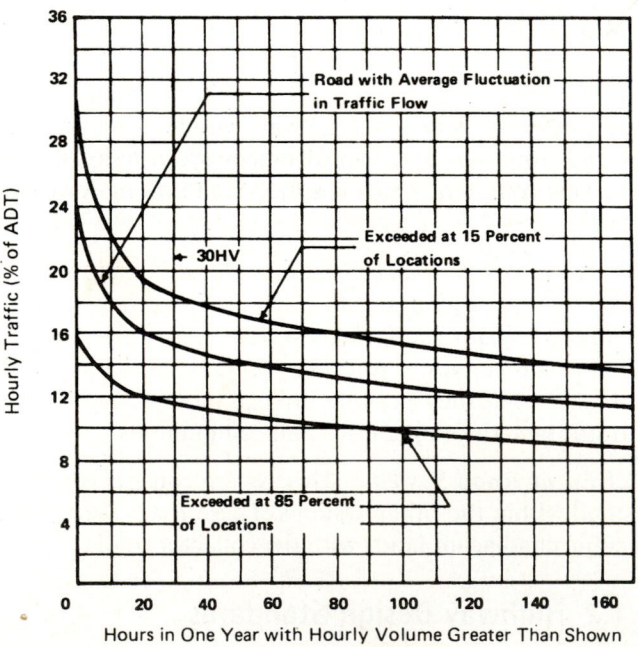

Figure 15.3   Relationship between Peak Hour and Annual Average Daily Traffic on Rural Arterials

SOURCE: From *A Policy on Geometric Design of Highways and Streets*, 2004, AASHTO, Washington, D.C. Used by permission.

The 30th-highest hourly volume should not be indiscriminately used as the DHV, particularly on highways with unusual or high seasonal fluctuation in the traffic flow. Although the percentage of annual average daily traffic (AADT) represented by the 30th-highest hourly volume on such highways may not be significantly different from those on most rural roads, this criterion may not be applicable, since the seasonal fluctuation results in a high percentage of high-volume hours and a low percentage of low-volume hours. For example, economic consideration may not permit the design to be carried out for the 30th-highest hourly volume, but at the same time, the design should not be such that severe congestion occurs during peak hours. A compromise is to select a DHV that will result in traffic operating at a somewhat slightly lower level of service than that which normally exists on rural roads with normal fluctuations. For this type of rural road, it is desirable to select 50 percent of the volume that occurs for only a few peak hours during the design year as the DHV, even though this may not be equal to the 30th-highest hourly volume. This may result in some congestion during the peak hour, but the capacity of the highway normally will not be exceeded.

In most cases, the 30th-highest hourly volume can be a reasonable estimate of the DHV for urban highways. It is usually determined by applying between 8 and 12 percent to the ADT. However, other relationships may be used for highways with seasonal fluctuation in traffic flow much different from that on rural roads. One alternative is to use the average of the highest afternoon peak hour volume for each week in the year as the DHV. Another approach is to determine AADT using methods of urban transportation forecasting described in Chapter 12 and applying an appropriate peak hour factor based on field data.

Design Speed

Design speed is defined as a selected speed to determine the various geometric features of the roadway. Design speed depends on the functional classification of the highway, the topography of the area in which the highway is located, traffic volume, and the land use of the adjacent area. For highway design, topography is generally classified into three groups: level, rolling, and mountainous terrain.

Level terrain is relatively flat. Horizontal and vertical sight distances are generally long or can be achieved without much construction difficulty or major expense.

Rolling terrain has natural slopes that often rise above and fall below the highway grade with occasional steep slopes that restrict the normal vertical and horizontal alignments.

Mountainous terrain has sudden changes in ground elevation in both the longitudinal and transverse directions, thereby requiring frequent hillside excavations to achieve acceptable horizontal and vertical alignments.

The design speed selected should be consistent with the speed that motorists will expect to drive. For example, a low design speed should not be selected for a rural collector road solely because the road is located in an area of flat topography, since motorists will tend to drive at higher speeds. The average trip length on the highway is another factor that should be considered in selecting the design speed. In general, highways with longer average trips should be designed for higher speeds.

Design speeds range from 32 km/h to 112 km/h, with intermediate values of 8 km/h increments. Design elements show little difference when increments are less than 16 km/h but exhibit very large differences with increments of 24 km/h or higher. In general, however, freeways are designed for 96 to 112 km/h, whereas design speeds for other arterial roads range from 48 km/h to 96 km/h. Table 15.1 provides design speed ranges for rural

Table 15.1   Minimum Design Speeds for Rural Collector Roads

Type of Terrain	Design Speed (mi/h) for Specified Design Volume (veh/day)		
	0 to 400	400 to 2000	Over 2000
Level	40	50	60
Rolling	30	40	50
Mountainous	20	30	40

Note: Where practical, design speeds higher than those shown should be considered. (1 m = 1.60 km)
SOURCE: From *A Policy on Geometric Design of Highways and Streets*, 2004, AASHTO, Washington, D.C. Used by permission.

collector roads based on design volume, and Table 15.2 provides recommended values for minimum design speed ranges for various functional classifications.

A design speed is selected to achieve a desired level of operation and safety on the highway. It is one of the first parameters selected in the design process because of its influence on other design variables.

Design Vehicle

A design vehicle is selected to represent all vehicles on the highway. Its weight, dimensions, and operating characteristics are used to establish the design standards of the highway. The different classes of vehicles and their dimensions were discussed in Chapter 3. The vehicle

Table 15.2   Minimum Design Speeds for Various Functional Classifications

Class		Speed (mi/h)					
		20	30	40	50	60	70
Rural principal arterial	Min 50 mi/h for freeways			x	x	x	x
Rural minor arterial				x	x	x	x
	DHV over 400			x	x	x	
Rural	DHV 20–400			x	x	x	
Collector	DHV 100–200		x	x	x		
Road	Current ADT over 400		x	x	x		
	Current ADT under 400	x	x	x			
	DHV over 400			x	x	x	
	DHV 200–400			x	x	x	
Rural	DHV 100–200			x	x	x	
Local	Current ADT over 400			x	x	x	
Road	Current ADT 250–400	x	x	x			
	Current ADT 50–250	x	x				
	Current ADT under 50	x	x				
Urban principal arterial	Minimum 50 mi/h for freeways			x	x	x	x
Urban minor arterial			x	x	x	x	
Urban collector street			x	x	x		
Urban local street		x	x				

SOURCE: *Road Design Manual*, Virginia Department of Transportation, Richmond, VA. (See www.virginiadot.org/business/locdes/rdmanual-index.asp for most current version.)

type selected as the design vehicle is the largest that is likely to use the highway with considerable frequency. The selected design vehicle is used to determine critical design features such as radii at intersections and turning roadways as well as highway grades.

In addition, the following guidelines apply when selecting a design vehicle:

- When a parking lot or a series of parking lots are the main traffic generators, the passenger car may be used.
- For the design of intersections of residential and park roads, a two-axle, single-unit truck may be used.
- At intersections of state highways and city streets that serve buses with relatively few large trucks, a city transit bus may be used.
- For the design of collector streets and other facilities where other single trucks are likely, a three-axle truck may be used.

Cross-Section Elements

The principal elements of a highway cross section consist of the travel lanes, shoulders, and medians (for some multilane highways). Marginal elements include median and roadside barriers, curbs, gutters, guardrails, sidewalks, and sideslopes. Figure 15.4 shows a typical cross section for a two-lane highway, and Figure 15.5 shows that for a multilane highway.

Width of Travel Lanes.   Travel lane widths usually vary from 2.7 to 3.6 m. Most arterials have 3.6 m travel lanes since the extra cost for constructing 3.6 m lanes over 3.0 m lanes is usually offset by the lower maintenance cost for shoulders and pavement surface, resulting

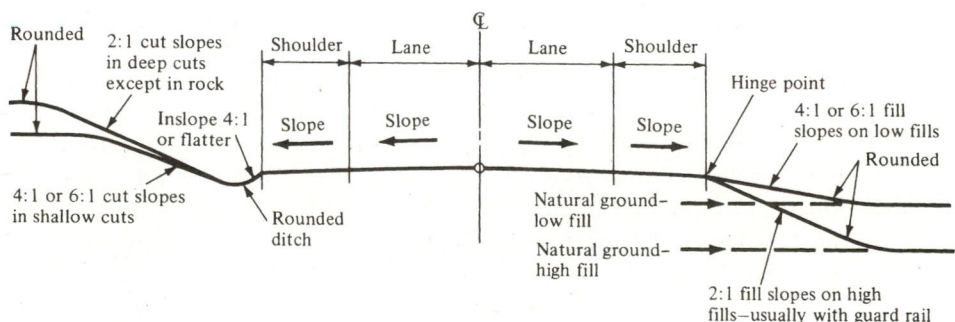

Figure 15.4   Typical Cross Section for Two-Lane Highways

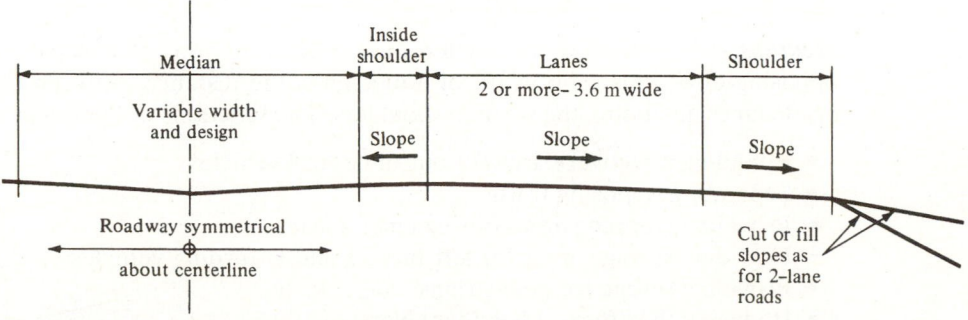

Figure 15.5   Typical Cross Section for Multilane Highways (half section)

in a reduction of wheel concentrations at the pavement edges. On two-lane, two-way rural roads, lane widths of 3 m or 3.3 m may be used, but two factors must be considered when selecting a lane width less than 3.6 m wide. When pavement surfaces are less than 6.7 m, the crash rates for large trucks tend to increase and, as the lane width is reduced from 3.6 m, the capacity of a highway significantly decreases. Lane widths of 3 m are therefore used only on low-speed facilities. Lanes that are 2.7 m wide are used occasionally in urban areas if traffic volume is low and there are extreme right-of-way constraints.

Shoulders.   The shoulder of a pavement cross section is always contiguous with the traveled lane so as to provide an area along the highway for vehicles to stop when necessary. In some cases, bicycles are permitted to use a highway shoulder, particularly on rural and collector roads. Shoulder surfaces range in width from 0.6 m on minor roads to 3.6 m on major arterials. Shoulders are also used to laterally support the pavement structure. The shoulder width is known as either *graded* or *usable*, depending on the section of the shoulder being considered. The graded shoulder width is the whole width of the shoulder measured from the edge of the traveled pavement to the intersection of the shoulder slope and the plane of the side slope. The usable shoulder width is that part of the graded shoulder that can be used to accommodate parked vehicles. The usable width is the same as the graded width when the side slope is equal to or flatter than 4:1 (horizontal:vertical), as the shoulder break is usually rounded to a width between 1.2 m and 1.8 m, thereby increasing the usable width.

When a vehicle stops on the shoulder, it is desirable for it to be at least 0.3 m and preferably 0.6 m from the edge of the pavement. Based on this consideration, usable shoulder widths should be at least 3 m on highways having a large number of trucks and on highways with heavy traffic volumes and high speeds. However, it may not always be feasible to provide this minimum width, particularly when the terrain is difficult or when traffic volume is low. A minimum shoulder width of 0.6 m may therefore be used on the lowest type of highways, but 1.8 to 2.4-m widths should preferably be used. When pedestrians and bicyclists are permitted, the minimum shoulder width should be 1.2 m. The width for usable shoulders within the median for divided arterials having two lanes in each direction may be reduced to 0.9 m, since drivers rarely use the median shoulder for stopping on these roads. The usable median shoulder width for divided arterials with three or more lanes in each direction should be at least 2.4 m, since drivers in the lane next to the median often find it difficult to maneuver to the outside shoulder when there is a need to stop.

All shoulders should be flush with the edge of the traveled lane and sloped to facilitate drainage of surface water on the traveled lanes. Recommended slopes are 2 to 6 percent for bituminous and concrete-surfaced shoulders, and 4 to 6 percent for gravel or crushed-rock shoulders. Rumble strips may be used on paved shoulders along arterials as a safety measure to warn motorists that they are leaving the traffic lane.

Medians.   A median is the section of a divided highway that separates the lanes in opposing directions. The width of a median is the distance between the edges of the inside lanes, including the median shoulders. The functions of a median include:

- Providing a recovery area for out-of-control vehicles
- Separating opposing traffic
- Providing stopping areas during emergencies
- Providing storage areas for left-turning and U-turning vehicles
- Providing refuge for pedestrians
- Reducing the effect of headlight glare
- Providing temporary lanes and crossovers during maintenance operations

Medians can either be raised, flush, or depressed. Raised medians are frequently used in urban arterial streets because they facilitate the control of left-turn traffic at intersections by using part of the median width for left-turn-only lanes. Some disadvantages associated with raised medians include possible loss of control of the vehicle by the driver if the median is accidentally struck, and they cast a shadow from oncoming headlights, which results in drivers finding it difficult to see the curb.

Flush medians are commonly used on urban arterials. They can also be used on freeways, but with a median barrier. To facilitate drainage of surface water, the flush median should be crowned. The practice in urban areas of converting flush medians into two-way left-turn lanes is common, since the capacity of the urban highway is increased while maintaining some features of a median.

Depressed medians are generally used on freeways and are more effective in draining surface water. A side slope of 6:1 is suggested for depressed medians, although a slope of 4:1 may be adequate.

Median widths vary from a minimum of 1.2 to 24 m or more. Median widths should be as wide as possible but should be balanced with other elements of the cross section and the cost involved. In general, the wider the median, the more effective it is in providing safe operating conditions and a recovery area for out-of-control vehicles. A minimum width of 3 m is recommended for use on four-lane urban freeways, which is adequate for two 1.2 m shoulders and a 0.6 m median barrier. A minimum of 6.6 m, preferably 7.8 m, is recommended for six or more lanes of freeway.

Median widths for urban collector streets vary from 0.6 to 12 m, depending on the median treatment. For example, when the median is a paint-striped separation, 0.6 to 1.2 m medians are required. For narrow raised or curbed areas, 0.6 to 1.8 m medians are required, and for curbed sections, 4.8 to 12 m. The larger width is necessary for curbed sections because it provides space for protecting vehicles crossing an intersection and also can be used for landscape treatment.

Roadside and Median Barriers.   A median barrier is defined as a longitudinal system used to prevent an errant vehicle from crossing the portion of a divided highway separating the traveled ways for traffic in opposite directions. Roadside barriers, on the other hand, protect vehicles from obstacles or slopes on the roadside. They also may be used to shield pedestrians and property from the traffic stream. The provision of median barriers must be considered when traffic volumes are high and when access to multilane highways and other highways is only partially controlled. However, when the median of a divided highway has physical characteristics that may create unsafe conditions, such as a sudden lateral drop-off or obstacles, the provision of a median barrier should be considered regardless of the traffic volume or the median width.

Roadside barriers should be provided whenever conditions exist requiring the protection for vehicles along the side of the road. For example, when the slope of an embankment is high or when traveling under an overhead bridge, the provision of a roadside barrier is warranted. There are a wide variety of roadside barriers, and the selection of the most desirable system should provide the required degree of shielding at the lowest cost for the specific application. Figure 15.6 illustrates a backed timber guardrail developed as an aesthetic alternative to conventional guardrail systems.

Median barriers can be composed of cable or post-and-beam systems or concrete. For major arterials, concrete barriers are commonly used because of their low life-cycle cost, effective performance, and easy maintenance. Figure 15.7 depicts a single-slope concrete median barrier. The primary advantage to this barrier shape is that adjacent pavement can be overlaid several times without affecting its performance.

Figure 15.6   Backed Timber Guardrail

SOURCE: From *Roadside Design Guide*, 2006, AASHTO, Washington, D.C. Used by permission.

Additional information on selecting, locating, and designing roadside and median barriers can be obtained from the AASHTO *Roadside Design Guide*.

Curbs and Gutters.   Curbs are raised structures made of either Portland cement concrete or bituminous concrete (rolled asphalt curbs) that are used mainly on urban highways to delineate pavement edges and pedestrian walkways. Curbs are also used to control drainage, improve aesthetics, and reduce right-of-way. Curbs can be generally classified as either vertical or sloping. Vertical curbs (which may be vertical or nearly vertical) range in height from 150 to 200, have steep sides, and are designed to prevent vehicles from leaving the highway. Sloping curbs are designed so that vehicles can cross them if necessary. Figure 15.8 illustrates typical highway curbs. Both vertical and sloping curbs may be designed separately or as integral parts of the pavement. In general, vertical curbs should not be used in conjunction with traffic barriers, such as bridge railings or median and roadside barriers, because they could contribute to vehicles rolling over the traffic barriers. Vertical curbs should also be avoided on highways with design speeds greater than 64 km/h, because at such speeds it is usually difficult for drivers to retain control of the vehicle after an impact with the curb.

Figure 15.7   Single-Slope Concrete Median Barrier

SOURCE: From *Roadside Design Guide*, 2006, AASHTO, Washington, D.C. Used by permission.

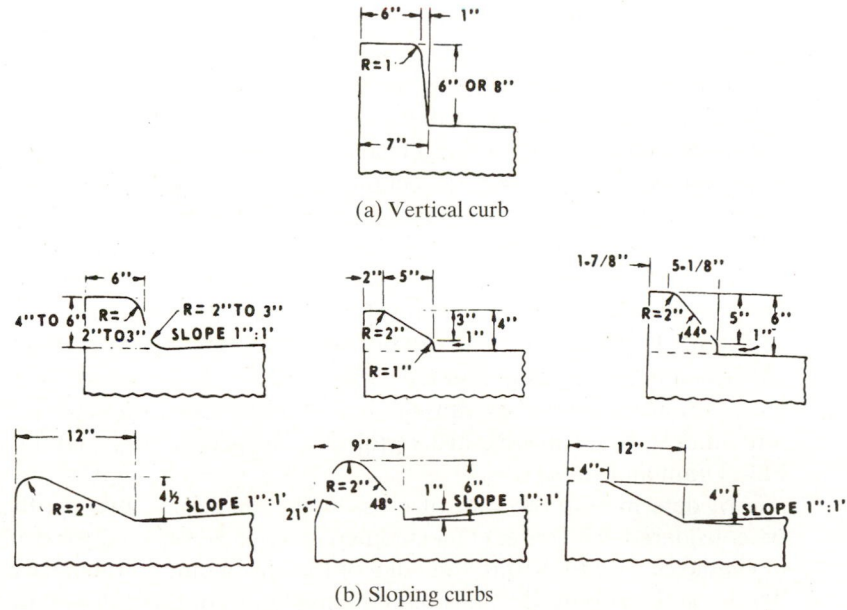

Figure 15.8   Typical Highway Curbs

SOURCE: From *A Policy on Geometric Design of Highways and Streets*, 2004, AASHTO, Washington, D.C. Used by permission.

Gutters or drainage ditches are usually located on the pavement side of a curb to provide the principal drainage facility for the highway. They are sloped to prevent any hazard to traffic, and they usually have cross slopes of 5 to 8 percent and are 0.3 to 1.8 m wide. Gutters can be designed as V-type sections or as broad, flat, rounded sections.

Guardrails.   Guardrails are longitudinal barriers placed on the outside of sharp curves and at sections with high fills. Their main function is to prevent vehicles from leaving the roadway. They are installed at embankments higher than 2.4 m and when shoulder slopes are greater than 4:1. Shapes commonly used include the W beam and the box beam. The weak post system provides for the post to collapse on impact, with the rail deflecting and absorbing the energy due to impact.

Sidewalks.   Sidewalks are usually provided on roads in urban areas, but are uncommon in rural areas. Nevertheless, the provision of sidewalks in rural areas should be evaluated during the planning process to determine sections of the road where they are required. For example, rural principal arterials may require sidewalks in areas with high pedestrian concentrations, such as adjacent to schools, industrial plants, and local businesses. Generally, sidewalks should be provided when pedestrian traffic is high along main or high-speed roads in either rural or urban areas. When shoulders are not provided on arterials, sidewalks are necessary even when pedestrian traffic is low. In urban areas, sidewalks should also be provided along both sides of collector streets that serve as pedestrian access to schools, parks, shopping centers, and transit stops, and along collector streets in commercial areas. Sidewalks should have a minimum clear width of 1.2 m in residential areas and a range of 1.2 to 2.4 m in commercial areas. To encourage pedestrians to use sidewalks, they should have all-weather surfaces since pedestrians will tend to use traffic lanes rather than unpaved sidewalks.

Cross Slopes.   Pavements on straight sections of two-lane and multilane highways without medians are sloped from the middle downward to both sides of the highway, resulting in a transverse or cross slope, with a cross-section shape that can be curved, plane, or a combination of the two. A parabola is generally used for curved cross sections, and the highest point of the pavement (called the crown) is slightly rounded, with the cross slope increasing toward the pavement edge. Plane cross slopes consist of uniform slopes at both sides of the crown. The curved cross section has one advantage: the slope increases outward to the pavement edge, thereby enhancing the flow of surface water away from the pavement. A disadvantage is that curved cross sections are difficult to construct.

Cross slopes on divided highways are provided by sloping each pavement two ways, as shown in Figure 15.9a (A–C) or by sloping each pavement one way, as shown in Figure 15.9b (D–G). The advantage of sloping the pavement in two ways is that surface water is quickly drained away from the traveled roadway during heavy rain storms, whereas the disadvantage is that additional drainage facilities, such as inlets and underground drains, are required. This method is mainly used at areas with heavy rain and snow.

In determining the rate of cross slope for design, two conflicting factors should be considered. They are: (1) a steep cross slope is required for drainage purposes, and (2) vehicles tend to drift to the edge of the pavement, particularly under icy conditions. To accomodate both factors, recommended rates of cross slopes are 1.5 to 2 percent for high-type pavements and 2 to 6 percent for low-type pavements. Cross slopes greater than 2 percent are sometimes used in areas with high-intensity rainfall.

High-type pavements have wearing surfaces that can adequately support the expected traffic load without visible distress due to fatigue and are not susceptible to

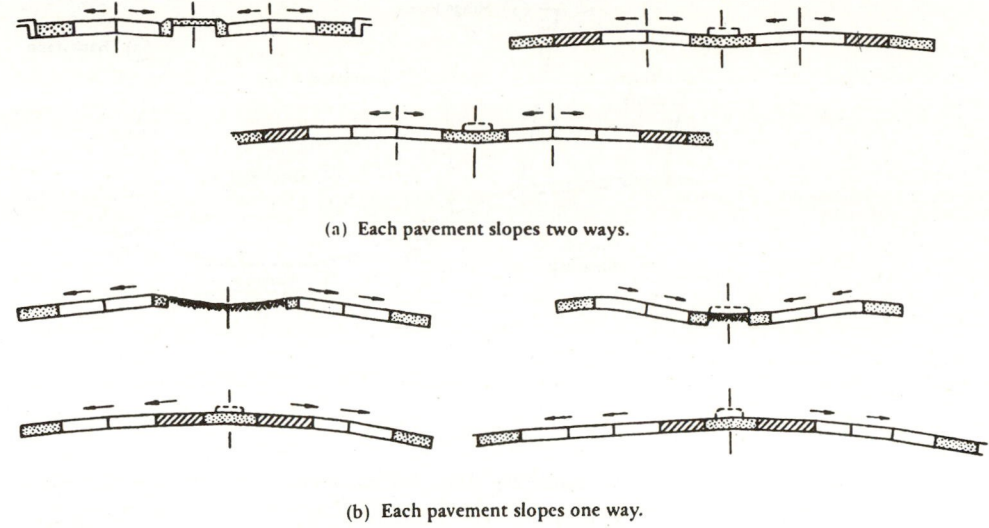

(a) **Each pavement slopes two ways.**

(b) **Each pavement slopes one way.**

Figure 15.9   Roadway Sections for Divided Highways (Basic Cross Slope Arrangements)

SOURCE: From *A Policy on Geometric Design of Highways and Streets*, 2004, AASHTO, Washington, D.C. Used by permission.

weather conditions. Low-type pavements are used mainly for low-cost roads and have wearing surfaces ranging from untreated loose material to surface-treated earth.

Side Slopes.   Side slopes are provided on embankments and fills to provide stability for earthworks. They also serve as a safety feature by providing a recovery area for out-of-control vehicles. When being considered as a safety feature, the important sections of the cross slope are the hinge point, the foreslope, and the toe of the slope, as shown in Figure 15.10. The hinge point should be rounded since it is potentially hazardous and may cause vehicles to become airborne while crossing it, resulting in loss of control of the vehicle. The foreslope serves principally as a recovery area, where vehicle speeds can be reduced and other recovery maneuvers taken to regain control of the vehicle. The gradient of the foreslope should therefore not be high. Slopes of 3:1 (horizontal:vertical) or flatter are generally used for high embankments. This can be increased based on the conditions at the site. As illustrated in Figure 15.10, the toe of slope is rounded up in order to facilitate the safe movement of vehicles from the foreslope to the backslope.

Right-of-Way.   The right-of-way is the total land area acquired for the construction of a highway. The width should be sufficient to accommodate all the elements of the highway cross section, any planned widening of the highway, and public-utility facilities that will be installed along the highway. In some cases, the side slopes may be located outside the right-of-way on easement areas. The right-of-way for two-lane urban collector streets should be between 12 to 18 m, whereas the desirable minimum for two-lane arterials is 25 m. Right-of-way widths for undivided four-lane arterials vary from 19 to 33 m, whereas for divided arterials, they range from about 36 to 90 m, depending on the numbers of lanes and whether frontage roads are included. The minimum right-of-way widths for freeways depend on the number of lanes and the existence of a frontage road.

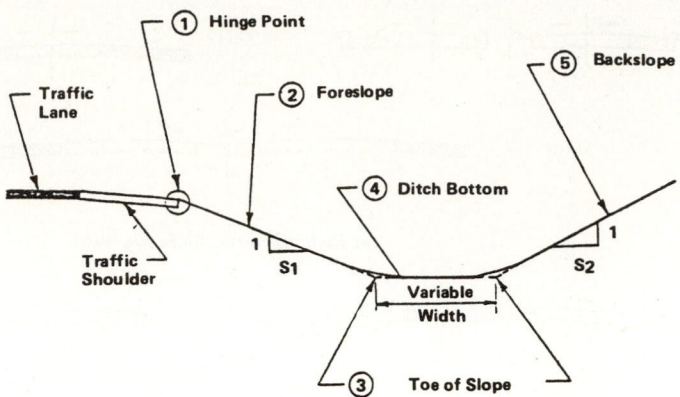

Figure 15.10   Designation of Roadside Regions

SOURCE: From *A Policy on Geometric Design of Highways and Streets*, 2004, AASHTO, Washington, D.C. Used by permission.

Maximum Highway Grades

The effect of grade on the performance of heavy vehicles was discussed in Chapter 3, where it was shown that the speed of a heavy vehicle is significantly reduced if the grade is steep and/or long. In Chapter 9, it was noted that steep grades affect not only the performance of heavy vehicles but also the performance of passenger cars. In order to limit the effect of grades on vehicular operation, the maximum grade on any highway should be selected judiciously.

The selection of maximum grades for a highway depends on the design speed and the design vehicle. It is generally accepted that grades of 4 to 5 percent have little or no effect on passenger cars, except for those with high weight/horsepower ratios, such as those found in compact and subcompact cars. As the grade increases above 5 percent, however, speeds of passenger cars decrease on upgrades and increase on downgrades.

Grade has a greater impact on trucks than on passenger cars. Extensive studies have been conducted, and results (some of which were presented in Chapter 9) have shown that truck speed may increase up to 5 percent on downgrades and decrease by 7 percent on upgrades, depending on the percent and length of the grade.

The impact of grades on recreational vehicles is more significant than that for passenger cars, but it is not as critical as that for trucks. However, it is very difficult to establish maximum grades for recreational routes, and it may be necessary to provide climbing lanes on steep grades when the percentage of recreational vehicles is high.

Maximum grades have been established based on the operating characteristics of the design vehicle on the highway. These vary from 5 percent for a design speed of 112 km/h to between 7 and 12 percent for a design speed of 48 km/h, depending on the type of highway. Table 15.3 gives recommended values of maximum grades. Note that these recommended maximum grades should not be used frequently, particularly when grades are long and the traffic includes a high percentage of trucks. On the other hand, when grade lengths are less than 150 m and roads are one-way in the downgrade direction, maximum grades may be increased by up to 2 percent, particularly on low-volume rural highways.

Table 15.3(a–e)   Recommended Maximum Grades

(a) Rural Collectors[a]
Design Speed (mi/h)

Type of Terrain	20	25	30	35	40	45	50	55	60
Grades (%)									
Level	7	7	7	7	7	7	6	6	5
Rolling	10	10	9	9	8	8	7	7	6
Mountainous	12	11	10	10	10	10	9	9	8

(b) Urban Collectors[a]
Design Speed (mi/h)

Type of Terrain	20	25	30	35	40	45	50	55	60
Grades (%)									
Level	9	9	9	9	9	8	7	7	6
Rolling	12	12	11	10	10	9	8	8	7
Mountainous	14	13	12	12	12	11	10	10	9

(c) Rural Arterials
Design Speed (mi/h)

Type of Terrain	40	45	50	55	60	65	70	75	80
Grades (%)									
Level	5	5	4	4	3	3	3	3	3
Rolling	6	6	5	5	4	4	4	4	4
Mountainous	8	7	7	6	6	5	5	5	5

(d) Rural and Urban Freeways[b]
Design Speed (mi/h)

Type of Terrain	50	55	60	65	70	75	80
Grades (%)							
Level	4	4	3	3	3	3	3
Rolling	5	5	4	4	4	4	4
Mountainous	6	6	6	5	5	—	—

(e) Urban Arterials
Design Speed (mi/h)

Types of Terrain	30	35	40	45	50	55	60
Grades (%)							
Level	8	7	7	6	6	5	5
Rolling	9	8	8	7	7	6	6
Mountainous	11	10	10	9	9	8	8

[a]Maximum grades shown for rural and urban conditions of short lengths (less than 150 m) and on one-way downgrades may be up to 2 percent steeper.
[b]Grades that are 1 percent steeper than the value shown may be used for extreme cases in urban areas where development precludes the use of flatter grades and for one-way downgrades, except in mountainous terrain.
SOURCE: Based on *A Policy on Geometric Design of Highways and Streets*, 2004, AASHTO, Washington, D.C.

Minimum grades depend on the drainage conditions of the highway. Zero-percent grades may be used on uncurbed pavements with adequate cross slopes to laterally drain the surface water. When pavements are curbed, however, a longitudinal grade should be provided to facilitate the longitudinal flow of the surface water. It is customary to use a minimum of 0.5 percent in such cases, although this may be reduced to 0.3 percent on high-type pavement constructed on suitably crowned, firm ground.

15.2 DESIGN OF THE ALIGNMENT

The alignment of a highway is composed of vertical and horizontal elements. The vertical alignment includes straight (tangent) highway grades and the parabolic curves that connect these grades. The horizontal alignment includes the straight (tangent) sections of the roadway and the circular curves that connect their change in direction. The design of the alignment depends primarily on the design speed selected for the highway. The least costly alignment is one that takes the form of the natural topography. It is not always possible to select the lowest cost alternative because the designer must adhere to certain standards that may not exist on the natural topography. It is important that the alignment of a given section has consistent standards to avoid sudden changes in the vertical and horizontal layout of the highway. It is also important that both horizontal and vertical alignments be designed to complement each other, since this will result in a safer and more attractive highway. One factor that should be considered to achieve compatibility is the proper balancing of the grades of tangents with curvatures of horizontal curves and the location of horizontal and vertical curves with respect to each other. For example, a design that achieves horizontal curves with large radii at the expense of steep or long grades is a poor design. Similarly, if sharp horizontal curves are placed at or near the top of pronounced crest vertical curves at or near the bottom of a pronounced sag vertical curve, a hazardous condition will be created at these sections of the highway. Thus, it is important that coordination of the vertical and horizontal alignments be considered at the early stages of preliminary design.

15.2.1 Vertical Alignment

The vertical alignment of a highway consists of straight sections known as grades (or tangents) connected by vertical curves. The design of the vertical alignment therefore involves the selection of suitable grades for the tangent sections and the appropriate length of vertical curves. The topography of the area through which the road traverses has a significant impact on the design of the vertical alignment.

Vertical curves are used to provide a gradual change from one tangent grade to another so that vehicles may run smoothly as they traverse the highway. These curves are usually parabolic in shape. The expressions developed for minimum lengths of vertical curves are therefore based on the properties of a parabola. Figure 15.11 illustrates vertical curves that are classified as crest or sag.

Length of Crest Vertical Curves

Provision of a minimum stopping sight distance (*SSD*) is the only criterion used for design of a crest vertical curve. As illustrated in Figures 15.12 and 15.13, there are two possible scenarios that could control the design length: (1) the *SSD* is greater than the length of the vertical curve, and (2) the *SSD* is less than the length of the vertical curve.

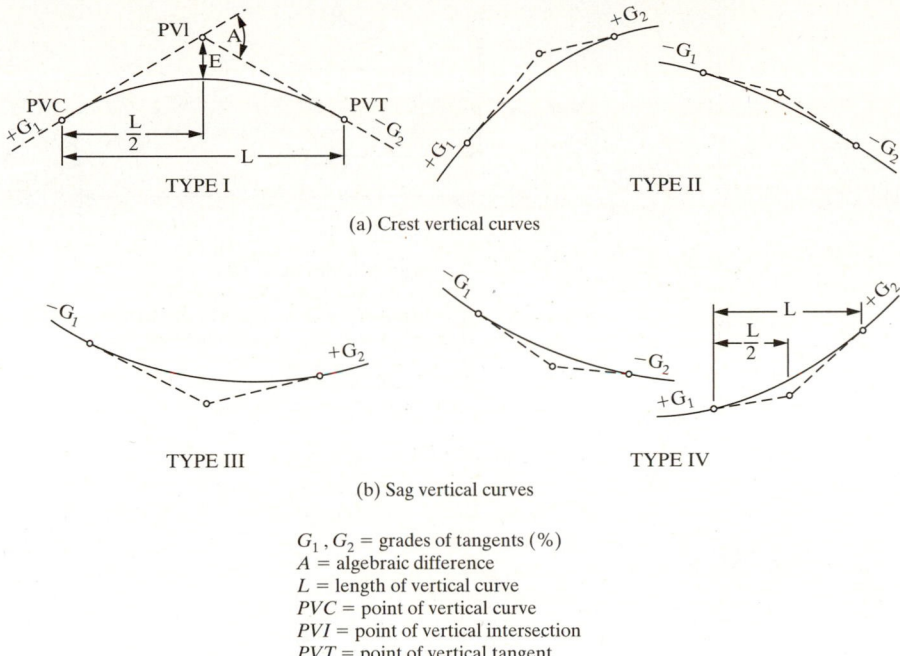

Figure 15.11 Types of Vertical Curves

Consider the case when the *SSD* is greater than the length of the vertical curve, as shown in Figure 15.12, where the driver eye height in a vehicle on the grade at point *C* is H_1 m and the object at point *D* seen by the driver is H_2 m. The driver's line of sight is *PN* ft and the *SSD* is *S* m. The line of sight, *PN*, may not necessarily be horizontal, but the value used in calculations for *SSD* considers the horizontal projection.

From the properties of the parabola,

$$X_3 = \frac{L}{2}$$

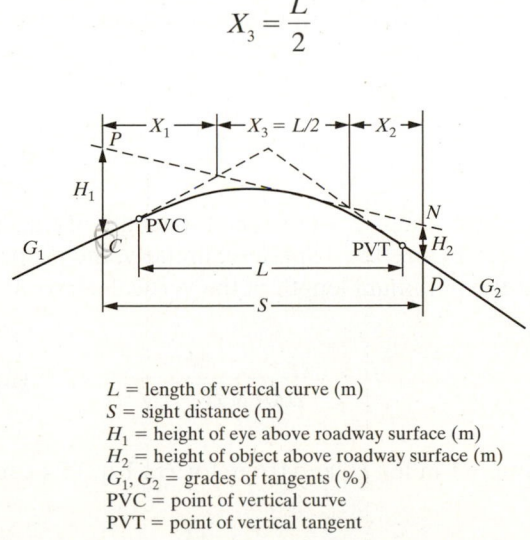

L = length of vertical curve (m)
S = sight distance (m)
H_1 = height of eye above roadway surface (m)
H_2 = height of object above roadway surface (m)
G_1, G_2 = grades of tangents (%)
PVC = point of vertical curve
PVT = point of vertical tangent

Figure 15.12 Sight Distance on Crest Vertical Curve (*S > L*)

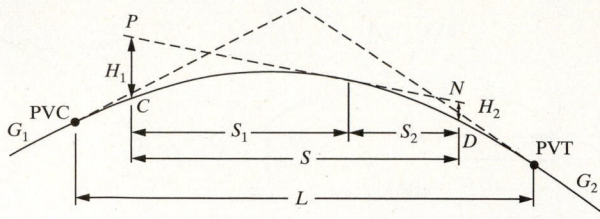

L = length of vertical curve (m)
S = sight distance (m)
H_1 = height of eye above roadway surface (m)
H_2 = height of object above roadway surface (m)
G_1, G_2 = grades of tangents (%)
PVC = point of vertical curve
PVT = point of vertical tangent

Figure 15.13   Sight Distance on Crest Vertical Curve $(S < L)$

The *SSD* is

$$S = X_1 + \frac{L}{2} + X_2 \tag{15.1}$$

X_1 and X_2 can be determined from grades G_1 and G_2 and their algebraic difference A. The minimum length of the vertical curve for the required sight distance is obtained as

$$L_{min} = 2S - \frac{200(\sqrt{H_1} + \sqrt{H_2})^2}{A} \qquad \text{(for } S > L) \tag{15.2}$$

It had been the practice to assume that the height of the driver's eye is 1.2 m, and the height of the object is 0.15 m. Due to the increasing number of compact automobiles on the nation's highways, the height of the driver's eye is now assumed to be 1.1 m, and the object height, considered to be the taillight of a passenger car, is 0.6 m. Under these assumptions, Eq. 15.2 can be written as

$$L_{min} = 2S - \frac{658}{A} \qquad \text{(for } S > L) \tag{15.3}$$

When the sight distance is less than the length of the crest vertical curve, the configuration shown in Figure 15.13 applies. Similarly, the properties of a parabola can be used to show that the minimum length of the vertical curve is

$$L_{min} = \frac{AS^2}{200(\sqrt{H_1} + \sqrt{H_2})^2} \qquad \text{(for } S < L) \tag{15.4}$$

Substituting 1.1 m for H_1 and 0.6 m for H_2, Eq. 15.4 can be written as

$$L_{min} = \frac{AS^2}{658} \qquad \text{(for } S < L) \tag{15.5}$$

Example 15.1 Minimum Length of a Crest Vertical Curve

A crest vertical curve is to be designed to join a +3% grade with a −2% grade at a section of a two-lane highway. Determine the minimum length of the curve if the design speed of the highway is 96 km/h, $S < L$, and a perception-reaction time of 2.5 sec. The deceleration rate for braking (a) is 3,41 m/sec².

Solution:

- Use the equation derived in Chapter 3 to determine the *SSD* required for the design conditions. (Since the grade changes constantly on a vertical curve, the worst-case value for *G* of 3 percent is used to determine the braking distance.)

$$SSD = 0.278ut + \frac{u^2}{254\left\{\left(\dfrac{a}{9.81}\right) - G\right\}}$$

$$= 0.278 \times 96 \times 2.5 + \frac{96}{254\left\{\dfrac{3.41}{9.81} - 0.03\right\}}$$

$$= 66.72 + 114.24$$

$$= 180.96 \text{ m}$$

- Use Eq. 15.5 to obtain the minimum length of vertical curve:

$$L_{min} = \frac{AS^2}{658}$$

$$= \frac{5 \times (180.96)^2}{658}$$

$$= 248.83 \text{ m}$$

Example 15.2 Maximum Safe Speed on a Crest Vertical Curve

An existing vertical curve on a highway joins a +4.4% grade with a −4.4% grade. If the length of the curve is 82 m, what is the maximum safe speed on this curve?
What speed should be posted if 8 km/h increments are used?
    Assume $a = 3.41$ m/sec², perception-reaction time = 2.5 sec, and $S < L$.

Solution:

- Determine the *SSD* using the length of the curve and Eq. 15.5.

$$L_{min} = \frac{AS^2}{658}$$

$$82 = \frac{8.8 \times S^2}{658}$$

$$S = 78.30 \text{ m}$$

- Determine the maximum safe speed for this sight distance using the equation for *SSD* from Chapter 3.

$$78.30 = 0.278 \times 2.5u + \frac{u^2}{254\left\{\dfrac{3.41}{9.81} - 0.044\right\}}$$

which yields the quadratic equation

$$u^2 + 53.6u - 6038 = 0$$

- Solve the quadratic equation to find the *u*, the maximum safe speed.

$$u = 55.4 \text{ km/h}$$

The maximum safe speed for an *SSD* of 78.30 m is therefore 55.4 km/h.

If a speed limit is to be posted to satisfy this condition, a conservative value of 48 km/h will be used.

Length of Sag Vertical Curves

The selection of the minimum length of a sag vertical curve is controlled by the following four criteria: (1) *SSD* provided by the headlight, (2) comfort while driving on the curve, (3) general appearance of the curve, and (4) adequate control of drainage at the low point of the curve.

Minimum Length Based on SSD Criterion.   The headlight *SSD* requirement is based on the fact that sight distance will be restricted during periods of darkness whereas during daylight periods, sight distance is unaffected by the sag curve. As a vehicle is driven on a sag vertical curve at night, the position of the headlight and the direction of the headlight beam will dictate the stretch of highway ahead that is lighted. Therefore, the headlight beam controls the distance that can be seen by the driver.

Figure 15.14 is a schematic of the case when $S > L$. The headlight is located at a height H above the ground, and the headlight beam is inclined upward at an angle b to the horizontal. The headlight beam intersects the road at point D, thereby restricting the available *SSD* S. The values used by AASHTO for H and β are 2 ft and 1 degree, respectively. Using the properties of the parabola, it can be shown that

$$L_{min} = 2S - \frac{200(H + S\tan\beta)}{A} \qquad (\text{for } S > L) \qquad (15.6)$$

Substituting 0.6 m for H and 1 degree for b in Eq. 15.6 yields the following:

$$L_{min} = 2S - \frac{(120 + 3.5S)}{A} \qquad (\text{for } S > L) \qquad (15.7)$$

Similarly, for the condition when $S < L$, it can be shown that

$$L_{min} = \frac{AS^2}{200(H + S\tan\beta)} \qquad (\text{for } S < L) \qquad (15.8)$$

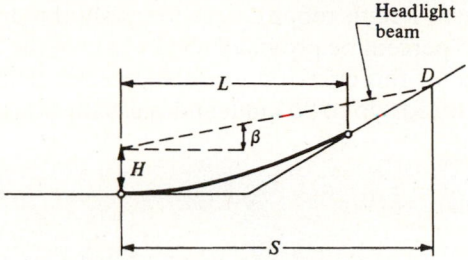

Figure 15.14   Headlight Sight Distance on Sag Vertical Curves ($S > L$)

and substituting 0.6 m for H and 1 degree for β in Eq. 15.8 yields

$$L_{min} = \frac{AS^2}{120 + 3.5S} \qquad \text{(for } S < L) \tag{15.9}$$

To provide a safe condition, the minimum length of the sag vertical curve should assure a light beam sight distance S at least equal to the SSD. The SSD for the appropriate design speed is the value for S when Eqs. 15.8 and 15.9 are used to compute minimum lengths of sag vertical curves.

Minimum Length Based on Comfort Criterion.   The comfort criterion is based on the fact that when a vehicle travels on a sag vertical curve, both the gravitational and centrifugal forces act in combination, resulting in a greater effect than on a crest vertical curve, where these forces act in opposition to each other. Several factors, such as weight carried, body suspension of the vehicle, and tire flexibility, affect comfort due to change in vertical direction, making it difficult for comfort to be measured directly. It is generally accepted that a comfortable ride will be provided if the radial acceleration is not greater than 0.3 m/sec². The following expression is used for the comfort criterion.

$$L_{min} = \frac{Au^2}{395} \tag{15.10}$$

where u is the design speed in km/h, L the minimum length based on comfort, and A the algebraic difference in grades. The length obtained from Eq. 15.10 is typically about 75 percent of that obtained from the headlight sight distance requirement.

Minimum Length of Curve Based on Appearance Criterion.   The criterion for acceptable appearance is usually satisfied by assuring that the minimum length of the sag curve is not less than expressed by the following equation:

$$L_{min} = 30\,A \tag{15.11}$$

where L is the minimum length of the sag vertical curve. Longer curves are frequently necessary for major arterials if the general appearance of these highways is to be considered to be satisfactory.

Minimum Length Based on Drainage Criterion.   The drainage criterion for sag vertical curves must be considered when the road is curbed. This criterion is different from the others in that there is a maximum length requirement rather than a minimum length.

The maximum length requirement to satisfy the drainage criterion is that a minimum slope of 0.35 percent be provided within 15 m of the lowest point of the curve. The maximum length for this criterion is usually greater than the minimum length for the other criteria for speeds up to 96 km/h and is usually equal for a speed of 112 km/h.

Example 15.3   Minimum Length of a Sag Vertical Curve

A sag vertical curve is to be designed to join a −5% grade to a +2% grade. If the design speed is 64 km/h, determine the minimum length of the curve that will satisfy all criteria. Assume $a = 3.41$ m/sec and perception-reaction time $= 2.5$ sec.

Solution:

- Find the stopping sight distance.

$$SSD = 0.278ut + \frac{u^2}{254\left(\dfrac{3.41}{9.81} - G\right)}$$

$$= 0.278 \times 64 \times 2.5 + \left[\frac{64}{254(0.35 - .05)}\right] = 44.48 + 53.75$$

$$= 98.35 \text{ m}$$

- Determine whether $S < L$ or $S > L$ for the headlight sight distance criterion. For $S > L$,

$$L_{min} = 2S - \frac{(120 + 3.5S)}{A}$$

$$= 2 \times 98.23 - \left[\frac{(120 + 3.5 \times 98.23)}{7}\right]$$

$$= 130.2 \text{ m}$$

(This condition is not appropriate since $324.78 < 430.03$. Therefore, $S \not> L$.) For $S < L$,

$$L_{min} = \frac{AS^2}{120 + 3.5S}$$

$$= \frac{7 \times (98.23)^2}{(120 + 3.5 \times 98.23)}$$

$$= 145.63 \text{ m}$$

This condition is satisfied since $98.23 < 145.63$.
- Determine minimum length for the comfort criterion.

$$L_{min} = \frac{Au^2}{395}$$

$$= \frac{7 \times 64}{395} = 72.60 \text{ m}$$

- Determine minimum length for the general appearance criterion.

$$L_{min} = 30A$$

$$= 30 \times 7 = 210 \text{ m}$$

The minimum length to satisfy all criteria is 210 m.

(*Note*: In order to check the maximum length drainage requirement, it is necessary to use procedures for calculating curve elevations that are discussed later in this chapter.)

The solution of Example 15.3 demonstrates a complete analysis to determine the minimum length of a sag vertical curve that satisfies each of the criteria discussed. The headlight sight distance criterion is the one commonly used for establishing minimum lengths of sag vertical curves.

15.2.2 Length of Crest and Sag Vertical Curves Based on *K* Factors

The expressions for the minimum lengths of the crest vertical curves are given in Eqs. 15.3 and 15.5 as follows:

$$L_{min} = 2S - \frac{658}{A} \qquad (\text{for } S > L) \tag{15.3}$$

and

$$L_{min} = \frac{AS^2}{658} \qquad (\text{for } S < L) \tag{15.5a}$$

Equation 15.5(a) can be written as:

$$L = KA \tag{15.5b}$$

where K is the length of the vertical curve per percent change in A. Since K is a function of design speed, it can be used as a convenient "shortcut" to compute the minimum length for a crest vertical curve.

Table 15.4 values for K are based on level roadway stopping sight distance requirements. The use of K as a design control is convenient, since the value for any design speed will represent all combinations of A and L for that speed. Similarly, K values can be computed for the case where the sight distance is less than the vertical curve.

In using Eq. 15.3, it has been found that the minimum lengths obtained for the case of S greater than L do not produce practical design values and generally are not used. The common practice for this condition is to set minimum limits, which range from 30 m to 100 m. Another method to establish minimum lengths when $S > L$ is to use three times the design speed. This has appeal as design lengths are directly related to the design speed.

Table 15.4   Values of *K* for Crest Vertical Curves Based on Stopping Sight Distance

Design Speed (mi/h)	Stopping Sight Distance (ft)	Rate of Vertical Curvature, *K*[a]	
		Calculated	*Design*
15	80	3.0	3
20	115	6.1	7
25	155	11.1	12
30	200	18.5	19
35	250	29.0	29
40	305	43.1	44
45	360	60.1	61
50	425	83.7	84
55	495	113.5	114
60	570	150.6	151
65	645	192.8	193
70	730	246.9	247
75	820	311.6	312
80	910	383.7	384

[a]Rate of vertical curvature, *K*, is the length of curve per percent algebraic difference in intersecting grades (*A*).

$$K = L/A$$

SOURCE: Based on *A Policy on Geometric Design of Highways and Streets*, 2004, AASHTO, Washington, D.C.

The design procedure of using *K* values for sag vertical curves is similar to that for crest vertical curves. The headlight sight distance requirement is used for design purposes since stopping sight distances based on this requirement are generally within the limits of practical accepted values. The expressions developed earlier for minimum length of sag vertical curves based on this requirement are:

$$L_{min} = 2S - \frac{(120 + 3.5S)}{A} \qquad \text{(for } S > L) \qquad (15.7)$$

$$L_{min} = \frac{AS^2}{120 + 3.5S} \qquad \text{(for } S < L) \qquad (15.9)$$

Equation 15.9 can be written as $L_{min} = KA$. The values of *K* for design where *S* is for a level road are shown in Table 15.5.

Elevation of Crest and Sag Vertical Curves

The minimum length of a crest and sag vertical curve must be known if the elevations are to be determined. As was the case for formulas to compute length, the method for computing elevations relies on the properties of the parabola. Consider a crest vertical curve as illustrated in Figure 15.15. (A similar diagram if inverted would apply to a sag vertical curve.) The beginning of the curve is the BVC, and the end of the curve is the EVC. The intersection of the grade lines (tangents) is the PVI, which is equidistant from the BVC and EVC.

Table 15.5 Values of *K* for Sag Vertical Curves Based on Stopping Sight Distance

Design Speed (mi/h)	Stopping Sight Distance (ft)	Rate of Vertical Curvature, K^a	
		Calculated	*Design*
15	80	9.4	10
20	115	16.5	17
25	155	25.5	26
30	200	36.4	37
35	250	49.0	49
40	305	63.4	64
45	360	78.1	79
50	425	95.7	96
55	495	114.9	115
60	570	135.7	136
65	645	156.5	157
70	730	180.3	181
75	820	205.6	206
80	910	231.0	231

^{a}Rate for vertical curvature, *K*, is the length of curve (ft) per percent algebraic difference intersecting grades (*A*).

$$K = L/A$$

SOURCE: Based on *A Policy on Geometric Design of Highways and Streets*, 2004, AASHTO, Washington, D.C.

Using the properties of a parabola, $Y = ax^2 + bx + c$, where *a* is constant and *b* and *c* are 0, the locations for the minimum and maximum points and the rate of change of slope are determined from the first and second derivative:

$$Y'_{\text{max/min}} = 2ax$$

$$\frac{d^2Y}{dx^2} = 2a$$

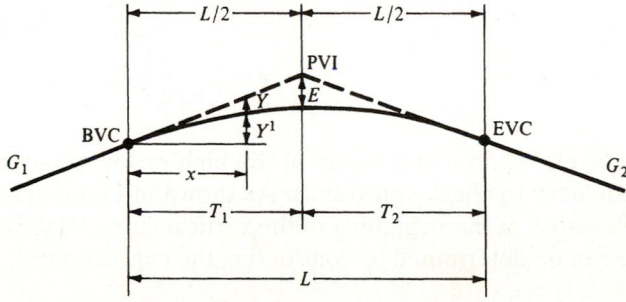

PVI = point of vertical intersection

BVC = beginning of vertical curve (same point as PVC)

EVC = end of vertical curve (same point as PVT)

E = external distance

G_1, G_2 = grades of tangents (%)

L = length of curve

A = algebraic difference of grades, $G_1 - G_2$

Figure 15.15 Layout of a Crest Vertical Curve for Design

The horizontal distance between the BVC or EVC and the PVI is $T_1 = T_2 = T_2$ and the length of the curve $L = 2T$, where L is in meters. (Recall that the length of the vertical curve is the horizontal projection of the curve and not the length along the curve.) If the total change in slope is A, then

$$2a = \frac{A}{100L}$$

and

$$a = \frac{A}{200L}$$

The equation of the curve, or tangent offset, is:

$$Y = \frac{A}{200L} x^2 \qquad (15.12)$$

When $x = L/2$, the external distance E from the point of vertical intersection (PVI) to the curve is determined by substituting $L/2$ for x in Eq. 15.12.

$$E = \frac{A}{200L}\left(\frac{L}{2}\right)^2 = \frac{AL}{800} \qquad (15.13)$$

Since stations are given in 30 m intervals, E can be written as

$$E = \frac{AN}{26.66} \qquad (15.14)$$

where N is the length of the curve in stations. The vertical offset Y at any point on the curve can also be given in terms of E. Substituting $800E/L$ for A in Eq. 15.15 will give

$$y = \left(\frac{x}{L/2}\right)^2 E \qquad (15.15)$$

The elevation and location of the high or low point of a vertical curve is used by the engineer to check constraints. As shown in Figure 15.15, Y^1, the distance between the elevation at the beginning of the vertical curve (BVC) and the highest point on the curve can be determined by considering the expression:

$$Y^1 = \frac{G_1 x}{100} - Y$$

$$= \frac{G_1 x}{100} - \frac{A}{200L} x^2 \qquad (15.16)$$

$$= \frac{G_1 x}{100} - \left(\frac{G_1 - G_2}{200L}\right) x^2$$

Differentiating Eq. 15.16 and equating it to 0 will give the value of x at the highest point on the curve:

$$\frac{dY^1}{dx} = \frac{G_1}{100} - \left(\frac{G_1 - G_2}{100L}\right) x = 0 \qquad (15.17)$$

Therefore:

$$X_{high} = \frac{100L}{(G_1 - G_2)} \frac{G_1}{100} = \frac{LG_1}{(G_1 - G_2)} \qquad (15.18)$$

where X_{high} = distance in feet from BVC to the turning point—that is, the point with the highest elevation on the curve. Equation 15.8 is also valid for sag curves.

Similarly, it can be shown that the difference in elevation between the BVC and the turning point (Y^1_{high}) can be obtained by substituting the expression X_{high} for x in Eq. 15.16. The result, Eq. 15.19, is also valid for sag curves.

$$Y^1_{high} = \frac{LG_1^2}{200(G_1 - G_2)} \qquad (15.19)$$

Design Procedure for Crest and Sag Vertical Curves

The design of a crest or sag vertical curve will generally proceed in the following manner:

Step 1. Determine the minimum length of curve to satisfy sight distance requirements and other criteria for sag curves (comfort, appearance, drainage).

Step 2. Determine from the layout plans the station and elevation of the point where the grades intersect (PVI).

Step 3. Compute the elevations of the beginning of vertical curve (BVC) and the end of vertical curve (EVC).

Step 4. Compute the offsets, Y (Eq. 15.12), as the distance between the tangent and the curve. Usually equal distances of 30 m (1 station) are used, beginning with the first whole station after the BVC.

Step 5. Compute elevations on the curve for each station as elevation of the tangent 6 offset from the tangent, ±. For crest curves the offset is (−), and for sag curves the offset is (+).

Step 6. Compute the location and elevation of the highest (crest) or lowest (sag) point on the curve using Eqs. 15.18 and 15.19.

Example 15.4 Design of Crest Vertical Curve

A crest vertical curve joining a +3 percent and a −4 percent grade is to be designed for 120 km/h. If the tangents intersect at station (345 + 18) at an elevation of 76.2 m, determine the stations and elevations of the BVC and EVC. Also, calculate the elevations of intermediate points on the curve at the whole stations. A sketch of the curve is shown in Figure 15.16.

Solution: For a design speed of 120 km/h, $K = 93.5$. From Table 15.4,

$$\text{Minimum length} = 93.5 \times [3 - (-4)] = 654.5 \text{ m}$$

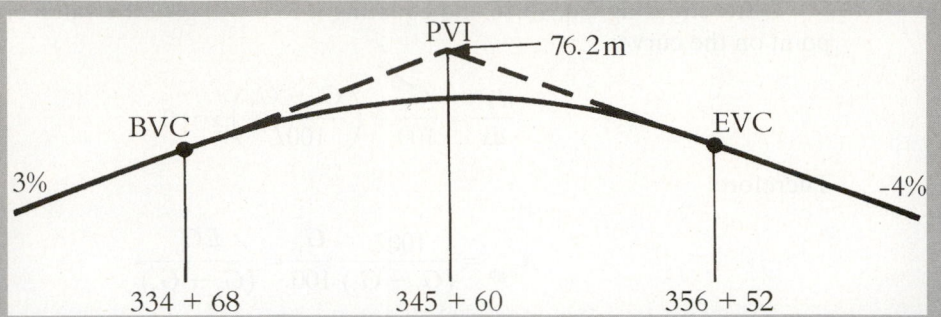

Figure 15.16   Layout of a Vertical Curve for Example 15.4

$$\text{Station of BVC} = (345 + 18) - \left(\frac{21 + 25}{2}\right) = 334 + 21$$

$$\text{Station of EVC} = (334 + 21) + (21 + 25) = 356 + 16$$

$$\text{Elevation of BVC} = 76.2 - \left(0.03 \times \frac{664.5}{2}\right) = 66.38$$

The remainder of the computation is efficiently done using the format shown in Table 15.6.

Table 15.6   Elevation Computations for Example 15.4

Station	Distance from BVC (x) (m)	Tangent Elevation (m)	Offset $\left[y = \dfrac{Ax^2}{200L}\right](m)$	Curve Elevation (Tangent Elevation – Offset) (m)
BVC 334 + 21	0	66.38	0.00	66.38
BVC 335 + 00	10	$217.24 + \dfrac{32}{100} \times 3 = 218.20$	0.01	66.67
BVC 336 + 00	40	67.58	0.09	67.49
BVC 337 + 00	70	68.48	0.26	68.22
BVC 338 + 00	100	69.38	0.53	68.85
BVC 339 + 00	130	70.28	0.90	69.38
BVC 340 + 00	160	71.18	1.37	69.81
BVC 341 + 00	190	72.08	1.93	70.15
BVC 342 + 00	220	72.98	2.59	70.39
BVC 343 + 00	250	73.88	3.34	70.54
BVC 344 + 00	280	74.78	4.19	70.59
BVC 345 + 00	310	75.68	5.14	70.54
BVC 346 + 00	340	76.58	6.18	70.40
BVC 347 + 00	370	77.48	7.32	70.16
BVC 348 + 00	400	78.38	8.56	69.82
BVC 349 + 00	430	79.28	9.89	69.39
BVC 350 + 00	460	80.18	11.32	68.86
BVC 351 + 00	490	81.08	12.84	69.76

Station	Distance from BVC (x) (m)	Tangent Elevation (m)	Offset $\left[\gamma = \dfrac{Ax^2}{200L}\right]$ (m)	Curve Elevation (Tangent Elevation − Offset) (m)
BVC 352 + 00	520	81.98	14.46	67.52
BVC 353 + 00	550	82.88	16.18	66.70
BVC 354 + 00	580	83.78	17.99	65.79
BVC 355 + 00	610	84.68	19.90	64.78
BVC 356 + 00	640	84.58	21.90	63.68
EVC 356 + 16	654.5	85.85	22.91	62.94

Computation of the elevations at different points on a sag curve follows the same procedure as that for the crest vertical curve. For sag curves, the offset Y is added to the appropriate tangent elevation to obtain the curve elevation since the elevation of the curve is higher than the elevation of the tangent.

Example 15.5 Design of Sag Vertical Curve

A sag vertical curve joins a −3 percent grade and a +3 percent grade. If the PVI of the grades is at station (435 + 50) and has an elevation of 70 m, determine the station and elevation of the BVC and EVC for a design speed of 112 km/h. Also, compute the elevation on the curve at 30-m intervals. Figure 15.17 shows a layout of the curve.

Solution: For a design speed of 112 km/h, $K = 54.3$, using the higher rounded value in Table 15.6.

$$\text{Length of curve} = 54.3 \times 6 = 325.8 \text{ m}$$

$$\text{Station of BVC} = (435 + 15) - (5 + 43) = 430 + 02$$

$$\text{Station of EVC} = (435 + 15) - (5 + 43) = 440 + 28$$

$$\text{Elevation of BVC} = 70 + 0.03 \times 163 = 74.89$$

$$\text{Elevation of EVC} = 70 + 0.03 \times 163 = 74.89$$

The computation of the elevations is shown in Table 15.7.

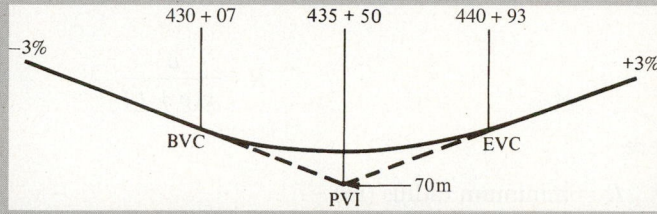

Figure 15.17 Layout for a Sag Vertical Curve for Example 15.5

Table 15.7   Elevation Computations for Example 15.5

Station	Distance from BVC (x) (m)	Tangent Elevation (m)	Offset $\left[\gamma = \dfrac{Ax^2}{200L}\right](m)$	Curve Elevation (Tangent Elevation + Offset) (m)
BVC 430 + 02	0	74.89	0.00	74.89
BCV 431 + 00	93	74.05	0.08	74.13
BCV 432 + 00	58	73.15	0.32	73.47
BCV 433 + 00	88	72.25	0.73	72.98
BCV 434 + 00	118	71.35	1.19	72.54
BVC 435 + 00	148	70.45	2.02	72.47
BCV 436 + 00	178	69.55	2.92	72.47
BCV 437 + 00	208	68.65	3.99	72.54
BCV 438 + 00	238	67.75	5.23	72.98
BCV 439 + 00	268	66.85	6.62	73.47
BCV 440 + 00	298	65.95	8.18	74.13
EVC 440 + 28	325.8	65.11	9.79	74.90

15.2.3 Horizontal Alignment

The horizontal alignment consists of straight sections of the road (known as tangents) connected by curves. The curves are usually segments of circles, which have radii that will provide for a smooth flow of traffic. The design of the horizontal alignment entails determination of the minimum radius, determination of the length of the curve, and computation of the horizontal offsets from the tangents to the curve to facilitate locating the curve in the field. In some cases, to avoid a sudden change from a tangent with infinite radius to a curve of finite radius, a curve with radii varying from infinite to the radius of the circular curve is placed between the circular curve and the tangent. Such a curve is known as a *spiral* or *transition curve*. There are four types of horizontal curves: simple, compound, reversed, and spiral. Computations required for each type are presented in the following sections.

Simple Curves

Figure 15.18 is a layout of a simple horizontal curve. The curve is a segment of a circle with radius R, which is discussed in Chapter 3 for the case when *SSD* is unobstructed. The relationship was shown to be

$$R = \frac{u^2}{g(e + f_s)}$$

where

R = minimum radius (m)
u = design speed (km/h)
e = superelevation (m/m)
f_s = coefficient of side friction

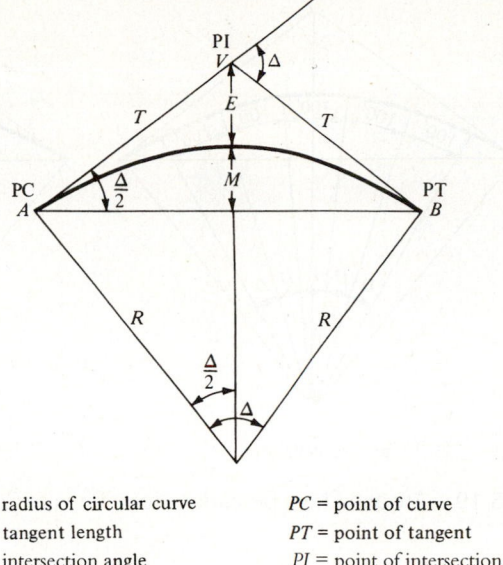

R = radius of circular curve PC = point of curve
T = tangent length PT = point of tangent
Δ = intersection angle PI = point of intersection
M = middle ordinate E = external distance

Figure 15.18   Layout of a Simple Horizontal Curve

When there are obstructions in the roadway that limit *SSD* on the curve, R is determined as described later in this section.

As Figure 15.18 illustrates, the point at which the curve begins (A) is known as the *point of curve* (PC), and the point at which it ends (B) is known as the *point of tangent* (PT). The point at which the two tangents intersect is known as the *point of intersection* (PI) or vertex (V). A simple circular curve is described either by its radius (for example, 60 m-radius curve) or by the degree of the curve (for example, a 4-degree curve). There are two ways to define degree of the curve, which is based on 30 m of arc length or on 30 m of chord length. Highway practice (which is the focus of this chapter) uses arc definition, whereas railroad practice uses chord definition.

The angle subtended at the center of a circular arc 30 m in length as shown in Figure 15.19a is the degree of curve as used in highway work. For example, a 2-degree curve subtends an arc of 30 m if the central angle is 2 degrees.

If θ is the angle in radians subtended at the center by an arc of a circle, the length of that arc is given by $R\theta = L$. If D_a° is the angle in degrees subtended at the center by an arc of length L, then

$$\theta = \frac{\pi D_a^\circ}{180}$$

since

$$\frac{\pi R}{180} = \frac{30}{D_a^\circ}$$

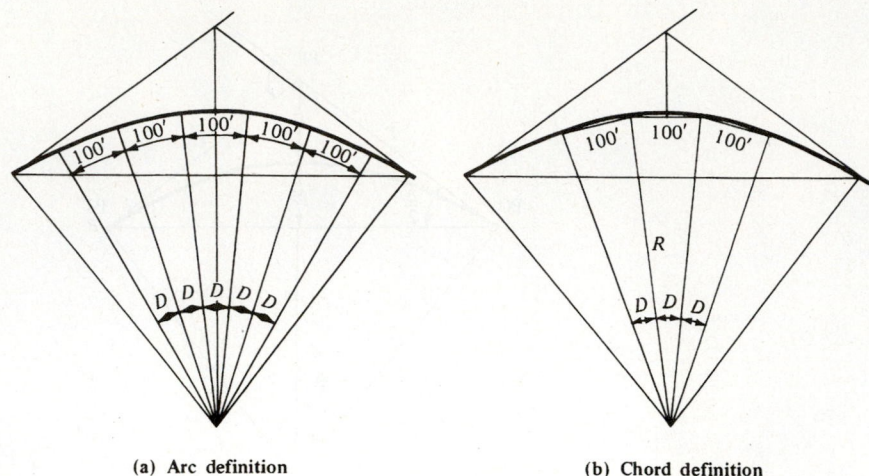

(a) Arc definition   (b) Chord definition

Figure 15.19   Arc and Chord Definitions for a Circular Curve

Solving for R yields

$$R = \frac{180 \times 30}{\pi D_a^\circ}$$

or

$$R = \frac{1719}{D_a^\circ} \tag{15.20}$$

Radius and degree of curve are not independent of each other when one value has been selected; the other has been defined as well. The degree of curve can thus be determined if the radius is known, or the radius can be determined if the degree of curve is known. For CADD programs such as InRoads, the radius is specified, whereas in traditional highway practice, the degree of curve is stated.

Chord definition for R in terms of D_c° is based on a chord of 30 m, as illustrated in Figure 15.19b. For this case, the radius of curve is as follows.

$$R = \frac{15}{\sin D_c^\circ/2} \tag{15.21}$$

Since the arc definition is used for highway work, all further references to degree of curve in this chapter are to D_c.

Formulas for Simple Curves.   Figure 15.18 illustrates the elements of a simple curve. Using the properties of a circle, the two tangent lengths AV and BV are equal and designated as T. The angle Δ that is formed by the two tangents is known as the *intersection* angle because it is the angle that is formed by the intersection of the tangents. The expression for the tangent length is

$$T = R \tan \frac{\Delta}{2} \tag{15.22}$$

The expression for the length C of the chord AB, which is known as the *long chord*, is

$$C = 2R \sin \frac{\Delta}{2} \tag{15.23}$$

The expression for the external distance E, which is the distance from the point of intersection to the curve on a radial line is

$$E = R \sec \frac{\Delta}{2} - R$$

$$E = R \left(\frac{1}{\cos \dfrac{\Delta}{2} - 1} \right) \tag{15.24}$$

The expression for the middle ordinate M, which is the distance between the midpoint of the long chord and the midpoint of the curve is:

$$M = R - R \cos \frac{\Delta}{2}$$

$$= R \left(1 - \cos \frac{\Delta}{2} \right) \tag{15.25}$$

The expression for the length of the curve L is:

$$L = \frac{R \Delta \pi}{180} \tag{15.26}$$

Field Location of a Simple Horizontal Curve.   Simple horizontal curves are usually located in the field by staking out points on the curve using angles measured from the tangent at the point of curve (PC) and the lengths of the chords joining consecutive whole stations. The angles are also called "deflection angles" because they are the angle that is "deflected" when the direction of the tangent changes direction to that of the chord. Figure 15.20 is a schematic of the procedure involved.

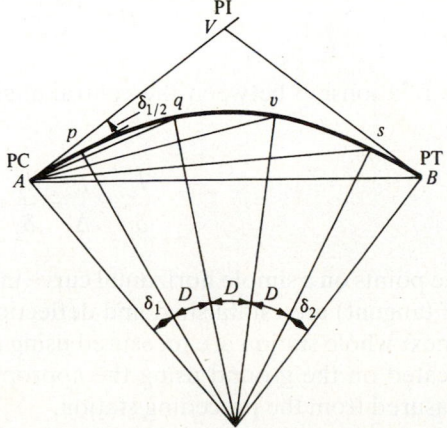

Figure 15.20   Deflection Angles on a Simple Circular Curve

The first deflection angle VAp to the first whole station on the curve, which is usually less than a station away from the PC, is equal to $\dfrac{\delta_1}{2}$ based on the properties of a circle.

The next deflection angle VAq is

$$\frac{\delta_1}{2}+\frac{D}{2}$$

and the next deflection angle VAv is

$$\frac{\delta_1}{2}+\frac{D}{2}+\frac{D}{2}=\frac{\delta_1}{2}+D$$

The next deflection angle VAs is

$$\frac{\delta_1}{2}+\frac{D}{2}+\frac{D}{2}+\frac{D}{2}+\frac{\delta}{2}+\frac{3D}{2}$$

and the last deflection angle VAB is

$$\frac{\delta_1}{2}+\frac{D}{2}+\frac{D}{2}+\frac{D}{2}+\frac{\delta_2}{2}=\frac{\delta_1}{2}+\frac{\delta_2}{2}+\frac{3D}{2}=\frac{\Delta}{2}$$

To set out the horizontal curve, it is necessary to determine δ_1 and δ_2. The length of the first arc, l_1, is related to δ_1 as

$$l_1=\frac{R\pi}{180}\delta_1 \tag{15.27}$$

Solving for R provided the following expression.

$$R=\frac{l_1\times180}{\delta_1\pi}$$

Equating R from Eq. 15.27,

$$R=\frac{180L}{\Delta\pi}$$

provides the relationship between the central angle that subtends the length of arc as follows.

$$\frac{l_1}{\delta_1}=\frac{L}{\Delta}=\frac{l_2}{\delta_2}$$

To locate points on a simple horizontal curve in the field, the PC (point of curve) and PT (point of tangent) are established and deflection angles between the tangent at the PC and the next whole station are measured using a transit set over the PC. Each whole station is located on the ground using the appropriate deflection angle and the chord distance measured from the preceding station.

Lengths l_1 and l_2 are the actual distance along the curve. Thus, to locate endpoints of these curves, chord lengths corresponding to the arc length must be computed.

The expression relating chord lengths to the corresponding arc length l_1 and l_2 and 100 ft are:

$$C_1 = 2R \sin \frac{\delta_1}{2}$$

$$C_D = 2R \sin \frac{D}{2}$$

$$C_2 = 2R \sin \frac{\delta_2}{2} \qquad (15.28)$$

where C_1, C_D, and C_2 are the first, intermediate, and last chords, respectively.

Figure 15.20 and the following formulas summarize the relationships for deflection angles and chord lengths required to set out a simple curve.

Chord: $Ap = 2R \sin \dfrac{\delta_1}{2}$ Deflection angle: $VAp = \dfrac{\delta_1}{2}$

Chord: $pq = 2R \sin \dfrac{D_a}{2}$ Deflection angle: $VAq = \dfrac{\delta_1 + D}{2}$

Chord: $sB = 2R \sin \dfrac{\delta_2}{2}$ Deflection angle: $VAB = \dfrac{\delta_1 + D + D + D + \delta_2}{2} = \dfrac{\Delta}{2}$

Example 15.6 Design of a Simple Horizontal Curve

The intersection angle of a 4° curve is 55°25′, and the PC is located at station 238 + 13.43. Determine the length of the curve, the station of the PT, the deflection angles, and the chord lengths for setting out the curve at whole stations from the PC. Figure 15.21 illustrates a layout of the curve.

Solution:

$$\text{Radius of curve} = \frac{1719}{D} = \frac{1719}{4}$$

$$= 429.8 \text{ m}$$

$$\text{Length of curve} = \frac{R\Delta\pi}{180} = \frac{429.8 \times 55.4167\pi}{180}$$

$$= 415.7 \text{ m}$$

The station at PT is equal to station $(238 + 13.43) + (13 + 25.63) = 252 + 9.05$ stations. The distance between the PC and the first station is $239 - (238 + 13.43) = 16.57$ m.

$$\frac{\delta_1}{\Delta} = \frac{l_1}{L}$$

$$\frac{\delta_1}{55.4167} = \frac{16.57}{415.7}$$

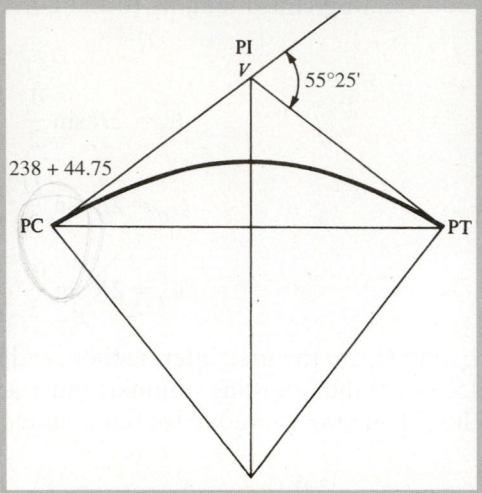

Figure 15.21 Layout of Curve for Example 15.6

Therefore,

$$\delta_1 = 2.210°$$

$$C_1 = 2 \times 429.8 \sin\left(\frac{2.210}{2}\right) = 16.58 \text{ m}$$

The first deflection angle to station 239 is $\delta_1/2 = 1.105° = 1°6'18''$.
   Similarly,

$$l_2 = (252 + 9.05) - (252) = 9.05 \text{ m}$$

$$\frac{\delta_2}{2} = \frac{9.05}{415.7} \times \frac{55.4167}{2} = 0.6034°$$

$$= 36'12''$$

$$C_2 = 2 \times 1432.4 \sin(0.6034°)$$

$$= 9.05 \text{ m}$$

$$D = 4°$$

$$C_D = 2 \times 429.8 \sin\left(\frac{4}{2}\right)$$

$$= 30 \text{ m}$$

   Note that the deflection angle to PT is half the intersection angle Δ of the tangents. This relationship serves as a check of the computation. Since highway curves are relatively flat, the chord lengths are approximately equal to the arc lengths.
   The other deflection angles are computed in Table 15.8.

Table 15.8   Computations of Deflection Angles and Chord Lengths for Example 15.6

Station	Deflection Angle	Chord Length (m)
PC 238 + 13.43	0	0
PC 239	1°6′18″	16.58
PC 240	3°6′18″	30
PC 241	5°6′18″	30
PC 242	7°6′18″	30
PC 243	9°6′18″	30
PC 244	11°6′18″	30
PC 245	13°6′18″	30
PC 246	15°6′18″	30
PC 247	17°6′18″	30
PC 248	19°6′18″	30
PC 249	21°6′18″	30
PC 250	23°6′18″	30
PC 251	25°6′18″	30
PC 252	27°6′18″	30
PT 252 + 9.06	27°42′30″	9.05

Compound Curves

Compound curves consist of two or more simple curves in succession, turning in the same direction, with any two successive curves having a common tangent point. Figure 15.22 shows a typical layout of a compound curve, consisting of two simple curves. These curves are used mainly to obtain desirable shapes of the horizontal alignment, particularly for at-grade intersections, ramps of interchanges, and highway sections in difficult topographic areas. To avoid abrupt changes in the alignment, the radii of any two consecutive simple curves that form a compound curve should not be widely different. AASHTO recommends that the ratio of the flatter radius to the sharper radius at intersections should not be greater than 2:1 so drivers can adjust to sudden changes in curvature and speed. The maximum desirable ratio recommended for interchanges is 1.75:1, although 2:1 may be used.

To provide a smooth transition from a flat curve to a sharp curve, and to facilitate a reasonable deceleration rate on a series of curves of decreasing radii, the length of each curve should observe minimum length requirements, based on the radius of each curve as recommended by AASHTO and given in Table 15.9, which are developed based on the premise that travel is in the direction of the sharper curve. The 2:1 ratio of the flatter radius should preferably not be exceeded but is not critical for the acceleration condition.

Figure 15.22 shows seven variables, R_1, R_2, Δ_1, Δ_2, Δ, T_1, and T_2, six of which are independent, since $\Delta = \Delta_1 + \Delta_2$. Several solutions can be developed for the compound curve. The vertex triangle method is presented since this method is frequently used in highway design. In Figure 15.22, R_1 and R_2 are usually known. The following equations can be used to determine the remaining variables.

$$\Delta = \Delta_1 + \Delta_2 \tag{15.29}$$

$$t_1 = R_1 \tan \frac{\Delta_1}{2} \tag{15.30}$$

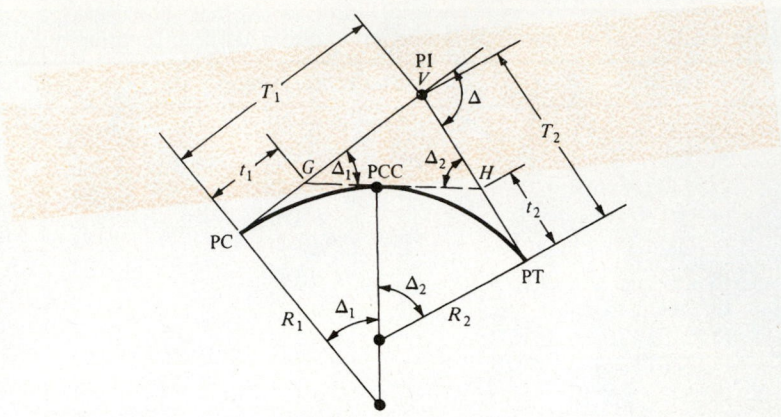

R_1, R_2 = radii of simple curves forming compound curve
Δ_1, Δ_2 = intersection angles of simple curves
Δ = intersection angle of compound curve
t_1, t_2 = tangent lengths of simple curves
T_1, T_2 = tangent lengths of compound curve
PCC = point of compound curve
PI = point of intersection
PC = point of curve
PT = point of tangent

Figure 15.22   Layout of a Compound Curve

$$t_2 = R_2 \tan \frac{\Delta_2}{2} \tag{15.31}$$

$$\frac{\overline{VG}}{\sin \Delta_2} = \frac{\overline{VH}}{\sin \Delta_1} = \frac{t_1 + t_2}{\sin (180 - \Delta)} = \frac{t_1 + t_2}{\sin \Delta} \tag{15.32}$$

$$T_1 = \overline{VG} + t_1 \tag{15.33}$$

$$T_1 = \overline{VH} + t_2 \tag{15.34}$$

where

R_1 and R_2 = radii of simple curves forming the compound curve
Δ_1 and Δ_2 = intersection angles of simple curves
t_1 and t_2 = tangent lengths of simple curves

Table 15.9   Lengths of Circular Arc for Different Compound Curve Radii

Minimum Length of Circular Arc (ft)	Radius (ft)						
	100	*150*	*200*	*250*	*300*	*400*	*500 or more*
Acceptable	40	50	60	80	100	120	140
Desirable	60	70	90	120	140	180	200

SOURCE: Based on *A Policy on Geometric Design of Highways and Streets*, 2004, AASHTO, Washington, D.C.

T_1 and T_2 = tangent lengths of compound curves
Δ = intersection angle of compound curve

In order to set out the curve, the intersection angles and chord lengths for both curves must be determined. Usually, Δ_1 or Δ_2 can be obtained from the layout plans and Eqs. 15.29 through 15.34 can be used to solve for Δ_1 or Δ_2, VG, VH, t_1, t_2, T_1, and T_2.

Example 15.7 Design of a Compound Curve

Figure 15.23 illustrates a compound curve that is to be set out at a highway intersection. If the point of compound curve (PCC) is located at station (565 + 10.5), determine the deflection angles for setting out the curve.

Solution:

$$t_1 = 150 \tan \frac{34}{2} = 45.86 \text{ m}$$

$$t_2 = 105 \tan \frac{26}{2} = 24.24 \text{ m}$$

For length of horizontal curve of 150 - m radius,

$$L = R\Delta_1 \frac{\pi}{180} = 105 \times \frac{34\pi}{180}$$

$$= 89.01 \text{ m}$$

For length of horizontal curve of 105 m radius,

$$L = R\Delta_2 \frac{\pi}{180} = 180 \times \frac{26\pi}{180}$$

$$= 47.65 \text{ m}$$

Therefore, station of the PC is equal to $(565 + 10.50) - (2 + 29.01) = 562 + 11.49$.

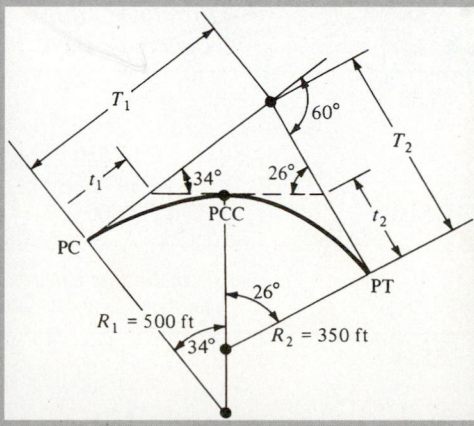

Figure 15.23 Compound Curve for Example 15.7

812 Part 4 Location, Geometrics, and Drainage

The station of the PT is equal to $(565 + 10.50) + (2 + 29.01) = 566 + 11.49$.
For curve of 150 m radius,

$$\frac{D}{2} = \frac{1719}{2 \times 150} = 5°43'47'' \qquad \text{(from Eq. 15.20)}$$

$$l_1 = (563 + 00) - (562 + 11.49) = 18.51 \text{ m}$$

$$\frac{\delta_1}{l_1} = \frac{\Delta}{L}$$

$$\frac{\delta_1}{2} = \frac{18.51 \times 34}{2 \times 89.01} = 3°32'8''$$

$$l_2 = (565 + 10.50) - (565 + 00) = 10.5 \text{ m}$$

$$\frac{\delta_2}{2} = \frac{35 \times 34}{2 \times 296.71} = 2°0'19''$$

For curve of 105 -m radius,

$$\frac{D}{2} = \frac{17.19}{2 \times 105} = 8°11'7''$$

$$l_1 = (566 + 00) - (565 + 10.50) = 19.5 \text{ m}$$

$$\frac{\delta_1}{2} = \frac{65 \times 26}{2 \times 158.82} = 5°19'16''$$

$$l_2 = (566 + 28.15) - (566 + 00) = 28.15 \text{ m}$$

$$\frac{\delta_2}{2} = \frac{28.15 \times 26}{2 \times 47.65} = 7°40'44''$$

Computations of the deflection angles are shown in Table 15.10.

The deflection angles for the 105 m radius curve are turned from the common tangent with the transit located at PCC. Since each simple curve is relatively flat, calculated lengths of the chords are almost equal to the corresponding arc lengths.

Table 15.10 Computations for Example 15.7

Station	150 m Radius Curve Deflection Angle	Chord Length (m)
PC 562 + 38.29	0	0
563	3°32'8''	61.66'
564	9°15'55''	99.84'
565	14°59'42''	99.84'
PCC 565 + 35.00	17°00'00''	35.00'

Station	105 m Radius Curve Deflection Angle	Chord Length (m)
PCC 565 + 35.00	0	0
566	5°19'16''	64.9'
PT 566 + 93.82	13°00'00''	93.5'

Reverse Curves

Reverse curves usually consist of two simple curves with equal radii turning in opposite directions with a common tangent. They are generally used to change the alignment of a highway. Figure 15.24 shows a reverse curve with parallel tangents.

If d and D are known, the following variables can be determined.

$$\Delta = \Delta_1 = \Delta_2$$

$$\text{Angle } OWX = \frac{\Delta_1}{2} = \frac{\Delta_2}{2}$$

$$\text{Angle } OYZ = \frac{\Delta_1}{2} = \frac{\Delta_2}{2}$$

Therefore, WOY is a straight line, and hence,

$$\tan \frac{\Delta}{2} = \frac{d}{D}$$

$$d = R - R \cos \Delta_1 + R - R \cos \Delta_2$$

$$= 2R(1 - \cos \Delta)$$

$$R = \frac{d}{2(1 - \cos \Delta)} \tag{15.35}$$

If d and R are known,

$$\cos \Delta = 1 - \frac{d}{2R}$$

$$D = d \cos \frac{\Delta}{2} \tag{15.36}$$

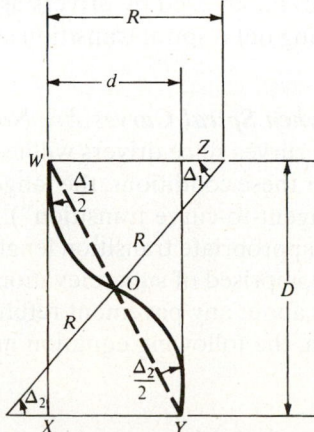

R = radius of simple curves
Δ_1, Δ_2 = intersection angles of simple curves
d = distance between parallel tangents
D = distance between tangent points

Figure 15.24   Geometry of a Reverse Curve with Parallel Tangents

Reverse curves are seldom recommended because sudden changes to the alignment may result in drivers finding it difficult to keep in their lanes. When it is necessary to reverse the alignment, a preferable design consists of two simple horizontal curves, separated by a tangent between them, to achieve full superelevation. Alternatively, an equivalent length of spiral, which is described in the next section, may separate the simple curves.

Transition Curves

Transition curves are placed between tangents and circular curves or between two adjacent circular curves having substantially different radii. The use of transition curves provides a vehicle path that gradually increases or decreases the radial force as the vehicle enters or leaves a circular curve.

Length of Spiral Curves.   If the transition curve is a spiral, the degree of curve between the tangent and the circular curve varies from 0 at the tangent end to the degree of the circular curve D_a at the curve end. When the transition is placed between two circular curves, the degree of curve varies from that of the first circular curve to that of the second circular curve.

The expression given in Eqs. 15.37 and 15.38 is used by some highway agencies to compute the minimum length of a spiral transition curve. The minimum length should be the larger of the values obtained from these equations.

$$L_{s,\min} = \frac{3.15u^3}{RC} \tag{15.37}$$

$$L_{s,\min} = \sqrt{24(p_{\min})R} \tag{15.38}$$

where

L_s = minimum length of curve (m)
u = speed (km/h)
R = radius of curve (m)
C = rate of increase of radial acceleration (m/sec^2/sec); values range from 1 to 3
$P_{\min}$ = minimum lateral offset between the tangent and the circular curve (0.2 m)

Under operational conditions, the most desirable length of a spiral curve is approximately the length of the natural spiral path used by drivers as they traverse the curve. The computations required for setting out a spiral transition curve is beyond the scope of this book.

Length of Superelevation Runoff when Spiral Curves Are Not Used.   Many highway agencies do not use spiral transition curves since drivers will usually guide their vehicles into circular curves gradually. Under these conditions, the tangent is joined directly with the main circular curve (called "tangent-to-curve transition"). However, if the curve is superelevated at a rate of e m/m, an appropriate transition length must be provided. This superelevation transition length is comprised of superelevation runoff and tangent runout. For highways where rotation is about any pavement reference line and the rotated width has a common superelevation, the following equation may be used to determine superelevation runoff.

$$L_r = \frac{(wn_1)e_d}{\Delta}(b_w) \tag{15.39}$$

where

L_r = minimum length of superelevation runoff
Δ = maximum relative gradient (%) (0.78% @ 24 kmph to 0.35% @ 128 kmph)
n_1 = number of lanes rotated
b_w = adjustment factor for number of lanes rotated (1 = 1.00, 2 = 0.75, 3 = 0.67)
w = width of one traffic lane (m) (typically 3.6 m)
e_d = design superelevation rate (%)

AASHTO recommends minimum superelevation runoff lengths where either one or two lanes are rotated about the pavement edge as shown in Table 15.11. These values are based on concerns for appearance and comfort and thus a maximum acceptable difference between the longitudinal grades of the centerline (or axis of rotation) and the edge of the pavement. In this case, superelevation runoff is defined as the distance over which the pavement cross slope on the outside lane changes from zero (flat) to full superelevation of the curve (e).

Typically, the runoff length is divided between the tangent and the curved section and avoids placing the runoff either entirely on the tangent or the curve. Theoretically, superelevation runoff should be placed entirely on the tangent section, thus providing full superelevation between the PC and PT. In practice, sharing the runoff between tangent and curve reduces peak lateral acceleration and its effect on side friction. Motorists tend to adjust their driving path by steering a "natural spiral," thus supporting the observation that some of the runoff length should be on the curve.

The length of the tangent runout consists of the length of roadway needed to accomplish a change on the outside-lane cross slope from normal (i.e., 2 percent) to zero, or vice versa. The sum of the superelevation runoff and tangent runout comprises the total distance over which transition from normal crown to full superelevation is achieved. The following equation can be used to determine the minimum runout length. The values

Table 15.11   Superelevation Runoff L_r (m) for Horizontal Curves

e (%)	Design Speed (km/h)													
	32		48		64		80		96		112		128	
	1	2	1	2	1	2	1	2	1	2	1	2	1	2
1.5	0	0	0	0	0	0	0	0	0	0	0	0	0	0
2.0	9.6	14.7	10.8	16.5	12.3	18.6	14.4	21.6	15.9	24.0	18.0	27.0	27.0	30.9
3.0	14.7	21.9	16.5	24.6	18.6	27.9	21.6	32.4	24.0	36.0	27.0	40.5	30.9	46.2
4.0	19.5	29.1	21.9	32.7	24.9	37.2	28.8	43.2	32.1	48.0	36.0	54.0	41.1	61.8
5.0	24.3	36.6	27.3	40.8	30.9	46.5	36.0	54.0	39.9	60.0	45.0	67.5	51.3	77.1
6.0	29.1	43.8	32.7	49.2	37.2	55.8	43.2	64.8	48.0	72.0	54.0	81.0	62.1	92.7
7.0	34.2	51.0	38.1	57.3	43.5	65.1	50.4	75.6	56.1	84.0	63.0	94.5	72.0	108.0
8.0	39.0	58.5	43.5	65.4	49.8	74.4	57.6	86.4	63.9	96.0	72.0	108.0	82.2	123.3
9.0	43.8	65.7	49.2	73.5	55.8	83.7	64.8	97.2	72.0	108.0	81.0	121.5	92.7	138.9
10.0	48.6	72.9	54.6	81.9	62.1	93.0	72.0	108.0	80.1	120.0	90.0	225.0	102.9	154.2
11.0	53.4	80.4	60.0	90.0	68.4	102.3	79.2	118.8	87.9	132.0	99.0	148.5	113.1	169.8
12.0	58.5	87.6	65.4	98.1	74.4	111.6	86.4	129.6	96.0	144.0	108.0	162.0	123.3	185.1

Note: (1) Two-lane—3.6 m 2% cross slope
(2) Multilane—3.6 m each direction rotated separately

obtained by substituting appropriate values into the equation are similar to those shown in Table 15.11 for $e = 2$ percent.

$$L_r = \frac{e_{NC}}{e_d} L_r \tag{15.40}$$

where

L_t = minimum length of tangent runout (m)
e_{NC} = normal cross slope rate (%)
e_d = design superelevation rate (%)
L_r = minimum length of superelevation runoff (m)

AASHTO Lengths and Tangent Runouts for Spiral Curves.

AASHTO recommends that when spiral curves are used in transition design, the superelevation runoff should be achieved over the length of the spiral curve. Based on this, it is recommended that the length of the spiral curve should be the length of the superelevation runoff. The runout spiral length and runout length are very different when values of e increase beyond 2 percent. Recommended lengths for tangent runout are shown in Table 15.12. Since it is desirable to maintain runoff within the spiral curve, its length should be increased to that of the runoff values shown in Table 15.12, which are based on a 2.0 percent normal cross slope. The change in cross slope begins by introducing a tangent runout section just in advance of the spiral curve.

Table 15.12   Tangent Runout Length for Spiral Curve Transition Design

| | Tangent Runout Length (ft) | | | | |
| | Superelevation Rate (%) | | | | |
Design Speed (mi/h)	2	4	6	8	10
15	44	—	—	—	—
20	59	30	—	—	—
25	74	37	25	—	—
30	88	44	29	—	—
35	103	52	34	26	—
40	117	59	39	29	—
45	132	66	44	33	—
50	147	74	49	37	—
55	161	81	54	40	—
60	176	88	59	44	—
65	191	96	64	48	38
70	205	103	68	51	41
75	220	110	73	55	44
80	235	118	78	59	47

Note: (1) Values for $e = 2\%$ represent the desirable lengths of the spiral curve transition.
        (2) Values shown for tangent runout should also be used as the minimum length of the spiral transition curve.
SOURCE: Based on *A Policy on Geometric Design of Highways and Streets*, 2004, AASHTO, Washington, D.C.

Attainment of Superelevation.   It is essential that the change from a crowned cross section to a superelevated one be achieved without causing any discomfort to motorists or creating unsafe conditions. One of four methods can be used to achieve this change on undivided highways.

1. A crowned pavement is rotated about the profile of the centerline.
2. A crowned pavement is rotated about the profile of the inside edge.
3. A crowned pavement is rotated about the profile of the outside edge.
4. A straight cross-slope pavement is rotated about the profile of the outside edge.

Figure 15.25 is a schematic of Method 1. This is the most commonly used method since the distortion obtained is less than that obtained with other methods. The procedure used is first to raise the outside edge of the pavement relative to the centerline, until the outer half of the cross section is horizontal. The outer edge is then raised by an additional amount to obtain a straight cross section. The inside edge is still at its original elevation. The entire cross section is then rotated as a unit about the centerline profile until the full superelevation is achieved. Straight lines have been used to illustrate this method, but in practice, angular breaks are appropriately rounded by using short vertical curves.

In Method 2, the centerline profile is raised with respect to the inside pavement edge to obtain half the required change, while the remaining half is achieved by raising the outside pavement edge with respect to the profile of the centerline.

Method 3 is similar to Method 2, with the only difference being that a change is effected below the outside edge profile. Method 4 is used for sections of straight cross slopes.

Curve Radii Based on Stopping Sight Distance

It was shown in Chapter 3 that the minimum radius of a horizontal curve depends on the design speed u of the highway, the superelevation e, and the coefficient of side friction f_s. This relationship was shown to be

$$R = \frac{u^2}{15(e + f_s)}$$

Normally, this value for R is sufficient for design purposes. However, there are instances when a constraint may exist. For example, if an object is located near the inside edge of the road, the driver's view may be blocked. When this situation exists, the

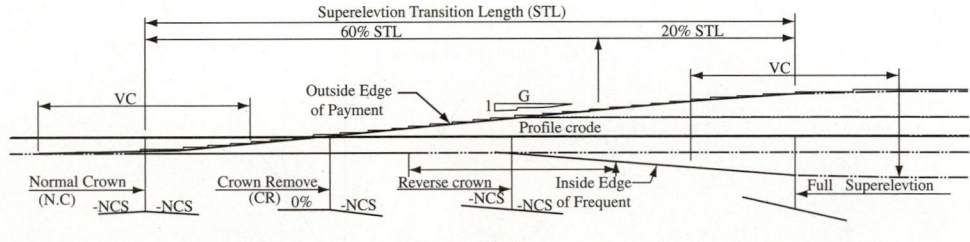

Figure 15.25   Diagrammatic Profiles Depicting Attaining Superelevation for a Crowned Pavement Rotated about the Profile of the Centerline

SOURCE: Alabama Department of Transportation, *ALDOT Standard Drawings,* 2011 Edition, SSEC-1 (Sheet 3 of 8)

design of the horizontal curve based on design speed is compromised and the designer has two options: change the radius of the curve to assure adequate *SSD* or post a lower speed limit on the curve. Figure 15.26a is a schematic diagram of a horizontal curve with sight distance restrictions due to an object located within the curve line of sight.

If a vehicle is located at point *A* on the curve and the object is at point *B*, the line of sight is the length of chord *AB*. The horizontal distance traversed by the vehicle when moving from point *A* to point *B* is the length of arc *AB*. The central angle for arc *AB* is defined as 2θ (in degrees). Thus, the expression for *SSD* is

$$\frac{2R\theta\pi}{180} = S$$

and

$$\theta = \frac{180(S)}{2\pi R} = \frac{28.65}{R}(S) \tag{15.41}$$

where

      R = radius of horizontal curve
      S = stopping sight distance
      θ = one half central angle

From Figure 15.26a, we can write

$$\frac{R - m}{R} = \cos\theta \tag{15.42}$$

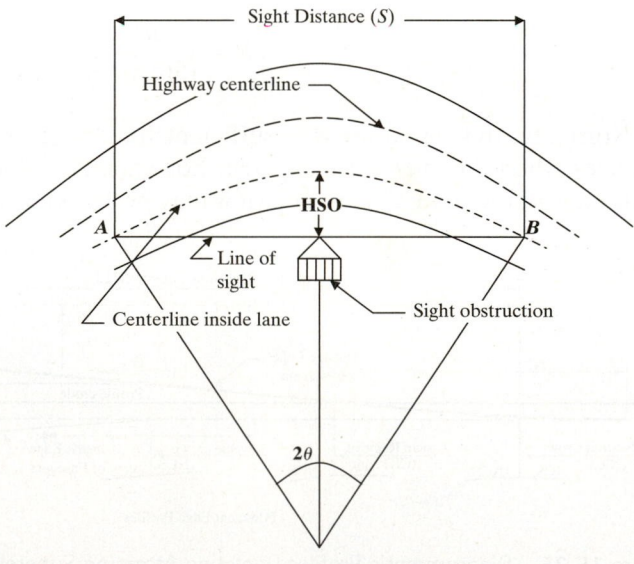

Figure 15.26(a)   Horizontal Curves with Sight-Distance Restrictions and Range of Lower Values for Stopping Sight Distances

SOURCE: Based on *A Policy on Geometric Design of Highways and Streets*, 2004, AASHTO, Washington, D.C.

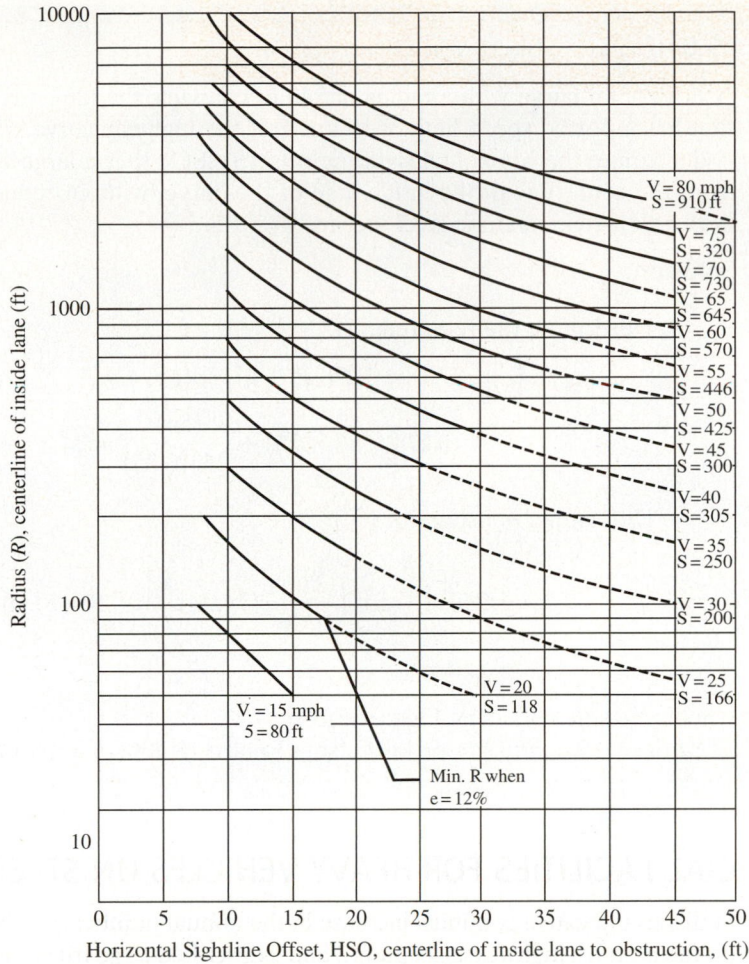

Figure 15.26(b)   Horizontal Curves with Sight-Distance Restrictions and Range of Lower Values for Stopping Sight Distances

SOURCE: Based on *A Policy on Geometric Design of Highways and Streets*, 2004, AASHTO, Washington, D.C.

Equating the cosines of θ from Eqs. 15.41 and 15.42 yields

$$\cos \frac{28.65}{R}(S) = \frac{R - m}{R}$$

where m = the horizontal sightline offset, HSO (m). Solving for m produces the following relationship:

$$m = R\left(1 - \cos \frac{28.65}{R} S\right) \tag{15.43}$$

Equation 15.43 can be used to determine m, R, or S, depending on which two of the three variables are known. Figure 15.26b is a graphical representation of Eq. 15.43 and can be used to determine the value of the unknown variable.

Example 15.8 Location of Object Near a Horizontal Curve

A horizontal curve with a radius of 240 m connects the tangents of a two-lane highway that has a posted speed limit of 56 km/h. If the highway curve is not superelevated, $e = 0$, determine the horizontal sightline offset (HSO) that a large billboard can be placed from the centerline of the inside lane of the curve, without reducing the required SSD. Perception-reaction time is 2.5 sec, and $f = 0.35$.

Solution:

- Determine the required SSD.

$$SSD = 0.278\,ut + u^2/30(f \pm G)$$

$$(0.278 \times 56 \times 25) + \frac{(56)^2}{254(0.35)} = 74.20 \text{ m}$$

- Determine m using Eq. 15.39.

$$m = 240\left[1 - \cos\left(\frac{28.65}{240}(74.20)\right)\right] = 240(1 - 0.988) \text{ m}$$

$$= 2.86 \text{ m}$$

Check the solution using Figure 15.26b.
    For $R = 240$ and $V = 56$ km/h from Figure 15.26b, m is estimated to be 2.85 m.

15.3 SPECIAL FACILITIES FOR HEAVY VEHICLES ON STEEP GRADES

Statistics indicate a continual increase in the annual number of vehicle-miles of large trucks on the nation's highways. Chapters 3 and 9 described large trucks as having different operating characteristics than those of passenger cars, a difference that increases with their weight and size. For example, as the grade of a highway section increases, the presence of trucks become more pronounced. Thus, it becomes necessary to consider the provision of special facilities on highways with steep grades where high volumes of heavy vehicles exist. The most common facilities that address this problem are climbing lanes and emergency escape ramps.

15.3.1 Climbing Lanes

A climbing lane is an extra lane in the upgrade direction for use by heavy vehicles whose speeds are significantly reduced by the grade. A climbing lane eliminates the need for drivers of light vehicles to reduce their speed when they encounter a heavy, slow-moving vehicle. Because of the increasing rate of crashes directly associated with the reduction in speed of heavy vehicles on steep sections of two-lane highways and the significant reduction of the capacity of these sections when heavy vehicles are present, the provision of climbing lanes should be considered.

    The need for a climbing lane is evident when a grade is longer than its critical length, defined as the length that will cause a speed reduction of the heavy vehicle by at least 16 km/h. The amount by which a truck's speed is reduced when climbing a steep grade depends on the length of the grade. For example, Figure 9.12 shows that the speed of a truck entering a grade of 5 percent at 88 km/h will be reduced to about 69 km/h for a grade length of 300 m and to about 43 km/h for a grade length of 1800 m.

The length of the climbing lane will depend on the physical characteristics of the grade, but, in general, the climbing lane should be long enough to facilitate the heavy vehicle's rejoining the main traffic stream without causing a hazardous condition. A climbing lane is provided only if (in addition to the critical length requirement) the upgrade traffic flow rate is greater than 200 veh/h and the upgrade truck flow is higher than 20 veh/h.

Climbing lanes are not typically used on multilane highways, since relatively faster moving vehicles can pass the slower moving vehicles by using a passing lane. Also, the provision of a climbing lane cannot be justified based on capacity criterion since multilane highways are designed to have capacity sufficient to carry the forecast traffic demand, which includes slow-moving vehicles.

15.3.2 Emergency Escape Ramps

An emergency escape ramp is provided on the downgrade of a highway for use by a truck that has lost control and cannot slow down. A lane is provided that diverges such that when a vehicle enters the escape ramp, its speed is gradually reduced, and eventually it stops. The four basic designs commonly used are shown in Figure 15.27. Design type A is an ascending-grade ramp illustrated in Figure 15.27a and combines the effect of gravity and the increased rolling resistance by providing an upgrade and an arresting bed. Design

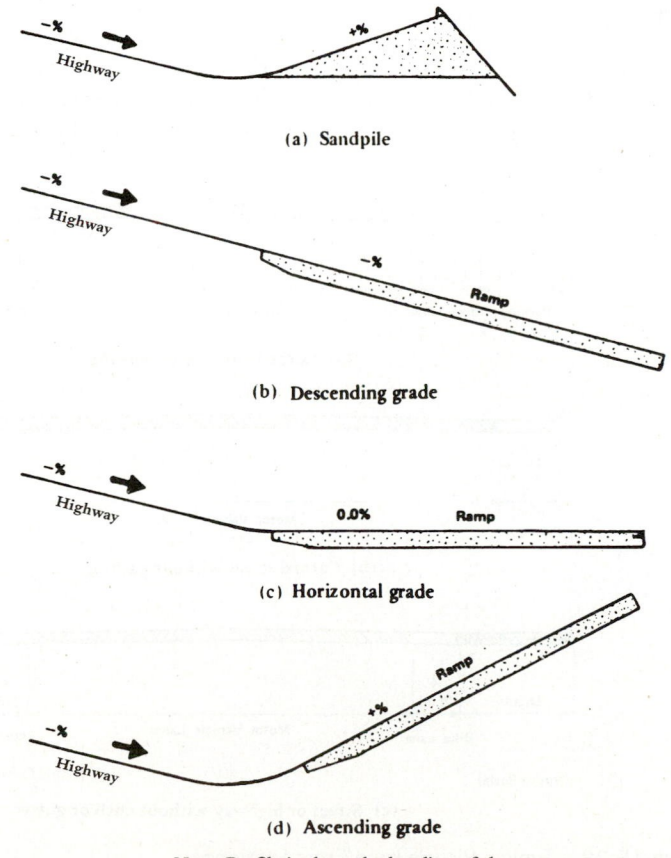

(a) Sandpile

(b) Descending grade

(c) Horizontal grade

(d) Ascending grade

Note: Profile is along the baseline of the ramp

Figure 15.27   Basic Types of Emergency Escape Ramps

SOURCE: Based on *A Policy on Geometric Design of Highways and Streets*, 2004, AASHTO, Washington, D.C.

types B and C use horizontal and descending grades, as illustrated in Figures 15.27b and 15.27c. These designs do not employ gravity in stopping the vehicle; rather they utilize the increased rolling resistance of the ramp surface that is comprised of loose aggregates. Design type D, the sandpile, shown in Figure 15.27d, is composed of loose dry sand that is positioned at the end of the escape ramp. The sandpile provides increased rolling resistance and is placed with an upgrade to assist stopping by gravity. Sandpiles are usually not greater than 400 ft in length. Ramp design A is the shortest of all the design types.

15.4 BICYCLE FACILITIES

The bicycle, a viable alternative mode of transportation, is popular, particularly in urban and suburban areas, as a means to travel within the community both for recreation and necessary travel. Thus, the bicycle is an important element in the design of highways. Almost all the existing highway and street systems can be used for bicycles, but improvements to these facilities are usually required to assure the safety of bicycle riders. A basis for the design of bicycle facilities is the AASHTO *Guide for Development of Bicycle Facilities*. Two types of bicycle facilities can be considered: (1) lanes that are contiguous with the existing street and highway system; and (2) paths that are constructed on a dedicated right-of-way for the exclusive use of bicycles.

15.4.1 Bicycle Lanes

A bicycle lane is that part of the street or highway specifically reserved for the exclusive or preferential use of bicycle riders. Striping, signing, or pavement markings can delineate bicycle lanes. These lanes should always be in the direction of traffic flow. Figure 15.28

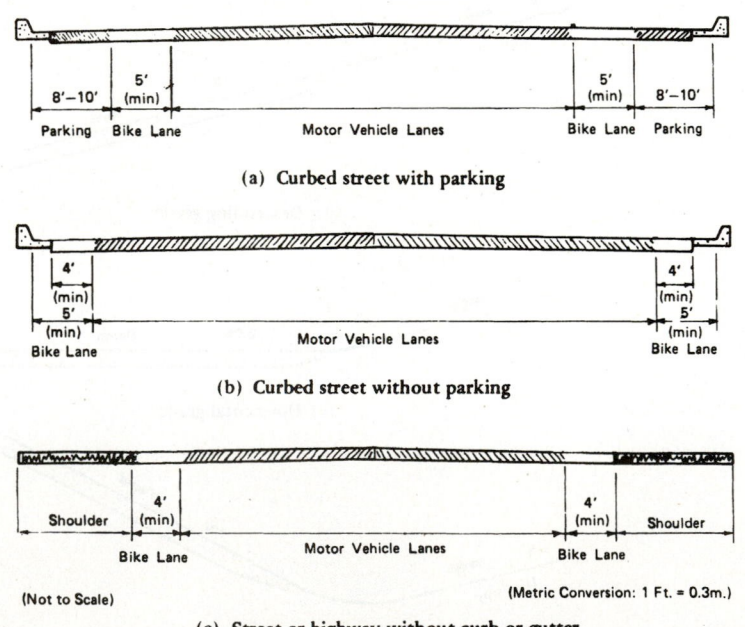

(a) Curbed street with parking

(b) Curbed street without parking

(c) Street or highway without curb or gutter

Figure 15.28 Typical Bicycle Lane Cross Sections

SOURCE: From *Guide for Development of Bicycle Facilities*, 1999, AASHTO, Washington, D.C. Used by permission.

shows typical bicycle-lane cross sections. The minimum lane width is 1.2 m but should be increased when possible. For example, when bicycle lanes are located on a curbed urban street with parking allowed at both sides, as shown in Figure 15.28a, a minimum width of 1.5 m is required. When parking is provided, the bicycle lane should always be located between the parking lane and the vehicle-traveled lanes, as shown in Figure 15.28a. When parking is prohibited, as in Figure 15.28b, a minimum lane width of 1.5 m should be provided, because cyclists tend to veer away from the curb. When the highway has no curb and gutter, a bicycle lane should be located between the shoulders and the vehicle travel lanes. A minimum bicycle lane of 1.2 m should be provided if the shoulders provide additional maneuvering width.

Special consideration should be given to pavement markings at intersections, since through movement of bicycles will tend to conflict with right-turning vehicle movements. Left-turning movements of bicycles will conflict with through vehicle movements, and pavements should be striped to encourage cyclists and motorists to merge to the left prior to arriving at the intersection. Figure 15.29 illustrates

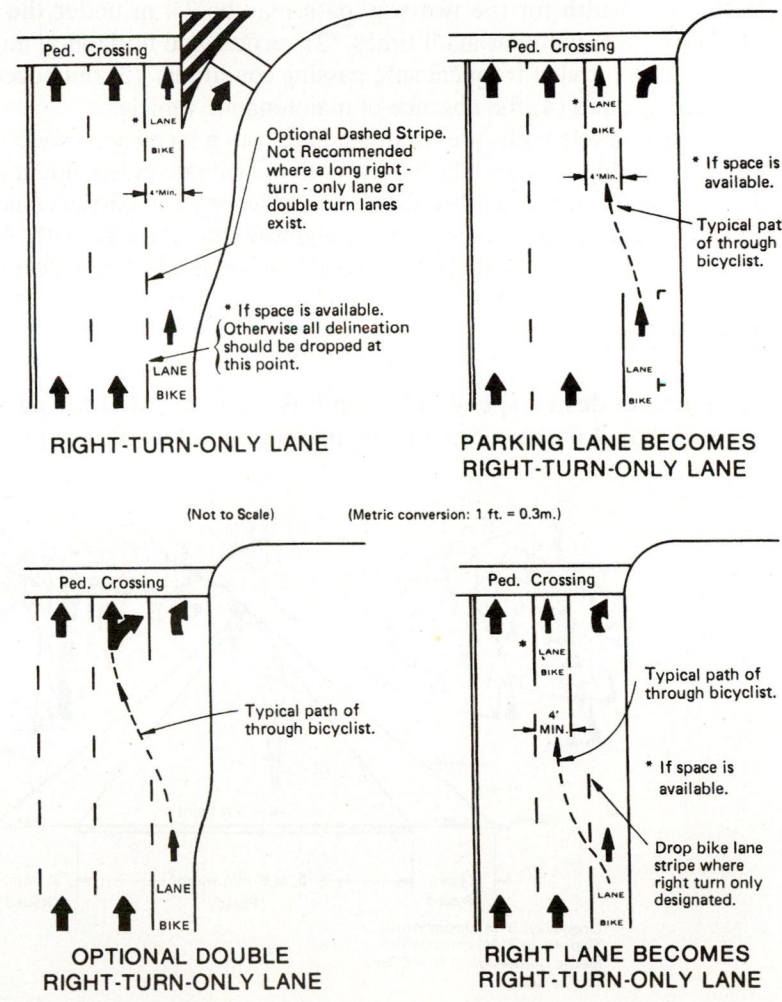

Figure 15.29   Channelization of Bicycle Lanes Approaching Motor Vehicle Right-Turn-Only Lanes

SOURCE: From *Guide for Development of Bicycle Facilities*, 1999, AASHTO, Washington, D.C. Used by permission.

alternative methods to achieve bicycle lane channelization that approaches vehicle right-turn-only lanes.

15.4.2 Bicycle Paths

Bicycle paths are physically separated from automobile traffic and may be used exclusively by cyclists or possibly with pedestrians. The design criteria for bicycle paths are similar to those for highways with criteria governed by bicycle operating characteristics. Important design considerations for a safe bicycle path include the path width and design speed, as well as horizontal and vertical alignment.

Width of Bicycle Paths

A typical layout for a bicycle path on a separate right-of-way is shown in Figure 15.30. The minimum recommended width is 3 m for two-way paths, and 1.5 m for one-way paths. The width for the two-way path may be 2.4 m under the following conditions: (1) low bicycle volumes at all times, (2) vertical and horizontal alignments are properly designed to provide frequent safe passing conditions, (3) only occasional use of path by pedestrians, and (4) the absence of maintenance vehicles.

When bicycle paths are located adjacent to a highway, a wide area must be provided between the highway and the bicycle path to facilitate independent use of the bicycle path. In cases where this area is less than 1.5 m wide, physical barriers (such as dense shrubs or a fence) should be used to separate the highway from the bike path. A uniform graded area, similar to a shoulder, should be provided that is at least 0.6 m wide on both sides of the path.

Design Speed

A minimum design speed of 32 km/h is recommended for paved paths. For grades greater than 4 percent or if strong prevailing tail winds are prevalent, design speeds

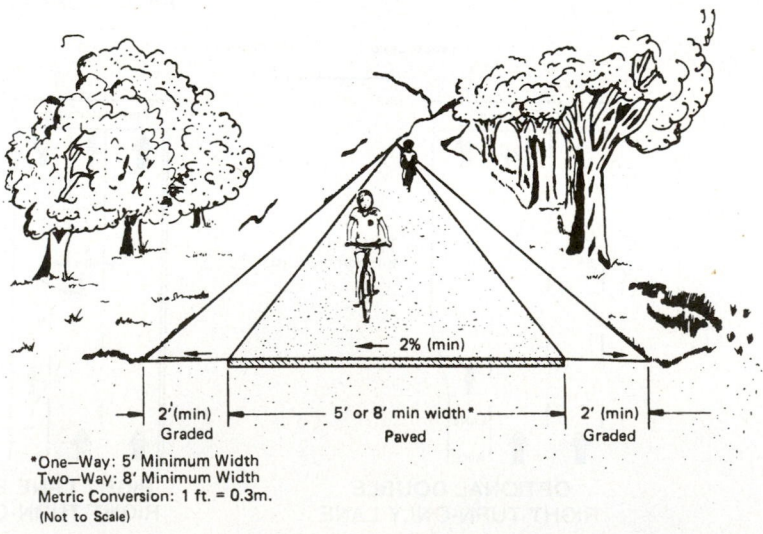

Figure 15.30   Bicycle Path on Separated Right-of-Way

SOURCE: From *Guide for Development of Bicycle Facilities*, 1999, AASHTO, Washington, D.C. Used by permission.

should be increased to 48 km/h. For unpaved bicycle paths, the minimum design speed is 24 km/h.

Horizontal Alignment

Criteria for bicycle paths are similar to those for highways. For example, the minimum radius, R, of bicycle path horizontal curves is a function of design speed, u, superelevation, e, and side friction factor, f_s, in the following equation:

$$R = \frac{u^2}{127(e + f_s)}$$

Superelevation rates for bicycle paths vary from a minimum of 2 percent to facilitate surface drainage, to a maximum of 5 percent. Higher superelevation rates may result in difficulty maneuvering around the curve. Coefficients of side friction used for design vary from 0.3 to 0.22. For unpaved bicycle paths, values vary from 0.15 to 0.11.

Vertical Alignment

The design of the vertical alignment for a bicycle path is also similar to that for a highway in that the selection of suitable grades and the design of appropriate vertical curves are required. Grades should not exceed 5 percent and should be lower when possible, particularly on long inclines, to enable cyclists to ascend the grade with little exertion and to prevent excessive speed while descending.

Vertical curves are designed to provide the minimum length as dictated by the stopping sight distance requirement. Figure 15.31 illustrates minimum stopping sight

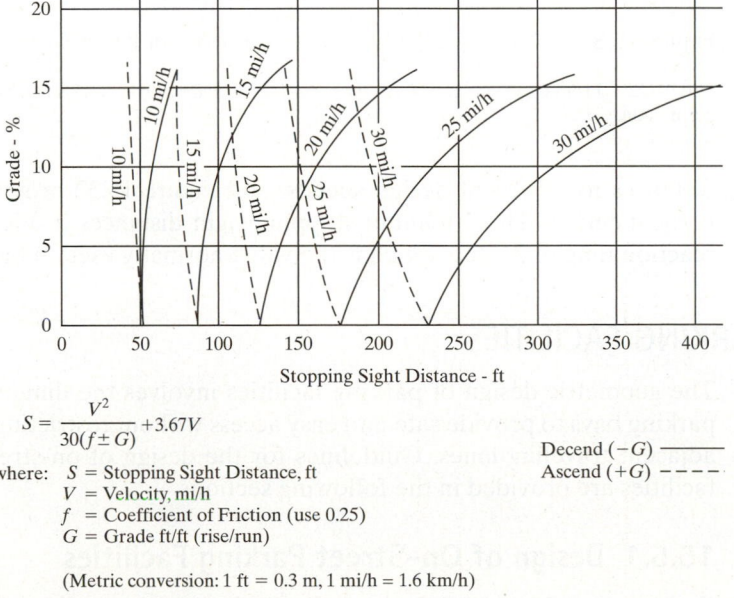

$$S = \frac{V^2}{30(f \pm G)} + 3.67V$$

where: S = Stopping Sight Distance, ft
V = Velocity, mi/h
f = Coefficient of Friction (use 0.25)
G = Grade ft/ft (rise/run)

Decend $(-G)$ ———
Ascend $(+G)$ – – – –

(Metric conversion: 1 ft = 0.3 m, 1 mi/h = 1.6 km/h)

Figure 15.31   Stopping Sight Distance for Bicycles

SOURCE: From *Guide for Development of Bicycle Facilities*, 1999, AASHTO, Washington, D.C. Used by permission.

$$L = 2S - \frac{200(\sqrt{h_1} + \sqrt{h_2})^2}{A} \text{ when } S > L$$

$$L = \frac{AS^2}{100(\sqrt{2h_1} + \sqrt{2h_2})^2} \text{ when } S > L$$

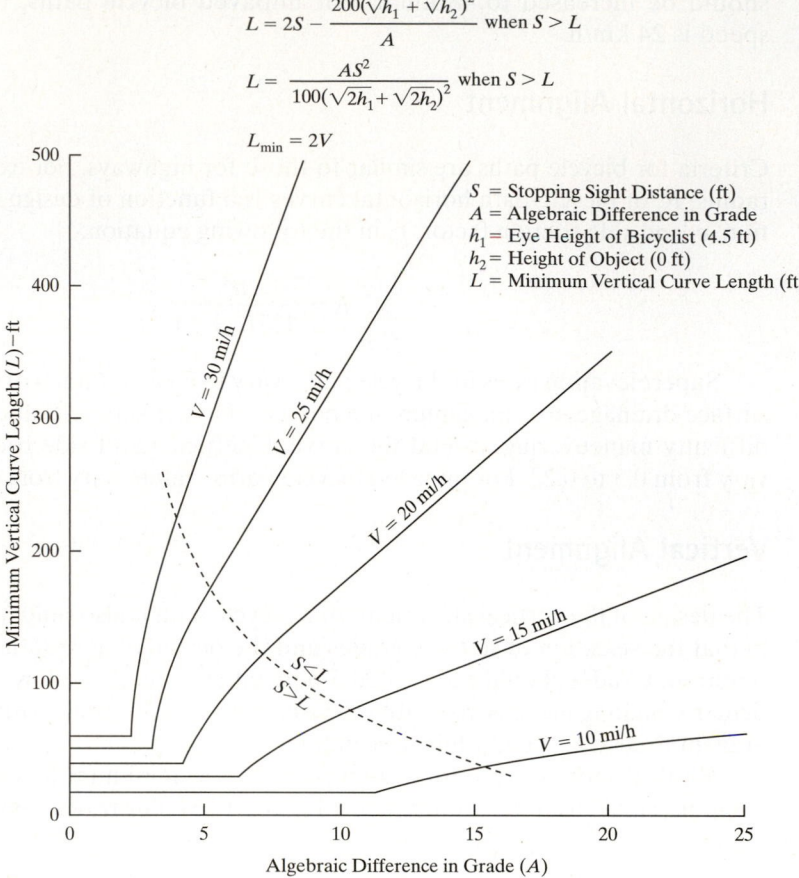

$L_{\min} = 2V$

S = Stopping Sight Distance (ft)
A = Algebraic Difference in Grade
h_1 = Eye Height of Bicyclist (4.5 ft)
h_2 = Height of Object (0 ft)
L = Minimum Vertical Curve Length (ft)

Figure 15.32   Minimum Lengths of Vertical Curves for Bicycles

SOURCE: From *Guide for Development of Bicycle Facilities*, 1999, AASHTO, Washington, D.C. Used by permission.

distances for different design speeds, and Figure 15.32 provides minimum lengths for vertical curves. The minimum stopping sight distances in Figure 15.31 are based on a reaction time of 2.5 sec, which is the value normally used in bicycle design.

15.5 PARKING FACILITIES

The geometric design of parking facilities involves the dimensioning and arranging of parking bays to provide safe and easy access without restricting the flow of traffic on the adjacent traveling lanes. Guidelines for the design of on-street and off-street parking facilities are provided in the following sections.

15.5.1 Design of On-Street Parking Facilities

On-street parking facilities may be designed with parking bays parallel or inclined to the curb. Figure 15.33 illustrates the angles of inclination commonly used for curb parking and the associated dimensions for automobiles. The number of parking bays that can be fitted along a given length of curb increases as the angle of inclination increases from parallel (0 degrees) to perpendicular (90 degrees). As the

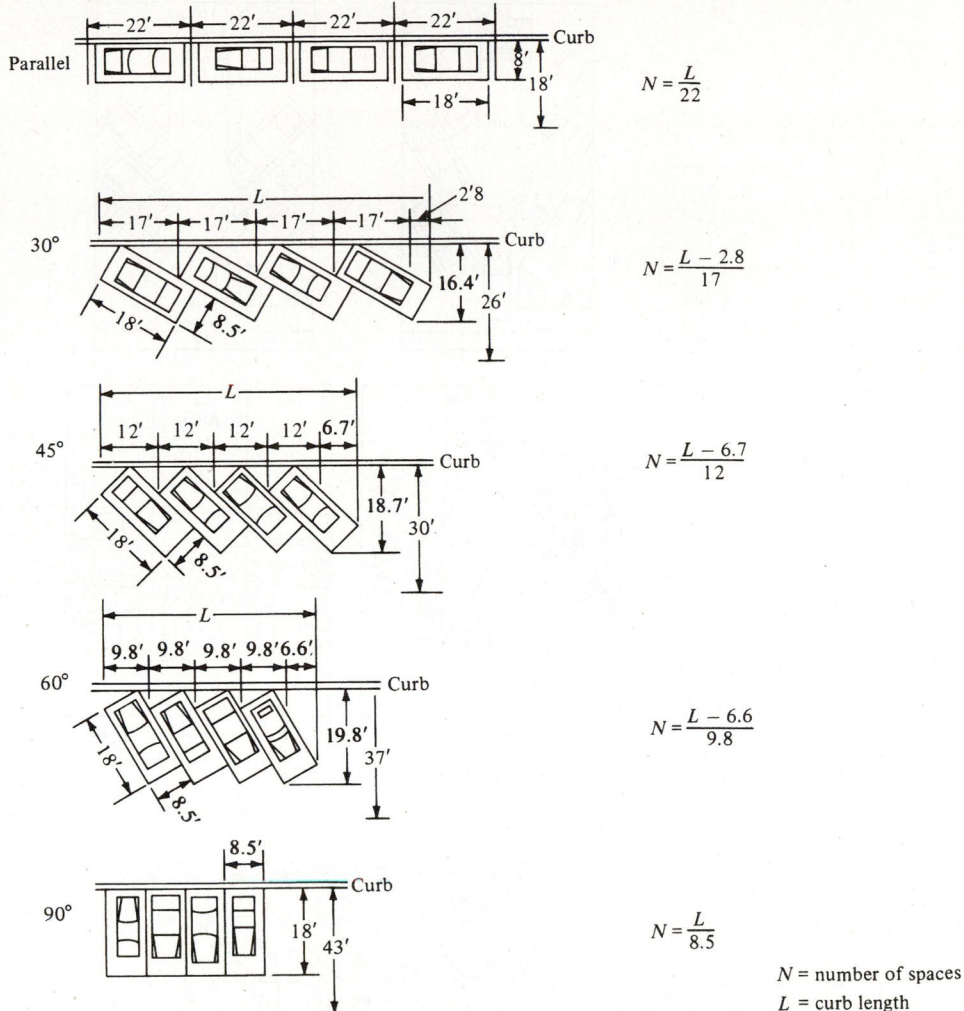

Figure 15.33 Street Space Used for Various Parking Configurations

SOURCE: From *Guide for Development of Bicycle Facilities*, 1999, AASHTO, Washington, D.C. Used by permission.

inclination angle increases, encroachment of the parking bays on the traveling pavement of the highway becomes more pronounced. Parking bays that are inclined at angles to the curb interfere with the movement of traffic, with the result that crash rates tend to be higher on sections of roads with angle parking than with parallel parking. Accordingly, public road agencies typically prefer parallel parking and prohibit perpendicular parking. When parking bays are to be provided for trucks and other types of vehicles, the dimensions should be based on characteristics of the design vehicle.

15.5.2 Design of Off-Street Parking Facilities—Surface Car Parks

The primary aim in designing off-street parking facilities is to obtain as many spaces as possible within the area provided. Figures 15.34 and 15.35 show different layouts that can be used to design a surface parking lot. The most important consideration in selecting

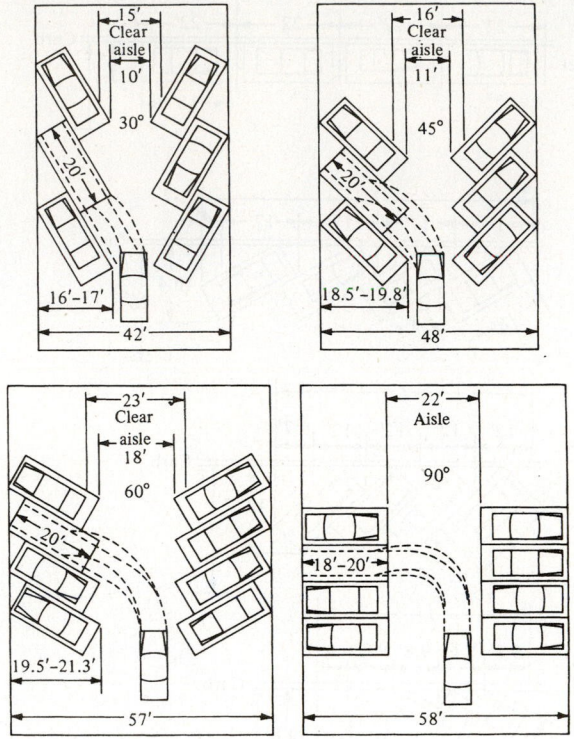

Figure 15.34   Parking Stall Layout (1 ft = 0.3 m)

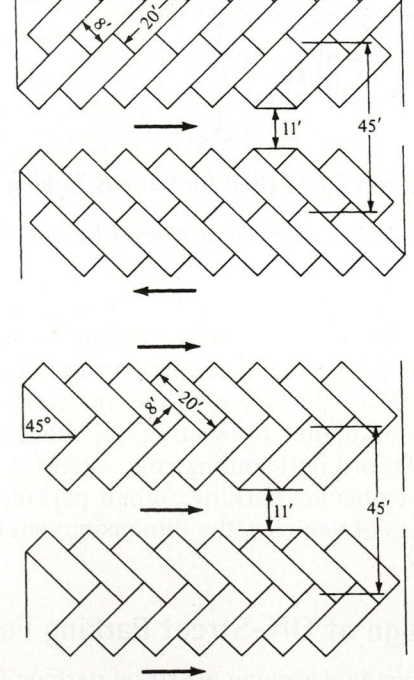

Figure 15.35   Herringbone Layout of Parking Stalls in an On-Surface Lot (1 ft = 0.3 m)

the layout is to assure that when parking that only one maneuver is necessary to reverse the vehicle. The layouts shown indicate that parking spaces are efficiently used when the parking bays are inclined at 90 degrees to the direction of traffic flow. The use of the herringbone layout as illustrated in Figure 15.35 facilitates traffic circulation because it provides for a one-way flow of traffic on each aisle.

15.5.3 Design of Off-Street Parking Facilities—Garages

Parking garages consist of several platforms, supported by columns, which are placed in such a way as to facilitate an efficient arrangement of parking bays and aisles. Access ramps connect each level with the one above. The gradient of these ramps is usually not greater than 1:10 on straight ramps and 1:12 on the centerline of curved ramps. The radius of curved ramps measured to the end of the outer curve should not be less than 21 m, and the maximum superelevation should be 0.15 m/m. The lane width should not be less than 4.8 m for curved ramps and 2.7 m for straight ramps. Ramps can be one-way or two-way, with one-way ramps preferred. When two-way ramps are used, the lanes must be clearly marked and, where possible, physically divided at curves and turning points to avoid head-on collisions, as drivers may cut corners or swing wide at bends.

Elevators into which cars are driven or placed mechanically may connect platforms. Elevators then lift the car to the appropriate level for parking. The vehicle is then either removed by an attendant, who parks it, or is mechanically removed and then parked by that attendant. Outside off-street attended parking lots with elevator platforms on which the cars are stored are used in NewYork city.

The size of the receiving area is an important factor in garage design and depends on whether the cars are owner-parked (self-parking) or attendant-parked. When cars are self-parked, very little or no reservoir space is required, since drivers need pause only for a short time to pick up a ticket from either a machine or an attendant. When attendants park cars, the driver must stop and leave the vehicle. The attendant then enters and drives the vehicle to the parking bay. Thus, a reservoir space must be provided that will accommodate temporary storage for entering vehicles. The size of reservoir space depends on the ratio of the rate of storage of vehicles to their rate of arrival. The rate of storage must consider the time required to transfer the vehicle from its driver to the attendant. The number of temporary storage bays can be determined by applying queuing theory, discussed in Chapter 6. Figure 15.36 is a chart that relates the size of reservoir space required for different ratios of storage rate to arrival rate, with overloading occurring only 1 percent of the time.

15.6 COMPUTER USE IN GEOMETRIC DESIGN

Advances in computer processing capabilities and data storage have allowed individual desktop and laptop computers to become the dominant means by which a set of highway construction plans is produced. The computer-aided design and drafting (CADD) programs in this field have matured within the last two decades or so. In 1988, only about half of the state DOTs were using CADD for roadway design. The CADD programs at that time relied on graphics terminals or workstations, with limited utility on individual computers. Today, these programs are fully functional on individual computers, and the costs of system operation have come down dramatically since then as well. Two programs commonly used for roadway design are InRoads, developed by Bentley Systems, and Civil 3D, from AutoDesk. These programs automate many of the calculations performed in the geometric design process in addition to developing plan and profile sheets and cross sections. However, the roadway designer still must make the decisions in

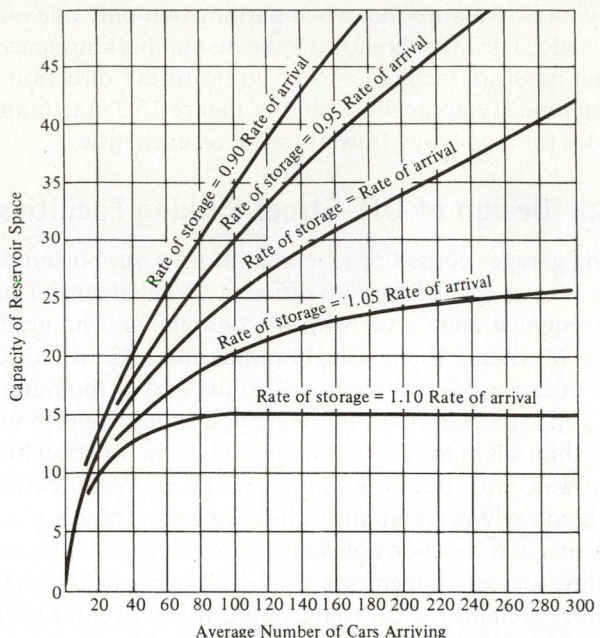

Figure 15.36 Reservoir Space Required If Facility Is Overloaded Less Than One Percent of Time

SOURCE: Reprinted with permission of the Eno Transportation Foundation, Washington, D.C. Redrawn from E. R. Ricker, *Traffic Design of Parking Garages.* Copyright 1957, Eno Transportation Foundation.

roadway design and supply the key inputs to the CADD program. Using base mapping and field-collected survey data, these programs can be used to create a digital record of the site, proposed horizontal and vertical alignments of the roadway, typical roadway sections, cross sections, and other elements that comprise a set of highway plans. The key contribution of these programs may simply lie in how they improve the efficiency of plan development, particularly when changes are made during the design process.

15.7 SUMMARY

This chapter presents the fundamental principles of highway geometric design, together with the necessary formulas required for the design of the vertical and horizontal alignments of the highway. The basic characteristics of the driver, pedestrian, vehicle, and road are used to determine the geometric characteristics of any highway or bikeway. The fundamental characteristic on which several design standards are based is the design speed, which depends on the type of facility being considered.

The material presented in this chapter provides the mathematical tools required by an engineer engaged in the geometric design of highways. Since any motor vehicle on a highway has to be parked at one time or another, the design of parking bays and garages is included.

A brief introduction to available computer programs for geometric design of highways is also presented. The intention is to let the reader know of the opportunities available in computer-aided design in highway engineering rather than simply enumerating an array of different computer programs. Many programs currently in use may become outdated as new programs become available, so the reader is advised to stay current about changes in computer hardware and applications software.

PROBLEMS

15-1   A rural collector highway located in a mountainous terrain is designed to carry a design volume of 800 veh/day. Determine the following: (a) a suitable design speed, (b) lane and usable shoulder widths, (c) maximum desirable grade.

15-2   Repeat Problem 15-1 for an urban freeway in rolling terrain.

15-3   A rural collector is to be constructed in rolling terrain with an ADT of 650 veh/day, determine:

  (a) minimum design speed
  (b) recommended lane width
  (c) preferable shoulder width
  (d) maximum grade

15-4   A +2 percent grade intersects with a –1 percent grade at station (535+24.25) at an elevation of 90 m. If the design speed is 104 km/h, determine:

  (a) the minimum length of vertical curve using the rate of vertical curvature
    Then, using the length found in part (a), find:
  (b) the stations and elevations of the BVC and EVC
  (c) the elevation of each 30 m station
  (d) the station and elevation of the highpoint

15-5   Determine the minimum length of a crest vertical curve, using the minimum length based on *SSD* criteria if the grades are +4 percent and –2 percent. Design speed is 112 km/h. State assumptions used.

15-6   Determine the minimum length of a sag vertical curve if the grades are –4 percent and +2 percent. Design speed is 112 km/h. State assumptions used. Consider the following criteria: stopping sight distance, comfort, and general appearance.

15-7   Show that the *offset y* of any point of the curve above the BVC is given as:

$$100y = g_1 x - \frac{(g_1 - g_2)x^2}{2L}$$

15-8   Given a sag vertical curve connecting a –1.5 percent grade with a +2.5 percent grade on a rural arterial highway, use the rate of vertical curvature and a design speed of 112 km/h to compute the elevation of the curve at 30 m stations if the grades intersect at station (475+00) at an elevation of 90 m. Identify the station and elevation of the low point.

15-9   A crest vertical curve connects a +4.44 percent grade and a –6.87 percent grade. The PVI is at station 43+50.00 at an elevation of 372 m. The design speed is 48 km/h. Determine:

  (a) The length of the vertical curve using the AASHTO method ("K" factors)
  (b) The station of the BVC
  (c) The elevation of the BVC
  (d) The station of the EVC
  (e) The elevation of the EVC
  (f) The station of the high point
  (g) The elevation of the high point
  (h) The elevation of station 44+23.23

15-10   A crest vertical curve connects a +4.90 percent grade and a –1.85 percent grade. The PVI is at station 77+00.00 at an elevation of 372 m. The design speed is 48 km/h.

Determine:

(a) The length of the vertical curve using the AASHTO method ("K" factors)
(b) The station of the BVC
(c) The elevation of the BVC
(d) The station of the EVC
(e) The elevation of the EVC
(f) The station of the high point
(g) The elevation of the high point
(h) The elevation of station 75+30.00

Also, provide a tabulation showing the elevations at each full station on the curve, including the distance from the PVC for each full station, the elevation on initial tangent, the offset, and the elevation on the vertical curve. This can be in the form of a computer printout from Microsoft Excel or another program.

15-11   A horizontal curve is to be designed for a two-lane road in mountainous terrain. The following data are known: Intersection angle: 40 degrees, tangent length = 131.03 m, station of PI: 2700+10.65, f_s = 0.12, e = 0.08.
Determine:

(a) design speed
(b) station of the PC
(c) station of the PT
(d) deflection angle and chord length to the first 30 m station

15-12   A proposed highway has two tangents of bearings N 45°54′36″ E and N 1°22′30″ W. The highway design engineer, attempting to obtain the best fit for the simple circular curve to join these tangents, decides that the external ordinate is to be 13 m. The PI is at station 65+43.21.
Determine:

(a) The central angle of the curve
(b) The radius of the curve
(c) The length of the tangent of the curve
(d) The station of the PC
(e) The length of the curve
(f) The station of the PT
(g) The deflection angle and chord from the PC to the first full station on the curve

15-13   Given a simple circular curve with the following properties: D = 11°, bearing on incoming (back) tangent is N 89°27′25" E, bearing on outgoing (forward) tangent is S 60°10′05" E, the station of the PI = 22+69.77.
Determine:

(a) The intersection angle (Δ)
(b) The radius (R)
(c) The tangent (T)
(d) The external distance (E),
(e) The middle ordinate (m),
(f) The long chord (C),
(g) The length of the curve (L)
(h) Station of the PC
(i) Station of the PT

15-14   A simple circular curve exists with a degree of curve $D = 12°$ and $e = 0.08$.
A structure is proposed on land on the inside of the curve. Assume the road is on level grade.
Determine:

(a) The radius of the curve
(b) The current maximum safe speed of the curve
(c) The minimum distance allowable between the proposed structure and the centerline of the curve such that the current maximum safe speed of the curve would not need to be reduced

15-15   A circular curve connecting two tangents intersect at an angle of 48 degrees. The PI is at station (948 + 67.32) and the design speed of the highway is 96 km/h.
Determine the point of the tangent and the deflection angles from the PC to full stations for laying out the curve.

15-16   Two chords xy and yz of 40.5 m each marked on an existing curve to determine the radius is given. The perpendicular distance between y and the chord xz is 4.5 m. Determine the radius of the curve and the angle set out from xz to get a line to the center of the curve.

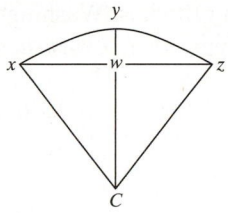

15-17   Given a compound circular curve with radii of 180 m and 135 m designed to connect two tangents that intersect at an angle of 75 degrees, determine the central angles and the corresponding chord lengths for setting out the curve if the central angle of the first curve is 45 degrees and the PCC is at station (675 + 35.25).

15-18   An arterial road is to be connected to a frontage road by a reverse curve with parallel tangents. The distance between the centerline of the two tangent sections is 18 m. The PC of the curve is located at station (38 + 25.31) and the central angle of each of the component curves is 25 degrees.
Determine the station of PT.

15-19   For the data provided in Problem 15-1, determine (a) minimum radius of horizontal curvature, assuming a superelevation rate of 8 percent, (b) minimum length of crest vertical curves, and (c) minimum length of sag vertical curves, using the rate of vertical curvature method when joining two segments at maximum grade.

15-20   Repeat Problem 15-19 for an urban freeway in rolling terrain.

15-21   For the data provided in Problem 15-3, determine (a) minimum length of crest vertical curves, (b) minimum length of sag vertical curves, using the rate of vertical curvature method when joining two segments at maximum grade, (c) maximum superelevation, and (d) maximum degree of curve (use $f_s = 0.15$) (deg-min-sec).

15-22   Determine the distance required to transition pavement cross-slope from a normal crown section with a normal crown cross-slope of 2 percent to superelevation of 6 percent on a two-lane highway with a design speed of 80 km/h.

15-23   A building is located 5.7 m from the centerline of the inside lane of a curved section of highway with a 120 m radius. The road is level; $e = 0.10$. Determine the appropriate speed limit (to the nearest 8 km/h) considering the following conditions: stopping sight distance and curve radius.

15-24   Describe the factors that must be taken into account in the design of bicycle paths.

15-25   For a bicycle path with an average speed of 32 km/h, a maximum superelevation of 2 percent, and a total change in grade of 10 percent, determine the minimum radius of horizontal curvature and the minimum length of the vertical curve.

15-26   Given an available area that is 120 m by 150 m, design a suitable parking lot layout for achieving each of the following objectives:

(a) Provide the maximum number of spaces.
(b) Provide the maximum number of spaces while facilitating traffic circulation by providing a one-way flow on each aisle.

REFERENCES

A Policy on Geometric Design of Highways and Streets, 6th ed. American Association of State Highway and Transportation Officials, Washington, D.C., 2011.

Guide for Development of Bicycle Facilities, 3rd ed. American Association of State Highway and Transportation Officials, Washington, D.C., 1999.

Kavanagh, Barry F., *Surveying with Construction Applications*, 6th ed. Prentice-Hall, Englewood Cliffs, New Jersey 2007.

Roadside Design Guide, 4th ed. American Association of State Highway and Transportation Officials, Washington, D.C., 2011.

Highway Drainage

Provision of sufficient drainage is an important factor in the location and geometric design of highways. Drainage facilities on any highway or street should adequately provide for the flow of water away from the surface of the pavement to properly designed channels. Inadequate drainage will eventually result in serious damage to the highway structure. In addition, traffic may be slowed by accumulated water on the pavement, and accidents may occur as a result of hydroplaning and loss of visibility from splash and spray. The importance of water management is recognized in the amount of highway construction dollars allocated to drainage facilities. About 25 percent of highway construction dollars are spent for erosion control and drainage structures, such as culverts, bridges, channels, and ditches.

The highway engineer is concerned primarily with two sources of water. The first, surface water, occurs as rain or snow. Some is absorbed into the soil, and the remainder remains on the surface of the ground and should be removed from the highway pavement. This source of water is referred to as *surface drainage*. The second source, ground water, flows in underground streams and may become important in highway cuts or at locations where a high water table exists near the pavement structure. This source of water is referred to as *subsurface drainage*.

The fundamental design principles for surface and subsurface drainage facilities are described in this chapter. The principles of hydrology necessary for understanding rainfall as a water source are also included, together with a brief discussion of erosion prevention.

CHAPTER OBJECTIVES:

- Understand the fundamental design principles for surface and subsurface drainage facilities.
- Apply the principles of hydrology necessary for understanding drainage design concepts.

- Explain the methods used to prevent erosion and control sediment from adjacent areas of the pavement.
- Apply hydraulic principles to the design of open channels and culverts.
- Design subsurface drainage facilities as an integral part of the total drainage system.

16.1 SURFACE DRAINAGE

Surface drainage encompasses all means by which surface water is removed from the pavement and right-of-way of the highway or street. A properly designed highway surface drainage system should effectively intercept all surface and watershed runoff and direct this water into channels and gutters for eventual discharge into the natural waterways. Water seeping through cracks in the highway riding surface and shoulder areas and contaminating underlying layers of the pavement may result in serious damage to the roadway. The major source of water for this type of intrusion is surface runoff. An adequately designed surface drainage system will therefore minimize this type of damage. The surface drainage system for rural highways should include sufficient transverse and longitudinal slopes on both the pavement and shoulder to ensure positive runoff and longitudinal channels (ditches), culverts, and bridges that provide for the discharge of the surface water to the natural waterways. Storm drains and inlets are also provided in the median of divided highways in rural areas. In urban areas, the surface drainage system also includes longitudinal and transverse slopes. The longitudinal drains are usually underground pipe drains designed to carry both surface runoff and ground water. Curbs and gutters also are used in urban and some rural areas to control surface runoff.

16.1.1 Transverse Slopes

The main objective for providing slopes in the transverse direction is to facilitate the removal of surface water from the pavement surface in the shortest possible time. This is achieved by crowning the surface at the center of the pavement, thereby providing cross slopes on either side of the centerline or providing a slope in one direction across the pavement width. Shoulders are usually sloped to drain away from the pavement, except on highways with raised narrow medians. The need for high cross slopes to facilitate drainage is in conflict with the need for relatively flat cross slopes for driver comfort. Selection of a suitable cross slope is therefore usually a compromise between the two requirements. Cross slopes of 2 percent or less do not significantly affect driver comfort, particularly with respect to the driver's effort in steering.

16.1.2 Longitudinal Slopes

A minimum gradient in the longitudinal direction of the highway is required to obtain adequate slope in the longitudinal channels, particularly at cut sections. Slopes in longitudinal channels should generally not be less than 0.2 percent for highways in very flat terrain. Although zero percent grades may be used on uncurbed pavements with adequate cross slopes, a minimum of 0.5 percent is recommended for curbed pavements. This may be reduced to 0.3 percent on suitably crowned high-type pavements constructed on firm ground.

16.1.3 Longitudinal Channels

Longitudinal channels or ditches are constructed along the sides of the highway to collect the surface water that runs off from the pavement surface, subsurface drains, and other areas of the highway right-of-way. When the highway pavement is located at a lower level than the adjacent ground, such as in cuts, water is prevented from flowing onto the pavement by constructing a longitudinal drain (intercepting drain) at the top of the cut to intercept the water. The water collected by the longitudinal ditches is then transported to a drainage channel and then to a natural waterway or retention pond.

16.1.4 Curbs and Gutters

Curbs and gutters can be used to control drainage as well as to prevent the encroachment of vehicles on adjacent areas and delineating pavement edges. Curbs and gutters are used more frequently in urban areas, particularly in residential areas, where they are used in conjunction with storm sewer systems to control street runoff. When it is necessary to provide relatively long continuous sections of curbs in urban areas, the inlets to the storm sewers must be adequately designed for both size and spacing so that the impounding of large amounts of water on the pavement surface is prevented.

16.2 HIGHWAY DRAINAGE STRUCTURES

Drainage structures are constructed to carry traffic over natural waterways that flow below the right-of-way of the highway. These structures also provide for the flow of water below the highway, along the natural channel, without significant alteration or disturbance to its normal course. An adequate size structure should be provided, such that the waterway opening is sufficiently large to discharge the expected flow of water. Inadequately sized structures can result in water impounding, which may lead to failure of the adjacent sections of the highway when embankments are submerged in water for long periods.

Drainage structures are categorized as either major or minor. Major structures are those with clear spans greater than 6 m, whereas minor structures are those with clear spans of 6 m or less. Major structures are usually large bridges, although multiple-span culverts also may be included in this class. Minor structures include small bridges and culverts.

16.2.1 Major Structures

It is beyond the scope of this book to discuss the different types of bridges. Emphasis is therefore placed on selecting the span and vertical clearance requirements for such structures. A necessary condition for the elevation of bridge superstructure is it must be located above the high water mark. The clearance above the high water mark depends on whether or not the waterway is navigable. If the waterway is navigable, the clearance above the high water mark should allow the largest ship using the channel to pass underneath the bridge without colliding with the bridge deck. The clearance height, type, and spacing of piers also depend on the probability of ice jams and the extent to which floating logs and debris appear on the waterway during high water.

The high water mark can be determined by an examination of the banks on either side of the waterway, as erosion and debris deposits are usually in evidence. Residents who have observed the waterway during flood stages over a number of years are also a source of information about the location of the high water mark. Stream gauges that have been installed in the waterway for many years can also provide similar data.

16.2.2 Minor Structures

Minor structures, consisting of short-span bridges and culverts, are the predominant type of drainage structures on highways. Although openings for these structures are not designed to be adequate for the worst flood conditions, they should be large enough to accommodate the flow conditions that might occur during the normal life expectancy of the structure. Provision also should be made for preventing clogging of the structure due to floating debris and large boulders rolling from the banks of steep channels.

Culverts are made of different materials and in different shapes. Materials used to construct culverts include concrete (reinforced and unreinforced), corrugated steel, and corrugated aluminum. Other materials also may be used to line the interior of the culvert to prevent corrosion and abrasion or to reduce hydraulic resistance. For example, asphaltic concrete may be used to line corrugated metal culverts. The different shapes normally used in culvert construction include circular, rectangular (box), elliptical, pipe arch, metal box, and arch. Figure 16.1a shows a corrugated metal arch culvert, and Figure 16.1b shows a rectangular (box) culvert.

(a) Corrugated metal arch culvert (ARMCO)

Figure 16.1   Different Types of Culverts

SOURCE: J.M. Normann, R.J. Houghtalen, W.J. Johnston, *Hydraulic Design of Highway Culverts*, Report No. FHWA-IP-85-15, US Department of Transportation, Office of Implementation, McLean, VA., September 1985

(b) Rectangular (box) culvert

Figure 16.1   Different Types of Culverts (*continued*)

SOURCE: Photograph by Lewis Woodson, Virginia Transportation Research Council, Charlottesville, VA. Used with permission.

16.3 SEDIMENT AND EROSION CONTROL

Soil erosion from adjacent areas of the pavement is the result of continuous flow of surface water over shoulders, side slopes, and unlined channels. Erosion can lead to conditions that are detrimental to the pavement structure and other adjacent facilities. For example, soil erosion of shoulders and side slopes can result in failure of embankment and cut sections, and soil erosion of highway channels often results in the pollution of nearby lakes and streams. Prevention of erosion is an important factor when highway drainage is being considered, both during construction and when the highway is completed. The methods used to prevent erosion and control sediment are discussed in the following sections.

16.3.1 Intercepting Drains

An intercepting drain at the top of a cut helps to prevent erosion of the side slopes of cut sections, since the water is intercepted before it can flow directly on the side slopes. The water is collected and transported in the intercepting drain to paved spillways that are placed at strategic locations on the side of the cut. The water is then transported through protected spillways to the longitudinal ditches alongside the highway.

16.3.2 Curbs and Gutters

Curbs and gutters can be used on rural highways to protect unsurfaced shoulders from eroding. They are placed along the edge of the pavement such that surface water is prevented from flowing over and eroding the unpaved shoulders. Curbs and gutters also can protect embankment slopes from erosion when paved shoulders are used. Curbs and gutters are placed on the outside edge of the paved shoulders, where surface water is directed to paved spillways located at strategic positions and then transported to the longitudinal drain at the bottom of the embankment.

16.3.3 Turf Cover

Turf cover on unpaved shoulders, ditches, embankments, and cut slopes is an efficient and economic method of preventing erosion when slopes are flatter than 3:1. Turf cover is developed by sowing suitable grasses immediately after grading. The disadvantages of using turf cover on unpaved shoulders are that it cannot resist continuous traffic and will eventually lose its firmness following heavy rains.

16.3.4 Slope and Channel Linings

When a highway is subjected to extensive erosion, a more effective preventive action is required. For example, when cut and embankment side slopes are steep and located in mountainous areas that are subjected to heavy rain or snow, a common method is to line the slope surface with rip-rap or hand-placed rock.

Channel linings are also used to protect longitudinal channels from eroding and the lining is placed both on the sides and bottom of the channel. Protective linings can either be flexible or rigid. Flexible linings include dense-graded bituminous mixtures and rock rip-rap, whereas rigid linings include Portland cement concrete and soil cement. Rigid linings are much more effective in preventing erosion under severe conditions, but they are more expensive and, because of their smoothness, can create unacceptably high velocities. When the use of rigid lining results in high velocities, a suitable energy dissipater must be placed at the lower end of the channel to prevent excessive erosion. The energy dissipater is not required if the water discharges into a rocky stream or into a deep pool. Detailed design procedures for linings are presented later in the chapter.

16.3.5 Erosion Control During Construction

Special precautions are required to control erosion and sediment accumulation during highway construction. Among the techniques used are: sediment basins, check dams, silt fence/filter barriers, brush barriers, diversion dikes, slope drains, and dewatering basins.

Sediment basins are required when runoff from drainage areas is greater than three acres flows across a disturbed area. The basin allows sediment-laden runoff to pond and settle. *Check dams* are used to slow the velocity of a concentrated flow of water and are made of local materials such as rock, logs, or straw bales. A *silt fence* is a fabric, often reinforced with wire mesh. *Brush barriers* are made of construction spoil material from the construction site, often combined with filter fabric. A *diversion dike* is an earthen berm that directs water to a sediment basin. *Slope drains* are used to convey water down a slope, thus avoiding erosion before a permanent drainage facilities are constructed. *Dewatering basins* are detention areas for the storage of sediment-laden water.

16.4 HYDROLOGIC CONSIDERATIONS

Hydrology is the science that deals with the characteristics and distribution of water in the atmosphere, on the earth's surface, and in the ground. The basic phenomenon in hydrology is the cycle that consists of precipitation in the form of rain, snow, and hail, and then returning to the atmosphere in the form of vapor. It is customary in hydrology to refer to all forms of precipitation as rainfall, measured in terms of the equivalent depth of water that is accumulated on the earth's surface.

The three properties of rainfall relevant to highway drainage design are: the precipitation, called *intensity;* the time over which the intensity is constant, called *duration;* and the average number of years between the occurrence of a given intensity and duration, called *frequency.* The National Weather Service has a network of automatic rainfall instruments that collect data on intensity and duration over the entire country. These data are used to draw rainfall-intensity curves from which rainfall intensity for a given frequency and duration can be obtained. Figure 16.2 is an example of a set of rainfall-intensity curves, and Figure 16.3 shows rainfall intensities for different parts of the United States for a storm of one-hour duration and one-year frequency. The use of the rainfall-intensity curves is illustrated in Figure 16.2. An intensity of 40 mm/h is obtained for a 10-year storm having a duration of 1.39 hours.

All estimate of rainfall intensity, duration, or frequency made from these data are based on probability. For example, if a culvert is designed to carry a "100-year" flood, then the probability is 1 in 100 that the culvert will flow full in any one year. Precipitation of the designed intensity and duration will not necessarily occur once every 100 years and precipitations of higher intensities can occur over a time period of 100 years. Thus, drainage facilities should be designed for very rare storms in order to reduce the chance of overflowing. Designing for a rare event requires large facilities that can result in large cost increases. The decision as to the frequency selected for design purposes must be based on a comparison of the capital cost for the drainage facility and the cost to the public in the event that the highway is severely damaged by storm runoff. Factors considered include the importance of the highway, the volume of traffic on the highway, and the population density of the area. Recommended storm frequencies, referred to as *return periods*, for various roadway classifications are shown in Table 16.1.

Other hydrologic variables that the engineer uses to determine surface runoff rates are the drainage area, the runoff coefficient, and the time of concentration.

The *drainage area, A,* is the area of land that contributes to the runoff at the point of channel capacity. Topographic maps are used to compute drainage areas.

The *runoff coefficient, C,* is the ratio of the amount of runoff and the total rainfall for a given drainage area. The runoff coefficient for a given drainage area depends on the type of ground cover, the slope, the storm duration, and the extent of saturation from prior rainfall. Constant values for C for any given storm are assumed and used in calculations.

Representative values for C for different runoff surfaces are shown in Table 16.2a for rural and urban areas and in Table 16.2b for additional urban land uses.

If the drainage area consists of different ground characteristics with different runoff coefficients, a weighted value C_w is computed using Eq. 16.1.

$$C_w = \frac{\sum_{i=1}^{n} C_i A_i}{\sum_{i=1}^{n} A_i}$$ (16.1)

where

C_w = weighted runoff coefficient for the whole drainage area
C_i = runoff coefficient for watershed i
A_i = area of watershed i (acres)
n = number of different watersheds in the drainage area

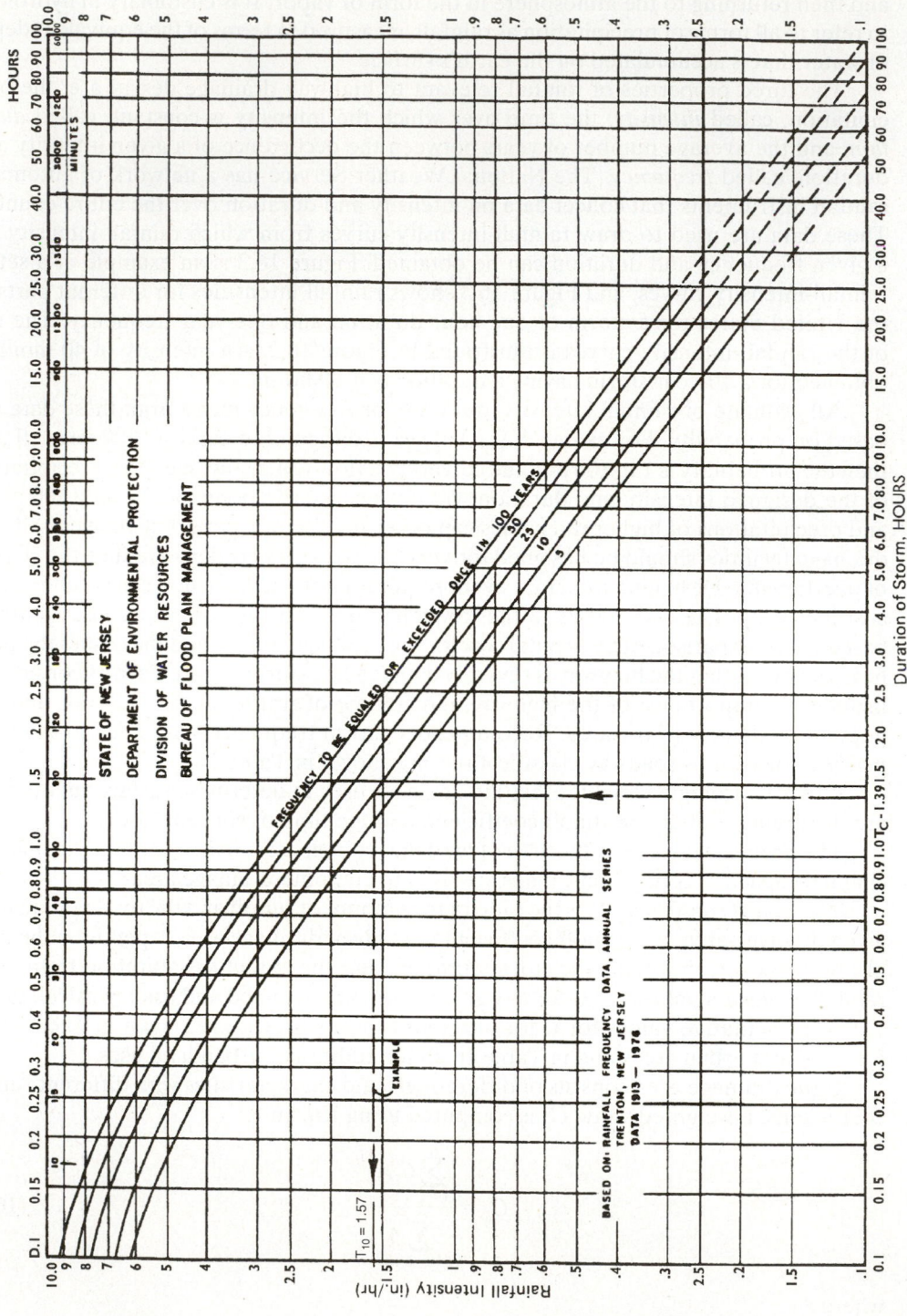

Figure 16.2 Rainfall Intensity Curves

SOURCE: G. S. Koslov, *Road Surface Drainage Design, Construction and Maintenance Guide for Pavements*, Report No. FHWA/NJ-82-004, State of New Jersey, Department of Transportation, Trenton, NJ, June 1981

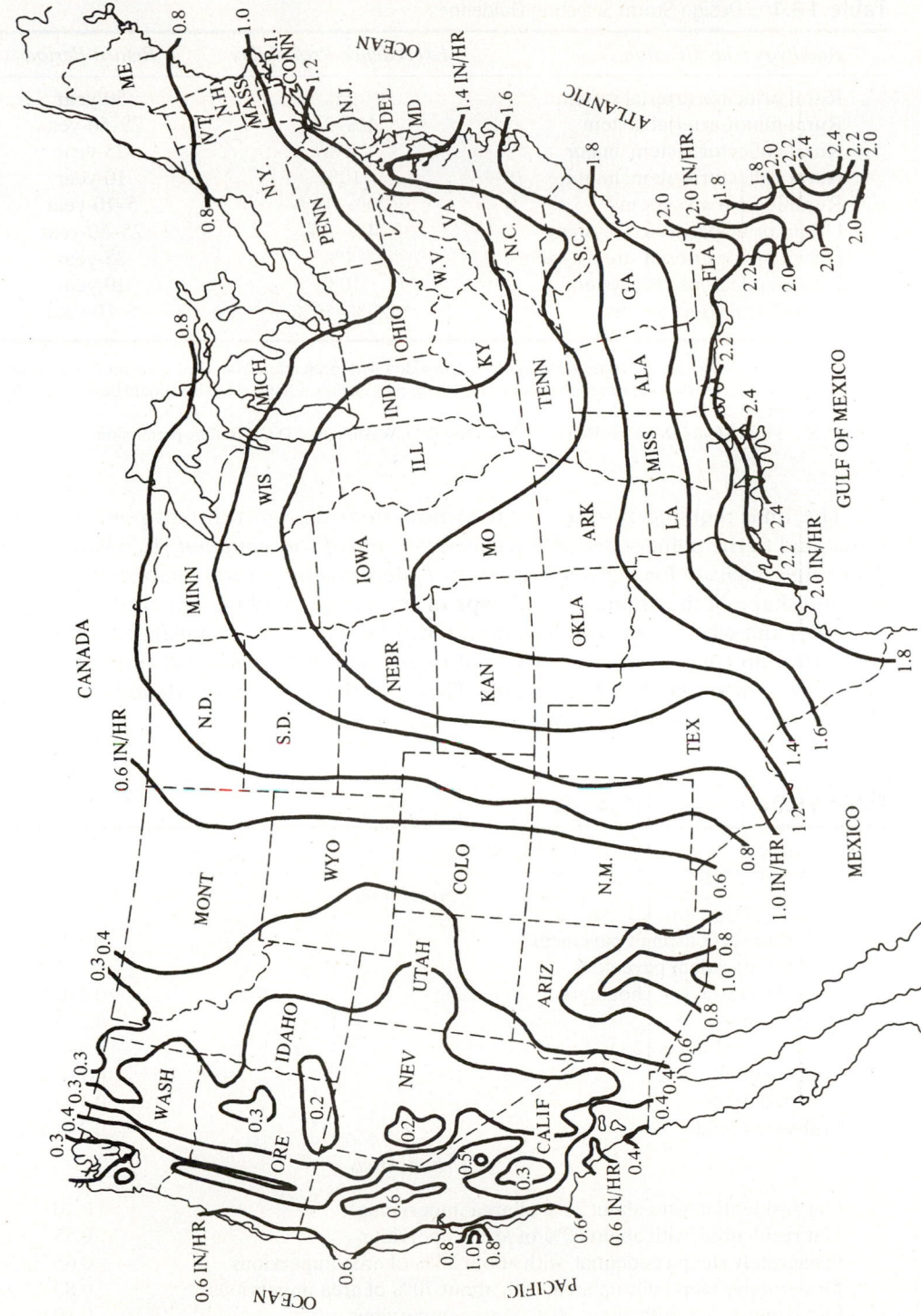

Figure 16.3   Rainfall Intensities for a One-Hour Duration Occurring at a Frequency of Once per Year

SOURCE: Redrawn from *Rainfall Frequency Atlas of the United States*, U.S. Department of Commerce, Washington, D.C., 1981

Table 16.1   Design Storm Selection Guidelines

Roadway Classification	Exceedence Probability	Return Period
Rural principal arterial system	2%	50-year
Rural minor arterial system	4%–2%	25–50-year
Rural collector system, major	4%	25-year
Rural collector system, minor	10%	10-year
Rural local road system	20%–10%	5–10-year
Urban principal arterial system	4%–2%	25–50-year
Urban minor arterial street system	4%	25-year
Urban collector street system	10%	10-year
Urban local street system	20%–10%	5–10-year

Note: Federal law requires interstate highways to be provided with protection from the 2 percent flood event, and facilities such as underpasses, depressed roadways, etc., where no overflow relief is available should be designed for the 2 percent event.
SOURCE: From *Model Drainage Manual*, 1991, AASHTO, Washington, D.C. Used by permission.

The time required for the runoff to flow from the most distant point within the watershed to the point of capacity is called *the time of concentration*, T_c, a value used to determine *intensity* for a given *frequency*. T_c depends on several factors, including the size and shape of the drainage basin, type of surface, slope of the drainage area, rainfall intensity, and whether the flow is entirely over land or partly channelized. The time of concentration consists of one or more of three components of travel times, depending on the location of the drainage systems. These are the times for overland flow, gutter or storm sewer flow (mainly in urban areas), and channel flow.

Table 16.2a   Values of Runoff Coefficients, *C*

Type of Surface	Coefficient, C*
Rural Areas	
Concrete sheet asphalt pavement	0.8–0.9
Asphalt macadam pavement	0.6–0.8
Gravel roadways or shoulders	0.4–0.6
Bare earth	0.2–0.9
Steep grassed areas (2:1)	0.5–0.7
Turf meadows	0.1–0.4
Forested areas	0.1–0.3
Cultivated fields	0.2–0.4
Urban Areas	
Flat residential, with about 30% of area impervious	0.40
Flat residential, with about 60% of area impervious	0.55
Moderately steep residential, with about 50% of area impervious	0.65
Moderately steep built-up area, with about 70% of area impervious	0.80
Flat commercial, with about 90% of area impervious	0.80

*For flat slopes or permeable soil, use the lower values. For steep slopes or impermeable soil, use the higher values.
SOURCE: Adapted from George Koslov, *Road Surface Drainage Design, Construction and Maintenance Guide for Pavements*, Report No. FHWA/NJ-82-004, State of New Jersey, Department of Transportation, Trenton, NJ, June 1981

Table 16.2b   Additional Runoff Coefficients for Urban Areas

Type of Drainage Area	Runoff Coefficient, C[1]
Business:	
Downtown areas	0.70–0.95
Neighborhood areas	0.50–0.70
Residential:	
Single-family areas	0.30–0.50
Multi-units, detached	0.40–0.60
Multi-units, attached	0.60–0.75
Suburban	0.25–0.40
Apartment dwelling areas	0.50–0.70
Industrial:	
Light areas	0.50–0.80
Heavy areas	0.60–0.90
Parks, cemeteries	0.10–0.25
Playgrounds	0.20–0.40
Railroad yard areas	0.20–0.40
Unimproved areas	0.10–0.30
Streets:	
Asphaltic	0.70–0.85
Concrete	0.80–0.95
Brick	0.75–0.85
Drives and walks	0.75–0.85
Roofs	0.75–0.95

[1]Higher values are usually appropriate for steeply sloped areas and longer return periods because infiltration and other losses have a proportionally smaller effect on runoff in these cases.
SOURCE: Adapted from Brown et al. *Urban Drainage Design manual*, Hydraulic Engineering Circular No. 22, 2nd. Ed., Department of Transportation, Federal Highway Administration Publication No. FHWA-NHI--01-021

The following procedure is used to calculate travel time and time of concentration. Water travels through a watershed as sheet flow, shallow concentrated flow, open-channel flow, or a combination of each flow type. The type of flow that occurs depends on the physical characteristics of the area. Travel time is the ratio of flow length to average flow velocity.

$$T_i = L/3600V \tag{16.2}$$

where

T_i = travel time for section i in watershed (h)
L = flow length (m)
V = average velocity (m/sec)

The time of concentration, T_c, is the sum of T_i for the various elements within the watershed. Thus,

$$T_c = \sum_{i}^{m} T_i \tag{16.3}$$

where

T_c = time of concentration (h)
T_i = travel time for segment i (h)
m = number of segments

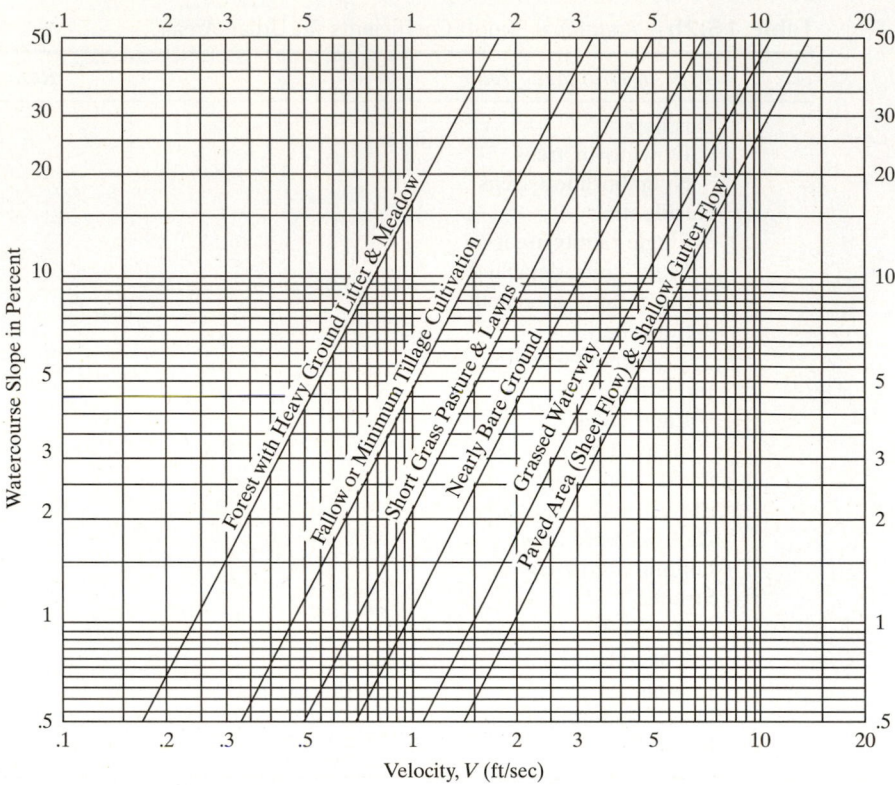

Figure 16.4   Average Velocities for Overland Flow

SOURCE: U.S. Department of Transportation, Federal Highway Administration, *Hydrology*, Hydraulic Engineering Circular, No. 19, 1984

Sheet-flow velocity depends on the slope and type of surface of the watershed area. Figure 16.4 provides average velocities for various surface types as a function of ground slope. For example, if a watercourse classified as "nearly bare ground" has an average slope of 5 percent, the average velocity is approximately 0.66 m/sec.

The travel time for flow in a gutter or storm sewer is the sum of the travel times in each component of the gutter and/or storm-sewer system between the farthest inlet and the outlet. Although velocities in the different components may be different, the use of the average velocity for the whole system does not usually result in large errors. When gutters are shallow, the curve for overland flow in paved areas in Figure 16.4 can be used to determine the average velocity.

The travel time in the open channel is determined in a manner similar to the procedure used for the flow in a storm sewer. For open channels, the velocity of flow is determined by using an equation such as Manning's formula, as explained later in this chapter.

16.4.1   Determination of Runoff

The amount of runoff for any combination of intensity and duration depends on the type of surface. For example, runoff will be much higher on rocky or bare impervious slopes, roofs, and pavements than on plowed land or heavy forest. Thus the proportion of rainfall that remains as runoff is needed for design purposes. Runoff rate for any given area during a single rainfall is not usually constant and several methods for estimating runoff are available. Two commonly used methods are described in the following sections.

The Rational Method

The rational method is based on the premise that the rate of runoff for any storm depends on the average storm intensity, the size of the drainage area, and the type of drainage area surface. For any given storm, the rainfall intensity is not constant over a large area, or during the entire storm's duration. The rational formula is based on the premise that, for a rainfall of average intensity (I) falling over an impervious area of size (A), the maximum rate of runoff at the outlet to the drainage area (Q) occurs when the whole drainage area is contributing to the runoff and this runoff rate is constant. This requires that the storm duration be at least equal to the time of concentration, which is the time required for the runoff to flow from the farthest point of the drainage area to the outlet. This condition is not always satisfied in practice, particularly in large drainage areas. It is therefore customary for the rational formula to be used for relatively small drainage areas not greater than 81 ha. The rational formula is shown in Eq. 16.4.

$$Q = CIA \qquad (16.4)$$

where

> Q = peak rate of runoff (m³/sec)
> A = drainage area (hectares)
> I = average intensity for a selected frequency and duration equal to at least the time of concentration (cm/h)
> C = a coefficient representing the fraction of rainfall that remains on the surface of the ground (runoff coefficient)

Although the formula is not dimensionally correct, it can be shown that a rainfall of 2.5 cm/h falling over an area of one acre produces a runoff of 0.0285 m³/sec if there are no losses. The runoff value Q is therefore almost exactly equal to the product of I and A. The losses due to infiltration and evaporation are accounted for by C. Values for C can be obtained from Tables 16.2 and 16.2a.

Example 16.1 Computing Rate of Runoff Using the Rational Formula

A 70.82 ha urban drainage area consists of three different watershed areas as follows.

> Apartment dwelling areas = 50%
>
> Parks = 30%
>
> Playgrounds = 20%

If the time of concentration for the drainage area is 1.5 hr, determine the runoff rate for a storm of 100-yr frequency. Assume that the rainfall-intensity curves in Figure 16.2 are applicable to this drainage area.

Solution: The weighted runoff coefficient should first be determined for the whole drainage area. From Table 16.2a, midpoint values for the different surface types are

Apartment dwelling areas = 0.6

Parks = 0.175

Playgrounds = 0.3

$$C_w = \frac{70.82(0.5 \times 0.6 + 0.3 \times 0.175 + 0.2 \times 0.3)}{70.82} = 0.413$$

The rainfall intensity for storm duration of at least 1.5 h (time of concentration) and frequency of 50-yr is obtained from Figure 16.2 as approximately 56 mm/h (2 in/h). From Eq. 16.4,

$$Q = 0.413 \times 55 \times 70.82 \times 10000/(1000 \times 3600) = 4.47 \text{ m}^3/\text{sec}$$

U.S. Soil Conservation Service (SCS) Method, TR–55

A method developed by the U.S. Soil Conservation Service (SCS) now known as the National Resources Conservation Service, (NRCS) is commonly used in determining surface runoffs. This method, referred to as TR-55, can be used to estimate runoff volumes and peak rate of discharge.

The TR-55 method consists of two parts; the first determines the runoff, h, in inches. The second part estimates the peak discharge using the value of h obtained from the first part and a graph that relates the time of concentration (hours) with the unit peak discharge (m³/sec/km²/cm).

The fundamental premise used in developing this method is that the depth of runoff (h) in inches depends on the rainfall (P) in inches. Some of the precipitation occurring at the early stage of the storm, known as initial abstraction (I_a), will not be part of the runoff. The potential maximum retention (S) of the surface (similar in concept to C in the rational method) is a measure of the imperviousness of the watershed area. The SCS equation is

$$h = \frac{(P - I_a)^2}{(P - I_a) + S} = \frac{(P - 0.2S)^2}{P + 0.8S} \tag{16.5}$$

where

h = runoff (mm)
P = rainfall (mm/24 h)
S = potential maximum retention after runoff begins (mm)
I_a = initial abstraction, including surface storage, interception, and infiltration prior to runoff. The relationship between I_a and S, developed empirically from watershed data, is $I_a = 0.2S$.

Use the following equation to calculate the peak discharge:

$$q_p = q_p' A h \tag{16.6}$$

where

q_p = peak discharge (m³/sec)
A = area of drainage basin (km²)
h = runoff (cm)
q_p' = peak discharge, for 24-h storm at time of concentration (m³/sec/km²/cm) (see Figure 16.5)

To determine h, first compute S, which is determined as

$$S = (2500/CN) - 25 \qquad (16.7)$$

where CN is the runoff curve number, which varies from 0 (pervious) to 100 (impervious).

High values of S indicate low retention of storm water, whereas low values of S indicate that the ground retains most water.

The runoff curve number accounts for the watershed characteristics, which include the soil type, land use, hydrologic condition of the cover, and soil moisture just prior to the storm (antecedent soil moisture). In order to determine values for CN, the SCS has developed classification systems for soil type, cover, and antecedent soil moisture.

Soils are divided into four groups as follows:

Group A: Deep sand, deep loess, aggregated silts

Group B: Shallow loess, sandy loam

Group C: Clay loams, shallow sandy loam, soils low in organic content, and soils usually high in clay

Group D: Soils that swell significantly when wet, heavy plastic clays, and certain saline soils

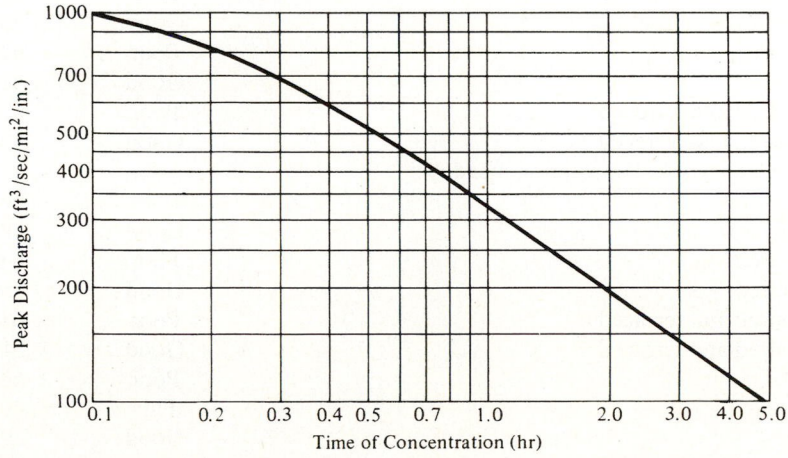

Figure 16.5   Peak Discharge in ft³/sec/mi²/in. of Runoff versus Time of Concentration, T_c, for 24-h Type II Storm Distribution

Note: This Type II storm distribution curve does not apply everywhere in the U.S. It is not valid for most coastal regions, including regions bordering the Gulf of Mexico. For more information, see Appendix B of the TR-55 manual.

Table 16.3 Runoff Curve Numbers for AMC II

Land-Use Description/Treatment/Hydrologic Condition		Hydrologic Soil Group			
		A	B	C	D
Residential[1]					
Average lot size:	Average Percent Impervious[2]				
⅛ acre or less	65	77	85	90	92
¼ acre	38	61	75	83	87
⅓ acre	30	57	72	81	86
½ acre	25	54	70	80	85
1 acre	20	51	68	79	84
Paved parking lots, roofs, driveways, etc.[3]		98	98	98	98
Streets and roads:					
Paved with curbs and storm sewers[3]		98	98	98	98
Gravel		76	85	89	91
Dirt		72	82	87	89
Commercial and business areas (85% impervious)		89	92	94	95
Industrial districts (72% impervious)		81	88	91	93
Open spaces, lawns, parks, golf courses, cemeteries, etc.:					
Good condition: grass cover on 75% or more of the area		39	61	74	80
Fair condition: grass cover on 50–75% of the area		49	69	79	84
Fallow:					
Straight row	—	77	86	91	94
Row crops:					
Straight row	Poor	72	81	88	91
Straight row	Good	67	78	85	89
Contoured	Poor	70	79	84	88
Contoured	Good	65	75	82	86
Contoured and terraced	Poor	66	74	80	82
Contoured and terraced	Good	62	71	78	81
Small grain:					
Straight row	Poor	65	76	84	88
Straight row	Good	63	75	83	87
Contoured	Poor	63	74	82	85
Contoured	Good	61	73	81	84
Contoured and terraced	Poor	61	72	79	82
Contoured and terraced	Good	59	70	78	81
Close-seeded legumes[4] or rotation meadow:					
Straight row	Poor	66	77	85	89
Straight row	Good	58	72	81	85
Contoured	Poor	64	75	83	85
Contoured	Good	55	69	78	83
Contoured and terraced	Poor	63	73	80	83
Contoured and terraced	Good	51	67	76	80
Pasture or range:	Poor	68	79	86	89
	Fair	49	69	79	84
	Good	39	61	74	80
Contoured	Poor	47	67	81	88
Contoured	Fair	25	59	75	83
Contoured	Good	6	35	70	79

(Continued)

Table 16.3   Runoff Curve Numbers for AMC II (*continued*)

Land-Use Description/Treatment/Hydrologic Condition		Hydrologic Soil Group			
		A	B	C	D
Meadow	Good	30	58	71	78
Woods or forest land	Poor	45	66	77	83
	Fair	36	60	73	79
	Good	25	55	70	77
Farmsteads	—	59	74	82	86

[1]Curve numbers are computed assuming the runoff from the house and driveway is directed toward the street with a minimum of roof water directed to lawns where additional infiltration could occur.
[2]The remaining pervious areas (lawn) are considered to be in good pasture condition for these curve numbers.
[3]In some warmer climates of the country, a curve number of 95 may be used.
[4]Close-drilled or broadcast.
SOURCE: Adapted from Richard H. McCuen, *A Guide to Hydrologic Analysis Using SCS Methods*, copyright © 1982.

Land use and hydrologic conditions of the cover are incorporated as shown in Table 16.3, which shows 14 different classifications for land use. In addition, agricultural land uses are usually subdivided into specific treatment or practices to account for the differences in hydrologic runoff potential. The hydrologic condition is considered in terms of the level of land management that is given in three classes: poor, fair, and good.

The effect of antecedent moisture condition (AMC) on rate of runoff is taken into consideration by classifying soil conditions into three categories:

Condition I: Soils are dry but not to wilting point; satisfactory cultivation has taken place.
Condition II: Average condition.
Condition III: Heavy rainfall, or light rainfall and low temperatures, have occurred within the last five days; saturated soils.

Table 16.4 provides seasonal rainfall limits as guidelines for determining the antecedent moisture condition (AMC). Note that the *CN* numbers in Table 16.3 are for AMC II. When antecedent moisture condition I or III exists, Table 16.3 is used to determine the condition II *CN* value, and then Table 16.5 is used to obtain the appropriate value for either condition I or III.

The first task in the second part of the procedure is to determine the unit peak discharge (q'_p) (m³/sec/km²/cm) from Figure 16.5, which relates the time of concentration (h) and the unit peak discharge for a 24-hour, type II storm which can be applied to most of the United States, excluding most of the coastal regions. A type II is the most intense short duration rainfall storm of the four synthetic 24-hour rainfall distributions (I, IA, II and III) developed by the NRCS.

Table 16.4   Total Five-Day Antecedent Rainfall (in.)

AMC	Dormant Season	Growing Season
I	< 0.5	> 1.4
II	0.5 to 1.1	1.4 to 2.1
III	< 1.1	> 2.1

Table 16.5   Corresponding Runoff Curve Numbers for Conditions I and III

	Corresponding CN for Condition	
CN for Condition II	I	III
100	100	100
95	87	99
90	78	98
85	70	97
80	63	94
75	57	91
70	51	87
65	45	83
60	40	79
55	35	75
50	31	70
45	27	65
40	23	60
35	19	55
30	15	50
25	12	45
20	9	39
15	7	33
10	4	26
5	2	17
0	0	0

SOURCE: Adapted from Richard H. McCuen, *A Guide to Hydrologic Analysis Using SCS Methods*, copyright ©1982.

Example 16.2 Computing Peak Discharge Using the TR-55 Method

A 1.28-km^2 drainage area consists of 20 percent residential ($\frac{1}{2}$-acre lots), 30 percent row crops with straight row treatment and good hydrologic condition, and 50 percent wooded area with good hydrologic condition. If the soil is classified as group C, with an AMC III, determine the peak discharge if the 24-hour precipitation is 15 cm and the time of concentration is 2 hours.

Solution: A weighted *CN* value should first be determined as follows.

$$\text{Weighted } CN = 0.2 \times 94 + 0.3 \times 97 + 0.5 \times 87 = 91.4$$

$$S = \frac{2500}{CN} - 25 = \frac{2500}{91.4} - 25 = 2.35$$

From Eq. 16.5,

$$h = \frac{(P - 0.2S)^2}{P + 0.8S} = \frac{(15 - 0.2 \times 2.35)^2}{15 + 0.8 \times 2.35} = \frac{211.12}{16.88} = 12.5 \text{ cm}$$

The peak unit discharge is determined from Figure 16.5 for a time of concentration of 2 h, $q'_p = 0.84$ m^3/sec/km^2/cm. From Eq. 16.6,

$$q_p = q'Ah$$
$$= 0.84 \times 1.28 \times 12.5$$
$$= 13.44 \text{ m}^3/\text{sec}$$

Land Use/Treatment/Hydrologic Condition	CN for AMC II	CN for AMC III
Residential ($\frac{1}{2}$-acre lots) (20%)	80	94
Row crops/straight row/good (30%)	85	97
Wooded/good (50%)	70	87

16.5 UNIT HYDROGRAPHS

A hydrograph is a plot that shows the relationship between stream flow and time. A unit hydrograph depicts runoff due to water of 1-in. depth uniformly distributed over the whole drainage area for a given storm with a specified duration. A unit hydrograph represents the effects of drainage area characteristics such as ground cover, slope, and soil type. Since these characteristics act concurrently, the runoff is representative of the drainage area.

Similar storms having the same duration will produce a unit hydrograph that is similar in shape. Thus, for a given type of rainfall, the timescale of the unit hydrograph is constant for a given drainage area, and its ordinates are approximately proportional to the runoff volumes. Accordingly, a unit hydrograph can be used to estimate the expected runoffs from similar storms having different intensities.

A unit hydrograph is valid for peak discharges from drainage areas as large as 5000 km². Other methods for quantifying hydrographs are the SCS (NRCS) Synthetic Unit Hydrograph and the USGS Nationwide Urban Hydrograph.

16.5.1 Computer Models for Highway Drainage

Computer models to generate hydrographs are in common use. Some models solve existing empirical formulas, whereas others use simulation techniques. Most simulation models provide for the drainage area to be divided into smaller sub-areas with similar characteristics. A design storm is then synthesized for each sub-area and the volume due to losses such as infiltration and interception deducted. The flow of the remaining water is then simulated using an overland flow routine. The overland flows from the subareas are collected by adjacent channels, which are eventually linked together to obtain the total response of the drainage area to the design storm.

Using measured historical data to calibrate the parameters of the model increases the validity of any simulation model. One major disadvantage of simulation models is that they usually require a large amount of input data and extensive user experience to obtain reliable results.

Highway drainage design utilizes software packages. These programs all have user-friendly graphic interfaces for easy use on laptop and notebook computers. The relevant computer packages are as follows.

- The FHWA Urban Storm Drainage Model consists of four modules: precipitation, hydraulic/quality, analysis, and cost estimation.
- FHWA and AASHTO have developed a microcomputer package called HYDRAIN, which consists of several programs.
- The HYDRO program generates design flows, or hydrographs, from rainfall and offers several hydrological analysis options, including the rational method.
- The PFP/HYDRA program is a storm and sanitary sewer-system analysis.
- The Culvert Design System (CDS) program can assist in the design or analysis of an existing or proposed culvert, and the Water Surface Profile (WSPRO) program is used for bridge waterway analysis to compute water surface profiles.
- SWMM is the Environmental Protection Agency (EPA) storm water management model for planning and design.
- STORM, the U.S. Army Corps of Engineers model, simulates the quality and quantity of overflows for combined sewer systems.

16.5.2 National Peak-Flow Data

The U.S. Geological Survey has compiled a comprehensive database for annual peak flow at more than 25,000 sites across the country with a total of 635,000 station years recorded. Such a database, available online at *http://waterdata.usgs.gov/sus/nwis/peak*, provides extremely useful information about the frequency and magnitude of flood flows for the design of hydraulic structures.

16.6 HYDRAULIC DESIGN OF HIGHWAY DRAINAGE STRUCTURES

The ultimate objective in determining the hydraulic requirements for any highway drainage structure is to provide a suitable structure size that will economically and efficiently dispose of the expected runoff. Certain hydraulic requirements also should be met to avoid erosion and/or sedimentation in the system.

16.6.1 Design of Open Channels

An important design consideration is that the velocity in a channel should not be so low as to cause deposits of transported material nor so high as to cause erosion. The velocity that will satisfy this condition usually depends on the shape and size of the channel, the type of lining in the channel, the quantity of water being transported, and the type of material suspended in the water. The most appropriate channel gradient range to produce the required velocity is between 1 and 5 percent. For most types of linings, sedimentation is usually a problem when slopes are less than 1 percent, and excessive erosion of the lining will occur when slopes are higher than 5 percent. Tables 16.6, 16.7, and 16.8 provide recommended maximum velocities in order to prevent erosion.

The point at which the channel discharges into the natural waterway requires special treatment. For example, if the drainage channel at the point of discharge is at a much higher elevation than the natural waterway, then the water should be discharged through a spillway or chute to prevent erosion.

Design Principles

The hydraulic design of a drainage ditch for a given storm entails the determination of the minimum cross-sectional area of the ditch that will accommodate the flow due to that storm

Table 16.6 Maximum Permissible Velocities in Erodable Channels, Based on Uniform Flow in Continuously Wet, Aged Channels

	Velocities		
Material	*Clear Water (m/sec)*	*Water Carrying Fine Silts (m/sec)*	*Water Carrying Sand and Gravel (m/sec)*
Fine sand (noncolloidal)	1.5	2.5	1.5
Sandy loam (noncolloidal)	1.7	2.5	2.0
Silt loam (noncolloidal)	2.0	3.0	2.0
Ordinary firm loam	2.5	3.5	2.2
Volcanic ash	2.5	3.5	2.0
Fine gravel	2.5	5.0	3.7
Stiff clay (very colloidal)	3.7	5.0	3.0
Graded, loam to cobbles (noncolloidal)	3.7	5.0	5.0
Graded, silt to cobbles (colloidal)	4.0	5.5	5.0
Alluvial silts (noncolloidal)	2.0	3.5	2.0
Alluvial silts (colloidal)	3.7	5.0	3.0
Coarse gravel (noncolloidal)	4.0	6.0	6.5
Cobbles and shingles	5.0	5.5	6.5
Shales and hard pans	6.0	6.0	5.0

Note: For sinuous (winding) channels, multiply allowable velocity by 0.95 for slightly sinuous channels by 0.9 for moderately sinuous channels and by 0.8 for highly sinuous channels.
SOURCE: Adapted from George S. Koslov, *Road Surface Drainage Design, Construction and Maintenance Guide for Pavements*, Report No. FHWA/NJ-82-004, State of New Jersey, Department of Transportation, Trenton, NJ, June 1981.

Table 16.7 Maximum Allowable Water Velocities for Different Types of Ditch Linings

	Maximum Velocity (m/sec)
Natural Soil Linings	
Bedrock or rip-rap sides and bottoms	15–18
Gravel bottom, rip-rap sides	8–10
Clean gravel	6–7
Silty gravel	2–5
Clayey gravel	5–7
Clean sand	1–2
Silty sand	2–3
Clayey sand	3–4
Silt	3–4
Light clay	2–3
Heavy clay	2–3
Vegetative Linings	
Average turf, erosion-resistant soil	4–5
Average turf, easily eroded soil	3–4
Dense turf, erosion-resistant soil	6–8
Gravel bottom, brushy sides	4–5
Dense weeds	5–6
Dense brush	4–5
Dense willows	8–9
Paved Linings	
Gravel bottom, concrete sides	10
Mortared rip-rap	8–10
Concrete or asphalt	18–20

SOURCE: Adapted from *Drainage Design for New York State*, U.S. Department of Commerce, Washington, D.C., November 1974.

Table 16.8   Maximum Permissible Velocities in Channels Lined with Uniform Strands of Various Well-Maintained Grass Covers

Cover	Slope Range (%)	Maximum Permissible Velocity on[a]	
		Erosion-Resistant Soils (m/sec)	*Easily Eroded Soils (m/sec)*
Bermuda grass	0–5	8	6
	5–10	7	5
	Over 10	6	4
Buffalo grass	0–5	7	5
Kentucky bluegrass	5–10	6	4
Smooth brome	Over 10	5	3
Blue grama			
Grass mixture	0–5[b]	5	4
	0–10[b]	4	3
Lespedeza sericea			
Weeping lovegrass			
Yellow bluestem			
Kudzu	0–5[c]	3.5	2.5
Alfalfa			
Crabgrass			
Common lespedeza[d]	0–5[c]	3.5	2.5
Sudangrass[d]			

[a]Use velocities over 5 ft/sec only where good covers and proper maintenance can be obtained.
[b]Do not use on slopes steeper than 10 percent.
[c]Not recommended for use on slopes steeper than 5 percent.
[d]Annuals, used on mild slopes or as temporary protection until permanent covers are established.
SOURCE: Adapted from *Drainage Design for New York State*, U.S. Department of Commerce, Washington, D.C., November 1974

and prevent water from overflowing its sides. The Mannings formula assumes uniform steady flow in the channel and estimates the mean velocity in the channel using Eq. 16.8.

$$v = \frac{1}{n} R^{2/3} S^{1/2} \tag{16.8}$$

where

v = average discharge velocity (m/sec)
R = mean hydraulic radius of flow in the channel (m)
   $= \dfrac{a}{P}$
a = channel cross-sectional area (m²)
P = wetted perimeter (m)
S = longitudinal slope in channel (m/m)
n = Manning's roughness coefficient

The Manning's roughness coefficient depends on the type of lining material used for the surface of the ditch. Table 16.9 provides recommended values for the roughness coefficients for different lining materials. The flow in the channel is then given as

$$Q = va = \frac{1}{n} a R^{2/3} S^{1/2} \tag{16.9}$$

where Q is the discharge (m³/sec).

Table 16.9 Manning's Roughness Coefficients

I. Closed conduits:	Manning's n range
A. Concrete pipe	0.011–0.013
B. Corrugated-metal pipe or pipe-arch:	
1. 2⅔ by ½-in. corrugation (riveted pipe):	
a. Plain or fully coated	0.024
b. Paved invert (range values are for 25 and 50 percent of circumference paved):	
(1) Flow full depth	0.021–0.018
(2) Flow 0.8 depth	0.021–0.016
(3) Flow 0.6 depth	0.019–0.013
2. 6 by 2-in. corrugation (field bolted)	0.03
C. Vitrified clay pipe	0.012–0.014
D. Cast-iron pipe, uncoated	0.013
E. Steel pipe	0.009–0.011
F. Brick	0.014–0.017
G. Monolithic concrete:	
1. Wood forms, rough	0.015–0.017
2. Wood forms, smooth	0.012–0.014
3. Steel forms	0.012–0.013
H. Cemented rubble masonry walls:	
1. Concrete floor and top	0.017–0.022
2. Natural floor	0.015–0.025
I. Laminated treated wood	0.015–0.017
J. Vitrified clay liner plates	0.015

II. Open channels, lined (straight alignment):	Manning's n range
A. Concrete, with surfaces as indicated:	
1. Formed, no finish	0.013–0.017
2. Trowel finish	0.012–0.014
3. Float finish	0.013–0.015
4. Float finish, some gravel on bottom	0.015–0.017
5. Gunite, good section	0.016–0.019
6. Gunite, wavy section	0.018–0.022
B. Concrete, bottom float finished, sides as indicated:	
1. Dressed stone in mortar	0.015–0.017
2. Random stone in mortar	0.017–0.020
3. Cement rubble masonry	0.020–0.025
4. Cement rubble masonry, plastered.	0.016–0.020
5. Dry rubble (riprap)	0.020–0.030
C. Gravel bottom, sides as indicated:	
1. Formed concrete	0.017–0.020
2. Random stone in mortar	0.020–0.023
3. Dry rubble (riprap)	0.023–0.033
D. Brick	0.014–0.017
E. Asphalt:	
1. Smooth	0.013
2. Rough	0.016
F. Wood, planed, clean	0.011–0.013
G. Concrete-lined excavated rock:	
1. Good section.	0.017–0.020
2. Irregular section	0.022–0.027

III. Open channels, excavated (straight, alinement, natural lining):	Manning's n range
A. Earth, uniform section:	
1. Clean, recently completed	0.016–0.018
2. Clean, after weathering	0.018–0.020
3. With short grass, few weeds	0.022–0.027
4. In gravelly soil, uniform section, clean	0.022–0.025
B. Earth, fairly uniform section:	
1. No vegetation	0.022–0.025
2. Grass, some weeds	0.025–0.030
3. Dense weeds or aquatic plants in deep channels	0.030–0.035
4. Sides clean, gravel bottom	0.025–0.030
5. Sides clean, cobble bottom	0.030–0.040
C. Dragline excavated or dredged:	
1. No vegetation	0.028–0.033
2. Light brush on banks	0.035–0.050
D. Rock:	
1. Based on design section	0.035
2. Based on actual mean section:	
a. Smooth and uniform	0.035–0.040
b. Jagged and irregular	0.040–0.045
E. Channels not maintained, weeds and brush uncut:	
1. Dense weeds, high as flow depth	0.08–0.12
2. Clean bottom, brush on sides	0.05–0.08
3. Clean bottom, brush on sides, highest stage of flow	0.07–0.11
4. Dense brush, high stage	0.10–0.14

IV. Highway channels and swales with maintained vegetation (values shown are for velocities of 2 and 6 f.p.s.):	Manning's n range
A. Depth of flow up to 0. 7 foot:	
1. Bermudagrass, Kentucky bluegrass, buffalograss:	
a. Mowed to 2 inches	0.07–0.045
b. Length 4–6 inches	0.09–0.05
2. Good stand, any grass:	
a. Length about 12 inches	0.18–0.09
b. Length about 24 inches	0.30–0.15
3. Fair stand, any grass:	
a. Length about 12 inches	0.14–0.08
b. Length about 24 inches	0.25–0.13
B. Depth of flow 0. 7–1. 5 feet:	
1. Bermudagrass, Kentucky bluegrass, buffalograss:	
a. Mowed to 2 inches	0.05–0.035
b. Length 4 to 6 inches	0.06–0.04
2. Good stand, any grass:	
a. Length about 12 inches	0.12–0.07
b. Length about 24 inches	0.20–0.10
3. Fair stand, any grass:	
a. Length about 12 inches	0.10–0.06
b. Length about 24 inches	0.17–0.09

(Continued)

Table 16.9   Manning's Roughness Coefficients (*continued*)

V. Street and expressway gutters:	*Manning's n range*
A. Concrete gutter, troweled finish..............	0.012
B. Asphalt pavement:	
1. Smooth texture	0.013
2. Rough texture..	0.016
C. Concrete gutter with asphalt pavement:	
1. Smooth...	0.013
2. Rough	0.015
D. Concrete pavement:	
1. Float finish..	0.014
2. Broom finish	0.016
E. For gutters with small slope, where sediment may accumulate, increase above values of *n* by...............................	0.002

VI. Natural stream channels:

A. Minor streams (surface width at flood stage less than 100 ft)	
1. Fairly regular section:	
a. Some grass and weeds, little or no brush	0.030–0.035
b. Dense growth of weeds, depth of flow materially greater than weed height	0.035–0.05
c. Some weeds, light brush on banks	0.035–0.05
d. Some weeds, heavy brush on banks	0.05–0.07
e. Some weeds, dense willows on banks	0.06–0.08
f. For trees within channel, with branches submerged at high stage, increase all above values by	0.01–0.02
2. Irregular sections, with pools, slight channel meander; increase values given in 1a–e about..................................	0.01–0.02

3. Mountain streams. No vegetation in channel, banks unusually steep, trees and brush along banks submerged at high stage:	
a. Bottom of gravel, cobbles, and few boulders...................................	0.04–0.05
b. Bottom of cobbles, with large boulders.......................................	0.05–0.07
B. Flood plains (adjacent to natural streams):	
1. Pasture, no brush:	
a. Short grass..	0.030–0.035
b. High grass..	0.035–0.05
2. Cultivated areas:	
a. No crop...	0.03–0.04
b. Mature row crops.............................	0.035–0.045
c. Mature field crops	0.04–0.05
3. Heavy weeds, scattered brush..............	0.05–0.07
4. Light brush and trees:	
a. Winter..	0.05–0.06
b. Summer	0.06–0.08
5. Medium to dense brush:	
a. Winter..	0.07–0.11
b. Summer	0.10–0.16
6. Dense willows, summer, not bent over by current..	0.15–0.20
7. Cleared land with tree stumps, 100–150 per acre:	
a. No sprouts	0.04–0.05
b. With heavy growth of sprouts...........	0.06–0.08
8. Heavy stand of timber, a few down trees, little undergrowth:	
a. Flood depth below branches............	0.10–0.12
b. Flood depth reaches branches..........	0.12–0.16
C. Major streams (surface width at flood stage more than 100 ft): Roughness coefficient is usually less than for minor streams of similar description on account of less effective resistance offered by irregular banks or vegetation banks. Values of *n* may be somewhat reduced. The values of *n* for larger streams of most regular section, with no boulders or brush, may be in range of..................................	0.028–0.033

SOURCE: Reproduced from *Design Charts for Open Channels*, U.S. Department of Transportation, Washington, D.C., 1980

The Federal Highway Administration (FHWA) has published a series of charts for channels of different cross sections that can be used to solve Eq. 16.9. Figures 16.6 and 16.7 are two examples of these charts.

Open-channel flows can be grouped into two general categories: steady and unsteady. Manning's formula assumes uniform steady flow. When the rate of discharge does not vary with time, the flow is steady; conversely, the flow is unsteady when the rate of discharge varies with time. Steady flow is further grouped into uniform and nonuniform, depending on the channel characteristics. Uniform flows are obtained when the channel properties (such as slope, roughness, and cross section) are constant along the length of the channel, whereas nonuniform flow is obtained when these properties vary along the length of the channel.

When uniform flow is achieved in the channel, the depth d and velocity v_n are defined as normal and the slope of the water surface is parallel to the slope of the channel. Since it is unusual for the exact same channel properties to exist throughout the entire length of

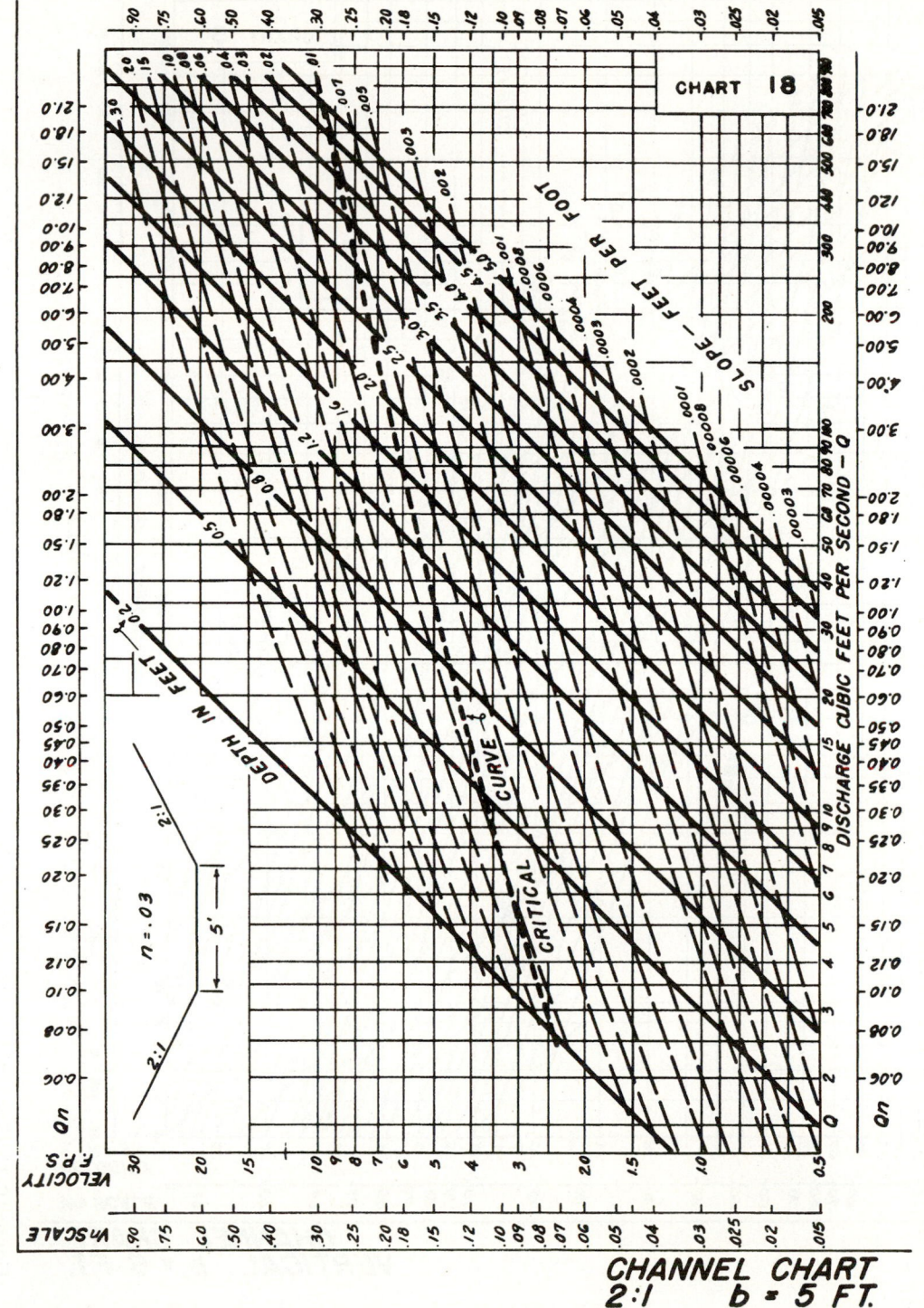

Figure **16.6** Graphical Solution of Manning's Equation for a 2:1 Side Slope Trapezoidal Channel

SOURCE: *Design Charts for Open Channels*, U.S. Department of Transportation, Washington, D.C., 1979

859

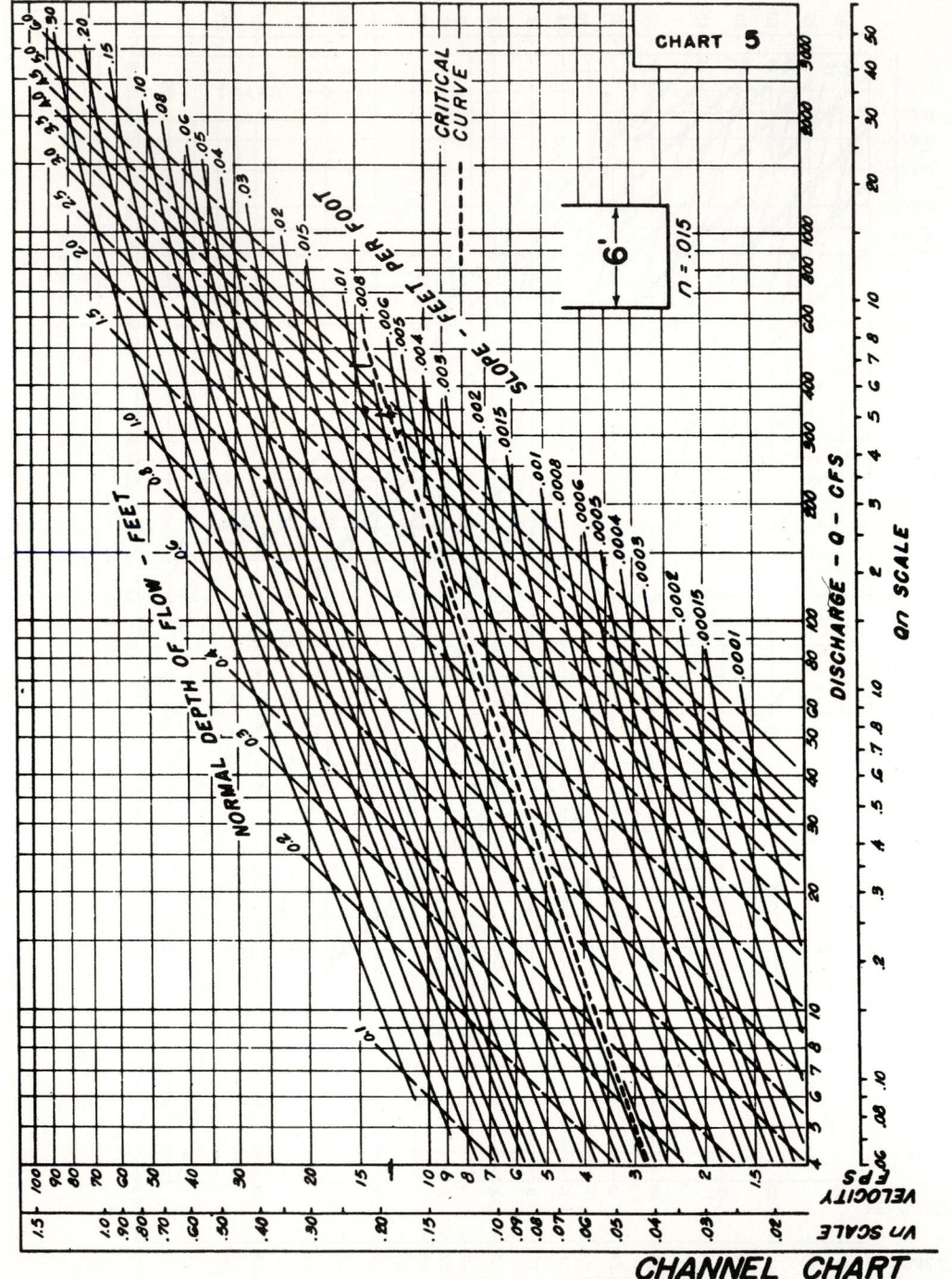

Figure 16.7 Graphical Solution of Manning's Equation for a Rectangular Channel, *b* = 6 ft

SOURCE: *Design Charts for Open Channels*, U.S. Department of Transportation, Washington, D.C., 1979

the channel, it is very difficult to obtain complete uniform flow conditions. Nevertheless, Manning's equation can be used to obtain practical solutions to stream flow problems in highway drainage design since the errors are usually insignificant.

Flows in channels also can be tranquil or rapid. Tranquil flow is similar to the flow of water in an open channel with a relatively flat longitudinal slope, whereas rapid flow is similar to water tumbling down a steep slope. The depth at which the flow in any channel changes from tranquil to rapid is known as the critical depth. When the flow depth is greater than the critical depth, the flow is known as subcritical. This type of flow often occurs in streams in the plains and broad valley regions. When the flow depth is less than the critical depth, the flow is known as supercritical. This type of flow is prevalent in steep flumes and mountain streams. The critical depth also may be defined as the depth of flow at which the specific energy is minimum. The critical depth depends only on the shape of the channel and the discharge. This implies that for any given channel cross section, there is only one critical depth for a given discharge.

The velocity and channel slope corresponding to uniform flow at critical depth are known as critical velocity and critical slope, respectively. The flow velocity and channel slope are therefore higher than the critical values when flow is supercritical and lower when flow is subcritical.

Consider a channel consisting of four sections, each with a different grade, as shown in Figure 16.8. The slopes of the first two sections are less than the critical slope (although the slope of the second section is less than that of the first section), resulting in a subcritical flow in both sections. The slope of the third section is higher than the critical slope, resulting in a supercritical flow. Flow in the fourth section is subcritical, since the slope is less than the critical slope. The depth of water changes as the water flows along all four sections of the channel.

As the water flows through section A of the channel, the depth of flow is greater than the critical depth, d_c, because the flow is subcritical. The slope of section B reduces, resulting in a lower velocity and a higher flow depth ($h_B > h_A$). This increase in depth takes place gradually and begins somewhere upstream in section A. The change from subcritical to supercritical flow in section C with the steep grade also takes place smoothly over some distance. The reduction in flow depth occurs gradually and begins somewhere upstream in the subcritical section. However, when the slope changes again to a value less than the critical slope, the flow changes abruptly from supercritical to subcritical, resulting in

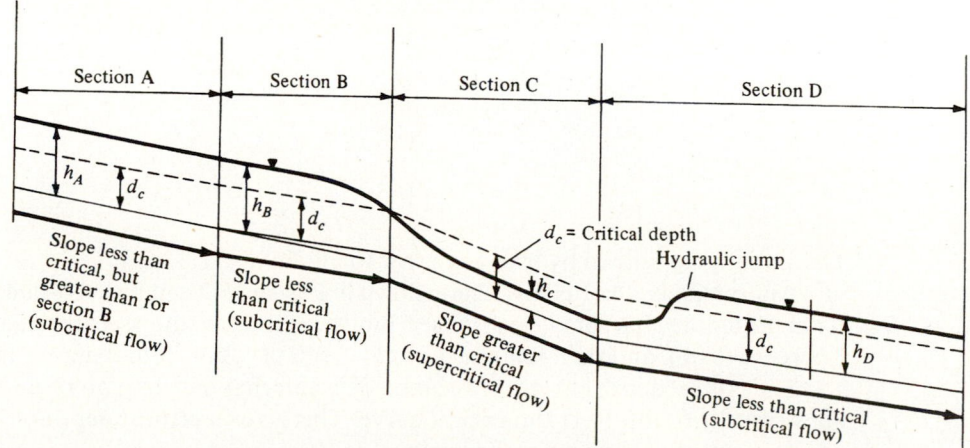

Figure 16.8   Schematic of the Effect of Critical Depth on Flow in Open Prismatic Channels

a *hydraulic jump*, in which some of the energy is absorbed by the resulting turbulence. This shows that downstream conditions may change the depth upstream of a subcritical flow, which means that *control is downstream*. Thus, when flow is subcritical, any downstream changes in slope, cross section, or intersection with another stream will result in the gradual change of depth upstream, known as *backwater curve*. Conversely, supercritical flow is normally not affected by downstream conditions and *control is upstream*.

Design Procedure

The design of a highway drainage channel is accomplished in two steps. The first step is to determine the channel cross section that will economically and effectively transport the expected runoff to a natural waterway, while the second step is to determine whether the channel requires any erosion protection and, if so, what type of lining is to be used.

Determination of Cross Section.   The Manning formula is solved for an assumed channel cross section to determine whether the channel is large enough for the runoff of the design storm. The solution may be obtained manually or by using the appropriate FHWA chart. Both methods are demonstrated in the following example. The FHWA curves for the solution of Manning's formula are only suitable for flows not affected by backwater.

Example 16.3 Design of an Open Channel

Determine a suitable cross section for a channel to carry an estimated runoff of 9.6 m³/sec if the slope of the channel is 1 percent and Manning's roughness coefficient, *n*, is 0.015.

Solution: Select a channel section and then use Manning's formula to determine the flow depth required for the estimated runoff. Assume a rectangular channel 6 ft wide.

$$\text{Flow depth} = d$$

$$\text{Cross-sectional area} = 6d$$

$$\text{Wetted perimeter} = 6 + 2d$$

$$\text{Hydraulic radius } R = 6d/(6 + 2d)$$

Using Eq. 16.9,

$$Q = \frac{1}{n} a R^{2/3} S^{1/2}$$

$$9.6 = \frac{1}{0.015}(6d)\left(\frac{6d}{6 + 2d}\right)^{2/3}(0.01)^{1/2}$$

This equation is solved by trial and error to obtain $d \approx 1.2$ m.
   Alternatively, the FHWA chart shown in Figure 16.7 can be used since the width is 1.8 m. Enter the chart at $Q = 9.6$ m³/sec, move vertically to intersect the channel slope of 1 percent (0.01), and then read the normal depth of 1.2 m from the normal depth lines.
   The critical depth also can be obtained by entering the chart at 9.6 m³/sec and moving vertically to intersect the critical curve. This gives a critical depth of about 1.4 m³, which means that the flow is supercritical.

The critical and flow velocities can be obtained similarly. The chart is entered at 9.6 m³/sec. Move vertically to the 0.01 slope and then horizontally to the velocity scale to read the flow velocity as approximately 4.2 m/sec. The critical velocity is obtained in a similar manner by moving vertically to the critical curve. The critical velocity is about 3.9 m/sec. The critical slope is about 0.007.

The solution indicates that if a rectangular channel 1.8 m wide is to be used to carry the runoff of 9.6 m³, the channel must be at least 1.2 m deep. However, it is necessary to provide a freeboard (distance from the water surface to the top of the channel) of at least 0.3 m, which makes the depth for this channel 1.5 m. Note that the formula for determining critical depth for a rectangular channel is

$$y_c = \left(\frac{q^2}{g}\right)^{1/3}$$

where q is the flow per foot of width, in m³/sec/m, and g is 9.81 m/sec². In this problem,

$$y_c = \left[\frac{\left(9.6/1.8\right)^2}{9.81}\right]^{1/3} = 1.46 \text{ m}$$

For nonrectangular sections, the critical depth occurs when

$$\frac{Q^2}{g} = \frac{A^3}{B}$$

where

B = width of the water surface (m)
A = area of cross section (m²)

This is a trial-and-error solution, and so charts and tables are typically utilized.

The same chart can be used for different values of n when all other conditions are the same. For example, if n is 0.02 in Example 16.3, Qn is first determined as $9.6 \times 0.02 = 0.192$. Then enter the chart at $qn = 0.192$ on the Qn scale, move vertically to the 0.01 slope, and read the flow depth as approximately 1.5 m. The velocity is also read from the Vn scale as $vn = 0.0675$, which gives V as 0.0675/0.02 = 3.38 m/sec.

Example 16.4 Computing Discharge Flow from a Trapezoidal Channel

A trapezoidal channel has 2:1 side slopes, a 1.5 m bottom width, and a depth of 1.2 m. If the channel is on a slope of 2 percent and $n = 0.030$, determine the discharge flow, velocity, and type of flow. The chart shown in Figure 16.6 is used to solve this problem.

Solution: The intersection of the 2 percent slope (0.02) and the 1.2 m depth line is located. Move vertically to the horizontal discharge scale to determine the discharge, $q = 17$ m³. Similarly, the velocity is 3.75 m/sec. The intersection of the 2 percent slope and 1.2 m depth lies above the critical curve, and thus the flow is supercritical.

Determination of Suitable Lining.   The traditional method for determining suitable lining material for a given channel is to select a lining material such that the flow velocity is less than the permissible velocity to prevent erosion of the lining. However, research results have shown that the maximum permissible depth of flow (d_{max}) for flexible linings should be the main criterion for selecting channel lining. Rigid channels, such as those made of concrete or soil cement, do not usually erode in normal highway work and therefore have no maximum permissible depth of flow to prevent erosion. Thus, maximum flow depth for rigid channels depends only on the freeboard required over the water surface.

The maximum permissible depth of flow for different flexible lining materials can be obtained from charts given in *Design of Stable Channels with Flexible Linings*. Figures 16.9 through 16.11 show some of these charts.

The design procedure for flexible linings is described in the following steps:

Step 1.   Determine channel cross section, as in Example 16.3 or Example 16.4.

Step 2.   Determine the maximum top width T of the selected cross section and select a suitable lining.

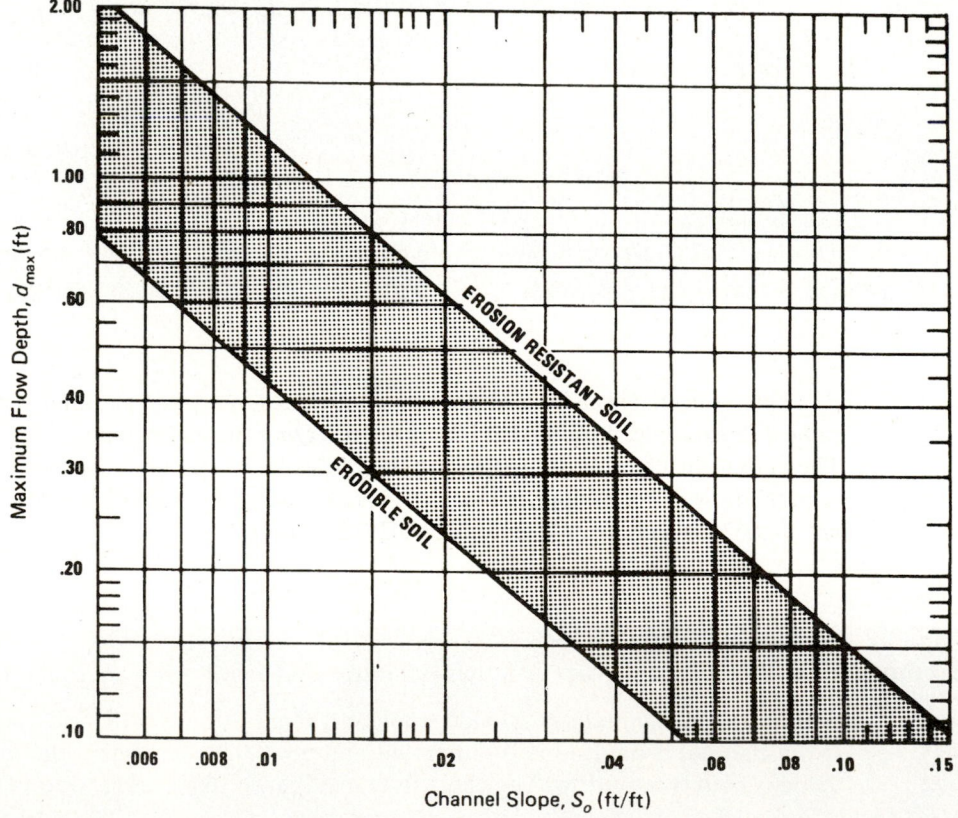

Figure 16.9   Maximum Permissible Depth of Flow, d_{max}, for Channels Lined with Jute Mesh

SOURCE: *Design of Stable Channels with Flexible Linings*, Hydraulic Engineering Circular No. 15, U.S. department of Transportation, Washington, D.C., October 1975

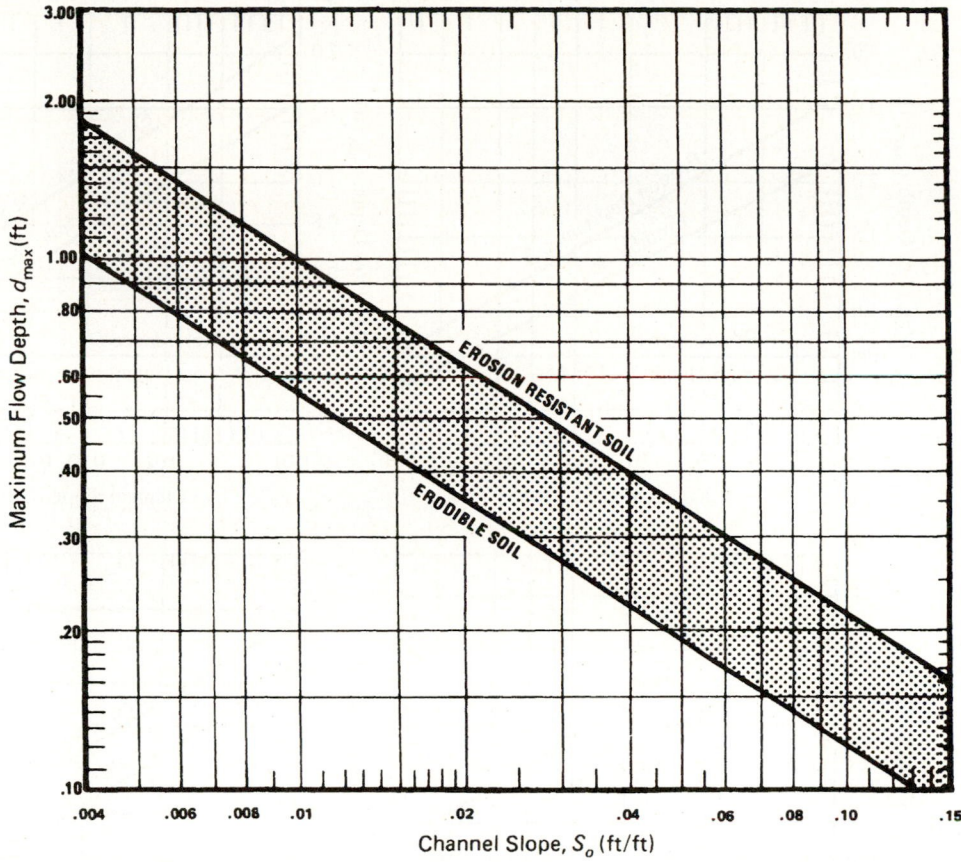

Figure 16.10 Maximum Permissible Depth of Flow, d_{max}, for Channels Lined with $^3/_8$-in. Fiberglass Mat

SOURCE: *Design of Stable Channels with Flexible Linings*, Hydraulic Engineering Circular No. 15, U.S. department of Transportation, Washington, D.C., October 1975

Step 3.   Using the appropriate d_{max} for the lining selected (obtained from the appropriate chart), determine the hydraulic radius R and the cross-sectional area a for the selected cross section. This can be done by calculation or by using the chart shown in Figure 16.12.

Step 4.   Determine the flow velocity V using the channel slope S and the hydraulic radius R determined in Step 3.

Step 5.   Determine the allowable flow in the channel using the velocity determined in Step 4 and the cross-sectional area a in Step 3—that is, $Q = aV$.

Step 6.   Compare the flow determined in Step 5 with the design flow. If there is a difference between the two flows, then the selected lining material is not suitable. For example, if the flow obtained in Step 5 is less than the design flow, then the lining is inadequate; if it is much greater, the channel is overdesigned. When the flows are much different, then another lining material should be selected and Steps 3 through 6 repeated.

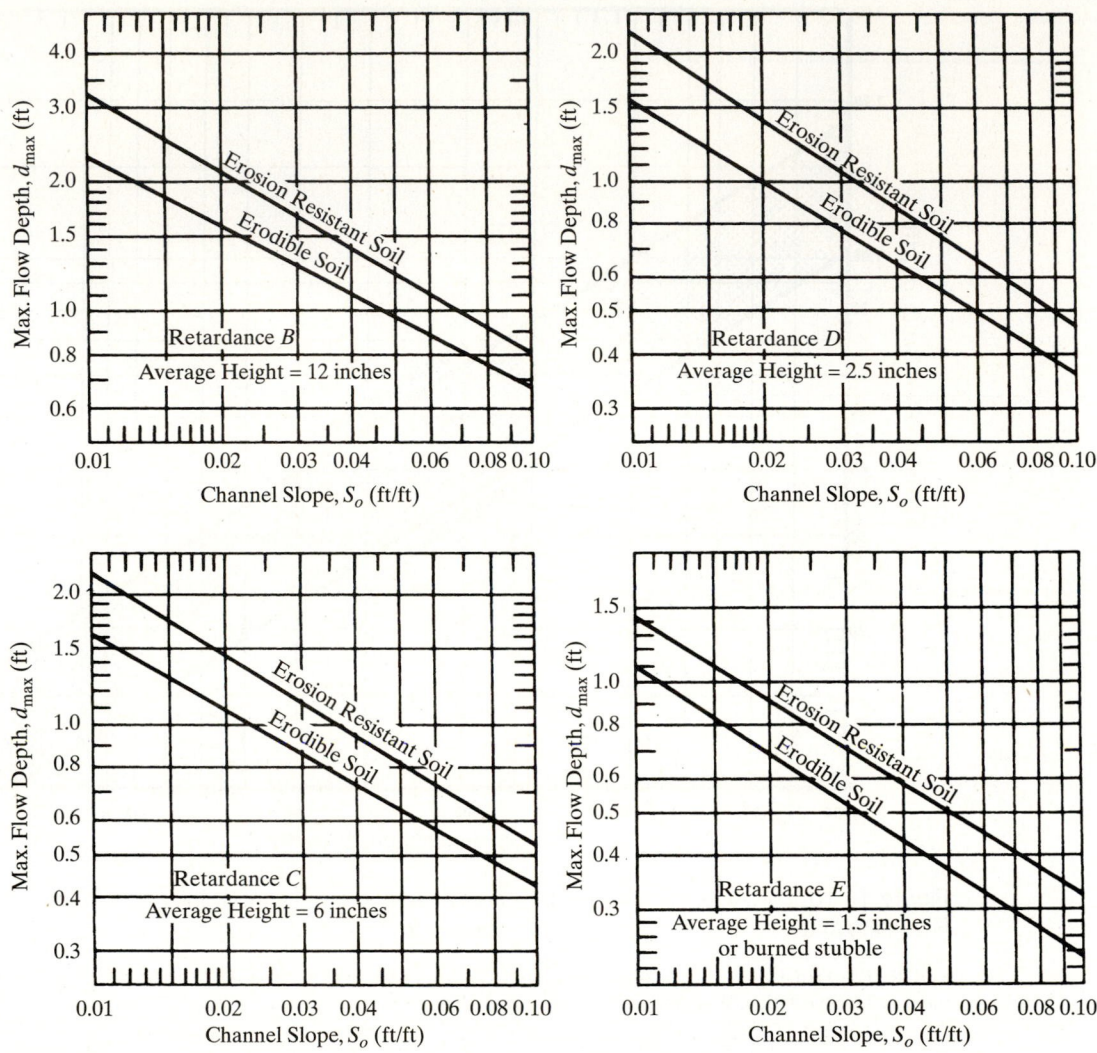

Figure 16.11 Maximum Permissible Depth of Flow, d_{max}, for Channels Lined with Bermuda Grass, Good Stand, Cut to Various Lengths

Note: Use on slopes steeper than 10 percent is not recommended.

SOURCE: *Design of Stable Channels with Flexible Linings*, Hydraulic Engineering Circular No. 15, U.S. department of Transportation, Washington, D.C., October 1975

Note that Manning's formula can be used to determine the allowable velocity in Step 4 if the Manning coefficient of the material selected is known. Since it is usually not easy to determine these coefficients, curves obtained from test results and provided in *Design of Stable Channels with Flexible Linings* can be used. Figures 16.13 and 16.14 are flow velocity curves for jute mesh and fiberglass mat, for which maximum permissible depth charts are shown in Figures 16.9 and 16.10.

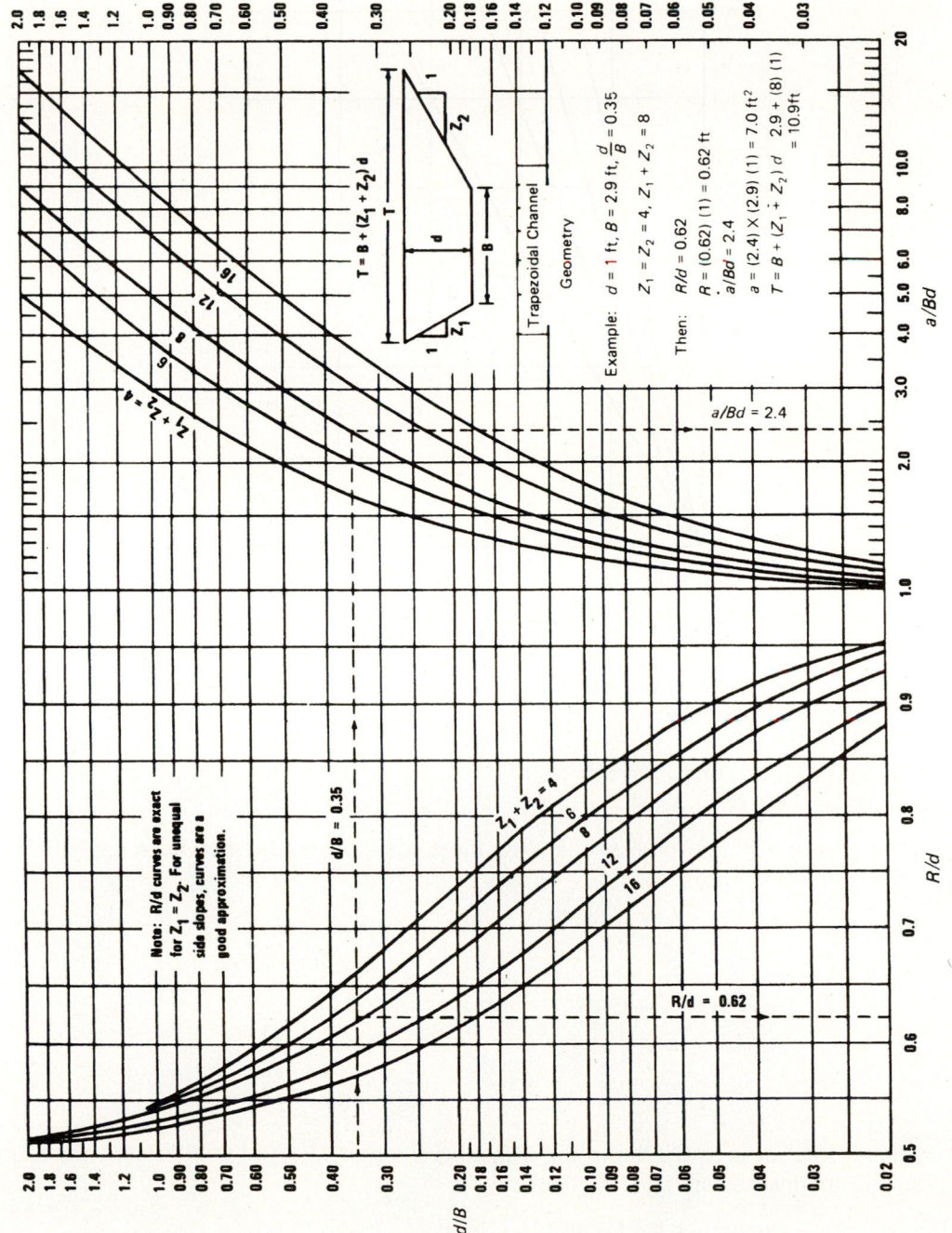

Figure 16.12 Trapezoidal Channel Geometry

SOURCE: *Design of Stable Channels with Flexible Linings*, Hydraulic Engineering Circular No. 15, U.S., department of Transportation, Washington, D.C., October 1975

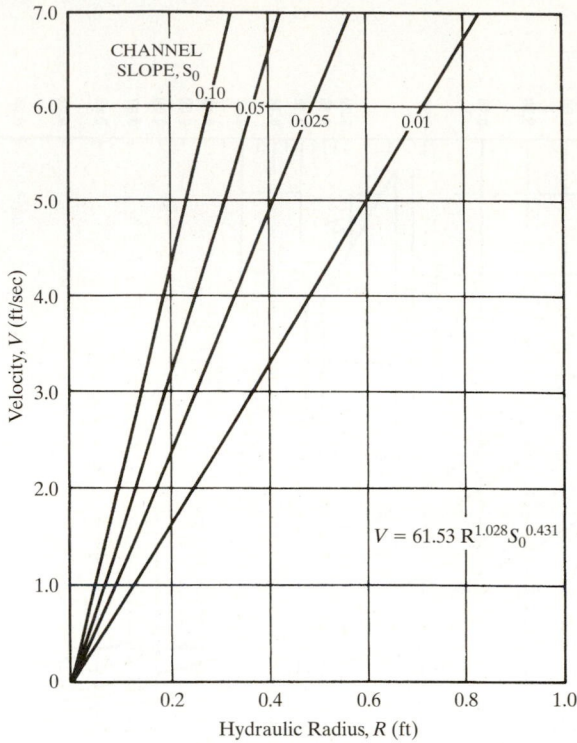

Figure 16.13 Flow Velocity for Channels Lined with Jute Mesh

SOURCE: *Design of Stable Channels with Flexible Linings*, Hydraulic Engineering Circular No. 15, U.S. department of Transportation, Washington, D.C., October 1975

Example 16.5 Checking for Lining Suitability

Determine whether jute mesh is suitable for the flow and channel section in Example 16.3. Assume erosion-resistant soil.

Solution:

$$d_{max} = 0.36 \text{ m} \qquad \text{(See Figure 16.9)}$$

$$\text{Wetted perimeter} = 2 \times 0.36 + 1.8 = 2.52 \text{ m}$$

$$\text{Cross-sectional area} = 0.36 \times 1.8 = 0.648 \text{ m}^2$$

$$R = \frac{0.648}{2.52} = 0.258$$

Selecting jute mesh as the lining and a channel slope of 0.01, the flow velocity for the lining is obtained from Figure 16.13 as

$$V = 61.53(0.86)^{1.028}(0.01)^{0.431} = 2.24 \text{ m/sec}$$

The maximum allowable flow is

$$Q = 2.24 \times 0.648 = 1.45 \text{ m}^3$$

Jute mesh is not suitable. A rigid channel is probably appropriate for this flow.

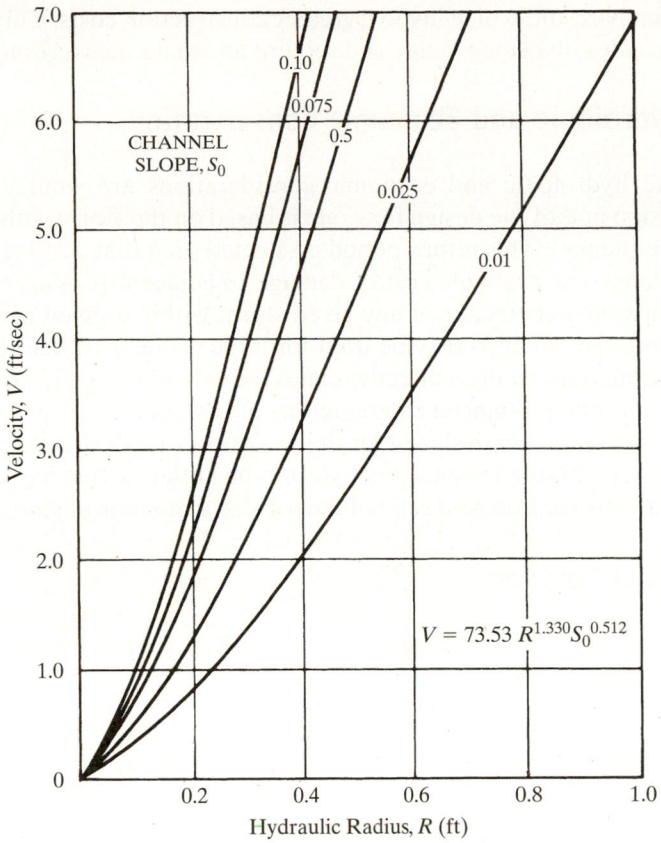

The figure shows a graph with y-axis labeled "Velocity, V (ft/sec)" ranging from 0 to 7.0, and x-axis labeled "Hydraulic Radius, R (ft)" ranging from 0 to 1.0. Curves are labeled with CHANNEL SLOPE, S_0 values: 0.10, 0.075, 0.5, 0.025, 0.01. The equation shown is:

$$V = 73.53\, R^{1.330} S_0^{0.512}$$

Figure 16.14   Flow Velocity for Channels Lined with $^3/_8$-in. Fiberglass Mat

SOURCE: *Design of Stable Channels with Flexible Linings*, Hydraulic Engineering Circular No. 15, U.S. department of Transportation, Washington, D.C., October 1975

16.6.2 Design of Culverts

Several hydraulic phenomena occur when water flows through highway culverts. The hydraulic design of culverts is therefore more complex than open channel design. The main factors considered in culvert design are the location of the culvert, the hydrologic characteristics of the watershed being served by the culvert, economy, and type of flow control.

Culvert Location

The most appropriate location of a culvert is in the existing channel bed, with the centerline and slope of the culvert coinciding with that of the channel. At this location, the cost associated with earth and channel work is achieved and stream-flow disturbance is minimized. However, other locations may have to be selected in some cases; for example, relocation of a stream channel may be necessary to avoid an extremely long culvert. The basic principle used in locating culverts is that abrupt stream changes at the inlet and outlet of the culvert should be avoided.

Special consideration should be given to culverts located in mountainous areas. When culverts are located in a natural channel in these areas, high fills and long channels usually result, which involve greater construction costs. Culverts are often located along the sides of a steep valley and require adequate measures to prevent erosion.

Hydrologic and Economic Considerations

The hydrologic and economic considerations are similar to those for open-channel design in that the design flow rate is based on the storm with an acceptable return period (frequency). This return period is selected such that construction and maintenance costs balance the probable cost of damage to adjacent properties if the storm should occur. Since the occurrence of any given storm within a given period is a stochastic phenomenon, risk analysis may be used for large projects to determine the appropriate return period. Risk analysis directly relates the culvert design to economic theory and identifies the probable financial consequences of both underdesign and overdesign.

Culverts are designed for the peak flow rate of the design storm. The peak flow rate may be obtained from a unit hydrograph at the culvert site; the hydrograph is developed from stream flow and rainfall records for a number of storm events.

Other Factors

Other factors that should be considered in culvert design are tailwater and upstream storage conditions.

Tailwater.   This is defined as the water depth at the outlet above the inside bottom of the culvert (culvert invert) as the water flows out of the culvert. Design of the culvert capacity must take tailwater into consideration, particularly when the design is based on the outlet conditions. High tailwater elevations may occur due to the hydraulic resistance of the channel or during flood events if the flow downstream is obstructed. Field observation and maps should be used to identify conditions that facilitate high tailwater elevations. These conditions include channel constrictions, intersections with other watercourses, downstream impoundments, channel obstructions, and tidal effects. If these conditions do not exist, the tailwater elevation is based on the elevation of the water surface in the natural channel.

Upstream Storage.   The ability of the channel to store large quantities of water upstream from the culvert may have some effect on the design of the culvert capacity. The storage capacity upstream is sometimes planned for retention and should therefore be checked using large-scale contour maps, from which topographic information is obtained.

Hydraulic Design of Culverts

As stated earlier, it is extremely difficult to carry out an exact theoretical analysis of culvert flow because of the many complex hydraulic phenomena that occur. For example, different flow types may exist at different times in the same culvert, depending on the tailwater elevation.

The design procedure presented here is that developed by FHWA and published in *Hydraulic Design of Highway Culverts*. The control section of the culvert is used to classify different culvert flows, which are then analyzed. The location at which a unique relationship exists between the flow rate and the depth of flow upstream from the culvert is the control section. When the flow is dictated by the inlet geometry, the control section is the culvert inlet (that is, the upstream end of the culvert), and the flow is *inlet controlled*. When the flow is governed by a combination of the tailwater, the culvert inlet, and the characteristics of the culvert barrel, the flow is *outlet controlled*. Although it is possible for the flow in a culvert to change from one control to the other and back, the design is based on the *minimum performance* concept, which provides for the culvert to perform at a level that is never lower than the designed level. This means that the culvert may perform at a more efficient level, which may not be desired as a higher flow may be obtained for a given headwater level. The design procedure uses several design charts and nomographs developed from a combination of theory and numerous hydraulic test results.

Inlet Control.   The flow in culverts operating under inlet control conditions is supercritical with high velocities and low depths. Four different flows under inlet control are shown in Figure 16.15. The flow type depends on whether the inlet and/or outlet of the culvert are submerged. In Figure 16.15a, both the inlet and outlet are above the water surface. In this case, the flow within the culvert is supercritical, the culvert is partly full throughout its length, and the flow depth approaches normal at the outlet end. In Figure 16.15b, only the downstream end (outlet) of the culvert is submerged, but this submergence does not result in outlet control. The flow in the culvert just beyond the culvert entrance (inlet) is supercritical, and a hydraulic jump occurs within the culvert. Figure 16.15c shows the inlet end of the culvert submerged, with the water flowing freely at the outlet. The culvert is partly full along its length, and the flow is supercritical within the culvert, since the critical depth is located just past the culvert inlet. Also, the flow depth at the culvert outlet approaches

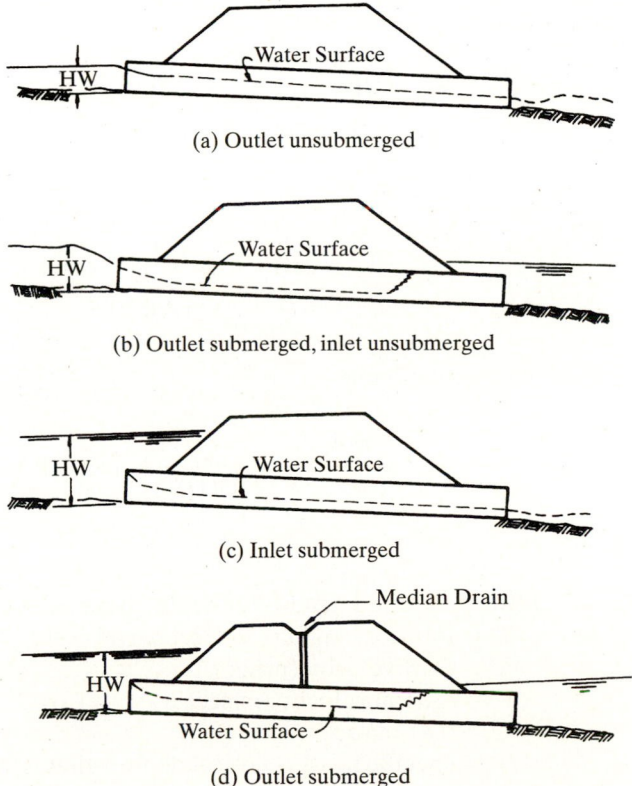

(a) Outlet unsubmerged

(b) Outlet submerged, inlet unsubmerged

(c) Inlet submerged

(d) Outlet submerged

Figure 16.15 Types of Inlet Control

SOURCE: J.M. Normann, R.J. Houghtalen, W.J. Johnston, *Hydraulic Design of Highway Culverts*, Report No. FHWA-IP-85-15, U.S. Department of Transportation, Office of Implementation, McLean, VA., September 1985

normal. This example of inlet control is more typical of design conditions. Figure 16.15d shows both the inlet and outlet of the culvert submerged, but the culvert is only partially full over part of its length. A hydraulic jump occurs within the culvert, resulting in a full culvert along its remaining length. Under these conditions, pressures less than atmospheric may develop, creating an unstable situation, with the culvert alternating between partly full flow and full flow. Providing the median inlet shown prevents this situation.

Several factors affect the performance of a culvert under inlet control conditions. These include the inlet area, inlet shape, inlet configuration, and the headwater depth. Several methods are available to increase the performance of culverts under inlet control. These include the use of special configurations for inlet edges and beveled edges at the culvert inlet. Detailed descriptions of these methods are given in *Hydraulic Design of Highway Culverts* and are briefly discussed later.

Model tests have been used to determine flow relationships between headwater (depth of water above the culvert inlet invert) and flow for culverts operating under inlet control conditions. The basic condition used to develop these equations is whether or not the inlet is submerged. The inlet performs as an orifice when it is submerged and as a weir when it is not. Two equations were developed for the unsubmerged condition. The first equation (Eq. 16.10) is based on the specific head at critical depth, and the second equation (Eq. 16.11) is exponential and similar to a weir equation. Equation 16.10 has more theoretical support, but Eq. 16.11 is simpler to use. Both equations will give adequate results. Equation 16.12 gives the relationship for the submerged condition.

For the *unsubmerged condition*,

$$\frac{HW_i}{D} = \frac{H_c}{D} + K\left[\frac{Q}{(A)(D)^{0.5}}\right]^M - 0.5S \tag{16.10}$$

and

$$\frac{HW_i}{D} = K\left[\frac{Q}{(A)(D)^{0.5}}\right]^M \tag{16.11}$$

For the *submerged condition*,

$$\frac{HW_i}{D} = c\left[\frac{Q}{(A)(D)^{0.5}}\right]^2 + Y - 0.5S \tag{16.12}$$

where

HW_i = required headwater depth above inlet control section invert (m)
D = interior height of culvert barrel (m)
V = flow velocity (m/sec)
V_c = critical velocity (m/sec)
g = 9.81 m/sec²
H_c = specific head at critical depth—that is, $d_c + (V_c^2/2g)$ (m)
d_c = critical depth (m)
Q = discharge (m³/sec)
A = full cross-sectional area of culvert barrel (m²)
S = culvert barrel slope (m/m)
K, M, c, Y = constants from Table 16.10

Table 16.10   Coefficients for Inlet Control Design Equations

Shape and Material	Inlet Edge Description	Form	Unsubmerged		Submerged	
			K	M	C	Y
Circular	Square edge w/headwall	1	0.0098	2.0	0.0398	0.67
Concrete	Groove end w/headwall		0.0078	2.0	0.292	0.74
	Groove end projecting		0.0045	2.0	0.0317	0.69
Circular	Headwall	1	0.0078	2.0	0.0379	0.69
CMP	Mitered to slope		0.0210	1.33	0.0463	0.75
	Projecting		0.0340	1.50	0.0553	0.54
Circular	Beveled ring, 45° bevels	1	0.0018	2.50	0.0300	0.74
	Beveled ring, 33.7° bevels		0.0018	2.50	0.0243	0.83
Rectangular	30° to 75° wingwall flares	1	0.026	1.0	0.0385	0.81
Box	90° and 15° wingwall flares		0.061	0.75	0.0400	0.80
	0° wingwall flares		0.061	0.75	0.0423	0.82
Rectangular	45° wingwall flare $d = .0430$	2	0.510	0.667	0.0309	0.80
Box	18° to 33.7° wingwall flare $d = .0830$		0.486	0.667	0.0249	0.83
Rectangular	90° headwall w/¾″ chamfers	2	0.515	0.667	0.0375	0.79
Box	90° headwall w/45° bevels		0.495	0.667	0.0314	0.82
	90° headwall w/33.7° bevels		0.486	0.667	0.0252	0.865
Rectangular	¾″ chamfers; 45° skewed headwall	2	0.522	0.667	0.0402	0.73
Box	¾″ chamfers; 30° skewed headwall		0.533	0.667	0.0425	0.705
	¾″ chamfers; 15° skewed headwall		0.545	0.667	0.04505	0.68
	45° bevels; 10° to 45° skewed headwall		0.498	0.667	0.0327	0.75
Rectangular	45° non-offset wingwall flares	2	0.497	0.667	0.0339	0.803
Box	18.4° non-offset wingwall flares		0.493	0.667	0.0361	0.806
¾″ Chamfers	18.4° non-offset wingwall flares 30° skewed barrel		0.495	0.667	0.0386	0.71
Rectangular	45° wingwall flares—offset	2	0.497	0.667	0.0302	0.835
Box	33.7° wingwall flares—offset		0.495	0.667	0.0252	0.881
Top Bevels	18.4° wingwall flares—offset		0.493	0.667	0.0227	0.887
C M Boxes	90° headwall	1	0.0083	2.0	0.0379	0.69
	Thick wall projecting		0.0145	1.75	0.0419	0.64
	Thin wall projecting		0.0340	1.5	0.0496	0.57
Horizontal	Square edge w/headwall	1	0.0100	2.0	0.0398	0.67
Ellipse	Groove end w/headwall		0.0018	2.5	0.0292	0.74
Concrete	Groove end projecting		0.0045	2.0	0.0317	0.69
Vertical	Square edge w/headwall	1	0.0100	2.0	0.0398	0.67
Ellipse	Groove end w/headwall		0.0018	2.5	0.0292	0.74
Concrete	Groove end projecting		0.0095	2.0	0.0317	0.69
Pipe Arch	90° headwall	1	0.0083	2.0	0.0379	0.69
18° Corner	Mitered to slope		0.0300	1.0	0.0463	0.75
Radius CM	Projecting		0.0340	1.5	0.0496	0.57
Pipe Arch	Projecting	1	0.0296	1.5	0.0487	0.55
18° Corner	No bevels		0.0087	2.0	0.0361	0.66
Radius CM	33.7° bevels		0.0030	2.0	0.0264	0.75
Pipe Arch	Projecting	1	0.0296	1.5	0.0487	0.55
31° Corner	No bevels		0.0087	2.0	0.0361	0.66
Radius CM	33.7° bevels		0.0030	2.0	0.0264	0.75
Arch CM	90° headwall	1	0.0083	2.0	0.0379	0.69
	Mitered to slope		0.0300	1.0	0.0463	0.75
	Thin wall projecting		0.0340	1.5	0.0496	0.57

(Continued)

Table 16.10   Coefficients for Inlet Control Design Equations (*continued*)

Shape and Material	Inlet Edge Description	Form	Unsubmerged		Submerged	
			K	M	C	Y
Circular	Smooth tapered inlet throat	2	0.534	0.555	0.0196	0.89
	Rough tapered inlet throat		0.519	0.64	0.0289	0.90
Elliptical	Tapered inlet—beveled edges	2	0.536	0.622	0.0368	0.83
Inlet Face	Tapered inlet—square edges		0.5035	0.719	0.0478	0.80
	Tapered inlet—thin edge projecting		0.547	0.80	0.0598	0.75
Rectangular	Tapered inlet throat	2	0.475	0.667	0.0179	0.97
Rectangular	Side tapered—less favorable edges	2	0.56	0.667	0.0466	0.85
Concrete	Side tapered—more favorable edges		0.56	0.667	0.0378	0.87
Rectangular	Slope tapered—less favorable edges	2	0.50	0.667	0.0466	0.65
Concrete	Slope tapered—more favorable edges		0.50	0.667	0.0378	0.71

SOURCE: Haestad Methods, Inc., *Computer Applications in Hydraulic Engineering*, Haestad Press, Waterbury, CT, 1997. Used with permission.

Note that the last term $(-0.5S)$ in Eqs. 16.10 and 16.12 should be replaced by $(+0.7S)$ when mitered corners are used. Equations 16.10 and 16.11 apply up to about $Q/(A)(D)^{0.5} = 3.5$. Equation 16.12 applies above about $Q/(A)(D)^{0.5} = 4.0$.

Several charts for different culvert shapes have been developed based on these equations and can be found in *Hydraulic Design for Highway Culverts*. Figure 16.16 is the chart for rectangular box culverts under inlet control, with flared wingwalls and beveled edge at top of inlet, and Figure 16.17 shows the chart for a circular pipe culvert under inlet control.

These charts are used to determine the depth of headwater required to accommodate the design flow through the selected culvert configuration under inlet control conditions. Alternatively, the iteration required to solve any of the equations may be carried out by using a computer. The use of the chart is demonstrated in Example 16.6.

Example 16.6   Computing Inlet Invert for a Box Culvert

Determine the required inlet invert for a 1.5 m × 1.5 m box culvert under inlet control with 45-degree flared wingwalls and beveled edge for the following flow conditions:

Peak flow = 6.75 m³/sec

Design headwater elevation (EL_{hd}) = 69.15 m (based on adjacent structures)

Stream bed elevation at face of inlet = 67.2 m

Solution: The chart shown in Figure 16.16 is applicable, and the solution is carried out to demonstrate the consecutive steps required.

Step 1.   Select the size of the culvert and locate the design flow rate on the appropriate scales (points *A* and *B*, respectively). Note that for rectangular box culverts, the flow rate per width of barrel width is used.

$$\frac{Q}{NB} = \frac{6.75}{1.5} = 4.5 \text{ m}^3/\text{sec/m} \qquad \text{(for point } B)$$

Step 2. Draw a straight line through points *A* and *B* and extend this line to the first headwater/culvert height (*HW/D*) scale. Read the value on this scale. (Note that the first line is a turning line and that alternate values of (*HW/D*) can be obtained by drawing a horizontal line from this point to the other scales as shown.) Using the first line in this example, (*HW/D*) = 1.41.

Step 3. The required headwater is determined by multiplying the reading obtained in Step 2; that is, the value for (*HW/D*) by the culvert depth, $HW = 1.41 \times 1.5 \sim 2.13$. This value is used for HW_i (required headwater depth above inlet control invert, ft) if the approach velocity head is neglected. When the approach velocity head is not neglected, then

$$HW_i = HW - \frac{V^2}{2g}$$

Neglecting the approach velocity head in this problem, $HW_i = 2.13$ m.

Step 4. The required depression (fall)—that is, the depth below the streambed at which the invert should be located—is obtained as follows.

$$HW_d = EL_{hd} - EL_{sf} \tag{16.13}$$

and

$$\text{Fall} = HW_i - HW_d \tag{16.14}$$

where

HW_d = design headwater depth (m)
EL_{hd} = design headwater elevation (m)
EL_{sf} = elevation at the stream bed at the face (m)

In this case,

$$HW_d = 69.15 - 67.2 = 1.95 \text{ m}$$

but required depth is 7.1 ft.

$$\text{Fall} = 2.13 - 1.95 = 0.18 \text{ m} \approx 18 \text{ cm}$$

The invert elevation is therefore 67.2 − 0.18 = 67.02 m. Note that the value obtained for the fall may be either negative, zero, or positive. When a negative or zero value is obtained, use zero. When a positive value that is regarded as being too large is obtained, another culvert configuration must be selected and the procedure repeated. In this case, a fall of 7 in. is acceptable, and the culvert is located with its inlet invert at 67.02 m.

For more sophisticated design, models such as HEC-RAS can be used to produce a transient response to rainfall. Thus, in addition to steady-state results as presented in Example 16.6, time-variable results the rising water surface in the culvert may also be generated by HEC-RAS, which is readily available from the Army Corps of Engineers (http://www.hec.usace.army.mil/software/hec-ras/). In addition, a software program, HY-8 (version 7.2), distributed by the Federal Highway Administration on August 8, 2009, may be used for culvert design.

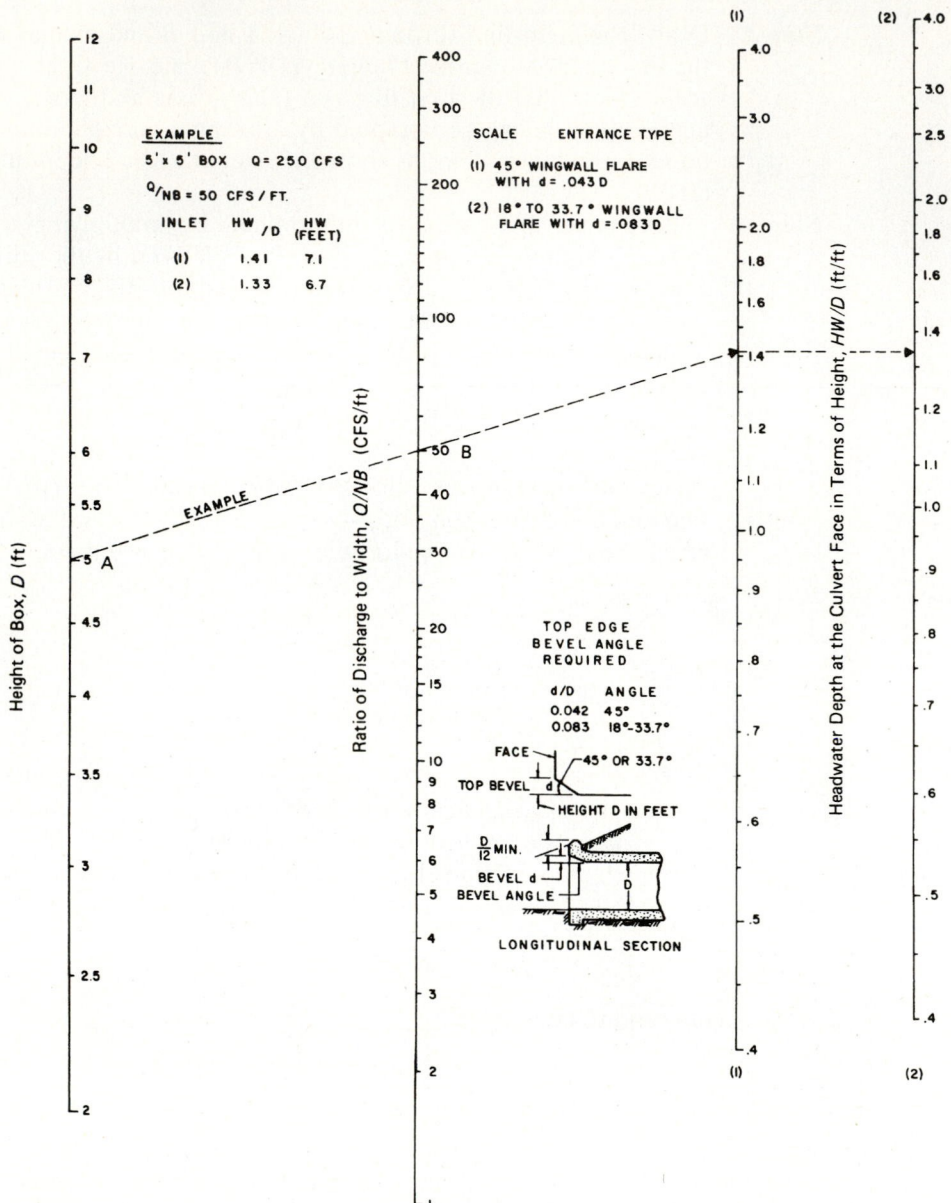

Figure 16.16   Headwater Depth for Inlet Control, Rectangular Box Culverts, Flared Wingwalls 18°
to 33.7°, and 45° with Beveled Edge at Top of Inlet

SOURCE: J.M. Normann, R.J. Houghtalen, W.J. Johnston, *Hydraulic Design of Highway Culverts*, Report No.
FHWA-IP-85-15, U.S. Department of Transportation, Office of Implementation, McLean, VA., September 1985

Outlet Control.   A culvert flows under outlet control when the barrel is incapable of
transporting as much flow as the inlet opening will receive. Figure 16.18 shows different
types of flows under outlet control conditions, where the control section is located at the
downstream end of the culvert or beyond. In Figure 16.18a, both the inlet and outlet of

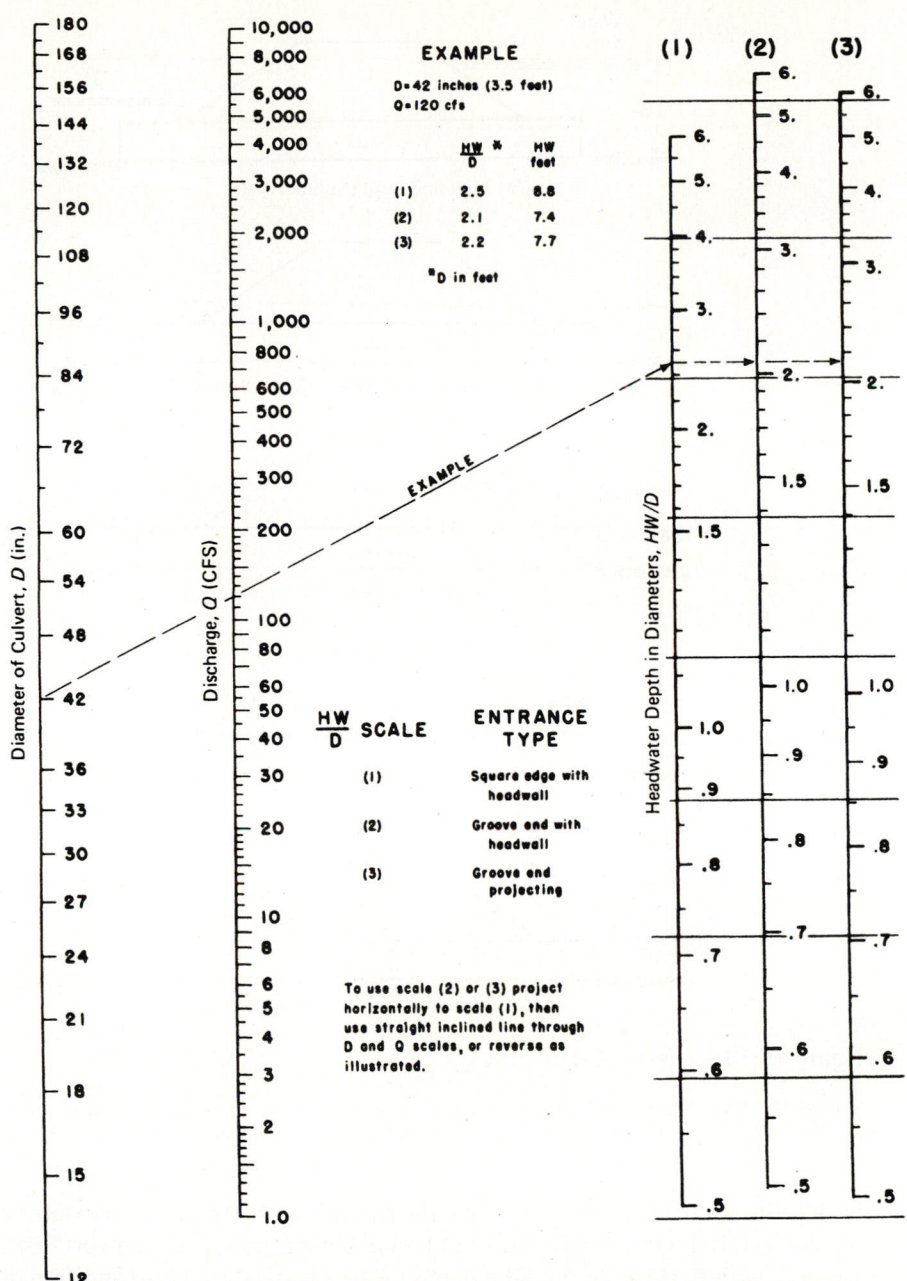

Figure 16.17 Headwater Depth for Concrete Pipe Culverts with Inlet Control

SOURCE: J.M. Normann, R.J. Houghtalen, W.J. Johnston, *Hydraulic Design of Highway Culverts*, Report No. FHWA-IP-85-15, U.S. Department of Transportation, Office of Implementation, McLean, VA., September 1985

the culvert are submerged, and the water flows under pressure along the whole length of the culvert with the culvert completely full. This is a common design assumption, although it does not often occur in practice. Figure 16.18b shows the inlet unsubmerged and the outlet submerged. This usually occurs when the headwater depth is low, resulting

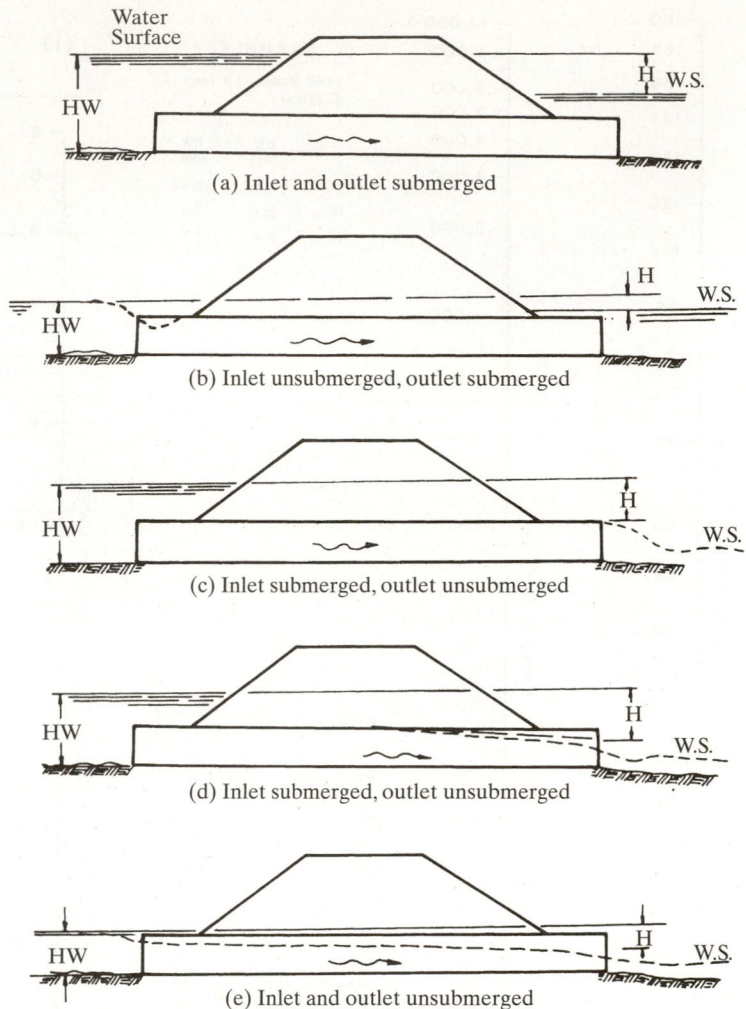

Figure 16.18   Types of Outlet Control

SOURCE: J.M. Normann, R.J. Houghtalen, W.J. Johnston, *Hydraulic Design of Highway Culverts*, Report No. FHWA-IP-85-15, U.S. Department of Transportation, Office of Implementation, McLean, VA., September 1985

in the top of the culvert being above the surface of the water as the water contracts into the culvert. In Figure 16.18c, the outlet is unsubmerged and the culvert flows full along its whole length because of the high depth of the headwater. This condition does not occur often, since it requires very high inlet heads. High outlet velocities are obtained under this condition. In Figure 16.18d, the culvert inlet is submerged and the outlet unsubmerged, with a low tailwater depth. The flow in the culvert is therefore partially full along part of its length. The flow is also subcritical along part of the culvert's length, but the critical depth is passed just upstream of the outlet. Figure 16.18e shows both the culvert's inlet and outlet unsubmerged with the culvert partly full along its entire length, with the flow being subcritical.

In addition to the factors that affect the performance of culverts under inlet control, the tailwater depth and certain culvert characteristics, which include the roughness, area, shape, slope, and length, also affect the performance of culverts under outlet control.

The hydraulic analysis of culverts flowing under outlet control is based on energy balance. The total energy loss through the culvert is given as

$$H_L = H_e + H_f + H_o + H_b + H_j + H_g \tag{16.15}$$

where

H_L = total energy required
H_e = energy loss at entrance
H_f = friction loss
H_o = energy loss at exit
H_b = bend loss
H_j = energy loss at junction
H_g = energy loss at safety grates

Losses due to bend, junction, and grates occur only when these features are incorporated in the culvert. For culverts without these features, the total head loss is given as

$$H_L = \left[1 + k_e + \frac{29n^2 L}{R^{1.33}}\right]\frac{V^2}{2g} \tag{16.16}$$

where

k_e = factor based on various inlet configurations (see Table 16.11)
n = Manning's coefficients for culverts (see Table 16.12)

R = hydraulic radius of the full culvert barrel = $\dfrac{a}{p}$ (m)

L = length of culvert barrel (m)
V = velocity in the barrel (m/s)

When special features such as grates, bends, and junctions are incorporated in the culvert, the appropriate additional losses may be determined from one or more of the following equations.

The bend loss H_b is given as

$$H_b = k_b \frac{v^2}{2g} \tag{16.17}$$

where

k_b = bend loss coefficient (see Table 16.13)
v = flow velocity in the culvert barrel (m/sec)
g = 9.8 m/sec²

The junction loss H_j is given as

$$H_j = y' + H_{v1} - H_{v2} \tag{16.18}$$

where

γ' = change in hydraulic grade line through the junction
= $(Q_2 v_2 - Q_1 v_1 - Q_3 v_3 \cos\theta_j)/[0.5(a_1 + a_2)g]$
Q_i = flow rate in barrel i (see Figure 16.19)
v_i = velocity in barrel i (m/sec)

Table 16.11   Entrance Loss Coefficients

Type of Structure and Design of Entrance	Coefficient, k_e
Pipe, Concrete	
Projecting from fill, socket end (groove-end)	0.2
Projecting from fill, square-cut end	0.5
Headwall or headwall and wingwalls:	
Socket end of pipe (groove-end)	0.2
Square-edged	0.5
Rounded (radius = $\frac{1}{12}D$)	0.2
Mitered to conform to fill slope	0.7
End-section conforming to fill slope	0.5
Beveled edges, 33.7° or 45° bevels	0.2
Side- or slope-tapered inlet	0.2
Pipe or Pipe-Arch, Corrugated Metal	
Projecting from fill (no headwall)	0.9
Headwall or headwall and wingwalls, square-edged	0.5
Mitered to conform to fill slope, paved or unpaved slope	0.7
End section conforming to fill slope	0.5
Beveled edges, 33.7° or 45° bevels	0.2
Side- or slope-tapered inlet	0.2
Box, Reinforced Concrete	
Headwall parallel to embankment (no wingwalls):	
Square edged on three edges	0.5
Rounded on three edges to radius of $\frac{1}{12}$ dimension,	0.2
or beveled edges on three sides	
Wingwalls at 30° to 75° to barrel:	
Square edged at crown	0.4
Crown edge rounded to radius of $\frac{1}{12}$ dimension,	0.2
or beveled top edge	
Wingwall at 10° to 25° to barrel, square edged at crown	0.5
Wingwalls parallel (extension of sides), square edged at crown	0.7
Side- or slope-tapered inlet	0.2

SOURCE: Adapted from J.M. Normann, R.J. Houghtalen, and W.J. Johnston, *Hydraulic Design of Highway Culverts*, Report to FHWA-IP-85-15, U.S. Department of Transportation, Office of Implementation, McLean, Va, September 1985

a_i = cross-sectional area of barrel i

θ_j = angle of the lateral with respect to the outlet conduit (degrees)

H_{v1} = velocity head in the upstream conduit (m)

H_{v2} = velocity head in the downstream conduit (m)

The head loss due to bar grate (H_g) is given as

$$H_g = k_g \frac{W}{x} \frac{v_u^2}{2g} \sin \theta_g \qquad (16.19)$$

where

x = minimum clear spacing between bars (m)

W = maximum cross-sectional width of the bars facing the flow (m)

θ_g = angle of grates with respect to the horizontal (degrees)

v_u = approach velocity (m/sec)

Table 16.12 Manning's Coefficients for Culverts

Type of Conduit	Wall and Joint Description	Manning n
Concrete pipe	Good joints, smooth walls	0.011–0.013
	Good joints, rough walls	0.014–0.016
	Poor joints, rough walls	0.016–0.017
Concrete box	Good joints, smooth finished walls	0.012–0.015
	Poor joints, rough, unfinished walls	0.014–0.018
Corrugated metal pipes and boxes, annular corrugations (Manning *n* varies with barrel size)	2⅔ by ½-in. corrugations	0.027–0.022
	6 by 1-in. corrugations	0.025–0.022
	5 by 1-in. corrugations	0.026–0.025
	3 by 1-in. corrugations	0.028–0.027
	6 by 2-in. structural plate corrugations	0.035–0.033
	9 by 2½-in. structural plate corrugations	0.037–0.033
Corrugated metal pipes, helical corrugations, full circular flow	2⅔ by ½-in. corrugations 24-in. plate width	0.012–0.024
Spiral rib metal pipe	¾ by ¾-in. recesses at 12-in. spacing, good joints	0.012–0.013

SOURCE: Adapted from J.M. Normann, R.J. Houghtalen, and W.J. Johnston, *Hydraulic Design of Highway Culverts*, Report to FHWA-IP-85-15, U.S. Department of Transportation, Office of Implementation, McLean, Va, September 1985

k_g = dimensionless bar shape factor
  = 2.42 for sharp-edged rectangular bars
  = 1.83 for rectangular bars with semicircular upstream face
k_g = 1.79 for circular bars
  = 1.67 for rectangular bars with semicircular upstream and downstream faces

Note that Eqs. 16.18 and 16.19 are both empirical, and caution must be exercised in using them.

Figure 16.20 is a schematic of the energy grade lines for a culvert flowing full. If the total energies at the inlet and outlet are equated, then

$$HW_o + \frac{v_u^2}{2g} = TW + \frac{v_d^2}{2g} + H_L \qquad (16.20)$$

Table 16.13 Loss Coefficients for Bends

Radius of Bend	Angle of Bend		
Equivalent Diameter	90°	45°	22.5°
1	0.50	0.37	0.25
2	0.30	0.22	0.15
4	0.25	0.19	0.12
6	0.15	0.11	0.08
8	0.15	0.11	0.08

SOURCE: Adapted from Ray F. Linsley and Joseph B. Franzini, *Water Resources Engineering*, McGraw Hill Book Company, copyright © 1992. Reproduced with permission of The McGraw Hill Companies.

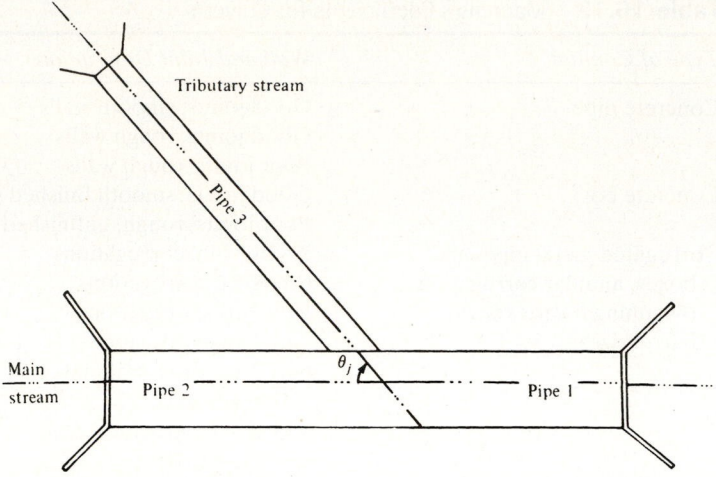

Figure 16.19   Culvert Junction

SOURCE: J.M. Normann, R.J. Houghtalen, W.J. Johnston, *Hydraulic Design of Highway Culverts*, Report No. FHWA-IP-85-15, U.S. Department of Transportation, Office of Implementation, McLean, VA., September 1985

where

HW_o = headwater depth above the outlet invert (m)
v_u = approach velocity
TW = tailwater depth above the outlet invert (m)
v_d = downstream velocity (m/sec)
H_L = sum of all losses
g = 9.81 m/sec²

When the approach and downstream velocity heads are both neglected, we obtain

$$HW_o = TW + H_L \qquad\qquad (16.21)$$

Note that Eqs. 16.15, 16.16, 16.20, and 16.21 were developed for the culvert flowing full and therefore apply to the conditions shown in Figures 16.18a, b, and c. Additional

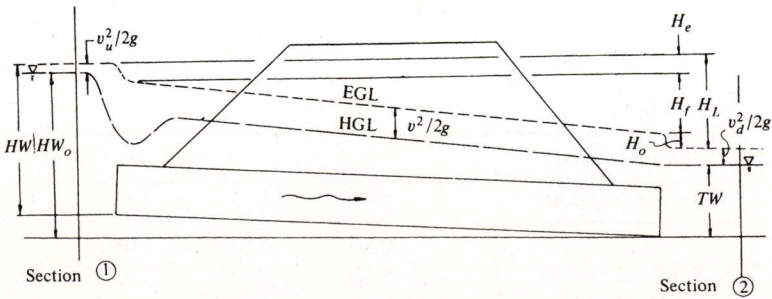

Figure 16.20   Full Flow Energy Grade Line (EGL) and Hydraulic Grade Line (HGL)

SOURCE: J.M. Normann, R.J. Houghtalen, W.J. Johnston, *Hydraulic Design of Highway Culverts*, Report No. FHWA-IP-85-15, U.S. Department of Transportation, Office of Implementation, McLean, VA., September 1985

calculations may be required for the conditions shown in Figures 16.18d and e. These additional calculations are beyond the scope of this book but are discussed in detail in *Hydraulic Design of Highway Culverts*.

Nomographs also have been developed for solving Eq. 16.20 for different configurations of culverts flowing full and performing under outlet control. Only entrance, friction, and exit losses are considered in the nomographs. Figures 16.21 and 16.22

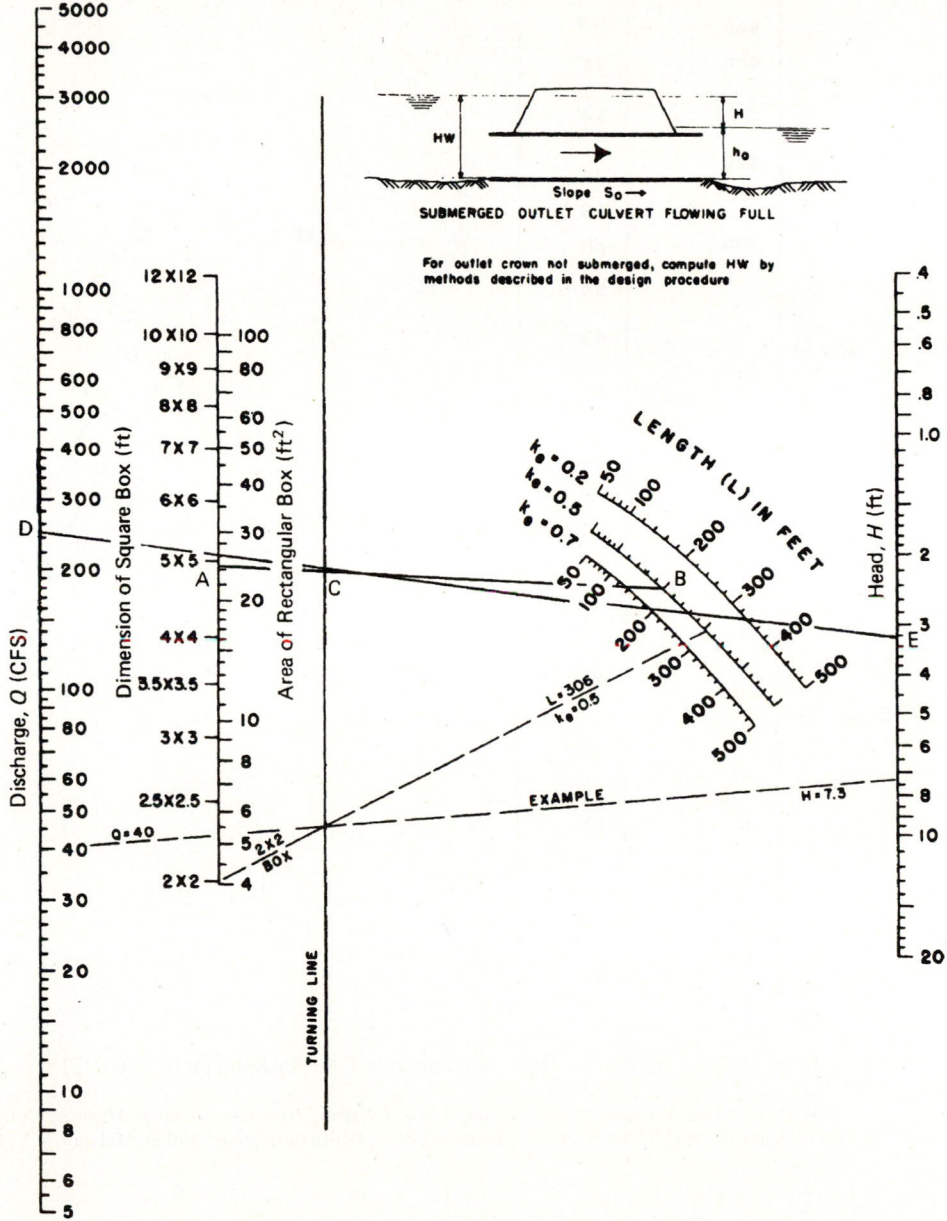

Figure 16.21   Headwater Depth for Concrete Box Culverts Flowing Full ($n = 0.012$)

SOURCE: J.M. Normann, R.J. Houghtalen, W.J. Johnston, *Hydraulic Design of Highway Culverts*, Report No. FHWA-IP-85-15, U.S. Department of Transportation, Office of Implementation, McLean, VA., September 1985

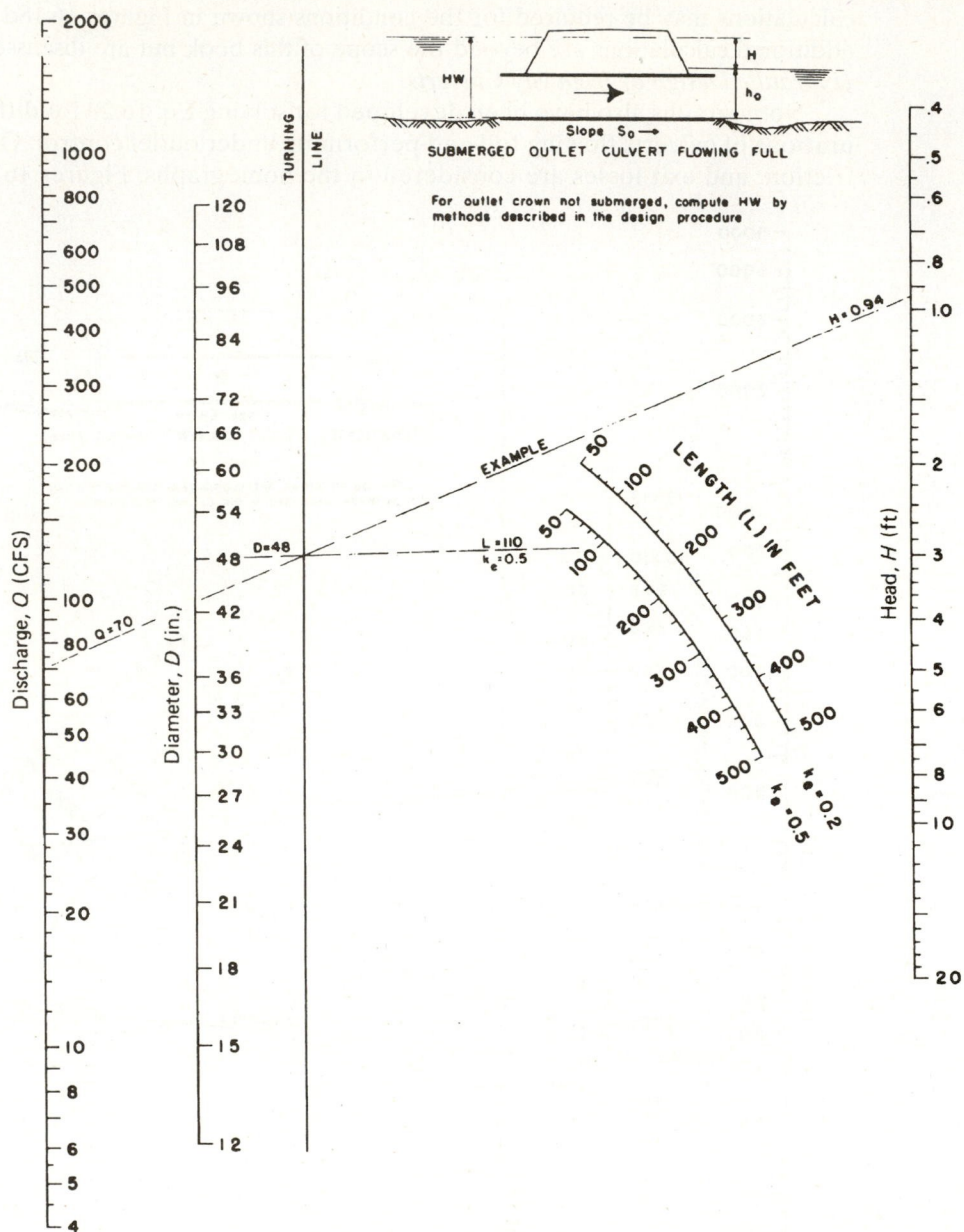

Figure 16.22   Headwater Depth for Concrete Pipe Flowing Full ($n = 0.012$)

SOURCE: J.M. Normann, R.J. Houghtalen, W.J. Johnston, *Hydraulic Design of Highway Culverts*, Report No. FHWA-IP-85-15, U.S. Department of Transportation, Office of Implementation, McLean, VA., September 1985

show examples of these nomographs for a concrete box culvert and a circular concrete pipe culvert. Figures 16.23 and 16.24 show the critical depth charts for these culverts that are also used in the design.

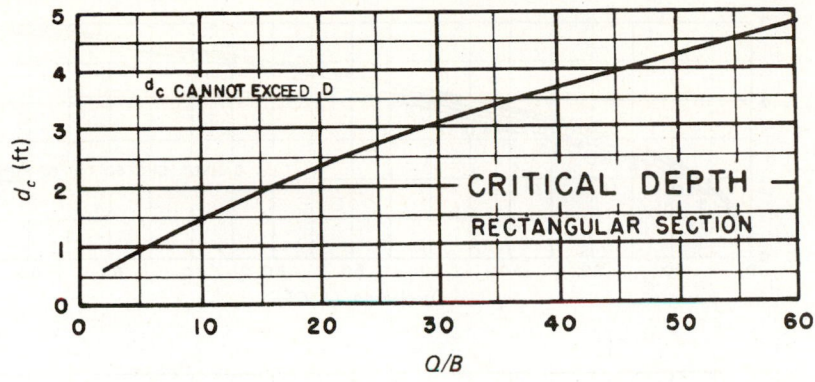

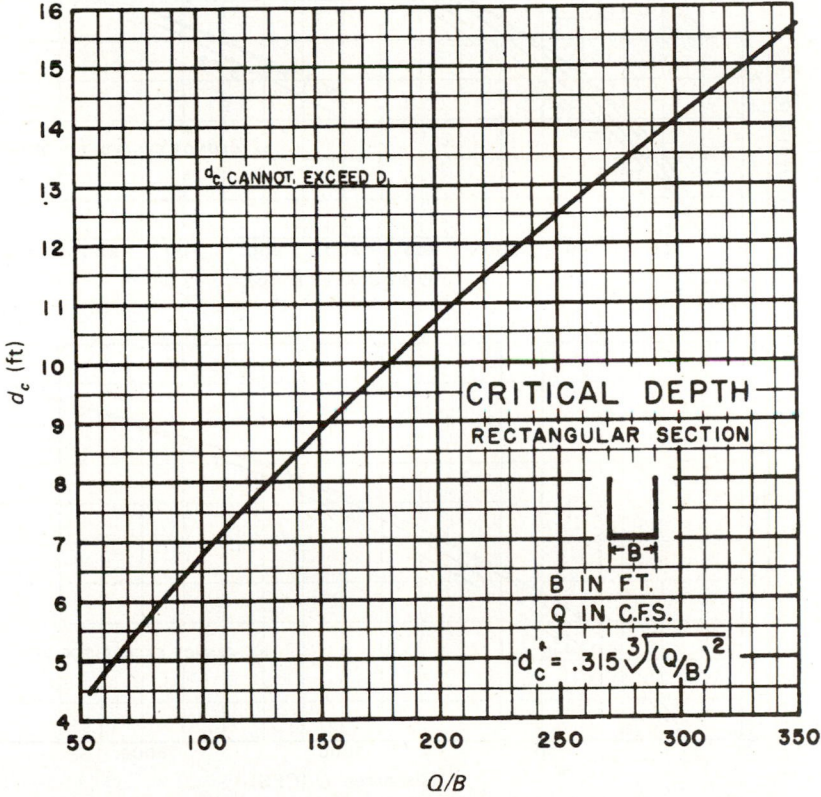

Figure 16.23   Critical Depth for Rectangular Sections

SOURCE: J.M. Normann, R.J. Houghtalen, W.J. Johnston, *Hydraulic Design of Highway Culverts*, Report No. FHWA-IP-85-15, U.S. Department of Transportation, Office of Implementation, McLean, VA., September 1985

    The nomographs for outlet control conditions are also used to determine the depth of the headwater required to accommodate the design flow through the selected culvert configuration under outlet control. The procedure is demonstrated in Example 16.7.

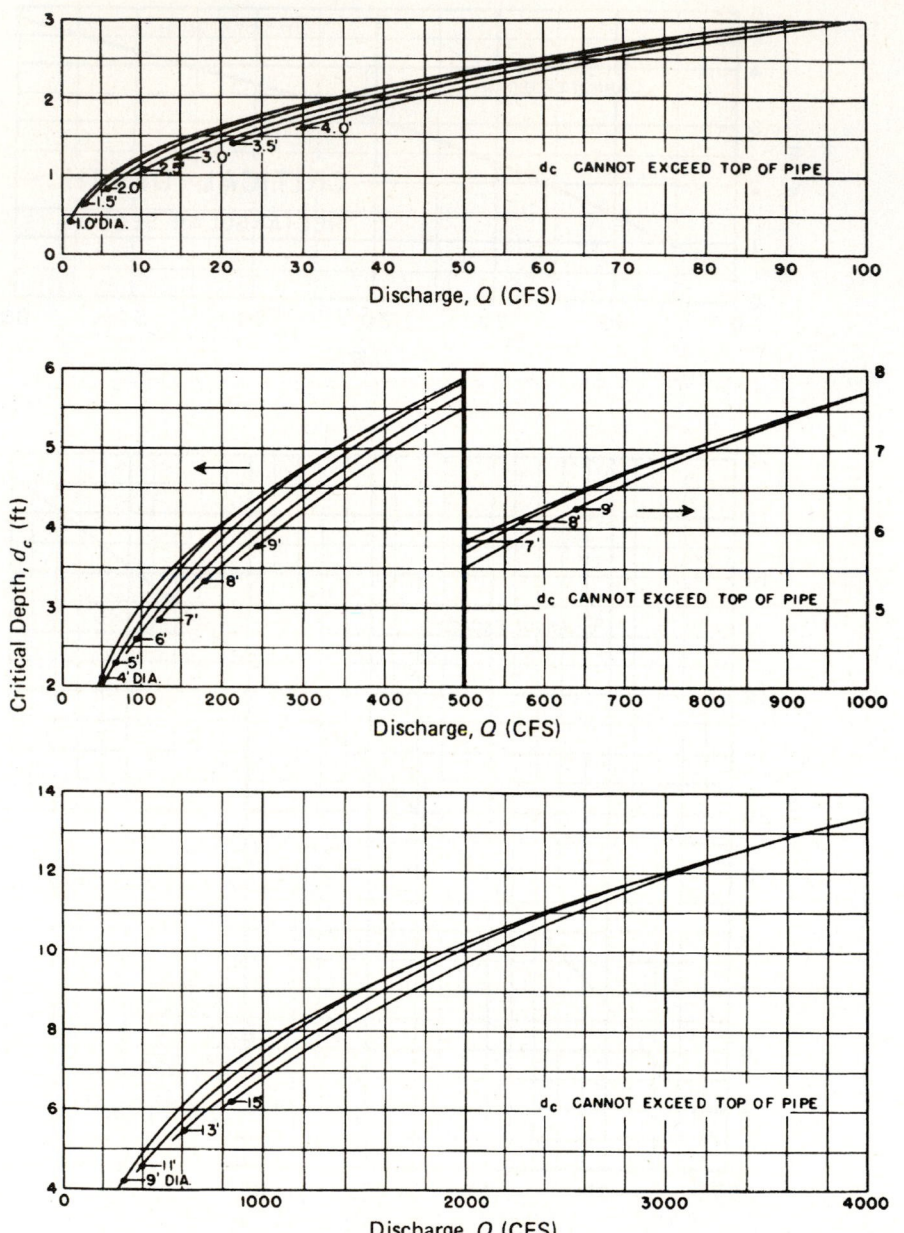

Figure 16.24   Critical Depth for Circular Pipes

SOURCE: J.M. Normann, R.J. Houghtalen, W.J. Johnston, *Hydraulic Design of Highway Culverts*, Report No. FHWA-IP-85-15, U.S. Department of Transportation, Office of Implementation, McLean, VA., September 1985

Example 16.7 Computing Required Headwater Elevation for a Culvert Flowing Full under Outlet Control

Determine the headwater elevation for Example 16.6 if the culvert is flowing full under outlet control, the tailwater depth above the outlet invert at the design flow rate is 1.98 m, the length of culvert is 61 m, and the natural stream slopes at 2 percent. The tailwater depth is determined using normal depth or backwater depth calculations or from on-site inspection. Assume $n = 0.012$.

Solution: The charts shown in Figures 16.21 and 16.23 are used in the following steps.

Step 1. From Figure 16.23, determine the critical depth.

$$Q/B = 50 \qquad d_c = 1.3 \text{ m}$$

Alternatively, d_c may be obtained from the equation

$$d_c = 0.315\sqrt[3]{(Q/B)^2}$$

Step 2. The depth (h_o) from the outlet invert to the hydraulic grade line is then determined. This is taken as $(d_c + D)/2$ or the tailwater depth TW, whichever is greater. In this case,

$$TW = 1.98 \text{ m}$$

$$(d_c + D)/2 = \frac{1.3 + 1.5}{2} = 1.40 \text{ m}$$

$$h_o = 1.98 \text{ m}$$

Step 3. The inlet coefficient k_e is obtained from Table 16.11 as 0.5.

Step 4. Locate the size, length, and k_e of the culvert as shown at A and B in Figure 16.21. Draw a straight line through A and B and locate the intersection C of this line with the turning line. Note that when the culvert material has a different n value than that given in the nomograph, an adjusted length L_1 of the culvert is determined as

$$L_1 = L\left(\frac{n_1}{n}\right)^2 \tag{16.22}$$

where

L = length of culvert (m)
n_1 = desired Manning coefficient
n = Manning n value for outlet control chart

Step 5. Locate the discharge D on the discharge scale and draw a straight line joining C and D. Extend this line to the head loss scale at E and determine the energy loss through the culvert. The total head loss (H) is obtained as 3.3 ft.

Step 6. The required outlet headwater elevation EL_{ho} is computed as

$$EL_{ho} = EL_o + H + h_o \tag{16.23}$$

where EL_o is the invert elevation at outlet. In this case,

$$EL_o = 223.4 - 0.017 \times 61 = 66 \text{ m}$$

(where 223.4 is the inlet invert from Example 16.6) and

$$EL_{ho} = 65.82 + 1 + 1.25 = 68.77 \text{ m}$$

If this computed outlet-control headwater elevation is higher than the design headwater elevation, another culvert should be selected and the procedure repeated.

The design headwater elevation is 230.5 ft (69 m) (from Example 16.6). The 1.5 m × 1.5 m culvert is therefore acceptable.

In the design of a culvert, the headwater elevations are computed for inlet and outlet controls and the control with the higher headwater elevation is selected as the controlling condition. In the above case, for example, the headwater elevation for inlet control = 67.02 + 2.13 = 69.15 m and the headwater elevation for outlet control is 68.77 m, which means that the inlet control governs.

The outlet velocity is then determined for the governing control. When the inlet control governs, the normal depth velocity is taken as the outlet velocity. When the outlet control governs, the outlet velocity is determined by using the area of flow at the outlet, which is based on the culvert's geometry and the following conditions.

1. If the tailwater depth is less than critical depth, use the critical depth.
2. If the level of the tailwater is between the critical depth and the top of the culvert barrel, use the tailwater depth.
3. If the tailwater is above the top of the barrel, use the height of the culvert barrel.

In this example, since the inlet control governs, the outlet velocity should be determined based on the normal depth.

The whole design procedure can be carried out using a table similar to the one in Figure 16.25, which facilitates the trial of different pipe configurations.

Computer Programs for Culvert Design and Analysis

Examples 16.6 and 16.7 indicate how tedious the analysis or design of a culvert can be, even with the use of available charts. Several computer and hand calculator programs are presently available that can be used to increase the accuracy of the results and to significantly reduce the time it takes for design or analysis.

Programs available for hand calculators include the Calculator Design Series (CDS) 1,2,3 and the Calculator Design Series (CDS) 4, developed by FHWA. The programs in CDS 1,2,3 are suitable for the Compucorp 325 Scientist, the HP-65, and the TI-59 programmable calculators, whereas the CDS 4 is suitable only for the TX-54. The CDS 1,2,3 consists of a set of subroutines that are run sequentially, some of which provide inputs for subsequent ones. The program can be used to analyze culvert sizes with different inlet configurations. The outputs include the dimensions of the barrel, the performance data, and the outlet velocities.

The CDS 4 program can be used to analyze corrugated metal and concrete culverts. This program also consists of a series of subroutines.

FHWA also has produced various software packages, some of which are suitable for use on mainframe computers and others on personal computers. The program H4-2 can be used to design pipe-arch culverts on a mainframe computer, whereas H4-6 (which is

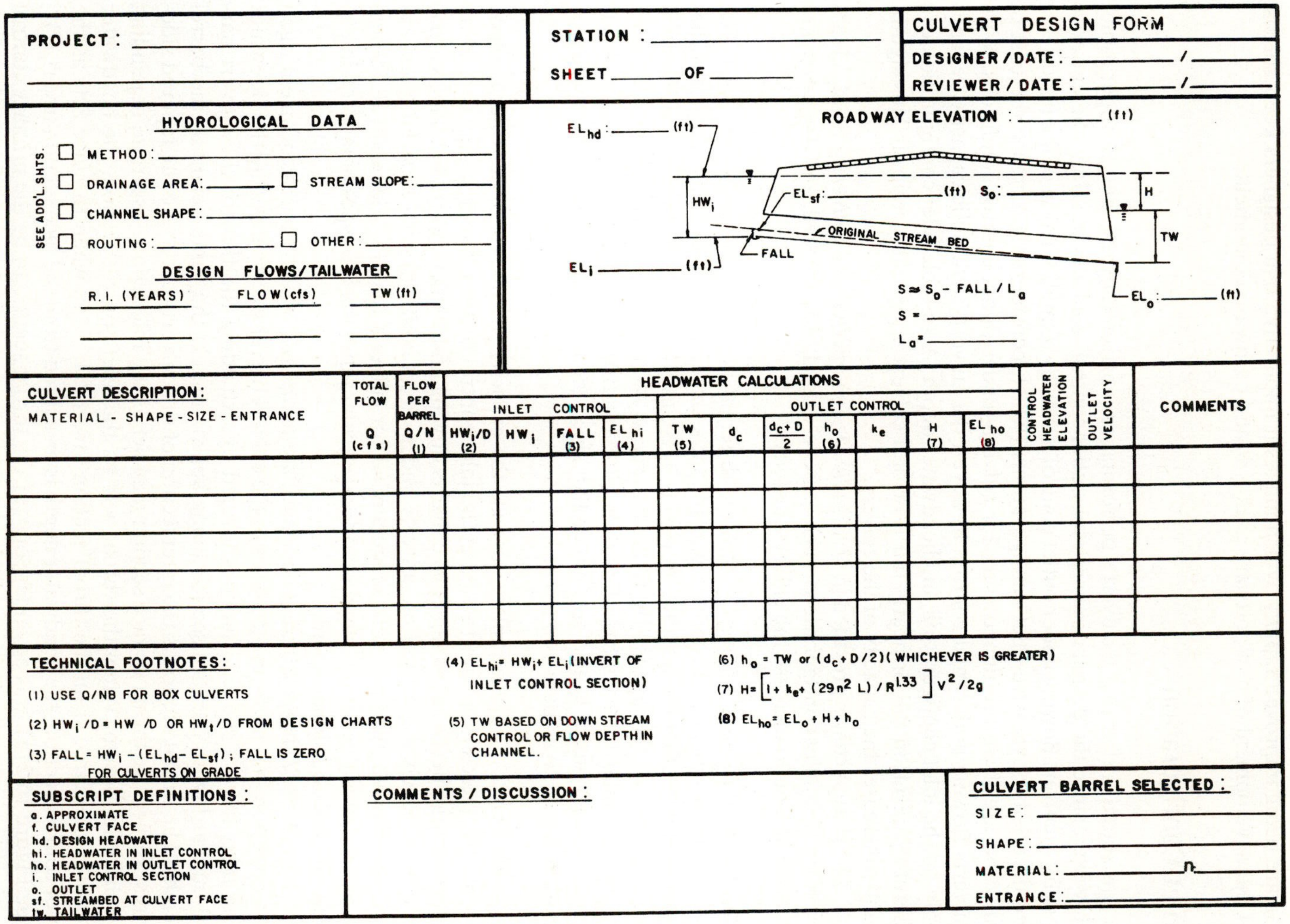

Figure 16.25 Work Sheet for Culvert Design

SOURCE: J.M. Normann, R.J. Houghtalen, W.J. Johnston, *Hydraulic Design of Highway Culverts*, Report No. FHWA-IP-85-15, U.S. Department of Transportation, Office of Implementation, McLean, VA., September 1985

also run on a mainframe) can be used to obtain a list of optional circular and box culverts for the specified site and hydrologic conditions.

Note, however, that both computer hardware and software in all fields of engineering are being improved rapidly, which makes it imperative that designers keep abreast of the development of new software packages and hardware equipment.

Inlet Configuration

It was stated earlier that a culvert's performance is affected by the inlet configurations. However, since the design discharge for outlet control results in full flow, the entrance loss for a culvert flowing under outlet control is usually a small fraction of the headwater requirements. Extensive inlet configurations (which considerably increase the cost of culverts) are therefore unnecessary for culverts under outlet-control conditions. With culverts flowing under inlet control, however, the culverts' hydraulic capacity depends only on the inlet configuration and headwater depth. A suitable inlet configuration can therefore result in full or nearly full flow of a culvert under inlet control. This significantly increases the capacity of the culvert.

The design charts provided in *Hydraulic Design of Highway Culverts* include charts for improved inlet configurations, such as bevel-edged, side-tapered, and slope-tapered inlets that help to increase the culvert capacity.

Bevel-Edged Inlets.   Figure 16.26 shows different beveled-edged configurations. The bevel edge is similar to a chamfer, except that a chamfer is usually much smaller. It has been estimated that the addition of bevels to a culvert having a square-edged inlet will increase the culvert capacity by 15 to 20 percent. As a minimum, therefore, all culverts operating under inlet control should be fitted with bevels. Use of bevels on culverts under outlet control is also recommended, since the entrance loss coefficient may be reduced by up to 40 percent.

Tapered Inlets.   Tapered inlets increase the capacity of culverts under inlet control, mainly by reducing the contraction at the inlet control section. Tapered inlets are more effective than bevel-edged inlets on culverts flowing under inlet control but have similar results as bevels when used on culverts under outlet control. Since tapered inlets are more expensive, they are not recommended for culverts under outlet control. Design charts are available for rectangular box and circular pipes for two types of tapered inlets: side-tapered and slope-tapered.

Figure 16.27 shows different designs of a *side-tapered inlet*. It consists of an enlarged faced section that uniformly reduces to the culvert barrel size by tapering the sidewalls. The inlet flow of the side taper is formed by extending the flow of the culvert barrel outward, with the face section being approximately the same height as the barrel height. The throat section is the intersection of the culvert barrel and the tapered sidewalls. Either the throat or face section may act as the inlet control section, depending on the inlet design. When the throat acts on the inlet control section, the headwater depth is measured from the throat section invert HW_t, and when the face acts as the inlet control section, the headwater depth is measured from the face section invert HW_f (see Figure 16.27a). It is advantageous for the throat section to be the primary control section, since the throat is usually lower than the face, resulting in a higher head on the throat for a specified headwater elevation. Figures 16.27b and c show two ways of increasing the effectiveness of the side-tapered inlet. In Figure 16.27b, the throat

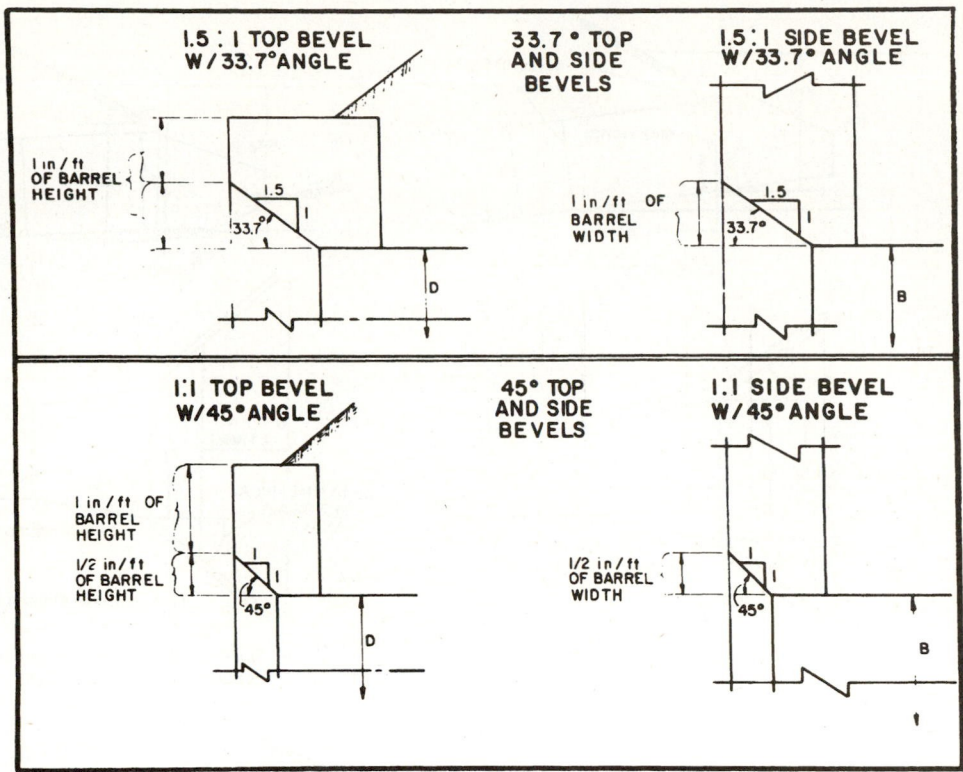

Figure 16.26   Beveled Edges

SOURCE: J.M. Normann, R.J. Houghtalen, W.J. Johnston, *Hydraulic Design of Highway Culverts*, Report No. FHWA-IP-85-15, U.S. Department of Transportation, Office of Implementation, McLean, VA., September 1985

section is depressed below the streambed, and a depression is constructed between the two wing walls. When this type of construction is used, it is recommended that the culvert barrel floor be extended upstream a minimum distance of $D/2$, before the steep upward slope begins. In Figure 16.27c, a sump is constructed upstream from the face section with the dimensional requirements given. When the side-tapered inlet is constructed as in Figures 16.27b or c, a crest is formed upstream at the intersection of the streambed and the depression slope. This crest may act as a weir if its length is too short. It should therefore be ascertained that the crest does not control the flow at the design flow and headwater.

The *slope-tapered inlet* is similar to the side-tapered inlet in that it also has an enlarged face section, which is gradually reduced to the barrel size at the throat section by sloping the sidewalls. The slope-tapered inlet, however, also has a uniform vertical drop (fall) between the face and throat sections (see Figure 16.28). A third section, known as the bend, is also placed at the intersection of the inlet slope and barrel slope, as shown in Figure 16.28.

Any one of three sections may act as the primary control section of the slope-tapered section. These are the face, the bend, and the throat. Design procedures (which are beyond the scope of this book) for the dimensions of the throat and face sections

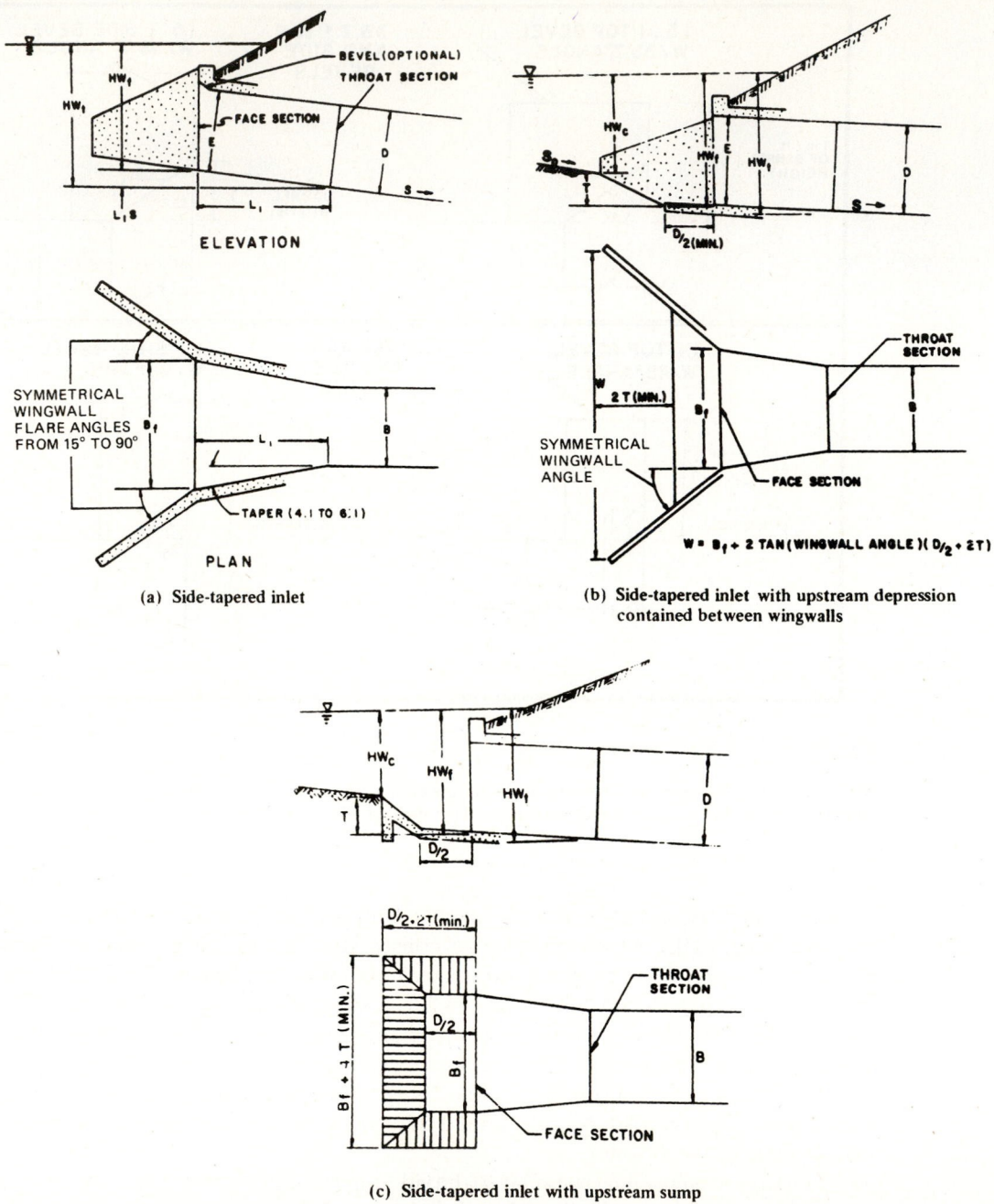

Figure 16.27   Side-Tapered Inlets

SOURCE: J.M. Normann, R.J. Houghtalen, W.J. Johnston, *Hydraulic Design of Highway Culverts*, Report No. FHWA-IP-85-15, U.S. Department of Transportation, Office of Implementation, McLean, VA., September 1985

are given in *Hydraulic Design of Highway Culverts*. The only criterion given for the size of the bend section is that it should be located a minimum distance from the throat. Again, the slope-tapered inlet is most efficient when the throat acts as the primary control section.

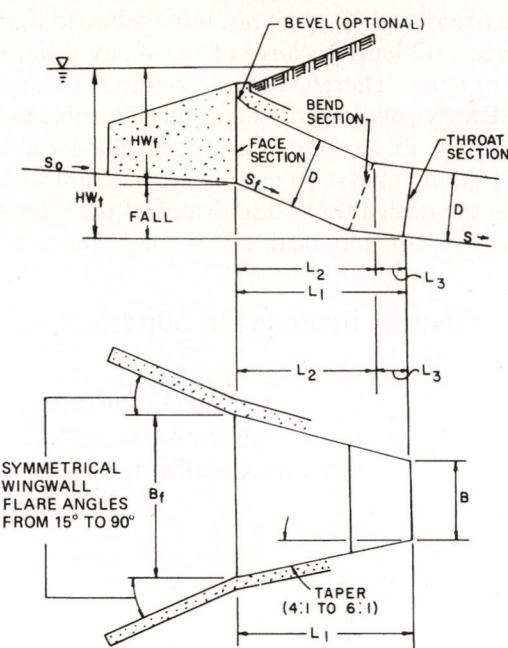

Figure 16.28   Slope-Tapered Inlet with Vertical Face

SOURCE: J.M. Normann, R.J. Houghtalen, W.J. Johnston, *Hydraulic Design of Highway Culverts*, Report No. FHWA-IP-85-15, U.S. Department of Transportation, Office of Implementation, McLean, VA., September 1985

16.7 SUBSURFACE DRAINAGE

Subsurface drainage systems are provided within the pavement structure to drain water in one or more of the following forms:

- Water that has permeated through cracks and joints in the pavement to the underlying strata.
- Water that has moved upward through the underlying soil strata as a result of capillary action.
- Water that exists in the natural ground below the water table, usually referred to as ground water.

The subsurface drainage system must be an integral part of the total drainage system, since the subsurface drains must operate in consonance with the surface drainage system to obtain an efficient overall drainage system.

The design of subsurface drainage should be carried out as an integral part of the complete design of the highway, since inadequate subsurface drainage also may have detrimental effects on the stability of slopes and pavement performance. However, certain design elements of the highway, such as geometry and material properties, are required for the design of the subdrainage system. Thus, the procedure usually adopted for subdrainage design is first to determine the geometric and structural requirements of the highway based on standard design practice, and then to subject these to a subsurface drainage analysis to determine the subdrainage requirements. In some cases, the subdrainage requirements determined from this analysis will require some changes in the original design.

It is extremely difficult, if not impossible, to develop standard solutions for solving subdrainage problems because of the many different situations that engineers come across in practice. Therefore, basic methods of analysis are given that can be used as tools to identify solutions for subdrainage problems. The experience gained from field and laboratory observations for a particular location, coupled with good engineering judgment, should always be used in conjunction with the design tools provided. Before presenting the design tools, discussions of the effects on the highway of an inadequate subdrainage system and the different subdrainage systems are presented.

16.7.1 Effect of Inadequate Subdrainage

Inadequate subdrainage on a highway will result in the accumulation of uncontrolled subsurface water within the pavement structure and/or right-of-way, which can result in poor performance of the highway or outright failure of sections of the highway. The effects of inadequate subdrainage fall into two classes: poor pavement performance and instability of slopes.

Pavement Performance

If the pavement structure and subgrade are saturated with underground water, the pavement's ability to resist traffic load is considerably reduced, resulting in one or more of several problems, which can lead to premature destruction of the pavement if remedial actions are not taken in time. In Portland cement concrete pavement, for example, inadequate subdrainage can result in excessive repeated deflections of the pavement (see "Pumping of Rigid Pavements" in Chapter 20), which will eventually lead to cracking.

When asphaltic concrete pavements are subjected to excessive uncontrolled subsurface water due to inadequate subdrainage, very high pore pressures are developed within the untreated base and subbase layers (see Chapter 19 for base and subbase definitions), resulting in a reduction of the pavement strength and thereby its ability to resist traffic load.

Another common effect of poor pavement performance due to inadequate subdrainage is frost action. As described later, this phenomenon requires that the base and/or subbase material be a frost-susceptible soil and that an adequate amount of subsurface water is present in the pavement structure. Under these conditions, during the active freezing period, subsurface water will move upward by capillary action toward the freezing zone and subsequently freeze to form lenses of ice. Continuous growth of the ice lenses due to the capillary action of the subsurface water can result in considerable heaving of the overlying pavement. This eventually leads to serious pavement damage, particularly if differential frost heaving occurs. Frost action also has a detrimental effect on pavement performance during the spring thaw period. During this period, the ice lenses formed during the active freeze period gradually thaw from the top down, resulting in the saturation of the subgrade soil, which results in a substantial reduction of pavement strength.

Slope Stability

The presence of subsurface water in an embankment or cut can cause an increase of the stress to be resisted and a reduction of the shear strength of the soil forming the embankment or cut. This can lead to a condition where the stress to be resisted is greater than the strength of the soil, resulting in sections of the slope crumbling down or a complete failure of the slope.

16.7.2 Highway Subdrainage Systems

Subsurface drainage systems are usually classified into five general categories:

- Longitudinal drains
- Transverse drains
- Horizontal drains
- Drainage blankets
- Well systems

Longitudinal Drains

Subsurface longitudinal drains usually consist of pipes laid in trenches within the pavement structure and parallel to the centerline of the highway. These drains can be used to lower the water table below the pavement structure, as shown in Figure 16.29, or to remove any water that is seeping into the pavement structure, as shown in Figure 16.30. In some cases, when the water table is very high and the highway is very wide, it may be necessary to use more than two rows of longitudinal drains to achieve the required reduction of the water table below the pavement structure (see Figure 16.31).

Transverse Drains

Transverse drains are placed below the pavement, usually in a direction perpendicular to the centerline, although they may be skewed to form a herringbone configuration. An example of the use of transverse drains is shown in Figure 16.32, where they are used to drain ground water that has infiltrated through the joints of the pavement. One disadvantage of transverse drains is that they can cause unevenness of the pavement when used in areas susceptible to frost action, where general frost

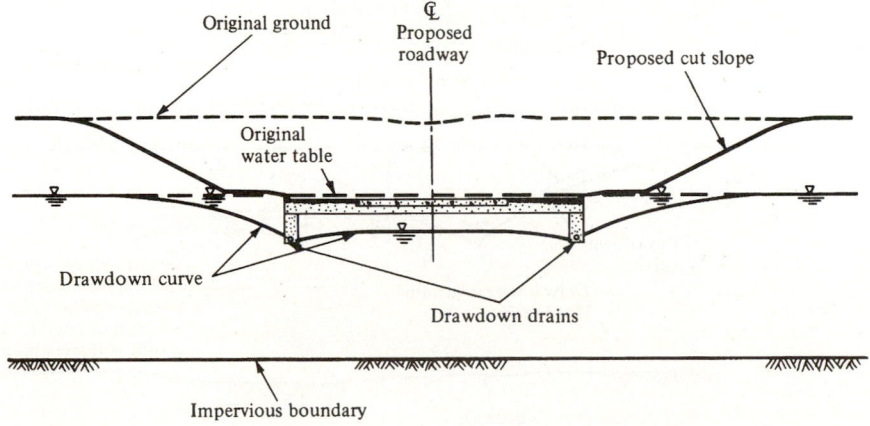

Figure 16.29   Symmetrical Longitudinal Drains Used to Lower Water Table

SOURCE: Redrawn from *Highway Subdrainage Design*, Report No. FHWA-TS-80-224, U.S. Department of Transportation, Washington, D.C., August 1980

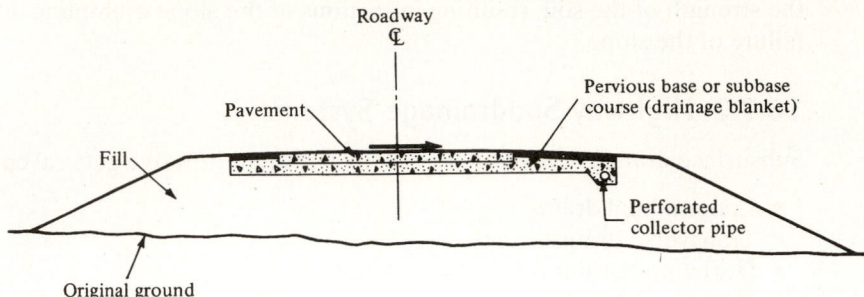

Figure 16.30   Longitudinal Collector Drain Used to Remove Water Seeping into Pavement Structural Section

SOURCE: Redrawn from *Highway Subdrainage Design*, Report No. FHWA-TS-80-224, U.S. Department of Transportation, Washington, D.C., August 1980

heaving occurs. The unevenness is due to the general heaving of the whole pavement, except at the transverse drains.

Horizontal Drains

Horizontal drains are used to relieve pore pressures at slopes of cuts and embankments on the highway. They usually consist of small diameter, perforated pipes inserted into the slopes of the cut or fill. The subsurface water is collected by the pipes and is then discharged at the face of the slope through paved spillways to longitudinal ditches.

Drainage Blankets

A drainage blanket is a layer of material that has a very high coefficient of permeability, usually greater than 30 ft/day, and is laid beneath or within the pavement structure such that its width and length in the flow direction are much greater than its thickness. The coefficient of

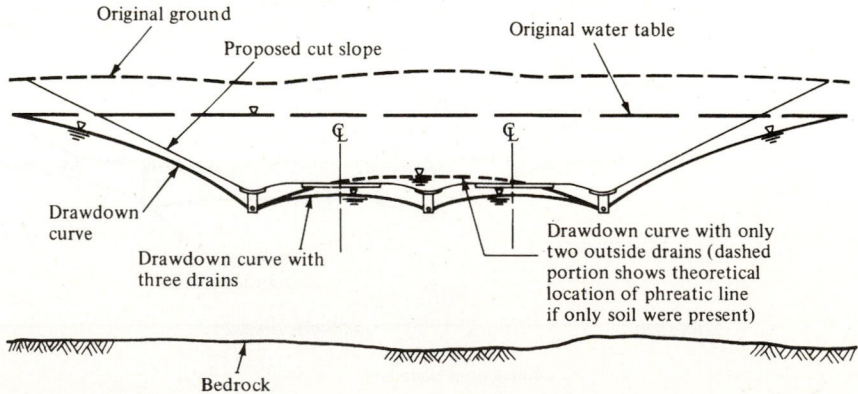

Figure 16.31   Multiple Longitudinal Drawdown Drain Installation

SOURCE: Redrawn from *Highway Subdrainage Design*, Report No. FHWA-TS-80-224, U.S. Department of Transportation, Washington, D.C., August 1980

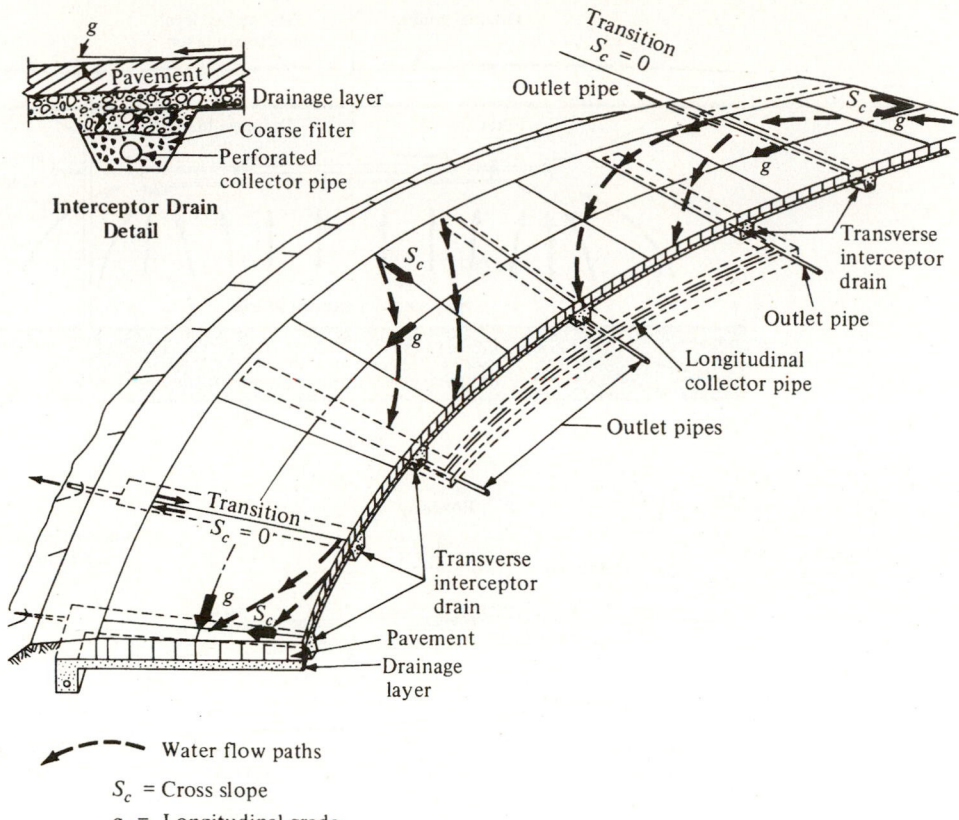

Figure 16.32   Transverse Drains on Superelevated Curves

SOURCE: Redrawn from *Highway Subdrainage Design*, Report No. FHWA-TS-80-224, U.S. Department of Transportation, Washington, D.C., August 1980

permeability is the constant of proportionality of the relationship between the flow velocity and the hydraulic gradient between two points in the material (see Chapter 17). Drainage blankets can be used to facilitate the flow of subsurface water away from the pavement, as well as to facilitate the flow of ground water that has seeped through cracks into the pavement structure or subsurface water from artesian sources. A drainage blanket also can be used in conjunction with longitudinal drains to improve the stability of cut slopes by controlling the flow of water on the slopes, thereby preventing the formation of a slip surface. However, drainage blankets must be properly designed to be effective. Figure 16.33 shows two drainage blanket systems.

Well Systems

A well system consists of a series of vertical wells, drilled into the ground, into which ground water flows, thereby reducing the water table and releasing the pore pressure. When used as a temporary measure for construction, the water collected in the wells is continuously pumped out, or else it may be left to overflow. A more common construction, however, includes a drainage layer either at the top or bottom of the wells to facilitate the flow of water collected.

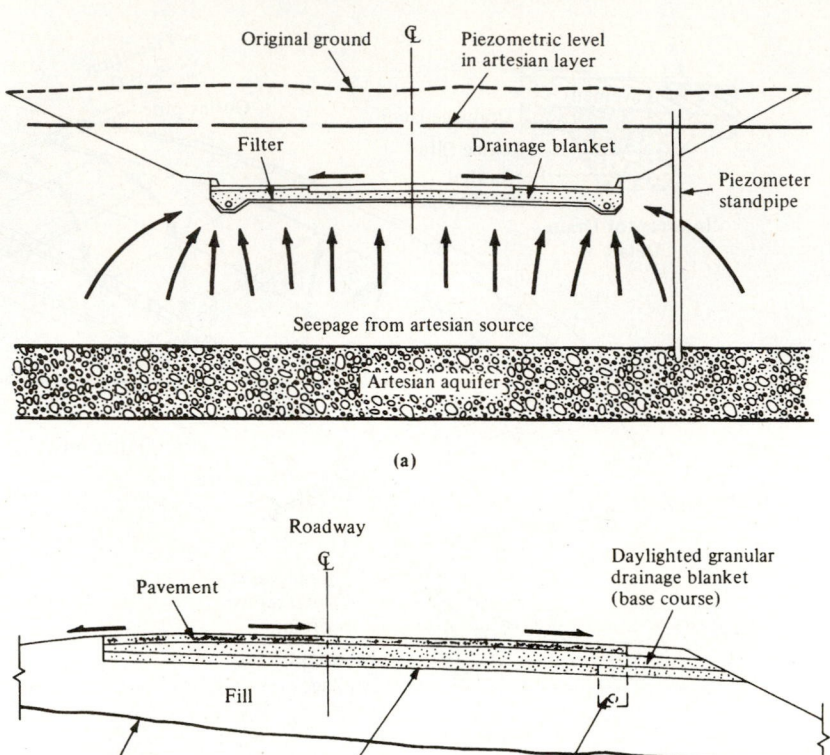

(a)

(b)

Figure 16.33   Applications of Horizontal Drainage Blankets

SOURCE: Redrawn from *Highway Subdrainage Design*, Report No. FHWA-TS-80-224, U.S. Department of Transportation, Washington, D.C., August 1980

16.7.3 Design of Subsurface Drainage

The design procedure for subsurface drainage involves the following.

1. Summarize the available data.
2. Determine the quantity of water for which the subdrainage system is being designed.
3. Determine the drainage system required.
4. Determine the capacity and spacing of longitudinal and transverse drains and select filter material, if necessary.
5. Evaluate the design with respect to economic feasibility and long-term performance.

Summarize Available Data

The data that should be identified and summarized can be divided into the following four classes:

- Flow geometry
- Materials' properties

- Hydrologic and climatic characteristics
- Miscellaneous information

The flow geometry is given by the existing subsurface characteristics of the area in which the highway is located and by the geometric characteristics of the highway. These are used to determine whether any special subdrainage problems exist and what conditions must be considered in developing solutions for these problems.

The main material property required is the permeability, since this is the property that indicates the extent to which water will flow through the material.

Hydrologic and climatic characteristics will indicate precipitation rates, the sources of subsurface water, and the possibility of frost action.

Miscellaneous information includes all other information that will aid in the design of an effective and economic subdrainage system, including information of any impact the subdrainage system may have on future construction and any difficulties identified that may preclude the construction of a subdrainage system.

Determine the Discharge Quantity

The net amount of water to be discharged consists of the following components:

- Water due to infiltration
- Ground water
- Water due to thawing of ice lenses
- Water flowing vertically from the pavement structure

Water Due to Infiltration, q_i.   This is the amount of surface water that infiltrates into the pavement structure through cracks in the pavement surface. It is extremely difficult to calculate this amount of water exactly, since the rate of infiltration depends on the intensity of the design storm, the frequency and size of the cracks and/or joints in the pavement, the moisture conditions of the atmosphere, and the permeability characteristics of the materials below the pavement surface.

FHWA recommends the use of the following empirical relationship to estimate the infiltration rate.

$$q_i = I_c\left(\frac{N_c}{W} + \frac{W_c}{WC_s}\right) + K_p \qquad (16.24)$$

where

q_i = design infiltration rate (m³/day/m² of drainage layer)
I_c = crack infiltration rate (m³/day/m of crack)
  (0.22 m³/day/m is recommended for most designs)
N_c = number of contributing longitudinal cracks or joints
W_c = length of contributing transverse cracks (m)
W = width of granular base or subbase subjected to infiltration (m)
C_s = spacing of the transverse cracks or joints (m)
K_p = rate of infiltration (m³/day/m²)

The suggested value of 2.4 for I_c normally should be used, but local experience also should be relied on to increase or decrease this value as necessary.

The value of N_c is usually taken as $N + 1$ for new pavements, where N is the number of traffic lanes. Local experience should be used to determine a value for C_s, although a value of 40 has been suggested for new bituminous concrete pavements. The rate of infiltration for Portland cement concrete and well-compacted, dense, graded asphaltic concrete pavements is usually very low and therefore can be taken as zero. However, when there is evidence of high infiltration rates, these should be determined from laboratory tests.

Example 16.8 Computing Infiltration Rate of a Flexible Pavement

Determine the infiltration rate for a new two-lane flexible pavement with the following characteristics.

Lane width = 3.3 m

Shoulder width = 2.4 m

Number of contributing longitudinal cracks $(N_c) = (N + 1) = 3$

Length of contributing transverse cracks $(W_c) = 6$ m

K_p (from laboratory tests) = 0.009

$W = 11.4$ m

Spacing of transverse cracks $(C_s) = 10.5$ m

Solution: Assuming $I_c = 0.22$ m³/day/m², then, from Eq. 16.24,

$$q_i = 0.22 \left[\frac{3}{11.4} + \frac{6}{11.4(10.5)} \right] + 0.009$$

$$= 0.0689 + 0.009$$

$$= 0.078 \text{ m}^3/\text{day}/\text{m}^2$$

Ground Water. When it is not possible to intercept the flow of ground water or lower the water table sufficiently before the water reaches the pavement, it is necessary to determine the amount of ground-water seepage that will occur. Figures 16.29 and 16.33a illustrate the two possible sources of ground water of interest in this case. Figure 16.29 shows a case of gravity drainage, whereas Figure 16.33a shows a case of artesian flow. A simple procedure to estimate the ground-water flow rate due to gravity drainage is to use the chart shown in Figure 16.34. In this case, the radius of influence L_i is first determined as

$$L_i = 1.14 \, (H - H_o) \tag{16.25}$$

where

H_o = depth of subgrade below the drainage pipe (m)
H = depth of subgrade below the natural water table (m)
$H - H_o$ = amount of drawdown (m)

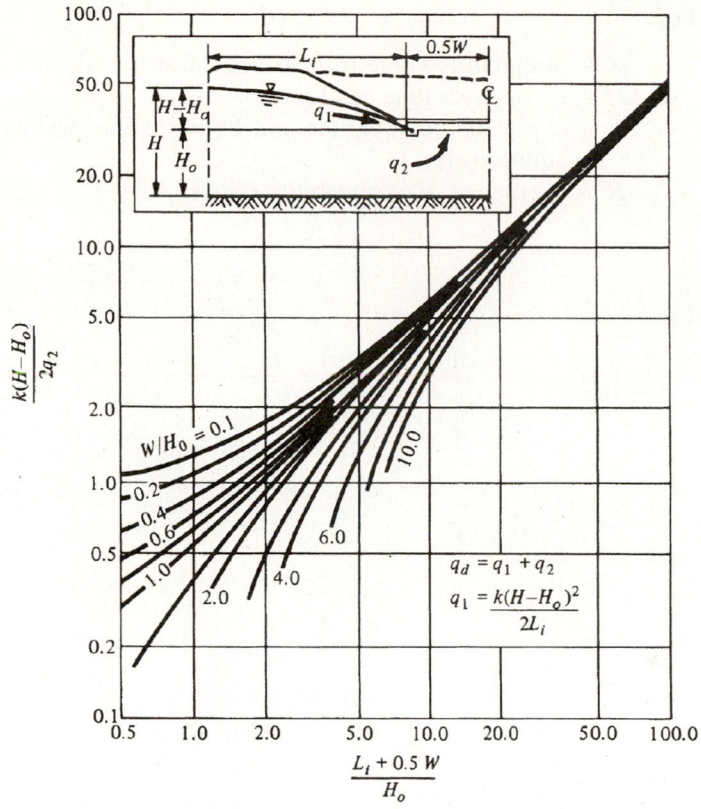

Figure 16.34 Chart for Determining Flow Rate in Horizontal Drainage Blanket

SOURCE: Redrawn from *Highway Subdrainage Design*, Report No. FHWA-TS-80-224, U.S. Department of Transportation, Washington, D.C., August 1980

The chart shown in Figure 16.34 is then used to determine the total quantity of upward flow (q_2) from which the average inflow rate q_g is determined.

$$q_g = \frac{q_2}{0.5W} \tag{16.26}$$

where

q_g = design inflow rate for gravity drainage (m^3/day/m^2 of drainage layer)
q_2 = total upward flow into one-half of the drainage blanket (m^3/day/linear m of roadway)
W = width of drainage layer (m)

For the case of the artesian flow, the average inflow rate is estimated using Darcy's law.

$$q_a = K \frac{\Delta H}{H_o} \tag{16.27}$$

where

q_a = design inflow rate from artesian flow (m³/day/m² of drainage area)

ΔH = excess hydraulic head (m)

H_o = depth of the subgrade soil between the drainage layer and the artesian aquifer (m)

K = coefficient of permeability (m/day)

Example 16.9   Computing Average Inflow Rate Due to Gravity Drainage

Using the chart shown in Figure 16.34, determine the average inflow rate (q_g) due to gravity drainage, as shown in Figure 16.35, for the following data.

$$\text{Depth of subgrade below drainage pipe } (H_o) = 4.5 \text{ m}$$

$$\text{Coefficient of permeability of native soil } (K) = 0.12 \text{ m/day}$$

$$\text{Width of drainage layer 12 m drawdown } (H - H_o) = 2.4 \text{ m}$$

$$\text{Radius of influence } (L_i) = 3.8 \times 2.4 \text{ (from Eq.16.25)} = 9.12 \text{ m}$$

Solution:

$$\frac{L_i + 0.5W}{H_o} = \frac{9.12 + 0.5 \times 12}{4.5} = 3.36$$

$$\frac{W}{H_o} = \frac{12}{4.5} = 2.67$$

Entering the chart at

$$\frac{L_i + 0.5W}{H_o} = 3.36$$

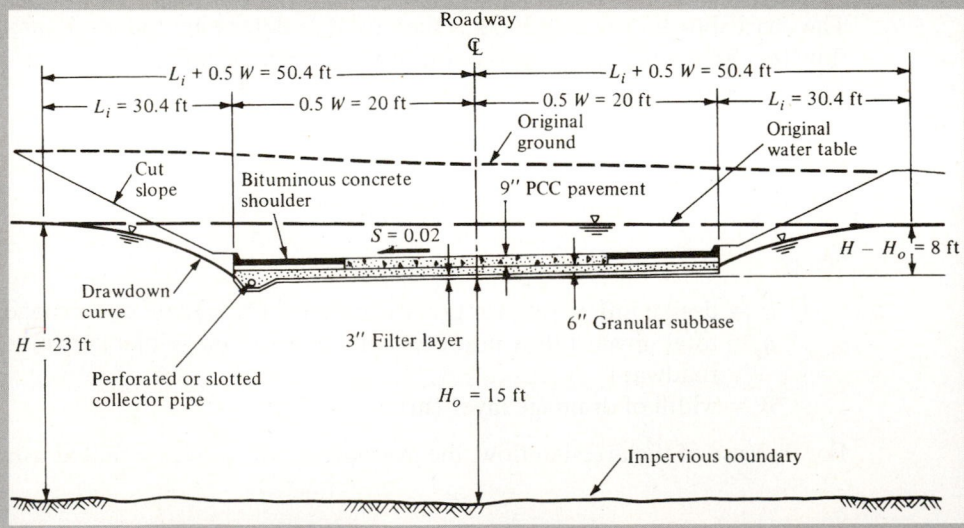

Figure 16.35   Rigid Pavement Section in Cut Dimensions and Details for Example 16.9

and

$$W/H_o = 2.67$$

to obtain

$$\frac{K(H - H_o)}{2q_2} \approx 13$$

$$q_2 = \frac{0.12 \times 2.4}{2 \times 13} = 0.0111 \text{ m}^3/\text{day/m}$$

$$q_g = \frac{0.0111}{0.5 \times 12} = 0.00185 \text{ m}^3/\text{day/m}^2$$

Example 16.10 Computing Average Inflow Rate Due to Artesian Flow

Determine the average flow rate of ground water into a pavement drainage layer due to artesian flow constructed on a subgrade soil having a coefficient of permeability of 0.015 m/day. A piezometer installed at the site indicates an excess hydraulic head of 3 m. The depth of the subgrade soil between the drainage layer and the artesian aquifer is 6 m (see Figure 16.36).

Solution: In this case,

$$q_a = \frac{0.015 \times 3}{6} = 0.0075 \text{ m}^3/\text{day/m}^2$$

Note that flow net analysis may be used to estimate the flow rates due to both gravity flow and artesian flow. Flow net analysis is beyond the scope of this book, but good

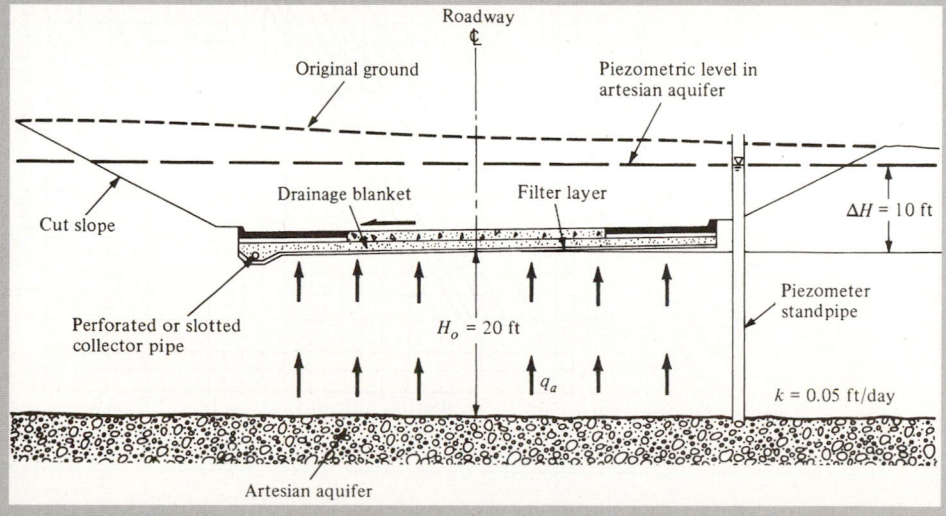

Figure 16.36   Artesian Flow of Groundwater into a Pavement Drainage Layer—Dimensions and Details for Example 16.10

estimates of these flows can be obtained by the methods presented. The amount of ground-water flow due to any one source is usually small, but ground-water flow should not be automatically neglected, as the sum of ground water flow from all sources may be significant. It is therefore essential that estimates be made of the inflow from all sources of ground water that influence the pavement structure.

Water from Ice Lenses.   As explained earlier, frost action results when ice lenses form within the pavement subgrade. The extent to which this occurs during the active freeze period is highly dependent on the frost susceptibility of the subgrade material. When ice lenses thaw during the spring thaw period, it is necessary to properly drain this water from the pavement environment. The rate of seepage of this water through the soil depends on several factors, including the permeability of the soil, the thawing rate, and the stresses imposed on the soil. It is very difficult to determine the extent to which each of these factors affects the flow rate, which makes it extremely difficult to develop an exact method for determining the flow rate. However, an empirical method for determining the design flow rate q_m has been developed using the chart shown in Figure 16.37. This requires determination from laboratory tests of the average rate of heave of the soil due to frost action or the classification of the soil with respect to its susceptibility to frost action and the stress σ_p imposed on the subgrade soil. The stress imposed on the

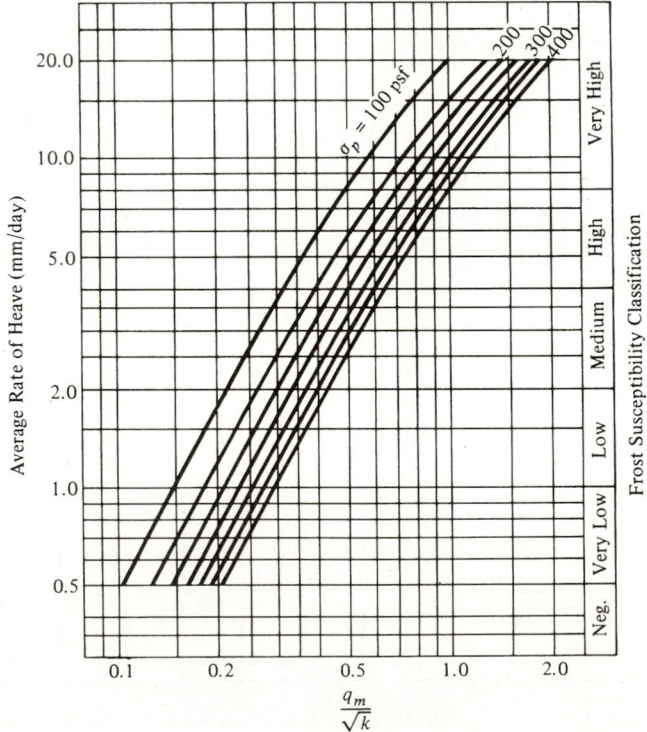

Figure 16.37   Chart for Estimating Design Inflow Rate of Melt Water from Ice Lenses

SOURCE: Redrawn from *Highway Subdrainage Design*, Report No. FHWA-TS-80-224, U.S. Department of Transportation, Washington, D.C., August 1980.

subgrade in pounds per square foot (kg/m²) due to the weight of a 1 m-square column of the pavement structure is usually taken as the value of σ_p. The design flow rate q_m obtained from the chart is the average flow during the first full day of thawing. This rate is higher than those for subsequent days because the rate of seepage decreases with time. Although the use of such a high value of flow is rather conservative, such a value may also cause the soil to be saturated for a period of up to 6 hours after thawing. In cases where saturation for this period of time is unacceptable, measures to increase the rate of drainage of the thawed water from the soil must be used. For example, a thick enough drainage layer that consists of a material with a suitable permeability that will never allow saturation to occur can be provided.

Example 16.11 Computing Flow Rate Due to Thawing of Ice Lenses

Determine the flow rate due to water thawing from ice lenses below a pavement structure that consists of a 15 cm concrete surface and a 22.5 cm granular base. The subgrade material is silty sand with a high frost susceptibility classification and a coefficient of permeability of 0.0225 m/day.

Solution: Assuming that concrete weighs 2520 kg/m³ and the granular material weighs 2184 kg/m³, then

$$\sigma_p = 0.15 \times 2520 + 0.225 \times 2184 = 869.4 \text{ kg/m}^2$$

Entering the chart in Figure 16.37 at the midlevel of the high range of frost susceptibility and projecting to a stress σ_p of 172.5 gives

$$q_m/\sqrt{k} \approx 0.3$$

$$q_m = [(\sqrt{0.0225})2]0.3 = 0.045 \text{ m}^3/\text{day/m}^2$$

Note that if the average rate of heave of the subgrade soil is determined from either laboratory tests or local experience based on observations of frost action, then this value may be used instead of the frost susceptibility classification.

Vertical Outflow, q_v. In some cases, the total amount of water accumulated within the pavement structure can be reduced because of the vertical seepage of some of the accumulated water through the subgrade. When this occurs, it is necessary to estimate the amount of this outflow in order to determine the net inflow for which the subdrainage system is to be provided. The procedure for estimating this flow involves the use of flow net diagrams, which is beyond the scope of this book. This procedure is discussed in detail in *Highway Subdrainage Design*. As will be seen later, however, the use of the vertical outflow to reduce the inflow is applicable only when there is neither groundwater inflow nor frost action.

Net Inflow. The net inflow is the sum of inflow rates from all sources less any amount attributed to vertical outflow through the underlying soil. However, note that all the different flows discussed earlier do not necessarily occur at the same time. For example, it is unlikely that flows from thawed water and ground water will occur at the same time, since soils susceptible to frost action will have very low permeability when frozen.

Similarly, downward vertical outflows will never occur at the same time as upward inflow from any other source. Downward vertical outflow will therefore only occur when there is no inflow due to ground water. A set of relationships for estimating the net inflow rate (q_n) has been developed, taking into consideration the different flows that occur concurrently, and is given in Eqs. 16.28 through 16.32.

$$q_n = q_i \qquad (16.28)$$

$$q_n = q_i + q_g \qquad (16.29)$$

$$q_n = q_i + q_a \qquad (16.30)$$

$$q_n = q_i + q_m \qquad (16.31)$$

$$q_n = q_i - q_v \qquad (16.32)$$

Guidelines for using these equations are given in Table 16.14.

16.7.4 Design of Drainage Layer

The design of the drainage layer involves either the determination of the maximum depth of flow H_m when the permeability of the material k_d is known or the determination of the required permeability of the drainage material when the maximum flow depth is stipulated. In each case, however, both the slope S of the drainage layer along the flow path and the length L of the flow path must be known. The flow through a drainage layer at full depth is directly related to the *coefficient of transmissibility*, which is the product of k_d and the depth H_d of the drainage layer. This relationship may be used to determine the characteristics of the drainage layer required, or, alternatively, the graphical solution presented in Figure 16.38 may be used. The chart shown in Figure 16.38 is based on steady inflow uniformly distributed across the surface of the pavement section. This condition does not normally occur in practice, but a conservative result is usually obtained when the chart is used in combination with the procedure presented herein for determining the net inflow rate q_n.

Table 16.14   Guidelines for Using Eqs. 16.28 through 16.32 to Compute Net Inflow, q_n, for Design of Pavement Drainage

Highway Cross Section	Ground Water Inflow	Frost Action	Net Inflow Rate, q_n, Recommended for Design
Cut	Gravity	Yes	Max. of Eqs. 16.29 and 16.31
		No	Eq. 16.29
Cut	Artesian	Yes	Max. of Eqs. 16.30 and 16.31
		No	Eq. 16.30
Cut	None	Yes	Eq. 16.31
		No	Eq. 16.28
Cut	None	Yes	Eq. 16.31
		No	Eq. 16.32
Fill	None	Yes	Eq. 16.31
		No	Eq. 16.32

SOURCE: Adapted from *Highway Subdrainage Design*, Report No. FHWA-TS-80-224, U.S. Department of Transportation, Washington, D.C., August 1980

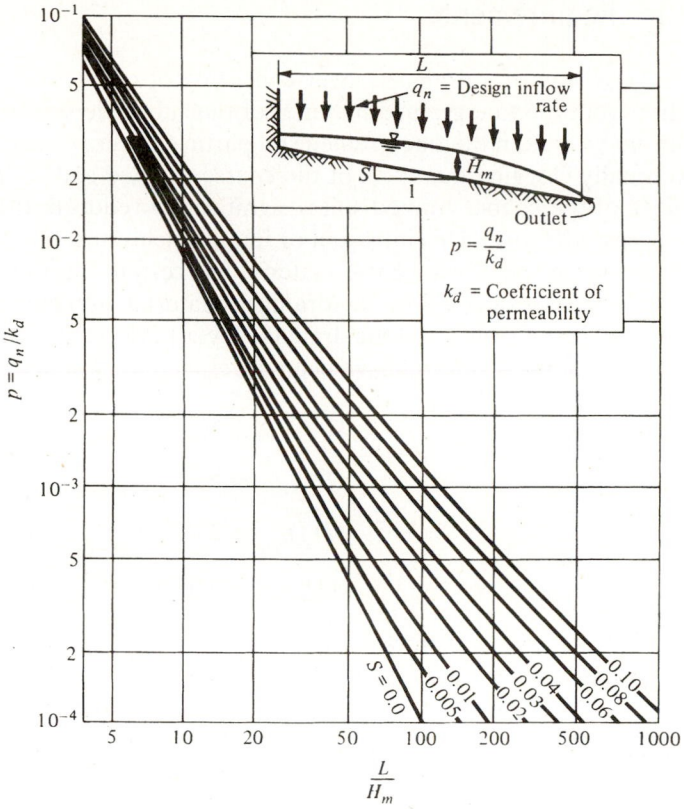

Figure 16.38   Chart for Estimating Maximum Depth of Flow Caused by Steady Inflow

SOURCE: Redrawn from *Highway Subdrainage Design*, Report No. FHWA-TS-80-224, U.S. Department of Transportation, Washington, D.C., August 1980

Example 16.12 Computing Required Depth for a Drainage Layer

Determine the depth required for a drainage layer to carry a net inflow of 0.15 m³/day/ m² if the permeability of the drainage material is 600 m/day. The drainage layer will be laid at a slope of 2 percent, and the length of the flow path is 12 m.

$$P = q_n/k_d \text{ (from Figure 16.38)}$$

$$= \frac{0.15}{600} = 2.50 \times 10^{-4}$$

Solution: Entering the chart at $p = 2.50 \times 10^{-4}$ and projecting horizontally to the slope of 0.02, we determine L/H_m as 130.

$$H_m = \frac{12}{130} \text{ (required depth of drainage layer)}$$

$$\approx 0.092 \text{ m}$$

$$= 9.2 \text{ cm (say, 10 cm —i.e., } H_d)$$

Note that $H_d > H_m$.

Filter Requirements

The provision of a drainage layer consisting of coarse material allows for the flow of water from the fine-grained material of the subgrade soil to the coarse drainage layer. This may result in the fine-grained soil particles being transmitted to the coarse soil and eventually clogging the voids of the coarse-grained soil. When this occurs, the permeability of the coarse-grained soil is significantly reduced, thereby making the drainage layer less effective. This intrusion of fine particles into the voids of the coarse material can be minimized if the coarse material has certain filter criteria. In cases where these criteria are not satisfied by the drainage material, a protective filter must be provided between the subgrade and the drainage layer to prevent clogging of the drainage layer.

The following criteria have been developed for soil materials used as filters.

$$(D_{15})_{\text{filter}} \leq 5(D_{85})_{\text{protected soil}}$$

$$(D_{15})_{\text{filter}} \geq 5(D_{15})_{\text{protected soil}}$$

$$(D_{50})_{\text{filter}} \leq 25(D_{50})_{\text{protectes soil}}$$

$$(D_{5})_{\text{filter}} \geq 0.074 \text{ mm}$$

where D_i is the grain diameter that is larger than the ith percent of the soil grains—that is, the ith percent size on the grain-size distribution curve (see Chapter 17).

16.7.5 Design of Longitudinal Collectors

Circular pipes are generally used for longitudinal collectors and are usually constructed of either porous concrete, perforated corrugated metal, or vitrified clay. The pipes are laid in trenches located at depths that will allow the drainage of the subsurface water from the pavement structure. The trenches are then backfilled with porous granular material to facilitate free flow of the subsurface water into the drains.

Design of the longitudinal collectors involves the determination of the pipe location and pipe diameter and the identification of a suitable backfill material.

Pipe Location

Shallow trenches within the subbase layer may be used in locations where the depth of frost penetration is insignificant and where the drawdown of the water table is low, as shown in Figure 16.39. In cases where the depth of frost penetration is high or the water table is high (thereby requiring a high drawdown), it is necessary to locate the pipe in a deeper trench below the subbase layer, as shown in Figure 16.40. Note, however, that the deeper the trench, the higher the construction cost of the system. The lateral location of the pipe depends on whether the shoulder is also to be drained. If the shoulder is to be drained, the pipe should be located close to the edge of the shoulder, as shown in Figures 16.39b and 16.40b, but if shoulder drainage is not required, the pipe is located just outside the pavement surface, as shown in Figures 16.39a and 16.40a.

Pipe Diameter

The diameter D_p of the collector pipe depends on the gradient g, the amount of water per running foot (q_d) that should be transmitted through the pipe, Manning's roughness

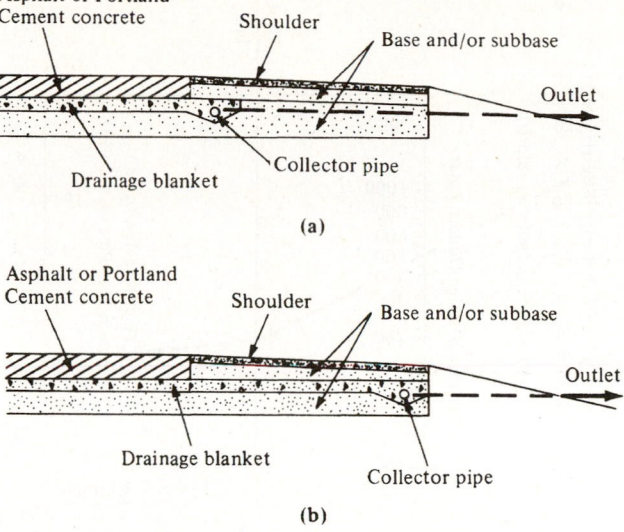

Figure 16.39 Typical Location of Shallow Longitudinal Collector Pipes

SOURCE: Redrawn from *Highway Subdrainage Design*, Report No. FHWA-TS-80-224, U.S. Department of Transportation, Washington, D.C., August 1980

coefficient of the pipe material, and the distance between the outlets L_o. The chart shown in Figure 16.41 can be used either to determine the minimum pipe diameter when the flow depth, the distance between the outlets, and the gradient are specified or to determine the maximum spacing between outlets for different combinations of gradient and pipe diameters.

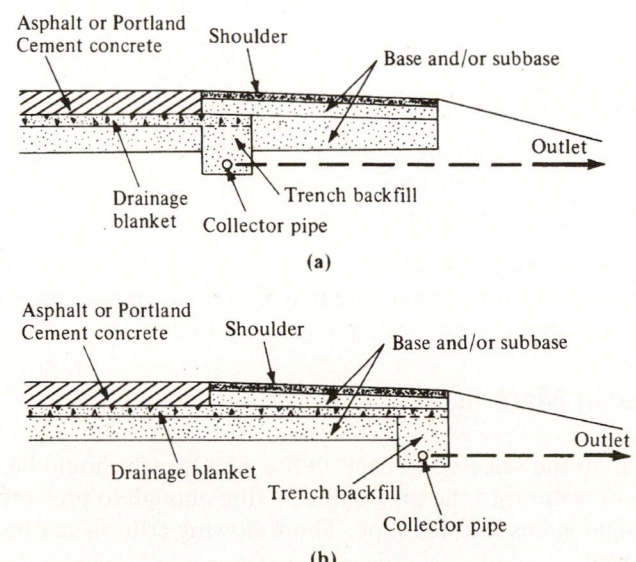

Figure 16.40 Typical Location of Deep Longitudinal Collector Pipes

SOURCE: Redrawn from *Highway Subdrainage Design*, Report No. FHWA-TS-80-224, U.S. Department of Transportation, Washington, D.C., August 1980

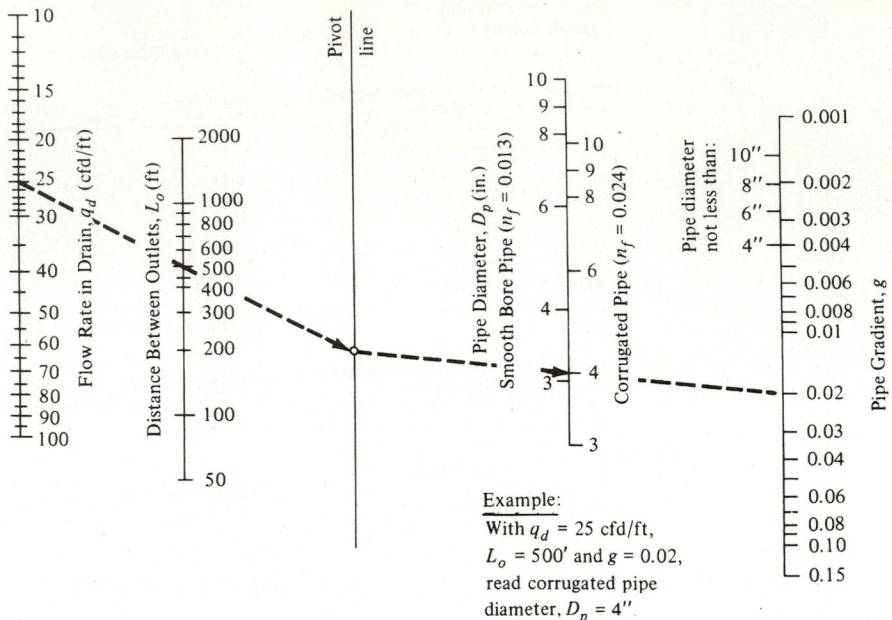

Figure 16.41 Nomogram Relating Collector Pipe Size with Flow Rate, Outlet Spacing, and Pipe Gradient

SOURCE: Redrawn from *Highway Subdrainage Design*, Report No. FHWA-TS-80-224, U.S. Department of Transportation, Washington, D.C., August 1980

In using the chart, it is first necessary to determine the amount of flow q_d from q_n as

$$q_d = q_n L \qquad (16.33)$$

where

q_d = flow rate in drain (m³/day/m)
q_n = net inflow (m³/day/m²)
L = the length of the flow path (m)

Note that some variation of L may occur along the highway. The average of all values of L associated with a given pipe may therefore be used for that pipe. The use of the chart is demonstrated in the example given in Figure 16.41.

Backfill Material

The material selected to backfill the pipe trench should be coarse enough to permit the flow of water into the pipe and also fine enough to prevent the infiltration of the drainage aggregates into the pipe. The following criteria can be used to select suitable filter material.

For slotted pipes $(D_{85})_{\text{filter}} > \frac{1}{2}$ slot width

For circular holes $(D_{85})_{\text{filter}} >$ hole diameter

16.7.6 ECONOMIC ANALYSIS

The economic analysis normally carried out is similar to that carried out for open-channel and culvert designs. In this case, however, the cost of the subdrainage system is highly dependent on the cost and availability of suitable drainage materials and the cost of the pipes used as longitudinal and transverse drains.

16.8 SUMMARY

The provision of adequate drainage facilities on a highway is fundamental and essential to the effective performance of the highway. The operation of any drainage system consists of complex hydraulic phenomena, which make it difficult to develop exact mathematical equations for design or analysis. The analysis and design of drainage facilities are therefore mainly based on empirical relations that have been developed from extensive test results. The material presented in this chapter provides the reader with the basic principles of analysis and design currently in use. However, the use of any methods presented should go hand in hand with experience that has been gained from local conditions.

The use of these procedures coupled with sound judgment will result in drainage facilities that effectively serve the highway.

PROBLEMS

16-1   What are the two sources of water a highway engineer is primarily concerned with? Briefly describe each.

16-2   Briefly describe the main differences between surface drainage and subsurface drainage.

16-3   What are the two main disadvantages of using turf cover on unpaved shoulders?

16-4   Briefly describe the three properties of rainfall that primarily concern highway engineers.

16-5   What is meant by a: (a) 10-year storm, (b) 50-year storm, (c) 100-year storm, and (d) 500-year storm?

16-6   Define the following: (a) drainage area, (b) run-off coefficient (C), (c) travel time (t_i), and (d) time of concentration (t_c).

16-7   A 196-acre rural drainage area consists of four different watershed areas as follows:

> Steep grass covered area = 35%
> Cultivated area = 15%
> Forested area = 40%
> Turf meadows = 10%

Using the rational formula, determine the runoff rate for a storm of 100-year frequency. Use Table 16.2 for runoff coefficients. Assume that the rainfall intensity curves in Figure 16.2 are applicable to this drainage area and the following land characteristics apply. Use Figure 16.4 to calculate average velocity using "fallow or minimum tillage cultivation" ground cover. Overland flow length = 0.5 mi. Average slope of overland area = 2 percent.

16-8   A 210-acre urban drainage area consists of three different watershed areas as follows:

> Flat residential (30% impervious area) = 58%
> Moderately steep residential (50% impervious area) = 28%
> Flat commercial (90% impervious area) = 14%

Using the rational formula, determine the runoff rate for a storm of 100-year frequency. Use Table 16.2 for runoff coefficients. Assume that the rainfall intensity curves in Figure 16.2 are applicable to this drainage area and the following land characteristics apply. Use Figure 16.4 to calculate average velocity using "fallow or minimum tillage cultivation" ground cover. Overland flow length = 0.4 mi. Average slope of overland area = 3 percent.

16-9   Compute rate of runoff using the rational formula for a 256-acre rural drainage area consisting of two different watershed areas as follows:

Steep grass area = 15%
Forested area = 30%
Cultivated fields = 55%

If the time of concentration for this area is 2.4 hours, determine the runoff rate for a storm of 50-year frequency. Use the rainfall intensity curves in Figure 16.2. Use Table 16.2 for runoff coefficients.

16-10   Using the TR-55 method, determine the depth of runoff for a 24-hour, 100-year precipitation event of 9 inches if the soil can be classified as group B and the watershed is contoured pasture with good hydrologic condition and an antecedent soil condition III.

16-11   Determine the depth of runoff by the TR-55 method for a 24-hour, 100-year precipitation of 9 inches for an antecedent moisture condition III if the following land uses and soil conditions exist.

Area Fraction	Land/Use Condition	Soil Group
0.33	Wooded/fair condition	D
0.27	Small grain/straight row/good condition	D
0.12	Pasture/contoured/fair condition	D
0.23	Meadow/good condition	D
0.05	Farmstead	D

16-12   Determine the peak discharge that will occur for the conditions indicated in Problem 16-11 if the drainage area is 1.28 km^2, and the time of concentration is 1.4 h.

16-13   What is the difference between supercritical and subcritical flow? Under what conditions will either of these occur?

16-14   A trapezoidal channel of 2:1 side slope and 1.5 m bottom width, discharges a flow of 7.425 m^3/sec. If the channel slope is 2.5 percent and the Manning coefficient is 0.03, determine (a) flow velocity, (b) flow depth, and (c) type of flow.

16-15   A 1,8 m-wide rectangular channel lined with rubble masonry is required to carry a flow of 8.1 m^3/sec. If the slope of the channel is 2 percent and $n = 0.015$, determine (a) flow depth, (b) flow velocity, and (c) type of flow.

16-16   For the conditions given in Problem 16-15, determine the critical depth and the maximum channel slope at which subcritical flow can occur.

16-17   Determine a suitable rectangular flexible lined channel to resist erosion for a maximum flow of 5.4 m^3/sec if the channel slope is 2 percent. Use channel dimensions given in Problem 16-15.

16-18   A trapezoidal channel of 2:1 side slope and 1.5 m bottom width is to be used to discharge a flow of 5.4 m^3/sec. If the channel slope is 1.5 percent and the Manning coefficient is 0.015, determine the minimum depth required for the channel. Is the flow supercritical or subcritical?

16-19   Determine whether a 1.5 m × 1.5 m reinforced concrete box culvert with 45° flared wing-walls and beveled edge at top of inlet carrying a 50-year flow rate of 5.4 m³/s will operate under inlet or outlet control for the following conditions. Assume $k_e = 0.5$.

>   Design headwater elevation (EL_{hd}) = 31.5 m
>   Elevation of stream bed at face of invert = 29.87 m
>   Tailwater depth = 1.43 m
>   Approximate length of culvert = 60 m
>   Slope of stream = 1.5%
> $n = 0.012$

16-20   Repeat Problem 16-19 using a 1.95 m diameter circular pipe culvert with $k_e = 0.5$.

16-21   Determine the ground water infiltration rate for a new two-lane pavement with the following characteristics:

>   Lane width = 3.6 m
>   Shoulder width = 3 m
>   Length of contributing transverse cracks (W_c) = 6 m
>   Rate of infiltration (K_p) = 0.015 m/day/m²
>   Spacing of transverse cracks = 9 m

16-22   Determine the ground water infiltration rate for a new four-lane pavement with the following characteristics:

>   Lane width = 3.6 m
>   Shoulder width = 3 m
>   Length of contributing transverse cracks (W_c) = 12 m
>   Rate of infiltration (K_p) = 0.015 m/day/m²
>   Spacing of transverse cracks = 12 m

16-23   In addition to the infiltration determined in Problem 16-21, ground water seepage due to gravity also occurs. Determine the thickness of a suitable drainage layer required to transmit the net inflow to a suitable outlet.

REFERENCES

Brown, S. A., S. M. Stein, and J. C. Warner, *Urban Drainage Design Manual*. Hydraulic Engineering Circular No. 22, 2nd ed., U.S. Department of Transportation, Federal Highway Administration Publication No. FHWA-NHI-01-021, Washington, D.C., 2001.

Brown, S. A., J. D. Schall, J. L. Morris, C. L. Doherty, S. M. Stein, and J. C. Warner, *Urban Drainage Design Manual*. Hydraulic Engineering Circular No. 22, 3rd ed., U.S. Department of Transportation, Federal Highway Administration Publication No. FHWA-NHI-01-021, Washington, D.C., September 2009.

Hayes, D. C. and R. L. Young, *Comparison of Peak Discharge and Runoff Characteristic Estimates from the Rational Method to Field Observations for Small Basins in Central Virginia*. Scientific Investigations Report 2005-5254, U.S. Geological Survey, Reston, VA, 2005.

Highway Drainage Guidelines. American Association of State Highway and Transportation Officials, Washington, D.C., 1992.

Model Drainage Manual. American Association of State Highway and Transportation Officials, Washington, D.C., 1991.

Norman, J. M., R. J. Houghtalen, and W. J. Johnston, *Hydraulic Design of Highway Culverts.* Report No. FHWA-NH1-01-020, U.S. Department of Transportation, Office of Implementation, September 2001 (revised May 2005).

Shaw, L. Yu, *Stormwater Management for Transportation Facilities.* Synthesis of Practice 174, Transportation Research Board, National Research Council, Washington, D.C., 1993.

Shaw, L. Yu and Robert J. Kaighn, Jr., *VDOT Manual of Practice for Planning Stormwater Management.* Virginia Transportation Research Council, Charlottesville, VA, January 1992.

Urban Hydrology for Small Watersheds, Technical Release No. 55, U.S. Department of Agriculture, Soil Conservation Service, Washington, D.C., 1975.

ADDITIONAL READING

Christopher, Barry R. and Verne G. McGuffey, *Pavement Subsurface Drainage Systems.* Synthesis of Highway Practice 239, Transportation Research Board, National Research Council, Washington, D.C., 1997.

Design Charts for Open Channel Flow, Hydraulic Design Series No. 3, U.S. Department of Transportation, Federal Highway Administration, Washington, D.C., December 1980.

Design and Construction of Urban Stormwater Management Systems, ASCE Manuals and Reports of Engineering Practices No. 77, American Society of Civil Engineers, Alexandria, VA, 1992.

Gray, Donald H. and Andrew T. Leiser, *Biotechnical Slope Protection and Erosion Control.* Krieger Publishing, Malabar, FL., 1989.

Haestad Methods, Inc., *Computer Applications in Hydraulic Engineering.* Haestad Press, Waterbury, CT., 1997.

Highway Drainage Guidance, Volume X: Evaluating Highway Effects on Surface Water Environment. Task Force on Hydrology and Hydraulics, American Association of State Highway and Transportation Officials, Washington, D.C., 1992.

Highway Drainage Manual, Chapter 8. Highway Drainage, Revision 61, U.S. Department of Transportation, Federal Highway Administration, Washington, D.C., March 25, 2011.

Highway Subdrainage Design. Report No. FHWA-TS-80-224, U.S. Department of Transportation, Federal Highway Administration, Washington, D.C., August 1980.

Kilgore, R. and G. C. Cotton, *Design of Roadside Channels with Flexible Linings.* Hydraulic Engineering Circular No. 15, 3rd ed., U.S. Department of Transportation, Federal Highway Administration, Publication No. FHWA-NHI-05-114, September 2005.

Thompson, P. L. and R. T. Kilgore, *Hydraulic Design of Energy Dissipators for Culverts and Channels.* Hydraulic Engineering Circular No. 14, 3rd ed., Publication No. FHWA-NHI-06-086, July 2006.

Wanielista, Martin, Robert Kersten, and Ron Eaglin, *Hydrology—Water Quantity and Quality Control.* 2nd ed., John Wiley & Sons, New York, 1997.

PART 5

Materials and Pavements

Highway pavements are constructed of either asphalt or concrete and ultimately rest on native soil. The engineer must be familiar with the properties and structural characteristics of materials that will be used in constructing or rehabilitating a roadway segment. The engineer must also be familiar with methods and theories for the design of heavy-duty asphaltic and concrete pavements, as well as various treatment strategies for low-volume roads.

CHAPTER 17

Soil Engineering for Highway Design

Highway engineers are interested in the basic engineering properties of soils because soils are used extensively in highway construction. Soil properties are of significant importance when a highway is to carry high traffic volumes with a large percentage of trucks. They are also of importance when high embankments are to be constructed as well as when the soil is to be strengthened and used as intermediate support for the highway pavement. Thus, several transportation agencies have developed detailed procedures for investigating soil materials used in highway construction.

This chapter presents a summary of current knowledge of the characteristics and engineering properties of soils that are important to highway engineers, including the origin and formation of soils, soil identification, and soil testing methods.

Procedures for improving the engineering properties of soils will be discussed in Chapter 19, "Design of Flexible Pavements".

CHAPTER OBJECTIVES:

- Become familiar with the basic characteristics of a soil sample.
- Learn the procedures for classifying soils for highway design purposes using their characteristics.
- Learn the procedures and equipment used in compacting soils for highway construction.
- Learn the procedures for testing soils for highway pavement design.
- Become familiar with the effect of frost action in soils.

17.1 SOIL CHARACTERISTICS

The basic characteristics of a soil may be described in terms of its origin, formation, grain size, and shape. It will be seen later in this chapter that the principal engineering properties of any soil are mainly related to the basic characteristics of that soil.

917

17.1.1 Origin and Formation of Soils

Soil can be defined from the civil engineering point of view as the loose mass of mineral and organic materials that cover the solid crust of granitic and basaltic rocks of the earth. Soil is mainly formed by weathering and other geologic processes that occur on the surface of the solid rock at or near the surface of the earth. Weathering is the result of physical and chemical actions, mainly due to atmospheric factors that change the structure and composition of the rocks. Weathering occurs through either physical or chemical means. Physical weathering, sometimes referred to as *mechanical weathering*, causes the disintegration of the rocks into smaller particle sizes by the action of forces exerted on the rock. These forces may be due to running water, wind, freezing and thawing, and the activity of plants and animals. *Chemical weathering* occurs as a result of oxidation, carbonation, and other chemical actions that decompose the minerals of the rocks.

Soils may be described as residual or transported. *Residual soils* are weathered in place and are located directly above the original material from which they were formed. *Transported soils* are those that have been moved by water, wind, glaciers, and so forth, and are located away from their parent materials.

The geologic history of any soil deposit has a significant effect on the engineering properties of the soils. For example, sedimentary soils, which are formed by the action of water, are usually particles that have settled from suspension in a lake, river, or ocean. These soils range from beach or river sands to marine clays. Soils that are formed by the action of wind are known as aeolian soils and are typically loess. Their voids are usually partially filled with water, and when submerged in water, the soil structure collapses.

Soils also may be described as organic when the particles are mainly composed of organic matter or as inorganic when the particles are mainly composed of mineral materials.

17.1.2 Surface Texture

The texture of a soil can be described in terms of its appearance, which depends mainly on the shapes and sizes of the soil particles and their distribution in the soil mass. For example, soils consisting mainly of silts and clays with very small particle sizes are known as *fine-textured soils*, whereas soils consisting mainly of sands and gravel with much larger particles are known as *coarse-textured soils*. The individual particles of fine-textured soils are usually invisible to the naked eye, whereas those of coarse-textured soils are visible to the naked eye.

It will be seen later in this chapter that the engineering properties of a soil are related to its texture. For example, the presence of water in fine-textured soils results in significant reduction in their strength, whereas this does not happen with coarse-textured soils. Soils can therefore be divided into two main categories based on their texture. Coarse-grained soils are sometimes defined as those with particle sizes greater than 0.05 mm, such as sands and gravel, and fine-grained soils are those with particle sizes less than 0.05 mm, such as silts and clays. The dividing line of 0.05 mm (0.075 mm has also been used) is selected because that is normally the smallest grain size that can be seen by the naked eye. Since there is a wide range of particle sizes in soils, both the coarse-grained soils and fine-grained soils may be further subdivided, as will be shown later under soil classification.

The distribution of particle size in soils can be determined by conducting a sieve analysis (sometimes known as mechanical analysis) on a soil sample if the particles are sufficiently large. This is done by shaking a sample of air-dried soil through a set of sieves with progressively smaller openings. The smallest practical opening of these sieves is 0.075 mm; this sieve is designated No. 200. Other sieves include No. 140 (0.106 mm), No. 100 (0.15 mm), No. 60 (0.25 mm), No. 40 (0.425 mm), No. 20 (0.85 mm), No. 10 (2.0 mm), No. 4 (4.75 mm), and several others, with openings increasing up to 125 mm or 5 in.

For soils containing particle sizes smaller than the lower limit, the hydrometer analysis is used. A representative sample of the air-dried soil is sieved through the No. 10 sieve, and a sieve analysis is carried out on the portion of soil retained. This will give a distribution of the coarse material. A portion of the material that passes through the No. 10 sieve is suspended in water, usually in the presence of a deflocculating agent, and is then left standing until the particles gradually settle to the bottom. A hydrometer is used to determine the specific gravity of the suspension at different times. The specific gravity of the suspension after any time t from the start of the test is used to determine the maximum particle sizes in the suspension as

$$D = \sqrt{\frac{18\eta}{\gamma_s - \gamma_w}\left(\frac{y}{t}\right)} \tag{17.1}$$

where

D = maximum diameter of particles in suspension at depth y; that is, all particles in suspension at depth y (cm) have diameters less than D (cm)

η = coefficient of viscosity of the suspending medium (in this case, water) in poises

γ_s = unit weight of soil particles (g/cm³)

γ_w = unit weight of water (g/cm³)

t = time from start of the test (sec)

The above expression is based on Stokes' law. Note, however, that corrections are usually made to the hydrometer readings to take into account the impact of factors that may affect the readings. These corrections can be found in most soil mechanics laboratory handbooks.

At the completion of this test, which lasts up to 24 hours, the sample of soil used for the hydrometer test is then washed over a No. 200 sieve. The portion retained in the No. 200 sieve is then oven-dried and sieved through Nos. 20, 40, 60, and 140 sieves. The combination of the results of the sieve analysis and the hydrometer test is then used to obtain the particle size distribution of the soil. This is usually plotted as the cumulative percentage by weight of the total sample less than a given sieve size or computed grain diameter, versus the logarithm of the sieve size or grain diameter. Figure 17.1 shows examples of particle size distributions of three different soil samples taken from different locations.

The natural shape of a soil particle is either round, angular, or flat. This natural shape is usually an indication of the strength of the soil, particularly for larger soil particles. Round particles are found in deposits of streams and rivers, have been subjected to extensive wear, and are therefore generally strong. Flat and flaky particles have not been subjected to similar action and are usually weak. Fine-grained soils generally have flat and flaky-shaped particles, whereas coarse-grained soils generally have round or angular-shaped particles. Soils with angular-shaped particles have more resistance to deformation than those with round particles, since the individual angular-shaped particles tend to lock together, whereas the rounded particles tend to roll over each other.

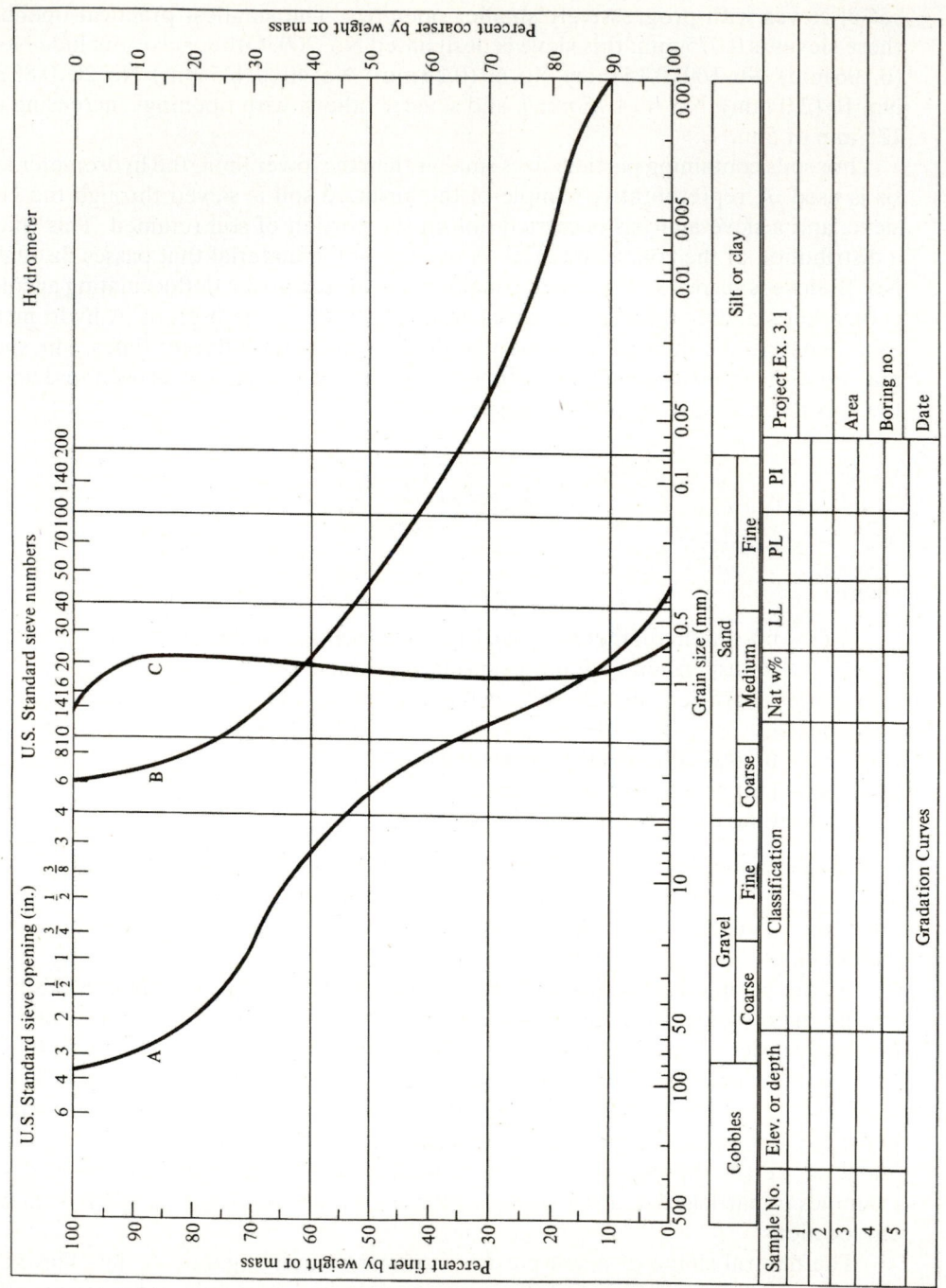

Figure 17.1 Distribution of Particle Size in Different Soils

17.2 BASIC ENGINEERING PROPERTIES OF SOILS

Highway engineers must be familiar with those basic engineering properties of soils that influence their behavior when subjected to external loads. The determination of how a specific soil deposit will behave when subjected to an external load is rather complicated because soil deposits may have heterogeneous properties. Highway engineers must always keep in mind that the behavior of any soil depends on the conditions of that soil at the time it is being tested.

17.2.1 Phase Relations

A soil mass generally consists of solid particles of different minerals with spaces between them. The spaces can be filled with air and/or water. Soils are therefore considered as three-phase systems that consist of air, water, and solids. Figure 17.2 schematically illustrates the three components of a soil mass of total volume V. The volumes of air, water, and solids are V_a, V_w, and V_s, respectively, and their weights are W_a, W_w, and W_s, respectively. The volume V_v is the total volume of the space occupied by air and water, generally referred to as a *void*.

Porosity

The relative amount of voids in any soil is an important quantity that influences some aspects of soil behavior. This amount can be measured in terms of the *porosity* of the soil, which is defined as the ratio of the volume of voids to the total volume of the soil and is designated as n, as shown in Eq. 17.2.

$$n = \frac{V_v}{V} \qquad (17.2)$$

Void Ratio

The amount of voids can also be measured in terms of the *void ratio*, which is defined as the ratio of the volume of voids to the volume of solids and is designated as e, as shown in Eq. 17.3.

$$e = \frac{V_v}{V_s} \qquad (17.3)$$

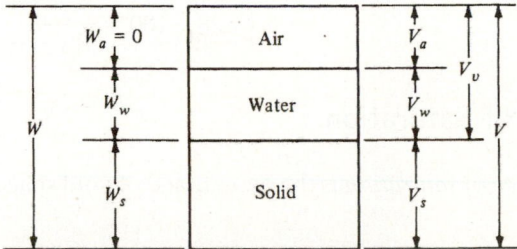

Figure 17.2   Schematic of the Three Phases of a Soil Mass

Combining Eqs. 17.2 and 17.3, we obtain

$$e = \frac{nV}{V_s} = \frac{n}{V_s}(V_s + V_v) = n(1+e)$$

and

$$n = \frac{e}{1+e} \qquad (17.4)$$

Similarly,

$$e = \frac{n}{1-n} \qquad (17.5)$$

Moisture Content

The quantity of water in a soil mass is expressed in terms of the *moisture content*, which is defined as the ratio of the weight of water, W_w, in the soil mass to the oven-dried weight of solids, W_s, expressed as a percentage. It is given as

$$w = \frac{W_w}{W_s}100 \qquad (17.6)$$

where w is the moisture content.

The moisture content of a soil mass can be determined in the laboratory by obtaining the weight of water in the soil and the weight of the dry solids after completely drying the soil in an oven at a temperature between 100 and 110°C. The weight of the water W_w is obtained by first obtaining the weight W_1 of the soil and container. The soil and container are then placed in the oven. The weight of the dry soil and container is again obtained after repeated drying until there is no further reduction in weight. This weight W_2 is that of the dry soil and container. The weight of water, W_w, is therefore given as

$$W_w = W_1 - W_2$$

The weight of the empty container W_c is then obtained, and the weight of the dry soil is given as

$$W_s = W_2 - W_c$$

and the moisture content is obtained as

$$w(\%) = \frac{W_w}{W_s}100 = \frac{W_1 - W_2}{W_2 - W_c}100 \qquad (17.7)$$

Degree of Saturation

The *degree of saturation* is the percentage of void space occupied by water and is given as

$$S = \frac{V_w}{V_v}100 \qquad (17.8)$$

where S is the degree of saturation.

The soil is saturated when the void is fully occupied with water (that is, when $S = 100\%$) and partially saturated when the voids are only partially occupied with water.

Density of Soil

A very useful soil property for highway engineers is the *density* of the soil. The density is the ratio that relates the mass side of the phase diagram to the volumetric side. Three densities are commonly used in soil engineering: total or bulk density γ, dry density γ_d, and submerged or buoyant density γ'.

Total Density.   The *total* (or *bulk*) *density* is the ratio of the weight of a given sample of soil to the volume or

$$\gamma = \frac{W}{V} = \frac{W_s + W_w}{V_s + V_w + V_a} \quad \text{(weight of air is negligible)} \tag{17.9}$$

The total density for saturated soils is the saturated density and is given as

$$\gamma_{\text{sat}} = \frac{W}{V} = \frac{W_s + W_w}{V_s + V_w} \tag{17.10}$$

Dry Density.   The *dry density* is the density of the soil with the water removed. It is given as

$$\gamma_d = \frac{W_s}{V} = \frac{W_s}{V_s + V_w + V_a} = \frac{\gamma}{1 + w} \tag{17.11}$$

The dry density is often used to evaluate how well earth embankments have been compacted and is therefore an important quantity in highway engineering.

Submerged Density.   The *submerged density* is the density of the soil when submerged in water and is the difference between the saturated density and the density of water, or

$$\gamma = \gamma_{\text{sat}} - \gamma_w \tag{17.12}$$

where γ_w is the density of water.

Specific Gravity of Soil Particles

The specific gravity of soil particles is the ratio of density of the soil particles to the density of distilled water.

Other Useful Relationships

The basic definitions presented above can be used to derive other useful relationships. For example, the bulk density can be given as

$$\gamma = \frac{G_s + Se}{1 + e} \gamma_w \tag{17.13}$$

where

γ = total or bulk density
S = degree of saturation
γ_w = density of water
G_s = specific gravity of the soil particles
e = void ratio

Example 17.1 Determining Soil Characteristics Using the Three-Phase Principle

The wet weight of a specimen of soil is 340 g and the dried weight is 230 g. The volume of the soil before drying is 210 cc. If the specific gravity of the soil particles is 2.75, determine the void ratio, porosity, degree of saturation, and dry density. Figure 17.3 is a schematic of the soil mass.

Solution: Since the weight of the air can be taken as zero, the weight of the water is

$$W_w = 340 - 230 = 110 \text{ g}$$

Therefore, the volume of water is

$$V_w = (110/1.00) = 110 \text{ cc (since density of water} = 1 \text{ gm/cc)}$$

For the volume of solids,

$$V_s = \frac{W_s}{\gamma_s} = \frac{230}{2.75} \quad (\gamma_s = \text{specific gravity} \times \text{density of water} = 2.75 \times 1)$$

$$= 83.64 \text{ cc}$$

For the volume of air,

$$V_a = 210 - 110 - 83.64 = 16.36 \text{ cc}$$

For the void ratio,

$$e = \frac{V_w + V_a}{V_s} = \frac{110 + 16.36}{83.64} = 1.51$$

For the porosity,

$$n = \frac{V_v}{V} = \frac{110 + 16.36}{210} = 0.60$$

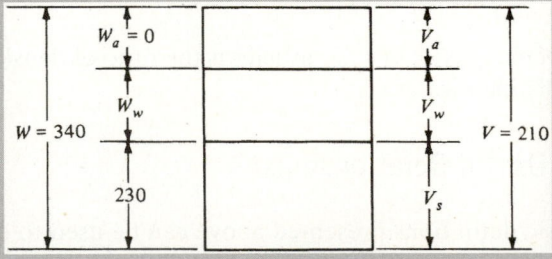

Figure 17.3   Schematic of Soil Mass for Example 17.1

For the degree of saturation,

$$S = \frac{V_w}{V_v} = \frac{110}{110 + 16.36} = 0.87 \text{ or } 87\%$$

For the dry density,

$$\gamma_d = \frac{W_s}{V} = \frac{230}{210} = 1.095 \text{ g/cc}$$

Example 17.2 Determining Soil Characteristics Using the Three-Phase Principle

The moisture content of a specimen of soil is 26 percent, and the bulk density is 1.86 g/cc. If the specific gravity of the soil particles is 2.76, determine the void ratio and the degree of saturation.

Solution: The weight of 1 cm³ of the soil is 1.86 g—that is,

$$W = 1.86 \text{ g} = W_s + W_w = W_s + wW_s = W_s(1 + 0.26)$$

Therefore,

$$W_s = \frac{1.86 \text{ g}}{1.26} = 1.476 \text{ g}$$

$$W_w = 0.26 \times 1.476 = 0.384 \text{ g}$$

$$V_s = \frac{W_s}{\gamma_s} \quad (\gamma_s = \text{specific gravity} \times \text{density of water} = 2.76 \times 62.4)$$

$$= \frac{1.476}{2.76 \times 1} = 0.535 \text{ cm}^3$$

$$V_w = \frac{W_w}{\gamma_w} = \frac{0.384}{62.4} = 0.384 \text{ cm}^3$$

$$V_a = V - V_s - V_w = 1 - 0.53 - 0.38 = 0.09 \text{ cm}^3$$

Therefore, the void ratio is

$$e = \frac{V_w + V_a}{V_s} = \frac{0.38 + 0.09}{0.53} = 0.89$$

and the degree of saturation is

$$S = \frac{V_w}{V_v} 100 = \frac{0.38}{0.38 + 0.09} 100 = 80.9\%$$

17.2.2 Atterberg Limits

Clay soils with very low moisture content will be in the form of solids. As the water content increases, however, the solid soil gradually becomes plastic—that is, the soil can easily be molded into different shapes without breaking up. Continuous increase of the water content will eventually bring the soil to a state where it can flow as a viscous liquid. The stiffness or consistency of the soil at any time therefore depends on the state in which the soil is, which in turn depends on the amount of water present in the soil. The water content levels at which the soil changes from one state to the other are the *Atterberg limits*. They are the shrinkage limit (SL), plastic limit (PL), and liquid limit (LL), as illustrated in Figure 17.4.

These are important limits of engineering behavior, because they facilitate the comparison of the water content of the soil with those at which the soil changes from one state to another. They are used in the classification of fine-grained soils and are extremely useful, since they correlate with the engineering behaviors of such soils.

Shrinkage Limit (SL)

When a saturated soil is slowly dried, the volume shrinks, but the soil continues to contain moisture. Continuous drying of the soil, however, will lead to a moisture content at which further drying will not result in additional shrinkage. The volume of the soil will stay constant, and further drying will be accompanied by air entering the voids. The moisture content at which this occurs is the *shrinkage limit*, or SL, of the soil.

Plastic Limit (PL)

The *plastic limit*, or PL, is defined as the moisture content at which the soil crumbles when it is rolled down to a diameter of one-eighth of an inch. The moisture content is higher than the PL if the soil can be rolled down to diameters less than one-eighth of an inch, and the moisture content is lower than the PL if the soil crumbles before it can be rolled to one-eighth of an inch diameter.

Liquid Limit (LL)

The *liquid limit*, or LL, is defined as the minimum moisture content at which the soil will flow when a small shearing force is applied. It is determined as the moisture content at which the soil will flow and close a groove of one-half inch within it after the standard liquid limit equipment has been dropped 25 times. The equipment and related appliances used for LL determination is shown in Figure 17.5. This device was developed by Casagrande, who worked to standardize the Atterberg limits tests. It is difficult in

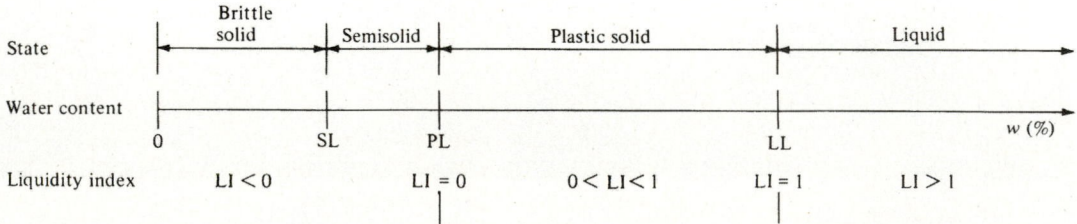

Figure 17.4   Consistency Limits

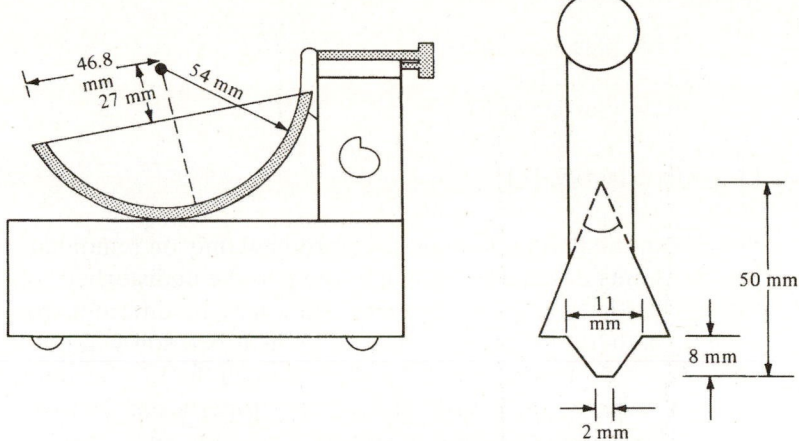

(a)   Schematic of the Casagrande Liquid Limit Apparatus

(b)   Casagrande Liquid Limit Apparatus

SOURCE: Science and Society / SuperStock

Figure 17.5   Casagrande Liquid Limit Equipment

practice to obtain the exact moisture content at which the groove will close at exactly 25 blows. The test is therefore conducted for different moisture contents, and the number of blows required to close the groove for each level of moisture content is recorded. A graph of moisture content versus the logarithm of the number of blows (usually a straight line known as the flow curve) is then drawn. The level of moisture content at which the flow curve crosses 25 blows is the LL.

The range of moisture content over which the soil is in the plastic state is the difference between the LL and the PL and is known as the *plasticity index* (PI).

$$PI = LL - PL \qquad (17.14)$$

where

PI = plasticity index
LL = liquid limit
PL = plastic limit

Liquidity Index (LI)

Since both the PL and LL can be determined only on remolded soils, it is quite possible that the limits determined may not apply to the undisturbed soil, since the structure of the soil particles in the undisturbed state may be different from that in the disturbed state. It is therefore possible that an undisturbed soil will not be in the liquid state if its moisture content is 35 percent, even though the LL was found to be 35 percent. The *liquidity index*, or LI, is used to reflect the properties of the natural soil and is defined as

$$LI = \frac{w_n - PL}{PI} \tag{17.15}$$

where

LI = liquidity index
W_n = natural moisture content of the soil

A soil with an LI less than zero will have a brittle fracture when sheared, and a soil with an LI between zero and one will be in a plastic state. When LI is greater than one, the soil will be in a state of viscous liquid if sheared.

Soils with LIs greater than one are known as *quick clays*. They stay relatively strong if undisturbed but become very unstable and can even flow like liquid if they are sheared.

Permeability

The *permeability* of a soil is the property that describes how water flows through the soil. It is usually given in terms of the coefficient of permeability (K), which is the constant of proportionality of the relationship between the flow velocity and the hydraulic gradient between two points in the soil. This relationship was first determined by the French engineer Henry D'Arcy and is given as

$$u = Ki \tag{17.16}$$

where

u = velocity of water in the soil
i = hydraulic gradient
  = $\frac{h}{l}$ (head loss h per unit length l)
K = coefficient of permeability

The coefficient of permeability of a soil can be determined in the laboratory by conducting either a constant head or falling head test, or in the field by pumping tests. Clays and fine-grained soils have very low permeability; thus hardly any flow of water occurs in these soils. Coarse-grained soils, such as gravel and sands, have high permeability, which allows for water to flow easily in them. Soils with high permeability are therefore generally stable, both in the dry and saturated states. Thus, coarse-grained soils make excellent subgrade (natural material along the alignment of a highway or compacted borrow material serving as the pavement foundation) materials. Note, however, that capillary action may occur in some permeable soils such as "dirty" gravel, which may cause serious stability problems. Capillary action is the movement of free moisture by capillary forces through small-diameter openings in the soil mass into pores that are not full of water. Although the moisture can move in any direction, the upward movement usually causes the most serious problems, since this may cause weakness or lead to frost heave. This is discussed further in Section 17.7, "Frost Action in Soils".

Shear Strength

The *shear strength* of soils is of particular importance to the highway engineer, because soil masses will usually fail in shear under highway loads. The shear strength of a soil depends on the cohesion and the angle of internal friction and is expressed as

$$S = C + \sigma \tan \phi \tag{17.17}$$

where

S = shear strength (N/m^2)
C = cohesion (N/m^2)
ϕ = angle of internal friction
σ = normal stress on the shear plane (N/m^2)

The degree of importance of either the cohesion or the angle of internal friction depends on the type of soil. In fine-grained soils such as clays, the cohesion component is the major contributor to the shear strength. In fact, it usually is assumed that the angle of internal friction of saturated clays is zero, which makes the shearing resistance on any plane of these soils equal to the cohesion C. Factors that affect the shear strength of cohesive soils include the geologic deposit, moisture content, drainage conditions, and density.

In coarse-grained soils such as sands, the shear strength is achieved mainly through the internal resistance to sliding as the particles roll over each other. The angle of internal friction is therefore important. The value of the angle of internal friction depends on the density of the soil mass, the shape of individual soil particles, and the surface texture. In general, the angle of internal friction is high when the density is high. Similarly, soils with rough particles such as angular sand grains will have a high angle of internal friction.

The shearing strength of a soil deposit may be obtained in the laboratory by conducting either the triaxial test, the unconfined compression test, or the direct shear test. These tests may be conducted either on the undisturbed soil or on remolded soils. Note, however, that in using remolded samples, the remolding should represent conditions similar to those in the field. The *in situ* shearing strengths of soils also can be obtained directly by conducting either the plate-bearing test or cone penetration test. Details of each of these tests can be found in AASHTO's *Standard Specifications for Transportation Materials and Methods of Sampling and Testing*.

17.3 CLASSIFICATION OF SOILS FOR HIGHWAY USE

Soil classification is a method by which soils are systematically categorized according to their probable engineering characteristics. It therefore serves as a means of identifying suitable subbase materials and predicting the probable behavior of a soil when used as subgrade material (see definitions of *subgrade* and *subbase* in Chapter 19.) The classification of a given soil is determined by conducting relatively simple tests on disturbed samples of the soil; the results are then correlated with field experience.

Note, however, that although the engineering properties of a given soil to be used in highway construction can in some cases be predicted reliably from its classification, this should not be regarded as a substitute for the detailed investigation of the soil properties. Classifying the soil should be considered as a means of obtaining a general idea of how the soil will behave if used as a subgrade or subbase material.

The most commonly used classification system for highway purposes is the American Association of State Highway and Transportation Officials (AASHTO) Classification System. The Unified Soil Classification System (USCS) is also used to a lesser extent. A slightly modified version of the USCS is used fairly extensively in the United Kingdom.

17.3.1 AASHTO Soil Classification System

The AASHTO Classification System is based on the Public Roads Classification System that was developed from the results of extensive research conducted by the Bureau of Public Roads, now known as the Federal Highway Administration. Several revisions have been made to the system since it was first published. The system has been described by AASHTO as a means for determining the relative quality of soils for use in embankments, subgrades, subbases, and bases.

In the current publication, soils are classified into seven groups, A-1 through A-7, with several subgroups, as shown in Table 17.1. The classification of a given soil is based on its particle size distribution, LL, and PI. Soils are evaluated within each group by using an empirical formula to determine the group index (GI) of the soils, given as

$$GI = (F - 35)[0.2 + 0.005(LL - 40)] + 0.01(F - 15)(PI - 10) \qquad (17.18)$$

where

GI = group index
F = percent of soil particles passing 0.075 mm (No. 200) sieve in whole number based on material passing 75 mm (3 in.) sieve
LL = liquid limit expressed in whole number
PI = plasticity index expressed in whole number

The GI is determined to the nearest whole number. A value of zero should be recorded when a negative value is obtained for the GI. Also, in determining the GI for A-2-6 and A-2-7 subgroups, the LL part of Eq. 17.18 is not used—that is, only the second term of the equation is used.

Under the AASHTO system, granular soils fall into classes A-1 to A-3. A-1 soils consist of well-graded granular materials, A-2 soils contain significant amounts of silts and clays, and A-3 soils are clean but poorly graded sands.

Classifying soils under the AASHTO system will consist of first determining the particle size distribution and Atterberg limits of the soil and then reading Table 17.1 from left to right to find the correct group. The correct group is the first one from the left that fits the particle size distribution and Atterberg limits and should be expressed in terms of group designation followed by the GI in parentheses. Examples are A-2-6(4) and A-6(10).

Table 17.1 AASHTO Classification of Soils and Soil Aggregate Mixtures

General Classification	Granular Materials (35% or Less Passing No. 200)								Silt-Clay Materials (More than 35% Passing No. 200)				
	A-1		A-3	A-2					A-4	A-5	A-6	A-7	
Group Classification	A-1-a	A-1-b		A-2-4	A-2-5	A-2-6	A-2-7					A-7-5,	A-7-6
Sieve analysis Percent passing:													
No. 10	50 max.	—											
No. 40	30 max.	50 max.	51 min.										
No. 200	15 max.	25 max.	10 max.	35 max.	35 max.	35 max.	35 max.		36 min.	36 min.	36 min.	36 min.	
Characteristics of fraction passing No. 40:													
Liquid limit	—		—	40 max.	41 min.	40 max.	41 min.		40 max.	41 min.	40 max.	41 min.	
Plasticity index	6 max.		N.P.	10 max.	10 max.	11 min.	11 min.		10 max.	10 max.	11 min.	11 min.*	
Usual types of significant constituent materials	Stone fragments, gravel and sand		Fine sand	Silty or clayey gravel and sand					Silty soils		Clayey soils		
General rating as subgrade	Excellent to good								Fair to poor				

*Plasticity index of A-7-5 subgroup $\leq$ LL − 30. Plasticity index of A-7-6 subgroup > LL − 30.

SOURCE: Based on *Standard Specifications for Transportation Materials and Methods of Sampling and Testing*, 27th ed., 2007, AASHTO, Washington, D.C.

931

In general, the suitability of a soil deposit for use in highway construction can be summarized as follows.

1. Soils classified as A-1-a, A-1-b, A-2-4, A-2-5, and A-3 can be used satisfactorily as subgrade or subbase material if properly drained (see definitions of *subgrade* and *subbase* in Chapter 19.) In addition, such soils must be properly compacted and covered with an adequate thickness of pavement (base and/or surface cover) for the surface load to be carried.
2. Materials classified as A-2-6, A-2-7, A-4, A-5, A-6, A-7-5, and A-7-6 will require a layer of subbase material if used as subgrade. If these are to be used as embankment materials, special attention must be given to the design of the embankment.
3. When soils are properly drained and compacted, their value as subgrade material decreases as the GI increases. For example, a soil with a GI of zero (an indication of a good subgrade material) will be better as a subgrade material than one with a GI of 20 (an indication of a poor subgrade material).

Example 17.3 Classifying a Soil Sample Using the AASHTO Method

The following data were obtained for a soil sample.

Mechanical Analysis

Sieve No.	Percent Finer	Plasticity Tests
4	97	LL = 48%
10	93	PL = 26%
40	88	
100	78	
200	70	

Using the AASHTO method for classifying soils, determine the classification of the soil and state whether this material is suitable in its natural state for use as a subbase material.

Solution:

- Since more than 35% of the material passes the No. 200 sieve, the soil is either A-4, A-5, A-6, or A-7.
- LL > 40%, and therefore the soil cannot be in group A-4 or A-6. Thus, it is either A-5 or A-7.
- The PI is 22% (48 − 26), which is greater than 10%, thus eliminating group A-5. The soil is A-7-5 or A-7-6.
- (LL − 30) = 18 < PI (22%). Therefore, the soil is A-7-6, since the plasticity index of A-7-5 soil subgroup is less than (LL−30). The GI is given as:

$$(70 - 35)[0.2 + 0.005(48 - 40)] + 0.01(70 - 15)(22 - 10) = 8.4 + 6.6 = 15$$

The soil is A-7-6 (15) and is therefore unsuitable as a subbase material in its natural state.

17.3.2 Unified Soil Classification System (USCS)

The original USCS system was developed during World War II for use in airfield construction. That system has been modified several times to obtain the current version, which also can be applied to other types of construction, such as dams and foundations. The fundamental premise used in the USCS system is that the engineering properties of any coarse-grained soil depend on its particle size distribution, whereas those for a fine-grained soil depend on its plasticity. Thus, the system classifies coarse-grained soils on the basis of grain size characteristics and fine-grained soils according to plasticity characteristics.

Table 17.2 lists the USCS definitions for the four major groups of materials, consisting of coarse-grained soils, fine-grained soils, organic soils, and peat. Material that is retained in the 75 mm (3 in.) sieve is recorded, but only that which passes is used for the classification of the sample. Soils with more than 50 percent of their particles being retained on the No. 200 sieve are coarse-grained, and those with less than 50 percent of their particles retained are fine-grained soils (see Table 17.3). The coarse-grained soils are subdivided into gravels (G) and sands (S). Soils having more than 50 percent of their particles larger than 4.75 mm—that is, retained on the No. 4 sieve—are gravels and those with more than 50 percent of their particles smaller than 4.75 mm—that is, passed through the No. 4 sieve—are sands. The gravels and sands are further divided into four subgroups—each based on grain-size distribution and the nature of the fine particles in them. They therefore can be classified as either well graded (W), poorly graded (P), silty (M), or clayey (C). Gravels can be described as either well-graded gravel (GW), poorly graded gravel (GP), silty gravel (GM), or clayey gravels (GC), and sands can be described as well-graded sand (SW), poorly graded sand (SP), silty sand (SM), or clayey

Table 17.2 USCS Definition of Particle Sizes

Soil Fraction or Component	Symbol	Size Range
1. Coarse-grained soils		
Gravel	G	75 mm to No. 4 sieve (4.75 mm)
Coarse		75 mm to 19 mm
Fine		19 mm to No. 4 sieve (4.75 mm)
Sand	S	No. 4 (4.75 mm) to No. 200 (0.075 mm)
Coarse		No. 4 (4.75 mm) to No. 10 (2.0 mm)
Medium		No. 10 (2.0 mm) to No. 40 (0.425 mm)
Fine		No. 40 (0.425 mm) to No. 200 (0.075 mm)
2. Fine-grained soils		
Fine		Less than No. 200 sieve (0.075 mm)
Silt	M	–(No specific grain size—use Atterberg limits)
Clay	C	–(No specific grain size—use Atterberg limits)
3. Organic soils	O	(No specific grain size)
4. Peat	Pt	(No specific grain size)

Gradation symbols	Liquid Limit Symbols
Well graded, W	High LL, H
Poorly graded, P	Low LL, L

SOURCE: Adapted from *The Unified Soil Classification System*, Annual Book of ASTM Standards, Vol. 04.08, American Society for Testing and Materials, West Conshohocken, PA., 2012

sand (SC). A gravel or sandy soil is described as well graded or poorly graded, depending on the values of two shape parameters known as the coefficient of uniformity, C_u, and the coefficient of curvature, C_c, given as

$$C_u = \frac{D_{60}}{D_{10}}$$

(17.19)

and

$$C_c = \frac{(D_{30})^2}{D_{10} \times D_{60}}$$

(17.20)

where

D_{60} = grain diameter at 60% passing
D_{30} = grain diameter at 30% passing
D_{10} = grain diameter at 10% passing

Gravels are described as well graded if C_u is greater than four and C_c is between one and three. Sands are described as well graded if C_u greater than six and C_c is between one and three.

The fine-grained soils, which are defined as those having more than 50 percent of their particles passing the No. 200 sieve, are subdivided into clays (C) or silt (M), depending on the PI and LL of the soil. A plasticity chart, shown in Table 17.3, is used to determine whether a soil is silty or clayey. The chart is a plot of PI versus LL, from which a dividing line known as the "A" line, which generally separates the more clayey materials from the silty materials, was developed. Soils with plots of LLs and PIs below the "A" line are silty soils, whereas those with plots above the "A" line are clayey soils. Organic clays are an exception to this general rule, since they plot below the "A" line. Organic clays, however, generally behave similarly to soils of lower plasticity.

Classification of coarse-grained soils as silty or clayey also depends on their LL plots. Only coarse-grained soils with more than 12 percent fines (that is, passes the No. 200 sieve) are so classified (see Table 17.3). Those soils with plots below the "A" line or with a PI less than four are silty gravel (CM) or silty sand (SM), and those with plots above the "A" line with a PI greater than seven are classified as clayey gravels (GC) or clayey sands (SC).

The organic, silty, and clayey soils are further divided into two groups, one having a relatively low LL (L) and the other having a relatively high LL (H). The dividing line between high LL soils and low LL soils is arbitrarily set at 50 percent.

Fine-grained soils can be classified as either silt with low plasticity (ML), silt with high plasticity (MH), clays with high plasticity (CH), clays with low plasticity (CL), or organic silt with high plasticity (OH).

Table 17.3 gives the complete layout of the USCS, and Table 17.4 shows an approximate correlation between the AASHTO system and USCS.

Table 17.3 Unified Soil Classification System

Major Divisions			Group Symbols	Typical Names	Laboratory Classification Criteria		
Coarse-grained soils (More than half of material is larger than No. 200 sieve size)	Gravels (More than half of coarse fraction is larger than No. 4 sieve size)	Clean gravels (Little or no fines)	GW	Well-graded gravels, gravel-sand mixtures, little or no fines	$C_u = \dfrac{D_{60}}{D_{10}}$ greater than 4; $C_c = \dfrac{(D_{30})^2}{D_{10} \times D_{60}}$ between 1 and 3		
			GP	Poorly graded gravels, gravel-sand mixtures, little or no fines	Not meeting all gradation requirements for GW		
		Gravels with fines (Appreciable amount of fines)	GM[a] d / u	Silty gravels, gravel-sand-silt mixtures	Atterberg limits below "A" line or P.I. less than 4	Above "A" line with P.I. between 4 and 7 are *borderline* cases requiring use of dual symbols	
			GC	Clayey gravels, gravel-sand-clay mixtures	Atterberg limits below "A" line with P.I. greater than 7		
	Sands (More than half of coarse fraction is smaller than No. 4 sieve size)	Clean sands (Little or no fines)	SW	Well-graded sands, gravelly sands, little or no fines	$C_u = \dfrac{D_{60}}{D_{10}}$ greater than 6; $C_c = \dfrac{(D_{30})^2}{D_{10} \times D_{60}}$ between 1 and 3		
			SP	Poorly graded sands, gravelly sands, little or no fines	Not meeting all gradation requirements for SW		
		Sands with fines (Appreciable amount of fines)	SM[a] d / u	Silty sands, sand-silt mixtures	Atterberg limits above "A" line or P.I. less than 4	Limits plotting in hatched zone with P.I. between 4 and 7 are *borderline* cases requiring use of dual symbols	
			SC	Clayey sands, sand-clay mixtures	Atterberg limits above "A" line with P.I. greater than 7		

Determine percentages of sand and gravel from grain-size curve. Depending on percentage of fines (fraction smaller than No. 200 sieve size), coarse-grained soils are classified as follows:
Less than 5 per cent — GW, GP, SW, SP
More than 12 per cent — GM, GC, SM, SC
5 to 12 per cent — *Borderline* cases requiring dual symbols[b]

Major Divisions			Group Symbols	Typical Names
Fine-grained soils (More than half of material is smaller than No. 200 sieve)	Silts and clays (Liquid limit less than 50)		ML	Inorganic silts and very fine sands, rock flour, silty or clayey fine sands, or clayey silts with slight plasticity
			CL	Inorganic clays of low to medium plasticity, gravelly clays, sandy clays, silty clays, lean clays
			OL	Organic silts and organic silty clays of low plasticity
	Silts and clays (Liquid limit greater than 50)		MH	Inorganic silts, micaceous or diatomaceous fine sandy or silty soils, elastic silts
			CH	Inorganic clays of high plasticity, fat clays
			OH	Organic clays of medium to high plasticity, organic silts
	Highly organic soils		Pt	Peat and other highly organic soils

Plasticity Chart

[a]Division of GM and SM groups into subdivisions of d and u are for roads and airfields only. Subdivision is based on Atterberg limits; suffix d used when L.L. is 28 or less and the P.I. is 6 or less; the suffix u used when L.L. is greater than 28.
[b]Borderline classifications, used for soils possessing characteristics of two groups, are designated by combinations of group symbols. For example: GW-GC, well-graded gravel-sand mixture with clay binder.

SOURCE: *The Unified Soil Classification System*, Annual Book of ASTM Standards, Vol. 4.08, American Society for Testing and Materials, West Conshohocken, PA, 2012

Table 17.4   Comparable Soil Groups in the AASHTO and USCS Systems

Soil Group in Unified System	Comparable Soil Groups in AASHTO System		
	Most Probable	Possible	Possible but Improbable
GW	A-1-a	—	A-2-4, A-2-5, A-2-6, A-2-7
GP	A-1-a	A-1-b	A-3, A-2-4 A-2-5, A-2-6, A-2-7
GM	A-1-b, A-2-4, A-2-5, A-2-7	A-2-6	A-4, A-5, A-6, A-7-5, A-7-6, A-1-a
GC	A-2-6, A-2-7	A-2-4, A-6	A-4, A-7-6, A-7-5
SW	A-1-b	A-1-a	A-3, A-2-4, A-2-5, A-2-6, A-2-7
SP	A-3, A-1-b	A-1-a	A-2-4, A-2-5, A-2-6, A-2-7
SM	A-1-b, A-2-4, A-2-5, A-2-7	A-2-6, A-4, A-5	A-6, A-7-5, A-7-6, A-1-a
SC	A-2-6, A-2-7	A-2-4, A-6 A-4, A-7-6	A-7-5
ML	A-4, A-5	A-6, A-7-5	—
CL	A-6, A-7-6	A-4	—
OL	A-4, A-5	A-6, A-7-5, A-7-6	—
MH	A-7-5, A-5	—	A-7-6
CH	A-7-6	A-7-5	—
OH	A-7-5, A-5	—	A-7-6
Pt	—	—	—

SOURCE: Adapted from *Comparison of AASHTO & USCS Soils Classification Systems,* Soils 2011, www.virginiadot.org/business/resources/.../bu-mat-Chapter1Soils.pdf

Example 17.4 Classifying a Soil Sample Using the Unified Soil Classification System

The results obtained from a mechanical analysis and a plasticity test on a soil sample are shown below. Classify the soil using the USCS and state whether or not it can be used in the natural state as a subbase material.

	Mechanical Analysis	
Sieve No.	*Percent Passing (by weight)*	*Plasticity Tests*
4	98	LL = 40%
10	93	PL = 30%
40	85	
100	73	
200	62	

Solution: The grain-size distribution curve is first plotted as shown in Figure 17.6. Since more than 50 percent (62 percent) of the soil passes the No. 200 sieve, the soil is fine grained. The plot of the limits on the plasticity chart is below the "A" line (PI = 40 − 30 = 10); therefore, it is either silt or organic clay. However, the LL is less than 50 percent (40 percent); therefore, it is low LL. This soil can be classified as ML or OL, which is probably equivalent to an A-4 or A-5 soil in the AASHTO classification system based on Table 17.4. It therefore is not useful as a subbase material based on the guidelines given in Section 17.3.1.

Example 17.5 Classifying a Soil Sample Using the Unified Soil Classification System

Repeat Example 17.4 for the data shown below.

	Mechanical Analysis	
Sieve No.	*Percent Passing (by weight)*	*Plasticity Tests:*
4	95	LL = nonplastic
10	30	PL = nonplastic
40	15	
100	8	
200	3	

Solution: The grain-size distribution curve is also plotted in Figure 17.6. Since only 3 percent of the particles pass through the No. 200 sieve, the soil is coarse grained. Since more than 50 percent pass through the No. 4 sieve, the soil is classified as sand. Because the soil is nonplastic, it is necessary to determine its coefficient of uniformity C_u and coefficient of curvature C_c.

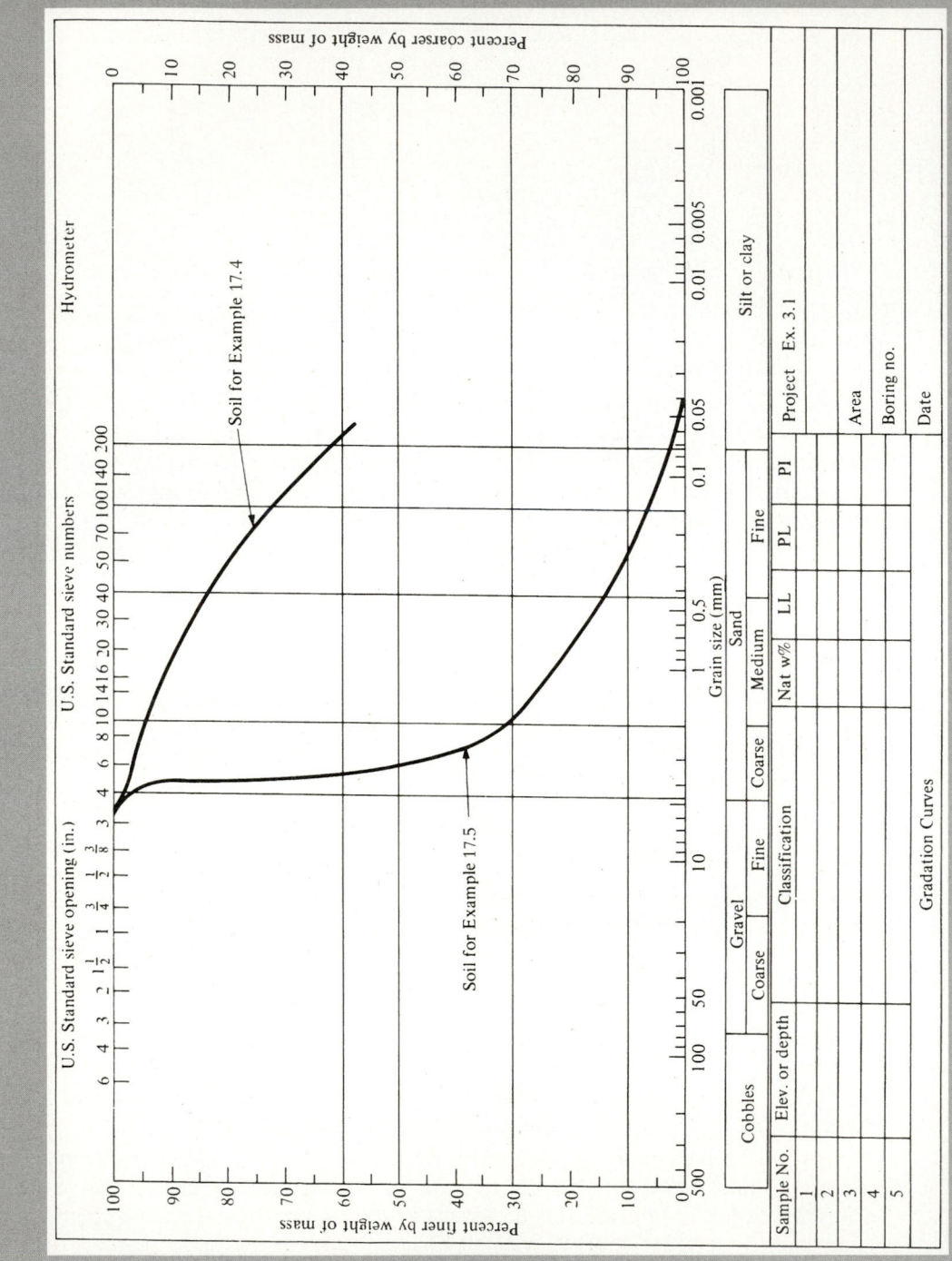

Figure 17.6 Grain-Size Distribution Curve for Examples 17.4 and 17.5

From the particle size distribution curve,

$$C_u = \frac{D_{60}}{D_{10}} = \frac{3.8}{0.25} = 15.2 > 6$$

$$C_c = \frac{(D_{30})^2}{D_{10} \times D_{60}} = \frac{2^2}{0.25 \times 3.8} = 4.2$$

This sand is not well graded, is classified as SP, and therefore can be used as a subbase material if properly drained and compacted.

17.4 SOIL SURVEYS FOR HIGHWAY CONSTRUCTION

Soil surveys for highway construction entail the investigation of the soil characteristics on the highway route and the identification of suitable soils for use as subbase and fill materials. Soil surveys are therefore normally an integral part of preliminary location surveys, since the soil conditions may significantly affect the location of the highway. A detailed soil survey is always carried out on the final highway location.

The first step in any soil survey is the collection of existing information on the soil characteristics of the area in which the highway is to be located. Such information can be obtained from geologic and agricultural soil maps, existing aerial photographs, and an examination of excavations and existing roadway cuts. It is also usually helpful to review the design and construction of other roads in the area. The information obtained from these sources can be used to develop a general understanding of the soil conditions in the area and to identify any unique problems that may exist. The extent of additional investigation usually depends on the amount of existing information that can be obtained.

The next step is to obtain and investigate enough soil samples along the highway route to identify the boundaries of the different types of soils so that a soil profile can be drawn. Samples of each type of soil along the route location are obtained by auger boring or from test pits for laboratory testing. Samples are usually taken at different depths down to about 1.50 m. In cases where rock locations are required, depths may be increased. The engineering properties of the samples are then determined and used to classify the soils. It is important that the characteristics of the soils in each hole be systematically recorded, including the depth, location, thickness, texture, and so forth. It is also important that the location of the water level be noted. These data are then used to plot a detailed soil profile along the highway (see Figure 17.7).

17.4.1 Geophysical Methods of Soil Exploration

Soil profiles can also be obtained from one of two geophysical methods of soil exploration known as the resistivity and seismic methods.

Resistivity Method

The *resistivity method* is based on the difference in electrical conductivity or resistivity of different types of soils. An electrical field is produced in the ground by means of two current electrodes, as shown in Figure 17.8, and the potential drop between

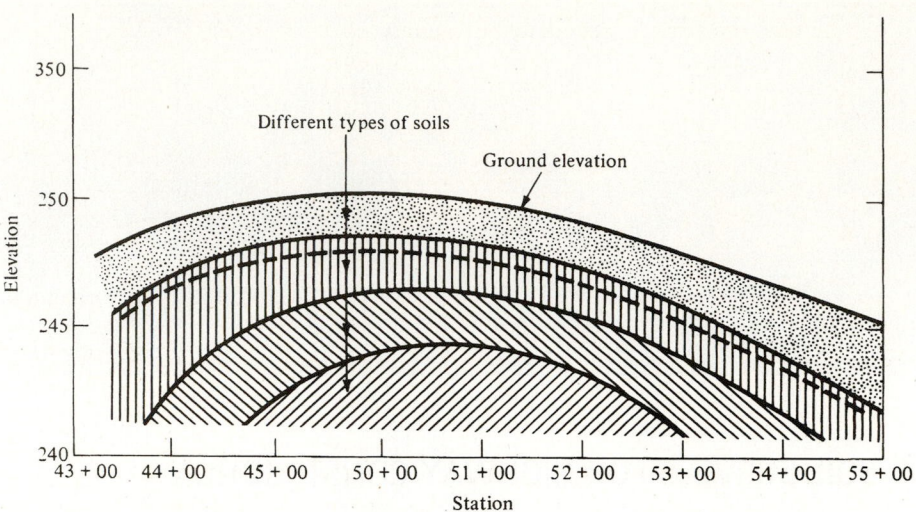

Figure 17.7 Soil Profile along a Section of Highway

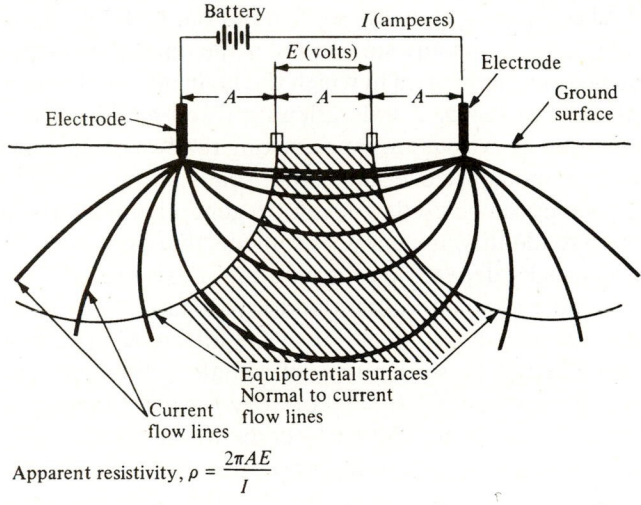

Apparent resistivity, $\rho = \dfrac{2\pi A E}{I}$

Figure 17.8 Electrical Resistivity Method of Soil Exploration

SOURCE: Redrawn from Hsai-Yang Fang, ed., *Foundation Engineering Handbook*, Routledge, Chapman & Hall, NY, 1990 with permission of Springer Science and Business Media

the two intermediate or potential electrodes is then recorded. The apparent resistivity of the soil to a depth approximately equal to the spacing A is then computed. The resistivity equipment used usually is designed such that the apparent resistivity can be read directly on the potentiometer. Data for the soil profile are obtained by moving the electrode along the center line of the proposed highway without changing the spacing. The apparent resistivity is then determined along the highway within a depth equal to the spacing A. The resistivities obtained are then compared with known values of different soils by calibrating the instrument using locally exposed materials.

Seismic Method

The *seismic method* is used to identify the location of rock profiles or dense strata underlying softer materials. (Figure 17.9 shows the layout for the seismic method.) It is conducted by inducing impact or shock waves into the soil by either striking a plate located on the surface with a hammer or exploding small charges in the soil. Listening devices known as geophones then pick up the shock waves. The time lapse of the wave traveling to the geophone then is used to calculate the velocity of the wave in the surface soil. Some of the shock waves can be made to pass from the surface stratum into underlying layers and then back into the surface stratum by moving the shock point away from the geophone. This permits the computation of the wave velocity in the underlying material.

The seismic test is conducted by moving the shock-producing device along the proposed centerline of the highway. The shock is produced at known distances from the geophone. A graph of the time it takes the shock waves to arrive at the geophone versus the distance of the geophone from the shock point is then drawn. These points will be on a straight line as long as the shock wave travels through the same soil material. When the distance of shock point from the geophone becomes large enough, the waves travel through the denser material and a break is observed, as shown in Figure 17.9. The inverse of the slope of each straight line will give the velocity of the wave within each layer. It can be shown that for three layers, as shown in Figure 17.9, the depths of the first and second soil layers, H_1 and H_2, are

$$H_1 = \frac{\overline{OP}u_1}{2\cos\alpha} \tag{17.21}$$

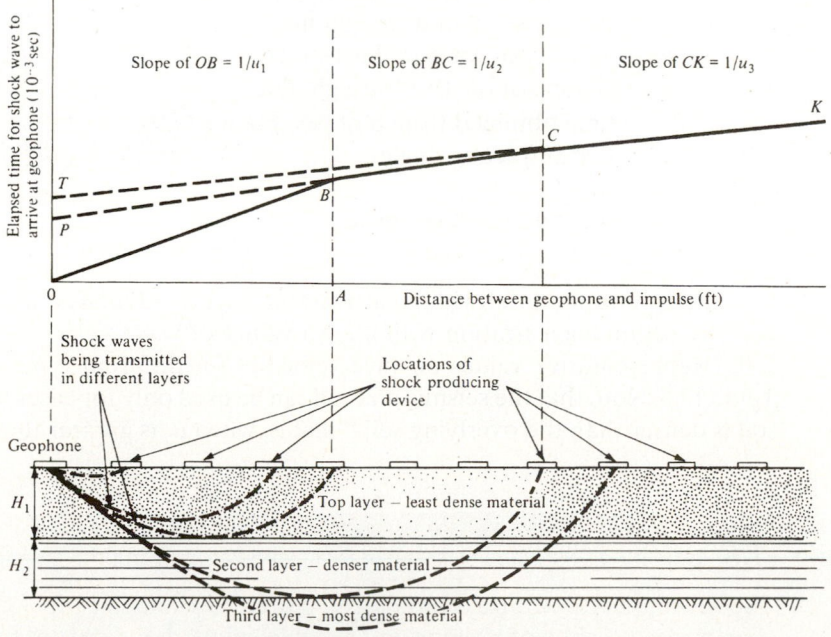

Figure 17.9   Soil Exploration by the Seismic Method

Table 17.5   Representative Values of Wave Velocities for Different Types of Soils

Material	Velocity (ft/sec)
Soil	
Sand, dry silt, and fine-grained top soil	650–3300
Alluvium	1650–6600
Compacted clays, clayey gravel, and dense clayey sand	3300–8200
Loess	800–2450
Rock	
Slate and shale	8200–16,400
Sandstone	4900–16,400
Granite	13,100–19,700
Sound limestone	16,400–32,800

SOURCE: From Das, *Principles of Foundation Engineering*, 7E. © 2011 Cengage Learning.

and

$$H_2 = \frac{\overline{PT}u_2}{2\cos\beta} \tag{17.22}$$

where

$$\overline{OP} = \text{time obtained from plot (see Figure 17.9)}$$
$$u_1 = \text{velocity of wave in upper stratum}$$
$$= 1/\text{slope of first straight line}$$
$$u_2 = \text{velocity of wave in underlying stratum}$$
$$= 1/\text{slope of second straight line}$$
$$u_3 = \text{velocity of wave in third stratum}$$
$$= 1/\text{slope of the third straight line}$$
$$\overline{PT} = \text{time obtained from plot (see Figure 17.9)}$$
$$\alpha = \text{first refraction angle}$$
$$\sin\alpha = u_1/u_2$$
$$\beta = \text{second refraction angle}$$
$$\sin\beta = u_2/u_3$$

The type of material within each stratum can be identified by comparing the wave velocity within each stratum with known values of wave velocity for different types of soils. Representative values of wave velocities for different types of soils are given in Table 17.5. Note that the seismic method can be used only for cases where the underlying soil is denser than the overlying soil—that is, when u_2 is greater than u_1.

Example 17.6 Estimating Depth and Soil Type of Each Soil Stratum Using the Seismic Method

The seismic method of exploration was used to establish the soil profile along the proposed centerline of a highway. The table below shows part of the results obtained.

Estimate the depth of each stratum of soil at this section and suggest the type of soil in each.

Distance of Impulse to Geophone (m)	Time for Wave Arrival (10^{-3} sec)
6	32
12	60
18	88
24	94
30	100
36	106
42	112
48	116
54	117
60	118.5
66	120
75	122

Solution: The plot of the data is shown in Figure 17.10. From Figure 17.10,

$$u_1 = \frac{OA}{AB} = \frac{18}{88 \times 10^{-3}} = 205 \text{ m/sec}$$

From Table 17.5, the soil in the first stratum is possibly sand.

$$u_2 = \frac{BD}{CD} = \frac{42 - 18}{(112 - 88) \times 10^{-3}} = 1000 \text{ m/sec}$$

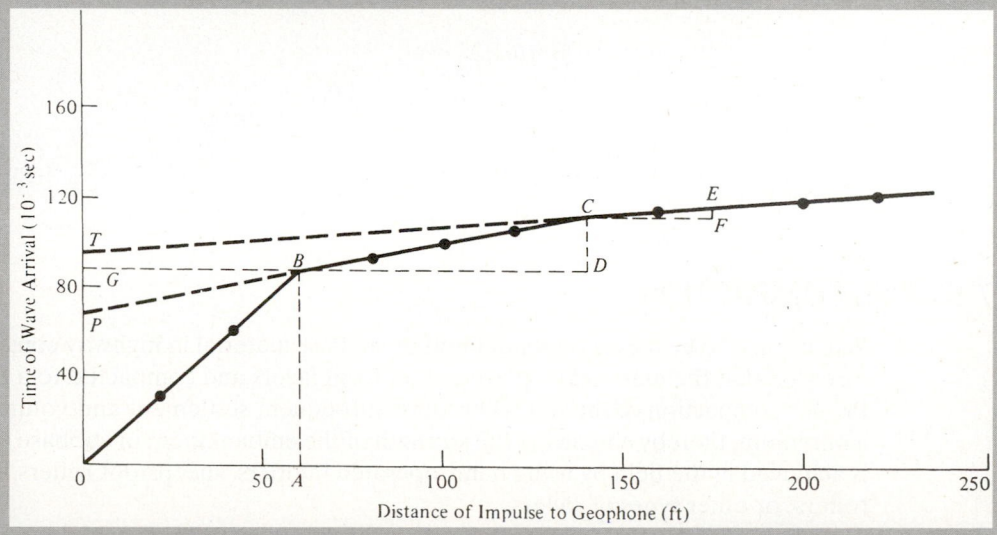

Figure 17.10   Solution for Example 17.6

From Table 17.5, the soil in the second stratum can be either alluvium, compacted clay, clayey gravel, or dense clayey sand.

$$u_3 = \frac{CF}{EF} = \frac{52.50 - 42}{(116 - 112) \times 10^{-3}} = 2625 \text{ m/sec}$$

The wave velocity in this material is high, which indicates some type of rock, such as sandstone or slate and shale.

$$H_1 = \frac{\overline{OP}u_1}{2 \cos \alpha}$$

$$\sin \alpha = \frac{u_1}{u_2} = \frac{681.8}{3333}$$

$$= 0.205, \ \alpha = 11.8°$$

$$\cos \alpha = 0.979$$

$$\overline{OP} = 68 \times 10^{-3} \text{ sec (from Figure 17.10)}$$

$$H_1 = \frac{(68 \times 10^{-3})681.8}{2 \times 0.979} = 7.10 \text{ m}$$

$$H_2 = \frac{\overline{PT}u_2}{2 \cos \beta}$$

$$\overline{PT} = 26 \times 10^{-3} \text{ sec (from Figure 17.10)}$$

$$\sin \beta = \frac{u_2}{u_3} = \frac{3333}{8750} = 0.381$$

$$\beta = 22.39°$$

$$\cos \beta = 0.925$$

$$H_2 = \frac{26 \times 10^{-3} \times 3333}{2 \times 0.925} = 14.05 \text{ m}$$

17.5 SOIL COMPACTION

When soil is to be used as embankment or subbase material in highway construction, it is essential that the material be placed in uniform layers and compacted to a high density. Proper compaction of the soil will reduce subsequent settlement and volume change to a minimum, thereby enhancing the strength of the embankment or subbase. Compaction is achieved in the field by using hand-operated tampers, sheepsfoot rollers, rubber-tired rollers, or other types of rollers.

The strength of the compacted soil is directly related to the maximum dry density achieved through compaction. The relationship between dry density and moisture content for practically all soils takes the form shown in Figure 17.11. It can be seen from this relationship that for a given compactive effort, the dry density attained is low at low moisture contents. The dry density increases with increase in moisture content to a maximum value when an optimum moisture content is reached. Further increase in moisture content results in a decrease in the dry density attained. This phenomenon is due to the effect of moisture on the soil particles. At low moisture content, the soil particles are not lubricated, and friction between adjacent particles prevents the densification of the particles. As the moisture content is increased, larger films of water develop on the particles, making the soil more plastic and easier for the particles to be moved and densified. When the optimum moisture content is reached, however, the maximum practical degree of saturation (where $S < 100\%$) is attained. The degree of saturation at the optimum moisture content cannot be increased by further compaction because of the presence of entrapped air in the void spaces and around the particles. Further addition of moisture therefore results in the voids being overfilled with water, with no accompanying reduction in the air. The soil particles are separated, resulting in a reduction in the dry density. The zero-air void curve shown in Figure 17.11 is the theoretical moisture-density curve for a saturated soil and zero-air voids, where the degree of saturation is 100 percent. This curve usually is not attained in the field, since the zero-air void cannot be attained as explained earlier. Points on the curve may be calculated from Eq. 17.23 as

$$\gamma_d = \frac{\gamma_w G_s}{1 + wG_s}$$

(17.23)

where

γ_w = density of water (g/cc)
G_s = specific gravity of soil particles
w = moisture content of soil
γ_d = dry density of soil (g/cc)

Although this curve is theoretical, the distance between it and the test moisture-density curve is of importance, since this distance is an indication of the amount of air voids remaining in the soil at different moisture contents. The farther away a point on

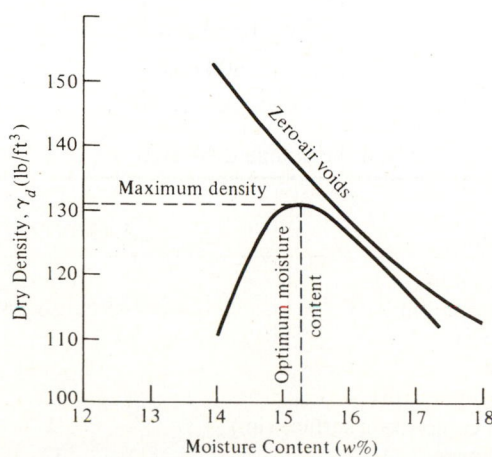

Figure 17.11   Typical Moisture-Density Relationship for Soils

the moisture-density curve is from the zero-air void curve, the more air voids remain in the soil and the higher the likelihood of expansion or swelling if the soil is subjected to flooding. It is therefore better to compact at the higher moisture content—that is, the wet side of optimum moisture content—if a given dry density other than the optimum is required.

17.5.1 Optimum Moisture Content

The determination of the optimum moisture content of any soil to be used as embankment or subgrade material is necessary before any field work is commenced. Most highway agencies now use dynamic or impact tests to determine the optimum moisture content and maximum dry density. In each of these tests, samples of the soil to be tested are compacted in layers to fill a specified size mold. Compacting effort is obtained by dropping a hammer of known weight and dimensions from a specified height a specified number of times for each layer. The moisture content of the compacted material is then obtained and the dry density determined from the measured weight of the compacted soil and the known volume of the mold. The soil is then broken down or another sample of the same soil is obtained. The moisture content is then increased and the test repeated. The process is repeated until a reduction in the density is observed. Usually a minimum of four or five individual compaction tests are required. A plot of dry density versus moisture content is then drawn from which the optimum moisture content is obtained. The two types of tests commonly used are the standard AASHTO or the modified AASHTO tests.

Table 17.6 shows details for the standard AASHTO (designated T99) and the modified AASHTO (designated T180) tests. Most transportation agencies use the standard AASHTO test.

Effect of Compacting Effort

Compacting effort is a measure of the mechanical energy per unit volume imposed on the soil mass during compaction. In the laboratory, it is given in units of cm-kg/mm^3 or cm-kg/cm^3, whereas in the field it is given in terms of the number of passes of a roller of known weight and type. The compactive effort in the standard AASHTO test, for example, is approximately calculated as

$$\frac{(2.5 \text{ kg})(0.30 \text{ m})(3)(25)}{1/900 \text{ cm}^3} = 5.5 \text{ cm-kg/cm}^3$$

Table 17.6   Details of the Standard AASHTO and Modified AASHTO Tests

Test Details	Standard AASHTO (T99)	Modified AASHTO (T180)
Diameter of mold (in.)	4 or 6	4 or 6
Height of sample (in.)	5 cut to 4.58	5 cut to 4.58
Number of lifts	3	5
Blows per lift	25 or 56	25 or 56
Weight of hammer (lb)	5.5	10
Diameter of compacting surface (in.)	2	2
Free-fall distance (in.)	12	18
Net volume (ft^3)	1/30 or 1/13.33	1/30 or 1/13.33

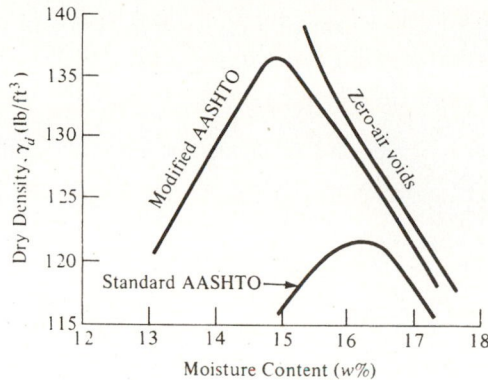

Figure 17.12   Effect of Compactive Effort in Dry Density

Note that the optimum moisture content and maximum dry density attained depend on the compactive effort used. Figure 17.12 shows that as the compactive effort increases, so does the maximum dry density. Also, the compactive effort required to obtain a given density increases as the moisture content of the soil decreases.

Example 17.7   Determining Maximum Dry Density and Optimum Moisture Content

The table shows results obtained from a standard AASHTO compaction test on six samples, 100 mm in diameter, of a soil to be used as fill for a highway. Determine the maximum dry density and the optimum moisture content of the soil.

Sample No.	Weight Compacted Soil, W (g)	Moisture Content, w (%)
1	1851	4.0
2	1953	6.1
3	2047	7.8
4	2082	10.1
5	2033	12.1
6	1989	14.0

Solution: Since we are using the standard AASHTO test, 100 mm in diameter, the volume of each sample is 1/900 cc. The dry densities are calculated as shown.

Sample No.	Bulk Density, γ 30W (g/cc)	Moisture Content, w (%)	Dry Density, γ_d g/cc $\left(\dfrac{\gamma}{1+w}\right)$
1	2.057	4.0	1.977
2	2.170	6.1	2.045
3	2.274	7.8	2.109
4	2.313	10.1	2.101
5	2.259	12.0	2.017
6	2.210	14.0	1.938

Figure 17.13 shows the plot of dry density versus moisture content, from which it is determined that maximum dry density is 2.109 g/cc and the optimum moisture content is 8 percent.

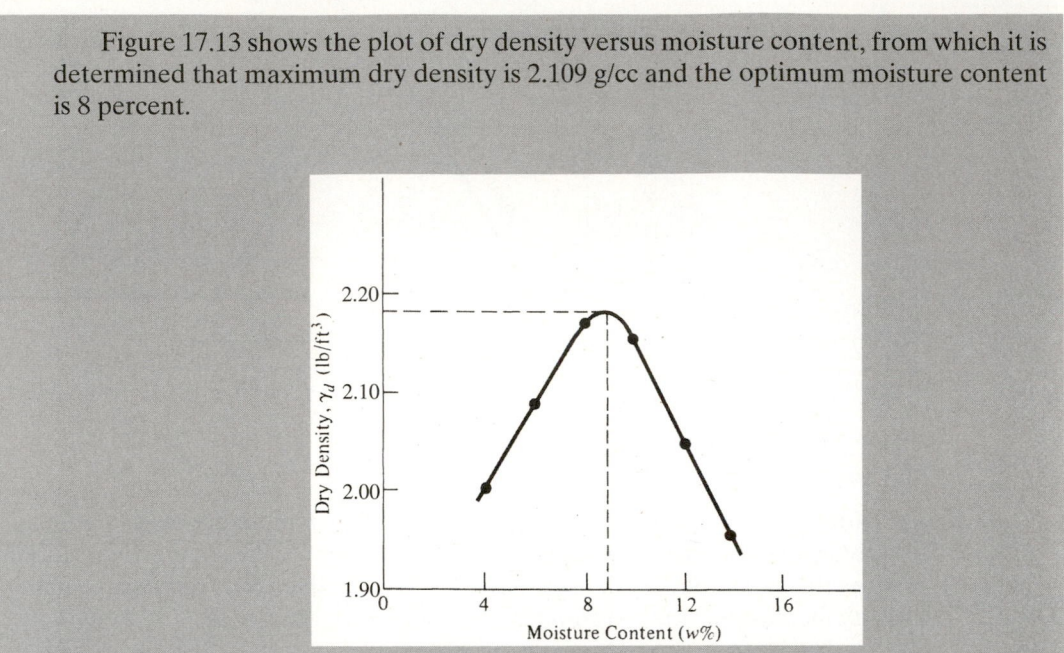

Figure 17.13   Moisture-Density Relationship for Example 17.7

17.5.2 Field Compaction Procedures and Equipment

A brief description of the field compaction procedures and the equipment used for embankment compaction is useful at this point, to show how the theory of compaction is applied in the field.

Field Compaction Procedures

The first step in the construction of a highway embankment is the identification and selection of a suitable material. This is done by obtaining samples from economically feasible borrow pits or borrow areas and testing them in the laboratory to determine the group of each. It has been shown earlier that, based on the AASHTO system of classification, materials classified as A-1, A-2-4, A-2-5, and A-3 are usually suitable embankment materials. In cases where it is necessary to use materials in other groups, special consideration should be given to design and construction. For example, soils in groups A-4 and A-6 can be used for embankment construction if the embankment height is low, the field compaction process is carefully controlled, and the embankment is located where the moisture content is not expected to exceed that at which the construction was undertaken. A factor that also significantly influences the selection of any material is whether that material can be economically transported to the construction site. Having identified suitable materials, their optimum moisture contents and maximum dry densities are determined.

Embankment Formation.   Highway embankments are formed by spreading thin layers of uniform thickness of the material and compacting each layer at or near the optimum moisture content. End dumping of the material from trucks is not recommended. The process of constructing one layer at a time facilitates obtaining uniform strength and moisture content in the embankment. End dumping or compaction of thick layers may result in variable strengths within the embankment, which could lead to differential settlement between adjacent areas.

Most states stipulate a thickness of 15 to 30 cm for each layer, although the thickness may be increased to 61 cm when the lower portion of an embankment consists mainly of large boulders.

All transportation agencies have their own requirements for the minimum density in the field. Some of these are based on the AASHTO specifications for transportation materials. Table 17.7 gives commonly used relative density values for different embankment heights. The relative density is given as a percentage of the maximum dry density obtained from the standard AASHTO (T99) test. Some agencies base their requirements on the maximum dry density obtained in the laboratory. For example, when the maximum dry density obtained in the laboratory is less than 1.65 g/cc, the required field density is 100 percent of the laboratory density. When the maximum dry density obtained in the laboratory is 1.65 g/cc or greater but less than 1.98 g/cc, 95 percent is required in the field, and so forth. The former practice of specifying the number of passes for different types of compacting equipment is not widely used at the present time.

Some transportation agencies also have specifications for the moisture content to be used during compaction. These specifications are usually given in general terms, although limits above and below the optimum moisture content have been given.

Control of Embankment Construction

The construction control of an embankment entails frequent and regular checks of the dry density and the moisture content of materials being compacted. The bulk density is obtained directly from measurements obtained in the field, and the dry density is then calculated from the bulk density and the moisture content. The laboratory moisture-density curve is then used to determine whether the dry density obtained in the field is in accordance with the laboratory results for the compactive effort used. These tests are conducted by using either a destructive method or a nondestructive method.

Table 17.7   Commonly Used Minimum Requirements for Compaction of Embankments and Subgrades

AASHTO Class of Soil	Minimum Relative Density		
	Embankments		
	Height Less Than 15 m	Height Greater Than 15 m	Subgrade
A-1, A-3	≥ 95	≥ 95	100
A-2-4, A-2-5	≥ 95	≥ 95	100
A-2-6, A-2-7	> 95	—[a]	≥ 95[b]
A-4, A-5, A-6, A-7	> 95	—[a]	≥ 95[b]

[a]Use of these materials requires special attention to design and construction.
[b]Compaction at 95 percent of T99 moisture content.

Destructive Methods.   In determining the bulk density by the destructive method, a cylindrical hole of about a 10 cm. diameter and a depth equal to that of the layer is excavated. The material obtained from the hole is immediately sealed in a container. Care should be taken not to lose any of the excavated material. The total weight of the excavated material is obtained, usually in the field laboratory, and the moisture content determined. The compacted volume of the excavated material is then measured by determining the volume of the excavated hole.

The moisture content is determined by either rapidly drying the soil in a field oven or by facilitating evaporation of the moisture by adding some volatile solvent material, such as alcohol, and igniting it. The volume of the excavated hole may be obtained by one of three methods: sand replacement, oil, or balloon. In the sand-replacement method, the excavated hole is carefully filled with standard sand from a jar originally filled with the standard sand. The jar can be opened and closed by a valve. When the hole is completely filled with sand, the valve is closed, and the weight of the remaining sand in the jar is determined. The weight of the quantity of sand used to fill the excavated hole is then obtained by subtracting the weight of the sand remaining in the jar from the weight of the sand required to fill the jar. The volume of the quantity of standard sand used to fill the excavated hole (that is, volume of hole) is then obtained from a previously established relationship between the weight and volume of the standard sand.

In the oil method, the volume of the hole is obtained by filling the excavated hole with a heavy oil of known specific gravity.

In the balloon method, a balloon is placed in the excavated hole and then filled with water. The volume of water required to fill the hole is the volume of the excavated hole.

The destructive methods are all subject to errors. For example, in the sand-replacement method, adjacent vibration will increase the density of the sand in the excavated hole, thereby indicating a larger volume hole. Large errors in the volume of the hole will be obtained if the balloon method is used in holes having uneven walls, and large errors may be obtained if the heavy oil method is used in coarse sand or gravel material.

Nondestructive Method.   The nondestructive method involves the direct measurement of the *in situ* density and moisture content of the compacted soil, using nuclear equipment. The density is obtained by measuring the scatter of gamma radiation by the soil particles since the amount of scatter is proportional to the bulk density of the soil. A calibration curve for the particular equipment is then used to determine the bulk density of the soil. A plot of the amount of scatter in materials of known density measured by the equipment versus the density of the materials is generally used in the calibration curve. The moisture content is also obtained by measuring the scatter of neutrons emitted in the soil. This scatter is due mainly to the presence of hydrogen atoms. The assumption is made that most of the hydrogen is in the form of water, which allows the amount of scatter to be related to the amount of water in the soil. The moisture content is then obtained directly from a calibrated gauge.

One advantage of the nondestructive method is that results are obtained speedily, which is essential if corrective actions are necessary. Another advantage is that more tests can be carried out, which facilitates the use of statistical methods in the control process. The main disadvantages are that a relatively high capital expenditure is required to obtain the equipment, and that field personnel are exposed to dangerous radioactive material, making it imperative that strict safety standards be enforced when nuclear equipment is used.

Field Compaction Equipment

Compaction equipment used in the field can be divided into two main categories. The first category includes the equipment used for spreading the material to the desired layer or lift thickness, and the second category includes the equipment used to compact each layer of material.

Spreading Equipment.   Spreading of the material to the required thickness is done by bulldozers and motor graders. Several types and sizes of graders and dozers are now available on the market. The equipment used for any specific project will depend on the size of the project. A typical motor grader is shown in Figure 17.14.

Compacting Equipment.   Rollers are used for field compaction and apply either a vibrating force or an impact force on the soil. The type of roller used for any particular job depends on the type of soil to be compacted.

A smooth wheel or drum roller applies contact pressure of up to 380 kN/m^2. This type of roller is generally used for finish rolling of subgrade material and can be used for all types of soil material except rocky soils. Figure 17.15a shows a typical smooth wheel roller. The rubber-tired roller is another type of contact roller, consisting of a heavily loaded wagon with rows of 3 to 6 tires placed close to each other. The pressure in the tires may be up to 690 kN/m^2. Rubber-tired rollers are used for both granular and cohesive materials. Figure 17.15b shows a typical rubber-tired roller.

One of the most frequently used rollers is the sheepsfoot. This roller has a drum wheel that can be filled with water. The drum wheel has several protrusions, which may be round or rectangular in shape, ranging from 31.5 to 75 cm^2 in area. The protrusions penetrate the loose soil and compact from the bottom to the top of each layer of soil, as the number of passes increases. Contact pressures ranging from 1380 to 6900 kN/m^2 can be

Figure 17.14   Typical Motor Grader

SOURCE: Robert J. Beyers II / Shutterstock.com

(a) Smooth wheel roller

(b) Rubber-tired roller

Figure 17.15   Typical Smooth Wheel and Rubber-Tired Rollers

SOURCE: (a) Michael Dechev / Shutterstock.com, (b) Vadim Ratnikov / Shutterstock.com

Figure 17.16   Typical Sheepsfoot Roller

SOURCE: Anatoliy Kosolapov / Shutterstock.com

obtained from sheepsfoot rollers, depending on the size of the drum and whether or not it is filled with water. The sheepsfoot roller is used mainly for cohesive soils. Figure 17.16 shows a typical sheepsfoot roller.

Tamping foot rollers are similar to sheepsfoot rollers in that they also have protrusions that are used to obtain high contact pressures, ranging from 1380 to 8280 kN/m^2. The feet of the tamping foot rollers are specially hinged to obtain a kneading action while compacting the soil. As with sheepsfoot rollers, tamping foot rollers compact from the bottom of the soil layer. Tamping foot rollers are used mainly for compacting fine-grained soils.

The smooth wheel and tamping foot rollers can be altered to vibrating rollers by attaching a vertical vibrator to the drum, producing a vibrating effect that makes the smooth wheel and tamping foot rollers more effective on granular soils. Vibrating plates and hammers are also available for use in areas where the larger drum rollers cannot be used.

17.6 SPECIAL SOIL TESTS FOR PAVEMENT DESIGN

Apart from the tests discussed so far, there are a few special soil tests that are sometimes undertaken to determine the strength or supporting value of a given soil if used as a subgrade or subbase material. The results obtained from these tests are used individually in the design of some pavements, depending on the pavement design method used (see Chapter 19). The two most commonly used tests under this category are the California Bearing Ratio Test and Hveem Stabilometer Test.

17.6.1 California Bearing Ratio (CBR) Test

This test is commonly known as the CBR test and involves the determination of the load-deformation curve of the soil in the laboratory using the standard CBR testing equipment shown in Figure 17.17. It was originally developed by the California Division of Highways prior to World War II and was used in the design of some highway pavements.

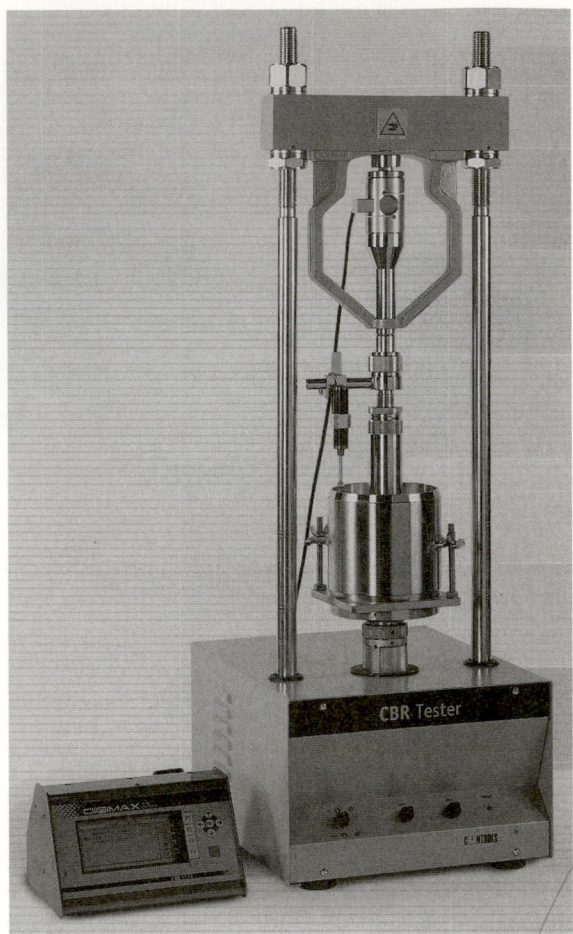

Figure 17.17   CBR Testing Equipment

SOURCE :   CONTROLS S.R.L. Used with permission.

The test is conducted on samples of soil compacted to required standards and immersed in water for four days, during which time the samples are loaded with a surcharge that simulates the estimated weight of pavement material the soil will support. The objective of the test is to determine the relative strength of a soil with respect to crushed rock, which is considered an excellent coarse base material. This is obtained by conducting a penetration test on the samples still carrying the simulated load and using standard CBR equipment. The CBR is defined as

$$\text{CBR} = \frac{(\text{unit load for 0.1 piston penetration in test specimen})(\text{kN}/\text{m}^2)}{(\text{unit load for 0.1 piston penetration in standard crushed roack})(\text{kN}/\text{m}^2)} \quad (17.24)$$

The unit load for 0.1 piston in standard crushed rock is usually taken as 6890 kN/m², which gives the CBR as

$$\text{CBR} = \frac{(\text{unit load for 0.1 piston penetration in test sample})}{6890} \times 100 \quad (17.25)$$

The test is fully described in *Standard Method of Test for the California Bearing Ratio* by AASHTO and is standardized under the AASHTO designation of T193.

The main criticism of the CBR test is that it does not simulate correctly the shearing forces imposed on subbase and subgrade materials as they support highway pavements. For example, it is possible to obtain a relatively high CBR value for a soil containing rough or angular coarse material and some amount of troublesome clay if the coarse material resists penetration of the piston by keeping together in the mold. When such a material is used in highway construction, however, the performance of the soil may be poor due to the lubrication of the soil mass by the clay, which reduces the shearing strength of the soil mass.

17.6.2 Hveem Stabilometer Test

This test is used to determine the resistance value R of the soil to the horizontal pressure obtained by imposing a vertical stress of 1100 kPa. on a sample of the soil. The value of R may then be used to determine the pavement thickness above the soil to carry the estimated traffic load (see Chapter 19). It now has been modified and made suitable for subgrade materials and is designated as T190 by AASHTO and fully described in *Standard Specifications for Transportation Materials and Methods of Sampling and Testing*. A brief description of the test is given here. The procedure consists of three phases: determination of the *exudation pressure*, determination of the *expansion pressure*, and determination of the *resistance value R* (stabilometer test). In order to carry out the test, four cylindrical specimens of 10.0-cm diameter and 6.25 cm height are prepared at different moisture contents by compacting samples of the soils in steel molds. The compaction is achieved by kneading the soil.

Exudation Pressure

Exudation pressure is the compressive stress that will exude water from the compacted specimen. Each specimen is pressed in the steel mold by applying a vertical load until water exudes from the soil. The base of the exudation equipment has several electrical circuits wired into it in parallel, and the exuded water completes these circuits. The pressure that exudes enough water to activate five or six of these circuits is the exudation pressure. Several tests in California have indicated that soils supporting highway pavements will exude moisture under pressure of about 2067 kPa. The moisture content of the stabilometer test (R value) samples is therefore set to that moisture content that exudes water at a pressure of 2067 kPa.

Expansion Pressure

The *expansion pressure* indicates the load and therefore the thickness of material required above the soil to prevent any swelling if the soil is inundated with water when used as a subgrade material. It is obtained by first allowing the test specimen to rebound after the exudation pressure has been determined, then by placing the specimen in a covered mold for about 30 min. A perforated brass plate is placed on each sample in the steel mold and the mold is placed in the expansion-pressure device and covered with water. The samples are then left to stand for a period of 16 to 24 h, during which time the expansion of the soil is measured. The expansion pressure is obtained as the product of the spring constant of the steel bar of the expansion-pressure device apparatus and expansion shown by the deflection gauge.

Resistance Value, R

A schematic of the Hveem stabilometer used in this phase of the test is shown in Figure 17.18. At the completion of the expansion test, the specimen is put into a flexible sleeve and placed in the stabilometer as shown in the figure. Vertical pressure is applied gradually on the specimen at a speed of 0.127 cm/min until a pressure of 1100 kPa is attained. The corresponding horizontal pressure is immediately recorded. To correct for any distortion of the results due to the surface roughness of the sample, the penetration of the flexible diaphragm into the sample is measured. This is done by reducing the vertical load on the specimen by half and also reducing the horizontal pressure to 1.46 g/m², using the screw-type pump. The number of turns of the pump required to increase the horizontal pressure to 29.26 g/m² is then recorded. The soil's resistance value is given as

$$R = 100 - \frac{100}{\dfrac{2.5}{D}\left(\dfrac{P_v}{P_h} - 1\right) + 1}$$

(17.26)

where

R = resistance value
P_v = vertical pressure (1100 kPa)
P_h = horizontal pressure at P_v at 1100 kPa (kPa)
D = number of turns of displacement pump

17.7 FROST ACTION IN SOILS

When the ambient temperature falls below freezing for several days, it is quite likely that the water in soil pores will freeze. Since the volume of water increases by about 10 percent when it freezes, the first problem is the increase in volume of the soil. The second problem is that the freezing can cause ice crystals and lenses that are several centimeters thick to form in the soil. These two problems can result in heaving of the subgrade (frost heave), which may result in significant structural damage to the pavement.

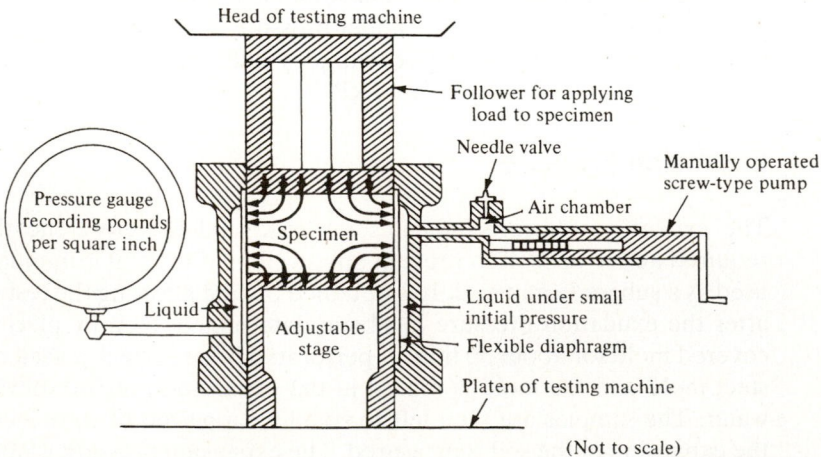

Figure 17.18 Hveem Stabilometer

In addition, the ice lenses melt during the spring (spring thaw), resulting in a considerable increase in the water content of the soil. This increase in water significantly reduces the strength of the soil, causing structural damage of the highway pavement known as "spring break-up."

In general, three conditions must exist for severe frost action to occur:

1. Ambient temperature must be lower than freezing for several days.
2. The shallow water table that provides capillary water to the frost line must be available.
3. The soil must be susceptible to frost action.

The first condition is a natural phenomenon and cannot be controlled by humans. Frost action therefore will be more common in cold areas than in warm areas if all other conditions are the same. The second condition requires that the ground-water table be within the height of the capillary rise, so that water will be continuously fed to the growing ice lenses. The third condition requires that the soil material be of such quality that relatively high capillary pressures can be developed, but at the same time that the flow of water through its pores is restricted.

Granular soils are therefore not susceptible to frost action because they have a relatively high coefficient of permeability. Clay soils also are not highly susceptible to frost action because they have very low permeability, so not enough water can flow during a freezing period to allow the formulation of ice lenses. Sandy or silty clays or cracked clay soils near the surface, however, may be susceptible to frost action. Silty soils are most susceptible to frost action. It has been determined that 0.02 mm is the critical grain size for frost susceptibility. For example, gravels with 5 percent of 0.02 mm particles are in general susceptible to frost action, whereas well-graded soils with only 3 percent by weight of their material finer than 0.02 mm are susceptible, and fairly uniform soils must contain at least 10 percent of 0.02 mm particles to be frost-susceptible. Soils with less than 1 percent of their material finer than the critical size are rarely affected by frost action.

Current measures taken to prevent frost action, as discussed in Chapter 16, include removing frost-susceptible soils to the depth of the frost line and replacing them with gravel material, lowering the water table by installing adequate drainage facilities, using impervious membranes or chemical additives, and restricting truck traffic on some roads during the spring thaw.

17.8 SUMMARY

Selection of suitable soils to be used as the foundation for the highway pavement surface is of primary importance in the design and construction of any highway. Use of unsuitable material will often result in premature failure of the pavement surface and reduction of the ability of the pavement to carry the design traffic load. Chapters 19 and 20 will show that the types of material used for the base, subbase, and/or subgrade of a highway significantly influence the depth of these materials used and also the thickness of the pavement surface.

It is therefore important that transportation engineers involved in the design and/or maintenance of highway pavements be familiar with the engineering properties of soils and the procedures through which the suitability of any soil for highway construction can be determined. A summary of some of the current procedures and techniques used for soil testing and identification is presented in this chapter. The techniques presented are those currently used in highway pavement design.

One of the most important tasks in highway pavement construction is the control of embankment compaction, and discussion of the different methods used in the control of embankment compaction is presented. A brief discussion of the different types of equipment used in the compaction of embankments is also presented to cover all aspects of embankment compaction.

PROBLEMS

17-1   Determine the void ratio of a soil if its bulk density is 2.06 g/cc and it has a moisture content of 24 percent. The specific gravity of the soil particles is 2.75. Also, determine its dry density and degree of saturation.

17-2   A soil has a bulk density of 2.29 g/cc and a dry density of 2.01 g/cc, and the specific gravity of the soil particles is 2.70. Determine: (a) moisture content, (b) degree of saturation, (c) void ratio, and (d) porosity.

17-3   The weight of a sample of saturated soil before drying is 1.246 kg and after drying is 0.934 kg. If the specific gravity of the soil particles is 2.6, determine: (a) moisture content, (b) void ratio, (c) porosity, (d) bulk density, and (e) dry density.

17-4   A moist soil has a moisture content of 12.5 percent, weighs 19.22 kg, and occupies a volume of 9450 cc. The specific gravity of the soil particles is 2.6. Find: (a) bulk density, (b) dry density, (c) void ratio, (d) porosity, (e) degree of saturation, and (f) volume occupied by water (ft^3).

17-5   The moist weight of 4050 cc of soil is 8.277 kg. If the moisture content is 17 percent and the specific gravity of soil solids is 2.62, find the following: (a) bulk density, (b) dry density, (c) void ratio, (d) porosity, (e) degree of saturation, and (f) volume occupied by water (cc).

17-6   The moist weight of 5130 cc of soil is 9.612 kg. If the moisture content is 18 percent and the specific gravity of soil solids is 2.55, find the following: (a) bulk density, (b) dry density, (c) void ratio, (d) porosity, (e) degree of saturation, and (f) volume occupied by water (cc).

17-7   A liquid limit test conducted in the laboratory on a sample of soil gave the following results. Determine the liquid limit of this soil from a plot of the flow curve.

Number of Blows (N)	Moisture Content (%)
20	45.0
28	43.6
30	43.2
35	42.8
40	42.0

17-8   A plastic limit test for a soil showed that moisture content was 19.2 percent. Data from a liquid limit test were as follows:

Number of Blows (N)	Moisture Content (%)
14	42.0
19	40.8
27	39.1

(a) Draw the flow curve and obtain the liquid limit.
(b) What is the plasticity index of the soil?

17-9   A plastic limit test for a soil showed that moisture content was 18.0 percent. Data from a liquid limit test were as follows:

Number of Blows (N)	Moisture Content (%)
16	39.8
21	39.2
26	38.4
30	36.9

(a) Draw the flow curve and obtain the liquid limit.
(b) What is the plasticity index of the soil?

17-10   The following results were obtained by a mechanical analysis. Classify the soil using the AASHTO classification system and give the group index.

Sieve Analysis, % Finer				
No. 10	No. 40	No. 200	Liquid Limit	Plastic Limit
98	81	38	42	23

17-11   The following results were obtained by a mechanical analysis. Classify the soil using the AASHTO classification system and give the group index.

Sieve Analysis, % Finer				
No. 10	No. 40	No. 200	Liquid Limit	Plastic Limit
84	58	8	—	N.P.

17-12   The following results were obtained by a mechanical analysis. Classify the soil using the AASHTO classification system and give the group index.

Sieve Analysis, % Finer				
No. 10	No. 40	No. 200	Liquid Limit	Plastic Limit
99	85	71	55	21

17-13   The following results were obtained by a mechanical sieve analysis. Classify the soil using the Unified Soil Classification System (USCS).

Sieve Analysis, % Passing by Weight				
No. 4	No. 40	No. 200	Liquid Limit	Plastic Limit
30	40	30	33	12

17-14   The following results were obtained by a mechanical sieve analysis. Classify the soil using the Unified Soil Classification System (USCS).

Sieve Analysis, % Passing by Weight				
No. 4	No. 40	No. 200	Liquid Limit	Plastic Limit
4	44	52	29	11

17-15   The following results were obtained by a mechanical sieve analysis. Classify the soil using the Unified Soil Classification System (USCS).

Sieve Analysis, % Passing by Weight				
No. 4	No. 40	No. 200	Liquid Limit	Plastic Limit
11	24	65	44	23

17-16   The following results were obtained by a mechanical sieve analysis. Classify the soil using the Unified Soil Classification System (USCS).

Sieve Analysis, % Passing by Weight				
No. 4	No. 40	No. 200	Liquid Limit	Plastic Limit
24	72	4	26	8

17-17   Following are the results of a sieve analysis:

Sieve No.	Percent Finer
4	100
10	92
20	82
40	67
60	58
80	38
100	22
200	6
Pan	—

(a) Plot the grain-size distribution curve for this sample.
(b) Determine D_{10}, D_{30}, and D_{60}.
(c) Calculate the uniformity coefficient, C_u.

17-18   Following are the results of a sieve analysis:

Sieve No.	Percent Finer
4	100
10	90
20	80
40	70
60	40

Sieve No.	Percent Finer
80	29
100	19
200	10
Pan	—

(a) Plot the grain-size distribution curve for this sample.
(b) Determine D_{10}, D_{30}, and D_{60}.
(c) Calculate the uniformity coefficient, C_u.
(d) Calculate the coefficient of gradation, C_c.

17-19   Following are the results of a sieve analysis:

Sieve No.	Percent Finer
4	99
10	87
20	77
40	65
60	32
80	23
100	14
200	7
Pan	—

(a) Plot the grain-size distribution curve for this sample.
(b) Determine D_{10}, D_{30}, and D_{60}.
(c) Calculate the uniformity coefficient, C_u.
(d) Calculate the coefficient of gradation, C_c.

17-20   The results of a compaction test on samples of soil that are to be used for an embankment on a highway project are listed below. Determine the maximum dry unit weight of compaction and the optimum moisture content.

Sample No.	Moisture Content	Bulk Density (g/cc)
1	4.8	2.164
2	7.5	2.322 g/cc
3	7.8	2.351 g/cc
4	8.9	2.345 g/cc
5	9.7	2.327 g/cc

17-21   The results of a compaction test on samples of soil that are to be used for an embankment on a highway project are listed below. Determine the maximum dry unit weight of compaction and the optimum moisture content.

Sample No.	Moisture Content	Bulk Density (g/cc)
1	15	1.837
2	16	1.90
3	19	1.93
4	22	1.923
5	25	1.894
6	26	1.875

17-22   The results obtained from a seismic study along a section of the centerline of a highway are shown below. Estimate the depths of the different strata of soil and suggest the type of soil in each stratum.

Distance of Impulse to Geophone (m)	Time for Wave Arrival (10⁻³ sec)
25	20
15	40
22.50	60
30	68
37	74
45	82
52.50	84
60	86
67.50	88
75	90

REFERENCES

Bowles, J. E., *Foundation Analysis and Design*, McGraw-Hill Publishing Company, New York, NY, 1988.

Das, B., *Principles of Foundation Engineering*, 4th ed., PWS Publishing, 1998.

Das, B., *Soil Mechanics Laboratory Manual*, Engineering Press, 1997.

Derucher, K. N. and G. P. Korflatis, *Materials for Civil and Highway Engineers*, Prentice Hall, Englewood Cliffs, NJ., 1994.

Fang, H.Y., *Foundation Engineering Handbook*, Routledge, Chapman & Hall, New York, NY, 1990.

Standard Specifications for Transportation Materials and Methods of Sampling and Testing, 31st ed., American Association of State Highway and Transportation Officials, Washington, D.C., 2011.

CHAPTER 18

Bituminous Materials

B ituminous materials are used widely all over the world in highway construction. These hydrocarbons are found in natural deposits or are obtained as a product of the distillation of crude petroleum. The bituminous materials used in highway construction are mainly asphalts.

All bituminous materials consist primarily of bitumen and have strong adhesive properties, with colors ranging from dark brown to black. They vary in consistency from liquid to solid; thus, they are divided into liquids, semisolids, and solids. The solid form is usually hard and brittle at normal temperatures but will flow when subjected to long, continuous loading.

The liquid form is obtained from the semisolid or solid forms by heating, dissolving in solvents, or breaking the material into minute particles and dispersing them in water with an emulsifier to form an asphalt emulsion.

This chapter presents a description of the different types of bituminous materials used in highway construction, the process by which they are obtained, and the tests required to determine those properties that are pertinent to highway engineering. It also includes a description of methods of mix design to obtain a paving material known as asphalt concrete.

CHAPTER OBJECTIVES:

- Become familiar with the sources of asphalt.
- Become familiar with the descriptions and uses of different bituminous binders.
- Learn about the properties of asphalt materials.
- Learn about the different tests of asphalt materials to ascertain their suitability for use in highway construction.
- Understand the *superior performing asphalt pavements (superpave)* procedure for specifying asphalt materials in asphalt concrete.

963

18.1 SOURCES OF ASPHALT

Asphalt is obtained from seeps or pools of natural deposits in different parts of the world or as a product of the distillation of crude petroleum.

18.1.1 Natural Deposits

Natural deposits of asphalt occur as either native asphalt or rock asphalt. The largest deposit of native asphalt is known to have existed in Iraq several thousand years ago. Native asphalts also have been found in Trinidad, Bermuda, and the La Brea asphalt pits in Los Angeles, California.

Native asphalt (after being softened with petroleum fluxes) was at one time used extensively as binders in highway construction. The properties of native asphalt vary from one deposit to another, particularly with respect to the amount of insoluble material the asphalt contains. The Trinidad deposit, for example, contains about 40 percent insoluble organic and inorganic materials, whereas the Bermuda material contains about 6 percent of such material.

Rock asphalt is a natural deposit of sandstone or limestone rocks filled with asphalt. Deposits have been found in California, Texas, Oklahoma, and Alabama. The amount of asphalt varies from one deposit to another and can be as low as 4.5 percent and as high as 18 percent. Rock asphalt can be used to surface roads after the mined or quarried material has been suitably processed. This process includes adding suitable mineral aggregates, asphalt binder, and oil, which facilitates the flowing of the material. Rock asphalt is not used widely because of its high transportation costs.

18.1.2 Petroleum Asphalt

The asphalt materials obtained from the distillation of petroleum are in the form of different types of asphalts, which include asphalt cements; cutbacks, which can be further classified into three types based on their curing time—slow-curing cutbacks, medium-curing cutbacks, rapid-curing cutbacks; and asphalt emulsions, which also exist in either anionic or cationic forms based on the charge of the bitumen droplets. They can also be further classified as slow-setting, medium-setting, and rapid-setting emulsions. The quantity of asphalt obtained from crude petroleum is dependent on the American Petroleum Institute (API) gravity of the petroleum. In general, large quantities of asphalt are obtained from crude petroleum with low API gravity. Before discussing the properties and uses of the different types of petroleum asphalt, we first describe the refining processes used to obtain them.

Refining Processes

The refining processes used to obtain petroleum asphalts can be divided into two main groups: *fractional distillation* and *destructive distillation* (cracking). The fractional distillation processes involve the separation of the different materials in the crude petroleum without significant changes in the chemical composition of each material. The destructive distillation processes involve the application of high temperature and pressure, resulting in chemical changes.

Fractional Distillation. The fractional distillation process removes the different volatile materials in the crude oil at successively higher temperatures until the petroleum asphalt is obtained as residue. Steam or a vacuum is used to gradually increase the temperature.

Steam distillation is a continuous flow process in which the crude petroleum is pumped through tube stills or stored in batches, and the temperature is increased gradually to facilitate the evaporation of different materials at different temperatures. Tube stills are more efficient than batches and are therefore preferred in modern refineries.

Immediately after increasing the temperature of the crude petroleum in the tube still, it is injected into a bubble tower, which consists of a vertical cylinder into which are built several trays or platforms stacked one above the other. The first separation of materials occurs in this tower. The lighter fractions of the evaporated materials collect on the top tray, and the heavier fractions collect in successive trays, with the heaviest residue containing asphalt remaining at the bottom of the distillation tower. The products obtained during this first phase of separation are gasoline, kerosene distillate, diesel fuel, lubricating oils, and the heavy residual material that contains the asphalt (see Figure 18.1). The various fractions

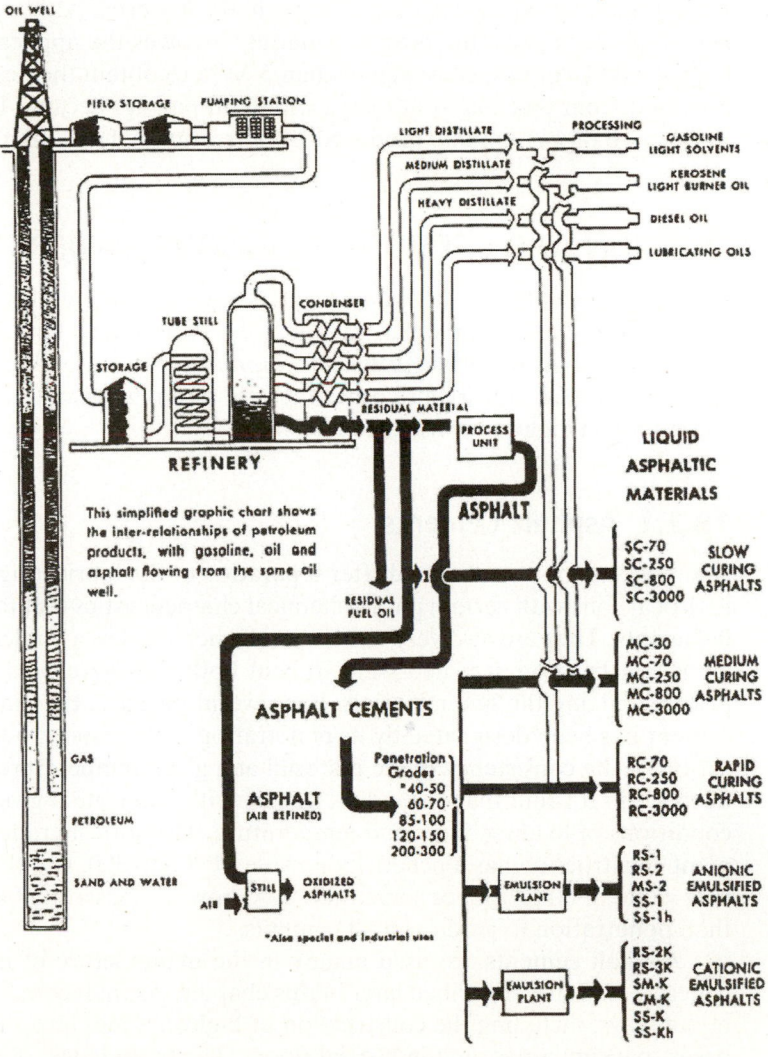

Figure 18.1   A Schematic Example of a Petroleum Distilling Plant

SOURCE: © 2007 Integrated Publishing, Inc. Navy Engine Mechanics CD-Rom - 14080 Equipment Operator, Advanced Petroleum Asphalt Flow Chart, Figure No.: 8-19, p. 183

collected are stored and refined further into specific grades of petroleum products. Note that a desired consistency of residue can be obtained by continuing the distillation process. Attainment of the desired consistency is checked by measuring the temperature of the residue or by observing the character of the distillate. The residue becomes harder the longer the distillation process is continued. Figure 18.1 shows an example of a petroleum distilling plant.

Further processing of the heavy residue obtained after the first separation will give asphalt cement of different penetration grades—or different grades of cutback—depending on the additional processing carried out. Emulsified asphalts also can be obtained, and a description of each of these different types of asphalt materials will be given later.

Destructive Distillation.   *Cracking processes* are used when larger amounts of the light fractions of materials (such as motor fuels) are required. Intense heat and high pressures are applied to produce chemical changes in the material. Although several specific methods of cracking exist, the process generally involves the application of temperatures as high as 600 C and pressure higher than 5 MPa to obtain the desired effect. The asphalt obtained from cracking is not used widely in paving, because it is more susceptible to weather changes than that produced from fractional distillation.

18.2 DESCRIPTION AND USES OF BITUMINOUS BINDERS

It is necessary at this point to describe the different bituminous binders and identify the type of construction for which each is used. Bituminous binders can be classified into three general groups: *asphalt cement, asphalt cutbacks*, and *emulsified asphalt*. Blown asphalt and road tars are also other types of bituminous material that now are not used commonly in highway construction.

18.2.1 Asphalt Cements

Asphalt cements are obtained after separation of the lubricating oils. They are semisolid hydrocarbons with certain physiochemical characteristics that make them good cementing agents. They are also very viscous, and when used as a binder for aggregates in pavement construction, it is necessary to heat both the aggregates and the asphalt cement prior to mixing the two materials. For several decades, the particular grade of asphalt cement has been designated by its penetration and viscosity, both of which give an indication of the consistency of the material at a given temperature. The penetration is the distance in 0.1 mm that a standard needle will penetrate a given sample under specific conditions of loading, time, and temperature. The softest grade used for highway pavement construction has a penetration value of 200 to 300, and the hardest has a penetration value of 60 to 70. For some time now, however, viscosity has been used more often than penetration to grade asphalt cements.

Asphalt cements are used mainly in the manufacture of hot-mix, hot-laid asphalt concrete, which is described later in this chapter. Asphalt concrete can be used in a variety of ways, including the construction of highways and airport pavement surfaces and bases, parking areas, and industrial floors. The specific use of a given sample depends on its grade. The procedures for determining the grades of asphalt cements through standard penetration and viscosity tests will be described in subsequent sections of this chapter.

18.2.2 Asphalt Cutbacks

The asphalt cutbacks are slow-curing asphalts, medium-curing cutback asphalts, and rapid-curing cutback asphalts. They are used mainly in cold-laid plant mixes, in road mixes (mixed-in-place), and as surface treatments.

Slow–Curing Asphalts

Slow-curing (SC) asphalts can be obtained directly as *slow-curing straight run asphalts* through the distillation of crude petroleum or as *slow-curing cutback asphalts* by "cutting back" asphalt cement with a heavy distillate, such as diesel oil. They have lower viscosities than asphalt cement and are very slow to harden. Slow-curing asphalts usually are designated as SC-70, SC-250, SC-800, or SC-3000, where the numbers relate to the approximate kinematic viscosity in centistokes at 60°C (140°F). Specifications for the use of these asphalts are no longer included in the American Association of State Highway and Transportation Officials (AASHTO) *Standard Specifications for Transportation Materials*.

Medium–Curing Cutback Asphalts

Medium-curing (MC) asphalts are produced by *fluxing*, or cutting back, the residual asphalt (usually 120 to 150 penetration) with light fuel oil or kerosene. The term *medium* refers to the medium volatility of the kerosene-type diluter used. Medium-curing cutback asphalts harden faster than slow-curing liquid asphalts, although consistencies of the different grades are similar to those of the slow-curing asphalts. However, the MC-30 is a unique grade in this series as it is very fluid and has no counterpart in the SC and RC series.

The fluidity of medium-curing asphalts depends on the amount of solvent in the material. MC-3000, for example, may have only 20 percent of the solvent by volume, whereas MC-70 may have up to 45 percent. These medium-curing asphalts can be used for the construction of pavement bases, surfaces, and surface treatments.

Rapid–Curing Cutback Asphalts

Rapid-curing (RC) cutback asphalts are produced by blending asphalt cement with a petroleum distillate that will evaporate easily, thereby facilitating a quick change from the liquid form at the time of application to the consistency of the original asphalt cement. Gasoline or naphtha generally is used as the solvent for this series of asphalts.

The grade of rapid-curing asphalt required dictates the amount of solvent to be added to the residual asphalt cement. For example, RC-3000 requires about 15 percent of distillate, whereas RC-70 requires about 40 percent. These grades of asphalt can be used for jobs similar to those for which the MC series is used. Specifications for the use of these asphalts are given in AASHTO's *Standard Specifications for Transportation Materials*.

18.2.3 Emulsified Asphalts

Emulsified asphalts are produced by breaking asphalt cement, usually of 100 to 250 penetration range, into minute particles and dispersing them in water with an emulsifier. These minute particles have like-electrical charges and therefore do not coalesce. They remain in suspension in the liquid phase as long as the water does not evaporate or the emulsifier does not break. Asphalt emulsions therefore consist of asphalt, which makes up about 55 to 70 percent by weight, water, and an emulsifying agent, which in some cases may also contain a stabilizer.

Asphalt emulsions generally are classified as anionic, cationic, or nonionic. The first two types have electrical charges surrounding the particles, whereas the third type is neutral. Classification as anionic or cationic is based on the electrical charges that surround the asphalt particles. Emulsions containing negatively charged particles of asphalt are classified as anionic, and those having positively charged particles of asphalt are classified as cationic. The anionic and cationic asphalts generally are used in highway maintenance and construction, although it is likely that the nonionic asphalts may be used more frequently in the future as emulsion technology advances.

Each of these categories is further divided into three subgroups based on how rapidly the asphalt emulsion returns to the state of the original asphalt cement. These subgroups are rapid-setting (RS), medium-setting (MS), and slow-setting (SS). A cationic emulsion is identified by placing the letter "C" in front of the emulsion type; no letter is placed in front of anionic and nonionic emulsions. For example, CRS-2 denotes a cationic emulsion, and RS-2 denotes either an anionic or nonionic emulsion.

Emulsified asphalts are used in cold-laid plant mixes and road mixes (mixed-in-place) for several purposes, including the construction of highway pavement surfaces and bases and in surface treatments. Note, however, that since anionic emulsions contain negative charges, they are more effective in treating aggregates containing electropositive charges (such as limestone), whereas cationic emulsions are more effective with electronegative aggregates (such as those containing a high percentage of siliceous material). Also note that ordinary emulsions must be protected during very cold spells because they will break down if frozen. Three grades of high-float, medium-setting anionic emulsions designated as HFMS have been developed and are used mainly in cold and hot plant mixes and coarse aggregate seal coats. These high-float emulsions have one significant property: They can be laid in relatively thicker films without a high probability of runoff.

Specifications for the use of emulsified asphalts are given in AASHTO M140 and ASTM D977.

18.2.4 Air-Blown Asphalts

Air-blown asphalt is obtained by blowing air through the semisolid residue obtained during the latter stages of the distillation process. The process involves stopping the regular distillation while the residue is in the liquid form and then transferring it into a tank known as a *converter*. The material is maintained at a high temperature while air is blown through it. This is continued until the required properties are achieved. Blown asphalts are relatively stiff compared to other types of asphalts and can maintain a firm consistency at the maximum temperature normally experienced when exposed to the environment.

Blown asphalt generally is not used as a paving material. However, it is very useful as a roofing material, for automobile undercoating, and as a joint filler for concrete pavements. If a catalyst is added during the air-blowing process, the material obtained usually will maintain its plastic characteristics, even at temperatures much lower than that at which ordinary asphalt cement will become brittle. The elasticity of catalytically blown asphalt is similar to that of rubber, and it is used for canal lining.

18.2.5 Road Tars

Tars are obtained from the destructive distillation of such organic materials as coal. Their properties are significantly different from petroleum asphalts. In general, they are more susceptible to weather conditions than similar grades of asphalts, and they set more quickly when exposed to the atmosphere. Because tars now are used rarely for highway pavements, this text does not include further discussion of the subject.

18.3 PROPERTIES OF ASPHALT MATERIALS

The properties of asphalt materials pertinent to pavement construction can be classified into four main categories:

- Consistency
- Aging and temperature sustainability
- Rate of curing
- Resistance to water action

18.3.1 Consistency

The consistency properties of an asphalt material usually are considered under two conditions: (1) variation of consistency with temperature and (2) consistency at a specified temperature.

Variation of Consistency with Temperature

The consistency of any asphalt material changes as the temperature varies. The change in consistency of different asphalt materials may differ considerably even for the same amount of temperature change. For example, if a sample of blown semisolid asphalt and a sample of semisolid regular paving-grade asphalt with the same consistency at a given temperature are heated to a high enough temperature, the consistencies of the two materials will be different at the high temperatures, with the regular paving-grade asphalt being much softer than the blown asphalt. Further increase in temperature eventually will result in the liquefaction of the paving asphalt at a temperature much lower than that at which the blown asphalt liquefies. If these two asphalts then are cooled down gradually to about the freezing temperature of water, the blown asphalt will be much softer than the paving-grade asphalt. Thus, the consistency of the blown asphalt is affected less by temperature changes than the consistency of regular paving-grade asphalt. This property of asphalt materials is known as *temperature susceptibility*. The temperature susceptibility of a given asphalt depends on the crude oil from which the asphalt is obtained, although variation in temperature susceptibility of paving-grade asphalts from different crudes is not as high as that between regular paving-grade asphalt and blown asphalt.

Consistency at a Specified Temperature

As stated earlier, the consistency of an asphalt material will vary from solid to liquid depending on the temperature of the material. It is therefore essential that when the consistency of an asphalt material is given, the associated temperature should also be given. Different ways for measuring consistency are presented later in this chapter.

18.3.2 Aging and Temperature Sustainability

When asphaltic materials are exposed to environmental elements, natural deterioration gradually takes place, and the materials eventually lose their plasticity and become brittle. This change is caused primarily by chemical and physical reactions that take place in the material.

This natural deterioration of the asphalt material is known as *weathering*. For paving asphalt to act successfully as a binder, the weathering must be minimized as much as possible. The ability of an asphalt material to resist weathering is described as the *durability* of the material. Some of the factors that influence weathering are oxidation, volatilization, temperature, and exposed surface area. These factors are discussed briefly in the following sections.

Oxidation

Oxidation is the chemical reaction that takes place when the asphalt material is attacked by oxygen in the air. This chemical reaction causes gradual hardening (eventually permanent hardening) and considerable loss of the plastic characteristics of the material.

Volatilization

Volatilization is the evaporation of the lighter hydrocarbons from the asphalt material. The loss of these lighter hydrocarbons also causes the loss of the plastic characteristics of the asphalt material.

Temperature

It has been shown that temperature has a significant effect on the rate of oxidation and volatilization. The higher the temperature, the higher the rates of oxidation and volatilization. The relationship between temperature increase and increases in rates of oxidation and volatilization is not linear; however, the percentage increase in rate of oxidation and volatilization is usually much greater than the percentage increase in temperature that causes the increase in oxidation and volatilization. It has been postulated that the rate of organic and physical reactions in the asphalt material approximately doubles for each 10°C (50°F) increase in temperature.

Surface Area

The exposed surface of the material also influences its rate of oxidation and volatilization. There is a direct relationship between surface area and rate of oxygen absorption and loss due to evaporation in g/cm³/min. An inverse relationship, however, exists between volume and rate of oxidation and volatilization. This means that the rate of hardening is directly proportional to the ratio of the surface area to the volume.

This fact is taken into consideration when asphalt concrete mixes are designed for pavement construction in that the air voids are kept to the practicable minimum required for stability to reduce the area exposed to oxidation.

18.3.3 Rate of Curing

Curing is defined as the process through which an asphalt material increases its consistency as it loses solvent by evaporation.

Rate of Curing of Cutbacks

As discussed earlier, the rate of curing of any cutback asphalt material depends on the distillate used in the cutting-back process. This is an important characteristic of cutback materials, since the rate of curing indicates the time that should elapse before a cutback will attain a consistency that is thick enough for the binder to perform satisfactorily. The rate of curing is affected by both inherent and external factors. The important inherent factors are

- Volatility of the solvent
- Quantity of solvent in the cutback
- Consistency of the base material

The more volatile the solvent is, the faster it can evaporate from the asphalt material, and therefore, the higher the curing rate of the material. This is why gasoline and naphtha are used for rapid-curing cutbacks, whereas light fuel oil and kerosene are used for medium-curing cutbacks.

For any given type of solvent, the smaller the quantity used, the less time is required for it to evaporate, and therefore, the faster the asphalt material will cure. Also, the higher the penetration of the base asphalt, the longer it takes for the asphalt cutback to cure.

The important external factors that affect curing rate are

- Temperature
- Ratio of surface area to volume
- Wind velocity across exposed surface

These three external forces are related directly to the rate of curing in that the higher these factors are, the higher the rate of curing. Unfortunately, these factors cannot be controlled or predicted in the field, which makes it extremely difficult to predict the expected curing time. The curing rates of different asphalt materials usually are compared with the assumption that the external factors are held constant.

Rate of Curing for Emulsified Asphalts

The curing and adhesion characteristics of emulsions (anionic and cationic) used for pavement construction depend on the rate at which the water evaporates from the mixture. When weather conditions are favorable, the water is displaced relatively rapidly, and so curing progresses rapidly. When weather conditions include high humidity, low temperature, or rainfall immediately following the application of the emulsion, its ability to properly cure is affected adversely. Although the effect of surface and weather conditions on proper curing is more critical for anionic emulsions, favorable weather conditions are also required to obtain optimum results for cationic emulsions. A major advantage of cationic emulsions is that they release their water more readily.

18.3.4 Resistance to Water Action

When asphalt materials are used in pavement construction, it is important that the asphalt continues to adhere to the aggregates even with the presence of water. If this bond between the asphalt and the aggregates is lost, the asphalt will strip from the aggregates, resulting in the deterioration of the pavement. The asphalt therefore must sustain its ability to adhere to the aggregates even in the presence of water. In hot-mix, hot-laid asphalt concrete, where the aggregates are thoroughly dried before mixing, stripping does not normally occur and so no preventive action is usually taken. However, when water is added to a hot-mix, cold-laid asphalt concrete, commercial anti-strip additives usually are added to improve the asphalt's ability to adhere to the aggregates.

18.3.5 Temperature Effect on Volume of Asphaltic Materials

The volume of asphalt is affected by changes in temperature significantly. The volume increases with an increase in temperature and decreases with a decrease in temperature. The rate of change in volume is given as the coefficient of expansion, which is the volume change in a unit volume of the material for a unit change in temperature. Because of this variation of volume with temperature, the volumes of asphalt materials usually are given for a temperature of 60°F (15.6°C). Volumes measured at other temperatures are converted to the equivalent volumes at 60°F by using appropriate multiplication factors published by the ASTM in their Petroleum Measurement Tables (ASTM D-1250).

18.4 TESTS FOR ASPHALT MATERIALS

Several tests are conducted on asphalt materials to determine both their consistency and quality to ascertain whether materials used in highway construction meet the prescribed specifications. Some of these specifications given by AASHTO and ASTM have been referred to earlier, and some are listed in Tables 18.1 and 18.2. Standard specifications also have been published by the Asphalt Institute for the types of asphalts used in pavement construction. Procedures for selecting representative samples of asphalt for testing have been standardized and are given in MS-18 by the Asphalt Institute and in D140 by the ASTM.

Some of the tests used to identify the quality of a given sample of asphalt on the basis of the properties discussed are described briefly, followed by a brief description of some general tests.

18.4.1 Consistency Tests

The consistency of asphalt materials is important in pavement construction because the consistency at a specified temperature will indicate the grade of the material. It is important that the temperature at which the consistency is determined be specified, since temperature significantly affects consistency of asphalt materials. As stated earlier, asphalt materials can exist in either liquid, semisolid, or solid states. This wide range dictates the necessity for more than one method for determining the consistency of asphalt materials. The property used to describe the consistency of asphalt materials in the liquid state is the viscosity, which can be determined by conducting either the Saybolt Furol viscosity test or the kinematic viscosity test. Tests used for asphalt materials in the semisolid and solid states include the penetration test and the float test. The ring-and-ball softening point test, which is not used often in highway specifications, also may be used for blown asphalt.

Saybolt Furol Viscosity Test

Figure 18.2 shows an example of the Saybolt Furol Viscometer. The principal part of the equipment is the standard viscometer tube, which is 7.5 cm long and about 2.5 cm in diameter. An orifice of specified shape and dimensions is provided at the bottom of the tube. The orifice is closed with a stopper, and the tube is filled with a quantity of the material to be tested. The standard tube then is placed in a larger oil or water bath and fitted with an electric heater and a stirring device. The material in the tube is brought to the specified temperature by heating the bath. Immediately upon reaching the prescribed temperature, the stopper is removed, and the time in seconds for exactly 60 ml of the asphalt material to flow through the orifice is recorded. This time is the Saybolt Furol viscosity in units of seconds at the specified temperature. Temperatures at which asphalt

Table 18.1 Specifications for Rapid–Curing Cutback Asphalts

	RC-70		RC-250		RC-800		RC-3000	
	Min.	Max.	Min.	Max.	Min.	Max.	Min.	Max.
Kinematic viscosity at 60°C (140°F) (See Note 1) centistokes	70	140	250	500	800	1600	3000	6000
Flash point (Tag, open-cup), °C (°F)	...	...	27 (80)	...	27 (80)	...	27 (80)	...
Water, percent	...	0.2	...	0.2	...	0.2	...	0.2
Distillation test:								
Distillate, percentage by volume of total distillate to 360°C (680°F)								
to 190°C (374°F)	10	...	...	...	...	...	...	...
to 225°C (437°F)	50	...	35	...	15	...	...	...
to 260°C (500°F)	70	...	60	...	45	...	25	...
to 315°C (600°F)	85	...	80	...	75	...	70	...
Residue from distillation to 360°C (680°F) volume percentage of sample by difference	55	...	65	...	75	...	80	...
Tests on residue from distillation:								
Absolute viscosity at 60°C (140°F) (See Note 3) poises	600	2400	600	2400	600	2400	600	2400
Ductility, 5 cm/min at 25°C (77°F) cm	100	...	100	...	100	...	100	...
Solubility in trichloroethylene, percent	99.0	...	99.0	...	99.0	...	99.0	...
Spot test (See Note 2) with:								
Standard naphtha			Negative for all grades					
Naphtha-xylene solvent, -percent xylene			Negative for all grades					
Heptane-xylene solvent, -percent xylene			Negative for all grades					

Note 1:   As an alternate, Saybolt Furol viscosities may be specified as follows:

Grade RC-70—Furol viscosity at 50°C (122°F)—60 to 120 sec
Grade RC-250—Furol viscosity at 60°C (140°F)—125 to 250 sec
Grade RC-800—Furol viscosity at 82.2°C (180°F)—100 to 200 sec
Grade RC-3000—Furol viscosity at 82.2°C (180°F)—300 to 600 sec

Note 2:   The use of the spot test is optional. When specified, the engineer shall indicate whether the standard naphtha solvent, the naphtha-xylene solvent, or the heptane-xylene solvent will be used in determining compliance with the requirement, and also, in the case of the xylene solvents, the percentage of xylene to be used.

Note 3:   In lieu of viscosity of the residue, the specifying agency, at its option, can specify penetration at 100 g; 5 sec at 25°C (77°F) of 80 to 120 for Grades RC-70, RC-250, RC-800, and RC-3000. However, in no case will both be required.

SOURCE: Photograph courtesy of STANHOPE-SETA, Ltd., Surrey, England. Used with permission

Table 18.2   Specifications for Medium-Curing Cutback Asphalts

	MC-30		MC-70		MC-250		MC-800		MC-3000	
	Min.	Max.	Min.	Max.	Min.	Max.	Min.	Max.	Min.	Max.
Kinematic Viscosity at 60°C (140°F) (See Note 1) mm²/s	30	60	70	140	250	500	800	1600	3000	6000
Flash point (Tag, open-cup), °C (F)	38 (100)	—	38 (100)	—	66 (150)	—	66 (150)	—	66 (150)	—
Water percent	—	0.2	—	0.2	—	0.2	—	0.2	—	0.2
Distillation test										
Distillate, percentage by volume of total distillate to 360°C (680°F)										
to 225°C (437°F)	—	25	0	20	0	10	—	—	—	—
to 260°C (500°F)	40	70	20	60	15	55	0	35	0	15
to 315°C (600°F)	75	93	65	90	60	87	45	80	15	75
Residue from distillation to 360°C (680°F) volume percentage of sample by difference	50	—	55	—	67	—	75	—	80	—
Tests on residue from distillation:										
Absolute viscosity at 60°C (140°F) (See Note 4) Pa · s (poises)	30 (300)	120 (1200)	30 (300)	120 (1200)	30 (300)	120 (1200)	30 (300)	120 (1200)	30 (300)	120 (1200)
Ductility, 5 cm/min, cm (See Note 2)	100	—	100	—	100	—	100	—	100	—
Solubility in Trichloroethylene, percent	99.0	—	99.0	—	99.0	—	99.0	—	99.0	—
Spot test (See Note 3) with:										
Standard naphtha				Negative for all grades						
Naphtha-xylene solvent, percent xylene				Negative for all grades						
Heptane-xylene solvent, percent xylene				Negative for all grades						

Note 1:   As an alternate, Saybolt-Furol viscosities may be specified as follows:

Grade MC-30—Furol viscosity at 25°C (77°F)—75 to 150 sec
Grade MC-70—Furol viscosity at 50°C (122°F)—60 to 120 sec
Grade MC-250—Furol viscosity at 60°C (140°F)—125 to 250 sec
Grade MC-800—Furol viscosity at 82.2°C (180°F)—100 to 200 sec
Grade MC-3000—Furol viscosity at 82.2°C (180°F)—300 to 600 sec

Note 2:   If the ductility at 25°C (77°F) is less than 100, the material will be acceptable if its ductility at 15.5°C (60°F) is more than 100.
Note 3:   The use of the spot test is optional. When specified, the engineer shall indicate whether the standard naphtha solvent, the naphtha-xylene solvent, or the heptane-xylene solvent will be used in determining compliance with the requirement and also (in the case of the xylene solvents) the percentage of xylene to be used.
Note 4:   In lieu of viscosity of the residue, the specifying agency, at its option, can specify penetration at 100 g: 5 sec at 25°C (77°F) of 120 to 250 for Grades MC-30, MC-70, MC-250, MC-800, and MC-3000. However, in no case will both be required.
SOURCE: From *Standard Specifications for Transportation Materials and Methods of Sampling and Testing*, 31st ed., 2011, AASHTO, Washington, D.C. Used by permission.

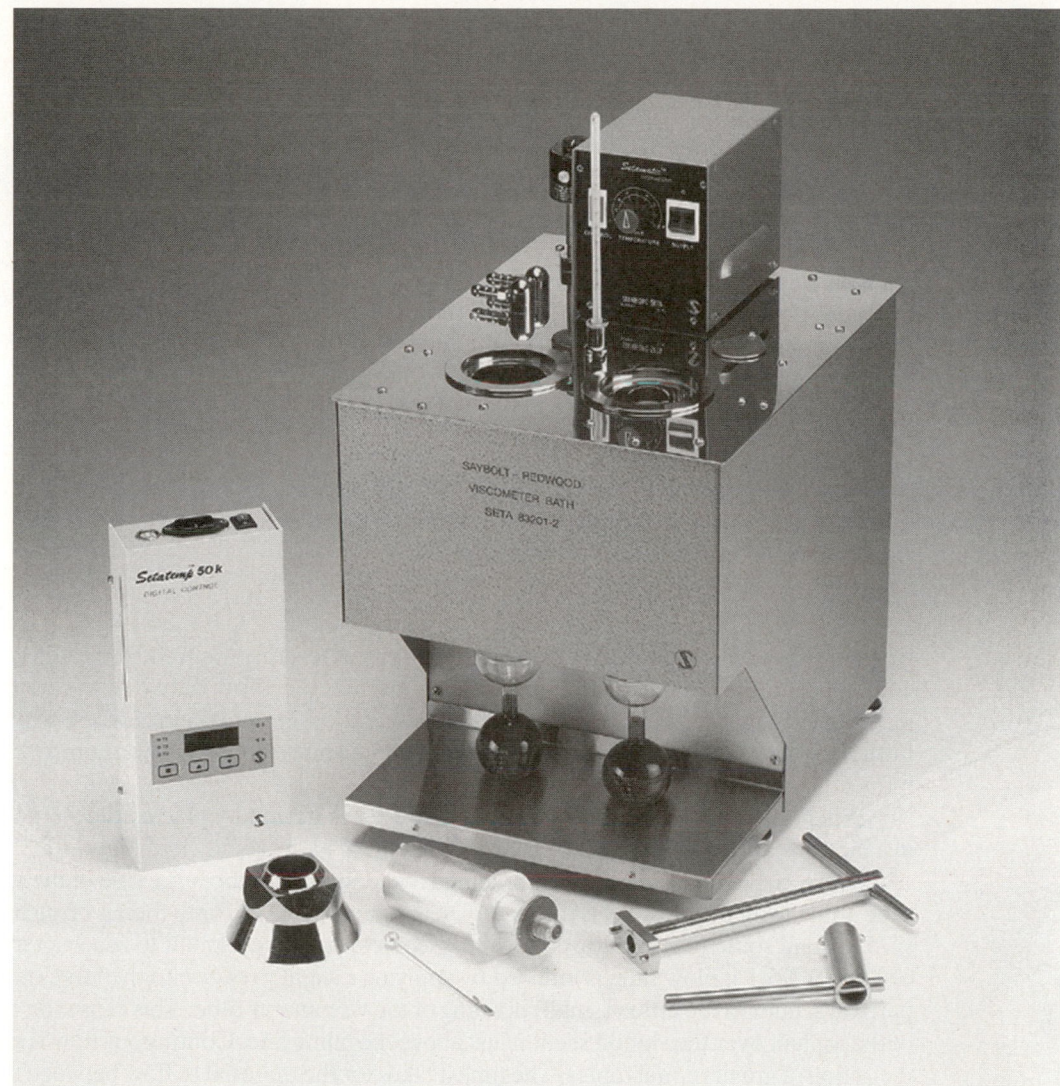

Figure 18.2 Saybolt Furol Viscometer

SOURCE: Photograph courtesy of STANHOPE-SETA, Ltd., Surrey, England. Used with permission.

materials for highway construction are tested include 25°C (77°F), 50°C (122°F), and 60°C (140°F). It is apparent that the higher the viscosity of the material, the longer it takes for a given quantity to flow through the orifice. Details of the equipment and procedures for conducting the Saybolt Furol test are given in AASHTO T72-2010.

Kinematic Viscosity Test

Figure 18.3 shows a schematic of the equipment used to determine the *kinematic viscosity*, which is defined as the absolute viscosity divided by the density. The test uses a capillary viscometer tube to measure the time it takes the asphalt sample to flow at a specified temperature between timing marks on the tube. One of three types of viscometer tubes can

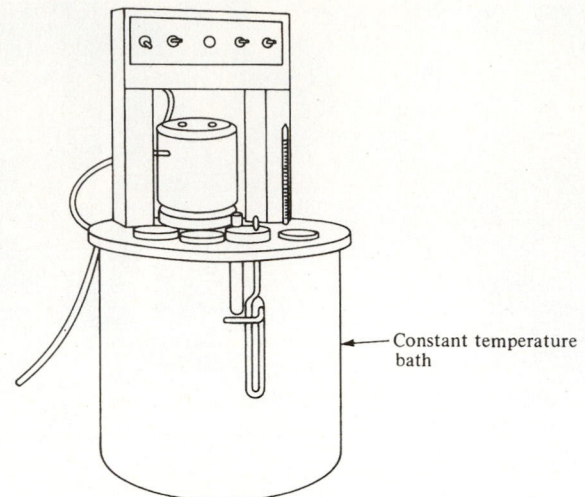

Figure 18.3   Kinematic Viscosity Apparatus

be used. These are the Zeitfuch's cross-arm viscometer, the Asphalt Institute vacuum viscometer, and the Cannon-Manning vacuum viscometer. Flow between the timing marks in the Zeitfuch's cross-arm viscometer is induced by gravitational forces, whereas flow in the Asphalt Institute vacuum viscometer and Cannon-Manning vacuum viscometer is induced by creating a partial vacuum.

When the cross-arm viscometer is used, the test is started by placing the viscometer tube in a thermostatically controlled constant-temperature bath, as shown in Figure 18.3. A sample of the material to be tested then is preheated and poured into the large side of the viscometer tube until the filling line level is reached. The temperature of the bath then is brought to 135°C (275°F), and some time is allowed for the viscometer and the asphalt to reach a temperature of 135°C (275°F). Flow then is induced by applying a slight pressure to the large opening or a partial vacuum to the efflux (small) opening of the viscometer tube. This causes an initial flow of the asphalt over the siphon section just above the filling line. Continuous flow is induced by the action of gravitational forces. The time it takes for the material to flow between two timing marks is recorded. The kinematic viscosity of the material in units of centistokes is obtained by multiplying the time in seconds by a calibration factor for the viscometer used. The calibration of each viscometer is carried out by using standard calibrating oils with known viscosity characteristics. The factor for each viscometer usually is furnished by the manufacturer. This test is described in detail in AASHTO Designation T201.

The test may also be conducted at a temperature of 60°C (140°F), as described in AASHTO Designation T202, by using either the Asphalt Institute vacuum viscometer or the Cannon-Manning vacuum viscometer. In this case, flow is induced by applying a prescribed vacuum through a vacuum control device attached to a vacuum pump. The product of the time interval and the calibration factor in this test gives the absolute viscosity of the material in poises.

Rotational Viscosity Test

This test is used to determine the viscosity of asphalt binders at elevated temperatures of 60°C to over 200°C. It is conducted in a rotational viscometer that can measure the torque required to rotate a cylinder submerged in a heated sample of the asphalt binder at a required speed. It is performed by submerging the cylindrical spindle of the viscometer in a specified amount of asphalt sample at the required temperature. The viscometer speed is set at 20 rpm. The torque required to maintain the speed of 20 rpm is then determined. The torque and speed are used to determine the viscosity of the binder in Pascal seconds. This test is described in detail in AASHTO Designation T316-11.

18.4.2 Penetration Test

The penetration test gives an empirical measurement of the consistency of a material in terms of the distance a standard needle sinks into that material under a prescribed loading and time. Although more fundamental tests are being substituted for this test, it still may be included in specifications for viscosity of asphalt cements to ensure the exclusion of materials with very low penetration values at 25°C (77°F).

Figure 18.4 shows a typical penetrometer and a schematic of the standard penetration test. A sample of the asphalt cement to be tested is placed in a container, which in

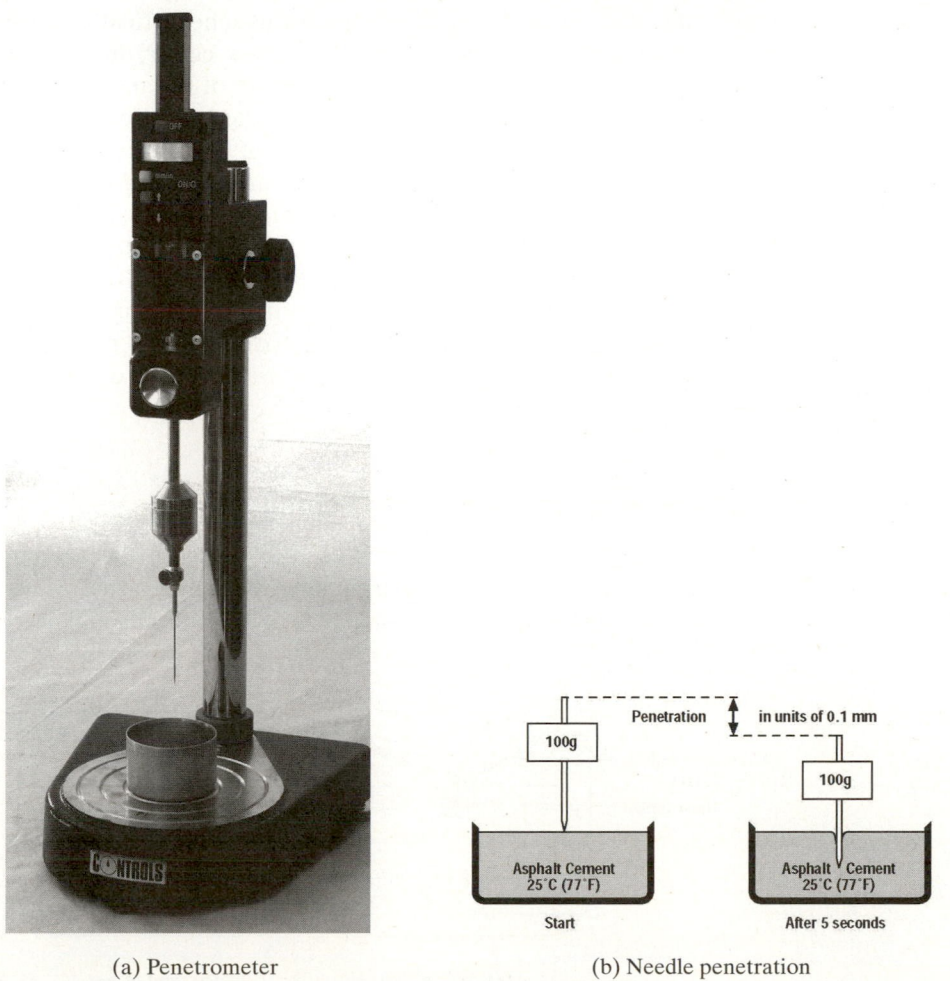

(a) Penetrometer   (b) Needle penetration

Figure 18.4 Standard Penetration Test and Equipment

SOURCE: CONTROLS S.R.L. Used with permission.

turn is placed in a temperature-controlled water bath. The sample is then brought to the prescribed temperature of 25°C (77°F) and the standard needle (loaded to a total weight of 100 g) is left to penetrate the sample of asphalt for the prescribed time of exactly 5 sec. The penetration is given as the distance in units of 0.1 mm that the needle penetrates the sample. For example, if the needle penetrates a distance of exactly 20 mm, the penetration is 200.

The penetration test also can be conducted at 0°C (32°F) or at 4°C (39.2°F), with the needle loaded to a total weight of 200 g and with penetrations allowed for 60 sec. Details of the penetration test are given in the AASHTO Designation T49 and the ASTM Test D5.

18.4.3 Float Test

The float test is used to determine the consistency of semisolid asphalt materials that are more viscous than grade 3000 or have penetration higher than 300, since these materials cannot be tested conveniently using either the Saybolt Furol viscosity test or the penetration test.

The float test is conducted with the apparatus schematically shown in Figure 18.5. It consists of an aluminum saucer (float), a brass collar open at both ends, and a water bath. The brass collar is filled with a sample of the material to be tested and is then attached to the bottom of the float and chilled to a temperature of 5°C (41°F) by immersing it in ice water. The temperature of the water bath is brought to 50°C (122°F), and the collar (still attached to the float) is placed in the water bath, which is kept at 50°C (122°F). The head gradually softens the sample of asphalt material in the collar until the water eventually forces its way through the plug into the aluminum float. The time in seconds that expires between the instant the collar is placed in the water bath and that at which the water forces its way through the bituminous plug is the *float-test value* and is a measure of the consistency. It is readily apparent that the higher the float-test value, the stiffer the material. Details of the float test are given in the ASTM D139.

18.4.4 Ring-and-Ball Softening Point Test

The ring-and-ball softening point test is used to measure the susceptibility of blown asphalt to temperature changes by determining the temperature at which the material will be adequately softened to allow a standard ball to sink through it. Figure 18.6 shows

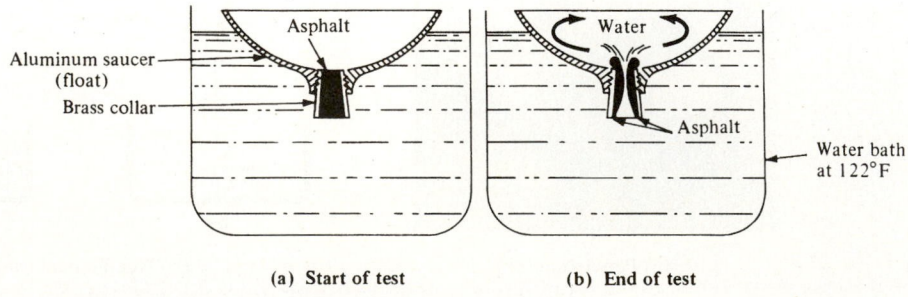

(a) **Start of test**   (b) **End of test**

Figure 18.5   Float Test

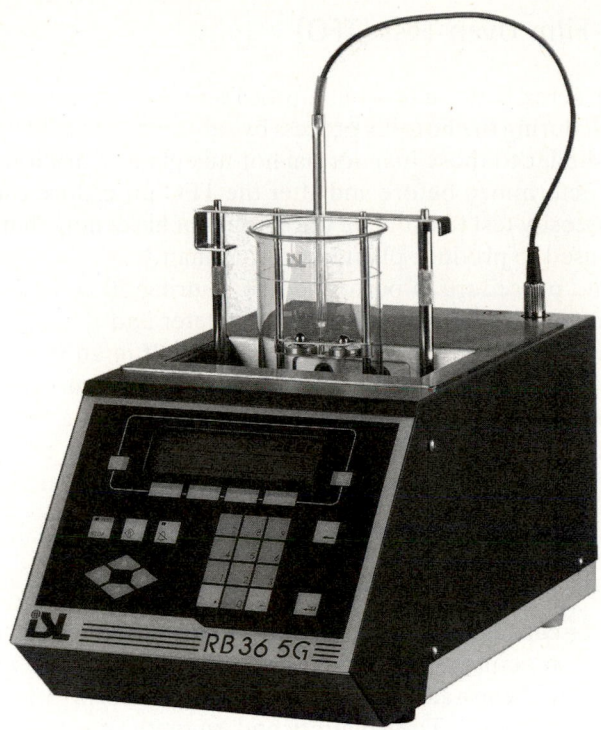

Figure 18.6   The RB36 5G Automated Ring-and-Ball Tester

SOURCE: Courtesy of PAC

an example of the apparatus commonly used for this test. It consists principally of a small brass ring of 16 mm inside diameter and 6.3 mm high, a steel ball 9.5 mm in diameter, and a water or glycerin bath. The test is conducted by first placing a sample of the material to be tested in the brass ring, which is cooled and immersed in the water or glycerin bath that is maintained at a temperature of 5°C (41°F). The ring is immersed to a depth such that its bottom is exactly 25 mm above the bottom of the bath. The temperature of the bath then is increased gradually, causing the asphalt to soften and permitting the ball to eventually sink to the bottom of the bath. The temperature in °F at which the asphalt material touches the bottom of the bath is recorded as the softening point. The resistance characteristic to temperature shear can also be evaluated using this test. Details of this test are given in the ASTM Designation D36.

18.4.5 Durability Tests

When asphalt materials are used in the construction of roadway pavements, they are subjected to changes in temperature (freezing and thawing) and other weather conditions over a period of time. These changes cause natural weathering of the material, which may lead to loss of plasticity, cracking, abnormal surface abrasion, and eventual failure of the pavement. One test used to evaluate the susceptibility characteristics of asphalt materials to changes in temperature and other atmospheric factors is the thin-film oven test.

Thin-Film Oven Test (TFO)

This is actually not a test but a procedure that measures the changes that take place in asphalt during the hot-mix process by subjecting the asphalt material to hardening conditions similar to those in a normal hot-mix plant operation. The consistency of the material is determined before and after the TFO procedure using either the penetration test or a viscosity test to estimate the amount of hardening that will take place in the material when used to produce plant hot-mix asphalt.

The procedure is performed by pouring 50 cc of material into a cylindrical flat-bottom pan, 5.5-in. (14 cm) inside diameter and $\frac{3}{8}$ in. (1 cm) high. The pan containing the sample is then placed on a rotating shelf in an oven and rotated for 5 h while the temperature is kept at 163°C (325°F). The amount of penetration after the TFO test is then expressed as a percentage of that before the test to determine percent of penetration retained. The minimum allowable percent of penetration retained is usually specified for different grades of asphalt cement. Details of this test are given in the AASHTO Designation T179.

18.4.6 Rate of Curing

Tests for curing rates of cutbacks are based on inherent factors which can be controlled. These tests compare different asphalt materials on the assumption that the external factors are held constant. Volatility and quantity of solvent are used commonly to indicate the rate of curing. The volatility and quantity of solvent may be determined from the distillation test; tests for consistency were described earlier.

Distillation Test for Cutbacks

Figure 18.7 is an example of the apparatus used in the distillation test. The apparatus consists principally of a flask in which the material is heated, a condenser, and a graduated cylinder for collecting the condensed material. A sample of 200 cc of the material to be tested is measured and poured into the flask and the apparatus is set up as shown in Figure 18.7. The material is then brought to boiling point by heating it with the burner. The evaporated solvent is condensed and collected in the graduated cylinder. The temperature in the flask is monitored continuously and the amount of solvent collected in the graduated cylinder is recorded when the temperature in the flask reaches 190°C (374°F), 225°C (437°F), 260°C (500°F), and 316°C (600°F). The amount of condensate collected at the different specified temperatures gives an indication of the volatility characteristics of the solvent. The residual in the flask is the base asphalt used in preparing the cutback. Details of this test are given in the AASHTO Designation T78.

Distillation Test for Emulsions

The distillation test for emulsions is similar to that described for cutbacks. A major difference, however, is that the glass flask and Bunsen burner are replaced with an aluminum-alloy still and a ring burner. This equipment prevents potential problems that may arise from the foaming of the emulsified asphalt as it is being heated to a maximum of 260°C (500°F).

Note, however, that the results obtained from the use of this method to recover the asphalt residue and to determine the properties of the asphalt base stock used in the

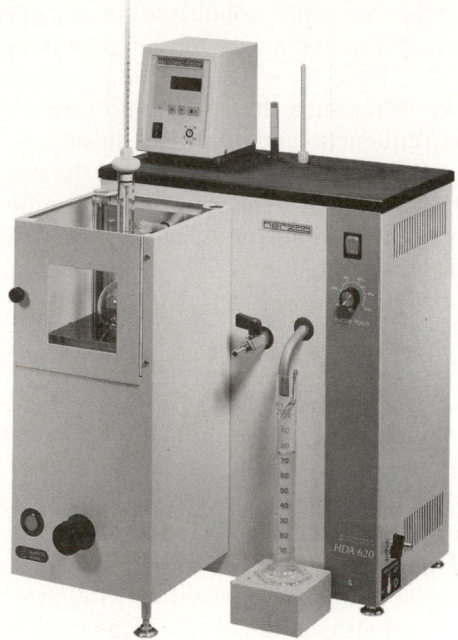

Figure 18.7 Herzog HAD 620 Manual Distillation Analyzer

SOURCE: Courtesy of PAC

emulsion may not always be accurate because of significant changes in the properties of the asphalt. These changes are due to

- Inorganic salts from an aqueous phase concentrating in the asphalt residue
- Emulsifying agent and stabilizer concentrating in the asphalt residue

These changes, mainly due to an increase in temperature, do not occur in field application of the emulsion, since the temperature in the field is usually much less than that used in the distillation test. The emulsion in the field, therefore, breaks either electrochemically or by evaporation of the water. An alternative method to determine the properties of the asphalt after it is cured on the pavement surface is to evaporate the water at subatmospheric pressure and lower temperatures. Such a test is designated AASHTO T59 or ASTM D244.

18.4.7 Rheological Tests

The dynamic shear test and the bending creep test are presented next. It should be noted that these tests do not lead to a full characterization of the viscoelastic properties of asphalt binders.

Dynamic Shear Test

This test is used to determine the dynamic (oscillatory) shear modulus and phase an of a sample of asphalt binder when tested in an oscillatory mode. A *Dynamic Sr Rheometer* (DSR) test system is used to conduct the test. The DSR test systerata sists of parallel metal plates, an environmental chamber, a loading device, an acquisition unit. The test is conducted on test specimens with dimensions of either 1 mm thick and 25 mm in diameter or 2 mm thick and 8 mm in diameter. The test specimen is placed between the parallel metal plates, and one of the plates is oscillated with respect to the other such that strain and stress are controlled. This is done by preselecting the frequencies and rotational deformation amplitudes that are dependent on the value of the complex shear modulus of the asphalt binder being tested. The manufacturer of the DSR provides a proprietary computer software package that automatically computes the complex modulus and the phase angle. This test is fully described in the AASHTO Designation T315.

Bending Creep Test

Indirect loading techniques are used in this test to determine the tensile creep compliance of hot-mix asphalt under different loading times, tensile strength, and Poisson's ratio. The test uses an indirect tensile test system consisting of several components including an axial loading device, a load measuring device, devices for measuring specimen deformation, an environmental chamber, and a control and data acquisition system. A static load of fixed magnitude is applied along the diametrical axis of the test specimen, having dimensions of 38 to 50 mm in height and 150 ± 9 mm in diameter. The horizontal and vertical deformations near the center of the system are recorded by the data acquisition system and then are used to determine the Poisson's ratio and a tensile creep compliance as a function of time. The tensile strength can be determined immediately after the creep test, or later by inducing a constant rate of vertical deformation to failure. This test is fully described in the AASHTO Designation T322.

18.4.8 Other General Tests

Several other tests are conducted routinely on asphalt materials used for pavement construction either to obtain specific characteristics for design purposes (for example, specific gravity) or to obtain additional information that aids in determining the quality of the material. Some of the more common routine tests are described briefly.

Specific Gravity Test

The specific gravity of asphalt materials is used mainly to determine the weight of a given volume of material (or vice versa), to determine the amount of voids in compacted mixes, and to correct volumes measured at high temperatures. Specific gravity is defined as the ratio of the weight of a given volume of the material to the weight of the same volume of water. The specific gravity of bituminous materials, however, changes with temperature, which dictates that the temperature at which the test is conducted should be indicated. For example, if the test is conducted at 20°C (68°F) and the specific gravity is determined to be 1.41, this should be recorded as 1.41/20°C. Note that both the asphalt material and the water should be at the same temperature. The usual temperature at which the specific gravities of asphalt materials are determined is 25°C (77°F).

The test normally is conducted with a pycnometer, examples of which are shown in Figure 18.8. The dry mass (m_1) of the pycnometer and stopper is obtained, and the pycnometer is filled with distilled water at the prescribed temperature. The mass (m_2) of the water and pycnometer together is determined. If the material to be tested can flow easily into the pycnometer, then the pycnometer must be filled completely with the material at

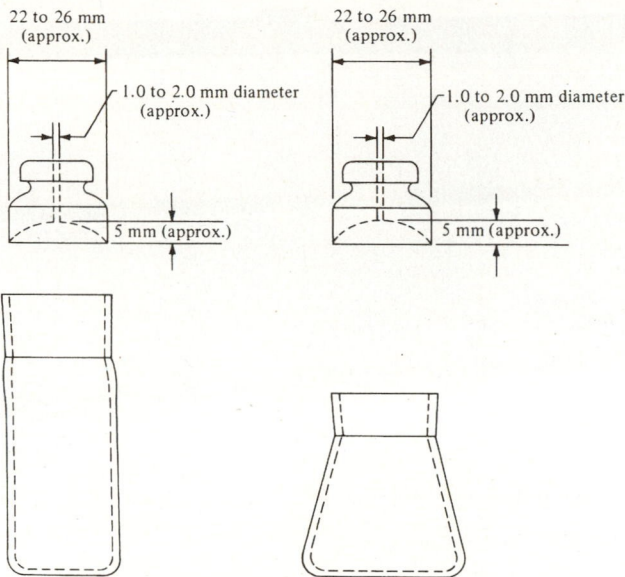

Figure 18.8   Pycnometers for Determining Specific Gravity of Asphalt Materials

the specified temperature after pouring out the water. The weight m_3 then is obtained. The specific gravity of the asphalt material is given as

$$G_b = \frac{m_3 - m_1}{m_2 - m_1} \tag{18.1}$$

where G_b is the specific gravity of the asphalt material, and m_1, m_2, and m_3 are in grams.

If the asphalt material cannot flow easily, a small sample of the material is heated gradually to facilitate flow and then poured into the pycnometer and left to cool to the specified temperature. The mass m_4 of pycnometer and material then is obtained. Water is poured into the pycnometer to completely fill the remaining space not occupied by the material. The mass m_5 of the filled pycnometer is obtained. The specific gravity is then given as

$$G_b = \frac{m_4 - m_1}{(m_2 - m_1) - (m_5 - m_4)} \tag{18.2}$$

Ductility Test

Ductility is the distance in centimeters a standard sample of asphalt material will stretch before breaking when tested on standard ductility test equipment at 25°C (77°F). The result of this test indicates the extent to which the material can be deformed without breaking. This is an important characteristic for asphalt materials, although the exact value of ductility is not as important as the existence or nonexistence of the property in the material.

Figure 18.9 shows the ductility test apparatus. The test is used mainly for semisolid or solid materials, which first are heated gently to facilitate flow and then poured into a standard mold to form a briquette of at least 1 cm² in cross section. The material then

Figure 18.9   Apparatus for Ductility Test

SOURCE: CONTROLS S.R.L. Used with permission.

is allowed to cool to 25°C (77°F) in a water bath. The prepared sample is placed in the ductility machine shown in Figure 18.9 and extended at a specified rate of speed until the thread of material joining the two ends breaks. The distance (in centimeters) moved by the machine is the ductility of the material. The test is described fully in the AASHTO Designation T51 and ASTM D113.

Solubility Test

The solubility test is used to measure the amount of impurities in the asphalt material. Since asphalt is nearly 100 percent soluble in certain solvents, the portion of any asphalt material that will be effective in cementing aggregates together can be determined from the solubility test. Insoluble materials include free carbon, salts, and other inorganic impurities.

This test is conducted by dissolving a known quantity of the material in a solvent (such as trichloroethylene) and then filtering it through a Gooch crucible. The material retained in the filter is dried and weighed. The test results are given in terms of the percent of the asphalt material that dissolved in the solvent. This test is fully described in the AASHTO Designation T44.

Flash-Point Test

The flash point of an asphalt material is the temperature at which its vapors will ignite instantaneously in the presence of an open flame. Note that the flash point normally is lower than the temperature at which the material will burn.

The test can be conducted by using either the Tagliabue open-cup apparatus or the Cleveland open-cup apparatus, which is shown in Figure 18.10. The Cleveland open-cup test is more suitable for materials with higher flash points whereas the Tagliabue open-cup is more suitable for materials with relatively low flash points, such as cutback asphalts.

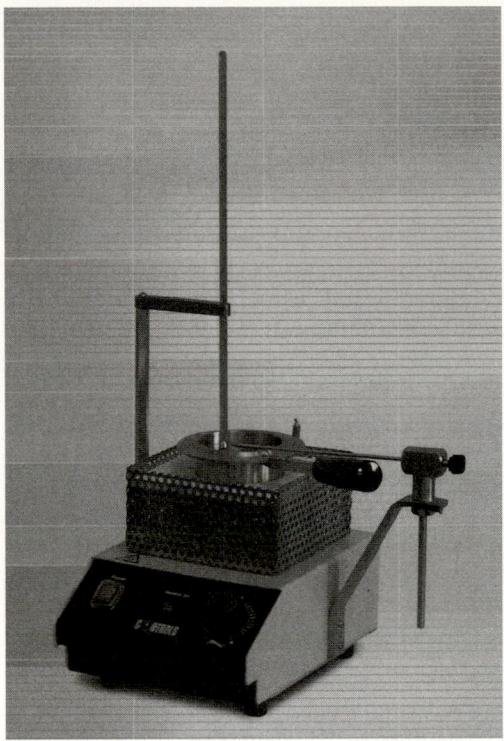

Figure 18.10   Apparatus for Cleveland Open-Cup Test

SOURCE: CONTROLS S.R.L. Used with permission.

The test is conducted by partly filling the cup with the asphalt material and gradually increasing its temperature at a specified rate. A small open flame is passed over the surface of the sample at regular intervals as the temperature increases. The increase in temperature will cause evaporation of volatile materials from the material being tested until a sufficient quantity of volatile materials is present to cause an instantaneous flash when the open flame is passed over the surface. The minimum temperature at which this occurs is the flash point. It can be seen that this temperature gives an indication of the temperature limit at which extreme care should be taken, particularly when heating is done over open flames in open containers. This test is fully described in the AASHTO Designation T48.

Loss-on-Heating Test

The loss-on-heating test is used to determine the amount of material that evaporates from a sample of asphalt under a specified temperature and time. The result indicates whether asphalt material has been contaminated with lighter materials. The test is conducted by pouring 50 g of the material to be tested into a standard cylindrical tin and leaving it in an oven for 5 h at a temperature of 163°C (325°F). The weight of the material remaining in the tin is determined and the loss in weight is expressed as a percentage of the original weight. The penetration of the sample also may be determined before and after the test in order to determine the loss of penetration due to the evaporation of the volatile material. This loss in penetration may be used as an indication of the weathering characteristics of the asphalt. Details of the test procedures are given in the AASHTO Designation T47.

Water Content Test

The presence of large amounts of water in asphalt materials used in pavement construction is undesirable, and to ensure that only a limited quantity of water is present, specifications for these materials usually include the maximum percentage of water by volume that is allowable. A quantity of the sample to be tested is mixed with an equal quantity of a suitable distillate in a distillation flask that is connected with a condenser and a trap for collecting the water. The sample is then heated gradually in the flask, eventually causing all of the water to evaporate and be collected. The quantity of water in the sample is then expressed as a percentage of the total sample volume. This test is described fully in the AASHTO Designation T55.

Demulsibility Test for Emulsion

The demulsibility test is used to indicate the relative susceptibility of asphalt emulsions to breaking down (coalescing) when in contact with aggregates. Asphalt emulsions are expected to break immediately when they come in contact with the aggregate, so that the material is prevented from washing away with rain that may occur soon after application. A high degree of demulsibility is required for emulsions used for surface treatments, such as RS1 and RS2. A relatively low degree of demulsibility is required for emulsions used for mixing coarse aggregates to avoid having the materials peel off before placing, and a very low degree is required for materials produced for mixing fine aggregates. Since calcium chloride will coagulate minute particles of asphalt, it is used in the test for anionic rapid-setting (RS) emulsions; however, dioctyl sodium sulfosuccinate is used instead of calcium chloride for testing cationic rapid-setting (CRS) emulsions. The test is conducted by thoroughly mixing the required standard solution with the asphalt emulsion and then passing the mixture through a No. 14 wire cloth sieve, which will retain the asphalt particles that have coalesced. The quantity of asphalt retained in the sieve is a measure of the breakdown that has occurred. Demulsibility is expressed in percentage as $(A/B)100$, where A is the average weight of demulsibility residue from three tests of each sample of emulsified asphalt, and B is the weight of residue by distillation in 100 g of the emulsified asphalt.

The strength of the test solution used and the minimum value of demulsibility required are prescribed in the relevant specifications. Details of this test are described in the AASHTO T59 and the ASTM D244.

Sieve Test for Emulsions

The sieve test is conducted on asphalt emulsions to determine to what extent the material has emulsified and the suitability of the material for application through pressure distributors. This test is conducted by passing a sample of the material through a No. 20 sieve and determining what percentage by weight of the material is retained in the sieve. A maximum value of 0.1 percent usually is specified. This test is described in the ASTM Designation D244 and the AASHTO T59.

Particle-Charge Test for Emulsions

The particle-charge test is used to identify CRS and CMS grades of emulsions. This test is conducted by immersing an anode electrode and a cathode electrode in a sample of the material to be tested and then passing an electric current through the system, as shown in Figure 18.11. The electrodes then are examined (after some time) to identify which one contains an asphalt deposit. If a deposit occurs on the cathode electrode, the emulsion is cationic. This test is described in the AASHTO Designation T59 and the ASTM D244.

18.5 ASPHALT MIXTURES

Asphalt mixtures are a uniformly mixed combination of asphalt cement, coarse aggregate, fine aggregate, and other materials, depending on the type of asphalt mixture. The different types of asphalt mixtures commonly used in pavement construction are hot-mix, hot-laid; hot-mix, cold-laid; and cold-mix cold-laid. Warm-mix asphalt is a fourth type that is increasingly receiving attention in the United States. It is produced by using a lower mixing temperature than that used for hot-mix asphalts. Reductions of 10 C to 38 C have been documented. This relatively lower mixing temperature reduces fuel consumption and the amount of greenhouse gases emitted. Asphalt mixtures are the most popular paving material used in the United States. When used in the construction of highway pavements, they must resist deformation from imposed traffic loads, be skid resistant even when wet, and not be affected easily by weathering forces. The degree to which an asphalt mixture achieves these characteristics mainly is dependent on the design of the mix used in producing the material. The three commonly used categories of asphalt mixtures will be described next, together with an appropriate mix design procedure.

18.5.1 Hot–Mix, Hot–Laid Asphalt Mixture

Hot-mix, hot-laid asphalt mixture is produced by properly blending asphalt cement, coarse aggregate, fine aggregate, and filler (dust) at temperatures ranging from about 80 C to 163 C, depending on the type of asphalt cement used. Suitable types of asphalt materials include AC-20, AC-10, and AR-8000, with penetration grades of usually 60 to 70 and 85 to 100. Hot-mix, hot-laid asphalt mixture is normally used for high-type pavement construction, and the mixture can be described as open-graded, coarse-graded,

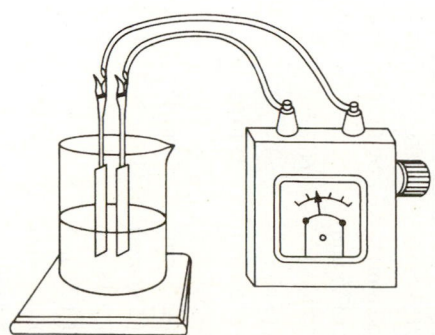

Figure 18.11   Particle-Charge Test for Emulsions

Table 18.3 Examples of Composition of Asphalt Paving Mixtures

Dense Mixtures

Mix Designation and Nominal Size of Aggregate

Grading of Total Aggregate (Coarse Plus Fine, Plus Filler if Required)
Amounts Finer Than Each Laboratory Sieve (Square Opening), Weight %

Sieve Size	2 in. (50 mm)	1½ in. (37.5 mm)	1 in. (25 mm)	¾ in. (19 mm)	½ in. (12.5 mm)	⅜ in. (9.5 mm)	No. 4 (4.75 mm) (Sand Asphalt)	No. 8 (2.36 mm)	No. 16 (1.18 mm) (Sheet Asphalt)
2½ in. (63 mm)	100	...	...	...	...	...	...	...	...
2 in. (50 mm)	90 to 100	100	...	...	...	...	...	...	...
1½ in. (37.5 mm)	...	90 to 100	100	...	...	...	...	...	...
1 in. (25 mm)	60 to 80	...	90 to 100	...	...	...	...	...	...
¾ in. (19 mm)	...	56 to 80	...	90 to 100	100	...	...	...	...
½ in. (12.5 mm)	35 to 65	...	56 to 80	...	90 to 100	100	...	...	...
⅜ in. (9.5 mm)	...	...	...	56 to 80	...	90 to 100	100	...	...
No. 4 (4.75 mm)[a]	17 to 47	23 to 53	29 to 59	35 to 65	44 to 74	55 to 85	80 to 100	...	100
No. 8 (2.36 mm)[a]	10 to 36	15 to 41	19 to 45	23 to 49	28 to 58	32 to 67	65 to 100	...	95 to 100
No. 16 (1.18 mm)	...	...	...	...	...	...	40 to 80	...	85 to 100
No. 30 (600 μm)	...	...	...	...	...	...	25 to 65	...	70 to 95
No. 50 (300 μm)	3 to 15	4 to 16	5 to 17	5 to 19	5 to 21	7 to 23	7 to 40	...	45 to 75
No. 100 (150 μm)	...	...	...	...	...	...	3 to 20	...	20 to 40
No. 200 (75 μm)[b]	0 to 5	0 to 6	1 to 7	2 to 8	2 to 10	2 to 10	2 to 10	...	9 to 20

Open Mixtures

Mix Designation and Nominal Maximum Size of Aggregate

Sieve Size	2 in. (50 mm)	1½ in. (37.5 mm)	1 in. (25 mm)	¾ in (19 mm)	½ in. (12.5 mm)	⅜ in. (9.5 mm)	No. 4 (4.75 mm) (Sand Asphalt)	No. 8 (2.36 mm)	No. 16 (1.18 mm) (Sheet Asphalt)
	Base and Binder Courses						Surface and Leveling Courses		
2½ in. (63 mm)	100	...	...	...	...	...	...	...	...
2 in. (50 mm)	90 to 100	100	...	...	...	...	...	...	...
1½ in. (37.5 mm)	...	90 to 100	100	...	...	...	...	...	...
1 in. (25 mm)	40 to 70	...	90 to 100	100	...	...	...	...	...

Table 18.3 Examples of Composition of Asphalt Paving Mixtures

	Open Mixtures								
	Mix Designation and Nominal Maximum Size of Aggregate								
	Base and Binder Courses				Surface and Leveling Courses				
Sieve Size	2 in. (50 mm)	1½ in. (37.5 mm)	1 in. (25 mm)	¾ in (19 mm)	½ in. (12.5 mm)	⅜ in. (9.5 mm)	No. 4 (4.75 mm) (Sand Asphalt)	No. 8 (2.36 mm)	No. 16 (1.18 mm) (Sheet Asphalt)
¾ in. (19 mm)	⋯	40 to 70	90 to 100	100	⋯	⋯	⋯	⋯	⋯
½ in. (12.5 mm)	18 to 48	⋯	40 to 70	85 to 100	100	⋯	⋯	⋯	⋯
⅜ in. (9.5 mm)	⋯	18 to 48	⋯	40 to 70	85 to 100	100	⋯	⋯	⋯
No. 4 (4.75 mm)	5 to 25	6 to 29	10 to 34	15 to 39	20 to 50	40 to 70	100	⋯	⋯
No. 8 (2.36 mm)[a]	0 to 12	0 to 14	1 to 17	2 to 18	5 to 25	10 to 35	75 to 100	100	⋯
No. 16 (1.18 mm)	⋯	⋯	⋯	⋯	3 to 19	5 to 25	50 to 75	75 to 100	⋯
No. 30 (600 μm)	0 to 8	0 to 8	0 to 10	0 to 10	⋯	⋯	28 to 53	50 to 75	⋯
No. 50 (300 μm)	⋯	⋯	⋯	⋯	0 to 10	0 to 12	8 to 30	28 to 53	⋯
No. 100 (150 μm)	⋯	⋯	⋯	⋯	⋯	⋯	0 to 12	8 to 30	⋯
No. 200 (75 μm)[b]	⋯	⋯	⋯	⋯	⋯	⋯	0 to 5	0 to 12	⋯
Bitumen, Weight % of Total Mixture[c]									
	2 to 7	3 to 8	3 to 9	4 to 10	4 to 11	5 to 12	6 to 12	7 to 12	8 to 12
Suggested Coarse Aggregate Sizes									
	3 and 57	4 and 67 or 4 and 68	5 and 7 or 57	67 or 68 or 6 and 8	7 or 78	8			

[a]In considering the total grading characteristics of a bituminous paving mixture, the amount passing the No. 8 (2.36 mm) sieve is a significant and convenient field control point between fine and coarse aggregate. Gradings approaching the maximum amount permitted to pass the No. 8 sieve will result in pavement surfaces having comparatively fine texture, while coarse gradings approaching the minimum amount passing the No. 8 sieve will result in surfaces with comparatively coarse texture.

[b]The material passing the No. 200 (75 μm) sieve may consist of fine particles of the aggregates or mineral filler, or both, but shall be free of organic matter and clay particles. The blend of aggregates and filler, when tested in accordance with Test Method D4318, shall have a plasticity index of not greater than 4, except that this plasticity requirement shall not apply when the filler material is hydrated lime or hydraulic cement.

[c]The quantity of bitumen is given in terms of weight % of the total mixture. The wide difference in terms of weight % of the total mixture. The wide difference in the specific gravity of various aggregates, as well as a considerable difference in absorption, results in a comparatively wide range in the limiting amount of bitumen specified. The amount of bitumen required for a given mixture should be determined by appropriate laboratory testing, on the basis of past experience with similar mixtures, or by a combination of both.

SOURCE: *Annual Book of ASTM Standards, Section 4, Construction,* Vol. 04.03, Road and Paving Materials; Pavement Management Technologies, American Society for Testing and Materials, West Conshohocken, PA., 2007

dense-graded, or fine-graded. When produced for high-type surfacing, maximum sizes of aggregates normally range from 9.5 mm to 19 mm for open-graded mixtures, 12.5 mm to 19 mm for coarse-graded mixtures, 12.5 mm to 25.4 mm for dense-graded mixtures, and 12.5 mm to 2 cm for fine-graded mixtures. When used as a base, maximum sizes of aggregates are usually 19 mm to 38 mm for open- and coarse-graded mixtures, 25 mm to 38 mm for dense-graded, and 19 mm for fine-graded mixtures. As stated earlier, the extent to which an asphalt mixture meets the desired characteristics for highway pavement construction is dependent mainly on the mix design, which involves the selection and proportioning of the different material components. However, note that when designing a hot-mix asphalt mixture, a favorable balance must be found between a highly stable product and a durable one. Therefore, the overall objective of the mix design is to determine an optimum blend of the different components that will satisfy the requirements of the given specifications.

Aggregate Gradation

Aggregates usually are categorized as crushed rock, sand, and filler. The rock material is predominantly coarse aggregate retained in a No. 8 sieve, sand is predominantly fine aggregate passing the No. 8 sieve and retained in a No. 200 sieve, and filler is predominantly mineral dust that passes the No. 200 sieve. It is customary for gradations of the combined aggregate and the individual fractions to be specified. Table 18.3 gives suggested grading requirements of aggregate materials. Note that the requirements given in Table 18.3 are no longer considered as required specifications, as local agencies tend to give their own specifications. The first phase in any mix design is the selection and combination of aggregates to obtain a gradation within the limits prescribed.

The procedure used to select and combine aggregates will be illustrated in the following example.

Example 18.1 Determining Proportions of Different Aggregates to Obtain a Required Gradation

Table 18.4 gives the specifications for the aggregates and mix composition for highway pavement asphaltic concrete, and Table 18.5 shows the results of a sieve analysis of samples from the materials available. We must determine the proportions of the separate aggregates that will give a gradation within the specified limits.

Solution: It can be seen that the amount of the different sizes selected should not only give a mix that meets the prescribed limits but also should be such that allowance is made for some variation during actual production of the mix. It also can be seen from Table 18.4 that to obtain the required specified gradation, some combination of all three materials is required, since the coarse and fine aggregates do not together meet the requirement of 2 to 6 percent by weight of filler material. Therefore, a trial mix is selected arbitrarily within the prescribed limits. Let this mix be

Coarse aggregates = 55% (48 − 65% specified)

Fine aggregates = 39% (35 − 50% specified)

Filler = 6% (5 − 8% specified)

Table 18.4   Required Limits for Mineral Aggregates Gradation and Mix Composition for an Asphalt Mixture for Example 18.1

Passive Sieve Designation	Retained on Sieve Designation	Percent by Weight
¾ in. (19 mm)	12.5 mm	0–5
½ in. (12.5 mm)	9.5 mm	8–42
⅜ in. (9.5 mm)	No. 4	8–48
No. 4 (4.75 mm)	No. 10	6–28
Total coarse aggregates	No. 10	48–65
No. 10 (2 mm)	No. 40	5–20
No. 40 (0.425 mm)	No. 80	9–30
No. 80 (0.180 mm)	No. 200	3–20
No. 200 (0.075 mm)	—	2–6
Total fine aggregate and filler	Passing No. 10	35–50
Total mineral aggregate in asphalt concrete		90–95
Asphalt cement in asphalt concrete		5–7
Total mix		100

Table 18.5   Sieve Analysis of Available Materials for Example 18.1

Passing Sieve Designation	Retained on Sieve Designation	Percent by Weight		
		Coarse Aggregate	Fine Aggregate	Mineral Filler
¾ in. (19 mm)	12.5 mm	5	—	—
½ in. (12.5 mm)	9.5 mm	35	—	—
⅜ in. (9.5 mm)	No. 4	38	—	—
No. 4 (4.75 mm)	No. 10	17	8	—
No. 10 (2 mm)	No. 40	5	30	—
No. 40 (0.425 mm)	No. 80	—	35	5
No. 80 (0.180 mm)	No. 200	—	26	35
No. 200 (0.075 mm)	—	—	1	60
Total		100	100	100

The selected proportions are then used to determine the combination of the different sizes, as shown in Table 18.6. The calculation is based on the fundamental equation for the percentage of material P passing a given sieve for the aggregates 1, 2, 3 and is given as

$$P = aA_1 + bA_2 + cA_3 + \cdots \qquad (18.3)$$

where

A_1, A_2, A_3 = the percentages of material passing a given sieve for aggregates 1, 2, 3

a, b, c = the proportions of aggregates 1, 2, 3 used in the combination

$a + b + c + \cdots = 100$

Note that this is true for any number of aggregates combined.

It can be seen that the combination obtained, as shown in the last column of Table 18.6, meets the specified limits as shown in the last column of Table 18.4. The trial combination is therefore acceptable. Note, however, that the first trial may not always meet the specified limits. In such cases, other combinations must be tried until a satisfactory one is obtained.

Table 18.6   Computation of Percentages of Different Aggregate Sizes for Example 18.1

Passing Sieve Size	Retained on Sieve Size	*Percent by Weight*			
		Coarse Aggregate	Fine Aggregate	Mineral Filler	Total Aggregate
19 mm	12.5 mm	$0.55 \times 5 = 2.75$	—	—	2.75
12.5 mm	9.5 mm	$0.55 \times 35 = 19.25$	—	—	19.25
9.5 mm	No. 4	$0.55 \times 38 = 20.90$	—	—	20.90
No. 4	No. 10	$0.55 \times 17 = 9.35$	$0.39 \times 8 = 3.12$	—	12.47
No. 10	No. 40	$0.55 \times 5 = 2.75$	$0.39 \times 30 = 11.70$	—	14.45
No. 40	No. 80	—	$0.39 \times 35 = 13.65$	$0.06 \times 5 = 0.3$	13.95
No. 80	No. 200	—	$0.39 \times 26 = 10.14$	$0.06 \times 35 = 2.10$	12.24
No. 200	—	—	$0.39 \times 1 = 0.39$	$0.06 \times 60 = 3.60$	3.99
Total		55.0	39.0	6.0	100.00

Several graphical methods have been developed for obtaining a suitable mixture of different aggregates to obtain a desired gradation. These methods tend to be rather complicated when the number of batches of aggregates is high. They generally can be of advantage over the trial-and-error method described here when the number of aggregates is not more than two. When the number of aggregates is three or greater, the trial-and-error method is preferable, although the graphical methods also can be used.

Asphalt Content

Having determined a suitable mix of aggregates, the next step is to determine the optimum percentage of asphalt that should be used in the asphalt mixture. This percentage should be within the prescribed limits (of course). The gradation of the aggregates determined earlier and the optimum amount of asphalt cement determined combine to give the proportions of the different materials to be used in producing the hot-mix, hot-laid mixture for the project under consideration. These determined proportions usually are referred to as the *job-mix formula*.

Two commonly used methods to determine the optimum asphalt content are the *Marshall method* and the *Hveem method*. The Marshall method is described here, since it has been more widely used.

Marshall Method Procedure.   The original concepts of this method were developed by Bruce Marshall, a bituminous engineer with the Mississippi State Highway Department. The original features have been improved by the U.S. Army Corps of Engineers, and the test now is standardized and described in detail in the ASTM Designation D1559. Test specimens of 100 mm diameter and 63.5 mm height are used in this method. They are prepared by a specified procedure of heating, mixing, and compacting the mixture of asphalt and aggregates, which is then subjected to a density-voids analysis and a stability-flow test. The *stability* is defined as the maximum load resistance N in pounds or kilo-Newtons (KN) that the specimen will achieve at 140°F (60°C) under specified conditions. The *flow* is the total movement of the specimen in units of 0.01 in. (0.25 mm) during the stability test as the load is increased from zero to the maximum.

Test specimens for the Marshall method are prepared for a range of asphalt contents within the prescribed limits. Usually the asphalt content is measured by 0.5 percent increments from the minimum prescribed, ensuring that at least two are below the optimum and two above the optimum so that the curves obtained from the result will indicate a well-defined optimum. For example, for a specified amount of 5 to 7 percent, mixtures of 5, 5.5, 6, 6.5, and 7 are prepared. At least three specimens are provided for each asphalt content, to facilitate the provision of adequate data. For this example of five different asphalt contents, therefore, a total minimum of 15 specimens is required. The amount of aggregates required for each specimen is about 1.2 kg.

A quantity of the aggregates having the designed gradation is dried at a temperature between 105°C (221°F) and 110°C (230°F) until a constant weight is obtained. The mixing temperature for this procedure is set as the temperature that will produce a kinematic viscosity of 170 ± 20 centistokes, or a Saybolt Furol viscosity of 85 ± 10 sec, in the asphalt. The compacting temperature is that which will produce a kinematic viscosity of 280 ± 30 centistokes, or a Saybolt Furol viscosity of 160 ± 15 sec. These temperatures are determined and recorded.

The specimens containing the appropriate amounts of aggregates and asphalt are prepared by thoroughly mixing and compacting each mixture. The compactive effort used is either 35, 50, or 75 blows of the hammer falling a distance of 450 mm, depending on the design traffic category. After the application on one face, the sample mold is reversed and the same number of blows is applied to the other face of the sample. The specimen then is cooled and tested for stability and flow after determining its bulk density.

The bulk density of the sample usually is determined by weighing the sample in air and in water. It may be necessary to coat samples made from open-graded mixtures with paraffin before determining the density. The bulk specific gravity G_{mb} of the compacted sample is given as

$$G_{mb} = \frac{m_a}{m_a - m_w}$$

(18.4)

where

m_a = mass of sample in air (g)
m_w = mass of sample in water (g)

Stability Test

In conducting the stability test, the specimen is immersed in a bath of water at a temperature of 60 ± 1°C (140 ± 1.8°F) for a period of 30 to 40 min. It is then placed in the Marshall stability testing machine, as shown in Figure 18.12, and loaded at a constant rate of deformation of 2 in. (5 mm) per minute until failure occurs. The total load N in Newtons or pounds that causes failure of the specimen at 60°C (140°F) is noted as the Marshall stability value of the specimen. The total amount of deformation in units of 0.01 in. (0.25 mm) that occurs up to the point the load starts decreasing is recorded as the flow value. The total time between removing the specimen from the bath and completion of the test should not exceed 30 sec.

Analysis of Results from Marshall Test.   The first step in the analysis of the results is the determination of the average bulk specific gravity for all test specimens having the same asphalt content. The average unit weight of each mixture is then obtained by multiplying its average specific gravity by the density of water γ_w. A smooth curve that represents the best fit of plots of bulk density versus percentage of asphalt is determined, as shown in Figure 18.13a. This curve is used to obtain the bulk specific gravity values that are used in further computations, as in Example 18.2 on page 1001.

Figure 18.12   Marshall Stability Equipment

SOURCE: CONTROLS S.R.L. Used with permission.

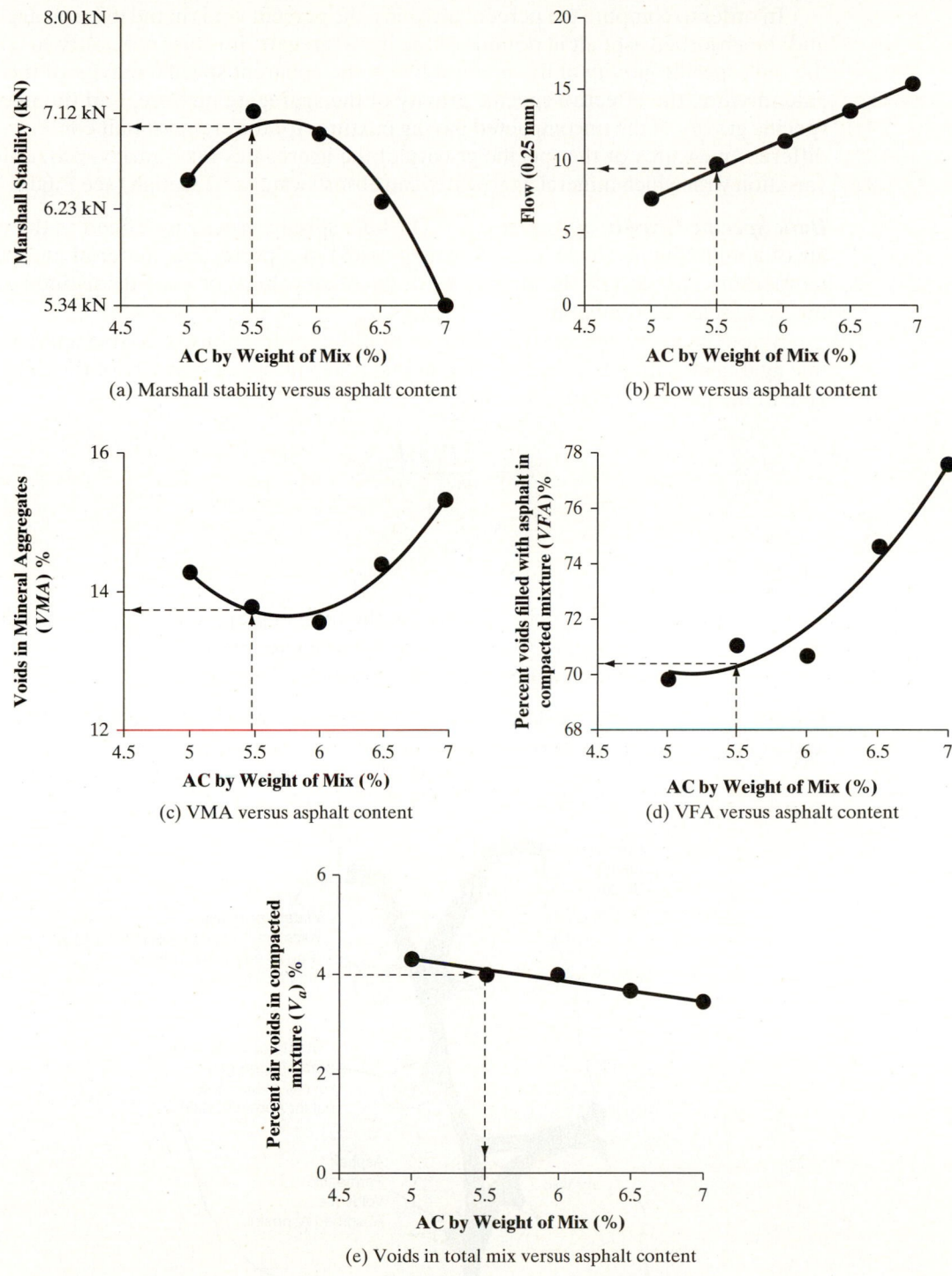

(a) Marshall stability versus asphalt content

(b) Flow versus asphalt content

(c) VMA versus asphalt content

(d) VFA versus asphalt content

(e) Voids in total mix versus asphalt content

Figure 18.13   Marshall Test Property Curves for Example 18.2 on page 999

In order to compute the percent air voids, the percent voids in the mineral aggregate, and the absorbed asphalt in pounds of the dry aggregate, it is first necessary to compute the bulk specific gravity of the mineral fillers, the apparent specific gravity of the aggregate mixture, the effective specific gravity of the aggregate mixture, and the maximum specific gravity of the uncompacted paving mixtures for different asphalt contents. These different measures of the specific gravity of the aggregates take into consideration the variation with which mineral aggregates can absorb water and asphalt (see Figure 18.14).

Bulk Specific Gravity of Aggregate.   The bulk specific gravity is defined as the mass in air of a unit volume (including all normal voids) of a permeable material at a selected temperature, divided by the mass in air of the same volume of gas-free distilled water at the same selected temperature.

Since the aggregate mixture consists of different fractions of coarse aggregate and fine aggregate with different specific gravities, the bulk specific gravity of the total aggregate in the paving mixture is given as

$$G_{sb} = \frac{P_1 + P_2 + \ldots + P_n}{\dfrac{P_1}{G_1} + \dfrac{P_2}{G_2} + \ldots + \dfrac{P_n}{G_n}} \tag{18.5}$$

where

G_{sb} = bulk specific gravity for the total aggregate $P_1 \, P_2 \ldots P_n$ = individual
percentages by mass of coarse aggregate
G_1, G_2, G_n = individual bulk specific gravities of aggregate

It is not easy to accurately determine the bulk specific gravity of the mineral filler. The apparent specific gravity may therefore be used with very little error.

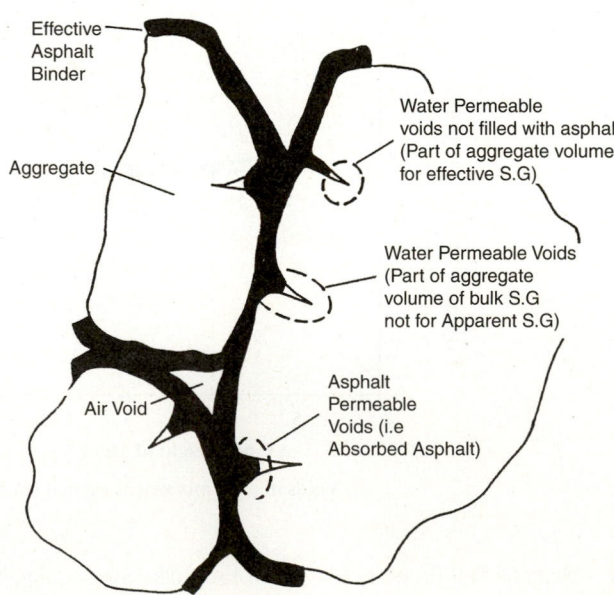

Figure 18.14   Bulk, Effective, and Apparent Specific Gravities; Air Voids; and Effective Asphalt Content in Compacted Asphalt Paving Mixture

Apparent Specific Gravity of Aggregates.   The apparent specific gravity is defined as the ratio of the weight in air of an impermeable material to the weight of an equal volume of distilled water at a specified temperature. The apparent specific gravity of the aggregate mix is therefore obtained as

$$G_{asb} = \frac{P_1 + P_2 + \ldots + P_n}{\dfrac{P_1}{G_1} + \dfrac{P_2}{G_2} + \ldots + \dfrac{P_n}{G_n}} \tag{18.6}$$

where

$$G_{asb} = \text{apparent specific gravity of the aggregate mixture}$$
$$P_1, P_2, P_n = \text{individual percentages by mass of aggregates}$$
$$G_1, G_2, G_n = \text{individual apparent specific gravities of aggregate}$$

Effective Specific Gravity of Aggregate.   The effective specific gravity of the aggregates is normally based on the maximum theoretical specific gravity of the paving mixture. Therefore, effective specific gravity of aggregates is defined as the specific gravity of the aggregates when all void spaces in the aggregate particles are included, with the exception of those that are filled with asphalt (see Figure 18.14.) It is given as

$$G_{se} = \frac{100 - P_b}{(100/G_{mm}) - (P_b/G_b)} \tag{18.7}$$

where

$$G_{se} = \text{effective specific gravity of the aggregates}$$
$$G_{mm} = \text{maximum theoretical specific gravity of paving mixture (no air voids)}$$
$$P_b = \text{asphalt percent by total weight of paving mixture (thus } P_s = 100 - P_b \text{ is the percent by mass of the base mixture that is not asphalt)}$$
$$G_b = \text{specific gravity of the asphalt}$$

Maximum Theoretical Specific Gravity of the Paving Mixture.   The maximum theoretical specific gravity of the paving mixture G_{mm} assumes that there are no air voids in the asphalt concrete. Although the G_{mm} can be determined in the laboratory by conducting the standard test (ASTM Designation D2041), the best accuracy is attained at mixtures near the optimum asphalt content. Since it is necessary to determine the G_{mm} for all samples, some of which contain much lower or much higher quantities than the optimum asphalt content, the following procedure can be used to determine the G_{mm} for each sample.

The ASTM Designation D2041 test is conducted on all specimens containing a selected asphalt cement content and the mean of these is determined. This value is then used to determine the effective specific gravity of the aggregates using Eq. 18.7. The effective specific gravity of the aggregates can be considered constant, since varying the asphalt content in the paving mixture does not significantly vary the asphalt absorption. The effective specific gravity obtained then is used to determine the maximum specific gravity of the paving mixtures with different asphalt cement contents using Eq. 18.8.

$$G_{mm} = \frac{100}{(P_s/G_{se}) + (P_b/G_b)} \tag{18.8}$$

where

G_{mm} = maximum specific gravity of paving mixture (no air voids)
P_s = percent by mass of aggregates in paving mixture
P_b = percent by mass of asphalt in paving mixture
G_{se} = effective specific gravity of the aggregates (assumed to be constant for different asphalt cement contents)
G_b = specific gravity of asphalt

Once these different specific gravities have been determined, the asphalt absorption, the effective asphalt content, the percent voids in mineral aggregates (*VMA*), and the percent air voids in the compacted mixture all can be determined.

Asphalt absorption is the percent by mass of the asphalt that is absorbed by the aggregates based on the total mass of the aggregates. This is given as

$$P_{ba} = 100\,\frac{G_{se} - G_{sb}}{G_{sb}G_{se}}\,G_b \qquad (18.9)$$

where

P_{ba} = amount of asphalt absorbed as a percentage of the total mass of aggregates
G_{se} = effective specific gravity of the aggregates
G_{sb} = bulk specific gravity of the aggregates
G_b = specific gravity of asphalt

Effective Asphalt Content.   The effective asphalt content is the difference between the total amount of asphalt in the mixture and that absorbed into the aggregate particles. The effective asphalt content is therefore that which coats the outside of the aggregate particles and influences the pavement performance. It is given as

$$P_{be} = P_b - \frac{P_{ba}}{100}\,P_s \qquad (18.10)$$

where

P_{be} = effective asphalt content in paving mixture (percent by total mass of the mix)
P_b = percent by mass of asphalt in paving mixture
P_s = percent of the aggregate by mass of the paving mixture
P_{ba} = amount of absorbed asphalt as a percentage of the total mass of aggregates

Percent Voids in Compacted Mineral Aggregates.   The percent voids in compacted mineral aggregates (*VMA*) is the percentage of void spaces between the granular particles in the compacted paving mixture, including the air voids and the volume occupied by the effective asphalt content. It usually is calculated as a percentage of the bulk volume of the compacted mixture based on the bulk specific gravity of the aggregates. It is given as

$$VMA = 100 - \frac{G_{mb}P_s}{G_{sb}} \qquad (18.11)$$

where

VMA = percent voids in compacted mineral aggregates (percent of bulk volume)
G_{mb} = bulk specific gravity of compacted mixture
G_{sb} = bulk specific gravity of aggregate
P_s = aggregate percent by weight of total paving mixture

Percent Air Voids in Compacted Mixture (V_a).   This is the ratio (expressed as a percentage) between the volume of the small air voids between the coated particles and the total volume of the mixture. It can be obtained from

$$V_a = 100 \, \frac{G_{mm} - G_{mb}}{G_{mm}} \qquad (18.12)$$

where

V_a = percent air voids in compacted paving mixture
G_{mm} = maximum theoretical specific gravity of the paving mixture
G_{mb} = bulk specific gravity of the compacted paving mixture

Percent of Voids Filled with Asphalt in Compacted Mixture (VFA).   This is the percent of the effective volume of asphalt to the volume of voids in mineral aggregates. This parameter expresses how rich or how dry the asphalt mix is and therefore how durable the asphalt mix is.

$$VFA = \frac{VMA - V_a}{VMA} \times 100 \qquad (18.13)$$

where

VFA = voids filled with asphalt, percent of VMA
VMA = voids in mineral aggregate, percent of bulk volume
V_a = air voids in compacted mixture, percent of total volume

Four additional separate smooth curves are drawn: percent voids in total mix versus percent of asphalt, percent voids in mineral aggregate versus percent of asphalt, Marshall stability versus percent of asphalt, and flow versus percent of asphalt. All of these graphs are then employed in the selection of the design asphalt content, as it is difficult to obtain a single asphalt content that will satisfy all the required criteria for the Marshall mix design as given in Table 18.7. Therefore, the optimum binder content is determined on the basis of optimizing percentage of air voids, voids in mineral aggregates, voids filled with asphalt, stability, and flow. AASHTO suggests that a suitable starting point is to select the asphalt content that corresponds to 4 percent air voids, as this is the median of the air voids limits (see Table 18.7). A comparison of each of the mix properties for this obtained asphalt content from the appropriate graph is then compared with the criteria given in Table 18.7. If all criteria are satisfied, then the asphalt content that corresponds to 4 percent air voids is selected as the preliminary asphalt content. If one or more criteria are not satisfied, it is necessary to compromise by making adjustments to the asphalt content or to redesign the mix. It should be noted that all criteria should be satisfied and not just the criterion for stability.

This analysis is illustrated further in Example 18.2.

Table 18.7 Suggested Test Limits

(a) Maximum and Minimum Values

Marshall Method Mix Criteria	*Light Traffic* *ESAL < 10⁴* *(see Chapter 19)*	*Medium Traffic* *10⁴ < ESAL < 10⁶* *(see Chapter 19)*	*Heavy Traffic* *ESAL > 10⁶* *(see Chapter 19)*
Compaction (no. of blows each end of specimen)	35	50	75
Stability N (lb)	3336 (750)	5338 (1200)	8006 (1800)
Flow, 0.25 mm (0.1 in.)	8 to 18	8 to 16	8 to 14
Air Voids (%)	3 to 5	3 to 5	3 to 5

(b) Mineral Percent Voids in Mineral Aggregates

Standard Sieve Designation	*Percent*
No. 16	23.5
No. 4	21
No. 8	18
⅜ in.	16
½ in.	15
¾ in.	14
1 in.	13
1½ in.	12
2 in.	11.5
2½ in.	11

SOURCE: Federal Highway Administration, U.S. Department of Transportation

Evaluation and Adjustment of Mix Design

As stated earlier, the overall objective of the mix design is to determine an optimum blend of different components that will satisfy the requirements of the given specifications. The mixture should have

- An adequate amount of asphalt to ensure a durable pavement
- An adequate mix stability to prevent unacceptable distortion and displacement when traffic load is applied
- Adequate voids in the total compacted mixture to permit a small amount of compaction when traffic load is applied without loss of stability, blushing, and bleeding, but, at the same time, with insufficient voids to prevent harmful penetration of air and moisture into the compacted mixture
- Adequate workability to facilitate placement of the mix without segregation

When the mix design for the optimum asphalt content does not satisfy all of the requirements given in Table 18.7, it is necessary to adjust the original blend of aggregates. Trial mixes can be adjusted by using the following general guidelines.

Low Voids and Low Stability.   In this situation, the voids in the mineral aggregates can be increased by adding more coarse aggregates. Alternatively, the asphalt content can be reduced, but only if the asphalt content is higher than that normally used and if the excess asphalt is not required as a replacement for the amount absorbed by the aggregates. Care should be taken when the asphalt content is reduced because this can lead to both a decrease in durability and an increase in permeability of the pavement.

Low Voids and Satisfactory Stability.   This mix can cause reorientation of particles and additional compaction of the pavement with time as continued traffic load is imposed on the pavement. This in turn may lead to instability or flushing of the pavement. Mixes with low voids should be altered by adding more aggregates.

High Voids and Satisfactory Stability.   When voids are high, it is likely that the permeability of the pavement also will be high, which will allow water and air to circulate through the pavement and result in premature hardening of the asphalt. High voids should be reduced to acceptable limits, even though the stability is satisfactory. This can be achieved by increasing the amount of mineral dust filler in the mix.

Satisfactory Voids and Low Stability.   This condition suggests low quality aggregates; the quality should be improved.

High Voids and Low Stability.   It may be necessary to carry out two steps in this case. The first step is to adjust the voids as discussed earlier. If this adjustment does not simultaneously improve the stability, the second step is to consider the improvement of the aggregate quality.

Example 18.2 Designing an Asphalt Concrete Mixture

In designing an asphalt concrete mixture for a highway pavement to support medium traffic, data in Table 18.8, showing the aggregate characteristics, and Table 18.9, showing data obtained using the Marshall method, were used. Determine the optimum asphalt content for this mix for the specified limits given in Table 18.7.

Solution: The bulk specific gravity of the mix for each asphalt cement content is determined by calculating the average value for the specimens with the same asphalt cement content using Eq. 18.4.

For example, for 5 percent asphalt content, the average bulk specific gravity is given as

$$G_{mb} = \frac{1}{3}\left(\frac{1325.6}{1325.6 - 780.1} + \frac{1325.4}{1325.4 - 780.3} + \frac{1325.0}{1325.0 - 779.8}\right)$$

$$= \frac{1}{3}(2.43 + 2.43 + 2.43)$$

$$= 2.43$$

Therefore, the bulk density is 2.43 × 1 = 2.43 gm/cm³. These results are shown in Table 18.9.

Table 18.8   Aggregate Characteristics for Example 18.2

Aggregate Type	Percent by Weight of Total Paving Mixture	Bulk Specific Gravity
Coarse	52.3	2.65
Fine	39.6	2.75
Filler	8.1	2.70

Note: The nominal maximum particle size in the aggregate mixture is 1 in.

Average bulk density is then plotted against asphalt content percent as shown in Figure 18.13a. Similarly, the average stability and flow for each asphalt cement content is determined, as shown in Table 18.9, and these values are plotted against asphalt content as shown in Figures 18.13b and c. For each specimen, we now

Table 18.9   Marshall Test Data and Results for Example 18.2

Asphalt % by Mass of Total Mix	Specimen Number	Mass of Specimen (g)		Volumetric Parameters						
		In Air	In Water	G_{mm}	G_{mb}	% VA	% VMA	% VFA	Stability	Flow
	a	1325.6	780.1	2.54	2.43	4.33	14.27	69.67	1460	7
5	b	1325.4	780.3		2.43	4.27	14.22	69.95	1450	7.5
	c	1325	779.8		2.43	4.32	14.26	69.72	1465	7
Average					**2.43**	**4.31**	**14.25**	**69.78**	**1458.33**	**7.17**
	a	1331.3	789.6	2.56	2.46	4.00	13.75	70.93	1600	10
5.5	b	1330.9	789.3		2.46	4.01	13.76	70.87	1610	9
	c	1331.8	790		2.46	3.98	13.74	71.02	1595	9.5
Average					**2.46**	**4.00**	**13.75**	**70.94**	**1601.67**	**9.50**
	a	1338.2	798.6	2.58	2.48	3.88	13.43	71.13	1560	11
6	b	1338.5	798.3		2.48	3.96	13.51	70.67	1540	11.5
	c	1338.1	797.3		2.47	4.10	13.63	69.94	1550	11
Average					**2.48**	**3.98**	**13.52**	**70.58**	**1550.00**	**11.17**
	a	1343.8	799.8	2.56	2.47	3.51	14.23	75.35	1400	13
6.5	b	1344	797.3		2.46	3.97	14.64	72.89	1420	13
	c	1343.9	799.9		2.47	3.50	14.22	75.39	1415	13.5
Average					**2.47**	**3.66**	**14.36**	**74.54**	**1411.67**	**13.17**
		1349	798.4	2.54	2.45	3.54	15.38	76.98	1200	16
7		1349.3	799		2.45	3.47	15.32	77.37	1190	15
		1349.8	800.1		2.46	3.33	15.20	78.11	1210	16
Average					**2.45**	**3.44**	**15.30**	**77.49**	**1200.00**	**15.67**

compute the average percent voids in the mineral aggregate (VMA) using Eq. 18.11, the percent air voids in the compacted mixture using (V_a) using Eq. 18.12, and the compacted mixture filled with asphalt (VFA) using Eq. 18.13. These are also plotted against asphalt content percent, as shown in Figures 18.13d-e.

$$VMA = 100 - \frac{G_{mb}P_s}{G_{sb}}$$

For example, for 5 percent asphalt content,

$$G_{mb} = 2.43$$

$$P_{ta} = 95.0 \quad \text{(total aggregate percent)}$$

Note: $P_{ta} = 100\%$ of asphalt

Use Eq. 18.5 to calculate G_{sb}.

$$G_{sb} = \frac{P_{ca} + P_{fa} + P_{mf}}{(P_{ca}/G_{bca}) + (P_{fa}/G_{bfa}) + (P_{mf}/G_{bmf})}$$

Determine P_{ca}, P_{fa}, and P_{mf} in terms of total aggregates.

$$P_{ca} = 0.523 \times 95.0 = 49.7$$

$$P_{fa} = 0.396 \times 95.0 = 37.6$$

$$P_{mf} = 0.081 \times 95.0 = 7.7$$

Therefore,

$$G_{sb} = \frac{49.7 + 37.6 + 7.7}{(49.7/2.65) + (37.6/2.75) + (7.7/2.70)} = 2.69$$

$$P_{ta} = (100 - 5) = 95$$

and

$$VMA = 100 - \frac{2.43 \times 95}{2.69} = 14.18$$

A plot of VMA versus asphalt content based on these calculations is shown in Figure 18.13d.

We now have to determine the percentage of air voids in each of the paving mixtures using Eq. 18.12.

$$V_a = 100 \frac{G_{mm} - G_{mb}}{G_{mm}}$$

For example, for 5 percent asphalt content,

$$V_a = \left(1 - \frac{G_{mb}}{G_{mm}}\right) \times 100 = \left(1 - \frac{2.43}{2.54}\right) \times 100 = 4.33\%$$

A plot of V_a versus asphalt content based on these calculations is shown in Figure 18.13e. Similarly, for 5 percent asphalt content,

$$VFA = \frac{VMA - V_a}{VMA} \times 100 = \frac{14.28 - 4.33}{14.28} \times 100 = 69.68\%$$

A plot of VFA versus asphalt content based on these calculations is shown in Figure 18.13d.

The asphalt content corresponding to the median binder content (4 percent) is determined as shown in Figure 18.13e, and this will be considered the initial optimum asphalt content; this value is 5.4 percent. The stability, flow, VMA, and VFA values corresponding to 5.4 percent are determined from figures 18.13 a, c and d respectively and checked against Marshall criteria given in Table 18.7. If all values are within the specified Marshall criteria, then the initial asphalt content is the optimum asphalt content. A plot of V_a versus asphalt content based on these calculations is shown in Figure 18.13e.

The values for the 4 percent mixture are

Unit weight = 2.46 gm/cc

Stability = 7.02 kN

Flow = 8.5 units of 0.25 mm

Percent voids in mineral aggregates = 13.75

This mixture meets all the criteria given in Table 18.7.

Example 18.3 Computing the Percent of Asphalt Absorbed

Using the information given in Example 18.2, determine the asphalt absorbed for the optimum mix. The maximum specific gravity for this mixture is 2.57 and the specific gravity of the asphalt cement is 1.02. From Eq. 18.9, the absorbed asphalt is given as

$$P_{ba} = 100 \frac{G_{se} - G_{sb}}{G_{sb} \times G_{se}} G_b$$

Solution: It is first necessary to determine the effective specific gravity of aggregates using Eq. 18.7.

$$G_{se} = \frac{100 - P_b}{(100/G_{mm}) - (P_b/G_b)}$$

$$G_{se} = \frac{100 - 5.6}{(100/2.57) - (5.6/1.02)} = 2.82$$

The bulk specific gravity of the aggregates in this mixture is

$$G_{sb} = \frac{0.523 \times 94.4 + 0.396 \times 94.4 + 0.081 \times 94.4}{\dfrac{0.523 \times 94.4}{2.65} + \dfrac{0.396 \times 94.4}{2.75} + \dfrac{0.081 \times 94.4}{2.7}} = 2.69$$

The asphalt absorbed is obtained from Eq. 18.9.

$$P_{ba} = 100 \frac{2.82 - 2.69}{2.60 \times 2.82} 1.02 \cong 1.81\%$$

18.5.2 Hot–Mix, Cold–Laid Asphalt Mixture

Asphalt mixtures in this category are manufactured hot and then shipped and laid immediately or they can be stockpiled for use at a future date. Thus, they are suitable for small jobs for which it may be uneconomical to set up a plant. They are also a suitable material for patching high-type pavements. The Marshall method of mix design can be used for this type of asphalt concrete, but high-penetration asphalt normally is used. The most suitable asphalt cements have been found to have penetrations within the lower limits of the 200 to 300 penetration grade.

Hot-mix, cold-laid asphalt mixtures are produced by first thoroughly drying the different aggregates in a central hot-mix plant and then separating them into several bins containing different specified sizes. One important factor in the production of this type of asphalt mixture is that the manufactured product should be discharged at a temperature of 77 c ± 10°. To achieve this, the aggregates are cooled to approximately 82 c after they are dried but before they are placed into the mixer. Based on the job-mix formula, the exact amount of the aggregates from each bin is weighed and placed in the mixer. The different sizes of aggregates are thoroughly mixed together and dried. About 0.75 percent by weight of a medium-curing cutback asphalt (MC-30), to which a wetting agent has been added, is mixed with the aggregates for another 10 sec. The high-penetration asphalt cement and water are then added simultaneously to the mixture. The addition of water is necessary to ensure that the material remains workable after it has cooled down to normal temperatures. The amount of asphalt cement added is the optimum amount obtained from the mix design, but the amount of water depends on whether the material is to be used within a few days or stockpiled for periods up to several months. When the material is to be used within a few days, 2 percent of water by weight is used; if it is to be stockpiled for a long period, 3 percent of water is used. The mixture is then mixed thoroughly for about 45 sec to produce a uniform mix.

18.5.3 Cold–Mix, Cold–Laid Asphalt Mixture

Emulsified asphalts and low-viscosity cutback asphalts are used to produce cold-mix asphalt mixtures. They also can be used immediately after production or stockpiled for use at a later date. The production process is similar to that of the hot-mix asphalts, except that the mixing is done at normal temperatures and it is not always necessary to dry the aggregates. However, saturated aggregates and aggregates with surface moisture should be dried before mixing. The type and grade of asphalt material used depends on whether the material is to be stockpiled for a long time, the use of the material, and the gradation of the aggregates. For example, MS-1 and MS-2 emulsified asphalts are used commonly in pavement bases and surfaces for open-graded aggregates, while CS-1 and CSH-1 are used commonly for well-graded aggregates.

Seal Coats

Seal coats are usually single applications of asphalt material that may or may not contain aggregates. The three types of seal coats commonly used in pavement maintenance are fog seals, slurry seals, and aggregate seals.

Fog Seal

Fog seal is a thin application of emulsified asphalt, usually with no aggregates added. Slow-setting emulsions, such as SS-1, SS-1H, CSS-1, and CSS-1H, are normally used for fog seals. The emulsion is sprayed at a rate of 0.46 to 0.92 l/m^2 after it has been diluted with clean water. Fog seals are used mainly to

- Reduce the infiltration of air and water into the pavement
- Prevent the progressive separation of aggregate particles from the surface downward or from the edges inward (raveling) in a pavement (raveling is mainly caused by insufficient compaction during construction carried out in wet or cold weather conditions)
- Bring the surface of the pavement to its original state

Slurry Seal

Slurry seal is a uniformly mixed combination of a slow-setting asphalt emulsion (usually SS-1), fine aggregate, mineral filler, and water. Mixing can be carried out in a conventional plastic mixer or in a wheelbarrow if the quantity required is small. It is usually applied with an average thickness of 1.6 mm to 3.2 mm.

Slurry seal is used as a low-cost maintenance material for pavements carrying light traffic. Note, however, that although the application of a properly manufactured slurry seal coat will fill cracks of about 6.4 mm or more and provide a fine-textured surface, existing cracks will appear through the slurry seal in a short time.

Aggregate Seals

Aggregate seals, also known as *chip seals,* are obtained by spraying asphalt, immediately covering it with aggregates, and then rolling the aggregates into the asphalt. Asphalts used for aggregate seals are usually the softer grades of paving asphalt and the heavier grades of liquid asphalts. Aggregate seals can be used to restore the surface of old pavements.

Prime Coats

Prime coats are obtained by spraying asphalt binder materials onto nonasphalt base courses. Prime coats are used mainly to

- Provide a waterproof surface on the base
- Fill capillary voids in the base
- Facilitate the bonding of loose mineral particles
- Facilitate the adhesion of the surface treatment to the base

Medium-curing cutbacks normally are used for prime coating with MC-30 recommended for priming a dense flexible base and MC-70 for more granular-type base materials. The rate of spray is usually between 0.92 and 1.6 l/m^2 for the MC-30 and between 1.42 and 2.8 l/m^2 for the MC-70. The amount of asphalt binder used, however, should be the maximum that can be absorbed completely by the base within 24 hours of application under favorable weather conditions. The base course must contain a nominal amount of water to facilitate the penetration of the asphalt material into the base. It is therefore necessary to lightly spray the surface of the base course with water just before the application of the prime coat if its surface has become dry and dusty.

Tack Coats

A *tack coat* is a thin layer of asphalt material sprayed over an old pavement to facilitate the bonding of the old pavement and a new course which is to be placed over the old pavement. In this case, the rate of application of the asphalt material should be limited, since none of this material is expected to penetrate the old pavement. Asphalt emulsions such as SS-1, SS-1H, CSS-1, and CSS-1H normally are used for tack coats after they have been thinned with an equal amount of water. The rate of application varies from 0.23 to 0.69 l/m^2 of the thinned material. Rapid-curing cutback asphalts such as RC-70 also may be used as tack coats.

Sufficient time must elapse between the application of the tack coat and the application of the new course to allow for adequate curing of the material through the evaporation of most of the diluter in the asphalt emulsion. This curing process usually takes several hours in hot weather but can take more than 24 h in cooler weather. When the material is satisfactorily cured, it becomes a highly viscous, tacky film.

Surface Treatments

Asphalt surface treatments are obtained by applying a quantity of asphalt material and suitable aggregates on a properly constructed flexible base course to provide a suitable wearing surface for traffic. Surface treatments are used to protect the base course and to eliminate the problem of dust on the wearing surface. They can be applied as a single course with thicknesses varying from 12.5 mm to 19 mm, or a multiple course with thicknesses varying from 22 mm to 51 mm.

A single-course asphalt treatment is obtained by applying a single course of asphalt material and a single course of aggregates. The rate of application of the asphalt material for a single course varies from 0.6 to 1.9 l/m^2 depending on the gradation of the aggregates used; the rate of application of the aggregates varies from 0.0037 to 0.017 m^3/m^2. Multiple-course asphalt surface treatments can be obtained either as a double asphalt surface treatment consisting of two courses of asphalt material and aggregates or as a triple asphalt treatment consisting of three layers. The multiple-course surface treatments are constructed by first placing a uniform layer of coarse aggregates over an initial application of the bituminous materials and then applying one or more layers of bituminous materials and smaller aggregates, with each layer having a thickness that is approximately equal to the nominal maximum size of the aggregates used for that layer. The maximum aggregate size of each layer subsequent to the initial layer usually is taken as one-half that of the aggregates used in the preceding layer. The recommended rates of application of the asphalt material and the aggregates are shown in Table 18.10.

18.6 SUPERPAVE SYSTEMS

As part of the Strategic Highway Research Program (SHRP), another system for specifying the asphalt materials in asphalt concrete has been developed. This system is known as *Superpave*, which is a shortened form for *su*perior *per*forming asphalt *pave*ments. The research leading to the development of this new system was initiated because prior to Superpave, it was difficult to relate the results obtained from laboratory analysis in

Table 18.10 Quantities of Materials for Bituminous Surface Treatments

Surface Treatment			Aggregate		Bituminous Material[a]
Type	Application	Size No.[b]	Nominal Size (Square Openings)	Typical Rate of Application, ft³/yd²	Typical Rate of Application, gal/yd²
Single	Initial	5	1 to ½ in.	0.50	0.42
		6	¾ to ⅜ in.	0.36	0.37
		7	½ in. to No. 4	0.23	0.23
		8	⅜ in. to No. 8	0.17	0.19
		9	No. 4 to No. 16	0.11	0.13
Double	Initial	5	1 to ½ in.	0.50	0.42
	Second	7	½ in. to No. 4	0.25	0.26
Double	Initial	6	¾ to ⅜ in.	0.36	0.37
	Second	8	⅜ in. to No. 8	0.18	0.20
Triple	Initial	5	1 to ½ in.	0.50	0.42
	Second	7	½ in. to No. 4	0.25	0.26
	Third	9	No. 4 to No. 16	0.13	0.14
Triple	Initial	6	¾ to ⅜ in.	0.36	0.37
	Second	8	¾ in. to No. 8	0.18	0.20
	Third	9	No. 4 to No. 16	0.13	0.14

Note: The values are typical design or target values and are not necessarily obtainable to the precision indicated.
[a]Experience has shown that these quantities should be increased slightly (5 to 10 percent) when the bituminous material to be used was manufactured for application with little or no heating.
[b]According to Specification D448.
SOURCE: *Annual Book of ASTM Standards, Section 4, Construction,* Vol. 04.03, Road and Paving Materials; Pavement Management Technologies, American Society for Testing and Materials, Philadelphia, PA, 2012

the existing systems to the performance of the pavement without field experience. For example, the old systems used the results of tests performed at standard test temperatures to determine whether the materials satisfied the specifications. However, these tests are mainly empirical, and field experience is required to determine whether the results obtained have meaningful information. The Superpave system includes a method for specifying asphalt binders and mineral aggregates, an asphalt mixing design, and a procedure for analyzing and predicting pavement performance. The material presented here is based on several reports published by the National Research Council on research conducted under the SHRP as noted in *The Superpave Mix Design Manual for New Construction and Overlays*. The objective of this mix design is to obtain a mixture of asphalt and aggregates that has the following characteristics:

- Sufficient asphalt binder
- Sufficient voids in the mineral aggregates (VMA) and air voids
- Sufficient workability
- Satisfactory performance characteristics over the service life of the pavement

A major difference between the Superpave mix design and other design methods, such as the Marshal and Hveem methods, is that the Superpave mix design method mainly uses performance-based and performance-related characteristics as the selection criteria for the mix design. Figure 18.15 illustrates the structure of the Superpave mix design system. An associated Superpave software is available for use by designers.

The Superpave system is unique in that it is performance-based and engineering principles can be used to relate the results obtained from its tests and analyses to field performance.

The system consists of the following parts:

- Selection of materials
- Volumetric trial mixture design
- Selection of final mixture design

18.6.1 Selection of Materials

Selection of materials includes the selection of asphalt binder and suitable mineral aggregates.

Selection of Asphalt Binder

The binder selected for a given project is based on the range of temperatures to which the pavement will be exposed and the traffic it will carry during its lifetime. The binders are classified with respect to the range of temperatures at which their physical property requirements must be met. For example, a binder that is classified as PG52-34 must satisfy all the high-temperature physical property requirements at temperatures up to at least 52°C and all the low-temperature physical requirements down to at least 34°C. The high-pavement design temperature is defined to be at a depth of 20 mm below the pavement surface and the low-pavement design temperature is defined to be at the surface of the pavement. Table 18.11 lists the more commonly used asphalt binder grades with their associated physical properties. The selection can be made in one of three ways.

1. The designer may select a binder based on the geographic location of the pavement.
2. The designer may determine the design pavement temperatures.

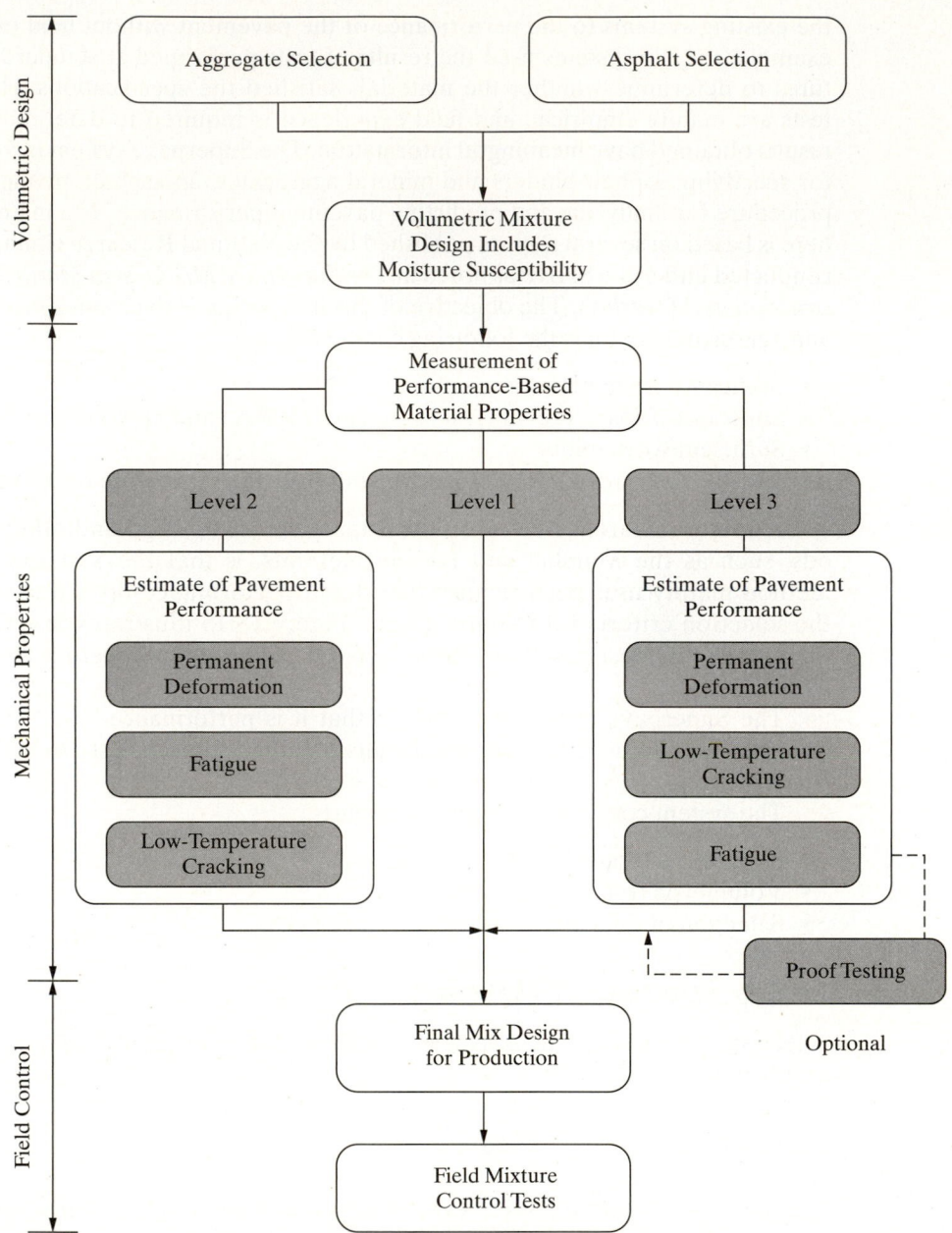

Figure 18.15   Structure of the Superpave Mix Design System

SOURCE: *The Superpave Mix Design Manual for New Construction and Overlays,* Strategic Highway Research Program, National Research Council, Washington, D.C., 1994

3. The designer may determine the design air temperatures that are then converted to design pavement temperatures.

In the first method, the designer selects a binder grade from a map that has been prepared by her agency indicating the binder that should be used for different locations. These maps usually are based on weather characteristics and/or policy

Table 18.11 Performance-Graded Asphalt Binder Specification

Performance Grade	PG 46			PG 52							PG 58					PG 64					
	-34	-40	-46	-10	-16	-22	-28	-34	-40	-46	-16	-22	-28	-34	-40	-10	-16	-22	-28	-34	-40
Average 7-day Maximum Pavement-Design Temperature, °C[a]	<46			<52							<58					<64					
Minimum Pavement-Design Temperature, °C[a]	>-34	>-40	>-46	>-10	>-16	>-22	>-28	>-34	>-40	>-46	>-16	>-22	>-28	>-34	>-40	>-10	>-16	>-22	>-28	>-34	>-40
Original Binder																					
Flash-Point Temp, T48: Minimum °C	230																				
Viscosity, ASTM D 4402:[b] Maximum, 3 Pa·s (300 cP), Test Temp, °C	135																				
Dynamic Shear, TP5:[c] G*/sin δ, Minimum, 1.00 kPa, Test Temperature @ 10 rad/s, °C	46			52							58					64					
Rolling Thin-Film Oven (T240) or Thin-Film Oven (T179) Residue																					
Mass Loss, Maximum, %	1.00																				
Dynamic Shear, TP5: G*/sin δ, Minimum, 2.20 kPa, Test Temp @ 10 rad/s, °C	46			52							58					64					
Pressure Aging Vessel Residue (PPI)																					
PAV Aging Temperature, °C[d]	90			90							100					100					
Dynamic Shear, TP5: G*sin δ, Maximum, 500 kPa, Test Temp @ 10 rad/s, °C	10	7	4	25	22	19	16	13	10	7	25	22	19	16	13	31	28	25	22	19	16
Physical Hardening[e]	Report																				
Creep Stiffness, TP1:[f] S, Maximum, 300 MPa, m-value, Minimum, 0.300, Test Temp, @ 60 s, °C	-24	-30	-36	0	-6	-12	-18	-24	-30	-36	-6	-12	-18	-24	-30	0	-6	-12	-18	-24	-30
Direct Tension, TP3:[f] Failure Strain, Minimum, 1.0%, Test Temp @ 1.0 mm/min, °C	-24	-30	-36	0	-6	-12	-18	-24	-30	-36	-6	-12	-18	-24	-30	0	-6	-12	-18	-24	-30

(Continued)

Table 18.11 Performance–Graded Asphalt Binder Specification (continued)

Performance Grade	PG 70						PG 76					PG 82				
	−10	−16	−22	−28	−34	−40	−10	−16	−22	−28	−34	−10	−16	−22	−28	−34
Average 7-day Maximum Pavement-Design Temperature, °C[a]	<70						<76					<82				
Minimum Pavement-Design Temperature, °C[a]	>−10	>−16	>−22	>−28	>−34	>−40	>−10	>−16	>−22	>−28	>−34	>−10	>−16	>−22	>−28	>−34
Original Binder																
Flash-Point Temperature, T48: Minimum °C	230															
Viscosity, ASTM D4402:[b] Maximum, 3 Pa · s (3000 cP), Test Temperature, °C	135															
Dynamic Shear, TP5:[c] G*/sin δ, Minimum, 1.00 kPa, Test Temperature @ 10 rad/s, °C	70						76					82				
Rolling Thin-Film Oven (T240) or Thin-Film Oven (T179) Residue																
Mass Loss, Maximum, %	1.00															
Dynamic Shear, TP5: G*/sin δ, Minimum, 2.20 kPa, Test Temp @ 10 rad/s, °C	70						76					82				
Pressure-Aging Vessel Residue (PPI)																
PAV Aging Temperature, °C[d]	100(110)						100(110)					100(110)				
Dynamic Shear, TP5: G*sin δ, Maximum, 5000 kPa, Test Temp @ 10 rad/s, °C	34	31	28	25	22	19	37	34	31	28	25	40	37	34	31	28
Physical Hardening[e]	Report															
Creep Stiffness, TP1:[f] S, Maximum, 300 MPa m-value, Minimum, 0.300, Test Temp, @ 60s, °C	0	−6	−12	−18	−24	−30	0	−6	−12	−18	−24	0	−6	−12	−18	−24
Direct Tension, TP3: Failure Strain, Minimum, 1.0%, Test Temp @ 1.0 mm/min, °C	0	−6	−12	−18	−24	−30	0	−6	−12	−18	−24	0	−6	−12	−18	−24

[a]Pavement temperatures can be estimated from air temperatures using an algorithm contained in the Superpave software program, or may be provided by the specifying agency.

[b]This requirement may be waived at the discretion of the specifying agency if the supplier warrants that the asphalt binder can be adequately pumped and mixed at temperatures that meet all applicable safety standards.

[c]For quality control of unmodified asphalt cement production, measurement of the viscosity of the original asphalt cement may be substituted for dynamic shear measurements of G*/sin δ at test temperatures where the asphalt is a Newtonian fluid. Any suitable standard means of viscosity measurement may be used, including capillary or viscometer (AASHTO T201 or T202).

[d]The PAV aging temperature is based on simulated climatic conditions and is one of three temperatures: 90°C, 100°C, or 110°C. The PAV aging temperature is 100°C for PG64 and above, except in desert climates, where it is 110°C.

[e]Physical Hardening-TP1 is performed on a set of asphalt beams according to Sections 13.1 of TP1, except the conditioning time is extended to 24 h ±10 min at 10°C above the minimum performance temperature. The 24-h stiffness and m-value are reported for information purposes only.

[f]If the creep stiffness is below 300 MPa, the direct tension test is not required. If the creep stiffness is between 300 and 600 MPa, the direct tension failure strain requirement can be used in lieu of the creep stiffness requirement. The m-value requirement must be satisfied in both cases.

SOURCE: *The Superpave Mix Design Manual for New Construction and Overlays*, Strategic Highway Research Program, National Research Council, Washington, D.C., 1994

decisions. In the second method, the designer determines the design pavement temperatures that should be used. In the third method, the designer uses the information given in the Superpave system on the maximum and minimum air temperatures of the project location and then converts these temperatures to the design pavement temperatures.

The Superpave system makes use of a temperature database consisting of data from 6092 reporting weather stations in the United States and Canada that have been in operation for 20 years or more. In determining the maximum temperature, the hottest seven-day period for each year was identified, and the average of the maximum air temperature during each of these periods was selected as the maximum temperature. The average of the minimum one-day air temperature for each year was selected as the minimum temperature. In addition, the standard deviations of the seven-day average maximum temperature and that for the one-day minimum air temperature for each year were determined. These standard deviations are determined to facilitate the use of reliability measurements in selecting the design pavement temperatures. For example, consider a location where the mean seven-day maximum temperature is 40°C with a standard deviation of ±1°C. There is a 50 percent chance that during an average year, the seven-day maximum air temperature will exceed 40°C, but only a 2 percent chance for it to exceed 42°C (mean plus two standard deviations), assuming a normal temperature distribution. Selecting a maximum air temperature of 42°C at this location will provide a reliability of 98 percent that the maximum air temperature will not be exceeded. Table 18.12 gives asphalt binder grades and reliability for selected cities.

The Superpave system also considers the fact that the pavement temperature and not the air temperature should be used as the design temperature. The system therefore uses the expression given in Eq. 18.14 to convert the maximum air temperature to the maximum design pavement temperature.

$$T_{20mm} = (T_{air} - 0.00618 Lat^2 + 0.2289 Lat + 42.2)(0.9545) - 17.78 \qquad (18.14)$$

where

T_{20mm} = high-pavement design temperature at a depth of 20 mm
T_{air} = seven-day average high air temperature (°C)
Lat = the geographical latitude of the project location (degrees)

The low-pavement design temperature can be selected as either the low air temperature, which is rather conservative, or can be determined from the low air temperature using the expression

$$T_{pav} = 1.56 + 0.72 T_{air} - 0.004 Lat^2 + 6.26 \log_{10}(H + 25) - Z(4.4 + 0.52\sigma_{air}^2)^{1/2} \quad (18.15)$$

where

T_{pav} = low AC-pavement temperature below surface (°C)
T_{air} = low air temperature (°C)
Lat = latitude of the project location (degrees)
H = depth of pavement surface (mm)
σ_{air} = standard deviation of the mean low air temperature (°C)
Z = from the standard normal distribution table, $Z = 2.055$ for 98 percent reliability

Table 18.12 Asphalt Binder Grades and Reliability for Selected Cities

ST	Station	Latitude	Min. 50% Grade	Actual Reliability High	Low	Min. 98% Grade	Actual Reliability High	Low
AL	Mobile	30.68	PG58-10	84	99	PG64-10	99.9	99
AK	Juneau 2	58.30	PG40-16	91	70	PG46-28	99.9	99
AZ	Phoenix WSFO AP	33.43	PG70-10	99.9	99.9	PG70-10	99.9	99.9
AR	Little Rock FAA AP	34.73	PG58-10	69	64	PG64-16	99.9	97
CA	Los Angeles WSO AP	33.93	PG52-10	66	99.9	PG58-10	99.9	99.9
	Sacramento WSO CI	38.58	PG58-10	61	99.9	PG64-10	99.9	99.9
	San Francisco WSO AP	37.62	PG52-10	98	99.9	PG52-10	98	99.9
CO	Denver WSFO AP	39.77	PG58-22	99.9	78	PG58-28	99.9	99
CT	Hartford WSO AP	41.93	PG52-22	54	89	PG58-28	99.9	99.7
DC	Wash Natl WSO AP	38.85	PG58-10	99.9	57	PG58-16	99.9	99
DE	Wilmington WSO AP	39.67	PG58-16	99	84	PG58-22	99	99.4
FL	Jacksonville WSO AP	30.50	PG58-10	91	98.6	PG64-10	99.9	98.6
	Miami	25.80	PG58-10	99	99.9	PG58-10	99	99.9
GA	Atlanta WSO AP	33.65	PG58-10	90	64	PG64-16	99.9	96.8
HI	Lahaina 361	20.88	PG58-10	99.9	99.9	PG58-10	99.9	99.9
IA	Des Moines WSFO AP	41.53	PG58-22	98	67	PG58-58	98	99.3
ID	Boise WSFO AP	43.57	PG58-16	93	61	PG64-28	99.9	99.6
IL	Chicago O'Hare WSO AP	41.98	PG52-22	58	67	PG58-28	99.9	99.3
	Peoria WSO AP	40.67	PG58-22	99.9	85	PG58-28	99.9	99.9
IN	Indianapolis SE Side	39.75	PG58-22	99	89	PG58-28	99	99.7
KS	Wichita WSO AP	37.65	PG64-16	99.9	57	PG64-22	99.9	98.5
KY	Lexington WSO AP	38.03	PG58-16	98	50	PG58-28	98	99.8
	Louisville WSFO	38.18	PG58-16	96	67	PG64-22	99.9	95
LA	New Orleans WSCMO AP	29.98	PG58-10	97	98.5	PG64-10	99.9	98.5
MA	Lowell	42.65	PG52-16	58	65	PG58-28	99.9	99.1
MD	Baltimore WSO AP	39.18	PG58-16	99.9	91	PG58-22	99.9	99.9
ME	Portland	43.67	PG52-16	97	59	PG58-28	99.9	98.7
MI	Detroit City AP	42.42	PG52-16	56	77	PG58-22	99.9	99.7
	Sault Ste. Marie WSO	46.47	PG52-28	99.9	90	PG52-34	99.9	99.9
MN	Duluth WSO AP	46.83	PG52-28	99.9	57	PG52-34	99.9	98.5
	Minn-St Paul WSO AP	44.88	PG52-28	71	90	PG58-34	99.9	99.9
MO	Kansas City FSS	39.12	PG58-16	80	57	PG64-22	99.9	98.5
	St. Louis WSCMO AP	38.75	PG58-16	91	57	PG64-22	99.9	98.5
MS	Jackson WSFO AP	32.32	PG58-10	55	77	PG64-16	99.9	99.7
MT	Great Falls	47.52	PG52-28	69	67	PG58-34	99.9	95
NC	Charlotte WSO AP	35.33	PG58-10	92	68	PG64-16	99.9	99.3
	Raleigh 4 SW	35.73	PG58-10	93	68	PG64-16	99.9	99.3

Table 18.12 Asphalt Binder Grades and Reliability for Selected Cities (*continued*)

ST	Station	Latitude	Min. 50% Grade	Actual Reliability High	Low	Min. 98% Grade	Actual Reliability High	Low
ND	Bismarck WSFO AP	46.77	PG58-34	99.9	88	PG58-40	99.9	99.6
NE	Omaha (North) WSFO	41.37	PG58-22	98	68	PG58-28	98	99.3
NH	Concord WSO AP	43.20	PG52-22	61	50	PG58-34	99.9	99.8
NJ	Atlantic City	39.38	PG52-10	77	68	PG58-16	99.9	99.3
NM	Laguna	35.03	PG58-16	81	64	PG64-22	99.9	96.7
NV	Reno WSFO AP	39.50	PG58-16	97	71	PG64-22	99.9	98
NY	Albany	42.65	PG52-22	58	94.5	PG58-22	99.9	94.5
	Buffalo WSFO AP	42.93	PG52-16	96	57	PG58-22	99.9	98.5
	New York Inter AP	40.65	PG52-16	61	97.1	PG58-16	99.9	97.1
OH	Cin Muni–Lunken Fld	39.10	PG58-16	99.9	64	PG58-22	99.9	96.7
	Cleveland WSO AP	41.42	PG52-16	51	50	PG58-28	99.9	99.8
	Columbus	39.98	PG58-16	99.9	50	PG58-28	99.9	99.8
OK	Oklahoma City WSFO AP	35.40	PG64-16	99.9	91	PG64-22	99.9	99.9
OR	Oregon City	45.35	PG52-10	55	89	PG58-16	99.9	99.7
PA	Philadelphia Drexel U	39.95	PG58-10	99.9	57	PG58-16	99.9	98.5
	Pittsburgh WSO CI	40.45	PG58-16	99.9	77	PG58-22	99.9	99.7
RI	Providence WSO AP	41.73	PG52-16	71	68	PG58-22	99.9	99.3
SC	Columbia WSFO AP	33.95	PG58-10	61	77	PG64-16	99.9	99.7
SD	Sioux Falls WSFO AP	43.57	PG58-28	99.9	87	PG58-34	99.9	99.9
TN	Knoxville U of Tenn	35.95	PG58-16	93	92	PG64-22	99.9	99.8
	Memphis FAA–AP	35.05	PG58-10	65	55	PG64-16	99.9	95
TX	Amarillo WSO AP	35.23	PG58-16	66	68	PG64-22	99.9	99.3
	Dallas FAA AP	32.85	PG64-10	99	77	PG64-16	99	99.7
	Houston FAA AP	29.65	PG64-10	99.9	99.3	PG64-10	99.9	99.3
UT	Salt Lake City WSO CI	40.77	PG58-16	98	84	PG58-22	98	99.4
VA	Norfolk WSO AP	36.90	PG58-10	98	85	PG58-16	98	99.9
	Richmond WSO AP	37.50	PG58-16	95	97.1	PG64-16	99.9	97.1
VT	Burlington WSO AP	44.47	PG52-22	93	50	PG58-28	99.9	97
WA	Seattle Tac WSCMO AP	47.45	PG52-10	99.9	83	PG52-16	99.9	98.5
	Spokane	47.67	PG58-16	98	59	PG58-28	98	98.7
WI	Milwaukee WSO AP	42.95	PG52-22	77	71	PG58-28	99.9	98
WV	Charleston WSFO AP	38.37	PG58-16	99	71	PG58-22	99	98
WY	Cheyenne WSFO AP	41.15	PG52-22	68	55	PG58-28	99.9	94.8
PR	San Juan WSFO	18.43	PG58-10	98	99.9	PG58-10	98	99.9

SOURCE: Adapted from *Weather Data Base for the Superpave Mix Design System*, Report ShRP A-648, Strategic Highway Research Program National Research Council, Washington, D.C., 1994.

Adjusting Binder Grade for Traffic Speed and Loading

The procedure described above for selecting the asphalt binder is based on an assumed traffic condition consisting of a designed number of fast transient loads. The selected binder should therefore be adjusted for traffic conditions that are different from that assumed in the procedure, as the speed of loading has an additional effect on the ability of the pavement to resist permanent deformation at the high temperature condition. When the design loads are moving slowly, the selected asphalt binder based on the procedure described earlier should be shifted higher one high-temperature grade. For example, when the standard procedure gives a PG52-28, a PG58-28 should be used for a slow-moving load. Also, when the design load is stationary, the binder selected from the procedure should be shifted higher two high-temperature grades. In addition to shifting the selected binder grade for speed, the designer should adjust the binder for the accumulated traffic load. For equivalent single-axle loads (ESAL)—see Chapter 19 for the definition of ESAL—of 10,000,000 to 30,000,000, the engineer should consider shifting the binder selected based on the procedure by one high-temperature binder grade, but for ESALs exceeding 30,000,000, a shift of one high-temperature grade is required.

The steps used to determine a suitable asphalt binder grade can be summarized as follows:

Step 1. Identify a weather station that is in the vicinity of the project location. If the project location is remote from the established stations, identify as many as three stations. Use the data at these stations to evaluate and estimate climate conditions at the project site. Note that weather data for a given site may be specified by a specific agency.

Step 2. Select a degree of design reliability for high- and low-temperature performance based on the agency's policy or use engineering judgment based on maintenance and rehabilitation costs.

Step 3. Using the selected design reliability, estimate the associated design pavement temperatures and associated risks.

Step 4. Determine the grade of asphalt binder that will just satisfy the minimum and maximum design pavement temperatures obtained from Steps 2 and 3.

Step 5. Select higher grades based on traffic characteristics, as shown in Table 18.13.

Note that the Superpave software can be used to conduct these steps.

Example 18.4 Determining a Suitable Binder Grade Using High and Low Air Temperatures

The latitude at a location where a high-speed rural road is to be located is 41°. The seven-day average high air temperature is 50°C and the low air temperature is −20°C. The standard deviation for both the high and low temperatures is ±1°C. Determine a suitable binder that could be used for the pavement of this highway if the depth of the pavement surface is 155 mm and the expected ESAL is 9×10^6.

Solution: Use Eq. 18.13 to determine the high-pavement temperature at a depth of 20 mm.

$$T_{20mm} = (T_{air} - 0.00618 Lat^2 + 0.2289 Lat + 42.2)(0.9545) - 17.78$$
$$= (50 - 0.00618(41)^2 + 0.2289(41) + 42.2)(0.9545) - 17.78$$
$$= 87.0 - 17.78$$
$$= 69.27°C$$

Use Eq. 18.14 to determine low-AC-pavement temperature.

$$T_{pav} = 1.56 + 0.72 T_{air} - 0.004 Lat^2 + 6.26 \log_{10}(155 + 25) - 2.055(4.4 + 0.52 \times 1)^{1/2}$$
$$= 1.56 + 0.72(-20) - 0.004(41)^2 + 6.26 \log_{10}(155 + 25) - 2.055(4.4 + 0.52 \times 1)^{1/2}$$
$$= -10.0°C$$

Select appropriate binder from Table 18.11. For minimum 50 percent reliability, binder is PG 70-10.

Note: No correction is needed for ESAL $< 10 \times 10^6$.

Determine seven-day maximum temperature for 98 percent reliability: Standard deviation $= \pm 1$.

There is a 2 percent chance that the seven-day maximum will exceed $(50 + 2 \times 1)°C$ (i.e., maximum air temperature for 98 percent reliability is 52°C).

Use Eq. 18.13 to determine the high-pavement temperature at a depth of 20 mm:

$$T_{20mm} = (T_{air} - 0.00618 Lat^2 + 0.2289 Lat + 42.2)(0.9545) - 17.78$$
$$= (52 - 0.00618(41)^2 + 0.2289 \times 41 + 42.2) - 17.78$$
$$= 17.18°C$$

Select the binder for 98 percent reliability from Table 18.13; use PG 76-10.

Example 18.5 Determining a Suitable Binder Grade

An urban interstate highway is being designed to carry an equivalent single-axle load (ESAL) of 31×10^6. It is anticipated that this road will be congested most of the time, resulting in slow-moving traffic. If the road is located in Richmond, Virginia, determine an appropriate asphalt binder grade for this project.

Solution: Use Table 18.12 to select binder based on latitude. For minimum 50 percent reliability, binder is PG 58-16. For minimum 98 percent reliability, binder is PG 64-16.

- Correct for slow-moving traffic: Move binder one grade higher. For minimum 50 percent reliability, binder is PG 64-16. For minimum 98 percent reliability, binder is PG 70-16.
- Correct for ESAL $> 30 \times 10^6$: Move binder one grade higher. For minimum 50 percent reliability, binder is PG 70-16. For minimum 98 percent reliability, binder is PG 76-16.

Table 18.13 Selection of Asphalt Binder Performance Grades on the Basis of Climate, Traffic Speed, and Traffic Volume

Recommendation for Selecting Binder Performance Grades							
Loads	*High-Pavement Design Temperature °C*						
Standing	*28 to 34*	*34 to 40*	*40 to 46*	*46 to 52*	*52 to 58*	*58 to 64*	*64 to 70*
(50 K/H) Slow Transient	*34 to 40*	*40 to 46*	*46 to 52*	*52 to 58*	*58 to 64*	*64 to 70*	*70 to 76*
(100 K/H) Fast Transient	*34 to 46*	*46 to 52*	*52 to 58*	*58 to 64*	*64 to 70*	*70 to 76*	*76 to 82*
>−10	PG46-10	PG52-10	PG58-10	PG64-10	PG70-10	PG76-10	PG82-10
−10 to −16	PG46-16	PG52-16	PG58-16	PG64-16	PG70-16	PG76-16	PG82-16
−16 to −22	PG46-22	PG52-22	PG58-22	PG64-22	PG70-22	PG76-22	PG82-22
−22 to −28	PG46-28	PG52-28	PG58-28	PG64-28	PG70-28	PG76-28	PG82-28
−28 to −34	PG46-34	PG52-34	PG58-34	PG64-34	PG70-34	PG76-34	PG82-34
−34 to −40	PG46-40	PG52-40	PG58-40	PG64-40	PG70-40		
−40 to −46	PG46-46	PG52-46	PG58-46	PG64-46			

Low-Pavement Design Temperature °C (row axis label)

Alaska-Canada Northern U.S.	*Canada North U.S.*	*Southern U.S.*	*Southwest U.S.-Desert Continental U.S.-Slow/ Heavy Traffic*

1. Select the type of loading.
2. Move horizontally to the high-pavement design temperature.
3. Move down the low-pavement design temperature.
4. Identify the binder grade.
5. ESALS $> 10^7$ consider increase of one high-temperature grade.
ESALS $> 3 \times 10^7$ increase one high-temperature grade.

Example:
Standing load, high design temperature 55°C
Low design temperature $= -25°C$
Grade = PG70-28
SOURCE: *The Superpave Mix Design Manual for New Construction and Overlays*, Strategic Highway Research Program National Research Council, Washington, D.C., 1994.

Selection of Mineral Aggregate

Based on the results obtained from surveying pavement experts, two categories of aggregate properties were identified by the developers of Superpave for use in the system. These are referred to as consensus standards and source (agency) standards. In addition to the consensus and source standards, Superpave uses a gradation system that is based on the 0.45 power gradation chart to determine the design aggregate structure. The aggregate characteristics that generally were accepted by the experts as critical for good performance of the hot-mix asphalt (HMA) are classified as *consensus proper-ties*. These properties include the angularity of the coarse aggregates, the angularity of the fine aggregates, the amount of flat and elongated particles in the coarse aggregates, and the clay content. The angularity of the coarse aggregate is defined as the percent by weight of coarse aggregates larger than 4.75 mm with one or more fractured faces. The angularity of coarse aggregates can be determined by conducting the Pennsylvania Department of Transportation Test No. 621, "Determining the Percentage of Crushed Pavements in Gravel." The criteria for angularity of coarse aggregates depend on the traffic level and are given in Table 18.14.

The *angularity of fine aggregates* is defined as the percent of air voids in loosely compacted aggregates smaller than 2.36 mm. The AASHTO designated test TP33, "Test Method for Uncompacted Void Content of Fine Aggregate (as Influenced by Particle, Shape, Surface Texture, and Grading) (Method A)," can be used to determine the angularity of fine aggregates. In this test, a standard cylinder of known volume (V) is filled with a washed sample of fine aggregates by pouring the aggregates through a standard funnel. The mass (W) of the fine aggregates filling the standard cylinder of volume V is then determined. The volume of the fine aggregates in the standard cylinder then is determined as (W/G_{bfa}), and the void content is determined as a percentage of the cylinder volume, as shown in Eq. 18.16.

$$\text{Uncompacted void} = \frac{V - W/G_{bfa}}{V} \times 100\% \qquad (18.16)$$

where G_{bfa} = bulk specific gravity of the fine aggregate. Table 18.15 gives the criteria for fine aggregate angularity.

A *flat and elongated particle* is defined as one that has its maximum dimension five times greater than its minimum dimension. The amount of fine and elongated particles in the coarse aggregate is obtained by conducting ASTM D4791 designated test "Flat or Elongated Particles in Coarse Aggregates" on coarse aggregates larger than 4.75 mm. This test involves the use of a proportional caliper device that automatically tells whether a particle is flat or elongated. The test is performed on a sample of the coarse aggregate

Table 18.14   Coarse Aggregate Angularity Criteria

Traffic, Million ESALs	Depth from Surface	
	< 100 mm	*> 100 mm*
< 0.3	55/–	–/–
< 1	65/–	–/–
< 3	75/–	50/–
< 10	85/80	60/–
< 30	95/90	80/75
< 100	100/100	95/90
> 100	100/100	100/100

Note: "85/80" indicates that 85 percent of the coarse aggregate has one or more fractured faces and 80 percent two or more fractured faces.
SOURCE: *The Superpave Mix Design Manual for New Construction and Overlays*, Strategic Highway Research Program National Research Council, Washington, D.C., 1994.

Table 18.15   Fine Aggregate Angularity Criteria

Traffic, Million ESALs	Percent Air Voids in Loosely Compacted Fine Aggregates Smaller than 2.36 mm	
	Depth from Surface	
	< 100 mm	> 100 mm
< 0.3	—	—
< 1	40	—
< 3	40	40
< 10	45	40
< 30	45	40
< 100	45	45
≥ 100	45	45

SOURCE: *Superpave Mix Design Manual for New Construction and Overlays*, Strategic Highway Research Program National Research Council, Washington, D.C., 1994.

and the percentage by mass determined. The maximum percentage allowed is given in Table 18.16 for different traffic loads.

The *clay content* is defined as the percentage of clayey material in the portion of aggregate passing through the 4.75-mm sieve. It is obtained by conducting the AASHTO T176 designated test "Plastic Fines in Graded Aggregates and Soils by Use of Sand Equivalent Test." This test is conducted by first mixing a sample of fine aggregates in a graduated cylinder with a flocculating solution. The clayey fine aggregates coating the aggregates are loosened from the aggregates by shaking the cylinder. The mixture is then allowed to stand for a period during which time the clayey material is suspended above the granular aggregates (sedimented sand). The heights of the suspended clay and the sedimented sand are measured. The ratio (expressed in percentage of the height of the sedimented sand to that of the suspended clay) is the sand equivalent value. Table 18.17 gives minimum allowable values for the sand equivalent for different traffic loads.

Table 18.16   Thin and Elongated Particles Criteria

Traffic, Million ESALs	Maximum, Percent
< 0.3	—
< 1	—
< 3	10
< 10	10
< 30	10
< 100	10
≥ 100	10

SOURCE: *Superpave Mix Design Manual for New Construction and Overlays*, Strategic Highway Research Program National Research Council, Washington, D.C., 1994.

Table 18.17   Clay Content Criteria

Traffic, Million ESALs	*Sand Equivalent Minimum, Percent*
< 0.3	40
< 1	40
< 3	40
< 10	45
< 30	45
< 100	50
≥ 100	50

SOURCE: *Superpave Mix Design Manual for New Construction and Overlays*, Strategic Highway Research Program National Research Council, Washington, D.C., 1994.

The other aggregate properties that were considered critical by the developers of the Superpave system (for which critical values were not determined by consensus) were classified as source aggregate properties. These properties are toughness, soundness, and maximum allowable percentage of deleterious materials. These properties are discussed in Chapter 20.

Gradation

The distribution of aggregate particle sizes for a given blend of aggregate mixture is known as the design aggregate structure. The gradation system used for Superpave is based on the 0.45 gradation plot. This is a plot of the percent passing a given sieve against the sieve size in mm raised to the 0.45 power. That is, the vertical axis of the graph is percent passing and the horizontal axis is the size of the sieve in millimeters raised to the 0.45 power. In order to understand the gradation system used, it is first necessary to define certain gradation terms that the Superpave system uses. These are *maximum size, nominal maximum size,* and *maximum density gradation.*

Maximum size is defined as one sieve larger than the nominal maximum size, and the *nominal maximum size* is one sieve larger than the first sieve that retains more than 10 percent of the aggregate. Five mixture gradations are specified in the Superpave system, as shown in Table 18.18.

Maximum density gradation is obtained when the aggregate particles fit together in their densest form. An important characteristic of the 0.45 power plot is that the maximum density gradation for a sample of soil is given by a straight line joining the maximum size and the origin, as shown in Figure 18.16. An acceptable aggregate gradation is defined by specifying control points on the maximum density gradation chart for the smallest sieve size (0.075 mm), the nominal maximum size, and an intermediate size sieve (2.36 mm). An acceptable soil blend therefore should have a maximum density gradation line that lies within these control points.

In addition, a restricted zone, also shown in Figure 18.16, is established along the maximum density curve between the 0.3 mm sieve and the intermediate sieve (2.36 mm). For a soil blend to be acceptable, its gradation must not pass within the restricted zone. Soils that have gradations that go through the restricted zone have been found to create compaction problems during construction and tend to have inadequate VMA. Also, the Superpave system recommends that the gradation pass below the restricted zone,

Table 18.18 Superpave Mixture Gradations

Superpave Designation	Nominal Maximum Size, mm	Maximum Size, mm
37.5 mm	37.5	50.0
25.0 mm	25.0	37.5
19.0 mm	19.0	25.0
12.5 mm	12.5	19.0
9.5 mm	9.5	12.5

SOURCE: *Superpave Mix Design Manual for New Construction and Overlays*, Strategic Highway Research Program, National Research Council, Washington, D.C., 1994.

although this is not a requirement. Table 18.19 gives control points (maximum and minimum limits) for different nominal sieve sizes. Aggregates that have maximum density gradation lying between the control points and outside the restricted zones are considered as having an acceptable design aggregate structure.

18.6.2 Volumetric Trial Mixture Design

Volumetric trial mixture design consists of:

- Selecting design aggregate structure
- Determining trial percentage of asphalt binder for each trial aggregate blend
- Evaluating trial mix designs
- Obtaining design asphalt binder content

Selecting Design Aggregate Structure

In selecting the design aggregate structure, the designer must first ascertain that the individual aggregate and asphalt materials meet the criteria discussed above. Trial blends should then be prepared by varying the percentages from the different available stockpiles.

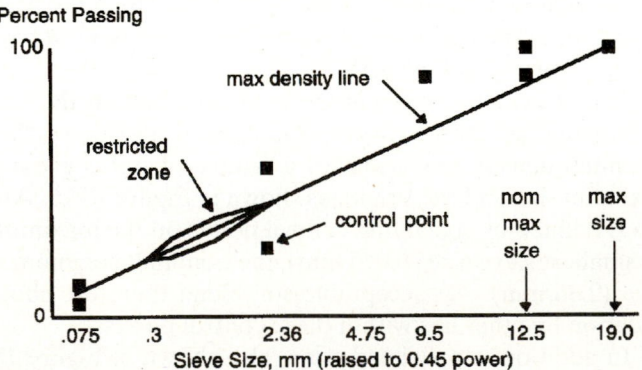

Figure 18.16 Superpave Gradation Control Points and Restricted Zone for 12.5 mm Nominal Maximum Size Gradation

SOURCE: *Superpave Mix Design Manual for New Construction and Overlays*, Strategic Highway Research Program, National Research Council, Washington, D.C., 1994.

Table 18.19   Superpave Aggregate Gradation Control Points and Restricted Zones

(a) 37.5 mm Nominal Size

	Control Points		Restricted Zone Boundary	
Sieve, mm	Minimum	Maximum	Minimum	Maximum
50	100			
37.5	90	100		
25		90		
19				
12.5				
9.5				
4.75			34.7	34.7
2.36	15	41	23.3	27.3
1.18			15.5	21.5
0.6			11.7	15.7
0.3			10	10
0.15				
0.075	0	6		

(b) 25 mm Nominal Size

	Control Points		Restricted Zone Boundary	
Sieve, mm	Minimum	Maximum	Minimum	Maximum
37.5		100		
25	90	100		
19		90		
12.5				
9.5				
4.75			39.5	39.5
2.36	19	45	26.8	30.8
1.18			18.1	24.1
0.60			13.6	17.6
0.30			11.4	11.4
0.15				
0.075	1	7		

(c) 19 mm Nominal Size

	Control Points		Restricted Zone Boundary	
Sieve, mm	Minimum	Maximum	Minimum	Maximum
25		100		
19	90	100		
12.5		90		
9.5				
4.75				
2.36	23	49	34.6	34.6
1.18			22.3	28.3
0.6			16.7	20.7

(Continued)

Table 18.19 Superpave Aggregate Gradation Control Points and Restricted Zones (*continued*)

Sieve, mm	Control Points		Restricted Zone Boundary	
	Minimum	Maximum	Minimum	Maximum
0.3			13.7	13.7
0.15				
0.075	2	8		

(d) 12.5 mm Nominal Size

Sieve, mm	Control Points		Restricted Zone Boundary	
	Minimum	Maximum	Minimum	Maximum
19		100		
12.5	90	100		
9.5		90		
4.75				
2.36	28	58	39.1	39.1
1.18			25.6	31.6
0.6			19.1	23.1
0.3			15.5	15.5
0.15				
0.075	2	10		

(e) 9.5 mm Nominal Size

Sieve, mm	Control Points		Restricted Zone Boundary	
	Minimum	Maximum	Minimum	Maximum
12.5		100		
9.5	90	100		
4.75		90		
2.36	32	67	47.2	47.2
1.18			31.6	37.6
0.6			23.5	27.5
0.3			18.7	18.7
0.15				
0.075	2	10		

SOURCE: *Superpave Mix Design Manual for New Construction and Overlays*, Strategic Highway Research Program National Research Council, Washington, D.C., 1994.

The mathematical procedure described in Section 18.5.1 of this chapter for determining the different proportions of different aggregates to obtain a required gradation may be used to determine trial blends that meet the control requirements. Although no specific number of trial blends is recommended, it is generally accepted that three trial blends with a range of gradation that will be adequate. The four consensus properties, the bulk and apparent specific gravities of the aggregates, and any source aggregate properties should be determined. At this stage, these properties can be determined by using mathematical expressions, but actual tests should be carried out on the aggregate blend finally selected. The bulk specific gravity may be obtained using Eq. 18.5, and the apparent specific gravity using Eq. 18.6.

Determining Trial Percentage of Asphalt Binder for Each Trial Aggregate Blend

This selection is based on certain volumetric characteristics of the mixture which are dependent on the accumulated traffic load. The Superpave volumetric mix design is proposed for three levels of traffic load. Level 1 is for accumulated traffic less than 10^6 ESAL. Level 2 is for accumulated traffic between 10^6 and 10^7 ESAL, and Level 3 is for accumulated traffic greater than 10^7 ESAL. The Level 1 design procedure primarily is based on the volumetric analysis of the mix, while Levels 2 and 3 mix design procedures incorporate performance tests that are used to measure fundamental properties and predict pavement performance. The difference between Levels 2 and 3 is that a more complete set of performance-based mixture properties is obtained and a more comprehensive set of models set is used to determine fatigue and permanent deformation. A brief discussion of the Level 1 procedure is given here. Interested readers may refer to the listed references for a more detailed description of the Level 1 procedure and the design procedures for Levels 2 and 3.

The basic assumption made in the Level 1 mix design process is that properties of the mix (such as air voids and voids in mineral aggregate) are suitable surrogates for mixture performance. It is suggested in the procedure that the designer could change the level of 10^6 ESAL if the policy for her jurisdiction allows it. The volumetric properties determined for the trial mixtures are the absorbed asphalt volume (V_{ba}), voids in mineral aggregates (VMA), air voids (V_a), asphalt content (P_b), effective asphalt volume (V_{be}), and voids filled with asphalt (VFA).

The first task is the computation of an initial trial asphalt content for a trial aggregate gradation that will meet VMA requirements shown in Table 18.20 through the following steps.

Step 1. Compute the bulk and apparent specific gravities of the total aggregates in the trial aggregate mix using Eqs. 18.5 and 18.6, respectively.

Step 2. Compute the effective specific gravity of the total aggregate in the trial gradation from Eq. 18.17.

$$G_{se} = G_{sb} + 0.8(G_{asb} - G_{sb}) \tag{18.17}$$

Table 18.20   Voids in Mineral Aggregate Criteria

Nominal Maximum Size (mm)	Minimum Voids in Mineral Aggregate (%)
9.5	15.0
12.5	14.0
19.0	13.0
25.0	12.0
37.5	11.0
50.0	10.5

SOURCE: Adapted from *The Superpave Mix Design Manual for New Construction and Overlays,* Strategic Highway Research Program, National Research Council, Washington, D.C., 1994

where

G_{se} = effective specific gravity of the aggregate blend
G_{sb} = bulk specific gravity of the aggregate blend
G_{asb} = apparent specific gravity of the aggregate blend

The designer may decide to change the implicit multiplication factor of 0.8, particularly when absorptive aggregates are used, as values close to 0.6 or 0.5 are more appropriate for these aggregates.

Step 3. The amount of asphalt binder absorbed by the aggregates is estimated from Eq. 18.18 as

$$V_{ba} = \frac{P_s(1 - V_a)}{\dfrac{P_b}{G_b} + \dfrac{P_s}{G_{se}}} \left[\frac{1}{G_{sb}} - \frac{1}{G_{se}} \right] \tag{18.18}$$

where

V_{ba} = volume of absorbed binder, cm³/cm³ of mix
P_b = percent of binder (assumed 0.05)
P_s = percent of aggregate (assumed 0.95)
G_b = specific gravity of binder (assumed 1.02)
V_a = volume of air voids (assumed 0.04 cm³/cm³ of mix)

Step 4. Estimate the percent of effective asphalt binder by volume using the empirical expression given in Eq. 18.19.

$$V_{be} = 0.176 - (0.0675) \log (S_n) \tag{18.19}$$

where

V_{be} = the volume of effective binder content
S_n = the nominal maximum sieve size of the total aggregate in the trial aggregate gradation (mm)

Step 5. A trial percentage of asphalt binder then is determined for each trial aggregate blend using the following equation.

$$P_{bi} = \frac{G_b(V_{be} + V_{ba})}{(G_b(V_{be} + V_{ba})) + W_s} \times 100 \tag{18.20}$$

where

P_{bi} = percent of binder by mass of mix i
V_{be} = volume of the effective binder and obtained from Eq. 18.21 as

$$V_{be} = 0.176 - 0.0675 \log (S_n) \tag{18.21}$$

S_n = the nominal maximum sieve size of the aggregate blend (mm)
V_{ba} = volume of absorbed binder, cm³/cm³ of mix

G_b = specific gravity of binder, assumed to be 1.02

W_s = mass of aggregate, $W_s = \dfrac{P_s(1 - V_a)}{\dfrac{P_b}{G_b} + \dfrac{P_s}{G_{se}}}$ grams

The Superpave software package can be used to carry out these computations. These computations will provide an initial trial asphalt content for each trial aggregate gradation.

Evaluating Trial Mix Designs

Two specimens of each of the trial asphalt mixes (using the computed trial asphalt contents) then will be prepared and compacted at the number of maximum gyrations (N_{max}) (see Table 18.21) using the Superpave gyratory compactor.

Figure 18.17 shows an example of a Superpave gyratory compactor. The main components of the compactor are

- A reaction frame, rotating base, and motor
- A loading system consisting of loading ram and pressure gauge
- A control and data acquisition system and a mold and base plate

The level of compaction in the Superpave system is given with respect to a design number of gyrations (N_{des}). The N_{des} depends on the average design high air temperature and the design ESAL. Two other levels of gyrations (maximum and initial) are important. The maximum number of gyrations, N_{max}, is used to compact the test specimens; and the initial number of gyrations, N_{ini}, is used to estimate the compactibility of the mixture. N_{max} and N_{ini} are obtained from N_{des} as shown.

$$\log N_{max} = 1.10 \log N_{des} \tag{18.22}$$

$$\log N_{ini} = 0.45 \log N_{des} \tag{18.23}$$

Table 18.21   Number of Initial (N_{ini}), Design (N_{des}) and Maximum (N_{max}) Gyrations Required for Various Traffic Levels and Maximum Temperature Environments

| Design ESALs (millions) | Average Design High Air Temperature | | | | | | | | | | | |
| | < 39°C | | | 39 to 40°C | | | 41 to 42°C | | | 43 to 45°C | | |
	N_{ini}	N_{des}	N_{max}	N_{ini}	N_{des}	N_{max}	N_{ini}	N_{des}	N_{max}	N_{ini}	N_{des}	N_{max}
< 0.3	7	68	104	7	74	114	7	78	121	7	82	127
0.3–1.0	7	76	117	7	83	129	7	88	138	8	93	146
1.0–3.0	7	86	134	8	95	150	8	100	158	8	105	167
3.0–10.0	8	96	152	8	106	169	8	113	181	9	119	192
10.0–30.0	8	109	174	9	121	195	9	128	208	9	135	220
30.0–100.0	9	126	204	9	139	228	9	146	240	10	153	253
> 100	9	143	235	10	158	262	10	165	275	10	172	288

SOURCE: *The Superpave Mix Design Manual for New Construction and Overlays*, Strategic Highway Research Program, National Research Council, Washington, D.C., 1994

Figure 18.17 Superpave Gyratory Compactors

SOURCE: Photograph by Lewis Woodson, Virginia Transportation Research Council. Used with permission.

Table 18.21 shows the values of N_{des} for different ESALs and average design air temperatures.

A detailed description of the procedures for aggregates and mixture preparation and the compaction of the volumetric specimens is beyond the scope of this book. Interested readers will find this description in *Superpave Mix Design System Manual of Specifications Test Methods and Practices*. The theoretical maximum specific gravity (G_{mm}) of the pavement mix is then determined using Eq. 18.8. Each trial aggregate gradation then is evaluated with respect to its *VMA* at N_{des} gyrations and the densities at N_{ini} and N_{max}. This involves determining the volumetric properties at the N_{des} gyration level of each paving mix and estimating the *VMA* at 4 percent air voids. As it is very likely that the air voids content will not be at the required 4 percent, the change in asphalt content to achieve a 4 percent air void content is obtained by a procedure that involves shifting the densification curves to determine the change in *VMA* that will occur as a result of this shift. This shifting procedure is described fully in the *Superpave Mix Design Manual for New Construction and Overlays*. The design parameters are estimated from the shifted curve and compared with the Level 1 criteria. The *VMA* criteria are given in Table 18.20, and the criteria for densities are that C_{ini} should be less than 89 percent and C_{max} should be less than 98 percent—where C is a correction factor as discussed later.

The following physical properties of the compacted specimens are determined as

- Estimated bulk specific gravity [G_{mbe} (estimated)]
- Corrected bulk specific gravity [G_{mbc} (corrected)]
- Corrected percentage of maximum theoretical specific gravity

The following equations are used to determine these properties.

$$G_{mbe} = \frac{\frac{W_m}{V_{mx}}}{\gamma_w} \tag{18.24}$$

where

W_m = mass of specimen (g)
γ_w = density of water = 1 g/cm
V_{mx} = volume of compaction mold (cm³)

$$V_{mx} = \frac{\pi d^2 h_x}{4} \times 0.01 \, \text{cm}^3/\text{mm}^3 \tag{18.25}$$

where

d = diameter of mold (150 mm)
h_x = height of specimen in mold during compaction (mm)

Since the cylinder in which the mold is compacted is not smooth and Eq. 18.24 is based on this assumption, the actual volume of the compacted mixture will be less than that obtained from the equation. The estimated bulk specific gravity therefore is corrected using the corrected factor obtained as the ratio of the measured bulk specific gravity to the estimated bulk specific gravity.

$$C = \frac{G_{mbm}(\text{measured})}{G_{mbe}(\text{measured})} \tag{18.26}$$

where

C = correction factor
G_{mbm} (measured) = measured bulk specific after N_{max}
G_{mbe} (estimated) = estimated bulk specific after N_{max}

The corrected bulk specific gravity [G_{mbc} (corrected)] at any other gyration level can then be obtained using Eq. 18.27.

$$G_{mbc}(\text{corrected}) = C \times G_{mbe}(\text{estimated}) \qquad (18.27)$$

where

G_{mbc} (corrected) = corrected bulk specific gravity for the specimen for a given gyration level
C = correction factor
G_{mbe} (estimated) = estimated bulk specific after N_{max}

The percentage maximum theoretical specific gravity G_{mm} for each desired gyration is then computed as the ratio of G_{mbc} (corrected) to G_{mm} (measured). The average G_{mm} value for each gyration level is obtained using the G_{mm} values for the samples compacted at that level.

The values for the properties, percent air voids at N_{des} (P_a), voids in mineral aggregate (VMA), and percent variation of VMA from 4 percent (VFA) at N_{des} are then determined using the following equations.

$$P_a = 100\left(\frac{G_{mm} - G_{mb}}{G_{nm}}\right) \qquad (18.28)$$

where

P_a = air voids at N_{des} percent of total volume
G_{mm} = maximum theoretical specific gravity at N_{des}
G_{mb} = bulk specific gravity of the compacted mixture

$$VMA = 100 - \left(\frac{G_{mb}P_s}{G_{sb}}\right) \qquad (18.29)$$

where

VMA = voids in mineral aggregate, percent in bulk volume
G_{mb} = bulk specific gravity of the compacted mixture
P_s = aggregate content cm^3/cm^3, by total mass of mixture
G_{sb} = bulk specific gravity of aggregates in the paving mixture

$$VFA = 100\left(\frac{VMA - P_s}{VMA}\right) \qquad (18.30)$$

If the percent of air voids is 4 percent, then the values obtained for the volumetric criteria are compared with the corresponding criteria values. For example, the values for VMA are shown in Table 18.20. If these criteria are met, then the blend is acceptable. When the V_a is not 4 percent, it is necessary to determine an estimated design asphalt content with a VMA of 4 percent using Eq. 18.31.

$$P_{b,\text{estimated}} = P_{b,i} - [0.4 \times (4 - V_a)] \qquad (18.31)$$

where

$$P_{b,\text{estimated}} = \text{estimated asphalt content, percent by mass of mixture}$$
$$P_{b,i} = \text{initial (trial) asphalt content, percent by mass of mixture}$$
$$V_a = \text{percent air voids at } N_{\text{des}} \text{ (trial)}$$

The *VMA* and *VFA* at N_{des}, G_{mm} at N_{max}, and N_{ini} are then estimated for the design asphalt content obtained using the following equations.

$$VMA_{\text{estimated}} = VMA_{\text{initial}} + C_1 \times (4 - V_a) \tag{18.32}$$

where

$$VMA_{\text{initial}} = VMA \text{ from trial asphalt binder content}$$
$$C_1 = \text{constant} = 0.1 \text{ if } V_a \text{ less than 4.0 percent}$$
$$= 0.2 \text{ if } V_a \text{ is greater than 4.0 percent}$$

This value is compared with the corresponding value given in Table 18.20.

$$VFA_{\text{estimated}} = 100 \frac{VMA_{\text{estimated}} - 4.0}{VMA_{\text{estimated}}} \tag{18.33}$$

This value is compared with the range of acceptable values given in Table 18.22. For G_{mm} (maximum specific gravity of paving mixture) at N_{max},

$$G_{mm,\text{estimated}} \text{ at } N_{\text{max}} = G_{mm,\text{trial}} \text{ at } N_{\text{max}} - (4 - V_a) \tag{18.34}$$
$$G_{mm,\text{trial}} = G_{mm} \text{ for the trial mix}$$

The maximum allowable mixture density at N_{max} is 98 percent of the maximum theoretical specific gravity (G_{mm}).
For the percent G_{mm} at N_{ini},

$$G_{mm,\text{estimated}} \text{ at } N_{\text{ini}} = G_{mm,\text{trial}} \text{ at } N_{\text{ini}} - (4 - P_{av}) \tag{18.35}$$

Table 18.22   VFA Criteria

Traffic, Million ESALs	*Design VFA, Percent*
< 0.3	70–80
< 1	65–78
< 3	65–78
< 10	65–75
< 30	65–75
< 100	65–75

SOURCE: *The Superpave Mix Design Manual for New Construction and Overlays,* Strategic Highway Research Program, National Research Council, Washington, D.C., 1994

The maximum allowable mixture density at N_{ini} is 89 percent of the maximum theoretical specific gravity (G_{mm}).

The effective binder content (P_{be}) is then determined from Eq. 18.36.

$$P_{be} = -(P_b \times G_b) \times \left(\frac{G_{se} - G_{sb}}{G_{se} - G_{sb}} \right) + P_{b,\text{estimated}} \qquad (18.36)$$

where

G_b = specific gravity of the asphalt
G_{se} = effective specific gravity of the aggregate
G_{sb} = bulk specific gravity of the aggregate
P_b = asphalt content, percent by total mass of the paving mixture

Obtaining Dust Percentage.   The dust percentage is determined as the proportion of the percentage by mass of the material passing the 0.075-mm sieve to the effective binder content by mass of the mix in percent as shown in Eq. 18.37.

$$DP = \frac{P_{0.75}}{P_{be}} \qquad (18.37)$$

where

DP = dust percentage
$P_{0.75}$ = aggregate content passing the 0.075-mm sieve, percent by mass of aggregate
P_{be} = effective asphalt content, percent by total mass of mixture

The range of acceptable dust proportion is from 0.6 to 1.2. Based on these results, the designer will accept one or more of the trial blends that meet the desired criteria.

Obtaining Design Asphalt Binder Content

After the selection of the design aggregate structure from the trial blends, a minimum of two specimens with an asphalt content of +0.5 and −5 percent of the estimated asphalt content and at least two with an asphalt content of +1.0 percent are compacted. These specimens are then tested using the same procedure described in the section on *select design aggregate structure*. The densification data at N_{ini}, N_{des}, and N_{max} are used to evaluate the properties of the selected aggregate blend for each of the asphalt binder contents. The volumetric properties (air voids, *VAM*, and *VFA*) at the N_{des} are calculated for each asphalt content and plotted against the asphalt content. The asphalt binder content at 4 percent air void is selected as the design asphalt binder content. The mixture with that asphalt binder content then is checked to ascertain that it meets all of the other mixture property criteria.

Plots of V_a, *VMA*, *VFA*, and C_{design} versus the percent asphalt content are drawn. The Superpave software can generate these plots automatically. These plots are used to ascertain that the design mix meets all the Superpave criteria given in Table 18.23. The use of these plots to select the design asphalt content is illustrated in Example 18.6, which is also given in the *Superpave Design Manual*.

Table 18.23 Summary of Volumetric Design Criteria

Volumetric Property	Superpave Limit
V_a at N_{des}	4%
VMA at N_{des}	See Table 18.20
C_{init}	$< 89.0\%$ of G_{mm}
C_{max}	$< 98.0\%$ of G_{mm}

SOURCE: *The Superpave Mix Design Manual for New Construction and Overlays,* Strategic Highway Research Program, National Research Council, Washington, D.C., 1994

Example 18.6 Check for Percent Voids Filled with Asphalt Criterion for a Superpave Mixture

Table 18.24 gives the volumetric design values obtained for an estimated design asphalt content of 4.8 percent. Figure 18.18 shows the four plots V_a, VMA, VFA, and density at N_{des} versus the asphalt percent. Note that these percentages are the estimated asphalt content of 4.8 percent, 0.5 percent below, 0.5 percent above, and 1 percent above the estimated asphalt content. Nominal maximum size of the aggregate mixture is 19 mm. If the estimated number of ESALs is 1.5×10^8, determine whether the percent voids meets the Superpave criteria.

From Table 18.20, the minimum requirement for VMA for the design aggregate structure (19-mm nominal maximum size) is 13. This condition is satisfied from Figure 18.18b. The asphalt content at 4 percent air void is 5.7 from Figure 18.18a. The VFA for an asphalt content of 5.7 is about 68 percent, which satisfies the Superpave VFA criteria as shown in Table 18.25.

Table 18.24 Sample Volumetric Design Data at N_{des} for Superpave Design Manual Example

P_b	V_a (percent)	VMA (percent)	VFA (percent)	Density (Kg/m³)
4.3	9.5	15.9	40.3	2320
4.8	7.0	14.7	52.4	2366
5.3	6.0	14.9	59.5	2372
5.8	3.7	13.9	73.5	2412

SOURCE: *The Superpave Mix Design Manual for New Construction and Overlays,* Strategic Highway Research Program, National Research Council, Washington, D.C., 1994

Table 18.25 Voids Filled with Asphalt Criteria

Traffic Level (ESALs)	Design Voids Filled with Asphalt (%)
$< 3 \times 10^5$	70–80
$> 3 \times 10^5$	65–78
$< 1 \times 10^8$	65–75
$> 1 \times 10^8$	65–75

SOURCE: *The Superpave Mix Design Manual for New Construction and Overlays,* Strategic Highway Research Program, National Research Council, Washington, D.C., 1994

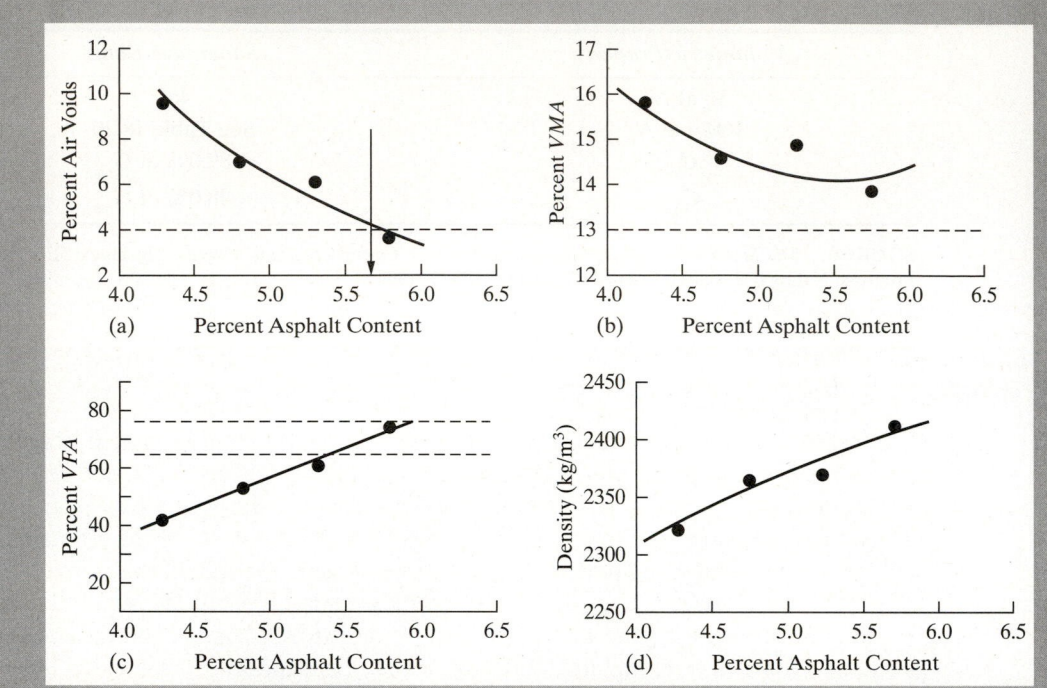

Figure 18.18 Plots of Air Voids, Percent VMA, Percent VFA, and Density versus Percent Asphalt Content

SOURCE: *The Superpave Mix Design Manual for New Construction and Overlays,* Strategic Highway Research Program, National Research Council, Washington, D.C., 1994

The results indicate that the paving mixture meets the Superpave requirements for *VFA.*

Moisture Sensitivity. Finally, the moisture sensitivity of the design mix may be established by conducting the AASHTO designated test T283 on specimens consisting of the selected aggregate blend and the design asphalt content. This test determines the indirect tensile strength of the specimens. A total of six specimens are prepared, three of which are considered as controlled, and the other three as conditioned. The conditioned specimens are first subjected to partial vacuum saturation, followed by an optional freeze cycle, and then a 24-hour thaw cycle at 60°C. The tensile tests of the specimens then are determined. The moisture sensitivity is given as the ratio of the average tensile strength of the conditioned specimens to that of the control specimens. The minimum acceptable value without remedial action is 0.80.

This section has described briefly the Superpave Level 1 mix design procedure as given in the *Superpave Mix Design Manual for New Construction and Overlays.* It should be emphasized that this description does not comprehensively cover the procedure but will enable the reader to have a general knowledge of it.

18.7 SUMMARY

Asphalt pavement is becoming the most popular type of pavement used in highway construction. It is envisaged that the use of asphalt in highway construction will continue to increase, particularly with the additional knowledge that will be obtained from research conducted over the next few years. The engineering properties of different asphalt materials are therefore of significant importance in highway engineering.

This chapter has presented information on the different types of asphalt materials, their physical characteristics, and some of the tests usually conducted on these materials when used in the maintenance and/or construction of highway pavements. Two mix design methods for asphalt mixtures (Marshall method and Superpave) also have been provided so the reader will understand the principles involved in determining the optimum mix for an asphalt mixture. The chapter contains sufficient material on the subject to enable the reader to become familiar with the fundamental engineering properties and characteristics of those asphalt materials used in pavement engineering.

PROBLEMS

18-1   Briefly describe the process of distillation by which asphalt cement is produced from crude petroleum. Also, describe in detail how to obtain asphalt binders that can be used to coat highly siliceous aggregates.

18-2   Describe both factors that influence the durability of asphalt materials and the effect each factor has on the materials.

18-3   Discuss the factors to consider in producing a rapid-curing cutback asphaltic material.

18-4   Results obtained from laboratory tests on a sample of RC-250 asphalt cement are given. Determine whether the properties of this material meet the Asphalt Institute specifications for this type of material; if not, note the differences.

Kinematic viscosity at 140°F (60°C) = 260 centistokes
Flash point (Tagliabue open cup) = 24 C
Distillation test where distillate percent by volume of total distillate percent by volume of total distillate to 680°F (360°C)
To 437°F (225°C) = 35
To 500°F (260°C) = 54
To 600°F (316°C) = 75
Residue from distillation to 680°F (360°C) by volume = 64 tests on residue from distillation
Ductility at 77°F (25°C) = 95 cm
Absolute viscosity at 140°F (60°C) = 750 poises
Solubility = 95%

18-5   Given the specifications for an asphaltic concrete mixture and the results of a sieve analysis, determine the proportion of different aggregates to obtain the required gradation.

Coarse aggregates: 60%
Fine aggregates: 35%
Filler: 5%

Passing Sieve Designation	Retained on Sieve Designation	Percent by Weight		
		Coarse Aggregate	Fine Aggregate	Mineral Filler
¾ in. (19 mm)	12.5 mm	4	—	—
½ in. (12.5 mm)	9.5 mm	36	—	—
³/₈ in. (9.5 mm)	No. 4	40	—	—
No. 4 (4.75 mm)	No. 10	15	6	—
No. 10 (2 mm)	No. 40	5	32	—
No. 40 (0.425 mm)	No. 80	—	33	5
No. 80 (0.180 mm)	No. 200	—	29	40
No. 200 (0.075 mm)	—	—	—	55
Total		100	100	100

18-6   Given the particle size distributions of two aggregates and the required limits of particle size distribution for the mix, determine a suitable ratio for blending the two aggregates to obtain an acceptable combined aggregate.

Sieve Size	Percent Passing by Weight		
	A	B	Required Mix
¾ in. (19 mm)	100	98	96 to 100
³/₈ in. (9.5 mm)	80	76	65 to 80
No. 4 (4.25 mm)	50	45	40 to 55
No. 10 (2 mm)	43	33	35 to 40
No. 40 (0.425 mm)	20	30	15 to 35
No. 200 (0.075 mm)	4	8	5 to 8

18-7   Given the particle size distributions of two aggregates and the required limits of particle size distribution for the mix, determine a suitable ratio for blending the two aggregates to obtain an acceptable combined aggregate.

Sieve Size	Aggregate A	Aggregate B	Required Mix (%)
19 mm	100	95	98–100
9.5 mm	80	74	75–85
No. 4	56	43	50–60
No. 10	42	32	37–47
No. 40	24	29	25–35
No. 200	5	12	7–10

18-8   Given four different types of aggregates to be used to produce a blended aggregate for use in the manufacture of asphaltic concrete, determine the bulk specific gravity of the aggregate mix.

Material	Percent by Weight	Bulk Specific Gravity
A	35	2.58
B	40	2.65
C	15	2.60
D	10	2.55

18-9   If the specific gravity of the asphalt cement used in a sample of asphalt concrete mix is 0.95, the maximum specific gravity of the mix is 2.58, and the mix contains 6.5 percent by weight of asphalt cement, determine the effective specific gravity of the mixture.

18-10   The table below lists data used in obtaining a mix design for an asphalt paving mixture. If the maximum specific gravity of the mixture is 2.41 and the bulk specific gravity is 2.35, determine:

(a) the bulk specific gravity of aggregates in the mix
(b) the asphalt absorbed
(c) the effective asphalt content of the paving mixture
(d) the percent voids in the mineral aggregate *VMA*

Material	Specific Gravity	Mix Composition by Weight of Total Mix
Asphalt cement	1.02	6.40
Coarse aggregate	2.51	52.35
Fine aggregate	2.74	33.45
Mineral filler	2.69	7.80

18-11   For hot-mix, hot-laid asphaltic concrete mixtures, if the asphaltic content is specified as 5 to 7 percent, how is the optimum percentage determined?

18-12   The aggregate mix used for the design of an asphalt paving concrete consists of 42 percent coarse aggregates, 51 percent fine aggregates, and 7 percent mineral fillers. If the respective bulk specific gravities of these materials are 2.60, 2.71, and 2.69, and the effective specific gravity of the aggregates is 2.82, determine the optimum asphalt content as a percentage of the total mix, using results obtained using the Marshall method as shown in the following table. The specific gravity of the asphalt cement is 1.02.

18-13   Determine the asphalt absorption of the optimum mix found in Problem 18-12.

18-14   The latitude at the location where a high-speed rural road is to be constructed is 35°. The expected ESAL is 32×10^6. The seven-day average high temperature is 53°C, and the low air temperature is -18°C. If the standard deviations for the high and low temperatures are ±2°C and ±1°C, respectively, and the depth of the pavement is 155 mm, determine an appropriate asphalt binder for a reliability of 98 percent.

18-15   An urban expressway is being designed for a congested area in Washington, D.C. It is expected that most of the time, traffic will be moving at a slow rate. If the anticipated ESAL is 8×10^6, determine an appropriate asphalt binder for this project.

18-16   The table below shows properties of three trial aggregate blends that are to be evaluated so as to determine their suitability for use in a Superpave mix. If the nominal maximum sieve of each aggregate blend is 19 mm, determine the initial trial asphalt content for each of the blends.

Property	Trial Blend 1	Trial Blend 2	Trial Blend 3
G_{mb}	2.698	2.696	2.711
G_{se}	2.765	2.766	2.764

Assume:

$P_b = 0.05$
$P_s = 0.95$
$P_a = 0.04$
$G_b = 1.02$

REFERENCES

AASHTO Guide for Design of Pavement Structures, American Association of State Highway and Transportation Officials, Washington, D.C., 2001.

Annual Book of ASTM Standards, Section 4, Construction, Vol. 04.03, *Road and Paving Materials: Pavement Management Technologies,* American Society for Testing and Materials, Philadelphia, PA, 2007.

Standard Specifications for Transportation Materials and Methods of Sampling and Testing, Part 2A and 2B, 27th ed., American Association of State Highway and Transportation Officials, Washington, D.C., 2001.

The Superpave Mix Design System Manual of Specifications, Test Methods and Practices, Strategic Highway Research Progam, National Research Council, Washington, D.C., 1994.

CHAPTER 19

Design of Flexible Highway Pavements

Highway pavements are divided into two main categories: rigid and flexible. The wearing surface of a rigid pavement usually is constructed of Portland cement concrete such that it acts like a beam over any irregularities in the underlying supporting material. The wearing surface of flexible pavements, on the other hand, usually is constructed of bituminous materials such that they remain in contact with the underlying material even when minor irregularities occur. Flexible pavements usually consist of a bituminous surface underlaid with a layer of granular material and a layer of a suitable mixture of coarse and fine materials. Traffic loads are transferred by the wearing surface to the underlying supporting materials through the interlocking of aggregates, the frictional effect of granular materials, and cohesion of fine materials.

Flexible pavements are further divided into three subgroups: high type, intermediate type, and low type. High-type pavements have wearing surfaces that adequately support the expected traffic load without visible distress due to fatigue and are not susceptible to weather conditions. Intermediate-type pavements have wearing surfaces that range from surface treated or chip seal to those with qualities just below that of high-type pavements. Low-type pavements are used mainly for low-cost roads and have wearing surfaces that range from untreated to loose natural materials to surface-treated earth.

This chapter deals with the design of new high-type pavements, although the methodologies presented also can be used for some intermediate-type pavements.

CHAPTER OBJECTIVES:

- Become familiar with the structural components of a flexible pavement.
- Learn about the different stabilization methods to improve soil properties for highway construction.
- Learn the AASHTO method for the design of flexible highway pavements.
- Learn the California (Hveem) method for the design of flexible highway pavements.
- Become familiar with the mechanistic-empirical method for the design of flexible highway pavements.

1039

19.1 STRUCTURAL COMPONENTS OF A FLEXIBLE PAVEMENT

Figure 19.1 shows the components of a flexible pavement: the subgrade or prepared roadbed, the subbase, the base, and the wearing surface. The performance of the pavement depends on the satisfactory performance of each component, which requires proper evaluation of the properties of each component separately.

19.1.1 Subgrade (Prepared Roadbed)

The subgrade is usually the natural material located along the horizontal alignment of the pavement and serves as the foundation of the pavement structure. It also may consist of a layer of selected borrow materials, well compacted to prescribed specifications discussed in Chapter 17. It may be necessary to treat the subgrade material to achieve certain strength properties required for the type of pavement being constructed. This will be discussed later.

19.1.2 Subbase Course

Located immediately above the subgrade, the subbase component consists of material of a superior quality to that which is generally used for subgrade construction. The requirements for subbase materials usually are given in terms of the gradation, plastic characteristics, and strength, as discussed in Chapter 17. When the quality of the subgrade material meets the requirements of the subbase material, the subbase component may be omitted. In cases where suitable subbase material is not readily available, the available material can be treated with other materials to achieve the necessary properties. This process of treating soils to improve their engineering properties is known as *stabilization*.

19.1.3 Base Course

The base course lies immediately above the subbase. It is placed immediately above the subgrade if a subbase course is not used. This course usually consists of granular materials such as crushed stone, crushed or uncrushed slag, crushed or uncrushed gravel, and sand. The specifications for base course materials usually include more strict requirements than those for subbase materials, particularly with respect to their plasticity, gradation, and strength. Materials that do not have the required properties can be used as base materials if they are properly stabilized with Portland cement, asphalt, or lime. In some cases, high-quality base course materials also may be treated with asphalt or Portland cement to improve the stiffness characteristics of heavy-duty pavements.

19.1.4 Surface Course

The surface course is the upper course of the road pavement and is constructed immediately above the base course. The surface course in a flexible pavement usually consists of a mixture of mineral aggregates and asphalt. It should be capable of withstanding

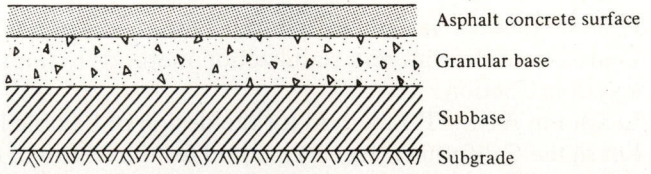

Figure 19.1 Schematic of a Flexible Pavement

high tire pressures, resisting abrasive forces due to traffic, providing a skid-resistant driving surface, and preventing the penetration of surface water into the underlying layers. The thickness of the wearing surface can vary from 75 mm to more than 150 mm, depending on the expected traffic on the pavement. It was shown in Chapter 18 that the quality of the surface course of a flexible pavement depends on the mix design of the asphalt concrete used.

19.2 SOIL STABILIZATION

Soil stabilization is the treatment of natural soil to improve its engineering properties. Soil stabilization methods can be divided into two categories, namely, mechanical and chemical. *Mechanical stabilization* is the blending of different grades of soils to obtain a required grade. This type of stabilization was discussed in Chapter 17. *Chemical stabilization* is the blending of the natural soil with chemical agents. Several blending agents have been used to obtain different effects. Table 19.1 shows stabilization methods that are most suitable for specific applications. The gradation triangle shown in Figure 19.2 and Table 19.2 can be used as a guide to determine the stabilizing additive for a particular soil type. The most commonly used agents are Portland cement, asphalt binders, and lime. It is necessary at this point to define some of the terms commonly used in the field of soil stabilization—particularly when Portland cement, lime, or asphalt is used—to help the reader fully understand Tables 19.1 and 19.2 and Figure 19.2.

1. ***Cement-stabilized soil (CSS)*** is a mixture of water, soil, and measured amounts of Portland cement—thoroughly mixed and compacted to a high density and then allowed to cure for a specific period, during which it is protected from loss of moisture.
2. ***Soil cement (SC)*** is a hardened material obtained by mechanically compacting a mixture of finely crushed soil, water, and a quantity of Portland cement that will make the mixture meet certain durability requirements.
3. ***Cement-modified soil (CMS)*** is a semihardened or unhardened mixture of water, Portland cement, and finely crushed soil. This mixture has less cement than the soil-cement mixture.
4. ***Plastic soil cement (PSC)*** is a hardened material obtained by mixing finely crushed soil, Portland cement, and a quantity of water, such that at the time of mixing and placing, a consistency similar to that of mortar is obtained.
5. ***Soil-lime (SL)*** is a mixture of lime, water, and fine-grained soil. If the soil contains silica and alumina, pozzolanic reaction occurs, resulting in the formation of a cementing-type material. Clay minerals, quartz, and feldspars are all possible sources of silica and alumina in typical fine-grained soils.
6. ***Soil-asphalt (SA)*** is obtained by mixing liquid asphalt with nonplastic or moderately plastic soils that exist naturally in order to improve the bearing qualities of the soil.

The stabilization process of each of the commonly used materials is briefly discussed. An in-depth discussion of soil stabilization is given in *Soil Stabilization in Pavement Structures.*

19.2.1 Cement Stabilization

Cement stabilization of soils usually involves the addition of 5 to 14 percent Portland cement by volume of the compacted mixture to the soil being stabilized. This type of stabilization is used mainly to obtain the required engineering properties of soils that are to be used as base course materials. Although the best results have been obtained

Table 19.1   Stabilization Methods Most Suitable for Specific Applications

Purpose	Soil Type	Method
Subgrade Stabilization	Fine-grained	SA, SC, MB, C
Improves load-carrying and stress-distribution characteristics	Coarse-grained	SA, SC, MB, C
	Clays of low PI	C, SC, CMS, LMS, SL
	Clays of high PI	SL, LMS
Reduces frost susceptibility	Fine-grained	CMS, SA, SC, LF
	Clays of low PI	CMS, SC, SL, LMS
Improves waterproofing and runoff	Clays of low PI	CMS, SA, LMS, SL
Controls shrinkage and swell	Clays of low PI	CMS, SC, C, LMS, SL
	Clays of high PI	SL
Reduces resiliency	Clays of high PI	SL, LMS
	Elastic silts or clays	SC, CMS
Base-Course Stabilization	Fine-grained	SC, SA, LF, MB
Improves substandard materials	Clays of low PI	SC, SL
Improves load-carrying and stress-distribution characteristics	Coarse-grained	SA, SC, MB, LF
	Fine-grained	SC, SA, LF, MB
Reduces pumping	Fine-grained	SC, SA, LF, MB, membranes
Dust Palliative	Fine-grained	CMS, SA, oil or bituminous surface spray, APSB
	Plastic soils	CMS, SL, LMS, APSB, DCA 70

C = Compaction   LMS = Lime-modified soil
CMS = Cement modified soil   MB = Mechanical blending
CL = Chlorides   PSC = Plastic soil cement
CS = Chemical solidifiers   SA = Soil asphalt
CW = Chemical waterproofers   SC = Soil cement
LF = Lime fly ash   SL = Soil lime

SOURCE: Soil Stabilization for Roads and Airfields, FM 5-410 Military Soils Engineering. www.Scribd.com/doc/487747/FM-5410-military-Soils-Engineering. Nov. 2007

when well-graded granular materials were stabilized with cement, the Portland Cement Association has indicated that nearly all types of soil can be stabilized with cement.

The procedure for stabilizing soils with cement involves:

- Pulverizing the soil
- Mixing the required quantity of cement with the pulverized soil
- Compacting the soil cement mixture
- Curing the compacted layer

Pulverizing the Soil

The soil to be stabilized first should be pulverized thoroughly to facilitate the mixing of the cement and soil. When the existing material on the roadway is to be used, the roadway is scarified to the required depth, using a scarifier attached to a grader. When the material to be stabilized is imported to the site, the soil is evenly spread to the required depth above the subgrade and then pulverized by using rotary mixers. Sieve analysis is

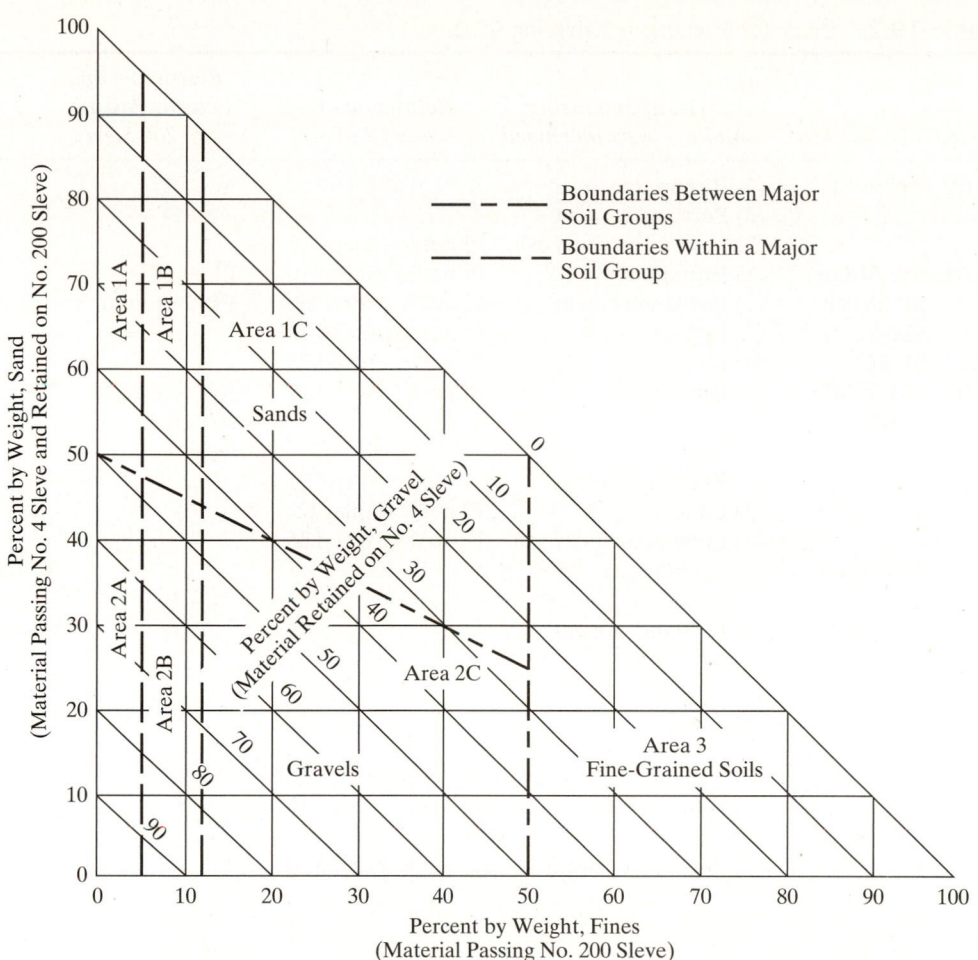

Figure 19.2   Graduation Triangle for Use in Selecting a Stabilizing Additive

SOURCE: *Soil Stabilization for Roads and Airfields, FM 5-410 Military Soils Engineering.*

conducted on the soil during pulverization, and pulverization is continued (except for gravel) until all material can pass through a 25 mm sieve and not more than 20 percent is retained on the No. 4 sieve. A suitable moisture content must be maintained during pulverization that may require air drying of wet soils or the addition of water to dry soils.

Mixing of Soil and Cement

The first step in the mixing process is to determine the amount of cement to be added to the soil by conducting laboratory experiments to determine the minimum quantity of cement required. The required tests are described by the Portland Cement Association. These tests generally include the classification of the soils to be stabilized, using the American Association of State Highway and Transportation Officials' (AASHTO) classification system, from which the range of cement content can be deduced based on past experience. Samples of soil-cement mixtures containing different quantities of cement within the range deduced are then prepared, and the American Society for Testing and Materials

Table 19.2   Guide for Selecting a Stabilizing Additive.

Area	Soils Class[a]	Type of Stabilizing Additive Recommended	Restriction on LL and PI of Soil	Restriction on Percent Passing No. 200 Sieve	Remarks
1A	SW or SP	(1) Bituminous (2) Portland cement (3) Lime-cement-fly ash	 PI not to exceed 25		
1B	SW-SM or SP-SM or SW-SC or SP-SC	(1) Bituminous (2) Portland cement (3) Lime (4) Lime-cement-fly ash	PI not to exceed 10 PI not to exceed 30 PI not to exceed 12 PI not to exceed 25	PI 30 or less PI 12 or greater	
1C	SM or SC or SM-SC	(1) Bituminous (2) Portland cement (3) Lime (4) Lime-cement-fly ash	PI not to exceed 10 PI not less than 12 PI not to exceed 25	Not to exceed 30 percent by weight	
2A	GW or GP	(1) Bituminous (2) Portland cement (3) Lime-cement-fly ash	 PI not to exceed 25		Well-graded material only Material should contain at least 45 percent by weight of material passing No. 4 sieve
2B	GW-GM or GP-GM or GW-GC or CP-GC	(1) Bituminous (2) Portland cement (3) Lime (4) Lime-cement-fly ash	PI not to exceed 10 PI not to exceed 30 PI not to exceed 12 PI not to exceed 25		Well-graded material only Material should contain at least 45 percent by weight of material passing No. 4 sieve
2C	GH or GC or GM-GC	(1) Bituminous (2) Portland cement (3) Lime (4) Lime-cement-fly ash	PI not to exceed 10 PI not less than 12 PI not to exceed 25	Not to exceed 30 percent by weight	Well-graded material only Material should contain at least 45 percent by weight of material passing No. 4 sieve
3	CH or CL or MH or ML or OH or OL or	(1) Portland cement	LL less than 40 and PI less than 20		Organic and strongly acid soils falling within this area are not susceptible to stabilization by ordinary means
	HL-CL	(2) Lime	PI not less than 12		

[a] See Table 17.3 for definitions of soil classes $PI = 10 + \dfrac{50 - percent\ passing\ No\ 200\ sieve}{\Delta}$

SOURCE: *Soil Stabilization for Roads and Airfields, FM 5-410 Military Soils Engineering*. www.Scribd.com/doc/487747/FM-5410-military-Soils-Engineering. Nov. 2007

moisture-density relationship test (ASTM Designation D558) is conducted on these samples. The results are used to determine the amount of cement to be used for preparing test specimens that are then used to conduct durability tests. These tests are described in "Standard Methods of Wetting-and-Drying Tests of Compacted Soil–Cement Mixtures," ASTM Designation D559, and "Standard Methods of Freezing-and-Thawing Tests of Compacted Soil–Cement Mixtures," ASTM Designation D560. The quantity of cement required to achieve satisfactory stabilization is obtained from the results of these tests. Table 19.3 and 19.4 give suggested cement contents for different types of soils.

Having determined the quantity of cement required, the mixing can be carried out either on the road (*road mixing*) or in a central plant (*plant mixing*). Plant mixing is used primarily when the material to be stabilized is borrowed from another site.

With road mixing, the cement usually is delivered in bulk in dump or hopper trucks and spread by a spreader box or some other type of equipment that will provide a uniform amount over the pulverized soil. Enough water is then added to achieve a moisture content that is 1 or 2 percent higher than the optimum required for compaction, and the soil, cement, and water are properly blended to obtain a uniform moisture of soil cement. Blending at a moisture content slightly higher than the compaction optimum moisture content allows for loss of water by evaporation during the mixing process.

In plant mixing, the borrowed soil is properly pulverized and mixed with the cement and water in either a continuous or batch mixer. It is not necessary to add more than the quantity of water that will bring the moisture content to the optimum for compaction, since moisture loss in plant mixing is relatively small. The soil-cement mixture is then delivered to the site in trucks and uniformly spread. The main advantage of plant mixing over road mixing is that the proportioning of cement, water, and soil can be controlled easily.

Compacting the Soil–Cement Mixture

It is essential that compaction of the mixture be carried out before the mixture begins to set. Specifications given by the Federal Highway Administration (FHWA) stipulate that the length of time between the addition of water and the compaction of the mix

Table 19.3   Average Cement Requirement for Granular and Sandy Soils

Material Retained on No. 4 Sieve, Percent	Material Smaller Than 0.05 mm, Percent	Cement Content, Percent by Weight Maximum Dry Density, lb/ft² (Treated Material)					
		116–120	121–126	127–131	132–137	138–142	143 or more
0–14	0–19	10	9	8	7	6	5
	20–39	9	8	7	7	5	5
	40–50	11	10	9	8	6	5
15–29	0–19	10	9	8	6	5	5
	20–39	9	8	7	6	6	5
	40–50	12	10	9	8	7	6
30–45	0–19	10	8	7	6	5	5
	20–39	11	9	8	7	6	5
	40–50	12	11	10	9	8	6

Note: Base course goes to 70 percent retained on the No. 4 sieve.
SOURCE: *Soil Stabilization for Roads and Airfields, FM 5-410 Military Soils Engineering.* www.Scribd.com/doc/487747/FM-5410-military-Soils-Engineering. Nov. 2007

Table 19.4.   Average Cement Requirement for Silty and Clayey Soils.

Group Index	Material Between 0.05 and 0.005 mm, Percent	Cement Content, Percent by Weight Maximum Dry Density, lb/ft² (Treated Material)						
		99–104	105–109	110–115	116–120	121–126	127–131	132 or more
0–3	0–19	12	11	10	8	8	7	7
	20–39	12	11	10	9	8	8	7
	40–59	13	12	11	9	9	8	8
	60 or more	—	—	—	—	—	—	—
3–7	0–19	13	12	11	9	8	7	7
	20–39	13	12	11	10	0	8	8
	40–59	14	13	12	10	10	9	8
	60 or more	15	14	12	11	10	9	9
7–11	0–19	14	13	11	10	9	8	8
	20–39	15	14	11	10	9	9	9
	40–59	16	14	12	11	10	10	9
	60 or more	17	15	13	11	10	10	10
11–15	0–19	15	14	13	12	11	9	9
	20–39	16	15	13	12	11	10	10
	40–59	17	16	14	12	12	11	10
	60 or more	18	16	14	13	12	11	11
15–20	0–19	17	16	14	13	12	11	10
	20–39	18	17	15	14	13	11	11
	40–59	19	18	15	14	14	12	12
	60 or more	20	19	16	15	14	13	12

SOURCE: *Soil Stabilization for Roads and Airfields, FM 5-410 Military Soils Engineering.* www.Scribd.com/doc/487747/FM-5410-military-Soils-Engineering. Nov. 2007

at the site should not be greater than 2 h for plant mixing and not more than 1 h for road mixing. The soil-cement mixture is initially compacted with sheepsfoot rollers, or pneumatic-tired rollers when the soil (for example, very sandy soil) cannot be effectively compacted with a sheepsfoot roller. The uppermost layer of 25 mm to 51 mm depth is usually compacted with a pneumatic-tired roller, and the final surface compaction is carried out with a smooth-wheeled roller.

Curing the Compacted Layer

Moisture loss in the compacted layer must be prevented before setting is completed because moisture is required for the hydration process. This is achieved by applying a thin layer of bituminous material such as RC-250 or MC-250. Tars and emulsions also can be used.

19.2.2 Asphalt Stabilization

Asphalt stabilization is carried out to achieve one or both of the following:

- Waterproofing of natural materials
- Binding of natural materials

Waterproofing the natural material through asphalt stabilization aids in maintaining the water content at a required level by providing a membrane that impedes the penetration of water, thereby reducing the effect of any surface water that may enter the soil when it is used as a base course. In addition, surface water is prevented from seeping into the subgrade, which protects the subgrade from failing due to increase in moisture content.

Binding improves the durability characteristics of the natural soil by providing an adhesive characteristic, whereby the soil particles adhere to each other, thus increasing cohesion.

Several types of soil can be stabilized with asphalt, although it is generally required that less than 25 percent of the material passes the No. 200 sieve. This is necessary because the smaller soil particles tend to have extremely large surface areas per unit volume and require a large amount of bituminous material for the soil surfaces to be adequately coated. It is also necessary to use soils that have a plasticity index (PI) of less than 10, because difficulty may be encountered in mixing soils with a high PI, which may result in the plastic fines swelling on contact with water and thereby losing strength.

The mixing of the soil and bituminous materials can also be done in a central or movable plant (plant mixing) or at the roadside (road mixing). In plant mixing, the desired amounts of water and bituminous material are automatically fed into the mixing hoppers, whereas in road mixing, the water and bituminous material are measured and applied separately using a pressure distributor. The materials then are mixed thoroughly in the plant when plant mixing is used or by rotary speed mixers or suitable alternative equipment when road mixing is used.

The material then is spread evenly in layers of uniform thickness, usually not greater than 150 mm and not less than 50 mm. Each layer is properly compacted until the required density is obtained using a sheepsfoot roller or a pneumatic-tired roller. The mixture must be aerated completely before compaction to ensure the removal of all volatile materials.

19.2.3 Lime Stabilization

Lime stabilization is one of the oldest processes of improving the engineering properties of soils and can be used for stabilizing both base and subbase materials. In general, the oxides and hydroxides of calcium and magnesium are considered as lime, but the materials most commonly used for lime stabilization are calcium hydroxide $Ca(OH)_2$ and dolomite $Ca(OH)_2 + MgO$. The dolomite, however, should not have more than 36 percent by weight of magnesium oxide (MgO) to be acceptable as a stabilizing agent.

Clayey materials are most suitable for lime stabilization, but these materials should also have PI values less than 10 for the lime stabilization to be most effective. When lime is added to fine-grained soil, cation exchange takes place, with the calcium and magnesium in the lime replacing the sodium and potassium in the soil. The tendency to swell as a result of an increase in moisture content is therefore immediately reduced. The PI value of the soil is also reduced. Pozzolanic reaction may also occur in some clays, resulting in the formation of cementing agents that increase the strength of the soil. When silica or alumina is present in the soil, a significant increase in strength may be observed over a long period of time. An additional effect is that lime causes flocculation of the fine particles, thereby increasing the effective grain size of the soil.

The percentage of lime used for any project depends on the type of soil being stabilized. The determination of the quantity of lime is usually based on an analysis of the effect that different lime percentages have on the reduction of plasticity and the increase in strength on the soil. The PI is most commonly used for testing the effect on plasticity, whereas the unconfined compression test, the Hveem Stabilometer test, or the California bearing-ratio (CBR) test can be used to test for the effect on strength. An alternative procedure to determine an initial design lime content is given in Figure 19.3. However, most fine-grained soil can be effectively stabilized with 3 to 10 percent lime, based on the dry weight of the soil.

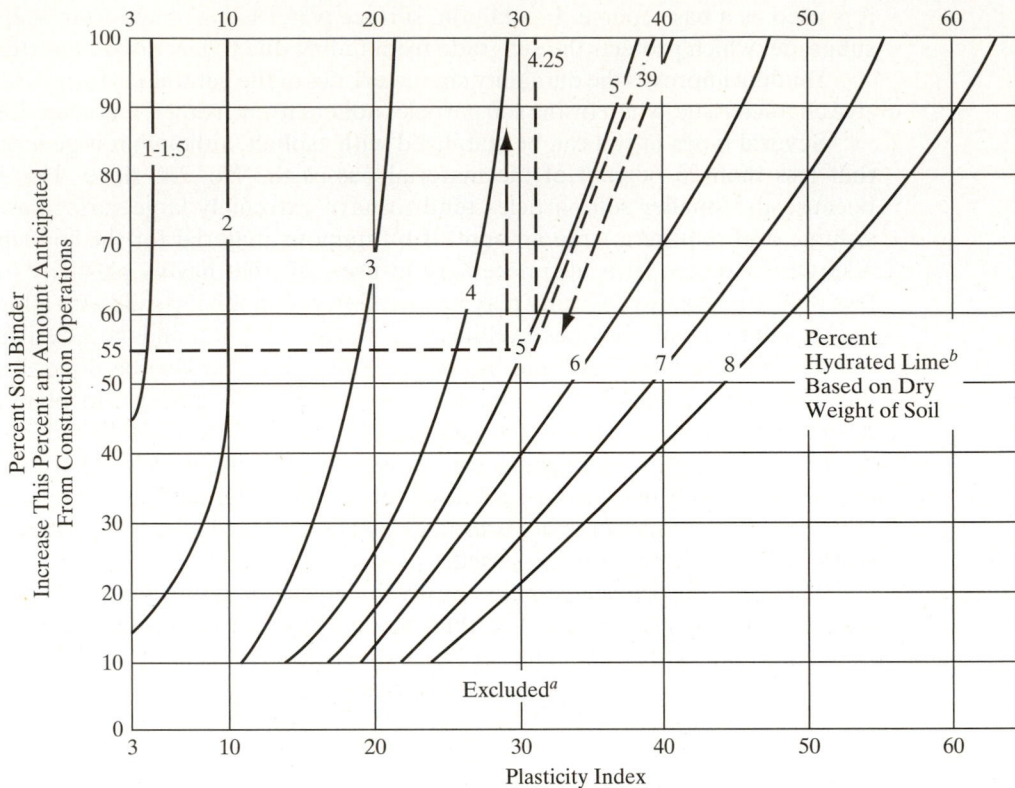

1. Enter the plot with the PI on the top scale.
2. Follow the curved line down to the percent of soil binder (percent passing No 40 sieve).
3. At the intersection with the percent of soil binder, move vertically upward to the 100 percent soil binder line.
4. Read the percent of lime represented at the intersection with the 100 percent soil binder line.
5. For soil having a PI of 39 and 55 percent soil binder, the lime required is +4.25 percent.

[a] Exclude use of chart for materials with less than 10 percent passing the No 40 sieve and for cohesionless materials (PI less than 3).
[b] The percent of relatively pure lime, usually 90 percent or more of calcium and/or magnesium hydroxides and 85 percent or more which pass the No 200 sieve. Percentages shown are for stabilizing subgrades and a base course where lasting effects are desired. Satisfactory temporary results are sometimes obtained by the use of as little as one-half of the above percentages. Reference to cementing strength is implied when such terms as "lasting effects" and "temporary results" are used.
Adapted from Soils Section, Materials and Test Division, Texas Highway Department *Courtesy of the National Lime Association, Washington, DC.*

Figure 19.3 Alternate Method of Determining Initial Design Lime Content

SOURCE: *Soil Stabilization for Roads and Airfields, FM 5-410 Military Soils Engineering.* www.Scribd.com/doc/487747/FM-5410-military-Soils-Engineering. Nov. 2007

19.3 GENERAL PRINCIPLES OF FLEXIBLE PAVEMENT DESIGN

In the design of flexible pavements, the pavement structure usually is considered as a multilayered elastic system, with the material in each layer characterized by certain physical properties that may include the modulus of elasticity, the resilient modulus, and the Poisson ratio. It is assumed initially that the subgrade layer is infinite in both the horizontal and vertical directions, whereas the other layers are finite in the vertical direction and infinite in the horizontal direction. The application of a wheel load causes a stress

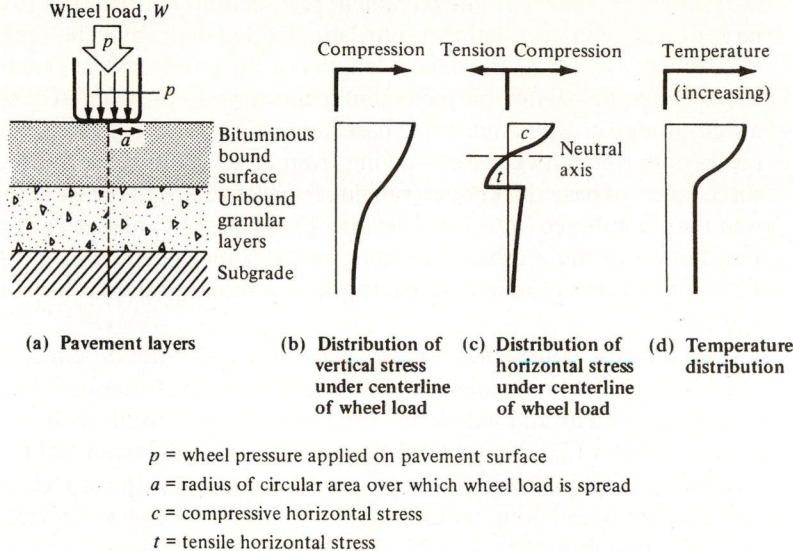

p = wheel pressure applied on pavement surface

a = radius of circular area over which wheel load is spread

c = compressive horizontal stress

t = tensile horizontal stress

Figure 19.4   Stress Distribution within a Flexible Pavement

distribution which can be represented as shown in Figure 19.4. The maximum vertical stresses are compressive and occur directly under the wheel load. These decrease with an increase in depth from the surface. The maximum horizontal stresses also occur directly under the wheel load but can be either tensile or compressive, as shown in Figure 19.4c. When the load and pavement thickness are within certain ranges, horizontal compressive stresses will occur above the neutral axis, whereas horizontal tensile stresses will occur below the neutral axis. The temperature distribution within the pavement structure, as shown in Figure 19.4d, will also have an effect on the magnitude of the stresses. The design of the pavement, therefore, is generally based on strain criteria that limit both the horizontal and vertical strains below those that will cause excessive cracking and excessive permanent deformation. These criteria are considered in terms of repeated load applications because the accumulated repetitions of traffic loads are of significant importance to the development of cracks and permanent deformation of the pavement.

The availability of highly sophisticated computerized solutions for multilayered systems, coupled with recent advances in materials evaluation, has led to the development of several design methods that are based wholly or partly on theoretical analysis. Several design methods have been used for flexible pavements. Those presented in this section are the *American Association of State Highway and Transportation Officials (AASHTO)* and the *California (Hveem)* methods. A brief description of the *Mechanistic-Empirical Pavement Design Guide (MEPDG)* method is also given. The Asphalt Institute method is also used by some agencies, but it is not discussed in this book because of copyright permission difficulties.

19.3.1 AASHTO Design Method

The AASHTO method for design of highway pavements is based primarily on the results of the AASHTO road test that was conducted in Ottawa, Illinois. It was a cooperative effort carried out under the auspices of 49 states, the District of Columbia, Puerto Rico, the Bureau of Public Roads, and several industry groups. Tests were conducted on short-span bridges and test sections of flexible and rigid pavements constructed on

A-6 subgrade material. The pavement test sections consisted of two small loops and four larger ones, with each being a four-lane divided highway. The tangent sections consisted of a successive set of pavement lengths of different designs, each length being at least 30 m. The principal flexible pavement sections were constructed of asphalt mixture surface, a well-graded crushed limestone base, and a uniformly graded sand-gravel subbase. Three levels of surface thicknesses ranging from 25 to 150 mm were used in combination with three levels of base thicknesses ranging from 0 to 230 mm. Each of these nine combinations was then combined with three levels of subbase thicknesses ranging from 0 to 150 mm. In addition to the crushed limestone bases, some special sections had bases constructed with either a well-graded uncrushed gravel, a bituminous plant mixture, or cement-treated aggregate.

Test traffic consisting of both single-axle and tandem-axle vehicles was then driven over the test sections until several thousand load repetitions had been made. Ten different axle-arrangement and axle-load combinations were used, with single-axle loads ranging from 8.9 kN to 133 kN and tandem-axle loads ranging from 107 kN to 214 kilo Newtons. Data were then collected on the pavement condition with respect to the extent of cracking and amount of patching required to maintain the section in service. The longitudinal and transverse profiles were also obtained to determine the extent of rutting, surface deflection caused by loaded vehicles moving at very slow speeds, pavement curvature at different vehicle speeds, stresses imposed on the subgrade surface, and temperature distribution in the pavement layers. These data then were analyzed thoroughly, and the results formed the basis for the AASHTO method of pavement design.

AASHTO initially published an interim guide for the design of pavement structures in 1961, which was revised in 1972. A further revision was published in 1986, consisting of two volumes: Volume 1 being the basic design guide and Volume 2 containing a series of appendices that provided documentation or further explanations for information contained in Volume 1. That edition incorporated new developments and specifically addressed pavement rehabilitation. Another edition was subsequently published in 1993, consisting of only one volume, which replaced Volume 1 of the 1986 guide. The major changes in the 1993 edition involved the overlay design procedure. Each edition of the guide specifically mentions that the design procedure presented cannot possibly include all conditions that relate to any one specific site. It is therefore recommended that, in using the guide, local experience be used to augment the procedures given in the guide.

Design Considerations

The factors considered in the AASHTO procedure for the design of flexible pavement as presented in the 1993 guide are:

- Pavement performance
- Traffic
- Roadbed soils (subgrade material)
- Materials of construction
- Environment
- Drainage
- Reliability

Pavement Performance.   The primary factors considered under pavement performance are the structural and functional performance of the pavement. Structural performance is related to the physical condition of the pavement with respect to factors that have a negative impact on the capability of the pavement to carry the traffic load. These factors include cracking, faulting, raveling, and so forth. Functional performance is an indication of how effectively the pavement serves the user. The main factor considered under functional performance is riding comfort.

To quantify pavement performance, a concept known as the *serviceability performance* was developed. Under this concept, a procedure was developed to determine the present serviceability index (PSI) of the pavement, based on its roughness and distress, which were measured in terms of extent of cracking, patching, and rut depth for flexible pavements. The original expression developed gave the PSI as a function of the extent and type of cracking and patching and the slope variance in the two wheel paths, which is a measure of the variations in the longitudinal profile. The mean of the ratings of individual engineers with wide experience in all facets of highway engineering was used to relate the PSI with the factors considered. The scale ranges from 0 to 5, where 0 is the lowest PSI and 5 is the highest. A detailed discussion of PSI is given in Chapter 21.

Two serviceability indices are used in the design procedure: the initial serviceability index (p_i), which is the serviceability index immediately after the construction of the pavement; and the terminal serviceability index (p_t), which is the minimum acceptable value before resurfacing or reconstruction is necessary. In the AASHTO road test, a value of 4.2 was used for p_i for flexible pavements. AASHTO, however, recommends that each agency determine more reliable levels for p_i based on existing conditions. Recommended values for the terminal serviceability index are 2.5 or 3.0 for major highways and 2.0 for highways with a lower classification. In cases where economic constraints restrict capital expenditures for construction, the p_t can be taken as 1.5, or the performance period may be reduced. However, this low value should be used only in special cases on selected classes of highways.

Traffic Load. In the AASHTO design method, the traffic load is determined in terms of the number of repetitions of an 18,000-lb (80 kilo-newtons [kN]) single-axle load applied to the pavement on two sets of dual tires. This is usually referred to as the *equivalent single-axle load* (ESAL). The dual tires are represented as two circular plates, each 114 mm radius, spaced 345 mm apart. This representation corresponds to a contact pressure of 12.25 kN/m The use of an 80 kN (18,000-lb) axle load is based on the results of experiments that have shown that the effect of any load on the performance of a pavement can be represented in terms of the number of single applications of an 80 kN single axle (ESALs). The equivalence factors used in this case are based on the terminal serviceability index to be used in the design and the structural number (SN) (see definition of SN under "Structural Design"). Table 19.5a gives traffic equivalence factors for p_t of 2.5 for single axles, and Table 19.5b gives the equivalence factors for p_t of 2.5 for tandem axles. To determine the ESAL, the number of different types of vehicles such as cars, buses, single-unit trucks, and multiple-unit trucks expected to use the facility during its lifetime must be known. The distribution of the different types of vehicles expected to use the proposed highway can be obtained from results of classification counts that are taken by state highway agencies at regular intervals. These can then be converted to equivalent 80 kN loads using the equivalency factors given in Table 19.5.

The total ESAL applied on the highway during its design period can be determined only after the design period and traffic growth factors are known. The design period is the number of years the pavement will effectively continue to carry the traffic load without requiring an overlay. Flexible highway pavements are usually designed for a 10 to 20-year period. Since traffic volume does not remain constant over the design period of the pavement, it is essential that the rate of growth be determined and applied

Table 19.5a Axle Load Equivalency Factors for Flexible Pavements, Single Axles, and p_t of 2.5

Axle Load (kips)	Pavement Structural Number (SN)					
	1	2	3	4	5	6
2	.0004	.0004	.0003	.0002	.0002	.0002
4	.003	.004	.004	.003	.002	.002
6	.011	.017	.017	.013	.010	.009
8	.032	.047	.051	.041	.034	.031
10	.078	.102	.118	.102	.088	.080
12	.168	.198	.229	.213	.189	.176
14	.328	.358	.399	.388	.360	.342
16	.591	.613	.646	.645	.623	.606
18	1.00	1.00	1.00	1.00	1.00	1.00
20	1.61	1.57	1.49	1.47	1.51	1.55
22	2.48	2.38	2.17	2.09	2.18	2.30
24	3.69	3.49	3.09	2.89	3.03	3.27
26	5.33	4.99	4.31	3.91	4.09	4.48
28	7.49	6.98	5.90	5.21	5.39	5.98
30	10.3	9.5	7.9	6.8	7.0	7.8
32	13.9	12.8	10.5	8.8	8.9	10.0
34	18.4	16.9	13.7	11.3	11.2	12.5
36	24.0	22.0	17.7	14.4	13.9	15.5
38	30.9	28.3	22.6	18.1	17.2	19.0
40	39.3	35.9	28.5	22.5	21.1	23.0
42	49.3	45.0	35.6	27.8	25.6	27.7
44	61.3	55.9	44.0	34.0	31.0	33.1
46	75.5	68.8	54.0	41.4	37.2	39.3
48	92.2	83.9	65.7	50.1	44.5	46.5
50	112.0	102.0	79.0	60.0	53.0	55.0

SOURCE: Based on *AASHTO Guide for Design of Pavement Structures*, 1993, AASHTO, Washington, D.C.

when calculating the total ESAL. Annual growth rates can be obtained from regional planning agencies or from state highway agencies. These usually are based on traffic volume counts over several years. It is also advisable to determine annual growth rates separately for trucks and passenger vehicles, since these may be significantly different in some cases. The overall growth rate in the United States is between 2 and 5 percent per year, although growth rates of up to 10 percent per year have been suggested for some interstate highways. The growth factors (G_{rn}) for different growth rates and design periods can be obtained from Eq. 19.1.

$$G_{rn} = [(1 + r)^n - 1]/r \qquad (19.1)$$

where

$r = \dfrac{i}{100}$ and is not zero. If annual growth is zero, growth factor = design period

i = growth rate

n = design life (years)

Table 19.5b   Axle Load Equivalency Factors for Flexible Pavements, Tandem Axles, and p_t of 2.5

Axle Load (kips)	Pavement Structural Number (SN)					
	1	2	3	4	5	6
2	.0001	.0001	.0001	.0000	.0000	.0000
4	.0005	.0005	.0004	.0003	.0003	.0002
6	.002	.002	.002	.001	.001	.001
8	.004	.006	.005	.004	.003	.003
10	.008	.013	.011	.009	.007	.006
12	.015	.024	.023	.018	.014	.013
14	.026	.041	.042	.033	.027	.024
16	.044	.065	.070	.057	.047	.043
18	.070	.097	.109	.092	.077	.070
20	.107	.141	.162	.141	.121	.110
22	.160	.198	.229	.207	.180	.166
24	.231	.273	.315	.292	.260	.242
26	.327	.370	.420	.401	.364	.342
28	.451	.493	.548	.534	.495	.470
30	.611	.648	.703	.695	.658	.633
32	.813	.843	.889	.887	.857	.834
34	1.06	1.08	1.11	1.11	1.09	1.08
36	1.38	1.38	1.38	1.38	1.38	1.38
38	1.75	1.73	1.69	1.68	1.70	1.73
40	2.21	2.16	2.06	2.03	2.08	2.14
42	2.76	2.67	2.49	2.43	2.51	2.61
44	3.41	3.27	2.99	2.88	3.00	3.16
46	4.18	3.98	3.58	3.40	3.55	3.79
48	5.08	4.80	4.25	3.98	4.17	4.49
50	6.12	5.76	5.03	4.64	4.86	5.28
52	7.33	6.87	5.93	5.38	5.63	6.17
54	8.72	8.14	6.95	6.22	6.47	7.15
56	10.3	9.6	8.1	7.2	7.4	8.2
58	12.1	11.3	9.4	8.2	8.4	9.4
60	14.2	13.1	10.9	9.4	9.6	10.7
62	16.5	15.3	12.6	10.7	10.8	12.1
64	19.1	17.6	14.5	12.2	12.2	13.7
66	22.1	20.3	16.6	13.8	13.7	15.4
68	25.3	23.3	18.9	15.6	15.4	17.2
70	29.0	26.6	21.5	17.6	17.2	19.2
72	33.0	30.3	24.4	19.8	19.2	21.3
74	37.5	34.4	27.6	22.2	21.3	23.6
76	42.5	38.9	31.1	24.8	23.7	26.1
78	48.0	43.9	35.0	27.8	26.2	28.8
80	54.0	49.4	39.2	30.9	29.0	31.7
82	60.6	55.4	43.9	34.4	32.0	34.8
84	67.8	61.9	49.0	38.2	35.3	38.1
86	75.7	69.1	54.5	42.3	38.8	41.7
88	84.3	76.9	60.6	46.8	42.6	45.6
90	93.7	85.4	67.1	51.7	46.8	49.7

SOURCE: Based on *AASHTO Guide for Design of Pavement Structures*, 1993, AASHTO, Washington, D.C.

Table 19.6 shows calculated growth factors (G_{rn}) for different growth rates (r) and design periods (n) that can be used to determine the total ESAL over the design period.

The portion of the total ESAL acting on the design lane (f_d) is used in the determination of pavement thickness. Either lane of a two-lane highway can be considered as the design lane, whereas for multilane highways, the outside lane is considered. The identification of the design lane is important because in some cases more trucks will travel in one direction than in the other or trucks may travel heavily loaded in one direction and empty in the other direction. Thus, it is necessary to determine the relevant proportion of trucks on the design lane. A general equation for the accumulated ESAL for each category of axle load is obtained as

$$ESAL_i = f_d \times G_{rn} \times AADT_i \times 365 \times N_i \times F_{Ei} \qquad (19.2)$$

where

$ESAL_i$ = equivalent accumulated 18,000-lb (80-kN) single-axle load for the axle category i

f_d = design lane factor

Table 19.6 Growth Factors

Design Period, Years (n)	No Growth	Annual Growth Rate, Percent (r)						
		2	4	5	6	7	8	10
1	1.0	1.0	1.0	1.0	1.0	1.0	1.0	1.0
2	2.0	2.02	2.04	2.05	2.06	2.07	2.08	2.10
3	3.0	3.06	3.12	3.15	3.18	3.21	3.25	3.31
4	4.0	4.12	4.25	4.31	4.37	4.44	4.51	4.64
5	5.0	5.20	5.42	5.53	5.64	5.75	5.87	6.11
6	6.0	6.31	6.63	6.80	6.98	7.15	7.34	7.72
7	7.0	7.43	7.90	8.14	8.39	8.65	8.92	9.49
8	8.0	8.58	9.21	9.55	9.90	10.26	10.64	11.44
9	9.0	9.75	10.58	11.03	11.49	11.98	12.49	13.58
10	10.0	10.95	12.01	12.58	13.18	13.82	14.49	15.94
11	11.0	12.17	13.49	14.21	14.97	15.78	16.65	18.53
12	12.0	13.41	15.03	15.92	16.87	17.89	18.98	21.38
13	13.0	14.68	16.63	17.71	18.88	20.14	21.50	24.52
14	14.0	15.97	18.29	19.16	21.01	22.55	24.21	27.97
15	15.0	17.29	20.02	21.58	23.28	25.13	27.15	31.77
16	16.0	18.64	21.82	23.66	25.67	27.89	30.32	35.95
17	17.0	20.01	23.70	25.84	28.21	30.84	33.75	40.55
18	18.0	21.41	25.65	28.13	30.91	34.00	37.45	45.60
19	19.0	22.84	27.67	30.54	33.76	37.38	41.45	51.16
20	20.0	24.30	29.78	33.06	36.79	41.00	45.76	57.28
25	25.0	32.03	41.65	47.73	54.86	63.25	73.11	98.35
30	30.0	40.57	56.08	66.44	79.06	94.46	113.28	164.49
35	35.0	49.99	73.65	90.32	111.43	138.24	172.32	271.02

SOURCE: *Thickness Design–Asphalt Pavements for Highways and Streets*, Manual Series No. 1, The Asphalt Institute, Lexington, KY, February 1991. Used with permission.

G_{rn} = growth factor for a given growth rate r and design period n
$AADT_i$ = first-year annual average daily traffic for axle category i
N_i = number of axles on each vehicle in category i
F_{Ei} = load equivalency factor for axle category i

Example 19.1 Computing Accumulated Equivalent Single-Axle Load for a Proposed Eight-Lane Highway Using Load Equivalency Factors

An eight-lane divided highway is to be constructed on a new alignment. Traffic volume forecasts indicate that the average annual daily traffic ($AADT$) in both directions during the first year of operation will be 12,000, with the following vehicle mix and axle loads.

Passenger cars 4.45 kN/axle(2000 lb/axle) = 50%

2-axle single-unit trucks 26.7 kN (6000 lb/axle) = 33%

3-axle single-unit trucks 44.5 kN (10,000 lb/axle) = 17%

The vehicle mix is expected to remain the same throughout the design life of the pavement. If the expected annual traffic growth rate is 4 percent for all vehicles, determine the design ESAL, given a design period of 20 years. The percent of traffic on the design lane is 45 percent, and the pavement has a terminal serviceability index (p_t) of 2.5 and SN of 5.
The following data apply:

Growth factor = 29.78 (from Table 19.6)

Percent truck volume on design lane = 45

Load equivalency factors (from Table 19.5a)

Passenger cars (2000 lb/axle) = 0.0002 (negligible)

2-axle single-unit trucks (6000 lb/axle) = 0.010

3-axle single-unit trucks (10,000 lb/axle) = 0.088

Solution: The ESAL for each class of vehicle is computed from Eq. 19.2.

$$ESAL = f_d \times G_{jt} \times AADT \times 365 \times N_i \times F_{Ei}$$

2-axle single-unit trucks = $0.45 \times 29.78 \times 12{,}000 \times 0.33 \times 365 \times 2 \times 0.010$

$$= 0.3874 \times 10^6$$

3-axle single-unit trucks = $0.45 \times 29.78 \times 12{,}000 \times 0.17 \times 365 \times 3 \times 0.088$

$$= 2.6343 \times 10^6$$

Passenger cars = $0.45 \times 29.78 \times 12{,}000 \times 0.5 \times 365 \times 2 \times 0.0002$

$$= .012 \times 10^6 \text{ (negligible)}$$

Thus,

$$\text{Total ESAL} = 3.0217 \times 10^6$$

It can be seen that the contribution of passenger cars to the ESAL is negligible. Passenger cars are therefore omitted when computing ESAL values. This example illustrates the conversion of axle loads to ESAL using axle load equivalency factors.

The equivalent 80 kN load can also be determined from the vehicle type, if the axle load is unknown, by using a truck factor for that vehicle type. The truck factor is defined as the number of 80 kN single-load applications caused by a single passage of a vehicle. These can be determined for each class of vehicle from the expression

$$\text{Truck factor} = \frac{\sum(\text{number of axles} \times \text{load equivalency factor})}{\text{number of vehicles}}$$

Roadbed Soils (Subgrade Material).   The 1993 AASHTO guide also uses the resilient modulus (M_r) of the soil to define its property. However, the method allows for the conversion of the CBR or R value of the soil to an equivalent M_r value using the following conversion factors:

$$M_r(\text{kN/m}^2) = 220\text{CBR} \text{ (for fine-grain soils with soaked CBR of 10 or less)} \quad (19.3)$$

$$M_r(\text{kN/m}^2) = 145 + 80.4R \text{ value(for } R \leq 20) \quad (19.4)$$

Figures 19.5 and 19.6 show average correlations among different test results obtained in several states that can be used when values fall outside the ranges specified in Eqs. 19.3 and 19.4.

Materials of Construction.   The materials used for construction can be classified under three general groups: those used for subbase construction, those used for base construction, and those used for surface construction.

Subbase Construction Materials. The quality of the material used is determined in terms of the layer coefficient, a_3, which is used to convert the actual thickness of the subbase to an equivalent SN. The sandy gravel subbase course material used in the AASHTO road test was assigned a value of 0.11. Layer coefficients are usually assigned, based on the description of the material used. Note, however, that due to the widely different environmental, traffic, and construction conditions, it is essential that each design agency develop layer coefficients appropriate to the conditions that exist in its own environment.

Charts correlating the layer coefficients with different soil engineering properties have been developed. Figure 19.5 shows one such chart for granular subbase materials.

Base Course Construction Materials. Materials selected should satisfy the general requirements for base course materials given earlier in this chapter and in Chapter 17. A structural layer coefficient, a_2, for the material used also should be determined. This can be done using Figure 19.6.

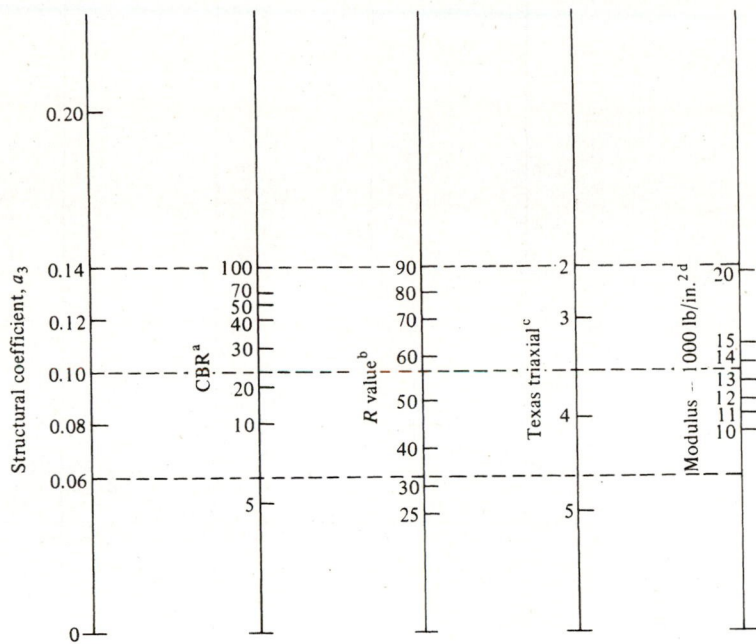

[a] Scale derived from correlations from Illinois.
[b] Scale derived from correlations obtained from The Asphalt
Institute, California, New Mexico, and Wyoming.
[c] Scale derived from correlations obtained from Texas.
[d] Scale derived on NCHRP project 128, 1972.

Figure 19.5 Variation in Granular Subbase Layer Coefficient, a_3, with Various Subbase Strength Parameters

SOURCE: From *AASHTO Guide for Design of Pavement Structures*, 1993, AASHTO, Washington, D.C. Used by permission.

Surface Course Construction Materials. The most commonly used material is a hot-plant mix of asphalt cement and dense-graded aggregates with a maximum size of 25 mm The procedure discussed in Chapter 18 for the design of asphalt mix can be used. The structural layer coefficient (a_1) for the surface course can be extracted from Figure 19.7, which relates the structural layer coefficient of a dense-grade asphalt concrete surface course with its resilient modulus at 20 C.

Environment. Temperature and rainfall are the two main environmental factors used in evaluating pavement performance in the AASHTO method. The effects of temperature on asphalt pavements include stresses induced by thermal action, changes in the creep properties, and the effect of freezing and thawing of the subgrade soil, as discussed in Chapters 16 and 17. The effect of rainfall is due mainly to the penetration of the surface water into the underlying material. If penetration occurs, the properties of the underlying materials may be altered significantly. In Chapter 16, different ways of preventing water penetration were discussed. However, this effect is taken into consideration in the design procedure, and the methodology used is presented later under "Drainage."

The effect of temperature, particularly with regard to the weakening of the underlying material during the thaw period, is considered a major factor in determining the strength of the underlying materials used in the design. Test results have shown that the

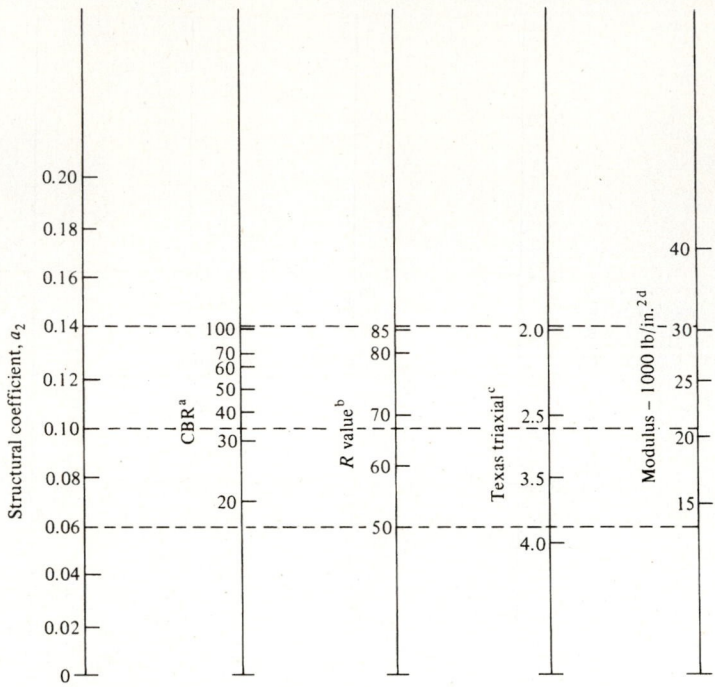

a Scale derived by averaging correlations obtained from Illinois.
b Scale derived by averaging correlations obtained from California,
New Mexico, and Wyoming.
c Scale derived by averaging correlations obtained from Texas.
d Scale derived on NCHRP project 128, 1972.

Figure 19.6   Variation in Granular Base Layer Coefficient, a_2, with Various base Strength Parameters

SOURCE: From *AASHTO Guide for Design of Pavement Structures*, 1993, AASHTO, Washington, D.C. Used by permission.

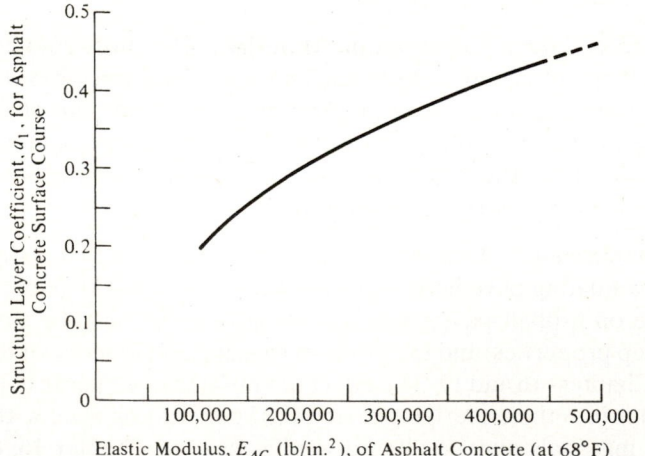

Figure 19.7   Chart for Estimating Structural Layer Coefficient of Dense-Graded/Asphalt Concrete Based on the Elastic (Resilient) Modulus

SOURCE: From *AASHTO Guide for Design of Pavement Structures*, 1993, AASHTO, Washington, D.C. Used by permission.

normal modulus (that is, modulus during summer and fall seasons) of materials suscep-tible to frost action can reduce by 50 percent to 80 percent during the thaw period. Also, the resilient modulus of a subgrade material may vary during the year, even when there is no specific thaw period. This occurs in areas subject to very heavy rains during specific periods of the year. It is likely that the strength of the material will be affected during the periods of heavy rains.

The procedure used to take into consideration the variation during the year of the resilient modulus of the roadbed soil is to determine an effective annual roadbed soil resilient modulus. The change in the PSI of the pavement during a full 12-month period will then be the same if the effective resilient modulus is used for the full period or if the appropriate resilient modulus for each season is used. This means that the effective resilient modulus is equivalent to the combined effect of the different seasonal moduli during the year.

The AASHTO guide suggests two methods for determining the effective resilient modulus. Only the first method is described here. In this method, a relationship between resilient modulus of the soil material and moisture content is developed using labora-tory test results. This relationship is then used to determine the resilient modulus for each season based on the estimated *in situ* moisture content during the season being considered. The whole year is then divided into different time intervals that correspond with the different seasonal resilient moduli. The AASHTO guide suggests that it is not necessary to use a time interval less than half a month. The relative damage u_f for each time period is then determined from the chart in Figure 19.8, using the vertical scale, or the equation given in the chart. The mean relative damage u_f is then computed, and the effective subgrade resilient modulus is determined using the chart and the value of u_f.

Example 19.2 Computing Effective Resilient Modulus

Figure 19.8 shows roadbed soil resilient modulus M_r for each month estimated from laboratory results correlating M_r with moisture content. Determine the effective resil-ient modulus of the subgrade.

Solution: Note that in this case, the moisture content does not vary within any one month. The solution of the problem is given in Figure 19.8. The value of u_f for each M_r is obtained directly from the chart. The mean relative damage $\bar{u}_f$ is 0.133, which in turn gives an effective resilient modulus of 50.025 kN/m².

Drainage.   The effect of drainage on the performance of flexible pavements is con-sidered in the 1993 guide with respect to the effect of water on the strength of the base material and roadbed soil. The approach used is to provide for the rapid drainage of the free water (noncapillary) from the pavement structure by providing a suitable drainage layer, as shown in Figure 19.9, and by modifying the structural layer coef-ficient. The modification is carried out by incorporating a factor m_i for the base and subbase layer coefficients (a_2 and a_3). The m_i factors are based both on the percent-age of time during which the pavement structure will be nearly saturated and on the quality of drainage, which is dependent on the time it takes to drain the base layer to 50 percent of saturation. Table 19.7 gives the general definitions of the different levels of drainage quality, while Table 19.8 gives recommended m_i values for different levels of drainage quality.

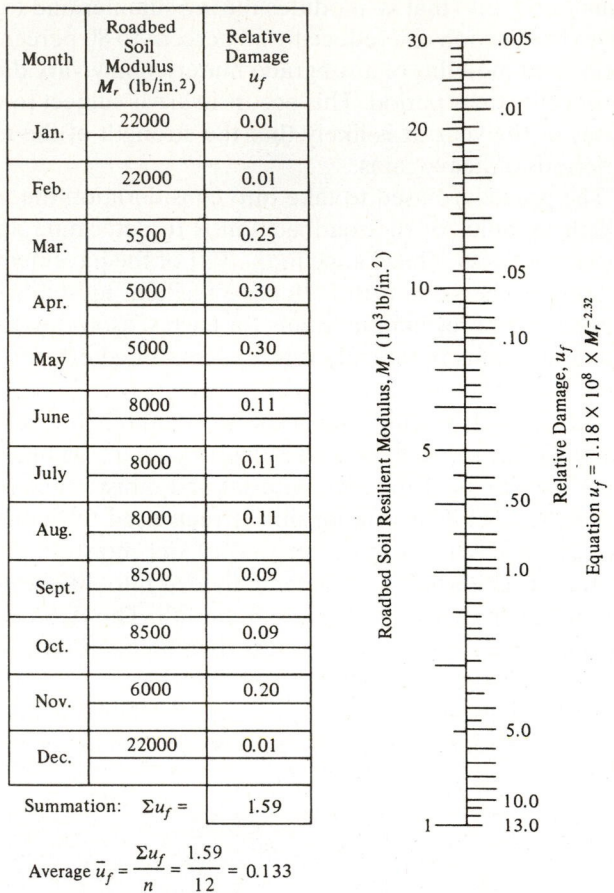

Month	Roadbed Soil Modulus M_r (lb/in.^{2})	Relative Damage u_f
Jan.	22000	0.01
Feb.	22000	0.01
Mar.	5500	0.25
Apr.	5000	0.30
May	5000	0.30
June	8000	0.11
July	8000	0.11
Aug.	8000	0.11
Sept.	8500	0.09
Oct.	8500	0.09
Nov.	6000	0.20
Dec.	22000	0.01
Summation: $\Sigma u_f =$		1.59

$$\text{Average } \bar{u}_f = \frac{\Sigma u_f}{n} = \frac{1.59}{12} = 0.133$$

Effective Roadbed Soil Resilient Modulus, M_r (lb/in.^{2}) = <u>7250</u> (corresponds to $\bar{u}_f$)

Figure 19.8   Chart for Estimating Effective Roadbed Soil Resilient Modulus for Flexible Pavements Designed Using the Serviceability Criteria

SOURCE: From *AASHTO Guide for Design of Pavement Structures*, 1993, AASHTO, Washington, D.C. Used by permission.

Reliability.   It has been noted that the cumulative ESAL is an important input to any pavement design method. However, the determination of this input is usually based on assumed growth rates, which may not be accurate. Most design methods do not consider this uncertainty, but the 1993 AASHTO guide proposes the use of a reliability factor that considers the possible uncertainties in traffic prediction and performance prediction. A detailed discussion of the development of the approach used is beyond the scope of this book; however, a general description of the methodology is presented to allow the incorporation of reliability in the design process, if so desired by a designer. Reliability design levels ($R\%$), which determine assurance levels that the pavement section designed using the procedure will survive for its design period, have been developed for different types of highways. For example, a 50 percent reliability design level implies a 50 percent chance for successful pavement performance—that is, the probability of design performance success is 50 percent. Table 19.9 shows suggested reliability levels based on a survey of the AASHTO pavement design task force. Reliability factors, $F_R \geq 1$, based on the reliability

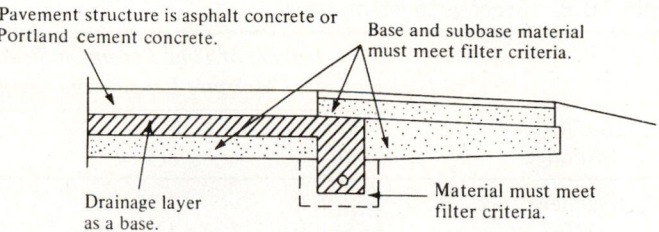

(a) **Base is used as the drainage layer.**

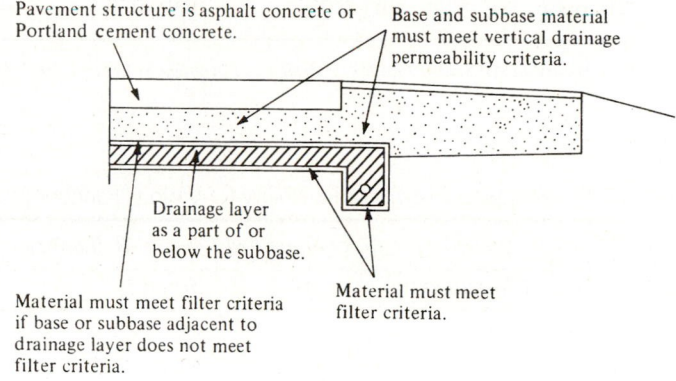

(b) **Drainage layer is part of or below the subbase.**

Note: Filter fabrics may be used in lieu of filter
        material, soil, or aggregate, depending on
        economic considerations.

Figure 19.9   Example of Drainage Layer in Pavement Structure

SOURCE: From *AASHTO Guide for Design of Pavement Structures*, 1993, AASHTO, Washington, D.C. Used by permission.

level selected and the overall variation, S_o^2, also have been developed. S_o^2 accounts for the chance variation in the traffic forecast and the chance variation in actual pavement performance for a given design period traffic, W_{18}.

The reliability factor F_R is given as

$$\log_{10} F_R = -Z_R S_o \tag{19.5}$$

Table 19.7   Definition of Drainage Quality

Quality of Drainage	Water Removed Within*
Excellent	2 hours
Good	1 day
Fair	1 week
Poor	1 month
Very poor	(water will not drain)

*Time required to drain the base layer to 50% saturation.
SOURCE: Based on *AASHTO Guide for Design of Pavement Structures*, 1993, AASHTO, Washington, D.C.

Table 19.8   Recommended m_i Values

Quality of Drainage	Percent of Time Pavement Structure Is Exposed to Moisture Levels Approaching Saturation			
	Less Than 1%	1 to 5%	5 to 25%	Greater Than 25%
Excellent	1.40–1.35	1.35–1.30	1.30–1.20	1.20
Good	1.35–1.25	1.25–1.15	1.15–1.00	1.00
Fair	1.25–1.15	1.15–1.05	1.00–0.80	0.80
Poor	1.15–1.05	1.05–0.80	0.80–0.60	0.60
Very poor	1.05–0.95	0.95–0.75	0.75–0.40	0.40

SOURCE: Based on *AASHTO Guide for Design of Pavement Structures*, 1993, AASHTO, Washington, D.C.

Table 19.9   Suggested Levels of Reliability for Various Functional Classifications

Functional Classification	Recommended Level of Reliability	
	Urban	Rural
Interstate and other freeways	85–99.9	80–99.9
Other principal arterials	80–99	75–95
Collectors	80–95	75–95
Local	50–80	50–80

Note: Results based on a survey of the AASHTO Pavement Design Task Force.
SOURCE: Based on *AASHTO Guide for Design of Pavement Structures*, 1993, AASHTO, Washington, D.C.

where

Z_R = standard normal variate for a given reliability ($R\%$)
S_o = estimated overall standard deviation

Table 19.10 gives values of Z_R for different reliability levels R.

Overall standard deviation (S_o) ranges have been identified as 0.40 to 0.50 for flexible pavements and 0.30 to 0.40 for rigid pavements. These values were derived through a detailed analysis of existing data. However, very little data currently exists for certain design components, such as drainage. A methodology for improving these estimates is presented in the 1993 AASHTO guide, which may be used when additional data are available.

Structural Design

The objective of the design using the AASHTO method is to determine a flexible pavement structural number (SN) adequate to carry the projected design ESAL. It is left to the designer to select the type of surface used, which can be either asphalt concrete, a single surface treatment such as chip seal, or a double surface treatment. This design procedure is used for ESALs greater than 50,000 for the performance period. The design for ESALs less than this is usually considered under low-volume roads.

The 1993 AASHTO guide gives the expression for SN as

$$\text{SN} = a_1 D_1 + a_2 D_2 m_2 + a_3 D_3 m_3 \tag{19.6}$$

Table 19.10   Standard Normal Deviation (Z_R) Values Corresponding to Selected Levels of Reliability

Reliability (R%)	Standard Normal Deviation, Z_R
50	−0.000
60	−0.253
70	−0.524
75	−0.674
80	−0.841
85	−1.037
90	−1.282
91	−1.340
92	−1.405
93	−1.476
94	−1.555
95	−1.645
96	−1.751
97	−1.881
98	−2.054
99	−2.327
99.9	−3.090
99.99	−3.750

SOURCE: Based on *AASHTO Guide for Design of Pavement Structures*, 1993, AASHTO, Washington, D.C.

where

m_i = drainage coefficient for layer i

a_1, a_2, a_3 = layer coefficients representative of surface, base, and subbase course, respectively

D_1, D_2, D_3 = actual thickness in inches of surface, base, and subbase courses, respectively

The basic design equation given in the 1993 guide is

$$\log_{10}W_{18} = Z_R S_o + 9.36\ \log_{10}(SN + 1) - 0.20 + \frac{\log_{10}[\Delta PSI/(4.2 - 1.5)]}{0.40 + [1094/(SN + 1)^{5.19}]} \quad (19.7)$$
$$+ 2.32 \log_{10}M_r - 8.07$$

where

W_{18} = predicted number of 18,000-lb (80-kN) single-axle load applications

Z_R = standard normal deviation for a given reliability

S_o = overall standard deviation

SN = structural number indicative of the total pavement thickness

$\Delta PSI = p_i - p_t$

p_i = initial serviceability index

p_t = terminal serviceability index

M_r = effective resilient modulus of roadbed soil (lb/in.^{2})

Equation 19.7 can be solved for SN using a computer program or the chart in Figure 19.10. The use of the chart is demonstrated by the example solved on the chart and in the solution of Example 19.3.

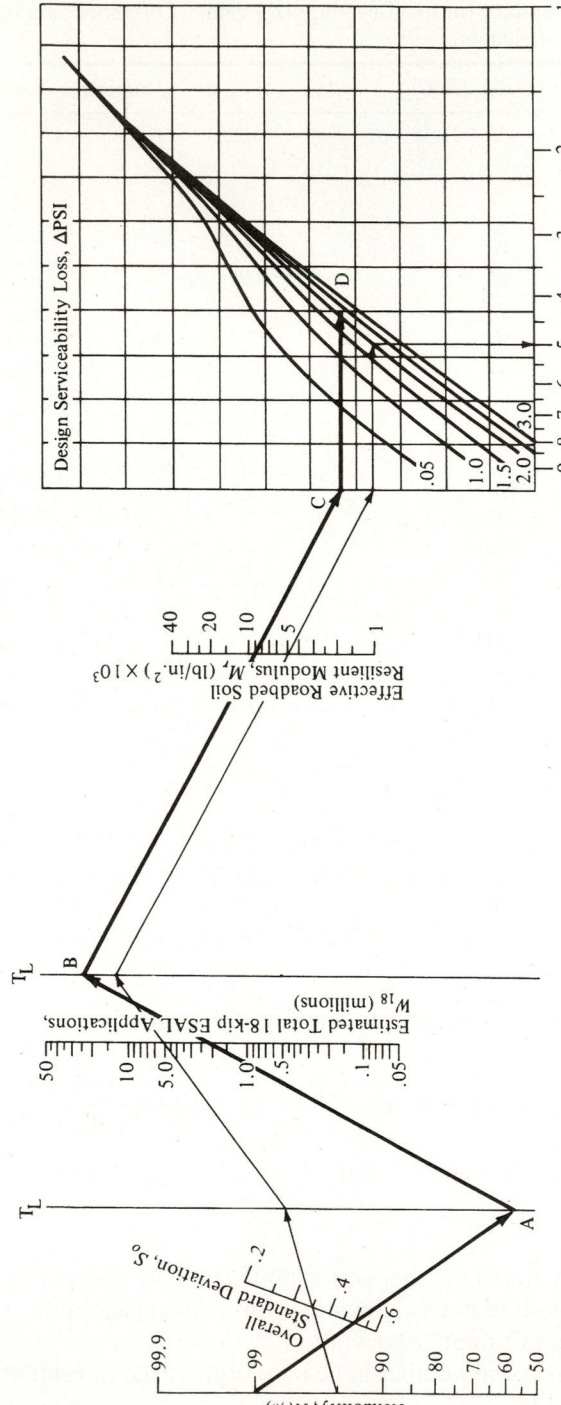

Figure 19.10 Design Chart for Flexible Pavements Based on Using Mean Values for Each Input

SOURCE: From *AASHTO Guide for Design of Pavement Structures*, 1993, AASHTO, Washington, D.C. Used by permission.

Example 19.3 Designing a Flexible Pavement Using the AASHTO Method

A flexible pavement for an urban interstate highway is to be designed using the 1993 AASHTO guide procedure to carry a design ESAL of 2×10^6. It is estimated that it takes about a week for water to be drained from within the pavement, and the pavement structure will be exposed to moisture levels approaching saturation for 30 percent of the time. The following additional information is available:

Resilient modulus of asphalt concrete at 20°C(68°F) = 3106 MPa (450000 lb/in²)

CBR value of base course material = 100, M_r = 213 MPa (31000 lb/in²)

CBR value of subbase course material = 22, M_r = 93 MPa (13,500 lb/in²)

CBR value of subgrade material = 6

Determine a suitable pavement structure, M_r, of subgrade = 6×1500 lb/in.² = 62 MPa (9000 lb/in²).

Solution: Since the pavement is to be designed for an interstate highway, the following assumptions are made.

Reliability level (R) = 99% (range is 85 to 99.9 from Table 19.9)

Standard deviation (S_o) = 0.49 (range is 0.4 to 0.5)

Initial serviceability index p_i = 4.5

Terminal serviceability index p_t = 2.5

The nomograph in Figure 19.10 is used to determine the design SN through the following steps.

Step 1.   Draw a line joining the reliability level of 99 percent and the overall standard deviation S_o of 0.49, and extend this line to intersect the first T_L line at point A.

Step 2.   Draw a line joining point A to the ESAL of 2×10^6, and extend this line to intersect the second T_L line at point B.

Step 3.   Draw a line joining point B and resilient modulus (M_r) of the roadbed soil (9,000 lb/in.²), and extend this line to intersect the design serviceability loss chart at point C.

Step 4.   Draw a horizontal line from point C to intersect the design serviceability loss (ΔPSI) curve at point D. In this problem, ΔPSI = 4.5 – 2.5 = 2.

Step 5.   Draw a vertical line to intersect the design SN, and read this value SN = 4.4.

Step 6.   Determine the appropriate structure layer coefficient for each construction material.

(a) Resilient value of asphalt = 3106 MPa (450000 lb/in²). From Figure 19.7, a_1 = 0.44.
(b) CBR of base course material = 100. From Figure 19.6, a_2 = 0.14.
(c) CBR of subbase course material = 22. From Figure 19.5, a_3 = 0.10.

Step 7.   Determine appropriate drainage coefficient m_i. Since only one set of conditions is given for both the base and subbase layers, the same value will be used for m_1 and m_2. The time required for water to drain from within pavement = 1 week, from Table 19.7, drainage quality is fair. The percentage of time pavement structure will be exposed to moisture levels approaching saturation = 30, and from Table 19.8, $m_i = 0.80$.

Step 8.   Determine appropriate layer thicknesses from Eq. 19.6:

$$= a_1 D_1 + a_2 D_2 m_2 + a_3 D_3 m_3$$

It can be seen that several values of D_1, D_2, and D_3 can be obtained to satisfy the SN value of 4.40. Layer thicknesses, however, are usually rounded up to the nearest 10 mm.

The selection of different layer thicknesses should also be based on constraints associated with maintenance and construction practices so that a practical design is obtained. For example, it is normally impractical and uneconomical to construct any layer with a thickness less than some minimum value. Table 19.11 lists minimum thicknesses suggested by AASHTO.

Taking into consideration that a flexible pavement structure is a layered system, the determination of the different thicknesses should be carried out as indicated in Figure 19.11. The required SN above the subgrade is first determined, and then the required SNs above the base and subbase layers are determined, using the appropriate strength of each layer. The minimum allowable thickness of each layer can then be determined using the differences of the computed SNs, as shown in Figure 19.11.

Using the appropriate values for M_r in Figure 19.10, we obtain $SN_3 = 4.4$ and $SN_2 = 3.8$. Note that when SN is assumed to compute ESAL, the assumed and computed SN_3 values

Table 19.11   AASHTO-Recommended Minimum Thicknesses of Highway Layers

Traffic, ESALs	Minimum Thickness (in.)	
	Asphalt Concrete	Aggregate Base
Less than 50,000	1.0 (or surface treatment)	4
50,001–150,000	2.0	4
150,001–500,000	2.5	4
500,001–2,000,000	3.0	6
2,000,001–7,000,000	3.5	6
Greater than 7,000,000	4.0	6

SOURCE: Based on *AASHTO Guide for Design of Pavement Structures*, 1993, AASHTO, Washington, D.C.

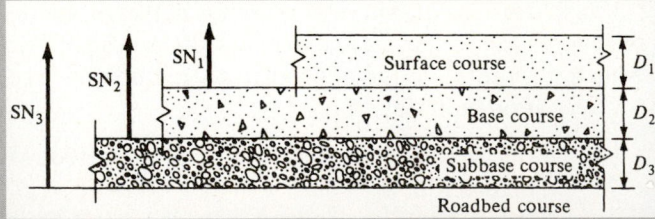

Figure 19.11   Procedure for Determining Thicknesses of Layers Using a Layered Analysis Approach

SOURCE: From *AASHTO Guide for Design of Pavement Structures*, 1993, AASHTO, Washington, D.C. Used by permission.

must be approximately equal. If these are significantly different, the computation must be repeated with a new assumed SN.

We know

$$M_r \text{ for base course} = 2139 \text{ MPa } (31{,}000 \text{ lb/in}^2)$$

Using this value in Figure 19.10, we obtain

$$SN_1 = 2.6$$

giving

$$D_1 = \frac{2.6}{0.44} = 150 \text{ mm } (5.9 \text{ in})$$

Using 6 in. for the thickness of the surface course,

$$D_1^* = 150 \text{ mm } (6 \text{ in})$$

$$SN_1^* = a_1 D_1^* = 0.44 \times 6 = 2.64$$

$$D_2^* \geq \frac{SN_2 - SN_1^*}{a_2 m_2} \geq \frac{3.8 - 2.64}{0.14 \times 0.8} \geq 10.36 \text{ in.} \quad (\text{use } 300 \text{ mm } (12 \text{ in}))$$

$$SN_2^* = 0.14 \times 0.8 \times 12 + 2.64 = 1.34 + 2.64$$

$$D_3^* = \frac{SN_3 - SN_2^*}{a_3 m_3} = \frac{4.4 - (2.64 + 1.34)}{0.1 \times 0.8} = 5.25 \text{ in.} \quad (\text{use } 150 \text{ mm } (6 \text{ in}))$$

$$SN_3^* = 2.64 + 1.34 + 6 \times 0.8 \times 0.1 = 4.46$$

The pavement will therefore consist of 6 in. asphalt concrete surface, 300 mm (12 in) granular base, and 150 mm (6 in) subbase.

Note that the design thicknesses (D^*) selected are not necessarily unique for a given set of input data. Different thicknesses may be selected by rounding up to take into consideration the cost of the different types of materials being considered. However, each selected D^* should be at least equal to the associated computed (D^*), and each SN should be at least the required associated SN.

---

*An asterisk with D or SN indicates that it represents the value actually used, which must be equal to or greater than the required value.

19.3.2 California (Hveem) Design Method

This method was originally developed in the 1940s based on a combination of information obtained from some test roads, theory, and experience. It is used widely in the western states and has been modified several times to take into consideration changes in traffic characteristics. The objective of the original design method was to avoid plastic deformation and the distortion of the pavement surface, but a later modification includes the reduction to a minimum of early fatigue cracking due to traffic load.

The factors considered are:

- Traffic load
- Strength of subgrade material
- Strength of construction materials

The traffic load is initially calculated as the ESAL, as discussed earlier—that is, the total number of 18,000-lb (80 kN) axle loads in one direction—and then converted to a traffic index (TI), where

$$TI = 9.0\left(\frac{ESAL}{10^6}\right)^{0.119} \tag{19.8}$$

In determining the ESAL, passenger vehicles and pickup trucks are not considered. The procedure was calibrated for traffic indices varying from 5.0 to 12 and for design lives varying from 10 to 20 years.

The strength of the subgrade material is given in terms of the resistance value R of the subgrade soil obtained from the Hveem Stabilometer test described in Chapter 17.

The strength characteristics of each construction material (asphalt concrete, base, and subbase materials) are given in terms of a gravel equivalent factor (G_f), which can be obtained from Eq. 19.9 or 19.10 for the asphalt cement.

$$G_f = 5.67/(TI)^{0.5} \qquad t < 0.5 \text{ ft (150 mm)} \tag{19.9}$$

$$G_f = (7.00)(t^{1/3})/(TI)^{0.5} \quad t > 0.5 \text{ ft (150 mm)} \tag{19.10}$$

where

G_f = gravel equivalent factor
TI = traffic index (see Eq. 19.8)
t = thickness of hot-mix asphalt layer

Representative G_f factors are given in Table 19.12, which also shows that the factors for asphalt concrete depend on the TI, while factors for base and subbase materials are independent of the TI.

Structural Design

The objective of the design is to determine the total thickness of material required above the subgrade to carry the projected traffic load. This thickness is determined in terms of gravel equivalent (GE) in feet, which is given as

$$GE = 0.00096(TI)(100 - R) \tag{19.11}$$

where

GE = thickness of material required above a given layer in terms of gravel equivalent in m
TI = traffic index
R = resistance value of the supporting layer material, normally determined at an exudation pressure of 2.07 MPa[2].

The actual depth of each layer can then be determined by dividing GE for that layer by the G_f factor for the material used in the layer. A check usually is made to ascertain that the thickness obtained is adequate for the expansion pressure requirements.

Table 19.12   Gravel Equivalent Factors for Different Types of Materials

Material	G_f
Cement-treated base	
Class A	1.7
Class B	1.2
Lime-treated base	1.2
Untreated aggregate base	1.1
Aggregate subbase	1.0
Asphalt concrete for TI of	
≤5.0	2.54
5.5–6.0	2.32
6.5–7.0	2.14
7.5–8.0	2.01
8.5–9.0	1.89
9.5–10.0	1.79
10.5–11.0	1.71
13.5–14.0	1.52

SOURCE: Adapted from *California Department of Transportation Design manual*, California Division of Highways, Sacramneto, CA, 2009

Example 19.4 Designing a Flexible Pavement Using the Hveem Method

A flexible pavement is to be designed to carry a 1.5×10^6 ESAL during its design period. Using the California (Hveem) method, determine appropriate thicknesses for an asphalt concrete surface and a granular base course over a subgrade soil for which stabilometer test data are shown in Table 19.13. The R value for the untreated granular base material is 70.

Solution: The design is carried out through the following steps.

Step 1.   Determine TI using Eq. 19.8.

$$TI = 9.0\left(\frac{ESAL}{10^6}\right)^{0.119}$$

$$= 9.0\left(\frac{1.5 \times 10^6}{10^6}\right)^{0.119} = 9.44 \quad (\text{say, } 9.5)$$

Step 2.   Determine the asphalt concrete surface thickness over the granular base. Use Eq. 19.11 to determine GE, with $R = 70$ and TI = 9.5.

$$GE = 0.00096(TI)(100 - R)$$

$$= 0.00096(9.5)(100 - 70) = 0.27 \text{ m}$$

The G_f factor from Table 19.12 is 1.79.

$$\text{Actual depth required} = 0.27/1.79 = 0.151 \text{ m}$$

$$= 151 \text{ mm (say 150 mm)}$$

Step 3.   Determine base thickness. The total thickness required above the subgrade should first be determined for an exudation pressure of 2.07 MPa[2]. This is obtained by plotting a graph of GE (given in Table 19.13) for all

Table 19.13   Stabilometer Test Data for Subgrade Soil of Example 19.4

Moisture Content (%)	R Value	Exudation Pressure (kN/m²)	Expansion Pressure (kN/m²)	GE (m) (from Eq. 20.15)	Expansion Pressure Thickness (m)
25.2	48	3795	7.11	0.49	0.34
25.8	37	2843	2.07	0.58	0.10
29.3	15	552	0.00	0.79	0.00

layers above the subgrade versus the exudation pressure, as shown in Figure 19.12a. The GE is obtained as 0.62 m and G_f for untreated aggregate base = 1.1. The base thickness is determined from the difference in total GE required over subgrade material and the GE provided by the asphalt surface.

$$\text{GE of base layer} = 0.62 - \frac{6.0}{12} \times 1.79$$

Therefore,

$$\text{Base thickness} = \left(0.62 - 0.16 \times 1.76\right)/1.1 \text{ m}$$

$$= 0.303 \text{ m (say 0.3 m)}$$

Step 4.   Check to see whether strength design satisfies expansion pressure requirements. The moisture content that corresponds with the strength design thickness (that is, an exudation pressure of 2.07 MPa²) is obtained by plotting the GE values for all layers above the subgrade versus the moisture content (given in Table 19.13), as shown in curve B in Figure 19.12b. This moisture content is determined as 27 percent. The thickness required to

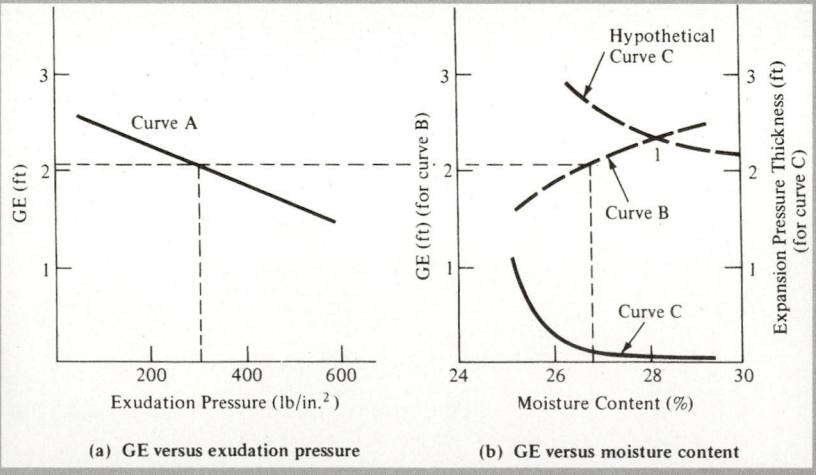

(a) GE versus exudation pressure   (b) GE versus moisture content

Figure 19.12   Plot of GE (given in Table 19.13) for all Layers above the Subgrade versus the Exudation Pressure

resist expansion at 27 percent moisture content is then determined from curve C, which is a plot of expansion pressure thickness against moisture content. Thickness required to prevent expansion at 27 percent moisture content is 0.06 m, which shows that strength design satisfies expansion pressure requirements since the total thickness of the materials above the subgrade is much higher than 0.06 m.

When the thickness based on strength design does not satisfy the expansion pressure requirements, a balanced design should be adopted where the design point is the intersection of the curves of GE (for strength) versus moisture content and GE (for expansion pressure requirement) versus moisture content. A hypothetical case is shown in Figure 19.12, where the balanced point is point 1, indicating that the total thickness GE must be increased to about 0.72 m and compaction of the embankment must take place at a moisture content of 28 percent rather than at 27 percent.

19.3.3 Mechanistic–Empirical Pavement Design

The Mechanistic-Empirical Pavement Design (MEPD) method uses empirical relationships between cumulative damage and pavement distress to determine the adequacy of a pavement structure to carry the expected traffic load on the pavement. The procedure is iterative in that the designer first selects a trial design and evaluates it for adequacy with respect to certain input performance criteria and reliability values based on predicted distresses and smoothness of the pavement. In selecting the trial pavement structure, the designer considers site conditions that include traffic, climate, and subgrade. If the predicted performance criteria do not satisfy the input performance criteria at the specified reliability, the trial design is revised and the evaluation repeated. Figure 19.13 shows a conceptual flow chart of the design process. The procedure briefly described here is for a new flexible pavement and is based on the Transportation Research Board's publication, *Interim Mechanistic-Empirical Pavement Design Guide (MEPDG) Manual of Practice.* The computational effort of this design process is enhanced by the use of the MEPDG software.

Design Procedure

The design procedure consists of three stages incorporating the following basic steps.

Step 1.   Select a trial design (trial pavement structure).
Step 2.   Select the appropriate performance indicator criteria (threshold values) and design reliability level for the design.
Step 3.   Obtain all inputs for the trial pavement structure under consideration.
Step 4.   Evaluate the trial design.
Step 5.   Revise the trial design as needed.

Select a Trial Design.   This can be achieved by either obtaining a pavement structure through the use of another design method, such as the AASHTO design method described in Section 19.3.1, or by using a specific method sanctioned by an agency.

Select the Appropriate Performance Indicator Criteria (Threshold Values) and Design Reliability Level.   In this step, the designer uses his or her agency's policies

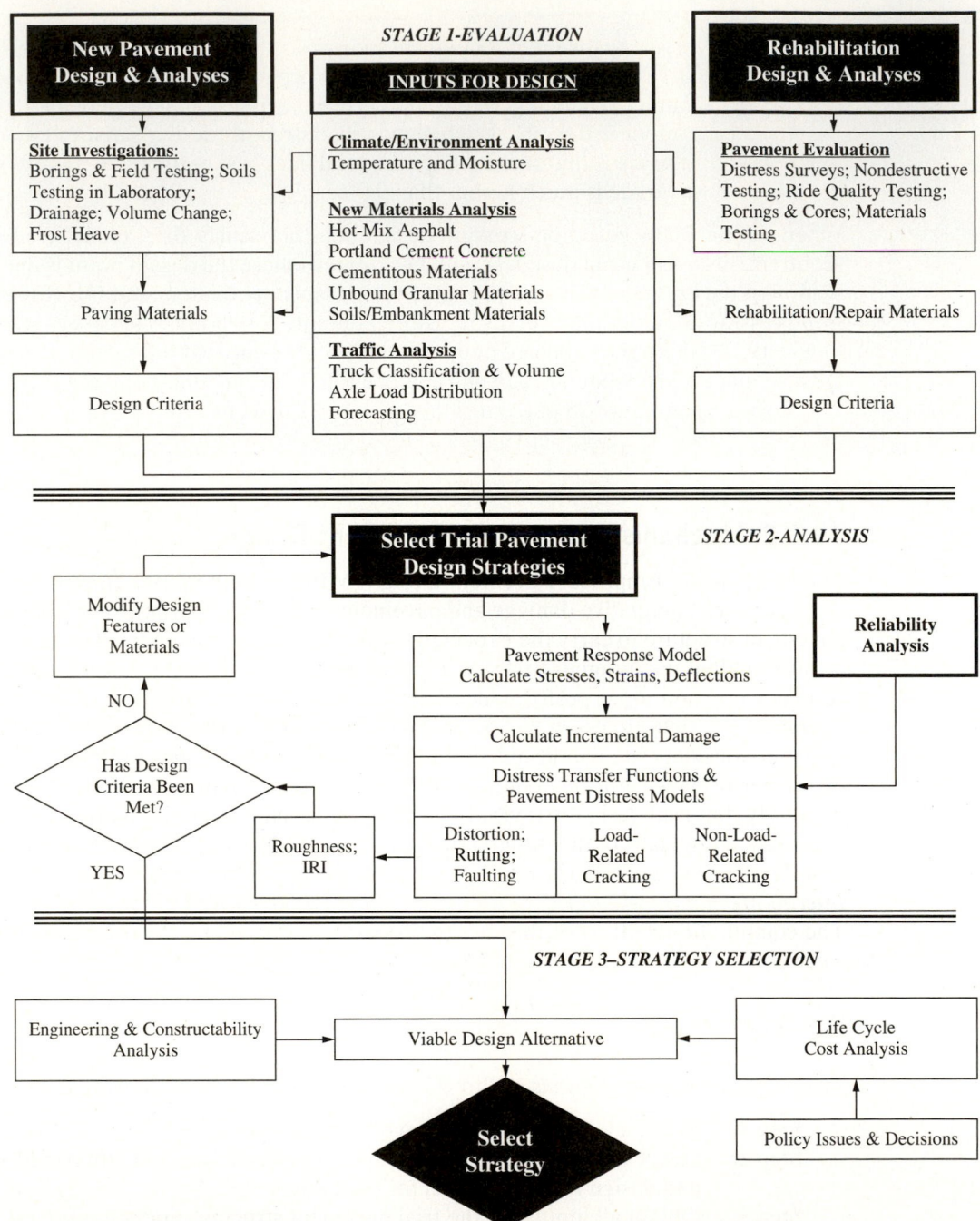

Figure 19.13   Conceptual Flow Chart of the Three-Stage Design/Analysis Process for the MEPDG

SOURCE: From *Mechanistic-Empirical Pavement Design Guide, Interim Edition*, 2008, AASHTO, Washington, D.C. Used by permission.

on rehabilitation or reconstruction to identify performance indicator criteria that will precipitate rehabilitation or reconstruction. Specific performance indicators for flexible pavements that can be predicted by the software include total rut depth, non-load-related cracking, load-related longitudinal cracking, load-related alligator cracking, and the pavement smoothness, which is given in terms of the International Roughness Index (IRI). Table 19.14 gives suggested critical values for hot-mix asphalt (HMA) pavements.

Obtain All Inputs for the Trial Pavement Structure under Consideration.   This step involves the identification of all the inputs required by the MEPDG software using a hierarchical structure that indicates the effort that should be placed in obtaining a given input based on the importance of the project, the importance of the input parameter, and the available resources. The hierarchical structure consists of the following three levels:

- **Input Level 1**—This level relates to parameters that are directly measured at the site and are therefore specific to the site or project. Input parameters obtained at this level provide the best information but are associated with the highest testing and data costs. This level is used for projects at unusual sites with materials and traffic characteristics for which inferences used to develop correlations and defaults obtained in Levels 2 and 3 cannot be used.
- **Input Level 2**—This level involves the use of correlations and regression analyses to obtain input parameters from data at other sites that have been obtained at Level 1, where testing and data costs are less expensive. Measured regional values of input parameters that may not be project-specific can also be used at this level.
- **Input Level 3**—Parameters obtained at this level are usually best-estimated or default values and are usually based on the median of a set of global or regional default values for sites or projects with similar characteristics. This level is associated with the least testing and data costs, but also gives the least knowledge of the input parameter for a specific site or project.

It should be noted that it is not necessary to use the same level for all input parameters. The input parameters can be classified into six broad areas: general project information, design criteria, traffic, climate, structural layering, and material properties.

Table 19.14   Recommended Design Criteria or Threshold Values for Flexible Pavements

Performance Criteria	*Maximum Value at End of Design Life*
Alligator cracking (HMA bottom-up cracking)	Interstate: 10% lane area
	Primary: 20% lane area
	Secondary: 35% lane area
Rut depth (permanent deformation in wheel paths)	Interstate: 0.40 in.
	Primary: 0.50 in.
	Others (< 45 mi/h): 0.65 in.
Traverse cracking length (thermal cracks)	Interstate: 500 ft/mi
	Primary: 700 ft/mi
	Secondary: 700 ft/mi
IRI (smoothness)	Interstate: 160 in./mi
	Primary: 200 in./mi
	Secondary: 200 in./mi

SOURCE: From *Mechanistic-Empirical Pavement Design Guide, Interim Edition*, 2008, AASHTO, Washington, D.C. Used by permission.

The general project information area includes information on design/analysis life, and construction and traffic opening dates. The design life is considered as the time from the initial construction to the time when the pavement has deteriorated to a specified condition which indicates significant rehabilitation or reconstruction of the pavement. The procedure given in the MEPDG can be used for design lives of 1 year to over 50 years; however, users of the procedure are cautioned that only a few pavements having design lives of over 30 years were included in the global calibration that was used in the design process. The prediction of the pavement distress is influenced by the construction and traffic opening dates. For example, the base/subgrade construction month is tied in with the monthly climatic inputs, which in turn affects the monthly layer and subgrade modulus values. Similarly, the traffic opening date is keyed to the monthly traffic loadings.

The design criteria selected for the project also serve as inputs. Suggested values are given in Table 19.14. Another input under the general project information area is the reliability value. The design reliability is defined by the MEPDG as the "probability that the predicted distress will be less than the critical level over the design period." The designer is free to select different reliability levels for different distress types and smoothness; however, probability levels should be higher than 50 percent. Table 19.15 gives suggested probability levels for different types of roadways.

The traffic data required for the MEPDG is much more detailed than that required for the 1993 AASHTO design method. In this case, rather than inputting the equivalent single-axle loads (ESALs), as is done for the 1993 AASHTO design guide, the full axle load spectrum data for each axle type is required. The MEPDG defines the axle load spectra as "a histogram or distribution of axle loads for a specific axle type (single, tandem, tridem, quad)"—that is, the number of axle applications within a specific axle load range. These are obtained through the analysis of weight-in-motion (WIM) data collected by the states. However, the MEPDG software can interface with other available software packages, such as the analysis software from the NCHRP Project 1-39, which can be used to obtain the required information from WIM data. The MEPDG also suggests that for cases where the required WIM data are not available, default values in the MEPDG software may be used, reflecting a Level 3 input. In addition to the axle load distributions, other inputs that can be obtained from the WIM data include axle load configurations (axle spacing and wheel base), monthly distribution factors that facilitate the distribution of the truck traffic within each class throughout the year, and hourly distribution factors that facilitate the distribution of truck traffic throughout a typical day. The axle load configurations (axle spacing) tend to be relatively constant for standard trucks, and the MEPDG suggests that 1425 mm, 1250 mm, and 1250 mm can be used for tandem axle

Table 19.15 Reliability Levels for Different Functional Classifications of the Roadway

Functional Classification	Reliability Level (%)	
	Urban	Rural
Interstate/Freeways	95	95
Principal arterials	90	85
Collectors	80	75
Local	75	70

SOURCE: From *Mechanistic-Empirical Pavement Design Guide, Interim Edition*, 2008, AASHTO, Washington, D.C. Used by permission.

spacing, tridem axle spacing, and quad axle spacing, respectively. In addition, roadway-specific traffic inputs include initial two-way average annual daily truck traffic (AADTT), percent trucks in design lane, percent trucks in design direction, truck operating speeds, and growth of truck traffic. Truck-related default values suggested in the MEPDG are 305 mm for dual tire spacing; 5.8 g/m^2 for tire pressure; and 305 mm, 254 mm, and 203 mm inches for lateral wander of axle loads for lane widths greater than 3.60 meters, lane widths between 3 and 3.60 meters, and lane widths less than 3 meters, respectively.

Input data related to climate include hourly temperature, precipitation, wind speed, relative humidity, and cloud cover and can be obtained from weather stations. Temperature and moisture contents in the pavement layers are determined from these climate-related data, and they also serve as inputs to the site factor parameter for the smoothness prediction models.

The most important inputs for the foundation soil are the in-place modulus and the resilient modulus of the in-place subgrade. Other required data for the foundation and subgrade soils include the vertical and horizontal variations of subsurface soil types (classification), moisture contents, densities, water table depth, and location of rock stratum, which are obtained as discussed in Chapter 17. These are used to identify suitable soils for use as a pavement material. Table 19.16 gives a summary of soil characteristics that may be used as a pavement material.

Evaluation of the Trial Design.   Evaluation involves running the MEPDG software to predict the pavement distresses and the smoothness of the trial pavement to determine whether the stipulated criteria are satisfied. The substeps carried out in this phase of the design process are checking the correctness of the input and assessing the acceptance or rejection of the trial design.

Checking for the Correctness of the Inputs.   At this stage of the analysis, the MEPDG gives all input values as an output. The designer checks that all necessary inputs are provided and that their values are correct and reflect the original design intention. The software also provides an assessment of the design reliability by reporting the distress predicted and the reliability predicted. This output is used to determine whether the reliability reported is greater than the target reliability. If the reported liability is less than the targeted reliability, the trial pavement fails. The MEPDG divides the asphalt surface into several layers. It then provides the material properties and other factors on a month-to-month basis for each layer over the design period. The designer should check these values to ascertain whether they are reasonable. These include the HMA dynamic modulus (E_{HMA}) for each sublayer, the unbound material resilient modulus (M_r), and the number of cumulative heavy trucks.

Assessing the acceptance or rejection of the trial design.   This part of the evaluation involves the comparison of the predicted performance at the design reliability with the stipulated design criteria. The performance indicators produced by the MEPDG software for a new flexible pavement are the **rut depth,** the **load-related cracking, non-load-related cracking (transverse cracking),** and **smoothness.**

Rut depth is an indication of the surface distortion that is caused by the plastic or permanent deformation of the hot-mix asphalt (HMA), unbound layers, and foundation soil. The MEPDG software computes incremental distortion or rutting within each sublayer (the HMA surface is divided into sublayers but not necessarily more than three sublayers) and determines the plastic deformation for each season by summing up the deformations within each layer. Laboratory triaxial tests are used to determine the rate or accumulation of plastic deformation of the HMA, unbound layers, and foundation

Table 19.16 Summary of Soil Characteristics as a Pavement Material

Major Divisions	Name	Strength When Not Subject to Frost Action	Potential Frost Action	Compressibility & Expansion	Drainage Characteristics
Gravel & Gravelly Soils	Well-graded gravels or gravel-sand mixes, little to no fines; GW	Excellent	None to very slight	Almost none	Excellent
	Poorly graded gravels or gravel-sand mixes little or no fines; GP	Good to excellent	None to very slight	Almost none	Excellent
	Silty gravels, gravel-sand silt mixes; GM	Good to excellent	Slight to medium	Very slight	Fair to poor
	Very Silty gravels, gravel-sand silt mixes; GM	Good	Slight to medium	Slight	Poor to practically impervious
	Clayey gravels, gravel-sand-clay mixes; GC	Good	Slight to medium	Slight	Poor to practically impervious
Sand and Sandy Soils	Well-graded sands or gravelly sands, little to no fines; SW	Good	None to very slight	Almost none	Excellent
	Poorly graded sands or gravelly sands; little or no fines; SP	Fair to good	None to very slight	Almost none	Excellent
	Silty sands, sand-silt mixes; SP	Fair to good	Slight to high	Very slight	Fair to poor
	Silty sands, sand-silt mixes; SM	Fair	Slight to high	Slight to medium	Poor to practically impervious
	Clayey sands, sand-clay mixes; SC	Poor to fair	Slight to high	Slight to medium	Poor to practically impervious

Table 19.16 Summary of Soil Characteristics as a Pavement Material (*continued*)

Major Divisions	Name	Strength When Not Subject to Frost Action	Potential Frost Action	Compressibility & Expansion	Drainage Characteristics
Silts & Clays with the Liquid Limit Less Than 50	Inorganic silts & very fine sand, rock flour, silty or clayey fine sand or clayey silts with slight plasticity; MG, MS, & ML	Poor to fair	Medium to very high	Slight to medium	Fair to poor
	Inorganic clays of low to medium plasticity, gravelly clays, sandy clays, silty clays, lean clays; CG, CL, & CS	Poor to fair	Medium to high	Slight to medium	Practically impervious
	Organic silts & organic silt-clays or low plasticity; MSO & CLO	Poor	Medium to high	Medium to high	Poor
Silts & Clays with Liquid Limit Greater Than 50	Inorganic silts, micaceous or diatomaceous fine sand or silty soils, elastic silts; MH	Poor	Medium to very high	High	Fair to poor
	Inorganic clays of high plasticity, fat clays; CH	Poor to fair	Medium to very high	High	Practically impervious
	Organic clays of medium to high plasticity, organic silts; MHO & CHO	Poor to very poor	Medium	High	Practically impervious
Highly Organic Soils	Peat & other highly organic soils	Not Suitable	Slight	Very high	Fair to poor

The information presented in this table is adopted after publications of the U.S. Army Corps of Engineers, Federal Aviation Administration, and the Federal Highway Administration.

SOURCE: From *Mechanistic-Empirical Pavement Design Guide, Interim Edition*, 2008, AASHTO, Washington, D.C. Used by permission.

soil. The results obtained are used to calibrate the relationship in the MEPDG model for rut depth in the HMA layer given as

$$\Delta_{p(HMA)} = \varepsilon_{p(HMA)}h_{(HMA)} = \beta_{1r}k_z\varepsilon_{r(HMA)}10^{k_{1r}}n^{k_{2r}\beta_{2r}}T^{k_{3r}\beta_{3r}} \tag{19.12}$$

where

$\Delta_{p(HMA)}$ = accumulated permanent or plastic vertical deformation in the HMA layer/sublayer (mm)

$\varepsilon_{p(HMA)}$ = accumulated permanent or plastic axial strain in the HMA layer/sublayer (mm/mm)

$\varepsilon_{r(HMA)}$ = resilient or elastic strain calculated by the structural response model at the mid-depth of each HMA sublayer (mm/mm)

$h_{(HMA)}$ = thickness of the HMA layer/sublayer, (mm)

n = number of axle load repetitions

T = mix or pavement temperature, (c)

k_z = depth confinement factor

$k_{1r, 2r, 3r}$ = global field calibration parameters (from the NCHRP I-40D recalibration)

$k_{1r} = -3.35412$

$k_{2r} = 0.4791$

$k_{3r} = 1.5606$

$\beta_{1r}, \beta_{2r}, \beta_3$ = local and or mixture field calibration constants; for the global calibration, these constants were all set to 1.0

$k_z = (C_1 + C_2 D)0.328196^D$

$C_1 = -0.1039(HMA)^2 + 2.4868H_{HMA} - 17.342$

$C_2 = 0.01712(H_{HMA})^2 - 1.7331H_{HMA} + 27.428$

D = depth below the surface, (in.)

H_{HMA} = total HMA thickness, (in.)

Example 19.5 Determining the Rut Depth in a HMA Layer

Determine the expected rut depth of the second layer of a 3 in (75 mm) thick HMA flexible pavement surface divided into three equal sublayers of 1 in (25 mm) each for the following conditions:

- Resilient or elastic strain calculated by the structural response model at the mid-depth of each HMA sublayer, in./in. = 40×10^6
- Number of axle load repetitions = 1×10^4
- Pavement temperature $(T) = 30°C$
- Mixture field calibration constants:

$$\beta_{1r} = 0.96$$
$$\beta_{2r} = 0.98$$
$$\beta_{3r} = 0.93$$

Solution: Use Eq. 19.12.

$$\Delta_{p(HMA)} = \varepsilon_{p(HMA)}h_{(HMA)} = \beta_{1r}k_z\varepsilon_{r(HMA)}10^{k_{1r}}n^{k_{2r}\beta_{2r}}T^{k_{3r}\beta_{3r}}$$

Compute $k_z = (C_1 + C_2 D) 0.328196^D$. (See Eq. 19.12.)

$$C_1 = -0.1039(H_{HMA})^2 + 2.4868 H_{HMA} - 17.342$$
$$= -0.1039(3)^2 + 2.4868 \times (3) - 17.342 = -10.8167$$

$$C_2 = 0.01712(H_{HMA})^2 - 1.7331 H_{HMA} + 27.428$$
$$= 0.01712(3)^2 - 1.7331 \times 3 + 27.428 = 22.3828$$

$$k_z = (C_1 + C_2 D) 0.0328196^D$$

$$k_z = (-10.8167 + 22.3828 \times 1.5)0.328196^{1.5} = 22.7575 \times 0.18802 = 4.27887$$

$$\Delta_{p(HMA)} = \beta_{1r} k_z \varepsilon_{r(HMA)} 10^{k_1} n^{k_2, \beta_2} T^{k_3, \beta_3}$$
$$= 0.96 \times 4.27887 \times 40 \times 10^{-6} \times 10^{-3.35412} \times 10^{4 \times 0.4791 \times 0.98} \times 85^{1.5606 \times 0.93}$$
$$= 3.47 \times 10^{-3} \text{ in.}$$
$$= 0.088 \text{ mm}$$

Note that Example 19.5 only illustrates the intensity of the computational effort involved in the MEPDG. This computation is repeated for each sublayer, for the full axle load spectrum data for each axle type, and for each season during the year. The procedure is therefore undertaken using the MEPDG software.

Equation 19.13 gives the expression in the MEPDG software for determining the plastic vertical deformation within the unbound pavement sublayers and the foundation or embankment soil.

$$\Delta_{p(soil)} = \beta_{s1} k_{s1} \varepsilon_v h_{soil} \left(\frac{\varepsilon_0}{\varepsilon_r}\right) e^{-\left(\frac{\rho}{n}\right)^\beta} \tag{19.13}$$

where

$\Delta_{p(soil)}$ = permanent or plastic deformation for the layer/sublayer (mm)
n = number of load applications
ε_0 = intercept determined from laboratory repeated load permanent deformation tests (mm/mm)
ε_r = resilient strain imposed in laboratory test to obtain material properties ε_0, β, and ρ (mm/mm)
ε_v = average vertical resilient or elastic strain in the layer/sublayer and calculated by the structural response model (mm/mm)
h_{soil} = thickness of the unbound layer/sublayer (mm)
k_{s1} = global calibration coefficients are 1.673 for granular materials and 1.35 for fine-grained materials
β_{s1} = local calibration constant for rutting in unbound layers; the local calibration constant was set to 1.0 for the global calibration effort

$$\log \beta = -0.61119 - 0.17638(W_c)$$

$$\rho = 10^9 \left(\frac{C_0}{1 - (10^9)^\beta}\right)^{\frac{1}{\beta}}$$

$$C_0 = \ln\left(\frac{a_1 M_r^{b_1}}{a_9 M_r^{b_9}}\right)$$

W_c = water content percent
M_r = resilient modulus of the unbound layer or sublayer (psi)
$a_{1,9}$ = regression constants; $a_1 = 0.15$ and $a_9 = 20.0$
$b_{1,9}$ = regression constants; $b_1 = 0.0$ and $b_9 = 0.0$

Equations 19.14, 19.15, and 19.16 give the standard error (s_e) for the HMA, the unbound layers for coarse, and for fine-grained materials and soils, respectively. The standard error for the total depth can then be found as the sum of the standard errors for the HMA, coarse, and fine-grained materials.

$$S_{e(\text{HMA})} = 0.1587(\Delta_{\text{HMA}})^{0.5479} + 0.001 \qquad (19.14)$$

$$S_{e(\text{gran})} = 0.1169(\Delta_{\text{gran}})^{0.5303} + 0.001 \qquad (19.15)$$

$$S_{e(\text{fine})} = 0.1724(\Delta_{\text{fine}})^{0.5516} + 0.001 \qquad (19.16)$$

where

Δ_{HMA} = plastic deformation in the HMA layers (in.)
Δ_{gran} = plastic deformation in the aggregate and coarse-grained layers (mm)
Δ_{fine} = plastic deformation in the fine-grained layers and soils (mm)

The ***load-related cracking*** computed by the MEPDG software consists of alligator cracking, which is assumed to start at the bottom of the HMA layers and extend upward to the surface of the HMA layers, and longitudinal cracking, which starts at the surface and extends to the bottom. The allowable number of axle load applications that is needed to compute the extent of each type of cracking is first determined using Eq. 19.17, which is known as Miner's hypothesis.

$$N_{f-\text{HMA}} = k_{f1}(C)(C_H)\beta_{f1}(\varepsilon_t)^{k_{f2}\beta_{f2}}(E_{\text{HMA}})^{k_{f3}\beta_{f3}} \qquad (19.17)$$

where

$N_{f-\text{HMA}}$ = allowable number of axle load applications for a flexible pavement and HMA overlays
ε_t = tensile strain at critical locations and calculated by the structural response model (mm/mm)
E_{HMA} = dynamic modulus of the HMA measured in compression (MPa²)
k_{f1}, k_{f2}, k_{f3} = global field-calibration parameters; $k_{f1} = 0.007566$, $k_{f2} = -3.49492$, and $k_{f3} = -1.281$
$\beta_{f1}, \beta_{f2}, \beta_{f3}$ = local or mixture-specific calibration constants; these were set to 1 in the global calibration effort
$C = 10^M$, with M calculated as

$$M = 4.84 \left(\frac{V_{be}}{V_a + V_{be}} - 0.69 \right)$$

V_{be} = effective asphalt content by volume, percent
V_a = percentage of air voids in the HMA mixture
C_H = thickness correction term, dependent on the type of cracking
Bottom-up or **alligator cracking**:

$$C_H = \frac{1}{0.000398 + \dfrac{0.003602}{1 + e^{(11.02 - 3.49H_{\text{HMA}})}}}$$

Top-down or **longitudinal cracking**:

$$C_H = \cfrac{1}{0.01 + \cfrac{12.00}{1 + e^{(15.676 - 2.8186 H_{HMA})}}}$$

H_{HMA} = total HMA thickness (mm)

After computing the allowable number of axle load applications, the incremental damage indices (ΔDI) are then computed at critical depths using a grid pattern throughout the HMA layers. This is done by first dividing the actual number of axle loads by the allowable obtained from Eq. 19.17 for each axle load interval and axle type and within each specific time interval. Equation 19.18 is then used to compute the cumulative damage index (DI) for each axle load.

$$DI = \sum (\Delta DI)_{j,m,l,p,T} = \sum \left(\frac{n}{N_{f-HMA}} \right)_{j,m,l,p,T} \tag{19.18}$$

where

n = actual number of axle load applications within a specific time period
j = axle load interval
m = axle load type (single, tandem, tridem, quad, or special axle configuration)
l = truck type using the truck classification groups included in the MEPDG
p = month
T = median temperature for the five temperature intervals or quantiles used to subdivide each month

The total damage computed from Eq. 19.18 is then used to determine the total area of alligator cracking and the length of longitudinal cracking using Eq. 19.19 and 19.20, respectively.

$$FC_{bottom} = \left(\frac{1}{60} \right) \left(\frac{C_4}{1 + e^{(c_1 c_1^* + c_2 c_2^* \log(DI_{bottom} * 100))}} \right) \tag{19.19}$$

where

FC_{bottom} = area of alligator cracking that starts at the bottom of the HMA layers, percent of total lane area
DI_{bottom} = cumulative damage index at the bottom of the HMA layers
$C_{1,2,4}$ = transfer function regression constants; C_4 = 6000; C_1 = 1.00, C_2 = 1.00
$C_1^* = -2C_2^*$
$C_2^* = -2.40874 - 39.748(1 + H_{HMA})^{-2.856}$
H_{HMA} = total HMA thickness (mm)

$$FC_{top} = 10.56 \left(\frac{C_4}{1 + e^{(C_1 - C_2 \log(DI_{top}))}} \right) \tag{19.20}$$

where

FC_{top} = length of longitudinal cracks that starts at the top of the HMA layer (m/km)
DI_{top} = cumulative damage index near the top of the HMA surface
$C_{1,2,4}$ = transfer function regression constants; C_4 = 1000, C_1 = 7.00, and C_2 = 3.5

The standard errors (s_e) (standard deviation of the residual errors) are given in Eqs. 19.21 and 19.22 for the alligator cracking and the longitudinal cracking, respectively.

$$S_{e(\text{alligator})} = 32.7 + \frac{995.1}{1 + e^{2 - 2\log(FC_{\text{bottom}} + 0.0001)}} \qquad (19.21)$$

where

$S_{e(\text{alligator})}$ = standard error for alligator cracking
FC_{bottom} = area of alligator cracking that starts at the bottom of the HMA layers, percent of total lane area

$$S_{e(\text{long})} = 200 + \frac{2300}{1 + e^{1.072 - 2.1654\log(FC_{\text{top}} + 0.0001)}} \qquad (19.22)$$

where

$S_{e(\text{long})}$ = standard error for longitudinal cracking
FC_{top} = area of longitudinal cracking that starts at the top of the HMA layers, percent of total lane area

The ***non-load-related cracking (transverse cracking)*** is obtained by using an enhanced version of a procedure that was developed under the Strategic Highway Research Program. The change in the thermal crack depth that occurred as a result of a given cooling cycle is obtained by Eq. 19.23.

$$\Delta C = A(\Delta K)^{\eta} \qquad (19.23)$$

where

ΔC = change in crack depth due to a cooling cycle
ΔK = change in the stress intensity factor due to a cooling cycle
A, η = fracture parameters for the HMA mixture

where

$A = 10^{k_t \beta_t (4.389 - 2.521\log(E_{\text{HMA}} \sigma_m \eta))}$

$\eta = 0.8 \left[1 + \dfrac{1}{m} \right]$

k_t = coefficient determined through global calibration for each input level (Level 1 = 5.0; Level 2 = 1.5, and Level 3 = 3.0)
E_{HMA} = HMA indirect tensile modulus (MPa^2)
σ_m = mixture tensile strength (MPa^2)
m = the m-value derived from the indirect tensile creep compliance curve measured in the laboratory
β_t = local or mixture calibration factor

The MEPDG software uses the expression given in Eq. 19.24 to determine the stress intensity factor K.

$$K = \sigma_{\text{tip}}(0.45 + 1.99(C_o)^{0.56}) \qquad (19.24)$$

where

σ_{tip} = far-field stress from pavement response model at depth of crack tip (MPa^2)
C_0 = current crack length (m)

The amount of thermal cracking is obtained from the expression

$$TC = \beta_{t1} N_{f\text{-}HMA} \left[\frac{1}{\sigma_d} \log \left(\frac{C_d}{H_{HMA}} \right) \right] \tag{19.25}$$

where

TC = observed amount of thermal cracking (m/km)
β_{t1} = regression coefficient determined through the global calibration (400)
σ_d = standard normal deviation of the log of the depth of cracks in the pavement = (0.769) (m)
C_d = crack depth (in.)
H_{HMA} = thickness of HMA layers (m)
$N_{f\text{-}HMA}$ = allowable number of axle load applications for a flexible pavement and HMA overlay.

Equations 19.26, 19.27, and 19.28 give the standard errors for the transverse cracking for input Levels 1, 2, and 3, respectively.

$$S_e(\text{Level 1}) = -0.0899(TC + 636.97) \tag{19.26}$$

$$S_e(\text{Level 2}) = -0.0169(TC + 654.86) \tag{19.27}$$

$$S_e(\text{Level 3}) = -0.0869(TC + 453.98) \tag{19.28}$$

The **smoothness** computation is based on the proposition that increased surface distress leads to increased roughness, which, in turn, results in an increased International Roughness Index (IRI). The MEPDG software uses Eq. 19.29 to determine the IRI.

$$IRI = IRI_0 + 0.0150(SF) + 0.400(FC_{\text{total}}) + 0.0080(TC) + 40.0(RD) \tag{19.29}$$

where

IRI_0 = initial IRI after construction (mm/km)
FC_{total} = area of fatigue cracking (combined alligator, longitudinal and reflection cracking in the wheel path), percent of total area. All load-related cracks are combined on an area basis—length of cracks is multiplied by 0.3 m to convert length into area basis
TC = length of transverse cracking (m/km)

$$SF = \text{Age}(0.02003(PI + 1) + 0.007947(Precip + 1) + 0.000636(FI + 1)) \tag{19.30}$$

where

Age = pavement age (years)
PI = percent plasticity index
FI = average annual freezing index, °C (days)
$Precip$ = average annual precipitation or rainfall (mm)
RD = average rut depth (mm)
SF = site factor

If any of the computed values do not satisfy the stipulated criteria, the trial design has to be changed and the analysis repeated for the new design.

The computational effort involved in the MEPDG is extremely intensive and cannot, therefore, be reasonably carried out manually. The use of the MEPDG software is therefore necessary when using the design process.

The short description given in this section is intended to give the reader a brief insight into the design process and does not cover all the requirements of the design process.

19.4 SUMMARY

The design of flexible pavements basically involves determining the strength characteristics of the materials of the pavement surface and underlying materials and then determining the respective thicknesses of the subbase (if any), base course, and pavement surface that should be placed over the native soil. The pavement is therefore usually considered as a multilayered elastic system. The thicknesses provided should be adequate to prevent excessive cracking and permanent deformation beyond certain limits. These limits are considered in terms of required load characteristics, which can be determined as the number of repetitions of 80 kN single-axle loads the pavement is expected to carry during its design life, as in the AASHTO design method, or in terms of the number of repetitions for a full axle load spectrum data for each axle type as in the MEPDG design guide.

The design methods described in this chapter are the AASHTO design method, which is commonly used by many state transportation agencies; the California (Hveem) design method; and the MEPDG design guide. The MEPDG design guide is the most comprehensive design method as, in addition to factors considered by the other methods, it also provides the opportunity for the designer to be fully involved in the design process with the flexibility to consider different material and design options.

Although the design steps involved in the AASHTO and Hveem methods may be carried out manually, it is highly unlikely that the MEPDG can be carried out manually. The MEPDG method therefore is described briefly to give the reader a general understanding of the design process.

PROBLEMS

19-1   Discuss the structural components of a flexible pavement.
19-2   Describe the purpose of soil stabilization and discuss at least three methods of achieving it.
19-3   Discuss the process of lime stabilization, and indicate the soil characteristics for which it is most effective.
19-4   An axle weight study on a section of highway gave the following data on axle load distribution. Determine the truck factor for this section of highway. Assume $SN = 4$ and $p_t = 2.5$.

Axle Load Group: Single (4.4 kN)	No. of Axles per 1000 Vehicles	Axle Load Group: Tandem (4.4 kN)	No. of Axles per 1000 Vehicles
<4	678	<6	18
4–8	775	6–12	236
8–12	500	12–18	170
12–16	150	18–24	120
16–18	60	24–30	152
18–20	40	30–32	66
20–22	7	32–34	30
22–24	4	34–36	12
24–26	3	36–38	4
		38–40	1

19-5   How does an ESAL differ from a truck factor?

19-6   A six-lane divided highway is to be designed to replace an existing highway. The present *AADT* (both directions) of 6000 vehicles is expected to grow at 5 percent per annum. Assume $SN = 4$ and $p_t = 2.5$. The percent of traffic on the design lane is 45 percent. Determine the design ESAL if the design life is 20 years and the vehicle mix is:

Passenger cars (1000 lb/axle) 4.45 kN/axle = 60%

2-axle single-unit trucks (5000 lb/axle) 22.25 kN/axle = 30%

3-axle single-unit trucks (7000 lb/axle) 31.35 kN/axle = 10%

19-7   A section of a two-lane rural highway is to be realigned and replaced by a four-lane highway with a full-depth asphalt pavement. The *AADT* (both ways) on the existing section can be represented by 500 ESAL. It is expected that construction will be completed 5 years from now. If the traffic growth rate is 5 percent and the effective CBR of the subgrade on the new alignment is 85, determine a suitable depth of the asphalt pavement using the AASHTO method. Take the design life of the pavement as 20 years. The resilient modulus of the asphalt (E_{AC}) is 2.76 MPa². Assume m_i for the subgrade is 1 and the percent of traffic on the design lane is 45 percent. Use a reliability level of 90 percent, a standard deviation of 0.45, and a design serviceability loss of 2.0.

19-8   A section of a two-lane rural highway is to be realigned and replaced by a four-lane highway with a full-depth asphalt pavement. The *AADT* (both ways) on the existing section can be represented by 750 ESAL. It is expected that construction will be completed 5 years from now. If the traffic growth rate is 2 percent and the effective CBR of the subgrade on the new alignment is 90, determine a suitable depth of the asphalt pavement using the AASHTO method. Take the design life of the pavement as 20 years. The resilient modulus of the asphalt (E_{AC}) is 3.11 MPa². Assume m_i for the subgrade is 1 and the percent of traffic on the design lane is 45 percent. Use a reliability level of 90 percent, a standard deviation of 0.45, and a design serviceability loss of 2.0.

19-9   The predicted traffic mix of a proposed four-lane urban non-interstate freeway is:

Passenger cars = 78%

Single-unit trucks:

2-axle, 22.2 KN/axle = 12%

2-axle, 39.9 KN/axle = 4%

3-axle or more, 102 KN/axle = 3%

Tractor semitrailers and combinations:

3-axle, 88.8 KN/axle = 3%

The projected *AADT* during the first year of operation is 3900 (both directions). If the traffic growth rate is estimated at 3 percent and the CBR of the subgrade is 75, determine the depth of a full-asphalt pavement using the AASHTO method and $n = 20$ years. The resilient modulus of the asphalt (E_{AC}) is 2.21 MPa. Assume m_i for the subgrade is 1 and the percent of traffic on the design lane is 42 percent, $p_t = 2.5$ and $SN = 4$. Use a reliability level of 90 percent, a standard deviation of 0.45, and a design serviceability loss of 2.0.

19-10   A rural principal arterial is expected to carry an ESAL of 0.188×10^6 during the first year of operation with an expected annual growth of 6 percent over the 20-year design life. If the subgrade has a resilient modulus of 1.04 MPa², design a suitable pavement consisting of a granular subbase with a layer coefficient of 0.13, a granular base layer with a layer coefficient of 0.14, and an asphalt concrete surface with an elastic modulus of 2.76 MPa^2. Assume all m_i values = 1, the percent of traffic on the design lane is 47 percent, and SN = 4. Use a reliability level of 85 percent, a standard deviation of 0.45, and a design serviceability loss of 2.0.

19-11 Using the information given in Problem 19-6, design a suitable pavement consisting of an asphalt mixture surface with an elastic modulus of 1.73 MPa^2, a granular base layer with a structural coefficient of 0.14 on a subgrade having a CBR of 10. Assume all m_i values = 1, and the percent of traffic on the design lane is 45 percent. Use a reliability level of 85 percent, a standard deviation of 0.45, and a design serviceability loss of 2.0.

19-12 Repeat Problem 19-11 for a pavement consisting of an asphalt mixture surface with an elastic modulus of 1.93 MPa^2, and 152 mm of granular subbase with a resilient modulus of 18×10^3 lb/in.^{2}.

19-13 Repeat Problem 19-7 using two different depths of untreated aggregate bases of 152 mm and 305 mm. Highway contractors in your area can furnish rates for providing and properly laying an asphalt concrete surface and untreated granular base. Assume a structural coefficient of 0.12 for the base course. If these rates are available, determine the cost for constructing the different pavement designs if the highway section is 5 miles long and the lane width is 3.7 m. Which design will you select for construction?

19-14 The traffic on the design lane of a proposed four-lane rural interstate highway consists of 40 percent trucks. If classification studies have shown that the truck factor can be taken as 0.45, design a suitable flexible pavement using the 1993 AASHTO procedure if the $AADT$ on the design lane during the first year of operation is 1000, p_i = 4.2, and p_t = 2.5.

$$\text{Growth rate} = 4\%$$
$$\text{Design life} = 20 \text{ years}$$
$$\text{Reliability level} = 95\%$$
$$\text{Standard deviation} = 0.45$$

The pavement structure will be exposed to moisture levels approaching saturation 20 percent of the time, and it will take about one week for drainage of water. Effective CBR of the subgrade material is 7. CBR of the base and subbase are 70 and 22, respectively, and M_r for the asphalt mixture is 3.11 MPa^2.

19-15 Repeat Problem 19-14 with the subgrade M_r values for each month from January through December being 0.12, 0.12, 0.08, 0.05, 0.05, 0.062, 0.062, 0.062, 0.065, 0.065, 0.08, and 0.14 kN/mm^2, respectively. The pavement structure will be exposed to moisture levels approaching saturation for 20 percent of the time, and it will take about 4 weeks for drainage of water from the pavement. Use untreated sandy gravel with M_r of 0.103 kN/mm^2 for the subbase and untreated granular material with M_r of 0.19 kN/mm^2 for the base course.

19-16 An existing two-lane rural highway is to be replaced by a four-lane divided highway on a new alignment. Construction of the new highway will commence two years from now and is expected to take three years to complete. The design life of the pavement is 20 years. The present ESAL is 150,000. Design a flexible pavement consisting of an asphalt concrete surface and lime-treated base using the California (Hveem) method. The results of a stabilometer test on the subgrade soil are as follows.

Moisture Content (%)	R Value	Exudation Pressure (kPa^2)	Expansion Pressure Thickness (m)
19.8	55	3967.5	0.30
22.1	45	3001.5	0.45
24.9	16	1138.5	0.30

19-17   Briefly describe the main steps in the Mechanistic-Empirical Pavement Design method as given in the MEPDG.

19-18   Determine the expected rut depth of the fourth layer of a 100 mm thick HMA flexible pavement surface divided into four equal sublayers for the following conditions:

Resilient or elastic strain calculated by the structural response model at the mid-depth of each HMA sublayer, mm/mm $= 45 \times 10^{-6}$

$$\text{Number of axle load repetitions} = 1.5 \times 10^4$$

Mixture field calibration constants:

$$\beta_{1r,} = 0.99$$
$$\beta_{2r,} = 0.98$$
$$\beta_3 = 0.95$$

19-19   Repeat Problem 19-18 for axle load repetitions of 2×10^4. Based on your answers for this problem and Problem 19-18, discuss the effect of axle load repetitions on the expected rut of HMA flexible pavements.

REFERENCES

AASHTO Guide for Design of Pavement Structures, American Association of State Highway and Transportation Officials, Washington, D.C., 1993.

Annual Book of ASTM Standards, Section 4, American Society for Testing and Materials, Philadelphia, PA, 2007.

Highway Design Manual: Design of the Pavement Standard Section, California Dept. of Transportation, 1995.

Interim Mechanistic-Empirical Pavement Design Guide Manual of Practice, Transportation Research Board, National Academies, Washington, D.C., 2008

Soil Cement Laboratory Handbook, Portland Cement Association, Skokie, IL, 1971.

Soil Stabilization in Pavement Structures, A User's Manual, Vols. I and II, U.S. Department of Transportation, Federal Highway Administration, Washington, D.C., October 1979.

CHAPTER 20

Design of Rigid Pavements

R igid highway pavements are normally constructed of Portland cement concrete and may or may not have a base course between the subgrade and the concrete surface. When a base course is used in rigid pavement construction, it is usually referred to as a subbase course. It is common, however, for only the concrete surface to be referred to as the rigid pavement, even where there is a base course. In this text, the terms *rigid pavement* and *concrete pavement* are synonymous. Rigid pavements have some flexural strength that permits them to sustain a beamlike action across minor irregularities in the underlying material. Thus, the minor irregularities may not be reflected in the concrete pavement. Properly designed and constructed rigid pavements have long service lives and usually are less expensive to maintain than flexible pavements but usually more expensive to construct.

Thickness of highway concrete pavements normally ranges from 150 to 330 mm. Different types of rigid pavements are described later in this chapter. These pavement types usually are constructed to carry heavy traffic loads, although they have been used for residential and local roads.

The topics discussed in this chapter cover the major aspects of rigid pavement design sufficient for an understanding of the design principles of the basic types: plain, simply reinforced, and continuous. The topics covered include a description of the basic types of rigid pavements, the materials used in their construction, and a discussion of the stresses imposed by traffic wheel loads and temperature differentials on the concrete pavement. The Alternative AASHTO Design Method and the American Concrete Pavement Association design method are described. The Alternative AASHTO Design Method is given in the AASHTO *Guide for Design of Pavement Methods,* and its supplement, *Design of Pavement Structures, Part II, Rigid Pavement Design and Rigid Pavement Joint Design.* In addition, a brief introduction is presented on the mechanistic-empirical design method.

CHAPTER OBJECTIVES:

- Become familiar with the characteristics of Portland cement used in rigid highway pavement construction.
- Learn about the different materials used in highway rigid pavement construction.
- Learn about the different joints in rigid highway pavements.
- Become familiar with the different wheel loads and stresses on rigid highway pavements.
- Learn the AASHTO method for the design of rigid highway pavements.
- Learn the PCA method for the design of highway rigid pavements.
- Become familiar with the mechanistic-empirical method for the design of rigid highway pavements

20.1 MATERIALS USED IN RIGID PAVEMENTS

The Portland cement concrete commonly used for rigid pavements consists of Portland cement, coarse aggregate, fine aggregate, and water. Steel reinforcing rods may or may not be used, depending on the type of pavement being constructed. A description of the quality requirements for each of the basic materials is presented in the following sections.

20.1.1 Portland Cement

Portland cement is manufactured by crushing and pulverizing a carefully prepared mix of limestone, marl, and clay or shale and by burning the mixture at a high temperature (about 1538 C) to form a clinker. The clinker is then allowed to cool, a small quantity of gypsum is added, and the mixture is then ground until more than 90 percent of the material passes the No. 200 sieve. The main chemical constituents of the material are tricalcium silicate (C_3S), dicalcium silicate (C_2S), and tetracalcium aluminoferrite (C_4AF).*

The material is usually transported in 0.028 m^3 bags, each weighing 42 kg, although it also can be transported in bulk for large projects.

Most highway agencies use either the American Society for Testing and Materials (ASTM) specifications (ASTM Designation C150) or the American Association of State Highway and Transportation Officials (AASHTO) specifications (AASHTO Designation M85) for specifying Portland cement quality requirements used in their projects. The AASHTO specifications list five main types of Portland cement.

- Type I is suitable for general concrete construction, where no special properties are required. A manufacturer will supply this type of cement when no specific type is requested.
- Type II is suitable for use in general concrete construction, where the concrete will be exposed to moderate action of sulphate or where moderate heat of hydration is required.
- Type III is suitable for concrete construction that requires a high concrete strength in a relatively short time. It is sometimes referred to as *high early strength cement*.
- Types IA, IIA, and IIIA are similar to Types I, II, and III, respectively, but contain a small amount (4 to 8 percent of total mix) of entrapped air. This is achieved during production by thoroughly mixing the cement with air-entraining agents and grinding the mixture. In addition to the properties listed for Types I, II, and III, Types IA, IIA,

---

*In expressing compounds, C = CaO, S = SiO_2, A = Al_2O_3, F = Fe_2O_3. For example, C_3A = 3CaO, Al_2O_3.

and IIIA are more resistant to calcium chloride and de-icing salts and are therefore more durable.

- Type IV is suitable for projects where low heat of hydration is necessary, while Type V is used in concrete construction projects where the concrete will be exposed to high sulphate action. Table 20.1 shows the proportions of the different chemical constituents for each of the five types of cement.

Table 20.1 Proportions of Chemical Constituents and Strength Characteristics for Different Types of Portland Cement

Cement Type[a]	I and IA	II and IIA	III and IIIA	IV	V
Silicon dioxide (SiO$_2$), min, percent	—	20.0[b,c]	—	—	—
Aluminum oxide (Al$_2$O$_3$), max, percent	—	6.0[b,c]	—	—	—
Ferric oxide (Fe$_2$O$_3$), max, percent	—	6.0[b,c]	—	6.5	—
Magnesium oxide (MgO), max, percent	6.0	6.0[b,c]	6.0	6.0[b]	6.0
Sulfur trioxide (SO$_3$),[d] max, percent					
When (C$_3$A)[e] is 8% or less	3.0	3.0	3.5	2.3	2.3
When (C$_3$A)[e] is more than 8%	3.5	f	4.5	f	f
Loss on ignition, max, percent	3.0	3.0	3.0	2.5	3.0
Insoluble residue, max, percent	0.75	0.75	0.75	0.75	0.75
Tricalcium silicate (C$_3$S)[e] max, percent	—	55	—	35[b]	—
Dicalcium silicate (C$_2$S)[e] min, percent	—	—	—	40[b]	—
Tricalcium aluminate (C$_3$A)[e] max, percent	—	8	15	7[b]	5[c]
Tetracalcium aluminoferrite plus twice the tricalcium aluminate[e] (C$_4$AF + 2 (C$_3$A)), or solid solution (C$_4$AF + C$_2$F), as applicable, max, percent	—	—	—	—	25[c]

[a]See source below.
[b]Does not apply when the heat of hydration limit is specified (see source below).
[c]Does not apply when the sulfate resistance limit is specified (see source below).
[d]There are cases where optimum SO$_3$ (using ASTM C563) for a particular cement is close to or in excess of the limit in this specification. In such cases where properties of a cement can be improved by exceeding the SO$_3$ limits stated in this table, it is permissible to exceed the values in the table, provided it has been demonstrated by ASTM C1038 that the cement with the increased SO$_3$ will not develop expansion in water exceeding 0.020% at 14 days. When the manufacturer supplies cement under this provision, he shall, upon request, supply supporting data to the purchaser.
[e]The expressing of chemical limitations by means of calculated assumed compounds does not necessarily mean that the oxides are actually or entirely present as such compounds.

When expressing compounds, C = CaO, S = SiO$_2$, A = Al$_2$O$_3$, F = Fe$_2$O$_3$. For example, C$_3$A = 3CaO. Al$_2$O$_3$. E Titanium dioxide and phosphorus pentoxide (TiO$_2$ and P$_2$O$_5$) shall not be included with the Al$_2$O$_3$ content.

When the ratio of percentages of aluminum oxide to ferric oxide is 0.64 or more, the percentages of tricalcium silicate, dicalcium silicate, tricalcium aluminate, and tetracalcium aluminoferrite shall be calculated from the chemical analysis as follows:

Tricalcium silicate = (4.071 × percent CaO) − (7.600 × percent SiO$_2$) − (6.718 × percent Al$_2$O$_3$) − (1.430 × percent Fe$_2$O$_3$) − (2.852 × percent SO$_3$)
Dicalcium silicate = (2.867 × percent SiO$_2$) − (0.7544 × percent C$_3$S)
Tricalcium aluminate = (2.650 × percent Al$_2$O$_3$) − (1.692 × percent Fe$_2$O$_3$)
Tetracalcium aluminoferrite = 3.043 × percent Fe$_2$O$_3$

When the alumina-ferric oxide ratio is less than 0.64, a calcium aluminoferrite solid solution (expressed as ss(C$_4$AF + C$_2$F)) is formed. Contents of this solid solution and of tricalcium silicate shall be calculated by the following formulas: E ss(C$_4$AF + C$_2$F) = (2.100 × percent Al$_2$O$_3$) + (1.702 × percent Fe$_2$O$_3$)
Tricalcium silicate = (4.071 × percent CaO) − (7.600 × percent SiO$_2$) − (4.479 × percent Al$_2$O$_3$) − (2.859 × percent Fe$_2$O$_3$) − (2.852 × percent SO$_3$). No tricalcium aluminate will be present in cements of this composition. Dicalcium silicate shall be calculated as previously shown.
[f]Not applicable.

SOURCE: Based on *Standard Specifications for Transportation Materials and Methods of Sampling and Testing*, 2011, AASHTO, Washington, D.C.

20.1.2 Coarse Aggregates

The coarse aggregates used in Portland cement concrete are inert materials that do not react with cement and usually are comprised of crushed gravel, stone, or blast furnace slag. The coarse aggregates may be any one of the three materials or else a combination of any two, or all three. One of the major requirements for coarse aggregates used in Portland cement concrete is the gradation of the material. The material is well graded, with the maximum size specified. Material retained in a No. 4 sieve is considered coarse aggregate. Table 20.2 shows gradation requirements for different maximum sizes as stipulated by ASTM.

Coarse aggregates must be clean. This is achieved by specifying the maximum percentage of deleterious substances allowed in the material. Other quality requirements include the ability of the aggregates to resist abrasion and the soundness of the aggregates.

A special test, known as the Los Angeles Rattler Test (AASHTO Designation T96), is used to determine the abrasive quality of the aggregates. In this test, a sample of the coarse aggregate retained in the No. 8 sieve and a specified number of standard steel spheres are placed in a hollow steel cylinder, with a diameter of 710 mm and a length of 500 mm, that is closed at both ends. The cylinder, which also contains a steel shelf that projects 89 mm radially inward, is mounted with its axis in the horizontal position. The cylinder is then rotated 500 times at a specific speed. The sample of coarse aggregate is then removed and sieved on a No. 12 sieve. The portion of the material retained on the No. 12 sieve is weighed, and the difference between this weight and the original weight is the weight loss. This loss is expressed as a percentage of the original weight. Maximum permissible loss in weight ranges from 30 to 60 percent, depending on the specifications used; however, a maximum of 40 to 50 percent has proved to be generally acceptable.

Soundness is defined as the ability of the aggregate to resist breaking up due to freezing and thawing. This property can be determined in the laboratory by first sieving a sample of the coarse aggregate through a No. 4 sieve and then immersing the portion retained in the sieve in water. The sample is frozen in the water for 2 hours and thawed for

Table 20.2   Gradation Requirements for Aggregates in Portland Cement Concrete (ASTM Designation C33)

Sieve Designation	Percent Passing by Weight		
	Aggregate Designation		
	2 in. to No. 4 (357)	*1 in. to No. 4 (467)*	*1 in. to No. 4 (57)*
2½ in. (63 mm)	100	—	—
2 in. (50 mm)	95–100	100	—
1½ in. (37.5 mm)	—	95–100	100
1 in. (25.0 mm)	35–70	—	95–100
¾ in. (19.0 mm)	—	35–70	—
½ in. (12.5 mm)	10–30	—	25–60
⅜ in. (9.5 mm)	—	10–30	—
No. 4 (4.75 mm)	0–5	0–5	0–10
No. 8 (2.36 mm)	—	—	0–5

SOURCE: Adapted from *ASTM Standards, Concrete and Aggregates*, Vol. 04.02, American Society for Testing and Materials, West Conshohocken, PA., October 2012.

one-half hour. This alternate freezing and thawing is repeated between 20 and 50 times. The sample is then dried and sieved again to determine the change in particle size. Sodium or magnesium sulphate may be used instead of water.

20.1.3 Fine Aggregates

Sand is mainly used as the fine aggregate in Portland cement concrete. Specifications for this material usually include grading requirements, soundness, and cleanliness. Standard specifications for the fine aggregates for Portland cement concrete (AASHTO Designation M6) give grading requirements normally adopted by state highway agencies (see Table 20.3).

The soundness requirement is usually given in terms of the maximum permitted loss in the material after five alternate cycles of wetting and drying in the soundness test. A maximum of 10 percent weight loss is usually specified.

Cleanliness is often specified in terms of the maximum amounts of different types of deleterious materials contained in the fine aggregates. For example, a maximum amount of silt (material passing No. 200 sieve) is usually specified within a range of 2 to 5 percent of the total fine aggregates. Since the presence of large amounts of organic material in the fine aggregates may reduce the hardening properties of the cement, a standard test (AASHTO Designation T21) is also usually specified as part of the cleanliness requirements. In this test, a sample of the fine aggregates is mixed with sodium hydroxide solution and then allowed to stand for 24 h. At the end of this period, the amount of light transmitted through the liquid floating above the test sample is compared with that transmitted through a standard color solution of reagent grade potassium dichromate ($K_2Cr_2O_2$) and concentrated sulfuric acid. The fine aggregate under test is considered to possibly contain injurious organic compounds if less light is transmitted through the liquid floating above the test sample. In such cases, additional tests should be carried out before using the fine aggregate in the concrete mix. For example, sand can be used only if the strength developed by 50 mm cubes made with this sand is at least 95 percent of that developed by similar cubes made with the same sand after washing it in a 3 percent hydroxide solution.

20.1.4 Water

The main water requirement stipulated is that the water used also should be suitable for drinking. This requires that the quantity of organic matter, oil, acids, and alkalis should not be greater than the allowable amount in drinking water.

Table 20.3   AASHTO-Recommended Particle Size Distribution for Fine Aggregates Used in Portland Cement Concrete

Sieve (M 92)	Mass Percent Passing
³⁄₈ in. (9.5 mm)	100
No. 4 (4.75 mm)	95 to 100
No. 8 (2.36 mm)	80 to 100
No. 16 (1.18 mm)	50 to 85
No. 30 (600 μm)	25 to 60
No. 50 (300 μm)	10 to 30
No. 100 (μm)	2 to 10

SOURCE: Based on *Standard Specifications for Transportation Materials and Methods of Sampling and Testing*, 2012, AASHTO, Washington, D.C.

20.1.5 Reinforcing Steel

Steel reinforcing may be used in concrete pavements to reduce the amount of cracking that occurs, as a load transfer mechanism at joints, or as a means of tying two slabs together. Steel reinforcement used to control cracking is usually referred to as *temperature steel,* whereas steel rods used as load transfer mechanisms are known as *dowel bars,* and those used to connect two slabs together are known as *tie bars.*

20.1.6 Temperature Steel

Temperature steel is provided in the form of a bar mat or wire mesh consisting of longitudinal and transverse steel wires welded at regular intervals. The mesh is usually placed about 76 mm below the slab surface. The cross-sectional area of the steel provided per foot width of the slab depends on the size and spacing of the steel wires forming the mesh. The amount of steel required depends on the length of the pavement between expansion joints, the maximum stress desired in the concrete pavement, the thickness of the pavement, and the moduli of elasticity of the concrete and steel. Equation 20.19 (developed later in this chapter) can be used to determine the area of steel required if the length of the slab is fixed. However, steel areas obtained by this equation for concrete slabs less than 13.5 m in length may be inadequate and therefore should be compared with the following general guidelines for the minimum cross-sectional area of the temperature steel.

1. Cross-sectional area of longitudinal steel should be at least equal to 0.1 percent of the cross-sectional area of the slab.
2. Longitudinal wires should not be less than No. 2 gauge, spaced at a maximum distance of 153 mm.
3. Transverse wires should not be less than No. 4 gauge, spaced at a maximum distance of 305 mm.

Temperature steel does not prevent cracking of the slab, but it does control the crack widths because the steel acts as a tie holding the edges of the cracks together. This helps to maintain the shearing resistance of the pavement, thereby maintaining its capacity to carry traffic load, even though the flexural strength is not improved.

20.1.7 Dowel Bars

Dowel bars are used mainly as load-transfer mechanisms across joints. They provide flexural, shearing, and bearing resistance. The dowel bars must be of a much larger diameter than the wires used in temperature steel. Size selection is based mainly on experience. Diameters of 25 to 38 mm and lengths of 0.60 to 0.90 m have been used, with the bars usually spaced at 0.30 m centers across the width of the slab. At least one end of the bar should be smooth and lubricated to facilitate free expansion.

20.1.8 Tie Bars

Tie bars are used to tie two sections of the pavement together, and therefore they should be either deformed bars or should contain hooks to facilitate the bonding of the two sections of the concrete pavement with the bar. These bars are usually much smaller in diameter than the dowel bars and are spaced at larger centers. Typical diameter and spacing for these bars are 19 mm and 0.90 m, respectively.

20.2 JOINTS IN CONCRETE PAVEMENTS

Different types of joints are placed in concrete pavements to limit the stresses induced by temperature changes and to facilitate proper bonding of two adjacent sections of pavement when there is a time lapse between their construction (for example, between the end of one day's work and the beginning of the next). These joints can be divided into four basic categories:

- Expansion joints
- Contraction joints
- Hinge joints
- Construction joints

20.2.1 Expansion Joints

When concrete pavement is subjected to an increase in temperature, it will expand, resulting in an increase in length of the slab. When the temperature is sufficiently high, the slab may buckle or "blow up" if it is long enough and if no provision is made to accommodate the increased length. Therefore, *expansion joints* are usually placed transversely, at regular intervals, to provide adequate space for the slab to expand. These joints are placed across the full width of the slab and are 19 to 25 mm. wide in the longitudinal direction. They must create a distinct break throughout the depth of the slab. The joint space is filled with a compressible filler material that permits the slab to expand. Filler materials can be cork, rubber, bituminous materials, or bituminous fabrics.

A means of transferring the load across the joint space must be provided since there are no aggregates that will develop an interlocking mechanism. The load-transfer mechanism is usually a smooth dowel bar that is lubricated on one side. An expansion cap is also usually installed, as shown in Figure 20.1, to provide a space for the dowel to occupy during expansion.

Some states no longer use expansion joints because of the inability of the load-transfer mechanism to adequately transfer the load. Other states continue to use expansion joints and may even use them in place of construction joints.

20.2.2 Contraction Joints

When concrete pavement is subjected to a decrease in temperature, the slab will contract if it is free to move. Prevention of this contraction movement will induce tensile stresses in the concrete pavement. *Contraction joints* therefore are placed

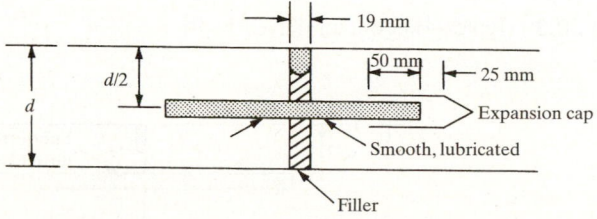

Figure 20.1 Typical Expansion Joint

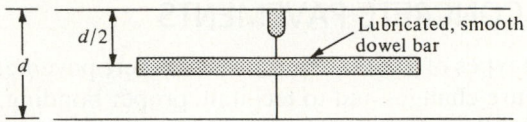

Figure 20.2   Typical Contraction Joint

transversely at regular intervals across the width of the pavement to release some of the tensile stresses that are so induced. A typical contraction joint is shown in Figure 20.2. It may be necessary in some cases to install a load-transfer mechanism in the form of a dowel bar when there is doubt about the ability of the interlocking gains to transfer the load.

20.2.3 Hinge Joints

Hinge joints are used mainly to reduce cracking along the center line of highway pavements. Figure 20.3 shows a typical hinge joint (keyed joint) suitable for single-lane-at-a-time construction.

20.2.4 Construction Joints

Construction joints are placed transversely across the pavement width to provide suitable transition between concrete laid at different times. For example, a construction joint is usually placed at the end of a day's pour to provide suitable bonding with the start of the next day's pour. A typical butt construction joint is shown in Figure 20.4. In some cases, as shown in Figure 20.3, a keyed construction joint may also be used in the longitudinal direction when only a single lane is constructed at a time. In this case, alternate lanes of the pavement are cast, and the key is formed by using metal formwork that has been cast with the shape of the groove or by attaching a piece of metal or wood to a wooden formwork. An expansion joint can be used in lieu of a transverse construction joint in cases where the construction joint falls at or near the same position as the expansion joint.

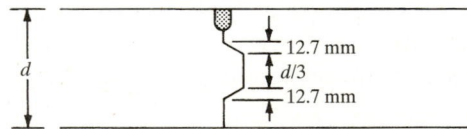

Figure 20.3   Typical Hinge Joint (Keyed Joint)

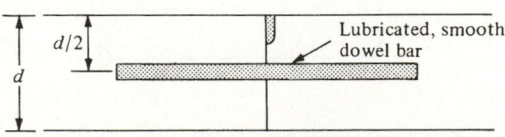

Figure 20.4   Typical Butt Joint

20.3 TYPES OF RIGID HIGHWAY PAVEMENTS

Rigid highway pavements can be divided into three general types: jointed plain concrete pavements, simply jointed reinforced concrete pavements, and continuously reinforced concrete pavements. The definition of each pavement type is related to the amount of reinforcement used.

20.3.1 Jointed Plain Concrete Pavement (JPCP)

Plain concrete pavement has no temperature steel or dowels for load transfer. However, steel tie bars often are used to provide a hinge effect at longitudinal joints and to prevent the opening of these joints. Jointed plain concrete pavements are used mainly on low-volume highways or when cement-stabilized soils are used as subbase. Joints are placed at relatively shorter distances (3 to 6 mm) than with other types of concrete pavements to reduce the amount of cracking. In some cases, the transverse joints of jointed plain concrete pavements are skewed about 1.2 to 1.50 m in plan, such that only one wheel of a vehicle passes through the joint at a time. This helps to provide a smoother ride.

20.3.2 Jointed Reinforced Concrete Pavement (JRCP)

Jointed reinforced concrete pavements have dowels for the transfer of traffic loads across joints, with these joints spaced at larger distances, ranging from 9 to 30 m. Temperature steel is used throughout the slab, with the amount dependent on the length of the slab. Tie bars also are used commonly at longitudinal joints.

20.3.3 Continuously Reinforced Concrete Pavement (CRCP)

Continuously reinforced concrete pavements have no transverse joints, except construction joints or expansion joints when they are necessary at specific positions, such as at bridges. These pavements have a relatively high percentage of steel, with the minimum usually at 0.6 percent of the cross section of the slab. They also contain tie bars across the longitudinal joints.

20.4 PUMPING OF RIGID PAVEMENTS

Pumping is an important phenomenon associated with rigid pavements. Pumping is the discharge of water and subgrade (or subbase) material through joints, cracks, and along the pavement edges. It primarily is caused by the repeated deflection of the pavement slab in the presence of accumulated water beneath it. The mechanics of pumping can best be explained by considering the sequence of events that lead to it.

The first event is the formation of void space beneath the pavement. This void forms from either the combination of plastic deformation of the soil, due to imposed loads and the elastic rebound of the pavement after it has been deflected by the imposed load, or warping of the pavement, which occurs as a result of temperature gradient within the slab. Water then accumulates in the space after many repetitions of traffic load. The water may be infiltrated from the surface through joints and the pavement edge. To a lesser extent, ground water may settle in the void. If the subgrade or base material is granular, the water will freely drain through the soil. If the material is fine-grained, however, the water is not easily discharged, and additional load repetitions will result in the soil going into suspension with the water to form a slurry. Further load repetitions and deflections of the slab will result in the slurry being ejected to the surface (pumping). Pumping action will then continue, with the result that a relatively large void space is formed underneath the concrete slab. This results in faulting of the joints and eventually the formation of transverse cracks or the breaking of the corners of the slab. Joint faulting and cracking is therefore progressive, since formation of a crack facilitates the pumping action.

Visual manifestations of pumping include:

- Discharge of water from cracks and joints
- Spalling near the centerline of the pavement and a transverse crack or joint
- Mud boils at the edge of the pavement
- Pavement surface discoloration (caused by the subgrade soil)
- Breaking of pavement at the corners

20.4.1 Design Considerations for Preventing Pumping

A major design consideration for preventing pumping is the reduction or elimination of expansion joints, since pumping is usually associated with these joints. This is the main reason why current design practices limit the number of expansion joints to a minimum. Since pumping is also associated with fine-grained soils, another design consideration is either to replace soils that are susceptible to pumping with a nominal thickness of granular or sandy soils, or to improve them by stabilization. Current design practices therefore usually include the use of 76 to 153 mm layers of granular subbase material at areas along the pavement alignment, where the subgrade material is susceptible to pumping, or stabilizing the susceptible soil with asphalt or Portland cement. The American Concrete Pavement Association method of rigid pavement design indirectly considers this phenomenon in the erosion analysis.

20.5 STRESSES IN RIGID PAVEMENTS

Stresses are developed in rigid pavements as a result of several factors, including the action of traffic wheel loads, the expansion and contraction of the concrete due to temperature changes, yielding of the subbase or subgrade supporting the concrete pavement, and volumetric changes. For example, traffic wheel loads will induce flexural stresses that are dependent on the location of the vehicle wheels relative to the edge of the pavement, whereas expansion and contraction may induce tensile and compressive stresses that are dependent on the range of temperature changes in the concrete pavement. These different factors that can induce stress in concrete pavement have made the theoretical determination of stresses rather complex, requiring the following simplifying assumptions.

1. Concrete pavement slabs are considered as unreinforced concrete beams. Any contribution made to the flexural strength by the inclusion of reinforcing steel is neglected.
2. The combination of flexural and direct tensile stresses will inevitably result in transverse and longitudinal cracks. The provision of suitable crack control in the form of joints, however, controls the occurrence of these cracks, thereby maintaining the beam action of large sections of the pavement.
3. The supporting subbase and/or subgrade layer acts as an elastic material in that it deflects at the application of the traffic load and recovers at the removal of the load.

20.5.1 Stresses Induced by Bending

The ability of rigid pavement to sustain a beamlike action across irregularities in the underlying materials suggests that the theory of bending is fundamental to the analysis of stresses in such pavements. The theory of a beam supported on an elastic foundation therefore can be used to analyze the stresses in the pavement when it is externally loaded. Figure 20.5 shows the deformation sustained by a beam on an elastic foundation when it is loaded externally. The stresses developed in the beam may be analyzed by assuming that a reactive pressure (p), which is proportional to the deflection, is developed as a result of the applied load. This pressure is given as

$$p = ky \tag{20.1}$$

where

p = reactive pressure at any point beneath the beam (kN/mm^2)
y = deflection at the point (mm)
k = modulus of subgrade reaction (kN/mm^3)

The modulus of subgrade reaction is the stress (kN/mm^2) that will cause an inch deflection of the underlying soil. Equation 20.1 assumes that k (kN/mm^3) is constant, which implies that the subgrade is elastic. However, this assumption is valid for only a limited range of different factors. Research has shown that the value of k depends on certain soil characteristics such as density, moisture, soil texture, and other factors that influence the strength of the soils. The k value of a particular soil will also vary with the size of the loaded area and the amount of deflection. The modulus of subgrade reaction is directly proportional to the loaded area and inversely proportional to the deflection. In pavement design, however, minor changes in k do not have significant impact on design results, and an average value is usually assumed. The plate-bearing test is commonly used for determining the value of k in the field.

A general relationship between the moment and the radius of curvature of a beam is given as

$$\frac{1}{R} = \frac{M}{EI} \tag{20.2}$$

where

R = radius of curvature
M = moment in beam
E = modulus of elasticity
I = moment of inertia

The general differential equation relating the moment at any section of a beam with the deflection at that section is given as

$$M = EI \frac{d^2y}{dx^2} \tag{20.3}$$

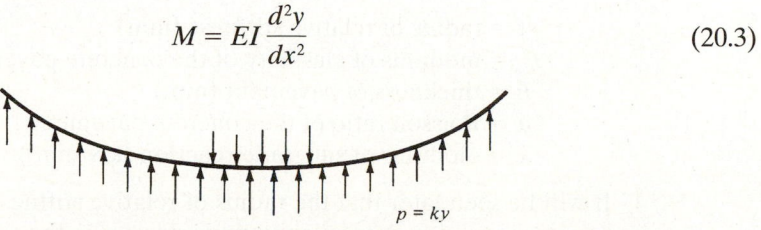

Figure 20.5   Deformation of a Beam on Elastic Foundation

whereas the basic differential equation for the deflection on an elastic foundation is given as

$$EI \frac{d^4y}{dx^4} = -ky \tag{20.4}$$

The basic differential equation for a slab is given as

$$M_x = \left[\frac{Eh^3}{12(1 - \mu^2)} \right] \frac{d^2w}{dx^2}$$

or

$$M_y = \left[\frac{Eh^3}{12(1 - \mu^2)} \right] \frac{d^2w}{dy^2} \tag{20.5}$$

where

h = thickness of slab
μ = Poisson ratio of concrete
w = deflection of the slab at a given point
M_x = bending moment at a point about the x-axis
M_y = bending moment at a point about the y-axis

The EI term in Eq. 20.4 is called the *stiffness* of the beam, whereas the stiffness of the slab is given by the term within the square brackets of Eq. 20.5. This term is usually denoted as D, where

$$D = \frac{E_c h^3}{12(1 - \mu^2)} \tag{20.6}$$

In developing expressions for the stresses in a concrete pavement, Westergaard made use of the radius of relative stiffness, which depends on the stiffness of the slab and the modulus of subgrade reaction of the soil. It is given as

$$\ell = \sqrt[4]{\frac{E_c h^3}{12(1 - \mu^2)k}} \tag{20.7}$$

where

ℓ = radius of relative stiffness (mm)
E_c = modulus of elasticity of the concrete pavement (kN/mm²)
h = thickness of pavement (mm)
μ = Poisson ratio of the concrete pavement
k = modulus of subgrade reaction (kN/mm³)

It will be seen later that the radius of relative stiffness is an important parameter in the equations used to determine various stresses in the concrete pavement.

20.5.2 Stresses Due to Traffic Wheel Loads

The basic equations for determining flexural stresses in concrete pavements due to traffic wheel loads were first developed by Westergaard. Although several theoretical developments have been made since then, the Westergaard equations are still considered a fundamental tool for evaluating stresses on concrete pavements. Westergaard considered three critical locations of the wheel load on the concrete pavement in developing the equations. These locations are shown in Figure 20.6 and are described as follows.

Case A. Load is applied at the corner of a rectangular slab. This provides for the cases when the wheel load is applied at the intersection of the pavement edge and a transverse joint. However, this condition is not common because pavements are generally much wider. Thus, no equation is presented for this case.

Case B. Load is applied at the interior of the slab at a considerable distance from its edges.

Case C. Load is applied at the edge of the slab at a considerable distance away from any corner.

The locations shown as Cases I, II, and III in Figure 20.6 are the critical locations presently used for the relatively wide pavements now being constructed.

The equations for determining these stresses were developed taking into consideration the different day and night temperature conditions that may exist. During the day, the temperature is higher at the surface of the slab than at the bottom. This temperature gradient through the depth of the slab will create a tendency for the slab edges to curl downward. During the night, however, the temperature at the bottom of the slab is higher than at the surface, thereby reversing the temperature gradient, which results in the tendency for the slab edges to curl upward. The equations for stresses due to traffic load reflect this phenomenon of concrete pavements.

The original equations developed by Westergaard were modified using the results of full-scale tests conducted by the Bureau of Public Roads. These modified equations for the different loading conditions are as follows.

1. Edge loading when the edges of the slab are warped upward at night

$$\sigma_e = \frac{0.572P}{h^2}\left[4\log_{10}\left(\frac{\ell}{b}\right) + \log_{10}b\right] \tag{20.8}$$

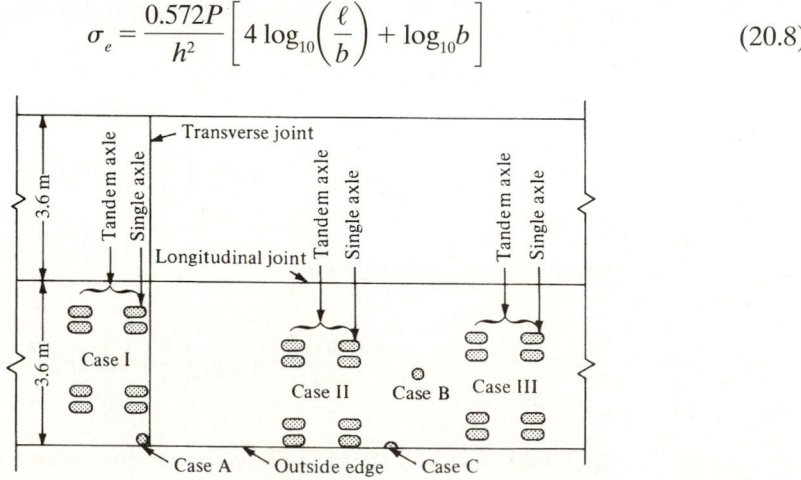

Figure 20.6   Critical Locations of Wheel Loads on Concrete Pavements

2. Edge loading when the slab is unwarped or when the edge of the slab is curled downward during the day

$$\sigma_e = \frac{0.572P}{h^2}\left[4\log_{10}\left(\frac{\ell}{b}\right) + 0.359\right] \tag{20.9}$$

3. Interior loading

$$\sigma_i = \frac{0.316P}{h^2}\left[4\log 10\left(\frac{\ell}{b}\right) + 1.069\right] \tag{20.10}$$

where

σ_e = maximum stress (lb/in.^{2}(kN/m^2)) induced in the bottom of the slab, directly under the load P and applied at the edge and in a direction parallel to the edge
σ_i = maximum tensile stress (lb/in.^{2}(kN/m^2)) induced at the bottom of the slab directly under the load P applied at the interior of the slab
P = applied load in pounds, including allowance for impact
h = thickness of slab (in.(mm))
ℓ = radius of relative stiffness
$\ell = \sqrt[4]{E_c h^3/[12(1 - \mu^2)k]}$
E_c = modulus of elasticity of concrete (lb/in.^{2} (kN/m^2))
μ = Poisson ratio for concrete = 0.15
k = subgrade modulus (lb/in.^{3})
b = radius of equivalent distribution of pressure (in.(mm))
  = $\sqrt{1.6a^2 + h^2} - 0.675h$ (for a < 1.724h)
  = a (for a > 1.724h)
a = radius of contact area of load (in.(mm)) (Contact area is usually assumed as a circle for interior and corner loadings and semicircle for edge loading.)

Revised equations for edge loadings have been developed by Ioannides et al. and are given as Eqs. 20.11 and 20.12.
For a circular loaded area,

$$\sigma_e = \frac{0.803P}{h^2}\left[4\log_{10}\left(\frac{\ell}{a}\right) + 0.666\frac{a}{\ell} - 0.034\right] \tag{20.11}$$

For a semicircular loaded area,

$$\sigma_e = \frac{0.803P}{h^2}\left[4\log_{10}\left(\frac{\ell}{a}\right) + 0.282\frac{a}{\ell} + 0.650\right] \tag{20.12}$$

It should be noted that the above equations assume a value of 0.15 for the Poisson ratio of the concrete pavement.
The results obtained from Ioannides' formulas differ significantly from those obtained from Westergaard's formulas, as illustrated in Example 20.1.

Example 20.1 Computing Tensile Stress Resulting from a Wheel Load on a Rigid Pavement

Determine the tensile stress imposed by a semicircular wheel load of 4 kN imposed during the day and located at the edge of a concrete pavement with the following dimensions and properties: (a) by using the Westergaard equation, and (b) by using the Ioannides equation.

$$\text{Pavement thickness} = 6 \text{ in. (153 mm)}$$

$$\mu = 0.15$$

$$E = 5 \times 10^6 \text{ lb/in.}^2 \text{ (34.47 kN/mm}^2\text{)}$$

$$k = 130 \text{ lb/in.}^3 \text{ (3.53 kN/mm}^3\text{)}$$

$$\text{Radius of loaded area} = 3 \text{ in. (76 mm)}$$

Solution:

(a) Use the Westergaard equation.
From Eq. 20.9,

$$\sigma_e = \frac{0.572P}{h^2}\left[4\log_{10}\left(\frac{\ell}{b}\right) + 0.359\right]$$

$$a = 3 \text{ in.} < 1.724h$$

Since $a < 1.724 \times 6$, we use the expressions

$$b = \sqrt{1.6a^2 + h^2} - 0.675h = \sqrt{1.6 \times 3^2 + 6^2} - 0.675 \times 6 = 7.1 - 4.05$$

$$= 3.05$$

$$\ell = \sqrt[4]{(5 \times 10^6 \times 6^3)/[12(1 - 0.15^2)130]} = 29.0$$

$$\sigma_2 = [(0.572 \times 900)/6^2][4\log_{10}(29/3.05) + 0.359 = 14.3(4 \times 0.978 + 0.359)]$$

$$= 61.07 \text{ lb/in.}^2 \text{ (425 kN/m}^2\text{)}$$

(b) Use the Ioannides equation.

From Eq. 20.12,

$$\sigma_e = \frac{0.803P}{h^2}\left[4\log\left(\frac{\ell}{a}\right) + 0.282\left(\frac{a}{\ell}\right) + 0.650\right]$$

$$= \left[\frac{0.803 \times 900}{6^2}\right]\left[4\log_{10}\left(\frac{29}{3.0}\right) + 0.282\frac{3.05}{29} + 0.650\right]$$

$$= 92.75 \text{ lb/in.}^2 \text{ (640 kN/m}^2\text{)}$$

20.5.3 Stresses Due to Temperature Effects

The tendency of the slab edges to curl downward during the day and upward during the night as a result of temperature gradients is resisted by the weight of the slab itself. This resistance tends to keep the slab in its original position, resulting in stresses being induced in the pavement. Compressive and tensile stresses therefore are induced at the top and bottom of the slab, respectively, during the day, whereas tensile stresses are induced at the top and compressive stresses at the bottom during the night.

Under certain conditions these curling stresses may have values high enough to cause cracking of the pavement. They also may reduce the subgrade support beneath some sections of the pavement, which can result in a considerable increase of stresses due to traffic loads over those pavements with uniform pavement support. Studies have shown that curling stresses can be higher than 1378 kN/m² for 3 m slabs and much higher for wider slabs. One of the main purposes of longitudinal joints is to limit the slab width by dividing the concrete pavement into individual slabs 3.3 or 3.6 m wide.

Tests have shown that maximum temperature differences between the top and bottom of the slab depend on the thickness of the slab, and that these differences are about 0.066 degrees Celsius/mm thickness for 153 to 229 mm-thick slabs. The temperature differential also depends on the season, with maximum differentials occurring during the day in the spring and summer months. Another factor that affects the temperature differential is the latitude of the location of the slab. The surface temperature of the pavement tends to be high if the angle of incidence of the sun's rays is high, as in areas near the equator.

These curling stresses can be determined from Eqs. 20.13 and 20.14.

$$\sigma_{xe} = \frac{C_x E_c e t}{2} \tag{20.13}$$

and

$$\sigma_{xi} = \frac{E_c e t}{2} \left(\frac{C_x + \mu C_y}{1 - \mu^2} \right) \tag{20.14}$$

where

σ_{xe} = maximum curling stress in kN/mm² at the edge of the slab in the direction of the slab length due to temperature difference between the top and bottom of the slab

σ_{xi} = maximum curling stress (kN/mm²) at the interior of the slab in the direction of the slab length due to temperature difference between the top and bottom of the slab

E_c = modulus of elasticity for concrete (kN/mm²)

μ = Poisson ratio for concrete

C_x, C_y = coefficients that are dependent on the radius of relative stiffness of the concrete and can be deduced from Figure 20.7

e = thermal coefficient of expansion and contraction of concrete °C

t = temperature difference between the top and bottom of the slab °C

Temperature changes in the slab will also result in expansion (for increased temperature) and contraction (for reduced temperature). The provision of suitable expansion/contraction joints in the slab reduces the magnitude of these stresses but does not entirely eliminate them, since considerable resistance to the free horizontal

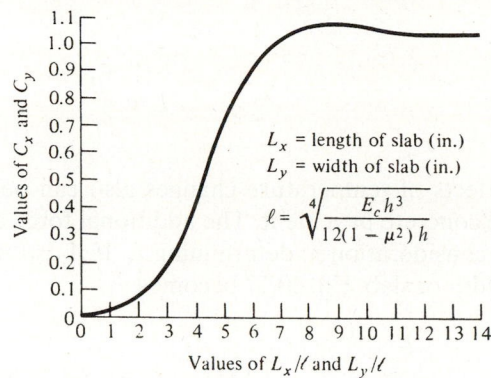

Figure 20.7   Values of C_x and C_y for Use in Formulas for Curling Stresses

movement of the pavement will still be offered by the subgrade, due to friction action between the bottom of the slab and the top of the subgrade. The magnitude of these stresses depends on the length of the slab, the type of concrete pavement, the magnitude of the temperature changes, and the coefficient of friction between the pavement and the subgrade. Tensile stresses greater than 690 kN/m^2 have been reported for an average temperature change of 4.4 C in a slab with a length of 30 m.

If only the effect of uniform temperature drop is being considered, the maximum spacing of contraction joints required to ensure that the maximum stress does not exceed a desired value p_c can be estimated by equating both the force in the slab and the force developed due to friction, as follows.

Considering a unit width of the pavement, the frictional force (F) developed due to a uniform drop in temperature is

$$F = f \frac{h}{12}(1)\frac{L}{2}\gamma_c \tag{20.15}$$

where

f = coefficient of friction between the bottom of concrete pavement and subgrade
h = thickness of concrete pavement (mm)
L = length of pavement between contraction joints (m)
γ_c = density of concrete (kN/mm^3)

The force (P) developed in the concrete due to stress p_c is given as

$$P = p_c(1)(1)h \tag{20.16}$$

where p_c is the maximum desired stress in the concrete in lb/in.^{2}.

Equating these two forces gives

$$fh(1)\frac{L}{2}\gamma_c = p_c(1)(h) \tag{20.17}$$

and

$$L = \frac{2p_c}{f\gamma_c} \tag{20.18}$$

The effects of temperature changes also can be reduced by including reinforcing steel in the concrete pavement. The additional force developed by the steel may also be taken into consideration in determining L. If A_s is the total cross-sectional area of steel per foot width of slab, Eq. 20.17 becomes

$$fh\,(1)\,\frac{L}{2}\,\gamma_c = p_c\,(h + nA_s) \tag{20.19}$$

or

$$L = \frac{2p_c}{h\gamma_c f}(h + nA_s) \tag{20.20}$$

where

$\quad E_s$ = modulus of elasticity of steel
$\quad n$ = modular ratio = E_s/E_c

20.6 THICKNESS DESIGN OF RIGID PAVEMENTS

The main objective in rigid pavement design is to determine the thickness of the concrete slab that will be adequate to carry the projected traffic load for the design period. Several design methods have been developed over the years, some of which are based on the results of full-scale road tests, others on theoretical development of stresses on layered systems, and others on the combination of the results of tests and theoretical development. However, two methods are used extensively: the AASHTO and American Concrete Pavement Association methods, which are presented in this chapter along with a brief introduction of the mechanistic-empirical design method given in the *Interim Mechanistic-Empirical Pavement Design Guide Manual of Practice* (MEPDG).

20.6.1 AASHTO Design Method as Given in AASHTO *Guide for Design of Pavement Structures*

The AASHTO method for rigid pavement design is mainly based on the results obtained from the AASHTO road test (Chapter 19). The design procedure was initially published in the early 1960s but was revised in the 1970s and 1980s. Further revisions have been carried out since then, incorporating new developments. For example, the *Supplement to the AASHTO Guide for Design of Pavement Structures, Part II, Rigid Pavement and Rigid Pavement Joint Design*, gives design equations that require an input of a "hypothetical" joint spacing, which is not required for the original AASHTO design equations. The design procedure provides for the determination of the pavement thickness and the amount of steel reinforcement when used. It is suitable for plain concrete, simply reinforced concrete, and continuously reinforced concrete pavements. The design procedure for the longitudinal reinforcing steel in continuously reinforced concrete pavements, however, is beyond the scope of this book. Interested readers should refer to the AASHTO guide for this procedure.

Design Considerations

The factors considered in the AASHTO procedure for the design of rigid pavements as presented in the 1993 guide are

- Pavement performance
- Subgrade strength
- Subbase strength
- Traffic
- Concrete properties
- Drainage
- Reliability

Pavement Performance. Pavement performance is considered in the same way as for flexible pavement, as presented in Chapter 19. The initial serviceability index (P_i) may be taken as 4.5, and the terminal serviceability index may also be selected by the designer.

Subbase Strength. The guide allows the use of either graded granular materials or suitably stabilized materials for the subbase layer. Table 20.4 gives recommended specifications for six types of subbase materials. AASHTO suggests that the first five types—A through E—can be used within the upper 102 mm layer of the subbase, whereas type F can be used below the uppermost 102 mm layer. Special precautions should be taken when certain conditions exist. For example, when A, B, and F materials are used in areas where the pavement may be subjected to frost action, the percentage of fines should be reduced to a minimum. Subbase thickness is usually not less than 153 mm and should be extended 0.30 to 0.90 m outside the edge of the pavement structure.

Subgrade Strength. The strength of the subgrade is given in terms of the Westergaard modulus of subgrade reaction k, which is defined as the load in kN/m² on a loaded area, divided by the deformation in inches. Values of k can be obtained by conducting a plate-bearing test in accordance with the AASHTO Test Designation T222 using a 762 mm-diameter plate. Estimates of k values can also be made either from experience or by correlating with other tests. Figure 20.8 shows an approximate interrelationship of soil classification and bearing values obtained from different types of tests.

Subbase Strength. The guide also provides for the determination of an effective modulus of subgrade reaction, which depends on (1) the seasonal effect on the resilient modulus of the subgrade, (2) the type and thickness of the subbase material used, (3) the effect of potential erosion of the subbase, and (4) whether bedrock lies within 10 ft of the subgrade surface. The seasonal effect on the resilient modulus of the subgrade was discussed in Chapter 19, and a procedure similar to that used in flexible pavement design is used here to take into consideration the variation of the resilient modulus during a 12-month period.

Table 20.4   Recommended Particle Size Distributions for Different Types of Subbase Materials

			Types of Subbase			
Sieve Designation	*Type A*	*Type B*	*Type C (Cement Treated)*	*Type D (Lime Treated)*	*Type E (Bituminous Treated)*	*Type F (Granular)*
Sieve analysis percent passing						
2 in.	100	100	—	—	—	—
1 in.	—	75–95	100	100	100	100
⅜ in.	30–65	40–75	50–85	60–100	—	—
No. 4	25–55	30–60	35–65	50–85	55–100	70–100
No. 10	15–40	20–45	25–50	40–70	40–100	55–100
No. 40	8–20	15–30	15–30	25–45	20–50	30–70
No. 200	2–8	5–20	5–15	5–20	6–20	8–25
(The minus No. 200 material should be held to a practical minimum.)						
Compressive strength lb/in.^{2} at 28 days			400–750	100		
Stability						
Hveem Stabilometer					20 min	
Hubbard field					1000 min	
Marshall stability					500 min	
Marshall flow					20 max	
Soil constants						
Liquid limit	25 max	25 max				25 max
Plasticity index[a]	N.P.	6 max	10 max[b]		6 max[b]	6 max

[a]As performed on samples prepared in accordance with AASHTO Designation T87.
[b]These values apply to the mineral aggregate prior to mixing with the stabilizing agent.
SOURCE: Based on *Standard Specifications for Transportation Materials and Methods of Sampling and Testing*, 2012, AASHTO, Washington, D.C.

Since different types of subbase materials have different strengths, the type of material used is an important input in the determination of the effective modulus of subgrade reaction. In estimating the composite modulus of subgrade reaction, the subbase material is defined in terms of its elastic modulus E_{SB}. It is also necessary to consider the combination of material types and the required thicknesses because this serves as a basis for determining the cost-effectiveness of the pavement. The chart in Figure 20.9 is used to estimate the composite modulus of subgrade reaction (k_∞), assuming an infinite depth for the type of subbase material, based on its elastic modulus, its resilient modulus, and the thickness of the subbase.

The effective k value also depends on the potential erosion of the subbase material. This effect is included by the use of a factor (see Table 20.5) for the loss of support (LS) in determining the effective k value. This factor is used to reduce the effective modulus of subgrade reaction, as shown in Figure 20.10.

The presence of bedrock, within a depth of 3 m of the subgrade surface and extending over a significant length along the highway alignment, may result in an increase of the overall modulus of subgrade reaction. This effect is taken into consideration by adjusting the effective modulus subgrade using the chart in Figure 20.11. The procedure is demonstrated in the solution of Example 20.2.

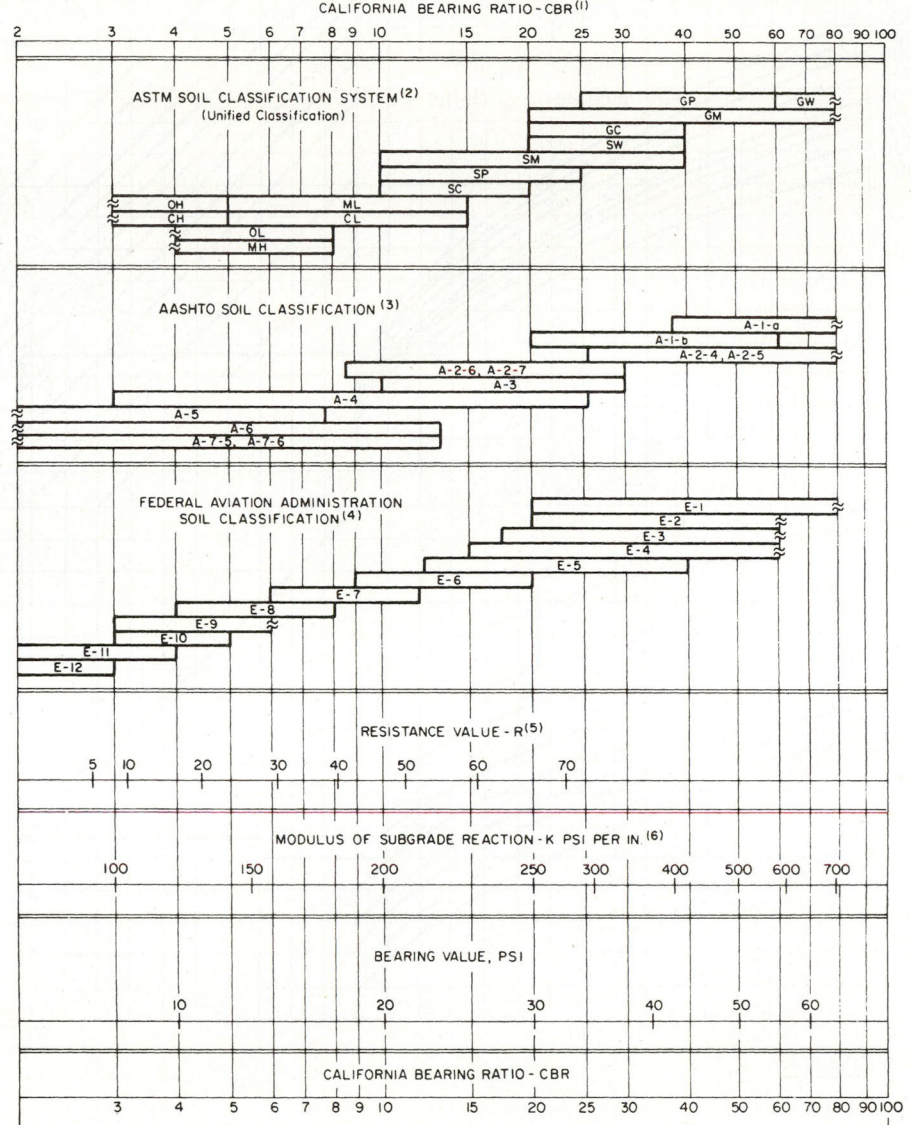

Figure 20.8   Approximate Interrelationship of Soil Classification and Bearing Values

(1) For the basic idea, see O. J. Porter, "Foundations for Flexible Pavements," Highway Research Board *Proceedings of the Twenty-second Annual Meeting*, 1942, Vol. 22, pages 100-136.
(2) ASTM Designation D2487.
(3) "Classification of Highway Subgrade Materials," Highway Research Board *Proceedings of the Twenty-fifth Annual Meeting*, 1945, Vol. 25, pages 376-392.
(4) *Airport Paving*, U.S. Department of Commerce, Federal Aviation Agency, May 1948, pages 11-16. Estimated using values given in FAA *Design Manual for Airport Pavements*.(Formerly used FAA Classification; Unified Classification now used.)
(5) C. E.Warnes, "Correlation Between *R* Value and *k* Value," unpublished report, Portland Cement Association, Rocky Mountain-Northwest Region, October 1971 (best-fit correlation with correction for saturation).
(6) See T. A. Middlebrooks and G. E. Bertram, "Soil Tests for Design of Runway Pavements," Highway Research Board *Proceedings of the Twenty-second Annual Meeting*, 1942, Vol. 22, page 152.

SOURCE: R.G. Packard, *Thickness Design for Concrete Highway and Street Pavements*, American Concrete Pavement Association, 1984.

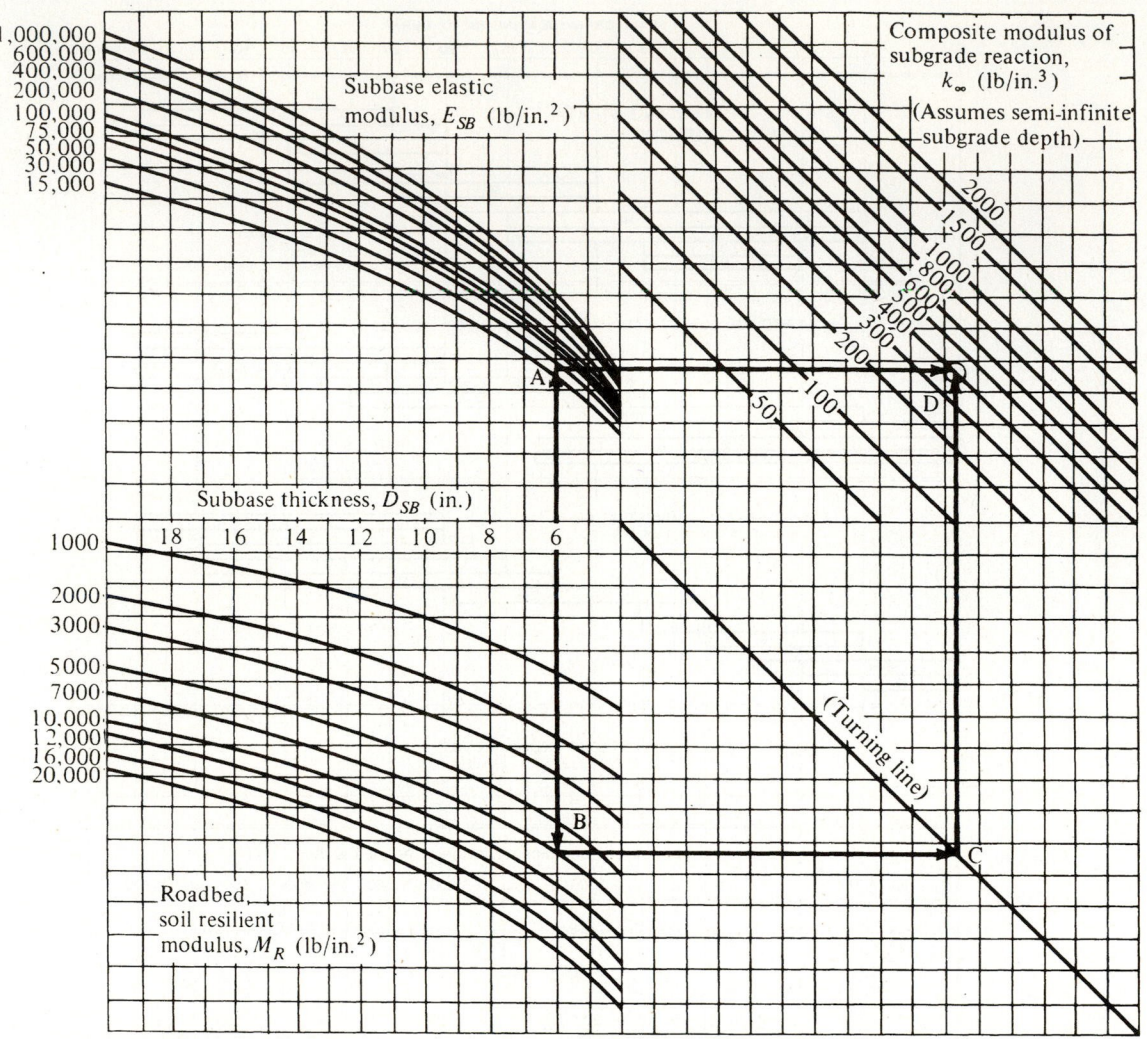

Example:

D_{SB} = 6 in.

E_{SB} = 20,000 lb/in.^{2}

M_R = 7,000 lb/in.^{2}

Solution: k_∞ = 400 lb/in.^{3}

Figure 20.9   Chart for Estimating Composite Modulus of Subgrade Reaction, K_∞, Assuming a Semi-Infinite Subgrade Depth*

*For practical purposes, a semi-infinite depth is considered to be greater than 10 ft below the surface of the subgrade.

SOURCE: From *AASHTO Guide for Design of Pavement Structures*, 1993, AASHTO, Washington, D.C. Used by permission.

Table 20.5 Typical Ranges of Loss of Support Factors for Various Types of Materials

Type of Material	Loss of Support (LS)
Cement-treated granular base ($E = 1,000,000$ to $2,000,000$ lb/in.²)	0.0 to 1.0
Cement aggregate mixtures ($E = 500,000$ to $1,000,000$ lb/in.²)	0.0 to 1.0
Asphalt-treated base ($E = 350,000$ to $1,000,000$ lb/in.²)	0.0 to 1.0
Bituminous stabilized mixtures ($E = 40,000$ to $300,000$ lb/in.²)	0.0 to 1.0
Lime-stabilized mixtures ($E = 20,000$ to $70,000$ lb/in.²)	1.0 to 3.0
Unbound granular materials ($E = 15,000$ to $45,000$ lb/in.²)	1.0 to 3.0
Fine-grained or natural subgrade materials ($E = 3000$ to $40,000$ lb/in.²)	2.0 to 3.0

Note: *E* in this table refers to the general symbol for elastic or resilient modulus of the material.
SOURCE: Adapted from B. F. McCullough and Gary E. Elkins, *CRC Pavement Design Manual,* Austin Research Engineers, Inc., Austin, TX, October 1979. Design of Continuously Reinforced Concrete for Highways, Associated Reinforcing Bar Producers—CRSI, 180 N. Lassalle Street, Chicago, IL, 1981.

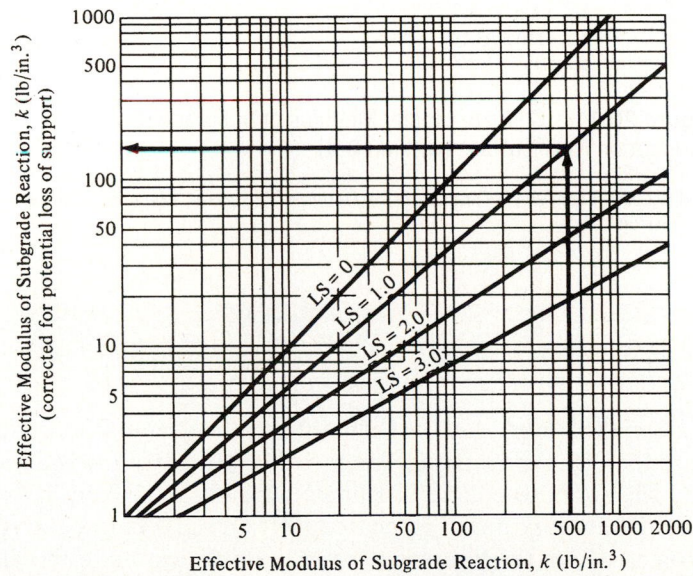

Figure 20.10   Correction of Effective Modulus of Subgrade Reaction for Potential Loss of Subbase Support

SOURCE: From *AASHTO Guide for Design of Pavement Structures*, 1993, AASHTO, Washington, D.C. Used by permission.

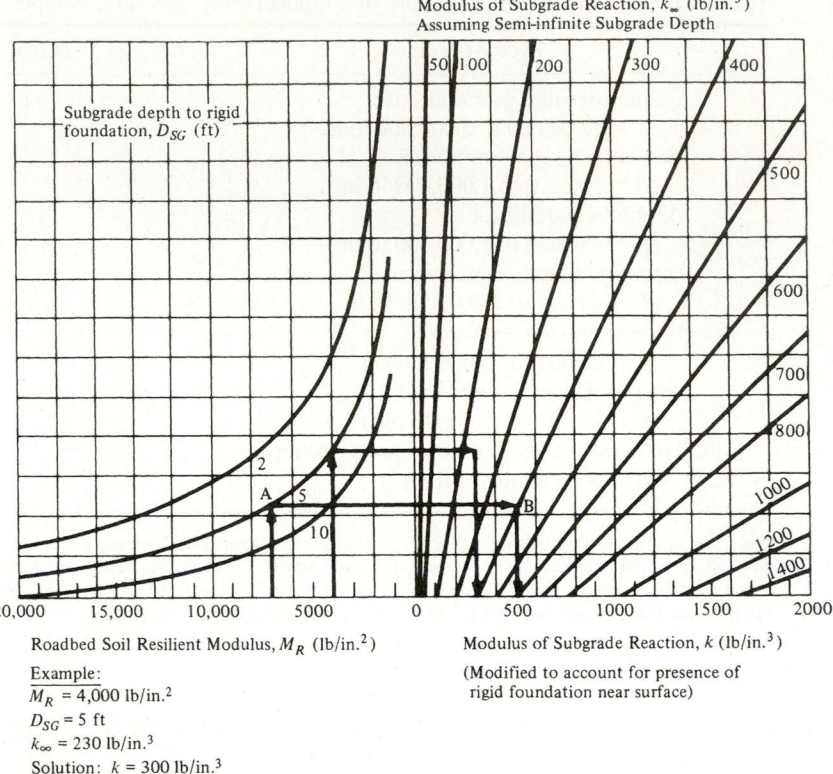

Modulus of Subgrade Reaction, k_∞ (lb/in.³)
Assuming Semi-infinite Subgrade Depth

Subgrade depth to rigid foundation, D_{SG} (ft)

Roadbed Soil Resilient Modulus, M_R (lb/in.²)

Modulus of Subgrade Reaction, k (lb/in.³)

Example:
$\overline{M_R}$ = 4,000 lb/in.²
D_{SG} = 5 ft
k_∞ = 230 lb/in.³
Solution: k = 300 lb/in.³

(Modified to account for presence of rigid foundation near surface)

Figure 20.11 Chart to Modify Modulus of Subgrade Reaction to Consider Effects of Rigid Foundation Near Surface (within 10 ft)

SOURCE: From *AASHTO Guide for Design of Pavement Structures*, 1993, AASHTO, Washington, D.C. Used by permission.

Example 20.2 Computing Effective Modulus of Subgrade Reaction for a Rigid Pavement Using the AASHTO *Guide for the Design of Pavement Structures*

A (6-in (152 mm)) layer of cement-treated granular material is to be used as a subbase for a rigid pavement. The monthly values for the roadbed soil resilient modulus and the subbase elastic (resilient) modulus are given in columns 2 and 3 of Table 20.6. If the rock depth is located (5 ft (1.5 m)) below the subgrade surface and the projected slab thickness is (9-in (228 mm)), estimate the effective modulus of subgrade reaction, using the AASHTO method.

Note that this is the example given in the 1993 AASHTO guide. Also note that the values for the modulus of the roadbed and subbase materials should be determined as discussed in Chapter 19, and the corresponding values shown in columns 2 and 3 of Table 20.6 should be for the same seasonal period.

Solution: Having determined the different moduli as given in columns 2 and 3 of Table 20.6, the next step is to estimate the composite subgrade modulus for each of the seasonal periods considered. At this stage, the effect of the existence of bedrock

Table 20.6   Data for and Solution to Example 20.2

(1) Month	(2) Roadbed Modulus M_R (lb/in.²)	(3) Subbase Modulus E_{SB} (lb/in.²)	(4) Composite k Value (lb/in.²) (Fig. 20.9)	(5) k Value (E_{SB}) on Rigid Foundation (Fig. 20.11)	(6) Relative Damage, u_r (Fig. 20.12)
January	20,000	50,000	1100	1350	0.35
February	20,000	50,000	1100	1350	0.35
March	2500	15,000	160	230	0.86
April	4000	15,000	230	300	0.78
May	4000	15,000	230	300	0.78
June	7000	20,000	400	500	0.60
July	7000	20,000	400	500	0.60
August	7000	20,000	400	500	0.60
September	7000	20,000	400	500	0.60
October	7000	20,000	400	500	0.60
November	4000	15,000	230	300	0.78
December	20,000	50,000	1100	1350	0.35
					$\Sigma u_r = 7.25$

Total Subbase
  Type: Granular
  Thickness (in.): 6
  Loss of Support, LS: 1.0
Depth to Rigid Foundation (ft): 5
Projected Slab Thickness (in.): 9

Average: $\bar{u}_r = \dfrac{\Sigma u_r}{n} = \dfrac{7.25}{12} = 0.60$

Effective modulus of subgrade reaction, k (lb/in.³) = 500
Corrected for loss of support: k (lb/in.³) = 170

within 10 ft from the subgrade surface is not considered, and it is assumed that the subgrade is of infinite depth. The composite subgrade k is then determined for each seasonal period, using the chart in Figure 20.9. The procedure involves the following steps:

Step 1.   Estimate k_∞. For example, in September, roadbed modulus $M_R = 7000$ lb/in.² and subbase modulus $E_{SB} = 20,000$ lb/in.².

  a) Enter the chart in Figure 20.9 at subbase thickness of 6 in. and draw a vertical line to intersect the subbase elastic modulus graph of 20,000 lb/in.² at A and the roadbed modulus graph of 7000 lb/in.² at B.
  b) Draw a horizontal line from B to intersect the turning line at C.
  c) Draw a vertical line upward from C to intersect the horizontal line drawn from A, as shown. This point of intersection D is the composite modulus of subgrade reaction.

  $k_\infty = 400$ lb/in.³. The computed k_∞ values are entered in column 4.

Step 2.   Adjust k_∞ for presence of rockbed within 10 ft of subgrade surface. This involves using the chart in Figure 20.11. This chart takes into account the depth below the subgrade surface at which the rockbed is located, the resilient modulus of the subgrade soil, and the composite modulus of subgrade reaction (k_∞) determined earlier. The adjustment is made as follows:

a) Enter Figure 20.11 at roadbed soil resilient modulus M_R of 7000 lb/in.² Draw a vertical line to intersect the graph corresponding to the depth of the rockbed below the subgrade surface, in this case, 5 ft. The intersection point is A.

b) From A, draw a horizontal line to intersect the appropriate k line. In this case, the intersection point is B.

c) From B, draw a vertical line downward to determine the adjusted modulus of subgrade reaction k. This is obtained as 500 lb/in.² Enter the k values in column 5.

Step 3. Determine the effective modulus of subgrade reaction by determining the average damage (u_r). The steps involved are similar to those for flexible pavements, but in this case, Figure 20.12 is used.

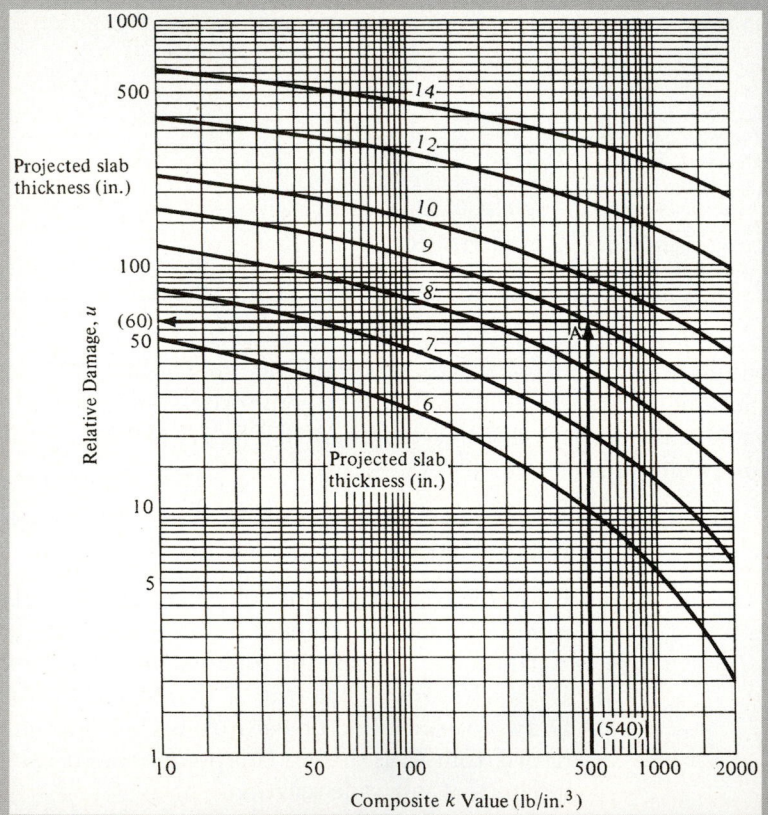

Figure 20.12   Chart for Estimating Relative Damage to Rigid Pavements Based on Slab Thickness and Underlying Support

a) Enter Figure 20.12 at the k value obtained in Step 2 (that is, 500 lb/in.^{2}) and project a vertical line to intersect the graph representing the projected slab thickness of 9 in. at point A.

b) From A, project a horizontal line to determine the appropriate u_r. In this case $u_r = 60$ percent or 0.60. These values are recorded in column 6 of Table 20.6.

c) Determine the mean u_r for all months by summing $u_r s$ for all months and dividing by the number of months as shown in Table 20.6; that is, $u_r = 0.6$.

d) Using the mean u_r of 0.6, obtain the effective modulus of subgrade reaction across all months as 500 lb/in.^{2} from Figure 20.12.

Step 4.   Adjust the effective modulus of subgrade reaction determined in Step 3 to account for the potential loss of subbase support due to erosion, using Figure 20.10. The LS factor given in Table 20.5 is used. Since the subbase consists of a cement-treated granular material, LS = 1 and $k = 500$ lb/in.^{2}. The corrected modulus of subgrade reaction is (46.1 MN/m^3).

Traffic.   The treatment of traffic load is similar to that presented for flexible pavements, in that the traffic load application is given in terms of the number of 18,000-lb (80-kN) equivalent single-axle loads (ESALs). ESAL factors depend on the slab thickness and the terminal serviceability index of the pavement. Tables 20.7 and 20.8 give ESAL factors for rigid pavements with a terminal serviceability index of 2.5. Since the ESAL factor depends on the thickness of the slab, it is necessary to assume the thickness of the slab at the start of the computation. This assumed value is used to compute the number of accumulated ESALs, which in turn is used to compute the required thickness. If the computed thickness is significantly different from the assumed thickness, the accumulated ESAL should be recomputed. This procedure should be repeated until the assumed and computed thicknesses are approximately the same.

Concrete Properties.   The concrete property is given in terms of its flexural strength (modulus of rupture) at 28 days. The flexural strength at 28 days of the concrete to be used in construction should be determined by conducting a three-point loading test as specified in AASHTO Designation T97.

Drainage.   The drainage quality of the pavement is considered by introducing a factor (C_d) into the performance equation. This factor depends on the quality of the drainage as described in Chapter 19 (see Table 19.5) and the percent of time the pavement structure is exposed to moisture levels approaching saturation. Table 20.9 gives AASHTO-recommended values for C_d.

Reliability.   Reliability considerations for rigid pavement are similar to those for flexible pavement as presented in Chapter 19. Reliability levels, $R\%$, and the overall standard deviation, S_o, are incorporated directly in the design charts.

Table 20.7 ESAL Factors for Rigid Pavements, Single Axles, and P_t of 2.5

Axle Load (kip)	Slab Thickness, D (in.)								
	6	7	8	9	10	11	12	13	14
2	.0002	.0002	.0002	.0002	.0002	.0002	.0002	.0002	.0002
4	.003	.002	.002	.002	.002	.002	.002	.002	.002
6	.012	.011	.010	.010	.010	.010	.010	.010	.010
8	.039	.035	.033	.032	.032	.032	.032	.032	.032
10	.097	.089	.084	.082	.081	.080	.080	.080	.080
12	.203	.189	.181	.176	.175	.174	.174	.173	.173
14	.376	.360	.347	.341	.338	.337	.336	.336	.336
16	.634	.623	.610	.604	.601	.599	.599	.599	.598
18	1.00	1.00	1.00	1.00	1.00	1.00	1.00	1.00	1.00
20	1.51	1.52	1.55	1.57	1.58	1.58	1.59	1.59	1.59
22	2.21	2.20	2.28	2.34	2.38	2.40	2.41	2.41	2.41
24	3.16	3.10	3.22	3.36	3.45	3.50	3.53	3.54	3.55
26	4.41	4.26	4.42	4.67	4.85	4.95	5.01	5.04	5.05
28	6.05	5.76	5.92	6.29	6.61	6.81	6.92	6.98	7.01
30	8.16	7.67	7.79	8.28	8.79	9.14	9.35	9.46	9.52
32	10.8	10.1	10.1	10.7	11.4	12.0	12.3	12.6	12.7
34	14.1	13.0	12.9	13.6	14.6	15.4	16.0	16.4	16.5
36	18.2	16.7	16.4	17.1	18.3	19.5	20.4	21.0	21.3
38	23.1	21.1	20.6	21.3	22.7	24.3	25.6	26.4	27.0
40	29.1	26.5	25.7	26.3	27.9	29.9	31.6	32.9	33.7
42	36.2	32.9	31.7	32.2	34.0	36.3	38.7	40.4	41.6
44	44.6	40.4	38.8	39.2	41.0	43.8	46.7	49.1	50.8
46	54.5	49.3	47.1	47.3	49.2	52.3	55.9	59.0	61.4
48	66.1	59.7	56.9	56.8	58.7	62.1	66.3	70.3	73.4
50	79.4	71.7	68.2	67.8	69.6	73.3	78.1	83.0	87.1

SOURCE: From *AASHTO Guide for Design of Pavement Structures*, 1993, AASHTO, Washington, D.C. Used by permission.

AASHTO Design Procedure

The objective of the design is to determine the thickness of the concrete pavement that is adequate to carry the projected design ESAL. The basic equation developed in the 1986 AASHTO design guide for the pavement thickness is given as

$$\log_{10} W_{18} = Z_R S_o + 7.35 \log_{10}(D + 1) - 0.06 + \frac{\log_{10}[\Delta PSI/(4.5 - 1.5)]}{1 + [(1.624 \times 10^7)/(D + 1)^{8.46}]}$$

$$+ (4.22 - 0.32 P_t) \log_{10}\left\{ \frac{S_c' C_d}{215.63 J}\left(\frac{D^{0.75} - 1.132}{D^{0.75} - [18.42/(E_c/k)^{25}]}\right)\right\} \tag{20.21}$$

where

Z_R = standard normal variant corresponding to the selected level of reliability
S_o = overall standard deviation (see Chapter 19)
W_{18} = predicted number of 18 kip ESAL applications that can be carried by the pavement structure after construction

Table 20.8   ESAL Factors for Rigid Pavements, Tandem Axles, and p_t of 2.5

Axle Load (kip)	Slab Thickness, D (in.)								
	6	7	8	9	10	11	12	13	14
2	.0001	.0001	.0001	.0001	.0001	.0001	.0001	.0001	.0001
4	.0006	.0006	.0005	.0005	.0005	.0005	.0005	.0005	.0005
6	.002	.002	.002	.002	.002	.002	.002	.002	.002
8	.007	.006	.006	.005	.005	.005	.005	.005	.005
10	.015	.014	.013	.013	.012	.012	.012	.012	.012
12	.031	.028	.026	.026	.025	.025	.025	.025	.025
14	.057	.052	.049	.048	.047	.047	.047	.047	.047
16	.097	.089	.084	.082	.081	.081	.080	.080	.080
18	.155	.143	.136	.133	.132	.131	.131	.131	.131
20	.234	.220	.211	.206	.204	.203	.203	.203	.203
22	.340	.325	.313	.308	.305	.304	.303	.303	.303
24	.475	.462	.450	.444	.441	.440	.439	.439	.439
26	.644	.637	.627	.622	.620	.619	.618	.618	.618
28	.855	.854	.852	.850	.850	.850	.849	.849	.849
30	1.11	1.12	1.13	1.14	1.14	1.14	1.14	1.14	1.14
32	1.43	1.44	1.47	1.49	1.50	1.51	1.51	1.51	1.51
34	1.82	1.82	1.87	1.92	1.95	1.96	1.97	1.97	1.97
36	2.29	2.27	2.35	2.43	2.48	2.51	2.52	2.52	2.53
38	2.85	2.80	2.91	3.03	3.12	3.16	3.18	3.20	3.20
40	3.52	3.42	3.55	3.74	3.87	3.94	3.98	4.00	4.01
42	4.32	4.16	4.30	4.55	4.74	4.86	4.91	4.95	4.96
44	5.26	5.01	5.16	5.48	5.75	5.92	6.01	6.06	6.09
46	6.36	6.01	6.14	6.53	6.90	7.14	7.28	7.36	7.40
48	7.64	7.16	7.27	7.73	8.21	8.55	8.75	8.86	8.92
50	9.11	8.50	8.55	9.07	9.68	10.14	10.42	10.58	10.66
52	10.8	10.0	10.0	10.6	11.3	11.9	12.3	12.5	12.7
54	12.8	11.8	11.7	12.3	13.2	13.9	14.5	14.8	14.9
56	15.0	13.8	13.6	14.2	15.2	16.2	16.8	17.3	17.5
58	17.5	16.0	15.7	16.3	17.5	18.6	19.5	20.1	20.4
60	20.3	18.5	18.1	18.7	20.0	21.4	22.5	23.2	23.6
62	23.5	21.4	20.8	21.4	22.8	24.4	25.7	26.7	27.3
64	27.0	24.6	23.8	24.4	25.8	27.7	29.3	30.5	31.3
66	31.0	28.1	27.1	27.6	29.2	31.3	33.2	34.7	35.7
68	35.4	32.1	30.9	31.3	32.9	35.2	37.5	39.3	40.5
70	40.3	36.5	35.0	35.3	37.0	39.5	42.1	44.3	45.9
72	45.7	41.4	39.6	39.8	41.5	44.2	47.2	49.8	51.7
74	51.7	46.7	44.6	44.7	46.4	49.3	52.7	55.7	58.0
76	58.3	52.6	50.2	50.1	51.8	54.9	58.6	62.1	64.8
78	65.5	59.1	56.3	56.1	57.7	60.9	65.0	69.0	72.3
80	73.4	66.2	62.9	62.5	64.2	67.5	71.9	76.4	80.2
82	82.0	73.9	70.2	69.6	71.2	74.7	79.4	84.4	88.8
84	91.4	82.4	78.1	77.3	78.9	82.4	87.4	93.0	98.1
86	102.0	92.0	87.0	86.0	87.0	91.0	96.0	102.0	108.0
88	113.0	102.0	96.0	95.0	96.0	100.0	105.0	112.0	119.0
90	125.0	112.0	106.0	105.0	106.0	110.0	115.0	123.0	130.0

SOURCE: From *AASHTO Guide for Design of Pavement Structures*, 1993, AASHTO, Washington, D.C. Used by permission.

Table 20.9 Recommended Values for Drainage Coefficient, C_d, for Rigid Pavements

Quality of Drainage	*Percent of Time Pavement Structure is Exposed to Moisture Levels Approaching Saturation*			
	Less Than 1%	*1–5%*	*5–25%*	*Greater Than 25%*
Excellent	1.25–1.20	1.20–1.15	1.15–1.10	1.10
Good	1.20–1.15	1.15–1.10	1.10–1.00	1.00
Fair	1.15–1.10	1.10–1.00	1.00–0.90	0.90
Poor	1.10–1.00	1.00–0.90	0.90–0.80	0.80
Very poor	1.00–0.90	0.90–0.80	0.80–0.70	0.70

SOURCE: From *AASHTO Guide for Design of Pavement Structures*, 1993, AASHTO, Washington, D.C. Used by permission.

D = thickness of concrete pavement to the nearest half-inch

ΔPSI = design serviceability loss = $p_i - p_t$

p_i = initial serviceability index

p_t = terminal serviceability index

E_c = elastic modulus of the concrete to be used in construction (lb/in.^{2})

S'_c = modulus of rupture of the concrete to be used in construction (lb/in.^{2})

J = load transfer coefficient = 3.2 (assumed)

C_d = drainage coefficient

Equation 20.21 can be solved for the thickness of the pavement (D) in inches by using either a computer program or the two charts in Figures 20.13 and 20.14. The use of a computer program facilitates the iteration necessary, since D has to be assumed to determine the effective modulus of subgrade reaction and the ESAL factors used in the design.

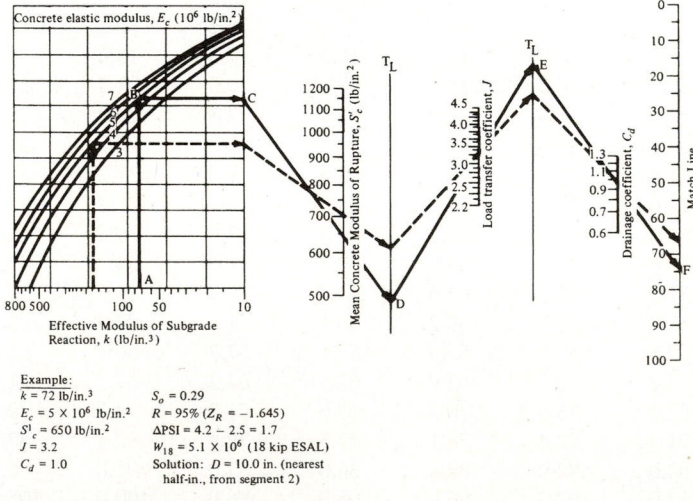

Figure 20.13 Design Chart for Rigid Pavements Based on Using Values for Each Input Variable (Segment 1)

SOURCE: From *AASHTO Guide for Design of Pavement Structures*, 1993, AASHTO, Washington, D.C. Used by permission.

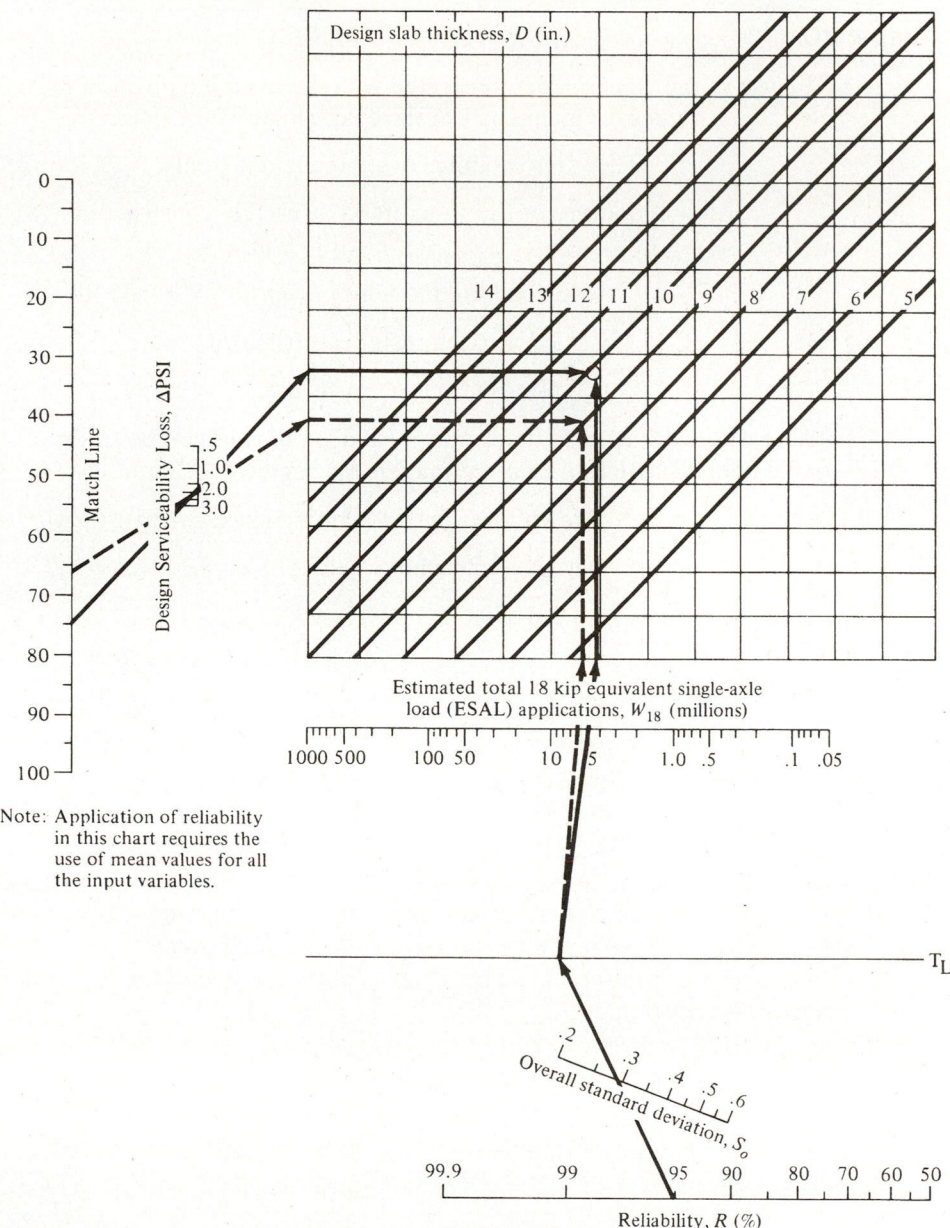

Figure 20.14   Design Chart for Rigid Pavements Based on Using Mean Values for Each Input Variable (Segment 2)

SOURCE: From *AASHTO Guide for Design of Pavement Structures*, 1993, AASHTO, Washington, D.C. Used by permission.

Example 20.3 Designing a Rigid Pavement Using the AASHTO Method

The use of the charts is demonstrated with the example given in Figure 20.13. In this case, input values for Segment 1 of the chart (Figure 20.13) are

Effective modulus of subgrade reaction, $k = 72$ lb/in.^{3}

Elastic modulus of the concrete to be used in construction (lb/in.^{2}) (E_c)
$$= 5 \times 10^6 \text{ lb/in.}^2$$

Mean concrete modulus of rupture, $S'_c = 650$ lb/in.^{2}

Load transfer coefficient, $J = 3.2$

Drainage coefficient, $C_d = 1.0$

These values are used to determine a value on the match line as shown in Figure 20.13 (solid line ABCDEF). Input parameters for Segment 2 (Figure 20.14) of the chart are

Match line value determined in segment 1 (74)

Design serviceability loss, $\Delta \text{PSI} = 4.2 - 2.5 = 1.7$

Reliability, $R\% = 95\%$ ($Z_R = 1.645$)

Overall standard deviation, $S_o = 0.29$

Cumulative 18 kip ESAL $= (5 \times 10^6)$

Solution: The required thickness of the concrete slab is then obtained, as shown in Figure 20.14, as 10 in (254 mm) (nearest half-inch).

Note that when the thickness obtained from solving Eq. 20.21 analytically or by use of Figures 20.13 and 20.14 is significantly different from that originally assumed to determine the effective subgrade modulus and to select the ESAL factors, the whole procedure has to be repeated until the assumed and designed values are approximately the same. This reinforces the importance of using a computer program to facilitate the necessary iteration.

Example 20.4 Evaluating the Adequacy of a Rigid Pavement Using the AASHTO Method

Using the data and effective subgrade modulus obtained in Example 20.2, determine whether the 9-in. pavement design of Example 20.2 will be adequate on a rural expressway for a 20-year analysis period and the following design criteria:

$$P_i = 4.5$$

$$P_t = 2.5$$

ESAL on design lane during first year of operation $= 0.2 \times 10^6$

Traffic growth rate $= 4\%$

Concrete elastic modulus, $E_c = 5 \times 10^6$ lb/in.^{2}

Mean concrete modulus of rupture = 700 lb/in.^{2}

Drainage conditions are such that $C_d = 1.0$

$R = 0.95$ ($Z_R = 1.645$)

Load transfer coefficient (J) = 3.2

$S_o = 0.30$ (for rigid pavements $S_o = 0.3 - 0.4$)

Growth factor = 29.78 (from Table 19.6)

$k = 170$ (from Example 20.2)

Assume $D = 9$ in. (from Example 20.2)

ESAL over design period = $0.2 \times 10^6 = 29.78 = 6 \times 10^6$

Solution: The depth of concrete required is obtained from Figures 20.13 and 20.14. The dashed lines represent the solution, and a depth of 9 in. is obtained. The pavement is therefore adequate.

20.6.2 Alternate AASHTO Design Method as Given in Supplement to AASHTO *Guide for Design of Pavement Structures, Part II, Rigid Pavement Design and Rigid Pavement Joint Design*

This section describes an alternate design procedure for rigid pavements, based on the recommendations given in NCHRP Project 1-30. These recommendations were derived from the results of a study that used the Long Term Pavement Performance (LTPP) database. The procedure is available for those who are interested in using the LTPP data in the design of highway rigid pavements. It requires a "hypothetical" joint spacing as one of the required inputs, in addition to those for the AASHTO design procedure described above.

The procedure requires the following inputs:

   (i) Estimated ESALs
  (ii) Design reliability, $R\%$
 (iii) Overall standard deviation S_o
  (iv) Design serviceability loss, $\Delta \text{PSI} = p_i - p_t$
   (v) Effective (seasonally adjusted) elastic k value of the subgrade (lb/in.^{2})
  (vi) Concrete modulus of rupture, S'_c (lb/in.^{2})
 (vii) Concrete elastic modulus, E_c (lb/in.^{2})
(viii) Joint spacing, L (in.)
  (ix) Base modulus, E_b (lb/in.^{2})
   (x) Slab/base friction coefficient, f
  (xi) Base thickness, H_b (in.)
 (xii) Effective positive temperature differential through concrete slab TD (°F)
(xiii) Lane edge support condition

   a. Conventional lane width, 12 ft with free edge
   b. Conventional lane of 12 ft width with tied concrete shoulder
   c. Wide slab; e.g., 14 ft with conventional traffic lane width of 12 ft

Input parameters (i), (ii), (iii), (iv), (vi), (vii), (ix), and (x) can be selected or computed using the procedures discussed in Section 20.6.2. The remaining input parameters are discussed in the following paragraphs.

Effective (seasonally adjusted) elastic k value of the subgrade.   In this procedure, the input k value is obtained by executing the following steps:

(1) Select a subgrade k value for each season.
(2) Determine a seasonally adjusted effective k value.
(3) Adjust the seasonally effective k to account for the presence of shallow rigid layer and/or an embankment above the natural grade.

Select a subgrade k value for each season.   The subgrade k value for each season can be obtained by one of three methods: (i) correlation of soil type and other properties or tests, (ii) deflection testing and back calculation, and (iii) plate-bearing tests. Only the correlation procedure is presented in this section. Readers interested in the other two methods may refer to the *Supplement to the Design of Pavement Structures, Part II, Rigid Pavement Design and Rigid Pavement Joint Design.* A season is a portion of the year when it can be assumed that the degree of saturation of the subgrade is approximately the same. It is often taken as three months, covering the fall, winter, spring, and summer seasons. However, this depends on the weather conditions at the location of the pavement. The length of the season and the number of seasons may be different for different locations. Figure 20.15 gives recommended k values for fine-grained soils for different degrees of saturation, and Tables 20.10 and 20.11 give recommended k value ranges for coarse- and fine-grained soils, respectively.

Determine a seasonally adjusted effective k value.   The effective seasonally adjusted k value reflects the k value that will result in the same pavement fatigue damage as that caused by the varying k values over the year. It is obtained by executing the following substeps:

  (i) Select a tentative value for the slab thickness (D) and the other input parameters listed above.
 (ii) Select a k value for each season of the year identified use Figure 20.15, Table 20.10 or 20.11.
(iii) Determine the number of allowable ESALs (W_{18}) for each seasonal k value selected in Step (ii), using the procedure discussed later under Design Equations for Rigid Pavements Using the Alternative Method.
 (iv) Determine the relative damage for each season as the inverse of the appropriate number of ESALs (W_{18}) determined in Step (iii).
  (v) Determine the mean annual average damage by dividing the sum of the individual seasonal damages by the number of seasons.
 (vi) Determine the corresponding number of ESALs (W_{18}) as the inverse of the mean annual average damage determined in Step (vi).
(vii) Determine the seasonally adjusted k value that will result in the number of ESALs computed in Step (vi) using the procedure discussed later under Design Equations for Rigid Pavements Using the Alternative Method.

The procedure can be executed by using the format shown in Table 20.12.
        Adjust the seasonally effective k value to account for the presence of a shallow rigid layer and/or an embankment above the natural grade.

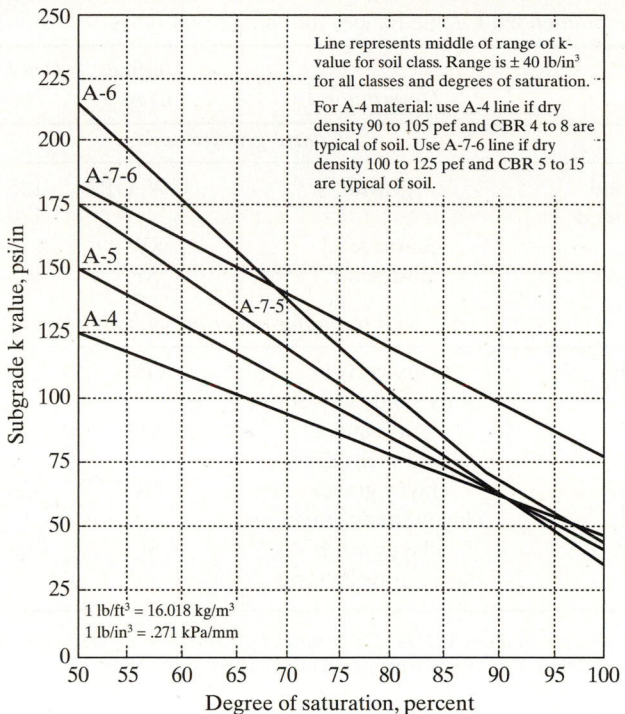

Figure 20.15   The *k* Value versus Degree of Saturation for Cohesive Soils

SOURCE: From Supplement to the *AASHTO Guide for Design of Pavement Structures, Part II, Rigid Pavement Design & Rigid Pavement Joint Design*, 1998, AASHTO, Washington, D.C. Used by permission.

The effective *k* value obtained from the procedure described above is adjusted for the presence of a fill material above the subgrade and/or a rigid layer such as bedrock or hardpan clay located at a depth of 10 ft or less below the surface of the existing subgrade. The monograph shown in Figure 20.16 is used to make this adjustment. Note that the adjustment for the presence of a rigid layer is made only when soil type or similar correlations were used to determine the *k* value of the subgrade as described above. This adjustment is not required when nondestructive deflection testing or plate-bearing tests were used to determine the *k* value of the subgrade, as the effect of the presence of a rigid layer less than 10 ft below the surface of the subgrade is already reflected in the test results.

Joint Spacing

The suggested hypothetical joint spacing for each rigid pavement type is given as:

Jointed plain concrete pavement (JPCP): Use actual joint spacing (ft).
Jointed reinforced concrete pavement (JRCP): Use actual joint spacing when it is less than 30 ft; otherwise, use 30 ft maximum.
Continuously reinforced concrete pavement: Use 15 ft.

Table 20.10   Recommended *k* Value Ranges for Various Soil Types

AASHTO Class	Description	Unified Class	Dry Density (lb/ft³)	CBR (percent)	k value (lb/in.³)
		Coarse-grained soils			
A-1-a, well graded	gravel	GW, GP	125–140	60–80	300–450
A-1-a, poorly graded			120–130	35–60	300–400
A-1-b	coarse sand	SW	110–130	20–40	200–400
A-3	fine sand	SP	105–120	15–25	150–300
		A-2 soils (granular materials with high fines)			
A-2-4, gravelly	silty gravel	GM	130–145	40–80	300–500
A-2-5, gravelly	silty sandy gravel				
A-2-4, sandy	silty sand	SM	120–135	20–40	300–400
A-2-5, sandy	silty gravelly sand				
A-2-6, gravelly	clayey gravel	GC	120–140	20–40	200–450
A-2-7, gravelly	clayey sandy gravel				
A-2-6, sandy	clayey sand	SC	105–130	10–20	150–350
A-2-7, sandy	clayey gravelly sand				

SOURCE: From Supplement to the *AASHTO Guide for Design of Pavement Structures, Part II, Rigid Pavement Design & Rigid Pavement Joint Design*, 1998, AASHTO, Washington, D.C. Used by permission.

Note that these hypothetical joint spacings are used only to obtain the slab design thickness. The methodology includes design considerations for different types of joints and a procedure to check the adequacy of the joint load transfer proposed.

Table 20.11   Recommended *k* values for Fine-Grained Soils

AASHTO class	Description	Unified Class	Dry Density (lb/ft³)	CBR (percent)	k value (lb/in³.)
A-4	silt	ML, OL	90–105	4–8	25–165*
	silt/sand/gravel mixture		100–125	5–15	40–220*
A-5	poorly graded silt	MH	80–100	4–8	25–190*
A-6	plastic clay	CL	100–125	5–15	25–255*
A-7-5	moderately plastic elastic clay	CL, OL	90–125	4–15	25–215*
A-7-6	highly plastic elastic clay	CH, OH	80–110	3–5	40–220*

*k value of fine-grained soil is highly dependent on degree of saturation. See Figure 20.15.
These recommended *k* value ranges apply to a homogeneous soil layer at least 10 ft (3 m) thick. If an embankment layer less than 10 ft (3 m) thick exists over a softer subgrade, the *k* value for the underlying soil should be estimated from this table and adjusted for the type and thickness of embankment material using Figure 20.16. If a layer of bedrock exists within 10 ft (3 m) of the top of the soil, the *k* should be adjusted using Figure 20.16.
1 lb/ft³ = 16.018 kg/m³, 1 psi/in. = 0.271 kPa/mm
SOURCE: From Supplement to the *AASHTO Guide for Design of Pavement Structures, Part II, Rigid Pavement Design & Rigid Pavement Joint Design*, 1998, AASHTO, Washington, D.C. Used by permission.

Table 20.12   Determination of Seasonally Adjusted Effective Subgrade k Value

Season	k Value (lb/in.²)	W_{18} (millions)	Relative Damage ($1/W_{18}$)
Spring			
Summer			
Fall			
Winter			
		Mean damage	
		W_{18}	
		Effective k value	

SOURCE: Supplement to the *AASHTO Guide for Design of Pavement Structures, Part II, Rigid Pavement Design and Rigid Pavement Joint Design,* American Association of State Highway and Transportation Officials, Washington, D.C., 1998. Used with permission.

Base Modulus, E_b and Slab/Base Friction Coefficient, f

The procedure also requires the selection of an appropriate elastic modulus and friction coefficient for the type of base used for the pavement. Table 20.13 gives suggested range of values for these input parameters.

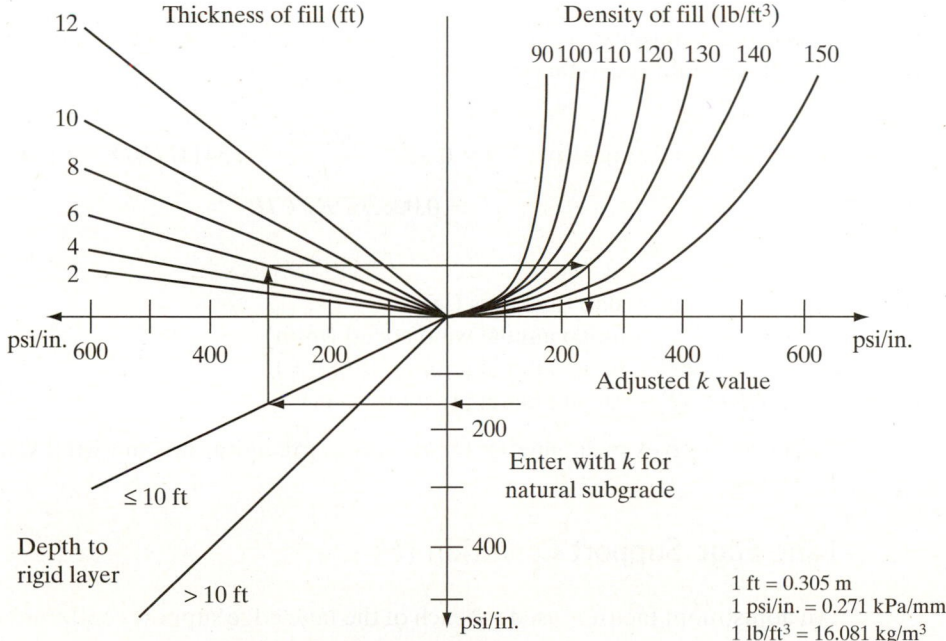

Figure 20.16   Adjustment to k for Fill and/or Rigid Layer

SOURCE: From Supplement to the *AASHTO Guide for Design of Pavement Structures, Part II, Rigid Pavement Design & Rigid Pavement Joint Design*, 1998, AASHTO, Washington, D.C. Used by permission.

Table 20.13   Modulus of Elasticity and Coefficient of Friction for Various Base Types

Base Type or Interface Treatment	Modulus of Elasticity (lb/in.²)	Peak Friction Coefficient		
		Low	Mean	High
Fine-grained soil	3,000–40,000	0.5	1.3	2.0
Sand	10,000–25,000	0.5	0.8	1.0
Aggregate	15,000–45,000	0.7	1.4	2.0
Polyethylene sheeting	NA	0.5	0.6	1.0
Lime-stabilized base	20,000–70,000	3.0	NA	5.3
Cement-treated gravel	$(500 + CS) \times 1000$	8.0	34	63
Asphalt-treated gravel	$300,000 - 600,000$	3.7	5.8	10
Lean concrete without curing compound	$(500 + CS) \times 1000$		> 36	
Lean concrete with single or double wax curing compound	$(500 + CS) \times 1000$	3.5		4.5

Notes: CS = compressive strength (lb/in.²)
Low, mean, and high measured peak coefficients of friction summarized from various references.
NA = Not available
SOURCE: From Supplement to the *AASHTO Guide for Design of Pavement Structures, Part II, Rigid Pavement Design & Rigid Pavement Joint Design*, 1998, AASHTO, Washington, D.C. Used by permission.

Effective Positive Temperature Differential through Concrete Slab, (°F)

This is the difference between the temperatures at the top of the slab and that at the bottom of the slab. It can be computed as:

$$\text{effective positive TD} = 0.962 - \frac{52.181}{h} + 0.341WIND + 0.184TEMP$$
$$- 0.00836PRECIP \qquad (20.22)$$

where

$$h = \text{slab thickness (in.)}$$
$$WIND = \text{mean annual wind speed (mi/h)}$$
$$TEMP = \text{mean annual temperature (°F)}$$
$$PRECIP = \text{mean annual precipitation (in.)}$$

Table 20.14 gives mean annual temperature, precipitation, and wind speed for selected U.S cities.

Lane Edge Support Condition (*E*)

An adjustment factor is used for each of the lane edge support conditions as given below:

E = 1.0 for conventional traffic lane width of 12 ft with free edge

= 0.94 for conventional traffic lane width of 12 ft with tied concrete shoulder

= 0.92 for wide slab (e.g., 14 ft with conventional traffic lane width of 12 ft)

Table 20.14   Mean Annual Temperature, Precipitation, and Wind Speed for Selected U.S. Cities

Location	Mean Annual Temperature, °F	Mean Annual Precipitation, in.	Mean Annual Wind Speed, mi/h	Location	Mean Annual Temperature, °F	Mean Annual Precipitation, in.	Mean Annual Wind Speed, mi/h	Location	Mean Annual Temperature, °F	Mean Annual Precipitation, in.	Mean Annual Wind Speed, mi/h
ALABAMA				KANSAS				OKLAHOMA			
Birmingham	62.2	52.2	7.2	Topeka	54.1	28.6	10.1	Oklahoma City	59.9	30.9	12.5
Mobile	67.5	64.6	9.0	Wichita	56.4	40.1	12.3	Tulsa	60.3	38.3	10.4
Montgomery	67.5	49.2	6.7	KENTUCKY				OREGON			
ALASKA				Lexington	54.9	45.7	7.1	Medford	53.6	19.3	4.8
Anchorage	35.3	15.2	6.9	Louisville	56.2	43.6	8.3	Portland	53.0	37.4	7.9
Fairbanks	25.9	10.4	5.5	LOUISIANA				Salem	52.0	40.4	7.0
King Salmon	32.8	19.3	10.8	Baton Rouge	67.5	55.8	7.7	PENNSYLVANIA			
ARIZONA				Lake Charles	68.0	53.0	8.6	Harrisburg	53.0	39.1	7.6
Flagstaff	45.4	20.9	7.1	New Orleans	68.2	59.7	8.2	Philadelphia	54.3	41.4	9.5
Phoenix	71.2	7.1	6.3	Shreveport	65.4	43.8	8.5	Pittsburgh	50.3	36.3	9.1
Tucson	68.0	11.1	8.2	MAINE				RHODE ISLAND			
ARKANSAS				Caribou	38.9	36.6	11.2	Providence	50.3	45.3	10.6
Little Rock	61.9	49.2	7.9	Portland	45.0	43.8	8.7	SOUTH CAROLINA			
CALIFORNIA				MARYLAND				Charleston	64.3	51.6	8.7
Bakersfield	65.6	5.7	6.4	Baltimore	55.1	41.8	9.2	Columbia	63.3	49.1	6.9
Fresno	62.5	10.5	6.4	MASSACHUSETTS				SOUTH DAKOTA			
Los Angeles	62.6	12.1	7.5	Boston	51.5	43.8	12.4	Huron	44.7	18.7	11.6
Sacramento	60.6	17.1	8.1	Worcester	46.8	47.6	12.4	Rapid City	46.7	16.3	11.3
San Diego	63.8	9.3	6.9	MICHIGAN				TENNESSEE			
San Francisco	56.6	19.7	10.5	Detroit	48.6	4.0	10.2	Chattanooga	59.4	52.6	6.1
Santa Barbara	58.9	16.2	6.1	Flint	46.8	29.2	10.6	Knoxville	58.9	47.3	7.1
COLORADO				Grand Rapids	47.5	34.4	9.7	Memphis	61.8	51.6	9.0
Colorado Springs	48.9	15.4	10.1	MINNESOTA				Nashville	59.2	48.5	8.0
Denver	50.3	15.3	8.8	Duluth	38.2	29.7	11.2	TEXAS			
CONNECTICUT				Minneapolis	44.7	26.4	10.6	Amarillo	57.2	19.1	13.6
Hartford	49.3	44.4	9.2	MISSISSIPPI				Brownsville	73.6	25.4	11.6
DC				Jackson	64.6	52.8	7.4	Corpus Christi	72.1	30.2	12.0
Washington	57.5	39.0	9.3	MISSOURI				Dallas	66.0	29.5	10.8
DELAWARE				Kansas City	56.3	35.2	10.7	El Paso	63.4	7.8	9.0
Wilmington	54.0	41.4	9.2	MONTANA				Galveston	69.6	40.2	11.0
FLORIDA				Great Falls	44.7	15.2	12.3	Houston	68.3	44.8	7.8
Jacksonville	68.0	52.8	8.1	NEBRASKA				Lubbock	59.9	17.8	12.4
Miami	75.6	57.6	9.2	Omaha	49.5	29.9	10.6	Midland	63.5	13.7	11.1
Orlando	72.4	47.8	8.6	NEVADA				San Antonio	68.7	29.2	9.4
Tallahassee	67.2	64.6	6.4	Las Vegas	66.3	4.2	9.2	Waco	67.0	31.0	11.3
Tampa	72.0	46.7	8.5	Reno	49.4	7.5	6.5	Wichita Falls	63.5	26.7	11.7
West Palm Beach	74.6	59.7	9.4	NEW JERSEY				UTAH			
GEORGIA				Atlantic City	53.1	41.9	10.1	Salt Lake City	51.7	15.3	8.8
Atlanta	61.2	48.6	9.1	NEW MEXICO				VERMONT			
Augusta	63.2	43.1	6.5	Albuquerque	56.2	8.1	9.0	Burlington	44.1	33.7	8.8
Macon	64.7	44.9	7.7	NEW YORK				VIRGINIA			
Savannah	65.9	49.7	7.9	Albany	47.3	35.7	8.9	Norfolk	59.5	45.2	10.6
HAWAII				Buffalo	47.6	37.5	12.1	Richmond	57.7	44.1	7.6
Hilo	73.6	128.2	7.1	New York City	54.5	44.1	12.1	Roanoke	56.1	39.2	8.2
Honolulu	77.0	23.5	11.5	Rochester	47.9	31.3	9.7	WASHINGTON			
IDAHO				Syracuse	47.7	39.1	9.7	Olympia	49.6	51.0	6.7
Boise	51.1	11.7	8.8	NORTH CAROLINA				Seattle	52.7	38.8	9.0
Pocatello	46.6	10.9	10.2	Charlotte	60.0	43.2	7.5	Spokane	47.2	16.7	8.8

(Continued)

Table 20.14 Mean Annual Temperature, Precipitation, and Wind Speed for Selected U.S. Cities (*continued*)

Location	Mean Annual Temperature, °F	Mean Annual Precipitation, in.	Mean Annual Wind Speed, mi/h	Location	Mean Annual Temperature, °F	Mean Annual Precipitation, in.	Mean Annual Wind Speed, mi/h	Location	Mean Annual Temperature, °F	Mean Annual Precipitation, in.	Mean Annual Wind Speed, mi/h
ILLINOIS				**NORTH CAROLINA**				**WEST VIRGINIA**			
Chicago	49.2	33.3	10.2	Greensboro	57.9	42.5	7.5	Charleston	54.8	42.4	6.4
Peoria	50.4	34.9	10.1	Raleigh	59.0	41.8	7.8	Huntington	55.2	40.7	6.5
Springfield	52.6	33.8	11.3	Wilmington	63.4	53.4	8.8	**WISCONSIN**			
INDIANA				**NORTH DAKOTA**				Green Bay	43.6	28.0	10.1
Evansville	55.7	41.6	8.2	Bismarck	41.3	15.4	10.3	Madison	45.2	30.8	9.8
Fort Wayne	49.7	34.4	10.1	Fargo	40.5	19.6	12.4	Milwaukee	46.1	30.9	11.6
Indianapolis	52.1	39.1	9.6	**OHIO**				**WYOMING**			
South Bend	49.4	38.2	10.4	Akron-Canton	49.5	35.9	9.8	Casper	45.2	11.4	13.0
IOWA				Cleveland	49.6	35.4	10.7	Cheyenne	45.7	13.3	12.9
Des Moines	49.7	30.8	10.9	Columbus	51.7	37.0	8.7				
Sioux City	48.4	25.4	11.0	Dayton	51.9	34.7	10.1				
Waterloo	46.1	33.1	10.7	Youngstown	48.3	37.3	10.0				

°C = (°F − 32)/1.8, 1 in. = 25.4 mm, 1 mi/h = 1.61 km/h
SOURCE: National Climatic Data Center, 1986

SOURCE: From Supplement to the *AASHTO Guide for Design of Pavement Structures, Part II, Rigid Pavement Design & Rigid Pavement Joint Design*, 1998, AASHTO, Washington, D.C. Used by permission.

Design Equations for Rigid Pavement Using the Alternate Method

The design equations are used to determine a slab thickness for the midslab loading position shown in Figure 20.17. When dowels are used at the transverse joints, the thickness obtained from the set of equations given below is the design thickness. When there are no dowels at the transverse joints, then a check is necessary to ascertain that a joint loading position does not result in a higher critical stress at the top of the slab. The rigid pavement design equation for the alternate method is given as:

$$\log W' = \log_{10} W + (5.065 - 0.03295 P_t^{2.4}) \left[\log_{10}\left(\frac{(S_c')'}{\sigma_t'} \right) - \log_{10}\left(\frac{690}{\sigma_t} \right) \right] \quad (20.23)$$

where

W' = number of 18-kip ESALs estimated for the design traffic lane
W = number of 18-kip ESALs computed from Eq. 20.24

$$\log_{10} W = \log_{10} R + \frac{G}{Y} \quad (20.24)$$

where

$$\log_{10} R = 5.85 + 7.35 \log_{10}(h + 1) - 4.62 \log_{10}(L1 + L2) + 3.28 \log L2 \quad (20.25)$$

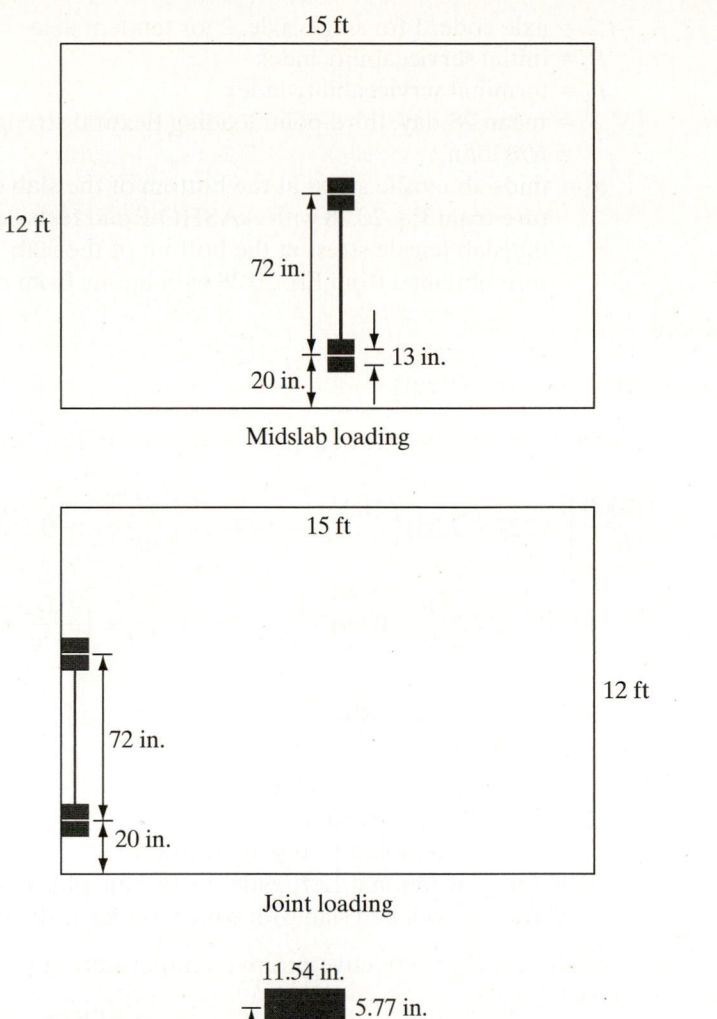

Figure 20.17   Midslab and Joint Loading Positions Defined

SOURCE: From Supplement to the *AASHTO Guide for Design of Pavement Structures, Part II, Rigid Pavement Design & Rigid Pavement Joint Design*, 1998, AASHTO, Washington, D.C. Used by permission.

$$Y = 1.00 + \frac{3.63(L1 + L2)^{5.2}}{(h + 1)^{8.46}(L2)^{3.52}} \tag{20.26}$$

$$G = \log_{10}\left(\frac{P_i - P_t}{P_i - 1.5}\right) \tag{20.27}$$

where

h = concrete slab thickness (in.)
$L1$ = load on single or tandem axle (kips)

$L2$ = axle code, 1 for single axle, 2 for tandem axle

P_i = initial serviceability index

P_t = terminal serviceability index

$(S'_c)'$ = mean 28-day, third-point loading flexural strength (lb/in.²)

= 690 lb/in.²

σ_t = midslab tensile stress at the bottom of the slab due to load and temperature from Eq. 20.28 with AASHO Road tests constants

σ'_t = midslab tensile stress at the bottom of the slab due to load and temperature obtained from Eq. 20.28 with inputs from new pavement design

where

$$\sigma_t = \sigma_l EF[1.0 + 10^{(\log b)}TD] \qquad (20.28)$$

σ_l = midslab tensile stress due to load only as given in Eq. 20.29

$$\sigma_l = \frac{18,000}{h^2}\left[4.227 - 2.381\left(\frac{180}{\ell}\right)^{0.2} - 0.0015\left[\frac{E_b H_b}{1.4k}\right]^{0.5} - 0.155\left[H_b\left(\frac{E_b}{E_c}\right)^{0.75}\right]^{0.5}\right] \quad (20.29)$$

$$\log b = -1.944 + 2.279\frac{h}{\ell} + 0.0917\frac{L}{\ell} - 433,080\frac{h^2}{kl^4} + \left(\frac{0.0614}{\ell}\right) \times \left(\frac{E_b H_b^{1.5}}{1.4k}\right)^{0.5}$$

$$- 438.642\frac{h^2}{kl^2} - 498,240\frac{h^3 L}{kl^6} \qquad (20.30)$$

E = edge support adjustment factor

= 1.00 for AASHO Road test

= 1.00 for conventional 12-ft-wide traffic lane

= 0.94 for conventional 12-ft-wide traffic lane plus tied concrete shoulder

= 0.92 for 2-ft widened slab with conventional 12-ft-wide lane width.

Effective positive TD_p = effective positive temperature differential, top of slab minus bottom of slab (°F)

$$TD_p = 0.962 - \frac{52.181}{h} + 0.341WIND + 0.184TEMP - 0.00836PRECIP \text{ see Eq. 20.22}$$

k = effective (seasonally adjusted) elastic value of the subgrade (lb/in.³) (110 lbs/in² for AASHO Road Test)

S'_c = concrete modulus of rupture (lb/in.²)

E_c = concrete elastic modulus (42,000 lb/in.² for AASHO Road Test)

L = joint spacing (in.)

E_b = base modulus (25,000 lb/in.²)

F = ratio between slab stress at a given coefficient of friction (f) between the slab and base, and slab stress at full friction (friction adjustment factor)

$$F = 1.177 - 4.3 \times 10^{-8} hE_b - 0.01155542h + 6.27 \times 10^{-7} E_b - 0.000315f \quad (20.31)$$

H_b = base thickness (in.)

E = edge support adjustment factor

f = friction coefficient between slab and base (see Table 20.12)

h = slab thickness (in.)

ℓ = radius of relative stiffness (see Eq. 20.7)

$\ell = \sqrt[4]{E_c h^3/[12(1 - \mu^2)k]}$

$WIND$ = mean annual wind speed (mi/h)
$TEMP$ = mean annual temperature (°F)
$PRECIP$ = mean annual precipitation (in.)

The design 18-kipESALs for the specified level design reliability is given as

$$W_{18R} = 10^{(\log W_{18} + ZS_o)} \qquad\qquad (20.32)$$

Where

W_{18R} = design 18-kip ESALs for a specified level of design reliability R
W_{18} = estimated design 18-kip ESALs over the design period in the design lane
Z = standard deviate from normal distribution table for given level of reliability (e.g., 1.28 = 90 percent
S_o = overall standard deviation

This alternate design procedure is similar to that described in Section 20.6.1, in that both of them are iterative as the depth of the pavement slab is required as an input to the design equation. However, because of the complicated nature of the equations in this alternative design procedure, there are no monographs available, and its use requires an appropriate spread sheet or a computer program. However, AASHTO has provided several tables that give the required slab thicknesses for a range of joint spacings, sub-grade k values, and temperature differentials for three different base types and three levels of design reliability. Table 20.15 shows the different base types and reliability levels used, and Tables 20.16 and 20.17 are two of the solution tables provided by AASHTO.

Table 20.15 Base Modulus for Different Base Types and Reliability Levels

Reliability	Base Type	Base Modulus (lb/in.²)
95	Granular	25,000
	Treated	500,000
	High strength	1,000,000
90	Granular	25,000
	Treated	500,000
	High strength	1,000,000
85	Granular	25,000
	Treated	500,000
	High strength	1,000,000

SOURCE: From Supplement to the *AASHTO Guide for Design of Pavement Structures, Part II, Rigid Pavement Design & Rigid Pavement Joint Design*, 1998, AASHTO, Washington, D.C. Used by permission.

Table 20.16 Slab Thickness Computed for Granular Base and 95 Percent Reliability

E_b = 25 ksi [172.25 MPa], R = 95 percent, S_o = 0.39, $P2$ = 2.5, 12-ft-wide [3.7-m-wide] lanes with AC shoulders. Computed thicknesses less than 6.0 in. [152 mm] or greater than 15.0 in. [381 mm] are not shown.

Joint Spacing (in.)	Flexural Strength (lb/in.²)	Subgrade k(lb/in.³)	Positive TD (°F)	Design ESALs, millions							
				5	10	20	30	40	50	75	100
144	600	100	5	10.7	11.8	12.9	13.6	14.0	14.4	—	—
144	600	100	7	10.6	11.7	12.8	13.5	13.9	14.3	14.9	—
144	600	100	9	10.5	11.7	12.8	13.4	13.9	14.3	14.9	—

(Continued)

Table 20.16 Slab Thickness Computed for Granular Base and 95 Percent Reliability (*continued*)

E_b = 25 ksi [172.25 MPa], R = 95 percent, S_o = 0.39, P2 = 2.5, 12-ft-wide [3.7-m-wide] lanes with AC shoulders. Computed thicknesses less than 6.0 in. [152 mm] or greater than 15.0 in. [381 mm] are not shown.

Joint Spacing (in.)	Flexural Strength (lb/in.²)	Subgrade k(lb/in.³)	Positive TD (°F)	Design ESALs, millions							
				5	10	20	30	40	50	75	100
144	600	100	11	10.5	11.6	12.7	13.4	13.8	14.2	14.9	—
144	600	100	13	10.4	11.5	12.6	13.3	13.8	14.1	14.8	—
144	600	250	5	10.1	11.4	12.7	13.4	13.9	14.4	—	—
144	600	250	7	10.2	11.5	12.8	13.6	14.1	14.6	—	—
144	600	250	9	10.3	11.7	13.0	13.8	14.4	14.8	—	—
144	600	250	11	10.4	11.8	13.1	13.9	14.5	14.9	—	—
144	600	250	13	10.5	11.9	13.3	14.1	14.7	—	—	—
144	600	500	5	9.1	10.8	12.5	13.4	14.1	14.7	—	—
144	600	500	7	9.5	11.3	13.1	14.1	14.9	—	—	—
144	600	500	9	9.9	11.8	13.7	14.8	—	—	—	—
144	600	500	11	10.3	12.3	14.3	—	—	—	—	—
144	600	500	13	10.7	12.8	14.8	—	—	—	—	—
144	700	100	5	9.7	10.8	11.8	12.4	12.9	13.2	13.8	14.3
144	700	100	7	9.6	10.7	11.7	12.3	12.8	13.1	13.7	14.2
144	700	100	9	9.6	10.7	11.7	12.4	12.8	13.1	13.8	14.2
144	700	100	11	9.5	10.6	11.7	12.3	12.7	13.1	13.7	14.1
144	700	100	13	9.5	10.5	11.6	12.2	12.7	13.0	13.6	14.1
144	700	250	5	9.0	10.2	11.4	12.2	12.7	13.1	13.8	14.3
144	700	250	7	9.2	10.4	11.7	12.4	12.9	13.3	14.0	14.5
144	700	250	9	9.3	10.6	11.8	12.6	13.1	13.5	14.2	14.7
144	700	250	11	9.4	10.7	11.9	12.7	13.2	13.6	14.3	14.8
144	700	250	13	9.5	10.8	12.1	12.8	13.4	13.8	14.5	15.1
144	700	500	5	8.1	9.6	11.1	12.0	12.6	13.1	14.0	14.6
144	700	500	7	8.5	10.0	11.6	12.6	13.2	13.7	14.7	—
144	700	500	9	8.7	10.4	12.2	13.2	13.9	14.4	—	—
144	700	500	11	9.1	10.9	12.6	13.7	14.4	15.0	—	—
144	700	500	13	9.4	11.3	13.1	14.1	14.9	—	—	—
144	800	100	5	8.9	9.9	10.9	11.5	11.9	12.2	12.8	13.2
144	800	100	7	8.9	9.9	10.8	11.4	11.8	12.1	12.7	13.1
144	800	100	9	8.8	9.8	10.8	11.4	11.8	12.1	12.6	13.0
144	800	100	11	8.8	9.8	10.7	11.3	11.7	12.0	12.6	13.0
144	800	100	13	8.8	9.7	10.7	11.3	11.7	12.0	12.5	12.9
144	800	250	5	8.2	9.3	10.5	11.1	11.6	12.0	12.6	13.1
144	800	250	7	8.3	9.5	10.7	11.3	11.8	12.2	12.9	13.4
144	800	250	9	8.5	9.6	10.8	11.5	12.0	12.4	13.1	13.6
144	800	250	11	8.6	9.8	11.0	11.7	12.2	12.6	13.3	13.8
144	800	250	13	8.7	9.9	11.1	11.9	12.4	12.8	13.5	14.0
144	800	500	5	7.2	8.6	10.0	10.8	11.4	11.9	12.7	13.3
144	800	500	7	7.6	9.1	10.5	11.4	12.0	12.5	13.4	14.0
144	800	500	9	7.9	9.5	11.0	11.9	12.6	13.0	14.0	14.6
144	800	500	11	8.2	9.8	11.4	12.3	13.0	13.5	14.4	—
144	800	500	13	8.6	10.2	11.8	12.7	13.4	13.9	14.8	—
192	600	100	5	10.8	12.0	13.1	13.8	14.2	14.6	—	—
192	600	100	7	10.8	11.9	13.1	13.7	14.2	14.5	—	—
192	600	100	9	10.8	11.9	13.0	13.7	14.1	14.5	—	—

Table 20.16   Slab Thickness Computed for Granular Base and 95 Percent Reliability (*continued*)

E_b = 25 ksi [172.25 MPa], R = 95 percent, S_o = 0.39, P2 = 2.5, 12-ft-wide [3.7-m-wide] lanes with AC shoulders. Computed thicknesses less than 6.0 in. [152 mm] or greater than 15.0 in. [381 mm] are not shown.

Joint Spacing (in.)	Flexural Strength (lb/in.²)	Subgrade k(lb/in.³)	Positive TD (°F)	Design ESALs, millions							
				5	10	20	30	40	50	75	100
192	600	100	11	10.7	11.9	13.0	13.6	14.1	14.5	—	—
192	600	100	13	10.7	11.8	12.9	13.6	14.1	14.4	—	—
192	600	250	5	10.4	11.7	13.0	13.7	14.3	14.7	—	—
192	600	250	7	10.6	11.9	13.2	14.0	14.6	15.0	—	—
192	600	250	9	10.8	12.2	13.5	14.3	14.9	—	—	—
192	600	250	11	10.9	12.3	13.7	14.5	—	—	—	—
192	600	250	13	11.1	12.5	14.0	14.8	—	—	—	—
192	600	500	5	9.5	11.3	13.0	14.1	—	—	—	—
192	600	500	7	10.1	12.0	13.9	15.0	—	—	—	—
192	600	500	9	10.7	12.7	14.7	—	—	—	—	—
192	600	500	11	11.3	13.4	—	—	—	—	—	—
192	600	500	13	11.8	14.0	—	—	—	—	—	—
192	700	100	5	9.9	10.9	12.0	12.6	13.1	13.4	14.0	14.5
192	700	100	7	9.9	10.9	12.0	12.6	13.0	13.4	14.0	14.4
192	700	100	9	9.8	10.9	12.0	12.6	13.0	13.4	14.0	14.4
192	700	100	11	9.8	10.9	11.9	12.6	13.0	13.3	14.0	14.4
192	700	100	13	9.8	10.9	11.9	12.5	13.0	13.3	13.9	14.4
192	700	250	5	9.4	10.6	11.8	12.5	13.0	13.4	14.1	14.6
192	700	250	7	9.6	10.8	12.0	12.8	13.3	13.7	14.4	14.9
192	700	250	9	9.8	11.1	12.3	13.1	13.6	14.0	14.8	—
192	700	250	11	10.0	11.3	12.6	13.4	13.9	14.3	—	—
192	700	250	13	10.2	11.5	12.8	13.3	14.1	14.5	—	—
192	700	500	5	8.5	10.1	11.6	12.6	13.2	13.7	—	—
192	700	500	7	9.0	10.7	12.4	13.4	14.1	14.7	—	—
192	700	500	9	9.6	11.3	13.1	14.1	14.9	—	—	—
192	700	500	11	10.1	11.9	13.7	14.8	—	—	—	—
192	700	500	13	10.6	12.4	14.3	—	—	—	—	—
192	800	100	5	9.1	10.1	11.1	11.7	12.1	12.4	13.0	13.4
192	800	100	7	9.1	10.1	11.1	11.7	12.1	12.4	13.0	13.4
192	800	100	9	9.1	10.1	11.1	11.6	12.1	12.4	13.0	13.4
192	800	100	11	9.1	10.1	11.1	11.6	12.1	12.4	12.9	13.4
192	800	100	13	9.1	10.1	11.1	11.6	12.0	12.4	12.9	13.3
192	800	250	5	8.5	9.7	10.8	11.5	12.0	12.4	13.0	13.5
192	800	250	7	8.7	9.9	11.1	11.8	12.3	12.7	13.4	13.9
192	800	250	9	8.9	10.2	11.4	12.1	12.6	13.0	13.7	14.2
192	800	250	11	9.2	10.4	11.6	12.3	12.8	13.2	13.9	14.4
192	800	250	13	9.4	10.6	11.8	12.5	13.0	13.4	14.1	14.6
192	800	500	5	7.7	9.2	10.6	11.5	12.1	12.5	13.4	14.0
192	800	500	7	8.3	9.8	11.3	12.2	12.8	13.3	14.2	14.8
192	800	500	9	8.7	10.3	11.9	12.8	13.5	14.0	14.9	—
192	800	500	11	9.2	10.8	12.5	13.4	14.1	14.6	—	—
192	800	500	13	9.6	11.3	13.0	13.9	14.6	—	—	—
240	600	100	5	11.0	12.1	13.2	13.9	14.3	14.7	—	—
240	600	100	7	11.0	12.1	13.2	13.9	14.4	14.7	—	—
240	600	100	9	11.0	12.1	13.3	13.9	14.4	14.7	—	—

(*Continued*)

Table 20.16   Slab Thickness Computed for Granular Base and 95 Percent Reliability (*continued*)

E_b = 25 ksi [172.25 MPa], R = 95 percent, S_o = 0.39, $P2$ = 2.5, 12-ft-wide [3.7-m-wide] lanes with AC shoulders. Computed thicknesses less than 6.0 in. [152 mm] or greater than 15.0 in. [381 mm] are not shown.

Joint Spacing (in.)	Flexural Strength (lb/in.²)	Subgrade k(lb/in.³)	Positive TD (°F)	Design ESALs, millions							
				5	10	20	30	40	50	75	100
240	600	100	11	11.0	12.1	13.3	13.9	14.4	14.8	—	—
240	600	100	13	11.0	12.2	13.3	13.9	14.4	14.8	—	—
240	600	250	5	10.7	12.0	13.3	14.1	14.6	15.0	—	—
240	600	250	7	11.0	12.3	13.7	14.5	15.0	—	—	—
240	600	250	9	11.3	12.7	14.1	14.9	—	—	—	—
240	600	250	11	11.6	13.0	14.5	—	—	—	—	—
240	600	250	13	11.8	13.2	14.7	—	—	—	—	—
240	600	500	5	10.0	11.9	13.7	—	—	—	—	—
240	600	500	7	10.8	12.8	14.8	—	—	—	—	—
240	600	500	9	11.7	13.8	—	—	—	—	—	—
240	600	500	11	12.4	14.6	—	—	—	—	—	—
240	600	500	13	13.2	—	—	—	—	—	—	—
240	700	100	5	10.1	11.1	12.2	12.8	13.3	13.6	14.2	14.7
240	700	100	7	10.1	11.1	12.2	12.8	13.2	13.6	14.2	14.6
240	700	100	9	10.1	11.2	12.2	12.8	13.3	13.6	14.2	14.7
240	700	100	11	10.1	11.2	12.2	12.9	13.3	13.6	14.2	14.7
240	700	100	13	10.2	11.2	12.3	12.9	13.3	13.7	14.3	14.7
240	700	250	5	9.7	10.9	12.1	12.9	13.4	13.8	14.5	15.0
240	700	250	7	10.0	11.3	12.6	13.3	13.8	14.3	15.0	—
240	700	250	9	10.3	11.6	12.9	13.7	14.2	14.6	—	—
240	700	250	11	10.6	11.9	13.2	14.0	14.5	14.9	—	—
240	700	250	13	10.9	12.2	13.5	14.3	14.9	—	—	—
240	700	500	5	9.0	10.7	12.3	13.3	14.0	—	—	—
240	700	500	7	9.8	11.6	13.3	14.3	15.0	—	—	—
240	700	500	9	10.6	12.4	14.2	—	—	—	—	—
240	700	500	11	11.2	13.1	14.9	—	—	—	—	—
240	700	500	13	11.9	13.7	—	—	—	—	—	—
240	800	100	5	9.3	10.3	11.3	11.9	12.3	12.6	13.2	13.6
240	800	100	7	9.3	10.3	11.3	11.9	12.3	12.7	13.2	13.7
240	800	100	9	9.4	10.4	11.4	12.0	12.4	12.7	13.3	13.7
240	800	100	11	9.4	10.4	11.4	12.0	12.4	12.8	13.4	13.8
240	800	100	13	9.5	10.5	11.5	12.1	12.5	12.8	13.4	13.8
240	800	250	5	8.8	10.0	11.2	11.9	12.4	12.8	13.5	14.0
240	800	250	7	9.3	10.4	11.6	12.3	12.8	13.2	13.8	14.3
240	800	250	9	9.6	10.8	12.0	12.7	13.2	13.6	14.3	14.8
240	800	250	11	9.8	11.0	12.2	13.0	13.5	13.8	14.5	15.0
240	800	250	13	10.1	11.3	12.5	13.3	13.8	14.2	14.9	15.4
240	800	500	5	8.3	9.8	11.3	12.1	12.7	13.2	14.1	14.7
240	800	500	7	9.0	10.6	12.2	13.1	13.7	14.2	—	—
240	800	500	9	9.7	11.3	12.9	13.9	14.5	15.0	—	—
240	800	500	11	10.4	12.0	13.6	14.6	—	—	—	—
240	800	500	13	10.9	12.5	14.2	—	—	—	—	—

1 in. = 25.4 mm, 1 lb/in.² = 6.89 kPa, 1 lb/in.³ = 0.271 kPa/mm, °C = (°F − 32)/1.8
SOURCE: From Supplement to the *AASHTO Guide for Design of Pavement Structures, Part II, Rigid Pavement Design & Rigid Pavement Joint Design*, 1998, AASHTO, Washington, D.C. Used by permission.

Table 20.17   Slab Thickness Computed for Treated Base and 95 Percent Reliability

E_b = 500 ksi [3445 MPa], R = 95 percent, S_o = 0.39, $P2$ = 2.5, 12-ft-wide [3.7-m-wide] lanes with AC shoulders. Computed thicknesses less than 6.0 in. [152 mm] or greater than 15.0 in. [381 mm] are not shown.

Joint Spacing (in.)	Flexural Strength (lb/in.²)	Subgrade k(lb/in.³)	Positive TD (°F)	Design ESALs, millions							
				5	10	20	30	40	50	75	100
144	600	100	5	9.2	10.3	11.3	12.0	12.4	12.7	13.4	13.8
144	600	100	7	9.4	10.4	11.4	12.1	12.5	12.8	13.4	13.9
144	600	100	9	9.5	10.5	11.5	12.1	12.6	12.9	13.5	13.9
144	600	100	11	9.6	10.6	11.6	12.2	12.6	12.0	13.6	14.0
144	600	100	13	9.6	10.7	11.7	12.3	12.7	13.0	13.6	14.1
144	600	250	5	8.6	10.0	11.3	12.1	12.6	13.0	13.8	14.3
144	600	250	7	9.1	10.4	11.7	12.5	13.0	13.4	14.2	14.7
144	600	250	9	9.4	10.7	12.1	12.8	13.4	13.8	14.6	—
144	600	250	11	9.7	11.0	12.4	13.1	13.7	14.1	14.9	—
144	600	250	13	10.0	11.3	12.6	13.4	14.0	14.4	—	—
144	600	500	5	7.5	9.2	10.9	11.9	12.6	13.1	—	—
144	600	500	7	8.4	10.1	11.8	12.8	13.5	14.0	—	—
144	600	500	9	9.1	10.8	12.5	13.5	14.2	14.8	—	—
144	600	500	11	9.4	11.3	13.2	14.3	—	—	—	—
144	600	500	13	10.1	11.9	13.8	14.9	—	—	—	—
144	700	100	5	8.5	9.5	10.4	11.0	11.4	11.8	12.3	12.7
144	700	100	7	8.6	9.6	10.6	11.1	11.5	11.9	12.4	12.8
144	700	100	9	8.8	9.7	10.7	11.2	11.6	12.0	12.5	12.9
144	700	100	11	8.9	9.8	10.8	11.3	11.7	12.0	12.6	13.0
144	700	100	13	9.0	9.9	10.9	11.4	11.8	12.1	12.7	13.0
144	700	250	5	8.0	9.2	10.3	11.0	11.5	11.9	12.6	13.1
144	700	250	7	8.3	9.5	10.7	11.5	12.0	12.4	13.1	13.6
144	700	250	9	8.7	9.9	11.1	11.8	12.3	12.7	13.4	13.9
144	700	250	11	9.0	10.2	11.4	12.1	12.6	13.0	13.7	14.2
144	700	250	13	9.2	10.4	11.7	12.4	12.9	13.3	14.0	14.5
144	700	500	5	6.6	8.2	9.8	10.8	11.4	12.0	12.9	13.6
144	700	500	7	7.6	9.1	10.7	11.6	12.3	12.8	13.7	14.3
144	700	500	9	8.3	9.3	11.4	12.3	12.9	13.4	14.4	15.0
144	700	500	11	8.6	10.3	12.0	12.9	13.6	14.2	—	—
144	700	500	13	9.2	10.9	12.5	13.4	14.1	14.6	—	—
144	800	100	5	7.9	8.8	9.7	10.3	10.7	11.0	11.5	11.9
144	800	100	7	8.0	8.9	9.9	10.4	10.8	11.1	11.6	12.0
144	800	100	9	8.2	9.1	10.0	10.5	10.9	11.2	11.7	12.1
144	800	100	11	8.3	9.2	10.1	10.6	11.0	11.3	11.8	12.1
144	800	100	13	8.4	9.3	10.2	10.7	11.1	11.3	11.8	12.2
144	800	250	5	7.3	8.5	9.6	10.2	10.7	11.1	11.7	12.2
144	800	250	7	7.7	8.9	10.0	10.6	11.1	11.4	12.1	12.6
144	800	250	9	8.1	9.2	10.3	11.0	11.4	11.8	12.5	12.9
144	800	250	11	8.4	9.5	10.6	11.3	11.7	12.1	12.7	13.2
144	800	250	13	8.6	9.7	10.9	11.5	12.0	12.3	13.0	13.5
144	800	500	5	6.4	7.7	9.1	9.9	10.5	11.0	11.8	12.4
144	800	500	7	6.9	8.4	9.8	10.7	11.3	11.7	12.6	13.2
144	800	500	9	7.6	9.1	10.5	11.3	11.9	12.4	13.2	13.8
144	800	500	11	7.9	9.5	11.0	11.9	12.5	13.0	13.9	14.5
144	800	500	13	8.5	10.0	11.5	12.4	13.0	13.5	14.3	14.9
192	600	100	5	9.5	10.5	11.6	12.2	12.6	13.0	13.6	14.0

(Continued)

Table 20.17 Slab Thickness Computed for Treated Base and 95 Percent Reliability (*continued*)

E_b = 500 ksi [3445 MPa], R = 95 percent, S_o = 0.39, P2 = 2.5, 12-ft-wide [3.7-m-wide] lanes with AC shoulders. Computed thicknesses less than 6.0 in. [152 mm] or greater than 15.0 in. [381 mm] are not shown.

Joint Spacing (in.)	Flexural Strength (lb/in.²)	Subgrade k(lb/in.³)	Positive TD (°F)	Design ESALs, millions							
				5	10	20	30	40	50	75	100
192	600	100	7	9.7	10.7	11.7	12.3	12.8	13.1	13.7	14.2
192	600	100	9	9.8	10.9	11.9	12.5	12.9	13.3	13.9	14.3
192	600	100	11	9.9	11.0	12.0	12.6	13.0	13.3	13.9	14.3
192	600	100	13	10.1	11.1	12.1	12.7	13.1	13.4	14.0	14.4
192	600	250	5	9.1	10.4	11.7	12.5	13.0	13.4	14.2	14.7
192	600	250	7	9.6	10.9	12.2	13.0	13.5	13.9	14.7	—
192	600	250	9	10.0	11.3	12.6	13.4	14.0	14.4	15.1	—
192	600	250	11	10.4	11.7	13.0	13.8	14.3	14.8	15.5	—
192	600	250	13	10.6	12.0	13.5	14.3	15.0	12.4	16.3	—
192	600	500	5	8.1	9.8	11.5	12.5	13.2	13.8	14.8	—
192	600	500	7	9.1	10.9	12.6	13.6	14.3	14.9	—	—
192	600	500	9	9.8	11.7	13.5	14.6	—	—	—	—
192	600	500	11	10.6	12.5	14.3	—	—	—	—	—
192	600	500	13	11.3	13.2	—	—	—	—	—	—
192	700	100	5	8.7	9.7	10.7	11.3	11.7	12.0	12.5	12.9
192	700	100	7	9.0	9.9	10.9	11.4	11.8	12.2	12.7	13.1
192	700	100	9	9.1	10.1	11.0	11.6	12.0	12.3	12.9	13.2
192	700	100	11	9.3	10.2	11.2	11.7	12.1	12.4	13.0	13.4
192	700	100	13	9.4	10.4	11.3	11.9	12.2	12.5	13.1	13.5
192	700	250	5	8.3	9.5	10.8	11.5	12.0	12.4	13.1	13.6
192	700	250	7	8.9	10.1	11.3	12.0	12.5	12.9	13.6	14.1
192	700	250	9	9.3	10.5	11.7	12.4	12.9	13.3	14.0	14.5
192	700	250	11	9.7	10.9	12.1	12.8	13.3	13.7	14.4	14.9
192	700	250	13	9.9	11.2	12.4	13.1	13.6	14.0	14.8	—
192	700	500	5	7.5	9.0	10.6	11.5	12.1	12.6	13.5	—
192	700	500	7	8.5	10.0	11.5	12.4	13.1	13.6	14.5	—
192	700	500	9	9.1	10.7	12.3	13.3	14.0	14.5	—	—
192	700	500	11	9.9	11.5	13.1	14.0	14.7	—	—	—
192	700	500	13	10.5	12.1	13.7	14.6	—	—	—	—
192	800	100	5	8.1	9.0	10.0	10.5	10.9	11.2	11.7	12.1
192	800	100	7	8.4	9.3	10.2	10.7	11.1	11.4	11.9	12.3
192	800	100	9	8.6	9.5	10.4	10.9	11.2	11.5	12.0	12.4
192	800	100	11	8.8	9.6	10.5	11.0	11.4	11.7	12.2	12.5
192	800	100	13	8.9	9.8	10.6	11.1	11.5	11.8	12.3	12.6
192	800	250	5	7.8	8.9	10.0	10.7	11.1	11.5	12.1	12.6
192	800	250	7	8.3	9.4	10.5	11.2	11.6	12.0	12.6	13.1
192	800	250	9	8.7	9.8	10.9	11.6	12.1	12.4	13.1	13.5
192	800	250	11	9.1	10.2	11.3	12.0	12.4	12.8	13.4	13.9
192	800	250	13	9.4	10.5	11.6	12.3	12.7	13.1	13.8	14.2
192	800	500	5	6.9	8.3	9.8	10.6	11.2	11.7	12.5	13.1
192	800	500	7	7.9	9.3	10.7	11.5	12.1	12.6	13.4	14.0
192	800	500	9	8.5	10.0	11.4	12.3	12.9	13.4	14.2	14.8
192	800	500	11	9.2	10.7	12.1	13.0	13.6	14.0	14.9	—
192	800	500	13	9.6	11.1	12.7	13.6	14.2	14.7	—	—
240	600	100	5	9.7	10.8	11.8	12.4	12.9	13.2	13.8	14.2
240	600	100	7	10.0	11.0	12.0	12.6	13.1	13.4	14.0	14.4

Table 20.17   Slab Thickness Computed for Treated Base and 95 Percent Reliability (*continued*)

E_b = 500 ksi [3445 MPa], R = 95 percent, S_o = 0.39, P2 = 2.5, 12-ft-wide [3.7-m-wide] lanes with AC shoulders. Computed thicknesses less than 6.0 in. [152 mm] or greater than 15.0 in. [381 mm] are not shown.

Joint Spacing (in.)	Flexural Strength (lb/in.²)	Subgrade k(lb/in.³)	Positive TD (°F)	Design ESALs, millions							
				5	10	20	30	40	50	75	100
240	600	100	9	10.2	11.2	12.2	12.8	13.3	13.6	14.2	14.6
240	600	100	11	10.4	11.4	12.4	13.0	13.4	13.7	14.3	14.8
240	600	100	13	10.5	11.5	12.5	13.1	13.5	13.8	14.4	14.8
240	600	250	5	9.5	10.8	12.1	12.9	13.4	13.9	14.6	—
240	600	250	7	10.1	11.4	12.8	13.5	14.1	14.5	—	—
240	600	250	9	10.7	12.0	13.3	14.0	14.6	15.0	—	—
240	600	250	11	11.0	12.4	13.8	14.6	—	—	—	—
240	600	250	13	11.5	12.8	14.2	15.0	—	—	—	—
240	600	500	5	8.8	10.6	12.3	13.3	14.0	—	—	—
240	600	500	7	9.8	11.7	13.6	14.7	—	—	—	—
240	600	500	9	10.9	12.8	14.6	—	—	—	—	—
240	600	500	11	11.8	13.7	—	—	—	—	—	—
240	600	500	13	12.5	14.5	—	—	—	—	—	—
240	700	100	5	9.0	10.0	11.0	11.5	11.9	12.2	12.8	13.2
240	700	100	7	9.3	10.3	11.2	11.8	12.2	12.5	13.0	13.4
240	700	100	9	9.5	10.5	11.4	12.0	12.4	12.7	13.2	13.6
240	700	100	11	9.7	10.7	11.6	12.2	12.6	12.9	13.4	13.8
240	700	100	13	9.9	10.8	11.8	12.3	12.7	13.0	13.6	14.0
240	700	250	5	8.8	10.0	11.2	11.9	12.4	12.8	13.5	14.0
240	700	250	7	9.5	10.7	11.8	12.5	13.0	13.4	14.1	14.6
240	700	250	9	9.9	11.1	12.4	13.1	13.6	14.0	14.7	—
240	700	250	11	10.4	11.6	12.8	13.5	14.0	14.4	—	—
240	700	250	13	10.8	12.0	13.2	13.9	14.4	14.7	—	—
240	700	500	5	8.3	9.8	11.3	12.2	12.9	13.4	14.3	14.9
240	700	500	7	9.2	10.8	12.5	13.4	14.1	14.6	—	—
240	700	500	9	10.3	11.8	13.4	14.3	15.0	—	—	—
240	700	500	11	11.0	12.7	14.4	—	—	—	—	—
240	700	500	13	11.7	13.4	—	—	—	—	—	—
240	800	100	5	8.5	9.4	10.3	10.8	11.1	11.4	12.0	12.3
240	800	100	7	8.8	9.6	10.5	11.0	11.4	11.7	12.2	12.6
240	800	100	9	9.0	9.9	10.8	11.3	11.6	11.9	12.4	12.8
240	800	100	11	9.2	10.1	10.9	11.5	11.8	12.1	12.6	13.0
240	800	100	13	9.4	10.2	11.1	11.6	12.0	12.3	12.8	13.1
240	800	250	5	8.3	9.4	10.5	11.1	11.6	12.0	12.6	13.1
240	800	250	7	8.9	10.0	11.1	11.8	12.2	12.6	13.2	13.7
240	800	250	9	9.4	10.5	11.6	12.3	12.7	13.1	13.7	14.2
240	800	250	11	9.8	10.9	12.0	12.7	13.1	13.5	14.1	14.6
240	800	250	13	10.2	11.3	12.4	13.0	13.5	13.8	14.4	14.9
240	800	500	5	7.8	9.2	10.6	11.4	12.0	12.4	13.2	13.8
240	800	500	7	8.8	10.3	11.6	12.5	13.1	13.5	14.4	14.9
240	800	500	9	9.7	11.1	12.5	13.3	13.9	14.4	—	—
240	800	500	11	10.4	11.8	13.2	14.1	14.7	—	—	—
240	800	500	13	11.0	12.4	13.9	14.7	—	—	—	—

1 in. = 25.4 mm, 1 lb/in.² = 6.89 kPa, 1 lb/in.³ = 0.271 kPa/mm, °C = (°F − 32)/1.8

SOURCE: From Supplement to the *AASHTO Guide for Design of Pavement Structures, Part II, Rigid Pavement Design & Rigid Pavement Joint Design*, 1998, AASHTO, Washington, D.C. Used by permission.

Design Check for Joint Load Position Cracking

This check is performed to ascertain that stresses developed at the top of the slab when an axle load is applied at the joint are below the maximum allowable. The test is usually not required when dowels or equivalent load transfer devices are placed in the traverse joints, as these result in stress levels that are much less than the stresses that occur at the midspan load condition, which is the basis for the design of the slab. However, when dowels or equivalent load transfer devices are not placed at the transverse joints, it is necessary to ascertain that stresses developed at the top of the slab due to axle load applications at the joint are not excessive and higher than those at the bottom of the slab developed by mid-slab loading. When critical stresses developed at the top of the slab due to axle loading at the joint are higher than those at the bottom of the slab developed by midslab loading, this will result in corner breaks, or diagonal cracks under certain load and climate conditions. The load and climate conditions that influence the occurrence of corner breaks or diagonal cracks are (i) location of axle load, (ii) negative temperature differential stress occurring at the top of the slab surface due to nighttime temperature differentials, (iii) construction curling stress, and (iv) moisture gradient stress. In addition to the material presented in Section 20.5.3 on the influence of these factors, it is suggested that because of the difficulty of quantifying construction curling stress and the moisture gradient stress separately, a combined effect can be considered in terms of a positive temperature differential. This positive temperature differential is considered as that which will be required to bring the slab to a flat position after curling without an actual temperature differential throughout the slab. AASHTO suggests the following approximations for the equivalent temperature differential, based on the climate conditions at the site and the conventional curing procedures (i.e., curing compound, no wet cure):

> For wet climate (annual precipitation $\geq$ 30 in (0.76 m) or Thornthwaite Moisture Index > 0),
> 0 to 2 F/in (0 to 0.044 C/mm) of slab thickness
>
> For dry climate (annual precipitation < 30 in (0.76 m) or Thornthwaite Moisture Index < 0),
> 1 to 3 F/in (0.021 to 0.066 C/mm) (0 to 0.044 C/mm) of slab thickness

Significant reduction of these values may be made when wet curing or night construction is used. Note that the Thornthwaite Moisture Index is an indicator of the precipitation in an area relative to the demand for water under prevailing climatic conditions.

The design check for joint load position cracking is carried out by executing the following steps:

1. Determine the required slab thickness using the midslab load position as the criterion for design as described in this section.
2. Determine the midslab tensile stress (σ_t) at the bottom of the slab using the required slab thickness and the climatic conditions existing at the site.
3. Estimate a total equivalent negative temperature differential taking into consideration the effective (weighted average annual) negative temperature differential, construction temperature differential, and moisture differential.
4. Estimate the critical stress at the top of the slab resulting from axle loading at the joint and negative temperature differential.
5. Compare the tensile stress determined for midslab loading at the bottom of the slab with that obtained at the top of the slab for joint loading.

Step 1. **Determine the required slab thickness using the midslab load position as the criterion for design as described in this section.** Use Equations 20.23 through

20.31 or the available Tables 20.16 or 20.17 to determine the required thickness, based on the midslab loading as the critical loading position.

Step 2. **Determine the midslab tensile stress (σ_{tb}) at the bottom of the slab using the required slab thickness and the climatic conditions existing at the site.** When Eqs. 20.23 through 20.31 are used to determine the required slab thickness in Step 1, the midslab stress may have been computed. Alternatively, the midslab stress can be obtained or computed as shown below:

(a) Obtain the midslab tensile stress (σ_t) at the bottom of the slab for full friction between the slab and base from the relevant charts provided by AASHTO. Figures 20.18 through 20.21 are examples of these charts. Interpolation can be carried out when necessary.

(b) Determine the friction adjustment factor from Eq. 20.31.

(c) Estimate the proper midslab tensile stress as the product of the full friction stress (σ_t) and the friction adjustment factor F obtained from Eq. 20.31.

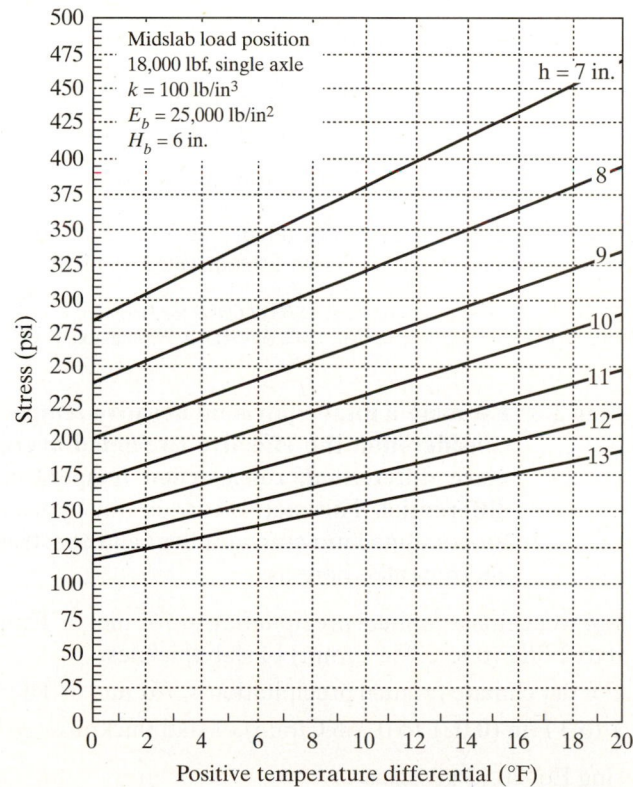

1 lbf = 4.45 N, 1 pci = 0.271 kPa/mm, 1 psi = 6.89 kPa, 1 in = 25.4 mm, °C = (°F – 32)/1.8

Figure 20.18   Tensile Stress at Bottom of Slab for Midslab Loading Position, Positive Temperature Differential, and Full Friction for Aggregate Base and Soft Subgrade

SOURCE: From Supplement to the *AASHTO Guide for Design of Pavement Structures, Part II, Rigid Pavement Design & Rigid Pavement Joint Design*, 1998, AASHTO, Washington, D.C. Used by permission.

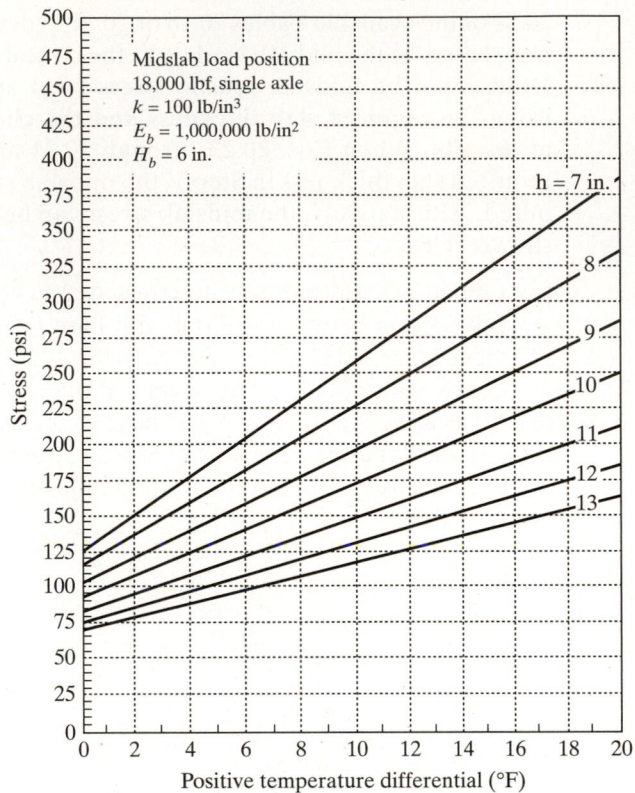

1 lbf = 4.45 N, 1 pci = 0.271 kPa/mm, 1 psi = 6.89 kPa, 1 in = 25.4 mm, °C = (°F − 32)/1.8

Figure 20.19   Tensile Stress at Bottom of Slab for Midslab Loading Position, Positive Temperature Differential, and Full Friction for High-Strength Base and Soft Subgrade

SOURCE: From Supplement to the *AASHTO Guide for Design of Pavement Structures, Part II, Rigid Pavement Design & Rigid Pavement Joint Design*, 1998, AASHTO, Washington, D.C. Used by permission.

Step 3.   Estimate a total equivalent negative temperature differential, taking into consideration the effective (weighted average annual) negative temperature differential, construction temperature differential, and moisture differential. This step is carried out by using the guidelines given above for combined moisture gradient and construction temperature differential and repeated here as:

For wet climate (annual precipitation ≥ 762 mm or Thornthwaite Moisture Index > 0) 0 to 2 F/in (0 to 0.044 C/mm) of slab thickness

For dry climate (annual precipitation < 762 mm or Thornthwaite Moisture Index < 0) 1 to 3 F/in (0.021 to 0.066 C/mm) of slab thickness

or using Eq. 20.33 given as:

$$\text{effective negative TD}_n = -18.14 + \frac{52.01}{h} + 0.394 WIND + 0.07 TEMP$$

$$+ 0.00407 PRECIP \tag{20.33}$$

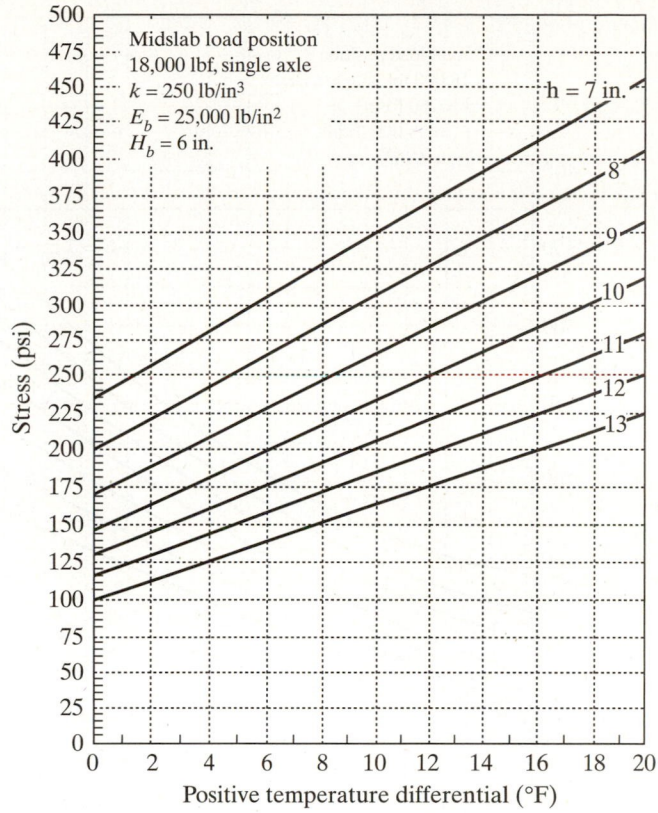

$$1 \text{ lbf} = 4.45 \text{ N}, 1 \text{ pci} = 0.271 \text{ kPa/mm}, 1 \text{ psi} = 6.89 \text{ kPa}, 1 \text{ in} = 25.4 \text{ mm}, °C = (°F - 32)/1.8$$

Figure 20.20   Tensile Stress at Bottom of Slab for Midslab Loading Position, Positive Temperature Differential, and Full Friction for Aggregate Base and Medium Subgrade

SOURCE: From Supplement to the *AASHTO Guide for Design of Pavement Structures, Part II, Rigid Pavement Design & Rigid Pavement Joint Design*, 1998, AASHTO, Washington, D.C. Used by permission.

where

$$\text{Effective negative TD}_n = \text{top temperature minus bottom temperature}$$
$$h = \text{slab thickness (mm)}$$
$$WIND = \text{mean annual wind speed (m/h)}$$
$$TEMP = \text{mean annual temperature (°C)}$$
$$PRECIP = \text{mean annual precipitation (mm)}$$

Step 4.   Estimate the critical stress at the top of the slab resulting from axle loading at the joint and negative temperature differential. AASHTO also provides several charts that give the critical stresses on top of the slab from joint loading. These are given for different negative temperature differentials, slab thickness, and base materials. Figures 20.22 through 20.25 are examples of these charts. These charts can be used to determine the tensile stress (σ_{tt}) at the top of the slab. The proper joint loading stress is obtained

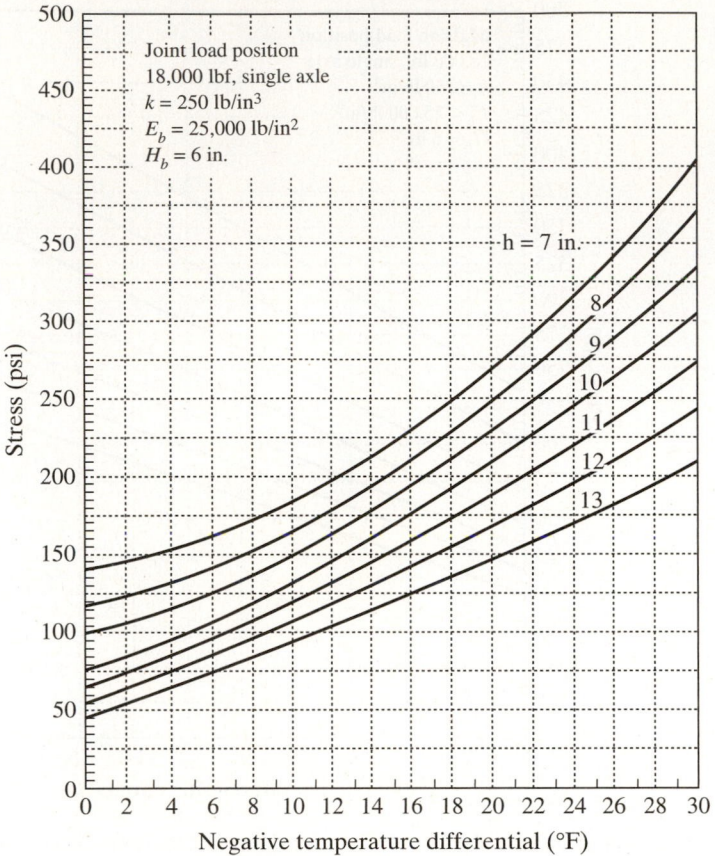

1 lbf = 4.45 N, 1 pci = 0.271 kPa/mm, 1 psi = 6.89 kPa, 1 in = 25.4 mm, °C = (°F − 32)/1.8

Figure 20.21   Tensile Stress at Bottom of Slab for Midslab Loading Position, Positive Temperature Differential, and Full Friction for High-Strength Base and Medium Subgrade

SOURCE: From Supplement to the *AASHTO Guide for Design of Pavement Structures, Part II, Rigid Pavement Design & Rigid Pavement Joint Design*, 1998, AASHTO, Washington, D.C. Used by permission.

as the product of the stress obtained from the appropriate chart and the friction adjustment factor.

Step 5.   **Compare the stress determined for midslab loading with that obtained for joint loading.** The value of tensile stress (σ_{tt}) at the top of the slab due to axle loading at the joint is compared with that obtained for the tensile stress (σ_{tb}) at the bottom of the slab due to midslab loading, and if

$\sigma_{tt} < \sigma_{tb}$: Use designed slab thickness.

$\sigma_{tt} > \sigma_{t}$: Consideration should be given to reduce joint loading stress by redesigning the joints.

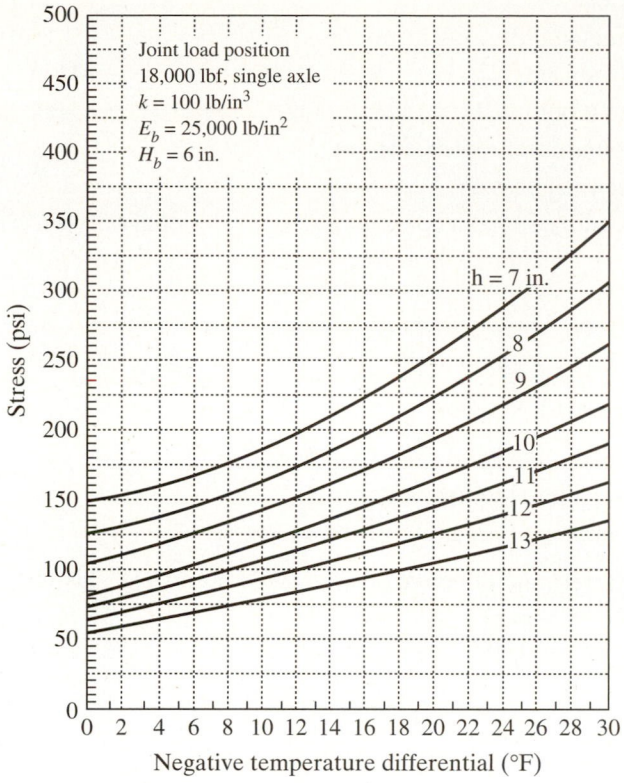

1 lbf = 4.45 N, 1 pci = 0.271 kPa/mm, 1 psi = 6.89 kPa, 1 in = 25.4 mm, °C = (°F − 32)/1.8

Figure 20.22   Tensile Stress at Top of Slab for Joint Loading Position, Negative Temperature Differential, and Full Friction for Aggregate Base and Soft Subgrade

SOURCE: From Supplement to the *AASHTO Guide for Design of Pavement Structures, Part II, Rigid Pavement Design & Rigid Pavement Joint Design*, 1998, AASHTO, Washington, D.C. Used by permission.

Example 20.5   Determine Required Slab Thickness Using the AASHTO Alternate Design Method

Determine the required slab thickness of a concrete rigid pavement located in Richmond, Virginia, using the AASHTO alternate design method for the following input data:

Estimated future traffic, W_{18}	5×10^6
Design reliability, R	95%
Overall standard deviation	0.39
Initial serviceability index, p_i	4.5
Terminal serviceability index, p_t	2.5
Effective subgrade k value, k	250 lb/in.³
Mean concrete modulus of rupture, S'_c	600 lb/in.³

Mean concrete elastic modulus, E_c 4.2×10^6 lb/in.^{2}

Joint spacing   144 in.

Base modulus, E_b 25,000 lb/in.^{2}

Slab/base friction coefficient, f 0.8

Base thickness, H_b 6 in.

Lane edge support condition   conventional slab width of 12 ft and AC shoulders

Solution:

- Determine the location's mean annual wind speed, mean annual temperature, and mean annual precipitation—Use Table 20.14.

For Richmond, Virginia:

Mean annual wind speed = 7.6 mi/h

Mean annual temperature = 57.7°F

Mean annual precipitation = 44.1 in.

- Determine effective positive differential TD_p. Use Eq. 20.22.

$$TD_p = 0.962 - \frac{52.181}{h} + 0.341WIND + 0.184TEMP - 0.00836PRECIP$$

$$= 0.962 - \frac{52.181}{6} + 0.341 \times 7.6 + 0.184 \times 57.7 - 0.00836 \times 44.1$$

$$= 5.1°F$$

- Determine slab thickness—Use Table 20.16, which is appropriate for the input data.

Depth of slab obtained = 10.1 in.; use 10.0 in (254 mm)

Example 20.6 Check for Joint Position Cracking

Determine whether the design slab thickness obtained for Example 20.5 is adequate for the tensile stress that will occur for joint loading.

Solution:

- Slab thickness = 10 in. from Example 20.5
- Compute the midslab load tensile stress at bottom of slab, for a slab of 10 in., assuming full friction between slab and base.

$$k = 250 \text{ lb/in.}^3$$

$$E_b = 25,000 \text{ lb/in.}^2$$

Effective positive temperature differential = 5.1°F (computed in Example 20.5 solution)

Use Figure 20.20, which is an appropriate chart for these input data.

Midslab loading tensile strength = 190 lb/in.^{2}

- Compute friction adjustment factor—Use Eq. 20.32.

$$F = 1.177 - 4.3 \times 10^{-8} h E_b - 0.01155542h + 6.27 \times 10^{-7} E_b - 0.000315f$$

$$= 1.177 - 4.3 \times 10^{-8} \times 10 \times 2.5 \times 10^4 - 0.01155542 \times 10 + 6.27 \times 10^{-7}$$

$$\times 2.5 \times 10^4 - 0.000315 \times 0.8 = 1.066$$

Proper estimate of the midslab loading stress = $190 \times 1.066 = 202.54$ lb/in.^{2}

- Estimate the total equivalent negative temperature differential—Use Eq. 20.33.

$$\text{TD}_n = -18.14 + \frac{52.01}{h} + 0.394 WIND + 0.07 TEMP + 0.00407 PRECIP$$

$$= -18.14 + 52.01/10 + 0.394 \times 7.6 + 0.07 \times 57.7 + 0.00407 \times 44.1$$

$$= -18.14 + 5.201 + 2.994 + 4.039 + 0.1795$$

$$= -5.7°F$$

- Compute combined moisture gradient and construction temperature differential.

Mean annual precipitation at location = 44.1 in. > 30 in., therefore, the location is considered a wet climate. Assuming the mean for this location, then the approximate equivalent differential for moisture gradient is $-1 \times 10 = 10°F$. Therefore, total negative temperature differential = $-10 + -5.7 = 15.7$.

- Compute the joint loading tensile stress at top of slab, for a slab of 10 in., assuming full friction between slab and base. Use Figure 20.21.

Tensile stress = 175 lb/in.^{2}

Proper estimate of the joint loading stress = $175 \times 1.066 = 186.55$ lb/in^2 (1287 kN/m^2)

Therefore, the joint loading stress < midslab loading stress, and it is not necessary to modify the joint design.

Design Check for Joint Load Position Cracking Joint Load Transfer

As noted in Section 20.6.1, the 1993 AASHTO pavement design method is mainly based on the results of the AASHO road test. However, the joints of the pavements used in that test were adequately doweled, and therefore the resulting faulting was insignificant. If joints are not adequately doweled, significant faulting will take place, resulting in a cumulative ESAL that is lower than the original design cumulative ESAL on the pavement for the terminal serviceability. This alternate design therefore provides a procedure for checking the adequacy of load transfer at the joints in order to avoid joint faulting that exceeds the following recommended values:

Joint spacing less than 25 ft, critical mean joint faulting = 0.06 in.

Joint spacing greater than 25 ft, critical mean joint faulting = 0.13 in.

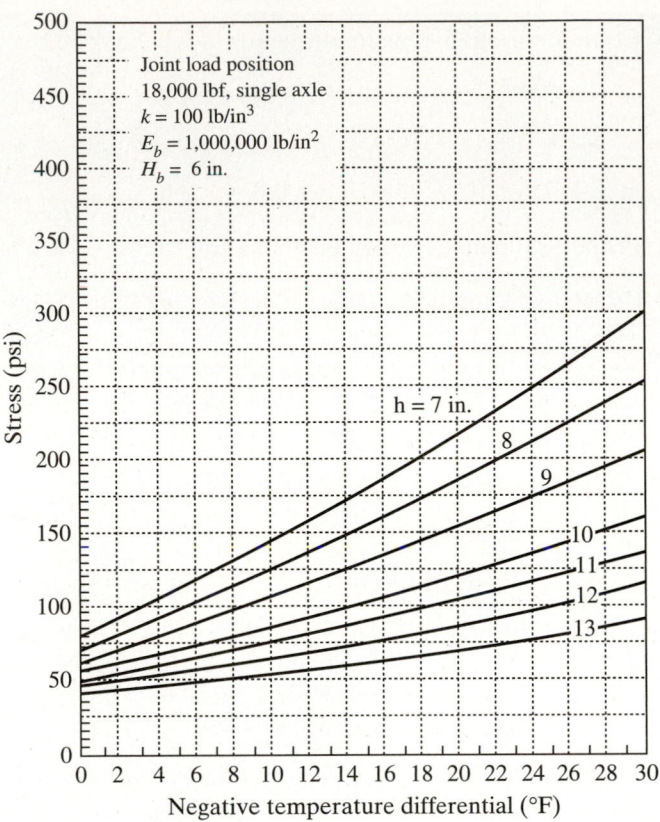

1 lbf = 4.45 N, 1 pci = 0.271 kPa/mm, 1 psi = 6.89 kPa, 1 in = 25.4 mm, °C = (°F − 32)/1.8

Figure 20.23   Tensile Stress at Top of Slab for Joint Loading Position, Negative Temperature Differential, and Full Friction for High-Strength Base and Soft Subgrade

SOURCE: From Supplement to the *AASHTO Guide for Design of Pavement Structures, Part II, Rigid Pavement Design & Rigid Pavement Joint Design*, 1998, AASHTO, Washington, D.C. Used by permission.

The procedure consists of the following steps:

1. Use the procedure described in this section for the alternate design. Determine the design thickness of the slab, including the comparison of tensile stresses for midslab and joint loadings, and note whether the pavement is doweled or not.
2. Use the appropriate model for doweled or undoweled pavement to estimate the expected mean joint faulting.
3. Compare the estimated joint faulting with the recommended mean joint faulting level to prevent significant loss in serviceability. If the estimated joint faulting is higher than the recommended value, the load transfer design should be altered.

   Step 1.   Determine required slab thickness. The required slab thickness of the pavement is determined using the AASHTO alternate design method described above, including checks for tensile stresses due to midslab and joint loading.

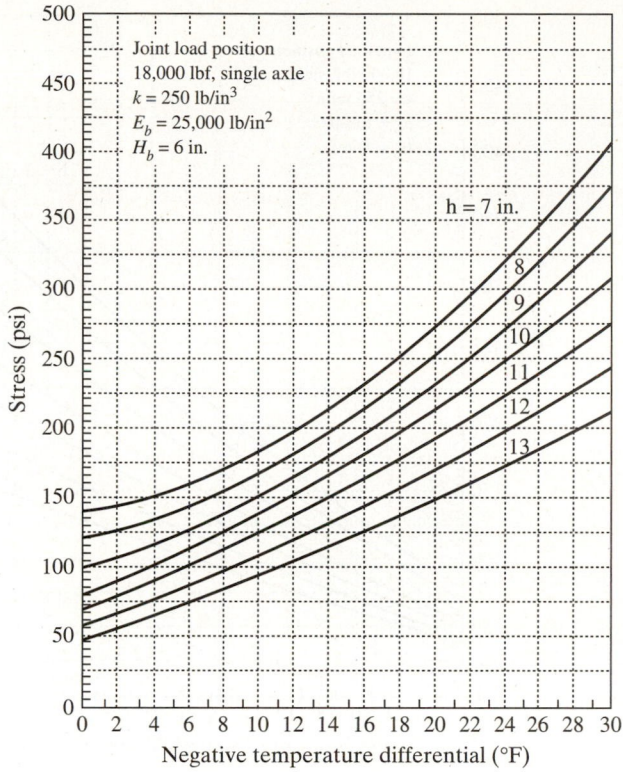

1 lbf = 4.45 N, 1 pci = 0.271 kPa/mm, 1 psi = 6.89 kPa, 1 in = 25.4 mm, °C = (°F − 32)/1.8

Figure 20.24   Tensile Stress at Top of Slab for Joint Loading Position, Negative Temperature Differential, and Full Friction for Aggregate Base and Medium Subgrade

SOURCE: From Supplement to the *AASHTO Guide for Design of Pavement Structures, Part II, Rigid Pavement Design & Rigid Pavement Joint Design*, 1998, AASHTO, Washington, D.C. Used by permission.

Step 2.   Use the appropriate model for doweled or undoweled pavement to estimate the expected mean joint faulting. The expected mean joint faulting is determined from the expressions given in Eqs. 20.34 to 20.39 for doweled joints and Eq. 20.40 for undoweled joints.

$$\text{FaultD} = \text{CESAL}^{0.25} \times$$

$$\begin{bmatrix} 0.0628 - 0.0628 \times C_d + 0.3673 \times 10^{-8} \times \text{Bstress}^2 \\ + 0.4116 \times 10^{-5} \times \text{Jtspace}^2 + 0.7466 \times 10^{-9} \times FI^2 \times \text{Precip}^{0.5} \\ - 0.009503 \times \text{Basetype} - 0.0197 \times \text{Widenlane} + 0.0009217 \times \text{Age} \end{bmatrix} \quad (20.34)$$

where

> FaultD = mean transverse doweled joint faulting (in.)
> CESAL = cumulative equivalent 18-kip single axle load (millions)
> C_d = modified AASHTO drainage coefficient (see Table 20.18)
> Bstress = maximum concrete bearing stress from closed form equation (lb/in.²)

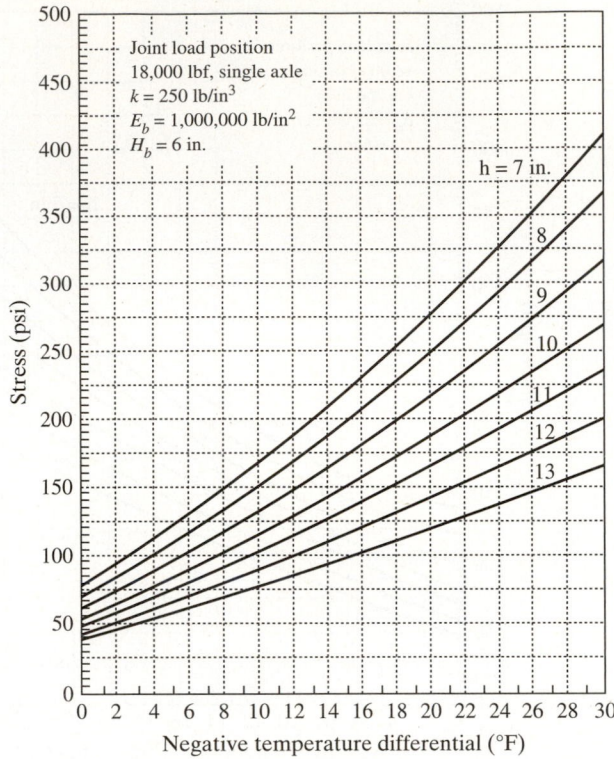

1 lbf = 4.45 N, 1 pci = 0.271 kPa/mm, 1 psi = 6.89 kPa, 1 in = 25.4 mm, °C = (°F − 32)/1.8

Figure 20.25   Tensile Stress at Top of Slab for Joint Loading Position, Negative Temperature Differential, and Full Friction for High-Strength Base and Medium Subgrade

SOURCE: From Supplement to the *AASHTO Guide for Design of Pavement Structures, Part II, Rigid Pavement Design & Rigid Pavement Joint Design*, 1998, AASHTO, Washington, D.C. Used by permission.

Table 20.18   Modified AASHTO Drainage Coefficients, C_d

| | | Fine-Grained Subgrade | | Coarse-Grained Subgrade | |
| | | Nonpermeable Base | Permeable Base | Nonpermeable Base | Permeable Base |
Edge Drains	Precip. Level				
No	Wet	0.70–0.90	0.85–0.95	0.75–0.95	0.90–1.00
	Dry	0.91–1.10	0.95–1.10	0.90–1.15	1.00–1.15
Yes	Wet	0.75–0.95	1.00–1.10	0.90–1.10	1.05–1.15
	Dry	0.95–1.15	1.10–1.20	1.10–1.20	1.15–1.20

Notes: 1. Fine subgrade = A-1 through A-3 classes;
Coarse subgrade = A-4 through A-8 classes.
2. Permeable base = k = 1000 ft/day (305 m/day) or uniformity coefficient (C_u) ≤ 6.
3. Wet climate = Precipitation > 25 in./year (635 mm/year);
Dry climate = Precipitation ≤ 25 in./year (635 mm/year).
4. Select midpoint of range and use other drainage features (adequacy of cross slopes, depth of ditches, presence of daylighting, relative drainability of base course, bathtub design, etc.) to adjust upward or downward.

SOURCE: From Supplement to the *AASHTO Guide for Design of Pavement Structures, Part II, Rigid Pavement Design & Rigid Pavement Joint Design*, 1998, AASHTO, Washington, D.C. Used by permission.

$$\text{Bstress} = f_d \times P \times T\left[\frac{k_d(2 + BETA \times OPENING)}{4 \times E_s \times I \times BETA^3}\right] \tag{20.35}$$

$$BETA = \sqrt[4]{\frac{k_d DOWEL}{4 \times E_s \times I}} \tag{20.36}$$

f_d = distribution factor = $2 \times (12/l + 12)$ (20.37)

l = radius of relative stiffness (in.)

I = moment of inertia of dowel bar cross section (in.⁴)

$$I = 0.25\pi\left(\frac{DOWEL}{2}\right)^4 \tag{20.38}$$

P = applied wheel load, set to 9000 lb

T = percent transferred load, set to 0.45

k_d = modulus of dowel support, set to 1,500,000 lb/in.²

$BETA$ = relative stiffness of the dowel-concrete system

$DOWEL$ = dowel diameter (in.)

E_s = modulus of elasticity of the dowel bar (lb/in.²)

k = modulus of subgrade reaction (lb/in.²)

$OPENING$ = average transfer joint spacing (lb/in.²)

$$OPENING = 12 \times CON \times \text{Jtspace} \times \left(\frac{ALPHA \times TRANGE}{2 + e}\right) \tag{20.39}$$

Jtspace = average transverse joint spacing (ft)

CON = adjustment factor due to base/slab frictional restraint:

= 0.65 (for stabilized bases)

= 0.80 (for aggregate base or lean concrete base with bond breaker)

$ALPHA$ = PCC thermal expansion coefficient, set to 0.000006/°F

$TRANGE$ = annual temperature range (°F)

e = PCC drying shrinkage coefficient, set to 0.00015 strain

FI = mean annual freezing index, Fahrenheit degree-days

Precip = mean annual precipitation (in.)

Basetype = 0 for unsaturated base, 1 for stabilized base

Widenlane = 0 if not widened, 1 if widened

Age = pavement age (yrs)

The expression for undoweled joints is given as:

$FaultND = CESAL^{0.25} \times$

$$\left[\begin{array}{l} 0.2347 - 0.1516 \times C_d - 0.000250 \times (h^2 / \text{Jtspace}^{0.25}) \\ - 0.0155 \times Basetype + 0.7784 \times 10^{-7} \times FI^{1.5} \times Precip^{0.25} \\ - 0.002478 \times Days90^{0.5} - 0.0415 \times Widenlane \end{array}\right] \tag{20.40}$$

where

$FaultND$ = mean transverse undoweled joint faulting (in.)

$CESAL$ = cumulative equivalent 18-kip single axle load (millions)

C_d = modified AASHTO drainage coefficient (see Table 20.18)

h = slab thickness (in.)

Jtspace = average transverse joint spacing (ft)

Basetype = 0 for unsaturated base, 1 for stabilized base

FI = mean annual freezing index, Fahrenheit degree-days

Precip = mean annual precipitation (in.)

Days90 = number of days with maximum temperature above 90°F

Widenlane = 0 if not widened, 1 if widened

AASHTO has also computed mean joint faulting predictions for 9-in.-thick pavements with doweled and undoweled joints using the procedure presented above. The computed predictions are shown in Tables 20.19 through 20.21.

Table 20.19 Mean Joint Faulting Predictions for Doweled Jointed Plain Concrete Pavement Using Equation 20.34.

ESALs, millions	Granular Base			Treated Base		
	Dowel Diameter 1.00 in.	Dowel Diameter 1.25 in.	Dowel Diameter 1.50 in.	Dowel Diameter 1.00 in.	Dowel Diameter 1.25 in.	Dowel Diameter 1.50 in.
1	0.03	0.01	0.01	0.02	0.00	0.00
2.5	0.04	0.02	0.01	0.03	0.01	0.00
5	0.05	0.03	0.02	0.04	0.01	0.00
10	0.07	0.04	0.03	0.05	0.02	0.01
20	0.10	0.07	0.06	0.08	0.05	0.04
30	0.13	0.10	0.08	0.11	0.07	0.06
40	0.16	0.13	0.11	0.14	0.10	0.09
50	0.20	0.16	0.14	0.17	0.13	0.12
75	0.29	0.24	0.23	0.26	0.22	0.20
100	0.38	0.33	0.32	0.35	0.30	0.29

Values shown in table are mean predicted joint faulting, inches [1 in. = 25.4 mm]
Joint spacing = 15 ft [4.6 m] Slab thickness = 9 in. [229 mm]
k value = 100 lb/in.^{3} [27 kPa/mm] E = 4,000,000 psi [27,580 MPa]
Precipitation = 30 in./year [762 mm/year] TRANGE = 85°F [29.4°C] (July max – January min)
FI = 200°F [93.3°C]-days E_s = 29,000,000 lb/in.^{2} [200,000 MPa]
Lane not widened Age = ESALs in millions
SOURCE: From Supplement to the *AASHTO Guide for Design of Pavement Structures, Part II, Rigid Pavement Design & Rigid Pavement Joint Design*, 1998, AASHTO, Washington, D.C. Used by permission.

Table 20.20 Mean Joint Faulting Predictions for Doweled Jointed Reinforced Concrete Pavement Using Eq. 20.34

ESALs, millions	Granular Base			Treated Base		
	Dowel Diameter 1.00 in.	Dowel Diameter 1.25 in.	Dowel Diameter 1.50 in.	Dowel Diameter 1.00 in.	Dowel Diameter 1.25 in.	Dowel Diameter 1.50 in.
1	0.04	0.02	0.02	0.03	0.01	0.01
2.5	0.05	0.03	0.02	0.04	0.02	0.01
5	0.06	0.04	0.03	0.05	0.02	0.02
10	0.08	0.05	0.04	0.07	0.04	0.03
20	0.12	0.08	0.07	0.10	0.06	0.05
30	0.15	0.12	0.10	0.13	0.09	0.08
40	0.19	0.15	0.13	0.16	0.12	0.11
50	0.22	0.18	0.16	0.20	0.15	0.14
75	0.31	0.27	0.25	0.28	0.24	0.22
100	0.41	0.36	0.34	0.38	0.33	0.31

Values shown in table are mean predicted joint faulting, inches [1 in. = 25.4 mm]
Joint spacing = 45 ft [13.7 m] Slab thickness = 9 in. [229 mm]
k value = 100 lb/in.^{3} [27 kPa/mm] E = 4,000,000 lb/in.^{2} [27,580 MPa]
Precipitation = 30 in./year [762 mm/yr] TRANGE = 85°F [29.4°C] (July max - January min)
FI = 200°C [93.3°C]-days E_s = 29,000,000 lb/in.^{2} [200,000 MPa]
Lane not widened Age = ESALs in millions
SOURCE: From Supplement to the *AASHTO Guide for Design of Pavement Structures, Part II, Rigid Pavement Design & Rigid Pavement Joint Design*, 1998, AASHTO, Washington, D.C. Used by permission.

Table 20.21   Mean Joint Faulting Predictions for Undoweled Jointed Plain Concrete Pavement Using Equation 20.40

ESAL, millions	$C_i = 0.80$		$C_d = 1.0$	
	Joint Spacing 15 ft	Joint Spacing 20 ft	Joint Spacing 15 ft	Joint Spacing 20 ft
1	0.09	0.09	0.05	0.05
2.5	0.12	0.12	0.06	0.06
5	0.14	0.14	0.08	0.08
10	0.16	0.17	0.09	0.09
20	0.20	0.20	0.11	0.11
30	0.22	0.23	0.12	0.12
40	0.23	0.23	0.13	0.13
50	0.25	0.25	0.13	0.14
75	0.27	0.27	0.15	0.15
100	0.29	0.29	0.16	0.16

Values shown in table are mean predicted joint faulting, inches [1 in. = 25.4 mm]
Joint spacing = 15 or 20 ft [4.6 or 6.1 m]
Slab thickness = 9 in. [229 mm]
Precipitation = 30 in./year [762 mm/year]
FI = 200°F [93.3°C]-days
Days90 = 20
Lane not widened
SOURCE: From Supplement to the *AASHTO Guide for Design of Pavement Structures, Part II, Rigid Pavement Design & Rigid Pavement Joint Design*, 1998, AASHTO, Washington, D.C. Used by permission.

20.6.3 PCA Design Method

The PCA method for concrete pavement design is based on a combination of theoretical studies, results of model and full-scale tests, and experience gained from the performance of concrete pavements normally constructed and carrying normal traffic loads. Tayabji and Colley have reported on some of these studies and tests. The design procedure was initially published in 1961 but was revised in 1984. The procedure provides for the determination of the pavement thickness for plain concrete, simply reinforced concrete, and continuously reinforced concrete pavements.

Design Considerations

The basic factors considered in the PCA design method are:

- Flexural strength of the concrete
- Subgrade and subbase support
- Traffic load

Flexural Strength of Concrete.   The flexural strength of the concrete used in this procedure is given in terms of the modulus of rupture obtained by the third-point method (ASTM Designation C78). The average of the 28-day test results is used as input by the designer. The design charts and tables, however, incorporate the variation of the concrete strength from one point to another in the concrete slab and the gain in strength with age.

Subgrade and Subbase Support.   The Westergaard modulus of subgrade reaction (k) is used to define the subgrade and subbase support. This can be determined by performing a plate-bearing test or by correlating with other test results, using the chart in Figure 20.8. No specific correction is made for the reduced value of k during the spring thaw period, but it is suggested that normal summer or fall k values should be used. The modulus of

Table 20.22   Design *k* Values for Untreated and Cement-Treated Subbases

(a) Untreated Granular Subbases

Subgrade *k* Value (lb/in.³)	Subbase *k* Value (lb/in.³)			
	4 in.	*6 in.*	*9 in.*	*12 in.*
50	65	75	85	110
100	130	140	160	190
200	220	230	270	320
300	320	330	370	430

(b) Cement-Treated Subbases

Subgrade *k* Value (lb/in.³)	Subbase *k* Value (lb/in.³)			
	4 in.	*6 in.*	*9 in.*	*12 in.*
50	170	230	310	390
100	280	400	520	640
200	470	640	830	—

SOURCE: R.G. Packard, *Thickness Design for Concrete Highway and Street Pavements*, American Concrete Pavement Association, 1984. Used with permission.

subgrade reaction can be increased by adding a layer of untreated granular material over the subgrade. An approximate value of the increased *k* can be obtained from Table 20.22.

Cement-stabilized soils can also be used as subbase material when the pavement is expected to carry very heavy traffic or when the subgrade material has a low value of modulus of subgrade reaction. Suitable soil materials for cement stabilization are those classified under the AASHTO Soil Classification System as A-1, A-2-4, A-2-5, and A-3. The amount of cement used should be based on standard ASTM laboratory freeze-thaw and wet-dry tests, ASTM Designation D560 and D559, respectively, and weight-loss criteria. It is permissible, however, to use other methods that will produce an equivalent quality material. Suggested *k* values for this type of stabilized material are given in Table 20.22b.

Traffic Load.   The traffic load is computed in terms of the cumulated number of single and tandem axles of different loads projected for the design period of the pavement. The information required to determine cumulated numbers are the average daily traffic (ADT), the average daily truck traffic (*ADTT*) (in both directions), and the axle load distribution of truck traffic. The truck factor is defined as the number of 80 kN single-axle load applications caused by a single passage of a vehicle. This can be determined for each class of vehicle for the expression

$$\text{Truck factor} = \frac{\sum (\textit{number of axles} \times \textit{load equivalency})}{\textit{Number of vehicles}}$$

Only trucks with six or more tires are included in this design. It can be assumed that truck volume is the same in each direction of travel. When there is reason to believe that truck volume varies in each direction, an adjustment factor can be used, as discussed in Chapter 19.

The design also incorporates a load safety factor (LSF), which is used to multiply each axle load. The recommended LSF values are:

- 1.2 for interstate and multilane projects with uninterrupted traffic flow and high truck volumes

- 1.1 for highways and arterials with moderate truck volume
- 1.0 for roads and residential streets with very low truck volume

The LSF can be increased to 1.3 if the objective is to maintain a higher-than-normal pavement serviceability level throughout the design life of the pavement. The design procedure also provides for a factor of safety of 1.1 or 1.2 in addition to the LSF to allow for unexpected truck traffic.

Design Procedure

The procedure is based on a detailed finite-element computer analysis of stresses and deflections of the pavement at edges, joints, and corners. It considers such factors as finite slab dimensions, position of axle load, load transfer at transverse joints or cracks, and load transfer at pavement and concrete shoulder joints. Load transfer characteristics are modeled using the diameter and modulus of elasticity of the dowels at dowel joints.

A spring stiffness value is used for keyway joints, aggregate interlocks, and cracks in continuously reinforced concrete pavements to represent the load deflection characteristics at such joints. Field and laboratory test results are used to determine the spring stiffness value.

Analysis of different axle load positions on the pavement reveals that the Case II position of Figure 20.6 is critical for flexural stresses, whereas Case I position is critical for deflection, but with one set of wheels at or near the corner of the free edge and the transverse joint.

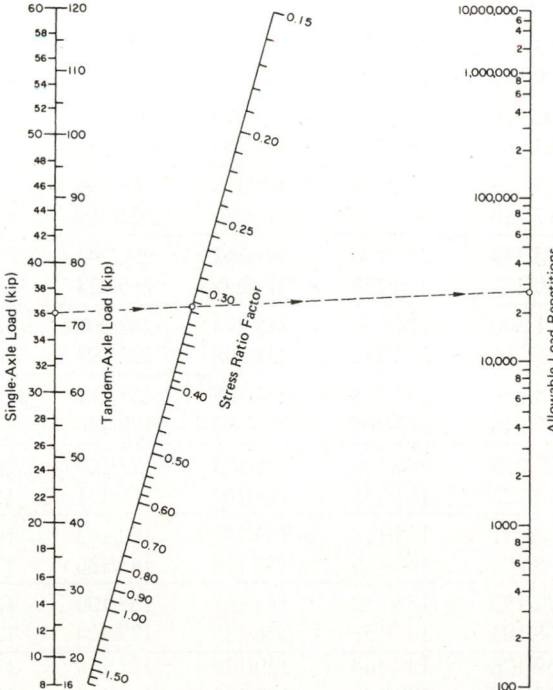

Figure 20.26   Allowable Load Repetitions for Fatigue Analysis Based on Stress Ratio

SOURCE: R.G. Packard, *Thickness Design for Concrete Highway and Street Pavements*, American Concrete Pavement Association, 1984. Used with permission.

The design procedure consists of two parts: fatigue analysis and erosion analysis. The objective of fatigue analysis is to determine the minimum thickness of the concrete required to control fatigue cracking. This is done by comparing the expected axle repetitions with the allowable repetitions for each axle load and ensuring that the cumulative repetitions are less than those allowable. Allowable axle repetitions depend on the stress ratio factor, which is the ratio of the equivalent stress of the pavement to the modulus of rupture of the concrete. The equivalent stress of the pavement depends on the thickness of the slab and the subbase-subgrade k. The chart in Figure 20.26 can be used to determine the allowable load repetitions based on the stress ratio factor. Tables 20.23 and 20.24 give equivalent stress values for pavements without concrete shoulders and with concrete shoulders, respectively.

The objective of the erosion analysis is to determine the minimum thickness of the pavement required to control foundation and shoulder erosion, pumping, and faulting. These pavement distresses are related more closely to deflection, as will be seen later. The erosion criterion is based mainly on the rate of work expended by an axle load in deflecting a slab, as it was determined that a useful correlation existed between pave-

Table 20.23   Equivalent Stress Values for Single Axles and Tandem Axles (without concrete shoulder)

| Slab Thickness (in.) | k of Subgrade-Subbase (lb/in.³) (Single Axle/Tandem Axle) | | | | | | |
	50	100	150	200	300	500	700
4	825/679	726/585	671/542	634/516	584/486	523/457	484/443
4.5	699/586	616/500	571/460	540/435	498/406	448/378	417/363
5	602/516	531/436	493/399	467/376	432/349	390/321	363/307
5.5	526/461	464/387	431/353	409/331	379/305	343/278	320/264
6	465/416	411/348	382/316	362/296	336/271	304/246	285/232
6.5	417/380	367/317	341/286	324/267	300/244	273/220	256/207
7	375/349	331/290	307/262	292/244	271/222	246/199	231/186
7.5	340/323	300/268	279/241	265/224	246/203	224/181	210/169
8	311/300	274/249	255/223	242/208	225/188	205/167	192/155
8.5	285/281	252/232	234/208	222/193	206/174	188/154	177/143
9	264/264	232/218	216/195	205/181	190/163	174/144	163/133
9.5	245/248	215/205	200/183	190/170	176/153	161/134	151/124
10	228/235	200/193	186/173	177/160	164/144	150/126	141/117
10.5	213/222	187/183	174/164	165/151	153/136	140/119	132/110
11	200/211	175/174	163/155	154/143	144/129	131/113	123/104
11.5	188/201	165/165	153/148	145/136	135/122	123/107	116/98
12	177/192	155/158	144/141	137/130	127/116	116/102	109/93
12.5	168/183	147/151	136/135	129/124	120/111	109/97	103/89
13	159/176	139/144	129/129	122/119	113/106	103/93	97/85
13.5	152/168	132/138	122/123	116/114	107/102	98/89	92/81
14	144/162	125/133	116/118	110/109	102/98	93/85	88/78

SOURCE: R.G. Packard, *Thickness Design for Concrete Highway and Street Pavements*, American Concrete Pavement Association, 1984. Used with permission.

Table 20.24 Equivalent Stress Values for Single Axles and Tandem Axles (with concrete shoulder)

Slab Thickness (in.)	k of Subgrade-Subbase (lb/in.³) (Single Axle/Tandem Axle)						
	50	100	150	200	300	500	700
4	640/534	559/468	517/439	489/422	452/403	409/388	383/384
4.5	547/461	479/400	444/372	421/356	390/338	355/322	333/316
5	475/404	417/349	387/323	367/308	341/290	311/274	294/267
5.5	418/360	368/309	342/285	324/271	302/254	276/238	261/231
6	372/325	327/277	304/255	289/241	270/225	247/210	234/203
6.5	334/295	294/251	274/230	260/218	243/203	223/188	212/180
7	302/270	266/230	248/210	236/198	220/184	203/170	192/162
7.5	275/250	243/211	226/193	215/182	201/168	185/155	176/148
8	252/232	222/196	207/179	197/168	185/155	170/142	162/135
8.5	232/216	205/182	191/166	182/156	170/144	157/131	150/125
9	215/202	190/171	177/155	169/146	158/134	146/122	139/116
9.5	200/190	176/160	164/146	157/137	147/126	136/114	129/108
10	186/179	164/151	153/137	146/129	137/118	127/107	121/101
10.5	174/170	154/143	144/130	137/121	128/111	119/101	113/95
11	164/161	144/135	135/123	129/115	120/105	112/95	106/90
11.5	154/153	136/128	127/117	121/109	113/100	105/90	100/85
12	145/146	128/122	120/111	114/104	107/95	99/86	95/81
12.5	137/139	121/117	113/106	108/99	101/91	94/82	90/77
13	130/133	115/112	107/101	102/95	96/86	89/78	85/73
13.5	124/127	109/107	102/97	97/91	91/83	85/74	81/70
14	118/122	104/103	97/93	93/87	87/79	81/71	77/67

SOURCE: R.G. Packard, *Thickness Design for Concrete Highway and Street Pavements*, American Concrete Pavement Association, 1984. Used with permission.

ment performance and the product of the corner deflection and the pressure at the slab-subgrade interface. It is suggested that engineers use the erosion criterion mainly as a guideline and modify it based on experience with local conditions of climate and drainage.

The erosion analysis is similar to that of fatigue analysis, except that an erosion factor is used instead of the stress factor. The erosion factor is also dependent on the thickness of the slab and the subgrade-subbase k. Tables 20.25 through 20.28 give erosion factors for different types of pavement construction. Figures 20.27 and 20.28 are charts that can be used to determine the allowable load repetitions based on erosion.

The minimum thickness that satisfies both analyses is the design thickness. Design thicknesses for pavements carrying light traffic and pavements with doweled joints carrying medium traffic will usually be based on fatigue analysis, whereas design thicknesses for pavements with undoweled joints carrying medium or heavy traffic and pavements with doweled joints carrying heavy traffic will normally be based on erosion analysis.

Table 20.25 Erosion Factors for Single Axles and Tandem Axles (doweled joints, without concrete shoulder)

Slab Thickness (in.)	_k of Subgrade-Subbase (lb/in.³) (Single Axle/Tandem Axle)_					
	_50_	_100_	_200_	_300_	_500_	_700_
4	3.74/3.83	3.73/3.79	3.72/3.75	3.71/3.73	3.70/3.70	3.68/3.67
4.5	3.59/3.70	3.57/3.65	3.56/3.61	3.55/3.58	3.54/3.55	3.52/3.53
5	3.45/3.58	3.43/3.52	3.42/3.48	3.41/3.45	3.40/3.42	3.38/3.40
5.5	3.33/3.47	3.31/3.41	3.29/3.36	3.28/3.33	3.27/3.30	3.26/3.28
6	3.22/3.38	3.19/3.31	3.18/3.26	3.17/3.23	3.15/3.20	3.14/3.17
6.5	3.11/3.29	3.09/3.22	3.07/3.16	3.06/3.13	3.05/3.10	3.03/3.07
7	3.02/3.21	2.99/3.14	2.97/3.08	2.96/3.05	2.95/3.01	2.94/2.98
7.5	2.93/3.14	2.91/3.06	2.88/3.00	2.87/2.97	2.86/2.93	2.84/2.90
8	2.85/3.07	2.82/2.99	2.80/2.93	2.79/2.89	2.77/2.85	2.76/2.82
8.5	2.77/3.01	2.74/2.93	2.72/2.86	2.71/2.82	2.69/2.78	2.68/2.75
9	2.70/2.96	2.67/2.87	2.65/2.80	2.63/2.76	2.62/2.71	2.61/2.68
9.5	2.63/2.90	2.60/2.81	2.58/2.74	2.56/2.70	2.55/2.65	2.54/2.62
10	2.56/2.85	2.54/2.76	2.51/2.68	2.50/2.64	2.48/2.59	2.47/2.56
10.5	2.50/2.81	2.47/2.71	2.45/2.63	2.44/2.59	2.42/2.54	2.41/2.51
11	2.44/2.76	2.42/2.67	2.39/2.58	2.38/2.54	2.36/2.49	2.35/2.45
11.5	2.38/2.72	2.36/2.62	2.33/2.54	2.32/2.49	2.30/2.44	2.29/2.40
12	2.33/2.68	2.30/2.58	2.28/2.49	2.26/2.44	2.25/2.39	2.23/2.36
12.5	2.28/2.64	2.25/2.54	2.23/2.45	2.21/2.40	2.19/2.35	2.18/2.31
13	2.23/2.61	2.20/2.50	2.18/2.41	2.16/2.36	2.14/2.30	2.13/2.27
13.5	2.18/2.57	2.15/2.47	2.13/2.37	2.11/2.32	2.09/2.26	2.08/2.23
14	2.13/2.54	2.11/2.43	2.08/2.34	2.07/2.29	2.05/2.23	2.03/2.19

SOURCE: R.G. Packard, _Thickness Design for Concrete Highway and Street Pavements_, American Concrete Pavement Association, 1984. Used with permission.

Table 20.26 Erosion Factors for Single Axles and Tandem Axles (aggregate interlock joints, without concrete shoulder)

Slab Thickness (in.)	_k of Subgrade-Subbase (lb/in.³) (Single Axle/Tandem Axle)_					
	_50_	_100_	_200_	_300_	_500_	_700_
4	3.94/4.03	3.91/3.95	3.88/3.89	3.86/3.86	3.82/3.83	3.77/3.80
4.5	3.79/3.91	3.76/3.82	3.73/3.75	3.71/3.72	3.68/3.68	3.64/3.65
5	3.66/3.81	3.63/3.72	3.60/3.64	3.58/3.60	3.55/3.55	3.52/3.52
5.5	3.54/3.72	3.51/3.62	3.48/3.53	3.46/3.49	3.43/3.44	3.41/3.40
6	3.44/3.64	3.40/3.53	3.37/3.44	3.35/3.40	3.32/3.34	3.30/3.30
6.5	3.34/3.56	3.30/3.46	3.26/3.36	3.25/3.31	3.22/3.25	3.20/3.21
7	3.26/3.49	3.21/3.39	3.17/3.29	3.15/3.24	3.13/3.17	3.11/3.13
7.5	3.18/3.43	3.13/3.32	3.09/3.22	3.07/3.17	3.04/3.10	3.02/3.06
8	3.11/3.37	3.05/3.26	3.01/3.16	2.99/3.10	2.96/3.03	2.94/2.99
8.5	3.04/3.32	2.98/3.21	2.93/3.10	2.91/3.04	2.88/2.97	2.87/2.93
9	2.98/3.27	2.91/3.16	2.86/3.05	2.84/2.99	2.81/2.92	2.79/2.87
9.5	2.92/3.22	2.85/3.11	2.80/3.00	2.77/2.94	2.75/2.86	2.73/2.81

Table 20.26   Erosion Factors for Single Axles and Tandem Axles (aggregate interlock joints, without concrete shoulder) (*continued*)

Slab Thickness (in.)	*k of Subgrade-Subbase (lb/in.³) (Single Axle/Tandem Axle)*					
	50	*100*	*200*	*300*	*500*	*700*
10	2.86/3.18	2.79/3.06	2.74/2.95	2.71/2.89	2.68/2.81	2.66/2.76
10.5	2.81/3.14	2.74/3.02	2.68/2.91	2.65/2.84	2.62/2.76	2.60/2.72
11	2.77/3.10	2.69/2.98	2.63/2.86	2.60/2.80	2.57/2.72	2.54/2.67
11.5	2.72/3.06	2.64/2.94	2.58/2.82	2.55/2.76	2.51/2.68	2.49/2.63
12	2.68/3.03	2.60/2.90	2.53/2.78	2.50/2.72	2.46/2.64	2.44/2.59
12.5	2.64/2.99	2.55/2.87	2.48/2.75	2.45/2.68	2.41/2.60	2.39/2.55
13	2.60/2.96	2.51/2.83	2.44/2.71	2.40/2.65	2.36/2.56	2.34/2.51
13.5	2.56/2.93	2.47/2.80	2.40/2.68	2.36/2.61	2.32/2.53	2.30/2.48
14	2.53/2.90	2.44/2.77	2.36/2.65	2.32/2.58	2.28/2.50	2.25/2.44

SOURCE: R.G. Packard, *Thickness Design for Concrete Highway and Street Pavements*, American Concrete Pavement Association, 1984. Used with permission.

Table 20.27   Erosion Factors for Single Axles and Tandem Axles (doweled joints, concrete shoulder)

Slab Thickness (in.)	*k of Subgrade-Subbase (lb/in.³) (Single Axle/Tandem Axle)*					
	50	*100*	*200*	*300*	*500*	*700*
4	3.28/3.30	3.24/3.20	3.21/3.13	3.19/3.10	3.15/3.09	3.12/3.08
4.5	3.13/3.19	3.09/3.08	3.06/3.00	3.04/2.96	3.01/2.93	2.98/2.91
5	3.01/3.09	2.97/2.98	2.93/2.89	2.90/2.84	2.87/2.79	2.85/2.77
5.5	2.90/3.01	2.85/2.89	2.81/2.79	2.79/2.74	2.76/2.68	2.73/2.65
6	2.79/2.93	2.75/2.82	2.70/2.71	2.68/2.65	2.65/2.58	2.62/2.54
6.5	2.70/2.86	2.65/2.75	2.61/2.63	2.58/2.57	2.55/2.50	2.52/2.45
7	2.61/2.79	2.56/2.68	2.52/2.56	2.49/2.50	2.46/2.42	2.43/2.38
7.5	2.53/2.73	2.48/2.62	2.44/2.50	2.41/2.44	2.38/2.36	2.35/2.31
8	2.46/2.68	2.41/2.56	2.36/2.44	2.33/2.38	2.30/2.30	2.27/2.24
8.5	2.39/2.62	2.34/2.51	2.29/2.39	2.26/2.32	2.22/2.24	2.20/2.18
9	2.32/2.57	2.27/2.46	2.22/2.34	2.19/2.27	2.16/2.19	2.13/2.13
9.5	2.26/2.52	2.21/2.41	2.16/2.29	2.13/2.22	2.09/2.14	2.07/2.08
10	2.20/2.47	2.15/2.36	2.10/2.25	2.07/2.18	2.03/2.09	2.01/2.03
10.5	2.15/2.43	2.09/2.32	2.04/2.20	2.01/2.14	1.97/2.05	1.95/1.99
11	2.10/2.39	2.04/2.28	1.99/2.16	1.95/2.09	1.92/2.01	1.89/1.95
11.5	2.05/2.35	1.99/2.24	1.93/2.12	1.90/2.05	1.87/1.97	1.84/1.91
12	2.00/2.31	1.94/2.20	1.88/2.09	1.85/2.02	1.82/1.93	1.79/1.87
12.5	1.95/2.27	1.89/2.16	1.84/2.05	1.81/1.98	1.77/1.89	1.74/1.84
13	1.91/2.23	1.85/2.13	1.79/2.01	1.76/1.95	1.72/1.86	1.70/1.80
13.5	1.86/2.20	1.81/2.09	1.75/1.98	1.72/1.91	1.68/1.83	1.65/1.77
14	1.82/2.17	1.76/2.06	1.71/1.95	1.67/1.88	1.64/1.80	1.61/1.74

SOURCE: R.G. Packard, *Thickness Design for Concrete Highway and Street Pavements*, American Concrete Pavement Association, 1984. Used with permission.

Table 20.28 Erosion Factors for Single Axles and Tandem Axles (aggregate interlock joints, concrete shoulder)

Slab Thickness (in.)	k of Subgrade-Subbase (lb/in.³) (Single Axle/Tandem Axle)					
	50	100	200	300	500	700
4	3.46/3.49	3.42/3.39	3.38/3.32	3.36/3.29	3.32/3.26	3.28/3.24
4.5	3.32/3.39	3.28/3.28	3.24/3.19	3.22/3.16	3.19/3.12	3.15/3.09
5	3.20/3.30	3.16/3.18	3.12/3.09	3.10/3.05	3.07/3.00	3.04/2.97
5.5	3.10/3.22	3.05/3.10	3.01/3.00	2.99/2.95	2.96/2.90	2.93/2.86
6	3.00/3.15	2.95/3.02	2.90/2.92	2.88/2.87	2.86/2.81	2.83/2.77
6.5	2.91/3.08	2.86/2.96	2.81/2.85	2.79/2.79	2.76/2.73	2.74/2.68
7	2.83/3.02	2.77/2.90	2.73/2.78	2.70/2.72	2.68/2.66	2.65/2.61
7.5	2.76/2.97	2.70/2.84	2.65/2.72	2.62/2.66	2.60/2.59	2.57/2.54
8	2.69/2.92	2.63/2.79	2.57/2.67	2.55/2.61	2.52/2.53	2.50/2.48
8.5	2.63/2.88	2.56/2.74	2.51/2.62	2.48/2.55	2.45/2.48	2.43/2.43
9	2.57/2.83	2.50/2.70	2.44/2.57	2.42/2.51	2.39/2.43	2.36/2.38
9.5	2.51/2.79	2.44/2.65	2.38/2.53	2.36/2.46	2.33/2.38	2.30/2.33
10	2.46/2.75	2.39/2.61	2.33/2.49	2.30/2.42	2.27/2.34	2.24/2.28
10.5	2.41/2.72	2.33/2.58	2.27/2.45	2.24/2.38	2.21/2.30	2.19/2.24
11	2.36/2.68	2.28/2.54	2.22/2.41	2.19/2.34	2.16/2.26	2.14/2.20
11.5	2.32/2.65	2.24/2.51	2.17/2.38	2.14/2.31	2.11/2.22	2.09/2.16
12	2.28/2.62	2.19/2.48	2.13/2.34	2.10/2.27	2.06/2.19	2.04/2.13
12.5	2.24/2.59	2.15/2.45	2.09/2.31	2.05/2.24	2.02/2.15	1.99/2.10
13	2.20/2.56	2.11/2.42	2.04/2.28	2.01/2.21	1.98/2.12	1.95/2.06
13.5	2.16/2.53	2.08/2.39	2.00/2.25	1.97/2.18	1.93/2.09	1.91/2.03
14.5	2.13/2.51	2.04/2.36	1.97/2.23	1.93/2.15	1.89/2.06	1.87/2.00

SOURCE: R.G. Packard, *Thickness Design for Concrete Highway and Street Pavements*, American Concrete Pavement Association, 1984. Used with permission.

Example 20.7 Designing a Rigid Pavement Using the PCA Method

The procedure is demonstrated by the following example, which is one of the many solutions provided by PCA. The following project and traffic data are available:

Four-lane interstate highway

Rolling terrain in rural location

Design period = 20 yr

Axle loads and expected repetitions are shown in Figure 20.29

Subbase-subgrade k = 35288 kN/m³

Concrete modulus of rupture = 4482 kN/m²

Determine minimum thickness of a pavement with doweled joints and without concrete shoulders.

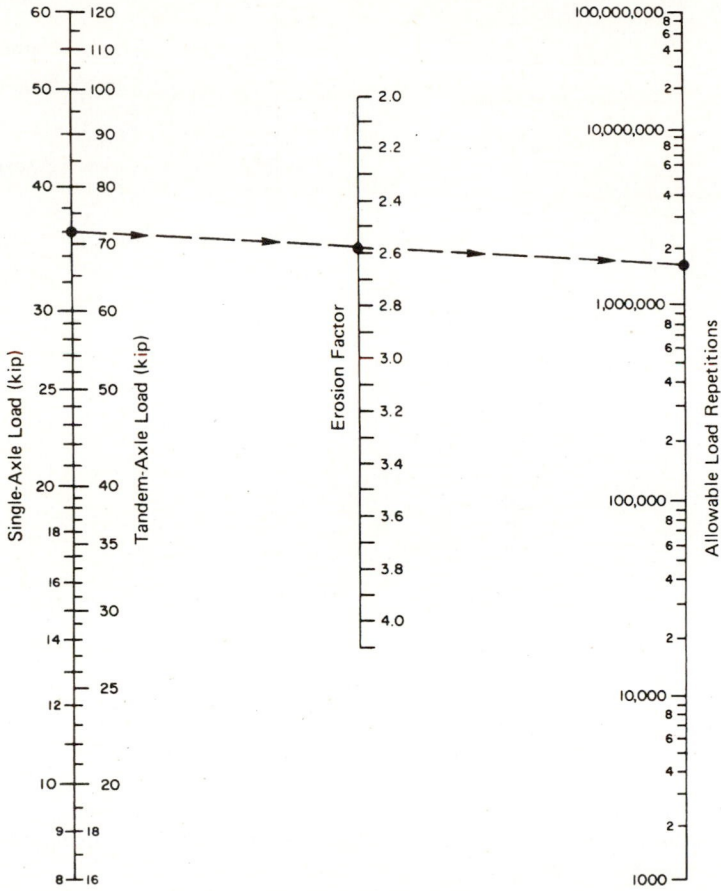

Figure 20.27   Allowable Load Repetitions for Erosion Analysis Based on Erosion Factors (without concrete shoulder)

SOURCE: R.G. Packard, *Thickness Design for Concrete Highway and Street Pavements*, American Concrete Pavement Association, 1984. Used with permission.

Solution: Figure 20.29 is used in working out the example through the following steps.

>   **Step 1.   Fatigue analysis.**
>
>   1. Select a trial thickness (241 mm).
>   2. Complete the information at the top of the form as shown.
>   3. Determine projected number of single-axle and tandem-axle repetitions in the different weight groups for the design period. This is done by determining the ADT, the percentage of truck traffic, the axle load distribution of the truck traffic, and the proportion of trucks on the design lane. The actual cumulative expected repetitions for each range of axle load (without converting to 80 kN equivalent axle load) is then computed using the procedure presented in Chapter 19.

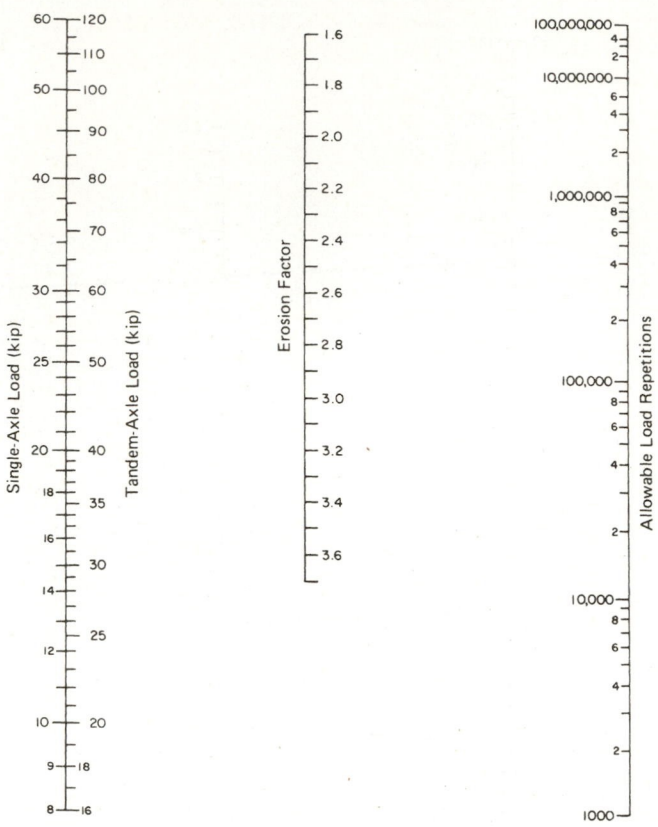

Figure 20.28   Allowable Load Repetitions for Erosion Analysis Based on Erosion Factors (with concrete shoulder)

SOURCE: R.G. Packard, *Thickness Design for Concrete Highway and Street Pavements*, American Concrete Pavement Association, 1984. Used with permission.

4. Complete columns 1, 2, and 3. Note that column 2 is column 1 multiplied by the LSF, which in this case is 1.2, since the design is for a four-lane interstate highway.

5. Determine the equivalent stresses for single axle and tandem axle. Table 20.23 is used in this case since there is no concrete shoulder. Interpolating for $k = 130$, for single axles and 241 mm-thick slab,

$$\text{Equivalent stress} = 215 - \frac{215 - 200}{50} \times 30$$
$$= 1420 \text{ kN/m}^2$$

Similarly, for the tandem axles,

$$\text{Equivalent stress} = 205 - \frac{205 - 183}{50} \times 30$$
$$= 1324 \text{ kN/m}^2$$

Record these values in spaces provided on lines 8 and 11, respectively.

Project _Design 1A, four-lane Interstate, Rural_

Trial thickness ___9.5___ in.   Doweled joints: yes ✔ no ___
Subbase-subgrade k ___130___ pci   Concrete shoulder: yes ___ no ✔
Modulus of rupture, MR ___650___ psi
Load safety factor, LSF ___1.2___   Design period __20__ years

4 in. untreated subbase

Axle load, kips	Multiplied by LSF 1.2	Expected repetitions	Fatigue analysis		Erosion analysis	
			Allowable repetitions	Fatigue, percent	Allowable repetitions	Damage, percent
1	2	3	4	5	6	7

8. Equivalent stress ___206___   10. Erosion factor ___2.59___

Single Axles   9. Stress ratio factor ___0.317___

30	36.0	6,310	27,000	23.3	1,500,000	0.4
28	33.6	14,690	77,000	19.1	2,200,000	0.7
26	31.2	30,140	230,000	13.1	3,500,000	0.9
24	28.8	64,410	1,200,000	5.4	5,900,000	1.1
22	26.4	106,900	Unlimited	0	11,000,000	1.0
20	24.0	235,800	"	0	23,000,000	1.0
18	21.6	307,200	"	0	64,000,000	0.5
16	19.2	422,500			Unlimited	0
14	16.8	586,900			"	0
12	14.4	1,837,000			"	0

11. Equivalent stress ___192___   13. Erosion factor ___2.79___

Tandem Axles   12. Stress ratio factor ___0.295___

52	62.4	21,320	1,100,000	1.9	920,000	2.3
48	57.6	42,870	Unlimited	0	1,500,000	2.9
44	52.8	124,900	"	0	2,500,000	5.0
40	48.0	372,900	"	0	4,600,000	8.1
36	43.2	885,800			9,500,000	9.3
32	38.4	930,700			24,000,000	3.9
28	33.6	1,656,000			92,000,000	1.8
24	28.8	984,900			Unlimited	0
20	24.0	1,227,000			"	0
16	19.2	1,356,000				
			Total	62.8	Total	38.9

Figure 20.29   Pavement Thickness Calculation

SOURCE: R.G. Packard, _Thickness Design for Concrete Highway and Street Pavements_, American Concrete Pavement Association, 1984. Used with permission.

6. Determine the stress ratio, which is the equivalent stress divided by the modulus of rupture.

   (a) For single axles:

   $$\text{Stress ratio} = \frac{206}{650} = 0.317$$

   (b) For tandem axles:

   $$\text{Stress ratio} = \frac{192}{650} = 0.295$$

   Record these values in spaces provided on lines 9 and 12, respectively.

7. Using Figure 20.26, determine the allowable load repetitions for each axle load based on fatigue analysis. Note that the axle loads used are those in column 2 (that is, after multiplying by the LSF). For example, for the first row of single axles, the axle load to be used is 36. Draw a line joining 36 kip on the single-axle load axis (left side) and 0.317 on the stress ratio factor line. Extend that line to the allowable load repetitions line and read allowable repetitions as 27,000. Repeat this for all axle loads, noting that for tandem axles the scale is on the right side of the axle load axis. Complete column 4 of Figure 20.18.

8. Determine the fatigue percentage for each axle load, which is an indication of the resistance consumed by the expected number of axle load repetitions:

   $$\text{Fatigue percentage} = \frac{\text{column 3}}{\text{column 4}} \times 100$$

   Complete column 5.

9. Determine total fatigue resistance consumed by summing up column 5 (single and tandem axles). If this total does not exceed 100 percent, the assumed thickness is adequate for fatigue resistance for the design period. The total in this example is 62.8 percent, which shows that 241 mm (9.5 in) is adequate for fatigue resistance.

Step 2.   Erosion analysis. Items 1 through 4 in the fatigue analysis are the same for the erosion analysis; therefore, they will not be repeated. The following additional procedures are necessary.

1. Determine the erosion factor for the single-axle loads and tandem-axle loads. Note that the erosion factors depend on the type of pavement construction—that is, doweled or undoweled joints, with or without concrete shoulders. The construction type in this project is doweled joints without shoulders. Table 20.25 is therefore used. For single axles, erosion factors are 2.59. For tandem axles, the erosion factor is 2.79. Record these values as items 10 and 13, respectively.

2. Determine the allowable axle repetitions for each axle load based on erosion analysis using either Figure 20.27 or Figure 20.28. In this

problem, Figure 20.27 will be used, as the pavement has no concrete shoulder. Enter these values under column 6.

3. Determine erosion damage percent for each axle load; that is, divide column 3 by column 6. Enter these values in column 7.
4. Determine the total erosion damage by summing column 7 (single and tandem axles). In this problem, total drainage is 38.9 percent.

The results obtained for Example 20.7 as shown in Figure 20.29 indicate that 9.5 in. is adequate for both fatigue and erosion analysis. Since the total damage for each analysis is much lower than 100 percent, the question may arise whether a 228 mm (9-in)-thick pavement would not be adequate. If the calculation is repeated for a 228 mm (9-in)-thick pavement, it will show that the fatigue condition will not be satisfied since total fatigue consumption will be about 245 percent. In order to achieve the most economic section for the design period, trial runs should be made until the minimum pavement thickness that satisfies both analyses is obtained.

20.6.4 Mechanistic–Empirical Design Guide (MEPDG) Method

The MEPDG design procedure presented is for jointed plain concrete pavement (JPCP), as the design of continuously reinforced concrete pavement (CRCP) is beyond the scope of this book. The procedure is similar to that for the flexible pavement described in Chapter 19, in that the designer starts with a trial design, which is then evaluated as to its adequacy based on certain criteria that have been selected by the designer. The design process also involves three major stages: Stage 1 involves determination of the input values for the trial design, Stage 2 involves analyses of the structural capability of the trial pavement through the prediction of the selected performance indicators, and Stage 3 is the evaluation of the structural viability of the trial design. The traffic, foundation, and subgrade soils and material characteristics for which input values are required are similar to those for the flexible pavement. For example, the traffic characterization involves estimating the number of axle load distributions that will be applied onto the pavement. These are given as the axle load spectra, defined by the MEPDG as the "distribution of axle loads for a specific axle type (single, tandem, tridem, quad). In other words, the number of axle applications within a specific axle load range." Similarly, the changes in horizontal and vertical characteristics of the foundation and subgrade soils, including soil types, moisture contents, densities, water-table depth, and location of rock strata, should be determined. In this case, since the surface material is Portland Cement Concrete (PCC), the inputs include the elastic modulus, Poisson's ratio, flexural strength, unit weight, coefficient of thermal expansion, thermal conductivity, and heat capacity. All of these can be obtained by conducting the appropriate tests described in the ASTM and/or AASHTO publications.

The predicted distresses for a new JPCP are the *Mean Transverse Joint Faulting,* the *Transverse Slab Cracking* (bottom-up and top-down), and the *Smoothness.* The design criteria or threshold values recommended in the MEPDG for these parameters are shown in Table 20.29. Suggested reliability levels for different functional classifications roadways were given in Chapter 19.

Transverse Slab Cracking

This may occur either in the bottom-up or top-down modes in a JPCP slab under typical service conditions. Therefore, a combined cracking that does not consider whether it is

Table 20.29   Design Criteria or Threshold Values Recommended for Use in Judging the Acceptability of a New JPCP, CRCP, and Overlay Trial Design

Mean joint faulting	Interstate: 0.15 in.
	Primary: 0.20 in.
	Secondary: 0.25 in.
Percent transverse slab cracking	Interstate: 10%
	Primary: 15%
	Secondary: 20%
IRI (smoothness)	Interstate: 160 in./mi
	Primary: 120 in./mi
	Secondary: 200 in./mi

SOURCE: From *Mechanistic-Empirical Pavement Design Guide, Interim Edition: A Manual of Practice*, 2008, AASHTO, Washington, D.C. Used by permission.

bottom-up or top-down is determined. It is based on the percentage of slabs with transverse cracks in a given traffic lane and is given as

$$CRK = \frac{1}{1 + (DI_F)^{-1.98}} \tag{20.41}$$

where

CRK = predicted amount of bottom-up or top-down cracking (fraction)
DI_F = fatigue damage calculated using the following procedure

Miner's hypothesis is used to determine the accumulated fatigue damage and is given as

$$DI_F = \sum \frac{n_{i,j,k,l,m,n,o}}{N_{i,j,k,l,m,n,o}} \tag{20.42}$$

where

DI_F = total fatigue damage (top-down or bottom-up)
$n_{i,j,k,l,m,n,o}$ = applied number of load applications at condition i,j,k,l,m,n,o
$N_{i,j,k,l,m,n}$ = allowable number of load applications at condition i,j,k,l,m,n
i = age (accounts for change in PCC modulus of rupture and elasticity, slab/base contact friction, and deterioration of shoulder load transfer efficiency (LTE))
j = month (accounts for change in base elastic modulus and effective dynamic modulus of subgrade reaction)
k = axle type (single, tandem, and tridem for bottom-up cracking; short, medium, and long wheel base for top-down cracking)
l = load level (incremental load for axle type)
m = equivalent temperature difference between top and bottom PCC surfaces
n = traffic offset path
o = hourly truck traffic

Note that the applied number of load applications ($n_{i,j,k,l,m,n,o}$) is the total number of type k axle passes carrying a load level l that passed through path n for each condition of age, season, and temperature difference. Also, the allowable number of load applications at condition i,j,k,l,m,n,o is the number of load cycles that will result in fatigue failure, usually taken as 50 percent slab cracking. The allowable number of load applications is obtained from Eq. 20.43:

$$\log (N_{i,j,k,l,m,n}) = C_1 \left(\frac{S_{c,i}^i}{\sigma_{i,j,k,l,m,n}} \right)^{C_2} \qquad (20.43)$$

where

$$
\begin{aligned}
N_{i,j,k,l,m,n} &= \text{allowable number of load applications at condition } i,j,k,l,m,n \\
S_{c,i}^i &= \text{PCC modulus rupture at age } i \text{ (kN/mm}^2\text{)} \\
\sigma_{i,j,k,l,m,n} &= \text{applied stress at condition } i,j,k,l,m,n \\
C_1 &= \text{calibration constant, 2.0} \\
C_2 &= \text{calibration constant, 1.22}
\end{aligned}
$$

Example 20.8 Determining the Transverse Cracking for a Set of Given Conditions.

Determine the expected predicted cracking for the three traffic levels and associated applied stresses given, respectively, in columns 1 and 2 in Table 20.30 for a given condition of age, season, and temperature difference if the modulus of rupture of the PCC is 4481 kN/m².

Solution:

- Determine the allowable number of load applications for each axle level load (use Eq. 20.43). See column 4 of Table 20.30.
- Determine the ratio of number of load applications to allowable number of load applications for each axle load level—see column 5 of Table 20.30.
- Determine total fatigue damage from Eq. 20.42.

$$DI_F = \sum \frac{n_{i,j,k,l,m,n,o}}{N_{i,j,k,l,m,n,o}} = 0.182 + 0.129 + 0.153 = 0.464$$

Table 20.30 Data and Solution for Example 20.8

Load Level (1)	Number of Applications (n) (2)	Applied Stress ($\sigma_{i,j,k,l,m,n}$) (kN/m²) Axle (3)	Allowable Number of Load Applications (N) $\log (N_{i,j,k,l,m,n}) = C_1 \left(\frac{S_{c,i}^i}{\sigma_{i,j,k,l,m,n}} \right)^{C_2}$ (4)	n/N (5)
1	0.025×10^6	2068	0.137×10^6	0.182
2	0.05×10^6	1930	0.387×10^6	0.129
3	0.079×10^6	1896	0.515×10^6	0.153
				$\Sigma = 0.464$

- Determine the predicted amount of transverse cracking. Use Eq. 20.41.

$$CRK = \frac{1}{1 + (DI_F)^{-1.98}}$$

$$CRK = \frac{1}{1 + (0.464)^{-1.98}} = 0.179$$

The *total transverse cracking* is obtained from the expression

$$TCrack = (CRK_{\text{bottom-up}} + CRK_{\text{top-down}} - CRK_{\text{bottom-up}}CRK_{\text{top-down}}) \bullet 100 \qquad (20.44)$$

where

$TCrack$ = total transverse cracking (percent, all severities)
$CRK_{\text{bottom-up}}$ = predicted amount of bottom-up transverse cracking (fraction)
$CRK_{\text{top-down}}$ = predicted amount of top-down transverse cracking (fraction)

The standard deviation of the residual error for the percentage of slabs cracked is given as

$$S_{e(CR)} = -0.00198(CRACK)^2 + 0.56857CRACK + 2.76825 \qquad (20.45)$$

where

$CRACK$ = predicted transverse cracking based on mean inputs (corresponding to 50 percent reliability), percentage of slabs
$S_{e(CR)}$ = standard error of the estimate of transverse cracking at the predicted level of mean cracking

Mean Transverse Joint Faulting

Determined for each month by first computing the incremental faulting during a given month from that of the previous month, the faulting at each month is then the sum of faulting increments from all the previous months, starting from the traffic opening date. Equations 20.46 through 20.49 are used for these computations.

$$Fault_m = \sum_{i=1}^{m} \Delta Fault_i \qquad (20.46)$$

$$\Delta Fault_i = C_{34}(Faultmax_{i-1} - Fault_{i-1})^2(DE_i) \qquad (20.47)$$

$$Faultmax_i = Faultmax_0 + C_7\left[\sum_{j=1}^{m} DE_j(\log(1 + 5.0^{EROD}C_5)^{C_6})\right] \qquad (20.48)$$

$$Faultmax_0 = \delta_{\text{curling}}C_{12}\left[Log(1 + 5.0^{EROD}(C_5))Log\left(\frac{WetDays(P_{200})}{P_s}\right)\right]^{C_6} \qquad (20.49)$$

where

$Fault_m$ = mean joint faulting at the end of month m (mm)
$\Delta Fault_i$ = incremental change (monthly) in mean transverse joint faulting during month i (mm)

Fault max$_i$ = maximum mean transverse joint faulting for month i (mm)

Fault max$_0$ = initial maximum mean transverse joint faulting i (mm)

$EROD$ = base/subbase erodability factor

DE_i = differential density of energy of subgrade deformation accumulated during month i

$\delta_{curling}$ = maximum mean monthly slab corner upward deflection of PCC due to temperature curling and moisture warping

P_s = overburden of subgrade (kN)

P_{200} = percent subgrade material passing No. 200 sieve

WetDays = average annual number of wet days (greater than 0.1 in. rainfall)

$C_{1,2,3,4,5,6,7,12,34}$ = global calibration constant; C_1 = 1.29; C_2 = 1.1; C_3 = 0.001725; C_4 = 0.0008; C_5 = 250; C_6 = 0.4; C_7 = 1.2; and

$$C_{12} = C_1 + C_2 \, (FR)^{0.25} \tag{20.50}$$

$$C_{34} = C_3 + C_4 \, (FR)^{0.25} \tag{20.51}$$

FR = base freezing index defined as percentage of time the top base temperature is below freezing (0 C) temperature.

Note that a finite element-based neural network response solution methodology that is embedded in the MEPDG software is used to determine the corner deflections due to curling and shrinkage. This computation is carried out for each month, using the effective temperature differential for the month and the corresponding k value and base modulus. The effective temperature differential for month m is given as

$$\Delta T_m = \Delta T_{t,m} - \Delta T_{b,m} + \Delta T_{sh,m} + \Delta T_{PCW} \tag{20.52}$$

where

ΔT_m = effective temperature differential for month m

$\Delta T_{t,m}$ = mean PCC top-surface nighttime temperature (from 8 p.m. to 8 a.m.) for month m

$\Delta T_{b,m}$ = mean PCC bottom-surface nighttime temperature (from 8 p.m. to 8 a.m.) for month m

$\Delta T_{sh,m}$ = equivalent temperature differential due to reversible shrinkage for month m for old concrete (i.e., shrinkage is fully developed)

ΔT_{PCW} = equivalent temperature differential due to permanent curl/warp

The design process also considers the load transfer efficiency (LTE) as it has a significant impact on the corner deflections of the loaded and unloaded slabs. This is checked monthly to ascertain that a high level is maintained. The MEPDG uses an equation relating the total transverse joint load transfer efficiency (LTE$_{joint}$) with that for the dowels, base or aggregate interlock, depending on the mechanism of load transfer used. The LTE for a given month is increased if the mean nighttime PCC temperature at the mid-depth of the slab is below 0 C. The MEPDG gives suggested LTE values for different base types. Another factor used in the faulting analysis is the differential energy of subgrade deformation (DE) caused by the difference between the loaded corner deflection and the unloaded corner deflection. The software computes the deflections at the loaded and unloaded corner of the slab using the neural networks, and then determines the DE from an expression that relates the DE with the coefficient of subgrade reaction and the difference between the loaded corner deflection and the unloaded corner deflection, as shown in Eq. 20.53. The shear stress at the slab corner (τ) and the maximum doweled bearing stress σ_b for doweled joints can also be determined from the difference between the loaded corner deflection and the unloaded corner deflection using Eqs. 20.53, 20.54, and 20.55.

$$DE = \frac{k}{2}(\delta_L^2 - \delta_U^2) \tag{20.53}$$

$$\tau = \frac{AGG(\delta_L - \delta_U)}{h_{PCC}} \tag{20.54}$$

$$\sigma_b = \frac{\xi_D(\delta_L - \delta_U)}{(d)(dsp)} \tag{20.55}$$

where

DE = differential energy (kN/mm)
δ_L = loaded corner deflection (mm)
δ_U = unloaded corner deflection (mm)
AGG = aggregate interlock stiffness factor
k = coefficient of subgrade reaction (kN/mm³)
h_{PCC} = PCC slab thickness (mm)
ξ_D = dowel stiffness factor = $J_d \times k \times l \times dsp$
d = dowel diameter (mm)
dsp = dowel spacing (mm)
J_d = nondimensional dowel stiffness at the time of load application
l = radius of relative stiffness (mm)

The standard error for the transverse joint faulting global prediction equation is given as

$$S_{e(F)} = (0.00761(Fault(t) + 0.00008099))^{0.445} \tag{20.56}$$

where

$S_{e(F)}$ = standard error for the transverse joint faulting global prediction equation
$Fault(t)$ = predicted mean transverse joint faulting at any given time t (mm)

The predicted value of the *Smoothness* depends on the initial as-constructed profile of the pavement, plus any changes of the longitudinal profile that occur as a result of traffic loads, plus distresses that result from movements of the foundation. The calibrated model for the IRI used by the MEPDG software is given as

$$IRI = IRI_1 + C1(CRK) + C2(spall) + C3(Tfault) + C4(SF) \tag{20.57}$$

where

IRI = predicted IRI (mm/m)
IRI_1 = initial smoothness measured as IRI (mm/m)
CRK = percent slabs with transverse cracks (all severities) (see Eq. 20.41)
$spall$ = percent of joints with spalling (medium and high severities)
$Tfault$ = total faulting cumulated (mm/m) (see Eq. 20.46)
$C1$ = 0.8203
$C2$ = 0.4417

$$C3 = 0.4929 \tag{20.58}$$
$$C4 = 25.24$$
$$SF = \text{site factor}$$
$$SF = AGE(1 + 0.5556FI(1 + P_{200})10^6$$

where

AGE = pavement age (yr)
FI = freezing index (°F-days)
P_{200} = percent subgrade material passing No. 200 sieve

The *transverse joint spalling* (spall) is obtained from Eq. 20.59.

$$Spall = \left[\frac{age}{age + 0.01}\right]\left[\frac{100}{1 + 1.005^{(-12age + scf)}}\right] \tag{20.59}$$

where

$Spall$ = percentage joints spalled (medium- and high-severities)
age = pavement age (yr)
scf = scaling factor based on site-, design-, and climate-related factors

$$scf = -1400 + 350(AC_{PCC})(0.5 + Preform) + 3.4(0.4f_c') - 0.2(Age)(FT_{cycles}) \tag{20.60}$$
$$+ 43H_{PCC} - 539(WC_{PCC})$$

where

AC_{PCC} = PCC air content (percent)
Age = pavement age (yr)
$Preform$ = 1 if reformed sealant is present; 0 if not
f_c' = PCC compressive strength (lb/in.^{2})
FT_{cycles} = average annual number of freeze-thaw cycles
H_{PCC} = PCC slab thickness (in.)
WC_{PCC} = PCC water/cement ratio

The standard error for the IRI prediction equation for JRCP is given as

$$s_{e(IRI)} = (Var_{IRIi} + C1^2(Var_{CRK}) + C2^2(Var_{spall}) + C3^2(Var_{fault}) + S_e^2)^{0.5} \tag{20.61}$$

where

$se_{(IRI)}$ = standard deviation of IRI at the predicted level of mean IRI
Var_{IRIi} = variance of initial IRI = 29.16 (in./mi)^2
Var_{CRK} = variance of cracking (percent slab)^2
Var_{spall} = variance of spalling (percent joints)^2 = 46.24 (percent joints)^2
Var_{fault} = variance of faulting (in./mi)^2
S_e^2 = variance of overall model error = 745.3 (in./mi)^2

The third stage is the evaluation of the structural viability of the trial design. This involves the comparison of the predicted values, including predicted probabilities, and the stipulated criteria to ascertain the adequacy of the trial design. If the predicted values do not meet the stipulated design criteria, the trial design is changed and the process repeated.

20.7 SUMMARY

This chapter presents the basic design principles for rigid highway pavements and the application of these principles in the AASHTO and PCA methods. A brief introduction of the MEPDG method is also presented. Rigid pavements have some flexural strength and can therefore sustain a beam-like action across minor irregularities in the underlying material. The flexural strength of the concrete used is therefore an important factor in the design of the pavement thickness. Although reinforcing steel sometimes may be used to reduce cracks in rigid pavement, it is not considered as contributing to the flexural strength of the pavement.

The AASHTO, PCA, and MEPDG methods of design are all iterative, indicating the importance of computers in carrying out the design. Design charts are included in this chapter for the AASHTO and PCA methods, which can be used if computer facilities are not readily available. A computer is, however, needed for the MEPDG method because of the extensive amount of calculations required. A trial design is required for each method and the required analysis is carried out to determine the adequacy of the trial design.

A phenomenon that may have significant effect on the life of a concrete pavement is pumping. The PCA method indirectly considers this phenomenon in the erosion consideration. When rigid pavements are to be constructed over fine-grain soils that are susceptible to pumping, effective means to reduce pumping, such as limiting expansion joints or stabilizing the subgrade, should be considered.

PROBLEMS

20-1   Portland cement concrete consists of what four primary elements?

20-2   List and briefly describe the five main types of Portland cement as specified by AASHTO.

20-3   What is the main requirement for the water used in Portland cement?

20-4   Describe the basic types of highway concrete pavements and give the conditions under which each type will be constructed.

20-5   List and briefly describe the four type of joints used in concrete pavements.

20-6   Define the phenomenon of pumping and its effects on rigid pavements.

20-7   List and describe the types of stresses that are developed in rigid pavements.

20-8   Determine the tensile stress imposed at night by a wheel load of 750 lb located at the edge of a concrete pavement with the following dimensions and properties.

Pavement thickness = 216 mm
μ = 0.15
E_c = 28.9 kN/mm^2
k = 5.43 kN/mm^3 × 10^{-5}
Radius of loaded area = 76 mm

20-9   Repeat Problem 20-8 for the load located at the interior of the slab.

20-10   Determine the tensile stress imposed at day by a wheel load of 3.5 kN located at the edge of a concrete pavement with the following dimensions and properties.

Pavement thickness = 254 mm
μ = 0.15
E_c = 31.7 kN/mm^2
k = 4.34 kN/mm^3 × 10^{-5}
Radius of loaded area = 76 mm

20-11   Repeat Problem 20-10 for the load located at the interior of the slab.

20-12   Determine the maximum distance of contraction joints for a plain concrete pavement if the maximum allowable tensile stress in the concrete is 50 lb/in.^{2} and the coefficient of friction between the slab and the subgrade is 1.7. Assume uniform drop in temperature. Weight of concrete is 144 lb/ft^3.

20-13   Repeat Problem 20-12, with the slab containing temperature steel in the form of welded wire mesh consisting of 0.125 in.^{2} steel/ft width. The modulus of steel, E_s, is 30×10^6 lb/in.^{2}, $E_c = 5 \times 10^6$ lb/in.^{2}, and h = 6 inches.

20-14   A concrete pavement is to be constructed for a 4-lane urban expressway on a subgrade with an effective modulus of subgrade reaction k of 100 lb/in.^{3}. The accumulated equivalent axle load for the design period is 3.25×10^6. The initial and terminal serviceability indices are 4.5 and 2.5, respectively. Using the AASHTO design method determine a suitable thickness of the concrete pavement if the working stress of the concrete is 600 lb/in.^{2} and the modulus of elasticity is 5×10^6 lb/in.^{2}. Take the overall standard deviation, S_o, as 0.30, the load transfer coefficient, J, as 3.2, the drainage coefficient as 0.9, and $R = 95\%$.

20-15   A six-lane concrete roadway is being designed for a metropolitan area. This roadway will be constructed on a subgrade with an effective modulus of subgrade reaction k of 170 lb/in.^{3}. The ESALs used for the design period is 2.5×10^6. Using the AASHTO design method, determine a suitable thickness of the concrete pavement (to the nearest 1/2 inch), provided that the working stress of the concrete to be used is 650 lb/in.^{2} and the modulus of elasticity is 5×10^6 lb/in.^{2}. Assume the initial serviceability is 4.75 and the terminal serviceability is 2.5. Assume the overall standard deviation, S_o, is 0.35, the load transfer coefficient J as 3.2, the drainage coefficient, C_d, is 1.15, and $R = 95\%$.

20-16   Revisit Problem 20-15 under the conditions in which a revised traffic analysis indicates that 3.5×10^6 ESALs are expected during the design life and the desired reliability has been revised to 99 percent.

20-17   Determine the required slab thickness, according to the AASHTO Alternate Design Method, for a rigid pavement being designed near Jacksonville, Florida, for the following input data:

Estimated future traffic load = 10,000,000 ESALs
Design reliability = 95%
Overall standard deviation = 0.39
Initial serviceability index = 4.5
Terminal serviceability index = 2.5
Effective subgrade modulus = 250 lb/in.^{2}
Mean concrete modulus of rupture = 700 lb/in.^{2}
Mean concrete elastic modulus = 4,200,000 lb/in.^{2}
Joint spacing = 20 ft
Base modulus = 25,000 lb/in.^{2}
Base thickness = 6 in.
Slab width = 12 ft supported by asphalt concrete shoulders

20-18   Determine whether the slab thickness obtained in Problem 20-17 is adequate for the tensile stress that will occur for joint loading.

20-19   An existing rural 4-lane highway is to be replaced by a 6-lane divided expressway (3 lanes in each direction). Traffic volume data on the highway indicate that the *AADT* (both directions) during the first year of operation is 24,000 with the following vehicle mix and axle loads.

Passenger cars = 50 percent
2-axle single-unit trucks (12,000 lb/axle) = 40 percent
3-axle single-unit trucks (16,000 lb/axle) = 10 percent

The vehicle mix is expected to remain the same throughout the design life of 20 years, although traffic is expected to grow at a rate of 3.5 percent annually. Using the AASHTO design procedure, determine the minimum depth of concrete pavement required for the design period of 20 years.

$P_i = 4.5$ $J = 3.2$
$P_t = 2.5$ $C_d = 1.0$
$S'_c = 650 \text{ lb/in.}^2$ $S_o = 0.3$
$E_c = 5 \times 10^6 \text{ lb/in.}^2$ $R = 95\%$
$k = 130 \text{ lb/in.}^3$

20-20   Repeat Problem 20-19 for a pavement containing doweled joints, 6-in. untreated subbase, and concrete shoulders, using the PCA design method.

20-21   Repeat Problem 20-19 using the PCA design method, for a pavement containing doweled joints and concrete shoulders. The modulus of rupture of the concrete used is 600 lb/in.².

20-22   Repeat Problem 20-19, using the PCA design method, if the subgrade k value is 50 and a 6-in. stabilized subbase is used. The modulus of rupture of the concrete is 600 lb/in.² and the pavement has aggregate interlock joints (no dowels) and a concrete shoulder.

20-23   Repeat Problem 20-22, using the PCA design method, assuming the pavement has doweled joints and no concrete shoulders.

20-24   Briefly describe the steps involved in the MEPDG method for JPCP.

20-25   What are the main input parameters used in the MEPDG for JPCP?

20-26   List and define each of the main criteria that are used for evaluating a trial JPCP structure in the MEPDG method.

20-27   Determine the expected predicted transverse cracking for the five traffic levels and associated applied stresses given on Table 20.31 for a given condition of age, season, and temperature difference if the modulus of rupture of the PCC is 650 lb/in.².

Table 20.31   Data for Problem 20-27

Load Level (1)	Number of Applications (n) (2)	Applied Stress $(\sigma_{i,j,k,l,m,n})$ $(lb/in.^2)$ Axle (3)
1	0.02×10^6	300
2	0.05×10^6	280
3	0.08×10^6	275
4	0.1×10^6	270
5	0.12×10^6	260

REFERENCES

AASHTO Guide for Design of Pavement Structures, American Association of State Highway and Transportation Officials, Washington, D.C., 1993.

Supplement to the *AASHTO Guide for the Design of Pavement Structures, Part II, Rigid Pavement Design and Rigid Pavement Joint Design*, American Association of State Highway and Transportation officials, Washington D.C., 1995.

Annual Book of ASTM Standards, Concrete and Mineral Aggregates, Vol. 04.02, Road and Paving Materials; Pavement Management Technologies, American Society for Testing and Materials, Philadelphia, PA., October 1992.

Interim Mechanistic-Empirical Pavement Design Guide Manual of Practice, Transportation Research Board, National Research Council, Washington, D.C., August 2007.

Ioannides, A. M., M. R. Thompson, and E. J. Barenberg, "Westergaard Solutions Reconsidered," *Transportation Research Record 1043*, Transportation Research Board, Washington, D.C., 1985.

Packard, R. G., *Thickness Design for Concrete Highway and Street Pavements*, American Concrete Pavement Association, Skokie, ILlo, 1984.

Standard Specifications for Transportation Materials and Methods of Sampling and Testing, 20th ed., American Association of State Highway and Transportation Officials, Washington, D.C., 2007.

CHAPTER 21

Pavement Management

The nation's highway system will be required to accommodate increasing numbers of motor vehicles during the coming decades. Accordingly, rehabilitation of the existing system has become a major activity for highway and transportation agencies. With the nation's interstate highway system completed, the focus is shifting from new construction to maintaining, preserving, and rehabilitating highway assets. Preserving and managing the nation's highways is a challenge, and transportation professionals are investigating tools and techniques to assist in this endeavor. This chapter discusses pavement management strategies and the data required to evaluate pavement condition.

CHAPTER OBJECTIVES:

- Understand the problems of pavement rehabilitation.
- Learn the methods used to establish roadway pavement condition.
- Predict pavement condition using mathematical models.
- Identify and apply strategies to rehabilitate pavements or to correct deficiencies.
- Analyze and select a program of pavement rehabilitation based on condition assessment, priority assessment, and optimization.

21.1 PROBLEMS OF HIGHWAY REHABILITATION

A major problem that faces highway and transportation agencies is that the funds they receive are usually insufficient to adequately repair and rehabilitate every roadway section that deteriorates. The problem is further complicated in that roads may be in poor condition but are still usable, making it easy to defer repair projects until conditions become unacceptable. Roadway deterioration is not usually the result of poor design and construction practices but is caused by the inevitable wear and tear that occurs over a period of years. The gradual deterioration of a pavement occurs due to many factors, including variations in climate, drainage, soil conditions, and truck traffic. Just as a piece of cloth eventually tears asunder if a small hole is not immediately repaired, so a roadway will unravel if its surface is allowed to deteriorate. Lack of funds often limits timely repair and rehabilitation of transportation facilities, causing a greater problem with more serious pavement defects and higher costs.

Because funds and personnel are often inadequate to address needs, the dilemma faced by many transportation agencies is to balance their work program between preventive maintenance activities and projects requiring immediate corrective action. When preventive maintenance has been neglected, roads will begin to deteriorate such that the basis for rehabilitation will be the extent of complaints by road users. The traveling public is unwilling to tolerate pavements that are extremely rough and cause vibration and severe damage to their vehicles. Poor-quality pavements may cause crashes to occur, and user costs will significantly increase. Ideally, preventive maintenance will be carried out in an orderly and systematic way and will be the least expensive approach in the long run. However, when funds are extremely limited, agencies often respond to either the most pressing and severe problems or the ones that generate the most vocal complaints.

21.1.1 Approaches to Pavement Management

The term *pavement management* refers to the various strategies that can be used to select a pavement restoration and rehabilitation policy. At one extreme, there is the "squeaky wheel" approach, wherein projects are selected because they have created the greatest attention. At the other extreme is a system wherein all roads are repaired on a regular schedule, for which money is no object. In realistic terms, pavement rehabilitation or repair strategies are plans that establish minimum standards for acceptable pavement condition and seek to establish the type of treatment required to maintain a minimum level of serviceability and the time frame for project completion. Rehabilitation management strategies include consideration of items such as pavement condition, first cost, annual maintenance, user costs, safety, physical, environmental, and economic constraints. Life-cycle cost analysis is the basis for pavement management and consists of the total cost of the project, including initial construction, routine maintenance, and major rehabilitation activities over the lifetime of the pavement.

Pavement management is a systematic process for maintaining, upgrading, and operating physical pavement assets in a cost-effective manner. The process combines applications of established engineering principles with sound business practices and economic theory, thus assuring an organized and scientific approach to decision making.

Pavement management involves the following steps: (1) Assess present pavement condition, (2) predict future condition, (3) conduct an alternatives analysis, and (4) select an appropriate rehabilitation strategy. The aforementioned process utilizes a computer-driven protocol called the pavement management system, which consists of the following elements: (1) an inventory and condition database, (2) mathematical models to forecast future pavement condition, (3) procedures for conducting alternative analyses, and (4) reporting and visualization tools to facilitate the interpretation and display of results.

21.1.2 Levels of Pavement Management

The total framework for pavement management illustrated in Figure 21.1 consists of two levels: the network level and the project level.

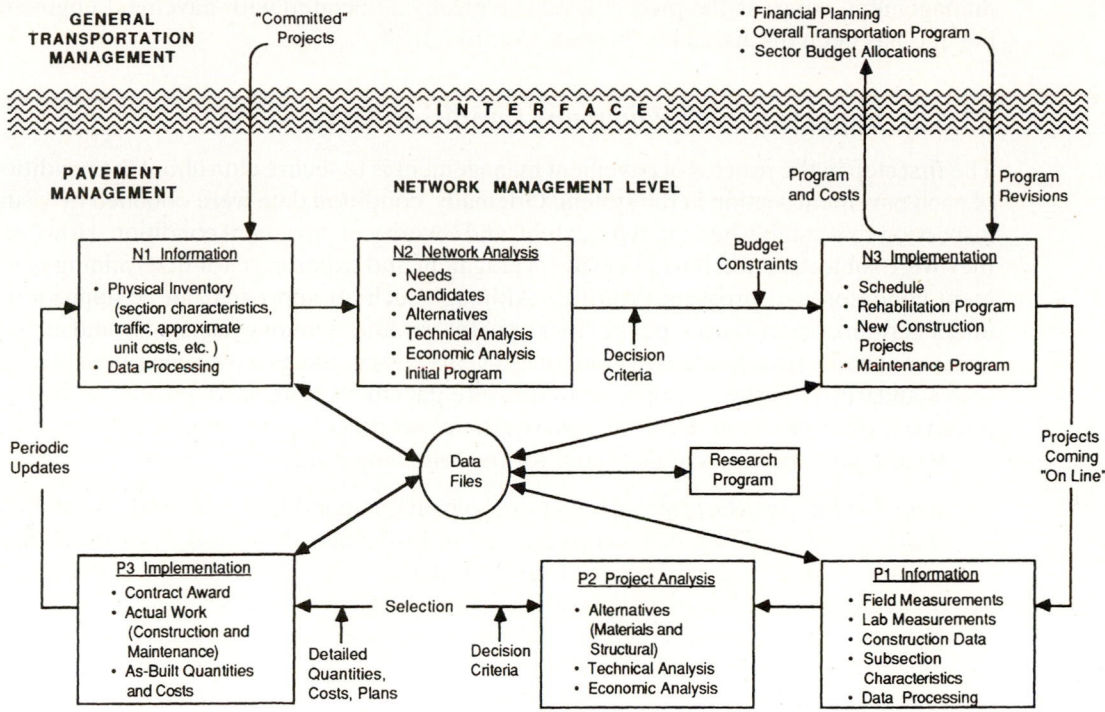

Figure 21.1   Framework for Total Pavement Management System

SOURCE: R. Haas, W. R. Hudson, and J. Zaniewski, *Modern Pavement Management*, Krieger Publishing, Malabar, FL, 1994, p. 38. Used with permission.

The network level is concerned with the entire highway network and all of the pavement sections that comprise the system. At the network level, the pavement management process is strategic and seeks answers to questions such as:

1. What is the current condition of the pavement network?
2. What resources in time, material, and personnel will be required to maintain the network at a specified performance standard?
3. What should be the annual work program to address the most critical needs that reflect available resources?

Since the scope is system-wide, appropriate data are less precise than required for a pavement segment analysis. The data used by the network-level management process are typically less detailed than those required by the project-level process.

At the project level, attention is directed toward determining the maintenance and/or rehabilitation action needed to preserve a specific element or project (i.e., a pavement segment) that has been selected by the network-level process. A detailed condition evaluation survey is typically conducted for each project, and the most cost-effective strategy is selected based upon life-cycle cost analysis. The level of detail of the data required is greater than at the network level. Furthermore, the project level does not consider resources that may be required to maintain other elements in the network. Network-level analysis is emphasized in this chapter, as this is the essence of pavement management, whereas the project level is typically associated with pavement engineering and has been discussed in Chapters 18 through 20.

21.1.3 Importance of Pavement Condition Data

The first step in the process of pavement management is to secure data about the condition of each pavement section in the system. Originally, condition data were obtained by visual inspection that established the type, extent, and severity of pavement condition. However, they were subjective and relied heavily on judgment and experience for determining pavement condition and program priorities. Although such an approach can be appropriate under certain circumstances, possibilities exist for variations among inspectors, and experience is not easily transferable. In more recent years, visual ratings have been supplemented with standardized testing equipment to measure pavement roughness, pavement distress, pavement structural condition, and skid resistance, as described in the next section.

Pavement condition data are used for the following purposes:

1. *Establishing project priorities*. Data on pavement condition are used to establish the relative condition of each pavement and to establish project priorities. There are several methods of data acquisition, and each state selects that combination of measures it considers most appropriate.
2. *Establishing options*. Pavement condition data can be used to develop a long-term rehabilitation program. Data about pavement condition in terms of type, extent, and severity are used to determine which available rehabilitation options should be selected.
3. *Forecasting performance*. By use of correlations between pavement performance indicators and variables such as traffic loadings, it is possible to predict the likely future condition of any given pavement section. This information is useful for preparing long-range budget estimates of the cost to maintain the highway system at a minimum standard of performance or to determine future consequences of various funding levels.

21.2 METHODS FOR DETERMINING ROADWAY CONDITION

Four characteristics of pavement condition are used in evaluating pavement rehabilitation needs: (1) pavement roughness (rideability), (2) pavement distress (surface condition), (3) pavement deflection (structural failure), and (4) skid resistance (safety). These characteristics are described in the following sections.

21.2.1 Pavement Roughness

Pavement roughness refers to irregularities in the pavement surface that affect the smoothness of the ride. The serviceability of a roadway was initially defined in the AASHTO road test: a national pavement research project. Two terms were defined: (1) present serviceability rating (PSR) and (2) present serviceability index (PSI).

PSR is a number grade given to a pavement section based on the ability of that pavement to serve its intended traffic. The PSR rating is established by observation and requires judgment on the part of the individual doing the rating. In the original AASHTO road test, a panel of raters drove on each test section and rated the performance of each section on the basis of how well the road section would serve if the rater were to drive his or her car over a similar road all day long. The ratings ranged between 0 and 5, with 5 being very good and 0 being impassable. Figure 21.2 illustrates the PSR

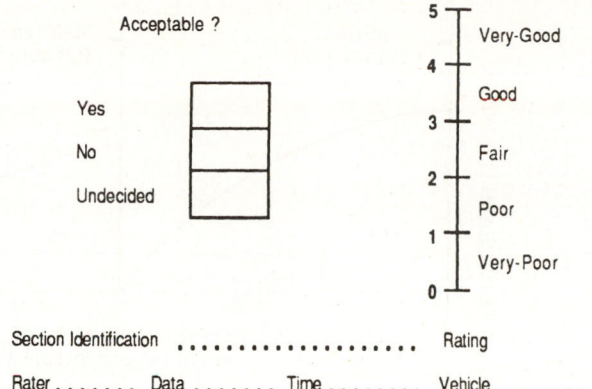

Figure 21.2   Individual Present Serviceability Rating

SOURCE: R. Haas, W. R. Hudson, and J. Zaniewski, *Modern Pavement Management*, Krieger Publishing, Malabar, FL, 1994, p. 80. Used with permission.

as used in the AASHTO road test. Serviceability ratings are based on the user's perception of pavement performance and are determined from the average rating of a panel of road users.

PSI is a value for pavement condition determined as a surrogate for PSR and is based on physical measurements. PSI is not based on panel ratings, as the primary measure of PSI is pavement roughness. The PSI is an objective means of estimating the PSR, which is subjective.

The performance of a pavement can be described in terms of its PSI and traffic loading over time, as illustrated in Figure 21.3. For example, when a pavement is originally constructed, it is in very good condition, with a PSI value of 4.5 (see Figure 21.4). Then

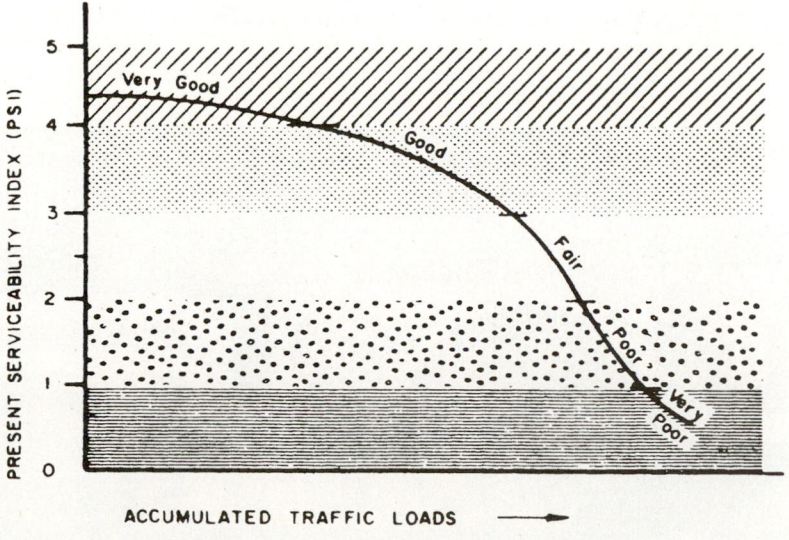

Figure 21.3   Typical Pavement Performance Curve

SOURCE: Modified from Peterson, D.E., NCHRP Synthesis 135: *Pavement Management Practices*, Transportation Research Board, National Research Council, Washington, D.C., 1987.

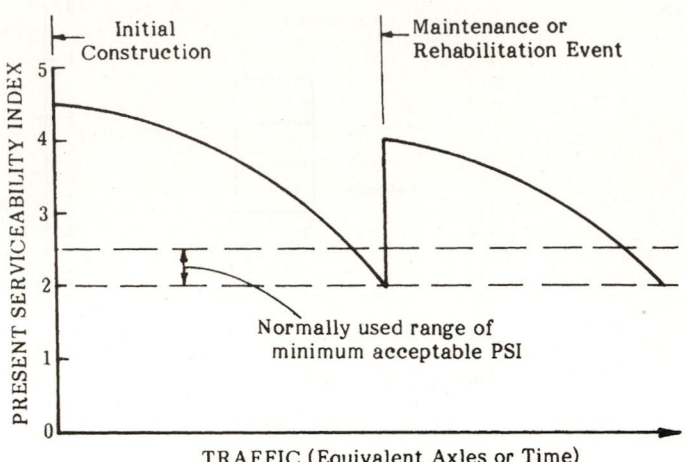

Figure 21.4   Performance History for Pavement Using PSI

SOURCE: Copas, T.L. and H.A. Pennock, *Collection and Use of Pavement Condition Data*, Transportation Research Board, National Research Council, Washington, D.C., 1981.

as the number of traffic loadings increases, the PSI declines to a value of 2, which can be the minimum acceptable PSI value. After the pavement section is rehabilitated, the PSI value increases to 4; as traffic loads increase, the PSI declines again until it reaches 2 and rehabilitation is again required.

Equipment for measuring roughness is classified into two basic categories: response type and profilometers. Response-type equipment does not measure the actual profile of the road but rather the response of the vehicle to surface roughness. Equipment used includes: (1) Mays Ride Meter (MRM), (2) Bureau of Public Roads (BPR) Roughmeter, and (3) Cox Road Meter. The Mays Ride Meter (see Figure 21.5) measures the number of mm of vertical movement per km: the greater the vertical movement, the rougher the road.

Figure 21.5   Mays Ride Meter Trailer Unit

SOURCE: Epps, J.A. and C.L. Monismith, NCHRP Synthesis of Highway Practice 126: *Equipment for Obtaining Pavement Condition and Traffic Loading Data*, Transportation Research Board, National Research Council, Washington, D.C., 1986.

Originally, some type of car-mounted meter was used. The advantages are low cost, simplicity, ease of operation, capability for acquisition of large amounts of data, and output correlated with PSI. However, the vehicle itself influences the measure obtained, and now meters are usually towed in trailers. Response-type equipment must be calibrated frequently to ensure reliable results. Accordingly, such equipment is not widely used.

Profilometers are devices that measure the true profile of the roadway and provide accurate and complete reproductions of the pavement profile. Profilometers also eliminate the need for the time-consuming and labor-intensive calibration required for response-type equipment performance.

There are several types of profilometers. Some equipment uses inertial profilometry that includes the following basic components.

- A device to measure the distance between the vehicle and the road surface
- An inertial referencing device to compensate for the vertical movement of the vehicle body
- A distance odometer to locate the profile points along the pavement
- An on-board processor for recording and analyzing data

Acoustic or optic and laser devices are used for measuring the distance between the pavement surface and the vehicle, d_p. Acoustic systems are more widely used because of their lower cost and the reliability and adequacy of results. Figure 21.6 illustrates a surface dynamics profilometer.

The inertial referencing device is usually either a mechanical or electronic accelerometer mounted to represent the vertical axis of the vehicle. The accelerometer measures the vertical acceleration of the vehicle body. These accelerations are then integrated twice to determine the vertical movement, d_v. These movements are added to the distance measurements d_p to obtain elevations of the pavement profile. Some profilometers record the actual profile, whereas others process the data on board and give only a roughness summary statistic.

The two most popular models of profilometers are the K. J. Law profilometer and the South Dakota profilometer. The K. J. Law profilometer measures and records the road surface profile in each of the vehicle's two wheel paths. An optical measuring system, based on reflectivity from the road surface, and an accelerometer are used together to measure both the distance between the vehicle and the road surface and the vehicle vertical movement. Operating speeds range between 16 and 88 km/hr. The South Dakota profilometer is used by most states. It costs less than the K. J. Law profilometer, although it is not as accurate. It has two additional sensors that can be used to automatically measure rut depth. Data can be collected at normal highway speeds, and about 1120 km of pavement can be measured within a single week.

A relationship exists among pavement roughness, distress condition, and crash frequency. The impact of rutting, road surface roughness, and the pavement PSI on crash frequency and crash type indicates that PSI is a pavement condition parameter that has a significant impact on crash frequency for all types of crashes considered (i.e., single-vehicle crashes, rear-end crashes, angle crashes, and sideswipe crashes). Lower values of the PSI are associated with a higher crash frequency. Moreover, improving the PSI by one unit (for example, from 3.0 to 4.0) can result in a decreased crash frequency.

21.2.2 Pavement Distress

The term *pavement distress* refers to the condition of a pavement surface in terms of its general appearance. A perfect pavement is level and has a continuous and unbroken surface. In contrast, a distressed pavement may be categorized as either fractured,

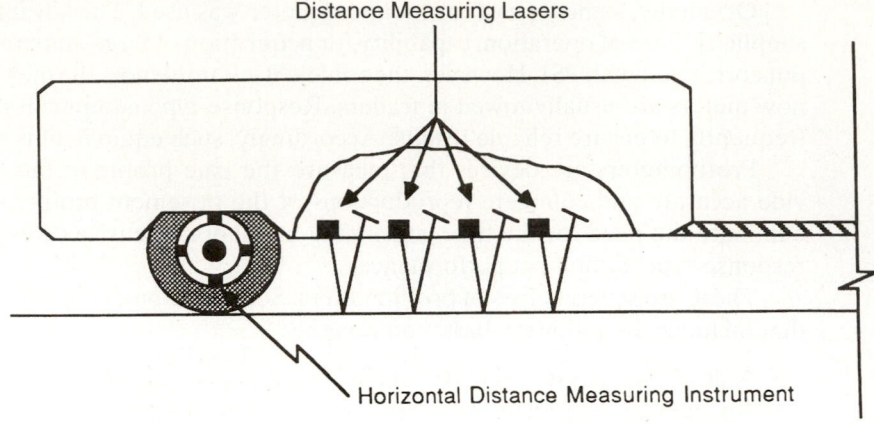

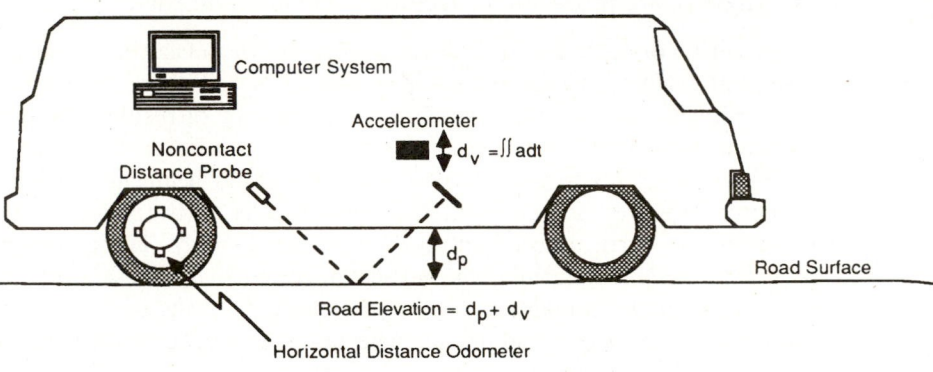

Figure 21.6 Example of Surface Dynamics Profilometer

SOURCE: R. Haas, W. R. Hudson, and J. Zaniewski, *Modern Pavement Management*, Krieger Publishing, Malabar, FL, 1994. p. 85. Used with permission.

distorted, or disintegrated, or a combibation of distress types. These categories can be further subdivided. For example, fractures can be seen as cracks or as spalling (chipping of the pavement surface). Cracks can be further described as generalized, transverse, longitudinal, alligator, and block. Ruts or corrugation of the surface may be evidence of pavement distortion.

Pavement disintegration can be observed as raveling (loosening of pavement structure), stripping of the pavement from the subbase, and surface polishing. The types of distress data collected for flexible and rigid pavements vary from one state to another. Figure 21.7 lists the three pavement distress groups, the measure of distress, and the probable causes.

Many highway agencies use some measure of cracking in evaluating the condition of flexible pavements. The most common measures are transverse, longitudinal, and alligator cracks. Distortion is usually measured by determining the extent of rutting. Disintegration is measured by the amount of raveling. Each state or federal agency has its own procedures for measuring pavement distress. Consequently, there are many methods used to conduct distress surveys. Typically, the agency has a procedural manual that defines each element of distress, with instructions as to how these are to be rated

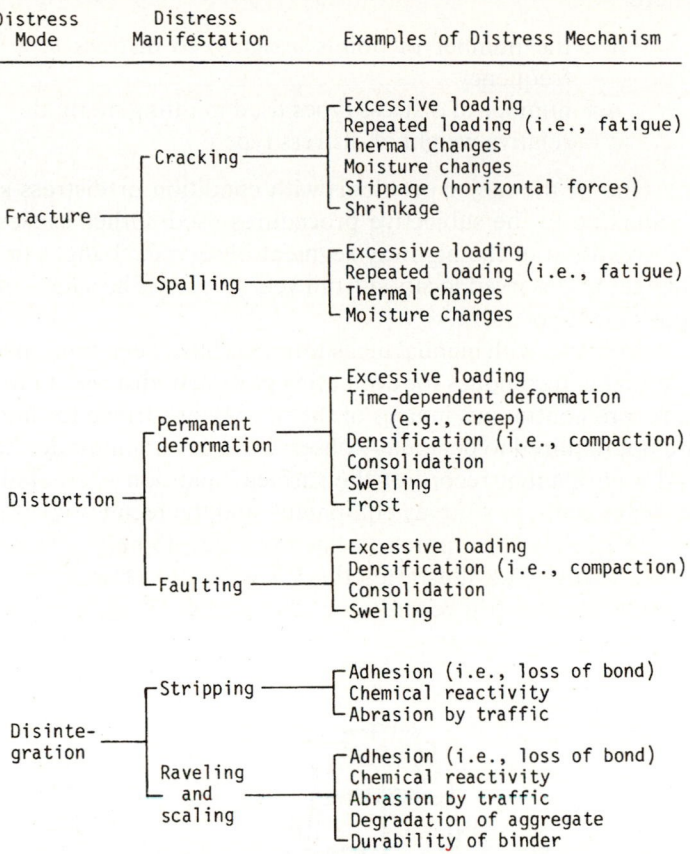

Distress Mode	Distress Manifestation	Examples of Distress Mechanism

Figure 21.7　Pavement Distress Groups and Their Causes

SOURCE:　Copas, T.L. and H.A. Pennock, *Collection and Use of Pavement Condition Data*, Transportation Research Board, National Research Council, Washington, D.C., 1981.

on a given point scale. The survey forms distinguish between bituminous and Portland cement concrete pavements. For bituminous pavements, the items observed are corrugations, alligator cracking, raveling, rutting, longitudinal cracking, transverse cracking, roughness, and patching. For Portland cement concrete, the measures are cracking, raveling, joint spalling, faulting, and patching.

Distress data may be obtained by employing trained observers to make subjective judgments about pavement condition based on predetermined factors. Often, photographs are used for making judgments. Some agencies use full sampling, while others randomly select pavement sections.

Measurements are usually made on a regular schedule of one to three years. After the data are recorded, the results are condensed into a single number called a *distress* (or *defect*) *rating* (DR). A perfect pavement is usually given a score of 100; if distress is observed, points are subtracted. The general equation is

$$\text{DR} = 100 - \sum_{i=1}^{n} d_i w_i \qquad (21.1)$$

where

d_i = the number of points assigned to distress type i for a given severity and frequency

n = number of distress types used in rating method

w_i = relative weight of distress type i

One of the major problems with condition or distress surveys is the variability in results due to the subjective procedures used. Other causes of error are variations in the condition of the highway segment observed, changes in evaluation procedure, and changes in observed location from year to year. The safety of pavement evaluators is of considerable concern.

Problems with manual measurements have been minimized through development of automated techniques for evaluating pavement distress. Film and video devices are used to record continuous images of the pavement surface for later evaluation; survey crews are not required to personally observe the pavement under hazardous traffic conditions, and a permanent record of the surface condition is created. There are many types of automated distress survey equipment, and the technology is constantly being improved.

The PASCO corporation has been developing a distress survey device since the 1960s. Figure 21.8 illustrates the PASCO ROADRECON system, which produces a continuous filmstrip recording of the pavement surface and a measure of roughness.

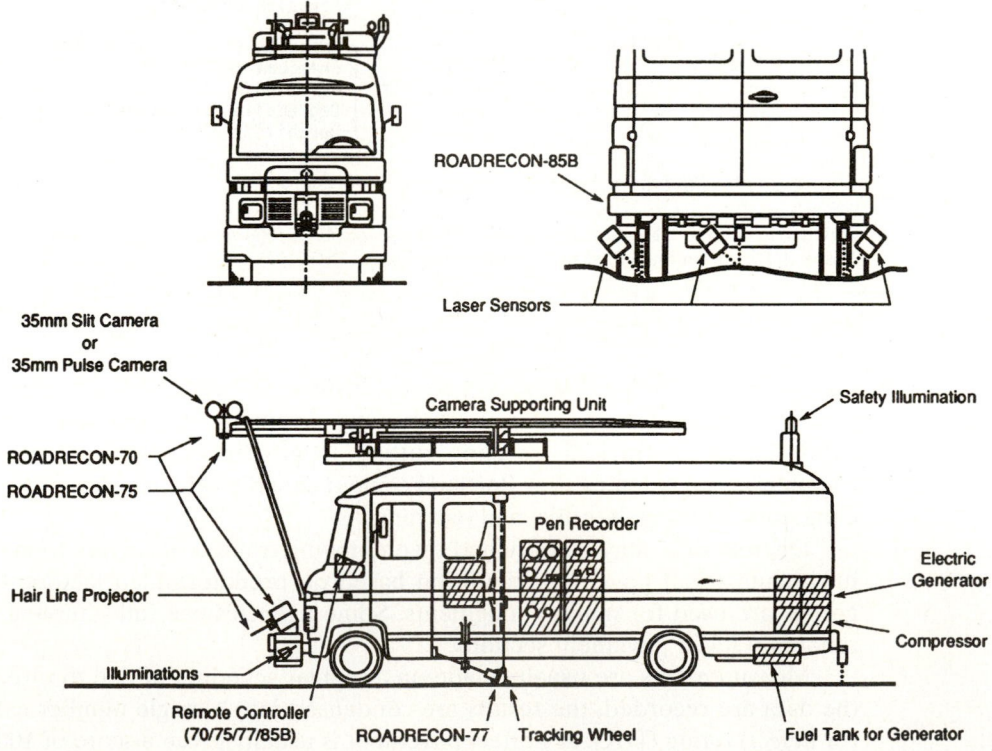

Figure 21.8   PASCO ROADRECON System for Distress Evaluation

SOURCE: R. Haas, W. R. Hudson, and J. Zaniewski, *Modern Pavement Management*, Krieger Publishing, Malabar, FL, 1994, p. 145. Used with permission.

Photographs are taken at night with a controlled amount and angle of lighting. The vehicle can operate at speeds up to 80 km/h. Manual interpretation of the photographs is still required for evaluating pavement distress. Computer vision technology is available to analyze images without human intervention.

Example 21.1 Computing Distress Rating of a Pavement Section

A pavement rating method for a certain state uses the following elements in its evaluation procedure: longitudinal or alligator cracking, rutting, bleeding, ravelling, and patching. The weighting factors are 2.4, 1.0, 1.0, 0.9, and 2.3, respectively. Each distress element is characterized by (1) its severity as not severe, severe, or very severe; and (2) its frequency as none, rare, occasional, or frequent. The categories for frequency are based on the percentage of area affected by a particular distress within the area of the section surveyed. For each combination of severity and distress, a rating factor is assigned, d_{id}, from 0 to 9, as shown in Table 21.1.

A one-km section of roadway was observed with results shown in Table 21.2. Calculate the distress rating for the section.

Solution: Using the data in Table 21.2 and the rating factors (d_i) in Table 21.1, each distress is categorized with factors as follows.

Distress Characteristic	Rating Factor, d_i	Weight, w_i
Cracking	1	2.4
Rutting	4	1.0
Bleeding	9	1.0
Raveling	0	0.9
Patching	2	2.3

Applying the weighting values for each characteristic, the distress rating for the section is determined using Eq. 21.1:

$$DR = 100 - \sum_{i=1}^{n} d_i w_i$$

$$= 100 - (1 \times 2.4 + 4 \times 1 + 9 \times 1 + 0 \times 0.9 + 2 \times 2.3)$$

$$= 100 - 20 = 80$$

Table 21.1   Rating Factor, d_{id}, Related to Severity and Frequency

| Frequency | Severity | | |
	Not Severe (NS)	Severe (S)	Very Severe (VS)
None (N)	0	0	0
Rare (R)	1	2	3
Occasional (O)	2	4	6
Frequent (F)	3	6	9

Table 21.2   Observed Distress Characteristics for Road Segment

Distress Characteristic	Frequency	Severity
Cracking	R	NS
Rutting	O	S
Bleeding	F	VS
Raveling	N	NS
Patching	R	S

Pavement Condition Index (PCI)

The Pavement Condition Index (PCI) is a widely used pavement distress index developed by the U.S. Army Corps of Engineers. PCI values range from 0 to 100 and are based on a visual condition survey that measures distress type, severity, and extent of pavement damage. The distress types that can be considered when using the PCI include the following.

- **Alligator cracking**. A series of interconnecting cracks that are caused by fatigue failure of the pavement surface under the repeated traffic loadings.
- **Bleeding**. A film of bituminous material on the pavement surface that becomes viscous when warm. It is caused by excessive amounts of bituminous material in the asphalt mix.
- **Block cracking**. Interconnected cracks that divide the pavement into rectangular pieces.
- **Corrugation**. A series of closely spaced ridges and valleys occurring at regular intervals.
- **Depressions**. Localized areas that are below the surrounding surface causing a "bowl-like" shape.
- **Longitudinal** and **tranverse cracking**. Cracks that are parallel or orthogonal to the centerline of the pavement.
- **Rutting**. A surface depression typically along the wheel paths of a road.
- **Raveling**. Wearing of the pavement surface caused by aggregate particles breaking loose and the loss of bituminous material binder.

For each distress type, the values to be deducted are determined by selecting a value from a chart or graph based on the survey results. Additional details regarding PCI calculations can be found in the references at the end of the chapter.

21.2.3 Pavement Structural Condition

The structural adequacy of a pavement is measured either by nondestructive means, which measure deflection under static or dynamic loadings, or by destructive tests, which involve removing sections of the pavement and testing these in the laboratory. Structural condition evaluations are rarely used by state agencies for monitoring network pavement condition due to the expense involved. However, nondestructive evaluations, which gather deflection data, are used by some agencies on a project basis for pavement design purposes and to develop rehabilitation strategies.

Nondestructive structural evaluation is based on the premise that measurements can be made on the surface of the pavement and *in situ* characteristics can be inferred from these measurements about the structural adequacy of the pavement. The four basic nondestructive test methods are (1) measurements of static deflection, (2) measurements of deflections due to dynamic or repeated loads, (3) measurements of deflections from a falling load (impulse load), and (4) measurements of density of pavement layers by nuclear radiation (used primarily to evaluate individual pavement layers during construction). Deflection data are primarily used for design purposes and not for pavement management. Some states use deflection equipment solely for research and special studies.

One method for measuring static deflections is the Benkelman beam, which is a simple hand-operated device designed to measure deflection responses of a flexible pavement to a standard wheel load. A probe point is placed between two dual tires and the motion of the beam is observed on a dial that records the maximum deflection. Other static devices that are used include the traveling deflectometer, the plate-bearing test, and the Lacroix Deflectograph. Most of these devices are based on the Benkelman beam principle, in which pavement deflections due to a static or a slowly moving load are measured manually or by automatic recording devices.

Another method for measuring pavement deflections is the Dynaflect. This device, shown in Figure 21.9, consists of a dynamic cyclical force generator mounted on a two-wheel trailer, a control unit, a sensor assembly, and a sensor calibration unit. The system provides rapid and precise measurements of roadway deflections,

Figure 21.9 Dynaflect Deflection Sensors

SOURCE: Epps, J.A. and C.L. Monismith, NCHRP Synthesis of Highway Practice 126: *Equipment for Obtaining Pavement Condition and Traffic Loading Data*, Transportation Research Board, National Research Council, Washington, D.C., 1986.

which in this test are caused by forces generated by unbalanced flywheels rotating in opposite directions. A vertical force of 4.45 kN is produced at the loading wheels, and deflections are measured at five points on the pavement surface located 0.3 m apart.

Most states use falling load-type equipment, referred to as "falling weight deflectometers" (FWDs), because force impulses created by a falling load more closely resemble the pulse created by a moving load than that created by either the vibratory or static load devices. Figure 21.10 illustrates the basic principle of a falling weight deflectometer.

Variations in the applied load may be achieved by altering either the magnitude of the mass or the height of drop. Vertical peak deflections are measured by the FWD in the center of the loading plate and at varying distances away from the plate. These data are used to draw what are known as "deflection basins."

21.2.4 Skid Resistance

Safety characteristics of a pavement are another measure of its condition, and highway agencies continually monitor this aspect to ensure that roadway sections are operating at the highest possible level of safety. The principal measure of pavement safety is its skid resistance. Other elements contributing to the extent to which pavements can perform safely are eliminating rutting (which causes water to collect that creates hydroplaning) and adequacy of visibility of pavement markings.

Skid resistance data are collected to monitor and evaluate the effectiveness of a pavement in preventing or reducing skid-related accidents. Skid data are used by highway

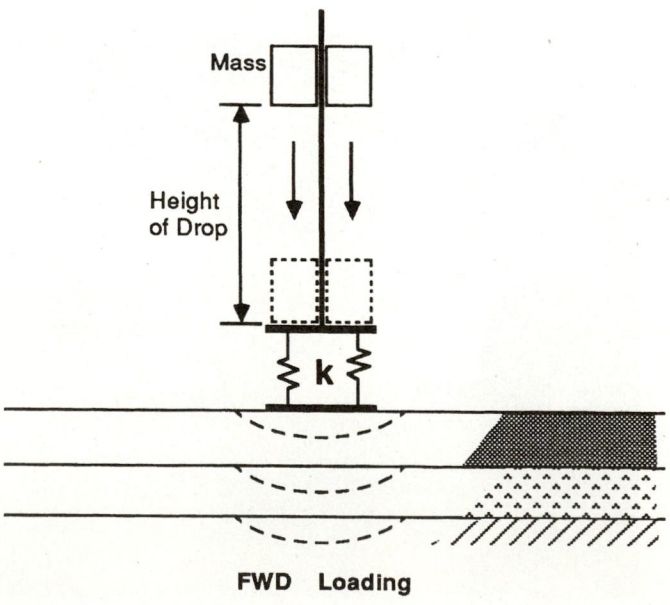

Figure 21.10 Principle of the Falling Weight Deflectometer

SOURCE: R. Haas, W. R. Hudson, and J. Zaniewski, *Modern Pavement Management*, Krieger Publishing, Malabar, FL, 1994, p. 120. Used with permission.

agencies to identify pavement sections with low skid resistance, to develop priorities for rehabilitation, and to evaluate the effectiveness of various pavement mixtures and surface types.

The coefficient of sliding friction between a tire and pavement depends on factors such as weather conditions, pavement texture, tire condition, and speed. Since skidding characteristics are not solely dependent on the pavement condition, it is necessary to standardize testing procedures and in this way eliminate all factors but the pavement. The basic formula for friction factor f is

$$f = \frac{L}{N} \tag{21.2}$$

where

L = lateral or frictional force required to cause two surfaces to move tangentially to each other
N = force perpendicular to the two surfaces

When skid tests are performed, they must conform to specified standards set by the American Society for Testing and Materials (ASTM). The test results produce a skid number SK, where

$$SK = 100f \tag{21.3}$$

The SK is usually obtained by measuring the forces obtained with a towed trailer riding on a wet pavement, equipped with standardized tires. The principal methods of testing are (1) locked-wheel trailers, (2) Yaw mode trailers, and (3) the British Portable Tester. Locked-wheel trailers, illustrated in Figure 21.11, are widely used skid-measuring devices. The test involves wetting the pavement surface and pulling a two-wheel trailer

Figure 21.11 Locked-Wheel Skid Trailer

SOURCE: Epps, J.A. and C.L. Monismith, NCHRP Synthesis of Highway Practice 126: *Equipment for Obtaining Pavement Condition and Traffic Loading Data*, Transportation Research Board, National Research Council, Washington, D.C., 1986.

whose wheels have been locked in place. The test is conducted at 64 km/h with standard tires each with seven grooves. The locking force is measured, and from this an SK value is obtained.

The Yaw mode test is done with the wheels turned at a specified angle to simulate the effects of cornering. The most common device for this test is a Mu-Meter, which uses two wheels turned at 7.5 degrees. The trailer is pulled in a straight line on a wetted surface with both wheels locked. Since both wheels cannot be in the wheel paths, friction values may be higher than those obtained by using a locked-wheel trailer.

The most common method for determining skid resistance is the locked-wheel trailer, and most state highway agencies own one or more of these devices. Several states use the Mu-Meter (a Yaw mode device). Figure 21.12 illustrates typical skid results for various pavement conditions. Skid resistance data are not typically used in developing rehabilitation programs. Rather, they are used to monitor the safety of the highway system and to assist in reducing potential crash locations.

High-skid-resistant pavements can help reduce the likelihood of skid-related crashes, whereas pavement surfaces with inadequate skid resistance can pose a safety risk. Skid resistance is a function of several factors, including pavement texture, tire condition, speed, and weather conditions such as rain or snow. Two parameters of pavement texture are: (1) micro-texture, which refers to fine-scale surface irregularities (≤ 1.0 mm depth) in the stone or aggregate particles making up the pavement mix; and (2) macro-texture, which refers to the large-scale roughness or variations in the road surface (in the range of 0.5 mm to 50 mm) at the pavement surface resulting from the way the aggregates are arranged. Both parameters have a direct impact on the pavement skid resistance, with micro-texture having the dominant impact at speeds less than 30 mi/h and macro-texture becoming more important at high speeds.

Mix design of asphalt can influence skid resistance and safety. However, the actual threshold value that may signal the need for pavement maintenance and rehabilitation on the basis of safety considerations is not fully understood. A skid number, SK, less than 35 suggests that a safety issue may exist.

Example 21.2 Measuring Skid Resistance

A 44.5 kN load is placed on two tires of a locked-wheel trailer. At a speed of 48 km/h, a force of 22.2 kN is required to move the device. Determine the SK and the surface type, assuming that treaded tires were used.

Solution:

$$SK = 100f = (100)\frac{L}{N}$$
$$= 100 \times \frac{22.2}{44.5} = 50$$

From Figure 21.12, at 48 km/h and SK = 50, the surface type is coarse-textured and gritty.

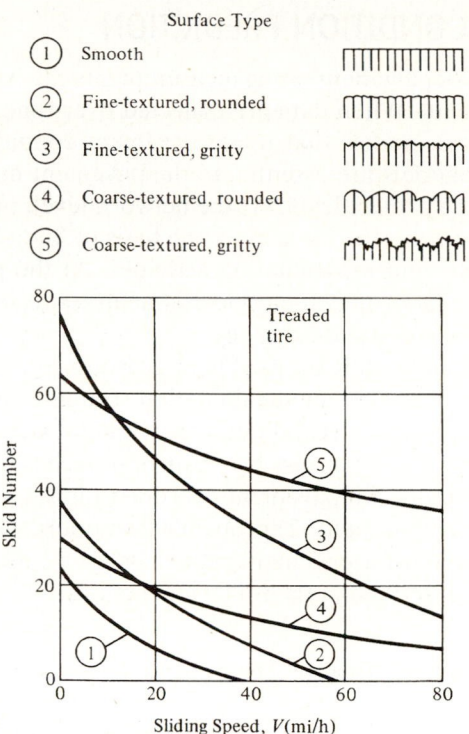

Figure 21.12 Skid Data for Various Pavement Surface Types

SOURCE: Redrawn from H.W. Kummer and W. E. Meyer, *Tentative Skid Resistance Requirements for Main Rural Highways*, NCHRP Report 37, Transportation Research Board, National Research Council, Washington, D.C., September 1967

21.2.5 Intelligent Transportation Systems and Pavement Condition Monitoring

In recent years, there has been increased interest in applying advances in information technologies (i.e., sensing, computing, communications, and control) to improve the efficiency, safety, resiliency, and environmental friendliness of the transportation system. This body of knowledge is based on Intelligent Transportation Systems (ITS), which envisions a totally networked environment supported by continuous wireless applications, such as Vehicle-to-Vehicle (V2V), Vehicle-to-Infrastructure (V2I), and Vehicle-to-Other (V2O) communications.

Previous research to develop a fully integrated transportation system has focused on applications to improve the system's safety and efficiency. The envisioned system provides opportunities for pavement condition and ride quality monitoring. This potential is being investigated involving vehicles equipped with accelerometers and Global Positioning System (GPS) receivers. These devices can detect potholes and other types of pavement distress, but it is more challenging to pinpoint the spatial location of the pavement distress and thus to allow direct averaging of the results from the different vehicles. The ITS connected-vehicle approach is a potential option that collects pavement distress and ride quality data, thus supplementing methods currently in use.

21.3 PAVEMENT CONDITION PREDICTION

The results of pavement rating measurements can be used to predict pavement condition in future years. When data are collected over time, the results can be used to develop mathematical models that relate pavement condition to age of pavement. Condition prediction models are essential to the pavement management process at both the network and the project level. At the network level, prediction models forecast the future condition of the network and are the basis of budget planning, inspection scheduling of maintenance, and rehabilitation activities. At the project level, prediction models are used in specific project life-cycle cost analyses to select the most cost-effective maintenance and rehabilitation strategy.

Prediction models are used in forecasting deterioration trends of pavement sections before and after a major rehabilitation strategy has been selected. For example, these models can predict the condition of newly constructed sections as well as the condition of overlaid sections. There are two main approaches that have been considered to develop prediction models for pavements: (1) deterministic models, which are developed through regression analysis, and (2) probabilistic models, which are based on tables that furnish the probability of a pavement rating change from one year to the next. Each of these methods is described in the following sections.

21.3.1 Deterministic Models

Many states accumulate pavement distress data for specific roadway sections. As is expected, with pavement aging, the distress rating will decrease. Figure 21.13 illustrates the variation in pavement condition (PCR) versus age for pavements in the state of Washington. There is a wide range of results, suggesting that a pavement model should only be used in circumstances similar to those existing when the data were collected (i.e., climate, subgrade strength, axle loads, etc.). Prediction models (such as the one shown in Figure 21.13) can be used to predict future pavement condition, information that is required to determine appropriate rehabilitation programs. Two regression-based prediction models are presented: (1) family based and (2) multiple regression.

Family–Based Prediction Models

This approach is referred to as "family-based" because different pavement sections are segmented into groups or families in such a manner that all pavement sections in a specific group or family have similar deterioration trend characteristics. Group selection is based on factors such as traffic volumes, climate, structural strength, and surface type.

Having defined each group or family of pavement sections, regression analysis is used to develop a separate prediction model for each family. The dependent variable is the pavement's condition index (PCI), and the independent variable is the corresponding number of years (age) since the pavement was constructed or resurfaced. The single independent variable—age—is sufficient to predict pavement condition since all other variables are classified and thus similar for this data set (or "family"). Pavement sections have already been grouped into families with similar deterioration trends, thus accounting for other factors that affect deterioration.

This technique is used for the prediction model in MicroPAVER, a pavement management software package developed by the U.S. Army Corps of Engineers. Pavements are divided into families based on factors such as pavement type (asphalt concrete, Portland cement concrete, or asphalt concrete overlay, etc.), roadway classification (primary,

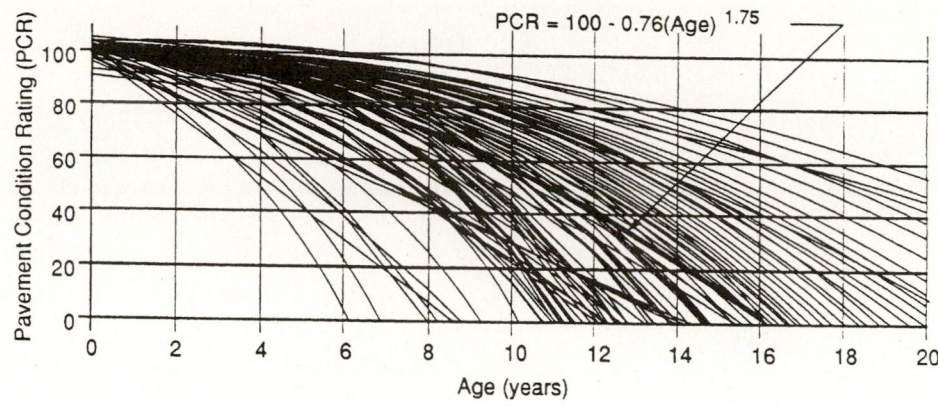

Figure 21.13 Pavement Condition versus Pavement Age

SOURCE: *An Advanced Course in Pavement Management Systems* (course text), Federal Highway Administration, Washington, D.C., 1990

secondary, etc.), and climatic conditions. For each family classification, MicroPAVER generates a data file that includes pavement section identification, age, and PCI. MicroPAVER contains a feature that allows the user to remove incorrect data points and outliers. The technique produces a mathematical fit to the data as a polynomial regression. This method for modeling curvature in the relationship between a response variable (y) and a predictor variable (x) extends a simple linear regression model by including higher order predictors, x^2 and x^3. Thus, the MicroPAVER model is expressed as follows.

$$P(x) = a_0 + a_1 + a_2x^2 + a_3x^3 + a_4x^4 \tag{21.4}$$

where

$$P(x) = \text{pavement condition index (PCI)}$$
$$x = \text{pavement age in years}$$
$$a_{0,1,2,3,4} = \text{constants}$$

By using the transformed variables (x^2, x^3, and x^4), a nonlinear model can be created using simple linear regression. The form of the model must be specified in advance. A characteristic of polynomial regression is that an increase in the slope of the curve can occur, which contradicts a physical reality that the pavement condition cannot improve as age increases. MicroPAVER recognizes this anomaly by imposing constraints on the regression to eliminate the contradiction.

Example 21.3 Developing a Family-Based Prediction Model

Table 21.3 lists inspection data that were recorded for a subset of pavement sections that are part of a statewide highway network. The PCI, the number of years since construction (age), the average annual daily traffic (*AADT*), and the structural number

(SN) are listed for each section. Develop a family-based prediction model based on the pavement sections data in Table 21.3.

Solution: Based on the data shown in Table 21.3, the eight pavement sections are divided into two families based solely on the $AADT$ values, since the values for the SN are not sufficiently distinctive to warrant further classification based on this independent variable. $AADT$ values are segmented into two categories: (1) sections with an $AADT < 10{,}000$ veh/day/ln, and sections with an $AADT \geq 10{,}000$ veh/day/ln.

Table 21.3   Pavement Section Data

ID	PCI	Age	AADT	SN
1	100	0	8700	4.3
1	98	0.5	8700	4.3
1	96	1.2	8700	4.3
1	93	3	8700	4.3
1	87	5	8700	4.3
2	100	0	14,000	4.7
2	95	1	14,000	4.7
2	89	3	14,000	4.7
2	83	5	14,000	4.7
2	78	7	14,000	4.7
3	100	0	9800	4.1
3	93	2	9800	4.1
3	88	4	9800	4.1
3	86	7	9800	4.1
4	100	0	18,000	4.9
4	93	1.5	18,000	4.9
4	87	3.5	18,000	4.9
4	79	6	18,000	4.9
5	100	0	8000	4.2
5	95	2	8000	4.2
5	90	4	8000	4.2
5	83	8	8000	4.2
6	100	0	12,000	4.5
6	90	3	12,000	4.5
6	78	7	12,000	4.5
6	75	9	12,000	4.5
7	100	0	9500	4.4
7	92	3	9500	4.4
7	85	6	9500	4.4
7	82	9	9500	4.4
7	81	10	9500	4.4
8	100	0	17,000	4.8
8	91	2.5	17,000	4.8
8	84	4.5	17,000	4.8
8	75	8	17,000	4.8
8	71	10	17,000	4.8

Polynomial regression is used to develop the prediction models using an x^2 term such that the regression model is

$$\text{Pavement Condition Index, PCI} = a_0 + a_1(\text{age}) + a_2(\text{age})^2 \qquad (21.5)$$

Sections belonging to each family are shown in Tables 21.4 and 21.5, with the new variable (age^2) added.

Table 21.4 Family 1 Sections ($AADT < 10,000$)

ID	PCI	Age	Age^2
1	100	0	0
1	98	0.5	0.25
1	96	1.2	1.44
1	93	3	9
3	100	0	0
3	93	2	4
3	88	4	16
3	86	7	49
5	100	0	0
5	95	2	4
5	90	4	16
5	83	8	64
7	100	0	0
7	92	3	9
7	85	6	36
7	82	9	81
7	81	10	100

Table 21.5 Family 2 Sections ($AADT \geq 10,000$)

ID	PCI	Age	Age^2
2	100	0	0
2	95	1	1
2	89	3	9
2	83	5	25
2	78	7	49
4	100	0	0
4	93	1.5	2.25
4	87	3.5	12.25
4	79	6	36
6	100	0	0
6	90	3	9
6	78	7	49
6	75	9	81
8	100	0	0
8	91	2.5	6.25
8	84	4.5	20.25
8	75	8	64
8	71	10	100

A spreadsheet or statistical analysis software is used to determine the constant values, a, in the prediction models based on the data in Tables 21.4 and Tables 21.5. The results are

Family 1: $AADT < 10{,}000$

PCI $= 99.84 - 3.04(\text{age}) + 0.12(\text{age})^2$ $R^2 = 0.99$ (21.6)

The value $R^2 = 0.99$ indicates a very good fit between the model and the data. This can be demonstrated in Figure 21.14, which shows a plot of the model results versus the actual data.

Family 2: $AADT \geq 10{,}000$

PCI $= 99.74 - 3.99(\text{age}) + 0.12(\text{age})^2$ $R^2 = 0.99$ (21.7)

Again, the value $R^2 = 0.99$ indicates a very good fit between the model and the data. This can be demonstrated in Figure 21.15, which shows a plot of the model results versus the actual data.

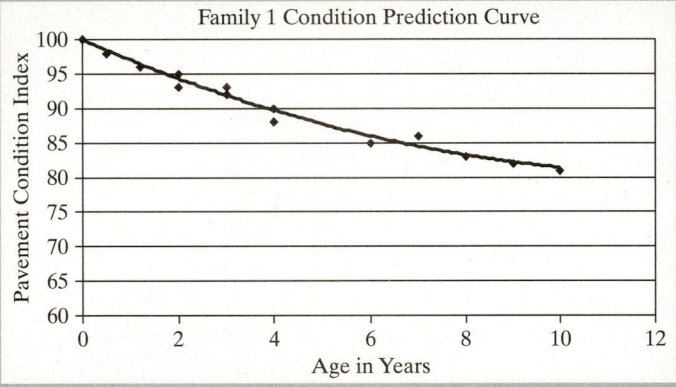

Figure 21.14   Condition Prediction Model: Pavement Sections $AADT < 10{,}000$

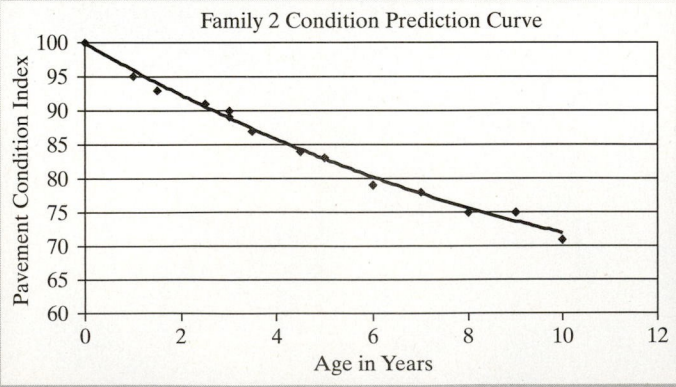

Figure 21.15   Condition Prediction Model: Pavement Sections $AADT > 10{,}000$

Example 21.4 Age When Pavement Resurfacing Is Required

Determine the number of years (age) before a pavement section belonging to Family 2 ($AADT \geq 10,000$) in Example 21.3 will need to be resurfaced if pavement overlays are necessary when the PCI value falls below 74.

Solution: Using the model previously developed as Eq. 21.7, determine the age at which the section will reach a PCI of 74 as follows:

$$74 = 99.74 - 3.99(\text{age}) + 0.12(\text{age})^2$$

Therefore,

$$0.12(\text{age})^2 - 3.99(\text{age}) + 25.74 = 0$$

Solve this quadratic equation as follows.

$$\text{Age} = \frac{+3.99 - \sqrt{3.99^2 - 4(0.12)(25.74)}}{2(0.12)} = 8.75 \text{ years}$$

Multiple-Regression Prediction Models

In contrast to family-based prediction models, which relate an element's condition index to a single independent variable, multiple-regression prediction models relate the PCI for a given highway section to several factors that affect deterioration, such as age, usage level, climatic and environmental conditions, structural strength, and construction materials. An example of a multiple-regression model follows.

Example 21.5 Developing a Multiple-Regression Prediction Model

Develop a multiple-regression-type prediction model using the data shown in Table 21.3. The form of the regression model is shown as Eq. 21.8.

$$\text{CI} = a + b(\text{age}) + c(AADT) + d(\text{SN}) \tag{21.8}$$

Solution: A spreadsheet or statistical analysis software is used to develop the prediction model based on the data in Table 21.3. The results are shown in Eq. 21.9.

$$\text{CI} = 101.48 - 2.46(\text{age}) - 0.5 \times 10^{-3}(AADT) + 0.81(\text{SN}) \quad R^2 = 0.97 \tag{21.9}$$

The value $R^2 = 0.97$ indicates a very good fit between the model and the data.

Table 21.6 Probabilities of Pavement Condition Changes

		\multicolumn{10}{c}{To PCR State}
		9 100 to 90	8 89 to 80	7 79 to 70	6 69 to 60	5 59 to 50	4 49 to 40	3 39 to 30	2 29 to 20	1 19 to 10	P
From PCR State	9 100 to 90	0.90	0.10								1.0
	8 89 to 80		0.70	0.30							1.0
	7 79 to 70			0.60	0.30	0.10					1.0
	6 69 to 60				0.50	0.30	0.15	0.05			1.0
	5 59 to 50					0.30	0.40	0.30			1.0
	4 49 to 40						0.30	0.70			1.0
	3 39 to 30							0.60	0.35	0.05	1.0
	2 29 to 20								0.20	0.80	1.0
	1 19 to 10									1.0	1.0

21.3.2 Probabilistic Models

Probability methods are based on the assumption that future conditions can be determined from the present state if the probabilities of given outcomes are known. An example of this approach is a Markovian model illustrated in Table 21.6, which shows the probability that the pavement condition in state i will change to a pavement condition in state j. For example, if in the current year the pavement condition (PCR) is within 70 to 79 (state 7), then the probability is 0.3 that in the next year the pavement condition will be within 60 to 69 (state 6). The probabilities shown in Table 21.6 can be used for successive years in a so-called Markovian chain. The assumption of Markovian models is that the future state is dependent on the current state regardless of how the pavement reached that current state.

Example 21.6 Predicting Future Condition Using a Markovian Model

A state DOT district office is responsible for maintaining a network of 1600 km. The results of the annual survey in a given year showed that the current network condition of these roads is

- Group 1: 960 km with a PCR between 90 and 100
- Group 2: 480 km with a PCR between 80 and 89
- Group 3: 160 km with a PCR between 70 and 79

Using the probability matrix given in Table 21.6, determine the condition of the network in year 1 and year 2.

Solution:

Step 1. Determine the current state of each of the three groups.

Group 1: state 9
Group 2: state 8
Group 3: state 7

Step 2. Determine probabilities of outcomes for each year, in one year.

For Group 1, state 9:
90% in state 9
10% in state 8

For Group 2, state 8:
70% in state 8
30% in state 7

For Group 3, state 7:
60% in state 7
30% in state 6
10% in state 5

Step 3. Determine the number of kilometers (km) involved.

Group 1:
$0.9 \times 960 = 864$ km in state 9
$0.1 \times 960 = 97$ km in state 8

Group 2:
$0.70 \times 483 = 336$ km in state 8
$0.30 \times 480 = 144$ km in state 7

Group 3:
$0.60 \times 160 = 96$ km in state 7
$0.30 \times 160 = 48$ km in state 6
$0.10 \times 160 = 16$ km in state 5

Step 4. Summarize the kilometers of highway in each PCR category.

$869 = 864$ km will be in state 9 (PCR between 90 & 100)
$96 + 336 = 432$ km will be in state 8 (PCR between 80 & 89)
$144 + 96 = 240$ km will be in state 7 (PCR between 70 & 79)
$48 = 48$ km will be in state 6 (PCR between 60 & 69)
$16 = 16$ km will be in state 5 (PCR between 50 & 59)

For year 2, the procedure is the same. Each PCR state is determined, followed by appropriate probabilities. Then calculations are made as shown in Step 3 and summarized in Step 4. The solution is left to the reader.

21.4 PAVEMENT REHABILITATION

A variety of methods can be used to rehabilitate pavements or to correct deficiencies in a given pavement section, including using overlays, sealing cracks, using seal coats, and repairing potholes.

21.4.1 Rehabilitation Techniques and Strategies

Rehabilitation techniques are classified as (1) corrective, which involves the permanent or temporary repair of deficiencies on an as-needed basis; or (2) preventive, which involves surface applications of either structural or nonstructural improvements intended to keep the quality of the pavement above a predetermined level. Corrective maintenance is analogous to repairing a small hole in a cloth whereas preventive maintenance can be thought of as sewing a large patch or replacing the lining in a suit. Just as with the sewing analogy, corrective measures can serve as prevention measures as well. To illustrate, sealing a crack is done to correct an existing problem, but it also prevents further deterioration that would occur if the crack were not repaired. Similarly, a chip seal coat (which is a layer of gravel placed on a thin coating of asphalt) is often used to correct a skid problem but also helps prevent further pavement deterioration.

Pavement rehabilitation strategies can be categorized in a variety of ways. One approach is in terms of the problem being solved, such as skid resistance, surface drainage, unevenness, roughness, or cracking. Another approach is in terms of the type of treatment used, such as surface treatment, overlay, or recycle. A third approach is in terms of the type of surface that will result from the process, such as asphalt overlay, rock seal coat, or liquid seal coat. The latter approach is the most commonly used because it enables the designer to consider each maintenance alternative in terms of a final product and then select the most appropriate one in terms of results desired and cost. Thus, the repair strategy information provides a means by which the problem at hand can be identified and the most economical alternative (based on annual cost and life expectancy) can be selected.

21.4.2 Alternatives for Repair and Rehabilitation

Figure 21.16 illustrates a variety of pavement repair and rehabilitation alternatives and differentiates between preventive and corrective approaches. For example, preventive strategies for pavement surfaces include fog-seal asphalt, rejuvenators, joint sealing, seal coats (with aggregate), and a thin blanket. This figure is provided to assist understanding the purpose for which a given treatment is intended.

There are a variety of pavement rehabilitation techniques for both flexible and rigid pavements, and they may be corrective or preventive depending on the circumstances. For example, patching is always considered to be corrective and can be effective if properly done. Many patching materials are available. At the other extreme, overlays are both corrective and preventive and are considered to be an effective technique. Surface treatments can be either preventive or corrective and are considered an effective means of maintaining roads on a regular basis.

Another method of describing pavement rehabilitation alternatives is to identify deficiencies and select the most appropriate treatment. If distress types are known in terms of severity and if the repair is considered temporary or permanent, then a repair strategy can be selected. For example, for flexible pavements, alligator cracking is repaired by removal and replacement of the surface course, by permanent patching or scarifying, and by mixing the materials with asphalt.

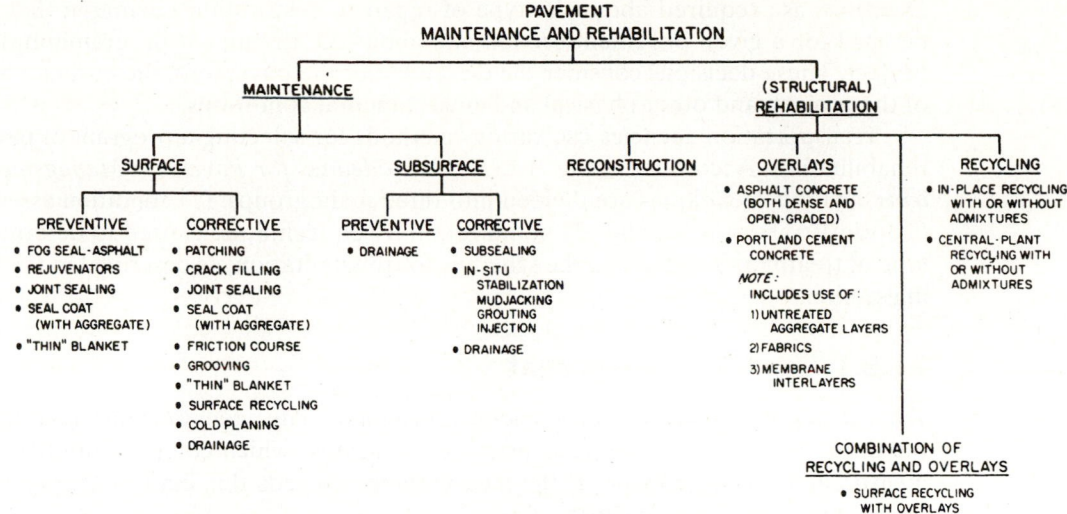

Figure 21.16   Pavement Maintenance and Rehabilitation Alternatives

SOURCE: Figure reproduced with permission of TRB, from C. L. Monismith, *Pavement Evaluation and Overlay Design Summary of Methods*, Transportation Research Record No. 700, Transportation Research Board, National Research Council, Washington, D.C., 1979

The state of the practice of pavement rehabilitation has been considerably advanced in recent years as a result of research on Long Term Pavement Performance (LTPP) supported by the Strategic Highway Research Program (SHRP). A list of the reports prepared through this program is available on the Federal Highway Administration Web site at www.fhwa.dot.gov. Additional reports produced by the National Cooperative Highway Research Program (NCHRP) of the Transportation Research Board are available on the TRB Web site at www.trb.org. Several relevant NCHRP synthesis reports are listed in the references section at the end of this chapter.

21.4.3 Expert Systems as a Tool to Select Maintenance and Rehabilitation Strategies

Expert systems (a branch of artificial intelligence) can be used in the selection of maintenance and rehabilitation strategies. Expert systems (ES) are computer models that exhibit, within a specific domain, a degree of expertise in problem solving that is comparable to that of a human being. The knowledge required to build the expert system is obtained by interviewing pavement engineers who have extensive experience and knowledge about pavement management. The acquired knowledge is stored in the expert system and then can be used to recommend appropriate maintenance or rehabilitation strategies. Since the information is stored in a computer knowledge base, diagnostic advice is available to inexperienced users.

21.5 PAVEMENT REHABILITATION PROGRAMMING

In previous sections, we described how pavement condition is measured and discussed the alternatives and strategies available to repair and rehabilitate these surfaces. In this section, we describe the process used to select a specific program for rehabilitation.

Decisions are required about the type of repair or restoration technique that should be used for a given pavement section and about the timing (or programming) of the project. These decisions consider the design life of the pavement, the cost and benefits of the project, and other physical and environmental conditions.

Transportation agencies use various methods for selecting a program of pavement rehabilitation. According to the *AASHTO Guidelines for Pavement Management Systems*, analysis techniques are divided into three main groups: (1) condition assessment, (2) priority assessment, and (3) optimization. Each technique is used to determine the type of treatment needed and the schedule for rehabilitation, as described in the following section.

21.5.1 Condition Assessment

This method is used to develop single-year programs. The agency establishes criteria for the different measures of pavement condition against which comparison of the actual measurements can be made. If the measurement exceeds this limit or "trigger point," then a deficiency or need exists.

Figure 21.17 illustrates this concept. For example, if a limit of PSI of 2.5 has been set as the minimum acceptable roughness level for a particular class of pavement, then any section with a PSI less than 2.5 will represent a current deficiency. The fixed trigger point thus resolves the timing issue in a simple manner. Whenever the condition index falls below the given trigger point (criterion), it is assumed that rehabilitation is needed. Therefore, by using the trigger criterion, all sections are separated into two groups: "now needs" and "later needs."

In this method, the concern is only with the "now needs." Having decided upon the sections that need action, the next step is to select a treatment for each of these "now needs." This could involve simple economic analysis methods such as the net present worth (NPW) or benefit–cost ratio (BCR) based on approximate estimates of the expected life of the different alternatives. After deciding on the treatment, three possible situations may arise: (1) the needs match the budget, (2) the needs exceed the budget, or (3) the needs are less than the budget.

Of the three situations, the most common occurs when needs exceed the available funds, and so a ranking of these projects is often needed. The purpose of such ranking

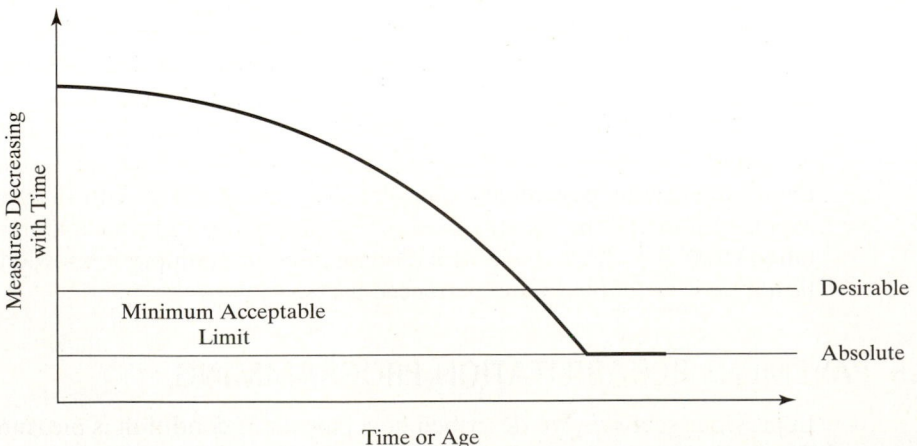

Figure 21.17   Determining Rehabilitation Needs Based on Established Criteria

is to determine which needs could be deferred to the following year. There are several possible alternatives for ranking. Projects may be ranked by distress, a combination of distress and traffic, the NPW, or the BCR. Sections are then selected until the budget is exhausted.

The rational factorial ranking method (FRM) is an example of a ranking method. It uses a priority index that combines climatic conditions, traffic, roughness, and distress. The priority index is expressed as

$$Y = 5.4 - (0.0263X_1) - (0.0132X_2) - \left[0.4 \log{(X_3)}\right]$$
$$+ (0.749X_4) + (1.66X_5) \tag{21.10}$$

where

Y = The priority index ranging from 1 to 10, with 1 representing very poor, and 10 representing excellent. Thus, a low value indicates a pavement that is a high priority for treatment.
X_1 = average rainfall (mm/yr)
X_2 = freeze and thaw (cycle/yr)
X_3 = traffic ($AADT$)
X_4 = present serviceability index
X_5 = distress (a subjective number between -1 and $+1$)

....

Example 21.7 Determining the Order of Priority for Rehabilitation

Three sections of highway have been measured for surface condition with results as shown in Table 21.7. Also shown are climatic and traffic conditions for each section. Use the rational factorial rating method to determine the order of priority for rehabilitation.

Solution: Using Eq. 21.10,

$$Y = 5.4 - (0.0263X_1) - (0.0132X_2) - \left[0.4 \log{(X_3)}\right] + (0.749X_4) + (1.66X_5)$$

The priority index (Y) for each section is:

- Section 1:

$$Y = 5.4 - (0.0263 \times 10) - (0.0132 \times 5) - \left[0.4 \log{(10,000)}\right]$$
$$+ (0.749 \times 2.4) + (1.66 \times 0.5) = 6.099$$

- Section 2:

$$Y = 5.4 - (0.0263 \times 30) - (0.0132 \times 15) - \left[0.4 \log{(5000)}\right]$$
$$+ (0.749 \times 3.2) + (1.66 \times (-0.2) = 4.998$$

- Section 3:

$$Y = 5.4 - (0.0263 \times 15) - (0.0132 \times 0) - \left[0.4 \log{(20,000)}\right]$$
$$+ (0.749 \times 3.0) + (1.66 \times 0.8 = 6.86$$

Since low index values indicate poor condition, section 2 should receive highest priority, followed by section 1, and lastly, section 3.

Table 21.7   Condition, Climatic, and Traffic Data for Highway Sections

Section	Rainfall	Freeze/Thaw	AADT	PSI	Distress
1	250 mm/yr	5 cycles/yr	10,000	2.4	+0.5
2	760 mm/yr	15 cycles/yr	5000	3.2	−0.2
3	380 mm/yr	0 cycles/yr	20,000	3.0	+0.8

21.5.2 Priority Assessment Models

Priority assessment is used to develop multiyear programs, an extension of the single-year model described. It addresses the question of when the "latter needs" are to be addressed and what action is required. To use this method, performance prediction models are required. For developing multiyear programs using ranking methods, either fixed-trigger-point or variable-trigger-point methods are used.

Trigger-Point Ranking

Models are used to predict when each road section will reach its trigger point. That is, instead of separating the network into two groups (present needs and later needs), this method separates the pavement sections into, say, six groups, according to the year when action will be needed, as illustrated in Figure 21.18. The process from this point is essentially the same as the previous method, except that more accurate economic analysis is possible as a result of the availability of prediction models. The use of a fixed-trigger-point method was illustrated in Example 21.4.

A variable-trigger-point ranking, as illustrated in Figure 21.19, is based on economic analysis models to establish the type of treatment as well as the timing. The advantage of using this method is that timing and treatment selection decisions are made simultaneously instead of being treated as two distinct stages in a sequential process, as was the case with the previous methods.

Near-Optimization Methods

Near-optimization methods are based on a heuristic approach that can usually yield good results. The marginal cost-effectiveness method illustrates near-optimization methods that have been widely applied. For this method, the cost effectiveness of various combinations of highway system project sections, maintenance/rehabilitation strategies, and their timing is computed.

Effectiveness is estimated by computing the area under the condition prediction model between the curve and the established critical value for highway condition. Effectiveness is calculated for a planning horizon (e.g., 10 years). Areas above the critical value are positive, and those below the critical value are negative. The section length and the average annual daily traffic (AADT) are multiplied by the net area. The result

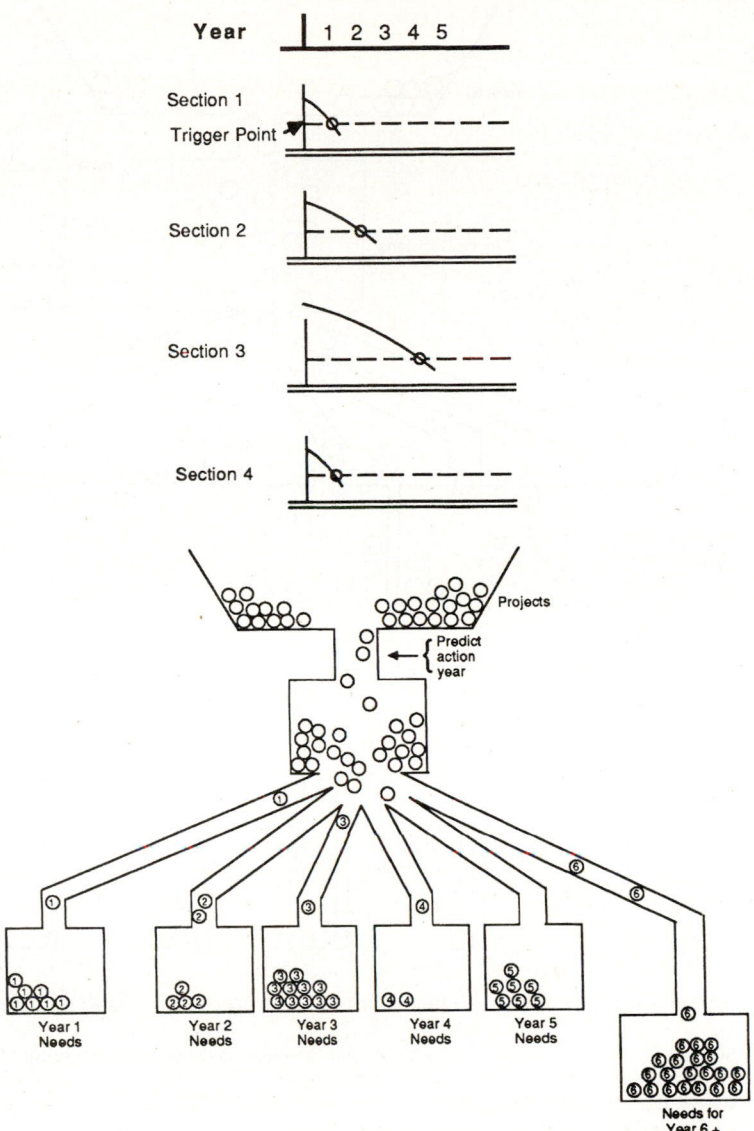

Figure 21.18   Fixed-Trigger-Point Ranking for Pavement Rehabilitation Programming

SOURCE: W. D. Cook and R. L. Lytton, "Recent Developments and Potential Future Directions in Ranking and Optimization Procedures for Pavement Management," *Second North American Conference for Managing Pavements*, vol. 2, 1987, p. 2.144

is a measure of the effectiveness of the maintenance or rehabilitation strategy being considered. Figure 21.20 illustrates the calculation procedure. The value of effectiveness is computed as the shaded area to the right of the rehabilitation strategy minus the shaded area to the left then multiplied by the traffic volume and by the length of the transportation element, as shown in Eq. 21.11.

$$\text{Effectiveness} = [\text{area 1} - \text{area 2}] \times [AADT] \times [\text{length of section}] \qquad (21.11)$$

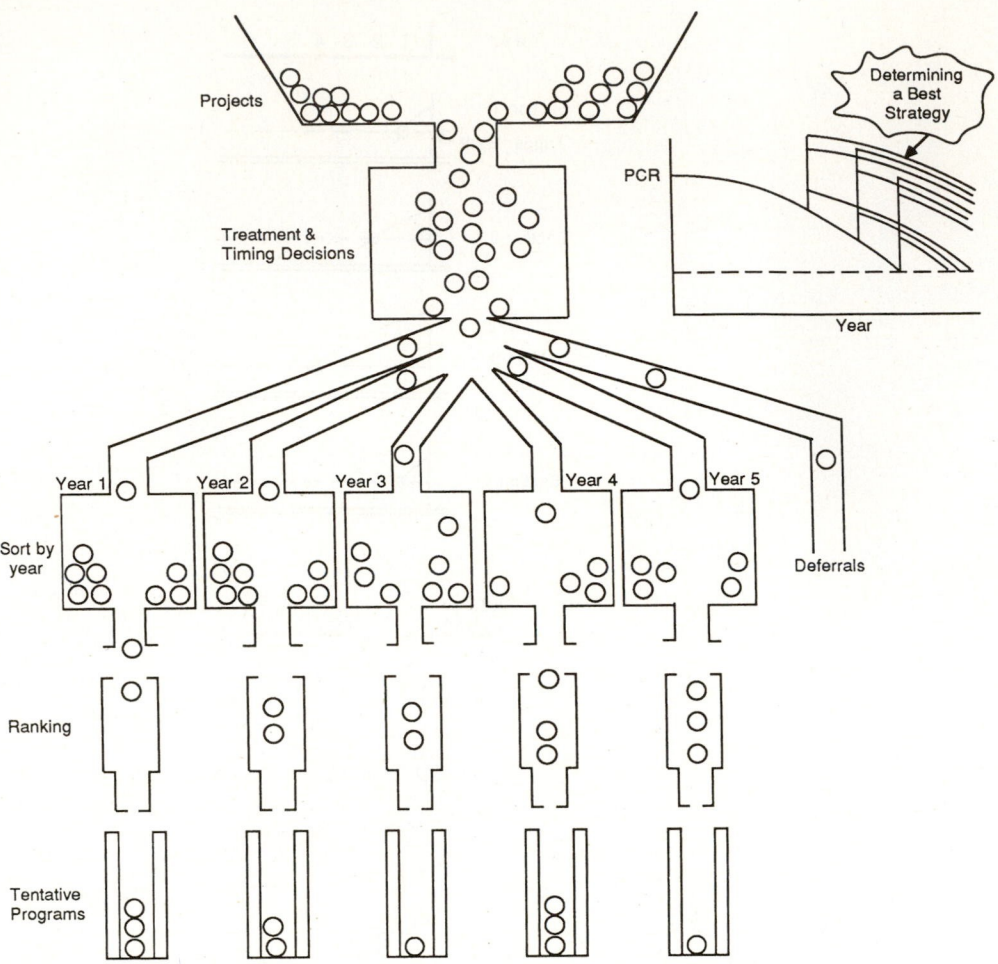

Figure 21.19   Variable-Trigger-Point Ranking Method for Pavement Rehabilitation Programming

SOURCE: W.D. Cook, R.L. Lytton, "Recent Developments and Potential Future Directions in Ranking and Optimization Procedures for Pavement Management," *Second North American Conference for managing Pavements*, vol. 2. 1987, p. 2.144.

The cost of the maintenance/rehabilitation project includes the cost of materials and construction on an in-place quantity basis, such as $/km, and user costs include delays and additional vehicle operating costs incurred during construction. Cost effectiveness of a given project is computed as

$$\text{Effectiveness ratio} = \frac{\text{effectiveness}}{\text{cost}}$$

The following steps are used to compute the marginal cost effectiveness.

Step 1.   For each highway section, select the combination of treatment alternative and year with the highest E/C ratio.

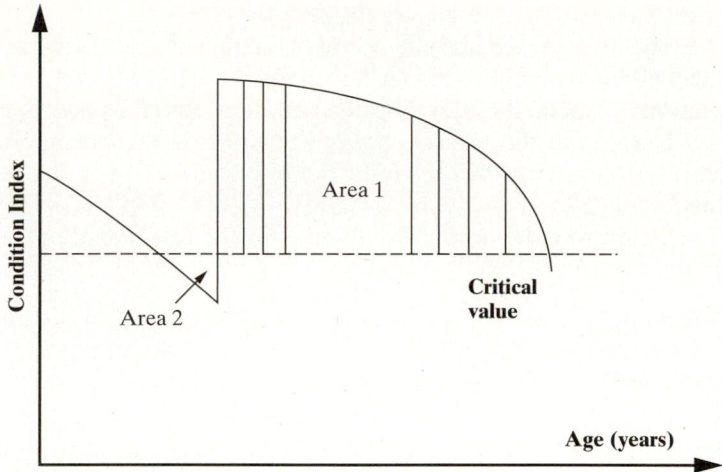

Figure 21.20   Calculating the Effectiveness of Maintenance and Rehabilitation Strategies

Step 2.   Compute the marginal cost effectiveness, MCE, of all other strategies (i.e., the combinations of other treatments and other timings) for each element, as

$$\text{MCE} = (E_s - E_r)/(C_s - C_r) \tag{21.12}$$

where

E_s = effectiveness of the selected combination
E_r = effectiveness of the combination being compared
C_s = cost of the combination selected
C_r = cost of the combination being compared

Step 3.   If MCE is negative, or if $E_r < E_s$, then the comparative strategy is removed from further consideration. If not, it replaces the selected combination.

Step 4.   Repeat the comparison process until the total cost of the selected projects utilizes the funding available.

Example 21.8 Effectiveness of Rehabilitation Strategies

The traffic volume of a 3.2 km pavement section is 10,000 veh/day. The deterioration relationship for sections in this family group is

$$\text{PQI} = 10 - 0.50(\text{age}) \tag{21.13}$$

where

PQI = pavement quality index with values between 0 and 10
age = pavement age in years

The highway agency policy is to maintain pavement sections with a PQI < 5. Resurfacing the pavement section will increase the PQI to a value of 9. After resurfacing, the section is expected to deteriorate at a rate of 1.50 PQI points per year.

Determine the effectiveness of resurfacing the pavement section after 12 years since construction. What would the effectiveness be if the resurfacing is postponed until year 13?

Determine the time it would take for the resurfaced pavement to deteriorate to a PQI value of 5.0.

Solution: Figure 21.21 illustrates the deterioration trend for the pavement section before and after the application of the rehabilitation strategy in year 12. The section condition deteriorates to a PQI value of 4.0 in year 12. After rehabilitation, the PQI value is restored to a value of 9. Following rehabilitation, the deterioration rate is much faster than the deterioration rate of the original pavement structure and is governed by the following equation.

$$PQI = 9 - 1.50 \times age \text{ (number of years after rehabilitation)} \qquad (21.14)$$

To determine when PQI will reach a value of 5.0 after rehabilitation, substitute 5.0 for PQI in Eq. 21.14 and solve for age (after rehabilitation) as follows.

$$5.0 = 9.0 - 1.50 \times age \text{(after rehabilitation)}$$

Therefore,

$$\text{Age after rehabilitation} = (9.0 - 5.0)/1.50 = 2.667 \text{ years}$$

Compute the area in Figure 21.22 to the right (area 1) and to the left (area 2) of the rehabilitation strategy as follows.

$$\text{Area 1} = \tfrac{1}{2} \times 2.667 \times 4.0 = 5.334$$
$$\text{Area 2} = \tfrac{1}{2} \times 2.0 \times 1.0 = 1.0$$

Use Eq. 21.11 to compute effectiveness.

$$\text{Effectiveness} = [\text{area 1} - \text{area 2}] \times [AADT] \times [\text{length of section}]$$
$$\text{Effectiveness} = (5.334 - 1.0) \times 10,000 \times 2 = 86,680$$

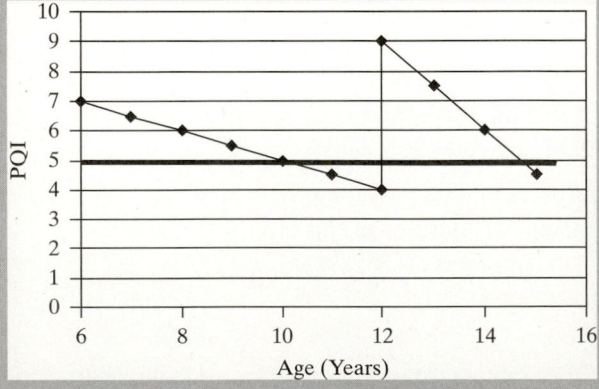

Figure 21.21   Deterioration Trend in Example 21.8: Rehabilitation in Year 12

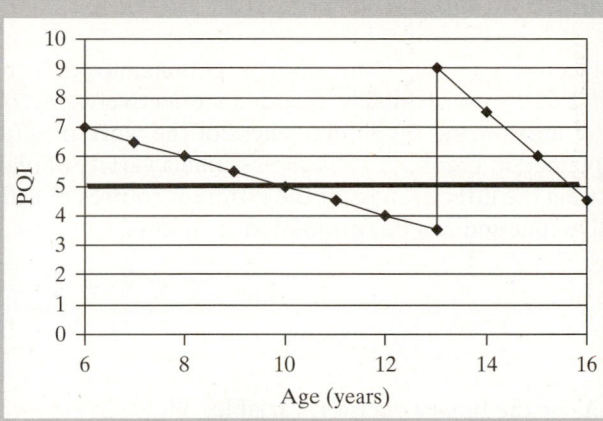

Figure 21.22   Deterioration Trend for the Pavement Section of Example 21.8: Rehabilitation Performed in Year 13

If resurfacing is postponed to year 13, the deterioration trend can be expressed as shown in Figure 21.22. Note that the PQI will be 5.0 in approximately 2.67 years following resurfacing (i.e., between years 15 and 16).

In this case, the area calculations are as follows.

$$\text{Area } 1 = \tfrac{1}{2} \times 2.67 \times 4.0 = 5.334$$
$$\text{Area } 2 = \tfrac{1}{2} \times 3.0 \times 1.5 = 2.25$$

Therefore,

$$\text{Effectiveness} = (5.334 - 2.25) \times 10{,}000 \times 2 = 61{,}680$$

21.5.3 Optimization Techniques

Optimization models provide the capability for the simultaneous evaluation of an entire pavement network, and as such, represent another method for scheduling rehabilitation programs in which alternative strategies satisfy an objective function subject to constraints. Optimization involves the following: (1) identifying the decision variables and determine optimal values, (2) formulating an objective function to be optimized, and (3) establishing the constraints of the model.

Identify Decision Variables

Decision variables reflect the selections regarding: (1) which pavement sections must be maintained, (2) the appropriate maintenance or rehabilitation strategy to select, and (3) the year when maintenance work is required. The following set of binary variables with values of 1 or 0 is used to represent these decisions.

$X_{ijt} = 1$, if section i is maintained using alternative j applied in year t

$X_{ijt} = 0$, if section i is not maintained using alternative j applied in year t

Formulate an Objective Function

The objective function is the value to be maximized or minimized. For example, if the objective is to use available resources effectively, the objective function is stated in terms of maximizing the effectiveness of the work program. Effectiveness calculations (as explained in Example 21.8) can assume a variety of planning horizons—for example, calculating the effectiveness of the different courses of action over the next 10 years. An objective function can be formulated as follows.

$$\text{Maximize} \sum_{i=1}^{m} \sum_{j=1}^{k} \sum_{t=1}^{T} X_{ijt} B_{ijt} \qquad (21.15)$$

where

X_{ijt} = the binary decision variable

B_{ijt} = effectiveness of maintenance strategy j for highway section i in year t, calculated using the appropriate prediction curve

Establish Constraints

Constraints are restrictions on the numerical value of a decision variable and reflect the limitations on available resources, including money, time, personnel, and materials. One inevitable constraint is that expenditures in a given year cannot exceed the available budget, which is formulated as follows.

$$\sum_{i=1}^{m} \sum_{j=1}^{k} X_{ijt} D_{ijt} \leq B_t \quad (\text{for } t = 1, 2, \ldots, T) \qquad (21.16)$$

where

D_{ijt} = the cost of maintenance strategy j for highway section i in year t

B_t = budget allocation in year t

An additional constraint is required to ensure that each section receives only one maintenance or rehabilitation strategy during the planning period which is formulated as follows.

$$\sum_{t=1}^{T} \sum_{j=1}^{k} X_{ijt} \leq 1 \quad (\text{for } i = 1, 2, \ldots, m) \qquad (21.17)$$

where X_{ijt} represents binary variables that take a value of 0 or 1.

Equations 21.15 through 21.17 can be combined to optimize the total present value of m highway improvement projects with k different maintenance strategies, and a planning horizon of T years which is formulated as follows.

$$\text{Maximize} \sum_{i=1}^{m} \sum_{j=1}^{k} \sum_{t=1}^{T} X_{ijt} B_{ijt}$$

subject to

$$\sum_{i=1}^{T} \sum_{j=1}^{k} X_{ijt} \leq 1 \quad (\text{for } i = 1, 2, \ldots, m)$$

$$\sum_{i=1}^{m} \sum_{j=1}^{k} X_{ijt} D_{ijt} \leq B_t \quad (\text{for } t = 1, 2, \ldots, T)$$

21.6 GIS AND PAVEMENT MANAGEMENT

Geographic Information Systems (GIS) are a set of computer software, hardware, data, and personnel that store, manipulate, analyze, and present geographically referenced (or spatial) data. GIS can link spatial information on maps (such as roadway alignment) with attribute or tabular data. For example, a GIS—a digital map of a road network—would be linked to an attribute table that stores pertinent information regarding each road section on the network. This information could include items such as the section ID number, length of section, number of lanes, condition of the pavement surface, and average daily traffic volume. By accessing a specific road segment, a complete array of relevant attribute data become available.

Figure 21.23 illustrates one application of GIS in infrastructure management. The computer-generated map shows the major roads and is color-coded based upon the values of the Ride Index (a measure of pavement surface condition). An attribute table is provided for pertinent information related to a specified road segment.

GIS includes procedures for: (1) data input; (2) data storage and retrieval; (3) data query, analysis, and modeling; and (4) data output. Data are accepted from a wide range of sources, including maps, aerial photographs, satellite image, and surveys. GIS also includes a comprehensive relational database management system (DBMS), which uses geo-references as the primary means of indexing information. GIS allows presentation of the data in a meaningful way, including map and textural/tabular reports. It provides

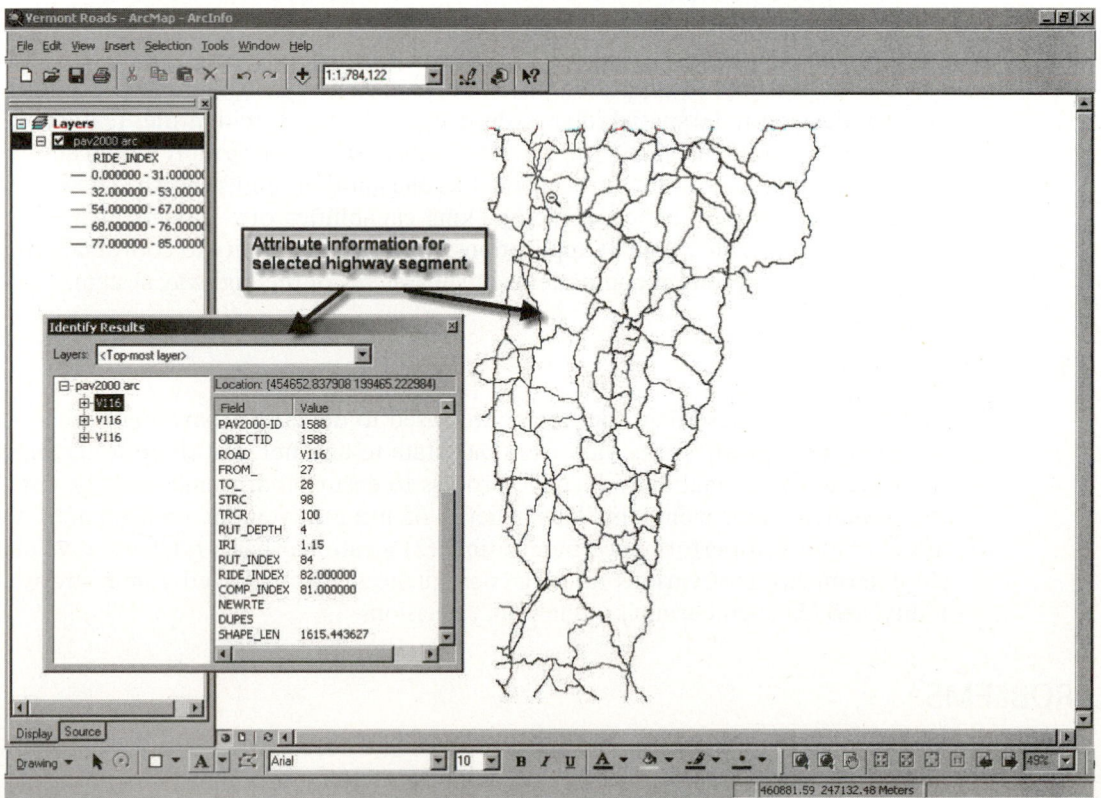

Figure 21.23   Use of GIS in Pavement Management

spatial analysis and modeling procedures that perform computations on data groups or layers and identifies patterns and relationships.

Due to the special capabilities of GIS (including data storage, retrieval, analysis, and presentation capabilities), the use of GIS for pavement management systems is increasing. Pavement management requires both spatial and attribute data; for example, location of pavement sections (i.e., spatial or location-type data) and the condition of those assets (i.e., attribute-type data). Other pertinent information is required regarding each highway section, such as year of construction, condition history, and maintenance history. For these reasons, GIS is an ideal application for pavement management.

There are several advantages of GIS over other forms of data presentation and analysis:

1. The ability to overlay different layers of information. Road surface condition can be stored in one layer, pavement construction information in another, and so forth. These different layers have a common coordinate system, allowing them to be overlaid on top of one another.

2. Provision of "intelligent maps," which access significant amounts of attribute data linked to the geographic features of the map. These "intelligent maps" can convey more information than may be immediately apparent. In addition, these maps can be managed, analyzed, queried, or presented in a more effective way than is possible with physical hard-copy maps, such as by identifying all pavement sections below a specified value for a condition index. Advanced queries and data analyses are also possible; for example, to identify sections whose condition is below a certain threshold and, at the same time, are subject to traffic volumes exceeding another threshold.

3. Use of topology to support decision making. Topology is a branch of mathematics that deals with the spatial relationships among features. It identifies features that are within a certain distance from another feature, or features totally within another feature, including ones that are adjacent to one another, and so forth. These functions can greatly support the decision-making capabilities of a pavement management system for a wide range of applications, including the ability to coordinate the work schedules of different components or subsystems of the highway system.

21.7 SUMMARY

This chapter has described the procedures used to develop a pavement rehabilitation program. The specifics may vary from one state to another, but there is agreement on the need for a rational and objective process to ensure that funds are efficiently used for pavement improvements. The benefits of using pavement management systems are (1) improved performance monitoring, (2) a rational basis for legislative support, (3) determination of various funding consequences, (4) improved administrative credibility, and (5) engineering input in policy decisions.

PROBLEMS

21-1   What is meant by the term *pavement management*? Describe three strategies used by public agencies to develop restoration and rehabilitation programs.

21-2   What are the three principal uses for pavement condition data?

21-3   What is the difference between PSI and PSR?

21-4 Draw a sketch showing the relationship between pavement condition (expressed as *PSI*) and time for a service life of 20 years, if the PSI values range from 4.5 to 2.5 in a six year period, and then the pavement is resurfaced such that the PSI is increased to 4.0. After another 6 years, the PSI has reached 2.0. With rehabilitation, PSI is increased to 4.5. At the end of its service life the PSI value is 3.4.

21-5 Describe the four characteristics of pavement condition used to evaluate whether a pavement should be rehabilitated, and if so, determine the appropriate treatment.

21-6 A given pavement rating method uses six distress types to establish the DR. These are: corrugation, alligator cracking, raveling, longitudinal cracking, rutting, and patching. For a section of highway, the number of points assigned to each category were 5, 3, 3, 3, 4, and 3. If the weighting factors are 1, 1, 1, 1, 2, and 1.5, respectively, determine the DR for the section.

21-7 What are the problems with distress surveys, and how can such problems be solved?

21-8 How are computers used in pavement management?

21-9 Describe the methods used to determine static and dynamic deflection of pavements. To what extent are these tests used in pavement rehabilitation management?

21-10 A 18 kN load is placed on two tires, which are then locked in place. A force of 9.4 kN is necessary to cause the trailer to move at a speed of 32 km/h. Determine the value of the skid number. If treaded tires were used, characterize the pavement type.

21-11 The PMS database has accumulated information regarding the performance of a pavement section before being overlaid as well as the performance of the overlay. The following data were recorded. Develop a linear prediction model (PCR=a + b*Age) for the two cases. If a criterion was established that maintenance or rehabilitation should occur when the PCR reaches a value of 81, when should such actions take place in the two cases?

Before Overlay		Overlay Performance	
PCR	Age (yr)	PCR	Age (yr)
100	0.5	98	0.5
98	1.5	96	1.5
94	3.5	92	3.5
91	4.5	86	5.5
89	5.5	83	6.5
86	6.5	80	7.5
83	8.0		

21-12 An agency uses the following formula to determine pavement condition rating:

PCR = 95.6 − 5.51 (5.0-ROUGH) − 1.59 LNALL − 0.221 AVGOUT − 0.0306 LONG − 0.531 TRAN

where:
   ROUGH = roughness measured by present serviceability rating (PSR), scale of 0 to 5
   LNALL = natural logarithm of alligator cracking, m per km
   AVGOUT = outer wheel path rutting (all locations averaged), 0.25 mm
   LONG = longitudinal cracking, m per km
   TRAN = transverse cracking, number of cracks per kilometer
And, present serviceability rating is determined as:

$$PSR = 5e^{-0.0051118\cdot IRI - 0.0016027}$$

A pavement section has the following distress data:

IRI = 1.66 m/km
Alligator cracking = 20 m/km
Average rutting = 5 mm
Longitudinal cracks = 25 m/km
Transverse cracks = 7.5/km

Determine the current PCR for this section. Then, assume a new roughness measurement is expected soon. When PCR falls below 60, the section will be scheduled for rehabilitation. Determine the new IRI value that will trigger the rehabilitation.

21-13   Referring to the Markovian transition matrix in Table 21.6, what is the probability that a section with a PCR value of 77 in the current year will have a PCR value between 60 and 69 (a) one year later, and (b) two years later.

21-14   Differentiate between corrective and preventive rehabilitation techniques. Cite three examples of surface treatments in each category. What is the best preventive maintenance technique for subsurface maintenance?

21-15   Describe the techniques used to repair flexible and rigid pavements and their effectiveness for the following treatment types: (a) patching, (b) crack maintenance, and (c) overlays.

21-16   What is the basic difference between an expert system and a conventional computer program?

21-17   Discuss the differences between condition and priority assessment models used in developing pavement improvement programs.

21-18   Describe six methods that can be used to select a program of pavement rehabilitation.

REFERENCES

AASHTO Guidelines for Pavement Management Systems, American Association of State Highway and Transportation Officials, Washington, D.C., 1990.

Barker, A. M., M. Ndoye, and D. M. Bullock, Opportunity to Leverage Vehicle Infrastructure Integration (VII) Data for Pavement Condition Monitoring, *Proceedings of the 87th Annual Transportation Research Board Annual Meeting*, Washington, D.C., 2008.

Chan, Y. C., B. Huang, X. Yan, and S. Richards, *Effects of Asphalt Pavement Condition on Traffic Accidents in Tennessee Utilizing Pavement Management Systems*, Final Report, Southeastern Transportation Center, University of Tennessee, Knoxville, TN, 2008.

Flintsch, G. W., R. Dymond, and J. Collura, *NCHRP Synthesis of Highway Practice, Report 335: Pavement Management Applications Using Geographic Information Systems*, Transportation Research Board, National Research Council, Washington, D.C., 2004.

Haas, R., W. R. Hudson, and J. Zaniewski, *Modern Pavement Management*, Krieger Publishing Co., Malabar, FL, 1994.

Longley, P., M. F. Goodchild, D. J. Macguire, and D. W. Rhind, *Geographic Information Systems and Science*, John Wiley & Sons, New York, NY, 2001.

McGhee, K. H., *NCHRP Synthesis of Highway Practice, Report 334: Automated Pavement Distress Collection Techniques*, Transportation Research Board, National Research Council, Washington, D.C., 2004.

National Cooperative Highway Research Program, Report 523: Optimal Timing of Pavement Preventive Maintenance Treatment Applications, Transportation Research Board, National Research Council, Washington, D.C., 2004.

National Cooperative Highway Research Program, Report 545: Analytical Tools for Asset Management, Transportation Research Board, National Research Council, Washington, D.C., 2005.

Noyce, D. A., H. U. Bahia, J. Yambo, J. Chapman, and A. Bill, *Incorporating Road Safety into Pavement Management: Maximizing Surface Friction for Road Safety Improvements*, Project 04–04, Final Report, Midwest Regional University Transportation Center, University of Wisconsin, Madison, WI, 2007.

Pavement Management Guide, 2nd Edition, American Association of State Highway and Transportation Officials, Washington D.C., 2012.

Tighe, S., L. C. Falls, and J. Morrall, "Integrating Safety with Asset Management Systems," presented at the *5th International Conference on Managing Pavements,* Seattle, Washington, 2001.

Critical Values for the Student's t and χ^2 Distributions

Table A.1 Student's t Distribution

	Level of Significance for One-Tailed Test							
	.250	.100	.050	.025	.010	.005	.0025	.0005
	Level of Significance for a Two-Tailed Test							
Degrees of Freedom	.500	.200	.100	.050	.020	.010	.005	.001
1	1.000	3.078	6.314	12.706	31.821	63.657	27.321	536.627
2	.816	1.886	2.920	4.303	6.965	9.925	14.089	31.599
3	.765	1.638	2.353	3.182	4.541	5.841	7.453	12.924
4	.741	1.533	2.132	2.776	3.747	4.604	5.598	8.610
5	.727	1.476	2.015	2.571	3.365	4.032	4.773	6.869
6	.718	1.440	1.943	2.447	3.143	3.707	4.317	5.959
7	.711	1.415	1.895	2.365	2.998	3.499	4.029	5.408
8	.706	1.397	1.860	2.306	2.896	3.355	3.833	5.041
9	.703	1.383	1.833	2.262	2.821	3.250	3.690	4.781
10	.700	1.372	1.812	2.228	2.764	3.169	3.581	4.587
11	.697	1.363	1.796	2.201	2.718	3.106	3.497	4.437
12	.695	1.356	1.782	2.179	2.681	3.055	3.428	4.318
13	.694	1.350	1.771	2.160	2.650	3.012	3.372	4.221
14	.692	1.345	1.761	2.145	2.624	2.977	3.326	4.140
15	.691	1.341	1.753	2.131	2.602	2.947	3.286	4.073
16	.690	1.337	1.746	2.120	2.583	2.921	3.252	4.015
17	.689	1.333	1.740	2.110	2.567	2.898	3.222	3.965
18	.688	1.330	1.734	2.101	2.552	2.878	3.197	3.922
19	.688	1.328	1.729	2.093	2.539	2.861	3.174	3.883
20	.687	1.325	1.725	2.086	2.528	2.845	3.153	3.850

(Continued)

Table A.1   Student's *t* Distribution (*continued*)

Degrees of Freedom	*Level of Significance for One-Tailed Test*							
	.250	.100	.050	.025	.010	.005	.0025	.0005
	Level of Significance for a Two-Tailed Test							
	.500	.200	.100	.050	.020	.010	.005	.001
21	.686	1.323	1.721	2.080	2.518	2.831	3.135	3.819
22	.686	1.321	1.717	2.074	2.508	2.819	3.119	3.792
23	.685	1.319	1.714	2.069	2.500	2.807	3.104	3.768
24	.685	1.318	1.711	2.064	2.492	2.797	3.091	3.745
25	.684	1.316	1.708	2.062	2.485	2.787	3.078	3.725
26	.684	1.315	1.706	2.056	2.479	2.779	3.067	3.707
27	.684	1.314	1.703	2.052	2.473	2.771	3.057	3.690
28	.683	1.313	1.701	2.048	2.467	2.763	3.047	3.674
29	.683	1.311	1.699	2.045	2.462	2.756	3.038	3.659
30	.683	1.310	1.697	2.042	2.457	2.750	3.030	3.646
35	.682	1.306	1.690	2.030	2.438	2.724	2.996	3.591
40	.681	1.303	1.684	2.021	2.423	2.704	2.971	3.551
45	.680	1.301	1.679	2.014	2.412	2.690	2.952	3.520
50	.679	1.299	1.676	2.009	2.403	2.678	2.937	3.496
55	.679	1.297	1.673	2.004	2.396	2.668	2.925	3.476
60	.679	1.296	1.671	2.000	2.390	2.660	2.915	3.460
65	.678	1.295	1.669	1.997	2.385	2.654	2.906	3.447
70	.678	1.294	1.667	1.994	2.381	2.648	2.899	3.435
80	.678	1.292	1.664	1.990	2.374	2.639	2.887	3.416
90	.677	1.291	1.662	1.987	2.368	2.632	2.878	3.402
100	.677	1.290	1.660	1.984	2.364	2.626	2.871	3.390
125	.676	1.288	1.657	1.979	2.357	2.616	2.858	3.370
150	.676	1.287	1.655	1.976	2.351	2.609	2.849	3.357
200	.676	1.286	1.653	1.972	2.345	2.601	2.839	3.340
∞	.6745	1.2816	1.6448	1.9600	2.3267	2.5758	2.8070	3.2905

SOURCE: Adapted from Richard H. McCuen, *Statistical Methods for Engineers*, copyright © 1985.

Table A.2   Critical Values of χ^2

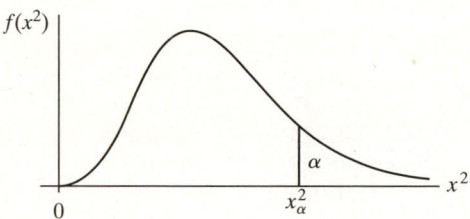

Degrees of Freedom	$\chi^2_{0.995}$	$\chi^2_{0.990}$	$\chi^2_{0.975}$	$\chi^2_{0.950}$	$\chi^2_{0.900}$
1	0.0000393	0.0001571	0.0009821	0.0039321	0.0157908
2	0.0100251	0.0201007	0.0506356	0.102587	0.210720
3	0.0717212	0.114832	0.215795	0.351846	0.584375
4	0.206990	0.297110	0.484419	0.710721	1.063623
5	0.411740	0.554300	0.831211	1.145476	1.61031
6	0.675727	0.872085	1.237347	1.63539	2.20413
7	0.989265	1.239043	1.68987	2.16735	2.83311
8	1.344419	1.646482	2.17973	2.73264	3.48954
9	1.734926	2.087912	2.70039	3.32511	4.16816
10	2.15585	2.55821	3.24697	3.94030	4.86518
11	2.60321	3.05347	3.81575	4.57481	5.57779
12	3.07382	3.57056	4.40379	5.22603	6.30380
13	3.56503	4.10691	5.00874	5.89186	7.04150
14	4.07468	4.66043	5.62872	6.57063	7.78953
15	4.60094	5.22935	6.26214	7.26094	8.54675
16	5.14224	5.81221	6.90766	7.96164	9.31223
17	5.69724	6.40776	7.56418	8.67176	10.0852
18	6.26481	7.01491	8.23075	9.39046	10.8649
19	6.84398	7.63273	8.90655	10.1170	11.6509
20	7.43386	8.26040	9.59083	10.8508	12.4426
21	8.03366	8.89720	10.28293	11.5913	13.2396
22	8.64272	9.54249	10.9823	12.3380	14.0415
23	9.26042	10.19567	11.6885	13.0905	14.8479
24	9.88623	10.8564	12.4011	13.8484	15.6587
25	10.5197	11.5240	13.1197	14.6114	16.4734
26	11.1603	12.1981	13.8439	15.3791	17.2919
27	11.8076	12.8786	14.5733	16.1513	18.1138
28	12.4613	13.5648	15.3079	16.9279	18.9392
29	13.1211	14.2565	16.0471	17.7083	19.7677
30	13.7867	14.9535	16.7908	18.4926	20.5992
40	20.7065	22.1643	24.4331	26.5093	29.0505
50	27.9907	29.7067	32.3574	34.7642	37.6886
60	35.5346	37.4848	40.4817	43.1879	46.4589
70	43.2752	45.4418	48.7576	51.7393	55.3290
80	51.1720	53.5400	57.1532	60.3915	64.2778
90	59.1963	61.7541	65.6466	69.1260	73.2912
100	67.3276	70.0648	74.2219	77.9295	82.3581

(Continued)

Table A.2   Critical Values of χ^2 (*continued*)

Degrees of Freedom	$\chi^2_{0.995}$	$\chi^2_{0.990}$	$\chi^2_{0.975}$	$\chi^2_{0.950}$	$\chi^2_{0.900}$
1	2.70554	3.84146	5.02389	6.63490	7.87944
2	4.60517	5.99147	7.37776	9.21034	10.5966
3	6.25139	7.81473	9.34840	11.3449	12.8381
4	7.77944	9.48773	11.1433	13.2767	14.8602
5	9.23635	11.0705	12.8325	15.0863	16.7496
6	10.6446	12.5916	14.4494	16.8119	18.5476
7	12.0170	14.0671	16.0128	18.4753	20.2777
8	13.3616	15.5073	17.5346	20.0902	21.9550
9	14.6837	16.9190	19.0228	21.6660	23.5893
10	15.9871	18.3070	20.4831	23.2093	25.1882
11	17.2750	19.6751	21.9200	24.7250	26.7569
12	18.5494	21.0261	23.3367	26.2170	28.2995
13	19.8119	22.3621	24.7356	27.6883	29.8194
14	21.0642	23.6848	26.1190	29.1413	31.3193
15	22.3072	24.9958	27.4884	30.5779	32.8013
16	23.5418	26.2962	28.8454	31.9999	34.2672
17	24.7690	27.5871	30.1910	33.4087	35.7185
18	25.9894	28.8693	31.5264	34.8053	37.1564
19	27.2036	30.1435	32.8523	36.1908	38.5822
20	28.4120	31.4104	34.1696	37.5662	39.9968
21	29.6151	32.6705	35.4789	38.9321	41.4010
22	30.8133	33.9244	36.7807	40.2894	42.7956
23	32.0069	35.1725	38.0757	41.6384	44.1813
24	33.1963	36.4151	39.3641	42.9798	45.5585
25	34.3816	37.6525	40.6465	44.3141	46.9278
26	35.5631	38.8852	41.9232	45.6417	48.2899
27	36.7412	40.1133	43.1944	46.9630	49.6449
28	37.9159	41.3372	44.4607	48.2782	50.9933
29	39.0875	42.5569	45.7222	49.5879	52.3356
30	40.2560	43.7729	46.9792	50.8922	53.6720
40	51.8050	55.7585	59.3417	63.6907	66.7659
50	63.1671	67.5048	71.4202	76.1539	79.4900
60	74.3970	79.0819	83.2976	88.3794	91.9517
70	85.5271	90.5312	95.0231	100.425	104.215
80	96.5782	101.879	106.629	112.329	116.321
90	107.565	113.145	118.136	124.116	128.299
100	118.498	124.342	129.561	135.807	140.169

SOURCE: From CATHERINE M. THOMPSON, TABLE OF PERCENTAGE POINTS OF THE χ^2 DISTRIBUTION, *Biometrika* 1941 32: 187–191, by permission of Oxford University Press.

Developing Equations for Computing Regression Coefficients

L et a dependent variable Y and an independent variable x be related by an estimated regression function

$$Y = a + bx \tag{B.1}$$

Let Y_i be an estimate and y_i be an observed value of Y for a corresponding value x_i for x. Estimates of a and b can be obtained by minimizing the sum of the squares of the differences (R) between Y_i and y_i for a set of observed values, where

$$R = \sum_{i=1}^{n} (y_i - Y_i)^2 \tag{B.2}$$

Substituting $(a + bx_i)$ for Y_i in Eq. B.2, we obtain

$$R = \sum_{i=1}^{n} (y_i - a - bx_i)^2 \tag{B.3}$$

Differentiating R partially with respect to a, then with respect to b, and equating each to zero, we obtain

$$\frac{\partial R}{\partial a} = -2 \sum_{i=1}^{n} (y_i - a - bx_i) = 0 \tag{B.4}$$

$$\frac{\partial R}{\partial b} = -2 \sum_{i=1}^{n} x_i (y_i - a - bx_i) = 0 \tag{B.5}$$

From Eq. B.4, we obtain

$$\sum_{i=1}^{n} y_i = na + b \sum_{i=1}^{n} x_i \tag{B.6}$$

giving

$$a = \frac{1}{n}\sum_{i=1}^{n} y_i - \frac{b}{n}\sum_{i=1}^{n} x_i \qquad (B.7)$$

From Eq. B.5, we obtain

$$\sum_{i=1}^{n} x_i y_i = a\sum_{i=1}^{n} x_i + b\sum_{i=1}^{n} x_i^2 \qquad (B.8)$$

Substituting for a, we obtain

$$b = \frac{\sum_{i=1}^{n} x_i y_i - \frac{1}{n}\left(\sum_{i=1}^{n} x_i\right)\left(\sum_{i=1}^{n} y_i\right)}{\sum_{i=1}^{n} x_i^2 - \frac{1}{n}\left(\sum_{i=1}^{n} x_i\right)} \qquad (B.9)$$

where

n = number of sets of observation
x_i = ith observation for x
y_i = ith observation for y

Equations B.8 and B.9 may be used to obtain estimated values for a and b in Eq. B.1. To test the suitability of the regression function obtained, the coefficient of determination, R^2 (which indicates to what extent values of Y_i obtained from the regression function agree with observed values y_i), is determined from the expression

$$R^2 = \frac{\sum_{i=1}^{n}(Y_i - \bar{y})^2}{\sum_{i=1}^{n}(y_i - \bar{y})^2} \qquad (B.10)$$

which is also written as

$$R^2 = \frac{\sum_{i=1}^{n}(x_i y_i - n\bar{x}\bar{y})^2}{\sum_{i=1}^{n} R(x_i^2 - n\bar{x})\left(\sum_{i=1}^{n} y_i^2 - n\bar{y}^2\right)} \qquad (B.11)$$

The closer the value of R^2 is to 1, the more suitable the estimated regression function is for the data.

APPENDIX C

Fitting Speed and Density Data for Example 6.3 to the Greenshields Model Using Excel

omparing the Greenshields expression to our estimated linear regression function, we see that the speed $\bar{u}_s$ in the Greenshields expression is represented by y in the estimated regression function, the mean free speed u_f is represented by a, and the value of the mean free speed u_f divided by the jam density k_j is represented by $-b$.

We must now perform a linear regression analysis in order to find the values of a and b in the estimated regression function.

Given the speed and density data shown in Table 6.2 (columns 1 and 2), first enter the data into columns A and B of your Excel Spreadsheet, as seen in Figure C.1.

Next, highlight all of the speed and density data that you have entered into your spreadsheet. Then select Data Analysis from under the Tools menu, as shown in Figure C.2.

Then, under the Data Analysis menu shown in Figure C.3, select Regression and click OK.

In the Regression menu, after ensuring that the speed data is selected for the Input <u>Y</u> Range and the density data is selected for the Input <u>X</u> Range, click OK. An example is shown in Figure C.4.

A Summary Output for the data analysis that is shown in Figure C.5 will be created in a new worksheet. Under the Coefficients column, it can be seen that the Intercept (which is the constant a) in the estimated regression function is 62.56, and the X Variable (which is the constant b) is -0.53.

This yields a regression formula for finding the speed depending on the density.

$$\bar{u}_s = 62.56 - 0.53k$$

Also seen in the Summary Output is the value of $R^2 = 0.95$, showing that this regression is a good fit for the data.

1223

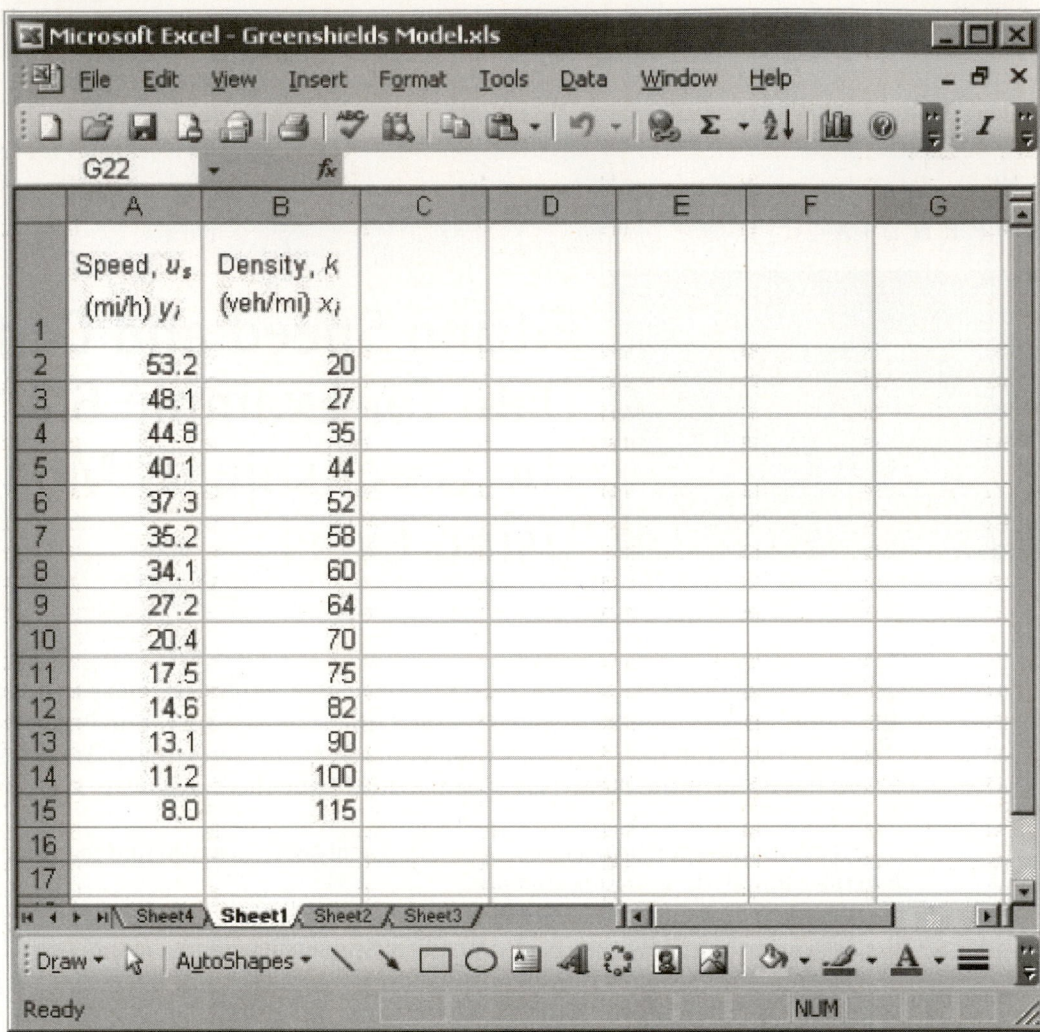

Figure C.1   Illustration of Input Data

SOURCE: Created using Microsoft Office

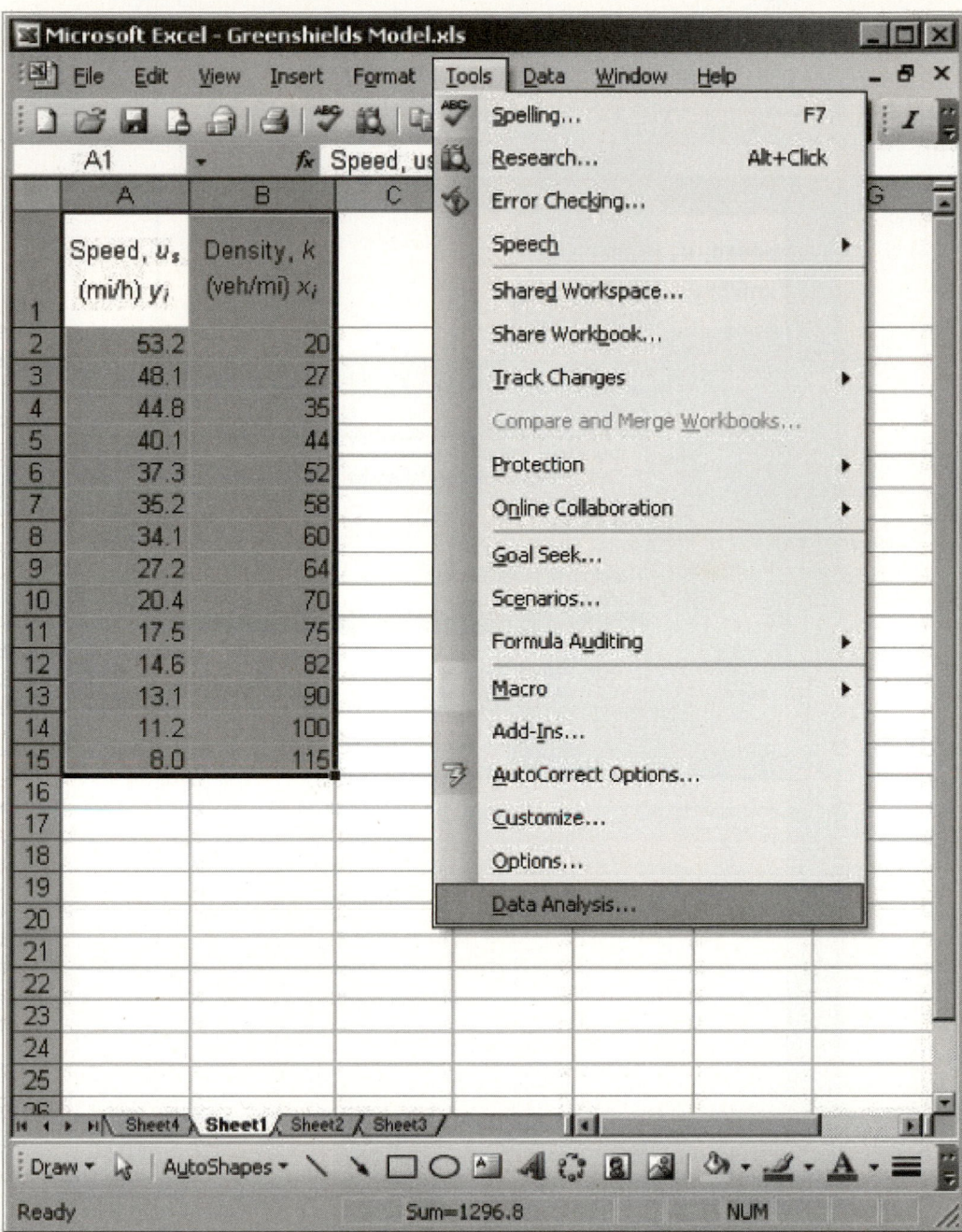

Figure C.2   Highlighted Data and Selection of Data Analysis under Tools Menu

SOURCE: Created using Microsoft Office

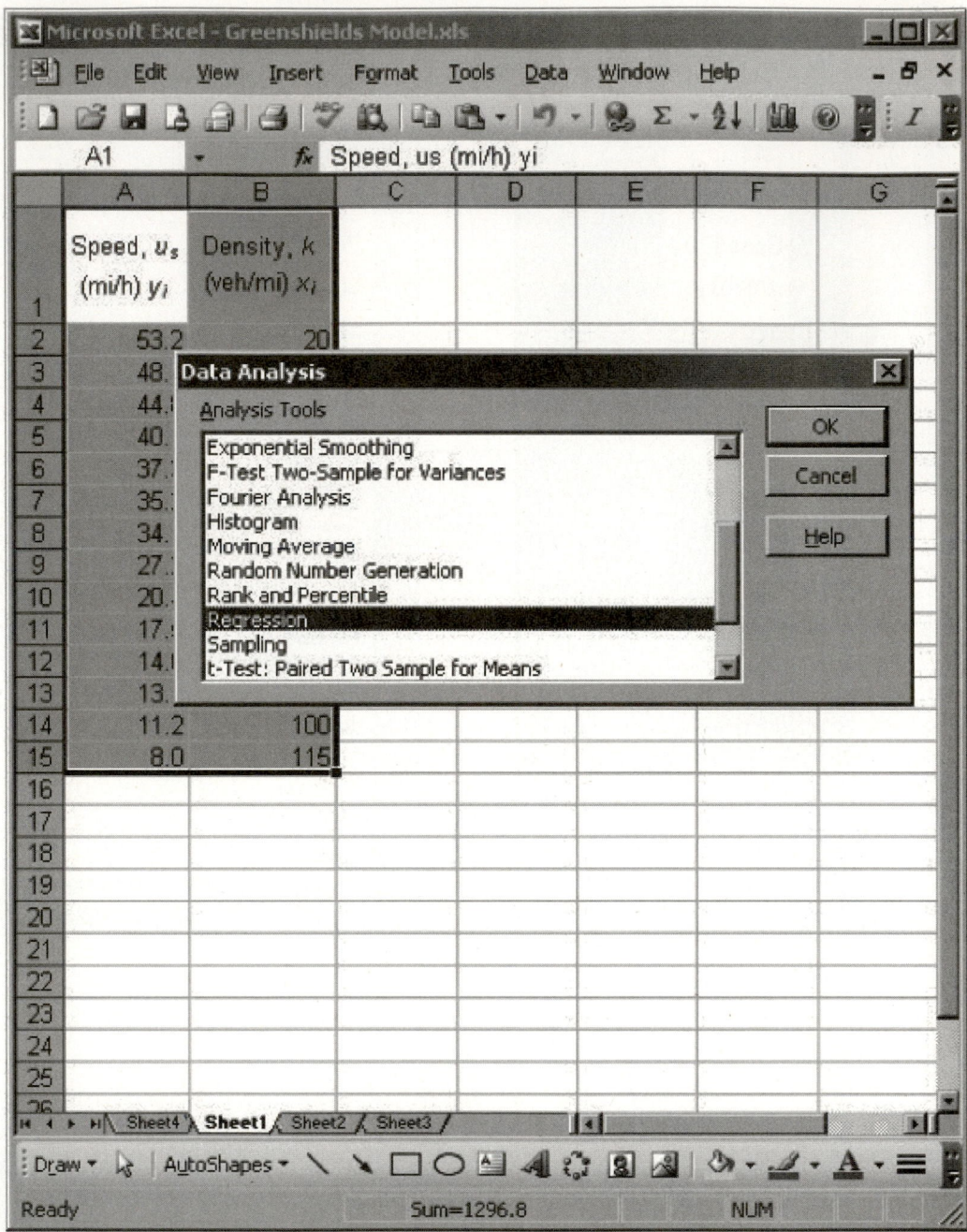

Figure C.3 Selection of Regression Analysis Procedure from Tools Menu

SOURCE: Created using Microsoft Office

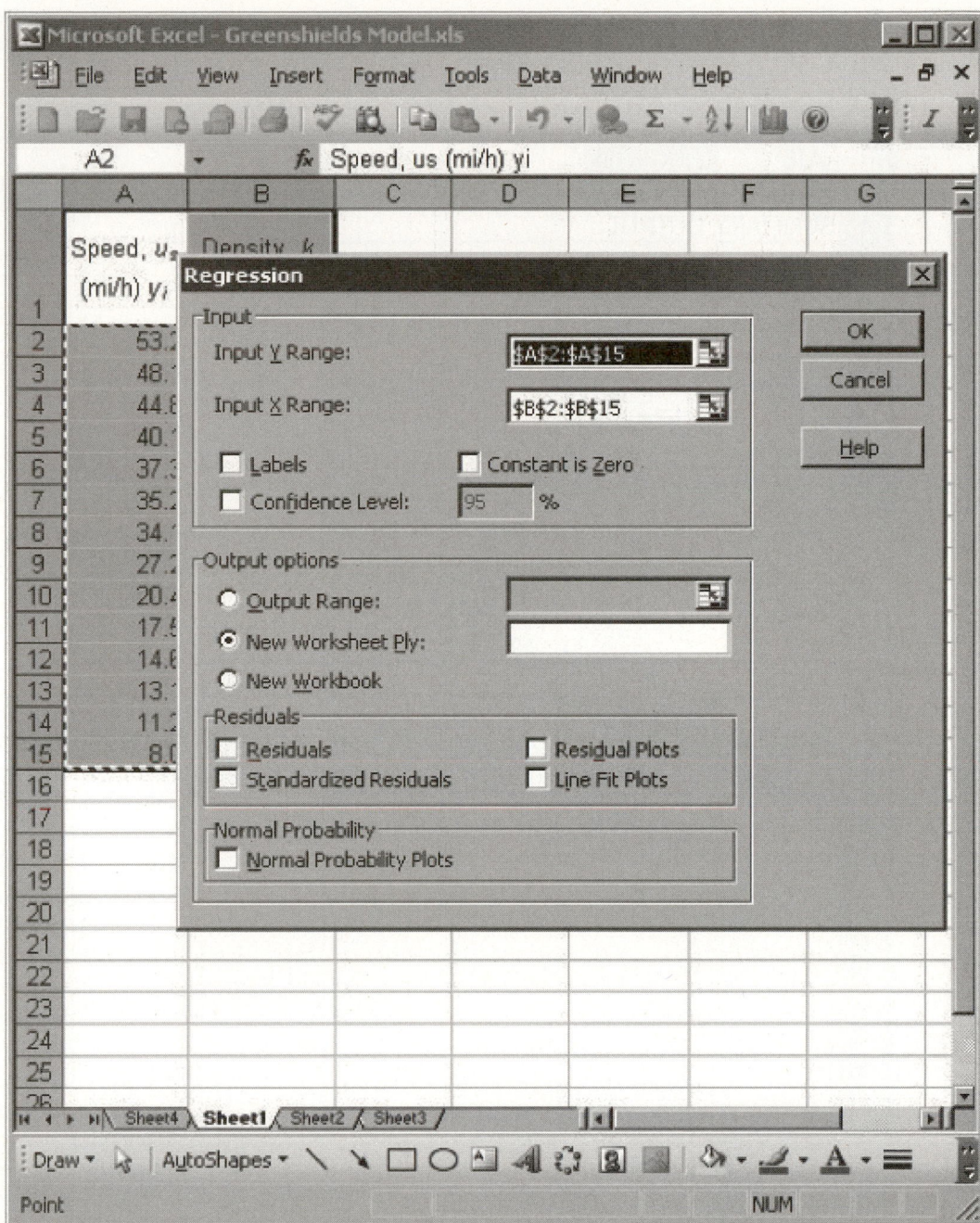

Figure C.4   Selecting Output Options

SOURCE: Created using Microsoft Office

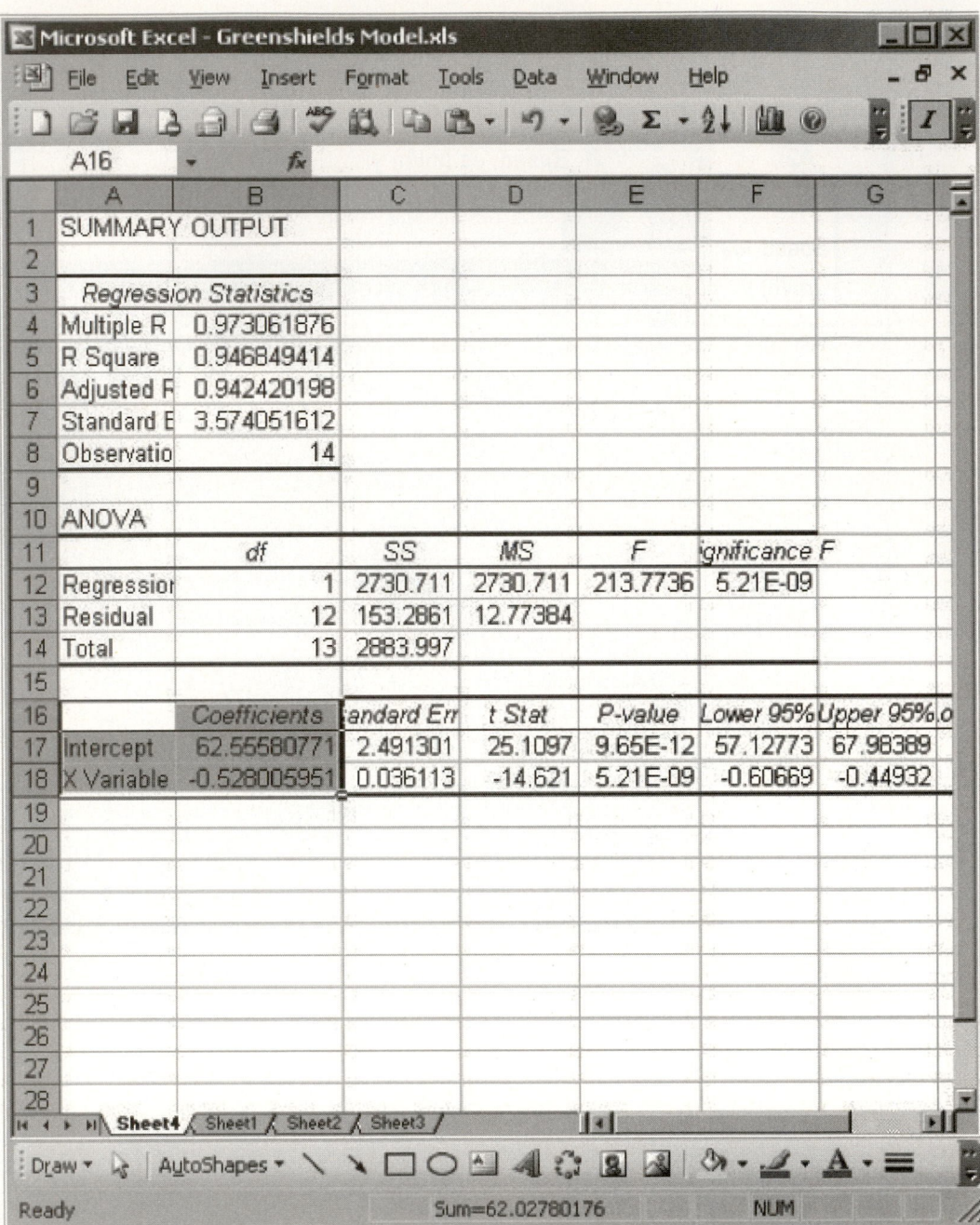

Figure C.5   Summary Output

SOURCE: Created using Microsoft Office

From the estimated regression function and the Greenshields expression, the jam density k_j can be found by dividing the mean free speed u_f by $-b$.

$$k_j = (62.56)/(0.53)$$

$$k_j = 118$$

Using u_f and k_j, we can determine the maximum flow from Eq. 6.25 as

$$q_{max} = (k_j u_f)/4$$

$$q_{max} = (118 \times 62.56)/4$$

$$q_{max} = 1846 \text{ veh/h}$$

Using Eq. 6.23, we also obtain the velocity at which flow is maximum; that is, $(62.56/2) = 31.3$ mi/h. From Eq. 6.24, we find the density at which flow is maximum at $(118/2) = 59$ veh/h.

An Example of Level of Service Determination Using HCS™ 2010

Source for all templates: Highway Capacity Software 6.1 (HCS™), McTrans Center, University of Florida, Gainesville, FL, 2010

Figure D.1   *HCS™* 2010 Main Interface

SOURCE: Highway Capacity Software 2010 6.50 (*HCS* 2010™). McTrans Center, University of Florida, Gainesville FL.

Signalized Intersection Primary Input Data

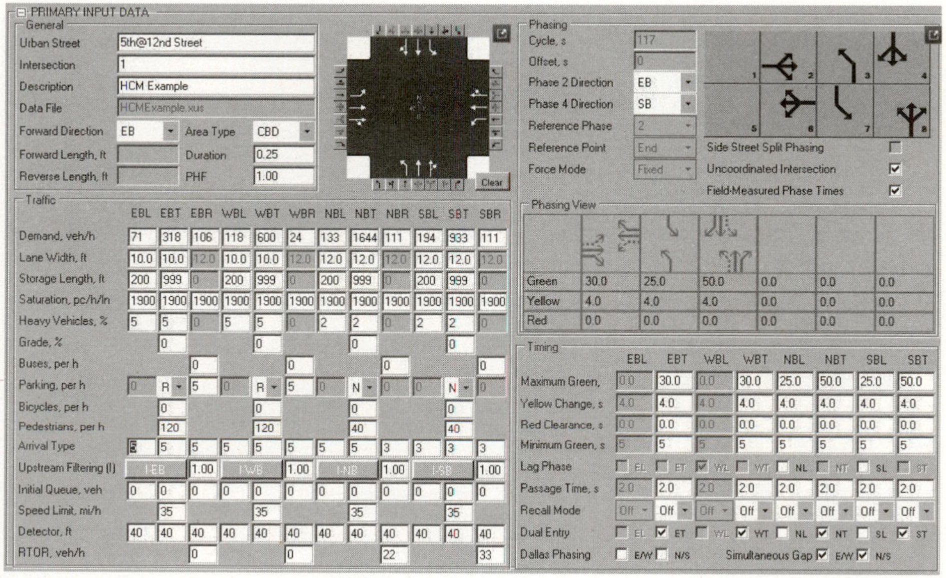

Figure D.2 Primary Input Data

SOURCE: Highway Capacity Software 2010 6.50 (*HCS* 2010™). McTrans Center, University of Florida, Gainesville FL.

Signalized Intersection Detailed Input Data

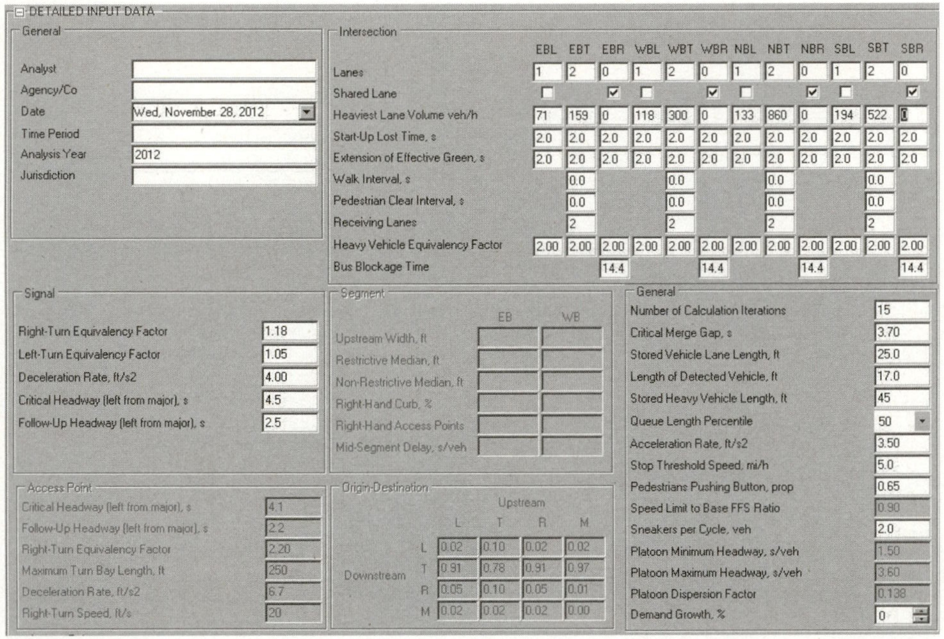

Figure D.3   Detailed Input Data

SOURCE: Highway Capacity Software 2010 6.50 (*HCS* 2010™). McTrans Center, University of Florida, Gainesville FL.

Signalized Intersection Summary Report

2010 HCS Signalized Intersection Results Summary

General Information

Agency			
Analyst		Analysis Date	Nov 28, 2012
Jurisdiction		Time Period	
Intersection	1	Analysis Year	2012
File Name	HCMExample.xus		
Project Description	HCM Example		

Intersection Information

Duration, h	0.25
Area Type	CBD
PHF	1.00
Analysis Period	1> 7:00

Demand Information

Approach Movement	EB			WB			NB			SB		
	L	T	R	L	T	R	L	T	R	L	T	R
Demand (v), veh/h	71	318	106	118	600	24	133	1644	89	194	933	78

Signal Information

Cycle, s	117.0	Reference Phase	2
Offset, s	0	Reference Point	End
Uncoordinated	Yes	Simult. Gap E/W	On
Force Mode	Fixed	Simult. Gap N/S	On

Green	30.0	25.0
Yellow	4.0	4.0
Red	0.0	0.0
Green	50.0	0.0
Yellow	4.0	0.0
Red	0.0	0.0
	0.0	
	0.0	
	0.0	

Timer Results

	EBL	EBT	WBL	WBT	NBL	NBT	SBL	SBT
Assigned Phase	2	2	6	6	3	8	7	4
Case Number	6.0	6.0	6.0	6.0	1.1	4.0	1.1	4.0
Phase Duration, s	34.0	34.0	34.0	34.0	29.0	54.0	29.0	54.0
Change Period, (Y+Rc), s	4.0	4.0	4.0	4.0	4.0	4.0	4.0	4.0
Max Allow Headway (MAH), s	3.4	3.4	3.4	3.4	3.1	3.0	3.1	3.0
Queue Clearance Time (gs), s	32.0	32.0	32.0	32.0	5.8	52.0	9.2	30.0
Green Extension Time (ge), s	0.0	0.0	0.0	0.0	0.2	0.0	0.3	6.9
Phase Call Probability	1.00	1.00	1.00	1.00	1.00	1.00	1.00	1.00
Max Out Probability	1.00	1.00	1.00	1.00	0.00	1.00	0.00	0.27

Figure D.4 Summary Report

SOURCE: Highway Capacity Software 2010 6.50 (*HCS* 2010™). McTrans Center, University of Florida, Gainesville FL.

Signalized Intersection Summary Report–Cont'd

Movement Group Results	EB L	EB T	EB R	WB L	WB T	WB R	NB L	NB T	NB R	SB L	SB T	SB R
Approach Movement												
Assigned Movement	5	2	12	1	6	16	3	8	18	7	4	14
Adjusted Flow Rate (V), veh/h	71	240	184	118	337	287	133	859	852	194	494	484
Adjusted Saturation Flow Rate (s), veh/h/ln	697	1629	1173	826	1629	1382	1597	1676	1649	1597	1676	1644
Queue Service time (g_s), s	7.2	15.1	16.1	13.9	22.7	22.8	3.8	50.0	50.0	7.2	28.0	28.0
Cycle Queue Clearance Time (g_c), s	30.0	15.1	16.1	30.0	22.7	22.8	3.8	50.0	50.0	7.2	28.0	28.0
Capacity (c), veh/h	104	418	301	159	418	354	500	716	705	407	716	702
Volume-to-Capacity Ratio (X)	0.680	0.576	0.610	0.740	0.807	0.810	0.266	1.198	1.210	0.477	0.689	0.689
Available Capacity (c_a), veh/h	104	418	301	159	418	354	500	716	705	407	716	702
Back of Queue (Q), veh/ln	2.5	6.1	4.8	4.2	10.2	8.9	1.3	40.0	40.3	2.7	11.5	11.2
Overflow Queue (Q_b), veh/ln	0.0	0.0	0.0	0.0	0.0	0.0	0.0	0.0	0.0	0.0	0.0	0.0
Queue Storage Ratio (RQ)	0.3	0.2	0.1	0.5	0.3	0.2	0.2	*1.0*	*1.0*	0.3	0.3	0.3
Uniform Delay (d_1), s/veh	56.3	38.0	38.3	52.8	40.8	40.8	12.7	33.5	33.5	23.5	27.2	27.2
Incremental Delay (d_2), s/veh	13.8	1.3	2.6	14.9	10.4	12.3	0.1	102.5	107.4	0.3	2.3	2.4
Initial Queue Delay (d_3), s/veh	0.0	0.0	0.0	0.0	0.0	0.0	0.0	0.0	0.0	0.0	0.0	0.0
Control Delay (d), s/veh	70.1	39.2	41.0	67.6	51.2	53.1	12.8	136.0	140.9	23.8	29.5	29.6
Level of Service (LOS)	E	D	D	E	D	D	B	F	F	C	C	C
Approach Delay, s/veh / LOS	44.3		D	54.5		D	129.4		F	28.6		C
Intersection Delay s/veh / LOS	78.6									E		

MultiModal Results	EB		WB		NB		SB	
Pedestrian LOS Score / LOS	2.9	C	2.9	C	2.8	C	2.8	C
Bicycle LOS Score / LOS	0.9	A	1.1	A	2.0	B	1.5	A

Figure D.4 Summary Report Cont'd

SOURCE: Highway Capacity Software 2010 6.50 (HCS 2010™). McTrans Center, University of Florida, Gainesville FL.

APPENDIX E

Metric Conversion Factors for Highway Geometric Design

This appendix provides information regarding areas critical to basic geometric design and provides the metric version of formulas that appear in Chapters 3 and 16, as given by the AASHTO publication, *A Policy on Geometric Design of Highways and Streets,* 6th edition, 2011, and the *Guide to Metric Conversion,* 1993.

There are nine areas critical to basic geometric design, for which AASHTO has provided metric values. These are speed, lane width, shoulders, vertical clearance, clear zone, curbs, sight distance, horizontal curvature, and structures. Recommended values are listed in Table E.1 of the following section.

Table E.2 provides conversion factors for length, area, and volume, and formulas converted to metric are listed together with the corresponding equation in the text.

E.1 METRIC FORMULAS USED IN HIGHWAY DESIGN

Braking Distance on Grade, *m*

$$d_B = \frac{v^2}{254\left[\left(\dfrac{a}{9.81}\right) \pm G\right]} \tag{3.24}$$

Stopping Sight Distance

$$S = 0.278vt + \frac{v^2}{254\left(\left(\dfrac{a}{9.81}\right) \pm G\right)} \tag{3.27}$$

where

d_B = braking distance on grade, m
S = stopping sight distance, m
t = brake reaction time, 2.5 sec

Table E.1   Selected Metric Values for Geometric Design

Metric	U.S. Customary	
Design Speed (km/h)	Corresponding Design Speed (mi/h)	
I. Design Speeds		
20	15	
30	20	
40	25	
50	30	
60	40	
70	45	
80	50	
90	55	
100	60	
110	70	
120	75	
130	80	
II. Lane Width		
2.7 m	(8.86 ft)	(1.56% less than 9′ lane)
3.0 m	(9.84 ft)	(1.60% less than 10′ lane)
3.3 m	(10.83 ft)	(1.55% less than 11′ lane)
3.6 m	(11.81 ft)	(1.58% less than 12′ lane)
III. Shoulders		
0.6 m	(1.97 ft)	
1.2 m	(3.94 ft)	
1.8 m	(5.91 ft)	
2.4 m	(7.87 ft)	
3.0 m	(9.84 ft)	
IV. Vertical Clearance		
3.8 m	(12.47 ft)	
4.3 m	(14.11 ft)	
4.9 m[a]	(16.08 ft)	
V. Clear Zone[b]		
Urban Conditions	0.5 m (1.64 ft)	
Locals/Collectors	3.0 m minimum (9.84 ft)	

[a]The 4.9 m value is seen to be the critical value since the federal legislation required Interstate design to have 16 ft vertical clearance. In view of the fact that the Interstate, now virtually complete, is based on this minimum clearance, the metric value should provide this clearance as a minimum.

[b]With two exceptions, the *Green Book* refers to the *Roadside Design Guide* for clear zone values. The two critical values are the clear zone for urban conditions and locals and collectors.

Table E.1 Selected Metric Values for Geometric Design (*continued*)

VI. Curbs

A. Curb Heights
  1. Mountable Curb, 150 mm max (5.91″)
  2. Barrier Curb, 225 mm max (8.86″)
B. The definition of high speed/low speed has an impact on where curb is used.
  Low speed: 60 km/h or less design speed
  High speed: 80 km/h or more design speed

VII. Sight Distance

Stopping Sight Distance	
Eye Height	1070 mm (3.51 ft)
Object Height	150 mm (5.91 in.)
Headlight Height	610 mm (2 ft.)
Passing Sight Distance	
Eye Height	1070 mm (3.51 ft)
Object Height	1300 mm (4.27 ft)

VIII. Horizontal Curvature

Radius definition should be used in lieu of degree of curve. Radius should be expressed in multiples of 5 m increments.

IX. Structures

Long bridges will be those over 60 m in length.

v = design speed, km/h
a = deceleration rate, m/sec^2

Minimum Radius of Curve

$$R_{\min} = \frac{v^2}{127(0.01e_{\max} + f_{s\max})} \tag{3.33}$$

where

v = initial speed (km/h)
$f_{s\max}$ = coefficient of friction (Table 3.3)
e = rate of superelevation

Length of Spiral

$$L = \frac{0.0214v^3}{RC} \tag{15.37}$$

where

C = the rate of increase of centripetal acceleration (m/sec^2); use 1 to 3 m/sec^2
L = minimum length of spiral (m)
v = speed (km/h)
R = curve radius (m)

Table E.2   Area, Length, and Volume Conversion Factors

Quantity	From Inch-Pound Units	To Metric Units	Multiply By
Length	mile	km	1.609344
	yard	m	0.9144
	foot	m	0.3048
		mm	304.8
	inch	mm	25.4
	square mile	km²	2.59000
Area	acre	m²	4046.856
		ha (10,000 m²)	0.4046856
	square yard	m²	0.83612736
	square foot	m²	0.09290304
	square inch	mm²	645.15
	acre foot	m³	1233.49
Volume	cubic yard	m³	0.764555
	cubic foot	m³	0.0283168
	cubic foot	cm³	28,316.85
	cubic foot	L (1000 cm³)	28.31685
	100 board feet	m³	0.235974
	gallon	L (1000 cm³)	3.78541
	cubic inch	cm³	16.387064
	cubic inch	mm³	16,387.064

Middle Ordinate (Horizontal Sight Line Offset)

$$M = R\left[1 - \cos\frac{28.65S}{R}\right] \tag{15.25}$$

where

M = middle ordinate (horizontal sight line offset) (m)
S = stopping sight distance (m)
R = radius of curve (m)

Length of Crest Vertical Curves

$$S < L : L = \frac{AS^2}{100\left(\sqrt{2h_1} + \sqrt{2h_2}\right)^2} = \frac{AS^2}{658} \tag{15.5}$$

$$S > L : L = 2S - \frac{200\left(\sqrt{2h_1} + \sqrt{2h_1}\right)}{A} = 2S - \frac{658}{A} \tag{15.3}$$

where

S = sight distance (m)
L = length of vertical curve (m)
A = algebraic difference in grades (percent)

h_1 = height of eye above roadway = 1.08 m
h_2 = height of object above roadway = 0.60 m

Sag Vertical Curves

$$H_e = \text{headlight height} = 600\,\text{mm}$$
$$\alpha = 1^*$$

$$S < L : L = \frac{AS^2}{120 + 3.5S} \qquad (15.9)$$

$$S > L : L = 2S - \left(\frac{120 + 3.5S}{A}\right) \qquad (15.7)$$

where

S = stopping sight distance (m)
L = length of vertical curve (m)
A = algebraic difference in grades (percent)

Comfort Criteria

$$L = \frac{AV^2}{395} \qquad (15.10)$$

where

v = design speed (km/h)
L = length of sag vertical curve (m)

Passing Sight Distance

Table E.3 shows the metric values for the components of safe passing sight distance on two-lane highways.

Table E.3   Passing Sight Distance for Design of Two-Lane Highways

Design Speed (km/h)	Assumed Speeds (km/h)		Passing Sight Distance (m)
	Passed Vehicle	Passing Vehicle	
30	11	30	120
40	21	40	140
50	31	50	160
60	41	60	180
70	51	70	210
80	61	80	245
90	71	90	280
100	81	100	320
110	91	110	355
120	101	120	395
130	111	130	440

SOURCE: Based on *A Policy on Geometric Design of Highways and Streets*, 2011, AASHTO, Washington, D.C.

REFERENCES

A Policy on Geometric Design of Highways and Streets, 6th ed., American Association of State Highway and Transportation Officials, Washington, D.C., 2011.

Guide to Metric Conversion, American Association of State Highway and Transportation Officials, Washington, D.C., 1993.

Index